Chi-Hai Ling

Oct., 1975

THEORETICAL HYDRODYNAMICS

THEORETICAL HYDRODYNAMICS

L. M. MILNE-THOMSON, C.B.E.

PROFESSOR OF MATHEMATICS EMERITUS IN THE UNIVERSITY OF ARIZONA
FORMERLY PROFESSOR OF APPLIED MATHEMATICS IN BROWN UNIVERSITY
AND THE MATHEMATICS RESEARCH CENTER, UNITED STATES ARMY
GRESHAM PROFESSOR OF GEOMETRY
PROFESSOR OF MATHEMATICS EMERITUS IN THE ROYAL NAVAL COLLEGE, GREENWICH
VISITING PROFESSOR AT THE UNIVERSITIES
OF ROME, QUEENSLAND, CALGARY, OTAGO

FIFTH EDITION

Revised

Distributed in the United States by
CRANE, RUSSAK & COMPANY, INC.
347 Madison Avenue
New York, New York 10017

MACMILLAN

quis crederet umquam
aërias hominem carpere posse vias?
OVID

First Edition 1938
Second Edition 1949
Third Edition 1955
Fourth Edition 1960
Reprinted 1962
Fifth Edition 1968
Reprinted 1972, 1974

Published by
THE MACMILLAN PRESS LTD
London and Basingstoke
Associated companies in New York Dublin Melbourne Johannesburg and Madras

SBN 333 07876 4

Printed in Great Britain by
ROBERT MACLEHOSE AND CO. LTD
The University Press, Glasgow

PREFACE

Not so many years ago the dynamics of a frictionless fluid had come to be regarded as an academic subject and incapable of practical application owing to the great discrepancy between calculated and observed results. The ultimate recognition, however, that Lanchester's theory of circulation in a perfect fluid could explain the lift on an aerofoil, and the adoption of Prandtl's hypothesis that outside a boundary layer the effect of viscosity is negligible, gave a fresh impetus to the subject which has always been necessary to the naval architect and which the advent of the modern aeroplane has placed in the front rank.

The investigation of fluid motion falls naturally into two parts; (i) the experimental or practical side; (ii) the theoretical side which attempts to explain why experimental results turn out as they do, and above all attempts to predict the course of experiments as yet untried. Thus the practical and theoretical sides supplement one another, and it is to the latter aspect that this book is devoted.

As a scientific theory becomes more exact, so does it of necessity tend to assume a more mathematical form. This statement must be construed to mean not that the form becomes more difficult or more abstruse, but rather that, when the fundamental laws have reached a stage of clear formulation, useful deductions can be made by the exact processes of mathematics. The object of this book, which is founded upon, and has grown out of, my lectures on the subject at Greenwich to junior members of the Royal Corps of Naval Constructors, is to give a thorough, clear and methodical introductory exposition of the mathematical theory of fluid motion which will be useful in applications to both hydrodynamics and aerodynamics.

I have ventured to depart radically from the traditional presentation of the subject by basing it consistently throughout on vector methods and notation with their natural consequence in two dimensions, the complex variable. It is not intended to imply that the application of the above methods to hydrodynamics is in itself a novelty, but their exclusive employment has not, so far as I know, been hitherto attempted. The previous mathematical knowledge required of the reader does not go beyond the elements of the infinitesimal calculus. The necessary additional mathematical apparatus is introduced as required and an attempt has been made to keep the book reasonably self-contained in this respect. As we are dealing with a real subject (even if in an idealised form) diagrams have been freely used. There are about 400 of these numbered in the decimal * notation with the number of the section in which they occur in order to facilitate reference.

The order of the chapters represents an attempt to give a rational classification to the topics treated. This is, of course, by no means the only possible order, but it seems to have some advantages. Chapter I is of an introductory character and is concerned mainly with inferences based on the famous theorem of Daniel Bernoulli who may justly be considered the father of Hydrodynamics.

Chapter II gives an account of such properties of vectors and tensors as are essential to the analysis of the motion of a fluid element and to the formulation of the hydrodynamical equations. Vectors are introduced here without any reference to systems of coordinates. The fundamental properties of vector operations are deduced by operational methods, which, in the form here explained, are easy to apply and lead directly to the theorems of Stokes, Gauss, and Green. As this is a book on hydrodynamics, not on vectors, the treatment is necessarily concise. On the other hand the subject matter has been arranged with a view to helping those to whom vector manipulation may be unfamiliar, and the reader is recommended to make himself thoroughly conversant with the contents of this chapter, if necessary, by frequent reference to it. Such a course will be amply rewarded by a physical insight into the phenomena under discussion which are, in general, made unnecessarily obscure by expression in particular coordinate systems. The proper function of coordinates is to perform the final step of algebraic interpretation. In Chapter III the general properties of fluid motion continuity, dynamical equation, pressure, energy, and vorticity are studied in the light of the vector formulation whose advantage is then clearly seen.

Chapter IV is occupied with intrinsic properties of two-dimensional motion in so far as they can be treated without the complex variable. Chapter V is a digression to introduce the complex variable, defined as a vector operator, and to prove such theorems as will be required in the sequel. In particular the properties of conformal mapping are treated in some detail in view of their subsequent fundamental importance.

Chapters VI to XV form a complete unit and embody an attempt to give a detailed discussion of two-dimensional motion from the unified standpoint of the complex variable, making full use of the circle theorem (6·21), the Area theorem (5·43), conformal mapping and the theorem of Blasius with its extensions. I have begun with a discussion of streaming motion in Chapter VI, followed by a consideration of simple Joukowski aerofoils in Chapter VII, while sources and

* It should be noted that the section numbers are decimals of which the integral part denotes the number of the chapter. Thus for example section 4·21 precedes section 4·5 and both belong to Chapter IV.

sinks are postponed to Chapter VIII. In Chapter IX the moving cylinder is treated in detail and a form of the theorem of Kutta and Joukowski, generalised to include the case of accelerated motion, is obtained (9·53). Chapter X contains a discussion of the mapping theorem of Schwarz and Christoffel with some immediate applications; in Chapters XI, XII further applications are made to the discontinuous motions of jets, currents, and the cavity behind a cylinder in a stream, including an account of the elegant method of Levi-Civita. Chapter XIII is devoted to the discussion of rectilinear vortices, Kármán's vortex street, and the drag due to a vortex wake. Chapter XIV is a new departure in that it brings together various exact treatments of two-dimensional motion with a free surface in the presence of a gravitational field. Chapter XV deals with approximations, generally linearised, to the subject of the preceding chapter with the emphasis on waves.

Chapter XVI introduces Stokes' stream function and the application of conformal mapping to three-dimensional problems with axial symmetry. The general motion of spheres and ellipsoids is treated in Chapter XVII. In Chapter XVIII partial differentiation with respect to a vector (2·71) is applied to obtain Kirchhoff's equations in vector form thus replacing six equations by two. It is believed that this method is still new and that it offers opportunities for research in stability problems. Chapter XIX discusses vortex motion in general with particular application to the aerofoil of finite span.

Chapter XX is intended as an introduction to the theory of the flow of a compressible fluid at subsonic and supersonic speeds. The source in a compressible fluid is discussed in 8·9, and the vortex in 13·8.

Chapter XXI introduces, in a cogent logical sequence, the application of tensor methods to the flow of viscous fluids, particularly liquids. It is interesting to note how simply the components of stress in a viscous liquid can be derived by vector methods for any system of orthogonal coordinates.

Chapter XXII, which is new, is concerned with flow at small Reynolds number and the approximations connected therewith, including an account of a novel application of the complex variable to Stokes flow.

Chapter XXIII, also new, gives an outline of the theory of two-dimensional laminar flow in a boundary layer.

There are 621 exercises in all collected into sets of examples at the end of each chapter. Many of these are taken with permission, for which I express my best thanks, from the Mathematical Tripos, the University of London's M.Sc. examination, and from the examination of Constructor Lieutenants at the Royal Naval College.* Apart from these I have included others of various origins, now unknown and so unacknowledged, used in my lectures and about 100 given me by the late Professor L. N. G. Filon. Some of the exercises are

* These sources are distinguished by the letters M.T., U.L., R.N.C., respectively.

very easy, others are quite difficult and may be regarded as supplementing the text.

In stating theorems I have, as far as possible, associated the name of the discoverer as sufficient indication of the origin, but it must not be assumed that the method of presentation here is in every case that in which the theorem was originally given. For example Gauss might well consider 2·60 as his theorem veiled in allegory and illustrated by symbols. Bibliographical references have occasionally been added where they appear to be useful or appropriate, but no systematic attempt has been made to give them. I have followed Lamb (1849–1934) in associating the negative sign with the gradient of the velocity potential. The preparation of the new edition has given me the opportunity to act upon suggestions made by a number of readers to whom I am very grateful.

The gratifying reception accorded to this work has encouraged me to continue to search for improvements. Apart from considerable rearrangements and new methods of presentation this fifth edition differs from the fourth by several important additions including chapters on gravity flows, flows at small Reynolds number and boundary layers.

The arduous task of proof reading has been greatly lightened by the help of my colleague Professor W. R. Dean and my former student Dr W. E. Conway both of whom have made many valuable suggestions. To these two friends I wish to give my lively thanks.

I take this opportunity of expressing my thanks to the officials of the Glasgow University Press not only for the ready way in which they have met my requirements but also for their careful attention to typographical detail which is so important in a work of this kind, and most of all for maintaining that standard of elegant mathematical printing for which they are justly renowned.

L. M. MILNE-THOMSON

MATHEMATICS DEPARTMENT
THE UNIVERSITY OF OTAGO
DUNEDIN, NEW ZEALAND
August 1971

CONTENTS

CHAPTER I
BERNOULLI'S EQUATION

CHAPTER II
VECTORS AND TENSORS

CHAPTER III

EQUATIONS OF MOTION

CHAPTER IV

TWO-DIMENSIONAL MOTION

CHAPTER V

COMPLEX VARIABLE

CHAPTER VI

STREAMING MOTIONS

CHAPTER VII
AEROFOILS

CHAPTER VIII
SOURCES AND SINKS

CHAPTER X

THEOREM OF SCHWARZ AND CHRISTOFFEL

CHAPTER XI

JETS AND CURRENTS

CHAPTER XII

HELMHOLTZ MOTIONS

CHAPTER XIII
RECTILINEAR VORTICES

CHAPTER XIV

FLOWS UNDER GRAVITY WITH A FREE SURFACE

CHAPTER XV

LINEARISED GRAVITY WAVES

CHAPTER XVI
STOKES' STREAM FUNCTION

CHAPTER XVII
SPHERES AND ELLIPSOIDS

CHAPTER XVIII
SOLID MOVING THROUGH A LIQUID

CHAPTER XIX

VORTEX MOTION

CHAPTER XX

SUBSONIC AND SUPERSONIC FLOW

CHAPTER XXI
VISCOSITY

CHAPTER XXII
STOKES AND OSEEN FLOWS

CHAPTER XXIII
BOUNDARY LAYERS

THE GREEK ALPHABET

alpha	α	A	nu	ν	N
beta	β	B	xi	ξ	Ξ
gamma	γ	Γ	omicron	o	O
delta	δ	Δ	pi	$\pi \quad \varpi$	Π
epsilon	ϵ	E	rho	ρ	P
zeta	ζ	Z	sigma	σ	Σ
eta	η	H	tau	τ	T
theta	θ	Θ	upsilon	υ	Υ
iota	ι	I	phi	ϕ	Φ
kappa	κ	K	chi	χ	X
lambda	λ	Λ	psi	ψ	Ψ
mu	μ	M	omega	ω	Ω

HISTORICAL NOTES

THE term *hydrodynamics* was introduced by Daniel Bernoulli (1700–1783) to comprise the two sciences of hydrostatics and hydraulics. He also discovered the famous theorem still known by his name.

d'Alembert (1717–1783) investigated resistance, discovered the paradox associated with his name, and introduced the principle of conservation of mass (equation of continuity) in a liquid.

Euler (1707–1783) formed the equations of motion of a perfect fluid and developed the mathematical theory. This work was continued by Lagrange (1736–1813).

Navier (1785–1836) derived the equations of motion of a viscous fluid from certain hypothesis of molecular interaction.

Stokes (1819–1903) also obtained the equations of motion of a viscous fluid. He may be regarded as having founded the modern theory of hydrodynamics.

Rankine (1820–1872) developed the theory of sources and sinks.

Helmholtz (1821–1894) introduced the term *velocity potential*, founded the theory of vortex motion, and discontinuous motion, making fundamental contributions to the subject.

Kirchhoff (1824–1887) and Rayleigh (1842–1919) continued the study of discontinuous motion and the resistance due to it.

Osborne Reynolds (1842–1912) studied the motion of viscous fluids, introduced the concepts of laminar and turbulent flow, and pointed out the abrupt transition from one to the other.

Joukowski (1847–1921) made outstanding contributions to aerofoil design and theory, and introduced the aerofoils known by his name.

Lanchester (1868–1945) made two fundamental contributions to the modern theory of flight; (i) the idea of circulation as the cause of lift, (ii) the idea of tip vortices as the cause of induced drag. He explained his theories to the Birmingham Natural History Society in 1894 but did not publish them till 1907 in his *Aerodynamics*.

Prandtl (1875–1953) introduced in 1904 his epoch making theory of the boundary layer thus reconciling the ideas of viscous and inviscid fluids.

PLATE 1

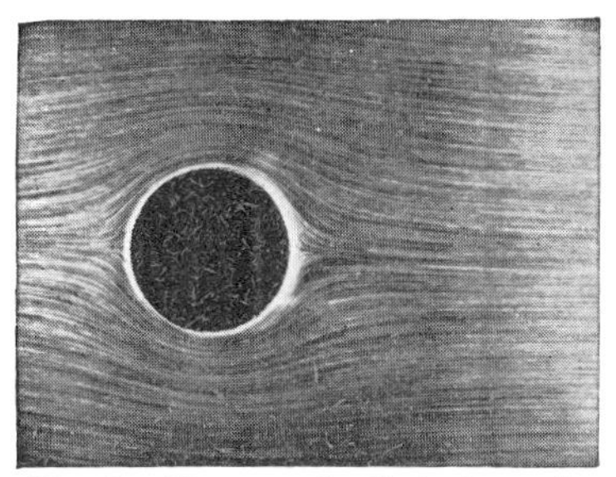

FIG. 1.—Flow round cylinder immediately after starting (potential flow).

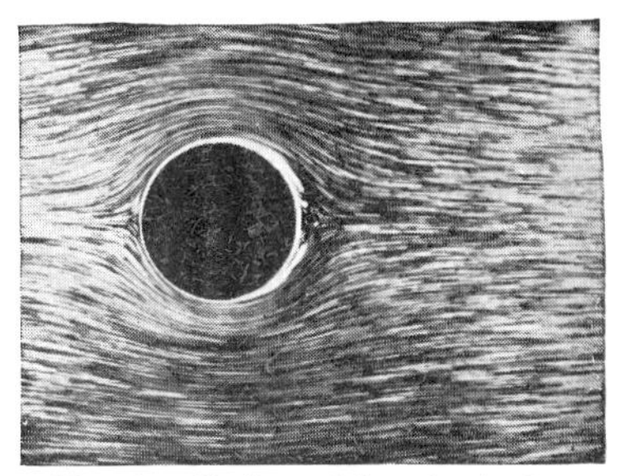

FIG. 2.—Backward flow in the boundary layer behind the cylinder; accumulation of boundary layer material.

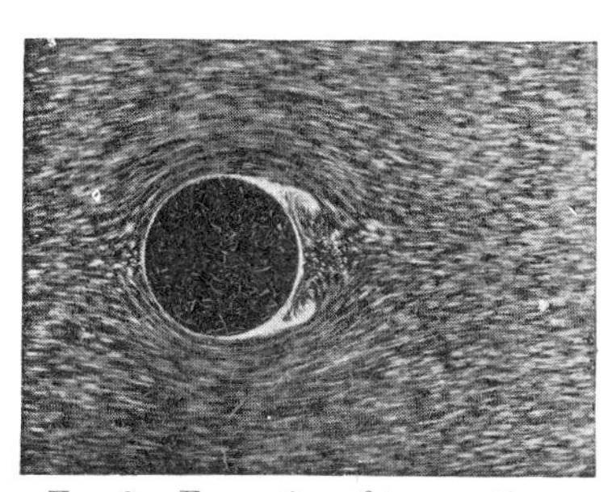

FIG. 3.—Formation of two vortices; flow breaking loose from cylinder.

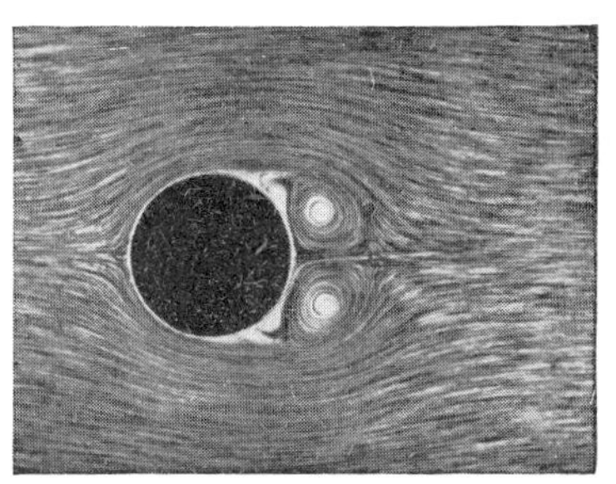

FIG. 4.—The eddies increase in size.

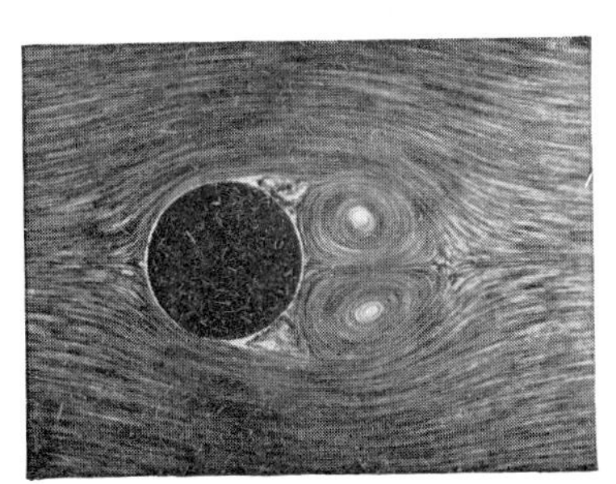

FIG. 5.—Final picture obtained a long time after starting.

FIG. 6.—The eddies grow still more; finally the picture becomes unsymmetrical and disintegrates.

The direction of flow in all photographs is from left to right.

Reprint from *Applied Hydro- and Aeromechanics*, by L. Prandtl, Ph.D., and O. G. Tietjens, Ph.D., through courtesy of United Engineering Trustees, Inc.

PLATE 2

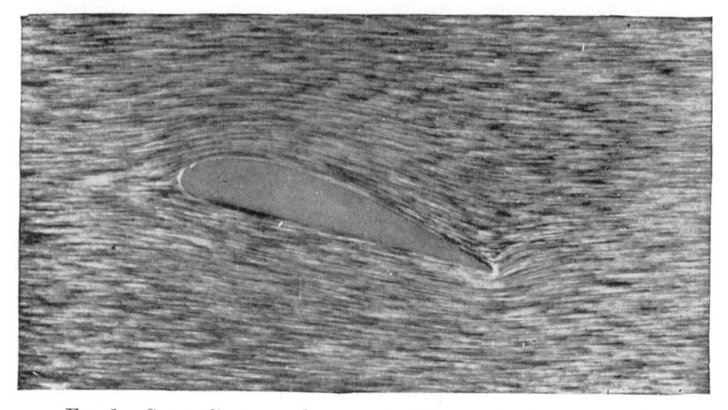

Fig. 1.—Streamlines round an aerofoil the very first moment after starting.

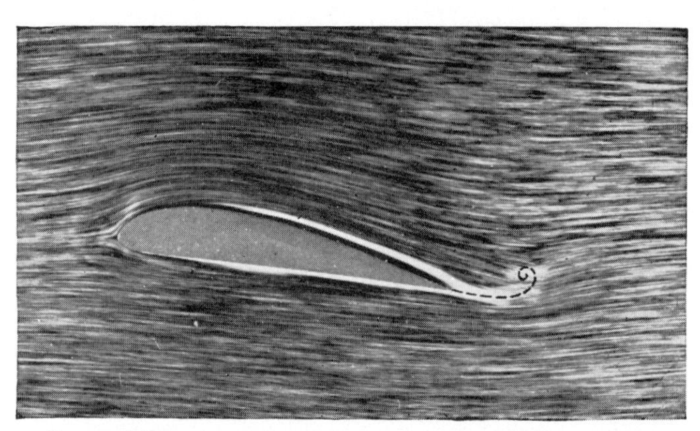

Fig. 2.—Formation of the starting vortex which is washed away with the fluid.

Reprint from *Applied Hydro- and Aeromechanics*, by L. Prandtl, Ph.D., and O. G. Tietjens, Ph.D., through courtesy of United Engineering Trustees, Inc.

PLATE 3

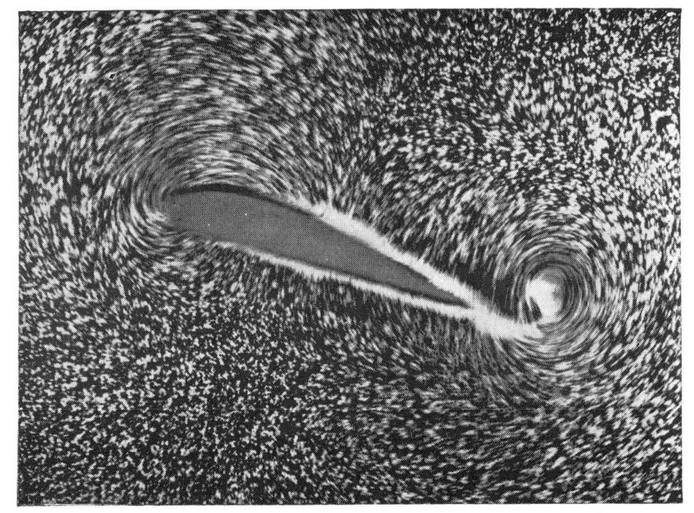

FIG. 1.—Like Fig. 1, Plate 2, but with the camera at rest with respect to the undisturbed fluid and a shorter exposure. Also with a greater angle of attack and consequently a greater starting vortex.

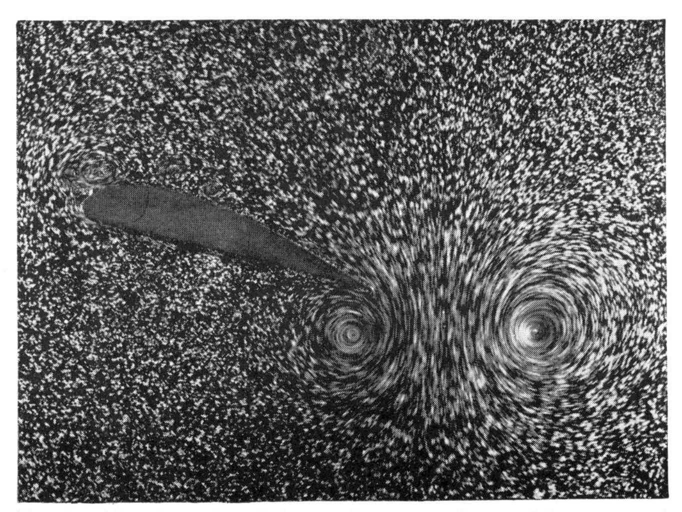

FIG. 2.—After formation of the starting vortex the aerofoil was stopped and then the picture was taken.

Reprint from *Applied Hydro- and Aeromechanics*, by L. Prandtl, Ph.D., and O. G. Tietjens, Ph.D. through courtesy of United Engineering Trustees, Inc.

PLATE 4

FIG. 1.—Kármán trail ; Reynolds number $wd/\nu = 250$.
The camera is at rest with respect to the cylinder.

FIG. 2.—Kármán trail ; Reynolds number $wd/\nu = 250$.
The camera is at rest with respect to the undisturbed fluid.

Reprint from *Applied Hydro- and Aeromechanics*, by L. Prandtl, Ph.D., and O. G. Tietjens, Ph.D.
through courtesy of United Engineering Trustees, Inc.

CHAPTER I

BERNOULLI'S EQUATION

1·0. The science of hydrodynamics is concerned with the behaviour of fluids in motion.

All materials * exhibit *deformation* under the action of forces; *elasticity* when a given force produces a definite deformation, which vanishes if the force is removed; *plasticity* if the removal of the forces leaves permanent deformation; *flow* if the deformation continually increases without limit under the action of forces, *however small.*

A *fluid* is material which flows.

Actual fluids fall into two categories, namely gases and liquids.

A *gas* will ultimately fill any closed space to which it has access and is therefore classified as a (highly) *compressible fluid.*

A liquid at constant temperature and pressure has a definite volume and when placed in an open vessel will take under the action of gravity the form of the lower part of the vessel and will be bounded above by a horizontal free surface. All known liquids are to some slight extent compressible. For most purposes it is, however, sufficient to regard liquids as *incompressible fluids.*

In this book we shall for the most part be concerned with the behaviour of fluids treated as incompressible and the term liquid will be used in this sense. But it is proper to observe that, for speeds which are not comparable with that of sound, the effect of compressibility on atmospheric air can be neglected, and in many experiments which are carried out in wind tunnels the air is considered to be a liquid, in the above sense, which may conveniently be called *incompressible air.*

Actual liquids (and gases) in common with solids exhibit *viscosity* arising from internal friction in the substance. Our definition of a fluid distinguishes a viscous fluid, such as treacle or pitch, from a plastic solid, such as putty or clay, since the former cannot permanently resist any shearing stress, however small, whilst in the case of the latter, stresses of a definite magnitude are required to produce deformation. Pitch is an example of a very viscous liquid; water is an example of a liquid which is but slightly viscous. A more precise definition

* In this summary description the materials are supposed to exhibit a macroscopic continuity, and the forces are not great enough to cause rupture. Thus a heap of sand is excluded, but the individual grains are not.

of viscosity will be given later. For the present, in order to render the subject amenable to exact mathematical treatment, we shall follow the course adopted in other branches of mechanics and make simplifying assumptions by defining an ideal substance known as an *inviscid* or *perfect* fluid.

Definition. An inviscid fluid is a continuous fluid substance which can exert no shearing stress *however small.*

The continuity is postulated in order to evade the difficulties inherent in the conception of a fluid as consisting of a granular structure of discrete molecules. The inability to exert any shearing stress, however small, will be shown later to imply that the pressure at any point is the same for all directions at that point.

Moreover, the absence of tangential stress between the fluid on the two sides of any small surface imagined as drawn in the fluid implies the entire absence of internal friction, so that no energy can be dissipated from this

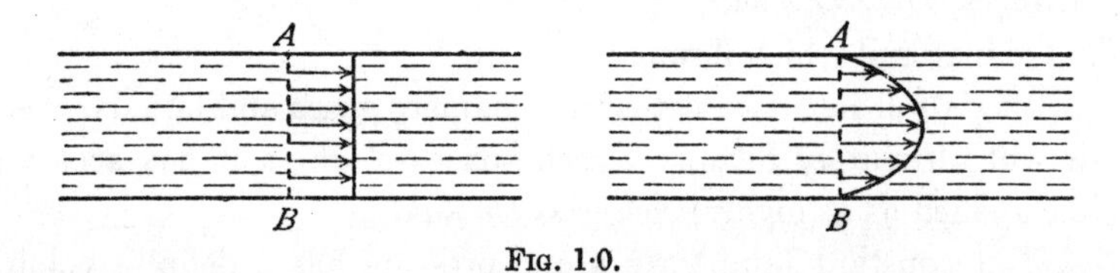

FIG. 1·0.

cause. A further implication is that, when a solid moves through the fluid or the fluid flows past a solid, the solid surface can exert no tangential action on the fluid, so that the fluid flows freely past the boundary and no energy can be dissipated there by friction. In this respect the ideal fluid departs widely from the actual fluid which, as experimental evidence tends to show, adheres to the surface of solid bodies immersed in it. The difference in behaviour is well illustrated by considering straight steady flow along a horizontal pipe. If we draw vectors to represent the velocity at points of AB, a diameter of the pipe, for an inviscid fluid their extremities will lie on another diameter, while for a viscous fluid the extremities will lie on a parabolic curve, passing through A and B. It might be thought that the study of the perfect fluid could throw but little light on the behaviour of actual fluids. As we shall see presently this is so far from being the case that the theory can, in important instances, explain not only qualitatively but also quantitatively the motion of actual fluids.

1·01. Physical dimensions. Physics deals with the measurable properties of physical quantities, certain of which, as for example, length, mass, time and temperature, are regarded as fundamental, since they are independent of one another, and others, such as velocity, acceleration, force, thermal conductivity, pressure, energy are regarded as derived quantities, since they are defined ultimately in terms of the fundamental quantities. Mathematical physics deals with the representation of the measures of these quantities by numbers and deductions therefrom. These measures are all of the nature of ratios of comparison of a measurable magnitude with a standard one of like kind, arbitrarily chosen as the unit, so that the number representing the measure depends on the choice of unit.

Consider a *dynamical system,* i.e. one in which the derived quantities depend only on length, mass and time, and change the fundamental units from, say, foot, pound, second, to mile, ton, hour. Let l_1, m_1, t_1 and l_2, m_2, t_2 be the measures of the same length, mass and time respectively in the two sets of units. Then we have

$$(1) \qquad l_1 = \frac{l_1}{l_2} \times l_2 = Ll_2, \quad m_1 = Mm_2, \quad t_1 = Tt_2,$$

where L, M, T are numbers independent of the particular length, mass or time measured, but depending only on the choice of the two sets of units. Thus in this case, we have $L = 5280$, $M = 2240$, $T = 3600$. These numbers L, M, T we call the respective *measure-ratios* of length, mass, time for the two sets of units, in the sense that measures of these quantities in the second set are converted into the corresponding measures in the first set by multiplication by L, M, T.

The measure-ratios V, A, F of the derived quantities, velocity v, acceleration a, and force f, are then readily obtained from the definitions of these quantities as

$$V = L/T, \quad A = V/T, \quad F = MA,$$

so that ultimately the measure ratio of a force is given by $F = ML/T^2$. And in general if n_1, n_2 are the measures of the same physical quantity n in the two sets of units, we arrive at the measure-ratio

$$(2) \qquad \frac{n_1}{n_2} = N = L^x M^y T^z.$$

and we express this conventionally by the statement that the quantity is of *dimensions* $L^x M^y T^z$ (or is of dimensions x in length, y in mass, and z in time). If $x = y = z = 0$, then $n_1 = n_2$, so the quantity in question is independent of any units which may be chosen, as for example, the quantity defined as the ratio of the mass of the engines to the mass of the ship. In such a case we say the quantity is dimensionless and is represented by *a pure number,* meaning that it does not change with units.

Now consider a definitive relation

$$(3) \qquad a = bc$$

between the measures a, b, c of physical quantities in a dynamical system, i.e. a relation which is to hold whatever the sets of units employed, and which is not merely an accidental relation between numbers arising from measurement in one particular set of units. Suppose the dimensions of a, b, c are respectively (p, q, r), (s, t, u), and (x, y, z), so that

$$(4) \qquad a_1 = a_2 L^p M^q T^r, \quad b_1 = b_2 L^s M^t T^u, \quad c_1 = c_2 L^x M^y T^z.$$

Then (3) would become $a_1 = b_1 c_1$, and (4) would then give by substitution

$$a_2 L^p M^q T^r = b_2 L^s M^t T^u c_2 L^x M^y T^z.$$

Now $a_2 = b_2 c_2$, since the form of (3) is independent of units, and therefore

$$L^p M^q T^r = L^{s+x} M^{t+y} T^{u+z}, \quad \text{or} \quad p = s+x, \quad q = t+y, \quad r = u+z.$$

In other words, each fundamental measure-ratio must occur with the same index on each side of (3), i.e. each side of (3) must be of the same physical dimensions.

In systems involving temperature as well as length, mass, and time as fundamental quantities (*thermodynamical systems*) a measure-ratio (say D) of temperature must be introduced.

1·1. Velocity. Since our fluid is continuous, we can define a *fluid particle* as consisting of the fluid contained within an infinitesimal volume, that is to say, a volume whose size may be considered so small that for the particular purpose in hand its linear dimensions are negligible. We can then treat a fluid particle as a geometrical point for the particular purpose of discussing its velocity and acceleration.

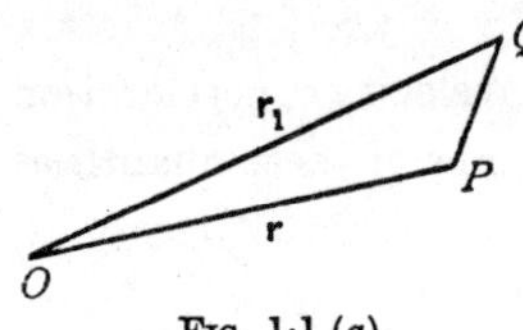

FIG. 1·1 (*a*).

If we consider, fig. 1·1 (*a*), the particle which at time t is at the point P, defined by the vector *

$$\mathbf{r} = \overrightarrow{OP},$$

at time t_1 this particle will have moved to the point Q, defined by the vector

$$\mathbf{r}_1 = \overrightarrow{OQ}.$$

The velocity of the particle at P is then defined by the vector †

$$\mathbf{q} = \lim_{t_1 \to t} \frac{\mathbf{r}_1 - \mathbf{r}}{t_1 - t} = \frac{d\mathbf{r}}{dt}.$$

Thus the velocity $\mathbf{q}$ is a function of $\mathbf{r}$ and t, say

$$\mathbf{q} = f(\mathbf{r}, t).$$

* The subject of vectors is explained at length in Chapter II.

† The symbol $\lim_{t_1 \to t}$ is to be read as " the limit when t_1 tends to the value t ". This is the usual method of defining differential coefficients, whose existence we shall infer on physical grounds. The symbol $\to$ alone is read " tends to ".

If the form of the function f is known, we know the motion of the fluid. At each point we can draw a short line to represent the vector $\mathbf{q}$, fig. 1·1 (*b*).

To obtain a physical conception of the velocity field defined by the vector $\mathbf{q}$, let us imagine the fluid to be filled with a large (but not infinitely large) number of luminous points moving with the fluid.

A photograph of the fluid taken with a short time exposure would reveal the tracks of the luminous points as short lines, each proportional to the distance moved by the point in the given time of the exposure and therefore proportional to its velocity. This is in fact the principle of one method of obtaining pictorial records of the motion of an actual fluid.* In an actual fluid the photograph may reveal a certain regularity of the velocity field in which the short tracks appear to form parts of a regular system of curves. The motion is then described as *streamline* motion. On the other hand, the tracks may be wildly irregular, crossing and recrossing, and the motion is then described as *turbulent*. The motions of our ideal inviscid fluid will always be supposed to be of the former character. An exact mathematical treatment of turbulent motion has not yet been achieved.

Fig. 1·1 (*b*).

1·11. Streamlines and paths of the particles. A line drawn in the fluid so that its tangent at each point is in the direction of the fluid velocity at that point is called a *streamline*.

When the fluid velocity at a given point depends not only on the position of the point but also on the time, the streamlines will alter from instant to instant. Thus photographs taken at different instants will reveal a different system of streamlines. The aggregate of all the streamlines at a given instant constitutes the *flow pattern* at that instant.

When the velocity at each point is independent of the time, the flow pattern will be the same at each instant and the motion is described as *steady*. In this connection it is useful to describe the type of motion which is *relatively steady*. Such a motion arises when the motion can be regarded as steady by imagining superposed on the whole system, including the observer, a constant velocity. Thus when a ship steams on a straight course with constant speed on an otherwise undisturbed sea, to an observer in the ship the flow pattern which accompanies him appears to be steady and could in fact be made so by superposing the reversed velocity of the ship on the whole system consisting of the ship and sea.

If we fix our attention on a particular particle of the fluid, the curve which this particle describes during its motion is called a *path line*. The direction of motion of the particle must necessarily be tangential to the path line, so that

* Plates 1–4 illustrate this.

the path line touches the streamline which passes through the instantaneous position of the particle as it describes its path.

Thus the streamlines show how each particle is moving at a given instant.

The path lines show how a given particle is moving at each instant.

When the motion is steady, the path lines coincide with the streamlines.

1·12. Stream tubes and filaments. If we draw the streamline through each point of a closed curve we obtain a *stream tube.*

A *stream filament* is a stream tube whose cross-section is a curve of infinitesimal dimensions.

When the motion is dependent on the time, the configuration of the stream tubes and filaments changes from instant to instant, but the most interesting applications of these concepts arise in the case of the steady motion of a liquid, which we shall now discuss.

In the steady motion of a liquid, a stream tube behaves like an actual tube through which the liquid is flowing, for there can be no flow into the tube across the walls since the flow is, by definition, always tangential to the walls. Moreover, these walls are fixed in space since the motion is steady, and therefore the motion of the liquid within the walls would be unaltered if we replaced the walls by a rigid substance.

Consider a stream filament of liquid in steady motion. We can suppose the cross-sectional area of the filament so small that the velocity is the same at each point of this area, which can be taken perpendicular to the direction of the velocity.

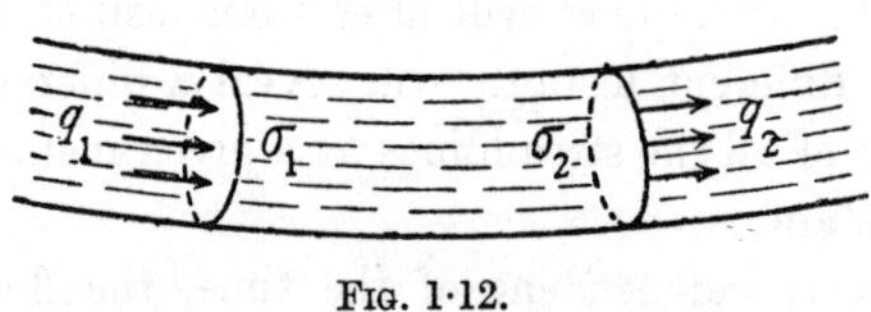

FIG. 1·12.

Now let q_1, q_2 be the speeds of the flow at places where the cross-sectional areas are σ_1 and σ_2. Since the liquid is incompressible, in a given time the same volume must flow out at one end as flows in at the other. Thus

$$q_1 \sigma_1 = q_2 \sigma_2 .$$

This is the simplest case of the equation of conservation of mass, or the *equation of continuity*, which asserts in the general case that the rate of generation of mass within a given volume must be balanced by an equal outflow of mass from the volume. The above result can be expressed in the following theorem.

The product of the speed and cross-sectional area is constant along a stream filament of a liquid in steady motion.

It follows from this that a stream filament is widest at places where the speed is least and is narrowest at places where the speed is greatest.

A further important consequence is that a stream filament cannot terminate at a point within the liquid unless the velocity becomes infinite at that point. Leaving this case out of consideration, it follows that in general stream filaments are either closed or terminate at the boundary of the liquid. The same is true of streamlines, for the cross-section of the filament may be considered as small as we please.

1·13. Fluid body. A surface is said to *move with the fluid* if the velocity of every point of the surface is the same as the velocity of the fluid at that point. It follows that a surface which moves with the fluid will, in general, deform or alter its shape as it moves about.

By continuous fluid *in continuous motion*, we shall mean that the velocity $\mathbf{q}$ is everywhere finite and continuous while its space derivatives of the first order are finite (but not necessarily continuous). It follows that any closed surface S, which moves with the fluid, permanently and completely separates the fluid matter inside S from that outside.

Definition. The fluid matter inside a closed surface which moves with the fluid constitutes a *fluid body* and the closed surface is the *boundary* of the fluid body.

The fluid body may be finite, as a raindrop, a drop of oil, or the whole of the water in a lake; or infinite as when the boundary surface is an infinite cylinder. Such an infinite fluid body cannot exist in nature but it is often advantageous to proceed as if it has real existence.

It is often convenient to isolate in thought the fluid body bounded by some imagined surface and to follow the motion of this body as time progresses and the bounding surface deforms.

1·2. Density. If M is the mass of the fluid within a closed volume V, we can write

$$M = V\rho_1, \tag{1}$$

and ρ_1 is then the average density of the fluid within the volume at that instant. In a hypothetical medium continuously distributed we can define the density ρ as the limit of ρ_1 when $V \to 0$.

In an actual fluid which consists of a large number of individual molecules we cannot let $V \to 0$, for at some stage there might be no molecules within the volume V. We must therefore be content with a definition of density given by (1) on the understanding that the dimensions of V are to be made very small, but not so small that V does not still contain a large number of molecules. In air at ordinary temperatures there are about 3×10^{19} molecules per cm.^{3}. A sphere of radius 0·001 cm. will then contain about 10^{11} molecules, and although small in the hydrodynamical sense will be reasonably large for the purposes of measuring average density.

1·3. Pressure. Consider a small plane of infinitesimal area $d\sigma$, whose centroid is P, drawn in the fluid, and draw the normal PN on one side of the area which we shall call the positive side. The other side will be called the negative side.

We shall make the hypothesis that the mutual action of the fluid particles on the two sides of the plane can, at a given instant, be represented by two equal but opposite forces $p\, d\sigma$ applied at P, each force being a push not a pull, that is to say, the fluid on the positive side pushes the fluid on the negative side with a force $p\, d\sigma$.

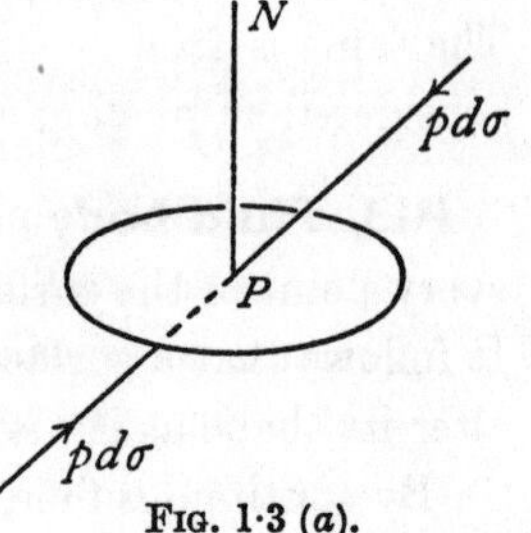

FIG. 1·3 (*a*).

Experiment shows that in a fluid at rest these forces act along the normal. In a real fluid in motion these forces make an angle ϵ with the normal (analogous to the angle of friction). When the viscosity is small, as in the case of air and water, ϵ is small. In an inviscid fluid which can exert no tangential stress $\epsilon = 0$, and in this case p is called the *pressure* at the point P.

In the above discussion there is nothing to show that the pressure p is independent of the orientation of the element $d\sigma$ used in defining p. That this independence does in fact exist is proved in the following theorem.

Theorem. The pressure at a point in an inviscid fluid is independent of direction.

Proof. Let $PQRS$ be representative of a family of homothetic tetrahedra of small dimensions, with a common centroid O, imagined drawn in the fluid.

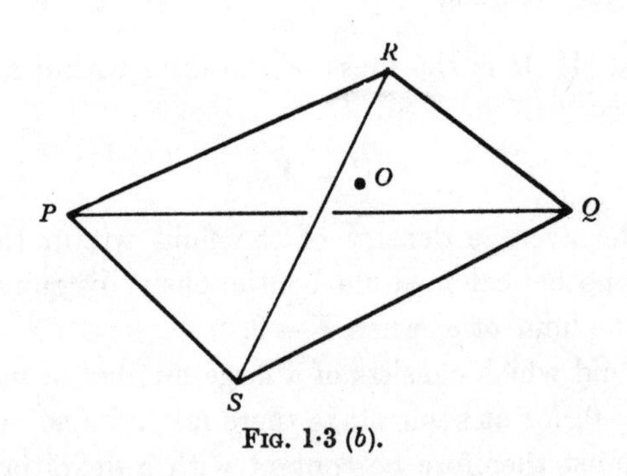

FIG. 1·3 (*b*).

Let p_1, p_2 be the average pressures defined by the faces PRS, QRS of areas σ_1, σ_2 respectively. Then the component in the direction PQ of the pressure thrust on the faces of $PQRS$ is $(p_1 - p_2)\sigma$, where σ is the common value of the projections of σ_1, σ_2 on a plane perpendicular to PQ. The volume of the fluid within $PQRS$ is $l\sigma$, where l is a small length of the same order as PQ. Let F be the component in the direction of PQ of the external force per unit mass of fluid, and let f be the acceleration of the fluid in the direction PQ. Then if ρ is the density, the second law of motion gives

$$(p_1-p_2)\sigma+Fl\rho\sigma=fl\rho\sigma \quad \text{or} \quad p_1-p_2=l\rho(f-F).$$

If we let $l\to 0$, we have, in the limit, $p_1=p_2$, where p_1 and p_2 are now the pressures defined at O by planes parallel to PRS and QRS respectively. Since the orientation of these planes is quite arbitrary, we conclude that the pressure at O is the same for all orientations of the defining element of area. Q.E.D.

Pressure is a scalar quantity, i.e. independent of direction. The dimensions of pressure in terms of measure-ratios (see 1·01) M, L, T of mass length and time are indicated by $ML^{-1}T^{-2}$.

The *thrust* on an area $d\sigma$ due to pressure is a force, that is a vector quantity, whose complete specification requires direction as well as magnitude.

Pressure in a fluid in motion is a function of the position of the point at which it is measured and of the time. When the motion is steady the pressure may vary from point to point, but at a given point it is independent of the time.

It should be noted that p is essentially positive.

1·4. Bernoulli's theorem. In the steady motion of an inviscid fluid the quantity

$$\frac{p}{\rho}+K$$

has the same value at every point of the same streamline where p and ρ are the pressure and density, and K is the energy per unit mass of the fluid.

Proof. Consider a stream filament bounded by cross-sections AB, CD of areas σ_1 and σ_2 and let p_1, q_1, K_1 refer to values at AB, while p_2, q_2, K_2 are the corresponding quantities at CD.

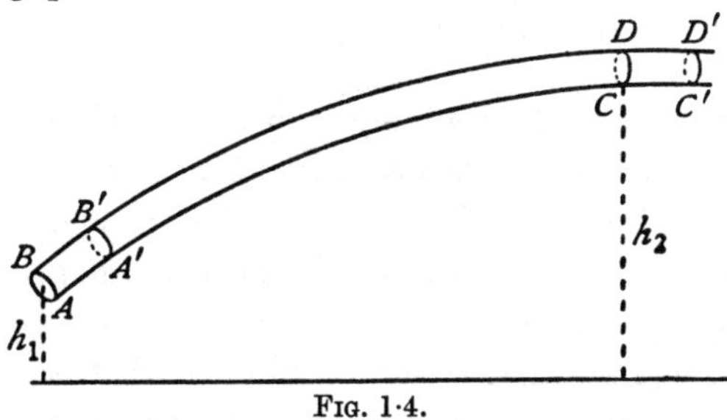

FIG. 1·4.

After a short time δt the fluid body $ABCD$ will have moved to the position $A'B'C'D'$ where

$$AA' = q_1\,\delta t,\quad CC' = q_2\,\delta t.$$

The mass m of the fluid between AB and $A'B'$, or between CD and $C'D'$ is

$$m = \sigma_1\,q_1\,\delta t\,\rho_1 = \sigma_2\,q_2\,\delta t\,\rho_2\,.$$

The work done by the pressure thrusts in moving the fluid body from $ABCD$ to $A'B'C'D'$ is

$$p_1 \sigma_1 q_1 \delta t - p_2 \sigma_2 q_2 \delta t = m\left(\frac{p_1}{\rho_1} - \frac{p_2}{\rho_2}\right)$$

since the thrusts on the walls of the filament, being perpendicular to the direction of motion, do no work.

The gain of energy of the fluid body is

$$mK_2 - mK_1 .$$

Equating the work done to the gain of energy we get

$$(1) \qquad \frac{p_1}{\rho_1} + K_1 = \frac{p_2}{\rho_2} + K_2 ,$$

which shows that $p/\rho + K$ is constant along the streamline to which the stream filament shrinks when its cross-section tends to zero. Q.E.D.

Corollary 1. In the case of a liquid in steady motion under gravity, the density ρ is constant and K is the sum of the kinetic and potential energies per unit mass

$$K = \tfrac{1}{2}q^2 + gh,$$

where h is the height above some fixed horizontal datum and g is the acceleration due to gravity. Bernoulli's theorem then becomes

$$(2) \qquad \frac{p}{\rho} + \tfrac{1}{2}q^2 + gh = \text{constant along a streamline.}$$

Corollary 2. For a liquid at rest under gravity every line is a streamline and Bernoulli's theorem becomes

$$(3) \qquad \frac{p}{\rho} + gh = \text{constant throughout the liquid.}$$

The field of gravitational force is a conservative field, meaning by this that the work done by the weight when a body moves from a point P to another point Q is independent of the path taken from P to Q and depends solely on the vertical height of Q above P. A conservative field of force gives rise to potential energy, which is measured by the work done in taking the body from one standard position to any other position. In order that potential energy of a unit mass at a point may have a definite meaning, it is obviously necessary that the work done by the forces of the field should be independent of the path by which that point was reached. The gravitational field is clearly the most important of conservative fields of force, but it is by no means the only conceivable field of this nature; for example, an electrostatic field has the conservative property. If more generally we denote by Ω the potential energy per unit mass in a conservative field, Bernoulli's theorem would take the more general form that

$$\frac{p}{\rho}+\tfrac{1}{2}q^2+\Omega \tag{4}$$

is constant along a streamline, and the same method of proof could be used.

1·41. Flow in a channel. Suppose water to flow steadily along a channel with a horizontal bottom and rectangular cross-section of breadth b. If h is the height of the free surface above the bottom, since the pressure at the free surface must be equal to that of the atmosphere, we shall have from Bernoulli's theorem u^2+2gh = constant, where u is the velocity supposed parallel to the walls and constant across the section. If the breadth of the channel varies slightly, there will be a small consequent change in u, and therefore by differentiation of the above

$$u\,du+g\,dh = 0.$$

Again, from the equation of continuity, ubh = constant, and therefore

$$\frac{du}{u}+\frac{db}{b}+\frac{dh}{h} = 0.$$

Elimination of du gives

$$\frac{dh}{db} = \frac{u^2h}{b(gh-u^2)}.$$

Thus the depth and breadth increase together if, and only if, $u^2<gh$, i.e. if u is less than the speed of propagation of long waves in the channel (cf. 15·62).

1·43. The constant in Bernoulli's theorem for a liquid. If we fix our attention on a particular streamline, 1, Bernoulli's theorem states that

$$\frac{p}{\rho}+\tfrac{1}{2}q^2+gh = C_1,$$

where C_1 is constant for that streamline. If we take a second streamline, 2, we get

$$\frac{p}{\rho}+\tfrac{1}{2}q^2+gh = C_2,$$

where C_2 is constant along the second streamline. We have not proved (and in the general case it is untrue) that $C_1 = C_2$. When, however, the motion is irrotational, a term which will be explained later (2·41), it is true that the constant is the same for all streamlines, so that

$$\frac{p}{\rho}+\tfrac{1}{2}q^2+gh = C,$$

where C has the same value at each point of the liquid. It will also be shown later (3·64) that this case arises whenever an inviscid liquid is set in motion by ordinary mechanical means, such as by moving the boundaries suddenly or slowly, by opening an aperture in a closed vessel, or by moving a body through the liquid.

1·44. Hydrodynamic pressure. In the steady motion of a liquid Bernoulli's theorem enables us to elucidate the nature of pressure still further. In a liquid at rest there exists at each point a hydrostatic pressure p_H, and the principle of Archimedes states that a body immersed in the fluid is buoyed up by a force equal to the weight of the liquid which it displaces. The particles of the liquid are themselves subject to this principle and are therefore in equilibrium under the hydrostatic pressure p_H and the force of gravity. It follows at once that $p_H/\rho+gh$ is constant throughout the liquid. When the liquid is in motion the buoyancy principle still operates, so that if we write

$$p = p_D+p_H,$$

Bernoulli's theorem gives

$$\frac{p_D}{\rho}+\tfrac{1}{2}q^2+\frac{p_H}{\rho}+gh = C,$$

and therefore

$$(1) \qquad \frac{p_D}{\rho}+\tfrac{1}{2}q^2 = C',$$

where $C' = C-(p_H/\rho+gh)$ is a new constant.

Now (1) is the form which Bernoulli's theorem would assume if the force of gravity were non-existent.

The quantity p_D may be called the *hydrodynamic pressure*, or the pressure due to motion. This pressure p_D measures the force with which two fluid particles are pressed together (for both are subject to the same force of buoyancy). It will be seen that the knowledge of the hydrodynamic pressure will enable us to calculate the *total* effect of the fluid pressure on an immersed body, for we have merely to work out the effect due to p_D and then add the effect due to p_H, which is known from the principles of hydrostatics. This is a very important result, for it enables us to neglect the external force of gravity in investigating many problems, due allowance being made for this force afterwards.

It is often felt that hydrodynamical problems in which external forces are neglected or ignored are of an artificial and unpractical nature. This is by no means the case. The omission of external forces is merely a device for avoiding unnecessary complications in our analysis.

It should therefore be borne in mind that when we neglect external forces we calculate in effect the hydrodynamic pressure.

We also see from (1) that the hydrodynamic pressure is greatest where the speed is least, and also that the greatest hydrodynamic pressure occurs at points of zero velocity.

It should be observed, however, that the device of introducing hydrodynamic pressure can be justified only when the boundaries of the fluid are fixed, for only in these conditions is the hydrostatic pressure constant at a given point. When the liquid has free surfaces which undulate, the hydrostatic pressure at a fixed point will vary, and we must consider the total pressure.

In the case of compressible fluids the pressure due to motion is usually called *aerodynamic pressure.*

1·5. The Pitot tube. Fig. 1·5 (*a*) shows a tube $ABCD$ open at A, where it is drawn to a fine point, and closed at D, containing mercury in the **U**-shaped part.

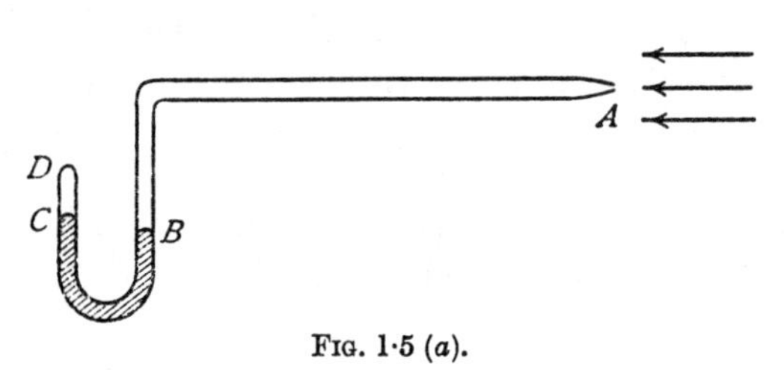

FIG. 1·5 (*a*).

If this apparatus is placed with the open end upstream in a steadily flowing liquid, the axis of the horizontal part in the figure will form part of the streamline which impinges at A. Hence if p_1 is the pressure just inside the tube at A, and p is the pressure ahead of A, we shall have, by Bernoulli's theorem,

$$\frac{p_1}{\rho} = \frac{p}{\rho} + \tfrac{1}{2}q^2,$$

since the fluid inside the tube is at rest. The pressure p_1 is measured by the difference in levels of the mercury at B and C, assuming a vacuum in the part CD. This is the simplest form of Pitot tube for determining the quantity $p + \frac{1}{2}\rho q^2$.

In applications it is often required to measure the speed q. In order to do this we must have a means of measuring p.

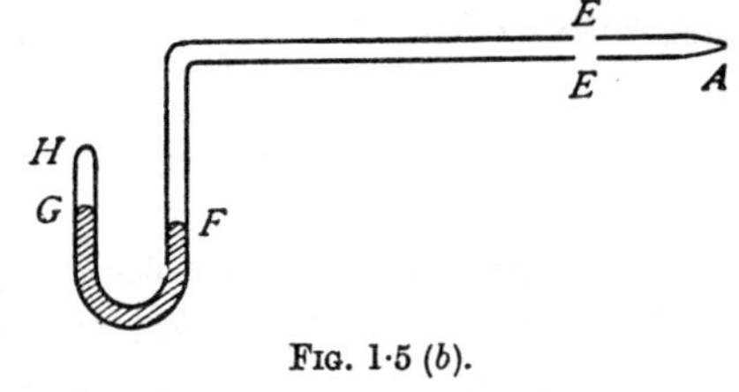

FIG. 1·5 (*b*).

This measurement can be made by means of the apparatus shown in fig. 1·5 (*b*), which differs from the former only in having the end A closed and holes in the walls of the tube at E slightly downstream of A. The streamlines now follow the walls of the tube from A, and the fluid within the tube being at rest and the pressure being necessarily continuous, the pressure just outside the tube at E is equal to

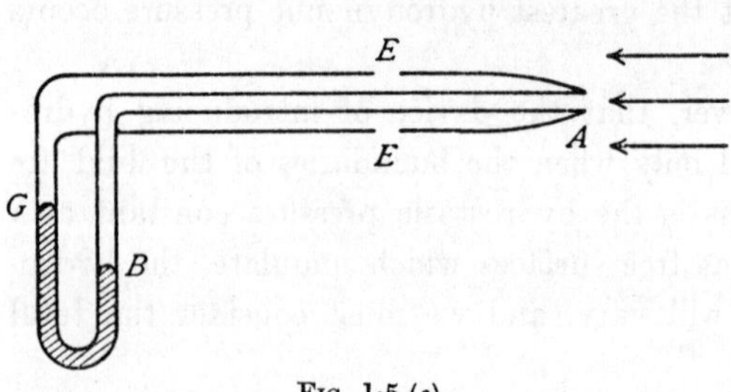

FIG. 1·5 (c).

the pressure just inside the tube at E, and this is measured by the difference in the levels of the mercury at G and F. In practice it is usual to combine both tubes into a single apparatus as shown in fig. 1·5 (c).

In this apparatus the difference in levels of the mercury at B and G measures $p_1 - p = \frac{1}{2}\rho q^2$.

The above description merely illustrates the principle of speed measurements with the Pitot tube. The actual apparatus has to be very carefully designed, to interfere as little as possible with the fluid motion. With proper design and precautions in use, the Pitot tube can give measurements within one per cent. of the correct values in an actual fluid, such as air or water.

1·6. The work done by a gas in expanding. Let S and S' be the surfaces of a unit mass of gas before and after a small expansion.

Let the normal displacement of the element dS of the surface S be dn.

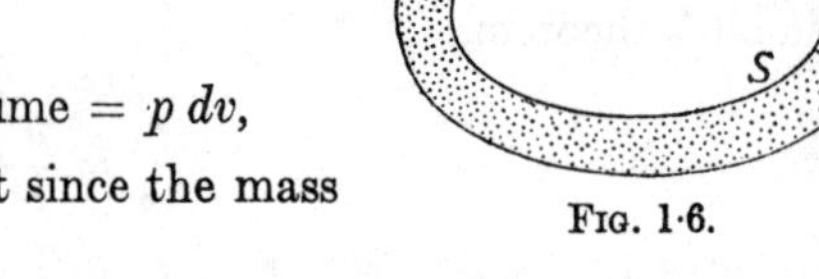

FIG. 1·6.

Suppose the pressure of the gas to be p. Then the work done by the gas is

$$p \,\Sigma\, dS \,.\, dn = p \times \text{increase in volume} = p\, dv,$$

where v is the volume within S. But since the mass is unity, $v\rho = 1$.

Hence the work done by the gas

$$= pd\left(\frac{1}{\rho}\right),$$

and if the expansion is from density ρ to density ρ_0,

$$\text{the work done} = \int_{\rho}^{\rho_0} pd\left(\frac{1}{\rho}\right).$$

We suppose that the pressure is a function of the density only.

We shall call *internal energy* per unit mass the work which a unit mass of the gas could do as it expands under the assumed relation between p and ρ from its actual state to some standard state in which the pressure and density are p_0 and ρ_0. Calling E the internal energy per unit mass, we get

$$E = \int_{\rho}^{\rho_0} pd\left(\frac{1}{\rho}\right) = \frac{p_0}{\rho_0} - \frac{p}{\rho} - \int_{p}^{p_0} \frac{dp}{\rho}$$

on integrating by parts. Thus

$$E = \frac{p_0}{\rho_0} - \frac{p}{\rho} + \int_{p_0}^{p} \frac{dp}{\rho}. \tag{1}$$

Note that internal energy is a form of strain energy analogous to that of a stretched elastic string.

1·61. Bernoulli's theorem for barotropic flow. The flow will be called *barotropic* when the pressure is a function of the density. This amounts to assuming that an equation of state $f(p, \rho, S) = 0$ exists wherein the entropy S has everywhere the same value, the homentropic case of 20·01.

Assuming steady barotropic flow and a conservative field of force for which the potential energy is Ω, the energy K per unit mass is

$$K = \tfrac{1}{2}q^2 + \Omega + E,$$

where E is the internal energy per unit mass. Thus Bernoulli's theorem 1·4 (1) becomes, using 1·6 (1),

$$\frac{p_1}{\rho_1} + \tfrac{1}{2}q_1^2 + \Omega_1 + \frac{p_0}{\rho_0} - \frac{p_1}{\rho_1} + \int_{p_0}^{p_1} \frac{dp}{\rho} = \frac{p_2}{\rho_2} + \tfrac{1}{2}q_2^2 + \Omega_2 + \frac{p_0}{\rho_0} - \frac{p_2}{\rho_2} + \int_{p_0}^{p_2} \frac{dp}{\rho},$$

or

$$\int_{p_0}^{p} \frac{dp}{\rho} + \tfrac{1}{2}q^2 + \Omega = \text{constant along a streamline.} \tag{1}$$

This agrees with 1·4 (4) when the fluid is incompressible so that ρ is constant. For the gravitational field $\Omega = gh$ and

$$\int_{p_0}^{p} \frac{dp}{\rho} + \tfrac{1}{2}q^2 + gh = \text{constant along a streamline.} \tag{2}$$

If we consider aerodynamic pressure (1·44), Bernoulli's theorem assumes the form

$$\int \frac{dp}{\rho} + \tfrac{1}{2}q^2 = \text{constant along a streamline,} \tag{3}$$

whence we get

$$dp = -\rho\, q\, dq. \tag{4}$$

1·62. Application of Bernoulli's theorem to adiabatic expansion. When a gas expands adiabatically (that is to say without gain or loss of heat), the pressure and the density are connected by the relation

$$p = \kappa \rho^{\gamma}, \tag{1}$$

where κ and γ are constants. For dry air, $\gamma = 1·405$. Therefore

$$\int_{p_0}^{p} \frac{dp}{\rho} = \kappa\gamma \int_{\rho_0}^{\rho} \rho^{\gamma-2}\, d\rho = \frac{\kappa\gamma}{\gamma-1}[\rho^{\gamma-1} - \rho_0^{\gamma-1}] = \frac{\gamma}{\gamma-1}\left(\frac{p}{\rho} - \frac{p_0}{\rho_0}\right).$$

Since p_0/ρ_0 refers to a standard state, this is constant, and therefore Bernoulli's theorem gives

$$\frac{\gamma}{\gamma-1}\frac{p}{\rho}+\tfrac{1}{2}q^2+gh = C.$$

If we take p_0 to be the pressure when the velocity is zero * and neglect the effect of gravity, we obtain

$$(2) \qquad \frac{\gamma}{\gamma-1}\frac{p}{\rho}+\tfrac{1}{2}q^2 = \frac{\gamma}{\gamma-1}\frac{p_0}{\rho_0},$$

so that

$$(3) \qquad q^2 = \frac{2\gamma}{\gamma-1}\frac{p_0}{\rho_0}\left(1-\frac{p\,\rho_0}{p_0\,\rho}\right).$$

Now $$\frac{p\,\rho_0}{p_0\,\rho} = \frac{\rho^{\gamma-1}}{\rho_0^{\gamma-1}} = \left(\frac{p}{p_0}\right)^{\frac{\gamma-1}{\gamma}} \quad \text{from (1).}$$

Also, from the theory of sound waves, it is known (15·86) that the speed of sound c_0 when the pressure is p_0 is given by

$$c_0^2 = \frac{\gamma\,p_0}{\rho_0}.$$

Therefore we obtain from (2)

$$\left(\frac{p}{p_0}\right)^{\frac{\gamma-1}{\gamma}} = 1-\frac{\gamma-1}{2}\left(\frac{q}{c_0}\right)^2,$$

and therefore

$$\frac{p}{p_0} = \left[1-\frac{\gamma-1}{2}\left(\frac{q}{c_0}\right)^2\right]^{\frac{\gamma}{\gamma-1}} = 1-\frac{\gamma}{2}\left(\frac{q}{c_0}\right)^2+\frac{\gamma}{8}\left(\frac{q}{c_0}\right)^4+\ldots$$

$$= 1-\frac{1}{2}\frac{\rho_0\,q^2}{p_0}+\frac{\gamma}{8}\left(\frac{q}{c_0}\right)^4+\ldots.$$

The ratio of the third term to the second in this expansion is $q^2/4c_0^2$, so that even when the speed q is equal to half the speed of sound this ratio is 1/16. Thus it appears that we may, to a good approximation, neglect the third term, unless q is a considerable fraction of c_0.

Bernoulli's theorem for air will then take the form

$$\frac{p}{\rho_0}+\tfrac{1}{2}q^2 = \frac{p_0}{\rho_0},$$

which means that the air may be treated as incompressible within a very considerable range of speeds. In particular, for air speeds of 300 miles per hour, the error in speed measurements made by the use of the Pitot tube (see 1·5) will be only about 2 per cent.

---

* It is not asserted that zero velocity is attained. The pressure p_0 is nevertheless uniquely defined by the equation which follows.

Again, the speed of flow in the neighbourhood of the wings of an aeroplane will be comparable with the forward speed, and therefore the effect of compressibility is small for small forward speeds. On the other hand, the compressibility cannot be neglected in the neighbourhood of the tips of the propeller blades.

1·63. Subsonic and supersonic flow. If c is the speed of sound when the pressure is p, we have (15·87) $c^2 = \gamma p/\rho$, and therefore 1·62 (2) gives

$$\frac{c^2}{\gamma-1}+\tfrac{1}{2}q^2 = \frac{c_0{}^2}{\gamma-1}, \tag{1}$$

which shows that c has the maximum value c_0 when $q = 0$, and that q has the maximum value $q_{\max}$ when $c = 0$, given by

$$q^2_{\max} = \frac{2c_0{}^2}{\gamma-1}. \tag{2}$$

The *critical speed* $q^\star$ occurs when sound speed and fluid speed are equal and therefore from (1),

$$q^\star = c^\star = c_0\sqrt{\frac{2}{\gamma+1}}. \tag{3}$$

The following forms of Bernoulli's equation (1) should be noted :

$$c^2 = \tfrac{1}{2}(\gamma-1)(q^2_{\max}-q^2), \tag{4}$$

$$\frac{q^2-c^2}{q^2_{\max}}+\frac{c^2}{c^{\star 2}} = 1. \tag{5}$$

The graph of q^2 as a function of c^2 is the straight line AB in fig. 1·63. This shows that along a streamline $c \leqslant c_0$, $q \leqslant q_{\max}$. The straight line $q^2 - c^2 = 0$ cuts AB at the point $C\,(c^{\star 2}, q^{\star 2})$, where $q^\star = c^\star$. The two portions AC, BC of this line correspond with two physically different *régimes*.

If we introduce the *Mach number*

$$M = q/c, \tag{6}$$

at any point of AC we have $q < q^\star = c^\star < c$, so that $M < 1$, provided that $q < c$. Flow for which $M < 1$ is called *subsonic*.

At any point of BC we have $q > q^\star = c^\star > c$, so that $M > 1$, and the flow is then said to be *supersonic*.

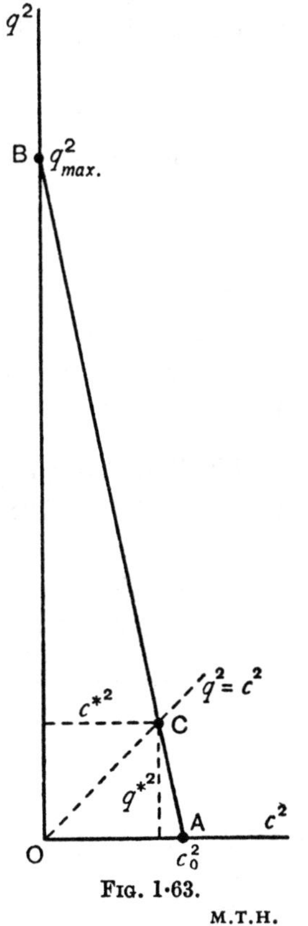

FIG. 1·63.

M.T.H.

We get from (1)

$$(7) \qquad 1+\tfrac{1}{2}(\gamma-1)M^2 = \frac{c_0^{\,2}}{c^2} = \left(\frac{\rho_0}{\rho}\right)^{\gamma-1} = \left(\frac{p_0}{p}\right)^{\frac{\gamma-1}{\gamma}}.$$

1·64. Flow of gas in a converging pipe. If ω is the area of the section, which is taken to be small, the pipe will converge if ω decreases as we go along the pipe, i.e. if $d\omega/ds<0$, where ds is an element of length of the pipe. The equation of continuity is $\omega\,\rho q$ = constant, which gives

$$(1) \qquad \frac{1}{\rho}\frac{d\rho}{ds}+\frac{1}{q}\frac{dq}{ds} = -\frac{1}{\omega}\frac{d\omega}{ds}.$$

Taking the adiabatic law, Bernoulli's theorem gives

$$\tfrac{1}{2}q^2+\frac{\gamma}{\gamma-1}\frac{p}{\rho} = \text{constant},$$

and therefore

$$q\frac{dq}{ds}+\gamma\kappa\,\rho^{\gamma-2}\frac{d\rho}{ds} = 0.$$

Let $c^2 = \gamma p/\rho$ denote the local speed of sound, i.e. the speed at the point we are considering. Then

$$\frac{1}{\rho}\frac{d\rho}{ds} = -\frac{q}{c^2}\frac{dq}{ds}.$$

Substitution in (1) then gives

$$q\frac{dq}{ds} = \frac{c^2q^2}{c^2-q^2}\left(-\frac{1}{\omega}\frac{d\omega}{ds}\right) = \frac{M^2c^2}{1-M^2}\left(-\frac{1}{\omega}\frac{d\omega}{ds}\right),$$

and so dq/ds is positive if $M<1$, i.e. if $q<c$.

Thus the speed increases as we go along the pipe in the direction in which it converges if the flow is subsonic ; for supersonic flow the speed decreases as the pipe gets narrower.

1·7. The Venturi tube. The principle of the Venturi tube is illustrated in fig. 1·7. The apparatus is used for measuring the flow in a pipe and

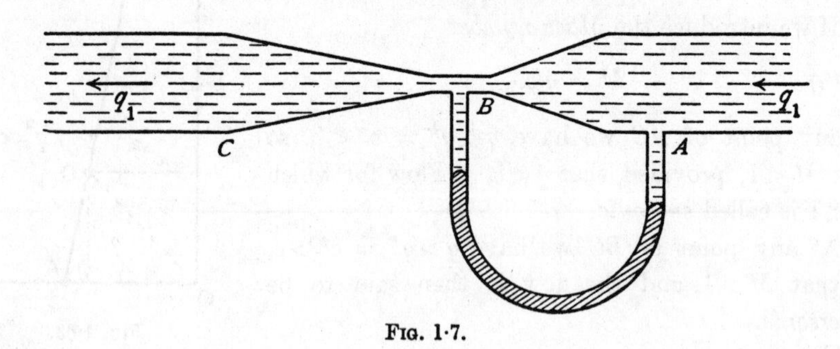

FIG. 1·7.

consists essentially of a conical contraction in the pipe from the full bore at A to a constriction at B, and a gradual widening of the pipe to full bore again at C. To preserve the streamline flow, the opening from B to C has to be very gradual. A **U**-tube manometer containing mercury joins openings at A and B, and the difference in level of the mercury measures the difference in pressures at A and B. Let p_1, q_1, p_2, q_2 be the pressures and speeds at A and B respectively. Then, for a liquid,

$$\frac{p_1}{\rho}+\tfrac{1}{2}q_1^2 = \frac{p_2}{\rho}+\tfrac{1}{2}q_2^2,$$

by Bernoulli's theorem.

Let S_1, S_2 be the areas of the cross-sections at A and B.

Then $$q_1 S_1 = q_2 S_2,$$

since the same volume of fluid crosses each section in a given time. Therefore

$$q_1 = \sqrt{\frac{2(p_1-p_2)}{\rho\left(\dfrac{S_1^2}{S_2^2}-1\right)}},$$

p_1-p_2 is given by observation and the value of q_1 follows.

If h is the difference in level of the mercury in the two limbs of the manometer and σ is the density of mercury, the formula becomes

$$q_1 = \sqrt{\frac{2gh\sigma}{\rho\left(\dfrac{S_1^2}{S_2^2}-1\right)}} = K\sqrt{h},$$

K being a constant for the apparatus.

1·71. Flow of a gas measured by the Venturi tube. Assuming adiabatic changes in the gas from the entrance to the throat, we obtain from Bernoulli's theorem and the equation of continuity

$$\frac{\gamma}{\gamma-1}\frac{p_1}{\rho_1}+\tfrac{1}{2}q_1^2 = \frac{\gamma}{\gamma-1}\frac{p_2}{\rho_2}+\tfrac{1}{2}q_2^2,$$

$$\rho_1 q_1 S_1 = \rho_2 q_2 S_2,$$

whence we easily obtain

$$q_1^2 = \frac{\dfrac{2\gamma}{\gamma-1}\left(\dfrac{p_1}{\rho_1}-\dfrac{p_2}{\rho_2}\right)}{\left(\dfrac{\rho_1}{\rho_2}\right)^2\dfrac{S_1^2}{S_2^2}-1}.$$

Now, $\dfrac{p_1}{p_2} = \left(\dfrac{\rho_1}{\rho_2}\right)^\gamma$, and therefore

$$q_1^2 = \frac{\dfrac{2\gamma}{\gamma-1}\dfrac{p_1}{\rho_1}\left[1-\left(\dfrac{p_2}{p_1}\right)^{\frac{\gamma-1}{\gamma}}\right]}{\left(\dfrac{p_1}{p_2}\right)^{\frac{2}{\gamma}}\left(\dfrac{S_1}{S_2}\right)^2-1}.$$

To use this formula we must know p_1, p_2 and ρ_1. The instrument must therefore be modified so that A and B in fig. 1·7 are connected to separate manometers, thereby obtaining measures of the actual pressures p_1, p_2 and not their difference, as in the case of a liquid. For speeds not comparable with that of sound, the ordinary formula and method for a liquid may be used (see 1·7).

1·8. Flow through an aperture. When a small hole is made in a wall of a large vessel which is kept full, it is found that the issuing jet of liquid

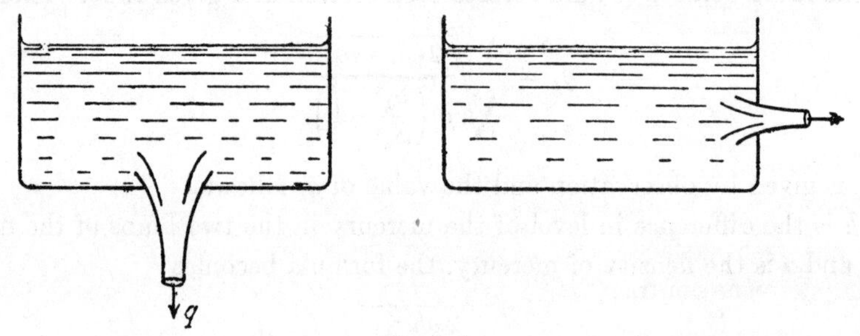

FIG. 1·8.

contracts at a short distance from the aperture to a minimum cross-section. At the contraction, called the *vena contracta*, the issuing jet is cylindrical in form and all the streamlines are parallel. If σ_1 is the area of the aperture and σ_2 the area of the cross-section of the jet, the ratio $\sigma_2 : \sigma_1$ is called the *coefficient of contraction*. The exact value α of the coefficient of contraction can only be rigorously evaluated in certain special cases, but plausible arguments can be adduced to show that $\alpha > \frac{1}{2}$. That $\alpha < 1$ follows experimentally from the existence of the contraction.

1·81. Torricelli's theorem. In fig. 1·8, let h be the depth of the vena contracta below the level of the upper surface of the water in a tank which is kept full, and let Π be the atmospheric pressure. If q is the speed of efflux at the vena contracta, Bernoulli's theorem gives

$$\frac{\Pi}{\rho}+gh = \frac{\Pi}{\rho}+\tfrac{1}{2}q^2,$$

since the velocity is practically zero at the free surface of the water in the tank, and the pressure is Π, both there and on the walls of the escaping jet.

Therefore $$q^2 = 2gh.$$

This is Torricelli's theorem, for the speed of efflux.

If σ_2 is the area of the cross-section of the jet at the vena contracta, the rate of efflux is

$$\sigma_2\sqrt{2gh}.$$

It is in most cases sufficient to take h as the depth of the orifice, for the vena contracta is at only a short distance from this. If σ_1 is the area of the orifice and α the coefficient of contraction, the rate of efflux is

$$\alpha\sigma_1\sqrt{2gh}.$$

1·82. The coefficient of contraction. Let there be a small hole AB in the wall of a vessel, which is kept full, and let h be the depth of the hole

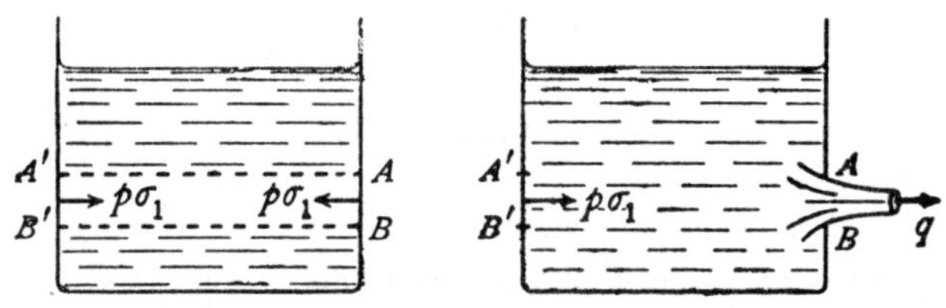

Fig. 1·82 (a).

below the free surface. Let Π be the atmospheric pressure, q the speed of efflux at the vena contracta. Let $A'B'$ be the projection of the area of the hole on the opposite wall, both walls being supposed vertical.

If p is the hydrostatic pressure at AB when the hole is closed, the action of AB and $A'B'$ on the fluid will consist of two equal but opposite forces $p\sigma_1$. When the hole is opened, the force $p\sigma_1$ at AB disappears and is replaced by a force $\Pi\sigma_1$. If we suppose, as a first approximation, that the hydrostatic pressure remains unaltered, except at the hole AB, the force accelerating the fluid is $(p-\Pi)\sigma_1$. The rate of outflow of momentum is $\rho\, q\, \sigma_2\, q$, where σ_2 is the area of the vena contracta. Thus *

$$(p-\Pi)\sigma_1 = \sigma_2\, \rho\, q^2.$$

By Bernoulli's theorem,

$$\frac{p}{\rho} = \frac{\Pi}{\rho} + \tfrac{1}{2}q^2.$$

Therefore $\sigma_2 = \frac{1}{2}\sigma_1$, and the coefficient of contraction is $\frac{1}{2}$.

Bernoulli's theorem also shows that when the hole is opened the pressure on the walls in the neighbourhood of the hole AB will fall below the hydrostatic pressure, so that the accelerating force is actually greater than $p-\Pi$, and therefore, in general, $\sigma_2/\sigma_1 > \frac{1}{2}$. (See 3·32.)

* From 3·40 it appears that when the motion is steady, the flux measures the rate of change of momentum.

If, however, we fit a small cylindrical nozzle projecting *inwards*, the original assumption is nearly exact and the coefficient of contraction is $\frac{1}{2}$. This arrangement is known as Borda's mouthpiece, fig. 1·82 (*b*).

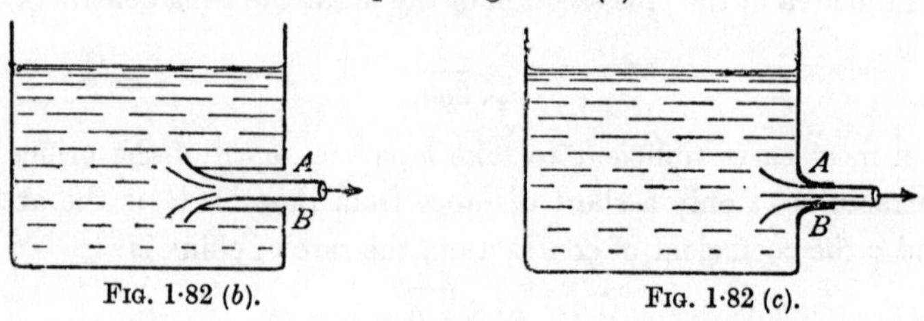

FIG. 1·82 (*b*). FIG. 1·82 (*c*).

On the other hand, a rounded nozzle projecting *outwards*, fig. 1·82 (*c*), will increase the flow, for the vena contracta will occur at the outlet and we shall get

$$p\sigma_1 - \Pi\sigma_2 = \sigma_2\,\rho\,q^2$$

and therefore

$$\frac{\sigma_2}{\sigma_1} = \frac{1}{2}\,\frac{p - \sigma_2\,\Pi/\sigma_1}{p - \Pi},$$

which is greater than the former value.

Torricelli's theorem shows that the rate of efflux increases with increasing coefficient of contraction so that this device increases the efflux. This fact was used by the Romans in the era of the Emperors, when the people were allowed as much water as they could draw in a given time from a supply flowing through an orifice.

1·9. Euler's momentum theorem. Consider a current filament bounded by cross-sections of areas σ_1, σ_2 at AB, CD respectively, in the steady motion of a liquid. If q_1, q_2 are the speeds at AB, CD, Euler's theorem states that, neglecting external forces, the resultant force due to pressure of the surrounding liquid on the walls and ends of the filament is equivalent to forces $\rho\sigma_1 q_1^2$ and $\rho\sigma_2 q_2^2$ normally outwards at the ends AB, CD respectively.

Proof. By Newton's second law of motion, the resultant force must produce the rate of change of the momentum of the fluid which occupies the portion of the filament between AB and CD in fig. 1·4 at a given instant t.

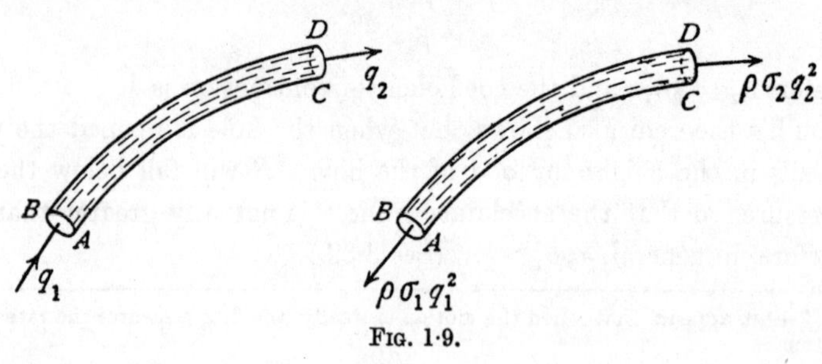

FIG. 1·9.

Now at time $t+\delta t$ the liquid in question will occupy the portion of the filament between $A'B'$, $C'D'$. Thus the momentum of the liquid in question has increased by the momentum of the fluid in between CD and $C'D'$ and has diminished by the momentum of the fluid between AB, $A'B'$.

Hence there has been a gain of momentum of amount $\rho\sigma_2 q_2 \delta t \times q_2$ at CD and a loss of amount $\rho\sigma_1 q_1 \delta t \times q_1$ at AB. Hence the rate of change is a gain of amount $\rho\sigma_2 q_2^2$ at CD and a loss of amount $\rho\sigma_1 q_1^2$ at AB. These rates of change are produced solely by the thrusts acting on the walls and ends of the filament. Hence these thrusts must be equivalent to the forces $\rho\sigma_1 q_1^2$ and $\rho\sigma_2 q_2^2$ normally *outwards* at AB, CD respectively. Q.E.D.

1·91. The force on the walls of a fine tube. Consider liquid flowing steadily through the portion AB of a tube whose cross-sectional area is so small that the liquid may be considered as part of a stream filament.

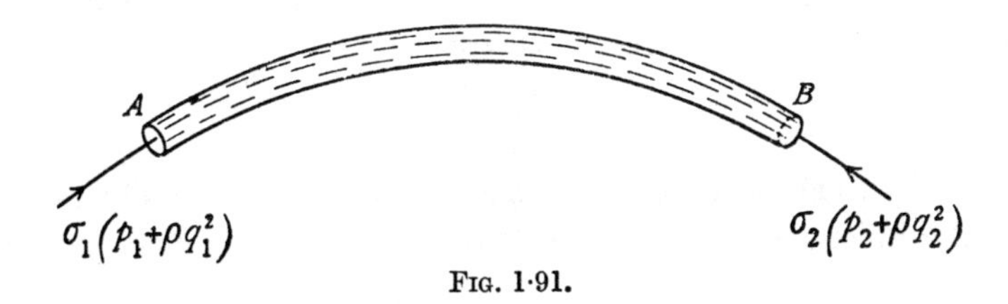

FIG. 1·91.

Let σ_1, p_1, q_1 denote the cross-sectional area, the pressure, and the speed at A, σ_2, p_2, q_2 the corresponding quantities at B. By Euler's momentum theorem, the total action of the pressures on the liquid in AB consists of normal forces $\rho\sigma_1 q_1^2$ at A and $\rho\sigma_2 q_2^2$ at B, both outwards. But the forces due to the pressures at A and B are $p_1\sigma_1$ and $p_2\sigma_2$, both normally inwards.

Hence the forces exerted by the walls on the liquid together with the normal inward forces $p_1\sigma_1$, $p_2\sigma_2$ are equivalent to the normal outward forces $\rho\sigma_1 q_1^2$, $\rho\sigma_2 q_2^2$.

Hence the forces exerted by the walls on the liquid are equivalent to normal outward forces $\sigma_1(p_1+\rho q_1^2)$ at A and $\sigma_2(p_2+\rho q_2^2)$ at B. By the principle of action and reaction, the forces exerted by the liquid on the tube are obtained by reversing these latter and are therefore equivalent to normal inward forces of the above amounts.

1·92. d'Alembert's paradox. Consider a long straight tube in which an inviscid liquid is flowing with constant speed U. If we place an obstacle A in the middle of the tube the flow in the immediate neighbourhood of A will be deranged, but at a great distance either upstream or downstream the flow will be undisturbed. To hold the obstacle at rest will in general require a force and a couple. Calling F the component of the force in the direction parallel to the current, we shall prove that $F = 0$. This is d'Alembert's paradox.

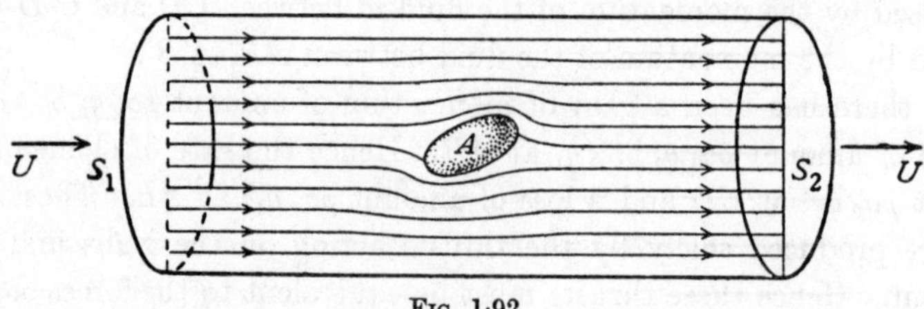

FIG. 1·92.

We shall neglect external forces such as gravity. Then F is the resultant in the direction of the flow of the pressure thrusts acting on the boundary of A.

Consider the two cross-sections S_1, S_2 at a great distance from A. The fluid between these sections can be split up into current filaments, to each of which Euler's momentum theorem is applicable. The outer filaments are bounded by the walls of the tube and on these the thrust components are perpendicular to the current. The walls of the filaments in contact with A are acted on by the solid by a force whose component in the direction of flow is $-F$. By Euler's theorem, the resultant of all the thrusts on the fluid considered is

$$-\rho\, S_1\, U^2 + \rho\, S_2\, U^2,$$

which vanishes since $S_1 = S_2$.

By Bernoulli's theorem, the pressure p_1 over S_1 is the same as the pressure p_2 over S_2. Thus

$$p_1\, S_1 - F - p_2\, S_2 = 0,$$

and therefore

$$F = 0.$$

If we suppose the walls of the tube to recede, we have the case of a body immersed in a current unbounded in every direction, and the above proof still shows that $F = 0$.

Finally, if we impose on the whole system a uniform velocity U in the direction opposite to that of the current, the liquid at a great distance is reduced to rest and A moves with uniform velocity U. Superposing a uniform velocity does not alter the dynamical conditions. Therefore the resistance to a body moving with uniform velocity through an unbounded inviscid fluid, otherwise at rest, is zero.

1·93. The flow past an obstacle. If we consider a sphere, fig. 1·93 (a), held in a stream which is otherwise uniform (uniform at a great distance from the sphere) and neglect external forces, the streamline flow must be symmetrical with respect to the diameter AC of the sphere which lies in the direction of the stream. The central streamline coming from upstream encounters the sphere at A and the fluid is there brought to rest. The point A is a point where the velocity is zero, usually called a *stagnation point*.

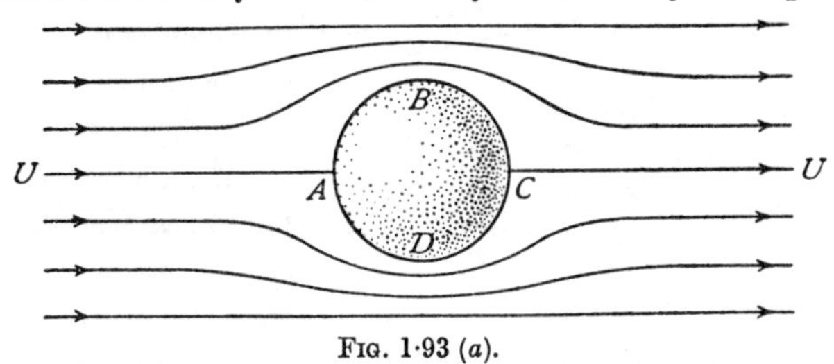

FIG. 1·93 (*a*).

This streamline then divides and passes round ABC, ADC, reuniting at C, which is a second stagnation point, and then proceeds downstream to infinity.* The streamlines adjacent to this are bent in the neighbourhood of the sphere and gradually straighten out. As we proceed further from the sphere the streamlines become less and less curved, so that at great distances laterally from AC their curvature becomes negligible. Photographs taken when the motion is in its *initial* stages confirm this qualitative description. (See Plate 1, fig. 1.)

In a real fluid, such as water, there is of necessity internal friction. Experimental evidence tends to show that the fluid in actual contact with the obstacle must be at rest. To reconcile the photographic evidence with this, the boundary layer hypothesis was introduced by Prandtl, namely, that in the immediate neighbourhood of the sphere there is a thin layer of fluid in which the tangential velocity component increases with great rapidity from zero to the velocity of the main stream as it passes the sphere, while the pressure is continuous as we pass normally outwards. As the velocity of the stream is increased, the boundary layer remains thin at A and on the anterior portion of the sphere but increases in thickness towards the rear, as illustrated in fig. 1·93 (*b*). (See also Plate 1, fig. 3.)

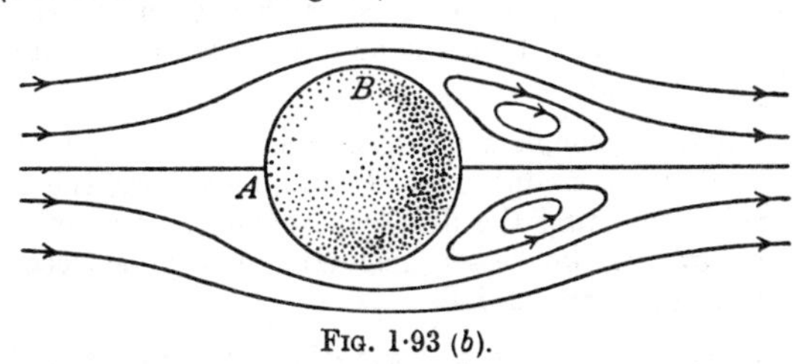

FIG. 1·93 (*b*).

Within this boundary layer there is reversal of the motion, forming eddies, while the theoretical motion subsists outside. The boundary layer thus separates from the sphere at a point in the neighbourhood of B.

* We shall use the term "infinity" as a convenient description of points so distant that the disturbing effect of the obstacle is negligible.

As the velocity of the stream is still further increased, the point of separation of the boundary layer moves further forward and the layer widens out behind into an eddying wake in which energy is continually washed away downstream with the eddies, fig. 1·93 (c).

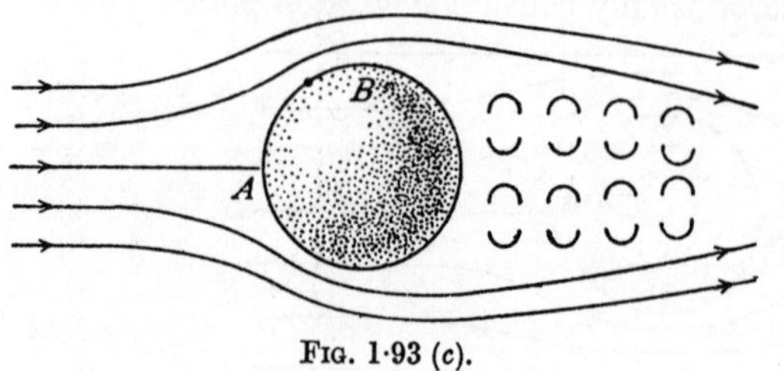

FIG. 1·93 (c).

The picture of the relative motion is the same when the sphere moves forward in otherwise still water with constant velocity and the sphere will undergo a resistance or *drag* to compensate for the loss of energy. To maintain the velocity, energy must be supplied to the sphere, and d'Alembert's paradox is avoided. The general validity of Prandtl's hypothesis is amply confirmed by photographs, and shows that the theoretical study of hydrodynamics can still fulfil a useful function, since the motion outside the wake is still a theoretical streamline motion. In another direction also we can apply the theory to the study of the behaviour of those bodies of " easy " shape in which the breaking away of the boundary layer is confined to a part near the rear with a consequent diminution in the breadth of the wake. Examples of these easy shapes occur in the forms of fish, in properly designed aerofoils, and in strut sections of small drag.

These are also the considerations on which we can repose our trust in the applications of Bernoulli's theorem to measurements made in actual fluids by the Pitot tube, and that for a twofold reason. In the first place, the apertures in a Pitot tube are on the anterior portion, where the boundary layer is thin, and in the second place, the pressure is transmitted with continuity through this thin layer.

EXAMPLES I

1. A water tap of diameter $\frac{1}{4}$ in. is 60 ft. below the level of the reservoir which supplies water to a town. Find the amount of water which can be delivered by the tap in gallons per hour.

2. Water is squirted through a small hole out of a large vessel in which a pressure of 51 atmospheres is maintained by compressed air, the external pressure being 1 atmosphere. Neglecting the difference of level between the hole and the free surface of the water in the vessel, calculate in feet per second the speed at which the water rushes through the hole.

3. Water flows steadily along a horizontal pipe of variable cross-section. If the pressure be 700 mm. of mercury (specific gravity 13·6) at a place where the speed is 150 cm./sec., find the pressure at a place where the cross-section of the pipe is twice as large, taking $g = 981$ cm./sec.^{2}.

4. A stream in a horizontal pipe, after passing a contraction in the pipe at which the sectional area is A, is delivered at atmospheric pressure at a place where the sectional area is B. Show that if a side tube is connected with the pipe at the former place, water will be sucked up through it into the pipe from a reservoir at a depth

$$\frac{S^2}{2g}\left(\frac{1}{A^2} - \frac{1}{B^2}\right)$$

below the pipe; S being the delivery per second.

5. An open rectangular vessel containing water is allowed to slide freely down a smooth plane inclined at an angle α to the horizontal. Find the inclination to the horizontal of the free surface of the water.

If the length and breadth of the vessel be a, b respectively and the mass of contained water be m, find the pressure on the base of the vessel, neglecting atmospheric pressure.

6. Liquid of density ρ is flowing along a horizontal pipe of variable cross-section, and the pipe is connected with a differential pressure gauge at two points A and B. Show that if $p_1 - p_2$ is the pressure indicated by the gauge, the mass m of liquid flowing through the pipe per second is given by

$$m = \sigma_1 \sigma_2 \sqrt{\frac{2\rho(p_1 - p_2)}{\sigma_1^2 - \sigma_2^2}},$$

where σ_1, σ_2 are the cross-sections at A, B respectively. (R.N.C.)

7. A vessel in the form of a hollow circular cone with axis vertical and vertex downwards, the top being open, is filled with water. A circular hole whose diameter is $1/n$th that of the top (n being large) is opened at the vertex. Show that the time taken for the depth of the water to fall to one-half of its original value (h) cannot be less than

$$\frac{(4\sqrt{2}-1)n^2\sqrt{h}}{20\sqrt{g}}.$$

8. If p/ρ^γ = constant, and the fluid flows out through a thin pipe leading out of a large closed vessel in which the pressure is n times the atmospheric pressure p, show that the speed V of efflux is given by

$$V^2 = \frac{2\gamma p}{(\gamma - 1)\rho}\left[n^{1-\frac{1}{\gamma}} - 1\right],$$

ρ being the density at the vena contracta. (R.N.C.)

9. A gas in which the pressure and the density are connected by the adiabatic relation $p = k\rho^\gamma$ flows along a pipe. Prove that

$$q^2 + \frac{2\gamma}{\gamma - 1}\frac{p}{\rho}$$

is constant, if the external forces are neglected, q being the speed. If the pipe converges in the direction of the flow, prove that q will increase and p/ρ will diminish in the direction of flow provided that $q^2\rho < \gamma p$. (R.N.C.)

10. Show that the speed q of gas flowing in a thin tube whose cross-section is σ at a point, of distance s in arc from a fixed cross-section, obeys the equation

$$\frac{d}{ds}(\log \sigma) + \left(1 - \frac{q^2}{c^2}\right)\frac{d}{ds}\log q = 0,$$

where c is the speed of sound in the gas at the point considered, the adiabatic law being followed throughout.

11. If gas flows from a vessel through a small orifice from a region where the pressure is p_1 to a region where the pressure is p_2, prove that the rate of efflux of mass is

$$c_2\,\rho_2\,\omega_2\left(\frac{2}{\gamma-1}\right)^{\frac{1}{2}}\left[\left(\frac{p_2}{p_1}\right)^{\frac{1}{\gamma}-1}-1\right]^{\frac{1}{2}},$$

where $p = k\rho^{\gamma}$, ω_2 is the area of the vena contracta, and $c_2{}^2 = \gamma p_2/\rho_2$ (cf. 1·64), ρ_2 being the density at the vena contracta.

12. If ω is the small cross-section of a tube of flow in a gas, prove that $q\rho\omega =$ constant along the tube and hence use the result of 1·64 to prove that $q\rho$ is a maximum when $q = c$, and that ω is then a minimum.

13. If c_m is the speed of sound at the minimum cross-section in Ex. 12, prove that there is an upper limit to the value of q given by

$$q_{\max} = c_m \times \sqrt{\frac{\gamma+1}{\gamma-1}} = 2{\cdot}45\,c_m.$$

14. Gas flows radially from a point symmetrically in all directions, the pressure and density being connected by the law $p = \kappa\rho$. If m is the rate of emission of mass, supposed constant, prove that

$$4\pi\,qr^2 = m\exp\left[\frac{q^2-q_1{}^2}{2\kappa}\right],$$

where q is the speed at distance r, and q_1 is the speed where $\rho = 1$.

CHAPTER II

VECTORS AND TENSORS

2·1. Scalars and vectors. Pure numbers and physical quantities which do not require direction in space for their complete specification are called *scalar quantities*, or simply *scalars*. Volume, density, mass and energy are familiar examples. Fluid pressure is also a scalar. The thrust on an infinitesimal plane area due to fluid pressure is, however, not a scalar, for to describe this thrust completely, the direction in which it acts must also be known.

A *vector quantity*, or simply a *vector*, is a quantity which needs for its complete specification both magnitude and direction, and which obeys the parallelogram law of composition (addition), and certain laws of multiplication which will be formulated later. Examples of vectors are readily furnished by velocity, linear momentum and force. Angular velocity and angular momentum are also vectors, as is proved in books on Mechanics.

A vector can be represented completely by a straight line drawn in the direction of the vector and of appropriate magnitude to some chosen scale. The sense of the vector in this straight line can be indicated by an arrow.

In some cases a vector must be considered as *localised* in a line. For instance, in calculating the moment of a force, it is clear that the position of the line of action of the force is relevant.

In many cases, however, we shall be concerned with *free vectors*, that is to say, vectors which are completely specified by their direction and magnitude, and which may therefore be drawn in any convenient positions. Thus if we wish to find only the magnitude and direction of the resultant of several given forces, we can use the polygon of forces irrespectively of the actual positions in space of the lines of action of the given forces.

We shall represent a vector by a single letter in clarendon (heavy) type and its magnitude by the corresponding letter in italic type. Thus if $\mathbf{q}$ is the velocity vector, its magnitude is q, the speed. Similarly the angular velocity $\boldsymbol{\omega}$ has the magnitude ω.

A *unit vector* is a vector whose magnitude is unity. Any vector can be represented by a numerical (scalar) multiple of a unit vector parallel to it. Thus if $\mathbf{i}_a$ is a unit vector parallel to the vector $\mathbf{a}$, we have

$$\mathbf{a} = a\mathbf{i}_a.$$

We proceed to develop some properties of vectors with a view to hydrodynamical applications.

In what follows, the magnitude of a vector will be supposed different from zero, unless the contrary is stated.

2·11. The scalar product of two vectors. Let $\mathbf{a}$, $\mathbf{b}$ be two vectors, of magnitudes a, b, represented by the lines OA, OB issuing from the point O.

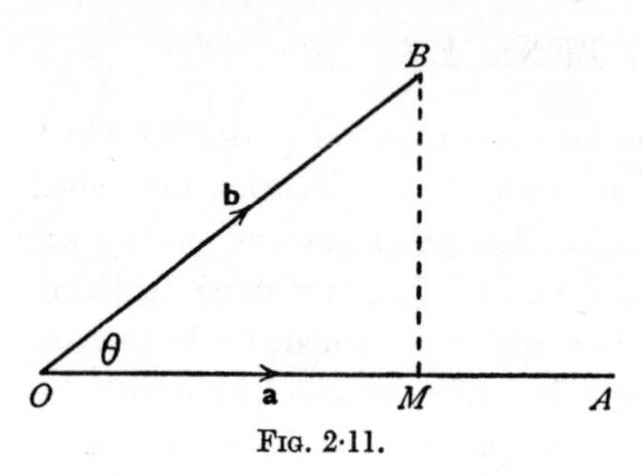

FIG. 2·11.

Let θ be the angle between the vectors, i.e. the angle AOB measured in the sense of minimum rotation from $\mathbf{a}$ to $\mathbf{b}$.

The *scalar product* of the vectors is then $\mathbf{ab}$ and is defined by the relation

$$\mathbf{ab} = ab \cos \theta.$$

The scalar product is a scalar and is measured by the product $OA \,.\, OM$, where M is the projection of B on OA, so that $OA = a$, $OM = b \cos \theta$. It is clear from the definition that

$$\mathbf{ba} = ba \cos(-\theta) = ab \cos \theta = \mathbf{ab},$$

so that the order of the two factors is irrelevant.

When the vectors are perpendicular, $\cos \theta = 0$, so that $\mathbf{ab} = 0$. Conversely this relation implies either that $\mathbf{a}$, $\mathbf{b}$ are perpendicular, or that $\mathbf{a} = 0$, or that $\mathbf{b} = 0$.

If $\mathbf{ab} = 0$, where $\mathbf{b}$ is an *arbitrary* vector, then $\mathbf{a} = 0$, for $\mathbf{a}$ cannot be perpendicular to every vector $\mathbf{b}$.

If θ is an obtuse angle, the scalar product is negative.

If $\mathbf{i}_a$ is a unit vector, then $\mathbf{i}_a \mathbf{b} = b \cos \theta$, which is the resolved part of the vector $\mathbf{b}$ along the direction of any vector which is parallel to $\mathbf{i}_a$.

If $\mathbf{i}_a$, $\mathbf{i}_b$ are both unit vectors, then $\mathbf{i}_a \mathbf{i}_b = \cos \theta$, which is the cosine of the angle between any two vectors parallel to $\mathbf{i}_a$ and $\mathbf{i}_b$.

If the point of application of a force $\mathbf{F}$ moves with velocity $\mathbf{v}$, the rate at which the force is doing work is the scalar product $\mathbf{Fv}$.

2·12. The vector product of two vectors. Let $\mathbf{a}$, $\mathbf{b}$ be two vectors of magnitudes a, b inclined at the angle θ measured as above from $\mathbf{a}$ to $\mathbf{b}$. We define the *vector product* $\mathbf{a}_\wedge \mathbf{b}$ as the vector of magnitude $ab \sin \theta$ which is perpendicular both to $\mathbf{a}$ and to $\mathbf{b}$ and whose sense is such that rotation from $\mathbf{a}$ to $\mathbf{b}$ is related to the sense of $\mathbf{a}_\wedge \mathbf{b}$ by the right-handed screw rule.

It follows from the definition that vector multiplication is not commutative, for $ba \sin(-\theta) = -ab \sin \theta$, and therefore

$$\mathbf{a}_\wedge \mathbf{b} = -\mathbf{b}_\wedge \mathbf{a}.$$

Also when the vectors are parallel ($\theta = 0$ or π) we have $\mathbf{a}_\wedge \mathbf{b} = 0$. Conversely this relation implies either that $\mathbf{a}$, $\mathbf{b}$ are parallel, or that one of them is zero.

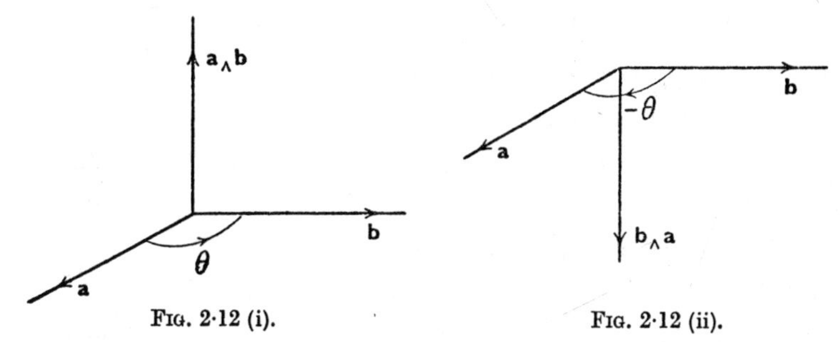

FIG. 2·12 (i).   FIG. 2·12 (ii).

As an example, let P be a point of a rigid body which is moving about the fixed point O with angular velocity $\boldsymbol{\omega}$. Let $\mathbf{r}$ be the position vector of P relative to O. Draw PN perpendicular to $\boldsymbol{\omega}$. Then the velocity of P is $\omega OP \sin\theta$ perpendicular to the plane PON and is therefore the vector $\boldsymbol{\omega}_\wedge \mathbf{r}$.

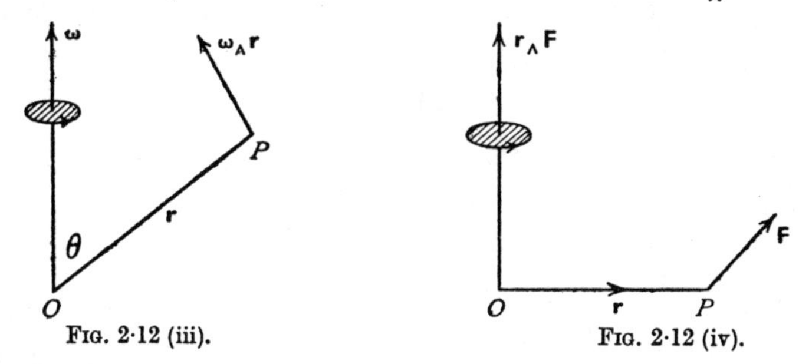

FIG. 2·12 (iii).   FIG. 2·12 (iv).

Similarly, the vector moment about O of a force $\mathbf{F}$ acting at P is $\mathbf{r}_\wedge \mathbf{F}$, fig. 2·12 (iv).

Since $ab \sin\theta$ measures the area of the parallelogram of which $\mathbf{a}$, $\mathbf{b}$ are adjacent sides, the vector product $\mathbf{a}_\wedge \mathbf{b}$ can be regarded as a *directed measure* of this area. It is the vector whose magnitude measures the area and whose direction is normal to the area.

2·121. The distributive law. Both scalar and vector multiplication are distributive, that is to say,

$$\mathbf{a}(\mathbf{b}+\mathbf{c}) = \mathbf{ab}+\mathbf{ac},$$

$$\mathbf{a}_\wedge(\mathbf{b}+\mathbf{c}) = \mathbf{a}_\wedge \mathbf{b}+\mathbf{a}_\wedge \mathbf{c}.$$

The proofs are left to the reader. See Ex. II, 27, 28.

2·13. Triple scalar product. If $\mathbf{a}$, $\mathbf{b}$, $\mathbf{c}$ are three vectors, the combination $\mathbf{a}(\mathbf{b}_\wedge \mathbf{c})$ is called their triple scalar product. This is the scalar product of the vectors $\mathbf{a}$ and $\mathbf{b}_\wedge \mathbf{c}$. The triple scalar product is measured by the volume of the parallelepiped whose conterminous edges are $\mathbf{a}$, $\mathbf{b}$, $\mathbf{c}$.

Proof. Since $\mathbf{b}_\wedge \mathbf{c}$ represents the area of the face whose edges are $\mathbf{b}$, $\mathbf{c}$ and in fig. 2·13 is directed along the normal on the same side as $\mathbf{a}$, the triple scalar product is measured by the volume.

Q.E.D.

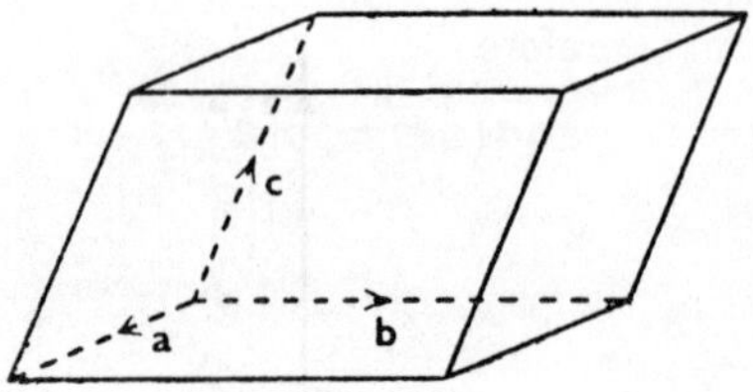

Thus $\mathbf{a}(\mathbf{b}_\wedge \mathbf{c}) = \mathbf{b}(\mathbf{c}_\wedge \mathbf{a}) = \mathbf{c}(\mathbf{a}_\wedge \mathbf{b})$.

But $\mathbf{a}(\mathbf{b}_\wedge \mathbf{c}) = -\mathbf{a}(\mathbf{c}_\wedge \mathbf{b})$,

since $\mathbf{b}_\wedge \mathbf{c} = -\mathbf{c}_\wedge \mathbf{b}$,

but note that

$$\mathbf{a}(\mathbf{b}_\wedge \mathbf{c}) = (\mathbf{b}_\wedge \mathbf{c})\,\mathbf{a}.$$

FIG. 2·13.

Hence the *cyclic rule*: the triple scalar product changes sign only with a change of cyclic order of the vectors. Note also that the actual position of the sign $_\wedge$ is unimportant, for

$$(\mathbf{a}_\wedge \mathbf{b})\,\mathbf{c} = \mathbf{a}(\mathbf{b}_\wedge \mathbf{c}) = [\mathbf{abc}],$$

the last being a convenient notation for the triple scalar product.

If two of the vectors are equal or parallel, or if all three are coplanar, the triple scalar product vanishes, e.g.

(1) $$[\mathbf{aab}] = 0.$$

2·14. Triple vector product. If $\mathbf{a}$, $\mathbf{b}$, $\mathbf{c}$ are three vectors, the combination $\mathbf{a}_\wedge(\mathbf{b}_\wedge \mathbf{c})$ is called a triple vector product.

This is the vector product of the vectors $\mathbf{a}$ and $\mathbf{b}_\wedge \mathbf{c}$.

Note that $\mathbf{a}_\wedge(\mathbf{b}_\wedge \mathbf{c}) = -\mathbf{a}_\wedge(\mathbf{c}_\wedge \mathbf{b}) = (\mathbf{c}_\wedge \mathbf{b})_\wedge \mathbf{a}$.

Hence the *centric rule*; the sign of the triple vector product changes only with a change of the centre vector.

The triple vector product has the very important property expressed by the relation

$$\mathbf{a}_\wedge(\mathbf{b}_\wedge \mathbf{c}) = -(\mathbf{ab})\,\mathbf{c} + (\mathbf{ac})\,\mathbf{b}.$$

Proof. The vector $\mathbf{a}_\wedge(\mathbf{b}_\wedge \mathbf{c})$ is perpendicular to the vector $(\mathbf{b}_\wedge \mathbf{c})$, which is itself perpendicular to the plane containing $\mathbf{b}$, $\mathbf{c}$. Thus $\mathbf{a}_\wedge(\mathbf{b}_\wedge \mathbf{c})$ is coplanar with $\mathbf{b}$, $\mathbf{c}$ and can therefore be compounded of scalar multiples of these latter. Therefore

$$\mathbf{a}_\wedge(\mathbf{b}_\wedge \mathbf{c}) = p\mathbf{b} - q\mathbf{c},$$

where p, q are scalars. Since $\mathbf{a}_\wedge(\mathbf{b}_\wedge \mathbf{c})$ is perpendicular to $\mathbf{a}$, the scalar product of these two vectors is zero. Therefore

$$0 = p\mathbf{ab} - q\mathbf{ac}.$$

Thus $$p = \lambda\mathbf{ac}, \quad q = \lambda\mathbf{ab},$$

where λ is a scalar. Hence

$$\mathbf{a}_\wedge(\mathbf{b}_\wedge \mathbf{c}) = -\lambda(\mathbf{ab})\,\mathbf{c} + \lambda(\mathbf{ac})\,\mathbf{b}.$$

To determine λ, take the scalar product with a vector $\mathbf{d}$ which is coplanar with $\mathbf{b}$, $\mathbf{c}$ and is perpendicular to $\mathbf{c}$, fig. 2·14.

Then $\mathbf{cd} = 0$,

and therefore

$$\lambda\mathbf{bd}(\mathbf{ac}) = \mathbf{d}[\mathbf{a}_\wedge(\mathbf{b}_\wedge\mathbf{c})]$$
$$= \mathbf{a}[(\mathbf{b}_\wedge\mathbf{c})_\wedge\mathbf{d}],$$

using the property of the triple scalar product.

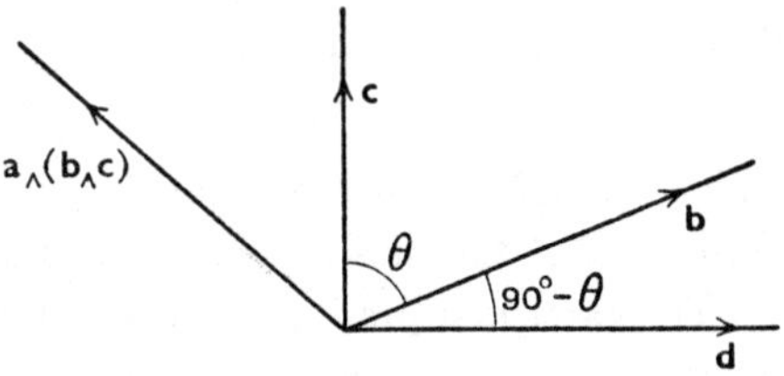

Fig. 2·14.

Now $(\mathbf{b}_\wedge\mathbf{c})_\wedge\mathbf{d}$ is a vector coplanar with $\mathbf{b}$, $\mathbf{c}$ and perpendicular to $\mathbf{d}$ and is therefore a vector along $\mathbf{c}$. If θ is the angle between $\mathbf{b}$, $\mathbf{c}$, the magnitude of this vector is

$$bcd \sin\theta = bd\cos(90° - \theta)c,$$

and therefore the vector $(\mathbf{b}_\wedge\mathbf{c})_\wedge\mathbf{d} = (\mathbf{bd})\mathbf{c}$.

Hence $\lambda(\mathbf{bd})(\mathbf{ac}) = (\mathbf{ac})(\mathbf{bd})$,

and therefore $\lambda = 1$. Q.E.D.

Note also the result $(\mathbf{a}_\wedge\mathbf{b})_\wedge\mathbf{c} = -\mathbf{a}(\mathbf{bc}) + \mathbf{b}(\mathbf{ac})$, and that as a mnemonic the term with the negative sign is always obtained by moving the brackets in the triple product but preserving the order.

2·15. Resolution of a vector. If $\mathbf{a}$, $\mathbf{b}$, $\mathbf{c}$ are given vectors, not all coplanar, and $\mathbf{x}$ is an arbitrary vector, then

$$\text{(i)}\quad \mathbf{x}[\mathbf{a}(\mathbf{b}_\wedge\mathbf{c})] = \mathbf{a}[(\mathbf{b}_\wedge\mathbf{c})\mathbf{x}] + \mathbf{b}[(\mathbf{c}_\wedge\mathbf{a})\mathbf{x}] + \mathbf{c}[(\mathbf{a}_\wedge\mathbf{b})\mathbf{x}].$$
$$\text{(ii)}\quad \mathbf{x}[\mathbf{a}(\mathbf{b}_\wedge\mathbf{c})] = (\mathbf{b}_\wedge\mathbf{c})(\mathbf{ax}) + (\mathbf{c}_\wedge\mathbf{a})(\mathbf{bx}) + (\mathbf{a}_\wedge\mathbf{b})(\mathbf{cx}).$$

The first resolves $\mathbf{x}$ along the given vectors, the second resolves $\mathbf{x}$ perpendicularly to the planes bc, ca, ab.

Proof of (i). Since $\mathbf{a}$, $\mathbf{b}$, $\mathbf{c}$ are not all coplanar, we can resolve $\mathbf{x}$ along them and so get

$$\mathbf{x} = p\mathbf{a} + q\mathbf{b} + r\mathbf{c},$$

where p, q, r are scalars. Form the scalar product with $(\mathbf{b}_\wedge\mathbf{c})$ which is perpendicular to $\mathbf{b}$ and $\mathbf{c}$. We then have

$$\mathbf{x}(\mathbf{b}_\wedge\mathbf{c}) = p\mathbf{a}(\mathbf{b}_\wedge\mathbf{c}),$$

which determines p. Q.E.D.

Proof of (ii). Let $\mathbf{x} = p(\mathbf{b}_\wedge\mathbf{c}) + q(\mathbf{c}_\wedge\mathbf{a}) + r(\mathbf{a}_\wedge\mathbf{b})$.

Form the scalar product with $\mathbf{a}$ which is perpendicular to $(\mathbf{c}_\wedge\mathbf{a})$ and $(\mathbf{a}_\wedge\mathbf{b})$. Then

$$\mathbf{ax} = p[\mathbf{a}(\mathbf{b}_\wedge\mathbf{c})]$$

which determines p. Q.E.D.

2·16. Tensors. In this section we look at a simple approach to tensor algebra.

Consider a scalar λ, vectors $\mathbf{a}$, $\mathbf{b}$, $\mathbf{c}$, ..., and an operation at present undefined, denoted by a semicolon (;) called *dyadic multiplication*. We can then write down the sequence of terms

(1) $$\lambda, \quad \mathbf{a}, \quad \mathbf{a};\mathbf{b}, \quad \mathbf{a};\mathbf{b};\mathbf{c}, \quad \mathbf{a};\mathbf{b};\mathbf{c};\mathbf{d}, \quad \ldots$$

which we shall call *continued dyadic products* of *ranks* 0, 1, 2, 3, 4, ... respectively. Observe that the rank is the same as the number of vectors in the product.

The particular product of two vectors $\mathbf{a};\mathbf{b}$ we shall call a *dyad*. We proceed to attach meaning to the operation (;) by assigning four laws of operation which must be obeyed. No interpretations will be admitted except those deducible from these laws.

I. *The associative law.* A continued dyadic product can be bracketed in any manner without change of meaning.

Thus for example

(2) $$\mathbf{a};\mathbf{b};\mathbf{c} = (\mathbf{a};\mathbf{b});\mathbf{c} = \mathbf{a};(\mathbf{b};\mathbf{c}),$$

(3) $$\mathbf{a};\mathbf{b};\mathbf{c};\mathbf{d} = (\mathbf{a};\mathbf{b});(\mathbf{c};\mathbf{d}) = \mathbf{a};(\mathbf{b};\mathbf{c});\mathbf{d}.$$

II. *The scalar law.* If in a continued dyadic product we replace a vector by a scalar, one semicolon adjacent to the scalar must be suppressed.

For example replacing $\mathbf{a}$ in (2) by the scalar λ we get

$$\lambda;\mathbf{b};\mathbf{c} = (\lambda;\mathbf{b});\mathbf{c} = \lambda;(\mathbf{b};\mathbf{c}),$$

or using the scalar law

(4) $$\lambda\,\mathbf{b};\mathbf{c} = (\lambda\,\mathbf{b});\mathbf{c} = \lambda(\mathbf{b};\mathbf{c}).$$

III. *The contraction law.* In an equality of continued dyadic products we can replace dyadic multiplication (;) by scalar multiplication ( ) in one and the same position on both sides of the equality.

Thus from (2), choosing the last semicolon, we get

(5) $$(\mathbf{a};\mathbf{b})\mathbf{c} = \mathbf{a};(\mathbf{bc}) = \mathbf{a}(\mathbf{bc})$$

on use of the scalar law. Thus the scalar product of a dyad and a vector is a vector.

This process, called *contraction*, lowers the rank of the original dyadic product by two units.

Again contracting (3) with respect to the middle semicolon we get

(6) $$(\mathbf{a};\mathbf{b})(\mathbf{c};\mathbf{d}) = \mathbf{a};(\mathbf{bc});\mathbf{d} = (\mathbf{a};\mathbf{d})(\mathbf{bc})$$

on using the scalar law. Thus the scalar product of two dyads is a dyad.

A second contraction of the right hand side of (6) leads to $(\mathbf{ad})(\mathbf{bc})$ which is known as the *twice contracted* or *double scalar product* of the dyads $(\mathbf{a};\mathbf{b})$ and $(\mathbf{c};\mathbf{d})$ written

(7) $$(\mathbf{a};\mathbf{b})\,..\,(\mathbf{c};\mathbf{d}) = (\mathbf{ad})(\mathbf{bc})$$

Since the scalar product of two vectors is independent of the order we have

$$(8) \qquad (\mathbf{ad})(\mathbf{bc}) = (\mathbf{cb})(\mathbf{da}) = (\mathbf{bc})(\mathbf{ad}) = (\mathbf{da})(\mathbf{cb}),$$

and therefore from (7)

$$(9) \quad (\mathbf{a};\mathbf{b})\,..\,(\mathbf{c};\mathbf{d}) = (\mathbf{c};\mathbf{d})\,..\,(\mathbf{a};\mathbf{b}) = (\mathbf{b};\mathbf{a})\,..\,(\mathbf{d};\mathbf{c}) = (\mathbf{d};\mathbf{c})\,..\,(\mathbf{b};\mathbf{a}).$$

If in a dyad we reverse the order of the vectors which compose it, we obtain the *conjugate* dyad, denoted by suffix c. We then see from (9) that

$$(10) \quad (\mathbf{a};\mathbf{b})..(\mathbf{c};\mathbf{d}) = (\mathbf{c};\mathbf{d})..(\mathbf{a};\mathbf{b}) = (\mathbf{a};\mathbf{b})_c..(\mathbf{c};\mathbf{d})_c = (\mathbf{c};\mathbf{d})_c..(\mathbf{a};\mathbf{b})_c,$$

that is to say double scalar multiplication of dyads is completely commutative whether we use both the originals or both their conjugates.

IV. *The distributive law.* Let $\mathbf{B}$ and $\mathbf{C}$ denote continued dyadic products of the same rank and let $\mathbf{A}$ be any continued dyadic product. Then

$$(11) \quad \mathbf{A};(\mathbf{B}+\mathbf{C}) = \mathbf{A};\mathbf{B}+\mathbf{A};\mathbf{C}, \quad (\mathbf{B}+\mathbf{C});\mathbf{A} = \mathbf{B};\mathbf{A}+\mathbf{C};\mathbf{A}.$$

Consider the triple vector product

$$(12) \qquad \begin{aligned}(\mathbf{a}_\wedge \mathbf{b})_\wedge \mathbf{c} &= -\mathbf{a}(\mathbf{bc})+\mathbf{b}(\mathbf{ac}) = \mathbf{c}(\mathbf{a};\mathbf{b})-\mathbf{c}(\mathbf{b};\mathbf{a})\\ &= \mathbf{c}[\mathbf{a};\mathbf{b}-\mathbf{b};\mathbf{a}] = \mathbf{c}[\mathbf{a};\mathbf{b}-(\mathbf{a};\mathbf{b})_c]\end{aligned}$$

by the distributive law, which expresses the triple vector product as the scalar product of a vector and the difference of two conjugate dyads.

It appears from (11) that any continued dyadic product $\mathbf{A}$ is a *linear vector operator* in the sense that $\mathbf{x}$ and $\mathbf{y}$ being arbitrary vectors.

$$(13) \qquad \mathbf{A}(\mathbf{x}+\mathbf{y}) = \mathbf{Ax}+\mathbf{Ay}, \quad (\mathbf{x}+\mathbf{y})\mathbf{A} = \mathbf{xA}+\mathbf{yA}.$$

*Definition of a tensor**

A linear vector operator $\mathbf{\Phi}^{(r)}$ is called a *tensor of rank* r, if $\mathbf{\Phi}^{(0)}$ is a scalar, and if for every positive integer $r \geqslant 1$ and for an arbitrary vector $\mathbf{x}$, $\mathbf{\Phi}^{(r)}\mathbf{x}$ is a tensor of rank $r-1$.

Thus a given vector $\mathbf{a}$ is a tensor of rank 1, for $\mathbf{ax}$ is a scalar, i.e. a tensor of rank zero.

A dyad $\mathbf{a};\mathbf{b}$ is a tensor of rank 2, for from (5) $(\mathbf{a};\mathbf{b})\mathbf{x} = \mathbf{a}(\mathbf{bx})$ which is a vector, i.e. a tensor of rank 1. This does not prove, and it is in fact false, that every tensor of rank 2 can be put in the form $\mathbf{a};\mathbf{b}$.

Proceeding as above we can show that tensors of every positive integral rank exist.

We shall confine our attention to tensors of the second rank, 2-tensors, in Euclidean space of 3-dimensions.

A particularly important 2-tensor is the *idemfactor* or *unit 2-tensor* $\mathbf{I}$ which has the property that for any vector $\mathbf{x}$

$$(14) \qquad \mathbf{Ix} = \mathbf{xI} = \mathbf{x}.$$

---

* Milne-Thomson, "Cálculo tensorial por métodos directos". *Rev. Mat. Hispano-Americana* 10 (1950), 1–27.

Its existence can be proved by means of the expression

(15) $$\mathbf{I} = \mathbf{i}_1 ; \mathbf{i}_1 + \mathbf{i}_2 ; \mathbf{i}_2 + \mathbf{i}_3 ; \mathbf{i}_3 ,$$

where $\mathbf{i}_1, \mathbf{i}_2, \mathbf{i}_3$ are mutually perpendicular unit vectors, for by resolution (2·15) we can write $\mathbf{x} = x_1\mathbf{i}_1 + x_2\mathbf{i}_2 + x_3\mathbf{i}_3$ and the truth of (14) is then easily verified by forming the products, using the distributive law.

Let $\mathbf{\Phi}$ be a 2-tensor in 3-space. Then

$$\begin{aligned}\mathbf{\Phi x} &= \mathbf{\Phi} x_1\mathbf{i}_1 + \mathbf{\Phi} x_2\mathbf{i}_2 + \mathbf{\Phi} x_3\mathbf{i}_3 \\ &= \mathbf{\Phi i}_1(\mathbf{i}_1\mathbf{x}) + \mathbf{\Phi i}_2(\mathbf{i}_2\mathbf{x}) + \mathbf{\Phi i}_3(\mathbf{i}_3\mathbf{x}) \\ &= [(\mathbf{\Phi i}_1) ; \mathbf{i}_1 + (\mathbf{\Phi i}_2) ; \mathbf{i}_2 + (\mathbf{\Phi i}_3) ; \mathbf{i}_3]\mathbf{x}.\end{aligned}$$

Now $\mathbf{x}$ is arbitrary. Therefore we must have

(16) $$\mathbf{\Phi} = (\mathbf{\Phi i}_1) ; \mathbf{i}_1 + (\mathbf{\Phi i}_2) ; \mathbf{i}_2 + (\mathbf{\Phi i}_3) ; \mathbf{i}_3 = \mathbf{\Phi}(\mathbf{i}_1 ; \mathbf{i}_1 + \mathbf{i}_2 ; \mathbf{i}_3 + \mathbf{i}_3 ; \mathbf{i}_3) = \mathbf{\Phi I}.$$

Thus the idemfactor operates also for 2-tensors (and indeed for tensors of any rank).

Also by the definition of a 2-tensor $\mathbf{\Phi i}_1, \mathbf{\Phi i}_2, \mathbf{\Phi i}_3$ are vectors.

Thus every 2-tensor in 3-space can be expressed as the sum of at most three dyads. Therefore we can always find vectors such that for a given $\mathbf{\Phi}$,

(17) $$\mathbf{\Phi} = \mathbf{a} ; \mathbf{b} + \mathbf{c} ; \mathbf{d} + \mathbf{e} ; \mathbf{f}$$

The *conjugate tensor* is defined by

(18) $$\mathbf{\Phi}_c = \mathbf{b} ; \mathbf{a} + \mathbf{d} ; \mathbf{c} + \mathbf{f} ; \mathbf{e}$$

just as in the above $(\mathbf{a} ; \mathbf{b})_c = \mathbf{b} ; \mathbf{a}$.

If $\mathbf{x}$ is any vector

(19) $$\mathbf{\Phi x} = \mathbf{a}(\mathbf{bx}) + \mathbf{c}(\mathbf{dx}) + \mathbf{e}(\mathbf{fx}) = \mathbf{x\Phi}_c ,$$

which could alternatively be taken to define the conjugate.

If $\mathbf{\Phi} = \mathbf{\Phi}_c$, the tensor $\mathbf{\Phi}$ is said to be *symmetric* and then

$$\mathbf{\Phi x} = \mathbf{x\Phi}_c = \mathbf{x\Phi}.$$

If $\mathbf{\Phi} = -\mathbf{\Phi}_c$, the tensor $\mathbf{\Phi}$ is said to be *antisymmetric* or *skew symmetric* and then $\mathbf{x\Phi} = -\mathbf{\Phi x}$.

Since we have identically

(20) $$\mathbf{\Phi} = \tfrac{1}{2}(\mathbf{\Phi} + \mathbf{\Phi}_c) + \tfrac{1}{2}(\mathbf{\Phi} - \mathbf{\Phi}_c) ,$$

any tensor can be expressed uniquely as the sum of a symmetric and an antisymmetric tensor.

If $\mathbf{\Phi}$ and $\mathbf{\Psi}$ are two 2-tensors, we can, as for dyads, distribute and form the double scalar product

(21) $$\mathbf{\Phi} \,..\, \mathbf{\Psi} = \mathbf{\Psi} \,..\, \mathbf{\Phi} = \mathbf{\Phi}_c \,..\, \mathbf{\Psi}_c = \mathbf{\Psi}_c \,..\, \mathbf{\Phi}_c .$$

If here $\mathbf{\Phi}$ is symmetric and $\mathbf{\Psi}$ is antisymmetric, we have from (21),

$$\mathbf{\Phi} \,..\, \mathbf{\Psi} = \mathbf{\Phi} \,..\, (-\mathbf{\Psi}) = -(\mathbf{\Phi} \,..\, \mathbf{\Psi}),$$

and therefore $\mathbf{\Phi} \,..\, \mathbf{\Psi} = 0$. Thus *the double scalar product of a symmetric and an antisymmetric 2-tensor is zero.*

A 2-tensor $\mathbf{\Phi}$ has two important invariants. The *first scalar invariant* can be formed from (17) by replacing dyadic by scalar multiplication to give the scalar

(22) $$\mathbf{\Phi}_{\mathrm{I}} = \mathbf{ab}+\mathbf{cd}+\mathbf{ef} = (\mathbf{\Phi}_c)_{\mathrm{I}}.$$

In particular from (15) the first scalar of the idemfactor is 3,

$$\mathbf{I}_{\mathrm{I}} = 3.$$

We also note that the first scalar invariant of a skew symmetric 2-tensor vanishes.

The *vector invariant* is formed by writing vector for dyadic multiplication in (15) to give

(23) $$\mathbf{\Phi}^{\times} = \mathbf{a}_{\wedge}\mathbf{b}+\mathbf{c}_{\wedge}\mathbf{d}+\mathbf{e}_{\wedge}\mathbf{f},$$

while from (18)

$$\mathbf{\Phi}_c^{\times} = \mathbf{b}\wedge\mathbf{a}+\mathbf{d}\wedge\mathbf{c}+\mathbf{f}\wedge\mathbf{e} = -\mathbf{\Phi}^{\times}.$$

It follows that for a symmetric 2-tensor the vector invariant vanishes.

2·19. Scalar and vector fields. If to each point of space there corresponds a scalar, then a *scalar field* is defined. Thus, for example, fluid pressure p and fluid density ρ constitute scalar fields.

If to each point of space there corresponds both a scalar and a direction, that is, if a vector is, as it were, tied to each point of space, then a *vector field* is defined. One of the most important vector fields in hydrodynamics is the field of fluid velocity $\mathbf{q}$. Another important field is that of vorticity (see 2·41).

2·20. Line, surface, and volume integrals. As we shall in the sequel have occasion to use these notions, this section will be devoted to explaining the sense in which the terms are to be understood. The object of this section is not to explain how the integrals may be calculated numerically, nor the conditions in which they exist, for these matters are fully treated in books on Analysis. When a particular case presents itself for numerical evaluation, that case will be dealt with as an individual instance.

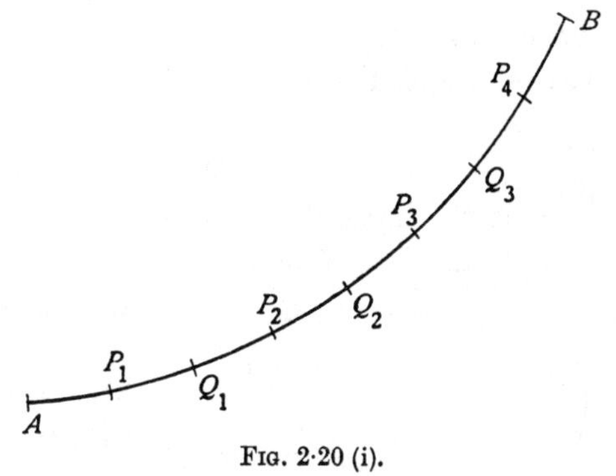

FIG. 2·20 (i).

Let AB be an arc of a given curve (not necessarily a plane curve). By marking points $Q_1, Q_2, \ldots, Q_{N-1}$, divide the arc AB into N sections $AQ_1, Q_1Q_2, \ldots, Q_{N-1}B$ of lengths δs_1,

$\delta s_2, \ldots, \delta s_N$ each less than ϵ, and take points $P_1, P_2, \ldots, P_N$, one in each section. Fig. 2·20 (i) illustrates the case $N = 4$. Let $f(P)$, or briefly f, be a function whose value is known at each point P of AB, and let $f_1, f_2, \ldots, f_N$ be the values of f at the points $P_1, P_2, \ldots, P_N$.

Then we can form the sum

$$(1) \qquad f_1\,\delta s_1 + f_2\,\delta s_2 + \ldots + f_N\,\delta s_N = \Sigma f\,\delta s.$$

If we now allow the number N to increase indefinitely, and at the same time let ϵ tend to zero, the *line integral* of f along AB, or the *curvilinear integral* of f along AB, is defined by

$$\int_{(AB)} f\,ds = \lim_{\substack{N\to\infty \\ \epsilon\to 0}} \Sigma f\,\delta s.$$

This definition applies whether f is a scalar or a vector.

If f is a vector, then the sum in (1) is a sum of vectors to be obtained by the law of addition of vectors, and the integral is then a vector quantity.

If f is constant, i.e. if f has the same value c at every point of AB, then it is clear from (1) that the sum is cl where l is the length of AB, and in this case the value of the integral is cl.

If f is a scalar function which obeys the inequality

$$(2) \qquad M > f > m,$$

where M and m are fixed numbers, then clearly

$$\Sigma(M-f)\,\delta s > 0, \quad \Sigma(f-m)\,\delta s > 0,$$

and therefore

$$\Sigma M\,\delta s > \Sigma f\,\delta s > \Sigma m\,\delta s.$$

so that

$$Ml > \int_{(AB)} f\,ds > ml.$$

Let $\mathbf{i}_s$ be a unit vector along the tangent to the element of arc ds. Then writing $d\mathbf{s} = \mathbf{i}_s\,ds$, so that $d\mathbf{s}$ is a *directed element of arc* of the curve AB, we have the equivalence

$$\int_{(AB)} X\,d\mathbf{s} = \int_{(AB)} X\mathbf{i}_s\,ds,$$

so that the integral on the left is defined in terms of the integrals already described. Here X may be a scalar or vector and the multiplication may be scalar, vector, or dyadic.

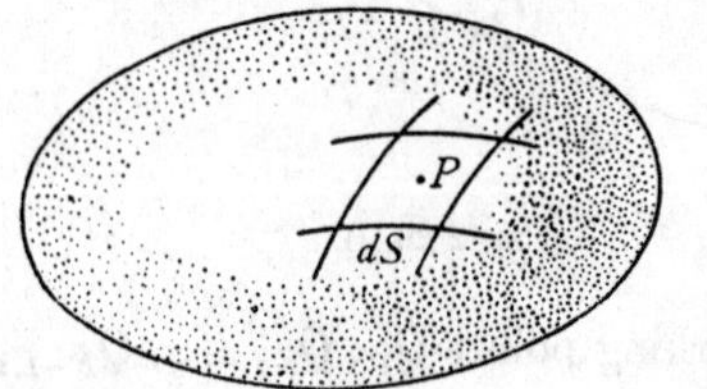

FIG. 2·20 (ii).

To define the *surface integral* of $f = f(P)$ over a surface S (not necessarily plane or closed), we divide the surface into elements of area $\delta S_1, \delta S_2, \ldots, \delta S_N$, each having its longest dimension less than ϵ. If $f_1, f_2, \ldots, f_N$ denote the values of f at points P_1, $P_2, \ldots$,

P_N, one within each element, we can form the sum

$$f_1\,\delta S_1+f_2\,\delta S_2+\ldots+f_N\,\delta S_N = \Sigma f\,\delta S.$$

The integral of f over the surface S is then defined by

$$\int_{(S)} f\,dS = \lim_{\substack{N\to\infty \\ \epsilon\to 0}} \Sigma f\,\delta S.$$

This definition applies to both scalar and vector functions.

If f has the constant value c over the surface, then the surface integral is cA, where A is the area of the surface S. Again, if f satisfies the inequality (2), then

$$MA > \int_{(S)} f\,dS > mA.$$

Again, if $\mathbf{n}$ is a unit vector drawn to the element dS in the direction of the outward normal to a *closed* surface S,

$$\int_{(S)} \mathbf{n}\,dS = 0, \tag{3}$$

for it is easily seen that the projection of this vector on any fixed plane is zero.

It is often convenient to replace $\mathbf{n}\,dS$ by the vector $d\mathbf{S}$ which represents an element of surface area directed along the normal (cf. 2·12). With this notation (3) becomes

$$\int_{(S)} d\mathbf{S} = 0.$$

More generally we are led to consider integrals of the type

$$\int_{(S)} X\,d\mathbf{S},$$

where X is a scalar or vector and the multiplication may be scalar, vector, or dyadic.

To define a *volume integral*, consider the volume V consisting of the region interior to a closed surface S. We divide V into elements of volume $\delta\tau_1, \delta\tau_2, \ldots, \delta\tau_N$, each having its longest dimension less than ϵ. If $f_1, f_2, \ldots, f_N$ denote the values of f at points $P_1, P_2, \ldots, P_N$, one within each element, we can form the sum

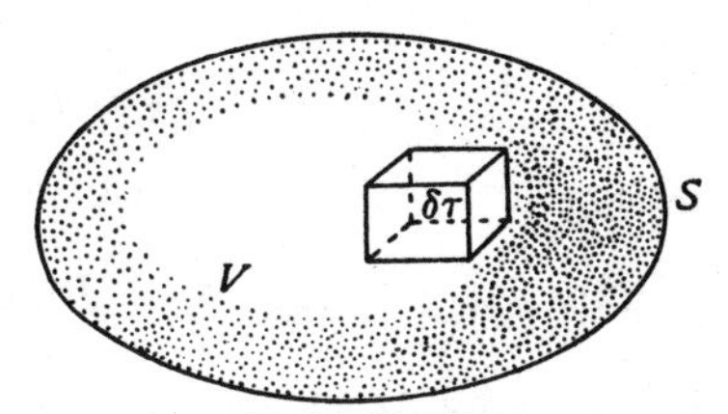

FIG. 2·20 (iii).

$$f_1\,\delta\tau_1+f_2\,\delta\tau_2+\ldots+f_N\,\delta\tau_N = \Sigma f\,\delta\tau.$$

The integral of f throughout the volume V is then defined by

$$\int_{(V)} f\,d\tau = \lim_{\substack{N\to\infty \\ \epsilon\to 0}} \Sigma f\,\delta\tau,$$

which again applies to scalar and vector functions.

If the function has the constant value c, then the integral is equal to cV, and if f satisfies the inequality (2), then

$$MV > \int_{(V)} f\,d\tau > mV.$$

Notation ; we use one sign of integration when there is only one differential ds, dS, or $d\tau$. When two differentials are used we shall use two integral signs. Thus if $dS = dx\,dy$, we write

$$\int_{(S)} f\,dS = \iint_{(S)} f\,dx\,dy.$$

2·22. Variation of a scalar function of position. Let ϕ be a scalar function of position so that the values of ϕ constitute a scalar field. We shall suppose ϕ to be a continuous function with continuous differential coefficients of the first order. Then there exists, in general, a family of surfaces on each of which ϕ is constant. We can call these *equi-ϕ surfaces*.

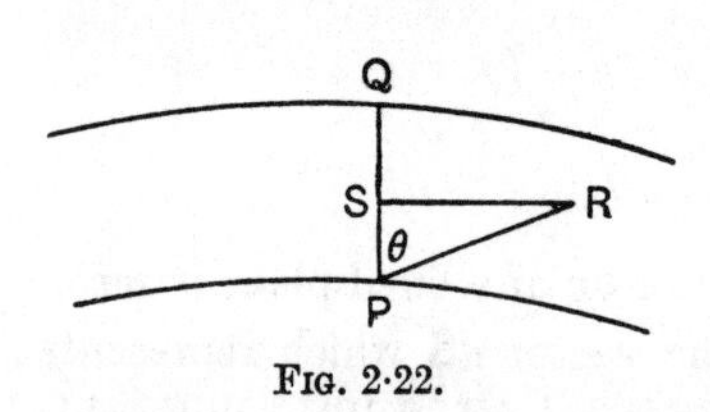

FIG. 2·22.

Let P be any point and let Q be a near point on the normal at P to the equi-ϕ surface $\phi = \phi_P$, where ϕ_P denotes the value of ϕ at P. Then if PQ is regarded as a small length of the first order, we can write

$$(1) \qquad \phi_Q - \phi_P = PQ\left(\frac{\partial\phi}{\partial s}\right)_{PQ},$$

where $\left(\frac{\partial\phi}{\partial s}\right)_{PQ}$ denotes the distance rate of change of ϕ for displacements in the direction of PQ.

Let R be any point near P, and let the equi-ϕ surface $\phi = \phi_R$ meet PQ in S, and assume that to the first order RS is perpendicular to PQ. Then

$$\phi_R = \phi_S = \phi_P + PS\left(\frac{\partial\phi}{\partial s}\right)_{PQ},$$

so that

$$(2) \qquad \phi_R - \phi_P = PR\cos\theta\left(\frac{\partial\phi}{\partial s}\right)_{PQ} = \overrightarrow{PR}\,(\text{grad}\,\phi),$$

where grad ϕ denotes a vector * whose direction is along PQ and whose magnitude is

$$|\,\text{grad}\,\phi\,| = \left(\frac{\partial\phi}{\partial s}\right)_{PQ}.$$

* grad ϕ is an abbreviation of the phrase " gradient of ϕ ".

It follows from this definition, by putting $\overrightarrow{PQ}=\mathbf{n}\,dn$, where $\mathbf{n}$ is the unit vector along the normal at P to the equi-ϕ surface through P, that

$$\text{(3)} \qquad \operatorname{grad}\phi = \mathbf{n}\,\frac{\partial\phi}{\partial n}.$$

Various alternative notations are available for grad ϕ.

$$\text{(4)} \qquad \operatorname{grad}\phi = \frac{\partial\phi}{\partial\mathbf{r}} = \frac{\partial\phi}{\partial\mathbf{P}} = \nabla\phi = \mathbf{n}\,\frac{\partial\phi}{\partial n}.$$

In the first of these we denote the change of position vector of P by $d\mathbf{r}$, in the second the change of position vector is denoted by $d\mathbf{P}$, the advantage here being that explicit attention is called to the point P by the notation. The notation $\partial\phi/\partial\mathbf{r}$ may be compared with the ordinary partial differential coefficient $\partial\phi/\partial x$, but it must be remembered that we cannot divide by a vector, so that $\partial\phi/\partial\mathbf{r}$ cannot be regarded as the limit of a quotient of two small quantities. The symbol ∇ (pronounced *nabla*) was introduced by Sir William Rowan Hamilton and so named from its fancied resemblance to a harp. The vector operator ∇ is analogous to the scalar operator $D = d/dx$, in that it does not call explicit attention to the independent variable. It is nevertheless convenient. We shall use in the sequel whichever of the notations indicated in (4) may appear most appropriate.

Returning to (2), the rate of change of ϕ when we proceed in the direction PR is

$$\left(\frac{\partial\phi}{\partial s}\right)_{PR} = \lim_{R\to P}\frac{\phi_R-\phi_P}{PR} = \cos\theta\left(\frac{\partial\phi}{\partial s}\right)_{PQ},$$

which is the component of grad ϕ in the direction PR.

Thus if in (2) we write $\overrightarrow{PR} = \mathbf{i}_s\,ds$, we have

$$\text{(5)} \qquad \frac{\partial\phi}{\partial s} = \mathbf{i}_s\operatorname{grad}\phi = \mathbf{i}_s\,\frac{\partial\phi}{\partial\mathbf{r}} = \mathbf{i}_s(\nabla\phi).$$

We must therefore regard ∇ as a *vector operator* which, applied to a scalar ϕ, gives a vector whose component in any direction is the rate of change of ϕ in that direction.

2·23. The operator $(\mathbf{a}\,\nabla)$. If $\mathbf{a} = a\mathbf{i}$, 2·22 (5) shows that $(\mathbf{a}\,\nabla)\phi$ is a times the rate of change of ϕ in the direction of $\mathbf{a}$.

Also, since $(\mathbf{a}\,\nabla)$ is a scalar operator, the ordinary rule for differentiating products gives

$$\text{(1)} \qquad (\mathbf{a}\,\nabla)(\mathbf{bc}) = \mathbf{b}[(\mathbf{a}\,\nabla)\mathbf{c}]+\mathbf{c}[(\mathbf{a}\,\nabla)\mathbf{b}].$$

$$\text{(2)} \qquad (\mathbf{a}\,\nabla)(\mathbf{b}_\wedge\mathbf{c}) = [(\mathbf{a}\,\nabla)\mathbf{b}]_\wedge\mathbf{c}+\mathbf{b}_\wedge[(\mathbf{a}\,\nabla)\mathbf{c}].$$

Note also that, for infinitesimal changes of position,

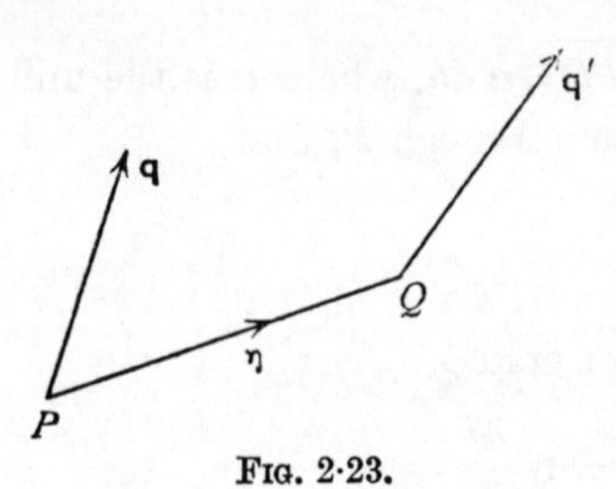

FIG. 2·23.

(3) $$d\phi = (d\mathbf{r}\,\nabla)\phi = d\mathbf{r}(\nabla\,\phi),$$
$$d\mathbf{q} = (d\mathbf{r}\,\nabla)\mathbf{q}.$$

As an important application let $\mathbf{q}$ be the fluid velocity at the point P, $\mathbf{q}'$ the velocity at a neighbouring point Q in the position measured from P by the infinitesimal vector $\boldsymbol{\eta}$.

Then $\mathbf{q}' = \mathbf{q}+(\boldsymbol{\eta}\,\nabla)\mathbf{q}$ to the first order.

2·24. Generalised definition of the operator ∇. Let $\mathbf{q}$ be the fluid velocity at a fixed point O and surround O by a convex surface S of small dimensions whose centroid is O.

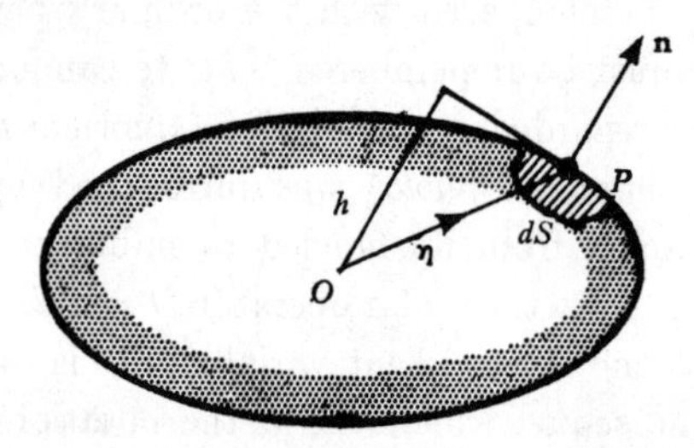

FIG. 2·24.

Then if P is any point on this surface $\mathbf{q}_P = \mathbf{q}+(\overrightarrow{OP}\,\nabla)\mathbf{q} = \mathbf{q}+(\boldsymbol{\eta}\,\nabla)\mathbf{q}$, where $\boldsymbol{\eta} = \overrightarrow{OP}$. Then, to the first order,

(1) $$\int_{(S)} \mathbf{n}\,dS\,;\,\mathbf{q}_P = \int_{(S)} \mathbf{n}\,dS\,;[\mathbf{q}+(\boldsymbol{\eta}\,\nabla)\mathbf{q}] = \left[\int_{(S)} (\mathbf{n}\,;\boldsymbol{\eta})\,dS\right](\nabla\,;\mathbf{q}).$$

Since $$\int_{(S)} \mathbf{n}\,dS\,;\mathbf{q} = \left[\int_{(S)} \mathbf{n}\,dS\right];\mathbf{q} = 0 \text{ from 2·20 (3).}$$

Now consider $$\int_{(S)} (\mathbf{n}\,;\boldsymbol{\eta})\,dS.$$

This is a symmetric tensor, for its vector invariant $\int_{(S)} (\mathbf{n}_\wedge\boldsymbol{\eta})\,dS$ vanishes since it is the moment about O of the thrusts due to uniform unit pressure acting on S, and such pressure thrusts are in equilibrium. Taking the first scalar invariant we have

$$\left[\int_{(S)} \mathbf{n}\,;\boldsymbol{\eta}\,dS\right]_I = \int_{(S)} \mathbf{n}\,\boldsymbol{\eta}\,dS = \int_{(S)} h dS = 3V = [\mathbf{I}V]_I,$$

where h is the perpendicular from O to the tangent plane at P and V is the volume enclosed by S. Thus

(2) $$\frac{1}{V}\int_{(S)} \mathbf{n}\,;\,\boldsymbol{\eta}\; dS = \mathbf{I} \quad \text{the idemfactor.}$$

Therefore from (1), to the first order,

$$\int_{(S)} \mathbf{n}\,;\,\mathbf{q}_P\, dS = \left[\int (\mathbf{n}\,;\,\boldsymbol{\eta})\, dS\right] \nabla\,;\,\mathbf{q}.$$

Dividing by V and taking the limit when $V \to 0$ we get the following theorem

(3) $$\lim_{V\to 0} \frac{1}{V}\int_{(S)} \mathbf{n}\,;\,\mathbf{q}_P\, dS = \mathbf{I}(\nabla\,;\,\mathbf{q}) = \nabla\,;\,\mathbf{q}.$$

Taking the first scalar and vector invariants we have

(4) $$\nabla\mathbf{q} = \lim_{V\to 0} \frac{1}{V}\int_{(S)} \mathbf{n}\mathbf{q}_P\, dS = \operatorname{div}\mathbf{q} \quad \text{(divergence)},$$

(5) $$\nabla_\wedge\mathbf{q} = \lim_{V\to 0} \frac{1}{V}\int_{(S)} \mathbf{n}_\wedge\mathbf{q}_P\, dS = \operatorname{curl}\mathbf{q} \quad \text{(curl or rotational)},$$

where div $\mathbf{q}$ and curl $\mathbf{q}$ are simply alternative respective names. The vector $\mathbf{q}$ can be any vector, not necessarily the velocity.

Lastly replace in (3) the vector $\mathbf{q}$ by the scalar ϕ and use the scalar law Then

(6) $$\nabla\phi = \lim_{V\to 0} \frac{1}{V}\int_{(S)} \mathbf{n}\phi_P\, dS = \operatorname{grad}\phi \quad \text{(gradient)}.$$

From the above results we may formulate the general definition

(7) $$\nabla \circ X = \lim_{V\to 0} \frac{1}{V}\int \mathbf{n} \circ X_P\, dS.$$

where X is a function of position, scalar, vector or tensor, and the small circle denotes multiplication, scalar, vector or dyadic, as the particular application may require.

A vector field whose divergence vanishes is called *solenoidal.*

A vector field whose curl vanishes is called *irrotational.*

From (2) and (3) we have for the position vector $\mathbf{r}$ the important theorem

(8) $$\nabla\,;\,\mathbf{r} = \lim_{V\to 0} \frac{1}{V}\int_{(S)} \mathbf{n}\,;\,(\mathbf{r}+\boldsymbol{\eta})\, dS = \mathbf{I} \quad \text{the idemfactor.}$$

To interpret $\mathbf{q}\,;\,\nabla$ we have

$$\mathbf{q}\,;\,\nabla = (\nabla\,;\,\mathbf{q})_c = \lim_{V\to 0} \frac{1}{V}\int (\mathbf{n}\,;\,\mathbf{q})_c\, dS.$$

If we use the expression for ∇ in terms of cartesian coordinates (x_1, x_2, x_3) and the corresponding unit vectors $\mathbf{i}_1, \mathbf{i}_2, \mathbf{i}_3$, we have from 2·70 (1)

$$\nabla = \mathbf{i}_1\frac{\partial}{\partial x_1} + \mathbf{i}_2\frac{\partial}{\partial x_2} + \mathbf{i}_3\frac{\partial}{\partial x_3},$$

(9) $$\nabla\,;\mathbf{q} = \left(\mathbf{i}_1\frac{\partial}{\partial x_1}+\mathbf{i}_2\frac{\partial}{\partial x_2}+\mathbf{i}_3\frac{\partial}{\partial x_3}\right);\,(\mathbf{i}_1\,\mathbf{q}_1+\mathbf{i}_2\,\mathbf{q}_2+\mathbf{i}_3\,\mathbf{q}_3) = \Sigma\,(\mathbf{i}_r\,;\,\mathbf{i}_s)\frac{\partial q_s}{\partial x_r}$$

and therefore taking the conjugate

(10) $$\mathbf{q}\,;\nabla = \Sigma\,(\mathbf{i}_s\,;\,\mathbf{i}_r)\frac{\partial q_s}{\partial x_r} = \Sigma\,(\mathbf{i}_r\,;\,\mathbf{i}_s)\frac{\partial q_r}{\partial x_s}.$$

2·32. Operations on a single vector or scalar.

(I) $$\operatorname{div}(\operatorname{grad}\phi) = \nabla(\nabla\,\phi) = (\nabla\,\nabla)\phi = \nabla^2\,\phi,$$

since ϕ is a scalar. The operator ∇^2 is called the Laplacian operator.

(II) $$\operatorname{div}(\operatorname{curl}\mathbf{q}) = \nabla(\nabla_\wedge\mathbf{q}) = [\nabla\,\nabla\,\mathbf{q}] = 0,$$

from 2·13 (1).

(III) $$\operatorname{curl}(\operatorname{grad}\phi) = \nabla_\wedge(\nabla\,\phi) = (\nabla_\wedge\nabla)\phi = 0,$$

since, from 2·12, $\mathbf{a}_\wedge\mathbf{a} = 0$.

(IV) $$\operatorname{curl}(\operatorname{curl}\mathbf{q}) = \nabla_\wedge(\nabla_\wedge\mathbf{q}).$$

Using the triple vector product, we get

$$\operatorname{curl}(\operatorname{curl}\mathbf{q}) = \nabla(\nabla\mathbf{q})-(\nabla\nabla)\mathbf{q} = \operatorname{grad}(\operatorname{div}\mathbf{q})-\nabla^2\,\mathbf{q}.$$

Thus

(V) $$\nabla^2\,\mathbf{q} = \nabla(\nabla\,\mathbf{q})-\nabla_\wedge(\nabla_\wedge\mathbf{q}).$$

The foregoing are all capable of direct proof. For example, to prove (II) we write, with an obvious notation,

$$\nabla_1(\nabla_{2\wedge}\mathbf{q}) = \lim_{V_1\to 0}\lim_{V_2\to 0}\frac{1}{V_1V_2}\iint\mathbf{n}_1(\mathbf{n}_{2\wedge}\mathbf{q})\,dS_2\,dS_1$$
$$= -\lim_{V_2\to 0}\lim_{V_1\to 0}\frac{1}{V_2V_1}\iint\mathbf{n}_2(\mathbf{n}_{1\wedge}\mathbf{q})\,dS_1\,dS_2,$$

using the triple scalar product cyclic rule, and assuming that the order of integrations can be inverted. Thus

$$\nabla_1(\nabla_{2\wedge}\mathbf{q}) = -\nabla_2(\nabla_{1\wedge}\mathbf{q}),$$

or

$$\nabla(\nabla_\wedge\mathbf{q}) = -\nabla(\nabla_\wedge\mathbf{q}) = 0.$$

Similarly for (V), we have

$$\nabla_\wedge(\nabla_\wedge\mathbf{q}) = \lim_{V_1\to 0}\lim_{V_2\to 0}\frac{1}{V_1V_2}\iint\mathbf{n}_{1\wedge}(\mathbf{n}_{2\wedge}\mathbf{q})\,dS_2\,dS_1$$
$$= \lim_{V_1\to 0}\lim_{V_2\to 0}\frac{1}{V_1V_2}\iint\{\mathbf{n}_2(\mathbf{n}_1\mathbf{q})-(\mathbf{n}_1\,\mathbf{n}_2)\mathbf{q}\}\,dS_2\,dS_1$$
$$= \nabla(\nabla\mathbf{q})-\nabla^2\,\mathbf{q}.$$

These specimen arguments show that manipulations with ∇ ultimately rest on the corresponding manipulations with $\mathbf{n}$.

2·33. Operations on a product.

To study operations on a product XY, we shall suppose X, Y to obey the following product law :

$$(X+X')(Y+Y') = XY+XY'+X'Y+X'Y',$$

the order of the factors in every product being, in general, relevant.

Let X, Y be the values of our symbols at the point P, and X', Y' their values at a point of a closed surface S surrounding P, and $\mathbf{n}$ a unit outward normal to the element dS of this surface. Then we have identically

$$\begin{aligned} X'Y' &= [X+(X'-X)]\,[Y+(Y'-Y)] \\ &= XY+X(Y'-Y)+(X'-X)Y+(X'-X)(Y'-Y), \end{aligned}$$

and therefore

$$\begin{aligned} \int \mathbf{n}\, X'Y'\, dS = \int \mathbf{n}\, XY\, dS + \int \mathbf{n}\, X(Y'-Y)\, dS \\ + \int \mathbf{n}\,(X'-X)\,Y\, dS + \int \mathbf{n}\,(X'-X)(Y'-Y)\,dS. \end{aligned}$$

If we let the surface surrounding P shrink to infinitesimal size, $X'-X$ $Y'-Y$ will also be infinitesimal, and therefore the last integral will itself be infinitesimal compared with the other integrals and may be neglected.

Also, X, Y, being calculated at the point P, are fixed, and $\int \mathbf{n}\, dS = 0$ when taken over a closed surface (2·20 (3)). Therefore

$$\int \mathbf{n}\, XY\, dS = 0, \tag{1}$$

and we get

$$\begin{aligned} \int \mathbf{n}\, X'Y'\, dS &= \int \mathbf{n}\, X(Y'-Y)\, dS + \int \mathbf{n}\,(X'-X)\,Y\, dS \\ &= \int \mathbf{n}\, XY'\, dS + \int \mathbf{n}\, X'Y\, dS \end{aligned}$$

on making a further application of (1).

Dividing by V the volume enclosed by the surface, we get

$$\frac{1}{V}\int \mathbf{n}\, X'Y'\, dS = \frac{1}{V}\int \mathbf{n}\, XY'\, dS + \frac{1}{V}\int \mathbf{n}\, X'Y\, dS.$$

If we now let $V \to 0$, this gives, by the definition of ∇,

$$\nabla(XY) = \nabla(X_0 Y) + \nabla(XY_0),$$

the suffix zero indicating that the corresponding quantity is not to be varied when applying the operator nabla.* This formula can be compared with the corresponding formula for the differentiation operator $D = d/dx$, namely:

$$D(XY) = D(X_0 Y) + D(XY_0) = X_0(DY) + (DX)\,Y_0 = X(DY) + (DX)\,Y,$$

the suffix zero being dropped, as it is no longer required. The above property,

* Note that this step is an essential preliminary to developing the result of operating on a product, cf. 2·34 (II), (III).

in conjunction with the gradient property (2·23), shows that ∇ is in the nature of a generalised differentiation operator.

2·34. Applications of ∇ to products. We shall now apply the result of the previous section to certain products of vectors and scalars. The triple scalar and vector products yield the following results which will be useful:

(A) $\quad \mathbf{p}(\mathbf{q}_\wedge \mathbf{r}) = \mathbf{r}(\mathbf{p}_\wedge \mathbf{q}) = -\mathbf{q}(\mathbf{p}_\wedge \mathbf{r}).$

(B) $\quad \mathbf{p}_\wedge(\mathbf{q}_\wedge \mathbf{r}) = (\mathbf{r}\mathbf{p})\mathbf{q} - \mathbf{r}(\mathbf{p}\mathbf{q}).$

(C) $\quad \mathbf{p}(\mathbf{q}\mathbf{r}) = \mathbf{q}_\wedge(\mathbf{p}_\wedge \mathbf{r}) + (\mathbf{q}\mathbf{p})\mathbf{r}.$

Observe that (C) is merely a rearrangement of (B).

(I)
$$\begin{aligned}\nabla(\mathbf{a}_\wedge \mathbf{b}) &= \nabla(\mathbf{a}_\wedge \mathbf{b}_0) + \nabla(\mathbf{a}_{0\wedge} \mathbf{b}) \\ &= \mathbf{b}_0(\nabla_\wedge \mathbf{a}) - \mathbf{a}_0(\nabla_\wedge \mathbf{b}), \text{ from (A).}\end{aligned}$$

The suffix zero is no longer required and we get

$$\nabla(\mathbf{a}_\wedge \mathbf{b}) = \mathbf{b}(\nabla_\wedge \mathbf{a}) - \mathbf{a}(\nabla_\wedge \mathbf{b}),$$

or

$$\operatorname{div}(\mathbf{a}_\wedge \mathbf{b}) = \mathbf{b} \operatorname{curl} \mathbf{a} - \mathbf{a} \operatorname{curl} \mathbf{b}.$$

(II)
$$\begin{aligned}\nabla_\wedge(\mathbf{a}_\wedge \mathbf{b}) &= \nabla_\wedge(\mathbf{a}_\wedge \mathbf{b}_0) + \nabla_\wedge(\mathbf{a}_{0\wedge} \mathbf{b}) \\ &= (\mathbf{b}_0 \nabla)\mathbf{a} - \mathbf{b}_0(\nabla \mathbf{a}) - (\mathbf{a}_0 \nabla)\mathbf{b} + \mathbf{a}_0(\nabla \mathbf{b}), \text{ from (B),} \\ \nabla_\wedge(\mathbf{a}_\wedge \mathbf{b}) &= (\mathbf{b} \nabla)\mathbf{a} - (\mathbf{a} \nabla)\mathbf{b} - \mathbf{b}(\nabla \mathbf{a}) + \mathbf{a}(\nabla \mathbf{b}).\end{aligned}$$

(III)
$$\begin{aligned}\nabla(\mathbf{a}\mathbf{b}) &= \nabla(\mathbf{a}_0 \mathbf{b}) + \nabla(\mathbf{a}\mathbf{b}_0) \\ &= \mathbf{a}_\wedge(\nabla_\wedge \mathbf{b}) + (\mathbf{a} \nabla)\mathbf{b} + \mathbf{b}_\wedge(\nabla_\wedge \mathbf{a}) + (\mathbf{b} \nabla)\mathbf{a}, \text{ from (C).}\end{aligned}$$

(IV) From (II) and (III), by subtraction,

$$\begin{aligned}(\mathbf{a} \nabla)\mathbf{b} = \tfrac{1}{2}\{\nabla(\mathbf{a}\mathbf{b}) - \nabla_\wedge(\mathbf{a}_\wedge \mathbf{b}) - \mathbf{b}_\wedge(\nabla_\wedge \mathbf{a}) \\ - \mathbf{a}_\wedge(\nabla_\wedge \mathbf{b}) - \mathbf{b}(\nabla \mathbf{a}) + \mathbf{a}(\nabla \mathbf{b})\}.\end{aligned}$$

In particular, since $\nabla_\wedge(\mathbf{q}_\wedge \mathbf{q}) = 0$, we get

$$(\mathbf{q} \nabla)\mathbf{q} = \tfrac{1}{2}\nabla q^2 - \mathbf{q}_\wedge(\nabla_\wedge \mathbf{q}).$$

(V) If $\boldsymbol{\eta}$ is a constant vector (unaffected by ∇), we get, from (II), (III),

$$\begin{aligned}(\boldsymbol{\eta} \nabla)\mathbf{b} &= -\nabla_\wedge(\boldsymbol{\eta}_\wedge \mathbf{b}) + \boldsymbol{\eta}(\nabla \mathbf{b}), \\ (\boldsymbol{\eta}\nabla)\mathbf{b} &= \nabla(\boldsymbol{\eta}\mathbf{b}) + (\nabla_\wedge \mathbf{b})_\wedge \boldsymbol{\eta}.\end{aligned}$$

(VI)
$$\begin{aligned}\nabla(\mathbf{a}\phi) &= \nabla(\phi_0\, \mathbf{a}) + \nabla(\phi \mathbf{a}_0) \\ &= \phi(\nabla \mathbf{a}) + \mathbf{a}(\nabla \phi), \text{ since } \phi \text{ is a scalar, or} \\ \operatorname{div}(\mathbf{a}\phi) &= \phi \operatorname{div} \mathbf{a} + \mathbf{a} \operatorname{grad} \phi\end{aligned}$$

(VII)
$$\nabla_{\wedge}(\phi\mathbf{a}) = \nabla_{\wedge}(\phi\mathbf{a}_0)+\nabla_{\wedge}(\phi_0\mathbf{a})$$
$$= \nabla\phi_{\wedge}\mathbf{a}+\phi(\nabla_{\wedge}\mathbf{a}),$$
$$\text{curl}\,(\mathbf{a}\phi) = \phi\,\text{curl}\,\mathbf{a}+\text{grad}\,\phi_{\wedge}\mathbf{a}.$$

(VIII)
$$\nabla(\phi\psi) = \nabla(\phi_0\psi)+\nabla(\phi\psi_0)$$
$$= \phi\nabla\psi+\psi\nabla\phi,$$

where ϕ, ψ are scalar functions.

(IX)
$$\nabla^2(\phi\psi) = \nabla(\nabla\phi\psi)$$
$$= \psi\nabla^2\phi+2(\nabla\phi)(\nabla\psi)+\phi\nabla^2\psi,$$

using (VIII) and then (VI).

(X)
$$\nabla(\mathbf{a}\,;\mathbf{b}) = \nabla(\mathbf{a}\,;\mathbf{b}_0)+\nabla(\mathbf{a}_0\,;\mathbf{b})$$
$$= \mathbf{b}(\nabla\mathbf{a})+(\mathbf{a}\nabla)\mathbf{b}.$$

In particular, $\nabla(\mathbf{q}\,;\mathbf{q}) = \mathbf{q}(\nabla\mathbf{q})+(\mathbf{q}\nabla)\mathbf{q}.$

(XI) If $\mathbf{T}$ is a 2-tensor, $\nabla(\mathbf{T}_{\wedge}\mathbf{r}) = (\nabla\mathbf{T})_{\wedge}\mathbf{r}-\mathbf{T}^{\times}$,

for $\nabla(\mathbf{T}_{\wedge}\mathbf{r}) = \nabla(\mathbf{T}_{\wedge}\mathbf{r}_0)+\nabla(\mathbf{T}_{0\wedge}\mathbf{r}) = (\nabla\mathbf{T})_{\wedge}\mathbf{r}+\nabla(\mathbf{T}_{0\wedge}\mathbf{r}).$

Now $\nabla(\mathbf{a}_0;\mathbf{b}_{0\wedge}\mathbf{r}) = (\mathbf{a}\nabla)(\mathbf{b}_{0\wedge}\mathbf{r}) = \mathbf{b}_{\wedge}\mathbf{a}(\nabla;\mathbf{r}) = \mathbf{b}_{\wedge}\mathbf{a} = -(\mathbf{a};\mathbf{b})^{\times}$,

since $\nabla;\mathbf{r} = \mathbf{I}$ from 2·24 (8). Therefore $\nabla(\mathbf{T}_{0\wedge}\mathbf{r}) = -\mathbf{T}^{\times}$

2·40. Analysis of the motion of a fluid element. Consider an infinitesimal element of fluid whose centroid is the point P.

Let $\boldsymbol{\eta}$ be the position vector of the point Q of the element relative to P. Then if $\mathbf{q}$ is the fluid velocity at P, the velocity at Q will be (2·23).

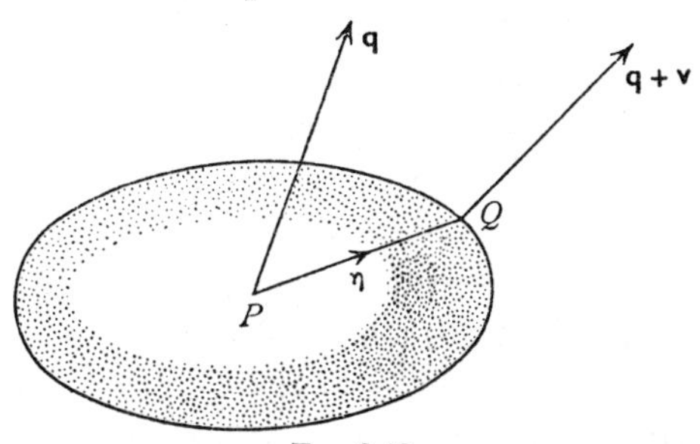

FIG. 2·40.

(1) $$\mathbf{q}+\mathbf{v} = \mathbf{q}+(\boldsymbol{\eta}\nabla)\mathbf{q}.$$

From the identity
$$(\boldsymbol{\eta}\nabla)\mathbf{q} = \tfrac{1}{2}\boldsymbol{\eta}(\nabla;\mathbf{q}-\mathbf{q};\nabla)+\tfrac{1}{2}\boldsymbol{\eta}(\nabla;\mathbf{q}+\mathbf{q};\nabla),$$
where the interpretation of $\mathbf{q}\,;\nabla$ is given in 2·24, we have, using 2·16 (12) with $\mathbf{a} = \nabla, \mathbf{b} = \mathbf{q}, \mathbf{c} = \boldsymbol{\eta}$,

(2) $$\mathbf{q}+\mathbf{v} = \mathbf{q}+\tfrac{1}{2}(\nabla_{\wedge}\mathbf{q})_{\wedge}\boldsymbol{\eta}+\boldsymbol{\eta}\mathbf{D}.$$

Here $\mathbf{D}$ is a symmetric tensor, known as the *rate of deformation tensor.*

(3) $$\mathbf{D} = \tfrac{1}{2}(\nabla;\mathbf{q}+\mathbf{q};\nabla).$$

Now consider the equation

(4) $$\tfrac{1}{2}\boldsymbol{\eta}\mathbf{D}\boldsymbol{\eta} = c,$$

where c is a constant. The left hand side is homogeneous and quadratic in the components of $\boldsymbol{\eta}$ and therefore represents a surface of the second degree. This surface is in fact a central quadric whose centre is P, for if the point $\boldsymbol{\eta}$ lies on it, so does the point $-\boldsymbol{\eta}$.

Also if $\boldsymbol{\eta}$ and $\boldsymbol{\eta}+d\boldsymbol{\eta}$ are adjacent points on (4) we have

$$0 = \tfrac{1}{2}(\boldsymbol{\eta}+d\boldsymbol{\eta})\,\mathbf{D}(\boldsymbol{\eta}+d\boldsymbol{\eta}) - \tfrac{1}{2}\boldsymbol{\eta}\,\mathbf{D}\boldsymbol{\eta} = \tfrac{1}{2}\boldsymbol{\eta}\,\mathbf{D}d\boldsymbol{\eta} + \tfrac{1}{2}d\boldsymbol{\eta}\,\mathbf{D}\boldsymbol{\eta} = \boldsymbol{\eta}\,\mathbf{D}d\boldsymbol{\eta}$$

to the first order, using the fact that $\mathbf{D}$ is a symmetric tensor.

Therefore the vector $\boldsymbol{\eta}\mathbf{D}$ is perpendicular to $d\boldsymbol{\eta}$ which is an arbitrary vector in the tangent plane to (4). Hence the vector $\boldsymbol{\eta}\mathbf{D}$ is in the direction of the normal to (4).

From this it appears that the velocity at Q is the sum of three parts, namely:

(i) The velocity $\mathbf{q}$ at P, which corresponds to a translation of the element as a whole.

(ii) The velocity $\tfrac{1}{2}(\nabla_\wedge \mathbf{q})_\wedge \boldsymbol{\eta}$, which is the velocity due to the rotation of the element as a whole with the angular velocity $\tfrac{1}{2}(\nabla_\wedge \mathbf{q})$. (See 2·12.)

(iii) A velocity $\boldsymbol{\eta}\mathbf{D}$ relative to P, which is in the direction of the normal to the quadric of the system of central quadrics (4), on which Q lies.

The first two of these motions are rigid body movements; they could still take place if the fluid element were frozen solid.

The third motion is called a velocity of *pure strain* and can only take place when the substance is deformable, as is the case with a fluid. This type of relative motion is characteristic of any deformable substance whether fluid or not.

To elucidate the nature of the pure strain, we observe that a central quadric has three perpendicular axes of symmetry which are normal to the tangent planes at their extremities. Lines parallel to these axes are being elongated at constant (though generally different) rates. Such a motion will distort an element originally spherical into an ellipsoid. We also note that lines in the direction of the axes of symmetry at time t will still be mutually perpendicular at time $t+\delta t$. Since the axes of symmetry are parallel to the normals at their extremities, the directions of these axes are given by the equation

$$\boldsymbol{\eta}_\wedge (\boldsymbol{\eta}\mathbf{D}) = 0.$$

The foregoing analysis shows that this description of the motion relates to an intrinsic property of the fluid, independent of any axes of reference.

2·41. Vorticity. The vector $\nabla_\wedge \mathbf{q} = \text{curl}\,\mathbf{q} = \boldsymbol{\zeta}$ say, is called the *vorticity vector*, or simply the *vorticity*. The angular velocity of an infinitesimal element, often but not very aptly called *molecular rotation*, is equal to half the vorticity.

If a spherical element of the fluid were suddenly solidified and the surrounding fluid simultaneously annihilated, this solid element would rotate with the above angular velocity. See Ex. II, 13.

A *vortex line* is a line drawn in the fluid such that the tangent to it at each point is in the direction of the vorticity vector at that point. It will be shown later (3·54) that vortex lines move with the fluid.

When the vorticity is different from zero the motion is said to be *rotational.*

A portion of the fluid at every point of which the vorticity is zero is said to be in *irrotational motion.* In such a portion of the fluid there are no vortex lines. Motions started from rest are always initially irrotational.

2·42. Circulation. Consider a closed curve C situated entirely in a moving fluid. Let $\mathbf{q}$ be the velocity at an arbitrary point P of the curve and $\mathbf{s}_1$ a unit vector drawn in the direction of the tangent at P, the direction being so chosen that an observer moving from P in the sense of $\mathbf{s}_1$ describes the curve in the sense chosen as positive. Take a point Q, on the curve, adjacent to P such that the arc PQ is of infinitesimal length δs. We can then form at P the scalar product $\mathbf{q}\mathbf{s}_1\,\delta s = \mathbf{q}\,\delta\mathbf{s}$, where $\delta\mathbf{s}$ is the directed element of arc at P (cf. 2·20).

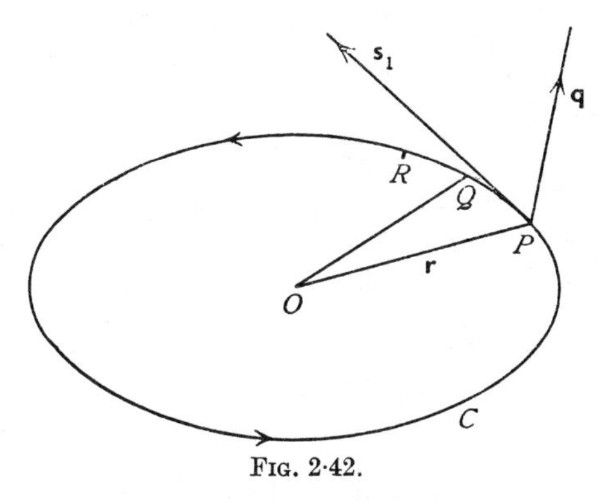

FIG. 2·42.

Forming the analogous products at Q, R, ..., and so on right round the curve back again to P, we define the *circulation* of the velocity vector round the curve C by the relation

$$\text{circulation} = \lim_{\delta s \to 0} \Sigma\,\mathbf{q}\,\delta\mathbf{s} = \int_{(C)} \mathbf{q}\,d\mathbf{s}.$$

The circulation may be written in the alternative forms

$$\text{circ } C = \int_{(C)} \mathbf{q}\,d\mathbf{s} = \int_{(C)} \mathbf{q}\,d\mathbf{r} = \int_{(C)} \mathbf{q}\,d\mathbf{P},$$

all of which mean the same thing.

We can form the circulation of any vector round a closed curve.

2·50. Stokes' theorem. Let S be a surface * having the closed curve C for boundary, and let $\mathbf{n}$ be a unit vector in that direction of the normal to the element of area dS which is related to the directions of circulation round dS and C by the right-handed screw rule. Then

$$\int_{(S)} \mathbf{n}(\nabla_\wedge \mathbf{q})\,dS = \int_{(C)} \mathbf{q}\,d\mathbf{s} = \text{circ } C.$$

This is Stokes' theorem.

Proof. If we join points of the curve C by sets of lines lying on the surface S so as to form a network, we see that every mesh of the network has lines in common with its neighbours, except those parts which belong to the curve C. Since a line which appears in two meshes is described twice in opposite senses, it follows that

circulation round C = sum of circulations in the meshes.

It is therefore sufficient to prove the theorem for a single mesh of an infinitesimal network covering S.

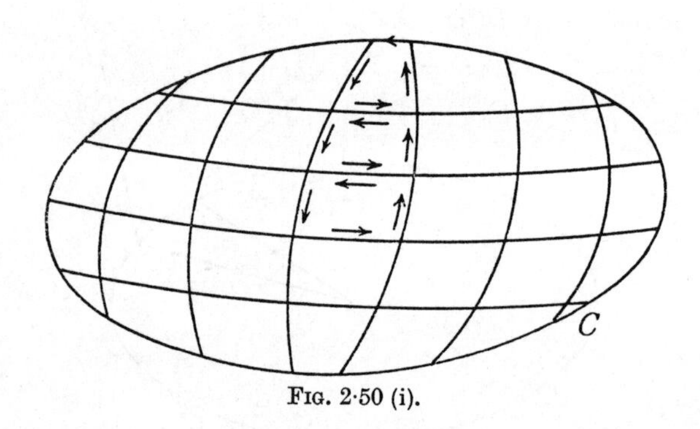

Fig. 2·50 (i).

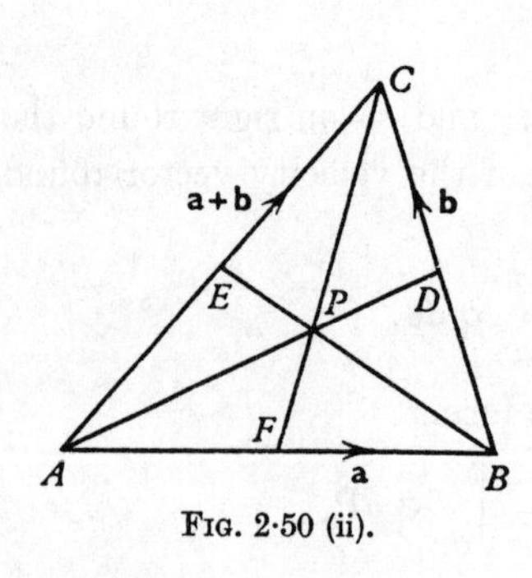

Fig. 2·50 (ii).

Since any mesh can be divided into triangles it is sufficient to prove the theorem for a single triangular mesh ABC whose sides are of infinitesimal length. Let D, E, F, be the mid-points of the sides, fig. 2·50 (ii), and let P be the centroid. Write

$$(1) \quad \overrightarrow{AB} = \mathbf{a}, \quad \overrightarrow{BC} = \mathbf{b}, \quad \overrightarrow{CA} = -(\mathbf{a}+\mathbf{b}).$$

Let $\mathbf{q}_M$ denote the value of $\mathbf{q}$ at any point M. Then by the definition of the integral we

* Such a surface may be conveniently described as a *diaphragm* closing C.

have, very nearly,

$$\int_{(ABC)} d\mathbf{s}\,\mathbf{q} = \overrightarrow{AB}\,\mathbf{q}_F + \overrightarrow{BC}\,\mathbf{q}_D + \overrightarrow{CA}\,\mathbf{q}_E$$

$$= \mathbf{a}(\mathbf{q}_F - \mathbf{q}_E) + \mathbf{b}(\mathbf{q}_D - \mathbf{q}_E).$$

Now from 2·31 (5)

$$\mathbf{q}_F = \mathbf{q}_P + (\overrightarrow{PF}\,\nabla)\mathbf{q}_P, \quad \mathbf{q}_E = \mathbf{q}_P + (\overrightarrow{PE}\,\nabla)\mathbf{q}_P.$$

Therefore by subtraction

$$(2) \qquad \mathbf{q}_F - \mathbf{q}_E = (\overrightarrow{EF}\,\nabla)\mathbf{q}_P = -\tfrac{1}{2}(\mathbf{b}\,\nabla)\mathbf{q}_P.$$

Similarly

$$\mathbf{q}_D - \mathbf{q}_E = \tfrac{1}{2}(\mathbf{a}\,\nabla)\mathbf{q}_P.$$

Therefore

$$(3) \qquad \int d\mathbf{s}\,\mathbf{q} = -\tfrac{1}{2}[\mathbf{a}(\mathbf{b}\,\nabla) - \mathbf{b}(\mathbf{a}\,\nabla)]\mathbf{q}_P = \tfrac{1}{2}[(\mathbf{a}_\wedge \mathbf{b})_\wedge \nabla]\mathbf{q}_P.$$

Now $\mathbf{n}\,dS = \frac{1}{2}(\mathbf{a}_\wedge \mathbf{b})$ if dS is the area of ABC and therefore to the same order of approximation

$$\int \mathbf{n}(\nabla_\wedge \mathbf{q})\,dS = \tfrac{1}{2}[\mathbf{a}_\wedge \mathbf{b}](\nabla_\wedge \mathbf{q}) = \tfrac{1}{2}[(\mathbf{a}_\wedge \mathbf{b})_\wedge \nabla]\mathbf{q}_P.$$

Comparing this with (3), we have proved the theorem for an infinitesimal triangle and therefore generally for any surface which can be regarded as the limit of a triangulation, bounded by a curve which can be regarded as the limit of an inscribed polygon. Q.E.D.

Stokes' theorem as stated above is a particular case of a more general theorem which, using directed areas, may be stated thus:

$$(1) \qquad \int_{(S)} \left(d\mathbf{S}_\wedge \frac{\partial}{\partial \mathbf{P}}\right) X = \int_{(C)} d\mathbf{s}\, X,$$

where X is any scalar or vector function of position and $d\mathbf{s}$ is the directed element of arc of C.

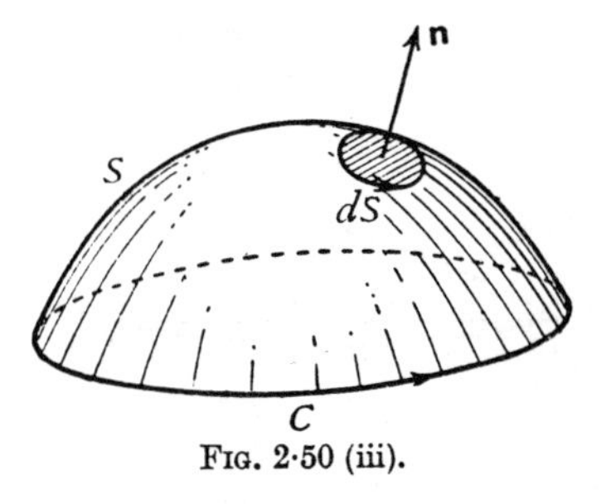

FIG. 2·50 (iii).

Proof. As before, it is sufficient to prove this for a single triangular mesh as follows. With the same steps,

$$\int d\mathbf{s}X = -\tfrac{1}{2}\mathbf{a}\left[\mathbf{b}\,\frac{\partial}{\partial \mathbf{P}}\right]X + \tfrac{1}{2}\mathbf{b}\left[\mathbf{a}\,\frac{\partial}{\partial \mathbf{P}}\right]X = \left[\tfrac{1}{2}(\mathbf{a}_\wedge \mathbf{b})_\wedge \frac{\partial}{\partial \mathbf{P}}\right]X,$$

and since $\frac{1}{2}\mathbf{a}_\wedge \mathbf{b}$ is the directed area of the mesh, the theorem follows for the mesh and therefore generally. Q.E.D.

A still more general form is

$$(4) \qquad \int_{(S)} \frac{\partial}{\partial \mathbf{P}}\left[\left(d\mathbf{S}\,_\wedge \frac{\partial}{\partial \mathbf{P}}\right) X\right] = \int_{(C)} \frac{\partial}{\partial \mathbf{P}}\,(d\mathbf{s}X),$$

the proof of which is an immediate inference from the method of proof given above; indeed we can even replace the first $\partial/\partial\mathbf{P}$ on each side by the same operation repeated n times.

2·51. Deductions from Stokes' theorem. Putting in turn $\mathbf{q}$, ϕ, $_\wedge\mathbf{q}$ for X in the general form of the theorem 2·50 (1), we get

$$(1) \qquad \int_{(C)} \mathbf{q}\, d\mathbf{s} = \int_{(S)} (\mathbf{n}\,_\wedge \nabla)\mathbf{q}\, dS = \int_{(S)} \mathbf{n}(\nabla\,_\wedge \mathbf{q}) dS$$

$$= \int_{(S)} \mathbf{n}\boldsymbol{\zeta}\, dS = \int_{(S)} \boldsymbol{\zeta}\, d\mathbf{S},$$

where $\boldsymbol{\zeta}$ is the vorticity. In words; the circulation of the velocity in any circuit is equal to the integral of the normal component of the vorticity over any diaphragm which closes the circuit.

$$(2) \qquad \int_{(C)} \phi\, d\mathbf{s} = \int_{(S)} (\mathbf{n}\,_\wedge \nabla)\phi\, dS.$$

Since $(\mathbf{n}_\wedge\ \nabla)_\wedge \mathbf{q} - \mathbf{n}_\wedge(\nabla_\wedge \mathbf{q}) = (\mathbf{n}\,\nabla)\mathbf{q} - \mathbf{n}(\nabla\mathbf{q})$ we have

$$(3) \qquad \int_{(C)} d\mathbf{s}\,_\wedge \mathbf{q} = \int_{(S)} (\mathbf{n}\,_\wedge \nabla)_\wedge \mathbf{q}\, dS = \int_{(S)} \left[(\mathbf{n}\nabla)\,\mathbf{q} - \mathbf{n}(\nabla\,\mathbf{q}) + \mathbf{n}\,_\wedge\boldsymbol{\zeta}\right] dS.$$

2·52. Irrotational motion. Let O be a fixed point, P an arbitrary point in a simply connected * region in which the motion of a fluid is irrotational. Join O to P by two paths OAP, OBP, each lying in the region in question.

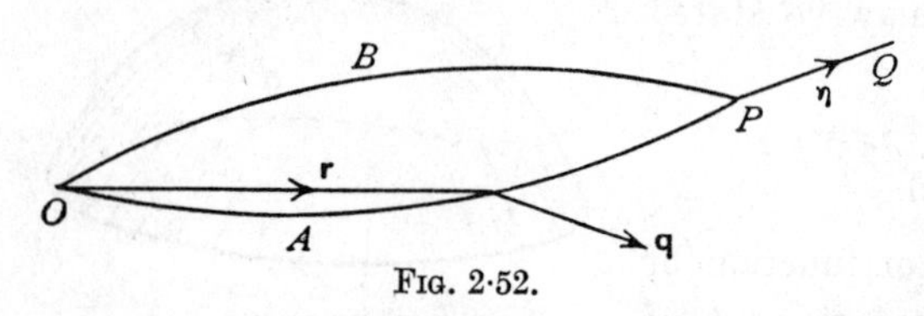

FIG. 2·52.

Then $OAPBO$ is a closed curve, and therefore, by Stokes' theorem,

$$\int_{(OAP)} \mathbf{q}\, d\mathbf{s} + \int_{(PBO)} \mathbf{q}\, d\mathbf{s} = \int_{(S)} \mathbf{n}(\nabla\,_\wedge \mathbf{q}) dS,$$

where S is any surface lying entirely in the fluid and having the curve $OAPBO$ for rim. Since the motion is irrotational, $\nabla\,_\wedge \mathbf{q} = 0$, and therefore

$$(1) \qquad \int_{(OAP)} \mathbf{q}\, d\mathbf{s} = \int_{(OBP)} \mathbf{q}\, d\mathbf{s} = -\phi_P,$$

* For the meaning of this term see 3·70.

say, and it is now clear that ϕ_P is a scalar function whose value depends solely on the position of P (and of the fixed point O) and not on the path from O to P.

Now take a point Q so near to P that the velocity vector $\mathbf{q}$ may be assumed nearly constant along PQ.

Let $\boldsymbol{\eta}$ be the position vector of Q with respect to P.

Then, approximately, if ϕ_P is denoted by ϕ,

$$-\boldsymbol{\eta}\,\nabla\,\phi = -\phi_Q+\phi_P = \int_{(PQ)} \mathbf{q}\,d\mathbf{s} = \mathbf{q}\boldsymbol{\eta}.$$

Since Q is arbitrary, provided it is near enough to P, the vector $\boldsymbol{\eta}$ is also arbitrary and therefore

$$\mathbf{q} = -\nabla\,\phi. \tag{2}$$

Thus when the motion is irrotational the velocity vector is the gradient of a scalar function * of position $-\phi$.

This scalar function is called the *velocity potential.* We have proved that the velocity potential necessarily exists when the motion is irrotational. Conversely, when the velocity potential exists, the motion is necessarily irrotational, for then

$$\nabla_{\wedge}\mathbf{q} = -\nabla_{\wedge}(\nabla\,\phi) = 0,$$

from 2·32 (III).

It also appears, from the meaning of $\nabla\,\phi$, that the fluid velocity at any point is normal to that member of the system of surfaces ϕ = constant, which passes through that point. In other words, the streamlines cut the equi-ϕ surfaces orthogonally.

2·53. Conservative field of force. In a conservative field of force (1·42), the work done by the force $\mathbf{F}$ of the field in taking a unit mass from O to P is independent of the path. Thus in fig. 2·52,

$$\int_{(OAP)} \mathbf{F}\,d\mathbf{r} = \int_{(OBP)} \mathbf{F}\,d\mathbf{r} = -\Omega_P,$$

where Ω_P is a scalar function whose value depends solely on the position of P (and of the fixed point O).

This equation is of the same form as (1) of section 2·52, and we can from that point repeat the same argument to show that

$$\mathbf{F} = -\nabla\,\Omega,$$

where Ω is a scalar function, known as the force potential. Physically, Ω measures the potential energy of the field, that is, the energy stored up in taking a unit mass from O to P.

---

* The negative sign for this scalar function is adopted by some writers and not by others. We have followed Lamb in adopting the negative sign, so that $\rho\phi$ is the impulsive pressure which will generate the motion from rest, cf. 3·64.

The negative sign in 2·52 (2) further brings out the mathematical (not physical) analogy between the velocity potential and the force potential.

2·60. Gauss's theorem. Let the closed surface S enclose the volume V, and let X be a scalar or vector function of position. Then, if $d\tau$ is an element of the volume V, and dS is an element of the surface S,

$$\int_{(V)} (\nabla X) d\tau = -\int_{(S)} \mathbf{n} X \, dS,$$

where $\mathbf{n}$ is a unit vector in the direction of the normal to dS drawn into the *interior* of the region enclosed by S. This is Gauss's theorem.*

Proof. By drawing three systems of surfaces, say parallel planes, the volume V will be divided into elements of volume. If $\delta\tau$ be such an element, we shall have approximately (2·24 (1)),

$$(\nabla X)\delta\tau = -\int_{(\delta\tau)} \mathbf{n} X \, dS,$$

the integral being taken over the surface of the volume $\delta\tau$, and by summation for all elements:

$$\int_{(V)} (\nabla X) d\tau = \lim_{\delta\tau \to 0} \Sigma \nabla X \, \delta\tau = -\Sigma \int_{(\delta\tau)} \mathbf{n} X \, dS.$$

Now at a point on the common boundary of two neighbouring elements the inward normals to each element are of opposite sign. Thus the surface integrals over boundaries which are shared by two elements of volume cancel out and we are left with the surface integral over S. Q.E.D.

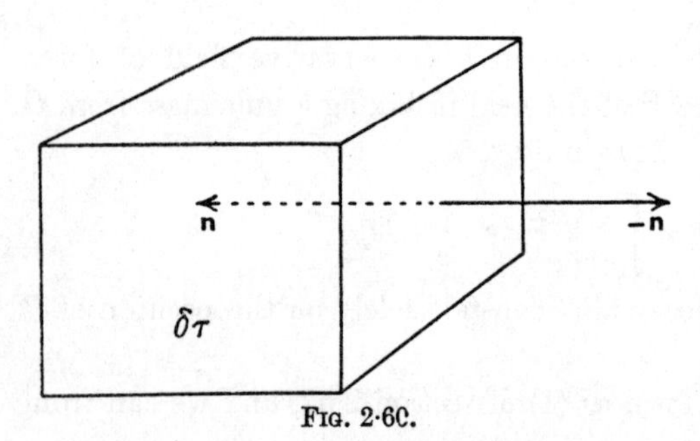

Fig. 2·6c.

Note that the minus sign in the above theorem arises from the fact that we have considered the normal drawn *into* the region enclosed by S. In applications to hydrodynamics we shall thus be considering the normal drawn into the fluid when S is the bounding surface.

It should be mentioned that the theorems of Stokes and Gauss, and the various deductions therefrom, depend for their complete validity on the existence and continuity of the partial derivatives implied in their enunciations. Discontinuity will manifest itself physically when it occurs in the motion of a

* C. F. Gauss, "Theoria attractionis", *Comm. soc. reg. Gott.*, Vol. II, Göttingen, 1813.

fluid, and we shall not therefore discuss conditions of validity, for that would lead us too far from the main theme.

If the region within S is m-ply connected (see 3·70), we modify it to become simply connected by inserting $m-1$ barriers $B_1, B_2, \ldots, B_{m-1}$; and reckon each face of a barrier as a separate boundary. Thus we have in the case of a doubly connected region, a single barrier B whose faces will be denoted by B^+ (the positive face) and B^- (the negative face). Then Gauss's theorem applied to the simply connected region so attained gives

$$\int_{(V)} \nabla X \, d\tau = -\int_{(S)} \mathbf{n} \, X \, dS - \int_{(B^+)} \mathbf{n}^+ X^+ \, dS - \int_{(B^-)} \mathbf{n}^- X^- \, dS$$

Since at any part of B, $\mathbf{n}^+ + \mathbf{n}^- = 0$, if we write

$$X^+ - X^- = [X]$$

for the jump in X when crossing B from the negative to the positive side Gauss's theorem for the doubly connected region in question is

$$\int_{(V)} \nabla X \, d\tau = -\int_{(S)} \mathbf{n} \, X \, dS - \int_{(B^+)} \mathbf{n}^+ [X] \, dS.$$

If $m > 2$ we simply add more terms on the right, one for each barrier.

2·61. Deductions from Gauss's theorem. If $\mathbf{a}$ is a vector and ϕ a scalar, let us write in the enunciation of Gauss's theorem the following forms instead of (∇X): $\nabla \mathbf{a}$, $\nabla_\wedge \mathbf{a}$, $\nabla \phi$, $(\nabla \nabla)\phi$, $(\nabla \nabla)\mathbf{a}$, $\nabla(\mathbf{q} \, ; \mathbf{a})$.

We then get the following theorems:

(1) $$\int_{(V)} \nabla \mathbf{a} \, d\tau = -\int_{(S)} \mathbf{n}\mathbf{a} \, dS.$$

(2) $$\int_{(V)} \nabla_\wedge \mathbf{a} \, d\tau = -\int_{(S)} \mathbf{n}_\wedge \mathbf{a} \, dS.$$

(3) $$\int_{(V)} \nabla \phi \, d\tau = -\int_{(S)} \mathbf{n}\phi \, dS.$$

(4) $$\int_{(V)} \nabla^2 \phi \, d\tau = -\int_{(S)} (\mathbf{n} \nabla)\phi \, dS = -\int_{(S)} \frac{\partial \phi}{\partial n} \, dS, \text{ from 2·22.}$$

(5) $$\int_{(V)} \nabla^2 \mathbf{a} \, d\tau = -\int_{(S)} (\mathbf{n} \nabla)\mathbf{a} \, dS.$$

(6) $$\int_{(V)} \nabla(\mathbf{q} \, ; \mathbf{a}) d\tau = -\int_{(S)} \mathbf{n}(\mathbf{q} \, ; \mathbf{a}) dS,$$

which may be called the *tensor form* of Gauss's theorem.

Using 2·34 (X), the last result leads to

(7) $$\int_{(S)} \mathbf{a}(\mathbf{n}\mathbf{q}) dS = -\int_{(V)} [\mathbf{a}(\nabla \mathbf{q}) + (\mathbf{q} \nabla) \mathbf{a}] \, d\tau.$$

Gauss's theorem may also be formulated thus:

$$(8) \qquad \int_{(V)} \frac{\partial X}{\partial \mathbf{P}}\, dV = -\int_{(S)} d\mathbf{S}\, X,$$

using dV for the element of volume and $d\mathbf{S}$ for the *inwardly* directed vector element of surface area.

2·615. A solenoidal vector forms tubes of constant intensity. If $\mathbf{a}$ is a vector field, an $\mathbf{a}$-line is a line whose tangent at every point is in the direction of the $\mathbf{a}$-vector through that point (cf. streamlines). An $\mathbf{a}$-tube results from drawing the $\mathbf{a}$-line through every point of a closed curve. Consider the portion of an $\mathbf{a}$-tube between two plane sections by *planes* S_1, S_2 whose outward normals are $\mathbf{n}_1$ and $-\mathbf{n}_2$. By Gauss's theorem

$$\int_{(S_1)} \mathbf{n}_1\mathbf{a}\, dS_1 - \int_{(S_2)} \mathbf{n}_2\mathbf{a}\, dS_2 = \int \nabla\cdot\mathbf{a}\, d\tau = 0\,;$$

since by definition* $\nabla\mathbf{a} = 0$ and since by the definition of an $\mathbf{a}$-tube, $\mathbf{na}\, dS = 0$ at the lateral surface.

Thus $A = \int \mathbf{na}\, dS$ is constant along the tube. We call A the *intensity* of the tube. We can therefore define a unit tube as one of unit intensity and we can speak of the number N of unit tubes which thread a given circuit C.

2·62. Green's theorem.† From 2.34 (VI) we have, for any vector $\mathbf{a}$,

$$\nabla(\phi\mathbf{a}) = \mathbf{a}(\nabla\,\phi) + \phi(\nabla\,\mathbf{a}).$$

Thus from 2·61 (1), we get

$$-\int_{(S)} \mathbf{n a}\phi\, dS = \int_{(V)} \mathbf{a}(\nabla\,\phi)\, d\tau + \int_{(V)} \phi(\nabla\,\mathbf{a})\, d\tau.$$

Putting instead of $\mathbf{a}$ the vector $\nabla\,\psi$, where ψ is a scalar function, and noticing that $\mathbf{n}\,\nabla\,\psi = \partial\psi/\partial n$ (2·22), we get

$$(1) \qquad \int_{(V)} (\nabla\,\psi\,\nabla\,\phi)\, d\tau = -\int_{(V)} \phi\,\nabla^2\psi\, d\tau - \int_{(S)} \phi\frac{\partial\psi}{\partial n}\, dS$$

$$= -\int_{(V)} \psi\,\nabla^2\phi\, d\tau - \int_{(S)} \psi\frac{\partial\phi}{\partial n}\, dS,$$

since the left-hand side is unaltered when ϕ and ψ are interchanged. The above relations constitute Green's theorem, or *Green's first identity*.

An immediate inference is *Green's second identity*

* See 2·24. † G. Green, *Essay on Electricity and Magnetism*, Nottingham, 1828.

$$(2)\qquad \int_{(V)} (\phi\nabla^2\psi - \psi\nabla^2\phi)\, d\tau = -\int_{(S)} \left(\phi\frac{\partial\psi}{\partial n} - \psi\frac{\partial\phi}{\partial n}\right) dS.$$

Put $\phi = \psi$ in (1). Then

$$(3)\qquad \int_{(V)} (\nabla\phi)(\nabla\phi)\, d\tau = -\int_{(V)} \phi\nabla^2\phi\, d\tau - \int_{(S)} \phi\frac{\partial\phi}{\partial n}\, dS.$$

Def. Any solution ϕ of Laplace's equation $\nabla^2 V = 0$ is called a *harmonic function.*

If ϕ is a harmonic function, it follows from (1) that

$$(4)\qquad \int_{(V)} (\nabla\phi)(\nabla\psi)\, d\tau = -\int_{(S)} \psi\frac{\partial\phi}{\partial n}\, dS.$$

Herein put $\psi = 1$. Then

$$(5)\qquad \int_{(S)} \frac{\partial\phi}{\partial n}\, dS = 0.$$

If ϕ and ψ are both harmonic functions, (2) gives

$$(6)\qquad \int_{(S)} \left(\phi\frac{\partial\psi}{\partial n} - \psi\frac{\partial\phi}{\partial n}\right) dS = 0.$$

In Green's theorem the functions ϕ and ψ must be one-valued, that is to say, to each point P of the region V there must correspond only one value of ϕ and one value of ψ. These functions will, in our applications, usually represent velocity potentials and, provided the region is simply connected,* the above condition will be satisfied. The same may be true in a multiply connected region, but here it is possible for the condition to be violated on account of the existence of circulations. When circulations exist Green's theorem requires modification.

Suppose for example that the region is doubly connected and that on crossing the barrier B which renders it simply connected, ϕ, ψ jump by constant quantities κ, λ, the *cyclic constants* of the barrier,

$$(7)\qquad \phi^+ - \phi^- = [\phi] = \kappa, \quad \psi^+ - \psi^- = [\psi] = \lambda.$$

Then the foregoing argument shows that

$$\begin{aligned}(8)\qquad \int_{(V)} (\nabla\phi\nabla\psi)\, d\tau &= -\int_{(V)} \phi\,\nabla^2\psi\, d\tau - \int_{(S)} \phi\frac{\partial\psi}{\partial n}\, dS - \int_{(B^+)} \kappa\frac{\partial\psi}{\partial n^+} dS \\ &= -\int_{(V)} \phi\,\nabla^2\psi\, d\tau - \int_{(S)} \phi\frac{\partial\psi}{\partial n} dS - \kappa\int_{(B^+)} \frac{\partial\psi}{\partial n^+} dS \\ &= -\int_{(V)} \psi\,\nabla^2\phi\, d\tau - \int_{(S)} \psi\frac{\partial\phi}{\partial n} dS - \lambda\int_{(B^+)} \frac{\partial\phi}{\partial n^+} dS,\end{aligned}$$

the last result being got by interchanging ϕ and ψ in (4), which is permissible since the left-hand side is unaltered.

---

* See 3·70.

The foregoing constitute Green's theorem for a doubly connected region. For an n-ply connected region we add one more term for each additional barrier. E.g. if $n=3$,

$$\int(\nabla\phi\,\nabla\psi)d\tau = -\int\phi\,\nabla^2\psi\,d\tau - \int\phi\frac{\partial\psi}{\partial n}dS - \kappa_1\int_{(B_1+)}\frac{\partial\psi}{\partial n^+}dS - \kappa_2\int_{(B_2}\frac{\partial\psi}{\partial n^+}dS.$$

2·63. An application of Green's theorem. Take a closed surface S at every point of whose interior $\nabla^2\phi = 0$, $\nabla^2\psi = 0$. Then, by Green's theorem,

$$\int_{(S)}\left(\phi\frac{\partial\psi}{\partial n} - \psi\frac{\partial\phi}{\partial n}\right)dS = 0. \tag{1}$$

Take a point P interior to S, and let r be the distance of P from the element of area dS. We shall prove that, if ϕ_P is the value of ϕ at P,

$$4\pi\,\phi_P = \int_{(S)}\left[\phi\frac{\partial}{\partial n}\left(\frac{1}{r}\right) - \frac{1}{r}\frac{\partial\phi}{\partial n}\right]dS, \tag{2}$$

which expresses the value of ϕ at any interior point in terms of its values on the boundary.

Proof. Take $\psi = 1/r$. It is easily verified that $\nabla^2 1/r = 0$. Draw a sphere Σ centre P, radius R, so small that Σ is entirely within S and apply 2·62 (2) to the region v between Σ and S. Since $dn = dR$ on Σ, we get

$$\int_{(v)}\frac{1}{r}\nabla^2\phi\,d\tau = \int_{(S)}\left[\phi\frac{\partial}{\partial n}\left(\frac{1}{r}\right) - \frac{1}{r}\frac{\partial\phi}{\partial n}\right]dS + \int_{(\Sigma)}\left[\phi\frac{\partial}{\partial R}\left(\frac{1}{R}\right) - \frac{1}{R}\frac{\partial\phi}{\partial R}\right]d\Sigma. \tag{3}$$

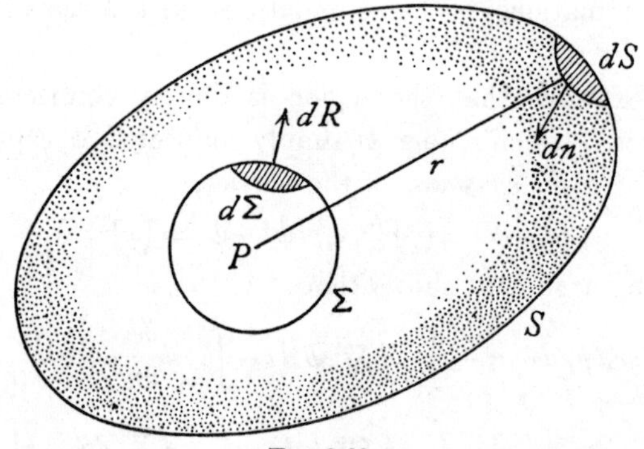

FIG. 2·63.

Now when $R\to 0$, $\lim v = V$ so that if we make R so small that $\phi = \phi_P$ nearly, over the whole surface, the limit of the last integral is

$$\lim_{R\to 0}\left[-\phi_P\frac{1}{R^2} - \frac{1}{R}\frac{\partial\phi}{\partial R}\right]4\pi R^2 = -4\pi\,\phi_P.$$

Therefore we obtain *Green's third identity*

$$(4)\qquad \phi_P = -\frac{1}{4\pi}\int_{(V)} \frac{1}{r}\nabla^2\phi\, d\tau - \frac{1}{4\pi}\int_{(S)} \frac{1}{r}\frac{\partial\phi}{\partial n} dS + \frac{1}{4\pi}\int_{(S)} \phi\frac{\partial}{\partial n}\left(\frac{1}{r}\right) dS,$$

and since $\nabla^2\phi = 0$ we have (2). Q.E.D.

It also follows from (1) that the left-hand side of (2) is zero for a point exterior to S.

2·70. Cartesian coordinates. If we take three mutually perpendicular axes of reference, Ox, Oy, Oz, and three unit vectors, $\mathbf{i}$, $\mathbf{j}$, $\mathbf{k}$, parallel to these axes, any vector $\mathbf{a}$ can be expressed in terms of its components,* a_x, a_y, a_z, along the axes in the form

$$\mathbf{a} = \mathbf{i}a_x + \mathbf{j}a_y + \mathbf{k}a_z.$$

The vectors $\mathbf{i}$, $\mathbf{j}$, $\mathbf{k}$ themselves combine according to the laws

$$\mathbf{i}^2 = \mathbf{j}^2 = \mathbf{k}^2 = 1, \quad \mathbf{ij} = \mathbf{jk} = \mathbf{ki} = 0$$

for their scalar products, since they are perpendicular.

For the vector products, we have

$$\mathbf{i}_\wedge\mathbf{i} = \mathbf{j}_\wedge\mathbf{j} = \mathbf{k}_\wedge\mathbf{k} = 0, \quad \mathbf{j}_\wedge\mathbf{k} = \mathbf{i}, \quad \mathbf{k}_\wedge\mathbf{i} = \mathbf{j}, \quad \mathbf{i}_\wedge\mathbf{j} = \mathbf{k},$$

for the same reason.

Taking a second vector $\mathbf{b}$, we have therefore the scalar product

$$\mathbf{ab} = (\mathbf{i}a_x + \mathbf{j}a_y + \mathbf{k}a_z)(\mathbf{i}b_x + \mathbf{j}b_y + \mathbf{k}b_z) = a_xb_x + a_yb_y + a_zb_z;$$

and the vector product

$$\begin{aligned}\mathbf{a}_\wedge\mathbf{b} &= (\mathbf{i}a_x + \mathbf{j}a_y + \mathbf{k}a_z)_\wedge(\mathbf{i}b_x + \mathbf{j}b_y + \mathbf{k}b_z)\\ &= a_x\mathbf{i}_\wedge(\mathbf{i}b_x + \mathbf{j}b_y + \mathbf{k}b_z) + a_y\mathbf{j}_\wedge(\mathbf{i}b_x + \mathbf{j}b_y + \mathbf{k}b_z) + a_z\mathbf{k}_\wedge(\mathbf{i}b_x + \mathbf{j}b_y + \mathbf{k}b_z)\\ &= \mathbf{i}(a_yb_z - a_zb_y) + \mathbf{j}(a_zb_x - a_xb_z) + \mathbf{k}(a_xb_y - a_yb_x).\end{aligned}$$

The vector product can be more conveniently written in the form of a determinant, thus:

$$\mathbf{a}_\wedge\mathbf{b} = \begin{vmatrix} \mathbf{i} & \mathbf{j} & \mathbf{k} \\ a_x & a_y & a_z \\ b_x & b_y & b_z \end{vmatrix}.$$

In this form it is clearly seen that $\mathbf{a}_\wedge\mathbf{b}$ and $\mathbf{b}_\wedge\mathbf{a}$ have opposite signs, for the second is obtained from the first by interchanging the last two rows of the determinant, thereby causing a change of sign, but not of absolute value.

If $\phi = \phi(x, y, z)$ is a scalar function, we have, from 2·22, $\mathbf{i}\nabla\phi = \partial\phi/\partial x$ and, from 2·15 (i),

$$\nabla\phi = \mathbf{i}(\mathbf{i}\nabla\phi) + \mathbf{j}(\mathbf{j}\nabla\phi) + \mathbf{k}(\mathbf{k}\nabla\phi),$$

* This notation for the components of a vector is very convenient. Thus the components of the velocity $\mathbf{q}$ would be (q_x, q_y, q_z), although they are more usually denoted by (u, v, w). We shall use both notations for $\mathbf{q}$.

and therefore $$\nabla \phi = \left(\mathbf{i}\frac{\partial}{\partial x}+\mathbf{j}\frac{\partial}{\partial y}+\mathbf{k}\frac{\partial}{\partial z}\right)\phi,$$

so that the vector operator ∇ is to be interpreted by

(1) $$\nabla = \mathbf{i}\frac{\partial}{\partial x}+\mathbf{j}\frac{\partial}{\partial y}+\mathbf{k}\frac{\partial}{\partial z}.$$

If we apply the operator to the vector $\mathbf{q}$ whose components are u, v, w parallel to the axes, we shall get

$$\nabla \mathbf{q} = \left(\mathbf{i}\frac{\partial}{\partial x}+\mathbf{j}\frac{\partial}{\partial y}+\mathbf{k}\frac{\partial}{\partial z}\right)(\mathbf{i}u+\mathbf{j}v+\mathbf{k}w),$$

which gives, on performing the multiplication,

(2) $$\nabla \mathbf{q} = \frac{\partial u}{\partial x}+\frac{\partial v}{\partial y}+\frac{\partial w}{\partial z}.$$

Repeating the operation on $\nabla \phi$, we obtain at once

$$\nabla^2 \phi = \frac{\partial^2 \phi}{\partial x^2}+\frac{\partial^2 \phi}{\partial y^2}+\frac{\partial^2 \phi}{\partial z^2}.$$

Again, $$\nabla_\wedge \mathbf{q} = \left(\mathbf{i}\frac{\partial}{\partial x}+\mathbf{j}\frac{\partial}{\partial y}+\mathbf{k}\frac{\partial}{\partial z}\right)_\wedge(\mathbf{i}u+\mathbf{j}v+\mathbf{k}w)$$
$$= \mathbf{i}\left(\frac{\partial w}{\partial y}-\frac{\partial v}{\partial z}\right)+\mathbf{j}\left(\frac{\partial u}{\partial z}-\frac{\partial w}{\partial x}\right)+\mathbf{k}\left(\frac{\partial v}{\partial x}-\frac{\partial u}{\partial y}\right).$$

We can also write symbolically

$$\nabla_\wedge \mathbf{q} = \begin{vmatrix} \mathbf{i} & \mathbf{j} & \mathbf{k} \\ \frac{\partial}{\partial x} & \frac{\partial}{\partial y} & \frac{\partial}{\partial z} \\ u & v & w \end{vmatrix}.$$

To interpret the expression $(\mathbf{a}\nabla)\mathbf{q}$, we observe that, if

$$\mathbf{a} = \mathbf{i}a_x+\mathbf{j}a_y+\mathbf{k}a_z,$$

then $$\mathbf{a}\nabla = a_x\frac{\partial}{\partial x}+a_y\frac{\partial}{\partial y}+a_z\frac{\partial}{\partial z},$$

and therefore

$$(\mathbf{a}\nabla)\mathbf{q} = \mathbf{i}\left(a_x\frac{\partial u}{\partial x}+a_y\frac{\partial u}{\partial y}+a_z\frac{\partial u}{\partial z}\right)$$
$$+\mathbf{j}\left(a_x\frac{\partial v}{\partial x}+a_y\frac{\partial v}{\partial y}+a_z\frac{\partial v}{\partial z}\right)+\mathbf{k}\left(a_x\frac{\partial w}{\partial x}+a_y\frac{\partial w}{\partial y}+a_z\frac{\partial w}{\partial z}\right).$$

Lastly,

$$\nabla^2 \mathbf{q} = \left(\frac{\partial^2}{\partial x^2}+\frac{\partial^2}{\partial y^2}+\frac{\partial^2}{\partial z^2}\right)(\mathbf{i}u+\mathbf{j}v+\mathbf{k}w)$$
$$= \mathbf{i}\left(\frac{\partial^2 u}{\partial x^2}+\frac{\partial^2 u}{\partial y^2}+\frac{\partial^2 u}{\partial z^2}\right)+\mathbf{j}\left(\frac{\partial^2 v}{\partial x^2}+\frac{\partial^2 v}{\partial y^2}+\frac{\partial^2 v}{\partial z^2}\right)+\mathbf{k}\left(\frac{\partial^2 w}{\partial x^2}+\frac{\partial^2 w}{\partial y^2}+\frac{\partial^2 w}{\partial z^2}\right).$$

The foregoing interpretations in cartesian coordinates serve to illustrate the economy of thought and writing arising from the use of vector notations independent of coordinates. The vector methods form a powerful tool for obtaining general theorems and afford immediate insight into their intrinsic character. In order to investigate particular problems which involve the carrying of calculations to a numerical conclusion it is nearly always necessary to introduce coordinates at some stage. It is clear that this stage may often be advantageously deferred as long as possible.

2·71. The alternative notation $\partial/\partial\mathbf{r}$. We have seen in 2·70 (1) that the gradient operator may be written

$$(1)\qquad \frac{\partial}{\partial\mathbf{r}} = \mathbf{i}\frac{\partial}{\partial x}+\mathbf{j}\frac{\partial}{\partial y}+\mathbf{k}\frac{\partial}{\partial z}, \quad \mathbf{r} = \mathbf{i}x+\mathbf{j}y+\mathbf{k}z,$$

and therefore the dyadic product

$$(2)\qquad \frac{\partial}{\partial\mathbf{r}};\mathbf{r} = \left(\mathbf{i}\frac{\partial}{\partial x}+\mathbf{j}\frac{\partial}{\partial y}+\mathbf{k}\frac{\partial}{\partial z}\right); (\mathbf{i}x+\mathbf{j}y+\mathbf{k}z)$$

$$= (\mathbf{i};\mathbf{i})+(\mathbf{j};\mathbf{j})+(\mathbf{k};\mathbf{k}) = \mathbf{I}, \text{ the idemfactor (2·16).}$$

Thus if $\mathbf{a}$ is a constant vector,

$$(3)\qquad \frac{\partial}{\partial\mathbf{r}}(\mathbf{ra}) = \left(\frac{\partial}{\partial\mathbf{r}};\mathbf{r}\right)\mathbf{a} = \mathbf{I}\,\mathbf{a} = \mathbf{a}.$$

Also $\dfrac{\partial\mathbf{r}}{\partial x} = \mathbf{i}$, and therefore

$$(4)\qquad \frac{\partial\phi}{\partial x} = \mathbf{i}\frac{\partial\phi}{\partial\mathbf{r}} = \frac{\partial\mathbf{r}}{\partial x}\frac{\partial\phi}{\partial\mathbf{r}}.$$

These results are capable of a simple generalisation.

Thus if $\mathbf{q} = \mathbf{i}u+\mathbf{j}v+\mathbf{k}w$, we can write

$$(5)\qquad \frac{\partial}{\partial\mathbf{q}} = \mathbf{i}\frac{\partial}{\partial u}+\mathbf{j}\frac{\partial}{\partial v}+\mathbf{k}\frac{\partial}{\partial w},$$

and therefore if $\mathbf{a}$ is a constant vector,

$$(6)\qquad \frac{\partial}{\partial\mathbf{q}}(\mathbf{qa}) = \mathbf{a},$$

and if T is a scalar function of u, v, w,

$$(7)\qquad \frac{\partial T}{\partial u} = \frac{\partial\mathbf{q}}{\partial u}\frac{\partial T}{\partial\mathbf{q}}.$$

Again, as in 2·33,

$$\frac{\partial q^2}{\partial\mathbf{q}} = \frac{\partial}{\partial\mathbf{q}}(\mathbf{qq}) = \frac{\partial}{\partial\mathbf{q}}(\mathbf{q}_0\mathbf{q})+\frac{\partial}{\partial\mathbf{q}}(\mathbf{qq}_0) = \mathbf{q}_0+\mathbf{q}_0 = 2\mathbf{q} \quad \text{from (6).}$$

If $\mathbf{r}_0$ and $\mathbf{r}$ are the position vectors of the same fluid particle at two different instants of time, it is easily verified that*

$$* \; d\phi = d\mathbf{r}\frac{\partial\phi}{\partial\mathbf{r}} = d\mathbf{r}_0\frac{\partial\phi}{\partial\mathbf{r}_0} = \left[\left(d\mathbf{r}\frac{\partial}{\partial\mathbf{r}}\right)\mathbf{r}_0\right]\frac{\partial\phi}{\partial\mathbf{r}_0} = d\mathbf{r}\left[\frac{\partial;\mathbf{r}_0}{\partial\mathbf{r}}\frac{\partial\phi}{\partial\mathbf{r}_0}\right].$$

(8) $$\frac{\partial}{\partial\mathbf{r}} = \frac{\partial\,;\,\mathbf{r}_0}{\partial\mathbf{r}}\cdot\frac{\partial}{\partial\mathbf{r}_0}, \quad\text{where}\quad \frac{\partial\,;\,\mathbf{r}_0}{\partial\mathbf{r}} \quad\text{means}\quad \frac{\partial}{\partial\mathbf{r}}\,;\,\mathbf{r}_0.$$

Thus in particular

(9) $$\frac{\partial\,;\,\mathbf{r}_0}{\partial\mathbf{r}}\cdot\frac{\partial\,;\,\mathbf{r}}{\partial\mathbf{r}_0} = \frac{\partial\,;\,\mathbf{r}}{\partial\mathbf{r}} = \mathbf{I} = \frac{\partial\,;\,\mathbf{r}}{\partial\mathbf{r}_0}\cdot\frac{\partial\,;\,\mathbf{r}_0}{\partial\mathbf{r}}.$$

Let T be a homogeneous scalar function of the second degree of two independent vectors $\mathbf{u}$, $\boldsymbol{\omega}$. By this we mean that if $T = T(\mathbf{u}, \boldsymbol{\omega})$, then if t is a scalar,

$$T(t\mathbf{u}, t\boldsymbol{\omega}) = t^2 T(\mathbf{u}, \boldsymbol{\omega}).$$

Write $\boldsymbol{\xi} = t\mathbf{u}$, $\boldsymbol{\eta} = t\boldsymbol{\omega}$, then

$$T(\boldsymbol{\xi}, \boldsymbol{\eta}) = t^2 T(\mathbf{u}, \boldsymbol{\omega}),$$

and therefore

$$\frac{\partial T(\boldsymbol{\xi}, \boldsymbol{\eta})}{\partial t} = 2tT(\mathbf{u}, \boldsymbol{\omega}).$$

But $$\frac{\partial T(\boldsymbol{\xi}, \boldsymbol{\eta})}{\partial t} = \frac{\partial T}{\partial\boldsymbol{\xi}}\frac{\partial\boldsymbol{\xi}}{\partial t} + \frac{\partial T}{\partial\boldsymbol{\eta}}\frac{\partial\boldsymbol{\eta}}{\partial t}$$
$$= \mathbf{u}\,\frac{\partial T}{\partial\boldsymbol{\xi}} + \boldsymbol{\omega}\,\frac{\partial T}{\partial\boldsymbol{\eta}}.$$

Thus $$\mathbf{u}\,\frac{\partial T}{\partial\boldsymbol{\xi}} + \boldsymbol{\omega}\,\frac{\partial T}{\partial\boldsymbol{\eta}} = 2tT(\mathbf{u}, \boldsymbol{\omega}).$$

Putting $t = 1$ we get

$$\mathbf{u}\,\frac{\partial T}{\partial\mathbf{u}} + \boldsymbol{\omega}\,\frac{\partial T}{\partial\boldsymbol{\omega}} = 2T,$$

which is the vector analogue of Euler's theorem on homogeneous functions (of degree 2).

The method of proof is quite general and applies to a homogeneous function of degree n, in which case 2 is replaced, in the above proof, by n.

2·72. Orthogonal curvilinear coordinates. In cartesian coordinates the position of a point is defined by the intersection of three mutually perpen-

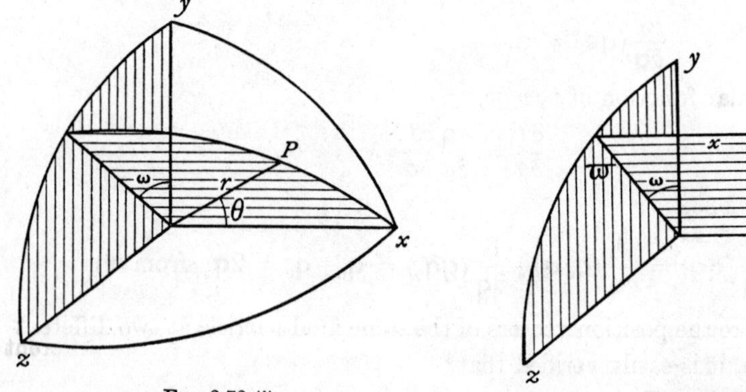

FIG. 2·72 (i). FIG. 2·72 (ii).

dicular planes, x = constant, y = constant, z = constant. For certain problems other systems of coordinates are convenient, for example, spherical polar coordinates, in which the position is defined by the intersection of a sphere, r = constant, a plane ω = constant, and a cone θ = constant, fig. 2·72 (i), or cylindrical coordinates, in which the position is defined by the intersection of two planes, x = constant, ω = constant, and a cylinder ϖ = constant, fig. 2·72 (ii).

To discuss the form taken by the nabla operator in such a system of orthogonal coordinates, suppose these to be defined by

$$x = f_1(u_1, u_2, u_3), \quad y = f_2(u_1, u_2, u_3), \quad z = f_3(u_1, u_2, u_3),$$

where the surfaces, u_1 = constant, u_2 = constant, u_3 = constant, intersect orthogonally. If we draw the surfaces corresponding to u_1, u_2, u_3 and $u_1+\delta u_1$, $u_2+\delta u_2$, $u_3+\delta u_3$, we obtain a figure which is to the first order a rectangular parallelepiped whose edges are $h_1\,\delta u_1$, $h_2\,\delta u_2$, $h_3\,\delta u_3$, fig. 2·72 (iii), where h_1, h_2, h_3 are functions of the coordinates obtained from the relation

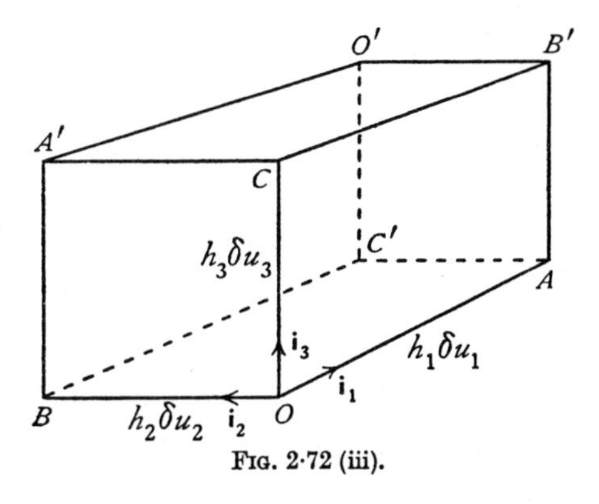

FIG. 2·72 (iii).

$$(ds)^2 = (dx)^2+(dy)^2+(dz)^2$$
$$= h_1^2(du_1)^2+h_2^2(du_2)^2+h_3^2(du_3)^2$$

where

$$dx = \frac{\partial x}{\partial u_1}du_1+\frac{\partial x}{\partial u_2}du_2+\frac{\partial x}{\partial u_3}du_3,$$

and so on, the product terms like du_1du_2 disappearing on account of the orthogonal property.

Let $\mathbf{i}_1, \mathbf{i}_2, \mathbf{i}_3$ denote unit vectors in the directions OA, OB, OC corresponding to increasing values of u_1, u_2, u_3. These vectors, being mutually perpendicular, satisfy the same relations among themselves as the vectors $\mathbf{i}, \mathbf{j}, \mathbf{k}$ of 2·70.

Then from 2·31 we have, for a scalar function ϕ,

$$\mathbf{i}_1 \nabla \phi = \frac{1}{h_1}\frac{\partial \phi}{\partial u_1},$$

and therefore, using 2·15 (i),

$$(1) \qquad \nabla \phi = \frac{1}{h_1}\frac{\partial \phi}{\partial u_1}\mathbf{i}_1+\frac{1}{h_2}\frac{\partial \phi}{\partial u_2}\mathbf{i}_2+\frac{1}{h_3}\frac{\partial \phi}{\partial u_3}\mathbf{i}_3.$$

Thus, in orthogonal curvilinear coordinates,

$$\nabla = \frac{\mathbf{i}_1}{h_1}\frac{\partial}{\partial u_1}+\frac{\mathbf{i}_2}{h_2}\frac{\partial}{\partial u_2}+\frac{\mathbf{i}_3}{h_3}\frac{\partial}{\partial u_3}.$$

Since the unit vectors are themselves functions of the coordinates, we must calculate expressions like $\nabla_\wedge \mathbf{i}_1$ and $\nabla\, \mathbf{i}_1$. To find the former we have, from 2·32 (III), $\nabla_\wedge (\nabla u_1) = 0$, and, from (1), $\nabla u_1 = \mathbf{i}_1/h_1$. Therefore 2·34 (VII) gives

$$\frac{1}{h_1}(\nabla_\wedge \mathbf{i}_1) = \mathbf{i}_{1\wedge}\nabla\left(\frac{1}{h_1}\right) = \mathbf{i}_{1\wedge}\left[-\frac{1}{h_1^3}\frac{\partial h_1}{\partial u_1}\mathbf{i}_1 - \frac{1}{h_1^2 h_2}\frac{\partial h_1}{\partial u_2}\mathbf{i}_2 - \frac{1}{h_1^2 h_3}\frac{\partial h_1}{\partial u_3}\mathbf{i}_3\right].$$

Hence
$$\nabla_\wedge \mathbf{i}_1 = \frac{\mathbf{i}_2}{h_1 h_3}\frac{\partial h_1}{\partial u_3} - \frac{\mathbf{i}_3}{h_1 h_2}\frac{\partial h_1}{\partial u_2}.$$

Again, $\nabla\, \mathbf{i}_1 = \nabla(\mathbf{i}_{2\wedge}\mathbf{i}_3) = \mathbf{i}_3(\nabla_\wedge \mathbf{i}_2) - \mathbf{i}_2(\nabla_\wedge \mathbf{i}_3)$, from 2·34 (I). Therefore

$$\nabla\, \mathbf{i}_1 = \frac{1}{h_1 h_2}\frac{\partial h_2}{\partial u_1} + \frac{1}{h_1 h_3}\frac{\partial h_3}{\partial u_1} = \frac{1}{h_1 h_2 h_3}\frac{\partial (h_2 h_3)}{\partial u_1} = \frac{1}{h_1}\frac{\partial \log (h_2 h_3)}{\partial u_1}.$$

Now let
$$\mathbf{q} = q_1\,\mathbf{i}_1 + q_2\,\mathbf{i}_2 + q_3\,\mathbf{i}_3.$$

Then
$$\nabla\, \mathbf{q} = \Sigma\, \nabla (q_1\,\mathbf{i}_1) = \Sigma(q_1 \nabla\, \mathbf{i}_1 + \mathbf{i}_1 \nabla q_1),$$

from 2·34 (VI), and this reduces to

$$(2)\quad \nabla\, \mathbf{q} = \frac{1}{h_1 h_2 h_3}\left\{\frac{\partial}{\partial u_1}(q_1 h_2 h_3) + \frac{\partial}{\partial u_2}(q_2 h_3 h_1) + \frac{\partial}{\partial u_3}(q_3 h_1 h_2)\right\}.$$

Thus if $\mathbf{q} = -\nabla\phi$, we get from (1) and (2)

$$(3)\quad \nabla^2\phi = \frac{1}{h_1 h_2 h_3}\left\{\frac{\partial}{\partial u_1}\left(\frac{h_2 h_3}{h_1}\frac{\partial\phi}{\partial u_1}\right) + \frac{\partial}{\partial u_2}\left(\frac{h_3 h_1}{h_2}\frac{\partial\phi}{\partial u_2}\right) + \frac{\partial}{\partial u_3}\left(\frac{h_1 h_2}{h_3}\frac{\partial\phi}{\partial u_3}\right)\right\}.$$

Again,
$$\nabla_\wedge \mathbf{q} = \Sigma\, \nabla_\wedge (q_1\,\mathbf{i}_1) = \Sigma[-\mathbf{i}_{1\wedge}\nabla q_1 + q_1(\nabla_\wedge \mathbf{i}_1)],$$

and therefore, after reduction, in determinantal form the vorticity is

$$(4)\quad \boldsymbol{\zeta} = \nabla_\wedge \mathbf{q} = \frac{1}{h_1 h_2 h_3}\begin{vmatrix} h_1\,\mathbf{i}_1 & h_2\,\mathbf{i}_2 & h_3\,\mathbf{i}_3 \\ \dfrac{\partial}{\partial u_1} & \dfrac{\partial}{\partial u_2} & \dfrac{\partial}{\partial u_3} \\ h_1 q_1 & h_2 q_2 & h_3 q_3 \end{vmatrix}.$$

$$(5)\quad \boldsymbol{\zeta} = \mathbf{i}_1\,\zeta_1 + \mathbf{i}_2\,\zeta_2 + \mathbf{i}_3\,\zeta_3.$$

For the acceleration * we have, from 3·10 (7) and 2·34 (IV).

$$\frac{d\mathbf{q}}{dt} = \frac{\partial \mathbf{q}}{\partial t} + \tfrac{1}{2}\nabla q^2 - \mathbf{q}_\wedge \boldsymbol{\zeta}.$$

Taking the component along $\mathbf{i}_1$, we get

$$\left(\frac{\partial \mathbf{q}}{\partial t}\right)_1 = \frac{\partial q_1}{\partial t},$$

$$(\tfrac{1}{2}\nabla q^2)_1 = \frac{1}{h_1}\left(q_1\frac{\partial q_1}{\partial u_1} + q_2\frac{\partial q_2}{\partial u_1} + q_3\frac{\partial q_3}{\partial u_1}\right), \quad \text{from (1),}$$

---

* This discussion is placed here for convenience, but section 3·10 of the next chapter should be read first.

$$(\mathbf{q}_\wedge \boldsymbol{\zeta})_1 = q_2 \zeta_3 - q_3 \zeta_2$$

$$= \frac{q_2}{h_1 h_2 h_3}\left(h_3 \frac{\partial q_2 h_2}{\partial u_1} - h_3 \frac{\partial q_1 h_1}{\partial u_2}\right) + \frac{q_3}{h_1 h_2 h_3}\left(h_2 \frac{\partial q_3 h_3}{\partial u_1} - h_2 \frac{\partial q_1 h_1}{\partial u_3}\right), \quad \text{from (4).}$$

Combining these, the component acceleration along $\mathbf{i}_1$ is

(6)
$$\frac{\partial q_1}{\partial t} + \frac{1}{h_1} q_1 \frac{\partial q_1}{\partial u_1} + \frac{1}{h_2} q_2 \frac{\partial q_1}{\partial u_2} + \frac{1}{h_3} q_3 \frac{\partial q_1}{\partial u_3}$$
$$+ \frac{1}{h_1} q_1 \left\{ \frac{q_1}{h_1}\left(\frac{\partial h_1}{\partial u_1}\right) + \frac{q_2}{h_2}\left(\frac{\partial h_1}{\partial u_2}\right) + \frac{q_3}{h_3}\left(\frac{\partial h_1}{\partial u_3}\right) \right\}$$
$$- \frac{1}{h_1} \left\{ \frac{q_1^2}{h_1}\left(\frac{\partial h_1}{\partial u_1}\right) + \frac{q_2^2}{h_2}\left(\frac{\partial h_2}{\partial u_1}\right) + \frac{q_3^2}{h_3}\left(\frac{\partial h_3}{\partial u_1}\right) \right\}.$$

The remaining components can be written down by symmetry.

To illustrate these results in the case of spherical polar coordinates, we have, fig. 2·72 (i), $x = r\cos\theta$, $y = r\sin\theta\cos\omega$, $z = r\sin\theta\sin\omega$, so that

$$(ds)^2 = (dx)^2 + (dy)^2 + (dz)^2 = (dr)^2 + r^2(d\theta)^2 + r^2\sin^2\theta\,(d\omega)^2.$$

Thus if $u_1 = r$, $u_2 = \theta$, $u_3 = \omega$, we have

(7) $h_1 = 1$, $h_2 = r$, $h_3 = r\sin\theta$, and therefore, from (3),

$$\nabla^2\phi = \frac{1}{r^2\sin\theta}\left\{ \frac{\partial}{\partial r}\left(r^2\sin\theta\frac{\partial\phi}{\partial r}\right) + \frac{\partial}{\partial\theta}\left(\sin\theta\frac{\partial\phi}{\partial\theta}\right) + \frac{\partial}{\partial\omega}\left(\frac{1}{\sin\theta}\frac{\partial\phi}{\partial\omega}\right)\right\}.$$

Again, with cylindrical coordinates,

$$x = x, \quad y = \varpi\cos\omega, \quad z = \varpi\sin\omega.$$

Taking $\quad u_1 = x, \quad u_2 = \varpi, \quad u_3 = \omega,$

we get $\quad h_1 = 1, \quad h_2 = 1, \quad h_3 = \varpi.$

Thus the vorticity, from (4), is given by

(8)
$$\boldsymbol{\zeta} = \frac{1}{\varpi}\begin{vmatrix} \mathbf{i}_x & \mathbf{i}_\varpi & \varpi\mathbf{i}_\omega \\ \dfrac{\partial}{\partial x} & \dfrac{\partial}{\partial\varpi} & \dfrac{\partial}{\partial\omega} \\ q_x & q_\varpi & \varpi q_\omega \end{vmatrix},$$

where the suffixes denote the direction of the corresponding unit vector or component. See also Ex. II (16), (17).

2·73. Rate of change of the unit vectors. In orthogonal curvilinear coordinates (2·72) we can calculate $\partial\mathbf{i}_r/\partial u_s$, $(r, s = 1, 2, 3)$ as follows. By Dupin's theorem,* that triply orthogonal surfaces intersect in lines of curvature, we see that the curves along which either u_1 or u_2 varies are lines of curvature of a surface $u_3 =$ constant. Now normals to a surface at adjacent points of a line of curvature intersect. Therefore as we go along OA in fig. 2·72 (iii), the normal $\mathbf{i}_3 + d\mathbf{i}_3$ intersects the normal $\mathbf{i}_3$ and therefore $d\mathbf{i}_3$ is perpendicular both

* R. J. T. Bell, *Coordinate geometry of three dimensions*, London (1926), pp. 334, 344.

to $\mathbf{i}_2$ and $\mathbf{i}_3$, and so is parallel to $\mathbf{i}_1$. Therefore $\partial \mathbf{i}_3/\partial u_1$ is parallel to $\mathbf{i}_1$, and similarly $\partial \mathbf{i}_3/\partial u_2$ is parallel to $\mathbf{i}_2$, and four similar results.

Let $\quad d\mathbf{s} = h_1\, du_1\, \mathbf{i}_1 + h_2\, du_2\, \mathbf{i}_2 + h_3\, du_3\, \mathbf{i}_3$. Thus

$$\frac{\partial \mathbf{s}}{\partial u_1} = h_1\, \mathbf{i}_1, \quad \frac{\partial \mathbf{s}}{\partial u_2} = h_2\, \mathbf{i}_2, \quad \frac{\partial \mathbf{s}}{\partial u_3} = h_3\, \mathbf{i}_3.$$

Therefore $\quad \partial(h_1\, \mathbf{i}_1)/\partial u_2 = \partial(h_2\, \mathbf{i}_2)/\partial u_1$, and so

$$h_1 \frac{\partial \mathbf{i}_1}{\partial u_2} - h_2 \frac{\partial \mathbf{i}_2}{\partial u_1} = \mathbf{i}_2 \frac{\partial h_2}{\partial u_1} - \mathbf{i}_1 \frac{\partial h_1}{\partial u_2}.$$

But $\partial \mathbf{i}_1/\partial u_2$ is parallel to $\mathbf{i}_2$, and $\partial \mathbf{i}_2/\partial u_1$ is parallel to $\mathbf{i}_1$. Therefore

$$(1) \qquad \frac{\partial \mathbf{i}_1}{\partial u_2} = \frac{\mathbf{i}_2}{h_1} \frac{\partial h_2}{\partial u_1}, \quad \frac{\partial \mathbf{i}_2}{\partial u_1} = \frac{\mathbf{i}_1}{h_2} \frac{\partial h_1}{\partial u_2}.$$

Also from $\mathbf{i}_1 = \mathbf{i}_2 \wedge \mathbf{i}_3$ we get

$$(2) \qquad \frac{\partial \mathbf{i}_1}{\partial u_1} = \frac{\partial \mathbf{i}_2}{\partial u_1} \wedge \mathbf{i}_3 + \mathbf{i}_2 \wedge \frac{\partial \mathbf{i}_3}{\partial u_1} = -\frac{\mathbf{i}_2}{h_2}\frac{\partial h_1}{\partial u_2} - \frac{\mathbf{i}_3}{h_3}\frac{\partial h_1}{\partial u_3} = \frac{\mathbf{i}_1}{h_1}\frac{\partial h_1}{\partial u_1} - \nabla h_1.$$

From (1) and (2) we can write down $\partial \mathbf{i}_r/\partial u_s$ for all values 1, 2, 3 of r, s. These results, together with

$$\nabla = \frac{\mathbf{i}_1}{h_1}\frac{\partial}{\partial u_1} + \frac{\mathbf{i}_2}{h_2}\frac{\partial}{\partial u_2} + \frac{\mathbf{i}_3}{h_3}\frac{\partial}{\partial u_3}$$

enable any nabla operation to be calculated with reasonable economy of effort.

EXAMPLES II

1. If masses m, n are at the extremities of the vectors $\mathbf{a}$, $\mathbf{b}$, prove that the centroid is $(m\mathbf{a} + n\mathbf{b})/(m+n)$.

2. Prove that

(i) $\mathbf{ab} = \mathbf{ba} = a_x b_x + a_y b_y + a_z b_z$,

(ii) $(\mathbf{a} + \mathbf{b})^2 = a^2 + b^2 + 2\mathbf{ab}$,

(iii) $\mathbf{a}(\mathbf{b} \wedge \mathbf{c}) = \begin{vmatrix} a_x & a_y & a_z \\ b_x & b_y & b_z \\ c_x & c_y & c_z \end{vmatrix} = [\mathbf{abc}]$

3. Prove that

(i) $(\mathbf{a} + \mathbf{b}) \wedge (\mathbf{a} - \mathbf{b}) = -2\mathbf{a} \wedge \mathbf{b}$,

(ii) $|\,\mathbf{a} \wedge \mathbf{b}\,|^2 = a^2 b^2 - (\mathbf{ab})^2 = (ab + \mathbf{ab})(ab - \mathbf{ab})$.

4. Prove that

(i) $\mathbf{a} \wedge (\mathbf{b} \wedge \mathbf{c}) + \mathbf{b} \wedge (\mathbf{c} \wedge \mathbf{a}) + \mathbf{c} \wedge (\mathbf{a} \wedge \mathbf{b}) = 0$,

(ii) $(\mathbf{a} \wedge \mathbf{b})(\mathbf{c} \wedge \mathbf{d}) = (\mathbf{ac})(\mathbf{bd}) - (\mathbf{ad})(\mathbf{bc})$ (Lagrange's identity),

(iii) $\mathbf{a}[\mathbf{b}(\mathbf{c} \wedge \mathbf{d})] - \mathbf{b}[\mathbf{a}(\mathbf{c} \wedge \mathbf{d})] + \mathbf{c}[\mathbf{d}(\mathbf{a} \wedge \mathbf{b})] - \mathbf{d}[\mathbf{c}(\mathbf{a} \wedge \mathbf{b})] = 0$.

5. Prove that the area of the triangle whose vertices are at $\mathbf{a}$, $\mathbf{b}$, $\mathbf{c}$ is the magnitude of the vector

$$\tfrac{1}{2}[(\mathbf{b} \wedge \mathbf{c}) + (\mathbf{c} \wedge \mathbf{a}) + (\mathbf{a} \wedge \mathbf{b})].$$

6. If λ is a scalar and $\mathbf{r}$, $\mathbf{s}$ vectors, all functions of t, prove that

(i) $\dfrac{d}{dt}(\lambda \mathbf{r}) = \lambda \dfrac{d\mathbf{r}}{dt} + \mathbf{r}\dfrac{d\lambda}{dt}$,

(ii) $\dfrac{d}{dt}(\mathbf{rs}) = \mathbf{r}\dfrac{d\mathbf{s}}{dt} + \mathbf{s}\dfrac{d\mathbf{r}}{dt}$,

$$\text{(iii)}\quad \frac{d}{dt}(\mathbf{r}_\wedge \mathbf{s}) = \mathbf{r}_\wedge \frac{d\mathbf{s}}{dt} + \frac{d\mathbf{r}}{dt}{}_\wedge \mathbf{s}.$$

7. If the surfaces $\phi = c$, where ϕ is the velocity potential, be drawn for equidistant infinitesimal values of the constant c, show that the velocity at any point is inversely proportional to the distance between consecutive surfaces in the neighbourhood of the point.

Prove also that if any surface of equal velocity potential intersects itself, the point of intersection is a stagnation point.

8. If $\phi(\mathbf{r}, t)$ denotes the velocity potential, prove that

$$d\phi = \frac{\partial \phi}{\partial t}\, dt - \mathbf{q}\, d\mathbf{r},$$

and show that the differential equations of the streamlines are given by

$$d\mathbf{r}_\wedge \nabla \phi = 0.$$

9. If ϕ, ϕ' are two distinct solutions of Laplace's equation (3·20) valid within the closed surface S, prove that

$$\int_{(S)} \phi \frac{\partial \phi'}{\partial n}\, dS = \int_{(S)} \phi' \frac{\partial \phi}{\partial n}\, dS.$$

10. If $\phi(x+h, y+k, z+l)$ be written in the form $\phi(\mathbf{r}+\mathbf{R})$, where

$$\mathbf{r} = \mathbf{i}x + \mathbf{j}y + \mathbf{k}z, \quad \mathbf{R} = \mathbf{i}h + \mathbf{j}k + \mathbf{k}l,$$

show that Taylor's theorem can be expressed in the symbolic form

$$\phi(\mathbf{r}+\mathbf{R}) = \phi(\mathbf{r}) + (\mathbf{R}\,\nabla)\phi(\mathbf{r}) + \frac{1}{2!}(\mathbf{R}\,\nabla)^2\phi(\mathbf{r}) + \dots .$$

11. If $\boldsymbol{\eta} = \mathbf{i}\,\delta x + \mathbf{j}\,\delta y + \mathbf{k}\,\delta z$, prove, with the notations of 2·40, that

$$\boldsymbol{\eta}\mathbf{D} = \mathbf{i}(a\,\delta x + h\,\delta y + g\,\delta z) + \mathbf{j}(h\,\delta x + b\,\delta y + f\,\delta z) + \mathbf{k}(g\,\delta x + f\,\delta y + c\,\delta z),$$

where

$$a = \frac{\partial u}{\partial x}, \quad b = \frac{\partial v}{\partial y}, \quad c = \frac{\partial w}{\partial z},$$

$$2f = \frac{\partial w}{\partial y} + \frac{\partial v}{\partial z}, \quad 2g = \frac{\partial u}{\partial z} + \frac{\partial w}{\partial x}, \quad 2h = \frac{\partial v}{\partial x} + \frac{\partial u}{\partial y},$$

and hence that the equation of the central quadric of 2·40 is

$$a(\delta x)^2 + b(\delta y)^2 + c(\delta z)^2 + 2f\,\delta y\,\delta z + 2g\,\delta z\,\delta x + 2h\,\delta x\,\delta y = \text{constant}.$$

12. If

$$\mathbf{q} = \mathbf{i}u + \mathbf{j}v + \mathbf{k}w, \quad \boldsymbol{\eta} = \mathbf{i}\,\delta x + \mathbf{j}\,\delta y + \mathbf{k}\,\delta z,$$

prove that

$$(\boldsymbol{\eta}\,\nabla)\mathbf{q} = \mathbf{i}\left(\frac{\partial u}{\partial x}\delta x + \frac{\partial u}{\partial y}\delta y + \frac{\partial u}{\partial z}\delta z\right) + \mathbf{j}\left(\frac{\partial v}{\partial x}\delta x + \frac{\partial v}{\partial y}\delta y + \frac{\partial v}{\partial z}\delta z\right)$$
$$+\mathbf{k}\left(\frac{\partial w}{\partial x}\delta x + \frac{\partial w}{\partial y}\delta y + \frac{\partial w}{\partial z}\delta z\right).$$

13. Prove that through any point P of a fluid in motion there is in general at any instant one set of three straight lines at right angles to each other such that, if the lines move with the fluid, then after a short time δt the angles between them remain right angles to the first order in δt, and that the angular velocity of this triad of lines, as it moves with the fluid, is $\frac{1}{2}$ curl $\mathbf{v}$, where $\mathbf{v}$ is the fluid velocity at P. Prove also that, if a small portion of the fluid with its mass centre at P be instantaneously solidified without change of angular momentum, then its angular velocity immediately after solidification is $\frac{1}{2}$ curl $\mathbf{v}$, if, and only if, the principal axes of inertia for the resulting solid lie along the above triad of lines. (U.L.)

14. Use the tensor form of Gauss's theorem to prove that

$$\text{(i)} \quad \int_{(S)} (\mathbf{nq})\mathbf{q}\, dS = \int_{(V)} [(\mathbf{q}\,\nabla)\mathbf{q} + \mathbf{q}(\nabla\,\mathbf{q})]\, d\tau,$$

$$\text{(ii)} \quad \tfrac{1}{2}\int_{(S)} [\mathbf{n}\mathbf{q}^2 - 2(\mathbf{nq})\mathbf{q}]\, dS = \int_{(V)} [\mathbf{q}(\nabla\mathbf{q}) - \mathbf{q}_\wedge \boldsymbol{\zeta}]\, d\tau.$$

15. If P, Q, R are finite, continuous, and single-valued functions of x, y, z throughout a space bounded by a closed surface S, prove that

$$\iiint \left(\frac{\partial P}{\partial x} + \frac{\partial Q}{\partial y} + \frac{\partial R}{\partial z}\right) dx\, dy\, dz = \int (lP + mQ + nR)\, dS,$$

where l, m, n are the direction-cosines of the outward normal at any point of S, and the integrals are taken throughout the volume and over the surface S.

Find the value of

$$\int \frac{dS}{p}$$

taken over the surface of an ellipsoid, where p is the perpendicular from the centre to the tangent plane. (R.N.C.)

16. For spherical polar coordinates, prove that the components of vorticity are given by

$$(\text{curl}\,\mathbf{q})_r = \frac{1}{r \sin\theta}\left[\frac{\partial(q_\omega \sin\theta)}{\partial\theta} - \frac{\partial q_\theta}{\partial\omega}\right],$$

$$(\text{curl}\,\mathbf{q})_\omega = \frac{1}{r}\left[\frac{\partial(q_\theta r)}{\partial r} - \frac{\partial q_r}{\partial\theta}\right],$$

$$(\text{curl}\,\mathbf{q})_\theta = \frac{1}{r}\left[\frac{1}{\sin\theta}\frac{\partial q_r}{\partial\omega} - \frac{\partial(q_\omega r)}{\partial r}\right].$$

17. Prove that in cylindrical coordinates

$$\nabla^2\phi = \frac{\partial^2\phi}{\partial x^2} + \frac{\partial^2\phi}{\partial\varpi^2} + \frac{1}{\varpi}\frac{\partial\phi}{\partial\varpi} + \frac{1}{\varpi^2}\frac{\partial^2\phi}{\partial\omega^2}.$$

18. Prove that the components of the vorticity in cylindrical coordinates are

$$(\text{curl}\,\mathbf{q})_x = \frac{1}{\varpi}\left[\frac{\partial(q_\omega\varpi)}{\partial\varpi} - \frac{\partial q_\varpi}{\partial\omega}\right],$$

$$(\text{curl}\,\mathbf{q})_\varpi = \frac{1}{\varpi}\left[\frac{\partial q_x}{\partial\omega} - \frac{\partial(q_\omega\varpi)}{\partial x}\right],$$

$$(\text{curl}\,\mathbf{q})_\omega = \frac{\partial q_\varpi}{\partial x} - \frac{\partial q_x}{\partial\varpi}.$$

19. Prove that if P lies on a straight line which passes through the extremity of $\mathbf{a}$ and is parallel to $\mathbf{b}$, then the equation of the line is $\boldsymbol{\eta} = \mathbf{a} + \mathbf{b}t$, where $\boldsymbol{\eta}$ is the position vector of P and t is a scalar.

20. Show that the equation of a plane whose normal is $\mathbf{n}$ and which passes through the extremity of $\mathbf{a}$ is $(\boldsymbol{\eta} - \mathbf{a})\mathbf{n} = 0$.

21. If $F = \mathbf{a}\,;\mathbf{a}'+\mathbf{b}\,;\mathbf{b}'+\mathbf{c}\,;\mathbf{c}'$, show that $\boldsymbol{\eta}F\boldsymbol{\eta}$ = constant represents a family of central quadrics.

22. Prove that $\nabla_\wedge \mathbf{r} = 0$, $\nabla\mathbf{r} = 3$, $\nabla r = \mathbf{r}/r$, $\nabla(1/r) = -\mathbf{r}/r^3$, $\nabla^2(1/r) = 0$.

23. Prove that in general orthogonal coordinates

$$(\mathbf{i}_1\nabla)\mathbf{i}_1 = -\frac{\mathbf{i}_2}{h_1h_2}\frac{\partial h_1}{\partial u_2} - \frac{\mathbf{i}_3}{h_1h_3}\frac{\partial h_1}{\partial u_3},$$

$$(\mathbf{i}_1\nabla)\mathbf{i}_2 = \frac{\mathbf{i}_1}{h_1h_2}\frac{\partial h_1}{\partial u_2}, \quad (\mathbf{i}_1\nabla)\mathbf{i}_3 = \frac{\mathbf{i}_1}{h_1h_3}\frac{\partial h_1}{\partial u_3}.$$

and deduce the expression for $(\mathbf{i}_1\nabla)\mathbf{q}$.

24. Show that typical terms in $(\nabla\,;\mathbf{q})$ are

$$\frac{\mathbf{i}_1\,;\mathbf{i}_1}{h_1}\left(\frac{\partial q_1}{\partial u_1} + \frac{q_2}{h_2}\frac{\partial h_1}{\partial u_2} + \frac{q_3}{h_3}\frac{\partial h_1}{\partial u_3}\right), \quad \frac{\mathbf{i}_2\,;\mathbf{i}_3}{h_2h_3}\left(h_3\frac{\partial q_3}{\partial u_2} - q_2\frac{\partial h_2}{\partial u_3}\right),$$

and hence write down the complete expression for $\nabla\,;\mathbf{q}$.

25. If Φ is any dyad, prove that

$$\text{(i) } \nabla(\mathbf{q}\,\Phi) = \mathbf{q}\,\nabla\Phi + (\Phi\,\nabla)\mathbf{q},$$

$$\text{(ii) } \Phi = [\Phi\mathbf{i}\,;\mathbf{i}] + [\Phi\mathbf{j}\,;\mathbf{j}] + [\Phi\mathbf{k}\,;\mathbf{k}].$$

26. If $\mathbf{a}$, $\mathbf{b}$, $\mathbf{c}$ are any three non-coplanar vectors, and if $\mathbf{a}^\star$, $\mathbf{b}^\star$, $\mathbf{c}^\star$ are so chosen that $\mathbf{a}^\star\mathbf{a} = \mathbf{b}^\star\mathbf{b} = \mathbf{c}^\star\mathbf{c} = 1$, while all cross-products such as $\mathbf{ab}^\star$, $\mathbf{ac}^\star$, etc., are zero, prove that $(\mathbf{a}\,;\mathbf{a}^\star)+(\mathbf{b}\,;\mathbf{b}^\star)+(\mathbf{c}\,;\mathbf{c}^\star)$ is the idemfactor.

27. If $\mathbf{u}$ is a unit vector, interpret geometrically the scalar product $\mathbf{ux}$, and prove geometrically that

$$\mathbf{u}(\mathbf{b}+\mathbf{c}) = \mathbf{ub}+\mathbf{uc}.$$

Deduce the distributive law for scalar products.

28. By considering the scalar product of an arbitrary unit vector $\mathbf{u}$ and the vector

$$\mathbf{x} = \mathbf{a}_\wedge(\mathbf{b}+\mathbf{c}) - \mathbf{a}_\wedge\mathbf{b} - \mathbf{a}_\wedge\mathbf{c},$$

prove, using the preceding example, that $\mathbf{x} = 0$, thus deducing the distributive law for vector products.

29. With the notation of 2·50 prove that

$$\int_{(S)} \nabla(\mathbf{n}\boldsymbol{\zeta})\,dS = \int_{(C)} \nabla(\mathbf{q}\,d\mathbf{s}).$$

30. With the notations of 2·72 show that

$$h_1^2 = (\partial x/\partial u_1)^2 + (\partial y/\partial u_1)^2 + (\partial z/\partial u_1)^2,$$

with similar expressions for h_2 and h_3.

31. Prove that $[\mathbf{abc}](\mathbf{p}_\wedge\mathbf{q}) = \begin{vmatrix} \mathbf{a} & \mathbf{b} & \mathbf{c} \\ \mathbf{pa} & \mathbf{pb} & \mathbf{pc} \\ \mathbf{qa} & \mathbf{qb} & \mathbf{qc} \end{vmatrix}$.

32. Prove that

$$\text{(i) } \nabla[(\mathbf{q}\,;\mathbf{q}) - \tfrac{1}{2}\mathbf{I}q^2] = \mathbf{q}(\nabla\mathbf{q}) - \mathbf{q}_\wedge\boldsymbol{\zeta}.$$

$$\text{(ii) } \nabla\{\mathbf{r}_\wedge[(\mathbf{q}\,;\mathbf{q}) - \tfrac{1}{2}\mathbf{I}q^2]\} = -\mathbf{r}_\wedge[\mathbf{q}(\nabla\mathbf{q}) - \mathbf{q}_\wedge\boldsymbol{\zeta}].$$

CHAPTER III

EQUATIONS OF MOTION

3·10. Differentiation with respect to the time. Fig. 3·10 shows the actual path of a fluid particle which at time t_0 was at the point A, whose position vector is $\mathbf{r}_0$ with respect to the fixed point O.

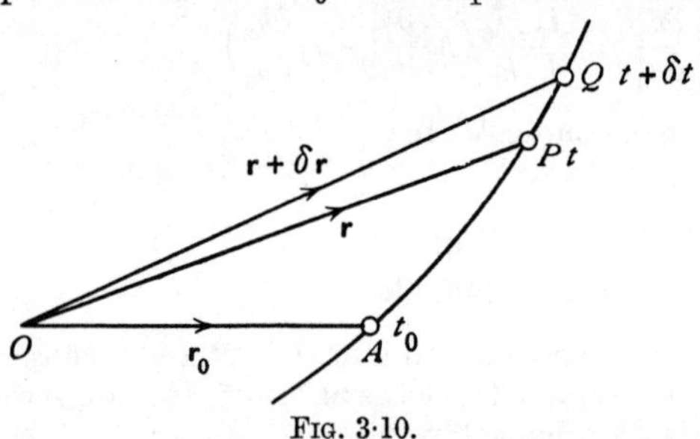

FIG. 3·10.

At time t the particle is at P, at time $t+\delta t$ it is at Q, position vectors $\mathbf{r}$ and $\mathbf{r}+\delta\mathbf{r}$ respectively.

When the particle is at P there are associated with it scalar functions, such as pressure and density at P, and vector functions, such as its velocity and acceleration at P.

We enquire how to form the differential coefficient with respect to the time of such scalar and vector functions. First note that the position vector $\mathbf{r}$ of this particular fluid particle is a function of t only, for it is clear that $\mathbf{r}$ can depend only on the time t and some position such as A considered as the initial position and therefore fixed.

We have seen in 1·1 that

(1) $$\mathbf{q} = \frac{d\mathbf{r}}{dt}.$$

Now consider, for example, the density ρ. If we fix our attention on the particle when at P, the density depends on the position vector $\mathbf{r}$ and the time t, so that

$$\rho = f(\mathbf{r}, t).$$

Since $\mathbf{r}$ is a function of t only, so is ρ, and therefore we can form a total differential coefficient $d\rho/dt$. To calculate this we have, in the notation of 2·71,

$$\frac{d\rho}{dt} = \frac{\partial f(\mathbf{r}, t)}{\partial t} + \frac{d\mathbf{r}}{dt}\frac{\partial f(\mathbf{r}, t)}{\partial \mathbf{r}} = \frac{\partial f(\mathbf{r}, t)}{\partial t} + \left(\mathbf{q}\,\frac{\partial}{\partial \mathbf{r}}\right) f(\mathbf{r}, t),$$

and therefore

(2) $$\frac{d\rho}{dt} = \frac{\partial \rho}{\partial t} + (\mathbf{q}\,\nabla)\rho.$$

The first term on the right represents the rate of change of ρ with respect to the time when P is regarded as fixed, the second the rate of change of ρ at a fixed time t due to the change of position from P to Q. Since ρ is a scalar, this can be written

$$\frac{d\rho}{dt} = \frac{\partial \rho}{\partial t} + \mathbf{q}(\nabla \rho). \tag{3}$$

This gives the rate of change of density as the particle moves about. If the fluid is incompressible, the density of the fluid particle does not change, so that

$$\frac{d\rho}{dt} = 0, \quad \text{or} \quad \frac{\partial \rho}{\partial t} + \mathbf{q}(\nabla \rho) = 0. \tag{4}$$

When ρ is constant, (4) is satisfied identically.

A similar argument can be applied to any scalar function α, giving

$$\frac{d\alpha}{dt} = \frac{\partial \alpha}{\partial t} + \mathbf{q}(\nabla \alpha). \tag{5}$$

To find the rate of change of a vector $\mathbf{a}$ associated with the particle, the steps of the argument are exactly the same down to (2), which now gives

$$\frac{d\mathbf{a}}{dt} = \frac{\partial \mathbf{a}}{\partial t} + (\mathbf{q}\,\nabla)\,\mathbf{a}, \tag{6}$$

which cannot be further reduced to the form (3).

The most important case here is that of the velocity vector $\mathbf{q}$ whose rate of change gives the *acceleration* of the particle, namely

$$\frac{d\mathbf{q}}{dt} = \frac{\partial \mathbf{q}}{\partial t} + (\mathbf{q}\,\nabla)\mathbf{q} = \frac{\partial \mathbf{q}}{\partial t} + \nabla(\tfrac{1}{2}q^2) - \mathbf{q}_\wedge \boldsymbol{\zeta}, \tag{7}$$

from 2·34 (IV).

Translating this into rectangular cartesian coordinates by means of 2·70, we see that

$$\mathbf{i}\frac{du}{dt} + \mathbf{j}\frac{dv}{dt} + \mathbf{k}\frac{dw}{dt} = \mathbf{i}\frac{\partial u}{\partial t} + \mathbf{j}\frac{\partial v}{\partial t} + \mathbf{k}\frac{\partial w}{\partial t}$$
$$+ \left(u\frac{\partial}{\partial x} + v\frac{\partial}{\partial y} + w\frac{\partial}{\partial z}\right)(\mathbf{i}u + \mathbf{j}v + \mathbf{k}w),$$

so that (7) is equivalent to the *three* equations:

$$\frac{du}{dt} = \frac{\partial u}{\partial t} + u\frac{\partial u}{\partial x} + v\frac{\partial u}{\partial y} + w\frac{\partial u}{\partial z}, \tag{8}$$

$$\frac{dv}{dt} = \frac{\partial v}{\partial t} + u\frac{\partial v}{\partial x} + v\frac{\partial v}{\partial y} + w\frac{\partial v}{\partial z},$$

$$\frac{dw}{dt} = \frac{\partial w}{\partial t} + u\frac{\partial w}{\partial x} + v\frac{\partial w}{\partial y} + w\frac{\partial w}{\partial z}.$$

Thus we have the equivalence of operators

$$\frac{d}{dt} = \frac{\partial}{\partial t} + u\frac{\partial}{\partial x} + v\frac{\partial}{\partial y} + w\frac{\partial}{\partial z}$$

in cartesian coordinates. In the vector form

$$(9) \qquad \frac{d}{dt} = \frac{\partial}{\partial t} + (\mathbf{q}\,\nabla).$$

The operation here implied is sometimes called *differentiation following the fluid*, implying that we are calculating the rate of change of some quantity associated with the same fluid particle as it moves about.*

3·20. The equation of continuity. If we consider a fluid particle of infinitesimal volume $d\tau$ and density ρ at time t, the mass of this fluid particle cannot change as it moves about and therefore

$$(1) \qquad \frac{d}{dt}(\rho\, d\tau) = 0.$$

This is one form of the *equation of continuity*, or *conservation of mass*. If the volume expands, the density diminishes, and vice versa, in such a way that (1) is always satisfied.

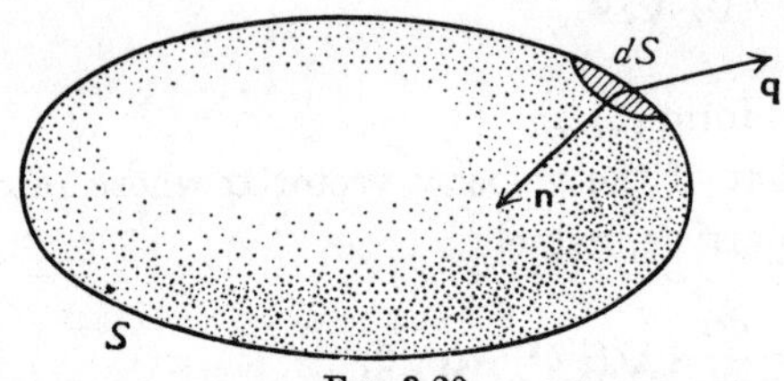

FIG. 3·20.

Let X denote any property per unit mass carried by a fluid particle as it moves. Then for a fluid body of volume V we have the *transport theorem*

$$(2) \qquad \frac{d}{dt}\int_{(V)} X\rho\, d\tau = \int_{(V)} \frac{dX}{dt}\rho\, d\tau + \int_{(V)} X\frac{d}{dt}(\rho\, d\tau) = \int_{(V)} \frac{dX}{dt}\rho\, d\tau.$$

Another point of view is the following:

Consider a fixed closed surface S lying entirely in the fluid. If $\mathbf{n}$ is a unit inward normal to the element dS, the rate at which mass flows into the surface through the boundary is

$$(3) \qquad \int_{(S)} \rho\mathbf{qn}\, dS.$$

The mass of fluid within the volume V enclosed by S is

$$\int_{(V)} \rho\, d\tau.$$

* Some writers use the notation D/Dt. The names *substantial* or *material differentiation* are also used.

Assuming that no fluid is created or annihilated within S, the mass can only increase by flow through the boundary. Equating the time rate of increase of the mass to (3), we get

$$\frac{\partial}{\partial t}\int_{(V)} \rho\, d\tau = \int_{(S)} \rho \mathbf{q}\mathbf{n}\, dS = -\int_{(V)} \nabla(\rho \mathbf{q}) d\tau$$

by Gauss's theorem. Thus

$$\int_{(V)} \left(\frac{\partial \rho}{\partial t} + \nabla(\rho \mathbf{q})\right) d\tau = 0.$$

Since the surface S can be replaced by any arbitrary closed surface drawn within it, we must have, at every point,*

$$\frac{\partial \rho}{\partial t} + \nabla\, (\rho \mathbf{q}) = 0, \tag{4}$$

which is another form of the equation of continuity.

Now, from 2·34 (VI) and 3·10 (9), we have successively

$$\frac{\partial \rho}{\partial t} + \mathbf{q}(\nabla\, \rho) + \rho\, \nabla\, \mathbf{q} = 0.$$

$$\frac{d\rho}{dt} + \rho\, \nabla\, \mathbf{q} = 0, \quad \nabla\, \mathbf{q} = \frac{d}{dt} \log\left(\frac{1}{\rho}\right). \tag{5}$$

In the case of an incompressible fluid, $d\rho/dt = 0$ (see 3·10 (4)), and therefore

$$\nabla\, \mathbf{q} = 0, \tag{6}$$

which is the equation of continuity for a liquid; the *expansion* $\nabla\, \mathbf{q}$ vanishes.

Using cartesian coordinates (2·70), this gives

$$\frac{\partial u}{\partial x} + \frac{\partial v}{\partial y} + \frac{\partial w}{\partial z} = 0. \tag{7}$$

In the extremely important case of irrotational motion, we have $\mathbf{q} = -\nabla\, \phi$, and therefore the equation of continuity (6) for a liquid in irrotational motion becomes

$$\nabla^2 \phi = 0, \tag{8}$$

or in cartesian coordinates,

$$\frac{\partial^2 \phi}{\partial x^2} + \frac{\partial^2 \phi}{\partial y^2} + \frac{\partial^2 \phi}{\partial z^2} = 0. \tag{9}$$

Equation (8) is known as Laplace's equation.

---

* If $\int_{(V)} A\, d\tau = 0$ for an arbitrary volume V, $\frac{1}{V}\int_{(V)} A\, d\tau = 0$, so that $\lim_{V \to 0} \frac{1}{V}\int_{(V)} A\, d\tau = 0$, i.e.

$$\lim \frac{1}{V} \cdot AV = A = 0.$$

From this investigation it appears that a fluid cannot move according to an arbitrarily assigned law of distribution of velocity. For the motion to be possible it is evidently necessary that the equation of continuity should be satisfied.

In particular, possible irrotational motions of a liquid are subject to the condition that the velocity potential ϕ shall satisfy Laplace's equation.

Equation (5) leads to another form of the transport theorem as follows

$$\frac{d}{dt}\left(\frac{X}{\rho}\right) = \frac{1}{\rho}\frac{dX}{dt} - \frac{X}{\rho^2}\frac{d\rho}{dt} = \frac{1}{\rho}\left(\frac{dX}{dt} + X\,\nabla\mathbf{q}\right) \quad \text{from (5)}$$

and therefore from (2) and 3·10 (9)

$$(10) \quad \frac{d}{dt}\int_{(V)} X\,d\tau = \int_{(V)} \frac{d}{dt}\left(\frac{X}{\rho}\right)\rho\,d\tau = \int_{(V)} \left(\frac{dX}{dt} + X\,\nabla\mathbf{q}\right)d\tau$$

$$= \int_{(V)} \left[\frac{\partial X}{\partial t} + \nabla(\mathbf{q}\,;X)\right]d\tau = \frac{\partial}{\partial t}\int_{(V)} X\,d\tau - \int_{(S)} (\mathbf{nq})X\,dS.$$

The last term gives the flux of X out of S through the boundary since $\mathbf{n}$ is the unit *inward* normal.

3·30. Boundary conditions (Kinematical).

When fluid is in contact with rigid surfaces or with other fluid with which it does not mix, a kinematical condition must be satisfied if contact is to be preserved, namely that the fluid and the surface with which contact is maintained must have the same velocity normal to the surface.

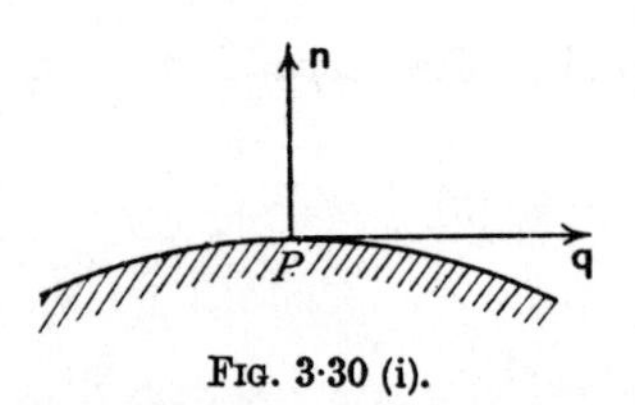

FIG. 3·30 (i).

If we denote by $\mathbf{n}$ the normal unit vector drawn at the point P of the surface of contact and by $\mathbf{q}$ the fluid velocity, we shall have, in the case of a fixed rigid surface, $\mathbf{qn} = 0$, which expresses the condition that the normal velocities are both zero, or, in other words, the fluid velocity is everywhere tangential to the fixed surface, fig. 3·30 (i).

When the rigid surface is in motion, if $\mathbf{U}$ is the velocity of the point P of the surface, we must have

$$\mathbf{qn} = \mathbf{Un},$$

or

$$(\mathbf{q} - \mathbf{U})\mathbf{n} = 0.$$

FIG. 3·30 (ii).

This equation points out that the velocity of the fluid relative to the surface is perpendicular to the normal, that is, tangential to the surface.

When two fluids which do not mix (such as air and water) are in contact along a common (geometrical) surface of separation S in order that contact may be maintained, it is clear that the relative velocity $\mathbf{q}-\mathbf{q}'$ must be again tangential to S. On the other hand, we note that in this case the form and movement of S are unknown until the problem of the motion has been solved.

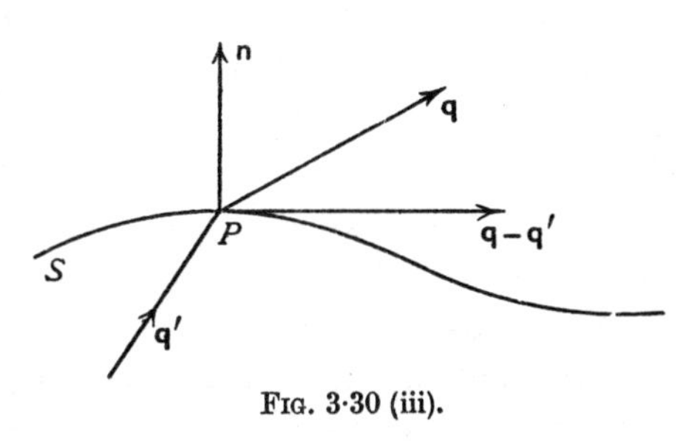

FIG. 3·30 (iii).

3·31. Boundary conditions (Physical). The kinematical boundary conditions just investigated must be satisfied independently of any special physical hypothesis.

In the case of an inviscid fluid in contact with rigid boundaries (fixed or moving), the additional condition to be satisfied is that the fluid thrust shall be normal to the boundary.

In the case of two inviscid fluids presenting a surface of separation S, the condition to be satisfied is that the pressure shall be continuous at the boundary as we pass from one side of S to the other.*

To prove this, take a cylinder whose generators are normal to S and whose cross-sections dS are small areas on either side of S. Then if p, p' are the pressures in the two fluids, we have, resolving along the normal,

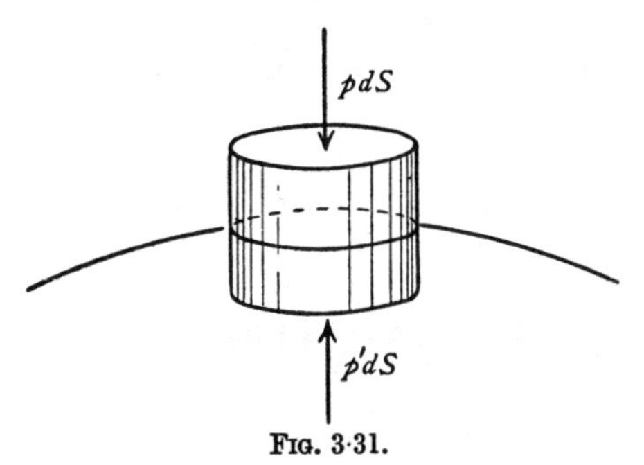

FIG. 3·31.

$$p\,dS-p'\,dS = 0, \text{ i.e. } p = p',$$

since, as in 1·3, the body forces and mass-accelerations are negligible compared with the terms retained.

Thus in the case of water in contact with the atmosphere, the pressure of the water at the free surface will be equal to that of the air, and if this latter is assumed to be constant, the water surface will be a surface of constant pressure.

Another important example of this principle occurs when the surface S separates not two different fluids, but two regions of the same fluid, there being a discontinuity of tangential velocity at the surface S which is then a vortex sheet (13·70). This may be conceived to occur in the case of air streaming past an aerofoil, where the two streams from the upper and lower faces glide

* This condition must be modified when surface tension is taken into account. See 15·50.

over one another along a surface of discontinuity, springing from the trailing edge. Bernoulli's equation then gives, when the motion is steady,

$$\frac{p'}{\rho}+\tfrac{1}{2}q'^2 = \frac{p}{\rho}+\tfrac{1}{2}q^2,$$

and since $p = p'$, we must have $q = q'$. Thus the surface will be a surface of discontinuity of *direction* of the velocity, not of speed.

In the case of a jet or current passing through fluid otherwise at rest where the pressure may be assumed constant, the continuity of pressure inside and outside the jet shows that the surface of the jet is one of constant speed.

In the case of a *viscous* fluid, experiment supports the view that at a rigid surface in contact with the fluid the relative velocity is zero, and this is the physical condition to be satisfied. This is the *adherence condition.*

The interface between a fluid and an immersed solid may be regarded as a vortex sheet, i.e. a surface of discontinuity of tangential velocity in passing from fluid to solid (13·70). Viscous contact is then distinguished by zero discontinuity.

3·32. Efflux. Returning to the subject of 1·82, consider the steady irrotational flow of liquid through an aperture of area σ_1 in the wall of a vessel, fig. 3·32.

Consider a horizontal plane section Σ of the vessel, so far removed from the aperture that all the stream filaments may be supposed to cross it with the same speed q_1, and let $\mathbf{m}$ be a unit normal drawn to Σ into the fluid below.

Let $\mathbf{l}$ be a unit normal drawn outwards at the section σ_2 of the vena contracta where the speed is q_2.

Let w denote the surface of the walls of the vessel below the section Σ, s the surface of the jet between σ_1 and σ_2.

Consider the fluid bounded by the total surface $\Sigma+w+s+\sigma_2$, and let $\mathbf{n}$ be a unit inward normal at any point of this surface.

Since $\nabla\,\mathbf{q} = 0$, Gauss's theorem, 2·61 (7), gives

$$\int \mathbf{q}(\mathbf{nq})\,dS = -\int (\mathbf{q}\,\nabla)\,\mathbf{q}\,d\tau = -\tfrac{1}{2}\int \nabla\,q^2\,d\tau = \tfrac{1}{2}\int \mathbf{n}q^2\,dS$$

using 2·34 (IV) and then Gauss's theorem 2·61 (3).

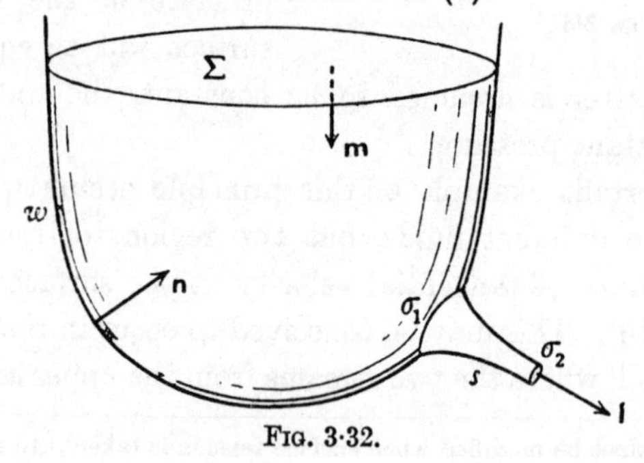

FIG. 3·32.

Now the values of $\mathbf{nq}$ on Σ, w, s, σ_2 are respectively q_1, 0, 0, $-q_2$, and the values of $\mathbf{n}$ are $\mathbf{m}$, $-\mathbf{l}$ on Σ, σ_2. Also, the speed is q_2 over the surface s by Bernoulli's theorem. Therefore

$$-\mathbf{l}q_2^2\,\sigma_2+\mathbf{m}q_1^2\,\Sigma = \int_{(w)} \mathbf{n}q^2\,dS + q_2^2\int_{(s)} \mathbf{n}\,dS.$$

Since the surface formed by $s+\sigma_1+\sigma_2$ is closed, 2·20 (3) gives

$$\int_{(s)} \mathbf{n}\,dS = -\int_{(\sigma_1)} \mathbf{n}\,dS - \int_{(\sigma_2)} \mathbf{n}\,dS = \mathbf{l}(\sigma_2-\sigma_1).$$

Therefore $$\mathbf{m}q_1^2\,\Sigma - \mathbf{l}(2\sigma_2-\sigma_1)q_2^2 = \int_{(w)} \mathbf{n}q^2\,dS.$$

Take the scalar product by $\mathbf{l}/(\sigma_1 q_2^2)$, and eliminate q_1 by the equation of continuity in the form $q_1\,\Sigma = q_2\,\sigma_2$. Then, if $\alpha = \sigma_2/\sigma_1$ is the coefficient of contraction, we get

$$\frac{\sigma_2}{\sigma_1} = \alpha = \frac{1-\dfrac{\mathbf{l}}{\sigma_1}\displaystyle\int_{(w)} \mathbf{n}\left(\frac{q}{q_2}\right)^2 dS}{2-\dfrac{\sigma_2}{\Sigma}\,\mathbf{lm}}.$$

When the plane of the orifice is vertical, $\mathbf{lm} = 0$ and the denominator can be replaced by 2. This is also the case when σ_2/Σ is negligible.

Thus for flow through a hole in an infinite plate $\mathbf{ln} = -1$, and

$$\alpha = \frac{1}{2}+\frac{1}{2\sigma_1}\int\left(\frac{q}{q_2}\right)^2 dS > \frac{1}{2}.$$

Again, when a vertical cylindrical nozzle, pointing inwards, is fitted to a hole in the horizontal bottom of a tank with vertical sides, $\mathbf{ln} = 0$ over the sides, and experiment shows that q is sensibly zero over the bottom. Thus

$$\alpha = \frac{1}{2-\dfrac{\sigma_2}{\Sigma}} > \frac{1}{2},$$

and when σ_2/Σ is negligible, $\alpha = 1/2$.

3·40. Rate of change of linear momentum. Consider the fluid body which at time t occupies the interior of a closed surface S.

Let $\mathbf{M}$ be the linear momentum at time t.

Then $$\mathbf{M} = \int \mathbf{q}\rho\,d\tau.$$

Therefore the required rate of change of momentum is, from 3·20 (10)

$$(1) \qquad \frac{\partial}{\partial t}\int_{(V)} \rho\mathbf{q}\,d\tau - \int_{(S)} \rho\mathbf{q}(\mathbf{qn})\,dS.$$

The second term gives the flux of linear momentum out of S through the boundary.

Using Gauss's theorem, 2·61 (7), (1) becomes

$$\int \frac{\partial}{\partial t}(\rho \mathbf{q})d\tau + \int [(\rho \mathbf{q})(\nabla \mathbf{q}) + (\mathbf{q}\,\nabla)(\rho \mathbf{q})]d\tau$$

$$= \int \left[\frac{d(\rho \mathbf{q})}{dt} + (\rho \mathbf{q})(\nabla \mathbf{q})\right] d\tau, \text{ from 3·10 (9),}$$

$$= \int \left[\rho \frac{d\mathbf{q}}{dt} + \mathbf{q}\left(\frac{d\rho}{dt} + \rho(\nabla \mathbf{q})\right)\right] d\tau.$$

(2) $$= \int \rho \frac{d\mathbf{q}}{dt}\, d\tau, \text{ using the equation of continuity.}$$

We can also look on this result as follows by using 3·20 (2).

The rate of change of the momentum within S as S moves about with the fluid is

$$\frac{d}{dt}\int \mathbf{q}\,\rho\, d\tau = \int \rho \frac{d\mathbf{q}}{dt}\, d\tau,$$

3·41. The equation of motion of an inviscid fluid. Consider the fluid body which at time t occupies the region interior to a closed surface S. By the second law of motion, the total force acting on this fluid body is equal to the rate of change of linear momentum.

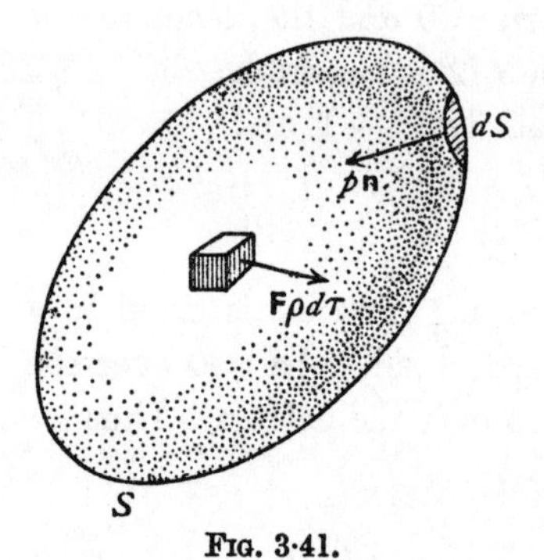

FIG. 3·41.

The force is due to

(i) the normal pressure thrusts on the boundary;

(ii) the external force (such as gravity), say $\mathbf{F}$ per unit mass.

Thus the total force is

$$\int p\mathbf{n}\, dS + \int \mathbf{F}\,\rho\, d\tau = -\int (\nabla p) d\tau + \int \mathbf{F}\,\rho\, d\tau,$$

using Gauss's theorem. Equating this to the rate of change of linear momentum calculated in 3·40 (2), we get

$$\int \left(\mathbf{F}\,\rho - \nabla p - \rho \frac{d\mathbf{q}}{dt}\right) d\tau = 0.$$

Since the shape of the fluid body and therefore the volume of integration is arbitrary we must have

$$\mathbf{F}\,\rho - \nabla p - \rho \frac{d\mathbf{q}}{dt} = 0, \quad \text{or}$$

(1) $$\frac{d\mathbf{q}}{dt} = \mathbf{F} - \frac{1}{\rho}\nabla p,$$

which is the equation of motion.

Again, from 3·10 (9) and 2·34 (IV),

$$\frac{d\mathbf{q}}{dt} = \frac{\partial \mathbf{q}}{\partial t} + (\mathbf{q}\,\nabla)\mathbf{q} = \frac{\partial \mathbf{q}}{\partial t} + \tfrac{1}{2}\nabla q^2 - \mathbf{q}_\wedge(\nabla_\wedge \mathbf{q}).$$

Therefore

$$\frac{\partial \mathbf{q}}{\partial t} - \mathbf{q}_\wedge(\nabla_\wedge \mathbf{q}) = \mathbf{F} - \frac{1}{\rho}\nabla p - \tfrac{1}{2}\nabla q^2 \tag{2}$$

which is another form of the equation of motion.

3·42. Euler's momentum theorem. We shall now obtain the general form of the theorem established in 1·9. From 3·40 (1), we have for the rate of change of momentum of the fluid body bounded by a closed surface S,

$$\frac{\partial}{\partial t}\int \rho \mathbf{q}\, d\tau - \int_{(S)} (\mathbf{n}\mathbf{q})\rho\mathbf{q}\, dS,$$

and therefore, from the second law of motion, using fig. 3·41,

$$\int_{(S)} \mathbf{n}\, p\, dS = -\int \rho \mathbf{F}\, d\tau + \frac{\partial}{\partial t}\int \rho\mathbf{q}\, d\tau - \int_{(S)} (\mathbf{n}\mathbf{q})\rho\mathbf{q}\, dS.$$

This formula states that the resultant thrust on the fluid contained within a closed surface S is equal to the reversed resultant of the body forces on the enclosed fluid, together with the rate of change $\partial/\partial t$ of the momentum of the fluid, and the rate of flow of momentum *outwards* across the boundary of S. This is the generalised form of the momentum theorem. It may also be regarded as a generalisation of the theorem known as the principle of Archimedes, to which it reduces when the fluid is at rest.

3·43. Conservative forces. For conservative forces derivable from a potential Ω we write $\mathbf{F} = -\nabla\Omega$. Also if the pressure is a function of the density so that $\int dp/\rho$ exists, we have, from 2·23 (3),

$$\frac{1}{\rho}(d\mathbf{r}\,\nabla)p = \frac{dp}{\rho} = d\int\frac{dp}{\rho} = (d\mathbf{r}\,\nabla)\int\frac{dp}{\rho}$$

and therefore, since $d\mathbf{r}$ is arbitrary,

$$\frac{1}{\rho}\nabla p = \nabla\int\frac{dp}{\rho}. \tag{1}$$

The equation of motion 3·41 (1) then assumes the form

$$\frac{d\mathbf{q}}{dt} = -\nabla\left(\int\frac{dp}{\rho} + \Omega\right), \tag{2}$$

which shows that the acceleration is derivable from the *acceleration potential* $\int dp/\rho + \Omega$.

Further, observing that the vorticity is $\boldsymbol{\zeta} = \nabla_\wedge \mathbf{q}$, equation 3·41 (2) can be written in the form

$$\text{(3)} \qquad \frac{\partial \mathbf{q}}{\partial t} - \mathbf{q}_\wedge \boldsymbol{\zeta} = -\nabla \chi, \quad \chi = \int \frac{dp}{\rho} + \Omega + \tfrac{1}{2}q^2$$

which puts the vorticity in evidence.

Again 3·41 (1) can be written

$$\rho \frac{\partial \mathbf{q}}{\partial t} + \rho(\mathbf{q}\,\nabla)\mathbf{q} = -\nabla p - \rho \nabla \Omega$$

while the equation of continuity 3·20 (4) gives

$$\mathbf{q}\,\frac{\partial \rho}{\partial t} + \mathbf{q}\,\nabla\,(\rho \mathbf{q}) = 0.$$

By addition, using 2·34 (X), we get

$$\text{(4)} \qquad \frac{\partial(\rho \mathbf{q})}{\partial t} + \nabla\,[\rho \mathbf{q}\,;\,\mathbf{q} + \mathbf{I}p] + \rho \nabla \Omega = 0,$$

where $\mathbf{I}$ is the idemfactor.

In cartesian coordinates, the equation (2) is equivalent to the following system of three equations :

$$\text{(5)} \qquad \begin{aligned} \frac{\partial u}{\partial t} + u\frac{\partial u}{\partial x} + v\frac{\partial u}{\partial y} + w\frac{\partial u}{\partial z} &= -\frac{\partial \Omega}{\partial x} - \frac{1}{\rho}\frac{\partial p}{\partial x}, \\ \frac{\partial v}{\partial t} + u\frac{\partial v}{\partial x} + v\frac{\partial v}{\partial y} + w\frac{\partial v}{\partial z} &= -\frac{\partial \Omega}{\partial y} - \frac{1}{\rho}\frac{\partial p}{\partial y}, \\ \frac{\partial w}{\partial t} + u\frac{\partial w}{\partial x} + v\frac{\partial w}{\partial y} + w\frac{\partial w}{\partial z} &= -\frac{\partial \Omega}{\partial z} - \frac{1}{\rho}\frac{\partial p}{\partial z}. \end{aligned}$$

If $\boldsymbol{\zeta} = \mathbf{i}\xi + \mathbf{j}\eta + \mathbf{k}\zeta$, so that ξ, η, ζ are the components of vorticity, equation (3) yields the following set :

$$\text{(6)} \qquad \begin{aligned} \frac{\partial u}{\partial t} + w\eta - v\zeta + \frac{\partial}{\partial x}(\tfrac{1}{2}q^2) &= -\frac{\partial \Omega}{\partial x} - \frac{1}{\rho}\frac{\partial p}{\partial x}, \\ \frac{\partial v}{\partial t} + u\zeta - w\xi + \frac{\partial}{\partial y}(\tfrac{1}{2}q^2) &= -\frac{\partial \Omega}{\partial y} - \frac{1}{\rho}\frac{\partial p}{\partial y}, \\ \frac{\partial w}{\partial t} + v\xi - u\eta + \frac{\partial}{\partial z}(\tfrac{1}{2}q^2) &= -\frac{\partial \Omega}{\partial z} - \frac{1}{\rho}\frac{\partial p}{\partial z}, \end{aligned}$$

where

$$q^2 = u^2 + v^2 + w^2,$$

$$\text{(7)} \qquad \xi = \frac{\partial w}{\partial y} - \frac{\partial v}{\partial z}, \quad \eta = \frac{\partial u}{\partial z} - \frac{\partial w}{\partial x}, \quad \zeta = \frac{\partial v}{\partial x} - \frac{\partial u}{\partial y}.$$

The reader should verify that the equations (5) and (6) are equivalent. The above results once more illustrate how effectively the vector notation condenses and illuminates the analysis.

Beltrami flows arise when $\mathbf{q}_\wedge \boldsymbol{\zeta} = 0$; the corresponding equation of motion is (3) in the form

(8) $$\frac{\partial \mathbf{q}}{\partial t} = -\nabla \chi, \quad \chi = \int \frac{dp}{\rho} + \Omega + \tfrac{1}{2}q^2.$$

If the vorticity is different from zero, the condition $\mathbf{q}_\wedge \boldsymbol{\zeta} = 0$ states that vortex lines and streamlines coincide.

If $\boldsymbol{\zeta}=0$ we have the important case of *irrotational motion* which is also a Beltrami flow and obeys equation (7).

In the case of a homogeneous liquid $\int dp/\rho$ is replaced in the above equations by p/ρ.

All the foregoing are known as *Eulerian* or *statistical* forms of the equation of motion. In them attention is directed to a particular point $\mathbf{r}$ of space. As time t elapses this point is occupied by a succession of fluid particles; $\mathbf{r}$ and t are independent variables.

3·44. Lagrangian form of the equation of motion. From the *Lagrangian* or *historical* point of view instead of fixing attention on a particular point of space we fix attention on a particular fluid particle and follow its progress. The independent variables are $\mathbf{r}_0$, the initial position vector of the particle, and t the time. If the particle occupies the position $\mathbf{r}$ at time t, we have $\mathbf{r}=\mathbf{r}(\mathbf{r}_0, t)$ so that the acceleration of the particle is $\partial^2\mathbf{r}/\partial t^2$, a partial derivative, and therefore from 3·41 (1) the equation of motion is

$$\frac{\partial^2 \mathbf{r}}{\partial t^2} = \mathbf{F} - \frac{1}{\rho}\frac{\partial p}{\partial \mathbf{r}} = \mathbf{F} - \frac{1}{\rho}\frac{\partial\,;\,\mathbf{r}_0}{\partial \mathbf{r}} \cdot \frac{\partial p}{\partial \mathbf{r}_0}$$

using 2·71 (8). Multiply in front by $\partial\,;\,\mathbf{r}/\partial\mathbf{r}_0$. Then from 2·71 (9) we have

(1) $$\frac{\partial\,;\,\mathbf{r}}{\partial \mathbf{r}_0} \cdot \left(\frac{\partial^2 \mathbf{r}}{\partial t^2} - \mathbf{F}\right) + \frac{1}{\rho}\frac{\partial p}{\partial \mathbf{r}_0} = 0,$$

and this is the Lagrangian form of the equation of motion, all differentiations being with respect to the independent variables $\mathbf{r}_0$, t.

If $\mathbf{F} = -\nabla\Omega$, integration from 0 to t gives Weber's transformation, namely

(2) $$\frac{\partial\,;\,\mathbf{r}}{\partial\,\mathbf{r}_0}\mathbf{q} - \mathbf{q}_0 = -\frac{\partial \chi}{\partial \mathbf{r}_0}, \quad \chi = \int_0^t \left\{\int \frac{dp}{\rho} + \Omega - \tfrac{1}{2}q^2\right\} dt.$$

The equation of continuity follows from 3·20 (1) in the form

(3) $$\rho\, d\tau = \rho_0\, d\tau_0$$

where suffix zero refers to the particle in its initial position, expressing the fact that the mass of the particle remains unaltered as it moves.

In cartesian coordinates we have $d\tau = dx\,dy\,dz$, $d\tau_0 = dx_0\,dy_0\,dz_0$ and

(4) $$dx\,dy\,dz = J\,dx_0\,dy_0\,dz_0, \quad J = \frac{\partial(x, y, z)}{\partial(x_0, y_0, z_0)},$$

J being the Jacobian of the coordinates (x, y, z) of $\mathbf{r}$ with respect to the coordinates (x_0, y_0, z_0) of $\mathbf{r}_0$. In this notation the equation of continuity becomes

$$\rho J = \rho_0. \tag{5}$$

A surface $\mathbf{F}(\mathbf{r}, t) = 0$ always consists of the same fluid particles if, and only if, $d\mathbf{F}/dt = 0$. For this condition means that $\mathbf{F}(\mathbf{r}, t)$ is independent of t and so when expressed in Lagrangian coordinates has the form $f(\mathbf{r}_0) = 0$. This occurs, in particular, in the case of the free surface of a liquid in continuous motion.

It is not essential that $\mathbf{r}_0$ should be the initial position vector. Any vector variable which serves to identify a particle and which varies continuously from one particle to another may be used; see for example 14·40

3·45. Steady motion. When the motion is steady, $\partial \mathbf{q}/\partial t = 0$, and we then get, from 3·43 (3),

$$\mathbf{q}_\wedge \boldsymbol{\zeta} = \nabla \chi \quad , \quad \chi = \int \frac{dp}{\rho} + \tfrac{1}{2}q^2 + \Omega. \tag{1}$$

From the meaning of ∇ applied to a scalar, this shows that the vector $\mathbf{q}_\wedge \boldsymbol{\zeta}$ is normal to the surfaces

$$\int \frac{dp}{\rho} + \tfrac{1}{2}q^2 + \Omega = c, \tag{2}$$

where c is a constant. Since $\mathbf{q}_\wedge \boldsymbol{\zeta}$ is perpendicular to $\mathbf{q}$ and to $\boldsymbol{\zeta}$, it follows that any particular surface of the system (2) contains both streamlines and vortex lines. Along every such streamline or vortex line the left-hand member of (2) has the same constant value. This is the general form of Bernoulli's equation for a fluid. For a liquid $\int dp/\rho$ in (1) and (2) is replaced by p/ρ.

The existence of the surfaces (2) is a necessary condition for steady motion to be possible.

When the motion is both steady and irrotational ($\boldsymbol{\zeta} = 0$), equation (1) shows that the constant in (2) is the same throughout the fluid.

3·50. The energy equation. When the forces are conservative, the equation of motion, 3·41 (1), after scalar multiplication by $\rho\mathbf{q}$, gives

$$\tfrac{1}{2}\rho \frac{d}{dt}(q^2) = -\mathbf{q} \nabla p - \rho \mathbf{q} \nabla \Omega.$$

If $\partial \Omega/\partial t = 0$, we have, from 3·10 (9),

$$\frac{d\Omega}{dt} = \mathbf{q} \nabla \Omega,$$

and therefore

$$\rho \frac{d}{dt}[\tfrac{1}{2}q^2 + \Omega] = -\mathbf{q} \nabla p.$$

Multiply by the volume element $d\tau$ and observe that, by the equation of continuity (3·20 (1)),

$$\frac{d}{dt}(\rho\, d\tau) = 0.$$

We then get, on integrating throughout the volume of the fluid,

$$\frac{d}{dt}\int \rho(\tfrac{1}{2}q^2+\Omega)\,d\tau = -\int \mathbf{q}\,\nabla\, p\, d\tau.$$

Now, if $\quad T = \int \tfrac{1}{2}\rho q^2 d\tau,\quad V = \int \rho\Omega\, d\tau,\quad J = \int \rho E\, d\tau,$

are the kinetic, potential and internal (1·6) energies respectively, we get, using 2·34 (VI),

$$\frac{d}{dt}(T+V) = -\int \nabla\,(p\mathbf{q})\,d\tau + \int p\,\nabla\,\mathbf{q}\,d\tau$$

$$= \int p\mathbf{q}\mathbf{n}\,dS + \int p\,\nabla\,\mathbf{q}\,d\tau,$$

by Gauss's theorem, the surface integral being taken over the bounding surface and $\mathbf{n}$ being the unit inward normal.

Now the last integral is $-dJ/dt$, see Ex. III, 31, and therefore

$$\frac{d}{dt}(T+V+J) = \int p\mathbf{q}\mathbf{n}\,dS,$$

which expresses that the rate of change of total energy of any portion of the fluid as it moves about is equal to the rate of working of the pressures on the boundary.

3·51 Rate of change of circulation. Let C be a closed circuit which moves with the fluid, i.e. a circuit which always consists of the same fluid particles. Let $\mathbf{a}$ be the acceleration of a fluid particle and $\mathbf{B}$ its curl:

(1) $$\mathbf{a} = d\mathbf{q}/dt,\quad \mathbf{B} = \nabla_\wedge\,\mathbf{a}.$$

Then for the rate of change of circulation in C as it moves we have

(2) $$\frac{d}{dt}\operatorname{circ} C = \frac{d}{dt}\int_{(C)} \mathbf{q}\,d\mathbf{r} = \int_{(C)} \frac{d\mathbf{q}}{dt}\,d\mathbf{r} + \int_{(C)} \mathbf{q}\,d\left(\frac{d\mathbf{r}}{dt}\right) = \int_{(C)} \mathbf{a}\,d\mathbf{r},$$

for $\mathbf{q}\,d(d\mathbf{r}/dt) = \mathbf{q}\,d\mathbf{q}$ and therefore its integral round C vanishes.

Also by Stokes' theorem

(3) $$\int_{(C)} \mathbf{a}\,d\mathbf{r} = \int_{(S)} \mathbf{n}\,(\nabla_\wedge\,\mathbf{a})\,dS$$

over any diaphragm S which closes C. Therefore

(4) $$\frac{d}{dt}\operatorname{circ} C = \int_{(S)} \mathbf{n}\,\mathbf{B}\,dS.$$

Now the vector field **B** is solenoidal (2·24), for by 2·32 (II)

(5) $$\nabla \mathbf{B} = \nabla (\nabla \wedge \mathbf{a}) = 0,$$

and therefore we can define unit **B**-tubes (2·615). Therefore from (4)

(6) $$\frac{d}{dt} \operatorname{circ} C = N,$$

where N is the number of unit **B**-tubes which thread the circuit C. This result holds for viscous and compressible as well as for inviscid or incompressible flows.

For an inviscid fluid under conservative forces we have

$$\mathbf{a} = -\frac{1}{\rho} \nabla p - \nabla \Omega,$$

and therefore

(7) $$\mathbf{B} = \nabla p \wedge \nabla \left(\frac{1}{\rho}\right).$$

If we call $1/\rho$ the *bulkiness* of the fluid, ∇p and $\nabla (1/\rho)$ are normals respectively to the surfaces of constant pressure and constant bulkiness, so that the vector **B** is tangential to the curve of intersection of these surfaces. The direction of **B** determines the sense of the circulation in C.

As an example; at given temperature and pressure water with greater salt content has higher density and therefore smaller bulkiness. Suppose in an ocean that the salinity decreases in a certain direction. Then the bulkiness increases in the same direction, and the pressure increases downwards. The result is that circulation is set up along the bottom in the sense of decreasing salinity and along the surface in the sense of increasing salinity. This explains the surface currents into the more saline Mediterranean from the Black Sea through the Bosporus and from the Atlantic through the Strait of Gibraltar.

From (6) it appears that the necessary and sufficient condition for constancy of the circulation in a circuit which moves with the fluid is $\nabla \wedge \mathbf{a} = 0$ or $\mathbf{B} = 0$.

A fundamental application of this result is Lord Kelvin's theorem concerning the *constancy of circulation in a circuit moving with the fluid in an inviscid fluid in which the density is either constant or is a function of the pressure* (*barotropic flow*).

Proof. If ρ is a constant $\nabla (1/\rho) = 0$ and therefore $\mathbf{B} = 0$, from (7).

If ρ is a function of p, $\nabla (1/\rho)$ and ∇p are parallel vectors and therefore from (7), $\mathbf{B} = 0$. In either case $d \operatorname{circ} C/dt = 0$ so that circ C is independent of time.

Q.E.D

3·52. Vortex motion. If $\boldsymbol{\zeta}$ is the vorticity vector, we have

$$\boldsymbol{\zeta} = \nabla \wedge \mathbf{q},$$

and therefore, from 2·32 (II),

$$\nabla \boldsymbol{\zeta} = 0,$$

so that the divergence of the vorticity is everywhere zero; vorticity is solenoidal.

Vortex lines have been defined already (2·41). If through every point of a closed curve we draw the vortex line, we obtain a *vortex tube.*

A *vortex filament* is a vortex tube whose cross-sectional area is of infinitesimal dimensions. By Gauss's theorem, applied to the volume enclosed between two cross-sections of areas $d\sigma_1$ and $d\sigma_2$ of a vortex filament, we get

$$\int \boldsymbol{\zeta}\mathbf{n}\, dS = -\int \nabla \boldsymbol{\zeta}\, d\tau = 0,$$

and since $\mathbf{n}\boldsymbol{\zeta} = 0$ on the walls of the filament,

$$\boldsymbol{\zeta}_1\, \mathbf{n}_1\, d\sigma_1 + \boldsymbol{\zeta}_2\, \mathbf{n}_2\, d\sigma_2 = 0,$$

where $\boldsymbol{\zeta}_1$, $\boldsymbol{\zeta}_2$ are the vorticities at the ends of the filament. Thus

$$\zeta_1\, d\sigma_1 = \zeta_2\, d\sigma_2,$$

which expresses that the magnitude of the vorticity multiplied by the cross-sectional area is constant along the filament.*

It therefore follows that a vortex filament cannot terminate at a point within the fluid. Vortex filaments must therefore be either closed (vortex rings) or terminate at the boundaries.

The analogy of the foregoing with the corresponding property of stream filaments in a liquid may be noted, for in the case of a liquid $\nabla\, \mathbf{q} = 0$, so that $\mathbf{q}$ like $\boldsymbol{\zeta}$ is solenoidal.

3·53. Permanence of vorticity. If $\mathbf{a}$ is the acceleration, we have

$$\mathbf{a} = \frac{d\mathbf{q}}{dt} = \frac{\partial \mathbf{q}}{\partial t} - \mathbf{q}_\wedge \boldsymbol{\zeta} + \nabla \tfrac{1}{2}q^2.$$

Taking the curl and using 2·32 (III) and 2·34 (II) we have

$$\nabla_\wedge \mathbf{a} = \frac{\partial \boldsymbol{\zeta}}{\partial t} - (\boldsymbol{\zeta}\nabla)\mathbf{q} + (\mathbf{q}\nabla)\boldsymbol{\zeta} + \boldsymbol{\zeta}(\nabla\mathbf{q}) - \mathbf{q}(\nabla\boldsymbol{\zeta}).$$

Now from 3·52 $\nabla \boldsymbol{\zeta} = 0$, and from 3·20 (5) $\rho \nabla \mathbf{q} = -d\rho/dt$. Thus using 3·10 (9) we find that

$$\nabla_\wedge \mathbf{a} = \frac{d\boldsymbol{\zeta}}{dt} - (\boldsymbol{\zeta}\nabla)\mathbf{q} - \frac{\boldsymbol{\zeta}}{\rho}\frac{d\rho}{dt} \quad \text{or}$$

$$\frac{d}{dt}\left(\frac{\boldsymbol{\zeta}}{\rho}\right) = \frac{1}{\rho}(\boldsymbol{\zeta}\nabla)\mathbf{q} + \frac{1}{\rho}\nabla_\wedge \mathbf{a}. \tag{1}$$

This *purely kinematical* relation gives the rate of change of $\boldsymbol{\zeta}/\rho$.

If *the forces are conservative and the pressure is a function of the density*, taking the curl of 3·43 (2) shows that $\nabla_\wedge \mathbf{a} = 0$, and in this case (1) becomes

* This result follows directly from the property of solenoidal vectors (2·615).

(2)
$$\frac{d}{dt}\left(\frac{\boldsymbol{\zeta}}{\rho}\right) = \left[\left(\frac{\boldsymbol{\zeta}}{\rho}\right)\nabla\right]\mathbf{q},$$

an equation due to Helmholtz.

To solve this equation we use the notation $\partial/\partial\mathbf{r}$ for ∇, and so, from 2·71 (8),

(3)
$$\frac{d}{dt}\left(\frac{\boldsymbol{\zeta}}{\rho}\right) = \frac{\boldsymbol{\zeta}}{\rho}\cdot\left(\frac{\partial\,;\,\mathbf{q}}{\partial\mathbf{r}}\right) = \frac{\boldsymbol{\zeta}}{\rho}\cdot\left(\frac{\partial\,;\,\mathbf{r}_0}{\partial\mathbf{r}}\right)\cdot\left(\frac{\partial\,;\,\mathbf{q}}{\partial\mathbf{r}_0}\right)$$

where $\mathbf{r}_0$ is the position vector of the particle at time t_0 as in fig. 3·10.

Differentiating 2·71 (9) with respect to t we get

(4)
$$\frac{d}{dt}\left(\frac{\partial\,;\,\mathbf{r}_0}{\partial\mathbf{r}}\right)\cdot\left(\frac{\partial\,;\,\mathbf{r}}{\partial\mathbf{r}_0}\right) + \frac{\partial\,;\,\mathbf{r}_0}{\partial\mathbf{r}}\cdot\frac{\partial\,;\,\mathbf{q}}{\partial\mathbf{r}_0} = 0,$$

since $d\mathbf{r}/dt = \mathbf{q}$. Thus (3) can be written

$$\frac{d}{dt}\left(\frac{\boldsymbol{\zeta}}{\rho}\right) + \frac{\boldsymbol{\zeta}}{\rho}\frac{d}{dt}\left(\frac{\partial\,;\,\mathbf{r}_0}{\partial\mathbf{r}}\right)\cdot\frac{\partial\,;\,\mathbf{r}}{\partial\mathbf{r}_0} = 0.$$

Multiply on the right by $\partial\,;\,\mathbf{r}_0/\partial\mathbf{r}$ and use 2·71 (9) again. Then we get

(5)
$$\frac{d}{dt}\left(\frac{\boldsymbol{\zeta}}{\rho}\cdot\frac{\partial\,;\,\mathbf{r}_0}{\partial\mathbf{r}}\right) = 0$$

and therefore

(6)
$$\frac{\boldsymbol{\zeta}}{\rho}\cdot\frac{\partial\,;\,\mathbf{r}_0}{\partial\mathbf{r}} = \text{constant} = \frac{\boldsymbol{\zeta}_0}{\rho_0},$$

where $\boldsymbol{\zeta}_0$ and ρ_0 are the values of $\boldsymbol{\zeta}$ and ρ at time t_0.

Multiply on the right by $\partial\,;\,\mathbf{r}/\partial\mathbf{r}_0$ and use 2·71 (9) once more. Then

(7)
$$\frac{\boldsymbol{\zeta}}{\rho} = \frac{\boldsymbol{\zeta}_0}{\rho_0}\cdot\frac{\partial\,;\,\mathbf{r}}{\partial\mathbf{r}_0}.$$

From (7) we see that if $\boldsymbol{\zeta}_0 = 0$, then $\boldsymbol{\zeta} = 0$ so that motion once irrotational remains so. Therefore a particle which has vorticity at any time continues to have vorticity. Thus rotational motion is permanent and so is irrotational motion.

Notice that this conclusion depends on the assumptions which led to equation (2); inviscid fluid, conservative forces, pressure a function of the density.

It also follows that if the flow is irrotational at infinity, non-zero vorticity can be possessed only by those particles whose path lines do not extend to infinity.

3·54. Permanence of vortex lines. When inviscid fluid moves under conservative forces and the pressure is a function of the density a vortex line consists always of the same fluid particles and therefore moves with the fluid.

Proof. Let a line of particles be specified by a Lagrangian parameter α so that at time t the position vector of a particle is $\mathbf{r} = \mathbf{r}(\alpha, t)$. Then at time t the tangent to the line is in the direction of the vector $\partial\mathbf{r}/\partial\alpha$.

By the definition of a vortex line the vorticity vector is tangent to the line and so if $\boldsymbol{\zeta}_0$ is the vorticity at time t_0, we have

$$\frac{\partial \mathbf{r}_0}{\partial \alpha} \wedge \boldsymbol{\zeta}_0 = 0 \quad \text{or} \quad \boldsymbol{\zeta}_0 = \lambda_0 \frac{\partial \mathbf{r}_0}{\partial \alpha}, \tag{1}$$

where λ_0 is a scalar, and these statements are equivalent. From 3·53 (7) we have at time t

$$\boldsymbol{\zeta} = \frac{\rho}{\rho_0} \lambda_0 \frac{\partial \mathbf{r}_0}{\partial \alpha} \cdot \frac{\partial ; \mathbf{r}}{\partial \mathbf{r}_0} = \frac{\rho}{\rho_0} \lambda_0 \frac{\partial \mathbf{r}}{\partial \alpha} = \lambda \frac{\partial \mathbf{r}}{\partial \alpha},$$

so that the same particle α is still on the vortex line. Q.E.D.

Such a line moves about with the fluid like a material substance. Moreover the line cannot disappear, for we have proved that rotational motion is permanent.

It follows that when in an actual fluid a vortex line does disappear, the internal friction must be the cause.

3·55. Relative motion. Velocity is a concept relative to a frame of reference which the observer sets up as his standard of " fixity ". Thus the velocity of a terrestrial body is usually measured with respect to a frame of reference fixed to the globe of the earth.

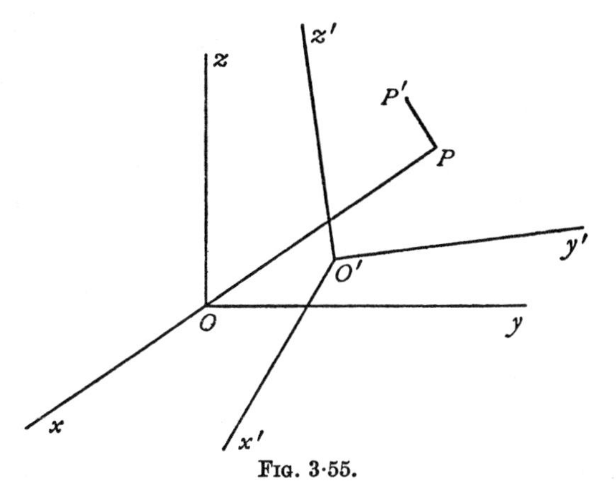

FIG. 3·55.

Now consider two cartesian frames of reference $Oxyz$ or F, and $O'x'y'z'$ or F'. Each frame may be imagined as identified by a set of wires rigidly connected and moving with the frame. Suppose that at time t the frames are *coincident* and that F' is moving relatively to F with a motion described by an observer in F as a velocity $\mathbf{U}$ of O' and an angular velocity $\boldsymbol{\omega}$. Then the position vector $\mathbf{r}$ of a particular fluid particle P at time t is the same for both frames.*

Let $\mathbf{q}$, $\mathbf{q}'$ be the velocities of the fluid particle P at time t as estimated by observers in F and F' respectively. Then $\mathbf{q} = \mathbf{q}' + \mathbf{U} + \boldsymbol{\omega} \wedge \mathbf{r}$, and therefore, for the vorticity,

$$\begin{aligned} \boldsymbol{\zeta} = \nabla \wedge \mathbf{q} &= \nabla \wedge \mathbf{q}' + \nabla \wedge (\boldsymbol{\omega} \wedge \mathbf{r}) \\ &= \boldsymbol{\zeta}' + \boldsymbol{\omega}(\nabla\, \mathbf{r}) - (\boldsymbol{\omega}\, \nabla)\,\mathbf{r} = \boldsymbol{\zeta}' + 3\boldsymbol{\omega} - \boldsymbol{\omega} = \boldsymbol{\zeta}' + 2\boldsymbol{\omega}, \end{aligned}$$

* Fig. 3·55 shows the relative positions of the frames at time $t + \delta t$, when they are no longer coincident. The fluid particle which was at P at time t is at P' at time $t + \delta t$.

so that, like velocity, vorticity is a concept relative to the frame of reference. If $\boldsymbol{\zeta} = 0$ in the frame F, the observer in that frame says that the motion is irrotational, and therefore also says that there is a velocity potential ϕ such that $\mathbf{q} = -\nabla\phi$, while the observer in the frame F' says that the motion is rotational with vorticity $-2\boldsymbol{\omega}$.

Similarly, circulation is a concept relative to the frame of reference, for if Γ is the circulation in a closed circuit C as measured by an observer in F and if Γ' is the circulation in the same circuit as measured by an observer in F', then

$$\Gamma = \Gamma' + 2\omega S,$$

where S is the area enclosed by the projection of C on a plane perpendicular to $\boldsymbol{\omega}$.

Proof. $\Gamma - \Gamma' = \int_{(C)} \mathbf{q}\,d\mathbf{r} - \int_{(C)} \mathbf{q}'\,d\mathbf{r} = \int_{(C)} (\mathbf{U} + \boldsymbol{\omega}_\wedge \mathbf{r})\,d\mathbf{r} = \boldsymbol{\omega}\int_{(C)} (\mathbf{r}_\wedge d\mathbf{r}).$

Take $\boldsymbol{\omega} = \omega\mathbf{k}$. Then

$$\boldsymbol{\omega}\int_{(C)} \mathbf{r}_\wedge d\mathbf{r} = \omega\int_{(C)} (x\,dy - y\,dx) = 2\omega S. \qquad \text{Q.E.D.}$$

These considerations are of importance in the hydrodynamics of meteorology on the rotating earth.

3·60. Irrotational motion. Pressure equation. When the pressure is a function of the density, $p = f(\rho)$, the equation of irrotational motion* under conservative forces is, 3·43 (8),

$$\frac{\partial \mathbf{q}}{\partial t} = -\nabla\left(\int \frac{dp}{\rho} + \tfrac{1}{2}q^2 + \Omega\right).$$

Since $\mathbf{q} = -\nabla\phi$ in irrotational motion, this gives

$$\nabla\left(\int \frac{dp}{\rho} + \tfrac{1}{2}q^2 + \Omega - \frac{\partial\phi}{\partial t}\right) = 0,$$

and therefore

$$\text{(1)} \qquad \int \frac{dp}{\rho} + \tfrac{1}{2}q^2 + \Omega - \frac{\partial\phi}{\partial t} = C(t),$$

where $C(t)$ denotes an instantaneous constant, that is to say, a function of t only, which therefore at a given instant has the same value throughout the liquid. This is the *pressure equation.* The function $C(t)$ may be replaced by an absolute constant by adding a suitable function of t to ϕ. The addition of such

* "With motion irrotational, in fluid incompressible,
A tiny little minnow swims along a line of flow,
And the greater its velocity—well cutting out verbosity—
The greater its velocity, the faster it will go."

Eureka, Cambridge.

a function to ϕ does not affect the relation $\mathbf{q} = -\nabla\phi$. When the motion is steady, $\partial\phi/\partial t = 0$, and we recover Bernoulli's equation, but with the same value of c throughout the fluid at all times.

The pressure equation is of paramount importance, for once we know the velocity potential ϕ, the velocity is determined by $\mathbf{q} = -\nabla\phi$, and the pressure is then found from the pressure equation and the relation $p = f(\rho)$.

Note that $\partial\phi/\partial t$ is calculated by varying t only and therefore refers to a point fixed in space.

When the fluid is incompressible the pressure equation is

$$\frac{p}{\rho} + \tfrac{1}{2}q^2 + \Omega - \frac{\partial\phi}{\partial t} = C(t).$$

It follows that in principle the solution of any problem of irrotational motion of a liquid is reduced to finding the velocity potential ϕ which satisfies Laplace's equation $\nabla^2\phi = 0$ and the other conditions of the problem. The calculation of fluid thrust on a surface is then reduced to an integration.

3·61. The pressure equation referred to moving axes. Consider as in 3·55 a moving rigid frame of reference F' whose motion referred to the instantaneous position F of the frame,* with O as base-point, is described by the linear velocity $\mathbf{U}$ and the angular velocity $\boldsymbol{\omega}$. The point P, whose position vector referred to O is $\mathbf{r}$, if rigidly attached to the frame F', has the velocity $\mathbf{V} = \mathbf{U} + \boldsymbol{\omega}_\wedge \mathbf{r}$. Thus if P is fixed in F instead of in the frame F', it will appear to an observer in the frame F' to move with velocity $-\mathbf{V}$. If the motion is irrotational when referred to the frame F, there exists a velocity potential ϕ such that $\mathbf{q} = -\nabla\phi$, and the rate of change of ϕ at a point fixed in F is now measured (cf. 3·10) by an observer in F' as

$$\left(\frac{\partial}{\partial t} - \mathbf{V}\nabla\right)\phi.$$

Hence the pressure equation for a liquid becomes

$$\frac{p}{\rho} + \tfrac{1}{2}q^2 + \Omega - \frac{\partial\phi}{\partial t} + \mathbf{V}\nabla\phi = C(t).$$

Let $\mathbf{q}_r$ be the velocity of the fluid relative to the moving frame. Then

$$\mathbf{q}_r = \mathbf{q} - \mathbf{V} = -\nabla\phi - \mathbf{V}.$$

Thus
$$\tfrac{1}{2}q^2 + \mathbf{V}(\nabla\phi) = \tfrac{1}{2}(\mathbf{q} - \mathbf{V})^2 - \tfrac{1}{2}\mathbf{V}^2,$$

and therefore the pressure equation with respect to the moving frame can be written

$$\frac{p}{\rho} + \tfrac{1}{2}q_r^{\,2} + \Omega - \frac{\partial\phi}{\partial t} - \tfrac{1}{2}V^2 = C(t),$$

* This instantaneous position F is taken as the standard of fixity referred to in 3·55.

where q_r is the magnitude of the fluid velocity at P relative to the moving frame, and V is the speed of the same point regarded as fixed to the moving frame.

3·62. The thrust on an obstacle. Consider the steady irrotational motion of a homogeneous liquid in the presence of a fixed obstacle S. Let $\mathbf{F}$ be the thrust on the obstacle due to the hydrodynamical pressure. Then if $\mathbf{n}$ is a unit outward normal to the element dS of the surface of the obstacle,

$$\mathbf{F} = -\int_{(S)} p\mathbf{n}\, dS.$$

Now, from the pressure equation, since the motion is steady,

$$p = \text{constant} - \tfrac{1}{2}\rho q^2,$$

and a constant pressure produces no resultant action on a closed surface. Therefore

$$\mathbf{F} = \tfrac{1}{2}\rho \int_{(S)} \mathbf{n} q^2\, dS.$$

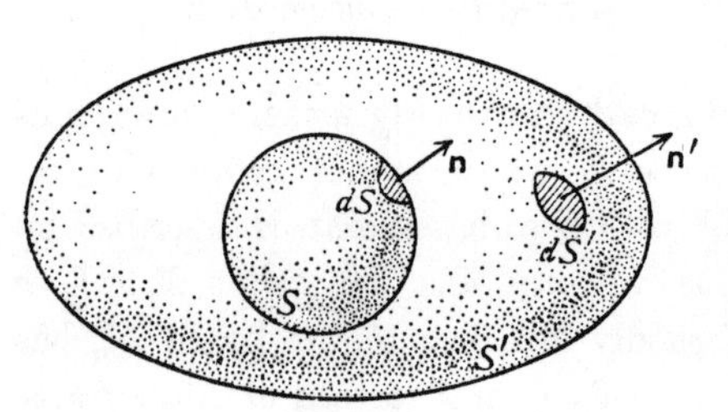

FIG. 3·62.

Now $\mathbf{n}\mathbf{q}$ is the component of the velocity of the fluid normal to the boundary, and therefore $\mathbf{n}\,\mathbf{q} = 0$ at points on the boundary. Therefore we can write

$$(1)\quad \mathbf{F} = \tfrac{1}{2}\rho \int_{(S)} [\mathbf{n} q^2 - 2\mathbf{q}\,(\mathbf{n}\,\mathbf{q})] dS,$$

the surface integral being taken over the surface of the obstacle. Let S' be a closed surface entirely surrounding the obstacle, and let $\mathbf{n}'$ be a unit normal (drawn outwards from the region between S and S') to the element dS'.

Then, if we integrate over the surface $S+S'$, we get from Gauss's theorem 2·61, (3), (7),

$$\int_{(S+S')} [\mathbf{n} q^2 - 2\mathbf{q}(\mathbf{n}\mathbf{q})] dS = -\int_{(V)} [\nabla q^2 - 2\mathbf{q}(\nabla\, \mathbf{q}) - 2(\mathbf{q}\, \nabla)\mathbf{q}] d\tau$$

$$= -\int_{(V)} [2\mathbf{q}_\wedge(\nabla_\wedge \mathbf{q}) - 2\mathbf{q}(\nabla\, \mathbf{q})] d\tau, \quad \text{from 2·34 (IV).}$$

Since the motion is irrotational, $\nabla_\wedge \mathbf{q} = 0$, and if the region (V) between S and S' encloses no points at which fluid is created or destroyed, $\nabla\, \mathbf{q} = 0$ from the equation of continuity, and therefore the volume integral vanishes. It follows that

$$\int_{(S)} [\mathbf{n} q^2 - 2\mathbf{q}(\mathbf{n}\mathbf{q})]\, dS = \int_{(S')} [\mathbf{n}' q^2 - 2\mathbf{q}(\mathbf{n}'\, \mathbf{q})] dS'.$$

Thus in (1) we can replace S by any enclosing surface, provided we cross no singularities in the fluid, i.e. by any reconcilable surface (see 3·70).

In the same way we prove that the moment about the origin is

$$\text{(2)} \qquad \mathbf{L} = \tfrac{1}{2}\rho \int_{(S')} \mathbf{r}_\wedge [\mathbf{n} q^2 - 2\mathbf{q}(\mathbf{n}\mathbf{q})] dS,$$

where S' is any surface reconcilable with S without crossing any singularities in the fluid.

3·64. Impulsive motion.

Let us suppose that a fluid in motion is subjected to external impulses and to impulsive pressure.

If $\mathbf{q}_1$ is the velocity generated in the element which was previously moving with velocity $\mathbf{q}$, $\mathbf{I}$ the external impulse per unit mass, and ϖ the impulsive pressure, by equating the impulse to the change of momentum of the fluid within a closed surface S, we get, as in 3·41,

$$\int \varpi \, \mathbf{n} \, dS + \int \mathbf{I} \rho \, d\tau = \int \rho \, (\mathbf{q}_1 - \mathbf{q}) \, d\tau.$$

Using Gauss's theorem this gives

$$\int \{\mathbf{I} \rho - \nabla \varpi - \rho(\mathbf{q}_1 - \mathbf{q})\} \, d\tau = 0.$$

Since the volume of integration is arbitrary we have

$$\mathbf{I} - \frac{1}{\rho} \nabla \varpi = \mathbf{q}_1 - \mathbf{q}.$$

This is the general equation of impulsive motion.

This equation provides a physical interpretation of the velocity potential as follows.

The external impulses being absent, let ϕ be the velocity potential of a motion generated from rest by impulsive pressure ϖ. Then in the above equation $\mathbf{I} = 0$, $\mathbf{q} = 0$, $\mathbf{q}_1 = -\nabla \phi$, and therefore

$$\nabla \varpi = \rho \nabla \phi,$$

which in the case of a homogeneous liquid gives

$$\varpi = \rho\phi + \text{constant}.$$

The constant can be ignored, for a pressure constant throughout the fluid produces no effect on the motion, and we see that $\rho\phi$ is the impulsive pressure which would instantaneously generate from rest the motion which actually exists (cf. Ex. III, 32).

Conversely, a motion generated from rest by impulsive pressure only is necessarily irrotational, the velocity potential being ϖ/ρ. This must necessarily be the case when a motion is, for example, started from rest by sudden motion of the boundaries. Thus when a body disturbs an irrotational flow, the disturbed flow is also irrotational.

The argument is true also for a viscous fluid as regards the *initial* motion (see Plates 1, 2, fig. 1), but vortex sheets (13·70) may form even in an inviscid fluid due to the bringing together of layers of fluid which were previously separated and are moving with different velocities. The presence of even slight viscosity may cause these sheets to roll up and form concentrated vortices (see Plates 1, 2, 3, 4).

3·70. Connectivity. *Definition.* A region of space is said to be *connected* if we can pass from any point of the region to any other point by moving along a path every point of which lies in the given region.

Thus the region interior to a sphere, fig. 3·70 (i), or the region between two coaxial infinitely long cylinders, fig. 3·70 (ii), are connected.

Definition. A closed circuit, all of whose points lie in the given region, is said to be *reducible*, if it can be contracted to a point of the region without ever passing out of the region.

The circuit $PRQS$ in figs. (i), (ii) is reducible; the circuit $P'R'Q'S'$ in fig. (ii) is *irreducible*, for it cannot be made smaller than the circumference of the inner cylinder.

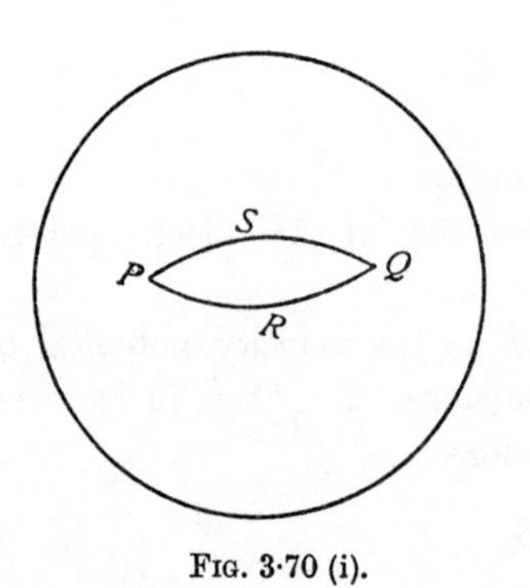

FIG. 3·70 (i).

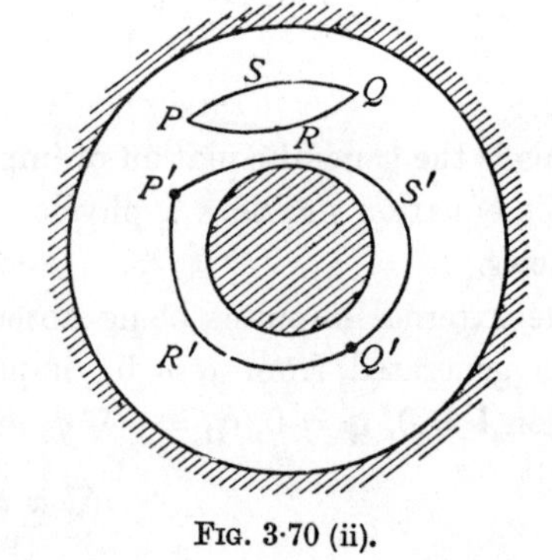

FIG. 3·70 (ii).

Definition. A region in which every circuit is reducible is said to be *simply connected.*

Examples of simply connected regions are: the region interior to a sphere; the region exterior to a sphere; the region exterior to any number of spheres; the region between two concentric spheres; unbounded space.

The region between the concentric cylinders in fig. (ii) is certainly not simply connected, for it contains irreducible circuits. We can, however, make this region simply connected by inserting one *barrier* or boundary which may not be crossed, such as the plane AB containing a generating line of each cylinder, fig. (iii).

When this barrier is inserted every circuit in the modified region is reducible and the modified region is therefore simply connected.

We also note that the insertion of an additional barrier between the inner and outer cylinders would break the region up into two parts which, although individually connected regions, would not form a connected region in their totality.

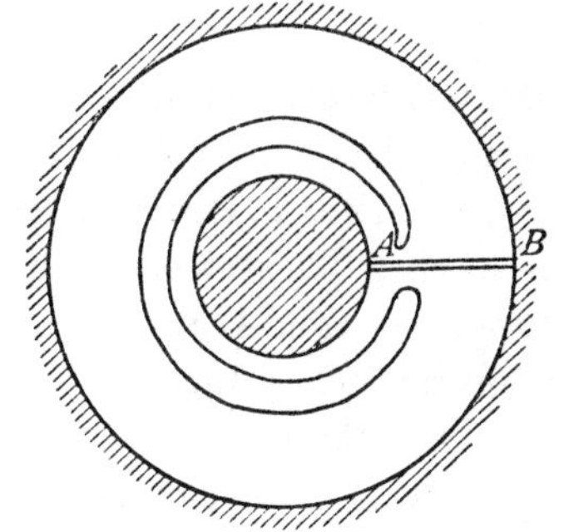

Fig. 3·70 (iii).

We thus arrive at the following definition.

Definition. A region is said to be *doubly connected*, if it can be made simply connected by the insertion of one barrier. A region is said to be n-ply connected, if it can be made simply connected by the insertion of $n-1$ barriers.

Examples of doubly connected regions are: the region between coaxial infinitely long cylinders; the region interior to an anchor ring; the region exterior to an anchor ring; the region exterior to an infinitely long cylinder.

Another useful idea is contained in the following definition.

Definition. The paths joining two points P and Q of a region are said to be *reconcilable*, if either can be continuously deformed into the other without ever passing out of the region.

Thus in figs. (i), (ii) the paths PRQ, PSQ are reconcilable. In fig. (ii) the paths $P'R'Q'$, $P'S'Q'$ are *irreconcilable.*

Two reconcilable paths taken together clearly constitute a reducible circuit.

Definition. Two closed circuits are said to be reconcilable, if either can be continuously deformed into the other without ever passing out of the region.

Reconcilable circuits are not necessarily reducible.

The term reconcilable can also be conveniently applied to surfaces (cf. 3·62). Thus the diaphragms referred to in the verbal enunciation of Stokes' theorem 2·51 (1) must all be reconcilable without passing out of the fluid.

The above properties of regions are termed *topological* rather than geometrical, for they do not essentially depend on the particular shapes of the boundaries mentioned. For example, the cross-sections of the cylinders could be ellipses or any other simple closed curves.

3·71. Acyclic and cyclic irrotational motion. When the region occupied by fluid moving irrotationally is simply connected, the velocity potential is one-valued, for the velocity potential at P is defined by (see 2·52)

$$(1) \qquad \phi_P = -\int_{(OAP)} \mathbf{q}\, d\mathbf{r},$$

and this integral is the same for all paths from O to P, for all such paths are reconcilable. Motion in which the velocity potential is one-valued is called *acyclic*. *Thus in a simply connected region the only possible irrotational motion is acyclic.* This result depends essentially on the possibility of joining any two paths from O to P by a surface lying entirely within the fluid and then applying Stokes' theorem (see 2·52).

When the region is not simply connected, two paths from O to P can be joined by a surface lying entirely within the fluid only when certain topological conditions are satisfied. When they are not, the inference from Stokes' theorem cannot be made, and the velocity potential *may* then have more than one value at P, according to the path taken from O to P.

When the velocity potential is not one-valued the motion is said to be *cyclic*.

In the continuous motion of a fluid the velocity at any point must be perfectly definite. Thus, even when ϕ has more than one value at a given point, $\nabla\phi$ must be one-valued. It follows that although two paths from O to P may lead to different values of ϕ_P, these values can differ only by a scalar κ, such that $\nabla\kappa = 0$, and κ is therefore independent of the coordinates of P. This scalar κ may be identified with the circulation in any one of a family of reconcilable irreducible circuits, for, if C be any circuit, (1) shows that

(2) circ C = *decrease in* ϕ *on describing the circuit once.*

We shall have occasion later to consider particular types of cyclic motion. For the present we shall consider only acyclic irrotational motion, and the general theorems which follow must be considered as applying to that type of motion only. In that sense the regions concerned may always be considered as simply connected, but it should be remembered that acyclic motion is also possible in multiply connected regions.

3·72. Kinetic energy of liquid. The kinetic energy is given by

$$T = \tfrac{1}{2}\int_{(V)} \rho q^2\, d\tau,$$

taken throughout the volume V occupied by the fluid.

When the motion is irrotational,

$$\mathbf{q} = -\nabla\phi,$$

and therefore by Green's theorem, if ϕ is single valued, and since $\nabla^2\phi = 0$,

$$T = \tfrac{1}{2}\rho\int_{(V)} (\nabla\phi)(\nabla\phi)\,d\tau = -\tfrac{1}{2}\rho\int_{(S)} \phi\frac{\partial\phi}{\partial n}\,dS,$$

taken over the bounding surface of the liquid, dn denoting an element of normal drawn into the liquid.

This result has a simple physical interpretation. Since the actual motion could be started from rest by the application of an impulsive pressure $\rho\phi$, and since $-\partial\phi/\partial n$ is the velocity of the liquid normal to the boundary, $\rho\phi\,\delta S \times -\tfrac{1}{2}\,\partial\phi/\partial n$ is the work done by the impulsive pressure on the element δS in accordance with the following dynamical theorem.

The work done by an impulse is equal to the product of the impulse into half the sum of the components in the direction of the impulse of the initial and final velocities of the point at which it is applied.

The surface integral therefore represents the work done by the impulsive pressure in starting the motion from rest.

3·73. Kelvin's minimum energy theorem. The irrotational motion of a liquid occupying a simply connected region has less kinetic energy than any other motion consistent with the same normal velocity of the boundary.

Proof. Let T be the kinetic energy of the irrotational motion, ϕ the velocity potential, and T_1 the kinetic energy of any other motion given by

$$\mathbf{q} = -\nabla\phi + \mathbf{q}_0, \quad \nabla\mathbf{q}_0 = 0, \quad \mathbf{n}\mathbf{q}_0 = 0 \text{ at the boundary,}$$

the second condition being the equation of continuity. Then

$$T_1 = \int \tfrac{1}{2}\rho(-\nabla\phi + \mathbf{q}_0)^2 d\tau = T + T_0 - \rho\int(\nabla\phi)\mathbf{q}_0\,d\tau,$$

where
$$T_0 = \int \tfrac{1}{2}\rho\,\mathbf{q}_0{}^2 d\tau.$$

Now, from 2·34 (VI),

$$\nabla(\phi\,\mathbf{q}_0) = \mathbf{q}_0(\nabla\phi) + \phi(\nabla\mathbf{q}_0) = \mathbf{q}_0(\nabla\phi).$$

Therefore
$$\int(\nabla\phi)\mathbf{q}_0\,d\tau = -\int \mathbf{n}\phi\mathbf{q}_0\,dS = 0,$$

by Gauss's theorem, and since $\mathbf{n}\mathbf{q}_0 = 0$.

Therefore
$$T_1 = T + T_0.$$

Since T_0 is positive, it follows that $T < T_1$. Q.E.D.

3·74. Mean value of the velocity potential. We shall prove the following theorem due to Gauss.

The mean value of ϕ *over any spherical surface, throughout whose interior* $\nabla^2\phi = 0$, *is equal to the value of* ϕ *at the centre of the sphere.*

Proof. Describe a sphere S of radius r about P. Then from 2·63 (2)

$$\phi_P = -\frac{1}{4\pi}\int_{(S)} \phi \frac{\partial}{\partial r}\left(\frac{1}{r}\right) dS + \frac{1}{4\pi r}\int_{(S)} \frac{\partial \phi}{\partial r} dS.$$

But the second integral vanishes by 2·62 (5). Therefore

$$\phi_P = \frac{1}{4\pi r^2}\int_{(S)} \phi \, dS.$$

Q.E.D.

Corollary. ϕ *cannot be a maximum or minimum in the interior of any region throughout which* $\nabla^2\phi = 0$.

For if ϕ_P were, say, a maximum, it would be greater than the value of ϕ at all points of a sufficiently small sphere, centre P, which contradicts the theorem just proved.

We can now prove the following theorem.

In irrotational motion the maximum values of the speed must occur on the boundary.

Proof. Take a point P *interior* to the fluid as origin, and take the axis of x in the direction of motion at P. Then if q_P, q_Q are the speeds at P and Q (a point near to P),

$$q_P^2 = \left(\frac{\partial \phi}{\partial x}\right)_P^2, \quad q_Q^2 = \left(\frac{\partial \phi}{\partial x}\right)_Q^2 + \left(\frac{\partial \phi}{\partial y}\right)_Q^2 + \left(\frac{\partial \phi}{\partial z}\right)_Q^2.$$

Now, $\dfrac{\partial \phi}{\partial x}$ satisfies Laplace's equation * and therefore cannot be a maximum or minimum at P. Therefore there are points such as Q in the immediate neighbourhood of P at which $\left(\dfrac{\partial \phi}{\partial x}\right)_Q^2 > \left(\dfrac{\partial \phi}{\partial x}\right)_P^2$, and therefore $q_Q^2 > q_P^2$.

Thus q_P cannot be a maximum in the interior of the fluid, and its maximum values, if any, must therefore occur on the boundary. Q.E.D.

It should be noted that q^2 may be a minimum in the interior of the fluid, for it is zero at a stagnation point.

From the above results we can deduce the following theorem.

In steady irrotational motion the hydrodynamical pressure has its minimum values on the boundary.

Proof. By Bernoulli's theorem,

$$\frac{p}{\rho} + \tfrac{1}{2}q^2 = \text{constant}.$$

---

* $\nabla^2 \dfrac{\partial \phi}{\partial x} = \dfrac{\partial}{\partial x}\nabla^2\phi = 0.$

Thus p is least where q^2 is greatest, and this cannot occur inside the fluid. Thus the minimum values of p must occur on the boundary. The maximum values of p occur at the stagnation points. Q.E.D.

3·75. Mean value of the velocity potential in a periphractic region.

A region is said to be *periphractic* * when it is bounded internally by one or more closed surfaces. For example, the region occupied by fluid in which a solid sphere is totally immersed is of this nature.

Consider liquid at rest at infinity bounded internally by a closed surface S and unbounded externally. With centre P, describe a sphere Σ, of radius R, large enough to enclose S. If the liquid is in irrotational motion, Gauss's theorem applied to the periphractic region between S and Σ gives

$$\int_{(S)} \frac{\partial \phi}{\partial n}\, dS + \int_{(\Sigma)} \frac{\partial \phi}{\partial n}\, d\Sigma = -\int \nabla^2 \phi \, d\tau = 0.$$

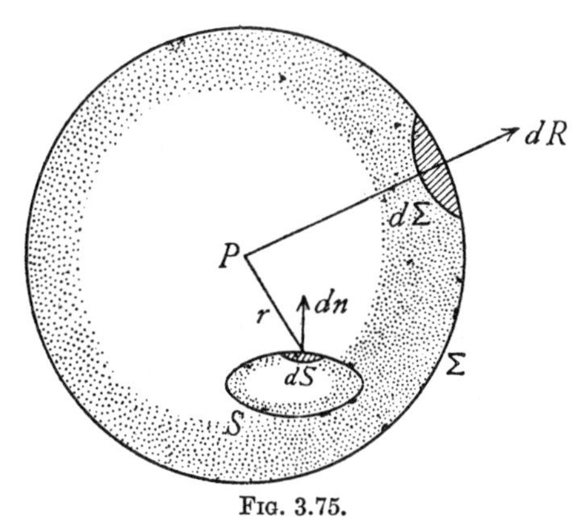

FIG. 3.75.

Therefore, since $dn = -dR$ on Σ,

$$(1) \qquad \int_{(\Sigma)} \frac{\partial \phi}{\partial R}\, d\Sigma = \int_{(S)} \frac{\partial \phi}{\partial n}\, dS = -F,$$

where F is the flux into the region considered across the internal boundary S.

Now, $d\Sigma = R^2\, d\omega$, where $d\omega$ is the solid angle subtended at P by $d\Sigma$. Therefore the above result can be written

$$\frac{\partial}{\partial R} \int \phi \, d\omega = -\frac{F}{R^2}.$$

But if $M(\phi)$ is the mean value of ϕ over the sphere Σ,

$$M(\phi) = \frac{1}{4\pi R^2} \int \phi \, d\Sigma = \frac{1}{4\pi} \int \phi \, d\omega,$$

* Greek, περιφρακτος = "fenced about".

and therefore

(2) $$\frac{\partial M(\phi)}{\partial R} = -\frac{F}{4\pi R^2}, \quad M(\phi) = \frac{F}{4\pi R} + C,$$

where C is independent of R. To show that C is also independent of the position of the centre of the sphere Σ, let us displace the centre through a distance δx, keeping R constant. Then

$$\frac{\partial C}{\partial x} = \frac{\partial M(\phi)}{\partial x} = \frac{1}{4\pi}\int \frac{\partial \phi}{\partial x}\, d\omega.$$

Since by hypothesis $\partial\phi/\partial x \to 0$ at infinity, by making R large enough we can make $\partial\phi/\partial x$ as small as we please, so that $\partial C/\partial x = 0$. Thus C is unaltered when the centre of the sphere Σ is displaced, provided that S is always within the sphere.

Thus it appears that at great distances R

$$\nabla\phi = O\left(\frac{1}{R^2}\right).$$

In the important case when S is the surface of a solid body, there is no flow across it, so that $F = 0$, and therefore the mean value of ϕ over any sphere enclosing the solid is constant and equal to C.

We now prove the important theorem that

$$\phi_P \to C \text{ when } P \to \infty.$$

Proof. Applying 2·63 to the region between S and Σ, we get

$$4\pi\phi_P = \int_{(S)} \left[\phi \frac{\partial}{\partial n}\left(\frac{1}{r}\right) - \frac{1}{r}\frac{\partial \phi}{\partial n}\right] dS$$
$$+ \int_{(\Sigma)} \left[\phi \frac{\partial}{\partial n}\left(\frac{1}{r}\right) - \frac{1}{r}\frac{\partial \phi}{\partial n}\right] d\Sigma.$$

Now, the latter integral is equal to

$$\frac{1}{R^2}\int_{(\Sigma)} \phi\, d\Sigma + \frac{1}{R}\int_{(\Sigma)} \frac{\partial \phi}{\partial R}\, d\Sigma = 4\pi C,$$

from (1) and (2). Therefore

(3) $$4\pi(\phi_P - C) = \int_{(S)} \left[\phi \frac{\partial}{\partial n}\left(\frac{1}{r}\right) - \frac{1}{r}\frac{\partial \phi}{\partial n}\right] dS.$$

If we now let $r \to \infty$, both $1/r$ and its differential coefficient tend to zero and therefore when $P \to \infty$,

$$\phi_P \to C.$$

3·76. Kinetic energy of infinite liquid. Taking liquid moving irrotationally, at rest at infinity, and bounded internally by a solid S, we shall suppose that the velocity potential ϕ is single valued. Applying the method of 3·72 to the region between the solid S and a large surface Σ, completely enclosing S, we get for the kinetic energy of the liquid occupying this region

$$T_1 = -\tfrac{1}{2}\rho \int_{(S)} \phi \frac{\partial \phi}{\partial n} dS - \tfrac{1}{2}\rho \int_{(\Sigma)} \phi \frac{\partial \phi}{\partial n} d\Sigma.$$

Since there is no flow into the region across S, the equation of continuity takes the form (cf. 3·20 (2))

$$(1) \qquad \int_{(S)} \frac{\partial \phi}{\partial n} dS + \int_{(\Sigma)} \frac{\partial \phi}{\partial n} d\Sigma = 0,$$

and therefore

$$T_1 = -\tfrac{1}{2}\rho \int_{(S)} (\phi - C) \frac{\partial \phi}{\partial n} dS - \tfrac{1}{2}\rho \int_{(\Sigma)} (\phi - C) \frac{\partial \phi}{\partial n} d\Sigma,$$

where C is any constant. It follows from (1) that $\int \partial\phi/\partial n \, d\Sigma$ is independent of Σ and is in fact zero since for a solid boundary $\int \partial\phi/\partial n \, dS = 0$. If therefore we take C to be the value to which ϕ tends at infinity (3·75) and then enlarge the surface Σ indefinitely in all directions, the second integral vanishes and we get the kinetic energy.

$$T = -\tfrac{1}{2}\rho \int_{(S)} (\phi - C) \frac{\partial \phi}{\partial n} dS = -\tfrac{1}{2}\rho \int_{(S)} \phi \frac{\partial \phi}{\partial n} dS.$$

3·77. Uniqueness theorems. We shall now prove some related theorems concerning acyclic irrotational motion of a liquid. The proofs are all based on the following equivalence of the expressions for the kinetic energy,

$$(1) \qquad \tfrac{1}{2}\rho \int q^2 d\tau = -\tfrac{1}{2}\rho \int \phi \frac{\partial \phi}{\partial n} dS.$$

where the volume integral is taken throughout the fluid and the surface integral is taken over the boundary.

(I) *Acyclic irrotational motion is impossible in a liquid bounded entirely by fixed rigid walls.*

For $\dfrac{\partial \phi}{\partial n} = 0$ at every point of the boundary, and therefore $\displaystyle\int q^2 d\tau = 0.$

Since q^2 cannot be negative, $q = 0$ everywhere and the liquid is at rest.

(II) *The acyclic irrotational motion of a liquid bounded by rigid walls will instantly cease if the boundaries are brought to rest.*

This is an immediate corollary to (I).

(III) *There cannot be two different forms of acyclic irrotational motion of a confined mass of liquid in which the boundaries have prescribed velocities.*

For, if possible, let ϕ_1, ϕ_2 be the velocity potentials of two different motions subject to the condition $\partial\phi_1/\partial n = \partial\phi_2/\partial n$ at each point of the boundary.

Then $\phi = \phi_1 - \phi_2$ is a solution of Laplace's equation and therefore represents a possible irrotational motion in which

$$\frac{\partial\phi}{\partial n} = \frac{\partial\phi_1}{\partial n} - \frac{\partial\phi_2}{\partial n} = 0.$$

Therefore, as in (I), $q = 0$ at every point, and therefore $\phi_1 - \phi_2 =$ constant, so that the motions are essentially the same.

This theorem shows that acyclic motion is uniquely determined when the boundary velocities are given.

(IV) *If given impulsive pressures are applied to the boundaries of a confined mass of liquid at rest, the resulting motion, if acyclic and irrotational, is uniquely determinate.*

If possible, let ϕ_1 and ϕ_2 be velocity potentials of two different motions. The impulsive pressure which would start the first motion is $\rho\phi_1$, that which would start the second is $\rho\phi_2$, and since the pressures are given at the boundaries

$$\rho\phi_1 = \rho\phi_2$$

at each point of the boundary.

Therefore $\phi = \phi_1 - \phi_2$ is the velocity potential of a possible irrotational motion such that $\phi = 0$ at each point of the boundary. Therefore, from (1), $q = 0$ at each point of the liquid. If follows that $\phi_1 - \phi_2$ is constant and the motions are essentially the same.

(V) *Acyclic irrotational motion is impossible in a liquid which is at rest at infinity and is bounded internally by fixed rigid walls.*

Since the liquid is at rest at infinity and there is no flow over the internal boundaries, the kinetic energy is still given by (1) (see 3·76) and the proof is therefore the same as in (I).

(VI) *The acyclic irrotational motion of a liquid at rest at infinity and bounded internally by rigid walls will instantly cease if the boundaries are brought to rest.*

This is an immediate corollary to (V).

(VII) *The acyclic irrotational motion of a liquid, at rest at infinity, due to the prescribed motion of an immersed solid, is uniquely determined by the motion of the solid.*

If possible, let ϕ_1, ϕ_2 be the velocity potentials of two different motions. The boundary conditions are

$$\frac{\partial\phi_1}{\partial n} = \frac{\partial\phi_2}{\partial n} \text{ at the surface of the solid,} \quad q_1 = q_2 = 0 \text{ at infinity.}$$

Thus $\phi = \phi_1 - \phi_2$ is the velocity potential of a possible motion, such that $\partial\phi/\partial n = 0$ at the surface of the solid, $q = 0$ at infinity. It then follows from

(1) that $q = 0$ everywhere, so that $\phi_1 - \phi_2$ = constant, and the motions are essentially the same.

(VIII) *If the liquid is in motion at infinity with uniform velocity, the acyclic irrotational motion, due to the prescribed motion of an immersed solid, is uniquely determined by the motion of the solid.*

For the relative kinematical conditions are unaltered if we superpose on the whole system of solid and liquid a velocity equal in magnitude and opposite in direction to the velocity at infinity. This brings the liquid to rest at infinity. The resulting motion is then determinate by (VII) and we return to the given motion by reimposing the velocity at infinity.

Theorem VIII shows that we can not prescribe more than the normal component of the fluid velocity at the bounding surfaces. In particular an adherence condition at a rigid surface is, in general, incompatible with inviscid irrotational motion.

EXAMPLES III

1. Establish the equation of continuity for an incompressible fluid in the form

$$\frac{\partial u}{\partial x} + \frac{\partial v}{\partial y} + \frac{\partial w}{\partial z} = 0.$$

Show that $u = -\dfrac{2xyz}{(x^2+y^2)^2}, \quad v = \dfrac{(x^2-y^2)z}{(x^2+y^2)^2}, \quad w = \dfrac{y}{x^2+y^2}$

are the velocity-components of a possible fluid motion. Is this motion irrotational? (R.N.C.)

2. If the fluid moves radially and the velocity u is a function of r, t only, where r is the distance from a fixed point, prove that the equation of continuity is

$$\frac{\partial \rho}{\partial t} + u\frac{\partial \rho}{\partial r} + \frac{\rho}{r^2}\frac{\partial}{\partial r}(r^2 u) = 0.$$

3. If every particle of fluid moves on the surface of a sphere, prove that the equation of continuity is

$$\cos\theta\frac{\partial \rho}{\partial t} + \frac{\partial}{\partial \theta}(\rho q_\theta \cos\theta) + \frac{\partial}{\partial \omega}(\rho q_\omega \cos\theta) = 0,$$

where θ, ω are the latitude and longitude, and q_θ, q_ω the angular velocities in latitude and longitude respectively.

4. If ω is the area of the cross-section of a stream filament, prove that the equation of continuity is

$$\frac{\partial}{\partial t}(\rho\omega) + \frac{\partial}{\partial s}(\rho\omega q) = 0,$$

where ds is an element of arc of the filament in the direction of flow and q is the speed.

5. If $F(\mathbf{r}, t) = 0$ is a surface which always consists of the same fluid particles, show that, after an infinitesimal time δt, $F(\mathbf{r}+\mathbf{q}\,\delta t, t+\delta t) = 0$, and deduce that

$$\frac{\partial F}{\partial t}+(\mathbf{q}\,\nabla)F = 0.$$

6. Explain the method of differentiation following the fluid, and find the condition that the surface $F(x, y, z, t) = 0$ may be a boundary surface.

Prove that the variable ellipsoid

$$\frac{x^2}{a^2k^2t^4}+kt^2\left(\frac{y^2}{b^2}+\frac{z^2}{c^2}\right) = 1$$

is a possible form of boundary surface of a liquid at time t.

7. A quantity of liquid occupies a length $2l$ of a straight tube of uniform small bore, under the action of a force towards a fixed point in the tube varying as the distance from the fixed point. Determine the pressure at a distance x from the fixed point when the nearer free surface is at a distance z.

8. For cylindrical coordinates (2·72), prove that the acceleration is

$$\frac{d\mathbf{q}}{dt} = \mathbf{i}_\varpi\left(\frac{d'q_\varpi}{dt}-\frac{q^2_\omega}{\varpi}\right)+\mathbf{i}_\omega\frac{1}{\varpi}\frac{d'}{dt}(\varpi q_\omega)+\mathbf{i}_x\frac{d'q_x}{dt},$$

where $$d'/dt = \partial/\partial t+q_\varpi\partial/\partial\varpi+\varpi^{-1}q_\omega\partial/\partial\omega+q_x\partial/\partial x.$$

9. Prove that the three equations of motion expressed in cylindrical coordinates are, (see Ex. 8),

$$F_x-\frac{1}{\rho}\frac{\partial p}{\partial x} = \frac{d'q_x}{dt},\quad F_\varpi-\frac{1}{\rho}\frac{\partial p}{\partial\varpi} = \frac{d'q_\varpi}{dt}-\frac{q_\omega{}^2}{\varpi},\quad \varpi F_\omega-\frac{1}{\rho}\frac{\partial p}{\partial\omega} = \frac{d'}{dt}(\varpi q_\omega).$$

10. If liquid rotates like a rigid body with constant angular speed ω about a vertical (z) axis and gravity is the only external force, prove that the pressure is given by

$$\frac{p}{\rho} = \tfrac{1}{2}\omega^2r^2-gz+\text{constant},$$

where r is the distance from the axis. Show that the surfaces of equal pressure are paraboloids with the same latus rectum.

11. If liquid contained within a closed circular cylinder rotates about the axis of the cylinder, prove that the equation of continuity and the boundary conditions are satisfied by $\mathbf{q} = \boldsymbol{\omega}\wedge\mathbf{r}$, where $\boldsymbol{\omega}$ is the angular velocity supposed dependent on the time only and $\mathbf{r}$ is the position vector measured from a point on the axis of rotation.

12. If the liquid in Ex. 11 starts from rest under the external forces whose components are $\alpha x+\beta y$, $\gamma x+\delta y$, 0 and the axis of the cylinder is the z-axis, write down the equations of motion and prove that

$$\frac{d\omega}{dt} = \tfrac{1}{2}(\gamma-\beta).$$

Prove also that the pressure is given by

$$\frac{p}{\rho} = \tfrac{1}{2}\omega^2r^2+\tfrac{1}{2}[\alpha x^2+(\beta+\gamma)xy+\delta y^2,]$$

where r is the distance from the z-axis.

13. If the motion of a fluid be referred to a moving frame of reference which rotates with angular velocity $\boldsymbol{\omega}$ and moves forward with velocity $\mathbf{u}$, show that the equation of motion is

$$\frac{\partial \mathbf{q}}{\partial t}+\boldsymbol{\omega}_\wedge\mathbf{q}+\left(\frac{d\mathbf{r}}{dt}\nabla\right)\mathbf{q} = \mathbf{F}-\frac{1}{\rho}\nabla p,$$

where $\dfrac{d\mathbf{r}}{dt} = \mathbf{q}-\mathbf{u}-\boldsymbol{\omega}_\wedge\mathbf{r}$, and that the equation of continuity is

$$\frac{\partial \rho}{\partial t}+\nabla\left(\rho\frac{d\mathbf{r}}{dt}\right) = 0,$$

where $\mathbf{q}$ is the fluid velocity and the position vector $\mathbf{r}$ is referred to the moving frame.

14. If the motion is referred to a moving frame which has velocity $\mathbf{u}$ and angular velocity $\boldsymbol{\omega}$, prove that the vorticity satisfies the equation

$$\frac{\partial \boldsymbol{\zeta}}{\partial t}+\boldsymbol{\omega}_\wedge\boldsymbol{\zeta}+(\mathbf{q}_r\nabla)\boldsymbol{\zeta} = (\boldsymbol{\zeta}\nabla)\mathbf{q},$$

where $\mathbf{q}_r = \mathbf{q}-\mathbf{u}-\boldsymbol{\omega}_\wedge\mathbf{r}$.

15. If $\mathbf{q}$ is the velocity, prove that

$$\frac{1}{2}\int_{(V)}\nabla q^2 d\tau = \int_{(V)}((\mathbf{q}\nabla)\mathbf{q}+\mathbf{q}_\wedge\boldsymbol{\zeta})d\tau,$$

and deduce that

$$\frac{1}{2}\int_{(S)}\mathbf{n}q^2dS = \int_{(S)}\mathbf{q}(\mathbf{n}\mathbf{q})dS+\int_{(V)}\mathbf{q}(\nabla\mathbf{q})d\tau-\int_{(V)}(\mathbf{q}_\wedge\boldsymbol{\zeta})d\tau,$$

where S is a closed surface and V the enclosed volume.

Use the above result to find the force on a body due to fluid pressure.

16. If Γ is the circulation around any closed circuit moving with the fluid, prove that

$$\frac{d\Gamma}{dt} = \int pd\left(\frac{1}{\rho}\right),$$

if the external forces have a potential, and the pressure is a function of the density alone. (U.L.)

17. A pulse travelling along a fine straight uniform tube filled with gas causes the density at time t and distance x from an origin where the velocity is u_0 to become $\rho_0\phi(Vt-x)$. Prove that the velocity u is given by

$$V+\frac{(u_0-V)\phi(Vt)}{\phi(Vt-x)}. \qquad \text{(U.L.)}$$

18. Every particle of a mass of liquid is revolving uniformly about a fixed axis, the angular speed varying as the nth power of the distance from the axis. Show that the motion is irrotational only if $n+2 = 0$.

If a very small spherical portion of the liquid be suddenly solidified, prove that it will begin to rotate about a diameter with an angular velocity $(n+2)/2$ of that with which it was revolving about the fixed axis.

19. An explosion takes place at a point O at some distance below the surface of deep water. If O' is the image of O in the free surface, show that the velocity-potential of the initial motion at any point P varies as

$$\frac{1}{OP}-\frac{1}{O'P}.$$

Determine the initial velocity of the free surface at any point. (R.N.C.)

20. Define irrotational motion and prove that under certain conditions the motion of a frictionless liquid, if once irrotational, is always so. Prove that this theorem remains true, if the motion of each particle be resisted by a force varying as its absolute velocity.

21. If ϕ is constant over the boundary of any simply connected region occupied by liquid in irrotational motion, prove that ϕ has the same constant value throughout the interior.

22. Prove that, if the normal velocity is zero at every point of the boundary of liquid occupying a simply connected region, and moving irrotationally, ϕ is constant throughout the interior of that region.

23. Liquid moving irrotationally occupies a simply connected region bounded partly by surfaces over which ϕ is constant, and partly by surfaces over which the normal velocity is zero. Prove that ϕ has the same constant value throughout the region.

24. A body moves in a given manner, without change of volume, in an inviscid liquid. T_0 denotes the kinetic energy of the fluid when it has no external boundary and is at rest at infinite distances; T_0' denotes the kinetic energy of that part of the fluid which is outside a closed surface S_0 which is external to the body; T denotes the kinetic energy of the fluid when S_0 is its external boundary and is fixed. Prove that, if the regions occupied by the fluid are simply connected,

$$T > T_0 + T_0'.$$

25. If α = constant, β = constant are the equations of a curve, show that the tangent is in the direction of the vector $\nabla \alpha_\wedge \nabla \beta$. Hence show that if the α and β surfaces are any two systems of surfaces which pass through the vortex lines, then $\boldsymbol{\zeta} = F \nabla \alpha_\wedge \nabla \beta$, where F is a scalar function.

26. In Ex. 25 use the fact that $\nabla \boldsymbol{\zeta} = 0$ to prove that $(\nabla F)(\nabla \alpha_\wedge \nabla \beta) = 0$, and hence show that this is equivalent to the vanishing of the Jacobian

$$\partial(F, \alpha, \beta)/\partial(x, y, z),$$

so that F is a function of α, β only.

27. Prove that $\nabla f(\alpha, \beta) = \dfrac{\partial f}{\partial \alpha} \nabla \alpha + \dfrac{\partial f}{\partial \beta} \nabla \beta$. With the notations of Ex. 25, 26, show that if the scalar function $f(\alpha, \beta)$ is so chosen that $\partial f/\partial \alpha = F$, then

(i) $\mathbf{q} = f(\alpha, \beta) \nabla \beta$ is a solution of the equation $\boldsymbol{\zeta} = \nabla_\wedge \mathbf{q}$.

(ii) $\boldsymbol{\zeta} = \dfrac{\partial f}{\partial \alpha} \nabla \alpha_\wedge \nabla \beta$.

28. Use Ex. 27 to prove that the general solution of $\boldsymbol{\zeta} = \nabla_\wedge \mathbf{q}$ is

$$\mathbf{q} = -\nabla \phi + f(\alpha, \beta) \nabla \beta,$$

where α = constant, β = constant are two systems of surfaces which pass through the vortex lines, and ϕ is a solution of Laplace's equation.

29. Obtain Clebsch's transformation that the velocity can be expressed in the form

$$\mathbf{q} = -\nabla \phi + \lambda \nabla \mu,$$

where the surfaces λ = constant, μ = constant move with the fluid, and the curves in which they intersect are vortex lines.

30. Prove the moment formula 3·62 (2).

31. If E is the internal energy per unit mass, prove that

$$\rho \frac{dE}{dt} = \frac{p}{\rho} \frac{d\rho}{dt} = -p \nabla \mathbf{q}.$$

32. Prove that for a compressible fluid moving irrotationally

$$\phi = \int \frac{d\varpi}{\rho}$$

where ϖ is the impulsive pressure which would generate the motion from rest.

CHAPTER IV

TWO-DIMENSIONAL MOTION

4·1. Motion in two dimensions. Two-dimensional motion is characterised by the fact that the streamlines are all parallel to a fixed plane and that the velocity at corresponding points of all planes parallel to the fixed plane has the same magnitude and direction. To explain this more fully, suppose that the fixed plane is the plane of xy and that P is any point in that plane. Draw PQ perpendicular to the plane xy (or parallel to Oz). Then points on the line PQ are said to correspond to P. Take any plane (in the fluid) parallel to xy and meeting PQ in R. Then, if the velocity at P is q in the xy plane in a direction making an angle θ with Oy, the velocity at R is equal in magnitude and parallel in direction to the velocity at P. The velocity at corresponding points is then a function of x, y and the time t, but not of z. It is therefore sufficient to consider the motion of fluid particles in a representative plane, say the xy plane, and we may properly speak of the velocity at the point P, which represents the other points on the line PQ at which the velocity is the same.

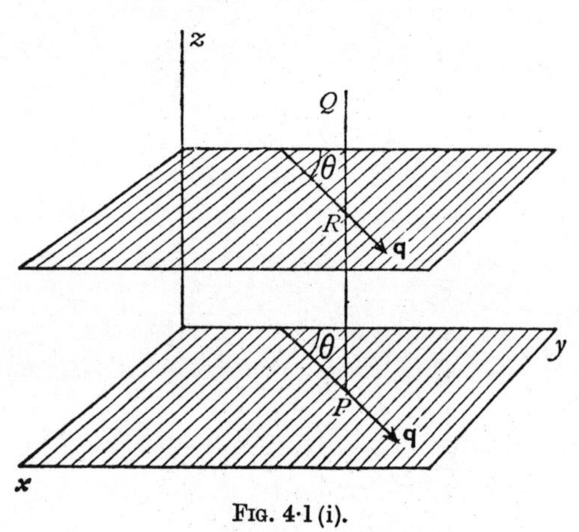

FIG. 4·1 (i).

In order to keep in touch with physical reality it is often useful to suppose the fluid in two-dimensional motion to be confined between two planes parallel to the plane of motion and at unit distance apart, the fluid being supposed to glide freely over these planes without encountering any resistance of a frictional nature. Thus in considering the problem of the flow of liquid past a cylinder in a two-dimensional motion in planes perpendicular to the axis of the cylinder, instead of considering a cylinder of infinite length, a more vivid picture is obtained by restricting attention to a unit length of cylinder confined between the said planes.

In considering the motion of a cylinder in a direction perpendicular to its axis, we can profitably suppose the cylinder to be of unit thickness * and to encounter no resistance from the barrier planes. This method of envisaging the phenomena in no way restricts the generality and does not affect the mathematical treatment.

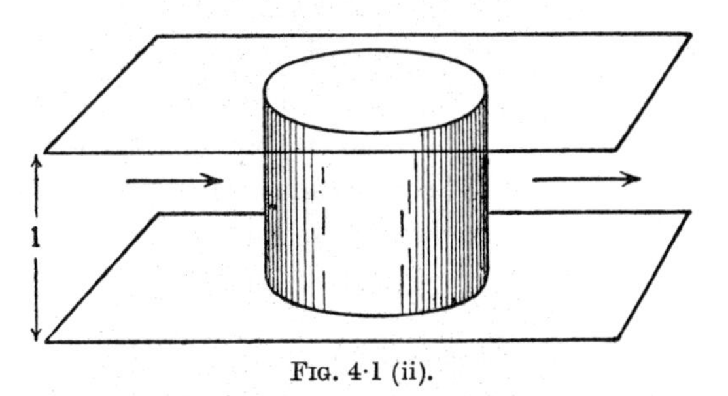

Fig. 4·1 (ii).

To complete the picture we shall adopt as our representative plane of the motion the plane which is parallel to our hypothetical fixed planes and midway between them.

Thus in the case of a circular cylinder moving in two dimensions the diagram will show the circle C which represents the cross-section of the cylinder by the aforesaid reference plane, and the centre A of this circle will be the point where the axis of the cylinder crosses the reference plane. This point may with propriety be called the *centre* of the cylinder. More generally any closed curve drawn in the reference plane represents a cross-section of a cylindrical surface bounded by the fixed planes.

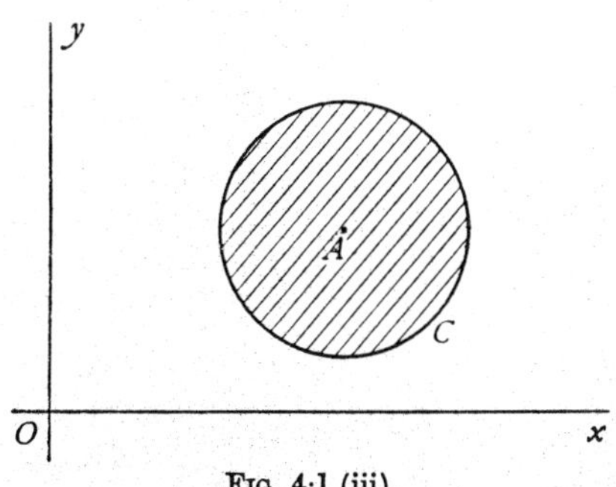

Fig. 4·1 (iii).

A clear understanding of the above conventional description will enable us to use the more familiar notation of ordinary two-dimensional geometry without confusion, and the reader is invited to form a mental picture of his results in the light of the diagram of fig. 4·1 (ii).

Two-dimensional motion, as will be seen in the sequel, presents opportunities for special mathematical treatment and enables us to investigate the nature of many phenomena which in their full three-dimensional form have so far proved intractable.

4·20. Intrinsic expression for the vorticity.

For two-dimensional flow in the xy plane the only non-zero component of the vorticity, see 3·43 (7), is

(1) $$\zeta = \frac{\partial v}{\partial x} - \frac{\partial u}{\partial y}.$$

* The term "thickness" will be used to denote dimensions perpendicular to the plane of the motion.

so that the vorticity vector is $\boldsymbol{\zeta} = \zeta\mathbf{k}$ and is everywhere perpendicular to the plane of the motion.

At a fixed instant t, let A and B be adjacent points of a streamline and denote by q the average fluid speed along AB. Let the normals at A and B meet at O, the centre of curvature at A of the streamline AB. Let θ be the infinitesimal angle between the normals, and let R be the radius of curvature at A, fig. 4·20. Let a second streamline DC meet OA at D and OB at C where $AD = BC = dn$ denotes an infinitesimal element of the normal at A.

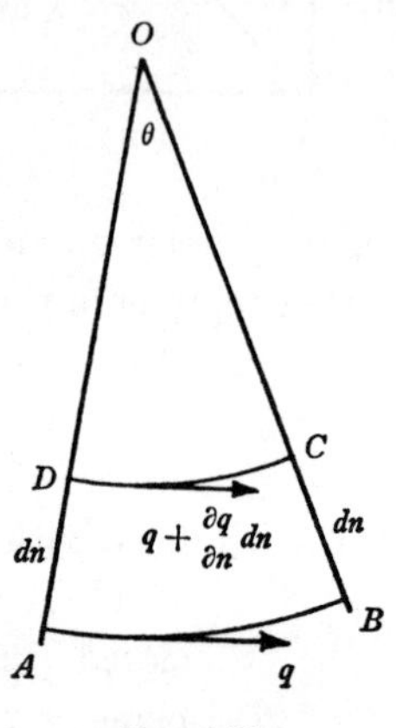

FIG. 4·20.

Then to the first order the average fluid speed on DC will be $q+(\partial q/\partial n)\,dn$, and since AB and CD are streamlines the fluid velocity is tangential to them.

Now apply Stokes' theorem 2·51 (1) to the circuit $ABCD$. Equating area times normal component of vorticity to the circulation and observing that AD and BC make no contributions we have, with sufficient approximation,

$$(R\theta\, dn)\zeta = (R\theta)q - [(R-dn)\theta]\left(q+\frac{\partial q}{\partial n}\, dn\right),$$

which in the limit gives the exact result

$$\zeta = \frac{q}{R} - \frac{\partial q}{\partial n}. \tag{2}$$

This is the intrinsic expression for the magnitude of the vorticity vector.
In cylindrical coordinates, from 2·72 (8)

$$\zeta = \frac{\partial q_{\varpi}}{\partial x} - \frac{\partial q_x}{\partial \varpi}, \tag{3}$$

while the cartesian expression is given by (1) above.

4·23. The rate of change of the vorticity. In two-dimensional motion the vorticity vector $\boldsymbol{\zeta}$ is necessarily at right angles to the plane of the motion and therefore the vector product $\boldsymbol{\zeta}_\wedge \mathbf{q}$ will represent a vector lying in the plane of the motion, at right angles to $\mathbf{q}$, and in such direction that the sense of rotation from $\mathbf{q}$ to $\boldsymbol{\zeta}_\wedge \mathbf{q}$ is counterclockwise.

From 3·53, we get the rate of change of vorticity

$$\frac{d\boldsymbol{\zeta}}{dt} = (\boldsymbol{\zeta}\,\nabla)\mathbf{q}. \tag{1}$$

Now, the right-hand member represents a rate of variation of $\mathbf{q}$ when we proceed in the direction $\boldsymbol{\zeta}$, i.e. at right angles to the plane of the motion. By the definition of two-dimensional motion this rate is zero. Thus

$$\frac{d\boldsymbol{\zeta}}{dt} = 0. \tag{2}$$

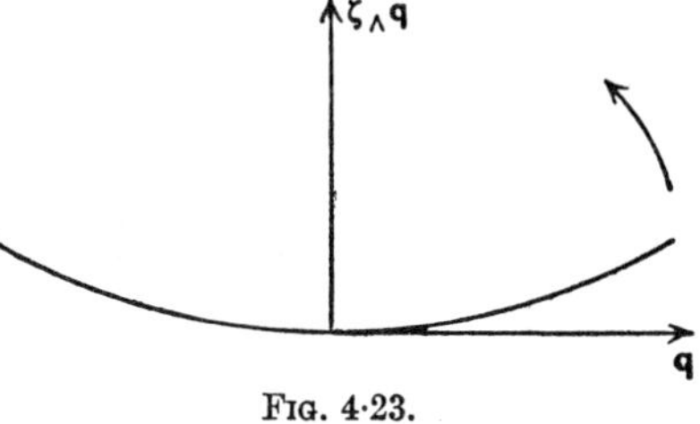

Fig. 4·23.

This means that the vorticity of a fluid particle does not change as the particle moves about.

This property of the vorticity is peculiar to two-dimensional motion, as is obvious from (1).

If the vorticity is constant at infinity, the vorticity is constant throughout the fluid except perhaps for those particles whose path lines do not extend to infinity.

In particular, if the flow is irrotational at infinity, it is irrotational everywhere except perhaps for those particles whose path lines do not extend to infinity.

In steady motion the paths of the particles are also the streamlines, and therefore the vorticity is constant along a streamline.

4·25. Intrinsic equations of steady motion. The accelerations of the element $ABCD$ in fig. 4·20 are

$$q\frac{\partial q}{\partial s}, \quad \frac{q^2}{R},$$

tangentially and normally. Equating the mass × acceleration to the force, we get

$$q\frac{\partial q}{\partial s} = -\frac{1}{\rho}\frac{\partial p}{\partial s} - \frac{\partial \Omega}{\partial s}, \tag{1}$$

$$\frac{q^2}{R} = -\frac{1}{\rho}\frac{\partial p}{\partial n} - \frac{\partial \Omega}{\partial n}, \tag{2}$$

where Ω is the potential per unit mass of the external forces. These equations can be written

$$\frac{\partial}{\partial s}\left(\frac{p}{\rho} + \tfrac{1}{2}q^2 + \Omega\right) = 0,$$

$$\frac{\partial}{\partial n}\left(\frac{p}{\rho} + \tfrac{1}{2}q^2 + \Omega\right) = -q\left(\frac{q}{R} - \frac{\partial q}{\partial n}\right) = -q\zeta,$$

where ζ is the magnitude of the vorticity. The first gives

$$\frac{p}{\rho}+\tfrac{1}{2}q^2+\Omega = C,$$

where C is constant along the streamline, which is Bernoulli's equation. The second then gives

$$\frac{\partial C}{\partial n} = -q\zeta,$$

which shows how the constant C changes as we move across the stream. When the motion is irrotational, $\zeta = 0$ and C is then constant throughout the fluid.

4·30. Stream function. In the two-dimensional motion of a liquid, let A be a fixed point in the plane of the motion, and ABP, ACP two curves also in the plane joining A to an arbitrary point P. We suppose that no fluid is created or destroyed within the region R bounded by these curves. Then the condition of continuity may be expressed in the following form.

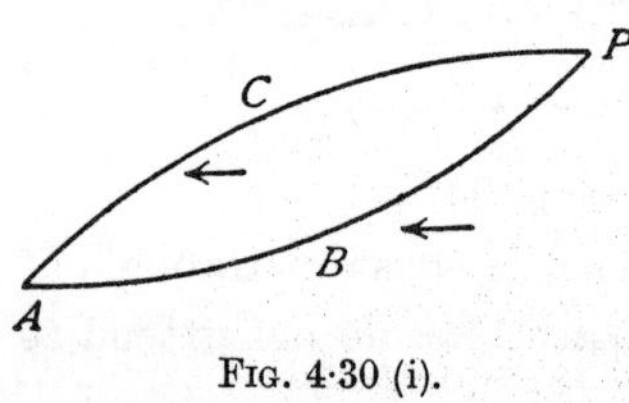

Fig. 4·30 (i).

The rate at which liquid flows into the region R from right to left across the curve ABP is equal to the rate at which it flows out from right to left across the curve ACP.

We shall use the convenient term *flux* for the rate of flow, and we shall assume the flux to be consistently reckoned in the same sense, here from right to left.

The term from right to left is relative to an observer who proceeds along the curve from the fixed point A in the direction in which the arc s of the curve measured from A is increasing.

Thus the flux across ACP is equal to the flux across any curve joining A to P.

Once the *base point* A has been fixed this flux therefore depends solely on the position of P, and the time t. If we denote this flux by ψ, ψ is a function of the position of P and the time. In cartesian coordinates, for example,

$$\psi = \psi(x, y, t).$$

The function ψ is called the *stream function*.

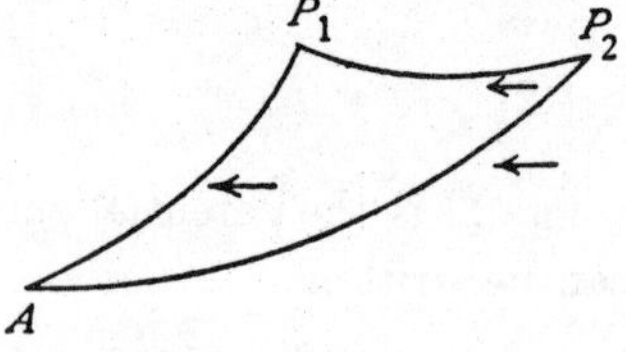

Fig. 4·30 (ii).

The existence of this function is merely a consequence of the assertion of the continuity and incompressibility of the fluid. Thus a stream function exists for a viscous liquid.

Now take two points P_1, P_2, and let ψ_1, ψ_2 be the corresponding values of the stream function.

Then, from the same principle, the flux across AP_2 is equal to the flux across AP_1 plus that across P_1P_2. Hence the flux across P_1P_2 is $\psi_2-\psi_1$.

It follows from this that if we take a different base point, A' say, the stream function merely changes by the flux across $A'A$.

Moreover, if P_1 and P_2 are points of the same streamline, the flux across P_1P_2 is equal to the flux across the streamline on which P_1 and P_2 lie. Thus $\psi_1-\psi_2=0$. Therefore

the stream function is constant along a streamline.

The equations of the streamlines are therefore obtained from $\psi = c$, by giving arbitrary values to the constant c.

When the motion is steady, the streamline pattern is fixed. When the motion is not steady, the pattern changes from instant to instant.

In terms of the measure-ratios L and T of length and time, the dimensions of the stream function are represented by L^2T^{-1}.

4·31. Velocity derived from the stream function. Let $P_1P_2 = \delta s$ be an infinitesimal arc of a curve. The fluid velocity across this arc can be resolved into components along and perpendicular to δs. The component along δs contributes nothing to the flux across. The component at right angles to δs

$$= \text{flux across divided by } \delta s$$

$$= (\psi_2-\psi_1)/\delta s,$$

Fig. 4·31 (i).

where ψ_1, ψ_2 are the values of the stream function at P_1, P_2.

Thus the velocity from right to left across δs becomes in the limit $\partial\psi/\partial s$.

In cartesian coordinates, by considering infinitesimal increments δx, δy, the components u, v of velocity parallel to the axes are given by

$$u = -\frac{\partial\psi}{\partial y},\quad v = \frac{\partial\psi}{\partial x}.$$

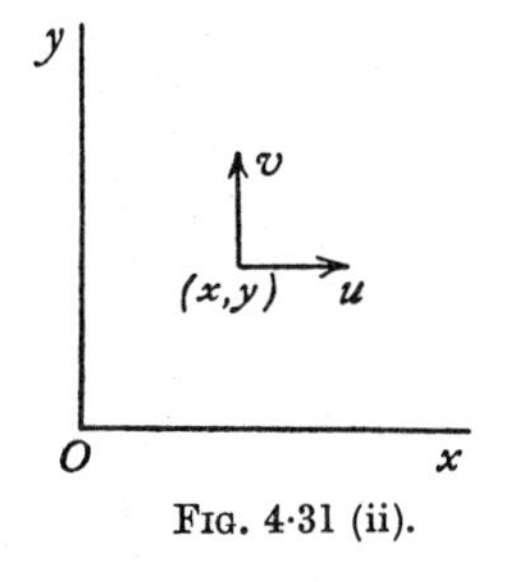

Fig. 4·31 (ii).

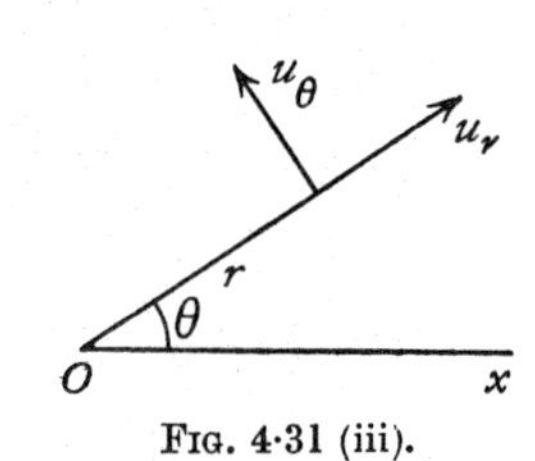

Fig. 4·31 (iii).

In polar coordinates, we get

$$u_r = -\frac{\partial\psi}{r\,\partial\theta}, \quad u_\theta = \frac{\partial\psi}{\partial r},$$

for the radial and transverse components, fig. 4·31 (iii).

4·32. Rankine's method. If the stream function ψ can be expressed as the sum of two functions in the form $\psi = \psi_1 + \psi_2$, the streamlines can be drawn when the curves $\psi_1 =$ constant, $\psi_2 =$ constant are known.

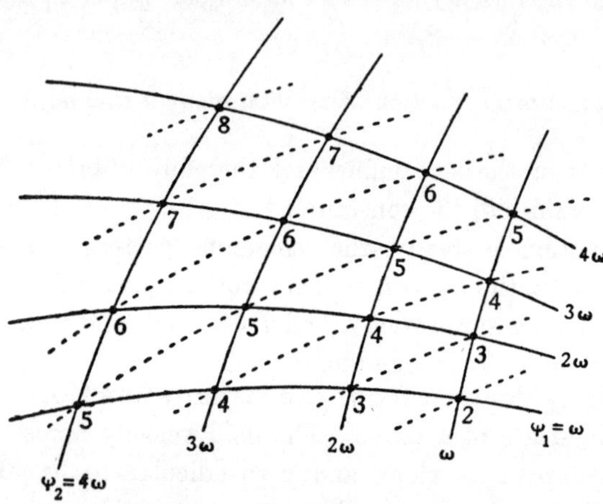

FIG. 4·32.

Taking a small constant ω, we draw the curves $\psi_1 = \omega, 2\omega, 3\omega, \ldots,$ $\psi_2 = \omega, 2\omega, 3\omega, \ldots,$ and so obtain a network as shown in fig. 4·32.

At the points marked 3, $\psi = 3\omega$, at the points marked 4, $\psi = 4\omega$, and so on. If we join the points with the same numeral we obtain lines along which $\psi =$ constant, the dotted lines in the figure.

The meshes of the network can be made as small as we please by taking ω small enough, and the meshes can be regarded as parallelograms (of different sizes). The streamlines are then obtained by drawing the diagonals of the meshes. The streamlines which pass through the corners of a mesh are approximately parallel in the neighbourhood of the mesh.

4·33. The stream function of a uniform stream. Suppose every fluid particle to move with the constant speed U parallel to the x-axis.

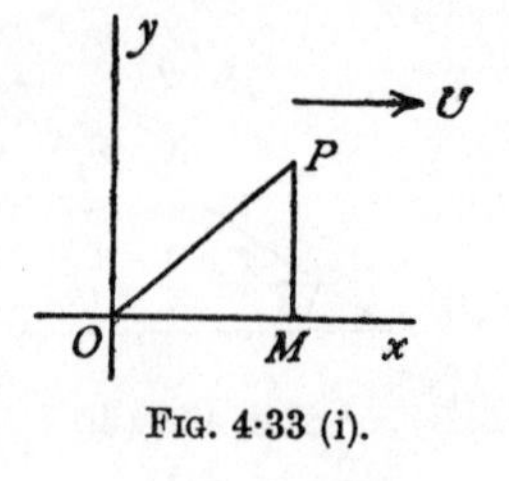

FIG. 4·33 (i).

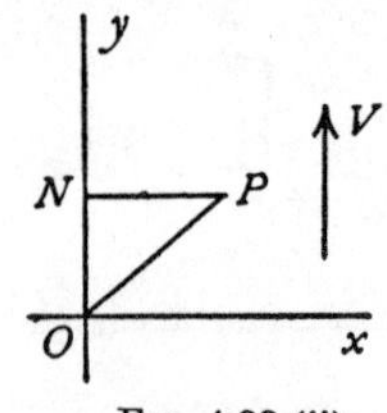

FIG. 4·33 (ii).

If P is the point (x, y), the flux across OP is the same as the flux across PM, where PM is perpendicular to Ox. Thus the flux is $-Uy$, and therefore

$$\psi = -Uy$$

is the stream function for this motion. In polar coordinates,

$$\psi = -Ur \sin\theta.$$

Similarly, for a uniform stream in the direction Oy of speed V, we get

$$\psi = Vx = Vr\cos\theta.$$

If we superpose the two streams, we get a stream of speed $\sqrt{U^2+V^2}$ inclined to the x-axis at the angle $\alpha = \tan^{-1} V/U$, and for this stream

$$\psi = -Uy + Vx.$$

Writing $U = Q\cos\alpha$, $V = Q\sin\alpha$, we obtain the stream function for a uniform stream, Q making an angle α with the x-axis, namely

$$\psi = Q(x\sin\alpha - y\cos\alpha),$$

or, in polar coordinates,

$$\psi = -Qr\sin(\theta - \alpha),$$

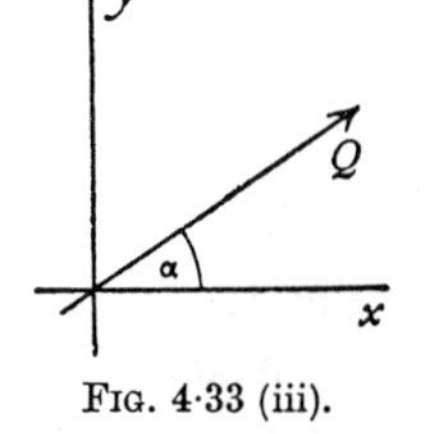

FIG. 4·33 (iii).

and in all these cases the streamlines are straight lines, as is indeed obvious.

The streamline which passes through the origin corresponds to $\psi = 0$ and is therefore the line $\theta = \alpha$.

4·40. Vector expressions for velocity and vorticity.

Let $\mathbf{s}_1$ be a unit vector tangent to the streamline ψ = constant, and in the direction of the velocity $\mathbf{q}$.

Let $\mathbf{n}$ be a unit vector in the direction of the normal to the streamline drawn in the direction towards which ψ diminishes, and let $\mathbf{k}$ be a unit vector

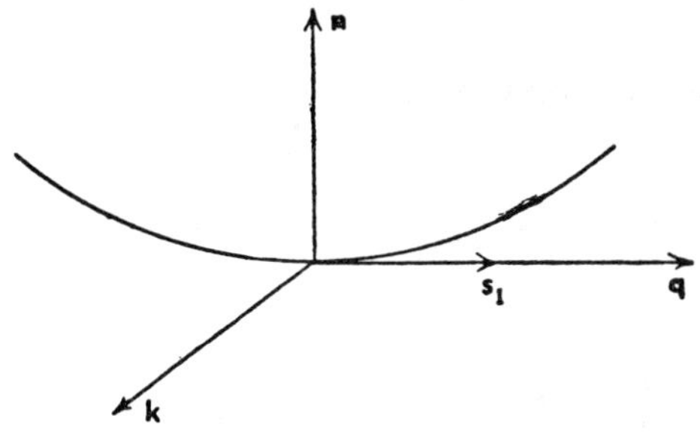

FIG. 4·40.

perpendicular to the plane of the motion in such sense that $\mathbf{k}$, $\mathbf{s}_1$, $\mathbf{n}$ form a right-handed system. Then $\mathbf{q} = q\mathbf{s}_1$, where q is the speed, and, from 4·31,

$$q = -\frac{\partial \psi}{\partial n} = -(\mathbf{n} \nabla \psi).$$

Since $\mathbf{n}$ and $-\nabla \psi$ are parallel and $\mathbf{n}$ is a unit vector, the magnitude of the velocity is that of $(-\nabla \psi)$. To obtain the velocity we must therefore turn this vector through a right angle from $\mathbf{n}$ to $\mathbf{s}_1$. Therefore

(1) $$\mathbf{q} = -\mathbf{k}_\wedge(-\nabla \psi) = \mathbf{k}_\wedge \nabla \psi.$$

Again, $$\boldsymbol{\zeta} = \nabla_\wedge \mathbf{q} = \nabla_\wedge(\mathbf{k}_\wedge \nabla \psi)$$
$$= \mathbf{k}\,[(\nabla)(\nabla \psi)] - (\mathbf{k}\,\nabla)\nabla \psi,$$

using the triple vector product. The second term represents a variation going along $\mathbf{k}$ and is therefore zero since the motion is two-dimensional. Hence

(2) $$\boldsymbol{\zeta} = \mathbf{k}\,\nabla^2 \psi.$$

Again, $$\mathbf{q}_\wedge \boldsymbol{\zeta} = (\mathbf{k}_\wedge \nabla \psi)_\wedge(\mathbf{k}\,\nabla^2 \psi), \text{ or}$$

(3) $$\mathbf{q}_\wedge \boldsymbol{\zeta} = (\nabla \psi)(\nabla^2 \psi),$$

using the triple vector product and observing that $\mathbf{kk} = 1$.

Finally, consider the operator $\mathbf{q}\,\nabla = (\mathbf{k}_\wedge \nabla \psi)\,\nabla$. Using the triple scalar product, we get

(4) $$\mathbf{q}\,\nabla = \mathbf{k}(\nabla \psi_\wedge \nabla).$$

It follows, from (2), that if ζ is the magnitude of the vorticity,

(5) $$\zeta = \nabla^2 \psi.$$

In cartesian coordinates, this becomes (2·70)

(6) $$\zeta = \frac{\partial^2 \psi}{\partial x^2} + \frac{\partial^2 \psi}{\partial y^2}.$$

In polar coordinates,

(7) $$\zeta = \frac{\partial^2 \psi}{\partial r^2} + \frac{1}{r}\frac{\partial \psi}{\partial r} + \frac{1}{r^2}\frac{\partial^2 \psi}{\partial \theta^2}.$$

4·41. Equation satisfied by ψ. We have proved, in 4·23, that $d\boldsymbol{\zeta}/dt = 0$. Therefore, using 3·10 (9),

$$\frac{\partial \boldsymbol{\zeta}}{\partial t} + (\mathbf{q}\,\nabla)\boldsymbol{\zeta} = 0.$$

Hence from 4·40 (2), (4), we get

$$\frac{\partial}{\partial t}(\mathbf{k}\,\nabla^2\psi) + (\mathbf{k}\,(\nabla \psi_\wedge \nabla))(\mathbf{k}\,\nabla^2\psi) = 0,$$

and since $\mathbf{kk} = 1$, it follows that

$$\mathbf{k}\frac{\partial}{\partial t}(\nabla^2\psi) + (\nabla\psi)_\wedge\nabla(\nabla^2\psi) = 0,$$

which is the equation satisfied by ψ.

When the motion is steady, this becomes

$$(\nabla\psi)_\wedge\nabla(\nabla^2\psi) = 0,$$

and therefore the vectors $\nabla\psi$, $\nabla(\nabla^2\psi)$ are parallel.

Since these vectors are normal respectively to the curves ψ = constant, $\nabla^2\psi$ = constant, it follows that ψ = constant implies $\nabla^2\psi$ = constant, and therefore that

$$\nabla^2\psi = f(\psi),$$

where $f(\psi)$ is a function of ψ only. This result also gives another proof that in steady motion the vorticity is constant along a streamline.

4·5. The pressure equation. If we put, with the usual notations,

$$\chi = \frac{p}{\rho} + \tfrac{1}{2}q^2 + \Omega,$$

the equation of motion

$$\frac{\partial \mathbf{q}}{\partial t} - \mathbf{q}_\wedge\boldsymbol{\zeta} = -\nabla\chi$$

becomes, using 4·40 (1), (3),

$$\mathbf{k}_\wedge\nabla\frac{\partial\psi}{\partial t} - (\nabla\psi)(\nabla^2\psi) = -\nabla\chi, \tag{1}$$

which is the equation of motion in terms of the stream function.

Now, let ds be the element of arc at P of a curve AP in the plane of the motion and $\mathbf{s}_1$ a unit vector along the tangent at P. Then (2·31),

$$\mathbf{s}_1\nabla\chi = \frac{\partial\chi}{\partial s}.$$

Taking the scalar product of (1) by $\mathbf{s}_1$, we get

$$\frac{\partial\chi}{\partial s} - \nabla^2\psi\frac{\partial\psi}{\partial s} + \left[\mathbf{k}_\wedge\nabla\left(\frac{\partial\psi}{\partial t}\right)\right]\mathbf{s}_1 = 0.$$

Now integrate along the arc AP. Then

$$\frac{p}{\rho} + \tfrac{1}{2}q^2 + \Omega - \int_{(AP)}\nabla^2\psi\, d\psi + \int_{(AP)}\left[\mathbf{k}_\wedge\nabla\frac{\partial\psi}{\partial t}\right]\mathbf{s}_1\, ds = F(t), \tag{2}$$

where $F(t)$ is an arbitrary function of the time t.

This is the pressure equation expressed in terms of the stream function.

The second integral on the left is

$$\frac{\partial}{\partial t}\int_{(AP)} \mathbf{q}\, d\mathbf{s},$$

where $d\mathbf{s}$ is the directed element of arc and this is the rate of change of circulation from A to P. We also note that, by the triple scalar product,

$$\left[\mathbf{k}_\wedge \nabla \frac{\partial \psi}{\partial t}\right]\mathbf{s}_1 = (\mathbf{s}_{1\,\wedge}\,\mathbf{k})\nabla \frac{\partial \psi}{\partial t} = -\mathbf{n}\,\nabla \frac{\partial \psi}{\partial t} = -\frac{\partial}{\partial t}\frac{\partial \psi}{\partial n},$$

where $\mathbf{n}$ is the unit normal to AP, drawn so that $\mathbf{k}$, $\mathbf{s}_1$, $\mathbf{n}$ form a right-handed system. Thus (2) can also be written

$$\frac{p}{\rho}+\tfrac{1}{2}q^2+\Omega-\int_{(AP)} \nabla^2\psi\, d\psi - \frac{\partial}{\partial t}\int_{(AP)} \frac{\partial \psi}{\partial n}\, ds = F(t).$$

In the case of steady motion, the terms involving the time disappear, and since in that case (4·41) $\nabla^2\psi$ is a function of ψ, we can write

$$\frac{p}{\rho}+\tfrac{1}{2}q^2+\Omega-f(\psi) = C,$$

where C is an absolute constant. This is Bernoulli's equation showing the dependence on the particular streamline chosen.

4·6. Stagnation points. Suppose that the origin is a stagnation point. Then the velocity vanishes there and hence at the origin

$$\frac{\partial \psi}{\partial x} = 0, \quad \frac{\partial \psi}{\partial y} = 0.$$

Without loss of generality we may suppose that $\psi = 0$ at the origin and therefore, by Maclaurin's theorem,

$$\begin{aligned}\psi &= \tfrac{1}{2}\left[x^2\left(\frac{\partial^2\psi}{\partial x^2}\right)_0 + 2xy\left(\frac{\partial^2\psi}{\partial x\,\partial y}\right)_0 + y^2\left(\frac{\partial^2\psi}{\partial y^2}\right)_0\right]+\dots \\ &= \tfrac{1}{2}(ax^2+2hxy+by^2)+\dots.\end{aligned}$$

Hence, when x and y are very small, the form of the streamline $\psi = 0$ is in general given approximately by the equation

$$(1) \qquad ax^2+2hxy+by^2 = 0,$$

which represents two straight lines. Thus at a stagnation point the streamline crosses itself, in other words it presents a double point. If the straight lines are imaginary, they define a real stagnation point around which the fluid in the immediate neighbourhood circulates.

When the motion is irrotational,

$$a+b = \left(\frac{\partial^2\psi}{\partial x^2}\right)_0 + \left(\frac{\partial^2\psi}{\partial y^2}\right)_0 = 0,$$

and therefore the lines (1) are perpendicular, so that the two branches of the streamline cut at right angles.

4·70. The velocity potential of a liquid. In irrotational motion the velocity is the negative gradient of a potential, namely $\mathbf{q} = -\nabla \phi$, and in cartesian coordinates its components are given by

$$u = -\frac{\partial \phi}{\partial x}, \quad v = -\frac{\partial \phi}{\partial y}.$$

Since the velocity components are also given in terms of the stream function, we have

$$\text{(1)} \qquad \frac{\partial \phi}{\partial x} = \frac{\partial \psi}{\partial y}, \quad \frac{\partial \phi}{\partial y} = -\frac{\partial \psi}{\partial x}.$$

In the notation of vectors,

$$\text{(2)} \qquad -\nabla \phi = \mathbf{k}_\wedge \nabla \psi.$$

Thus if $\mathbf{s}_1$ is a unit vector in any direction, and $\mathbf{n}$ a unit vector normal to $\mathbf{s}_1$ measured counterclockwise from $\mathbf{s}_1$, we get

$$-\mathbf{s}_1 \nabla \phi = \mathbf{s}_1 (\mathbf{k}_\wedge \nabla \psi) = (\mathbf{s}_{1\wedge} \mathbf{k}) \nabla \psi = -\mathbf{n} \nabla \psi,$$

or

$$\frac{\partial \phi}{\partial s} = \frac{\partial \psi}{\partial n},$$

which yields equations (1) if we take $ds = dx$, $ds = dy$ in turn, for the corresponding values of dn are dy, and $-dx$.

We also conclude from (2) that $\nabla \phi$ and $\nabla \psi$ are at right angles. This means that the curves, $\phi =$ constant, $\psi =$ constant, intersect at right angles. Thus the curves of constant velocity potential cut the streamlines orthogonally.

The following points should be observed :

(*a*) The stream function ψ exists whether the motion is irrotational or not.

(*b*) The velocity potential can exist only where the motion is irrotational.

(*c*) Where the motion is irrotational, the velocity potential does exist.

(*d*) Part of the fluid may be moving irrotationally and other parts rotationally. The velocity potential exists only and always in those parts of the fluid where the motion is irrotational.

(*e*) As the fluid moves about, the irrotational part may occupy different regions of space. The existence of the velocity potential is a property of those parts of the fluid which are moving irrotationally, not of the regions of space which they may temporarily occupy.

(*f*) The flow pattern in irrotational motion under conservative forces depends solely on the boundary conditions. In particular, when the fluid has no free surface, the flow pattern of acyclic irrotational motion depends solely on the motion of those boundaries, and not at all on the field of external force which merely affects the pressure.

Consider the stream function

$$\psi = xy.$$

We find that $\nabla^2 \psi = 0$, so that the motion is irrotational.

The velocity components are $-x$, y.

Hence, to find the velocity potential, we can write

$$(1) \qquad -\frac{\partial \phi}{\partial x} = -x, \quad -\frac{\partial \phi}{\partial y} = y,$$

so that $$d\phi = \frac{\partial \phi}{\partial x}\,dx + \frac{\partial \phi}{\partial y}\,dy = x\,dx - y\,dy = d\,\tfrac{1}{2}(x^2 - y^2).$$

Thus $$\phi = \tfrac{1}{2}(x^2 - y^2).$$

The streamlines are xy = constant, that is, rectangular hyperbolas having the axes of reference for asymptotes. The lines of constant velocity potential are also rectangular hyperbolas Thus this stream and velocity function give the flow round a rectangular corner, fig. 4·70, where the dotted lines correspond to constant values of ϕ.

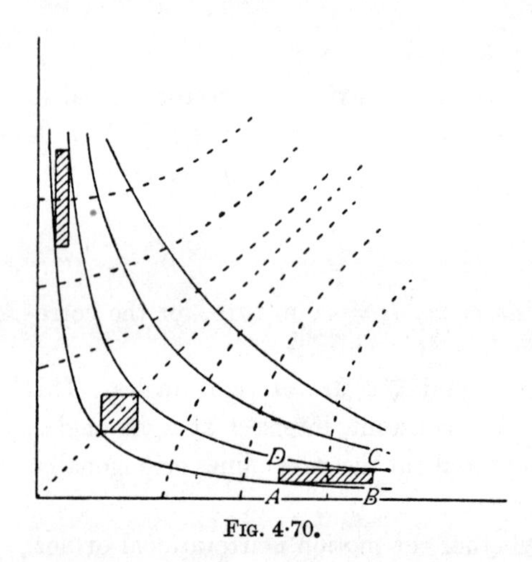

FIG. 4·70.

Consider a rectangular drop of fluid $ABCD$ with sides parallel to the axes. From (1) we see that u is the same for all points on BC and v is the same for all points on AB. Hence $ABCD$ remains a rectangle as AB moves up. Also the area $ABCD$ remains constant (equation of continuity) since the same particles are involved. Clearly AB continually decreases in length while BC continually increases. Hence the drop alters its shape, but the sides remain parallel to the axes. This illustrates the irrotational character of the motion, and the rate of pure strain referred to in 2·40.

4·71. The equation satisfied by the velocity potential. When the motion is irrotational the vorticity is zero and therefore

$$\nabla^2 \psi = 0.$$

Again, $\mathbf{q} = -\nabla \phi$, and $\nabla \mathbf{q} = 0$, from the equation of continuity. Therefore

$$\nabla^2 \phi = 0.$$

It follows that ϕ and ψ both satisfy Laplace's equation $\nabla^2 V = 0$, or, in cartesian coordinates,

$$\frac{\partial^2 V}{\partial x^2} + \frac{\partial^2 V}{\partial y^2} = 0.$$

We have now arrived at the point where two-dimensional irrotational motion can be most profitably investigated in terms of the complex variable. The next chapter will be devoted to a brief description of the necessary mathematical apparatus.

In Chapter VI we shall see that, in terms of the complex variable, irrotational motion in two dimensions admits of a special mathematical treatment which enables us to solve problems which in their full three-dimensional form cannot be attacked with the means at present at our disposal. By limiting ourselves to two dimensions, we are thus enabled to discuss many peculiarities of fluid motion which might otherwise elude treatment, and so to throw light on important physical properties of hydrodynamical problems.

EXAMPLES IV

1. Wind blows over the surface of water which is flowing in the direction of the wind, but with different velocity. Explain why, in general, any small departure of the water surface from a plane form will tend to increase.

2. Determine the condition that

$$u = ax+by, \quad v = cx+dy$$

may give the velocity components of an incompressible fluid. Show that the streamlines of this motion are conic sections in general, and rectangular hyperbolas when the motion is irrotational.

3. Prove that in a two-dimensional motion of a liquid the mean tangential fluid velocity round any *small* circle of radius r is ωr, where 2ω is the value of

$$\frac{\partial v}{\partial x} - \frac{\partial u}{\partial y}$$

at the centre of the circle, terms of order r^3 being neglected.

4. Show that $u = 2cxy$, $v = c(a^2+x^2-y^2)$ are the velocity components of a possible fluid motion. Determine the stream function and sketch the streamlines.

5. Obtain the equation of continuity for the two-dimensional motion of an incompressible fluid in the form

$$\frac{\partial (ur)}{\partial r} + \frac{\partial v}{\partial \theta} = 0,$$

where u, v are the velocities in the directions of increasing r and θ respectively r, θ having the usual meanings.

Show that this equation is satisfied by $u = akr^n e^{-k(n+1)\theta}$, $v = ar^n e^{-k(n+1)\theta}$, and determine the stream function. Show also that the fluid speed at any point is

$$(n+1)\,\psi\sqrt{1+k^2}/r,$$

where ψ is the stream function. (R.N.C.)

6. If u, v are the velocity components of a continuous two-dimensional motion of an incompressible fluid, show that

$$\frac{\partial u}{\partial x} + \frac{\partial v}{\partial y} = 0.$$

Deduce the existence of the stream function. If the circulation round any closed path is zero, prove that the stream function satisfies

$$\frac{\partial^2\psi}{\partial x^2}+\frac{\partial^2\psi}{\partial y^2} = 0.$$

7. The stream function in a two-dimensional motion is given by $\psi = Cr^2\theta$, r, θ being polar coordinates. Find the vorticity and the velocity at any point. Show further that this motion corresponds to the case of two plane boundaries hinged together along their intersection, opening out or closing in.

8. In two-dimensional irrotational motion, prove that, if the speed is everywhere the same, the streamlines are straight.

9. Determine the condition that the equation $\psi(x, y, c) = 0$ shall give the streamlines of an irrotational motion, c being a parameter which is constant along any one line of the system.

10. In a two-dimensional motion, show that a streamline cuts itself at a point of zero velocity, and that the two branches are at right angles when the motion is irrotational.

Sketch the streamline which passes through the stagnation point of the motion given by

$$\psi = U\left(y-a\tan^{-1}\frac{y}{x}\right),$$

and determine the velocity at the points where this line crosses the axis of y.

11. Show that the velocity potential

$$\phi = \tfrac{1}{2}\log\frac{(x+a)^2+y^2}{(x-a)^2+y^2}$$

gives a possible motion, and determine the form of the streamlines.

Show that the curves of equal speed are the ovals of Cassini given by

$$rr' = \text{constant}.$$

12. Liquid moves irrotationally in two dimensions under the action of conservative forces whose potential Ω satisfies $\nabla^2\Omega = 0$. Prove that the pressure satisfies the equation $\nabla^2(\log\nabla^2 p) = 0$.

13. In irrotational motion in two dimensions, prove that

$$\left(\frac{\partial q}{\partial x}\right)^2+\left(\frac{\partial q}{\partial y}\right)^2 = q\,\nabla^2 q.$$

14. Prove that
$$\frac{\partial v}{\partial x}-\frac{\partial u}{\partial y} = \frac{q}{R}-\frac{\partial q}{\partial n}.$$

CHAPTER V

COMPLEX VARIABLE

5·01. Complex numbers. Let $\mathbf{i}$, $\mathbf{j}$ be unit vectors along the axes of x, y, and let $\mathbf{k}$ be a unit vector perpendicular to each of these, the three forming a right-handed system.

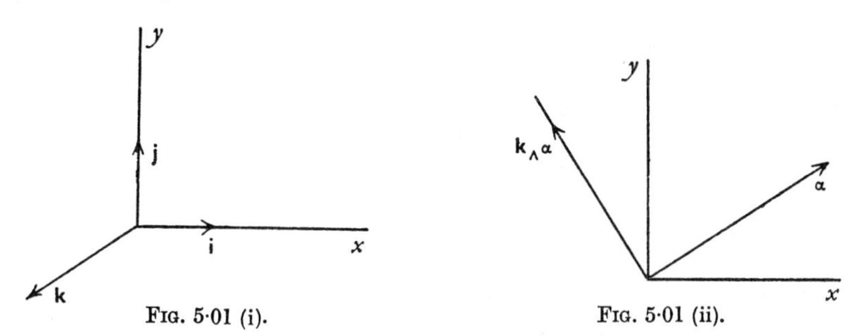

Fig. 5·01 (i). Fig. 5·01 (ii).

If we confine our attention to vectors in the xy plane, the vectors $\boldsymbol{\alpha}$ and $\mathbf{k}_\wedge \boldsymbol{\alpha}$ are perpendicular and lie in that plane. Thus the operation of multiplying a given vector $\boldsymbol{\alpha}$ in the xy plane by $\mathbf{k}_\wedge$ results in a rotation of that vector, without change of magnitude, through a right angle in the sense x to y, i.e. counterclockwise in the figure. If b is a scalar, then $b\mathbf{k}_\wedge \boldsymbol{\alpha}$ is the vector obtained by rotating $\boldsymbol{\alpha}$ through a right angle and multiplying its magnitude by b.

Thus as regards vectors in the xy plane we can look upon $\mathbf{k}_\wedge$ as an operator which turns a given vector through a right angle.

If we multiply a given vector $\boldsymbol{\alpha}$ by $a+b\mathbf{k}_\wedge$, we get the vector $a\boldsymbol{\alpha}+b(\mathbf{k}_\wedge \boldsymbol{\alpha})$, which is also in the xy plane. Thus the operator $a+b\mathbf{k}_\wedge$ applied to a vector in the xy plane changes it into another vector in that plane.

Definition. The operator $a+b\mathbf{k}_\wedge$ is called a complex number, where a and b are real scalars.

It is customary in mathematics to write i instead of $\mathbf{k}_\wedge$, and the complex number is then written in the form

$$a+ib.$$

5·10. Argand diagram. The complex number $x+iy$ applied to the vector $\mathbf{i}$ gives

$$(x+iy)\,\mathbf{i} = x\mathbf{i}+y\mathbf{k}_\wedge \mathbf{i} = x\mathbf{i}+y\mathbf{j},$$

that is, the position vector $\overrightarrow{OP}$ of the point $P(x, y)$.

Thus any complex number applied to the vector $\mathbf{i}$ yields the position vector of a definite point.

This point is termed the *representative point* of the complex number, and is taken as a geometrical representation of the complex number $z = x+iy$. In this sense we may refer to the point z, meaning thereby the representative point in the above geometrical description which is known as the Argand diagram.

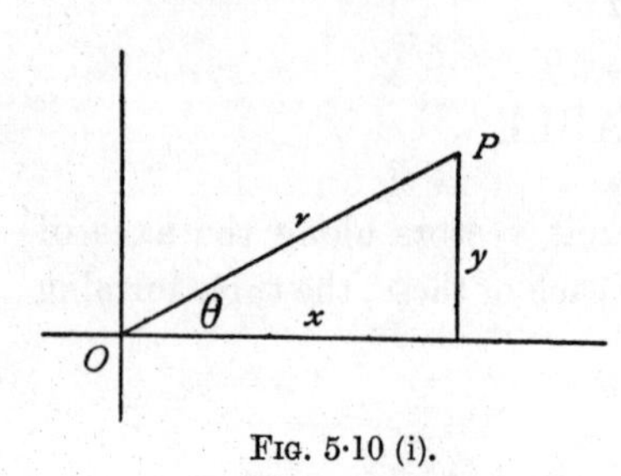

FIG. 5·10 (i).

The law of addition of complex numbers is now easy to obtain. Let

$$z_1 = x_1+iy_1, \quad z_2 = x_2+iy_2.$$

Then, operating on the vector $\mathbf{i}$,

$$(x_1+iy_1)\,\mathbf{i} = x_1\,\mathbf{i}+y_1\,\mathbf{j}, \quad (x_2+iy_2)\,\mathbf{i} = x_2\,\mathbf{i}+y_2\,\mathbf{j}.$$

Hence

$$(x_1+iy_1)\,\mathbf{i}+(x_2+iy_2)\,\mathbf{i} = (x_1+x_2)\,\mathbf{i}+(y_1+y_2)\,\mathbf{j} = [(x_1+x_2)+i(y_1+y_2)]\,\mathbf{i},$$

so that we may write

$$(x_1+iy_1)+(x_2+iy_2) = (x_1+x_2)+i(y_1+y_2),$$

and this means that the addition of complex numbers follows the same law as the addition of vectors.

Thus if A, B, C are the representative points of z_1, z_2, z_1+z_2, the four points O, A, C, B are at the vertices of a parallelogram. We also note that, since

$$\overrightarrow{OB} = \overrightarrow{OC}-\overrightarrow{OA},$$

the same method can be applied to obtaining the difference of two complex numbers marked on the Argand diagram.

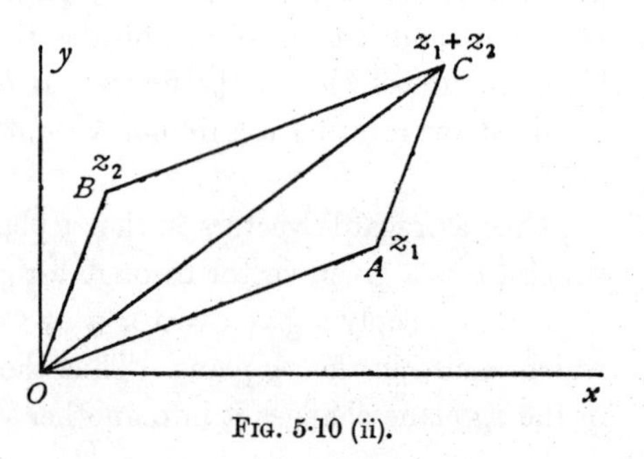

FIG. 5·10 (ii).

5·11. Multiplication. Let $z_1 = x_1+iy_1$, $z_2 = x_2+iy_2$. Then, operating on the vector $\mathbf{i}$,

$$\begin{aligned}(x_1+iy_1)(x_2+iy_2)\,\mathbf{i} &= (x_1+iy_1)(x_2\,\mathbf{i}+y_2\,\mathbf{j}) \\ &= x_1\,x_2\,\mathbf{i}+x_1\,y_2\,\mathbf{j}+y_1\,x_2\,\mathbf{j}-y_1\,y_2\,\mathbf{i},\end{aligned}$$

since, by definition, $i\,\mathbf{j} = -\,\mathbf{i}$.

Thus
$$\begin{aligned}(x_1+iy_1)(x_2+iy_2)\,\mathbf{i} &= (x_1\,x_2-y_1\,y_2)\,\mathbf{i}+(x_1\,y_2+x_2\,y_1)\,\mathbf{j} \\ &= [(x_1\,x_2-y_1\,y_2)+i(x_1\,y_2+x_2\,y_1)]\,\mathbf{i},\end{aligned}$$

and therefore

$$(1) \qquad (x_1+iy_1)(x_2+iy_2) = (x_1 x_2 - y_1 y_2) + i(x_1 y_2 + x_2 y_1).$$

It is easy to prove that the same result is obtained if the factors are taken in the order $(x_2+iy_2)(x_1+iy_1)$.

Thus the order of the factors may be interchanged without altering the product; multiplication is commutative. Moreover, if we multiply the factors in (1) according to the ordinary laws of algebra, we get

$$x_1 x_2 + i(x_1 y_2 + x_2 y_1) + i^2 y_1 y_2.$$

Comparison with (1) shows that the product is obtained by the ordinary laws of algebra, provided that we put

$$i^2 = -1,$$

an interpretation perfectly in accordance with the definition of i as the operator $\mathbf{k}_\wedge$, two successive applications of which to a vector reverse its direction and therefore multiply it by -1.

5·12. Equality of complex numbers. The equation

$$x_1 + iy_1 = x_2 + iy_2,$$

implies the equality of the vectors

$$(x_1+iy_1)\,\mathbf{i} = x_1\,\mathbf{i} + y_1\,\mathbf{j},$$

$$(x_2+iy_2)\,\mathbf{i} = x_2\,\mathbf{i} + y_2\,\mathbf{j},$$

and therefore $\qquad x_1 = x_2, \quad y_1 = y_2.$

Complex numbers are often called imaginary numbers, and in this terminology x is the real part of the complex number $z = x+iy$, and iy is the imaginary part.* The equality of two complex numbers therefore implies the equality of the real parts and the equality of the imaginary parts. In an equation between complex numbers we may therefore equate the real parts of the two sides of the equation, and equate the imaginary parts.

A complex number is said to be *zero* if both its real and its imaginary parts vanish.

We can apply the principle of equality to find p, q, such that

$$(x_1+iy_1) = (x_2+iy_2)(p+iq),$$

for $\qquad x_1 = px_2 - qy_2, \quad y_1 = py_2 + qx_2,$

which give $\qquad p = \dfrac{x_1 x_2 + y_1 y_2}{x_2^2 + y_2^2}, \quad q = \dfrac{x_2 y_1 - x_1 y_2}{x_2^2 + y_2^2}.$

---

* This name is usually applied to y itself. The x- and y-axes are called the real and imaginary axes respectively.

The number $p+iq$ is called the *quotient* of the numbers x_1+iy_1, x_2+iy_2, so that

$$\frac{x_1+iy_1}{x_2+iy_2} = \frac{x_1x_2+y_1y_2}{x_2^2+y_2^2}+i\frac{x_2y_1-x_1y_2}{x_2^2+y_2^2} = \frac{(x_1+iy_1)(x_2-iy_2)}{(x_2+iy_2)(x_2-iy_2)}.$$

We have thus found that the fundamental rules of algebra—addition, subtraction, multiplication and division—apply to complex numbers, subject to the interpretation

$$i^2 = -1,$$

and relying upon this, we shall manipulate complex numbers according to those rules.

5·13. Euler's theorem.

$$\cos\theta+i\sin\theta = e^{i\theta}.$$

We define $e^{i\theta}$ by putting $x = i\theta$ in the exponential series

$$e^x = 1+x+\frac{x^2}{2!}+\frac{x^3}{3!}+\ldots,$$

from which it follows at once that

$$\frac{de^{i\theta}}{d\theta} = ie^{i\theta}.$$

Also $$\frac{d(\cos\theta+i\sin\theta)}{d\theta} = -\sin\theta+i\cos\theta = i(\cos\theta+i\sin\theta).$$

Thus the linear differential equation

$$\frac{du}{d\theta} = iu$$

has the two solutions $u_1 = e^{i\theta}$, $u_2 = \cos\theta+i\sin\theta$,

both of which become equal to unity when $\theta = 0$. Therefore they are identical and thus

(1) $$e^{i\theta} = \cos\theta+i\sin\theta.$$ Q.E.D.

The complex number $z = x+iy$ can therefore be expressed in the form

$$z = r\cos\theta+ir\sin\theta = re^{i\theta},$$

where (r, θ) are the polar coordinates of the point (x, y), fig. 5·10 (i).

In this notation $r = (x^2+y^2)^{\frac{1}{2}}$ is called the *modulus* of z, written

$$r = |z|.$$

The modulus of a complex number measures the distance of the representative point from the origin. Thus the modulus is essentially a positive quantity. An important result is $|e^{i\theta}| = 1$, if θ is real. This follows at once from (1).

The angle θ is called the *argument* of z. Thus

$$\arg z = \theta.$$

Again, if $z_1 = r_1 e^{i\theta_1}$, $z_2 = r_2 e^{i\theta_2}$, then $z_1 z_2 = r_1 r_2 e^{i(\theta_1+\theta_2)}$

Therefore $$\arg (z_1 z_2) = \arg z_1 + \arg z_2.$$

In applying this result it is important to remember that $\arg z = \theta$ is indeterminate to an integral multiple of 2π, for

$$e^{i(\theta+2\pi)} = e^{i\theta} \times e^{2\pi i},$$

and $$e^{2\pi i} = \cos 2\pi + i \sin 2\pi = 1.$$

We also note that

$$e^{i\pi} = \cos \pi + i \sin \pi = -1,$$

$$e^{i\pi/2} = \cos \tfrac{1}{2}\pi + i \sin \tfrac{1}{2}\pi = i.$$

Thus $$\arg(-1) = \pi, \quad \arg(i) = \tfrac{1}{2}\pi.$$

5·14. Conjugate complex numbers. If in an expression involving i we change the sign of i throughout, the expression so formed is said to be the *conjugate complex* of the original expression.

Thus, if $$z = x + iy = re^{i\theta},$$

then the conjugate is $$\bar{z} = x - iy = re^{-i\theta}.$$

We shall express the conjugate complex by placing a bar above the original symbol. Observe that the conjugate of $\bar{z}$ is z, and that z, $\bar{z}$ have the same modulus.

We have, from the above,

$$z + \bar{z} = 2x, \qquad z - \bar{z} = 2iy, \qquad z\bar{z} = x^2 + y^2 = r^2.$$

Thus we have the following important theorems:

(1) The sum of two conjugate complex numbers is real.

(2) The difference of two conjugate complex numbers is purely imaginary (i.e. has zero for real part).

(3) The product of two conjugate complex numbers is real and equal to the square of the modulus of either.

(4) If a complex number is equal to its conjugate, the number is real (use (2)).

If $f(z)$ is a function of z, we denote the conjugate complex by $\bar{f}(\bar{z})$. Thus if $f(z) = 6z + 3iz^2$, $\bar{f}(\bar{z}) = 6\bar{z} - 3i\bar{z}^2$; replacing herein $\bar{z}$ by z we get $\bar{f}(z) = 6z - 3iz^2$.

5·15. The reciprocal of a complex number. If $z = re^{i\theta}$, the reciprocal of z is

$$\frac{1}{z} = \frac{e^{-i\theta}}{r}.$$

To represent z and its reciprocal on the Argand diagram, draw a circle of unit radius, centre O.

Then if P is the point z, and if on OP we take the point Q' such that

$$OQ' \,.\, OP = 1,$$

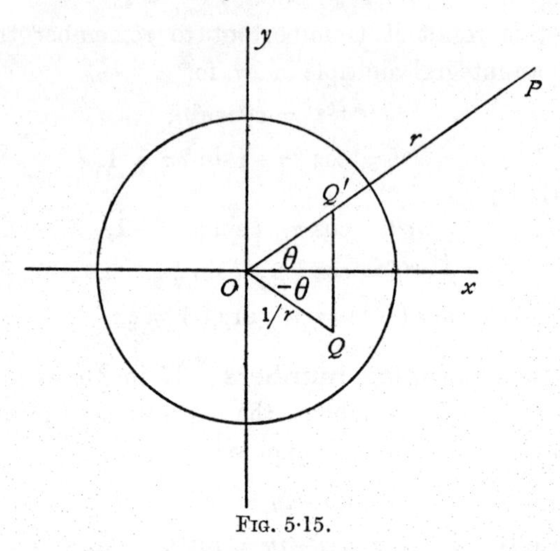

FIG. 5·15.

(so that Q' and P are *inverse points* with respect to the circle), the point Q which represents $1/z$ is the image of Q' in the x-axis, fig. 5·15.

5·16. Vector properties of complex numbers. We have already seen that complex numbers obey the vector law of addition when represented on the Argand diagram. If P_1, P_2 are the representative points of z_1, z_2, for purposes of *addition*, we can identify the vectors $\overrightarrow{OP_1}$, $\overrightarrow{OP_2}$ with z_1 and z_2 in the sense that if

$$\overrightarrow{OP_1} = z_1, \quad \overrightarrow{OP_2} = z_2, \quad \text{then} \quad \overrightarrow{OP_1} + \overrightarrow{OP_2} = z_1 + z_2.$$

On the other hand, the scalar product is not represented by $z_1 \,.\, z_2$. We note, however, that

$$z_1 \,.\, \bar{z}_2 = (x_1 + iy_1)(x_2 - iy_2) = x_1 x_2 + y_1 y_2 - i(x_1 y_2 - x_2 y_1)$$
$$= \overrightarrow{OP_1} \,.\, \overrightarrow{OP_2} - i \,|\, \overrightarrow{OP_1} {}_\wedge \overrightarrow{OP_2} \,|.$$

We therefore obtain the important and useful results

$$\overrightarrow{OP_1} \,.\, \overrightarrow{OP_2} = \text{real part of } z_1 \,.\, \bar{z}_2 = \tfrac{1}{2}(z_1 \bar{z}_2 + \bar{z}_1 z_2),$$

$$|\, \overrightarrow{OP_1} {}_\wedge \overrightarrow{OP_2} \,| = \text{real part of } iz_1 \,.\, \bar{z}_2 = \tfrac{1}{2} i (z_1 \bar{z}_2 - \bar{z}_1 z_2).$$

For example, the moment about the origin of the complex force $F = X + iY$, acting at the point z, is the real part of $iz\bar{F}$, that is of $iz(X - iY)$.

5·17. Rotation of axes of reference. If we wish to change from axes Ox, Oy to axes Ox', Oy', where Ox' makes an angle α with Ox, we have

$$x'+iy' = z' = re^{i\theta'} = re^{i(\theta-\alpha)} = ze^{-i\alpha}$$

and $$z = z'e^{i\alpha}.$$

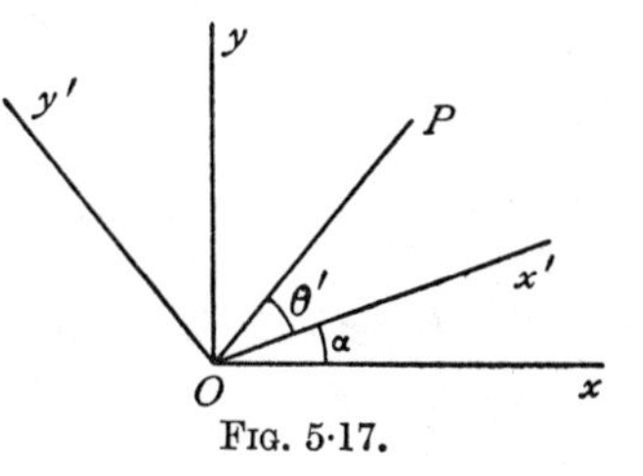

Fig. 5·17.

If in addition we change the origin to the point z_0 (referred to Ox, Oy), we get

$$z = z_0+z'e^{i\alpha},$$
$$z' = (z-z_0)e^{-i\alpha}.$$

5·20. Logarithms.

Let $$z = x+iy = re^{i\theta}.$$

Then $$\log z = \log r+i\theta$$
$$= \tfrac{1}{2}\log(x^2+y^2)+i\tan^{-1}\frac{y}{x}.$$

Thus the real part of $\log z$ is $\log r$, or $\frac{1}{2}\log(x^2+y^2)$.

The imaginary part of $\log z$ is θ, or $\tan^{-1}\dfrac{y}{x}$.

It is important to note that θ is not determined, except to an integral multiple of 2π, for the addition of 2π to θ does not alter the position of the point (r, θ).

Thus if we draw a circle, centre O, radius r, and, starting from A, describe the circle once in the counterclockwise or positive sense, when we return to A the argument has increased by 2π, assuming that it changes continuously. If we go round again, the argument increases again by 2π.

The argument therefore depends not only on the point A, but on the history of our movements in arriving at that point. The same applies if we move from A to A round any curve which encloses the origin.

Thus the imaginary part of $\log z$ may have the values

$$\theta,\quad \theta+2\pi,\quad \theta+4\pi,\ \ldots,$$

or $$\theta,\quad \theta-2\pi,\quad \theta-4\pi,\ \ldots.$$

5·21. Real and imaginary parts. We shall frequently require to separate a function of $z = x+iy$ into its real and imaginary parts. We have just seen that

$$\log z = \tfrac{1}{2}\log(x^2+y^2)+i\tan^{-1}\frac{y}{x},$$

and therefore, if X, Y are real functions,

$$\log(X+iY) = \tfrac{1}{2}\log(X^2+Y^2)+i\tan^{-1}\frac{Y}{X}.$$

Again, from Euler's theorem (5·13),

$$\cos\theta = \frac{e^{i\theta}+e^{-i\theta}}{2}, \quad \sin\theta = \frac{e^{i\theta}-e^{-i\theta}}{2i}.$$

Changing θ into $i\alpha$, we get

$$\cos i\alpha = \frac{e^{-\alpha}+e^{\alpha}}{2} \qquad \sin i\alpha = \frac{i}{2}(e^{\alpha}-e^{-\alpha}).$$

The *hyperbolic functions* $\cosh\alpha$, $\sinh\alpha$ are *defined* by the equations

$$\cosh\alpha = \tfrac{1}{2}(e^{\alpha}+e^{-\alpha}), \quad \sinh\alpha = \tfrac{1}{2}(e^{\alpha}-e^{-\alpha}),$$

so that $$\cosh\theta = \cos i\theta, \quad \sin i\theta = i\sinh\theta.$$

Hence $$\sin z = \sin x\cos iy + \cos x\sin iy$$
$$= \sin x\cosh y + i\cos x\sinh y,$$
$$\cos z = \cos x\cosh y - i\sin x\sinh y.$$

Similarly, $$\cosh z = \cos(iz) = \cos(ix-y)$$
$$= \cosh x\cos y + i\sinh x\sin y,$$
$$\sinh z = \sinh x\cos y + i\cosh x\sin y.$$

5·30. Definition of a holomorphic function of z. If $\phi = \phi(x, y)$, $\psi = \psi(x, y)$ are any functions whatever of x and y, the combination $\phi+i\psi$ is a function of the complex variable $z = x+iy$, in the sense that given z (i.e. x and y) there must correspond to this value of z one or more values of $\phi+i\psi$. This conception is far too general to be useful. We shall therefore restrict the functions which we shall consider to the class of holomorphic * functions of z which will now be defined.

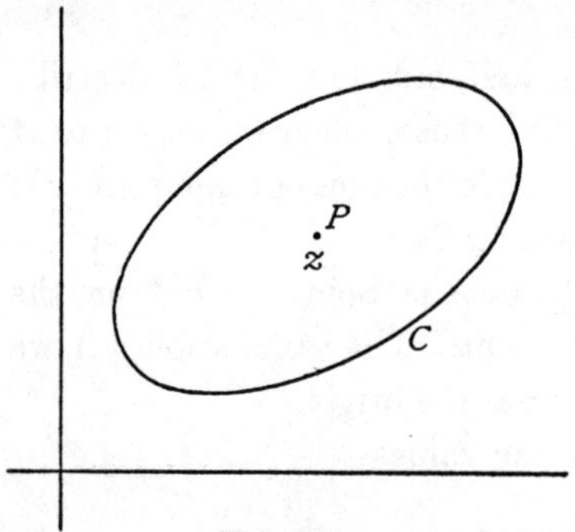

FIG. 5·30.

A *simple arc* is an arc which does not intersect itself and is rectifiable, i.e. has a definite length.

A *simple closed curve* is a closed curve which is separated by *every pair* of points on it into two simple arcs.

Let there be given a simple closed curve (or *contour*) C in the plane of the Argand diagram of z (briefly, the z-plane) and a function $f(z)$. The function $f(z)$ is said to be *holomorphic within the contour* C, if it satisfies the following conditions.

(*a*) To each value of z within C there corresponds one, and only one, value of $f(z)$, and that value is finite (i.e. its modulus is not infinite). Briefly, $f(z)$ is finite and one-valued in C.

---

* Greek ὅλος complete, μορφή form.

(b) For each value of z within C the function has a one-valued finite differential coefficient with respect to z.

We investigate requirement (b).

Since $x = \frac{1}{2}(z+\bar{z})$, $y = -\frac{1}{2}i(z-\bar{z})$ any function of x, y is a function of z and $\bar{z}$. Thus for example if $\phi(x, y)$, $\psi(x, y)$ are given functions,

$$\phi(x, y) + i\,\psi(x, y) = f(z, \bar{z}).$$

Now
$$df = \frac{\partial f}{\partial z}\,dz + \frac{\partial f}{\partial \bar{z}}\,d\bar{z}.$$

Therefore
$$\frac{df}{dz} = \frac{\partial f}{\partial z} + \frac{\partial f}{\partial \bar{z}} \lim_{|dz| \to 0} \frac{d\bar{z}}{dz}.$$

But
$$\lim \frac{d\bar{z}}{dz} = \lim \frac{\delta x - i\,\delta y}{\delta x + i\,\delta y} = \lim_{\delta x,\, \delta y \to 0} \frac{1 - i\dfrac{\delta y}{\delta x}}{1 + i\dfrac{\delta y}{\delta x}},$$

which is indeterminate, since δx and δy can tend to zero independently of one another.

Therefore a determinate derivative can exist only if $\partial f/\partial \bar{z} = 0$.

Therefore a holomorphic function of z is necessarily independent of $\bar{z}$, i.e. $\partial f/\partial \bar{z} = 0$.

Suppose then that $f = \phi(x, y) + i\,\psi(x, y) = \phi + i\,\psi$ and that $\partial f/\partial \bar{z} = 0$. Since

$$\frac{\partial f}{\partial \bar{z}} = \frac{\partial f}{\partial x}\frac{\partial x}{\partial \bar{z}} + \frac{\partial f}{\partial y}\frac{\partial y}{\partial \bar{z}} \quad \text{and} \quad x = \tfrac{1}{2}(z+\bar{z}),\ y = -\tfrac{1}{2}i(z-\bar{z}) \text{ we have}$$

$$0 = \tfrac{1}{2}\frac{\partial f}{\partial x} + \tfrac{1}{2}i\frac{\partial f}{\partial y} = \tfrac{1}{2}\left(\frac{\partial \phi}{\partial x} + i\frac{\partial \psi}{\partial x}\right) + \tfrac{1}{2}i\left(\frac{\partial \phi}{\partial y} + i\frac{\partial \psi}{\partial y}\right).$$

Therefore

$$\text{(1)} \qquad \frac{\partial \phi}{\partial x} = \frac{\partial \psi}{\partial y}, \quad \frac{\partial \psi}{\partial x} = -\frac{\partial \phi}{\partial y}.$$

These results are known as the Cauchy-Riemann equations. They are necessary but not sufficient. Sufficient conditions are obtained by adjoining to (1) the further conditions :

(2) All the partial derivatives $\dfrac{\partial \phi}{\partial x}, \dfrac{\partial \phi}{\partial y}, \dfrac{\partial \psi}{\partial x}, \dfrac{\partial \psi}{\partial y}$ are continuous.

Thus $\partial f/\partial \bar{z} = 0$, together with (2) are necessary and sufficient conditions that f shall be a holomorphic function of z.

Obvious examples of holomorphic functions are $\sin z$, e^z, $z^3 + 5z^2 - 3$, $(1+z)/(1-z^3)$. In the last case we must exclude the points at which $z^3 = 1$.

On the other hand, $|z|$ is not a holomorphic function of z, for $|z| = \sqrt{(z\bar{z})}$, and so involves $\bar{z}$.

5·31. Conjugate functions. The real and imaginary parts of a holomorphic function of z are called *conjugate functions*. Thus, if

$$f(z) = \phi(x, y)+i\psi(x, y) = \phi+i\psi,$$

ϕ and ψ are conjugate functions. As an example,

$$z^3 = (x^3-3xy^2)+i(3x^2y-y^3)$$

yields the conjugate functions x^3-3xy^2, $3x^2y-y^3$.

The Cauchy-Riemann conditions (5·30) give

$$\frac{\partial\phi}{\partial x} = \frac{\partial\psi}{\partial y}, \quad \frac{\partial\phi}{\partial y} = -\frac{\partial\psi}{\partial x}, \tag{1}$$

from which we deduce

$$\frac{\partial^2\phi}{\partial x^2}+\frac{\partial^2\phi}{\partial y^2} = 0, \quad \frac{\partial^2\psi}{\partial x^2}+\frac{\partial^2\psi}{\partial y^2} = 0.$$

Thus if $\nabla_1{}^2 = \dfrac{\partial^2}{\partial x^2}+\dfrac{\partial^2}{\partial y^2}$, the two-dimensional form of Laplace's operator, we see that conjugate functions are solutions of the equation $\nabla_1{}^2 V = 0$.

If we equate conjugate functions to constant values, say, $\phi(x, y) = c_1$, $\psi(x, y) = c_2$, we get two systems of curves. These curves are *orthogonal*, that is to say, their tangents at a point of intersection are at right angles. To prove this, observe that the gradient dy/dx of the curve $\phi(x, y) = c_1$ is given by

$$\frac{\partial\phi}{\partial x}+\frac{\partial\phi}{\partial y}\frac{dy}{dx} = 0.$$

Thus the gradient is $-\dfrac{\partial\phi}{\partial x}\Big/\dfrac{\partial\phi}{\partial y}$.

The gradient of $\psi(x, y) = c_2$ is $-\dfrac{\partial\psi}{\partial x}\Big/\dfrac{\partial\psi}{\partial y}$.

From (1), we see that the product of these gradients is -1, and therefore the tangents are at right angles.

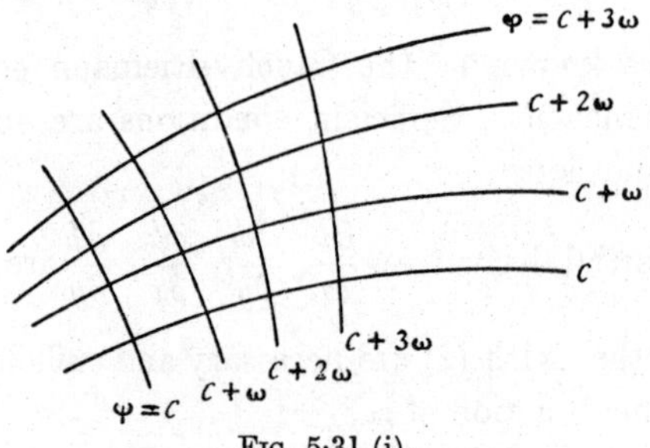

FIG. 5·31 (i).

Another proof is as follows. We have

$$f'(z)dz = d\phi+i\,d\psi.$$

Therefore $$(\arg dz)_{\phi=\text{constant}} = \tfrac{1}{2}\pi+(\arg dz)_{\psi=\text{constant}},$$

so that the elements of arc of the curves ϕ = constant, ψ = constant, are at right angles.

It follows that the curves $\phi = c_1$, $\psi = c_2$, if drawn at frequent small intervals of the constants c_1, c_2, divide the plane into infinitesimal rectangles (not all of the same size).

To illustrate this point, consider the conjugate functions defined by $\phi + i\psi = \log z$. Now, $\log z$ is not holomorphic in any curve which encircles the origin, for taking z once round the origin in the positive sense increases $\arg z$ by 2π and therefore $\log z$ by $2\pi i$, so that $\log z$ is not one-valued. If $f(z)$ is holomorphic, it must be continuous and one-valued in the region considered. This can be achieved by introducing suitable barriers. Let us then exclude the origin by drawing round it a circle of small radius ϵ and make a *cut* or impassable barrier along the positive part of the real axis, so that z may move in any manner outside the circle without crossing the positive part of the real axis. To fix the determination of the logarithm, let us agree that $\log z = i\pi$ when $z = -1$. We then get

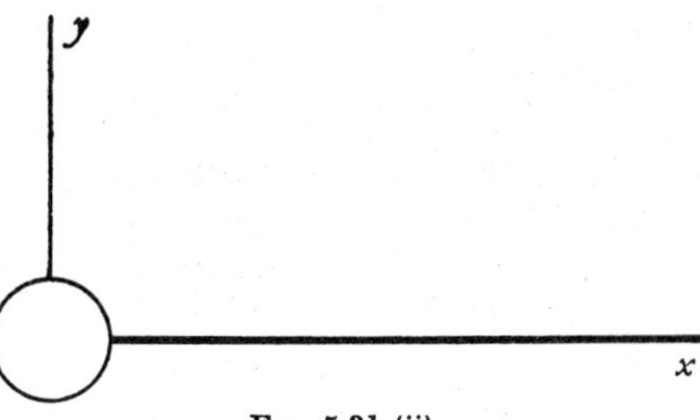

FIG. 5·31 (ii).

$$\phi = \tfrac{1}{2}\log(x^2+y^2), \quad \psi = \tan^{-1}\frac{y}{x},$$

where $\tan^{-1} y/x$ can now take all values between 0 and 2π, but no other values. The curves $\phi = c_1$ are circles, centre the origin, the curves $\psi = c_2$ are straight lines radiating from the origin.

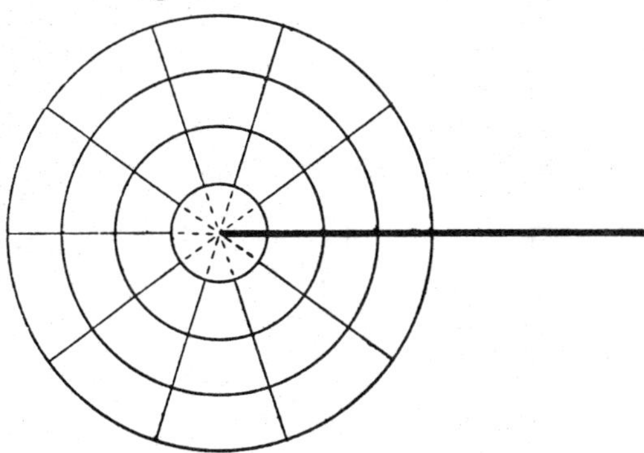
FIG. 5·31 (iii).

The resulting pattern is shown in fig. 5·31 (iii).

5·32. On the relation of conjugate functions to $f(z)$. A given holomorphic function $f(z)$ can be written in the form

(1) $$f(z) = f(x+iy) = \phi(x, y) + i\psi(x, y) = \phi + i\psi.$$

Therefore we have the identities

$$f(x+iy)+\bar{f}(x-iy)=2\phi(x,y),\quad f(x+iy)-\bar{f}(x-iy)=2i\,\psi(x,y).$$

Put $x=\frac{1}{2}z$, $y=-\frac{1}{2}iz$. Then these identities give

$$f(z)=2\phi(\tfrac{1}{2}z,\,-\tfrac{1}{2}iz)-\bar{f}(0),\quad f(z)=2i\,\psi(\tfrac{1}{2}z,\,-\tfrac{1}{2}iz)+\bar{f}(0).$$

Let $f(0)=\alpha+i\beta$, and $\bar{f}(0)=\alpha-i\beta$. Then

$$2\alpha=f(0)+\bar{f}(0)=2\phi(0,0),\quad 2i\beta=f(0)-\bar{f}(0)=2i\psi(0,0).$$

Therefore if $\phi(x,y)$ or $\psi(x,y)$ is given we determine $f(z)$ from

$$f(z)=2\phi(\tfrac{1}{2}z,\,-\tfrac{1}{2}iz)-\phi(0,0)+i\beta,\quad f(z)=2i\psi(\tfrac{1}{2}z,\,-\tfrac{1}{2}iz)-i\psi(0,0)+\alpha,$$

wherein β and α are arbitrary real constants.

Example. $\phi(x,y)=\sin x\cosh y+2\cos x\sinh y+x^2-y^2+4xy$,

$$f(z)=2\sin\tfrac{1}{2}z\cosh(-\tfrac{1}{2}iz)+4\cos\tfrac{1}{2}z\sinh(-\tfrac{1}{2}iz)+\tfrac{1}{2}z^2+\tfrac{1}{2}z^2-2iz^2,$$

Since

$$\cosh(i\theta)=\cos\theta,\quad \sinh i\theta=i\sin\theta,$$

$$f(z)=\sin z-2i\sin z+z^2-2iz^2.$$

5·33. The solution of Laplace's equation. To solve the equation

$$\nabla_1^2 V\equiv\frac{\partial^2 V}{\partial x^2}+\frac{\partial^2 V}{\partial y^2}=0,$$

put

$$z=x+iy,\quad \bar{z}=x-iy.$$

Then

$$\frac{\partial V}{\partial x}=\frac{\partial V}{\partial z}\frac{\partial z}{\partial x}+\frac{\partial V}{\partial \bar{z}}\frac{\partial \bar{z}}{\partial x}=\frac{\partial V}{\partial z}+\frac{\partial V}{\partial \bar{z}}$$

$$\frac{\partial V}{\partial y}=\frac{\partial V}{\partial z}\frac{\partial z}{\partial y}+\frac{\partial V}{\partial \bar{z}}\frac{\partial \bar{z}}{\partial y}=i\left(\frac{\partial V}{\partial z}-\frac{\partial V}{\partial \bar{z}}\right).$$

Thus we have the equivalence of operators

(1) $$2\frac{\partial}{\partial z}=\frac{\partial}{\partial x}-i\frac{\partial}{\partial y},\quad 2\frac{\partial}{\partial \bar{z}}=\frac{\partial}{\partial x}+i\frac{\partial}{\partial y}.$$

Therefore $$\frac{\partial^2 V}{\partial x^2}+\frac{\partial^2 V}{\partial y^2}=\left(\frac{\partial}{\partial x}-i\frac{\partial}{\partial y}\right)\left(\frac{\partial V}{\partial x}+i\frac{\partial V}{\partial y}\right)=4\frac{\partial^2 V}{\partial z\,\partial \bar{z}}=0.$$

It follows successively that

$$\frac{\partial V}{\partial \bar{z}}=f_1'(\bar{z}),\quad V=f_1(\bar{z})+f_2(z),$$

where $f_1(\bar{z})$, $f_2(z)$ are arbitrary functions, and this is the general solution. We thus see that any holomorphic function $f(z)$ satisfies Laplace's equation and therefore that this is the most general continuous solution involving z only. The most general real solution is $V=f(z)+\bar{f}(\bar{z})$.

The conjugate functions to which $f(z)$ gives rise must also be solutions, for the real and imaginary parts of $f(z)$ must separately satisfy the equation. This is in agreement with the result already obtained in 5·31. Solutions of Laplace's equation are often termed *harmonic functions*. Thus conjugate functions are also harmonic functions.

5·40. Sense of description of a contour. In calculating integrals taken round a contour C we can go round the contour in either sense : clockwise or anticlockwise. We shall make the convention that the sense of description shall be called *positive* if the contour is described, so as to leave the area regarded as bounded by it on the *left*, the region L in fig. 5·40.

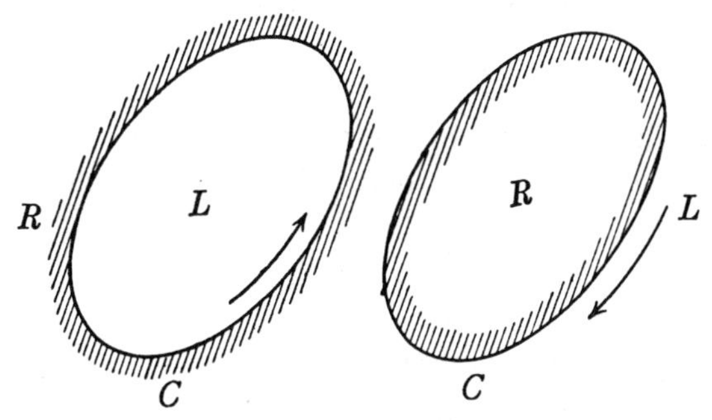

FIG. 5·40.

Fig. 5·40 shows the positive sense of description when the area is regarded as internal to the contour or external to it. The values obtained for the integral in the two cases will differ in sign.

5·43. The Area theorem. If $f(z, \bar{z})$ is a function of $z = x+iy$, $\bar{z} = x-iy$ which is continuous and differentiable in the area S enclosed by the contour C, then

$$(1) \qquad \int_{(C)} f(z, \bar{z})\,dz = 2i\int_{(S)} \frac{\partial f}{\partial \bar{z}}\,dS,$$

$$(2) \qquad \int_{(C)} f(z, \bar{z})\,d\bar{z} = -2i\int_{(S)} \frac{\partial f}{\partial z}\,dS.$$

Proof. By Stokes' theorem applied to the *plane* contour C closed by the *plane* diaphragm S

$$\int_{(C)} f\,d\mathbf{r} = \int_{(S)} (\mathbf{k}_\wedge \nabla) f\,dS = \mathbf{k}_\wedge \int_{(S)} \left(\mathbf{i}\frac{\partial f}{\partial x} + \mathbf{j}\frac{\partial f}{\partial y}\right) dS.$$

Now $d\mathbf{r} = \mathbf{i}\,dx + \mathbf{j}\,dy = (dx + i\,dy)\,\mathbf{i} = dz\,\mathbf{i}$, since $\mathbf{j} = \mathbf{k}_\wedge \mathbf{i} = i\,\mathbf{i}$.

Therefore, removing the factor **i**,

$$\int_{(C)} f\,dz = i\int_{(S)} \left(\frac{\partial f}{\partial x}+i\,\frac{\partial f}{\partial y}\right) dS = 2i\int_{(S)} \frac{\partial f}{\partial \bar{z}}\,dS,$$

from 5·33. Formula (2) follows by taking complex conjugates and then replacing $\bar{f}$ by f. Q.E.D.

Corollary. In (1) put $f = u - iv$. Then equating real and imaginary parts we get

$$\int_{(C)} (u\,dx+v\,dy) = \int_{(S)} \left(\frac{\partial v}{\partial x}-\frac{\partial u}{\partial y}\right) dS,\quad \int_{(C)} (u\,dy-v\,dx) = \int_{(S)} \left(\frac{\partial u}{\partial x}+\frac{\partial v}{\partial y}\right) dS.$$

The fundamental importance of the Area theorem was not perceived at the time when it appeared as a lemma in Chapter 9, of the first edition of this book.

5·50. Cauchy's integral theorem. Let C be a simple closed contour such that the function $f(z)$ is holomorphic at every point of C and in the interior of C.* Then

$$\int_{(C)} f(z)\,dz = 0.$$

This is Cauchy's integral theorem.

Proof. Since $f(z)$ is holomorphic, $\partial f/\partial \bar{z} = 0$.

Therefore, from 5·43 (1), $\int_{(C)} f(z)\,dz = 0.$ Q.E.D.

The proof here given is based on the assumption pointed out in 5·30 that sufficient conditions of holomorphy are satisfied. A complete proof would be long and difficult but the conditions here assumed are satisfied in the applications.

5·51. Morera's theorem. This is the converse of Cauchy's integral theorem, and states that, if

$$\int_{(C)} f(z)\,dz = 0,$$

for *every* simple closed contour within a region R, then $f(z)$ is a holomorphic function of z within that region.

Proof. From 5·43 (1) we get

$$\int_{(S)} \frac{\partial f}{\partial \bar{z}}\,dS = 0,$$

* This means that C and its interior lie wholly within a larger contour inside which the function is holomorphic.

where S is the region enclosed by C. Since this region is arbitrary, provided it lies within R, we must have

$$\frac{\partial f}{\partial \bar{z}} = 0,$$

so that $f(z)$ is a holomorphic function of z. Q.E.D.

The above argument requires a considerable amount of amplification to make it completely satisfactory. For a complete exposition the reader is referred to works on Analysis.

5·52. Analytical continuation. Let R_1, R_2 be two regions, separated by the line Σ, in which functions $f_1(z), f_2(z)$ are holomorphic, and such that

$$f_1(z) = f_2(z) \text{ on } \Sigma.$$

Then the function $f(z)$, which is equal to $f_1(z)$ if z is in R_1 and to $f_2(z)$ if z is in R_2, is a holomorphic function in the total region R_1+R_2. To prove this, we have only to show that

$$\int_{(C)} f(z)\,dz = 0$$

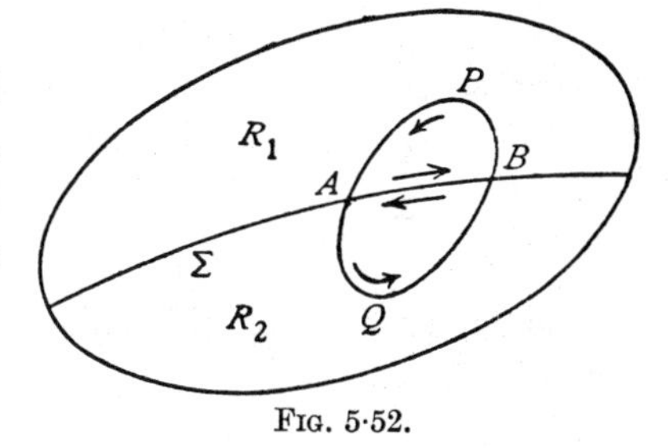

FIG. 5·52.

when C is a contour within R_1+R_2. Since $f_1(z)$, $f_2(z)$ are holomorphic, the only case for which this is not obvious is when the contour cuts Σ, see fig. 5·52.

For such a contour, we have

$$\int_{(C)} f(z)\,dz = \int_{(ABPA)} f_1(z)\,dz + \int_{(AQBA)} f_2(z)\,dz = 0,$$

since the integrals along AB and BA annul one another. Thus, by Morera's theorem, $f(z)$ is holomorphic in the total region R_1+R_2.

This situation is described by saying that $f_2(z)$ is the *analytical continuation of* $f_1(z)$ into the region R_2.

5·53. The principle of reflection. Let $f_1(z)$ be a holomorphic function defined within the region R_1 which is bounded by a straight line Σ on which $f_1(z)$ takes *real values*.

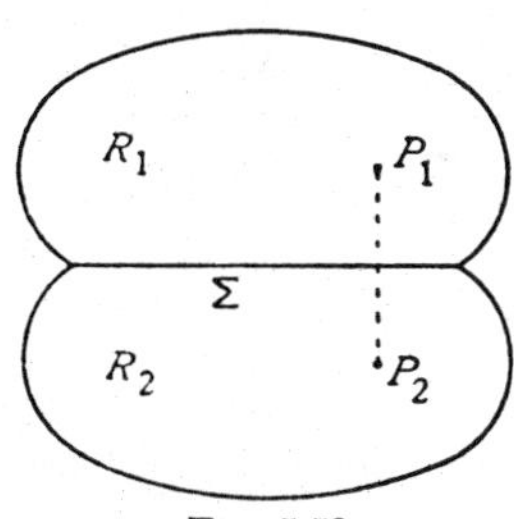

FIG. 5·53.

Let R_2 be the reflection of the region R_1 in the line Σ regarded as a mirror.

Let P_2 be the reflection of P_1 in Σ.

To continue $f_1(z)$ analytically into the region R_2, it is only necessary to take as $f_2(z)$ a function whose value at each point P_2 is the complex conjugate of the value of $f_1(z)$ at the corresponding point P_1.

5·54. Contraction or enlargement of a contour. Let us apply Cauchy's theorem to the contour consisting of two closed curves C, C' and the line AB joining two points on them, as shown in fig. 5·54.

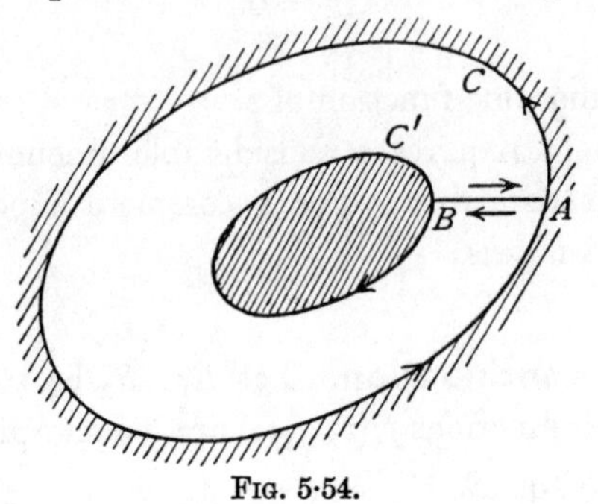

FIG. 5·54.

Then assuming $f(z)$ to be holomorphic on C and C' and at every point in the region between them, we have

$$\int_{(C)} f(z)\,dz + \int_{(AB)} f(z)\,dz - \int_{(C')} f(z)\,dz + \int_{(BA)} f(z)\,dz = 0.$$

The integrals along AB and BA cancel because $f(z)$ is one-valued and therefore

$$\int_{(C)} f(z)\,dz = \int_{(C')} f(z)\,dz,$$

both integrals being taken in the positive (anticlockwise) sense round the respective contours C and C'.

This means that, starting with the contour C, we can replace it by a contracted contour C', provided that $f(z)$ does not cease to be holomorphic at any point between C and C'. Similarly, under the same conditions the contour C' may be enlarged to C.

5·55. Case where the function ceases to be holomorphic. We can apply the method of argument of section 5·54 to obtain an important result.

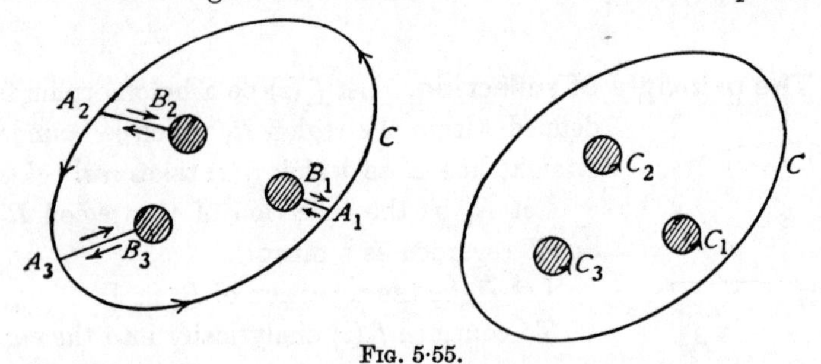

FIG. 5·55.

If the function ceases to be holomorphic at a finite number of points within a contour, we can draw small circles with centres at these points, such that each circle encloses only one point at which the function ceases to be holomorphic. We can join these circles by non-intersecting straight lines to the contour C. Fig. 5·55 illustrates the case where the function ceases to be holomorphic at three points. The circles are C_1, C_2, C_3, and the lines are A_1B_1, A_2B_2, A_3B_3. Then, by Cauchy's integral theorem,

$$\int_{(C)} f(z)\,dz + \int_{(A_1B_1)} - \int_{(C_1)} + \int_{(B_1A_1)} + \int_{(A_2B_2)} - \int_{(C_2)} + \int_{(B_2A_2)} + \int_{(A_3B_3)} - \int_{(C_3)} + \int_{(B_3A_3)} = 0,$$

where the integrand $f(z)\,dz$ is understood throughout. Thus

$$\int_{(C)} f(z)\,dz = \int_{(C_1)} f(z)\,dz + \int_{(C_2)} f(z)\,dz + \int_{(C_3)} f(z)\,dz.$$

This means that the integral round a contour can be replaced by the sum of the integrals round small circles centred at the points within the contour at which the function ceases to be holomorphic.

5·56. Singularities. A point at which a function ceases to be holomorphic is called a *singular point*, or *singularity* of the function.

Thus the function $f(z) = (z-a)^{-1}$ is holomorphic in any region from which the point $z = a$ is excluded (e.g. by drawing a small circle round it). At $z = a$ the function ceases to be finite and therefore does not satisfy the first part of the definition of holomorphy.

More generally, if near the point $z = a$ the function can be expanded in positive and negative powers of $z-a$, say

$$f(z) = \ldots + A_2(z-a)^2 + A_1(z-a) + A_0 + \frac{B_1}{z-a} + \frac{B_2}{(z-a)^2} + \ldots,$$

the point $z = a$ is a singular point.

If only a finite number of terms contain negative powers of $z-a$, the point $z = a$ is called a *pole*.

Again, consider the function $f(z) = \log z$. This function ceases to be holomorphic at $z = 0$. We have seen in 5·20 that $\log z$ is many-valued. If we choose one particular determination, say that which reduces to zero when $z = 1$, and allow z to describe a closed curve which does not encircle the point $z = 0$, $\log z$ will return to its starting value and will be holomorphic inside the curve.

5·57. Residues. We have seen that a function, which in the neighbourhood of $z = a$ has an expansion which contains negative powers of $z-a$, is singular at $z = a$.

In this case the coefficient of $(z-a)^{-1}$ is called the *residue* of the function at $z = a$.

Let us consider

$$\int (z-a)^n \, dz$$

taken round a circle of radius R whose centre is at the point $z = a$. On the circumference of this circle $z-a = Re^{i\theta}$, and therefore

$$\int (z-a)^n \, dz = \int_0^{2\pi} R^{n+1} e^{(n+1)i\theta} \, i \, d\theta = \frac{R^{n+1}}{(n+1)} \Big[e^{(n+1)i\theta} \Big]_0^{2\pi} = 0, \text{ if } n \neq -1.$$

If, however, $n = -1$, we get

$$\int \frac{dz}{z-a} = \int_0^{2\pi} i \, d\theta = 2\pi i.$$

Now, suppose that $f(z)$ has an expansion in the neighbourhood of $z = a$ of the form

$$\ldots + A_2(z-a)^2 + A_1(z-a) + A_0 + \frac{B_1}{(z-a)} + \frac{B_2}{(z-a)^2} + \ldots.$$

If we integrate round a small circle surrounding $z = a$, we get

$$\int f(z) \, dz = 2\pi i B_1 ,$$

for all the integrals vanish except that of $B_1(z-a)^{-1}$.

Thus we see the importance of the residues, for they form the only contributions to the integral of a function which is holomorphic at all points except singularities of the kind described above.

More generally we define the residue r_1 at an isolated singularity a_1 by

$$(1) \qquad r_1 = \text{residue} = \frac{1}{2\pi i} \int_{(C_1)} f(z) \, dz,$$

where C_1 is a small circle (or simple closed curve) about a_1 which encloses no other singularity.

5·58. Cauchy's residue theorem. Let C be a closed contour inside and upon which the function $f(z)$ is holomorphic, except at a finite number of singular points within C at which the residues are $r_1, r_2, \ldots, r_n$. Then

$$\int_{(C)} f(z) \, dz = 2\pi i (r_1 + r_2 + \ldots + r_n).$$

Proof. Suppose there are three singularities. Surround them by small circles, as in 5·55. Then

$$\int_{(C)} f(z) \, dz = \int_{(C_1)} f(z) \, dz + \int_{(C_2)} f(z) \, dz + \int_{(C_3)} f(z) \, dz$$
$$= 2\pi i \, r_1 + 2\pi i \, r_2 + 2\pi i \, r_3 ,$$

from 5·57 (1). This proves the theorem in the case of three singularities. The proof for any finite number is the same. Q.E.D.

5·59. Cauchy's formula. Let $f(\zeta)$ be a function of the complex variable ζ, holomorphic inside and on a closed contour C and let z be any point not on C. Then

$$\frac{1}{2\pi i}\oint_{(C)} \frac{f(\zeta)}{\zeta - z}\,d\zeta = f(z) \text{ or } 0,$$

according as z is inside or outside C.

Proof. Let $F(\zeta) = [f(\zeta) - f(z)]/(\zeta - z)$. Then $F(\zeta)$ is holomorphic everywhere within C except at $\zeta = z$, where it is undefined.

But since $f(\zeta)$ is holomorphic,

$$\lim_{\zeta \to z} F(\zeta) = f'(z).$$

Let us therefore define $F(\zeta)$ to be equal to $f'(z)$ when $\zeta = z$.

With this definition $F(\zeta)$ is holomorphic everywhere within C, and therefore by Cauchy's theorem $\int F(\zeta)\,d\zeta = 0$. Therefore

$$\frac{1}{2\pi i}\oint_{(C)} \frac{f(\zeta)}{\zeta - z}\,d\zeta = \frac{1}{2\pi i}\oint_{(C)} \frac{f(z)}{\zeta - z}\,d\zeta = f(z) \text{ or } 0$$

by Cauchy's residue theorem, according as z is inside or outside C. Q.E.D.

5·591. Cauchy's formula for the exterior domain. Let the function $f(\zeta)$ be holomorphic in the region L *exterior* to a closed contour C and continuous on C. Then

$$(1) \qquad \frac{1}{2\pi i}\oint_{(C)} \frac{f(\zeta)}{\zeta - z}\,d\zeta = f(z) - f(\infty) \text{ or } -f(\infty)$$

according as z is in the region exterior to C or inside C.

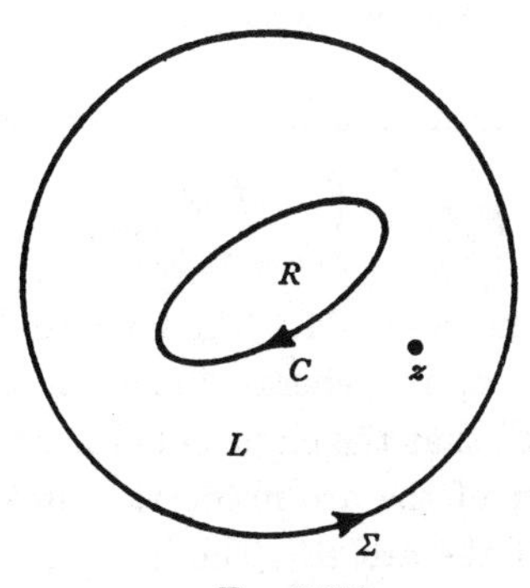

FIG. 5·591.

Proof. Surround C by a circle Σ, of radius ρ, about the origin. Let z be a point between C and Σ. Then Cauchy's formula applied to the region between C and Σ gives

(2) $$\frac{1}{2\pi i}\oint_{(C)}\frac{f(\zeta)}{\zeta-z}\,d\zeta-f(z)=-\frac{1}{2\pi i}\oint_{(\Sigma)}\frac{f(\zeta)}{\zeta-z}\,d\zeta=-\frac{1}{2\pi i}\int_0^{2\pi}\frac{f(\rho e^{i\theta})}{1-(z/\rho e^{i\theta})}\,id\theta$$

on putting $\zeta=\rho e^{i\theta}$ on Σ so that $d\zeta=i\rho e^{i\theta}\,d\theta$. But the left-hand side of (2) is independent of ρ and so therefore is the right-hand side. Putting $\rho=\infty$ the last integral becomes $-f(\infty)$.

This proves the first part of (1). The second part follows by noting that if z is inside C, $f(z)$ in (2) is replaced by zero.

More generally a given sense of description of the closed contour C determines regions L and R (5·40). Now suppose that $f(\zeta)$ is holomorphic everywhere in the region L except for a singularity in L where the principal part is $g(\zeta)$. Then

(3) $$\frac{1}{2\pi i}\oint\frac{f(\zeta)}{\zeta-z}\,d\zeta=f(z)-g(z)\text{ or }-g(z)$$

according as z is in L or R.

If L includes the point infinity it is here assumed that $f(z)-g(z)$ tends to zero at ∞.

5·595. Principal value of an integral.

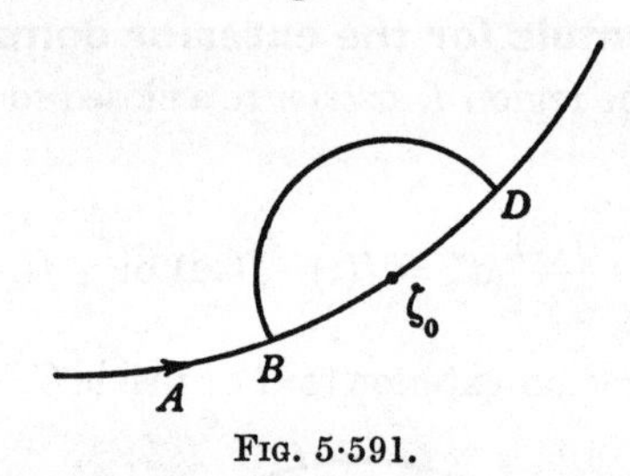

FIG. 5·591.

Let ζ_0 be a point on an arc A (which may be a closed contour) and consider

(1) $$F(\zeta_0)=\int_{(A)}\frac{f(\zeta)}{\zeta-\zeta_0}\,d\zeta,$$

where $f(\zeta)$ is given when ζ moves on A. The integrand becomes infinite when $\zeta=\zeta_0$ and so this integral is, in general, indeterminate. Describe a circle centre ζ_0 of radius ϵ so small that the circle cuts the arc A in two points B, D, say. Denote by a the part of the arc inside the circle, i.e. the arc BD, and denote by $A-a$ the rest of the arc A. The integral (1) is said to exist as a *Cauchy principal value* if

(2) $$\lim_{\epsilon\to 0}\int_{(A-a)}\frac{f(\zeta)}{\zeta-\zeta_0}\,d\zeta\quad\text{exists.}$$

Observe that an integral which exists in the ordinary sense, exists also as a

Cauchy principal value. The converse is false. Every integral can therefore be considered as a Cauchy principal value when this latter exists.

Consider in particular $\int_{(C)} \frac{d\zeta}{\zeta - \zeta_0}$ taken round a closed contour C. Here

$$\lim_{\epsilon \to 0} \int_{(C-a)} \frac{d\zeta}{\zeta - \zeta_0} = \lim_{\epsilon \to 0} \Big[\log(\zeta - \zeta_0) \Big]_{(C-a)} = \lim_{\epsilon \to 0} i \Big[\arg(\zeta - \zeta_0) \Big]_{(C-a)} = i\pi.$$

Therefore as a Cauchy principal value

$$\int_{(C)} \frac{d\zeta}{\zeta - \zeta_0} = i\pi.$$

5·596. The formulae of Plemelj. Let ζ_0 be a given point on a simple closed contour C and let $\phi(\zeta)$ be a function given on C such that

$$\frac{1}{2\pi i} \int_{(C)} \frac{\phi(\zeta) - \phi(\zeta_0)}{\zeta - \zeta_0} \, d\zeta \tag{1}$$

exists, at least as a Cauchy principal value.

Having fixed a positive sense of description, the curve C separates the plane into two regions, L on the left and R on the right. See fig. 5·40. We consider

$$\Phi(z) = \frac{1}{2\pi i} \int_{(C)} \frac{\phi(\zeta)}{\zeta - z} \, d\zeta \equiv \frac{1}{2\pi i} \int_{(C)} \frac{\phi(\zeta) - \phi(z)}{\zeta - z} \, d\zeta + \frac{1}{2\pi i} \int_{(C)} \frac{\phi(z) \, d\zeta}{\zeta - z}. \tag{2}$$

If z is in L, we write $\Phi^L(z)$ for $\Phi(z)$ giving

$$\Phi^L(z) = \frac{1}{2\pi i} \int_{(C)} \frac{\phi(\zeta) - \phi(z)}{\zeta - z} \, d\zeta + \phi(z) \tag{3}$$

by Cauchy's residue theorem.

Now let z, while remaining in L, tend to ζ_0. Then we write

$$\Phi^L(\zeta_0) = \frac{1}{2\pi i} \int_{(C)} \frac{\phi(\zeta) - \phi(\zeta_0)}{\zeta - \zeta_0} \, d\zeta + \phi(\zeta_0). \tag{4}$$

Again if z is in R, we have $\int_{(C)} d\zeta/(\zeta - z) = 0$ by Cauchy's residue theorem, and therefore from (2)

$$\Phi^R(z) = \frac{1}{2\pi i} \int_{(C)} \frac{\phi(\zeta) - \phi(z)}{\zeta - z} \, d\zeta .$$

Therefore if z, remaining in R, tends to ζ_0, we write

$$\Phi^R(\zeta_0) = \frac{1}{2\pi i} \int_{(C)} \frac{\phi(\zeta) - \phi(\zeta_0)}{\zeta - \zeta_0} \, d\zeta. \tag{5}$$

Subtracting (5) from (4) we get the *first Plemelj formula* *

* Plemelj, J., " Ein Ergänzungssatz . . . ", *Monatshefte für Math. & Phys.*, 19 (1908), 205–10.

(6) $$\Phi^L(\zeta_0) - \Phi^R(\zeta_0) = \phi(\zeta_0)$$

and adding (4) to (5) we get the *second Plemelj formula* *

(7) $$\Phi^L(\zeta_0) + \Phi^R(\zeta_0) = \frac{1}{\pi i}\int_{(C)} \frac{\phi(\zeta)}{\zeta - \zeta_0}\, d\zeta\,.$$

If instead of a closed contour C we have an open arc A the formulae still subsist, for we can close the arc by joining its ends and on this closure ascribe the value zero to $\phi(\zeta)$.

One of the most valuable conclusions from the first Plemelj formula is embodied in the following theorem.

The Plemelj theorem. The functional equation

(8) $$\Phi^L(\zeta_0) - \Phi^R(\zeta_0) = \phi(\zeta_0)$$

on an arc A has a particular solution

(9) $$\Phi(z) = \frac{1}{2\pi i}\int_{(A)} \frac{\phi(\zeta)\, d\zeta}{\zeta - z}\,.$$

This is the unique solution which is holomorphic in the whole plane except on the arc A and which tends to zero at infinity.

That (9) is a solution follows at once from (6). To see that it is unique let $\Psi(z)$ be the *difference* of two solutions which satisfy the given conditions. Then by suitably defining $\Psi(z)$ on A (where it is undefined) we find that $\Psi(z)$ is holomorphic in the whole plane including infinity, and therefore by Liouville's theorem reduces to a constant which must be zero for $\Phi(z)$ has to vanish at ∞.

5·60. Zeros. If a holomorphic function $f(z)$ can be expressed in the form $f(z) = (z-z_0)^n g(z)$, where n is a positive integer and $g(z)$ is not zero when $z = z_0$, the function $f(z)$ is said to have a *zero* of *multiplicity* n at $z = z_0$. If $n = 1$, z_0 is a *simple zero.*

Since $$f'(z) = n(z-z_0)^{n-1} g(z) + (z-z_0)^n g'(z),$$

$f'(z)$ will have a zero of multiplicity $n-1$ at $z = z_0$.

In the case of a simple zero, $f'(z_0) \neq 0$.

Thus if $f'(z) \neq 0$ inside a given contour in which $f(z)$ is holomorphic, $f(z)$ can have only simple zeros within this contour.

Again, since the argument of a product is equal to the sum of the arguments (5·13):

$$\arg f(z) = \arg (z-z_0)^n + \arg g(z) = n \arg (z-z_0) + \arg g(z)$$

for the same reason.

In counting the zeros of $f(z)$, it is convenient to regard a zero of multiplicity n as n zeros (all equal).

* Plemelj, J., "Ein Ergänzungssatz . . . ", *Monatshefte für Math. & Phys.*, 19 (1908), 205–10.

5·61. The principle of the argument. If C is a simple closed contour upon which $f(z)$ has no zeros, and within and upon which $f(z)$ is holomorphic, then the number N of zeros of $f(z)$ within the contour is given by

$$2\pi N = [\arg f(z)]_{(C)},$$

where the notation means the increase in $\arg f(z)$, when z describes the contour once in the positive sense.

Proof. For simplicity, suppose there are two zeros inside, say z_1 and z_2, of multiplicities n_1 and n_2. Then

$$f(z) = (z-z_1)^{n_1}(z-z_2)^{n_2}g(z),$$

where $g(z)$ has no zeros inside C. Thus (5·60),

$$\arg f(z) = n_1 \arg(z-z_1)+n_2 \arg(z-z_2)+\arg g(z).$$

When z describes C once in the positive sense, $\arg(z-z_1)$ and $\arg(z-z_2)$ each increase by 2π, while $\arg g(z)$ returns to its original value.

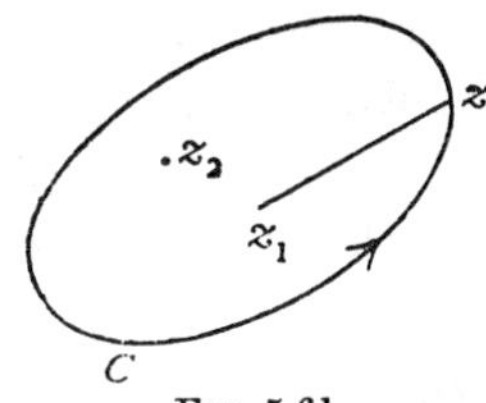

FIG. 5·61.

Therefore $[\arg f(z)]_{(C)} = 2\pi(n_1+n_2) = 2\pi N$. Q.E.D.

If in addition $f(z)$ has a zero, say z_3, *on* the contour C, when z describes the contour once in the positive sense, there will be an *increase* in $\arg(z-z_3)$. This increase will be π, if z_3 is an ordinary * point of C; it will be the angle between the tangents at z_3, if z_3 is a point at which there are two distinct tangents.† Thus in any case we shall have

$$[\arg f(z)]_{(C)} \geqslant 2\pi N,$$

where N is the number of zeros of $f(z)$ *within* the contour C.

5·62. Mapping. Let $f(z)$ be a function of $z = x+iy$, which is holomorphic inside and upon a simple closed contour C in the x, y plane,

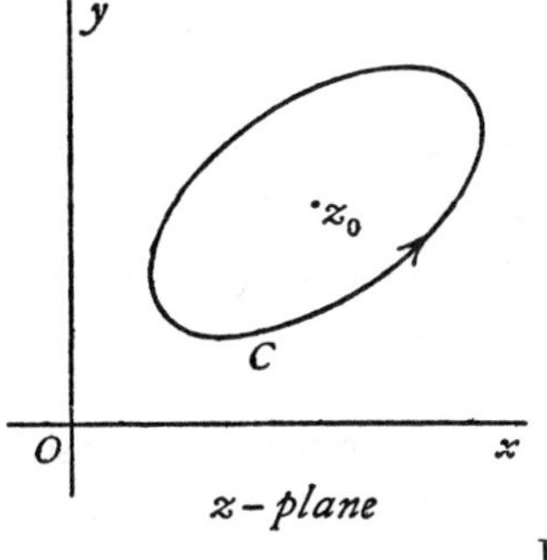

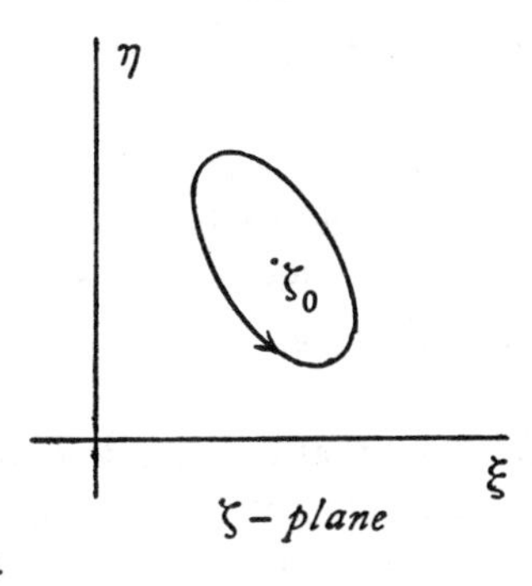

FIG. 5·62.

which we shall call the z-plane. We take a second complex variable $\zeta = \xi+i\eta$ and mark the representative points of ζ in a second Argand dia-

* See fig. 7·32 (i). † See fig. 7·32 (ii).

gram, axes $O\xi$, $O\eta$, which we shall call the ζ-plane. Now consider the relation

(1) $$\zeta = f(z).$$

By means of this relation, to each point within or upon C there corresponds one point in the ζ-plane, and, since $f(z)$ being holomorphic is one-valued, only one point. Thus the points of C and its interior are mapped into certain points in the ζ-plane. We shall enquire into the nature of the map on the following hypotheses :

(*a*) $f(z)$ never takes the same value at two different points of the contour C ; and

(*b*) the derivative $f'(z)$ has no zero on the contour C.

We shall now prove several properties of the mapping given by (1).

(i) When z describes C once, ζ describes a closed curve Γ in the ζ-plane, and this curve has no double points.

Proof. The function $f(z)$ being holomorphic on C varies continuously, and therefore ζ varies continuously, so that ζ describes a continuous curve Γ.

Since $f(z)$ being holomorphic is one-valued, when z describes C once, returning to the same point, $f(z)$ and therefore ζ returns to its initial value. Therefore Γ is a closed curve.

Since, by (*a*), $f(z)$ never takes the same value twice when z describes C, ζ never takes the same value twice when it describes Γ. This means that the curve Γ does not cross itself, that is, it has no double points.

(ii) Given the point z_0 inside C, the corresponding point ζ_0 is inside Γ.

Proof. Let

$$n = \frac{1}{2\pi}\,[\arg\{f(z)-f(z_0)\}]_{(C)} = \frac{1}{2\pi}\,[\arg(\zeta-\zeta_0)]_{(\Gamma)}.$$

Since $f(z)-f(z_0)$ has at least one zero inside C, namely z_0, section 5·61 shows that $n \geqslant 1$.

Now, when ζ describes Γ once, the increase in $\arg(\zeta-\zeta_0)$ is 0, $\pm\alpha$ (where * $\alpha<2\pi$), or $\pm 2\pi$ according as ζ_0 is outside, on, or inside Γ.

The corresponding values of n are 0, $\pm m\,(m<1)$, ± 1. But $n \geqslant 1$. Hence $n = 1$, and therefore

$$[\arg(\zeta-\zeta_0)]_{(\Gamma)} = 2\pi.$$

This shows that ζ_0 is inside Γ and that Γ is described in the positive sense. This means, 5·40, that ζ_0 is on the left of an observer who describes the contour in the positive sense.

(iii) If z describes C in the positive sense, ζ describes Γ in the positive sense.

---

* The curve Γ has no double point and therefore α must be less than 2π. See figs. 7·32 (i), (ii).

This is an immediate corollary to (ii), where it was proved that when z describes C positively, ζ_0 is inside Γ and Γ is described positively.

(iv) Given the point ζ_0 inside Γ, there is exactly one point z_0 inside C such that $\zeta_0 = f(z_0)$.

Proof. Since ζ_0 is inside Γ, $\zeta - \zeta_0$ has exactly one zero inside Γ, and therefore

$$1 = \frac{1}{2\pi}[\arg(\zeta - \zeta_0)]_{(\Gamma)} = \frac{1}{2\pi}[\arg\{f(z) - \zeta_0\}]_{(C)},$$

which shows that $f(z) - \zeta_0$ has exactly one zero inside C. Calling this z_0, we get $f(z_0) - \zeta_0 = 0$.

(v) The derivative $f'(z)$ cannot vanish inside or upon C.

Proof. If possible, let z_1 be a zero of $f'(z)$ inside C. Then $f(z) - f(z_1)$ has a zero of multiplicity greater than 1 since $f'(z_1) = 0$ (see 5·60).

Therefore the equation $f(z) - f(z_1) = 0$ has at least two roots at z_1, which is inside C. This contradicts (iv), and so the hypothesis that $f'(z)$ vanishes inside C is false. That $f'(z)$ cannot vanish upon C follows from (*b*).

(vi) When ζ moves inside Γ, z is a holomorphic function of ζ.

Proof. From (iv), we see that to each value of ζ within Γ there corresponds a single definite value of z within C, so that z is a one-valued function of ζ.

It remains to show that z has a unique finite derivative for each value of ζ within Γ. Now, if $f'(z)$ is not zero,

$$\frac{dz}{d\zeta} = \left(\frac{d\zeta}{dz}\right)^{-1} = \frac{1}{f'(z)},$$

and since $f'(z)$ is unique and never zero while z moves within C, the required result follows. On account of (*b*) the result is still true when ζ moves on Γ

The above results show that the relation (1), subject to the condition (*a*) constitutes a *bi-uniform mapping* whereby the region within C is mapped point by point on the region within Γ, and conversely, the region within Γ is mapped point by point on the region within C, in such a way that to the point z_0 within C there corresponds one, and only one, point ζ_0 within Γ, and to the point ζ_0 within Γ there corresponds one and the same point z_0 within C. The adjunction of condition (*b*) ensures that the bi-uniform character of the mapping extends to the boundaries C and Γ.

5·63. Indented contours. It may happen that we require to map a contour C on which there occurs a zero of $f'(z)$, say at P.

To do this we *indent* the contour C, that is to say, we replace it by a modified contour C' in which an infinitesimal arc of C containing P is replaced by

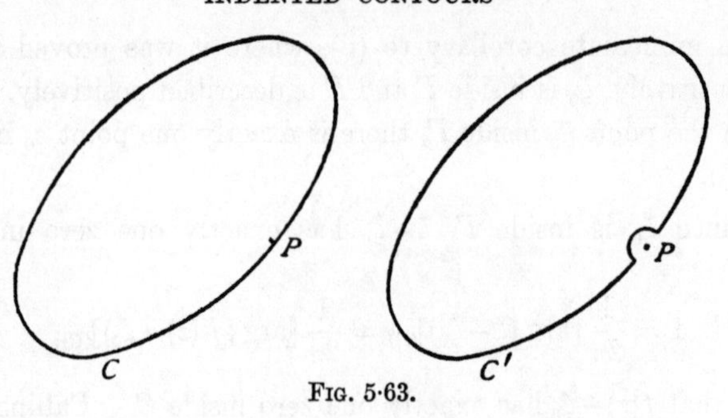

FIG. 5·63.

a circular arc, centre P, of infinitesimal radius, so that P is now *outside* the modified contour C', fig. 5·63.

To the modified contour C' the theorems of mapping now apply. We then let the radius of the indentation tend to zero. A contour may of course be indented at as many points as may be necessary.

5·70. Conformal representation. Let a bi-uniform mapping of a region of the z-plane on a region of the ζ-plane be defined by

(1) $$\zeta = f(z).$$

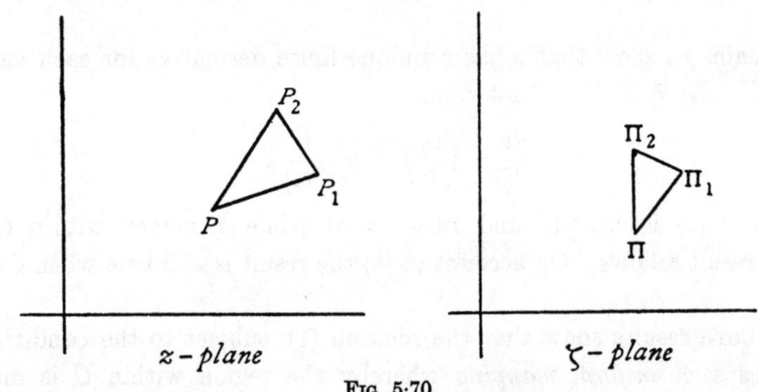

FIG. 5·70.

Let z, z_1, z_2 be represented by the points P, P_1, P_2 of the z-plane, and let the corresponding values ζ, ζ_1, ζ_2 be represented by the points Π, Π_1, Π_2 of the ζ-plane. Then

$$\frac{\zeta_1 - \zeta}{z_1 - z} = \frac{f(z_1) - f(z)}{z_1 - z}, \quad \frac{\zeta_2 - \zeta}{z_2 - z} = \frac{f(z_2) - f(z)}{z_2 - z}.$$

If we suppose $z_1 - z$, and $z_2 - z$ to be small, we then have

(2) $$\frac{\zeta_1 - \zeta}{z_1 - z} = f'(z), \quad \frac{\zeta_2 - \zeta}{z_2 - z} = f'(z)$$

very nearly, and hence

$$\frac{\zeta_1 - \zeta}{z_1 - z} = \frac{\zeta_2 - \zeta}{z_2 - z} = f'(z) = \frac{d\zeta}{dz}.$$

Thus, taking modulus and argument,

$$(3) \qquad \frac{\Pi\Pi_1}{PP_1} = \frac{\Pi\Pi_2}{PP_2} = |f'(z)| = \left|\frac{d\zeta}{dz}\right|,$$

$$\arg \Pi\Pi_1 - \arg PP_1 = \arg \Pi\Pi_2 - \arg PP_2 = \arg f'(z).$$

Hence $\arg \Pi\Pi_2 - \arg \Pi\Pi_1 = \arg PP_2 - \arg PP_1,$

and therefore

$$(4) \qquad \angle \Pi_1 \Pi \Pi_2 = \angle P_1 P P_2.$$

Equations (3) and (4) mean geometrically that the triangles P_1PP_2, $\Pi_1\Pi\Pi_2$ are similar, so that an infinitesimal triangle in the z-plane maps into a similar infinitesimal triangle in the ζ-plane. Thus the mapping preserves

(*a*) the angles ;

(*b*) the similarity of corresponding infinitesimal triangles.

From these properties we derive the name *conformal representation* of the mapping given by (1).

The relation (3) gives the scale of the mapping at the point Π. This scale since it is a function of z, varies from point to point. An illustration of conformal mapping is afforded by an ordinary map on Mercator's projection. It is well known that the angle between two lines as measured on the map is equal to the angle at which the two corresponding lines intersect on the earth's surface ; in fact, it is this property which renders the map useful in navigation.

In particular the lines on the map which represent the meridians and parallels of latitude are at right angles. If we confine our attention to a small portion of the map, we also know that distances measured on the map will represent to scale the corresponding distances on the globe, but that the scale changes as the latitude increases.

We also obtain from (3) the ratio of corresponding small areas. Thus

$$\frac{\Delta \Pi_1 \Pi \Pi_2}{\Delta P_1 P P_2} = |f'(z)|^2 = f'(z) \times \bar{f}'(\bar{z}) = \frac{d\zeta}{dz} \times \frac{d\bar{\zeta}}{d\bar{z}},$$

where $\bar{f}(\bar{z})$ is the conjugate complex of $f(z)$.

To illustrate this last point suppose that

$$f(z) = 6z + 3i\,z^2.$$

Then

$$f'(z) = 6 + 6i\,z = 6 + 6i\,(x + iy),$$

$$\bar{f}'(\bar{z}) = 6 - 6i\,\bar{z} = 6 - 6i\,(x - iy),$$

and

$$|f'(z)|^2 = (6 - 6y)^2 + (6x)^2.$$

5·71. The mapping of infinite regions. In most of the applications of conformal representation to hydrodynamics one or both of the regions concerned extends to infinity and it becomes of importance to have a clear idea as to what constitutes the " inside " of the boundary. To elucidate this point, consider the mapping given by

$$\zeta = z^{\alpha}, \quad \alpha > 1$$

applied to the region in the z-plane bounded by the circular arcs $r = a$, $r = b$ and the radii $\theta = 0$, $\theta = \pi/\alpha$, fig. 5·71 (i).

Put $z = re^{i\theta}$, $\zeta = Re^{i\gamma}$. Then

$$Re^{i\gamma} = r^{\alpha} e^{i\alpha\theta}.$$

Thus $\gamma = 0$ and ζ moves along $A'B'$ when z moves along AB $(\theta = 0)$, $\gamma = \pi$ and ζ moves along $C'D'$ when z moves along $CD(\theta = \pi/\alpha)$, while on the arc $AMD(r = a)$, $R = a^{\alpha}$, and on the arc $BLC(r = b)$, $R = b^{\alpha}$, so that the corresponding paths in the ζ-plane are the semicircles $D'M'A'$, $B'L'C'$.

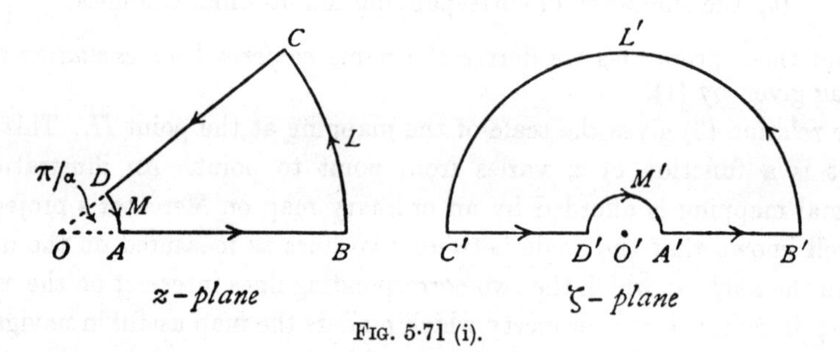

FIG. 5·71 (i).

It is clear that the conditions 5·62 (a), (b) are both fulfilled by the mapping function, for the origin, at which the derivative of z^{α} vanishes, is excluded from the region considered. Thus the mapping is bi-uniform and the inside of the sectorial region in the z-plane is conformally represented on the inside of the region between the semicircles in the ζ-plane. Moreover, the sense of description corresponds in the two diagrams, the area mapped in both being on the left when the contours are described in the senses shown. These statements are true however great b may be, and therefore letting $b \to \infty$ and denoting points at infinity by the suffix ∞, we obtain fig. 5·71 (ii), in which the hatching indicates the exterior. This shows that the interior of the infinite sectorial region is mapped on the upper half of the ζ-plane, and that now the term interior is to be inferred from the limiting form of the finite case and is related in the same way as before to the sense of description. The indentations at the origin may now be removed by supposing a to tend to zero.

Simple considerations on the above lines will generally be adequate to

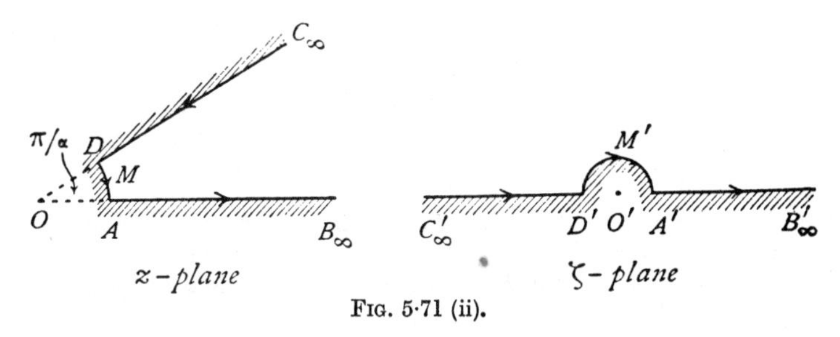

FIG. 5·71 (ii).

decide on the meaning which is to be assigned to the terms inside or interior when applied to mapping in which infinite regions are involved; indeed the sense of description alone will generally furnish the required information.

EXAMPLES V

1. If $\phi + i\psi = f(z)$, and $f(z)$ is real when $y = a$, show that $\psi = 0$ when $y = a$.

2. Find the function of z whose imaginary part is

$$2x(x^2 - 3y^2) + \tfrac{1}{2}(x^2 - y^2) + \alpha xy.$$

3. Taking the transformations

$$\text{(i) } \zeta = z + a, \quad \text{(ii) } \zeta = ze^{i\alpha}, \quad \text{(iii) } \zeta = bz, \quad \text{(iv) } \zeta = z^{-1},$$

prove that the first is a translation, the second a rotation, the third a rotation and a magnification, the fourth an inversion followed by a reflection; where α is real and a, b may be complex.

Prove that $\zeta = (\alpha z + \beta)/(\gamma z + \delta)$, where $\alpha\delta - \beta\gamma \neq 0$, may be compounded of a succession of the above transformations and hence gives a mapping in which circles and straight lines transform into circles or straight lines.

4. Prove that the transformation $\zeta = (z - i)/(z + i)$ maps the upper half of the z-plane on the interior of the circle $|\zeta| = 1$. Find the points corresponding to $z = \infty, -1, 0, 1$.

5. Show that the transformation $\zeta = z^2$ maps the half-plane $y \geqslant 0$ on the whole of the ζ-plane, provided the part of the real axis for which $\xi \geqslant 0$ in the ζ-plane is regarded as an impassable barrier, so that ζ may not be taken along any path which crosses this barrier.

6. Prove that the transformation $\zeta = -i(z-1)/(z+1)$ maps the region within the circle $|z| = 1$, indented at the points $z = 1$, $z = -1$, on the region in the ζ-plane within a semicircle of great radius indented at the origin. Find the relation between the radius of the semicircle in the ζ-plane and the radius of the indentation at $z = -1$, and hence show that when the latter tends to zero the whole of the upper half ζ-plane is mapped.

7. Show that the transformation $z = \cos\zeta$ maps the whole of the z-plane in which there is an impassable barrier along the real axis from $z = -\infty$ to $z = 1$, on the semi-infinite rectangle bounded by $\xi = -\pi$, $\xi = \pi$ for which $\eta \geqslant 0$. Show that the curves η = constant are confocal ellipses.

8. Show that the relation

$$z = \frac{2a}{\pi} \log \frac{1+\zeta}{1-\zeta}$$

maps the region between the lines $y = a$, $y = -a$ on the interior of a circle of unit radius and centre at the origin in the ζ-plane.

9. If OA is the line $y = x\tan(k\pi)$ from $x = 0$ to $x = l_1\cos(k\pi)$, where $k<1$, and OB is the line joining the origin to $x = -l_2$, show that the transformation

$$z = Fe^{ik\alpha}\zeta^{-1}(\zeta - e^{-i\alpha})^{1+k}(\zeta - e^{i\alpha})^{1-k}$$

maps the circumference of the unit circle in the ζ-plane on the broken line AOB described twice, the points $\zeta = e^{i\alpha}$, $\zeta = e^{-i\alpha}$ mapping into the origin and the points $\zeta = e^{i\beta}$, $\zeta = e^{i(\pi-\beta)}$ into A and B respectively, where $\sin\beta = k\sin\alpha$ and F is a suitably chosen constant.

10. If the circle $|\zeta|<r$ is mapped on a region B of the z-plane by the relation $z = \zeta + a_2\zeta^2 + a_3\zeta^3 + \dots$, prove that the area of B is

$$\pi\{r^2 + 2|a_2|^2 r^4 + 3|a_3|^2 r^6 + \dots\},$$

and is therefore greater than the area of the given circle. (Bieberbach)

11. Use the theorem of the preceding exercise to show that the problem of mapping a given region B in the ζ-plane on a circle in the z-plane is reducible to the problem of determining $a_2, a_3, \dots$, such that, if $z = \zeta + a_2\zeta^2 + \dots = f(\zeta)$,

$$\iint_{(B)} |f'(\zeta)|^2\, d\xi\, d\eta$$

is a minimum. Show that, by restricting the series to a few terms, the region B can be mapped on an almost circular area. (Bieberbach)

12. The transformation Ex. 3 is called a Möbius transformation. Prove that the inverse transformation $z = (-\delta\zeta+\beta)/(\gamma\zeta - \alpha)$ is also a Möbius transformation.

13. If successive Möbius transformations transform ζ into z_1 and z_1 into z, prove that ζ is transformed directly into z by a Möbius transformation. Deduce that all Möbius transformations form a group.

14. Prove that the Möbius transformation maps the whole z-plane (including $z = \infty$) on itself.

15. Prove that the Möbius transformation

$$\zeta = e^{i\alpha}\frac{z-c}{z-\bar{c}}, \quad c \text{ not real},$$

maps the half plane $y>0$ on the unit circle $|\zeta| < 1$ and maps $z=c$ on $\zeta = 0$.

16. Prove that the Möbius transformation

$$\zeta = e^{i\alpha}\frac{z-c}{1-\bar{c}z}, \quad |c|<1$$

maps the unit circle $|z| \leqslant 1$ on the unit circle $|\zeta| \leqslant 1$ and $z = c$ on $\zeta = 0$.

17. Prove that the transformation

$$\zeta = \frac{(z+1)^2 - i(z-1)^2}{(z+1)^2 + i(z-1)^2}$$

maps the semi-circle on the line joining $z = -1$ to $z = 1$ as diameter on the unit circle $|\zeta| \leqslant 1$.

18. Prove that the necessary and sufficient condition for the points z_1, z_2, z_3 to be collinear is that $(z_1 - z_3)/(z_2 - z_3)$ shall be real.

19. Prove that if the four points z_1, z_2, z_3, z_4 lie on a circumference, the double ratio

$$\frac{z_1 - z_3}{z_2 - z_3} : \frac{z_1 - z_4}{z_2 - z_4}$$

must be real.

20. Three points z_1, z_2, z_3 satisfy the conditions

$$z_1 + z_2 + z_3 = 0, \quad |z_1| = |z_2| = |z_3| = 1.$$

Prove that the points are the vertices of an equilateral triangle inscribed in the unit circle.

CHAPTER VI

STREAMING MOTIONS

6·0. Complex potential. Let ϕ, ψ be the velocity potential and stream function of the irrotational two-dimensional motion of an inviscid liquid. Then equating the velocity components,

$$\text{(1)} \qquad \frac{\partial \phi}{\partial x} = \frac{\partial \psi}{\partial y}, \quad \frac{\partial \phi}{\partial y} = -\frac{\partial \psi}{\partial x}.$$

We define the *complex potential* of the motion by the relation

$$w = \phi + i\psi.$$

We see from 5·30 that, on account of (1), w is a holomorphic function of the complex variable $z = x+iy$ in any region where ϕ and ψ are one-valued.

Conversely, if we assume for w any holomorphic function of z, the corresponding real and imaginary parts give the velocity potential and stream function of a possible two-dimensional irrotational motion, for they satisfy (1) and Laplace's equation.

Thus $\qquad w = z^2 \quad \text{gives} \quad \phi = x^2 - y^2, \quad \psi = 2xy,$

a motion which has already been discussed (4·70).

Since iw is likewise a function of z, it follows that $-\psi$ and ϕ are the velocity potential and stream function of another motion in which the streamlines and lines of equal velocity potential are interchanged.

It will be found that the mathematical analysis is very considerably simplified by working with the complex potential instead of ϕ and ψ separately. The simplification is of the same nature as that attained by using one vector equation instead of three cartesian equations. In two dimensions we work with one equation in z instead of two in x and y.

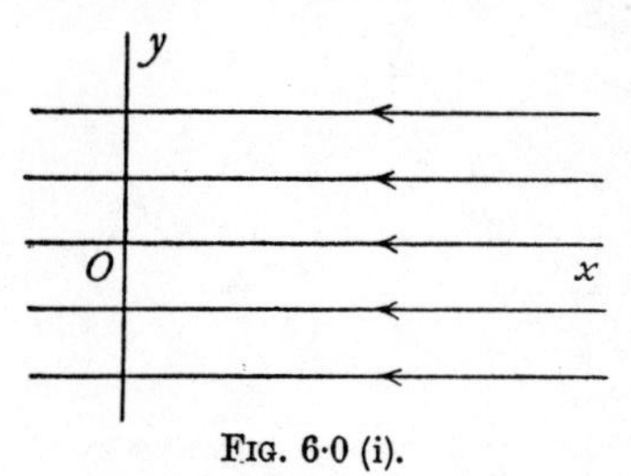

FIG. 6·0 (i).

The dimensions of the complex potential are those of a velocity multiplied by a length, i.e. $L^2 T^{-1}$.

We give a few illustrations in which U represents a velocity and a is a length both real.

$$\text{(i)} \qquad w = Uz.$$

Here $\psi = Uy$, and the motion is a uniform stream parallel to the negative direction of the x-axis.

(ii) $$w = \frac{Ua^2}{z},$$

$$\psi = -\frac{Ua^2}{r}\sin\theta = -\frac{Ua^2 y}{x^2+y^2}.$$

The streamlines ψ = constant are circles; all touch the x-axis at the origin. The motion is due to a doublet at the origin (see 8·23).

(iii) $$w = Ua\left(\frac{z}{a}\right)^{\pi/\alpha}.$$

Fig. 6·0 (ii).

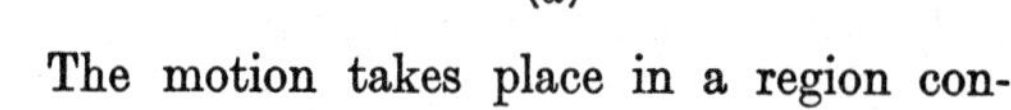

The motion takes place in a region contained by the arms of an angle α, and the streamlines are asymptotic to these arms. The special case $\alpha = \pi/2$ has been considered in 4·70.

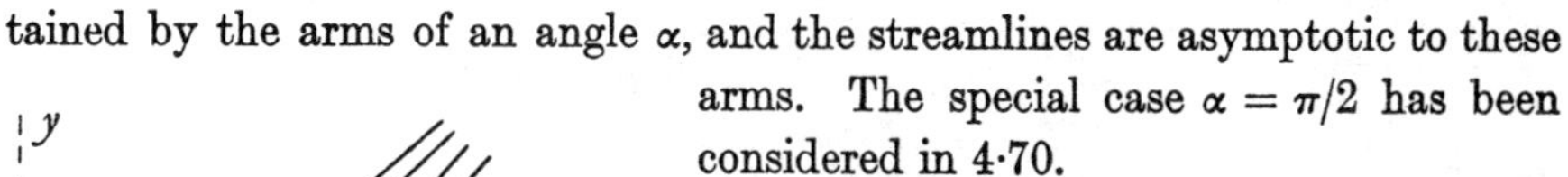

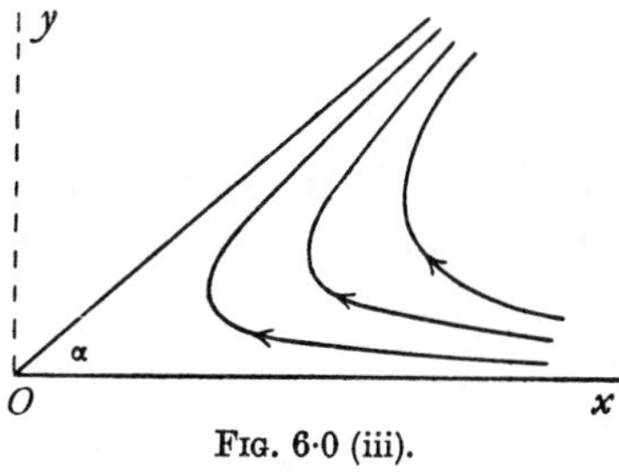

Fig. 6·0 (iii).

From the mathematical standpoint, the complex potential in the form $w = f(z)$ determines a mapping of the z-plane on the w-plane in which the streamlines of the motion in the z-plane map into the straight lines ψ = constant, parallel to the real axis in the w-plane. The determination of this mapping is the basic principle of the solution of hydrodynamical problems by means of the complex potential.

6·01. The complex velocity. From the complex potential $w = \phi + i\psi$ we get

$$\frac{\partial \phi}{\partial x} + i\frac{\partial \psi}{\partial x} = \frac{\partial w}{\partial x} = \frac{dw}{dz}\frac{\partial z}{\partial x} = \frac{dw}{dz}.$$

Now $$u = -\frac{\partial \phi}{\partial x}, \quad v = \frac{\partial \psi}{\partial x},$$

and therefore

(1) $$\upsilon = u - iv = -\frac{dw}{dz},$$

using the Greek letter υ (upsilon) to denote the combination $u - iv$, which we shall call the *complex velocity*. We note that the complex velocity is obtained directly from the complex potential as shown in (1). Graphically, the vector representing the complex velocity is the reflection, in the line through the point considered parallel to the x-axis, of the vector of the actual velocity.

The relation is shown in fig. 6·01.

It is very important to notice that $-dw/dz$ gives $u-iv = v$ and not $u+iv = \bar{v}$. If we want to obtain $u+iv$ we must change the sign of i throughout, so that $u+iv = -d\bar{w}/d\bar{z}$, where $\bar{w}$ is the conjugate complex function of $\bar{z}$. Thus, if $w = iz^2$, we shall have $\bar{w} = -i\bar{z}^2$, changing the sign of i throughout, and

$$v = u-iv = -2iz, \quad \bar{v} = u+iv = 2i\bar{z},$$

either of these leading to $u = 2y$, $v = 2x$.

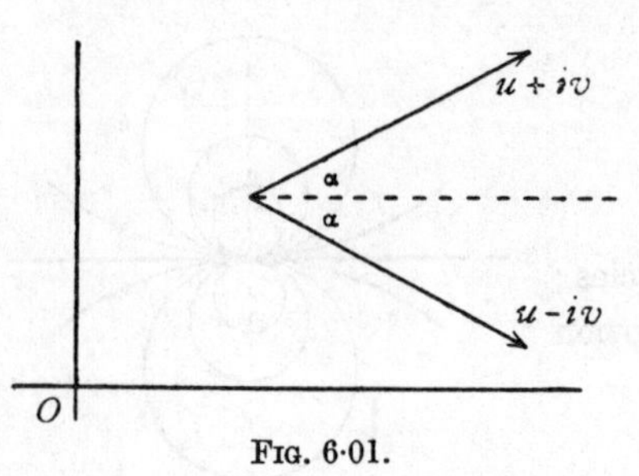

FIG. 6·01.

As a simple application, consider the uniform stream depicted in fig. 4·33 (iii). We have

$$-\frac{dw}{dz} = Q\cos\alpha - i\,Q\sin\alpha = Qe^{-i\alpha},$$

which gives $$w = -Qe^{-i\alpha}z.$$

6·02. Stagnation points. At a stagnation point the velocity is zero, and therefore the complex velocity vanishes. It follows that the stagnation points are given by

$$\frac{dw}{dz} = 0.$$

Hence, if $w = Ua\left(\frac{z}{a}\right)^{\pi/\alpha}$, the stagnation points are given by

$$z^{\frac{\pi}{\alpha}-1} = 0.$$

Thus, if $\pi < \alpha$, the stagnation point is at infinity. If $\pi > \alpha$, the stagnation point is at the origin. See fig. 6·0 (iii).

6·03. The speed. For the speed, we have

$$q = \sqrt{(u^2+v^2)} = \left|\frac{dw}{dz}\right|.$$

An alternative method is as follows:

$$q^2 = u^2+v^2 = (u-iv)(u+iv) = v\,.\,\bar{v} = \frac{dw}{dz}\cdot\frac{d\bar{w}}{d\bar{z}}.$$

To illustrate the calculation, suppose $w = 2z+3iz^2$, then $\bar{w} = 2\bar{z}-3i\bar{z}^2$,

$$q^2 = (2+6iz)(2-6i\bar{z}) = 4+36z\bar{z}+12i(z-\bar{z}) = 4+36(x^2+y^2)-24y.$$

There is a stagnation point given by $2+6iz = 0$, whence $z = i/3$, and the point is $(0, \frac{1}{3})$.

6·04. Inviscid flow past a wedge.

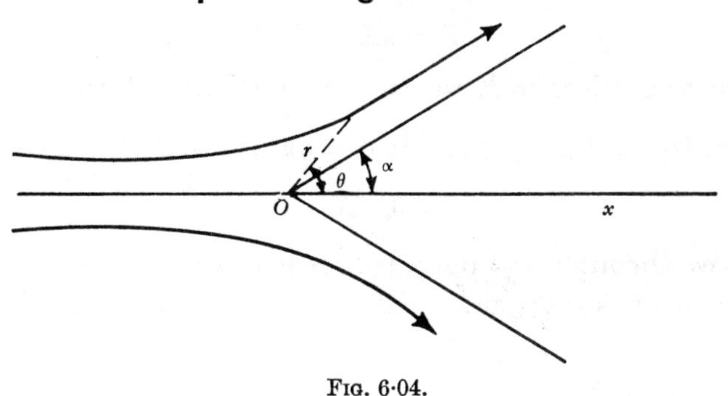

FIG. 6·04.

The complex potential

(1) $$w = -Ce^{-in\pi}z^{n+1}, \quad n = \alpha/(\pi-\alpha)$$

gives for the stream function

(2) $$\psi = -Cr^{n+1}\sin[n(\theta-\pi)+\theta]$$

which is zero when $\theta = \alpha$ and $\theta = 2\pi-\alpha$.

Thus w gives flow past a wedge of angle 2α whose section is symmetrically placed with respect to the x-axis.

The speed q is given by

$$q^2 = \frac{dw}{dz}\cdot\frac{d\bar{w}}{d\bar{z}} = C^2(n+1)^2r^{2n}.$$

If V is the speed when $r = 1$ we have $V = C(n+1)$ and

(3) $$q = Vr^n.$$

Since the velocity on the face of the wedge is along the face we see that (3) applies to give the velocity on the face at distance r from the vertex.

6·05. The equations of the streamlines. We here explain a method whereby the equations of the streamlines, namely ψ = constant, can often be deduced with the minimum labour. Let

$$w = \phi+i\psi.$$

Then $$\exp(\phi+i\psi) = \exp w.$$

This may be written

$$e^{\phi}\cos\psi+ie^{\phi}\sin\psi = X+iY,$$

where X, Y are the real and imaginary parts of $\exp w$.

Thus $$X = e^{\phi}\cos\psi, \quad Y = e^{\phi}\sin\psi.$$

Eliminating ϕ, we get

$$Y = X\tan\psi.$$

Thus when ψ = constant, we can write $\tan\psi = k$, and the streamlines are

$$Y = kX.$$

By attributing values to k, we get the individual streamlines. The lines corresponding to $k = 0$, $k = \infty$, i.e. to $\psi = n\pi$, $\psi = (2n+1)\frac{\pi}{2}$ are respectively

$$Y = 0, \quad X = 0.$$

6·10. Flow through an aperture. If w is a function of z, then z is a function of w, and it is sometimes useful to use this form of relationship between z and w.

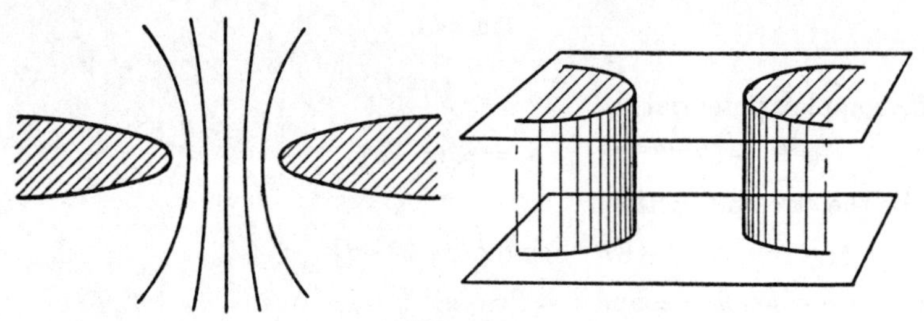

FIG. 6·10.

If we take $z = c\cosh w$, we get

$$x+iy = c\cosh\phi\cos\psi + ic\sinh\phi\sin\psi,$$

$$x = c\cosh\phi\cos\psi, \quad y = c\sinh\phi\sin\psi.$$

Eliminating ϕ, we get

$$\frac{x^2}{c^2\cos^2\psi} - \frac{y^2}{c^2\sin^2\psi} = 1,$$

so that the streamlines ψ = constant are confocal hyperbolas, whose real and transverse semi-axes are $c\cos\psi$, $c\sin\psi$, and whose foci are $(c, 0)$, $(-c, 0)$.

If we take the cylinder whose cross-section is one of these hyperbolas for a fixed boundary we obtain the pattern of liquid flow through the aperture so formed. As a limiting case, if we take the hyperbolic boundary quite flat ($\psi = 0, \pi$), we get the flow through an aperture of breadth $2c$ in a flat plate. This limiting case, however, cannot accord with physical reality since the speed is infinite at the edges.

To prove this, we have

$$\frac{1}{q^2} = \frac{dz}{dw}\frac{d\bar{z}}{d\bar{w}} = c^2\sinh w\sinh\bar{w} = \tfrac{1}{2}c^2(\cosh(w+\bar{w}) - \cosh(w-\bar{w}))$$

$$= \tfrac{1}{2}c^2(\cosh 2\phi - \cos 2\psi).$$

At the edges $(c, 0)$, $(-c, 0)$, we have $\phi = 0$, $\psi = 0$ or π. Thus $q^{-2} = 0$, and therefore the speed is infinite.

6·11. Circulation about an elliptic cylinder. Taking $z = c\cos w$, we get

$$x = c\cosh\psi\cos\phi, \quad y = -c\sinh\psi\sin\phi,$$

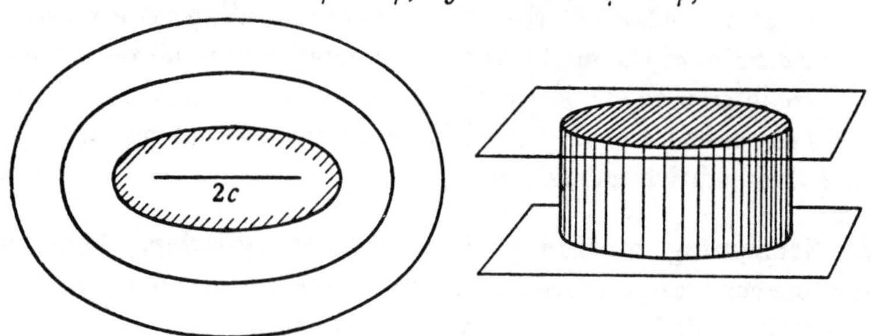

FIG. 6·11.

and eliminating ϕ,

$$\frac{x^2}{c^2\cosh^2\psi}+\frac{y^2}{c^2\sinh^2\psi}=1,$$

so that the streamlines are now confocal ellipses whose semi-axes are $c\cosh\psi$, $c\sinh\psi$.

If we take the cylinder represented by one of these ellipses as a fixed boundary, we have the case of liquid circulating about a fixed elliptic cylinder. As a limiting case, if we take as our fixed cylinder $\psi = 0$, the ellipse reduces to a line of length $2c$ and we get the case of liquid circulating about a flat plate of breadth $2c$, but here again the speed at the edges will be infinite, for

$$\frac{1}{q^2}=\frac{dz}{dw}\frac{d\bar{z}}{d\bar{w}}=\sqrt{c^2-z^2}\sqrt{c^2-\bar{z}^2},$$

which vanishes when $z = \pm c$.

Also when $|z|$ is large, we have approximately $q = 1/r$, where

$$r = |z| = |\bar{z}|.$$

Since $\cos iw = \cosh w$, we see that the formulae of this and the preceding section illustrate the interchange of streamlines and lines of equal velocity potential when we write iw for w (see 6·0).

6·21. The circle theorem. We now prove a general theorem * which will be of great use subsequently.

The circle theorem. Let there be irrotational two-dimensional flow of incompressible inviscid fluid in the z-plane. Let there be no rigid boundaries and let the complex potential of the flow be $f(z)$, where the singularities of $f(z)$ are all at a distance greater than a from the origin. If a circular cylinder, typified by its cross-section the circle C, $|z| = a$, be introduced into the field of flow, the complex potential becomes

$$(1) \qquad w = f(z)+\bar{f}\left(\frac{a^2}{z}\right).$$

* Milne-Thomson, *Proc. Camb. Phil. Soc.*, 36 (1940).

Proof. Since $\bar{z} = a^2/z$ on the circle, we see that w as given by (1) is purely real on the circle C and therefore $\psi = 0$. Thus C is a streamline.

If the point z is outside C, the point a^2/z is inside C, and vice-versa. Since all the singularities of $f(z)$ are by hypothesis exterior to C, all the singularities of $\bar{f}(a^2/z)$ are interior to C; in particular $\bar{f}(a^2/z)$ has no singularity at infinity, since $f(z)$ has none at $z = 0$. Thus w has exactly the same singularities as $f(z)$ and so all the conditions are satisfied. Q.E.D.

6·22. Streaming motion past a circular cylinder. Consider the stream whose complex potential is Uz. If we insert the cylinder $|z| = a$, by the circle theorem (6·21), the complex potential becomes

$$w = U\left(z+\frac{a^2}{z}\right), \tag{1}$$

which is therefore the complex potential of a circular cylinder placed in a stream whose velocity at infinity is U negatively along the x-axis. This system is generally referred to as a circular cylinder in a uniform stream. Actually the stream is disturbed by the presence of the cylinder and only remains uniform at a great distance from it. Accepting this conventional meaning, the terminology is convenient.

More generally, if we insert the cylinder in the uniform stream $Uze^{-i\alpha}$, the complex potential, again by the circle theorem, is

$$w = Uze^{-i\alpha}+\frac{Ua^2e^{i\alpha}}{z}. \tag{2}$$

If the centre of the cylinder is at the point z_0, a simple change of origin yields the complex potential

$$w = Uze^{-i\alpha}+\frac{Ua^2e^{i\alpha}}{z-z_0}. \tag{3}$$

Returning to (1), since $z = re^{i\theta}$, the stream function is

$$\psi = U\left(r\sin\theta-\frac{a^2}{r}\sin\theta\right) = Uy\left(1-\frac{a^2}{r^2}\right) = \psi_1+\psi_2$$

where $$\psi_1 = Uy, \quad \psi_2 = -\frac{a^2Uy}{x^2+y^2}.$$

Putting $$\psi_1 = mUa, \quad \psi_2 = -nUa,$$

we get $$y = ma, \quad x^2+\left(y-\frac{a}{2n}\right)^2 = \frac{a^2}{4n^2},$$

so that the lines corresponding to ψ_1 and ψ_2 are straight lines parallel to the x-axis and circles touching the x-axis at the origin. By giving m, n the values 0·1, 0·2, 0·3, ..., the streamlines can be readily plotted with the aid of Rankine's method (4·32).

The streamlines are symmetrical with respect to the y-axis, for changing the sign of x does not alter their equation. The streamlines above the x-axis are the reflections in that axis of the lines below it, as is obvious from symmetry.

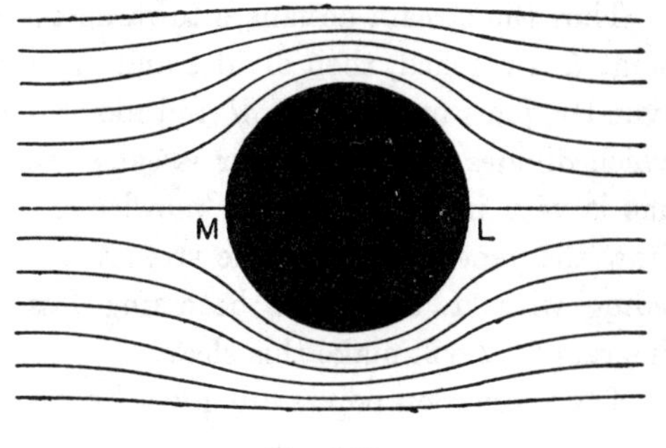

FIG. 6·22.

If the velocity U is reversed, the streamline pattern is unaltered.

Writing $\psi = kUa$, the equation of the streamlines is

$$ka = y\left(1-\frac{a^2}{r^2}\right),$$

so that, when $r\to\infty$, $y\to ka$, and therefore $y = ka$ is the asymptote of the streamline. Also if $k>0$, then $y>ka$, and therefore the streamline approaches its asymptote from above.

Again, consider the streamlines which are asymptotic to

$$y = ka, \quad y = (k+1)a.$$

Let y_1, y_2 be the respective ordinates of these lines as they pass over the cylinder, i.e. when $x = 0$.

Then

$$ka = y_1\left(1-\frac{a^2}{y_1^2}\right),$$

$$(k+1)a = y_2\left(1-\frac{a^2}{y_2^2}\right).$$

Subtracting and rearranging, we get

$$\frac{a}{y_2-y_1} = 1+\frac{a^2}{y_1 y_2}.$$

Since the term on the right is positive and greater than unity, we conclude that $y_2>y_1$, and that $y_2-y_1<a$. But at infinity the distance between these lines is a. Hence the lines come closer together as they pass over the cylinder. Since the same mass must cross every section of a stream tube, the velocity where a given line passes over the cylinder is greater than the velocity at infinity, and hence by Bernoulli's theorem, in the absence of extraneous forces, the pressure is less than the pressure at infinity.

6·23. The dividing streamline. In the flow past a cylinder the contour of the cylinder itself must form part of a streamline. Since the stream function is

$$\psi = Uy\left(1-\frac{a^2}{r^2}\right),$$

we see that the contour $r = a$ corresponds to $\psi = 0$.

The complete streamline $\psi = 0$ consists therefore of $y = 0$ and $r = a$, that is to say of the circle $r = a$ and that part of the x-axis which lies outside it, fig. 6·22.

Thus the stream advances towards the cylinder along the x-axis until the point L is reached, then divides and proceeds in opposite directions round the cylinder, joins up again at M and moves off along the x-axis. This streamline which divides on the contour is called the *dividing streamline.* The dividing line is very important, for a knowledge of its position at once enables us to draw the general form of the stream by successive lines at first nearly coinciding with it, and then becoming less and less curved. A study of the diagram 6·22 will make this clear.

The stagnation points are given by $dw/dz = 0$, that is, by

$$U\left(1-\frac{a^2}{z^2}\right) = 0,$$

whence $z = a$ or $z = -a$. These are the points L, M where the dividing line meets the cylinder, and we observe that, in accordance with the general property of intersecting streamlines (4·6), L, M are double points where the tangents are at right angles.

6·24. The pressure distribution on the cylinder.

To calculate the speed at the point $z = ae^{i\theta}$ on the cylinder, we have

$$\frac{dw}{dz} = U\left(1-\frac{a^2}{z^2}\right) = U(1-e^{-2i\theta}) = ie^{-i\theta}\,.\,2U\sin\theta,$$

and therefore

$$(1) \qquad q = 2U\sin\theta.$$

Thus q^2 is greatest when $\theta = \pm\pi/2$ and the speed at these points is $2U$ that is, twice the velocity of the stream at infinity.

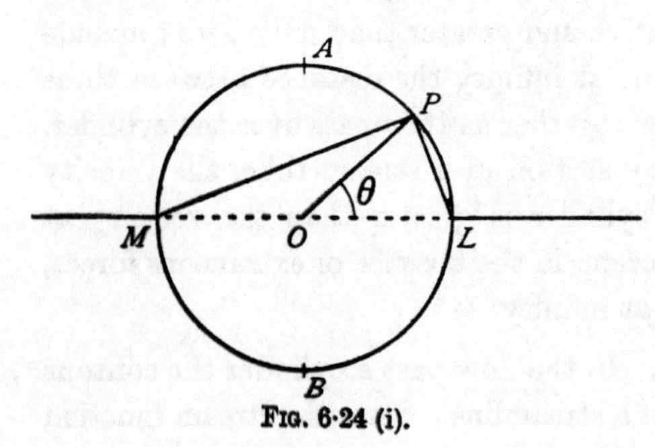

FIG. 6·24 (i).

Thus the speed is greatest at A and B where the diameter perpendicular to the stream at infinity meets the cylinder.

It also follows from (1) that the speed at P on the cylinder is proportional to the area of the triangle LPM.

If Π is the pressure at infinity, Bernoulli's theorem gives for the hydrodynamic pressure at a point of the cylinder

$$\frac{p}{\rho}+2U^2\sin^2\theta = \frac{\Pi}{\rho}+\tfrac{1}{2}U^2,$$

or
$$p - \Pi = \tfrac{1}{2}\rho U^2(1 - 4\sin^2\theta).$$

We can represent the pressure distribution on a polar diagram in which, taking the radius a to represent the pressure Π, the pressure at each point is measured by a length drawn along the radius through that point. With this representation, fig. 6·24 (ii), we see that at N_1, N_2, N_3, N_4, whose vectorial

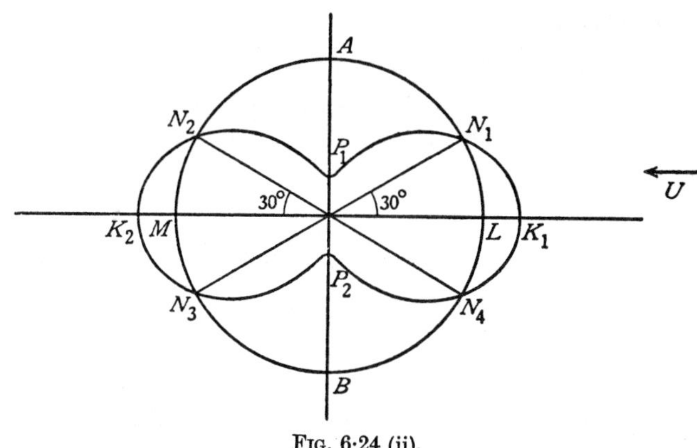

Fig. 6·24 (ii).

angles are 30°, 150°, 210°, 330°, the pressure is Π, along the arcs $N_4 L N_1$, $N_2 M N_3$ the pressure exceeds Π, the maximum excess being $\tfrac{1}{2}\rho U^2$ at L, M, while along the arcs $N_1 A N_2$, $N_3 B N_4$ the pressure is less than Π, the maximum defect being $\tfrac{3}{2}\rho U^2$. This pressure diagram is symmetrical, the pressure at the angles θ and $\theta + \pi$ being the same, so that there is no resultant force on the cylinder due to the hydrodynamical pressure. This result agrees with experiment only as regards the anterior portion $N_1 L N_4$, elsewhere the pressure is generally in defect (cf. fig. 1·93 (c)).

6·25. Cavitation. A fluid is presumed to be incapable of sustaining a negative pressure. In the relative motion of a solid boundary and fluid, the fluid will remain everywhere in contact with the boundary only so long as the pressure at every point of the boundary remains positive. Thus at points where the pressure vanishes a slight further diminution would render the pressure negative and a vacuum would tend to form. The formation of a vacuous space in a fluid is called *cavitation*. The phenomenon commonly occurs, for example, near the rapidly moving tips of propeller blades.

In the case of the flow past a circular cylinder, cavitation will tend to set in if the pressure is zero where it is least, i.e. at the sides ($\theta = \pm\tfrac{1}{2}\pi$). The condition for this is

$$\frac{\Pi}{\tfrac{1}{2}\rho U^2} = 3,$$

and if U exceeds the value given by this formula the liquid will cavitate at the sides of the cylinder.

6·26. Rigid boundaries and the circle theorem. The admissibility, in certain circumstances, of rigid boundaries when applying the circle theorem has been investigated by E. Levin.* His results may be summed up in the following theorem.

Theorem. Let B_1 be a given curve and B_2 its inverse with respect to the circumference $|z| = a$. Then if B_1 and B_2 are streamlines for the flow whose complex potential is $f(z)$, they are also streamlines for the flow whose complex potential is

$$(1) \qquad w = f(z) + \bar{f}\left(\frac{a^2}{z}\right).$$

Proof. Let z and ζ be inverse points so that

$$(2) \qquad z\bar{\zeta} = a^2.$$

Then if ζ lies on B_1, z lies on B_2, and if ζ lies on B_2, z lies on B_1. Since B_1 is a streamline (say $\psi = k_1$) and B_2 is a streamline (say $\psi = k_2$) for the flow whose complex potential is $f(z)$, when ζ is on B_1 we have

$$(3) \qquad k_1 = Im\, f(\zeta) = -Im\, \bar{f}(\bar{\zeta}) = -Im\, \bar{f}\left(\frac{a^2}{z}\right),$$

when z is on B_2, and when ζ is on B_2,

$$(4) \qquad k_2 = Im\, f(\zeta) = -Im\, \bar{f}(\bar{\zeta}) = -Im\, \bar{f}\left(\frac{a^2}{z}\right)$$

when z is on B_1.

Now for the flow (1)

$$\psi = Im\, f(z) + Im\, \bar{f}\left(\frac{a^2}{z}\right)$$

and therefore from (3) and (4)

$$(5) \qquad \psi = k_1 - k_2,\ z \text{ on } B_1, \text{ and } \psi = k_2 - k_1,\ z \text{ on } B_2.$$

Therefore B_1 and B_2 are streamlines for the flow (1). Q.E.D.

Observe that, in any case, as proved in 6·21, the circumference C of the circle $|z| \leqslant a$ is a streamline of the flow (1).

We shall assume that in the above theorem, as in 6·21, the singularities of $f(z)$ are all at a distance greater than a from the origin. We then have the following corollaries.

* *Q.A.M.* 12 (1954) 315, 316.

Corollary 1. When the flow whose complex potential is $f(z)$ is disturbed by introducing C the circle theorem still applies in the form (1) to give the flow when either B_1 or B_2 or both are replaced by rigid boundaries.

Corollary 2. If B_1 and B_2 coincide with a single curve B which is therefore its own inverse, the circle theorem still applies when C is introduced in the presence of the rigid boundary B.

Example. The circumference $B:(x-b)^2+y^2 = b^2-a^2$, $b>a$ is its own inverse with respect to the circumference $C: x^2+y^2 = a^2$, since if $P(x, y)$ lies on B, so does its inverse $(a^2x/(x^2+y^2),\ a^2y/(x^2+y^2))$.

If B disturbs the stream $Uze^{-i\alpha}$, the circle theorem gives for the flow the complex potential

$$f(z) = U(z-b)e^{-i\alpha}+U(b^2-a^2)e^{i\alpha}/(z-b).$$

Hence if we introduce the circle C, the complex potential becomes

$$w = f(z)+\bar{f}(a^2/z) \quad \text{or}$$

$$w = U(z-b)e^{-i\alpha}+\frac{U(b^2-a^2)e^{i\alpha}}{z-b}+U\left(\frac{a^2}{z}-b\right)e^{i\alpha}+\frac{U(b^2-a^2)e^{-i\alpha}}{(a^2/z)-b}$$

and this gives the flow when the cylinder whose cross-section is bounded by the orthogonal circumferences $x^2+y^2 = a^2$, $x^2+y^2-2bx+a^2 = 0$ disturbs the stream $Uze^{-i\alpha}$.

6·29. Application of conformal representation. Consider a mapping of the ζ-plane on the z-plane by

(1) $$z = f(\zeta),$$

such that the region R exterior to C in the ζ-plane maps into the region S exterior to A in the z-plane. Then the contour C maps into the contour A.

Let a fluid motion in the region R of the ζ-plane be given by the complex potential

(2) $$w(\zeta) = w = \phi+i\psi.$$

Then at corresponding points ζ and z given by (1), w and therefore ϕ and ψ take the same values.

Now C is a boundary and so a streamline, and therefore $\psi = k$, a constant,

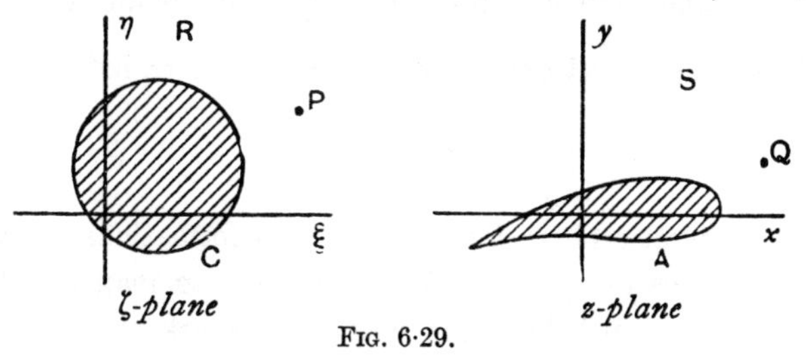

FIG. 6·29.

at all points of C. Since A corresponds point by point with C, $\psi = k$ at all points of A. Therefore A is a streamline in the motion given by (2) and (1) together in the z-plane.

The actual form of the complex potential in terms of z would be got by eliminating ζ between (1) and (2), but it is often preferable to look on ζ as a parameter and forgo the elimination. Thus to find the velocity at Q in the z-plane corresponding with P in the ζ-plane, we have

$$\frac{dw}{dz} = \frac{dw}{d\zeta} \times \frac{d\zeta}{dz},$$

and therefore

$$u_Q - iv_Q = (u_P - iv_P)/f'(\zeta). \tag{3}$$

Let q_1, q_2 be the speeds at P and Q respectively. Then

$$q_1^2 = q_2^2 \times \left|\frac{dz}{d\zeta}\right|^2. \tag{4}$$

Let dS_1, dS_2 be corresponding elements of area surrounding P and Q. Since the representation is conformal, we know that dS_1, dS_2 are similar and that the ratio of corresponding lengths in dS_1 and dS_2 is $|d\zeta/dz|$. Thus

$$\frac{dS_1}{dS_2} = \left|\frac{d\zeta}{dz}\right|^2,$$

and therefore $$q_1^2 dS_1 = q_2^2 dS_2,$$

and hence $$\int \tfrac{1}{2}\rho q_1^2 dS_1 = \int \tfrac{1}{2}\rho q_2^2 dS_2,$$

the integrals being taken over corresponding areas. But these integrals measure the kinetic energies of the liquid in the corresponding areas. Thus the kinetic energies of the two motions are the same.

We now see the application of these results. If we know the complex potential of a motion in the ζ-plane given by (2), and if we then transform to the z-plane by means of (1), we obtain the complex potential of a motion in the z-plane, the boundaries of the motion being the lines corresponding in the z-plane by means of (1) to the boundaries in the ζ-plane. The stream-lines correspond and the velocities at corresponding points are given by (3).

6·30. The Joukowski transformation. The transformation

$$z = Z + \frac{c^2}{4Z} \tag{1}$$

is one of the simplest and most important transformations of two-dimensional motions. By means of this transformation we can map the Z-plane on the z-plane, and vice versa. We begin with the remark that when $|z|$ is large, we have $Z = z$ nearly, so that the distant parts of the two planes correspond unaltered. Thus a uniform stream at infinity in the z-plane will correspond to a uniform stream of the same strength and direction in the Z-plane.

Let us now enquire into the transformation of circles in the Z-plane whose centres are at the origin.

We first note that if L, M are the points $Z = \frac{1}{2}c$, $Z = -\frac{1}{2}c$, the corresponding points in the z-plane are $z = c$, $z = -c$, say S, H.

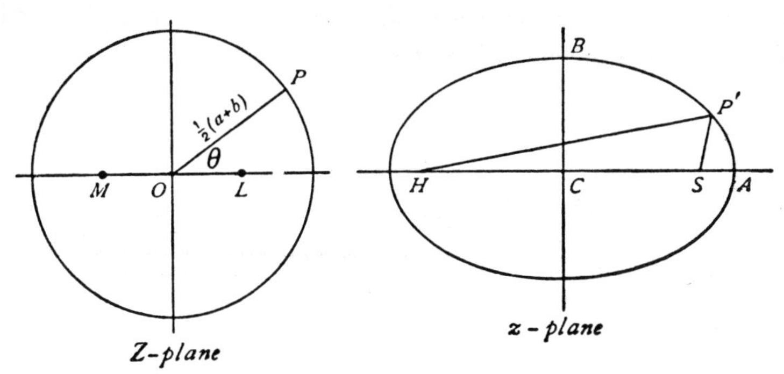

FIG. 6·30 (i).

Let the point P' in the z-plane correspond to the point P in the Z-plane on the circle $|Z| = \frac{1}{2}(a+b)$ and suppose $a^2-b^2 = c^2$. The transformation then gives

$$Z+\frac{c^2}{4Z}-c = z-c, \quad Z+\frac{c^2}{4Z}+c = z+c.$$

Hence
$$P'S = |z-c| = \left|\frac{(Z-\frac{1}{2}c)^2}{Z}\right| = \frac{2PL^2}{a+b},$$

$$P'H = |z+c| = \left|\frac{(Z+\frac{1}{2}c)^2}{Z}\right| = \frac{2PM^2}{a+b},$$

and therefore
$$SP'+HP' = 2(PL^2+PM^2)/(a+b).$$

But since OP is a median of the triangle MPL,

$$PL^2+PM^2 = 2(OL^2+OP^2)$$

and therefore
$$SP'+HP' = 2a,$$

so that P' describes an ellipse whose foci are S and H, and whose major axis is $2a$.

Now, in the ellipse, if B is an end of the minor axis and C the centre,

$$CB^2 = SB^2-CS^2 = b^2,$$

therefore the semi-axes of the ellipse are a, b.

Thus concentric circles with centre at the origin in the Z-plane map into confocal ellipses in the z-plane.

In particular, if we take $b = 0$, the circle $|Z| = \frac{1}{2}a$ maps * into the straight line SH joining the foci, for the minor axis of the corresponding ellipse is then zero, and $a = c$.

This result is easy to establish analytically, for any point on the circle is $Z = \frac{1}{2}ae^{i\theta}$, and therefore $z = \frac{1}{2}ae^{i\theta} + \frac{1}{2}ae^{-i\theta} = a\cos\theta$.

Hence as θ goes through the values $0, \pi/2, \pi, 3\pi/2, 2\pi$, z goes through the values $a, 0, -a, 0, a$, and as P describes the semicircle LDM, P' describes the line SH, and when P completes the circumference along the semicircle MEL, P' moves back along the line HS.

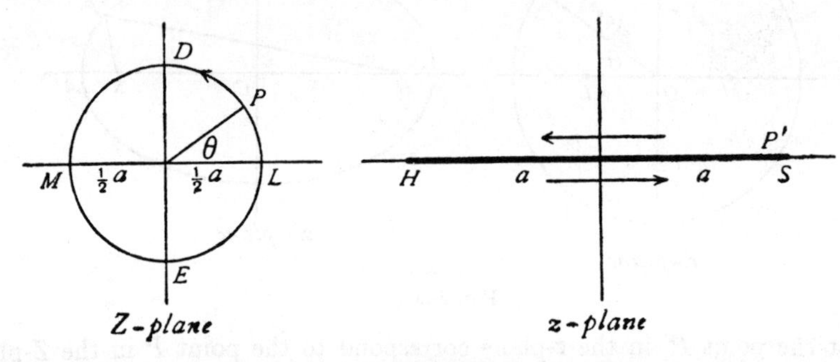

FIG. 6·30 (ii).

Now consider the inverse transformation which gives Z in terms of z. From (1),

$$Z^2 - zZ + \tfrac{1}{4}c^2 = 0,$$

$$Z = \tfrac{1}{2}(z \pm \sqrt{z^2 - c^2}). \tag{2}$$

The positive sign outside the square root means that the value of $\sqrt{(z^2 - c^2)}$ is to be taken which becomes real and positive when z is on the positive part of the real axis which lies *outside* the ellipse. When $|z|$ is very large, we have from (2), $Z = z$ or $Z = 0$, approximately, according as the positive or negative sign is taken. Therefore, if we take the positive sign for the square root, the points *outside* the ellipse in the z-plane will map into the points *outside* the circle in the Z-plane.

Therefore the transformation

$$Z = \tfrac{1}{2}(z + \sqrt{z^2 - c^2}), \quad c^2 = a^2 - b^2$$

maps the region outside the ellipse of semi-axes a, b in the z-plane on the region outside the circle of radius $\frac{1}{2}(a+b)$ in the Z-plane.

---

* In the present case $dz/dZ = 0$ at L, M. We must therefore suppose the circle to be indented at these points as explained in 5·63.

6·31. The flow past an elliptic cylinder. If we take in the Z-plane of fig. 6·30 (i) a stream U which makes an angle α with the real axis, the complex potential, from 6·22 (2), is

$$w = U\left(Ze^{-i\alpha} + \frac{(a+b)^2 e^{i\alpha}}{4Z}\right).$$

The region outside the circle is mapped on the region outside an ellipse in the z-plane whose semi-axes are a, b, centre at the origin and axis a along the x-axis, by the transformation

$$Z = \tfrac{1}{2}[z + \sqrt{(z^2 - c^2)}], \quad c^2 = a^2 - b^2,$$

and therefore the complex potential for the flow past an elliptic cylinder is

$$w = \tfrac{1}{2}U\left[e^{-i\alpha}\{z + \sqrt{(z^2-c^2)}\} + \frac{(a+b)^2 e^{i\alpha}}{z + \sqrt{(z^2-c^2)}}\right],$$

and since
$$[z + \sqrt{(z^2-c^2)}]^{-1} = \frac{1}{c^2}[z - \sqrt{(z^2-c^2)}],$$

$$w = \tfrac{1}{2}U(a+b)\left[\frac{e^{-i\alpha}(z + \sqrt{(z^2-c^2)})}{a+b} + \frac{e^{i\alpha}(z - \sqrt{(z^2-c^2)})}{a-b}\right],$$

and this is the solution of the problem.

In the above form the complex potential does not readily lend itself to detailed description of the flow. To simplify the treatment we shall now introduce elliptic coordinates.

6·32. Elliptic coordinates. Let

(1) $$z = c\cosh\zeta,$$

where
$$\zeta = \xi + i\eta.$$

Then
$$x + iy = c\cosh(\xi + i\eta) = c\cosh\xi\cos\eta + ic\sinh\xi\sin\eta,$$

so that

(2) $$x = c\cosh\xi\cos\eta, \quad y = c\sinh\xi\sin\eta,$$

and therefore

(3) $$\frac{x^2}{c^2\cosh^2\xi} + \frac{y^2}{c^2\sinh^2\xi} = 1,$$

(4) $$\frac{x^2}{c^2\cos^2\eta} - \frac{y^2}{c^2\sin^2\eta} = 1.$$

From (3), it appears that if ξ has the constant value ξ_0, the point (x, y) lies on an ellipse whose semi-axes a, b are given by

(5) $$a = c\cosh\xi_0, \quad b = c\sinh\xi_0,$$

and therefore $a^2 - b^2 = c^2$.

The ellipses corresponding to constant values * of ξ are therefore confocal, the distance between the foci being $2c$. The curves (4) corresponding to constant values of η are hyperbolas confocal with one another and with the ellipses.

Now, through any point of the plane we can draw two conics of a confocal system, one an ellipse and the other a hyperbola. On the ellipse ξ has a constant value, on the hyperbola η is constant. If we know these values of ξ, η, the conics can be drawn and by their intersections fix the point. For this reason the parameters ξ, η are called *elliptic coordinates*.

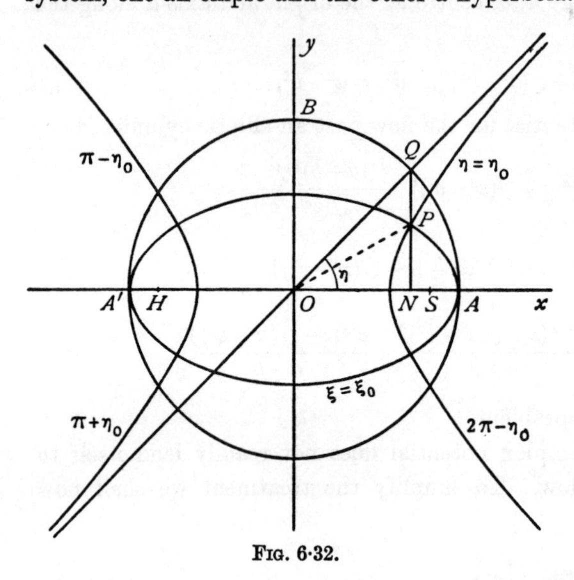

FIG. 6·32.

If we fix our attention on the ellipse $\xi = \xi_0$, we see from (3) that η is the eccentric angle of the point (x, y) on the ellipse. The geometrical interpretation is shown in fig. 6·32.

In this figure AA' is the major axis of the ellipse $\xi = \xi_0$, S, H are the foci. The confocal hyperbola which passes through P on the ellipse is also shown. On AA' as diameter, the auxiliary circle of the ellipse is drawn.

The ordinate PN meets this circle in Q. The angle $QON = \eta$.

For if a, b are the semi-axes of the ellipse, comparison of (2) and (5) shows that η is the eccentric angle of P. But

$$x = ON = OQ \cos QON = a \cos QON$$

and therefore the result follows.

We can now see that, if $\eta = \eta_0$ on the hyperbolic branch which lies in the first quadrant, the values of η on the branches of the same hyperbola which lie in the second, third, and fourth quadrants are $\pi - \eta_0$, $\pi + \eta_0$, $2\pi - \eta_0$ respectively.

* In what follows we shall let ξ_0 correspond to the whole ellipse, and take $0 \leqslant \xi < \infty$, $0 \leqslant \eta \leqslant 2\pi$. Another possible interpretation is $\xi > 0$ for $y > 0$ and $\xi < 0$ for $y < 0$. The corresponding ranges are then $-\infty < \xi < +\infty$, $0 \leqslant \eta \leqslant \pi$.

It also appears from (4) that the line

$$\frac{y}{x} = \tan \eta$$

is an asymptote of the hyperbola through P. This asymptote is the radius OQ.

In order to complete our description of the ellipse $\xi = \xi_0$, we use equations (5). These give

$$(6) \qquad a+b = c\,(\cosh \xi_0 + \sinh \xi_0) = ce^{\xi_0},$$

$$a-b = c\,(\cosh \xi_0 - \sinh \xi_0) = ce^{-\xi_0}.$$

By division, we obtain

$$e^{2\xi_0} = \frac{a+b}{a-b},$$

and therefore

$$\xi_0 = \tfrac{1}{2} \log \frac{a+b}{a-b}.$$

This equation determines the parameter ξ_0 in terms of the semi-axes a, b.

Lastly, we note that the foci $(c, 0)$, $(-c, 0)$ correspond to $\xi = 0$, $\eta = 0$; $\xi = 0$, $\eta = \pi$, as is clear from (2).

6·33. Application of elliptic coordinates to the streaming past an ellipse. The complex potential was found in 6·31. If we put $z = c \cosh \zeta$, we get $\sqrt{(z^2 - c^2)} = c \sinh \zeta$, and therefore

$$z + \sqrt{(z^2 - c^2)} = c\,(\cosh \zeta + \sinh \zeta) = ce^{\zeta},$$

$$z - \sqrt{(z^2 - c^2)} = c\,(\cosh \zeta - \sinh \zeta) = ce^{-\zeta}.$$

Also, on the ellipse $\xi = \xi_0$, we have, from 6·32 (6),

$$a+b = ce^{\xi_0}, \quad a-b = ce^{-\xi_0}.$$

Therefore

$$w = \tfrac{1}{2} U\,(a+b)\,[e^{-i\alpha+\zeta-\xi_0} + e^{i\alpha-\zeta+\xi_0}].$$

$$(1) \qquad w = U\,(a+b) \cosh (\zeta - \xi_0 - i\alpha).$$

This expresses the complex potential for the streaming motion past an ellipse in terms of the elliptic coordinates.

If we put $\xi = \xi_0$, we obtain

$$w = U\,(a+b) \cosh i\,(\eta - \alpha)$$

$$= U\,(a+b) \cos (\eta - \alpha),$$

so that $\psi = 0$. Hence the ellipse $\xi = \xi_0$ forms part of the streamline $\psi = 0$, which is therefore the dividing streamline.

The stream function, from (1), is

$$\psi = U\,(a+b) \sinh (\xi - \xi_0) \sin (\eta - \alpha).$$

Hence the complete dividing line is given by

$$\sinh (\xi - \xi_0) = 0 \quad \text{and} \quad \sin (\eta - \alpha) = 0,$$

that is, by

$$\xi = \xi_0, \quad \eta = \alpha, \quad \eta = \alpha + \pi.$$

These last values of η correspond to branches of a hyperbola confocal with the ellipse and therefore cutting it orthogonally.

The line through O in the direction of the stream at infinity is an asymptote of the hyperbola.

The general form of the streamlines is shown in fig. 6·33. The asymptote to the dividing line is shown dotted. The dividing line intersects the cylinder at L, M, which are therefore stagnation points and consequently points of

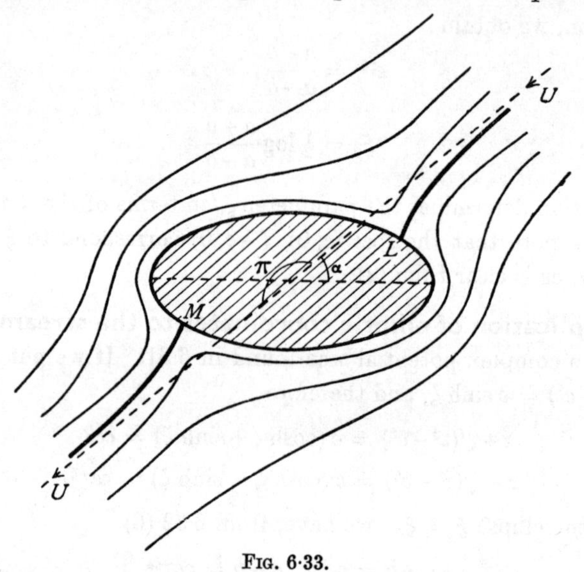

FIG. 6·33.

maximum pressure. It would therefore appear that the cylinder is subjected to a couple tending to set it broadside on to the stream. We shall calculate the magnitude of this couple in section 6·42.

For the velocity, we have

$$\frac{dw}{dz} = \frac{dw}{d\zeta}\frac{d\zeta}{dz} = \frac{U(a+b)\sinh(\zeta - \xi_0 - i\alpha)}{c \sinh \zeta},$$

and therefore at the stagnation points

$$\zeta - \xi_0 - i\alpha = 0 \text{ or } i\pi,$$

so that $\xi = \xi_0$, $\eta = \alpha$, or $\alpha + \pi$, giving the points L, M, already found from the dividing line.

Also $$q^2 = \frac{U(a+b)\sinh(\zeta - \xi_0 - i\alpha)}{c\sinh\zeta} \times \frac{U(a+b)\sinh(\bar{\zeta} - \xi_0 + i\alpha)}{c\sinh\bar{\zeta}}$$

$$= \frac{U^2(a+b)^2}{c^2}\frac{\cosh(\zeta + \bar{\zeta} - 2\xi_0) - \cosh(\zeta - \bar{\zeta} - 2i\alpha)}{\cosh(\zeta + \bar{\zeta}) - \cosh(\zeta - \bar{\zeta})}$$

$$= \frac{U^2(a+b)}{a-b}\frac{\cosh 2(\xi-\xi_0)-\cos 2(\eta-\alpha)}{\cosh 2\xi-\cos 2\eta}$$

$$= \frac{U^2(a+b)}{a-b}\frac{\sinh^2(\xi-\xi_0)+\sin^2(\eta-\alpha)}{\sinh^2\xi+\sin^2\eta}.$$

Since $\xi = 0$ only at a focus, it follows that the denominator cannot vanish and so the speed is never infinite.

The pressure distribution on the elliptic cylinder is found from Bernoulli's theorem which gives

$$\text{(2)} \qquad \frac{p}{\rho}+\frac{1}{2}\frac{U^2(a+b)}{a-b}\frac{1-\cos 2(\eta-\alpha)}{\cosh 2\xi_0-\cos 2\eta} = \frac{\Pi}{\rho}+\frac{1}{2}U^2,$$

where Π is the pressure at infinity. To find where the pressure is greatest and least, we have $dp/d\eta = 0$, which leads to

$$\sin 2(\eta-\alpha)\cosh 2\xi_0-\sin 2\eta+\sin 2\alpha = 0,$$

or

$$\sin(\eta-\alpha)\{\cos(\eta-\alpha)\cosh 2\xi_0-\cos(\eta+\alpha)\} = 0.$$

Now $\sin(\eta-\alpha) = 0$ gives the stagnation points where the pressure is greatest. The points of minimum pressure are therefore given by

$$\frac{\cos(\eta+\alpha)}{\cos(\eta-\alpha)} = \cosh 2\xi_0,$$

whence

$$\tan\eta\tan\alpha = \frac{1-\cosh 2\xi_0}{1+\cosh 2\xi_0} = -\tanh^2\xi_0 = -\frac{b^2}{a^2},$$

from 6·32. If P denotes a point given by this equation, the result means that the tangent at P is parallel to the normal at a stagnation point.

If we substitute $\tan\eta = -b^2\cot\alpha/a^2$ in (2), we get, after some reduction, for the minimum pressure the value

$$\Pi+\tfrac{1}{2}\rho U^2\left\{1-(a+b)^2\left(\frac{\cos^2\alpha}{a^2}+\frac{\sin^2\alpha}{b^2}\right)\right\}.$$

The condition that there shall be no cavitation is therefore

$$\Pi > \tfrac{1}{2}\rho U^2\left\{(a+b)^2\left(\frac{\cos^2\alpha}{a^2}+\frac{\sin^2\alpha}{b^2}\right)-1\right\}.$$

6·34. Flow past a plate. If $b = 0$, our ellipse degenerates into the line joining the foci, namely $\xi_0 = 0$, and therefore $a = c$. Hence for the flow past a plate inclined at angle α to the stream, we have

$$w = Ua\cosh(\zeta-i\alpha).$$

The stagnation points still lie on the hyperbolic branches

$$\eta = \alpha, \quad \eta = \pi+\alpha.$$

The speed becomes infinite at the edges of the plate, so that the solution cannot represent the complete motion past an actual plate.

In terms of z, we have

$$w = U(z \cos \alpha - i\sqrt{(z^2 - a^2)} \sin \alpha).$$

When the plate is perpendicular to the stream,

$$w = -iU\sqrt{(z^2 - a^2)}.$$

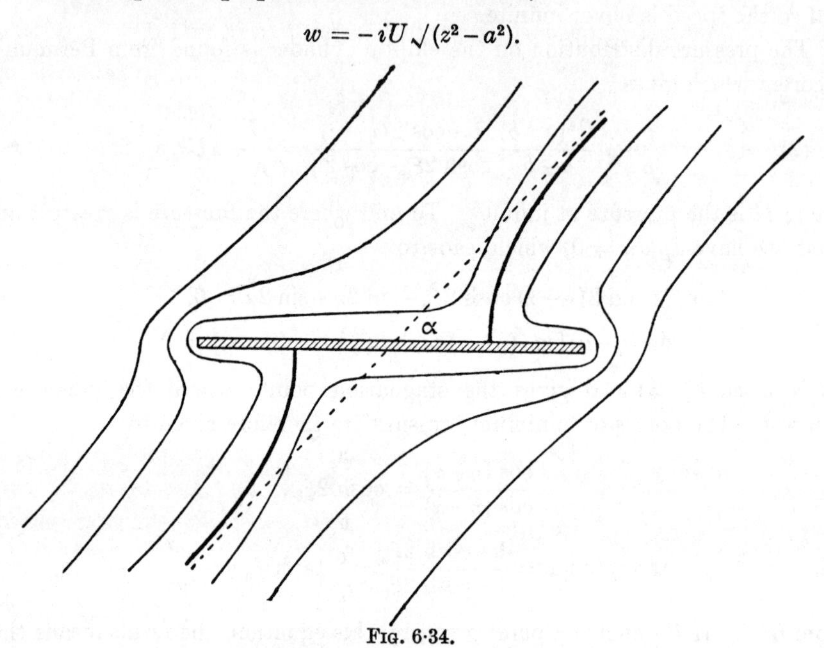

FIG. 6·34.

6·35. A general method. Consider a cylinder of cross-section C placed in the stream $Uze^{-i\alpha}$. Analogously to elliptic coordinates, let

$$z = f(\zeta) \tag{1}$$

define a system of coordinates (ξ, η) in which the curve C is given by $\xi = \xi_0$. Then on C, $\bar{\zeta} = 2\xi_0 - \zeta$. Thus the complex potential

$$w = F(\zeta) + \bar{F}(2\xi_0 - \zeta) \tag{2}$$

is purely real on C, which is therefore the streamline $\psi = 0$.

Now the complex potential of the uniform stream is

$$Uze^{-i\alpha} = Uf(\zeta)e^{-i\alpha} = F_1(\zeta) + F_2(\zeta), \tag{3}$$

where we suppose $F_2(\zeta)$ to contain only terms which tend to zero at infinity. If we can arrange this split into the sum of two functions in such a way that $\bar{F}_1(2\xi_0 - \zeta)$ also tends to zero at infinity and has no singularities in the flow, our streaming problem is solved by

$$w = F_1(\zeta) + \bar{F}_1(2\xi_0 - \zeta). \tag{4}$$

Thus for the ellipse we have $z = c \cosh \zeta$, and so (3) gives

$$F_1(\zeta) = \tfrac{1}{2}Uce^{\zeta - i\alpha}, \quad F_2(\zeta) = \tfrac{1}{2}Uce^{-\zeta - i\alpha},$$

and therefore

$$w = \tfrac{1}{2}Uce^{\zeta - i\alpha} + \tfrac{1}{2}Uce^{2\xi_0 - \zeta + i\alpha},$$

which is 6·33 (1).

To determine a coordinate system of the required type, let C be given by the parametric equations $x = f_1(t)$, $y = f_2(t)$. Writing $i(\xi_0 - \zeta)$ instead of t, we get

(5) $$z = f_1(i\xi_0 - i\zeta) + if_2(i\xi_0 - i\zeta),$$

which has the required property.

Thus in the case of the ellipse $x = a \cos t$, $y = b \sin t$, we get

$$z = (a \cosh \xi_0 - b \sinh \xi_0) \cosh \zeta + (b \cosh \xi_0 - a \sinh \xi_0) \sinh \zeta,$$

which reduces to the standard elliptic coordinates by taking $a = c \cosh \xi_0$ $b = c \sinh \xi_0$.

The foregoing remarks embody a principle whose general application is not confined to the particular mode of coordinate expression here used to illustrate it.

6·41. Theorem of Blasius. Let a fixed cylinder be placed in a liquid which is moving steadily and irrotationally. Let X, Y and M be the components along the axes and the moment about the origin of the pressure thrusts on the cylinder. Then, neglecting external forces,

$$X - iY = \tfrac{1}{2}i\rho \int_{(C)} \left(\frac{dw}{dz}\right)^2 dz,$$

$$M = \text{real part of } -\tfrac{1}{2}\rho \int_{(C)} z \left(\frac{dw}{dz}\right)^2 dz,$$

where w is the complex potential, ρ the density, and the integrals are taken round the contour of the cylinder.

Proof. For the action on the arc ds at P we have

$dX = -p\, dy$, $dY = p\, dx$,

$dM = p(x\, dx + y\, dy)$. Thus

(1) $d(X - iY) = -ip\, d\bar{z}$,

(2) $dM =$ real part of $pz\, d\bar{z}$.

From the pressure equation,

(3) $p = a - \frac{1}{2}\rho q^2$,

where a is a constant.

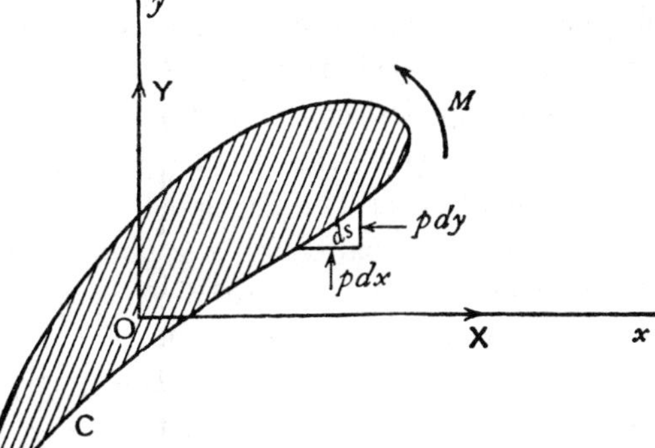

FIG. 6·41.

Since a constant pressure can have no resultant effect, we can take

$$p = -\tfrac{1}{2}\rho q^2 = -\tfrac{1}{2}\rho\,\frac{dw}{dz}\frac{d\bar{w}}{d\bar{z}}, \text{ so that}$$

$$d(X-iY) = \tfrac{1}{2}i\rho\,\frac{dw}{dz}\,d\bar{w}, \quad dM = \text{real part of } -\tfrac{1}{2}\rho z\,\frac{dw}{dz}\,d\bar{w}.$$

But on C, ψ = constant and therefore $d\bar{w} = dw$, so that

$$d(X-iY) = \tfrac{1}{2}i\rho\left(\frac{dw}{dz}\right)^2 dz, \quad dM = \text{real part of } -\tfrac{1}{2}\rho z\left(\frac{dw}{dz}\right)^2 dz,$$

and the theorem follows by integration round C. Q.E.D.

It is sometimes convenient to write

$$M+iN = -\tfrac{1}{2}\rho\int_{(C)} z\left(\frac{dw}{dz}\right)^2 dz,$$

where N is the imaginary part of the integral.

When the motion is not steady the pressure equation contains the term $\rho\,\partial\phi/\partial t$, and therefore to the expressions for the force and moment we must add

$$\int -i\rho\,\frac{\partial\phi}{\partial t}\,d\bar{z}, \quad \int z\rho\,\frac{\partial\phi}{\partial t}\,d\bar{z}.$$

Now on the cylinder ψ is an instantaneous constant, say $c(t)$, and therefore

$$\frac{\partial\phi}{\partial t} = \frac{\partial\bar{w}}{\partial t} + ic'(t).$$

Thus the above expressions for $X-iY$, $M+iN$ are increased by the respective terms,

$$-i\rho\,\frac{\partial}{\partial t}\int\bar{w}\,d\bar{z}, \quad \rho\,\frac{\partial}{\partial t}\int[\bar{w}+ic(t)]z\,d\bar{z}.$$

The theorem of Blasius in the form just enunciated refers to integrals round the contour of the cylinder. This contour can be enlarged to any extent, provided that we do not pass over any singularity of the integrand. Such singularities can only occur in hydrodynamics when the fluid contains sources or vortices. We shall deal with these matters later. At present we shall make some simple applications of the theorem to cases of streaming.

In the case where the cylinder is in uniform motion in liquid otherwise at rest the forces can still be calculated by the theorem of Blasius, for the *dynamical* conditions are unaltered if a *uniform* velocity equal and opposite to that of the cylinder is superposed in the whole system. The cylinder is then reduced to rest and the liquid streams past it.

We can also obtain formulae in terms of the stream function ψ, which exists even when the motion is rotational. In fact, using 5·33 (1),

$$u - iv = -\frac{\partial\psi}{\partial y} - i\,\frac{\partial\psi}{\partial x} = -2i\,\frac{\partial\psi}{\partial z} \tag{4}$$

and therefore, taking $a=0$, in (3) we have

$$p\,d\bar{z} = -2\rho\frac{\partial\psi}{\partial z}\frac{\partial\psi}{\partial\bar{z}}d\bar{z}. \tag{5}$$

Now $\psi=\psi(z,\bar{z})$ is constant on C and therefore

$$\frac{\partial\psi}{\partial z}dz+\frac{\partial\psi}{\partial\bar{z}}d\bar{z} = 0 \text{ on } C. \tag{6}$$

Combining (1), (2), (5), and (6) we have

$$X-iY = -2i\rho\int_{(C)}\left(\frac{\partial\psi}{\partial z}\right)^2 dz,\quad M = \text{real part of } 2\rho\int_{(C)} z\left(\frac{\partial\psi}{\partial z}\right)^2 dz. \tag{7}$$

Further observe that, although ψ is a function of both z and $\bar{z}$, on C the variable $\bar{z}$ is a function of z, and therefore by first eliminating $\bar{z}$, the residue theorem, and change of contour may be applied to (7).

6·42. The action of a uniform stream on an elliptic cylinder. Referring to 6·33, we see that the complex potential is of the form

$$w = cA\cosh(\zeta-\zeta_0),\quad z = c\cosh\zeta,$$

where

$$cA = U(a+b),\quad \zeta_0 = \xi_0+i\alpha.$$

The force and moment on the cylinder are given by the theorem of Blasius. Now

$$\frac{dw}{dz} = \frac{dw}{d\zeta}\div\frac{dz}{d\zeta} = \frac{A\sinh(\zeta-\zeta_0)}{\sinh\zeta}$$

$$= A\left(\cosh\zeta_0-\sinh\zeta_0\frac{z}{\sqrt{(z^2-c^2)}}\right).$$

We shall integrate round a circle, enclosing the cylinder, whose radius is so large that dw/dz can be expanded in a convergent series of powers of $1/z$. The only contributions to the integrals will then arise from the coefficient of $1/z$ in the integrand (5·57). Then

$$\frac{z}{\sqrt{(z^2-c^2)}} = 1+\frac{c^2}{2z^2}+\dots$$

and therefore

$$\frac{dw}{dz} = A\left(e^{-\zeta_0}-\frac{c^2\sinh\zeta_0}{2z^2}+\dots\right),$$

so that

$$\left(\frac{dw}{dz}\right)^2 = A^2\left(e^{-2\zeta_0}-\frac{c^2e^{-\zeta_0}\sinh\zeta_0}{z^2}+\dots\right).$$

Hence $X-iY = 0$, or $X=0$, $Y=0$, and

$$M+iN = -\tfrac{1}{2}\rho\times 2\pi i(-c^2A^2e^{-\zeta_0}\sinh\zeta_0) = \tfrac{1}{2}i\pi\rho c^2A^2(1-e^{-2\zeta_0})$$
$$= \tfrac{1}{2}i\pi\rho c^2A^2(1-e^{-2\xi_0}(\cos 2\alpha - i\sin 2\alpha)).$$

Therefore

$$M = -\pi\rho c^2A^2e^{-2\xi_0}\sin\alpha\cos\alpha.$$

Now, from 6·33,

$$A^2 = \frac{U^2(a+b)^2}{c^2} = U^2\left(\frac{a+b}{a-b}\right) = U^2e^{2\xi_0}.$$

Hence $$M = -\pi\rho(a^2-b^2)\,U^2 \sin\alpha\cos\alpha.$$

The negative value indicates that the cylinder is acted upon by a couple which tends to set it broadside on to the stream. An inspection of fig. 6·33 shows the reason for the existence of this couple, for the stagnation points or points of maximum pressure are unsymmetrically situated.

The result is typical of any elongated body in a stream, and affords an explanation of the behaviour of a boat drifting in a stream.

We note that the couple vanishes if $a = b$, the case of a circular cylinder, as is indeed obvious.

The couple also vanishes if $\alpha = 0$, that is, if the major axis of the ellipse points upstream. The slightest deviation from this orientation of the major axis calls the disturbing couple into play, with increasing moment until $\alpha = \frac{1}{4}\pi$. Thus an elliptic cylinder with its major axis pointing upstream is unstable. This phenomenon is well exemplified in the case of a ship which needs the continual attention of the helmsman to maintain the course.

The couple also vanishes when $\alpha = \frac{1}{2}\pi$, that is, when the cylinder is broadside on, but this position is stable, for a deviation from it calls into play a restoring couple whose moment increases with the deviation.

6·50. Coaxal coordinates. Let A, B be the points $(c, 0)$, $(-c, 0)$ respectively. Taking the x-axis as initial line and A and B for poles, the coordinates of any point P are (r_1, θ_1), (r_2, θ_2) respectively. The numbers r_1, r_2 are the *bipolar coordinates* of P.

If P describes a circle passing through A and B, then $\angle APB = \theta_1 - \theta_2$ is constant. Such circles form a coaxal system. The orthogonal system has A and B for limiting points, and when P describes a circle of this system r_2/r_1 is constant.

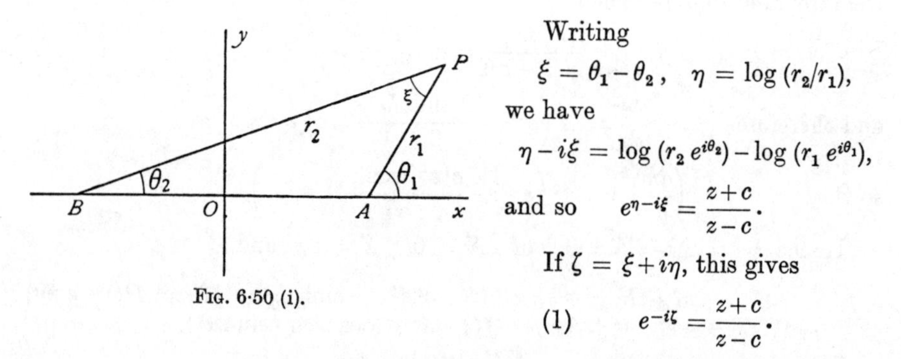

FIG. 6·50 (i).

Writing
$$\xi = \theta_1 - \theta_2, \quad \eta = \log(r_2/r_1),$$
we have
$$\eta - i\xi = \log(r_2\, e^{i\theta_2}) - \log(r_1\, e^{i\theta_1}),$$
and so $$e^{\eta - i\xi} = \frac{z+c}{z-c}.$$

If $\zeta = \xi + i\eta$, this gives

(1) $$e^{-i\zeta} = \frac{z+c}{z-c}.$$

The position of a point P is fixed if we know in which quadrant it lies, and the constant values of ξ, η on the circles which pass through it. Thus, just as we introduced elliptic coordinates, so we may appropriately call the ξ, η, defined by (1), *coaxal coordinates*.

From (1), we get

$$\frac{z}{c} = \frac{e^{-i\zeta}+1}{e^{-i\zeta}-1} = i \cot \tfrac{1}{2}\zeta,$$

so that

(2) $\quad z = ic \cot \tfrac{1}{2}\zeta,$

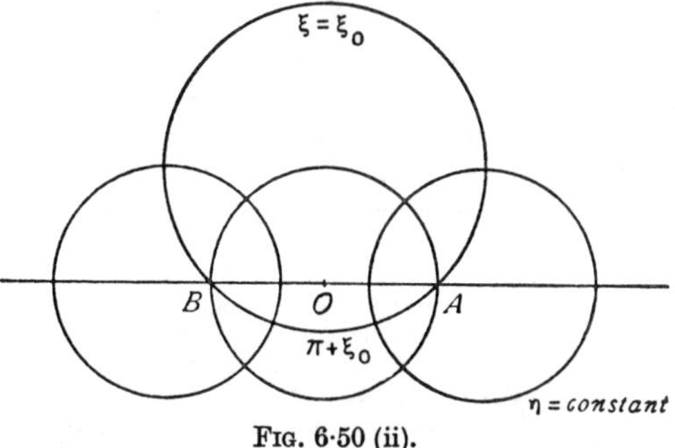

FIG. 6·50 (ii).

which may be compared with the equation defining elliptic coordinates.

The curve ξ=constant is a circle whose centre is the point $(0, c \cot \xi)$ and whose radius is $c \operatorname{cosec} \xi$.

The curve η=constant is a circle whose centre is the point $(c \coth \eta, 0)$ and whose radius is $c \operatorname{cosech} \eta$.

Since $\bar{z} = -ic \cot \tfrac{1}{2}\bar{\zeta}$, we get

$$\frac{x}{c} = \frac{i}{2}(\cot \tfrac{1}{2}\zeta - \cot \tfrac{1}{2}\bar{\zeta}), \quad \frac{y}{c} = \frac{1}{2}(\cot \tfrac{1}{2}\zeta + \cot \tfrac{1}{2}\bar{\zeta}).$$

Since $2 \sin \tfrac{1}{2}\zeta \sin \tfrac{1}{2}\bar{\zeta} = \cos i\eta - \cos \xi = \cosh \eta - \cos \xi$, we conclude that

$$\frac{x}{c} = \frac{\sinh \eta}{\cosh \eta - \cos \xi}, \quad \frac{y}{c} = \frac{\sin \xi}{\cosh \eta - \cos \xi}.$$

On the real axis $\xi = 0$, except for points between A and B for which $\xi = \pi$.

Observe that $\xi < \pi$ when $y > 0$ and that if $\xi = \xi_0$ on the arc of a circle through AB for which $y > 0$, then $\xi = \xi_0 + \pi$ on the arc of the same circle for which $y < 0$.

At infinitely distant points PA, PB are parallel, and $PA = PB$ nearly. Thus $\eta \to 0$, when $P \to \infty$, and $\xi \to 0$, or $\xi \to 2\pi$, according as $y > 0$ or $y < 0$.

At A and B, η is infinite.

6·51. Flow over a ditch or mound. The complex potential

$$w = U\frac{2ci}{n} \cot \frac{\zeta}{n}, \quad z = ic \cot \tfrac{1}{2}\zeta,$$

where n is real, makes $\psi = 0$, when $\xi = 0$ and when $\xi = \tfrac{1}{2}n\pi$, for in both these cases w is purely real.

Thus w gives the flow past a boundary which consists of the arc of a circle through A, B on which $\xi = n\pi/2$, and the part of the x-axis which lies outside this boundary.

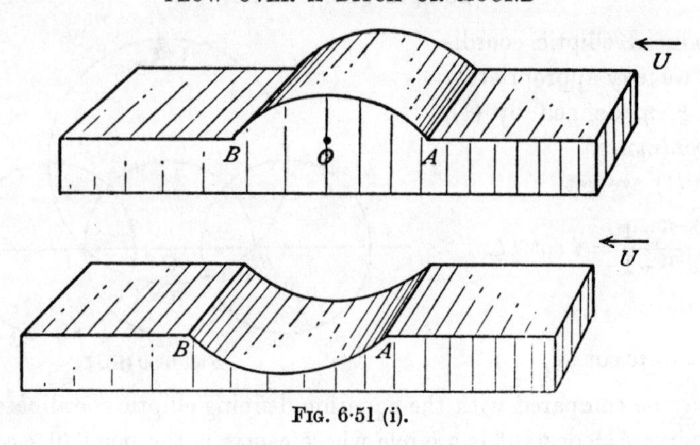

FIG. 6·51 (i).

To find the velocity, we have

$$u - iv = -\frac{dw}{dz} = -\frac{dw}{d\zeta}\frac{d\zeta}{dz}$$

$$= -\frac{4}{n^2}\left(\frac{\sin \frac{1}{2}\zeta}{\sin \frac{\zeta}{n}}\right)^2 U,$$

at infinity

$$\zeta \to 0,$$

$$u - iv \to -U,$$

so that there is a uniform stream. Again

$$q^2 = \frac{16U^2}{n^4}\left(\frac{\sin \frac{1}{2}\zeta \sin \frac{1}{2}\bar{\zeta}}{\sin \frac{\zeta}{n} \sin \frac{\bar{\zeta}}{n}}\right)^2 = \frac{16U^2}{n^4}\left(\frac{\cosh \eta - \cos \xi}{\cosh \frac{2\eta}{n} - \cos \frac{2\xi}{n}}\right)^2.$$

As we approach A, $\eta \to \infty$, and

$$q^2 = \frac{16U^2}{n^4} e^{2\left(1-\frac{2}{n}\right)\eta} \text{ nearly.}$$

Thus if $\quad n<2, \quad q \to 0, \quad$ and if $\quad n>2, \quad q \to \infty$.

If $n<2$, we have a mound and the velocity is everywhere finite. If $n = 1$ in particular, we have a semicircular mound and the problem is the same as that of the flow past a circular cylinder (see 6·22).

If $n = 3$, we have a semicircular ditch with infinite speed at A and B.

At the bottom of this ditch $\eta = 0$, $\xi = 3\pi/2$. Hence the speed is $2U/9$.

We may also observe that when $n \leqslant 2$ the same complex potential gives the flow past a cylinder whose cross-section consists of two equal circular segments on opposite sides of a common base, for such a flow is clearly symmetrical about the plane of the common base.

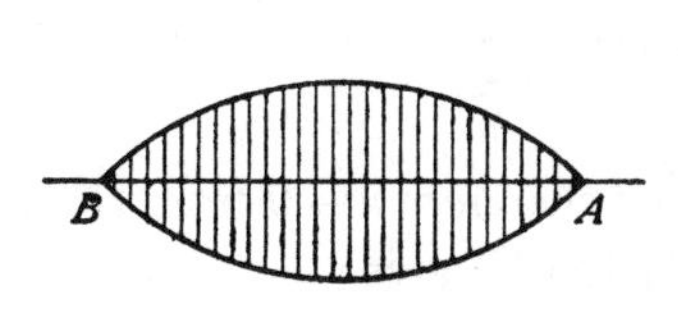

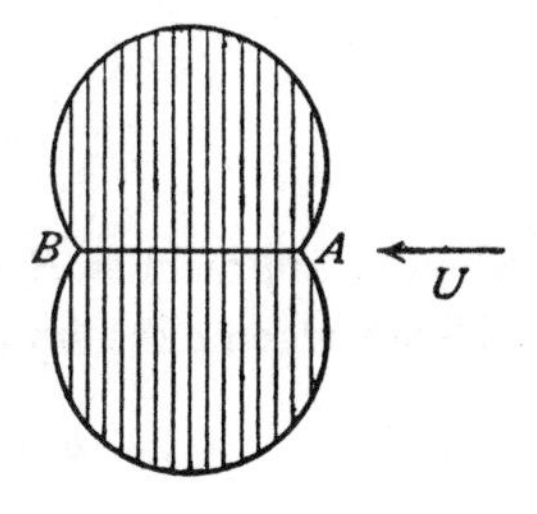

FIG. 6·51 (ii).

Fig. 6·51 (ii) illustrates two such cross-sections. The circular cylinder is intermediate between these cases.

If we impress on everything a uniform velocity $U(w = -Uz)$ from left to right, we get

$$w = U\frac{2ci}{n}\cot\frac{\zeta}{n} - Uic\cot\frac{\zeta}{2}$$

for the complex potential when a cylinder of such cross-section moves in the direction BA with velocity U.

The complex potential

$$w = \frac{2Uc}{n}\left(i\cos\alpha\cot\frac{\zeta}{n} + \sin\alpha\operatorname{cosec}\frac{\zeta}{n}\right)$$

gives the flow past the above solids when they disturb the stream

$$(-U\cos\alpha,\ -U\sin\alpha).$$

For $\xi = \pm n\pi/2$ makes w real and so $\psi = 0$, while

$$-dw/dz \to -U\cos\alpha + iU\sin\alpha \quad \text{when } \zeta \to 0.$$

6·52. Flow past a cylindrical log. The flow past a mound, discussed in the preceding section, may be made to yield the flow past a log by allowing the points A, B to come into coincidence.

The radius a of the circle at whose circumference AB subtends the angle ξ is given by

$$2a = \frac{2c}{\sin\xi}.$$

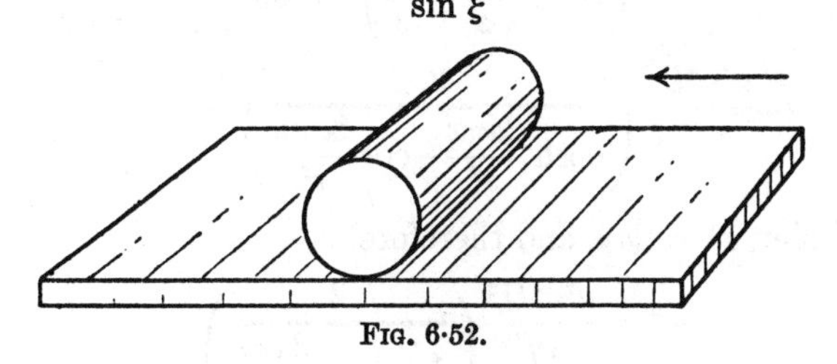

FIG. 6·52.

In the case of the mound, $\xi = n\pi/2$, and therefore

$$a = \frac{c}{\sin\frac{n\pi}{2}}, \tag{1}$$

(2) $$z = ic \cot \tfrac{1}{2}\zeta,$$

(3) $$w = \frac{2cUi}{n} \cot \frac{\zeta}{n}.$$

Now, when A approaches B, $c \to 0$, and therefore $n \to 0$ from (1), and $\zeta \to 0$ from (2). Hence when c is very small (1) and (2) give

$$a = \frac{2c}{n\pi}, \quad z = \frac{2ic}{\zeta}.$$

Substituting in (3), we get $w = ia\pi U \cot \dfrac{ia\pi}{z} = a\pi U \coth \dfrac{a\pi}{z}$

for the complex potential of the flow past a cylindrical log of radius a. We readily verify this result as follows. Since

$$w = \phi + i\psi = a\pi U \coth \left(\frac{a\pi x}{r^2} - \frac{a\pi i y}{r^2} \right), \quad r^2 = x^2 + y^2,$$

when $y = 0$, w is real, so that $\psi = 0$.

When $\dfrac{a\pi y}{r^2} = \dfrac{\pi}{2}$, w is also real, so that again $\psi = 0$.

Thus the streamline $\psi = 0$ consists of the real axis $y = 0$, and the circle $x^2 + y^2 = 2ay$.

Again, for large values of $|z|$,

$$w = a\pi U \times \frac{z}{a\pi} = Uz.$$

Hence there is a uniform stream at infinity parallel to the real axis and from right to left.

To find the speed, we have

$$q^2 = \frac{dw}{dz} \cdot \frac{d\bar{w}}{d\bar{z}} = a^2\pi^2U^2 \cdot \frac{a^2\pi^2}{z^2\bar{z}^2} \operatorname{cosech}^2 \frac{a\pi}{z} \operatorname{cosech}^2 \frac{a\pi}{\bar{z}}$$

$$= \frac{a^4\pi^4U^2}{r^4} \left(\frac{2}{\cosh\left(\dfrac{a\pi}{z} + \dfrac{a\pi}{\bar{z}}\right) - \cosh\left(\dfrac{a\pi}{z} - \dfrac{a\pi}{\bar{z}}\right)} \right)^2$$

$$= \frac{a^4\pi^4U^2}{r^4} \left(\frac{2}{\cosh \dfrac{2a\pi x}{r^2} - \cos \dfrac{2a\pi y}{r^2}} \right)^2.$$

On the cylinder, $r^2 = 2ay$, and therefore

$$q^2 = \frac{a^2\pi^4U^2}{4y^2} \left(\frac{2}{1 + \cosh \dfrac{2a\pi x}{r^2}} \right)^2$$

$$= \frac{a^2\pi^4U^2}{4y^2 \cosh^4 \dfrac{a\pi x}{r^2}}.$$

On the plane, $y = 0$,

$$q^2 = \frac{a^4\pi^4 U^2}{x^4} \frac{1}{\sinh^4 \dfrac{a\pi}{x}}.$$

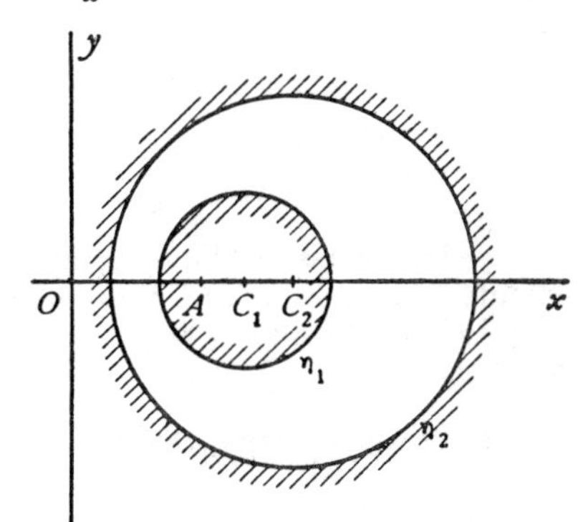

Fig. 6·53 (i).

6·53. Cylinder in a tunnel.

If

$$w = -\kappa\zeta, \quad z = ic \cot \tfrac{1}{2}\zeta,$$

then

$$\phi = -\kappa\xi, \quad \psi = -\kappa\eta,$$

and ψ is constant when η = constant, while ϕ decreases by $2\pi\kappa$ when we go round one of the circles η = constant. It follows from 3·71 (2) that the potential w represents the flow due to the circulation * $2\pi\kappa$ about a cylinder $\eta = \eta_1$ enclosed within a cylinder $\eta = \eta_2$ (cylinder in a tunnel). Eliminating ζ, we obtain

$$z = -ic \cot \frac{w}{2\kappa}, \quad w = 2\kappa \cot^{-1} \frac{iz}{c},$$

whence

$$\frac{dw}{dz} = \frac{2\kappa ic}{z^2 - c^2} = \frac{i\kappa}{z-c} - \frac{i\kappa}{z+c}.$$

Thus, by the theorem of Blasius, the force on the inner cylinder is given by

$$X - iY = \frac{i\rho}{2} \int -\kappa^2 \left(\frac{1}{(z-c)^2} - \frac{2}{(z-c)(z+c)} + \frac{1}{(z+c)^2} \right) dz$$

$$= \frac{i\kappa^2\rho}{2} \int \frac{1}{c} \left(\frac{1}{z-c} - \frac{1}{z+c} \right) dz = \frac{i\kappa^2\rho}{2c} \times 2\pi i = -\frac{\pi\kappa^2\rho}{c}.$$

Therefore

$$X = -\pi\kappa^2\rho/c, \quad Y = 0,$$

and the resultant fluid thrust therefore tends to increase the distance between the axis of the cylinder and the axis of the tunnel.

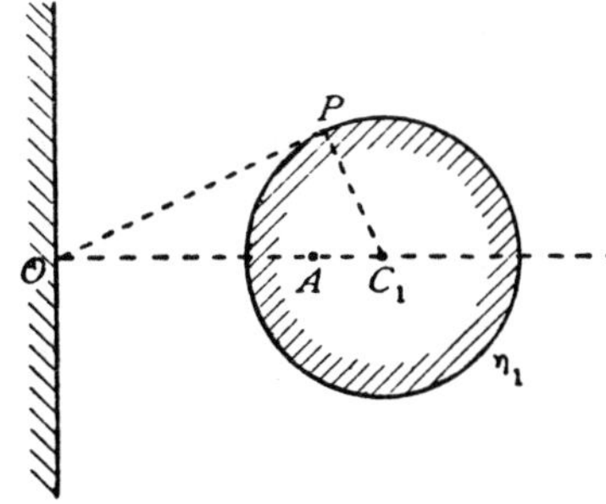

Fig. 6·53 (ii).

An interesting case occurs when the radius of the circle η_2 becomes infinite, so that this circle coincides with the radical axis. We have then the case of a cylinder whose axis is parallel to a wall. The cylinder is urged towards the wall with the force

$$\pi\kappa^2 \rho/c.$$

* Note that the region is doubly connected (3·70).

Drawing the tangent OP, we have, since A is a limiting point,

$$c^2 = OA^2 = OP^2 = OC_1^2 - a^2 = h^2 - a^2,$$

where h is the distance of the axis from the wall and a is the radius. Hence the force is

$$\frac{\pi\kappa^2\rho}{(h^2-a^2)^{\frac{1}{2}}}.$$

EXAMPLES VI

1. In the case of liquid streaming past a fixed circular disc, the velocity at infinity being u in a fixed direction where u is variable, show that the force necessary to hold the disc at rest is $2m\, du/dt$ where m is the mass of fluid displaced by the disc.

2. Prove, or verify, that the velocity potential

$$\phi = U\left(r + \frac{a^2}{r}\right)\cos\theta$$

represents a streaming motion past a fixed circular cylinder.

The pressure at infinity being given, calculate the resultant fluid action per unit length on half the cylinder lying on one side of a plane through the axis and parallel to the stream. (R.N.C.)

3. Liquid flows steadily and irrotationally in two dimensions in a space with fixed boundaries, the cross-section of which consists of the two lines $\theta = \pm\pi/10$ and the curve $r^5 \cos 5\theta = k^5$.

Prove that, if V is the speed of the liquid in contact with one of the plane boundaries at unit distance from their intersection, the volume of liquid which passes in unit time through a circular ring in the plane $\theta = 0$ is

$$\tfrac{1}{8}\pi V\, a^2(a^4 + 12a^2c^2 + 8c^4),$$

where a is the radius of the ring and c the distance of its centre from the intersection of the planes of the boundaries.

4. Sketch the streamlines represented by $\phi + i\psi = Az^2$, and show that the speed is everywhere proportional to the distance from the origin.

5. Discuss the motion represented by $w = \frac{1}{2}Ua^3/z^2$, and show that the streamlines are lemniscates.

6. If $w^2 = z^2 - 1$, prove that the streamline for which $\psi = 1$ is $y^2(1+x^2) = x^2$. Regarding this as a fixed boundary, show that the motion is that of a uniform stream flowing past the boundary.

7. Verify that the streaming motion past a solid bounded by

$$(x+1)^2 + y^2 = 2, \quad (x-1)^2 + y^2 = 2,$$

when the stream is asymptotic to the y-axis, is given by

$$\phi = -Uy\left[1 + \frac{1}{x^2+y^2} + \frac{2}{(x+1)^2+y^2} + \frac{2}{(x-1)^2+y^2}\right],$$

$$\psi = -U\left[-x + \frac{x}{x^2+y^2} + \frac{2(x+1)}{(x+1)^2+y^2} + \frac{2(x-1)}{(x-1)^2+y^2}\right].$$

8. Trace the streamline along which $\psi = 0$ and ϕ diminishes from $+\infty$ to $-\infty$ in the two cases:

$$\text{(i)}\ z^2 = 4w^3;$$

$$\text{(ii)}\ z = (w-1)^{\frac{3}{2}}+(w+1)^{\frac{3}{2}};$$

and indicate roughly the form of the streamlines for which ψ has a positive value.

9. If $w^2 = U^2(z^2+c^2)$, obtain the equation of the streamlines in the form

$$y^2 = \psi^2\,[U^2(x^2+c^2)+\psi^2]/U^2\,(U^2x^2+\psi^2),$$

and show that this gives the flow of a wide stream of velocity U past a thin obstacle of length c projecting perpendicularly from a straight boundary. (R.N.C.)

10. By considering the transformation $z = \zeta + a^2/\zeta$ applied to a stream flowing with velocity U past a circular island $r = c$ in the z-plane, obtain the corresponding solution for the stream of the same velocity flowing past an island of the shape given by a branch of the curve

$$(x^2+y^2)^2+a^4+2a^2(x^2-y^2)-c^2(x^2+y^2) = 0.$$

Trace this curve for various values of the ratio a/c and discuss the physical significance of the results obtained.

11. In the steady two-dimensional motion of an incompressible fluid given by $w = f(z)$, prove that, if p_0 is the pressure at a point where the speed is V, the resultant force (X, Y) exerted across any arc AB of a streamline by the fluid on the right of AB upon the fluid on the left is given by

$$\{X+(p_0+\tfrac{1}{2}\rho V^2)(y_B-y_A)\}-i\{Y-(p_0+\tfrac{1}{2}\rho V^2)(x_B-x_A)\} = \tfrac{1}{2}i\rho\int_A^B \left(\frac{dw}{dz}\right)^2 dz,$$

where the integral on the right-hand side is taken along any arc reducible to AB without passing over a singularity in the fluid.

A cylinder of radius a is placed in a stream of velocity V and pressure p_0 at infinity. Show that the resultant thrust (per unit thickness) on a quadrant of the cylinder between $\theta = 0$, $\theta = \pi/2$, where $\theta = 0$ points upstream, is given by

$$X = a[-p_0+\tfrac{1}{6}\rho V^2],\quad Y = a[-p_0+\tfrac{5}{6}\rho V^2].$$

(R.N.C.)

12. Apply to the motion of a uniform stream, given by the complex potential $w = Uz$, the successive transformations

$$z = z_1+\frac{(a+b)^2}{4z_1},\quad z_2 = z_1 e^{i\alpha},\quad z_3 = z_2+\frac{a^2-b^2}{4z_2},$$

and show that the two-dimensional motion in the z_3 plane corresponds to an elliptic cylinder fixed in a uniform stream making angle α with the major axis. Prove also that the argument of z_2 gives the eccentric angle of points on the ellipse.

Prove that the pressure on the surface of the elliptic cylinder is least where the eccentric angle θ is given by

$$a^2\tan\theta+b^2\cot\alpha = 0.$$

(R.N.C.)

13. If the two-dimensional motion of a fluid is given by $w = f(z)$, where $w = \phi+i\psi$, $z = x+iy$, obtain the form of the function f in the case of a stream of velocity V in a direction making an angle α with the axis of x, flowing past a fixed circular cylinder $x^2+y^2 = a^2$.

By the use of the transformation $z' = z+c^2/z$ $(c<a)$, or otherwise, find the solution for the same stream flowing past a fixed elliptic cylinder

$$x^2\operatorname{sech}^2\beta+y^2\operatorname{cosech}^2\beta = 4c^2,$$

where $\beta = \log a/c$, and calculate the torque tending to turn the cylinder about its axis. (U.L.)

14. With the usual notation, show that for liquid streaming past an elliptic cylinder in a direction parallel to the minor axis the stream function is

$$\psi = -Vc\, e^{\xi_0} \sinh(\xi - \xi_0) \cos\eta.$$

Hence show that for a stream of velocity Q in a direction making an angle θ with OX the stream function is

$$\psi = Qc\, e^{\xi_0} \sinh(\xi - \xi_0) \sin(\eta - \theta).$$

15. The elliptic cylinder $\xi = \xi_0$ is placed in a stream U parallel to the major axis. Prove that the speed q at any point is given by

$$q^2 = U^2 \frac{a+b}{a-b} \frac{\sinh^2(\xi-\xi_0) + \sin^2\eta}{\sinh^2\xi + \sin^2\eta},$$

and that it has the maximum value $U(a+b)/a$ at the end of the minor axis.

16. The velocity resolutes at distant points in an infinite liquid which streams past an elliptic cylinder are $-V\cos\beta$, $-V\sin\beta$ parallel to the major and minor axes respectively of the cross-section, and there is a circulation of amount κ about the cylinder. Find the force and couple resultants, per unit thickness, exerted by the fluid on the cylinder.

17. Show that

$$w = -Uz - Ub\sqrt{\frac{a+b}{a-b}}\, e^{-(\xi+i\eta)} + \frac{iI(\xi+i\eta)}{2\pi},$$

where

$$z = \sqrt{(a^2-b^2)} \cosh(\xi + i\eta),$$

gives the solution of the problem of flow of a stream of velocity U past a fixed elliptic cylinder of semi-axes a, b, whose major axis is parallel to the stream, there being a circulation I round the cylinder. Find also the resultant thrust on the cylinder. (R.N.C.)

18. Liquid of density ρ is circulating irrotationally between two confocal elliptic cylinders $\xi = \alpha$, $\xi = \beta$ where

$$x + iy = c\cosh(\xi + i\eta).$$

Prove that, if k is the circulation, the kinetic energy per unit thickness is

$$\rho k^2(\beta - \alpha)/8\pi.$$

19. If $x + iy = (\xi + i\eta)^2$, the streaming motion with velocity U parallel to the axis past the parabola $\xi = \xi_0$ is given by

$$\psi = 2U(\xi - \xi_0)\eta.$$

20. Prove that $w^2 = z$ gives the motion in the space bounded by two confocal and coaxial parabolic cylinders.

21. Homogeneous liquid streams past the infinite parabolic cylinder

$$r^{\frac{1}{2}} \cos(\theta/2) = a^{\frac{1}{2}},$$

the velocity at infinity being V in the positive direction of the axis of x. Prove that the velocity potential is

$$-Vr\cos\theta + 2Va^{\frac{1}{2}}r^{\frac{1}{2}}\cos(\theta/2)$$

and that the resultant thrust on the cylinder per unit length is $\pi\rho aV^2$, the pressure at infinity being taken to be zero.

22. Prove that the formula

$$\phi + i\psi = ik\log[(x+iy-c)/(x+iy+c)],$$

in which k is real, gives the irrotational motion of fluid circulating about two fixed circles, the circulations being $2\pi k$ for one and $-2\pi k$ for the other.

Determine the motion obtained by applying the transformation

$$x' + iy' = a^2/(x+iy-c),$$

where a is real, obtaining the boundaries of the region in which it takes place. (U.L.)

23. In the case of flow past a log of radius a lying on the bed of a deep stream, show that the difference in pressure at the highest and lowest points of the log is $\pi^4 \rho\, U^2/32$, where U is the velocity of the stream. (M.T.)

24. Using coaxal coordinates, verify that

$$w = U\frac{2ic}{n}\cot\frac{\zeta}{n}$$

gives a uniform stream if $n = 2$ and the flow past a circular cylinder if $n = 1$.

25. Homogeneous incompressible frictionless fluid occupies the region bounded by the plane $x = 0$ and the cylinder $(x-b)^2+y^2 = a^2$, where $b > a$. The fluid is streaming with the general velocity V in the negative direction of the axis of y. Prove that the motion is expressed by the equation

$$\phi + i\psi = -iVz + 2iVa^2z\left\{\frac{1}{z^2-b^2} + \sum_{n=1}^{\infty}\frac{a^{2n}}{\prod_{r=0}^{n-1}(b+x_r)^2\,.\,(z^2-x_n^{\,2})}\right\},$$

where $z = x+iy$, $x_0 = b$, $x_n = b-a^2/(b+x_{n-1})$. (U.L.)

CHAPTER VII

AEROFOILS

7·10. Circulation about a circular cylinder. Consider the complex potential

(1) $$w = i\kappa \log \frac{z}{a},$$

where κ is real.

On the cylinder $|z| = a$ we have $z = ae^{i\theta}$.

Hence $w = -\kappa\theta$, so that $\psi = 0$ and $\phi = -\kappa\theta$.

Thus the cylinder is the streamline $\psi = 0$.

Also, if we go once round the cylinder in the positive sense, θ changes into $\theta + 2\pi$ and therefore ϕ decreases by $2\pi\kappa$.

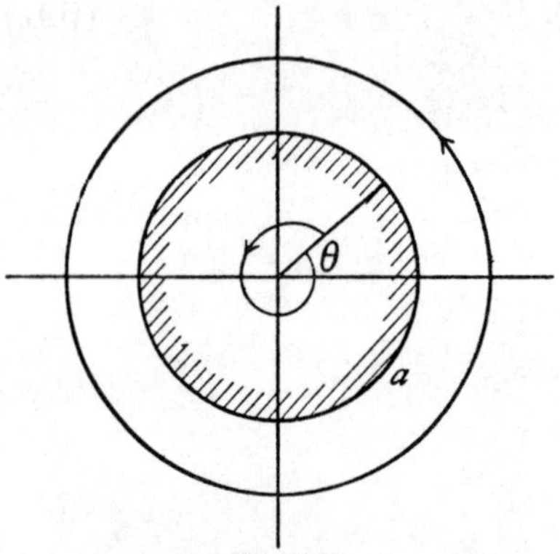

Thus, as appears from 3·71 (2), there is a circulation of amount $2\pi\kappa$ about the cylinder. More generally (1) gives

$$\phi = -\kappa\theta, \quad \psi = \kappa \log \frac{r}{a},$$

so that the circulation is $2\pi\kappa$ in every circuit which embraces the cylinder once (see 3·71), and the streamlines are concentric circles whose centres are on the axis of the cylinder.

FIG. 7·10.

Definition. When the circulation in a circuit is $2\pi\kappa$, we shall call κ the *strength* of the circulation.

The object of this definition is to avoid the constant occurrence of the factor 2π in the analysis.

In the present case

$$-\frac{\partial\phi}{\partial r} = 0, \quad -\frac{\partial\phi}{r\,\partial\theta} = \frac{\kappa}{r},$$

so that κ is the speed at unit distance from the origin.

Again, $$w = i\kappa \log z - i\kappa \log a,$$

and since the addition of a constant to the potential has no physical effect, we can, if we like, work with the complex potential $i\kappa \log z$, and indeed this is often convenient, in spite of the apparent lack of consistency in the physical dimensions, and we see that κ is still the speed at unit distance from the origin. The effect of dropping the constant is merely to make the boundary of the cylinder the streamline $\psi = \kappa \log a$ instead of $\psi = 0$.

It is very important to realise that the motion here described is indeed irrotational in the sense that the vorticity is zero. In fact the vorticity is (see 4·20) given by

$$\omega = \frac{q}{r} - \frac{\partial q}{\partial n} = \frac{q}{r} + \frac{\partial q}{\partial r} = \frac{\kappa}{r^2} - \frac{\kappa}{r^2} = 0.$$

7·11. Circulation between concentric cylinders.

The complex potential $w = i\kappa \log z$ will also apply to the circulatory motion of liquid between two concentric cylinders, for the stream function $\psi = \kappa \log r$ is constant on the cylinders $r = a$, $r = b$.

The possibility of cyclic motion in the case considered here and in the preceding section is due to the fact that the region occupied by the fluid is doubly connected (see 3·70).

7·12. Streaming and circulation for a circular cylinder.

The streaming motion past a circular cylinder of radius a is given by the complex potential

$$V\left(z + \frac{a^2}{z}\right).$$

The circulation of strength κ about the cylinder is given by

$$i\kappa \log \frac{z}{a}.$$

Combining these motions, we obtain the complex potential

$$w = V\left(z + \frac{a^2}{z}\right) + i\kappa \log \frac{z}{a}, \tag{1}$$

and the cylinder is still part of the streamline $\psi = 0$, for, putting $z = ae^{i\theta}$, we find that w is real and therefore $\psi = 0$.

To find the general form of the streamlines, we first investigate the stagnation points given by

$$\frac{dw}{dz} = V\left(1 - \frac{a^2}{z^2}\right) + \frac{i\kappa}{z} = 0, \quad \text{or} \quad \frac{z^2}{a^2} + \frac{z}{a}\frac{i\kappa}{aV} - 1 = 0,$$

whence

$$z = a\left(-\frac{i\kappa}{2aV} \pm \sqrt{\left(1 - \frac{\kappa^2}{4a^2V^2}\right)}\right).$$

We must now distinguish the cases

$$\kappa < 2aV, \quad \kappa = 2aV, \quad \kappa > 2aV.$$

Case I If $\kappa < 2aV$, put $\dfrac{\kappa}{2aV} = \sin\beta$. Then $z = a(-i\sin\beta \pm \cos\beta)$,

so that the stagnation points lie on the cylinder and on a line below the centre parallel to the real axis.

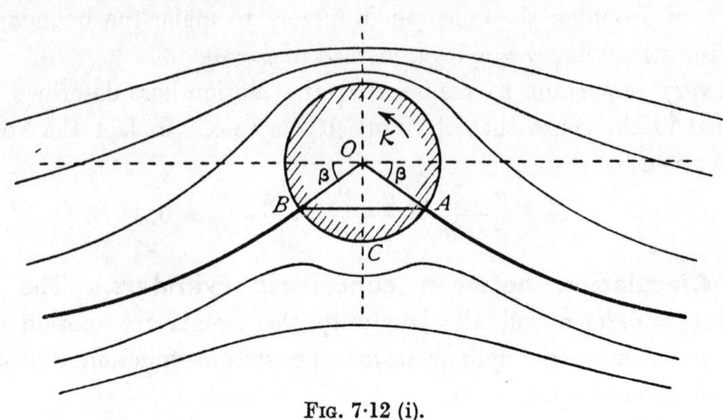

FIG. 7·12 (i).

Fig. 7·12 (i) shows the stagnation points A, B, the interpretation of the angle β, and the disposition of the streamlines.

The general effect of the circulation is to increase the speed of the fluid at points above the cylinder, and to diminish the speed at points below. Thus the pressure above is diminished and the pressure below is increased, and therefore there will be an upward force on the cylinder in the direction of the y-axis.

The stream function $\quad \psi = Vy\left(1 - \dfrac{a^2}{r^2}\right) + \kappa \log\dfrac{r}{a}$

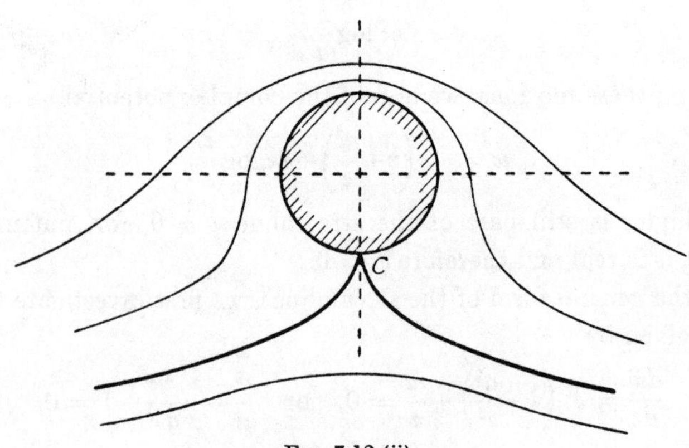

FIG. 7·12 (ii).

is unaltered when $-x$ is written for x and therefore the streamlines are symmetrical about the y-axis, so that there will be no resultant force in the direction of the x-axis, i.e. no resistance or drag.

When the circulation is zero we have seen that the stagnation points lie on the x-axis. Another effect of the circulation is thus to move these points downwards.

Case II $\kappa = 2aV$. In this case $\beta = \pi/2$ and the stagnation points coincide at the bottom, C, of the cylinder, fig. 7·12 (ii).

Case III $\kappa > 2aV$. Put $\dfrac{\kappa}{2aV} = \cosh\beta$. Then

$$z = ai(-\cosh\beta \pm \sinh\beta) = -ai\,e^{\beta} \quad \text{or} \quad -ai\,e^{-\beta}.$$

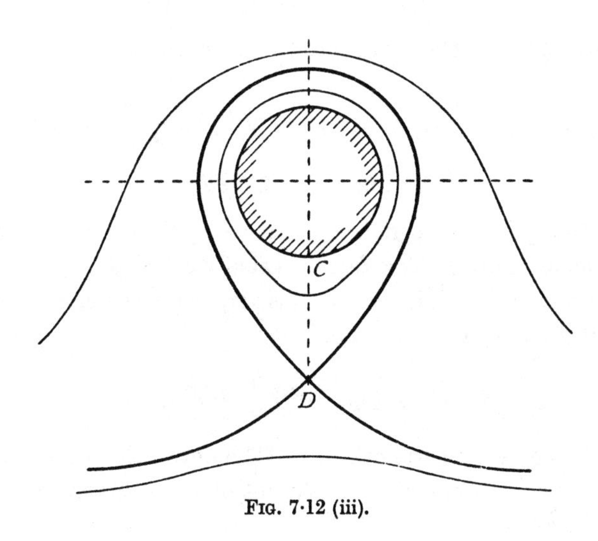

FIG. 7·12 (iii).

Calling these z_1, z_2, we have

$$|z_1 z_2| = a^2.$$

Thus the stagnation points are now inverse points on the imaginary axis, and one is therefore inside the cylinder and does not belong to the motion considered.

At the stagnation point the streamline cuts itself, necessarily at right angles (see 4·6), and the fluid within the loop thus formed circulates round the cylinder, never joining the main stream.

To find the pressure at points on the cylinder, we have

$$\frac{dw}{dz} = V(1-e^{-2i\theta}) + \frac{i\kappa}{a}e^{-i\theta} = e^{-i\theta}\left[2iV\sin\theta + \frac{i\kappa}{a}\right].$$

Hence
$$q^2 = \left(2V\sin\theta + \frac{\kappa}{a}\right)^2,$$

and therefore
$$\frac{p}{\rho} = \frac{\Pi}{\rho} + \tfrac{1}{2}V^2 - \tfrac{1}{2}\left(4V^2\sin^2\theta + \frac{\kappa^2}{a^2} + \frac{4\kappa V}{a}\sin\theta\right),$$

and the components of thrust on the cylinder due to the pressure are

$$X = -\int_0^{2\pi} p \cos\theta \, a d\theta, \quad Y = -\int_0^{2\pi} p \sin\theta \, a \, d\theta.$$

When θ is replaced by $\theta+\pi$, the only term in the expression for p which changes is the last. Thus the pressures at diametrically opposite points are

$$p_1 - \frac{2\kappa V}{a} \rho \sin\theta, \quad p_1 + \frac{2\kappa V}{a} \rho \sin\theta,$$

where
$$\frac{p_1}{\rho} = \frac{\Pi}{\rho} + \tfrac{1}{2} V^2 - 2V^2 \sin^2\theta - \frac{\kappa^2}{2a^2}.$$

The pressures p_1 have clearly no effect for they cancel one another. Therefore

$$X = 2\kappa V \rho \int_0^{2\pi} \sin\theta \cos\theta \, d\theta, \quad Y = 2\kappa V \rho \int_0^{2\pi} \sin^2\theta \, d\theta,$$

whence
$$X = 0, \quad Y = 2\pi\kappa\rho V,$$

and therefore the cylinder experiences a force $2\pi\kappa\rho V$, tending to lift it at right angles to the main stream. This force is usually called the *lift*.

The calculation of the lift is, even in this very simple case, greatly facilitated by using the theorem of Blasius. For

$$X - iY = \tfrac{1}{2} i\rho \int \left(V + \frac{i\kappa}{z} - \frac{Va^2}{z^2} \right)^2 dz,$$

taken round the contour of the cylinder. The only pole inside the contour is $z = 0$ and the residue there is the coefficient of $1/z$ in the integrand, which is at once seen to be $2iV\kappa$.

Hence, by Cauchy's residue theorem,

$$X - iY = \tfrac{1}{2} i\rho \times 2\pi i \times 2iV\kappa = -i \, . \, 2\pi\rho V\kappa,$$

so that $X = 0$, $Y = 2\pi\kappa\rho V$, as before.

The advantage of the theorem of Blasius lies in working with the single variable z and the elimination of irrelevant terms by the residue theorem.

7·13. Flow with constant vorticity. Let there be constant vorticity ω throughout the fluid. Since from 4·40 (5) and 5·33

(1)
$$\omega = \nabla^2 \psi = \frac{4 \, \partial^2 \psi}{\partial z \, \partial \bar{z}}$$

integration with respect to z and then $\bar{z}$ gives

(2)
$$\psi = f(z) + \bar{f}(\bar{z}) + \tfrac{1}{4} \omega z \bar{z}$$

the arbitrary functions $f(z)$, $\bar{f}(\bar{z})$ being necessarily conjugate complex since ψ is real.

Thus the most general two-dimensional flow with constant vorticity consists of the flow whose stream function is $\frac{1}{4}\omega z\bar{z}$ superposed on an irrotational flow whose complex potential is $2if(z)$.

7·14. The second circle theorem. The circle theorem of 6·21 applies to irrotational two-dimensional motion. The theorem about to be enunciated applies to two-dimensional motion in which the vorticity is constant.

The second circle theorem. Let there be two-dimensional flow with constant vorticity ω in the z-plane given by the stream function

$$(1) \qquad \psi_0(z, \bar{z}) = F(z) + \bar{F}(\bar{z}) + \tfrac{1}{4}\omega z\bar{z}.$$

Let there be no rigid boundaries and let all the singularities of $F(z)$, and therefore also of $\bar{F}(\bar{z})$, be at a distance greater than a from the origin. If a circular cylinder typified by its cross-section of circumference C, $|z| = a$, be introduced into the field of flow, the stream function of the perturbed flow becomes

$$(2) \qquad \psi(z, \bar{z}) = F(z) - F\left(\frac{a^2}{\bar{z}}\right) + \bar{F}(\bar{z}) - \bar{F}\left(\frac{a^2}{z}\right) + \tfrac{1}{4}\omega z\bar{z}.$$

Proof. Since on C, $z\bar{z} = a^2$ the stream function $\psi(z, \bar{z})$ assumes the constant value $\frac{1}{4}\omega a^2$ on C which is therefore a streamline for the flow given by (2).

Since all the singularities of $F(z)$, $\bar{F}(\bar{z})$ are outside the circumference C, all the singularities of $F(a^2/\bar{z})$ and $\bar{F}(a^2/z)$ are inside C. Therefore in the region of flow (1) and (2) have the same singularities. In particular, at infinity the flows given by (1) and (2) are the same.

Also the vorticity of the flow given by (2) is $4\,\partial^2\psi/\partial z\,\partial\bar{z} = \omega$.

Thus (2) satisfies all the conditions and is therefore the stream function of the perturbed motion. Q.E.D.

Corollary. If in (1) we replace $\frac{1}{4}\omega z\bar{z}$ by $\frac{1}{4}\omega z\bar{z} + \frac{1}{2}\kappa \log(z\bar{z})$ we get for the perturbed flow

$$(3) \quad \psi(z, \bar{z}) = F(z) - F\left(\frac{a^2}{\bar{z}}\right) + \bar{F}(\bar{z}) - \bar{F}\left(\frac{a^2}{z}\right) + \tfrac{1}{4}\omega z\bar{z} + \tfrac{1}{2}\kappa \log(z\bar{z}).$$

This allows for circulation $2\pi\kappa$ about C.

Notes. (i) Save for an added constant, (2) is the unique solution, for (2) solves the Dirichlet problem for the function $\psi(z, \bar{z}) - \frac{1}{4}\omega z\bar{z}$.

(ii) Formula (2) can also be written in the form

$$(4) \qquad \psi(z, \bar{z}) = \psi_0(z, \bar{z}) - \psi_0\left(\frac{a^2}{\bar{z}}, \frac{a^2}{z}\right) + \tfrac{1}{4}\frac{\omega a^4}{z\bar{z}}.$$

(iii) Another form of (2) is

$$(5) \qquad \psi(z, \bar{z}) = 2Re\left[F(z) - \bar{F}\left(\frac{a^2}{z}\right) + \tfrac{1}{8}\omega z\bar{z}\right].$$

(iv) Using polar coordinates we write

$$(6) \qquad F(z) + \bar{F}(\bar{z}) = S(r, \theta)$$

where $S(r, \theta)$ is a harmonic function (5·33). Therefore an alternative statement of the second circle theorem (2) is

(7) $$\psi(r, \theta) = S(r, \theta) - S\left(\frac{a^2}{r}, \theta\right) + \tfrac{1}{4}\omega r^2 .$$

This could also be written

(8) $$\psi(r, \theta) = \psi_0(r, \theta) - \psi_0\left(\frac{a^2}{r}, \theta\right) + \tfrac{1}{4}\omega \frac{a^4}{r^2} .$$

(v) The advantage of (2) is that it permits the full use of the complex variable.

7·15. Uniform shear flow.

Let the x-axis be horizontal, say on ground level, and let the y-axis be vertically upwards. The velocity distribution

(1) $$u = -\omega y, \quad v = 0, \quad \omega = \text{constant}$$

is one in which the speed is proportional to the distance from the ground and decreases to zero as the ground is approached.

This type of velocity distribution is frequently exhibited by natural wind and is known as *uniform shear flow.*

The stream function for the flow (1) is

$$\tfrac{1}{2}\omega y^2 = -\tfrac{1}{8}\omega(z - \bar{z})^2 .$$

Hence by a simple rotation of the axes through the angle β, we find that if the velocity of the shear flow is parallel to the line $y \cos\beta - x \sin\beta = 0$, the stream function is

(2) $$\psi_0 = -\tfrac{1}{8}\omega(ze^{-i\beta} - \bar{z}e^{i\beta})^2 .$$

The vorticity of this flow is $4\,\partial^2\psi_0/\partial z\,\partial\bar{z} = \omega$, so that the vorticity in uniform shear flow is constant.

We can also write

(3) $$\psi_0 = -\tfrac{1}{8}\omega z^2 e^{-2i\beta} - \tfrac{1}{8}\omega\bar{z}^2 e^{2i\beta} + \tfrac{1}{4}\omega z\bar{z}.$$

Comparing this with 7·14 (1) for the second circle theorem we see that

(4) $$F(z) = -\tfrac{1}{8}\omega z^2 e^{-2i\beta} .$$

If in addition to uniform shear flow, we wish to take account of circulation we simply add $\tfrac{1}{2}\kappa \log(z\bar{z})$ to the right hand side of (3).

7·16. Circular cylinder in uniform shear flow.

Consider the flow consisting of a uniform stream $(-V\cos\alpha, V\sin\alpha)$, uniform shear flow parallel to the line $y\cos\beta - x\sin\beta = 0$, and circulation $2\pi\kappa$. The stream function for this flow is

(1) $$\psi_0 = -\tfrac{1}{2}iVze^{i\alpha} + \tfrac{1}{2}iV\bar{z}e^{-i\alpha} - \tfrac{1}{8}\omega z^2 e^{-2i\beta} - \tfrac{1}{8}\omega\bar{z}^2 e^{2i\beta} + \tfrac{1}{4}\omega z\bar{z} + \tfrac{1}{2}\kappa \log z\bar{z}.$$

To this apply the second circle theorem 7·14 (3) with

(2) $$F(z) = -\tfrac{1}{2}iVze^{i\alpha} - \tfrac{1}{8}\omega z^2 e^{-2i\beta}$$

so that the stream function when the flow is disturbed by the circular cylinder whose cross-section is $z\bar{z} \leqslant a^2$ is

$$(3) \quad \psi = -\tfrac{1}{2}iVze^{i\alpha} - \tfrac{1}{8}\omega z^2 e^{-2i\beta} - \tfrac{1}{2}iV\frac{a^2}{z}e^{-i\alpha} + \tfrac{1}{8}\omega\frac{a^4}{z^2}e^{2i\beta} + \tfrac{1}{2}iV\bar{z}e^{-i\alpha}$$
$$- \tfrac{1}{8}\omega\bar{z}^2 e^{2i\beta} + \tfrac{1}{2}iV\frac{a^2}{\bar{z}}e^{i\alpha} + \tfrac{1}{8}\omega\frac{a^4}{\bar{z}^2}e^{-2i\beta} + \tfrac{1}{4}\omega z\bar{z} + \tfrac{1}{2}\kappa \log(z\bar{z}).$$

We note that this fulfils all the conditions and reduces to (1) when $z \to \infty$.

To find the force on the cylinder we use the theorem of Blasius. Thus from (3), putting $\bar{z} = a^2/z$ after differentiation, we get

$$(4) \quad \frac{\partial\psi}{\partial z} = -\tfrac{1}{4}\omega e^{-2i\beta}z - \tfrac{1}{2}iVe^{i\alpha} + \frac{a^2\omega + 2\kappa}{4z} + \tfrac{1}{2}i\frac{Va^2e^{-i\alpha}}{z^2} - \tfrac{1}{4}\frac{\omega a^4 e^{2i\beta}}{z^3}.$$

Therefore from 6·41 (7) and the residue theorem

$$(5) \qquad X - iY = -2\pi\rho iV\{\kappa e^{i\alpha} + a^2\omega e^{-i\beta}\cos(\alpha+\beta)\}.$$

This gives, as it should, the Kutta-Joukowski lift when $\omega = 0$, but we note that even when $\kappa = 0$ there can still be a lift.

7·17. Elliptic cylinder in uniform shear flow. Consider the ellipse $\xi = \xi_0$ in the net $z = c\cosh\zeta$, disturbing the uniform shear flow whose stream function is

$$\psi_0 = -\tfrac{1}{8}\omega(ze^{-i\beta} - \bar{z}e^{i\beta})^2 = -\tfrac{1}{8}\omega c^2(e^{-i\beta}\cosh\zeta - e^{i\beta}\cosh\bar{\zeta})^2.$$

After a simple reduction we get

$$(1) \quad \psi_0 = -\tfrac{1}{8}\omega c^2\{\cos 2\beta - \cosh 2\xi + \cos 2\eta\,(-1 + \cos 2\beta\cosh 2\xi)$$
$$+ \sin 2\eta \sin 2\beta \sinh 2\xi\}.$$

The required stream function is $\psi = \psi_0 + \psi_1$, where ψ_1 is a harmonic function which tends to zero at infinity and is such that ψ is constant on the boundary $\xi = \xi_0$ which must be a streamline. Looking at (1) we see that the appropriate form is

$$(2) \qquad \psi_1 = -\tfrac{1}{8}\omega c^2\{Ae^{-2\xi}\cos 2\eta + Be^{-2\xi}\sin 2\eta\}$$

where the constants A and B are to be chosen to make the coefficients of $\cos 2\eta$ and $\sin 2\eta$ in ψ vanish when $\xi = \xi_0$. This condition gives at once

$$Ae^{-2\xi_0} = 1 - \cos 2\beta\cosh 2\xi_0, \quad Be^{-2\xi_0} = -\sin 2\beta\sinh 2\xi_0$$

and therefore dropping the constant $-\tfrac{1}{8}\omega c^2\cos 2\beta$ we have

$$\psi = -\tfrac{1}{8}\omega c^2\{-\cosh 2\xi + \cos 2\eta\,[-1 + \cos 2\beta\cosh 2\xi - e^{2\xi_0 - 2\xi}$$
$$(-1 + \cos 2\beta\cosh 2\xi_0)] + \sin 2\eta\,[\sin 2\beta\sinh 2\xi - e^{2\xi_0-2\xi}\sin 2\beta\sinh 2\xi_0]\}.$$

This clearly becomes constant when $\xi = \xi_0$. It is left as an exercise to show that

$$\nabla^2\psi = \frac{2}{\cosh 2\xi - \cos 2\eta}\left(\frac{\partial^2\psi}{\partial\xi^2} + \frac{\partial^2\psi}{\partial\eta^2}\right) = \omega$$

so that all the conditions are satisfied.

If we wish to take account also of circulation $2\pi\kappa$ about the ellipse, we simply add $\kappa\xi$ to the above value of the stream function, for this is constant on $\xi = \xi_0$ and the corresponding velocity potential is $-\kappa\eta$ which decreases by $2\pi\kappa$ in making one circuit about the ellipse.

The case of a flat plate is obtained by putting $\xi_0 = 0$.

7·20. The aerofoil. The aerofoil used in modern aeroplanes has a profile of the " fish " type, indicated in fig. 7·20. Such an aerofoil has a blunt *leading edge* and a sharp *trailing edge.* The projection of the profile on the double tangent, as shown in the diagram, is the *chord.* The ratio of the span to the chord is the *aspect ratio.*

The *camber line* of a profile is the locus of the point midway between the points in which an ordinate perpendicular to the chord meets the profile.

The *camber* is the ratio of the maximum ordinate of the camber line to the chord.

It is proposed to outline the elements of the theory of the flow round such an aerofoil on the following assumptions :

(1) That the air behaves as an incompressible inviscid fluid.

(2) That the aerofoil is a cylinder whose cross-section is a curve of the above type.

(3) That the flow is two-dimensional irrotational cyclic motion.

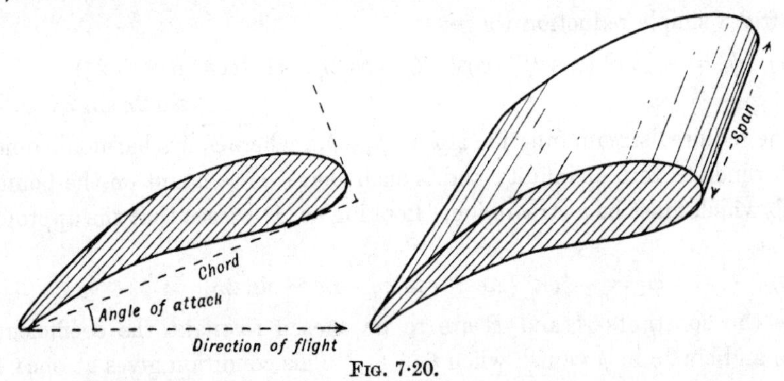

FIG. 7·20.

The above assumptions are of course only approximations to the actual state of affairs, but by making these simplifications it is possible to arrive at a general understanding of the principles involved. There is a considerable and increasing literature on this subject which cannot even be outlined here. Our purpose is merely to give an introductory view of the simplest aspects of the phenomenon.*

It has been found that profiles obtained by conformal transformation of a circle by the simple Joukowski transformation (see 6·30) make good wing

* For a more detailed treatment see Milne-Thomson, *Theoretical Aerodynamics*, London 1966.

shapes, and that the lift can be calculated from the known flow with respect to a circular cylinder. There are two ways of approaching this type of aerofoil design.

(*a*) By transforming a given circle.

(*b*) By enquiring what circle would give rise to a given profile previously drawn. Naturally, the inverse process of (*b*) is more difficult. We shall confine our investigations to the process (*a*). To this end we shall consider the transformation of 6·30 in more detail.

7·30. Further investigation of the Joukowski transformation. The transformation

$$z = \zeta + \frac{l^2}{\zeta},$$

regarded as mapping the ζ-plane on the z-plane, is equivalent to the successive transformations

$$\zeta_1 = \frac{l^2}{\zeta}, \quad z = \zeta + \zeta_1.$$

Given ζ and ζ_1, the second of these reduces to simple addition on the Argand diagram. Let us then consider how to obtain ζ_1 when ζ is given. Writing

$$\zeta = re^{i\theta},$$

we have

$$\zeta_1 = \frac{l^2}{r} e^{-i\theta}.$$

The points $P(\zeta)$ and $P_1(\zeta_1)$ are shown in fig. 7·30.

If we draw P_1P' perpendicular to the real axis to meet OP at P', we see that

$$OP' = OP_1 = l^2/r,$$

and therefore

$$OP \,.\, OP' = l^2.$$

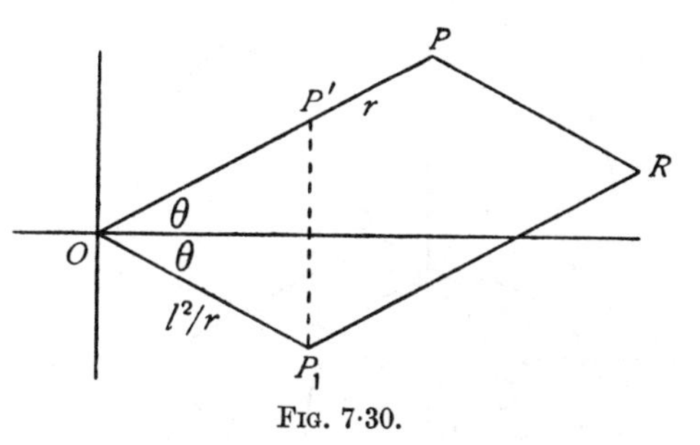

FIG. 7·30.

Thus P and P' are inverse points with respect to O, and to obtain P_1 we therefore first find the inverse point P' and then reflect OP' in the real axis, thereby obtaining the position of P_1.

Finally, to obtain the point z, we add the complex numbers represented by the points P, P_1 by completing the parallelogram OP_1RP. The fourth vertex R then represents z and the transformation is complete.

Now in the problem to be treated here P will be made to describe a circle. The point P' will then describe the inverse of this circle, which will be shown to be another circle, and P_1 will therefore describe a circle obtained by reflecting the locus of P' in the real axis. We shall now investigate a geometrical construction for finding the locus of P_1.

7·31. Geometrical construction for the transformation. Let C be the centre of the given circle, cutting the real axis at A, B where $OB = l$.

Let P be any point on the given circle, P' its inverse with respect to the circle centre O, radius l, that is to say the point on OP such that

(1) $$OP \,.\, OP' = l^2 = OB^2.$$

Let PO cut the circle again at Q and draw $P'C'$ parallel to CQ to meet CO at C'.

We shall first prove that the locus of P' is a circle whose centre is C'.

Proof. Since AOB, POQ are chords of a circle intersecting at O,

(2) $$OP \,.\, OQ = OA \,.\, OB.$$

Dividing (1) by (2),

$$\frac{OP'}{OQ} = \frac{OB}{OA}.$$

Now the triangles $OP'C'$, OQC are similar since $P'C'$ is parallel to QC. Therefore

$$\frac{OC'}{OC} = \frac{C'P'}{CQ} = \frac{OP'}{OQ} = \frac{OB}{OA}.$$

Since $OC' : OC$ is constant, C' is a fixed point.

Since $CQ = a$, the radius of the given circle, and $C'P' : CQ$ is constant, it follows that $C'P'$ is of constant length.

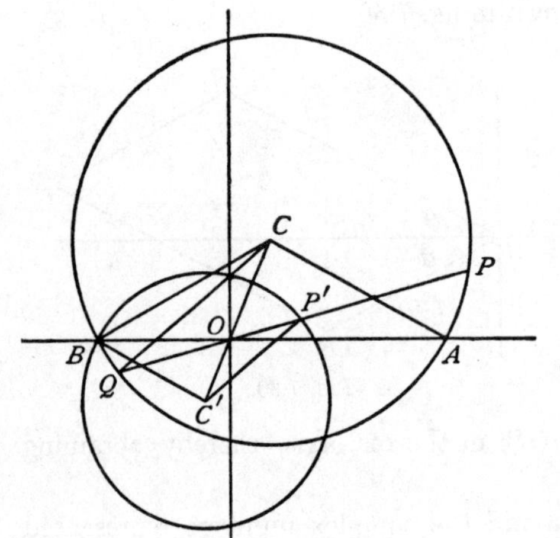

FIG. 7·31 (i).

Therefore P' describes a circle whose centre is C'. Q.E.D.

Since $OB = l$, the point B is its own inverse, and therefore the locus of P' passes also through B.

Since

$$\frac{OC'}{OC} = \frac{OB}{OA}$$

the triangles OAC, OBC' are similar and similarly situated. It follows that BC', CA are parallel, and therefore

$$\angle ABC' = \angle BAC = \angle CBA.$$

Hence BC' and BC are equally inclined to the real axis. If, therefore, we reflect the circle which is the locus of P' in the real axis, we shall obtain an equal circle whose centre D lies on BC and which passes through B. This is the required circle, the locus of P_1 the reflection of P'.

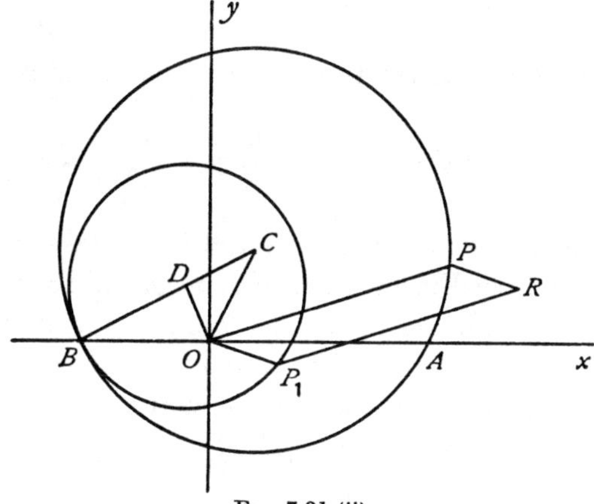

FIG. 7·31 (ii).

Since B lies on the line of centres CD, the two circles must touch at B.

Since OC' and OC are equally inclined to Oy, and since OD is the reflection of OC', it follows that OD and OC are equally inclined to Oy. This remark enables us to find D and draw the circle without any difficulty. The construction of the point R, representing $z=\zeta+l^2/\zeta$, is then finished as described in 7·30.

Fig. 7·31 (iii) shows an aerofoil section sketched through points obtained by drawing radii vectores at 30° intervals. Corresponding points on the aerofoil and the circle bear the same number.

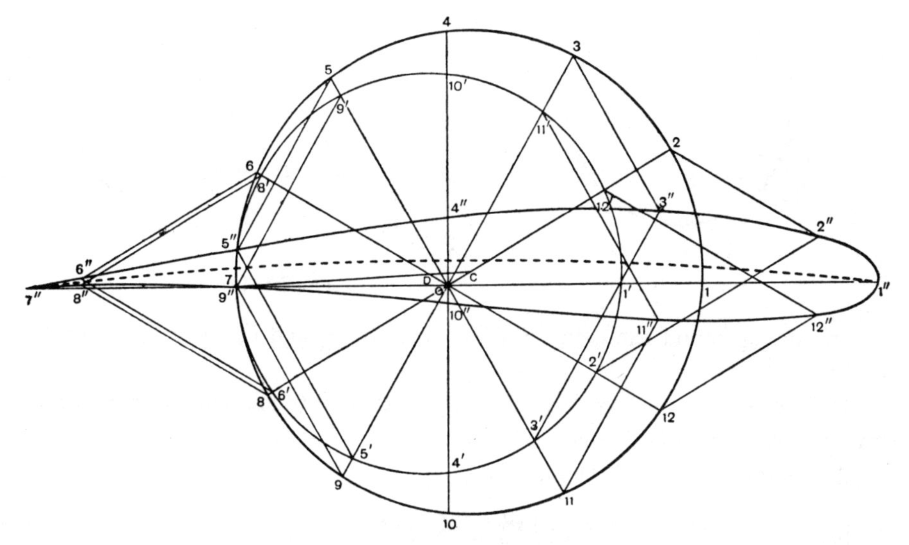

FIG. 7·31 (iii).

Aerofoils obtained by this construction are known as Joukowski aerofoils. They have a blunt nose and a sharp trailing edge corresponding to the point B on the circle.

7·32. The nature of the trailing edge.

The transformation

$$z = \zeta + \frac{l^2}{\zeta} \quad \text{gives} \quad \frac{dz}{d\zeta} = 1 - \frac{l^2}{\zeta^2},$$

so that $dz/d\zeta$ vanishes at the points $\zeta = -l$, $\zeta = l$, and therefore the representation ceases to be conformal in the immediate neighbourhood of these points. The point $\zeta = l$, being inside the circle, transforms to a point inside the aerofoil and need not be considered further. The point $\zeta = -l$ transforms into $z = -2l$, the trailing edge. The transformation can be written

$$\text{(1)} \qquad \frac{z+2l}{z-2l} = \left(\frac{\zeta+l}{\zeta-l}\right)^2$$

In the neighbourhood of $\zeta = -l$ and $z = -2l$, put

$$\zeta = -l + re^{i\theta}, \quad z = -2l + se^{i\chi},$$

where r and s are infinitesimal. Then, approximately,

$$\frac{s\, e^{i\chi}}{-4l} = \frac{r^2\, e^{2i\theta}}{4l^2},$$

and therefore

$$\chi + \pi = 2\theta.$$

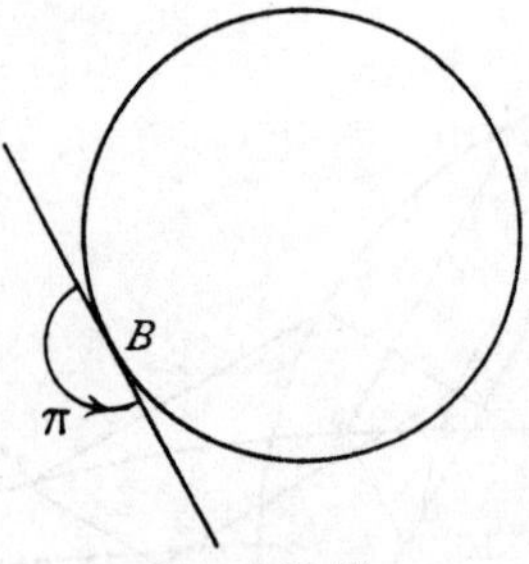

Fig. 7·32 (i).

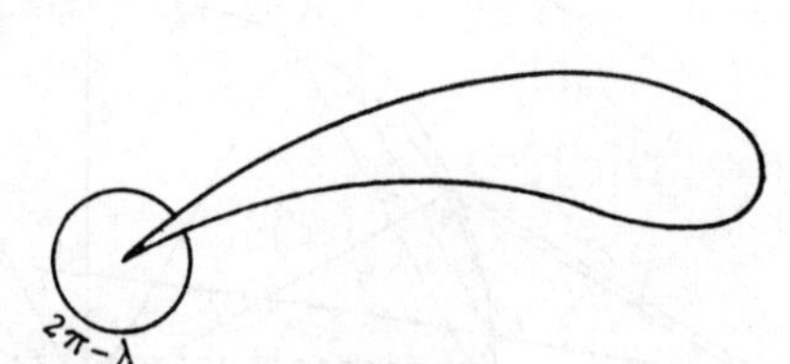

Fig. 7·32 (ii).

In moving round the point B, θ increases by π, and therefore χ increases by 2π.

It follows that the two branches of the aerofoil touch one another at the trailing edge, which is therefore a cusp.

A generalised form of (1), namely

$$\text{(2)} \qquad \frac{z+nl}{z-nl} = \left(\frac{\zeta+l}{\zeta-l}\right)^n,$$

is also used in design work and gives rise to a class of curves known as Kármán-Trefftz profiles. Using the same notations, we find

$$\chi+\pi = n\theta,$$

so that if $n = 2-\dfrac{\lambda}{\pi}$, the increase π of θ at B gives for χ the increase $2\pi-\lambda$ and the trailing edge has now two distinct branches intersecting at the angle λ.

The transformation (2) does not permit the simple geometrical treatment which is available when $n = 2$.

7·40. Joukowski's hypothesis. If q is the speed at the point B of the cylinder which transforms into the trailing edge of the aerofoil, and q' the corresponding speed at the trailing edge, we have, from 6·03,

$$q' = \left|\frac{dw}{dz}\right| = \left|\frac{dw}{d\zeta}\right| \times \left|\frac{d\zeta}{dz}\right| = q\left|\frac{d\zeta}{dz}\right|.$$

Now we have seen that at the trailing edge $dz/d\zeta = 0$, and therefore q' will be infinite. This could be avoided if B were a stagnation point so that $q = 0$, An inspection of the position of the stagnation points, discussed in 7·12, shows that, by proper choice of the strength of the circulation κ, the stagnation points can be placed anywhere on the lower half of the cylinder so that B can always be made a stagnation point.

Joukowski's hypothesis is that the circulation in the case of a properly designed aerofoil always adjusts itself so that B is a stagnation point and the velocity at the trailing edge is finite. This condition* appears to be satisfied with reasonable exactness within the working range of well designed aerofoils.

The physical explanation of the origin of the circulation is probably somewhat as follows. (See also Plates 2, 3.)

When the motion is just starting, i.e. for low velocities of the airstream, the flow is ordinary streamline flow, with the stagnation point just ahead of the trailing edge on the upper surface of the aerofoil. As the speed increases, even with small viscosity, the viscous forces increase and the air is no longer able to turn round the sharp edge and a vortex is formed.

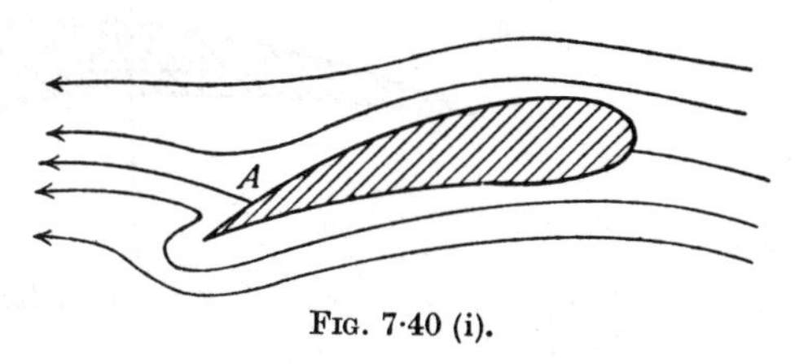

FIG. 7·40 (i).

Since the circulation in any circuit large enough to enclose the aerofoil and the vortex was zero to start with it must remain zero, and so a circulation now exists round the aerofoil equal and opposite to that of the vortex. The vortex

* Behind the aerofoil there exists a vortex wake which causes the measured circulation to be less than that given by Joukowski's hypothesis. The effect of the wake increases with increasing angle of attack.

gets washed away downstream, and when the steady state is reached the circulation round the aerofoil remains.

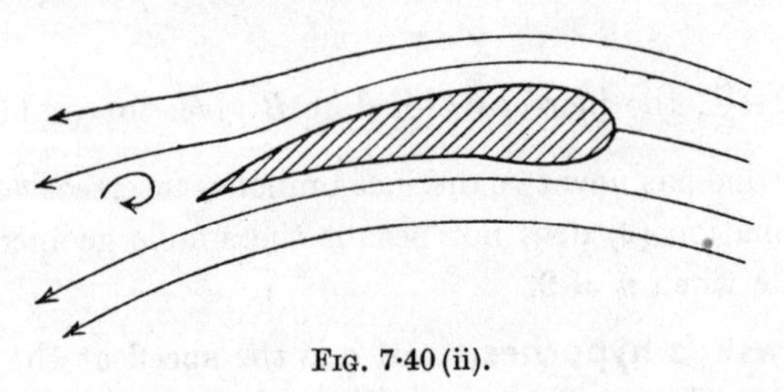

FIG. 7·40 (ii).

7·45. The theorem of Kutta and Joukowski.

An aerofoil at rest in a uniform wind of speed V, with circulation K round the aerofoil, undergoes a lift $K\rho V$ perpendicular to the wind. The direction of the lift vector is got by rotating the wind velocity vector through a right angle in the sense opposite to that of the circulation.

Proof. Since there is a uniform wind, the velocity at a great distance from the aerofoil must tend simply to the wind velocity, and therefore if $|z|$ is sufficiently large, we may write

$$(1) \qquad -\frac{dw}{dz} = -Ve^{i\alpha} + \frac{A}{z} + \frac{B}{z^2} + \ldots,$$

where α is the *incidence* or *angle of attack*, fig. 7·20. Thus

$$w = Ve^{i\alpha} z - A \log z + \frac{B}{z} + \ldots,$$

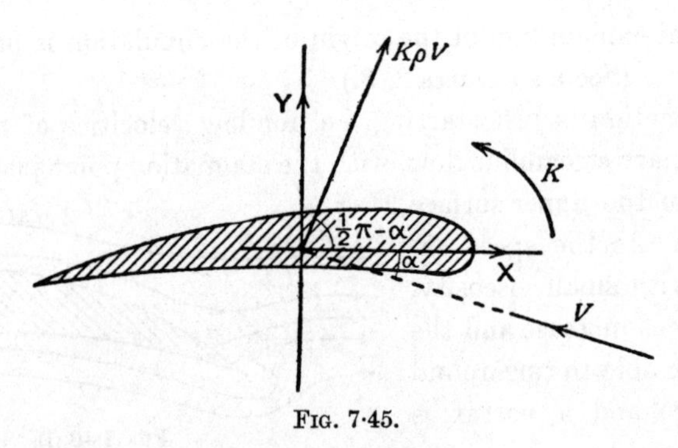

FIG. 7·45.

and since there is circulation K, we must have

$$(2) \qquad -A = \frac{iK}{2\pi},$$

for $\log z$ increases by $2\pi i$ when we go once round the aerofoil in the positive sense. From (1) and (2) we get

(3) $$\left(\frac{dw}{dz}\right)^2 = V^2 e^{2i\alpha} + \frac{iK\, Ve^{i\alpha}}{\pi z} - \frac{K^2 + 8\pi^2 BV\, e^{i\alpha}}{4\pi^2 z^2} - \ldots .$$

If we now integrate round a circle whose radius is sufficiently large for the expansion (3) to be valid, the theorem of Blasius gives (see 6·41)

$$X - iY = \tfrac{1}{2} i\rho \times 2\pi i \left(\frac{iK\, Ve^{i\alpha}}{\pi}\right) = -iK\rho\, Ve^{i\alpha},$$

so that, changing the sign of i,

(4) $$X + iY = iK\rho\, Ve^{-i\alpha} = K\rho\, Ve^{i(\frac{1}{2}\pi - \alpha)}.$$

Comparison with fig. 7·45 shows that this force has all the properties stated in the enunciation. Q.E.D.

Notes. (i) The theorem was discovered by Kutta (1902), and independently by Joukowski (1906).

(ii) *The lift is independent of the form of the profile.*

(iii) The theory gives no drag since we have taken no account of the wake or of viscosity. See 21·82, 22·76.

(iv) If the aerofoil is regarded as moving in air otherwise at rest, the lift is got by rotating the velocity vector *of the aerofoil* through a right angle in the *same sense* as the circulation.

(v) The theorem of Blasius applied to (3) gives the moment about the origin

(5) $$M = \text{real part of } 2\pi i\rho BVe^{i\alpha}.$$

7·50. The lift on an aerofoil in a uniform stream. The Joukowski transformation $z = \zeta + l^2/\zeta$ is a particular case of a more general type of transformation

(1) $$z = \zeta + \frac{a_1}{\zeta} + \frac{a_2}{\zeta^2} + \ldots ,$$

which applied to the circle of radius a with its centre C at the point $\zeta = s$ will yield an aerofoil profile.

The aerodynamic force on the aerofoil is due to the aerodynamic pressure thrusts on the elements of its surface. It is known that a system of forces acting on a rigid body (and we shall assume our aerofoil to be rigid) can be replaced at any chosen *base point* by a force acting at that base point and a couple. Moreover the magnitude and direction of the force are the same for all base points, whereas the moment of the couple depends upon the particular base point selected.

For the present investigation we shall take as base point the centre C of the circle. This point is called the *centre of the profile*; the actual position it occupies with respect to the profile is shown when the points of the circle and the corresponding points of the profile are marked in the *same* Argand diagram as in fig. 7·31 (iii). In the present case we shall take C as origin. This entails writing in (1) $z+s$ for z and $\zeta+s$ for ζ, which for sufficiently large values of $|\zeta|$ leads to the following convergent expansion

$$z = \zeta + \frac{a_1}{\zeta} + \frac{a_2 - a_1 s}{\zeta^2} + \ldots . \tag{2}$$

We also note that (2) can be *reversed* to give

$$\zeta = z - \frac{a_1}{z} - \ldots = z\left(1 - \frac{a_1}{z^2} - \ldots\right), \tag{3}$$

which can easily be verified to this degree of accuracy by substitution in (2).

Fig. 7·50 shows the disposition, C being the origin in both planes.

If α is the *incidence* or *angle of attack*, the circle theorem (6·21) gives for the complex potential for the flow past the circle $Ve^{i\alpha}\zeta + a^2Ve^{-i\alpha}/\zeta$, and if in addition we have a circulation $2\pi\kappa$, we get

$$w = Ve^{i\alpha}\zeta + \frac{a^2Ve^{-i\alpha}}{\zeta} + i\kappa \log \zeta. \tag{4}$$

If we now replace ζ by the function of z which defines the transformation, we get the complex potential for the flow past the aerofoil. In the present case

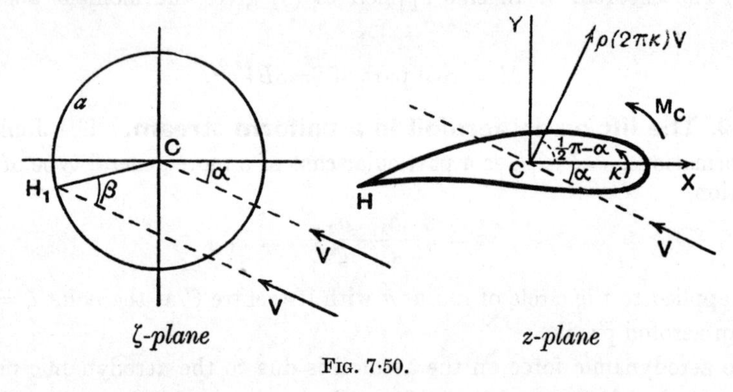

FIG. 7·50.

for sufficiently large values of $|z|$ this is defined by (3) and (4), which give in the z-plane

$$w = Ve^{i\alpha}\left(z - \frac{a_1}{z} - \ldots\right) + \frac{Ve^{-i\alpha}a^2}{z}\left(1 - \frac{a_1}{z^2} - \ldots\right)^{-1}$$
$$+ i\kappa\left[\log z + \log\left(1 - \frac{a_1}{z^2} - \ldots\right)\right].$$

$$= Ve^{i\alpha}z + i\kappa \log z + \frac{Ve^{-i\alpha}a^2 - Ve^{i\alpha}a_1}{z} + \ldots , \tag{5}$$

where the dots indicate omitted powers of $1/z$.

Comparison with 7·45 shows that here

$$(6) \qquad A = -i\kappa, \quad B = Ve^{-i\alpha}a^2 - Ve^{i\alpha}a_1,$$

and therefore from 7·45 (4) and (5)

$$(7) \qquad X+iY = 2\pi i\kappa\rho Ve^{-i\alpha} = 2\pi\kappa\rho Ve^{i\left(\frac{\pi}{2}-\alpha\right)},$$

$$(8) \qquad M_C = \text{real part of } (-2\pi\rho V^2 i a_1 e^{2i\alpha}),$$

where M_C is the moment with respect to the centre C.

If L is the lift, M the moment, and c the chord, the dimensionless numbers

$$C_L = \frac{L}{\frac{1}{2}\rho V^2 c}, \quad C_m = \frac{M}{\frac{1}{2}\rho V^2 c^2},$$

are called the *lift coefficient*, and *moment coefficient* respectively. The moment coefficient depends on the choice of point about which moments are taken.

Observe that

$$\textit{Lift} = (\textit{wind speed}) \times (\textit{air density}) \times (\textit{circulation}),$$

and is independent of shape.

7·51. Axes of a profile. If we draw the circle and the profile in one Argand diagram, the line joining the centre C to the rear stagnation point of the circle is called the *first axis* of the profile (Axis I), and then in the notation of fig. 7·50, $\kappa = 2aV \sin\beta$. Thus the lift L is proportional to $\sin\beta$ and vanishes when $\beta = 0$, i.e. when the wind stream is along Axis I (sense CH_1 in fig. 7·50). The first axis is therefore also known as the axis of zero lift.

Again, if in 7·50 (2) we put $a_1 = l^2 e^{i\mu}$, we get from 7·50 (8)

$$M_C = 2\pi\rho\, l^2 V^2 \sin(2\alpha+\mu),$$

so that the pitching moment M_C vanishes when the incidence is $-\frac{1}{2}\mu$. The corresponding wind direction through C is called the *second axis* of the profile (Axis II) or the axis of zero pitching moment. If we call γ the angle between Axes I and II, we get

$$M_C = 2\pi\rho l^2 V^2 \sin(2\beta - 2\gamma).$$

7·52. Focus of a profile. The *focus* is the point such that the moment of aerodynamic force about it is independent of the incidence.

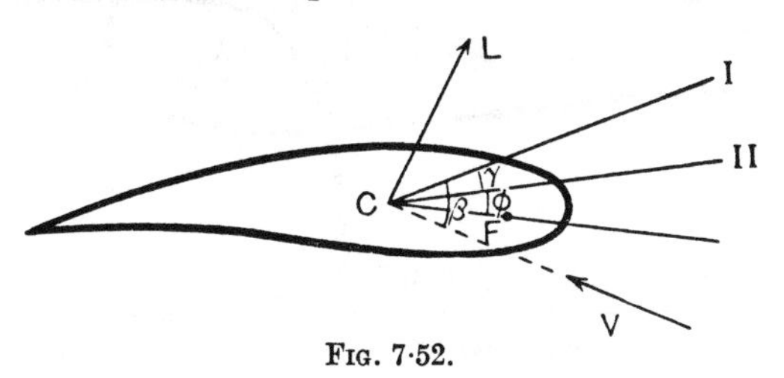

FIG. 7·52.

To establish the existence of the focus we note that if F is any point,

$$M_F = M_C - CF \cos(\beta-\gamma-\phi) \,.\, L,$$

where ϕ is the angle between CF and Axis II as shown in fig. 7·52 and L is the lift.

Using the values

$$M_C = 2\pi\rho V^2 l^2 \sin 2(\beta-\gamma), \quad L = 4\pi\rho a V^2 \sin\beta,$$

we have

$$\begin{aligned} M_F &= 2\pi\rho V^2\{l^2 \sin(2\beta-2\gamma) - 2a\, CF \,.\, \sin\beta \cos(\beta-\gamma-\phi)\} \\ &= 2\pi\rho V^2\{l^2 \sin(2\beta-2\gamma) - a \,.\, CF \,.\, \sin(2\beta-\gamma-\phi) - a \,.\, CF \,.\, \sin(\gamma+\phi)\}. \end{aligned}$$

This will be independent of β, the *absolute incidence*,* if we take

$$l^2 = a \,.\, CF, \quad \phi = \gamma.$$

This proves the existence of the focus F and gives its position as distant l^2/a from the centre on a line which is the reflection of Axis I in Axis II.

The moment about the focus is

$$M_F = -2\pi\rho V^2 l^2 \sin 2\gamma.$$

Our diagram has been drawn on the assumption that Axis I is above Axis II in the sense indicated in fig. 7·52. In this case the pitching moment *about the focus* is negative. If, however, Axis II were above Axis I, γ would change sign and the moment would become positive. The relative positions of Axes I and II therefore correspond with different dynamical properties of the profile.

Moreover, if $\gamma = 0$, we have $M_F = 0$ at all incidences and therefore the lift always passes through the focus. In this case the aerofoil is said to have a *centre of lift*.

For a flat aerofoil the focus is the quarter point midway between the centre and the leading edge.

7·53. Metacentric parabola.

Let L be the actual line of action of the lift $4\pi\rho a V^2 \sin\beta$. The direction of L is perpendicular to the wind. Let the line L meet the line KF, which is

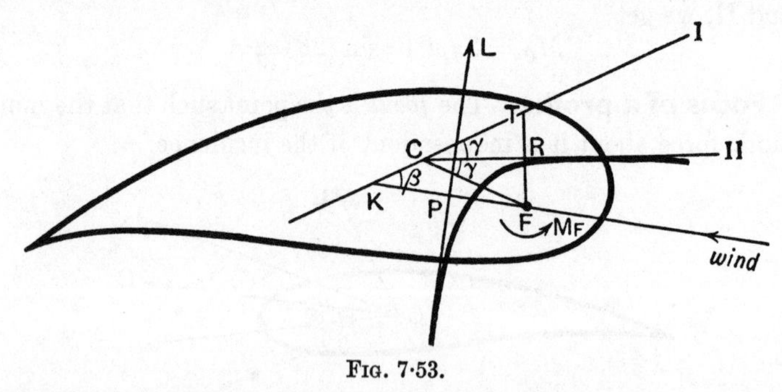

FIG. 7·53.

* Absolute incidence is the angle between the wind direction and the axis of zero lift.

drawn through the focus F parallel to the wind, at P, the point K being on Axis I. Taking moments about the focus F

$$M_F + FP \, . \, L = 0,$$

$$FP = -\frac{M_F}{L} = \frac{1}{2}\frac{l^2}{a}\frac{\sin 2\gamma}{\sin \beta} = \tfrac{1}{2}CF \cdot \frac{FK}{FC} = \tfrac{1}{2}FK,$$

using the sine formula for the triangle FKC.

Thus the locus of P is a straight line parallel to Axis I and midway between F and Axis I. From a known property of the parabola that the foot of the perpendicular to a tangent from the focus lies on the tangent at the vertex, it follows that the line of action of the lift touches a parabola whose focus is F and whose directrix is Axis I. This is called the *metacentric parabola.*

To find the resultant lift we draw that tangent to this parabola which is perpendicular to the wind direction.

Axis II touches the metacentric parabola, for if FRT is perpendicular to Axis II, $FR = RT$ and hence R lies on the tangent at the vertex.

Since perpendicular tangents intersect on the directrix the corresponding lift passes through C when the wind direction is along Axis II.

EXAMPLES VII

1. When an aerofoil is obtained by transforming a circle as in 7·31, prove that the moment of the forces, due to the pressures, about the centre of the circle is

$$2\pi\rho \, V^2 \, l^2 \sin 2\alpha.$$

2. If a circle whose centre is on the imaginary axis is transformed as in 7·31, show that the resulting aerofoil degenerates into a circular arc described twice, while if the centre is on the real axis, a symmetrical aerofoil is obtained.

3. Apply the geometrical construction for the simple Joukowski transformation to the following:

(*a*) A circle whose centre is the origin and whose radius is the radius l of the transformation.

(*b*) A circle, centre the origin, and radius greater than l.

4. Discuss the type of transformation that will convert the flow past a circular cylinder (with or without circulation) to the flow past a body of aerofoil shape in a perfect fluid. Explain in particular how the flow past a circular arc and past a strut may be found.

How is the transformation used to find the flow past an aerofoil of fixed shape at varying angles of attack? (U.L.)

5. In the usual notation for two-dimensional motion of a perfect liquid, determine w as a function of z for a stream of velocity (U, V) flowing past a fixed circular cylinder $| z - z_0 | = b$, when there is a circulation I round the cylinder.

By applying the transformation

$$z' = z + a^2/z,$$

where $| a - z_0 | = b$, z_0 being small and a real and positive, obtain the solution for a stream past an aerofoil with a Joukowski profile, and explain how the circulation can be chosen so as to make the velocity finite at the cusp. (U.L.)

6. Prove that the velocity potential and stream function when a stream of speed V impinges, at an angle α to the axis of x, on a stationary circular cylinder of radius b, whose centre is the point $(c, 0)$ and round which there is a circulation I, are given by

$$w = V\left\{(z-c)e^{-i\alpha}+\frac{b^2e^{i\alpha}}{z-c}\right\}+\frac{iI}{2\pi}\log(z-c).$$

Applying the transformation $z' = z+a^2/z$, where $a = b-c$, show that, if c is small, the transformation gives the flow of the same stream past an aerofoil having a symmetrical Joukowski profile, and that the condition that the velocity of the fluid remains finite at the cusp is

$$I+4\pi Vb\sin\alpha = 0.$$

Hence calculate the force on the aerofoil. (U.L.)

7. The boundary of an obstacle in the z-plane is mapped on the circle $|\zeta| = a$ in the ζ-plane by the transformation

$$\zeta = z+\frac{a_1}{z}+\frac{a_2}{z^2}+\ldots.$$

Show that the motion of the liquid past the obstacle is given by a complex potential of the form

$$\phi+i\psi = V(e^{i\alpha}\zeta+a^2e^{-i\alpha}\zeta^{-1})+(ik/2\pi)\log(\zeta/a).$$

Prove that the resultant force per unit length of the obstacle has a magnitude $k\rho V$ and that the resultant moment per unit length about the centre of the circle is

$$2\pi\rho\, b^2V^2\sin 2(\alpha+\mu),$$

where

$$a_1 = -b^2\exp(2i\mu).$$

(U.L.)

8. A circle $|\zeta| = a$ is transformed into a thin aerofoil section by the equation

$$z = \zeta\left(1+\Sigma a_n\frac{a^n}{\zeta^n}\right).$$

Show how to find the complex quantities a_n in terms of the thickness and camber of the aerofoil.

Obtain the lift formula

$$L = 4\pi\rho aV^2\sin(\alpha+\beta)$$

and show that the moment about the centre is

$$M = 2\pi\rho\, b^2V^2\sin 2(\alpha+\mu),$$

where α is the angle of attack and β, b, μ are constants of the transformation. (U.L.)

9. An aerofoil is derived from the circle $|\zeta-be^{i\beta}| = a$ by the conformal transformation

$$z = \zeta+\sum_{r=1}^{n}\frac{a_r}{\zeta^r},$$

which is such that the zeros of $dz/d\zeta$ all lie within the circle except one which falls on the circumference at $\zeta = -l = be^{i\beta}-ae^{i\alpha}$, where a, b, l are real and in general the coefficients a_r are complex. Show that, if the circulation about the aerofoil is chosen in accordance with Joukowski's hypothesis, then there is a lift at right angles to a steady stream in which the aerofoil is placed, vanishing for a certain angle of incidence, and find the form of a_1 if the moment about the centre of the circle vanishes with the lift. (U.L.)

10. The transformation $z' = z + l^2/z$, where l is real, transforms the circle $|z + l - ae^{i\beta}| = a$, where a, β are real, into a Joukowski profile in the z'-plane with a cusp at $z' = -2l$. Show that the tangent to the cusp makes an angle 2β with the axis of x'.

If the profile is fixed in a stream of incompressible inviscid fluid of density ρ, whose velocity at infinity is given by $-u + iv = Ve^{i\alpha}$, and the circulation about the profile is chosen so that no infinite velocities occur in the fluid, show that the fluid streams tangentially past the cusp at a rate $Vl \cos(\alpha + \beta)/a$ and that the turning moment M of the pressures on the profile about the point $z' = -l + ae^{i\beta}$ is given by $M = 2\pi\rho l^2 V^2 \sin 2\alpha$. (U.L.)

11. Show that the domain outside the circle $|Z| = a$ in the Z-plane is transformed into the domain outside a circular arc in the z-plane by the conformal relation

$$\frac{z - a\,e^{2i\alpha}}{z - a\,e^{-2i\alpha}} = \left(\frac{Z - ia\,e^{i\alpha}}{Z + ia\,e^{-i\alpha}}\right)^2,$$

where the circular arc subtends an angle 4α at its centre. Show also that z/Z tends to $\sin\alpha$ at infinity.

A cylinder whose section is the above circular arc is placed in a stream of fluid in which the velocity at a great distance from the cylinder is V. This velocity is perpendicular to the generators and makes a positive angle β with the radius from the centre to the middle point of the arc. If, in addition, there is a circulation κ round the cylinder in the positive sense, show that the complex potential w can be derived by eliminating Z from the above relation and the equation

$$w = -V \sin\alpha \left(Ze^{-i\beta} + \frac{a^2}{Ze^{-i\beta}}\right) + \frac{i\kappa}{2\pi} \log Z.$$

Prove that the velocity at the upper edge is finite when, and only when,

$$\kappa = 2\pi a V[\sin\beta + \sin(2\alpha - \beta)]. \qquad \text{(M.T.)}$$

12. The circle $|\zeta| = a$ is transformed into a flat profile by $z = \zeta + a^2/\zeta$. Prove that near $z = 2a$

$$\frac{d\zeta}{dz} = \tfrac{1}{2} \frac{\sqrt{a}}{\sqrt{(z - 2a)}} + \dots.$$

13. In the Joukowski profile, fig. 7·31 (iii), show that if the centre of the circle is the point $se^{i\mu}$, the chord $1''7''$ is

$$4l + \frac{4s^2 \cos^2\mu}{l + 2s\cos\mu},$$

and that for slender profiles of small camber the chord is approximately $4a$.

14. Apply the construction of 7·31 to draw the profiles obtained from the circles through B whose centres are the points

$$\frac{l}{10} e^{i\pi/6}, \quad \frac{l}{10} e^{i\pi/3}$$

respectively. Measure the camber and thickness ratio in each case.

15. The circle $|\zeta| = a$ in the ζ-plane is transformed into a thin aerofoil section in the z-plane by the equations

$$z' = \zeta\left[1 + \sum_{n=1}^{\infty} A_n (a/\zeta)^n\right], \quad A_n = B_n + iC_n, \quad z = z' + b^2/z',$$

where b is real and nearly equal to a, and is so chosen that the point in the ζ-plane corresponding to $z' = -b$ lies on the circle; and B_n, C_n are small. Show how the constants B_n, C_n are related to the thickness of the aerofoil and to the ordinate of its middle line.

Prove that, if the circulation is adjusted to make the velocity finite at the trailing edge, the lift coefficient of the aerofoil is $C_L = 2\pi(\alpha+\lambda)$ where α is its incidence, supposed small, and

$$\lambda = \sum_{n=1}^{\infty} (-1)^{n-1} C_n = \frac{1}{2\pi b}\int_0^{\pi} \frac{y\,d\theta}{1+\cos\theta},$$

y being the ordinate of the middle line at a distance $2b\cos\theta$ from the point half-way between the leading and trailing edges.

Show also that the moment coefficient about the leading edge is approximately $\frac{1}{4}C_L+\frac{1}{2}\pi\lambda-\frac{1}{4}\pi C_2$, and that, if the moment coefficient vanishes when the lift is zero, then the centre of pressure is at a quarter of the chord from the leading edge for all (small) values of the incidence. (M.T.)

16. Calculate the lift on an elliptic cylinder in shear flow.

CHAPTER VIII

SOURCES AND SINKS

8·10. Two-dimensional source. *Definition.* If the two-dimensional motion of a liquid consists of outward radial flow from a point, symmetrical in all directions in the reference plane, the point is called a *simple source.*

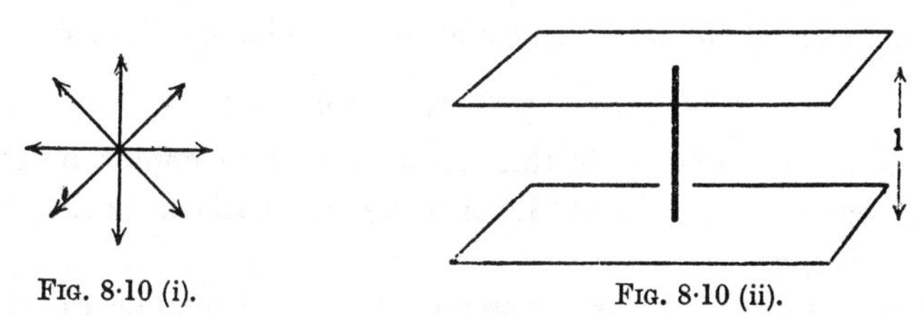

FIG. 8·10 (i). FIG. 8·10 (ii).

A two-dimensional source can be regarded as a straight axis (of unit length between two fixed planes) which emits fluid in the manner described.

Definition. If $2\pi m$ is the rate of emission of *volume* per unit time, m is called the *strength of the source.**

A source is a purely abstract conception which does not occur in nature. The idea is nevertheless useful, as will appear, for we can describe many fluid motions as due to sources which are outside the boundaries of the fluid which we consider.

A source is thus a point at which fluid is continuously created and distributed. Since the velocity near a source is very great, Bernoulli's theorem demands a great *negative* pressure. This fact alone shows that a source in the above sense can have no actual existence. An expanding bubble of gas pushes away the surrounding fluid and so imitates a source. When the rate of emission is constant, not intermittent, the source is said to be *steady.*

Definition. A *sink* is a negative source.

Thus a sink is a point of inward radial flow at which fluid is absorbed or annihilated continuously.

If u_r is the radial velocity at distance r from the source the flux out of the circle of radius r is

$$2\pi r u_r = 2\pi m.$$

* Some writers denote by m the rate of emission of volume. What we call the strength would then be $m/2\pi$. The object of our notation is to avoid the recurrence of the factor 2π in the subsequent analysis, cf. strength of circulation, 7·10.

Thus $$u_r = \frac{m}{r},$$

and this is the entire velocity for an isolated source.

8·12. The complex potential for a simple source. If the source of strength m is at the origin, the velocity at (r, θ) is m/r radially. Therefore

$$-\frac{dw}{dz} = u - iv = \frac{m}{r}(\cos\theta - i\sin\theta) = \frac{m}{z},$$

$$w = -m\log z.$$

The stream function is $\psi = -m\theta$.

If the source is at the point z_0, we have, by a change of origin,

$$w = -m\log(z - z_0).$$

It is interesting to compare this result with the complex potential for a vortex of strength κ given in 13·21. A vortex is (mathematically) a source of imaginary strength.

It will be observed that as r increases, the speed diminishes, so that at a great distance from the source the fluid is almost quiescent.

It is characteristic of a source (or sink) that the speed tends to infinity as we approach the source, and therefore in the immediate neighbourhood of the source the velocity is always radial, no matter how the fluid may be moving at distant points.

8·20. Combination of sources and streams. The motions due to a uniform stream, and any number of sources, can be obtained by addition of the corresponding complex potentials, when no boundaries occur in the liquid.

To prove this, consider the complex potential

$$w = -Uz - m_1\log z - m_2\log(z - z_0).$$

We show that this potential gives a uniform stream U at infinity, and sources of strengths m_1, m_2 at $z = 0$, $z = z_0$.

Since $$u - iv = -\frac{dw}{dz} = U + \frac{m_1}{z} + \frac{m_2}{z - z_0},$$

when $|z| \to \infty$, we have $u = U$, $v = 0$, so that there is a uniform stream. Again, near $z = z_0$, put $z = z_0 + re^{i\theta}$, where r is small. Then

$$u - iv = U + \frac{m_1}{z_0 + re^{i\theta}} + \frac{m_2}{r}e^{-i\theta},$$

the first two terms are negligible compared with the third, and therefore

$$u = \frac{m_2}{r}\cos\theta, \quad v = \frac{m_2}{r}\sin\theta,$$

so that there is an outward radial flow from z_0, due to a source of strength m_2 at that point.

In the same way we prove that there is a source of strength m_1 at the origin. The proof can clearly be extended to any number of sources and sinks.

We have insisted on proving this additive property, for it is not obvious for sources and is in general not applicable in other cases.

For example, the flow past a circular cylinder, centre at the origin, is given by the complex potential

$$U\left(z+\frac{a^2}{z}\right),$$

and the motion due to a source at z_0 is given by $-m \log(z-z_0)$.

If we add these, we get

$$U\left(z+\frac{a^2}{z}\right)-m \log(z-z_0),$$

which is indeed the complex potential of some motion, but not that of the stream past a cylinder in the presence of the source. The failure of the additive property is obvious, for the stream function does not become constant on the circle $r = a$, so that the cylinder is not a streamline.

8·21. Source in a uniform stream. Let us combine a source of strength m at the origin with a stream U parallel to the x-axis. Then we may add the potentials (8·20) and so obtain

$$w = -Uz - m \log z,$$

$$\frac{dw}{dz} = -U - \frac{m}{z}.$$

So $z = -\dfrac{m}{U}$ is the only stagnation point and lies on the real axis at the

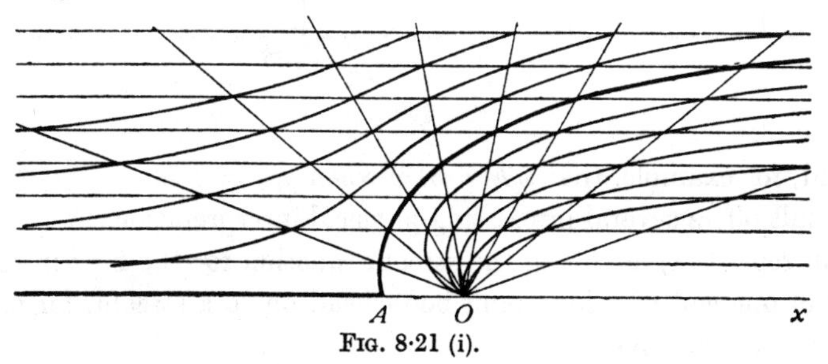

FIG. 8·21 (i).

point where the stream velocity and the velocity due to the source neutralise one another. The stream function is

$$\psi = -Uy - m \tan^{-1}\frac{y}{x} = -Uy - m\theta, \tag{1}$$

and the streamlines are easily drawn by Rankine's method, which gives fig. 8·21 (i).

We see that the streamlines are symmetrical about the x-axis, across which there is no flow. The dividing streamline passes through the stagnation point A, and this curve separates the flow into two parts.

We could therefore suppose this curve to be replaced by a solid wall. The stream function (1) would then give the disturbance in a uniform stream due to the presence of this wall, and the source would be outside the fluid, and thus we could have a representation of an actual motion.

Consider the part of the flow for which $y \geqslant 0$. If we measure θ counter-clockwise from the value zero for points on the positive part of the x-axis, on the negative part of the x-axis we have $y = 0$ and $\theta = \pi$. Therefore (1) gives, for this part, $\psi = -m\pi$ and the dividing streamline is

$$-m\pi = -Uy - m\theta,$$

which includes the wall and the negative part of the x-axis. When $\theta \to 0$, $y \to m\pi/U = h$, say, and therefore the asymptote is $y = h$. By symmetry there is a second asymptote $y = -h$. From (1) the equation of the wall is then

$$\frac{1}{x} = -\frac{1}{y}\tan\frac{\pi y}{h}, \quad \text{and} \quad OA = \frac{h}{\pi}.$$

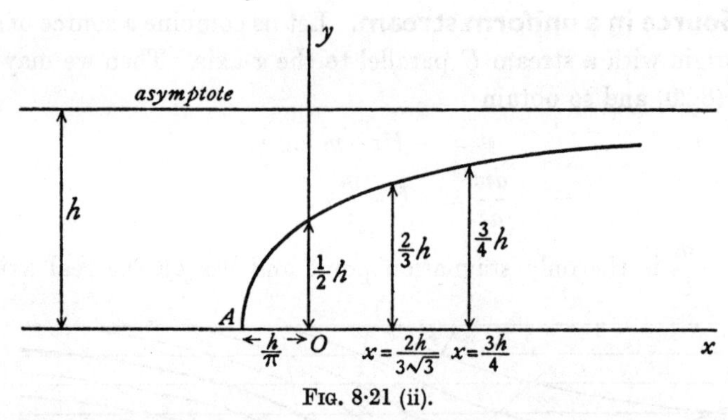

FIG. 8·21 (ii).

We find, for example, that $x/h = 31{\cdot}9$ when $y/h = 0{\cdot}99$.

The result offers several interesting physical interpretations.

We can regard (1) as giving the stream function for the flow in the neighbourhood of one end of a long bluff body head on to a stream, for example a long island in a wide river.

Again, if we confine our attention to the part above the x-axis, we have the flow pattern at the bottom of an ocean where the level changes from 0 to h by fairly gradual stages.

We can also regard the picture as representing the flow of wind meeting a cliff. In this connection it is interesting to note that the stagnation point A would be the most sheltered place.

We can moreover regard any streamline as a rigid wall, and the flow of wind over more gradually sloping ground would be so represented, but then there would be no stagnation point.

8·22. Source and sink of equal strengths. Let there be a source and sink, each of strength m, at the points B, A,

$$a\,e^{i\alpha}, \quad -a\,e^{i\alpha}$$

respectively. Then

$$w = -m \log (z - a\,e^{i\alpha}) + m \log (z + a\,e^{i\alpha}),$$

so that, if P is the point z,

$$\psi = -m \times (\text{angle } APB).$$

The streamlines are given by ψ = constant, or $\angle APB$ = constant, and are therefore coaxal circles passing through A and B.

The flow is directed from the source to the sink, so that the parts of a given circle on either side of AB are described in opposite senses. The line

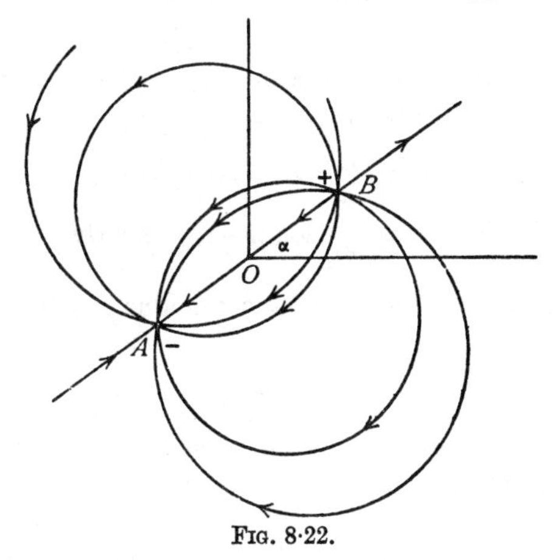

FIG. 8·22.

through AB is a limiting case of the circles. The directions of flow in this line are indicated in fig. 8·22.

8·23. Doublet, or double source. In the case of a source and equal sink just considered, suppose that A and B are very close together, so that a is small. Then

$$w = -m \log \left[z \left(1 - \frac{a\,e^{i\alpha}}{z} \right) \right] + m \log \left[z \left(1 + \frac{a\,e^{i\alpha}}{z} \right) \right]$$

$$= -m \log \left(1 - \frac{a\,e^{i\alpha}}{z} \right) + m \log \left(1 + \frac{a\,e^{i\alpha}}{z} \right)$$

$$= \frac{2ma\,e^{i\alpha}}{z} + \frac{2ma^3\,e^{3i\alpha}}{3z^3} + \ldots,$$

using the logarithmic series for the expansion of $\log(1+k)$. Let $2ma = \mu$. Then

$$w = \frac{\mu\, e^{i\alpha}}{z} + \frac{\mu a^2\, e^{3i\alpha}}{3z^3} + \dots .$$

Now let $a \to 0$, μ remaining constant so that $m \to \infty$. Then when A and B coincide, we get

$$w = \frac{\mu\, e^{i\alpha}}{z}.$$

This combination of an infinite source and sink at an infinitesimal distance apart is called a *doublet* of strength μ. The streamlines are still circles, fig. 6·0 (ii), having a common tangent which makes an angle α with the x-axis. This common tangent is called the *axis* of the doublet, the positive direction along the axis being reckoned from sink to source. The physical conception of a doublet may be helped by considering it as approximately represented by a short length of two-dimensional tube into one end of which fluid flows to emerge immediately from the other end, the direction of the tube being the axis of the doublet.

The complex potential can be obtained in another way which is instructive. Consider a sink at z_0 and a source at $z_0 + \delta z_0$.

Then $$w = -m \log(z - z_0 - \delta z_0) + m \log(z - z_0)$$

$$= -m\, \delta z_0 \frac{\partial}{\partial z_0} \log(z - z_0) \quad \text{nearly.}$$

Let $\delta z_0 = r\, e^{i\alpha}$. Then, if $mr = \mu$, and remains constant while $r \to 0$, we get

$$w = \frac{\mu\, e^{i\alpha}}{z - z_0}$$

for the complex potential of a doublet of strength μ at the point z_0, the axis of the doublet being in the direction α.

8·24. Green's equivalent stratum of doublets. Consider irrotational motion of a liquid, in the region L of fig. 5·40 bounded by the contour C, to be given by a complex potential $w(z)$ holomorphic throughout L. This condition excludes singularities and circulations. Then by Cauchy's formula (5·59)

$$(1) \qquad \frac{1}{2\pi i} \int_{(C)} \frac{w(\zeta)\, d\zeta}{\zeta - z} = w(z) \quad \text{or} \quad 0$$

according as z is in L or outside L.

If ds is an element of arc of C drawn in the positive direction of description, we can write $d\zeta = ds\, e^{i\theta}$, and therefore for a point z of L, (1) gives

$$(2) \qquad w(z) = \frac{1}{2\pi i} \int_{(C)} \frac{w(\zeta) e^{i\theta}\, ds}{\zeta - z} = \int_{(C)} \frac{\mu e^{i\chi}\, ds}{z - \zeta},$$

where

(3) $$\mu e^{i\chi} = w(\zeta)e^{i(\theta+\frac{1}{2}\pi)}/(2\pi),$$

and this equation defines a real positive μ and an angle χ. Now $\mu e^{i\chi}\, ds/(z-\zeta)$ is the complex potential at z of a doublet at ζ of strength $\mu\, ds$, whose axis is in the direction χ. Therefore (2) shows that the complex potential $w(z)$ could be imagined to arise from a continuous distribution of doublets ranged round the contour C whose density per unit length of arc is given by (3). This distribution is known as *Green's equivalent stratum of doublets*. For another type of stratum, also due to Green, see 13·64.

Note that if z is outside L, the velocity due to the distribution is zero, for as (1) shows, $w(z)$ is then constant, namely zero.

If the region L is doubly connected (as in fig. 9·11), we can modify it to be simply connected by inserting an imagined barrier AB. The foregoing considerations then apply if we also place a distribution on each face of AB.

8·30. Source and equal sink in a stream. Let there be a source of strength m at A $(a, 0)$, and a sink of strength m at $B\,(-a, 0)$, and a uniform stream

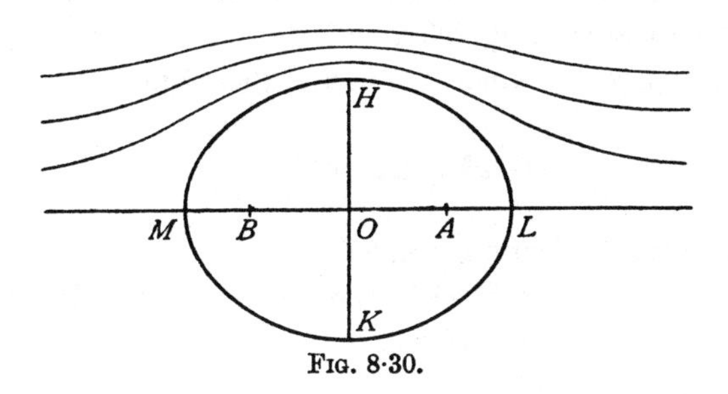

FIG. 8·30.

U parallel to the real axis. The interesting case arises when the stream is in the direction from source to sink, that is, in the direction of x negative.

Here $$w = Uz - m\log(z-a) + m\log(z+a).$$

The stagnation points are given by

$$U - \frac{m}{z-a} + \frac{m}{z+a} = 0,$$

and therefore $$z = \pm\sqrt{\left(a^2 + \frac{2am}{U}\right)}.$$

Let $$\frac{2am}{U} = b^2 - a^2.$$

Then $$z = \pm b,$$

so that the stagnation points are L, M, where $OL = OM = b$.

The stream function is

$$\psi = Uy - m\tan^{-1}\frac{y}{x-a} + m\tan^{-1}\frac{y}{x+a} = Uy - m\tan^{-1}\frac{2ay}{x^2+y^2-a^2}.$$

The streamline $\psi = 0$ contains the real axis $y = 0$, and therefore the dividing streamline is

$$\frac{2ay}{x^2+y^2-a^2} = \tan\frac{Uy}{m}.$$

Rearranging, we get

$$x^2+y^2-a^2 = 2ay\cot\frac{Uy}{m} = 2ay\cot\frac{2ay}{b^2-a^2}.$$

This equation represents a curve which is symmetrical about both axes, for if the point (x, y) lies on it so do the points $(\pm x, \pm y)$.

The value of y cannot become infinite on this curve, for as we go away from AB the stream becomes parallel to the x-axis. Therefore the curve is a closed oval of the type indicated in fig. 8·30.

Let $OH = c$, then $y = c$ when $x = 0$, and therefore

$$c^2 - a^2 = 2ac\cot\frac{2ac}{b^2-a^2},$$

and the value of c can be found graphically.

If we take this curve as a fixed boundary, we get the flow past a cylinder whose cross-section is the above curve.

When a is small,

$$\cot\frac{2ac}{b^2-a^2} = \frac{b^2-a^2}{2ac} \text{ approximately},$$

and therefore $$c^2 = b^2.$$

Thus as $a \to 0$, $c \to b$, and the oval becomes a circle. In this case the source and sink form a doublet, and we have again the flow round a circular cylinder of radius b. The strength of the doublet is $2am = \mu$, and therefore

$$\psi = Uy - \frac{b^2 Uy}{x^2+y^2},$$

which agrees with the result already obtained in 6·22.

The complex potential for the flow past a circular cylinder is

$$w = Uz + \frac{Ua^2}{z}.$$

The first term represents the stream, the second term the disturbance due to the cylinder. Thus a cylinder of radius a placed in a stream of velocity U behaves as would a doublet of strength Ua^2 on the axis of the cylinder.

8·31. Two equal sources. Equal sources of strength m at the points $A(a, 0)$, $B(-a, 0)$, give

(1) $w = -m \log(z-a) - m \log(z+a)$, $\phi + i\psi = -m \log(x^2 - y^2 - a^2 + 2ixy)$.

The stream function is therefore

$$\psi = -m \tan^{-1} \frac{2xy}{x^2 - y^2 - a^2},$$

which gives

$$x^2 + 2xy \cot \frac{\psi}{m} - y^2 = a^2. \tag{2}$$

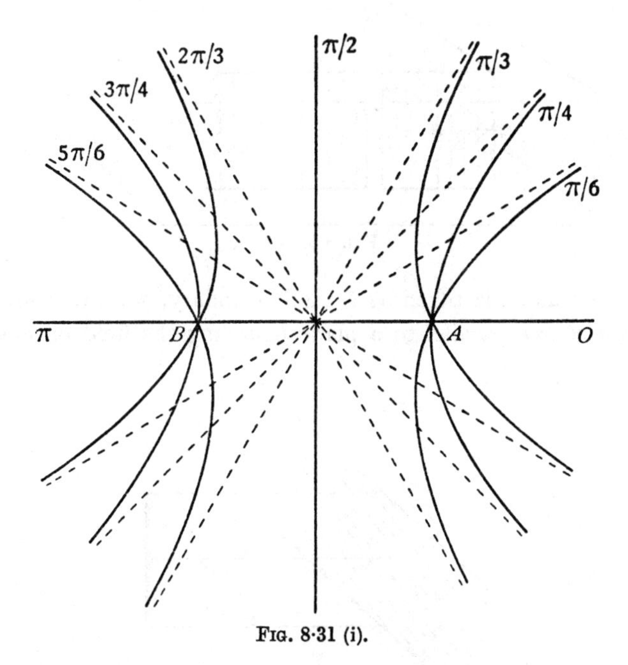

Fig. 8·31 (i).

Thus the streamlines are rectangular hyperbolas with centres at the origin. This is easily seen, for (2) can be written in the form

$$\left(x + y \cot \frac{\psi}{2m}\right)\left(x - y \tan \frac{\psi}{2m}\right) = a^2,$$

so that the asymptotes obtained by equating each factor of the left side to zero are at right angles.

By giving $\psi/(2m)$ successively the values

$$0, \frac{\pi}{6}, \frac{\pi}{4}, \frac{\pi}{3}, \frac{\pi}{2}, \frac{2\pi}{3}, \frac{3\pi}{4}, \frac{5\pi}{6}, \pi,$$

we obtain fig. 8·31 (i) in which the dotted lines are the asymptotes.

The axes of reference are streamlines, intersecting at right angles at the origin which is a stagnation point.

Since the flow is ultimately along the asymptotes, at a great distance the two sources behave like a single source of strength $2m$ placed midway between them.

If we replace the streamlines Ox, Oy by rigid walls, (1) then gives the flow into an infinite region bounded by two rectangular walls through a narrow slit in one of the walls as indicated in fig. 8·31 (ii).

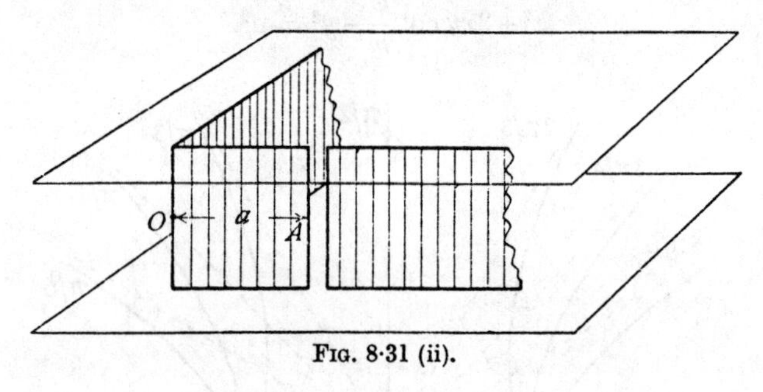

Fig. 8·31 (ii).

If the y-axis alone is taken as a rigid boundary, we obtain from (1) the flow due to a source parallel to a plane bounding the fluid on one side, fig.

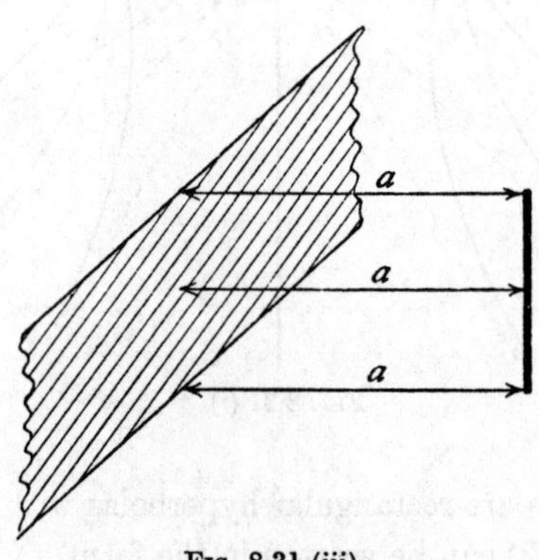

Fig. 8·31 (iii).

8·31 (iii), where we suppose as usual that the fluid is also bounded by parallel planes at unit distance apart.

This last result is of great theoretical importance, for it forms the foundation of the method of images which we proceed to discuss.

8·40. The method of images. We have seen in the preceding section that the flow due to a source m at $A(a, 0)$ in the presence of a plane represented by the y-axis is given by

$$w = -m \log(z-a) - m \log(z+a),$$

and this is the same complex potential as would be obtained if we placed a source m at the point $B(-a, 0)$ and imagined the fluid to have access to the whole region on both sides of the plane $x = 0$. The y-axis being a streamline for this system, the plane could be considered as removed. The source m at B is called the image in the given plane barrier of the source m at A. This is the simplest case of the method of images, which may be briefly described as follows.

Suppose that a system S of sources and sinks exists in fluid possessing one or more boundaries C. If by placing a system S' of sources and sinks in the region outside C and then allowing the fluid to have access to the whole region, we get C as a streamline, then the system S' is said to be the image in C of the system S.

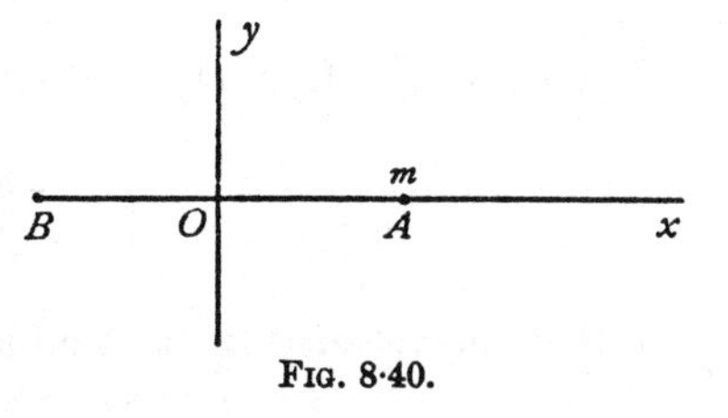

FIG. 8·40.

In the case of the source parallel to a plane, the system S consists of the single given source at A, the boundary C consists of the given plane, and the image system S' consists of the source at B.

We note that B is the optical image of A in the given plane regarded as a reflecting surface.

8·41. Effect on a wall of a source parallel to the wall. Let the source be at $A(a, 0)$ and the wall at $x = 0$. The image of the source in the wall is an equal source at $B(-a, 0)$, and therefore

$$w = -m \log(z-a) - m \log(z+a) = -m \log(z^2 - a^2),$$

$$\frac{dw}{dz} = \frac{-2mz}{z^2 - a^2}.$$

Now on the wall $x = 0$, and therefore

$$\frac{dw}{dz} = \frac{-2miy}{-y^2 - a^2}, \quad \frac{d\bar{w}}{d\bar{z}} = \frac{2miy}{-y^2 - a^2},$$

and

$$q^2 = \frac{4m^2 y^2}{(y^2 + a^2)^2}, \tag{1}$$

where q is the velocity, which is directed along the wall since this is a streamline.

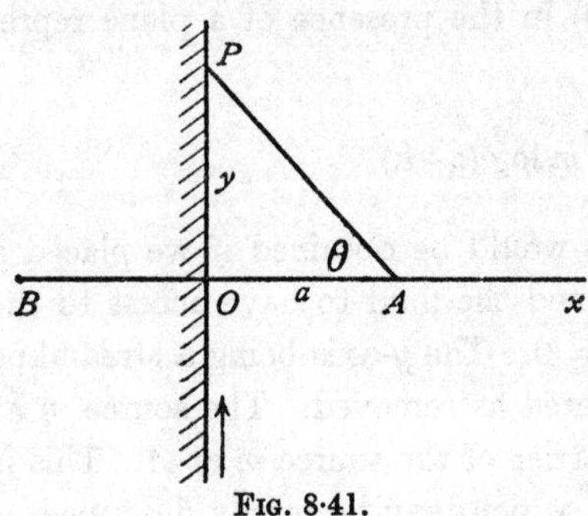

FIG. 8·41.

The pressure on the wall is therefore given by

$$\frac{p}{\rho} = \frac{\Pi}{\rho} - \tfrac{1}{2}q^2 = \frac{\Pi}{\rho} - \frac{2m^2y^2}{(y^2+a^2)^2},$$

where Π is the pressure at infinity.

If the liquid were at rest, the pressure would be Π everywhere. Thus the effect of the motion is to diminish the pressure on the wall. Hence the wall is urged towards the source with a force (per unit breadth of wall) given by

$$(2) \qquad F = \int_{-\infty}^{+\infty} \frac{\rho \, . \, 2m^2y^2}{(y^2+a^2)^2}\, dy = \frac{\pi \rho m^2}{a}.$$

We also see, from (1), that on the wall, putting $y = a \tan\theta$,

$$q = \frac{2ma\tan\theta}{a^2\sec^2\theta} = \frac{m}{a}\sin 2\theta,$$

so that the velocity on the wall is greatest when $\theta = \pm\pi/4$.

8·42. General method for images in a plane. We can proceed on much the same lines as for the circle theorem of 6·21. Thus if

$$f(z) = -\Sigma m_r \log(z - z_r)$$

is the complex potential of sources and sinks all of which lie in the half-plane $y>0$, the insertion of the plane barrier $y = 0$ leads to the complex potential

$$w = f(z) + \bar{f}(z) = -\Sigma m_r \log(z - z_r) - \Sigma m_r \log(z - \bar{z}_r),$$

since on $y = 0$, we have $z = \bar{z}$ so that w is purely real and $y = 0$ is the streamline $\psi = 0$. Moreover if z_r lies in the region $y>0$, then $\bar{z}_r$ lies in the region $y<0$ so that this process introduces no new singularities into the region $y>0$.

Similarly, if all the sources and sinks lie in the half-plane $x>0$ the complex potential when the plane barrier $x = 0$ is inserted is

$$w = f(z) + \bar{f}(-z) = -\Sigma m_r \log(z - z_r) - \Sigma m_r \log(-z - \bar{z}_r),$$

for here, on $x = 0$, we have $-z = \bar{z}$ and so $\psi = 0$. This method is applicable equally well to the rectilinear vortices of Chapter XIII.

8·43. Image of a doublet in a plane. Taking a two-dimensional doublet of strength μ and inclination α to the x-axis, we can regard this as the limit of a sink at A and a source at B, where AB makes an angle α with the x-axis.

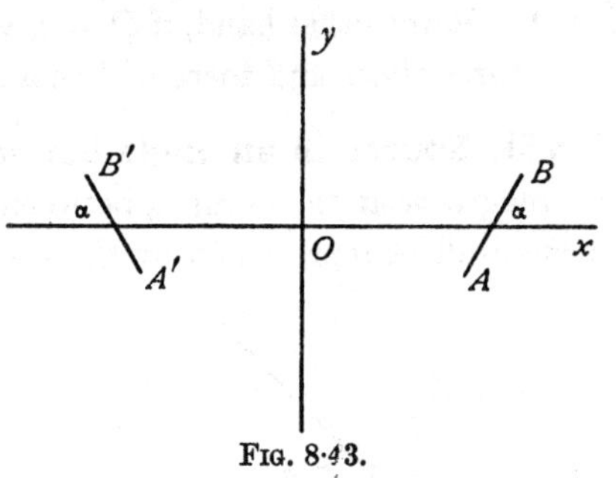

FIG. 8·43.

The images of the source and sink are at B', A', the optical images of B, A in the y-axis regarded as representing the given plane. Proceeding to the limit when $A \to B$, we have $A' \to B'$, and the image therefore is a doublet of equal strength, making an angle $\pi - \alpha$ with the x-axis, in fact, an equal antiparallel doublet.

Using the method of 8·42 we have for the isolated doublet at the point z_0, the potential $f(z) = \mu e^{i\alpha}/(z - z_0)$, and therefore, in the presence of the plane $x = 0$,

$$w = \frac{\mu e^{i\alpha}}{z - z_0} - \frac{\mu e^{-i\alpha}}{z + \bar{z}_0}.$$

8·50. Sources in conformal transformation. If we map the z-plane on the ζ-plane by a conformal transformation

$$\zeta = f(z),$$

a source in the z-plane will transform into a source at the corresponding point of the ζ-plane.

Proof. Let there be a source of strength m at the point P, z_0 in the z-plane, and let Π, ζ_0 in the ζ-plane correspond to P. Draw a small circle c, centre z_0, and let γ be the corresponding curve in the ζ-plane. This curve γ must enclose Π.

Since the stream function ψ has the same value at corresponding points in both planes,

$$-2\pi m = \int_{(c)} d\psi = \int_{(\gamma)} d\psi.$$

We can take c as small as we please and γ will also diminish, but the integral of $d\psi$ round γ will remain constant and therefore there will be a source at Π. If γ encircles Π once only (the usual case), the sources will be of the same strength in the two planes.

If γ encircles Π n times when c encircles P once, the source at Π will be of strength m/n.

As an example, suppose $\zeta = z^3$, and that there is a source of strength m at $z = 0$.

Since $$\arg \zeta = 3 \arg z$$

when $\arg z$ increases by 2π, $\arg \zeta$ will increase by 6π and therefore γ will encircle $\zeta = 0$ three times. Therefore there is a source of strength $m/3$ at $\zeta = 0$. On the other hand, if $\zeta^3 = z$, γ will encircle $\zeta = 0$ once when c encircles $z = 0$ three times and there will be a source of strength $3m$ at $\zeta = 0$.

8·51. Source in an angle between two walls. Let there be a source of strength m at the point z_0 between two walls inclined at an angle π/n, and let one wall be represented by the x-axis.

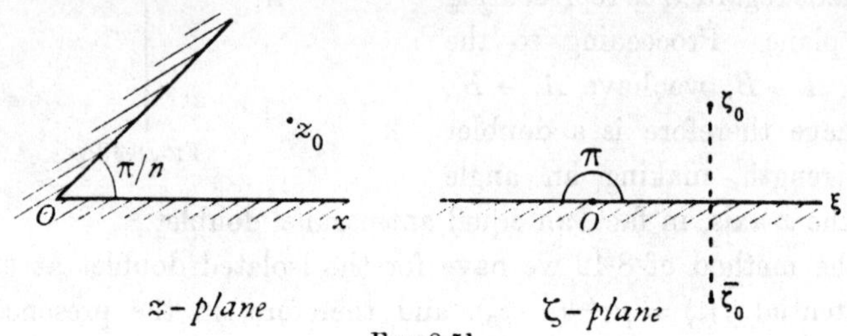

FIG. 8·51.

Consider the transformation

$$\zeta = z^n.$$

Then
$$\arg \zeta = n \arg z,$$

and therefore, as $\arg z$ increases from 0 to π/n, $\arg \zeta$ increases from 0 to π and the interior of the angle is mapped on the upper half of the ζ-plane.

A source at z_0 corresponds to an equal source at

$$\zeta_0 = z_0{}^n.$$

The image of this source is an equal source m at $\bar{\zeta}_0$, and therefore the complex potential is

$$w = -m \log(\zeta - \zeta_0) - m \log(\zeta - \bar{\zeta}_0).$$

Therefore in the z-plane

$$w = -m \log(z^n - z_0{}^n) - m \log(z^n - \bar{z}_0{}^n).$$

8·60. Source outside a circular cylinder. Let there be a source of strength m at $z = f$, where f is real, outside the cylinder radius a whose centre is at the origin. When the source is alone in the fluid the complex potential is $-m \log(z-f)$. Therefore by the circle theorem (6·21), when the cylinder is inserted,

$$w = -m \log(z-f) - m \log\left(\frac{a^2}{z} - f\right).$$

8·61. The image system for a source outside a circular cylinder. The complex potential, from the preceding section, if we add the constant $m \log(-f)$, is

$$w = -m \log(z-f) - m \log\left(z - \frac{a^2}{f}\right) + m \log z.$$

This is the complex potential of, fig. 8·61 (*a*),

(1) a source m at A, $z = f$;

(2) a source m at B, $z = a^2/f$;

(3) a sink $-m$ at the origin.

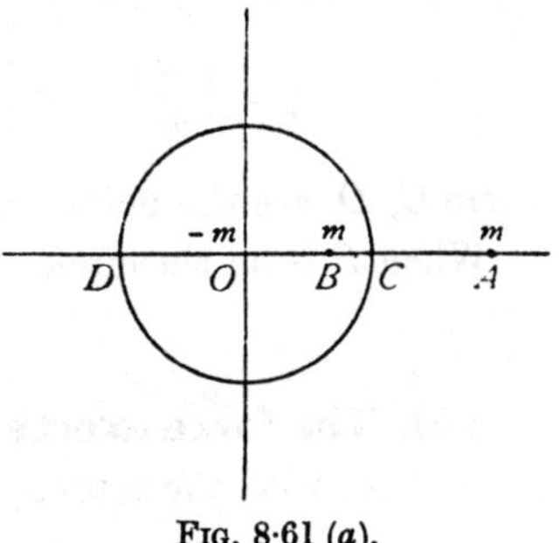

FIG. 8·61 (*a*).

Since $OA \,.\, OB = a^2$, A and B are inverse points with respect to the circular section of the cylinder, and therefore B is inside the cylinder.

Thus the image system for a source outside a circular cylinder consists of an equal source at the inverse point and an equal sink at the centre of the cylinder. The streamlines are shown in fig. 8·61 (*b*).

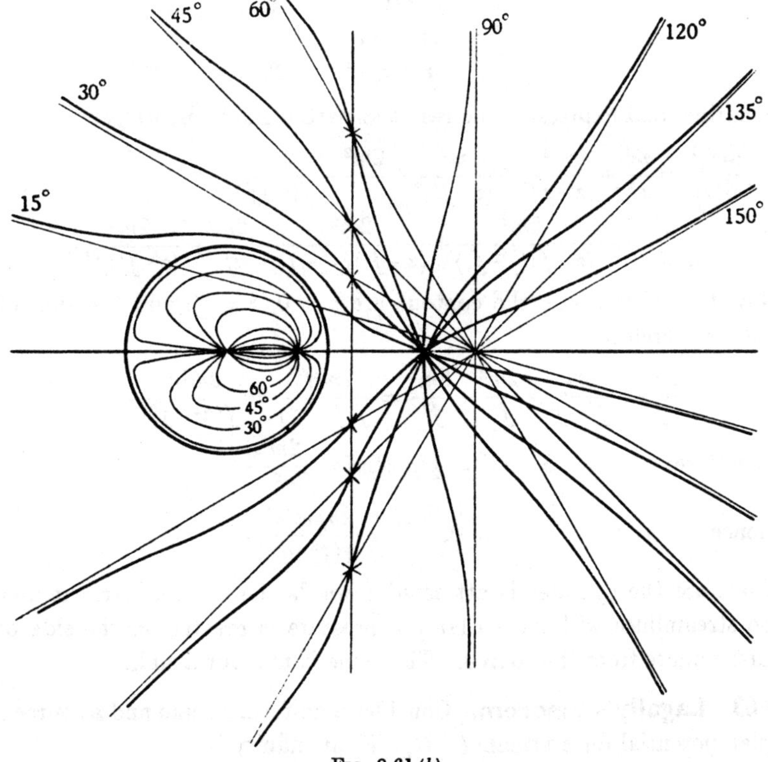

FIG. 8·61 (*b*).

As a corollary it follows that a source inside a cylinder and an equal sink at the centre has for image system an equal source at the inverse point of the given source.

The speed at any point P is given by

$$q = \left|\frac{dw}{dz}\right| = m\left|\frac{z^2-a^2}{(z-f)z(z-a^2/f)}\right| = m\frac{PC\,.\,PD}{PA\,.\,PO\,.\,PB},$$

where C, D, are the points in which AB cuts the circle.

When P is on the circle the triangles OBP, OPA are similar and

$$PB : PA = a : f.$$

8·62. The force exerted on a circular cylinder by a source. Taking fig. 8·61 (a) with the source at A on the x-axis, the theorem of Blasius gives

$$X - iY = \tfrac{1}{2}i\rho\int\left(\frac{dw}{dz}\right)^2 dz$$

taken round the contour of the cylinder. Now

$$w = m\log z - m\log(z-f) - m\log(z-f'),$$

where

$$OA = f, \quad OB = f' = a^2/f,$$

$$\frac{dw}{dz} = \frac{m}{z} - \frac{m}{z-f} - \frac{m}{z-f'}.$$

Squaring, and expressing the result in partial fractions, we get

$$\left(\frac{dw}{dz}\right)^2 = \frac{m^2}{z^2} + \frac{m^2}{(z-f)^2} + \frac{m^2}{(z-f')^2} + \frac{2m^2}{zf} - \frac{2m^2}{f(z-f)}$$
$$+ \frac{2m^2}{(z-f)(f-f')} + \frac{2m^2}{(z-f')(f'-f)} + \frac{2m^2}{zf'} - \frac{2m^2}{(z-f')f'}.$$

Now the poles inside the contour are $z = 0$, $z = f'$, and the sum of the residues is therefore

$$\frac{2m^2}{f} + \frac{2m^2}{f'-f} + \frac{2m^2}{f'} - \frac{2m^2}{f'} = \frac{2m^2 f'}{f(f'-f)} = \frac{2m^2a^2}{f(a^2-f^2)}.$$

Therefore
$$X - iY = \tfrac{1}{2}i\rho \times 2\pi i \times \frac{2m^2a^2}{f(a^2-f^2)}.$$

Hence
$$Y = 0, \quad X = \frac{2\pi\rho m^2a^2}{f(f^2-a^2)}.$$

Therefore the cylinder is attracted towards the source. An examination of the streamlines will show that the pressure is greater on the side of the cylinder remote from the source. The same is true for a sink.

8·63. Lagally's theorem. Consider a uniform stream and a source; the complex potential for a stream $(-U, -V)$ at infinity is

$$(U - iV)z - m\log(z-a),$$

the source of strength m being at $z = a$.

If we introduce a cylinder into the stream, the complex potential is "disturbed" by the addition of a function which must vanish at infinity, for the presence of the cylinder cannot affect the distant parts of the fluid. To make the case general, suppose there is a circulation of strength κ about the cylinder. Then the complete complex potential at great distances will be of the form

$$(1)\qquad w = (U-iV)z - m\log(z-a) + i\kappa\log z + \frac{A}{z} + \frac{B}{z^2} + \ldots,$$

the last terms giving the effect of the circulation and the disturbance introduced by the cylinder.

The complex velocity is then given by

$$(2)\qquad -\frac{dw}{dz} = -(U-iV) + \frac{m}{z-a} - \frac{i\kappa}{z} + \frac{A}{z^2} + \frac{2B}{z^3} + \ldots.$$

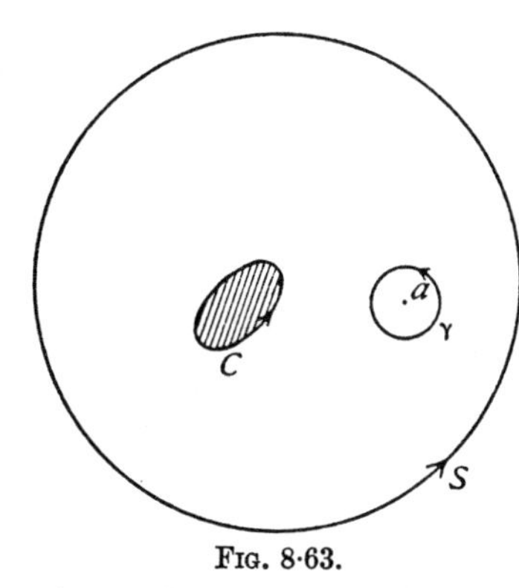

FIG. 8·63.

To find the force on the cylinder, we have, from the theorem of Blasius,

$$X - iY = \frac{i\rho}{2}\int_{(C)}\left(\frac{dw}{dz}\right)^2 dz.$$

Now let S be a circle of great radius which includes both the cylinder and the source. By the method of 5·54, we can enlarge the contour to S and so write

$$\int_{(S)}\left(\frac{dw}{dz}\right)^2 dz = \int_{(C)}\left(\frac{dw}{dz}\right)^2 dz + \int_{(\gamma)}\left(\frac{dw}{dz}\right)^2 dz,$$

where γ is a small contour drawn round the source. Thus

$$(3)\qquad X - iY = \frac{i\rho}{2}\int_{(S)}\left(\frac{dw}{dz}\right)^2 dz - \frac{i\rho}{2}\int_{(\gamma)}\left(\frac{dw}{dz}\right)^2 dz.$$

Now on the circle S, since $|z|$ is large, we can expand $1/(z-a)$ in powers of $1/z$, and therefore from (2),

$$-\frac{dw}{dz} = -(U-iV) + \frac{m}{z}\left(1 + \frac{a}{z} + \frac{a^2}{z^2} + \ldots\right) - \frac{i\kappa}{z} + \frac{A}{z^2} + \ldots,$$

and hence

$$\left(\frac{dw}{dz}\right)^2 = (U-iV)^2 - \frac{2(U-iV)(m-i\kappa)}{z} + \frac{A'}{z^2} + \frac{B'}{z^3} + \ldots,$$

where A', B', ... are certain constants. Thus, by the residue theorem,

$$(4)\qquad \tfrac{1}{2}i\rho\int_{(S)}\left(\frac{dw}{dz}\right)^2 dz = 2\pi\rho(U-iV)(m-i\kappa).$$

To calculate the second integral in (3), we see from (2) that

$$-\frac{dw}{dz} = f(z) + \frac{m}{z-a},$$

where

$$f(z) = -\frac{d}{dz}\left[w - (-m\log(z-a))\right].$$

The function $f(z)$ is therefore the complex velocity obtained by omitting the source from the original complex potential, and $f(z)$ is holomorphic within the contour γ. Now

$$(5) \qquad \left(\frac{dw}{dz}\right)^2 = (f(z))^2 + \frac{m^2}{(z-a)^2} + \frac{2mf(z)}{z-a}.$$

By Taylor's theorem

$$f(z) = f[(z-a)+a] = f(a) + (z-a)f'(a) + \ldots .$$

Hence the residue of $(dw/dz)^2$ at $z = a$ is $2mf(a)$.

Once more using the residue theorem, we get, from (3) and (4),

$$X - iY = 2\pi\rho(U-iV)(m-i\kappa) + 2\pi\rho m f(a).$$

$$(6) \qquad = -2\pi\rho i\kappa(U-iV) + 2\pi\rho m(U-iV+f(a)).$$

Now $f(a)$ is the complex velocity "*induced*" at a by the remaining part of the complex potential when we omit the source m. Thus, calling this induced velocity $u_m - iv_m$, we have finally

$$(7) \qquad X - iY = -2\pi\rho i\kappa(U-iV) + 2\pi\rho m(U-iV+u_m-iv_m),$$

and this is Lagally's theorem, which can clearly be extended to any number of sources by adding for each a term of the same type as the last term in the formula.

To find a corresponding expression for the moment M, we have

$$M + iN = -\tfrac{1}{2}\rho\int_{(S)}\left(\frac{dw}{dz}\right)^2 z\,dz + \tfrac{1}{2}\rho\int_{(\gamma)}\left(\frac{dw}{dz}\right)^2 z\,dz.$$

Calculating the residues by means of the preceding formulae, we get

$$M + iN = -\pi\rho i[(m-i\kappa)^2 - m^2 - 2(U-iV)(A+ma) - 2amf(a)]$$

$$(8) \qquad = 2\pi\rho iA(U-iV) + \pi\rho i\kappa(\kappa+2im) + 2\pi\rho ima(U-iV+u_m-iv_m)$$

and M is the real part of this expression.

Lagally's theorem assumes a striking form in the case where stream and circulation are absent, so that the cylinder disturbs the field of the source alone. In this case (7) gives

$$(9) \qquad X + iY = 2\pi\rho m(u_m + iv_m),$$

while the moment expression $M + iN = 2\pi\rho ima(u_m - iv_m)$ shows that M is the moment of the force (X, Y) acting at the point a.

It follows that the force on the cylinder is (X, Y) acting in a line through the source, and in the same direction as the velocity induced at the source.

We thus obtain the following theorem.

Theorem. A source of strength m in the presence of a cylinder exerts on the cylinder a complex force $2\pi\rho m(u_m + iv_m)$, where $u_m + iv_m$ is the velocity induced at the source by all causes except the source, and this force is localised in a line through the source.

When there are several sources, there is the appropriate force (9) localised at each.*

We can at once apply this theorem to find the force exerted on a circular cylinder by a source.

Thus, with the notations of 8·62,

$$w = m \log z - m \log (z-f) - m \log (z-f').$$

Hence

$$f(z) = -\frac{m}{z} + \frac{m}{z-f'},$$

$$u_m - iv_m = -\frac{m}{f} + \frac{m}{f-f'} = \frac{ma^2}{f(f^2-a^2)},$$

and therefore

$$X - iY = \frac{2\pi\rho m^2 a^2}{f(f^2-a^2)},$$

as already obtained.

8·64. Source outside an elliptic cylinder. In the same way that the Joukowski transformation was used in section 6·31 to deduce the flow past an elliptic cylinder when the corresponding flow past a circular cylinder had been obtained, the complex potential due to a source outside a circular cylinder can be made to yield the complex potential for a source outside an elliptic cylinder. Considering the circular cylinder of radius $(a+b)/2$ with a source at Z_0, we get for the Z-plane

$$w = +m \log Z - m \log (Z - Z_0) - m \log (Z - Z_0')$$

where Z_0' is the inverse of Z_0, and therefore if

$$Z_0 = r\,e^{i\alpha},$$

we get

$$Z_0' = \frac{(a+b)^2}{4r}\,e^{i\alpha}.$$

Making the transformation of section 6·31

$$Z = \tfrac{1}{2}[z + \sqrt{(z^2-c^2)}],$$

$$Z_0 = \tfrac{1}{2}[z_0 + \sqrt{(z_0^2-c^2)}],$$

we get the complex potential for a source m at the point z_0 in the presence of an elliptic cylinder whose semi-axes are a and b.

8·70. Mapping on a unit circle. Consider the circle of unit radius, centre the origin, in the Z-plane. The coordinates of any point on this circle can be expressed in the form $X = \cos\theta$, $Y = \sin\theta$, and as θ increases from 0 to 2π, the point (X, Y) describes the circle in the counterclockwise sense. As we shall be concerned with the region outside this circle it is convenient to

* A corresponding three-dimensional theorem is proved in 16·42.

write $\theta = -\xi$, so that, as ξ increases from 0 to 2π, the point (X, Y) describes the circle in the clockwise sense and therefore leaves the region outside the

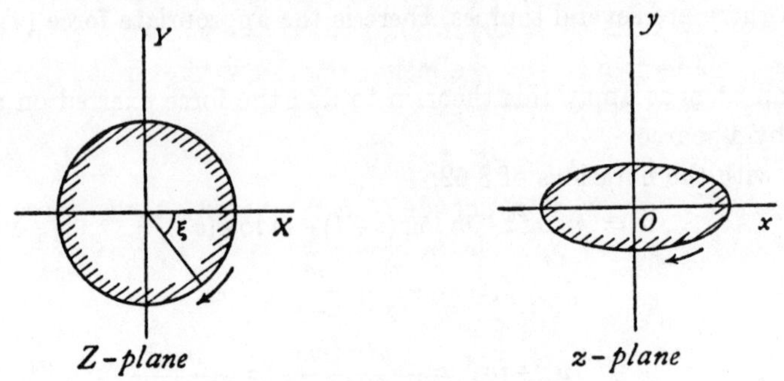

FIG. 8·70.

circle on the left. Any point on the circle can therefore be expressed in the form

(1) $$Z = X + iY = \cos\xi - i\sin\xi = e^{-i\xi}.$$

If the coordinates of a point on a given curve can be expressed in the form

(2) $$z = a_0\, e^{-i\xi} + b_0 + b_1\, e^{i\xi} + b_2\, e^{2i\xi} + \ldots,$$

where the curve is described clockwise as ξ increases from 0 to 2π, the region exterior to the given curve is mapped on the region exterior to the unit circle by

(3) $$z = f(Z) = a_0 Z + b_0 + \frac{b_1}{Z} + \frac{b_2}{Z^2} + \ldots.$$

This follows at once by eliminating ξ between (1) and (2), since the senses of description correspond.

For example in the case of the ellipse (a, b) we have

(4) $$z = a\cos\xi - ib\sin\xi = \tfrac{1}{2}(a+b)e^{-i\xi} + \tfrac{1}{2}(a-b)e^{i\xi},$$

so that the required mapping is

(5) $$z = \frac{a+b}{2} Z + \frac{a-b}{2Z},$$

a Joukowski transformation.

A class of curves having this property has been described by D. Wrinch * and are included in the equation

(6) $$z = \frac{a+b}{2}\, e^{-i\xi} + \frac{a-b}{2}\, e^{i\xi} + \tfrac{1}{2} b\, e^{2i\xi}, \quad 0 \leqslant b \leqslant a.$$

These curves are mapped on the unit circle $|Z| = 1$ by

(7) $$z = \frac{a+b}{2} Z + \frac{a-b}{2Z} + \frac{b}{2Z^2},$$

which should be compared with (5).

* *Phil. Mag.* (6), 48, 1924.

These curves range from the hypocycloid with three cusps when $a = b$ to symmetrical aerofoil shapes when $b < \frac{1}{2}a$.

8·71. Source outside a cylinder. Let the region exterior to the curve C in the z-plane be mapped on the region exterior to the circle $|Z| = 1$ in the Z-plane by

$$(1) \qquad z = f(Z).$$

If there is a source m at z_0 outside C in the z-plane, there is a source m at the corresponding point Z_0 outside the circle, and therefore by the circle theorem

$$(2) \qquad w = -m \log (Z - Z_0) - m \log \left(\frac{1}{Z} - \bar{Z}_0\right),$$

which with (1) determines w in the z-plane.

In the case of a doublet we have, as in 8·23,

$$w = -m\,\delta Z_0 \frac{\partial}{\partial Z_0} \log (Z - Z_0) - m\,\delta \bar{Z}_0 \frac{\partial}{\partial \bar{Z}_0} \log \left(\frac{1}{Z} - \bar{Z}_0\right).$$

Now if $\mu e^{i\alpha} = m\,\delta z_0$, we get from (1)

$$m\,\delta Z_0 = \mu e^{i\alpha}/f'(Z_0), \quad m\,\delta \bar{Z}_0 = \mu e^{-i\alpha}/\bar{f}'(\bar{Z}_0),$$

and so

$$w = \frac{\mu e^{i\alpha}}{(Z - Z_0) f'(Z_0)} + \frac{\mu e^{-i\alpha} Z}{(1 - Z\bar{Z}_0)\bar{f}'(\bar{Z}_0)}.$$

8·72. Force on the cylinder. We use Lagally's theorem, 8·63, which gives

$$X - iY = 2\pi\rho m (u_m - iv_m),$$

where the induced velocity is calculated by omitting the source from the complex potential. Thus $u_m - iv_m = F'(z_0)$ where $F(z) = -w - m \log (z - z_0)$.

Thus using 8·71 (2),

$$F(z) = m \log (Z - Z_0) + m \log \left(Z - \frac{1}{\bar{Z}_0}\right) - m \log Z - m \log \{f(Z) - f(Z_0)\},$$

together with $z = f(Z)$.

Thus

$$u_m - iv_m = \left\{\frac{m}{Z - Z_0} - \frac{m f'(Z)}{f(Z) - f(Z_0)} + \frac{m\bar{Z}_0}{Z\bar{Z}_0 - 1} - \frac{m}{Z}\right\}_{Z=Z_0} \times \left(\frac{dZ}{dz}\right)_{Z=Z_0}.$$

Now by Taylor's theorem

$$f(Z) = f(Z_0) + (Z - Z_0) f'(Z_0) + \tfrac{1}{2}(Z - Z_0)^2 f''(Z_0) + \dots,$$

whence after reduction

$$X - iY = \frac{2\pi\rho m^2}{f'(Z_0)} \left\{ -\frac{f''(Z_0)}{2f'(Z_0)} + \frac{1}{Z_0(Z_0\bar{Z}_0 - 1)} \right\},$$

and by the theorem proved in 8·63 this force is localised in a line through z_0, so that no separate calculation of moment is required.

A simple illustration of this result is its application to the elliptic cylinder given by the Joukowski mapping of 8·70 (5). This is left as an exercise.

8·80. Source and sink outside a circular cylinder. Consider a sink at S_1 and an equal source of strength m at S_2 outside a circular cylinder, centre O. If S_1', S_2' are the inverse points, the image system consists of

$$-m \text{ at } S_1', \quad m \text{ at } O, \quad m \text{ at } S_2', \quad -m \text{ at } O,$$

and therefore reduces to

$$-m \text{ at } S_1' \quad \text{and} \quad +m \text{ at } S_2',$$

for the source and sink at O neutralise one another.

Since $$OS_1 \,.\, OS_1' = a^2 = OS_2 \,.\, OS_2',$$

the points S_1, S_2, S_1', S_2' are concyclic.

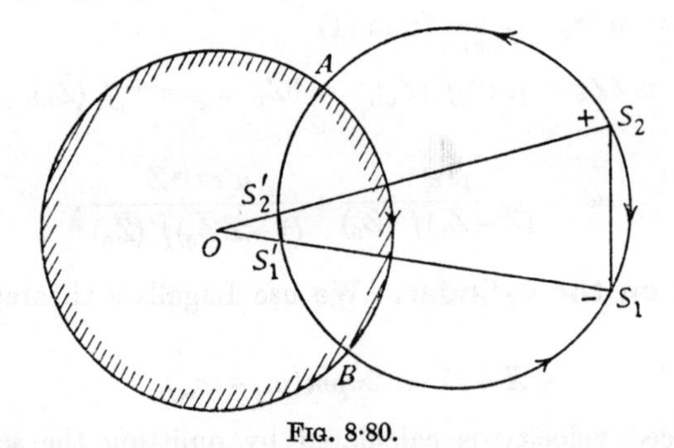

FIG. 8·80.

Since the streamlines for a source and equal sink are circles, the circle through the above four points is a streamline. The cross-section of the cylinder is also a streamline. Hence the circles intersect in stagnation points A and B.

Since $OA^2 = OS_1 \,.\, OS_1'$, OA is a tangent to the circle S_1, S_1', S_2, and therefore the two circles cut orthogonally at A and B (cf. 4·6).

The directions of flow on the dividing streamline are indicated in fig. 8·80.

8·81. The image of a doublet in a circular cylinder. If in fig. 8·80 we let S_1 approach S_2 while the product $m \times S_1S_2 = \mu$ remains finite, we get the image of the doublet μ.

Since the triangles $OS_1'S_2'$, OS_2S_1 are similar,

$$\frac{S_1'S_2'}{S_1S_2} = \frac{OS_1'}{OS_2},$$

and therefore the strength of the image doublet is

$$\mu' = \mu \times \lim_{s_1 \to s_2} \frac{OS_1'}{OS_2} = \mu \times \frac{OD}{OC},$$

where C is the position of the doublet μ and D is the position of μ'. Since C and D are inverse points, if $OC = f$, then $OD = a^2/f$, and therefore

$$\mu' = \mu a^2/f^2.$$

Again referring to fig. 8·80, we see that when S_1 comes to coincide with S_2 the circle through $S_1 S_2 S_1' S_2'$ touches the axes of the doublets at C and D.

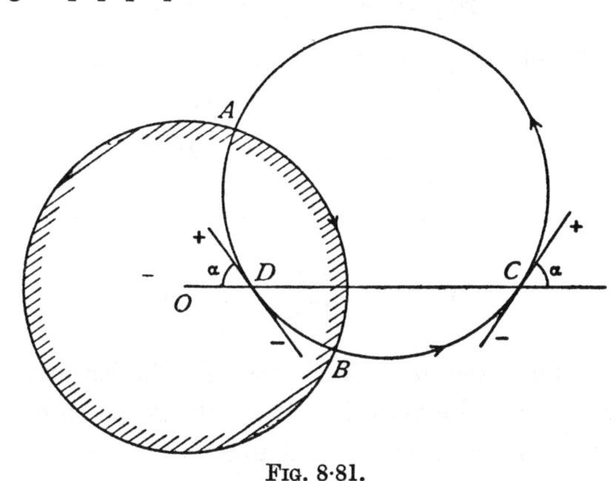

FIG. 8·81.

Therefore the image of a doublet of strength μ at distance f from the centre of a cylinder of radius a is a doublet at the inverse point of strength $\mu a^2/f^2$, and the axes of the doublet and its image are equally inclined to the line joining them, but are not parallel. (Such lines are conveniently called *antiparallel.*)

The above circle which touches the doublets at C and D is still a streamline and cuts the cylinder in stagnation points A and B.

8·82. The force on a cylinder due to a doublet. Let the doublet be at distance f from the centre of a cylinder of radius a and on the real axis, fig. 8·81.

Let μ and α be the strength and inclination of the doublet. Then if $ff' = a^2$, we have (8·23)

$$w = \frac{\mu e^{i\alpha}}{z-f} - \frac{\mu a^2}{f^2}\frac{e^{-i\alpha}}{z-f'},$$

the second term being due to the image at D of the doublet at C. Then

$$\left(\frac{dw}{dz}\right)^2 = \frac{\mu^2 e^{2i\alpha}}{(z-f)^2} + \frac{\mu^2 a^4 e^{-2i\alpha}}{f^4(z-f')^4} - \frac{2\mu^2 a^2}{f^2(z-f)^2(z-f')^2}.$$

Using the theorem of Blasius,

$$X - iY = \tfrac{1}{2} i\rho \int \left(\frac{dw}{dz}\right)^2 dz,$$

taken round the circle. Now, the only pole inside the contour is $z = f'$ and therefore the residue is the coefficient of $(z-f')^{-1}$ when the third term of the above expression is put into partial fractions or expanded in ascending powers of $z-f'$. We shall do the latter.

Writing $y = z-f'$, we have

$$\frac{1}{(z-f')^2(z-f)^2} = \frac{1}{y^2(y+f'-f)^2} = \frac{1}{y^2(f-f')^2}\left(1-\frac{y}{f-f'}\right)^{-2}$$
$$= \frac{1}{y^2(f-f')^2}\left(1+\frac{2y}{f-f'}+\frac{3y^2}{(f-f')^2}+\dots\right).$$

Hence the coefficient of $(z-f')^{-1}$ or y^{-1} is

$$\frac{2}{(f-f')^3} = \frac{2f^3}{(f^2-a^2)^3}.$$

Therefore $$X-iY = \tfrac{1}{2}i\rho\times 2\pi i\times\frac{-2\mu^2a^2}{f^2}\times\frac{2f^3}{(f^2-a^2)^3},$$

$$X = \frac{4\pi\rho\mu^2a^2f}{(f^2-a^2)^3},\quad Y = 0.$$

This gives the force per unit thickness of cylinder and shows that the cylinder is urged towards the doublet. It is interesting to note that the force is independent of the orientation of the doublet.

Since all the pressure thrusts on the boundary pass through the centre of the cylinder, their moment about the centre is zero.

8·83. Extension of Lagally's theorem to doublets. A doublet will give rise in the complex potential to a term $\mu\, e^{i\alpha}/(z-a)$. Thus with the notations of 8·63, we shall have

$$\int_{(S)}\left(\frac{dw}{dz}\right)^2 dz = -2\pi\rho i\kappa(U-iV);$$

the terms corresponding to m now disappear.

Again, (5) will become

$$\left(\frac{dw}{dz}\right)^2 = (f(z))^2+\frac{\mu^2e^{2i\alpha}}{(z-a)^4}+\frac{2\mu e^{i\alpha}f(z)}{(z-a)^2},$$

where $f(z)$ is still the part of the complex velocity obtained by omitting the doublet. By Taylor's theorem,

$$f(z) = f[(z-a)+a] = f(a)+(z-a)f'(a)+\dots,$$

and the residue at $z = a$ is now $2\mu e^{i\alpha}f'(a)$.

Thus (6) becomes

$$X-iY = -2\pi\rho i\kappa(U-iV)+2\pi\rho\mu e^{i\alpha}f'(a),$$

which is the required extension.

To apply it to the case of 8·82, we have, omitting the term due to the doublet at $z = f$ from w,

$$f(z) = -\frac{\mu a^2}{f^2} \frac{e^{-i\alpha}}{(z-f')^2},$$

and therefore
$$X - iY = 2\pi\rho\mu e^{i\alpha} \frac{\mu a^2}{f^2} \frac{2e^{-i\alpha}}{(f-f')^3} = \frac{4\pi\rho\mu^2 a^2 f}{(f^2-a^2)^3}$$

as before.

8·9. Source in compressible flow. Let there be a (two-dimensional) source whose output of mass is $2\pi m$ per unit time, and let q be the speed at distance r from the source. Then the equation of continuity gives

$$2\pi r\rho q = 2\pi m,$$

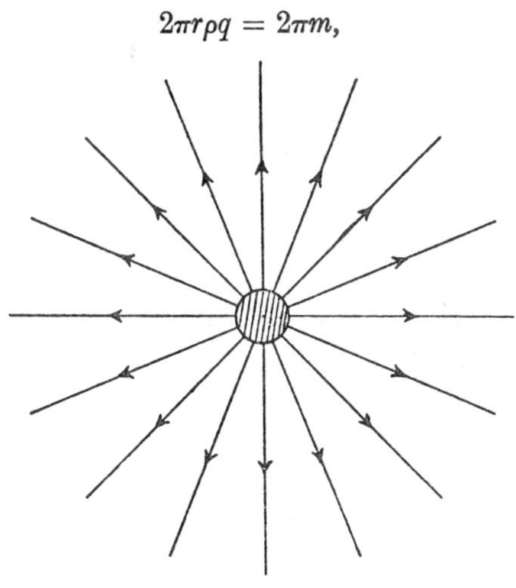

FIG. 8·9.

whence with the notations of 1·63, using 1·63 (7),

$$(1) \qquad r = \frac{m}{c_0\rho_0 M} \frac{c_0\rho_0}{c\rho} = \frac{m}{c_0\rho_0 M} [1 + \tfrac{1}{2}(\gamma-1) M^2]^{\frac{\gamma+1}{2(\gamma-1)}}$$

where M is the Mach number.

By differentiation with respect to M we easily show that r has a *minimum* value when $M = 1$, and therefore (1) gives

$$(2) \qquad \frac{r}{r_{\min}} = \left(\frac{2}{\gamma+1} + \frac{\gamma-1}{\gamma+1} M^2\right)^{\frac{(\gamma+1)}{2(\gamma-1)}} \Big/ M.$$

The motion thus exists only outside the circle whose radius is $r_{\min}$ and cannot be continued inside it. Thus there is a physical lower limit to the size of the source. It cannot be imagined contracted to a point. Outside the circle we have either pure subsonic flow in which q steadily decreases until $M = 0$ at infinity; or pure supersonic flow in which q steadily increases, till $M = \infty$ at infinity.

Such flows can take place between two rigid planes inclined at an angle as in fig. 8·51, the source being at the angle.

EXAMPLES VIII

1. Apply Rankine's method to drawing the streamlines for the flow due to two equal sources.

2. Draw the streamlines for the flow due to a source and equal sink

(i) alone in the fluid ;

(ii) in a stream perpendicular to the line joining source and sink.

3. A source and a sink of the same strength are placed at a given distance apart in an infinite fluid which is otherwise at rest. Show that the streamlines are circles, and that the fluid speed along any streamline is inversely as the distance from the line joining the source and sink.

4. Two sources of equal strength are situated respectively at the points $(\pm a, 0)$ in an unbounded fluid. Show that at any point on the circle $x^2+y^2 = a^2$ the fluid velocity is parallel to the axis of y, and inversely as the ordinate of the point. Determine also the point in the axis of y at which the velocity is greatest.

Hence show that, if a uniform stream parallel to the axis of y be combined with the two sources, there are necessarily two points at which the velocity vanishes. (R.N.C.)

5. If there is a source m at A and a sink $-m$ at B and a uniform stream U in the direction BA, find the stagnation points, and prove that they lie on AB or its perpendicular bisector according as the stream is relatively strong or feeble. Draw the streamlines in each case.

6. There is a source at A and an equal sink at B. AB is the direction of a uniform stream. Determine the form of the streamlines. If A is $(a, 0)$, B is $(-a, 0)$ and the ratio of the flow issuing from A in unit time to the speed of the stream is $2\pi b$, show that the stream function is

$$\psi = Vy - Vb\tan^{-1}\frac{2ay}{x^2+y^2-a^2},$$

and that the length, $2l$, and the breadth, $2d$, of the closed wall that forms part of the dividing streamline is given by

$$l = \sqrt{a^2+2ab}, \quad \tan\frac{d}{b} = \frac{2ad}{d^2-a^2},$$

and the locus of the point at which the speed is equal to that of the stream is

$$x^2-y^2 = a^2+ab.$$

7. A two-dimensional source of strength m is situated at the point $(a, 0)$, the axis of y being a fixed boundary. Find the points on the boundary at which the fluid velocity is a maximum.

Show that the resultant thrust on that part of the axis of y which lies between $y = \pm b$ is

$$2p_0 b - 2m^2\rho\left\{\frac{1}{a}\tan^{-1}\frac{b}{a} - \frac{b}{a^2+b^2}\right\}$$

where p_0 is the pressure at infinity. (R.N.C.)

8. Calculate the force on a wall due to a doublet of strength μ at distance a from the wall and inclined to it at an angle α. In what direction is the wall urged by this force?

9. Prove that in conformal transformation a doublet will transform into a doublet, but that the strengths will differ.

10. Two sources, each of strength m, are placed at the points $(-a, 0)$ and $(a, 0)$, and a sink of strength $2m$ is placed at the origin. Show that the stream-lines are the curves

$$(x^2+y^2)^2 = a^2(x^2-y^2+\lambda xy),$$

where λ is a variable parameter.

Show also that the fluid speed at any point is

$$\frac{2ma^2}{r_1 r_2 r_3},$$

where r_1, r_2, r_3 are respectively the distances of the point from the sources and the sink. (R.N.C.)

11. If there is a source at $(a, 0)$ and $(-a, 0)$ and sinks at $(0, a)$, $(0, -a)$, all of equal strength, show that the circle through these four points is a streamline.

12. OX, OY are fixed rigid boundaries and there is a source at (a, b). Find the form of the streamlines and show that the dividing line is

$$xy(x^2-y^2-a^2+b^2) = 0.$$

13. In liquid bounded by the axes of x and y in the first quadrant there is a source of strength m at distance a from the origin on the bisector of the angle xOy. Prove that the complex potential is $-m\log(a^4+z^4)$.

14. Between the fixed boundaries $\theta = \pi/4$, $\theta = -\pi/4$, the two-dimensional motion is due to a source of strength m at $r = a$, $\theta = 0$, and an equal sink at $r = b$, $\theta = 0$. Show that the stream function is

$$-m\tan^{-1}\frac{r^4(a^4-b^4)\sin 4\theta}{r^8-r^4(a^4+b^4)\cos 4\theta+a^4b^4}.$$

15. Show that the velocity components given by

$$u = U\left(1-\frac{ay}{x^2+y^2}+\frac{b^2(x^2-y^2)}{(x^2+y^2)^2}\right),$$

$$v = U\left(\frac{ax}{x^2+y^2}+\frac{2b^2xy}{(x^2+y^2)^2}\right),$$

represent a possible fluid motion in two dimensions.

Show that the motion is irrotational, and interpret the meaning of the terms in the complex potential. (R.N.C.)

16. A, B, C is an equilateral triangle. There is a source of strength 2 at A and sinks of strength 1 at B and C, and a stream in the direction from A perpendicular to BC. Determine the form of the streamlines when the relative strengths are such that the dividing streamline consists in part of a closed wall.

17. Show that $w = \kappa\log(z^2-l^2)$ gives the motion due to a two-dimensional source in the presence of a fixed wall, and, by using the transformation given by

$$\frac{d\zeta}{dz} = \frac{A}{(z^2-1)^{\frac{1}{2}}},$$

obtain the solution for such a source in a semi-infinite rectangle.

18. Use the transformation $z' = e^{\pi z/a}$ to find the streamlines of the motion in two dimensions due to a source midway between two infinite parallel boundaries, assuming the liquid drawn off equally by sinks at the end of the region. If the pressure tends to zero at the ends of the streams, prove that the planes are pressed apart with a force which varies inversely as their distance from each other.

19. A source is placed midway between two planes whose distance from one another is $2a$. Find the equation of the streamlines when the motion is in two dimensions, and show that those particles, which at an infinite distance are distant $\frac{1}{2}a$ from one of the boundaries, issued from the source in a direction making an angle $\pi/4$ with it.

20. The irrotational motion in two dimensions of a fluid bounded by the lines $y = \pm b$ is due to a doublet of strength μ at the origin, the axis of the doublet being in the positive direction of the axis of x. Prove that the motion is given by

$$\phi + i\psi = \frac{\pi\mu}{2b} \coth [\pi(x+iy)/2b].$$

Show also that the points where the fluid is moving parallel to the axis of y lie on the curve

$$\cosh (\pi x/b) = \sec (\pi y/b).$$

21. The space on one side of an infinite plane wall $y = 0$ is filled with inviscid incompressible fluid moving at infinity with velocity U in the direction of the axis of x. The motion of the fluid is wholly two-dimensional, in the (x, y) plane. A doublet of strength μ is at a distance a from the wall and points in the negative direction of the axis of x. Show that, if μ is less than $4a^2 U$, the pressure of the fluid at the wall is a maximum at points distant $a\sqrt{3}$ from O, the foot of the perpendicular from the doublet on to the wall, and is a minimum at O.

If $\mu = 4a^2U$, find the stagnation points, and show that the streamlines include the circle $x^2 + (y-a)^2 = 4a^2$, where the origin is taken at O. (M.T.)

22. In two-dimensional motion there is a uniform source along the real axis of total output $2\pi m$ stretching from $x = 0$ to $x = a$. Show that the complex potential is

$$w = -\frac{m}{a}\int_0^a \log (z-\xi)\, d\xi = -m\left\{\frac{z}{a}\log z - \frac{z-a}{a}\log (z-a)\right\}.$$

Combine this with a uniform stream U parallel to the x-axis, and show that the dividing streamline is

$$Uy + \frac{m}{a}\left\{x(\theta_2 - \theta_1) + a\theta_2 + y \log \frac{r_1}{r_2} - \pi a\right\} = 0,$$

where r_1, r_2 are the distances and θ_1, θ_2 the corresponding angles from a point on the line to the ends of the source. Trace the form of this line.

23. Along the x-axis there exists for each stretch from $x = 2na$ to $x = (2n+1)a$ a two-dimensional source of strength k per unit length, and from $x = (2n-1)a$ to $x = 2na$ a two-dimensional sink of equal strength when n takes all positive and negative integral values. If w is the complex potential, find $-dw/dz$.

If in a channel bounded by walls at $x = a$ and $x = -a$ a line source stretches from $x = 0$ to $x = a$, and an equal line sink from $x = 0$ to $x = -a$, find the velocity at any point along the walls. (U.L.)

24. Prove by direct calculation that the radial velocity on a circular cylinder due to a source and its image system vanishes.

25. Verify that a source and its image system in a circular cylinder do in fact make the section of the cylinder a streamline.

26. In the case of a source outside a circular cylinder, prove that the equation of the streamline ψ = constant is

$$(x^2+y^2)(cx+y) - (x^2+y^2)c(f+f') + a^2(cx-y) = 0,$$

where $c = \tan (\psi/m)$, $ff' = a^2$.

27. In Ex. 26 above, prove that

(i) the asymptote to the streamline is $cx+y-c(f+f') = 0$;

(ii) all the asymptotes pass through $(f+f', 0)$;

(iii) each streamline equation gives rise to a closed curve lying entirely inside the cylinder.

28. In the case of a source at a point A outside a circular disc, prove that the velocity of slip of the fluid in contact with the disc is greatest at the points where the circle is cut by the lines joining A to the ends of the diameter perpendicular to OA, and that its magnitude at these points is

$$\frac{2m \,.\, OA}{OA^2 - a^2},$$

where O is the centre and a the radius of the disc.

29. If the axis of y and the circle $x^2+y^2 = a^2$ are fixed boundaries, and there is a two-dimensional source at the point $(c, 0)$ where $c>a$, show that the radius, drawn from the origin to the point on the circle where the velocity is a maximum, makes with the axis of x an angle

$$\cos^{-1} \frac{a^2+c^2}{\sqrt{2(a^4+c^4)}}.$$

(R.N.C.)

30. A two-dimensional source I of strength m is outside a fixed circle, centre C. Prove that the value of q at any point P is

$$mr_3 r_4/rr_1 r_2,$$

where r, r_1, r_2, r_3, r_4 denote the distances of P from the points C, I, J, A, B respectively, J being the inverse point of I with respect to the circle and A, B the points in which CI cuts the circle. (U.L.)

31. If a circle be cut in half by the y-axis, forming a rigid boundary, and a source, of strength m, be on the x-axis at a distance a, equal to half the radius, from the centre, prove that the streamlines are given by

$$(16a^4+r^4)\cos 2\theta - 17a^2r^2 = (16a^4-r^4)\sin 2\theta \cot\left(\frac{\psi}{m}\right),$$

ψ being a suitably adjusted value of the stream function.

Show that the streamline $\psi = m\pi/2$ leaves the source in a direction perpendicular to Ox and enters the sink at an angle $\pi/4$ with Ox, and sketch the streamlines.

32. In the two-dimensional motion of an infinite liquid there is a rigid boundary consisting of that part of the circle $x^2+y^2 = a^2$ which lies in the first and fourth quadrants and the parts of the axis of y which lie outside the circle. A simple source of strength m is placed at the point $(f, 0)$ where $f>a$. Prove that the speed of the fluid at the point $(a\cos\theta, a\sin\theta)$ of the semicircular boundary is

$$\frac{4maf^2 \sin 2\theta}{a^4+f^4-2a^2f^2\cos 2\theta}.$$

Find at what points of the boundary the pressure is least. (R.N.C.)

33. Water enters a circular enclosure of radius a at the centre O and escapes by a small hole at the point A of the boundary into the region outside which is also occupied by water and is unbounded. The motion being considered two-dimensional, prove (i) that the asymptotes of the streamlines pass through a fixed point; (ii) that the tangent at O to a streamline and the corresponding asymptote are equally inclined to OA; (iii) that the streamline has a double point at A, the tangents at which are perpendicular. Sketch one of the streamlines. (U.L.)

34. Within a circular boundary of radius a there is a source of strength m at distance f from the centre and an equal sink at the centre. Find the complex potential, and show that the resultant thrust on the boundary is

$$\frac{2\pi\rho m^2 f^3}{a^2(a^2-f^2)}.$$

In what direction is the boundary urged by this thrust?

Deduce as a limiting case the velocity potential due to a doublet at the centre. (R.N.C.)

35. A source is situated at the point (c, c) in the region bounded by the axis of x and the circle $x^2+y^2 = a^2$, the source being outside the circle. Show that the fluid velocity vanishes at the points $(\pm a, 0)$ and that it will vanish at one other point on the circle provided that $2c<(2+\sqrt{2})a$.

36. The boundary of a semi-infinite liquid consists of an infinite plane surmounted by a cylindrical boss of semicircular cross-section of radius a, and the liquid contains a line source everywhere at a distance c from the plane and the axis of the boss, where $c = a\tan\lambda$. Show that the velocity at points on the boss is a maximum along the generators lying in the axial planes, making an angle θ with the axial plane containing the line source, given by $\tan\theta = \pm\cos 2\lambda$. (U.L.)

37. Show that $w = m\log\{(z-z_1)(z-z_2)/z\}$, where $z_2 : z_1$ is real, gives the motion for a simple two-dimensional sink of strength m at $z = z_1$, in the presence of a fixed circle, centre the origin, and of radius a, where $a^2 = |z_1 z_2|$.

Using the transformation

$$z' = z+c^2/z, \quad (c<a<|z_1|),$$

obtain the solution for the motion due to such a sink outside a fixed ellipse, and find the resultant force on the ellipse. (U.L.)

38. Sources of equal strength m are placed at the points $z = nia$ when

$$n = \ldots, \quad -2, \ -1, 0, 1, 2, 3, \ldots.$$

Prove that the complex potential is $w = -m\log\sinh(\pi z/a)$. Hence show that the complex potential for doublets, parallel to the x-axis, of strength μ at the same points, is given by

$$w = \mu\coth(\pi z/a).$$

39. If the row of doublets of Ex. 38 is placed in a uniform stream $-U$ parallel to the x-axis, prove that the streamline $\psi = 0$ is

$$\frac{ay}{\pi b^2} = \frac{\sin(2\pi y/a)}{\cosh(2\pi x/a)-\cos(2\pi y/a)},$$

and show that this consists in part of the x-axis and in part of an oval curve which is nearly circular (diameter $2b$) if b is small compared with a. Show that this solves the problem of a stream flowing through a set of parallel equidistant rails of approximately circular section.

40. Prove or verify that the complex potential defined by

$$\frac{z}{a} = \exp\left(-\frac{nw}{m}\right)+\exp\left(\frac{(1-n)w}{m}\right)$$

makes the streamlines $\psi = \pm m\pi$ straight and radiating from the origin. Prove that the flow is inwards towards the origin in one of the angles thus formed and outwards from the origin in the other (re-entrant) angle.

41. If $w = f(z)$ and $\mathbf{v} = -dz/dw$, show that $\mathbf{v}$ and $\bar{\mathbf{v}}$ are inverse points in the Argand diagram. Show that if $\mathbf{v}$ can be determined as a function of w, then

$$z = -\int \mathbf{v}\, dw.$$

Prove that the assumption of a simple source in the $\mathbf{v}$-plane in the form

$$w = -\frac{aU}{\pi}\log\left(\mathbf{v}-\frac{1}{U}\right)+\frac{aU}{\pi}\log\frac{1}{U} \quad \text{leads to}$$

$$Uz = -w+\frac{aU}{\pi}\exp\left(-\frac{\pi w}{aU}\right),$$

and that this represents the flow of water out of a canal of breadth $2a$, the asymptotic velocity in the canal being U.

42. Use the result of Ex. 41 to prove that the complex potential due to a stream U flowing against the mouth of an infinite canal of breadth $2a$ is given by

$$\frac{\pi z}{a} = \frac{\pi w}{aU}+\log\frac{\pi w}{aU}.$$

43. Two infinite planes converge at the angle $2\alpha = 2n\pi$, but do not meet, to form a spout into which liquid flows. Show that the diagram in the $\mathbf{v}$-plane ($\mathbf{v} = -dz/dw$) corresponds to the type of flow given in Ex. 40, and deduce that for flow into the spout

$$z = -\frac{m}{nU}\exp\left(-\frac{nw}{m}\right)+\frac{m}{(1-n)U}\exp\left(\frac{(1-n)w}{m}\right)+C.$$

Taking $C = m/(nU)$, show that the result of Ex. 42 can be deduced.

44. If liquid moves inside a thin shell between two plane laminae, show that a corresponding motion in a thin spherical shell can be obtained by inverting the streamlines in the first motion with regard to any origin, and find the factor by which the velocities must be multiplied to transform one motion into the other.

A source and an equal sink are placed at two points of a thin spherical shell. Show that the equipotential and streamlines on the sphere are small circles.

CHAPTER IX

MOVING CYLINDERS

9·10. Kinetic energy of acyclic irrotational motion. Consider two-dimensional acyclic irrotational motion of liquid bounded internally by a

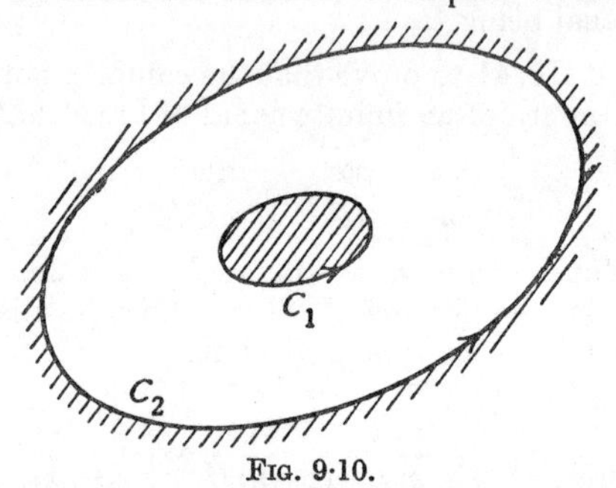

FIG. 9·10.

cylinder C_1 and externally by a cylinder C_2 and of unit thickness (i.e. comprised between two planes parallel to the plane of the motion and at unit distance apart). For such a motion to subsist, it follows from 3·77 (I) that either or both cylinders must be in motion.

Then if S is the region between C_1 and C_2, we have for the kinetic energy

$$T = \tfrac{1}{2}\rho \int_{(S)} q^2 dS = \tfrac{1}{2}\rho \int_{(S)} \frac{dw}{dz}\frac{d\bar{w}}{d\bar{z}}\, dS = \tfrac{1}{2}\rho \int_{(S)} \frac{\partial}{\partial z}\left(w \frac{d\bar{w}}{d\bar{z}}\right) dS.$$

Using the Area theorem, 5·43, we get

$$T = -\tfrac{1}{4} i\rho \int_{(C_1)} w\, d\bar{w} + \tfrac{1}{4} i\rho \int_{(C_2)} w\, d\bar{w},$$

each contour being described in the counterclockwise sense.

Observe that steady motion is not assumed.

9·11. Kinetic energy of cyclic motion. Consider cyclic irrotational motion taking place in liquid contained between *fixed* cylinders C_1 and C_2, the region being doubly connected.

Let w_0 denote the complex potential. By hypothesis, there is a circulation of strength κ, so that w_0 decreases to $w_0 - 2\pi\kappa$ when we encircle the cylinder C_1 in the anticlockwise sense.

Imagine a barrier AB between the cylinders, thus rendering the region occupied by the fluid simply connected. The barrier AB is a geometrical conception which does not interfere with the motion. This will be the case if AB consists always of the same fluid particles. This barrier gives us a simply connected region in which w_0 is one-valued.

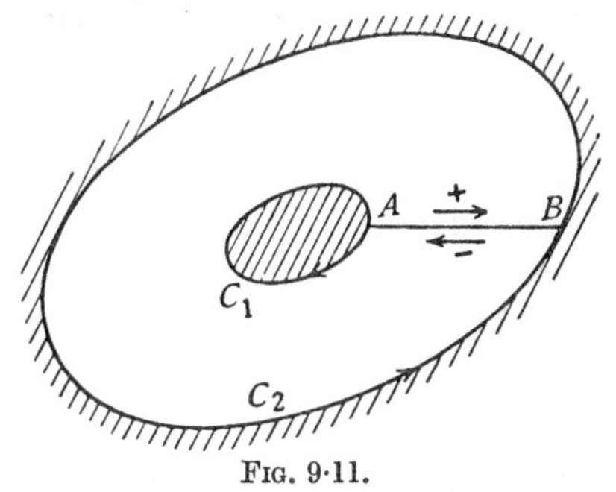

FIG. 9·11.

Let C denote the circuit C_2+BA+C_1+AB, where C_2 is described anti-clockwise and C_1 clockwise. The kinetic energy is given by

$$(1) \qquad T_0 = \tfrac{1}{4}i\rho \int_{(C)} w_0\, d\bar{w}_0 = \tfrac{1}{4}i\rho \int_{(C)} w_0\, dw_0 + \tfrac{1}{2}\rho \int_{(C)} w_0\, d\psi_0,$$

since $\bar{w}_0 = w_0 - 2i\psi$ and therefore $d\bar{w}_0 = dw_0 - 2i\, d\psi$. The first integral vanishes by Cauchy's theorem, since w_0 is one-valued in the region bounded by C. Since ψ_0 is constant on the streamlines C_1 and C_2 the last integral reduces to one along $AB+BA$. On AB w_0 has the value w_0, but on BA it has the value $w_0 - 2\pi\kappa$. Therefore

$$(2) \quad T_0 = \tfrac{1}{2}\rho \int_{AB} w_0\, d\psi_0 + \tfrac{1}{2}\rho \int_{BA} (w_0 - 2\pi\kappa)\, d\psi_0 = \pi\kappa\rho \Big[(\psi_0)_B - (\psi_0)_A \Big],$$

where $(\psi_0)_B$, $(\psi_0)_A$ are the values of ψ_0 at B and A. Thus

$$(3) \qquad T_0 = \pi\kappa\rho m,$$

where m is the flux from right to left across AB.

It follows also from (3) that

$$(4) \qquad T_0 = \tfrac{1}{2}\rho \int_{(AB)} 2\pi\kappa \frac{\partial \psi_0}{\partial s}\, ds = \int \pi\kappa\rho q_n\, ds,$$

where q_n is the velocity normal to the barrier AB. This is the work done by an impulsive pressure $2\pi\kappa\rho$ applied to the barrier AB, the liquid being initially at rest.

Thus the given cyclic motion could be generated from rest by this impulsive pressure applied to the barrier, the barrier being supposed to disappear immediately after the application of the impulses. Conversely, the cyclic motion being established, the liquid could only be brought to rest by the application of this impulsive pressure (but in the opposite sense) to a barrier such as AB. It follows that the cyclic motion cannot be either created or destroyed by impulsive pressures applied to the boundaries C_1 and C_2 alone.

We can therefore generalise theorem II of 3·77 as follows (at least for two-dimensional motion).

If liquid occupying doubly connected space is bounded by rigid walls, the motion, if acyclic, will instantly cease, if the boundaries are brought to rest, but, if cyclic, the cyclic part of the motion will persist.

Theorem VI admits of a similar generalisation, and generally it follows that, given the strength of the circulation, irrotational motions in doubly connected space are determinate.

The foregoing remarks, here justified in the case of doubly connected two-dimensional regions, are of general application to space of any connectivity in three dimensions.

We can now generalise to the case of any two-dimensional irrotational motion of liquid between two moving cylinders C_1 and C_2. The complex potential of any such motion can be expressed in the form $w+w_0$, where w applies to acyclic motion while w_0 applies to cyclic motion with the boundaries at rest. In this case the whole kinetic energy (per unit thickness) is

$$(5)\quad T=\tfrac{1}{4}i\rho\int_{(C)}(w+w_0)(d\bar{w}+d\bar{w}_0)=\tfrac{1}{4}i\rho\int_{(C)}w\,d\bar{w}+\tfrac{1}{4}i\rho\int_{(C)}w_0\,d\bar{w}_0+T',$$

where

$$T'=\tfrac{1}{4}i\rho\int_{(C)}(w\,d\bar{w}_0+w_0\,d\bar{w})=\tfrac{1}{4}i\rho\int_{(C)}\left\{w(dw_0-2i\,d\psi_0)+(\bar{w}_0+2i\psi_0)\,d\bar{w}\right\}.$$

Now by Cauchy's theorem $\int_{(C)}w\,dw_0=\int_{(C)}\bar{w}_0\,d\bar{w}=0$, and as before ψ_0 is constant on C_1 and C_2. Therefore

$$T'=\tfrac{1}{2}\rho\int_{AB+BA}(w\,d\psi_0-\psi_0\,d\bar{w})=0$$

since w and ψ_0 are one-valued and therefore the integrals along AB and BA cancel. Thus

$$(6)\qquad T=\tfrac{1}{4}i\rho\int_{(C)}w\,d\bar{w}+\pi\kappa\rho\,[(\psi_0)_B-(\psi_0)_A],$$

in other words, T is the direct sum of the kinetic energies of each motion considered independently. Since w is one-valued the barrier AB does not intervene in calculating the integral in (6).

As an example of the kinetic energy of cyclic motion, take the case of cyclic motion, with circulation of strength κ between two circular cylinders of radii a and b (7·11). Here

$$T=\pi\kappa\rho\,(\kappa\log b-\kappa\log a)=\pi\kappa^2\rho\log(b/a).$$

9·20. Circular cylinder moving forward. Taking the origin at the centre of the cylinder, radius a, when liquid streams past with velocity U at infinity in the negative direction of the x-axis, the complex potential, from 6·22, is

$$Uz+\frac{Ua^2}{z}.$$

If we impress on the whole system a velocity U in the positive direction of the x-axis, the cylinder moves forward with velocity U and the liquid is at rest at infinity, so that

$$(1) \qquad w = \frac{Ua^2}{z}.$$

Comparing this with 8·23, we see that the complex potential is the same as that of a two-dimensional doublet at the centre of the cylinder in the positive direction of the x-axis and of strength Ua^2.

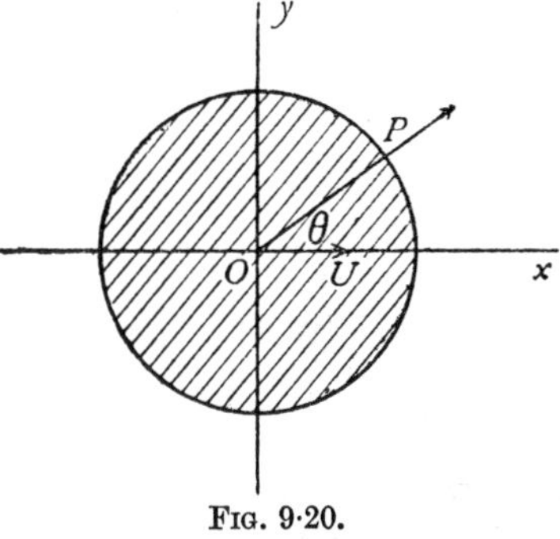

FIG. 9·20.

From (1), we get

$$u-iv = \frac{Ua^2}{z^2}, \quad u+iv = \frac{Ua^2e^{2i\theta}}{r^2}.$$

The radial and transverse components of the velocity at the point (r, θ) are

$$\frac{Ua^2\cos\theta}{r^2}, \quad \frac{Ua^2\sin\theta}{r^2}.$$

It must be emphasised that these are the components of the absolute velocity of the liquid at the fixed point in space whose coordinates are (r, θ) with respect to the moving axes at the instant considered. The only property required of the complex potential is that its derivative should yield the velocity. We also note that

$$q^2 = \frac{U^2a^4}{r^4},$$

so that the speed is the same at all points equidistant from the centre of the cylinder. In particular when $r = a$, $q = U$, and therefore the speed on the boundary of the cylinder is U.

9·21. Paths of the particles. Consider fixed axes Ox, Oy at the instant

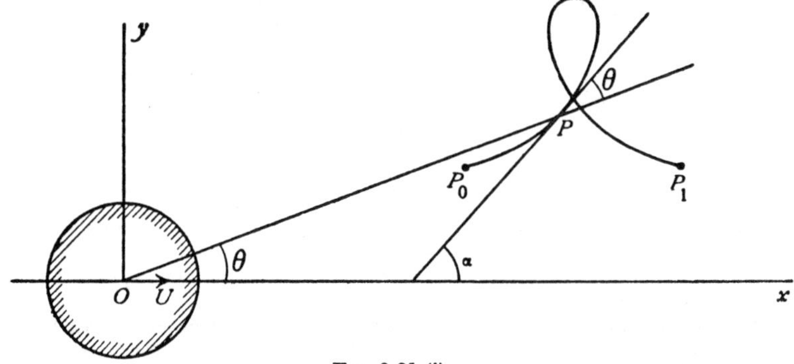

FIG. 9·21 (i).

when the centre of the cylinder is at O. The particle at the point $P(x, y)$ is moving with velocity Ua^2/r^2 at an angle θ with the radius vector (9·20), and therefore the tangent to the path of P makes an angle α with Ox where $\alpha = 2\theta$. Hence, if R is the radius of curvature of the path of P,

$$\frac{1}{R} = \frac{d\alpha}{ds} = \frac{d(2\theta)}{dy}\frac{dy}{ds} = 2\frac{d\theta}{dy}\cdot \sin 2\theta.$$

Now when the liquid is streaming past the fixed cylinder, the particle P moves along a streamline whose equation (6·22) is *

(1) $$\eta = y\left(1 - \frac{a^2}{r^2}\right) = y\left(1 - \frac{a^2 \sin^2\theta}{y^2}\right).$$

whence $$2\frac{d\theta}{dy}\sin 2\theta = \frac{4}{a^2}(y - \tfrac{1}{2}\eta).$$

Therefore $$\frac{1}{R} = \frac{4}{a^2}(y - \tfrac{1}{2}\eta).$$

This is the equation of the *elastica* or curve assumed by a perfectly flexible rod subjected to longitudinal thrusts. As the cylinder moves from $-\infty$ to $+\infty$, P moves from P_0 to P_1, the points of the elastica at which the tangents are parallel to the x-axis.

We now calculate the *drift*,† $\xi = P_0P_1$ in fig. 9·21 (i).

To this end we consider the motion of the liquid relative to the cylinder regarded as fixed, so that the liquid moves from right to left with general velocity U. Using the radial and transverse velocity components of 9·20 we have for the relative motion

(2) $$\frac{1}{U}\frac{dr}{dt} = \cos\theta\left(\frac{a^2}{r^2} - 1\right), \quad \frac{1}{U}\frac{r\,d\theta}{dt} = \sin\theta\left(\frac{a^2}{r^2} + 1\right).$$

One integral of these equations is the stream function (1), where the constant η gives the initial and final distance of the particle from the line of motion of the centre of the cylinder. For the drift we then have

(3) $$\xi = \int_{-\infty}^{\infty}\left(\frac{dx}{dt} + U\right)dt = \int_0^{\pi} \frac{a^2 \cos 2\theta\, d\theta}{(\eta^2 + 4a^2 \sin^2\theta)^{1/2}}$$

on the use of (1) and (2). The motion can be obtained in terms of elliptic functions by putting for the squared modulus

(4) $$m = k^2 = \frac{4a^2}{\eta^2 + 4a^2},$$

and writing $\cos\theta = -\operatorname{sn} v$, so that v ranges from $-K$ to K, where K is the

* The relation between y and θ is that appropriate to an observer moving with the cylinder, i.e. a streamline when the cylinder is fixed.

† The term and the treatment which follows are due to Darwin who seems to be the first to give a physically satisfactory intuitive picture of this phenomenon and of virtual mass. C. Darwin, *Proc. Camb. Phil. Soc.*, 49, (1953), 342–354.

complete elliptic integral of the first kind. Then the whole course of the motion is given in terms of the parameter v by

$$y(v) = \frac{a}{k}(k' + \operatorname{dn} v), \quad \xi(v) = \frac{a}{k}\{(1 - \tfrac{1}{2}k^2)v - E(v)\}$$

$$U\,t(v) = \frac{a}{k}\operatorname{sc} v\,(k' + \operatorname{dn} v) + \xi(v),$$

where $(\xi(v), y(v))$ are the cartesian coordinates of the particle, at time $t(v)$, relative to the initial undisturbed position. These equations enable us to plot the paths* and to calculate the drift

$$\xi = \frac{2a}{k}\{(1 - \tfrac{1}{2}k^2)K - E\}. \tag{5}$$

Some of these paths are shown in fig. 9·21 (ii) adapted from Darwin's paper. The origin of time has been taken at the moment of central passage of the cylin-

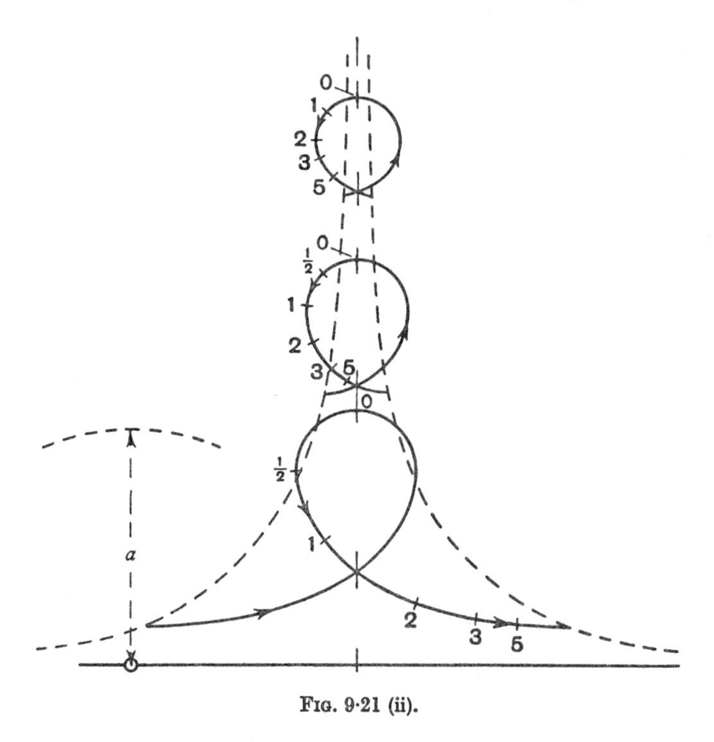

FIG. 9·21 (ii).

der. The numbers on the curves record the times at those points in a suitable unit; thus a point marked 2 gives the position of the liquid particle when the cylinder has moved forward 2 radii from the centre position. For this set of particles the broken curved line on the left of the figure shows the initial positions

* L. M. Milne-Thomson, *Jacobian elliptic function tables*, New York, (1960).

when the cylinder is at $-\infty$ and the broken line on the right the final positions when the cylinder has passed to $+\infty$. Thus there is indeed a drift of the liquid from left to right. The mass of liquid between the initial and final positions (taking unit thickness of liquid) may be called the *drift-mass* ρD where

$$\rho D = \rho \int_{-\infty}^{\infty} \xi \, d\eta. \tag{6}$$

It can be shown by performing the integration that $\rho D = \pi a^2 \rho = M'$ the mass of liquid displaced by the cylinder.

9·22. Kinetic energy. When a circular cylinder of radius a moves forward with velocity U, the kinetic energy of the fluid is given by

$$T_f = -\tfrac{1}{4} i\rho \int w \, d\bar{w}.$$

Also
$$w = \frac{Ua^2}{z}, \quad \bar{w} = \frac{Ua^2}{\bar{z}}, \quad d\bar{w} = -\frac{Ua^2}{\bar{z}^2} d\bar{z}.$$

Now on the cylinder, $z = a\, e^{i\theta}$, $\bar{z} = a\, e^{-i\theta}$, $d\bar{z} = -ia\, e^{-i\theta}\, d\theta$, so that

$$T_f = -\tfrac{1}{4} i\rho \int_0^{2\pi} \frac{U^2 a^4}{a^3\, e^{-i\theta}} ia\, e^{-i\theta} d\theta = \tfrac{1}{2}\pi\rho a^2 U^2.$$

Let $M' = \pi\rho a^2$. Then M' is the mass of liquid (per unit thickness) displaced by the cylinder.

If then M is the mass of the cylinder (per unit thickness), the total kinetic energy of the fluid and cylinder is

$$T = \tfrac{1}{2}(M+M')U^2.$$

Let F be the external force in the direction of motion of the cylinder necessary to maintain the motion. Then the rate at which F does work must be equal to the rate of increase of the total kinetic energy, and therefore

$$FU = \frac{dT}{dt} = (M+M')U\frac{dU}{dt},$$

$$F - M'\frac{dU}{dt} = M\frac{dU}{dt}.$$

Had the liquid been absent, the second term would have vanished. Thus the cylinder experiences a resistance to its motion of amount

$$M'\frac{dU}{dt}$$

per unit thickness, due to the presence of the liquid.

9·221. Virtual mass. An inspection of the final equations of 9·22 proves that the presence of the liquid effectively increases the mass of a moving circular cylinder from M to $M+M'$, where M' is the mass of liquid displaced. The mass $M+M'$ is called the *virtual mass* of the cylinder, and the virtual mass is obtained by increasing the mass M by the *added mass* or *hydrodynamic mass* which in the case of the circular cylinder is M'. Observe that this hydrodynamic mass M' is equal to the drift-mass ρD calculated in 9·21. It would appear that all moving bodies, in so far as the motion takes place in a medium, should be affected by added mass, so that in dynamical experiments the masses enter as virtual masses of the type $M+kM'$, where the coefficient k depends on the shape of the body and the nature of the motion. Darwin in the paper just cited has proved that for bodies moving in a straight line in unbounded liquid the hydrodynamic mass is given by the drift-mass, i.e. that

$$kM' = \rho D,$$

and in the case of a circular cylinder $k = 1$.

9·222. Virtual mass in two-dimensional motion. We consider a cylinder of any form moving two-dimensionally with velocity U in a straight line in unbounded liquid. In coordinates fixed with reference to the cylinder the flow will be represented by a complex potential $w=Uz+(f+ig)$ or in terms of velocity potential and stream function

$$(1) \qquad \phi = Ux+f, \quad \psi = Uy+g.$$

Since $f+ig$ represents the disturbance made to the stream by the presence of the cylinder $f+ig$ must tend to zero at infinity and must therefore be expressible in a series of negative powers of z, so that f and g will be expressible in series whose terms will tend to zero at infinity by involving negative powers of r. Thus we shall have

$$(2) \qquad \phi = Ux+\frac{Ax}{r^2}+\frac{By}{r^2}+\frac{C(x^2-y^2)}{r^4}+\dots.$$

$$(3) \qquad \psi = Uy-\frac{Ay}{r^2}+\frac{Bx}{r^2}-\frac{2Cxy}{r^4}+\dots.$$

The boundary condition can be expressed either as

$$(4) \qquad \psi = \text{constant on the boundary, or}$$

$$(5) \qquad lf_x+mf_y = -Ul \text{ on the boundary,}$$

where (l, m) are the direction cosines of the normal to the boundary drawn towards the fluid, and suffixes denote partial differentiation.

The motion of a particle is given by

$$(6) \qquad \frac{dx}{dt} = -U-f_x, \quad \frac{dy}{dt} = -f_y.$$

The stream function gives one integral of these equations say

$$(7) \qquad \psi = Uy+g = U\eta,$$

so that η defines the initial and final asymptotic line of flow. Again

$$\frac{d\phi}{dt} = -\phi_x^2 - \phi_y^2 = -\frac{\partial(\phi, \psi)}{\partial(x, y)},$$

the Jacobian, on account of the Cauchy-Riemann equations 6·0 (1). Thus the drift ξ is given by

$$(8) \qquad \xi = \int_{-\infty}^{\infty} \left(\frac{dx}{dt} + U\right) dt = \int_{-\infty}^{\infty} (-f_x)\, d\phi \Big/ \frac{\partial(\phi, \psi)}{\partial(x, y)}.$$

Here the integrand is to be transformed from its expression in x and y into terms of ϕ and ψ, and is then to be integrated with respect to ϕ with ψ kept constant.

The *drift-volume* D is then given by

$$(9) \quad D = \int_{-\infty}^{\infty} \xi\, d\eta = \frac{1}{U}\int d\psi \int d\phi(-f_x) \Big/ \frac{\partial(\phi, \psi)}{\partial(x, y)} = \frac{1}{U}\iint -f_x\, dx\, dy.$$

Here the field of integration extends over the whole plane of the motion except the cross-section of the cylinder. But it is important to notice that the integral is not absolutely convergent and can attain quite different values according to which integration is done first. In the present case there is no doubt of the order for the ϕ-integration was to be done before the ψ-integration. Since in the distant parts of the fluid $\phi \to Ux$, $\psi \to Uy$ as is clear from (2) and (3), in the last integral of (9) the x-integration is to be done first. An alternative way of stating this is that if the integrations are taken respectively between the infinite values $x = \pm\lambda$, $y = \pm\mu$ then λ is to be much greater than μ. Let us then consider the possible values of the integral

$$(10) \qquad J = \iint (-f_x)\, dx\, dy = \iint (U - \phi_x)\, dx\, dy.$$

Apply Stokes' theorem. Then in terms of the stream function ψ

$$(11) \quad J = \iint (U - \psi_y)\, dx\, dy = -\int^{(0)} (Uy - \psi)\, m\, ds - \int^{(\infty)} (Uy - \psi)\, m\, ds,$$

where (0) signifies integration over the body and (∞) over the surface at infinity. Now by (4) ψ is constant on the body and the associated integral vanishes while

$$(12) \qquad \int^{(0)} ym\, ds = V,$$

where V is the volume of the body. Thus the first term on the right of (11) always gives $-VU$. As regards the second term let us define the field of integration as a " box " $x = \pm\lambda$, $y = \pm\mu$ where both λ and μ are to be infinite. If we insert the expansion (3) into the last term of (11) we find that only one term contributes and that

$$(13) \qquad J = -VU + 4A \tan^{-1}(\lambda/\mu).$$

Thus the extreme values that J can assume are

(14) $$J = -VU, \text{ when } \mu \text{ is large compared with } \lambda.$$

(15) $$J = 2\pi A - VU, \text{ when } \mu \text{ is small compared with } \lambda.$$

We proceed to show that (15) is proportional to the hydrodynamic mass. We have already seen that when λ is much greater than μ (9) gives the drift-volume so that in this case $J = DU$. There are however other interpretations. In the system in which the body is moving the velocity of the fluid in the x-direction is $u = U - \phi_x$, and the total flux of fluid is $\int u\, dy$ across any transverse plane. The total transfer of fluid is the time integral of this; and the time integral multiplied by U is the x-integral which is J so that the time integral is J/U. Here the y-integration was done first so that we must have μ much greater than λ and the answer is (see (14)) $-V$, the reflux of fluid displaced by the body.

Again if ρ is the density

(16) $$\iint \rho u\, dx\, dy = \rho J$$

represents the total momentum of the fluid.

The kinetic energy of the fluid is

(17) $$\tfrac{1}{2} H U^2 = \iint \tfrac{1}{2}\rho\{(U - \phi_x)^2 + \phi_y^2\}\, dx\, dy$$

which is an absolutely convergent integral defining H the *hydrodynamic mass*. But if H can yield energy, it should also be able to yield momentum so (16) and (17) should be related. We have, since $u = U - \phi_x$, from (16) and (17)

$$\begin{aligned} HU^2 - J\rho U &= \rho \iint \{\phi_x(\phi_x - U) + \phi_y^2\}\, dx\, dy \\ &= \rho \iint \left\{\frac{\partial}{\partial x}(\phi - xU)\phi_x + \frac{\partial}{\partial y}(\phi - xU)\,\phi_y\right\} dx\, dy \\ &= -\rho \int^{(0)} (\phi - Ux)(l\phi_x + m\phi_y)\, ds - \rho \int^{(\infty)} (\phi - Ux)(l\phi_x + m\phi_y)\, ds \end{aligned}$$

by Stokes' theorem.

Now by (5) $l\phi_x + m\phi_y = 0$ on the body, while at infinity, from (2), $\phi_x \to U$, $\phi_y \to 0$. Thus only the leading term in (2) contributes and it yields $4A \tan^{-1}(\mu/\lambda)$. Thus if μ is small compared with λ

$$HU - J\rho = 0, \quad J = 2\pi A - VU$$

from (15) and so the virtual mass is

$$H = (2\pi A - VU)\rho/U.$$

This proves that for unbounded fluid the drift-volume measures the hydrodynamic mass. Thus the added mass really represents a mass of the liquid entrained by the cylinder.

9·23. Circular cylinder falling under gravity. Suppose the cylinder of radius a, density σ, to fall, the axis remaining horizontal, in fluid of density ρ.

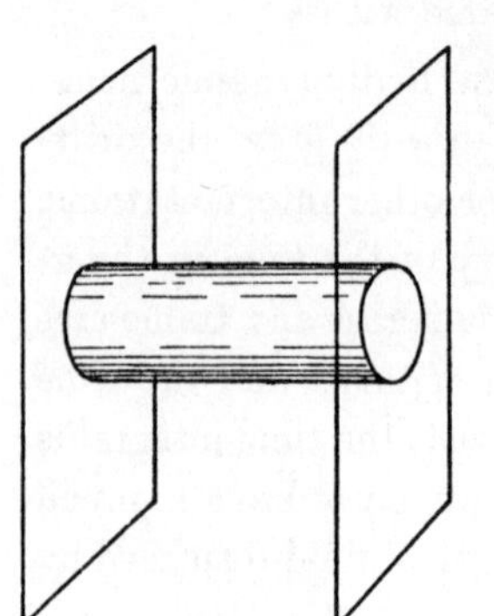

Fig. 9·23.

Consider a unit length of the cylinder limited by smooth vertical walls.

The weight of the cylinder is $\pi a^2 \sigma g$.

The upthrust of the liquid is $\pi a^2 \rho g$, by the principle of Archimedes. Hence the vertical downward force on the cylinder due to gravity is $\pi a^2(\sigma - \rho)g$.

If y is the vertical depth measured from the surface, we have, from 9·22,

$$\pi a^2 \sigma \frac{d^2 y}{dt^2} = \pi a^2 (\sigma - \rho)\, g - \pi a^2 \rho \frac{d^2 y}{dt^2}.$$

Thus
$$\frac{d^2 y}{dt^2} = \frac{\sigma - \rho}{\sigma + \rho} g,$$

and the cylinder descends with this constant acceleration, provided that y is large enough for the surface conditions to be negligible. When $\sigma < \rho$, as in the case of a balloon, the cylinder rises with an acceleration $(\rho - \sigma)\, g/(\rho + \sigma)$.

9·24. Circular cylinder with circulation. Let the centre C of the cross-section of a circular cylinder of radius a be moving with velocity $U + iV$ and let C be at the point z at time t. Then

$$(1) \qquad \dot{z} = U + iV,$$

where the dot denotes differentiation with respect to t, and if $2\pi\kappa$ is the circulation, the cylinder (7·45) experiences a lift $2\pi\kappa\rho i\dot{z}$. Also the acceleration of the centre is $\ddot{z}$ and therefore there is (9·22) a resistance $-M'\ddot{z}$.

Therefore the force exerted by the fluid on the cylinder is

$$X + iY = -M'\ddot{z} + 2\pi\kappa\rho i\dot{z}.$$

If there are no external forces, the equation of motion of the cylinder, mass M, is

$$M\ddot{z} = X + iY.$$

Therefore

$$(1) \qquad (M + M')\ddot{z} - 2\pi\kappa\rho i\dot{z} = 0.$$

The integral of this linear differential equation of the second order with constant coefficients can be found or verified to be

$$z = z_0 + A\, e^{i(\omega t + \epsilon)}, \quad \omega = \frac{2\pi\kappa\rho}{M + M'},$$

$A\,e^{i\epsilon}$, z_0 being arbitrary constants, and A being real. Thus

$$|\,z-z_0\,| = A,$$

so that the centre of the cylinder describes a circle whose centre is z_0. Again,

$$U+iV = \dot{z} = i\omega A\,e^{i(\omega t+\epsilon)}.$$

Thus
$$U^2+V^2 = \omega^2 A^2,$$

so that the circle is described with constant speed in time $2\pi/\omega$, and its radius is

$$A = \frac{(M+M')(U^2+V^2)^{\frac{1}{2}}}{2\pi\kappa\rho}.$$

9·25. Cylinder moving under gravity. If the cylinder considered in 9·24 moves under gravity with its axis horizontal, we take the y-axis vertically upwards. The effect of gravity on the cylinder is to cause a vertically down-

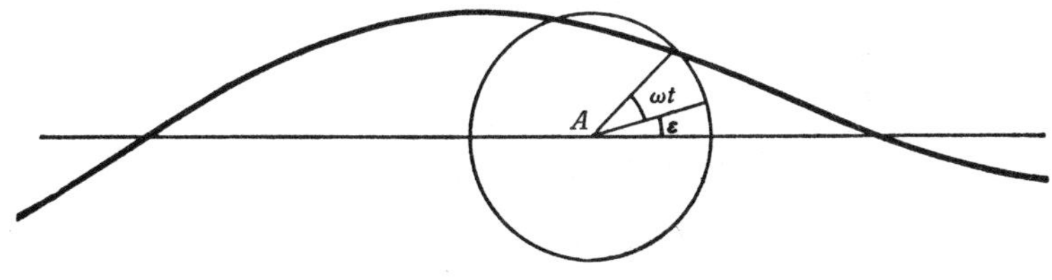

FIG. 9·25.

ward force Mg, the weight, and a vertically upward force $M'g$, the buoyancy (principle of Archimedes). Therefore equation (1) of 9·24 is replaced by

$$(M+M')\ddot{z} - 2\pi\kappa\rho i\dot{z} = -i(M-M')g,$$

or
$$\ddot{z} - i\omega\dot{z} = -ig_0\,,$$

where
$$\omega = \frac{2\pi\kappa\rho}{M+M'}, \quad g_0 = \frac{M-M'}{M+M'}g.$$

A particular integral is clearly $z = g_0\,t/\omega$, and therefore

$$z = z_0 + A\,e^{i(\omega t+\epsilon)} + \frac{g_0 t}{\omega}.$$

Therefore $\quad x = x_0 + \dfrac{g_0 t}{\omega} + A\cos(\omega t+\epsilon), \quad y = y_0 + A\sin(\omega t+\epsilon).$

Thus the path of the centre of the cylinder is the trochoid described by a point on the circumference of a circle of radius A which rotates with angular velocity ω, while its centre moves on a straight horizontal line with constant velocity g_0/ω.

The precise value of A will depend on the initial speed and direction of motion of the centre of the cylinder. If these are so arranged that $A = 0$, then the path of the centre of the cylinder will be a straight line. Also, when the path is a trochoid, the mean direction of progress is horizontal; in other words, the cylinder does not tend to descend under the action of gravity. This phenomenon has been advanced as a partial explanation of the observed behaviour of a tennis ball.

9·30. Pressure equation referred to moving axes. Let the origin have velocities U and V along the instantaneous position of the axes, and let ω be the angular velocity. The pressure equation is given in 3·61. To adapt it to this case we have to calculate the square of the speed at the point z. Now, with the notations of 3·61 and 5·10, we get, for the velocity of z,

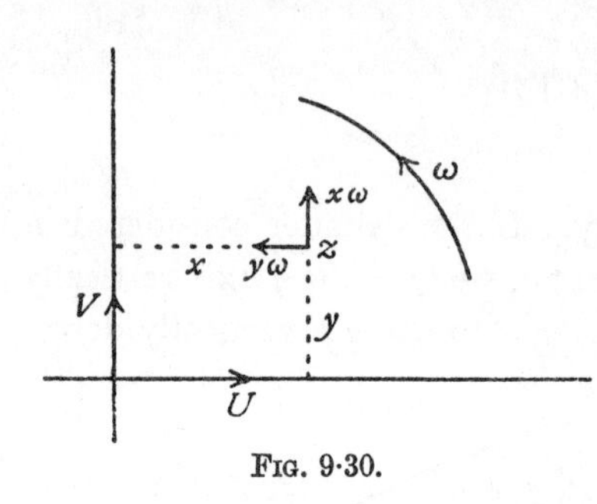

FIG. 9·30.

$$\mathbf{U}+\boldsymbol{\omega}_\wedge \mathbf{r} = \mathbf{i}U+\mathbf{j}V+\omega\mathbf{k}_\wedge(x\mathbf{i}+y\mathbf{j})$$
$$= [(U-y\omega)+(V+x\omega)\,\mathbf{k}_\wedge]\,\mathbf{i}$$
$$= (W+i\omega z)\,\mathbf{i},$$

where $W = U+iV$. Thus the square of the speed of z is

$$(W+i\omega z)(\overline{W}-i\omega\bar{z}),$$

and the pressure equation is

$$\frac{p}{\rho}+\tfrac{1}{2}q_r^2+\Omega-\frac{\partial\phi}{\partial t}-\tfrac{1}{2}(W+i\omega z)(\overline{W}-i\omega\bar{z}) = C(t),$$

where q_r is the speed of the fluid relative to the moving axes.

When the relative motion is steady, as for example in the case of an observer in a ship moving with constant course and speed, we get

$$\frac{p}{\rho}+\tfrac{1}{2}q_r^2+\Omega-\tfrac{1}{2}(W+i\omega z)(\overline{W}-i\omega\bar{z}) = C,$$

where C is now an absolute constant.

9·40. The stream function on the boundary. Consider axes fixed in a cylinder which is moving with a velocity of translation and rotation.

Let U, V be the components of the velocity of the origin O, and let ω be the angular velocity. Then the components of velocity of the point $P(x, y)$ of the boundary are $U-y\omega$, $V+x\omega$. Resolving along the outward normal to the boundary at P, we get

$$(U-y\omega)\sin\theta-(V+x\omega)\cos\theta,$$

where θ is the inclination of the tangent to the axis of x.

Now $\sin\theta = dy/ds$, $\cos\theta = dx/ds$, and the normal velocity of the fluid is $-\partial\psi/\partial s$. Equating the normal velocities, we get

$$-\frac{\partial\psi}{\partial s} = (U - y\omega)\frac{dy}{ds} - (V + x\omega)\frac{dx}{ds}.$$

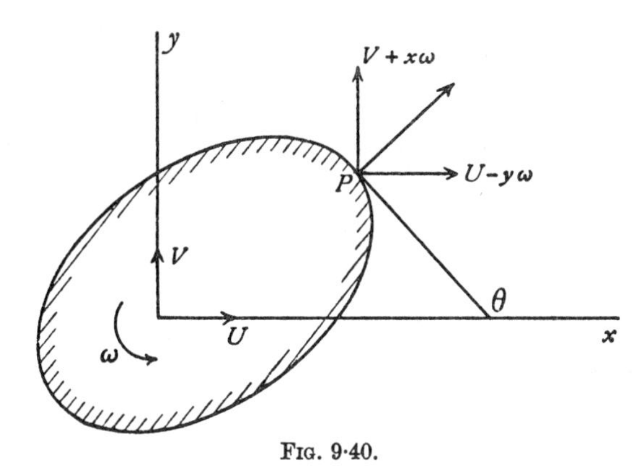

FIG. 9·40.

Integrating along the boundary,

$$\psi = Vx - Uy + \tfrac{1}{2}\omega(x^2 + y^2) + B,$$

where B is an arbitrary constant, and therefore we have the value of the stream function on the boundary. We now see that, ignoring an added constant, ψ is equal to the imaginary part of the function

(1) $$f(z, \bar{z}) = -(U - iV)z + \tfrac{1}{2}i\omega z\bar{z}.$$

If we put $U\cos\alpha$, $U\sin\alpha$ for U, V, so that the resultant velocity is U at angle α to Ox, we get

(2) $$f(z, \bar{z}) = -Uz\,e^{-i\alpha} + \tfrac{1}{2}i\omega z\bar{z}.$$

The conjugate complex function is

(3) $$\bar{f}(\bar{z}, z) = -U\bar{z}\,e^{i\alpha} - \tfrac{1}{2}i\omega\bar{z}z.$$

Since ψ is the imaginary part of (2), $-\psi$ is the imaginary part of (3). Therefore, by subtraction,

(4) $$2i\psi = -Uz\,e^{-i\alpha} + U\bar{z}\,e^{i\alpha} + i\omega z\bar{z}.$$

9·50. Force on a moving cylinder. Referring to 6·41, we have for the action on the element ds of the boundary

$$dX - i\,dY = -ip\,d\bar{z}, \quad dM + i\,dN = pz\,d\bar{z}.$$

Now let $dz = ds\,e^{i\alpha}$ and therefore $d\bar{z} = dz\,e^{-2i\alpha}$. Therefore

(1) $$X - iY = -i\int_{(C)} p\,e^{-2i\alpha}\,dz, \quad M + iN = \int_{(C)} pz\,e^{-2i\alpha}\,dz,$$

where the integrals are taken round the contour C of the cross-section of the cylinder. Now, if $W = U+iV$ is the velocity of the origin with respect to axes fixed in the cylinder, and ω is the angular velocity, the pressure equation is (9·30),

$$\text{(2)} \qquad \frac{p}{\rho} = \frac{\partial \phi}{\partial t} - \tfrac{1}{2}q_r^2 + \tfrac{1}{2}(W+i\omega z)(\overline{W}-i\omega\bar{z}),$$

where q_r is the relative speed. Now, on the boundary the fluid is moving tangentially relatively to the cylinder, and therefore the relative complex velocity is

$$q_r\, e^{-i\alpha} = -\frac{dw}{dz} - \overline{W} + i\omega\bar{z},$$

for the left side expresses that the relative velocity is tangential and the right side measures the relative velocity.

Substitute the value of q_r given by this equation in (2) and then substitute for p in (1). We thus get, making use of the relation $d\bar{z} = e^{-2i\alpha}\, dz$ at points on the cylinder,

$$\text{(3)} \qquad X - iY = \tfrac{1}{2}i\rho \int_{(C)} \left(\frac{dw}{dz} + \overline{W} - i\omega\bar{z}\right)^2 dz$$

$$- \tfrac{1}{2}i\rho \int_{(C)} (W+i\omega z)(\overline{W}-i\omega\bar{z})\, d\bar{z} - i\rho \int_{(C)} \frac{\partial \phi}{\partial t}\, d\bar{z}.$$

$$\text{(4)} \qquad M + iN = -\tfrac{1}{2}\rho \int_{(C)} z \left(\frac{dw}{dz} + \overline{W} - i\omega\bar{z}\right)^2 dz$$

$$+ \tfrac{1}{2}\rho \int_{(C)} z(W+i\omega z)(\overline{W}-i\omega\bar{z})\, d\bar{z} + \rho \int z \frac{\partial \phi}{\partial t}\, d\bar{z}.$$

These equations constitute a generalisation of the theorem of Blasius, to which they reduce when the motion is steady and the cylinder is at rest. In their present form they are unwieldy. The simplification of these results is most rapidly effected by the use of the Area theorem (5·43), which gives

$$\text{(5)} \qquad \int_{(C)} (\overline{W}-i\omega\bar{z})^2 dz = 2i \int_{(S)} -2i\omega(\overline{W}-i\omega\bar{z})\, dS = 4\omega A(\overline{W}-i\omega\bar{z}_c),$$

where A is the area enclosed by the contour and $z_c = x_c + iy_c$ is the position of the centroid of this area.

$$\text{(6)} \qquad \int_{(C)} (W+i\omega z)(\overline{W}-i\omega\bar{z})\, d\bar{z} = -2i \int_{(S)} i\omega(\overline{W}-i\omega\bar{z})\, dS = 2\omega A(\overline{W}-i\omega\bar{z}_c).$$

$$\text{(7)} \qquad \int_{(C)} (-\overline{W}z + W\bar{z} + i\omega z\bar{z})\, d\bar{z} = -2i \int_{(S)} (-\overline{W}+i\omega\bar{z})\, dS = 2iA(\overline{W}-i\omega\bar{z}_c).$$

9·52. Extension of the theorem of Blasius. Equation (3) of 9·50 gives the force on a moving cylinder. This can be written

$$X-iY = \tfrac{1}{2}i\rho\int\left(\frac{dw}{dz}\right)^2 dz+i\rho\int\left(\overline{W}\frac{dw}{dz}-i\omega\bar{z}\frac{dw}{dz}\right)dz+\tfrac{1}{2}i\rho\int(\overline{W}-i\omega\bar{z})^2dz$$

$$-\tfrac{1}{2}i\rho\int(W+i\omega z)(\overline{W}-i\omega\bar{z})\,d\bar{z}-i\rho\frac{\partial}{\partial t}\int\phi\,d\bar{z},$$

where the integrals are all taken round the contour of the cylinder. Using (5) and (6) of the preceding section, we get

$$(1)\quad X-iY = \tfrac{1}{2}i\rho\int\left(\frac{dw}{dz}\right)^2 dz+i\rho\overline{W}\int dw+\omega\rho\int\bar{z}\,dw$$

$$+2i\rho\omega A(\overline{W}-i\omega\bar{z}_c)-i\rho\omega A(\overline{W}-i\omega\bar{z}_c)-i\rho\frac{\partial}{\partial t}\int\phi\,d\bar{z}.$$

Now $\int dw$ is the increase in w as we pass round the cylinder and is therefore equal to $-2\pi\kappa$, where κ is the strength of the circulation (which may be zero).

Also, $\int\bar{z}\,dw = \int\bar{z}(d\bar{w}+2i\,d\psi)$, since $w = \bar{w}+2i\psi$, and, integrating by parts,

$$\int\bar{z}\,d\psi = [\bar{z}\psi]_C-\int\psi\,d\bar{z}.$$

Since the product $\bar{z}\psi$ returns to its initial value on going once round the cylinder $[\bar{z}\psi]_C = 0$, and hence

$$\int\bar{z}\,dw = \int\bar{z}\,d\bar{w}-2i\int\psi\,d\bar{z}.$$

Now, from 9·40 (4), we have on the cylinder

$$2i\psi = -\overline{W}z+W\bar{z}+i\omega z\bar{z}.$$

Therefore, from 9·50 (7),

$$(2)\qquad \int\psi\,d\bar{z} = A(\overline{W}-i\omega\bar{z}_c).$$

(3) Hence $\int\bar{z}\,dw = \int\bar{z}\,d\bar{w}-2iA(\overline{W}-i\omega\bar{z}_c).$

(4) Again, $\int\phi\,d\bar{z} = \int(\bar{w}+i\psi)\,d\bar{z} = \int\bar{w}\,d\bar{z}+Ai(\overline{W}-i\omega\bar{z}_c),$

from (2). Substituting (3) and (4) in (1), we get

$$(5)\qquad X-iY = \tfrac{1}{2}i\rho\int\left(\frac{dw}{dz}\right)^2 dz+\omega\rho\int\bar{z}\,d\bar{w}-i\rho\frac{\partial}{\partial t}\int\bar{w}\,d\bar{z}$$

$$-2\pi\kappa\rho i\overline{W}-i\rho A\left\{\omega(\overline{W}-i\omega\bar{z}_c)+i\left(\frac{d\overline{W}}{dt}-i\bar{z}_c\frac{d\omega}{dt}\right)\right\}.$$

This may be regarded as the extended form of the theorem of Blasius for the force on a moving cylinder, its advantage being that the integrals are all taken round the contour of the cylinder or any larger contour reconcilable with this without passing over singularities such as sources, sinks, or vortices. A similar calculation will show that the moment about the origin of the forces due to the pressure is the real part of

$$(6) \qquad -\tfrac{1}{2}\rho\int z\left(\frac{dw}{dz}\right)^2 dz-\rho\overline{W}\int z\,dw+\omega\rho A\overline{W}z_c$$
$$+\rho\,\frac{\partial}{\partial t}\int zw\,d\bar{z}-A\rho\left\{3iz_c\frac{d\overline{W}}{dt}+2k^2\frac{d\omega}{dt}\right\},$$

where k is the radius of gyration of the section with respect to O.

The reduction of (4) of 9·50 to the above form by use of 5·43 is left as an exercise to the reader.

It will be seen that it is advantageous to take the origin at the centroid of the section, for then $z_c = 0$.

The interest of the above results lies in their complete generality, since they apply equally whether the motion is steady or not.

In the case of relatively steady motion, the terms involving differentiations with respect to the time disappear.

9·53. Cylinder moving in unbounded fluid.

When a cylinder moves in unbounded liquid which is at rest at infinity, the disturbance due to the motion of the cylinder must be negligible at great distances from the cylinder. Thus, for large values of z, we must have $dw/dz = 0$.

The most general form of w consistent with this condition and with continuity of the motion of the fluid and of the potential is, for large values of $|z|$,

$$(1) \qquad w = i\kappa\log z+\frac{a_1}{z}+\frac{a_2}{z^2}+\dots,$$

κ being the strength of the circulation.

We then obtain

$$(2) \qquad \frac{dw}{dz}=\frac{i\kappa}{z}-\frac{a_1}{z^2}-\frac{2a_2}{z^3}-\dots,$$
$$\left(\frac{dw}{dz}\right)^2=-\frac{\kappa^2}{z^2}-\frac{2a_1 i\kappa}{z^3}-\dots,$$

and it follows that the first integral in the Blasius formula 9·52 (5) vanishes.

Changing the sign of i throughout, and taking the origin at the centroid of the section, we get

$$(3) \quad X+iY=\omega\rho\int z\,dw+i\rho\,\frac{\partial}{\partial t}\int w\,dz+2\pi\kappa\rho iW+i\rho A\left(\omega W-i\,\frac{dW}{dt}\right).$$

From (2), $$\int z\,dw = \int \left(i\kappa - \frac{a_1}{z} - \frac{2a_2}{z^2} \ldots\right) dz = -2\pi i a_1,$$

by the residue theorem.

Again, $$\int w\,dz = i\kappa\,[z \log z] + 2\pi i a_1,$$

where $[z \log z]$ represents the variation of $z \log z$ when we go once round the contour. If the circulation remains constant, the differential coefficient of this term with respect to t vanishes. We then get, from (3),

$$(4) \quad X + iY = -2\pi\rho\omega i a_1 + 2\pi\kappa\rho i W + i\rho A\left(\omega W - i\frac{dW}{dt}\right) - 2\pi\rho\frac{da_1}{dt}.$$

This formula is very convenient for it contains no integrations. In the complex potential (1), let $a_1 = a + ib$. Then remembering that $W = U + iV$, where (U, V) are the velocity components of the motion of the origin, we get

$$X = 2\pi\rho b\omega - 2\pi\kappa\rho V - A\rho\omega V + A\rho\frac{dU}{dt} - 2\pi\rho\frac{da}{dt},$$

$$Y = -2\pi\rho a\omega + 2\pi\kappa\rho U + A\rho\omega U + A\rho\frac{dV}{dt} - 2\pi\rho\frac{db}{dt}.$$

We may also note that $A\rho = M'$, the mass of fluid displaced by the cylinder (per unit thickness), and that if $\omega = 0$, the last two terms of (4) measure the hydrodynamic mass for linear motion.

The above formulae may be applied to give the results of 9·24, 9·25. This is left as an exercise.

The theorem of Kutta and Joukowski (7·45) follows as a special case of (4), for taking $\omega = 0$, and $W =$ constant, we get $X + iY = 2\pi\kappa\rho i W$, which is a force at right angles to the direction of W, of intensity $2\pi\kappa\rho\sqrt{(U^2 + V^2)}$, and independent of the shape or area of the cross-section of the cylinder. Equation (4) may therefore be regarded as an extension of the theorem of Kutta and Joukowski.

The corresponding extension on the lines of Lagally's theorem, when sources and sinks are present, offers no difficulty.

9·62. Cylinder moving in a general manner. The complex potential in the case of a circular cylinder moving transversely was derived in 9·20 from the corresponding case of liquid streaming past a fixed cylinder by superposing on the whole system a velocity opposite to that of the stream. The case of transverse motion of an elliptic cylinder could be similarly derived from the streaming past applied to the result of 6·33. We shall now, however, explain a method of more general application, whereby a direct attack can be made on the problem of a cylinder moving with translation and rotation in a fluid at rest at infinity.

The method consists essentially of mapping the region exterior to the cross-section of the cylinder in the z-plane on the region exterior to the unit circle $|\zeta| = 1$ in the ζ-plane combined with a particular application of 9·40 (4).

9·63. The complex potential for a moving cylinder. Let C be the contour of the cross-section of a cylinder moving two-dimensionally in infinite liquid at rest at infinity, with no circulation about it. The motion of the cylinder is described by the velocity of translation U of a point O of the cross-section at an angle α with the x-axis, and an angular velocity ω. We suppose the domain outside the cylinder C in the z-plane (referred to axes at O fixed in the cylinder) to be mapped conformally on the outside of the unit circle $|\zeta| = 1$ in a complex ζ-plane, by a relation

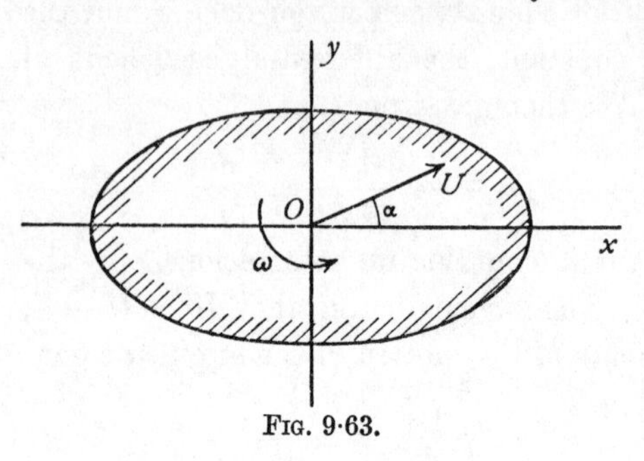

FIG. 9·63.

(1) $$z = f(\zeta),$$

the points at infinity in the z- and ζ-planes corresponding.

Then for the liquid to be at rest at infinity, the complex potential w cannot contain positive powers of z (or ζ) when expanded in a power series in z (or ζ).

Also, on the boundary C of the cylinder, the stream function ψ is such that (see 9·40 (4))

(2) $$2i\psi = -Uz\,e^{-i\alpha} + U\bar{z}\,e^{i\alpha} + i\omega z\bar{z}.$$

We shall denote a general point on the unit circle by σ. Then

(3) $$\sigma = e^{i\theta}, \quad \bar{\sigma} = e^{-i\theta} = 1/\sigma.$$

Therefore on the unit circle (2) gives

(4) $$2i\psi = B(\sigma) = -Uf(\sigma)\,e^{-i\alpha} + U\bar{f}(1/\sigma)\,e^{i\alpha} + i\omega\, f(\sigma)\,\bar{f}(1/\sigma).$$

The function $B(\sigma)$ may be conveniently called the *boundary function.* If this is expanded in powers of σ, we can write

(5) $$B(\sigma) = B_1(\sigma) + B_2(\sigma),$$

where $B_1(\sigma)$ contains all the negative powers of σ and no non-negative powers. Thus $B_1(\zeta)$ is holomorphic outside the unit circle and vanishes at infinity.

We can now write the boundary condition (4) in the form

(6) $$w(\sigma) - \bar{w}(1/\sigma) = B_1(\sigma) + B_2(\sigma).$$

Multiply by $d\sigma/\{2\pi i(\sigma - \zeta)\}$ and integrate round γ the circumference of the unit circle. Then

$$(7)\quad \frac{1}{2\pi i}\int_{(\gamma)}\frac{w(\sigma)\,d\sigma}{\sigma-\zeta}-\frac{1}{2\pi i}\int_{(\gamma)}\frac{\bar{w}(1/\sigma)\,d\sigma}{\sigma-\zeta}=\frac{1}{2\pi i}\int_{(\gamma)}\frac{B_1(\sigma)\,d\sigma}{\sigma-\zeta}+\frac{1}{2\pi i}\int_{(\gamma)}\frac{B_2(\sigma)\,d\sigma}{\sigma-\zeta}$$

Now $w(\zeta)$ and $B_1(\zeta)$ are holomorphic outside γ, while $\bar{w}(1/\zeta)$ and $B_2(\zeta)$ are holomorphic inside γ. Therefore if ζ is outside γ, the application of Cauchy's formula (5·59) shows that the second and fourth integrals vanish while the first and third give

$$(8)\qquad w = B_1(\zeta),$$

and since $B_1(\zeta)$ contains only negative powers of ζ the condition of vanishing velocity at infinity is also satisfied.

To see that the velocity is physically admissible everywhere in the liquid, we have

$$-u+iv = w'(z) = B_1'(\zeta)/f'(\zeta),$$

and since the transformation (1) is conformal at all points in the exterior domain, there are no zeros of $f'(\zeta)$ in the liquid, hence the velocity is finite everywhere in the liquid.

To sum up, by means of (1) we form the boundary function (4), separate out the negative powers of ζ which tend to zero as $|\zeta|$ tends to infinity, and these give immediately the complex potential (8) as a function of ζ.

Elimination of ζ between (1) and (8) would, of course, yield the complex potential as a function of z. In many cases it is impossible or undesirable to effect the elimination.

Finally, we can deduce w for the streaming motion past the cylinder by putting $\omega = 0$ in (2) and superposing the stream U reversed. This gives

$$w = B_1(\zeta)+Uf(\zeta)e^{-i\alpha}.$$

9·64. Circular cylinder (general method). The simplest illustration of the general method is afforded by the circular cylinder of radius a moving with velocity U at an angle α with the real axis. Taking the origin at the centre of the circular cross-section, the mapping function of 9·63 (1) is

$$z = a\zeta.$$

The boundary function is

$$B(\zeta) = -Ua\zeta\, e^{-i\alpha}+Ua\zeta^{-1}\,e^{i\alpha}+i\omega a^2,$$

so that

$$B_1(\zeta) = Ua\, e^{i\alpha}\zeta^{-1};$$

hence

$$w = Ua\, e^{i\alpha}\zeta^{-1} = Ua^2\, e^{i\alpha}z^{-1}.$$

As we should expect, this does not involve the angular velocity.

9·65. Elliptic cylinder. If the unit circle in the complex ζ-plane is given by $\zeta = e^{i\eta}$, the transformation

$$(1) \qquad z = c(\zeta + \lambda\zeta^{-1}), \quad 0 \leqslant \lambda \leqslant 1,$$

maps the region outside the boundary C given by

$$(2) \qquad z = a \cos \eta + ib \sin \eta,$$

where

$$(3) \qquad a = c(1+\lambda), \quad b = c(1-\lambda),$$

conformally on the region outside the unit circle and the points at infinity correspond. Clearly C is an ellipse of axes $2a$, $2b$, and the eccentric angle of the point z is η. Note that $f'(\zeta) = 0$ only for

$$\zeta = \lambda^{\frac{1}{2}} = \left(\frac{a-b}{a+b}\right)^{\frac{1}{2}} < 1,$$

which lies within the unit circle, so that the transformation of the exterior domain is everywhere conformal.

The boundary function is

$$B(\zeta) = -Uc\, e^{-i\alpha}(\zeta + \lambda\zeta^{-1}) + Uc\, e^{i\alpha}(\zeta^{-1} + \lambda\zeta) + i\omega c^2\{1 + \lambda^2 + \lambda(\zeta^2 + \zeta^{-2})\},$$

so that

$$B_1(\zeta) = -U\lambda c\, e^{-i\alpha}\zeta^{-1} + Uc\, e^{i\alpha}\zeta^{-1} + i\omega\lambda c^2\zeta^{-2}.$$

Hence from 9·63 (8) the complex potential is

$$(4) \qquad w = A\zeta^{-1} + B\zeta^{-2},$$

where, from (3),

$$(5) \qquad A = U(b \cos \alpha + ia \sin \alpha), \quad B = \frac{i}{4}\omega(a^2 - b^2).$$

When $a = b$, we again have the results for the circular cylinder.

The kinetic energy of the liquid (per unit thickness) is given by (9·10)

$$T = -\tfrac{1}{4}i\rho \int_{(C)} w\, d\bar{w},$$

taken round the elliptic boundary C, hence

$$T = -\tfrac{1}{4}i\rho \int_{(C)} (A\zeta^{-1} + B\zeta^{-2})(\bar{A} + 2\bar{B}\zeta)\, d\zeta = -\tfrac{1}{4}i\rho\, 2\pi i (A\bar{A} + 2B\bar{B}),$$

or $\quad T = \tfrac{1}{2}\rho\pi U^2(b^2 \cos^2\alpha + a^2 \sin^2\alpha) + \tfrac{1}{16}\rho\pi\omega^2(a^2 - b^2)^2.$

When $U = 0$ so that the cylinder rotates without translation,

$$T = \tfrac{1}{16}\pi\rho\omega^2(a^2 - b^2)^2,$$

which is the same for all confocal ellipses. In particular, this gives the kinetic energy when the ellipse reduces to the straight line of length $2c$ joining the foci. We then have the case of the flat plate rotating, but the velocity at the edges is then infinite, so that the solution cannot apply without modification to a real fluid.

The case of the rotating plate offers some other features of interest.* For the plate we have $b=0$, and so $\lambda=1$, $a=2c$ and the length of the plate is $4c$ from (3). Thus from (4) the stream function is $\psi=\frac{1}{2}\omega c^2(\zeta^{-2}+\bar{\zeta}^{-2})$. To find the streamlines relative to plate we superpose the angular velocity $-\omega$ on the whole system by adding to ψ the stream function $-\frac{1}{2}\omega(x^2+y^2) = -\frac{1}{2}\omega z\bar{z}$. The resulting stream function of the relative motion is then

$$\Psi = \tfrac{1}{2}\omega c^2\left\{\frac{1}{\zeta^2}+\frac{1}{\bar{\zeta}^2}-\left(\zeta+\frac{1}{\zeta}\right)\left(\bar{\zeta}+\frac{1}{\bar{\zeta}}\right)\right\} \tag{6}$$

and the streamlines relative to the plate are the lines Ψ=constant. On the plate itself $\zeta\bar{\zeta}=1$ since the plate maps into the circumference of the unit circle, and then $\Psi=-\omega c^2$. Therefore the relative dividing streamline is $\Psi+\omega c^2=0$, or after reduction

$$(\zeta\bar{\zeta}-1)(\zeta^2\bar{\zeta}^2-\zeta\bar{\zeta}+\zeta^2+\bar{\zeta}^2)=0.$$

The first factor gives the circle i.e. the plate and the remaining part of the dividing line is

$$\zeta^2\bar{\zeta}^2-\zeta\bar{\zeta}+\zeta^2+\bar{\zeta}^2=0. \tag{7}$$

This meets the plate where $\zeta\bar{\zeta}=1$ or $\zeta^4=-1$ and so $z=\pm 2c/\sqrt{2}$, the points L and L' in fig. 9·65 in which AA' is the plate which is rotating *anticlockwise*.

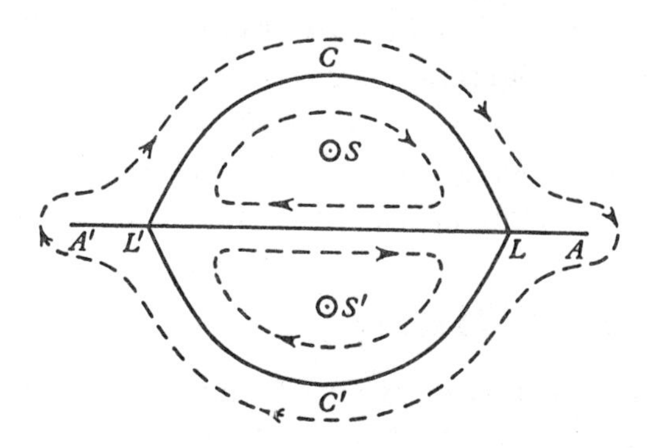

Fig. 9·65.

The curve (7) meets the y-axis where ζ is imaginary i.e. $\zeta=-\bar{\zeta}$, whence $\zeta^2=-3$ and $z=\pm 2ic/\sqrt{3}$ the points C, C' in fig. 9·65. Thus with the plate the dividing streamline forms two closed loops as indicated in the figure by $L'CL$, $L'C'L$. The liquid within these is trapped and must perforce move round with the rotating plate, always with a velocity distribution consistent with irrotational

* C. Darwin, *loc. cit.* p. 244.

motion. Within these loops there are *relative* stagnation points S and S' (given by $\partial\Psi/\partial\zeta = 0$) which lie on the y-axis at distance $c(3^{1/4} - 3^{-1/4}) = 0{\cdot}556c$ from the centre of the plate. The particles at these points move as if rigidly attached to the plate. In fig. 9·65 the dotted lines show other relative paths. The relative motion in these is *clockwise*, that is to say, against the sense of rotation of the plate. In fact the relative angular velocity of the radius from the centre of the plate to a fluid particle is less than ω and so there is a general anticlockwise drift of the fluid, leading to rotational added mass (cf. Ex. IX, 8).

Problems relating to elliptic cylinders can also be solved by the direct method of 6·35. Thus when the cylinder moves forward with velocity $U e^{i\alpha}$, we have on the boundary ψ = imaginary part of

$$-Uz\,e^{-i\alpha} = -Uc\,e^{-i\alpha}\cosh\zeta$$

in elliptic coordinates. Therefore we must have

$$w = -Uc\,e^{-i\alpha}\cosh\zeta + F(\zeta),$$

where $F(\zeta)$ is to be chosen so as to be real on the boundary and to make $w \to 0$ when $|\zeta| \to \infty$. If the ellipse is defined by $\xi = \xi_0$ so that $\bar{\zeta} = 2\xi_0 - \zeta$ on the boundary, we see that the suitable form for $F(\zeta)$ is

$$\tfrac{1}{2}Uc\,e^{-i\alpha}\,e^{\zeta} + \tfrac{1}{2}Uc\,e^{i\alpha}\,e^{2\xi_0-\zeta},$$

whence

$$w = U(a+b)\sinh(\xi_0 + i\alpha)e^{-\zeta}.$$

Similarly for the rotating elliptic cylinder, on the boundary

$$\psi = \tfrac{1}{2}\omega c^2\cosh\zeta\cosh\bar{\zeta} = \tfrac{1}{4}\omega c^2\cosh(\zeta - \bar{\zeta}) + \text{constant},$$

and a similar argument leads to

$$w = \tfrac{1}{4}i\omega c^2\cosh(2\zeta - 2\xi_0) - \tfrac{1}{4}i\omega c^2\sinh(2\zeta - 2\xi_0) = \tfrac{1}{4}i\omega(a+b)^2 e^{-2\zeta}.$$

More generally, if ψ on the boundary is the imaginary part of the complex potential $F_1(\zeta) + F_2(\zeta)$, where $F_1(\zeta) \to \infty$, $F_2(\zeta) \to 0$, at infinity, then

$$w = -\bar{F}_1(2\xi_0 - \zeta) + F_2(\zeta),$$

provided that $\bar{F}_1(2\xi_0 - \zeta) \to 0$ at infinity.

9·66. Cylinder with circulation. To allow for the circulation about a cylinder of any form, we observe that the complex potential

$$(1) \qquad w = i\kappa\log\zeta$$

makes $\psi = \kappa\log|\zeta| = 0$ along the unit circle $\zeta = e^{i\eta}$, i.e. the boundary C is a streamline, also $\phi = -\kappa\eta$, so that ϕ decreases by $2\pi\kappa$ when we go once round the cylinder in the positive (counter-clockwise) direction. Thus (1) gives circulation $2\pi\kappa$ round a cylinder of any form which can be mapped on the unit circle. In particular for the elliptic cylinder of 9·65, we have

$$(2) \qquad w = i\kappa\log\zeta + U(b\cos\alpha + ia\sin\alpha)\zeta^{-1} + i\frac{\omega}{4}(a^2 - b^2)\zeta^{-2}.$$

9·70. Rotating cylinder. When a cylinder containing liquid rotates about an axis through the origin parallel to the generators the following considerations may be used.

If the equation of the boundary of the cross-section can be written in the form

$$(1) \qquad z\bar{z} = f(z) + \bar{f}(\bar{z}),$$

where $f'(z)$ has no singularities within the cross-section, then the problem is solved by the complex potential

$$(2) \qquad w = i\omega f(z),$$

for then $\psi = \frac{1}{2}\omega z\bar{z}$ on the boundary.

If all the singularities of $f'(z)$ are inside the contour, then (2) is the solution when the cylinder rotates in fluid external to it.

More generally if $z = F(\zeta)$ defines some system of coordinates, e.g. elliptic, such that on the boundary

$$z\bar{z} = f(\zeta) + \bar{f}(\bar{\zeta}),$$

then $w = i\omega f(\zeta)$ is the complex potential when the fluid is inside or outside the cylinder according as the singularities of dw/dz, that is of $f'(\zeta)/F'(\zeta)$, are outside or inside the cylinder.

9·71. Rotating elliptic cylinder containing liquid. Taking the cross-section to be the ellipse,

$$\frac{x^2}{a^2} + \frac{y^2}{b^2} = 1, \quad \text{or} \quad \frac{(z+\bar{z})^2}{4a^2} - \frac{(z-\bar{z})^2}{4b^2} = 1,$$

comparison with 9·70 (1) gives

$$f(z) = \frac{1}{2}\frac{a^2-b^2}{a^2+b^2}z^2 + \frac{a^2b^2}{a^2+b^2}.$$

The constant is irrelevant, so that

$$w = \frac{1}{2}i\omega\left(\frac{a^2-b^2}{a^2+b^2}\right)z^2.$$

To find the paths of the particles relative to the cylinder, we can superpose the angular velocity $-\omega$ by adding to ψ the stream function $-\frac{1}{2}\omega(x^2+y^2)$, which gives

$$\Psi = -\frac{a^2b^2\omega}{a^2+b^2}\left(\frac{x^2}{a^2} + \frac{y^2}{b^2} - 1\right) + A,$$

so that when Ψ is constant the relative paths are the ellipses

$$\frac{x^2}{a^2} + \frac{y^2}{b^2} = \text{constant},$$

which are similar to the cross-section of the cylinder.

For the relative velocity we have

$$\frac{dx}{dt} = -\Psi_y = \frac{2a^2\omega y}{a^2+b^2}, \quad \frac{dy}{dt} = \Psi_x = -\frac{2b^2\omega x}{a^2+b^2}.$$

Consider the particle which at time $t=0$ lies at the point $(ka, 0)$ on the major axis. Then at time t

$$x = ka \cos \Omega t, \quad y = -kb \sin \Omega t, \quad \Omega = \frac{2ab}{a^2+b^2}\omega.$$

Thus at time t the particle is at the point of its ellipse whose eccentric angle is $-\Omega t$. This refers to the moving axes. Referred to fixed axes the particle will have a mean angular velocity of drift $\omega - \Omega = (a-b)^2\omega/(a^2+b^2)$, superposed on its oscillatory motion.

Suppose, for example, that the positive half of the major axis is initially marked by dye. This line will remain a radius of the ellipse, and periodically it will coincide with the major axis again. In the course of $(a^2+b^2)/(a-b)^2$ complete turns of the cylinder it will have rotated completely once right round; nevertheless the motion remains irrotational.

9·72. Rotating equilateral prism containing liquid. The lines

$$x-a = 0, \quad x-y\surd 3+2a = 0, \quad x+y\surd 3+2a = 0$$

bound the equilateral triangle ABC, whose centroid is the origin. The length of the side of the triangle is then $2a\surd 3$. Combining these into one equation, we get for the equation of the boundary

$$F(x, y) = x^3 - 3xy^2 + 3a(x^2+y^2) - 4a^3 = 0, \quad \text{or} \quad \tfrac{1}{2}(z^3+\bar{z}^3) + 3az\bar{z} - 4a^3 = 0,$$

whence by comparison with 9·70 (1)

$$f(z) = -\tfrac{1}{6}\frac{z^3}{a} + \frac{2a^2}{3},$$

and therefore

$$w = -\frac{i\omega z^3}{6a}.$$

Superposing the stream function $\psi = -\frac{1}{2}\omega(x^2+y^2)$, we get the equations of the relative streamlines $F(x, y) =$ constant, or

$$(x-a)(x-y\surd 3+2a)(x+y\surd 3+2a) = c^3,$$

where c is a constant. These are cubic curves having the sides of the triangle as asymptotes and loops within the triangle ABC formed by these asymptotes. In particular when $c = 0$, we see that the line ABC is a relative streamline.

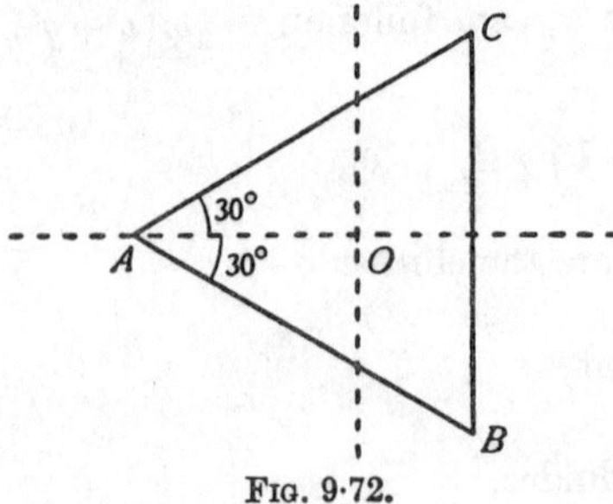

FIG. 9·72.

9·73. Slotted circular cylinder. The cross-section is the lune comprised between the circles

$$x^2+y^2-b^2 = 0, \quad x^2+y^2-2ax = 0,$$

as shown in fig. 9·73, and the fluid is inside. The centre of the first circle is on the circumference of the second.

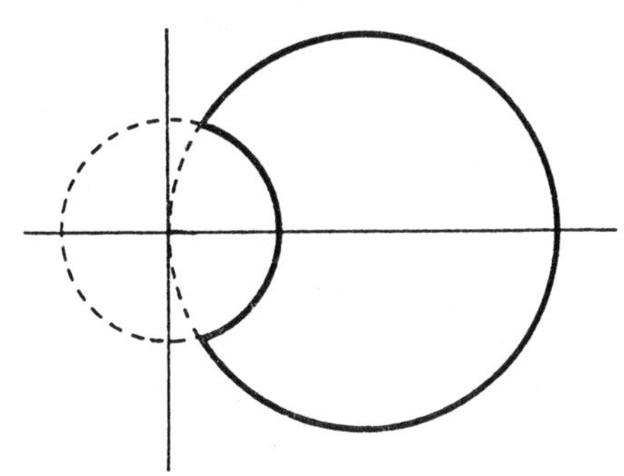

FIG. 9·73.

Multiplying the above equations we get

$$z\bar{z} = a(z+\bar{z})+b^2-ab^2\left(\frac{1}{z}+\frac{1}{\bar{z}}\right),$$

whence
$$w = i\omega\left(az-\frac{ab^2}{z}\right).$$

Note that the singularity $z = 0$ is external to the cross-section.

9·74. Cross-section Booth's lemniscate.* This curve is the inverse of an ellipse with respect to its centre and has for equation

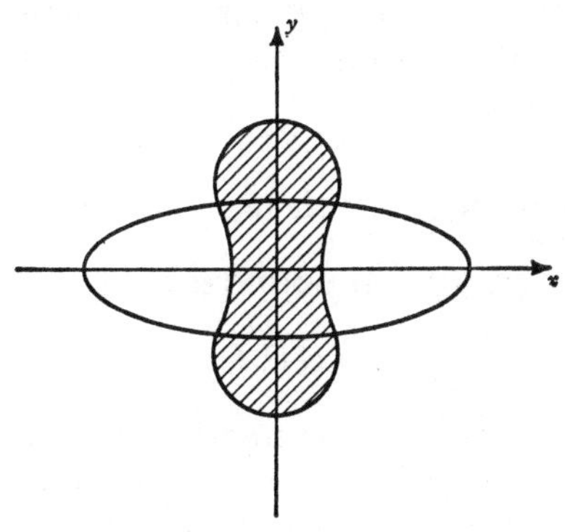

FIG. 9·74.

* See Milne-Thomson, *Antiplane elastic systems*, Springer, Berlin, 1962.

(1) $$z = \frac{k\zeta}{\zeta^2+a^2}, \quad \zeta = e^{i\theta}$$

whereby it is mapped on the unit circumference. Then $\bar{\zeta} = e^{-i\theta} = 1/\zeta$ and therefore on the periphery, where $\zeta\bar{\zeta} = 1$,

$$z\bar{z} = \frac{k^2\zeta\bar{\zeta}}{(\zeta^2+a^2)(\bar{\zeta}^2+a^2)} = \frac{k^2\zeta^2}{(\zeta^2+a^2)(1+a^2\zeta^2)} = \frac{-k^2a^2}{(\zeta^2+a^2)(1-a^4)} + \frac{k^2}{(1-a^4)(1+a^2\zeta^2)}.$$

Now

$$\frac{1}{1+a^2\zeta^2} = \frac{\bar{\zeta}^2}{\bar{\zeta}^2+a^2\zeta^2\bar{\zeta}^2} = \frac{\bar{\zeta}^2}{\bar{\zeta}^2+a^2} = 1-\frac{a^2}{\bar{\zeta}^2+a^2}.$$

Therefore

$$z\bar{z} = -\frac{k^2a^2}{(\zeta^2+a^2)(1-a^4)} - \frac{k^2a^2}{(\bar{\zeta}^2+a^2)(1-a^4)} + \frac{k^2}{1-a^4}.$$

Therefore the complex potential is

(2) $$w = \frac{i\omega a^2k^2}{(a^4-1)(\zeta^2+a^2)}, \quad z = \frac{k\zeta}{\zeta^2+a^2}.$$

9·75. Mapping method for the complex potential. When the contour of the cross-section of the cylinder containing liquid and rotating about a point O of the cross-section with angular velocity ω is a curve C, such that the domain *inside* C can be mapped conformally upon the *interior* of the unit circle in a complex ζ-plane by a relation

(1) $$z = f(\zeta)$$

we may proceed by the method of 9·63 (7) which gives

(2) $$w = B_2(\zeta),$$

and since this contains only positive powers the solution gives finite velocities at the origin, and indeed everywhere, for

$$-u+iv = dw/dz = B_2'(\zeta)/f'(\zeta),$$

which cannot become infinite, since there are no zeros of $f'(\zeta)$ in the liquid.

9·76. Curvilinear polygonal boundary. The transformation

(1) $$z = c\zeta(1+\lambda\zeta^n),$$

where c and n are real positive constants, maps the space inside the unit circle in the ζ-plane conformally upon the space inside a regular curvilinear polygon* of n " sides " so that the cylinder is a grooved or fluted column of a

* The curve is an epitrochoid.

special type. The transformation is conformal at all points inside the unit circle, if $f'(\zeta)$ does not vanish or become infinite inside the unit circle, which is so if

$$(2) \qquad 0 \leqslant \lambda(n+1) \leqslant 1.$$

Now if $\zeta = e^{i\eta}$ and $z = r\,e^{i\theta}$, it is readily shown that

$$(3) \qquad r^2 = z\bar{z} = c^2(1+\lambda^2+2\lambda\cos n\eta)$$

and

$$(4) \qquad \tan\theta = \frac{\sin\eta+\lambda\sin(n+1)\eta}{\cos\eta+\lambda\cos(n+1)\eta}.$$

Hence the boundary curve C is such that r is stationary for $\sin n\eta = 0$ or when

$$\eta = \theta = s\pi/n, \quad s = 0,\ 1,\ 2\ldots 2n-1.$$

and so

$$1-\lambda \leqslant r/c \leqslant 1+\lambda.$$

The curve C has n axes of symmetry if n is odd, $2n$ if n is even. In the case $n = 1$, a simple change of origin given by $z' = z+\lambda c = r'e^{i\theta'}$ allows us to recognise C as the kidney-shaped elliptic limaçon

$$(5) \qquad r' = a+b\cos\theta', \quad b \leqslant a, \quad (a = c,\ b/a = 2\lambda).$$

For a cylinder of cross-section given by (3) and (4), rotating with liquid inside it, the boundary function is given by

$$B(\zeta) = i\omega c^2\{1+\lambda^2+\lambda(\zeta^n+\zeta^{-n})\};$$

hence 9·75 (2) gives

$$(6) \qquad w = i\omega c^2\lambda\zeta^n.$$

For the kinetic energy T of the liquid we have

$$T = \tfrac{1}{4}i\rho\int_{(C)} w\,d\bar{w} = \tfrac{1}{4}i\rho\int_{(C)} i\omega c^2\lambda\zeta^n(ni\omega c^2\lambda\zeta^{-n-1}d\zeta),$$

or

$$T = -\frac{i}{4}\rho\omega^2c^4\lambda^2 n\int_{(C)} d\zeta/\zeta = \tfrac{1}{2}\rho\pi\omega^2c^4\lambda^2 n.$$

9·77. Rotation about an eccentric point. If the axis of rotation passes through the point z_0 instead of through the origin, taking the origin as base point, the motion is equivalent to rotation with angular velocity ω about the origin together with a complex velocity $i\bar{z}_0\omega$ of the origin. The new boundary condition (9·40) is then satisfied by the complex potential

$$w - i\omega\bar{z}_0 z,$$

where w is the complex potential when rotation takes place about the origin.

EXAMPLES IX

1. A circular cylinder of radius a moves transversely through an infinite incompressible fluid of density ρ with velocity U, and there is also a circulation I about the cylinder. Show that, if (u, v) is the velocity at any point z $(= x+iy)$,

$$-u+iv = -\frac{a^2 U}{z^2}+i\frac{I}{2\pi z}.$$

If the cylinder rotates with spin ω, and I is chosen so as to make the mean square velocity of slip at the boundary of the cylinder a minimum, prove that $I = 2\pi a^2\omega$, and find the force exerted by the fluid on the cylinder. (R.N.C.)

2. In the case of a fluid streaming past a fixed circular disc, the velocity at infinity is V. Find the velocity function. Show that the maximum velocity at any point of the fluid is $2V$. Show that, in the case of a cylinder moving forward in a fluid otherwise at rest, the speed of the fluid varies inversely as the square of the distance from the centre.

3. If the complete boundary of a region occupied by liquid is at rest, there can be no purely irrotational motion. Prove this theorem, introducing and explaining the necessary restriction on the nature of the region.

The space between two fixed coaxial circular cylinders of radii a and b and between two planes perpendicular to the axis and distant c apart is occupied by liquid of density ρ. Find the velocity potential of a motion whose kinetic energy shall equal a given quantity T.

4. If ϕ, ψ are the velocity and stream functions for an elliptic cylinder moving forward in the direction of the major axis, show that

$$x = \tfrac{1}{2}c\phi\left\{\frac{1}{k}+\frac{k}{\phi^2+\psi^2}\right\}, \quad y = \tfrac{1}{2}c\psi\left\{\frac{1}{k}-\frac{k}{\phi^2+\psi^2}\right\},$$

where $k = Vc\,e^{\xi}\sinh\xi$. Hence plot the curves ϕ = constant, ψ = constant.

5. A very long thin rigid plank of breadth $2c$ is floating on the surface of deep water and receives a normal downwards blow of impulse I at its centre. Show that the upwards velocity of the water at a distance x from the axis of the plank is given by

$$2I/[\{\pi\,\rho\surd(x^2-c^2)\}\{x+\surd(x^2-c^2)\}],$$

where ρ is the density of the water.

6. An elliptic cylinder, mass M, semi-axes a, b, surrounded by a fluid, is acted on by an impulse. Show that the initial motion is given by

$$u(M+\pi\rho\, b^2) = I, \quad v(M+\pi\rho\, a^2) = J,$$
$$\omega[Mk^2+\tfrac{1}{8}\pi\rho\,(a^2-b^2)^2] = G,$$

where I, J, G are the components of impulse.

7. An elliptic cylinder, the semi-axes of whose cross-section are a and b, rotates about its axis with angular velocity ω in a liquid which is at rest at infinity. Find the velocity potential and the stream function, and calculate the kinetic energy of the liquid per unit layer. Find also at what points of the boundary the fluid velocity is greatest and least. State, with reasons, whether it can be inferred without further examination that these are points of least and greatest pressure respectively. (R.N.C.)

8. Prove that the square of the radius of gyration about its axis of an infinitely long cylinder of density σ, whose cross-section is an ellipse of semi-axes a, b, is effectively increased by the quantity

$$\frac{\rho}{8\sigma}\frac{(a^2-b^2)^2}{ab}$$

when the cylinder is rotating in an infinite liquid of density ρ.

9. An infinite flat plate of zero mass and of breadth $2l$ is rotating. Prove that the couple (per unit thickness) necessary to maintain the rotation is

$$\tfrac{1}{8}\pi\rho\, l^4 \frac{d\omega}{dt},$$

where ω is the angular velocity and ρ is the density of the fluid. (R.N.C.)

10. A hollow cylinder, bounded by the ellipse $b^2x^2+a^2y^2 = a^2b^2$, contains fluid and is rotating with angular velocity ω about its axis. Show that the stream function of the fluid motion is given by

$$\psi = \frac{\omega}{2}\frac{a^2-b^2}{a^2+b^2}(x^2-y^2).$$

Prove that, relatively to the cylinder, the fluid particles describe ellipses in a common period

$$\frac{\pi(a^2+b^2)}{\omega ab}.$$

(R.N.C.)

11. A liquid of density ρ completely fills a vessel in the form of a long elliptic cylinder; the semi-axes of cross-section are a and b, and its mass may be neglected. The cylinder is caused to rotate about its axis with spin ω. Find the kinetic energy of the fluid per unit length of the cylinder, and express the result in terms of the effective moment of inertia.

12. In the case of a rotating elliptic cylinder, prove that the kinetic energy of the contained fluid is less than if the fluid were moving round like a solid in the ratio

$$\left(\frac{a^2-b^2}{a^2+b^2}\right)^2.$$

13. An elliptic cylinder of semi-axes a and b is filled with incompressible fluid and rotates about its axis with angular velocity ω. Prove that the velocity components (u, v) parallel to the axes Ox, Oy of the ellipse are given by

$$u = \omega y\frac{a^2-b^2}{a^2+b^2}, \quad v = \omega x\frac{a^2-b^2}{a^2+b^2}.$$

Show that the coordinates X, Y (relative to axes through O fixed in space) of a given particle at time t can be written

$$X = \lambda\left\{(a+b)\cos\left[\frac{(a-b)^2\omega t}{a^2+b^2}\right]+(a-b)\cos\left[\frac{(a+b)^2\omega t}{a^2+b^2}\right]\right\},$$

$$Y = \lambda\left\{(a+b)\sin\left[\frac{(a-b)^2\omega t}{a^2+b^2}\right]+(a-b)\sin\left[\frac{(a+b)^2\omega t}{a^2+b^2}\right]\right\},$$

where λ is a constant depending on the particle and $t = 0$ when the particle crosses the axis OX. (R.N.C.)

14. A thin shell in the form of an elliptic cylinder, the axes of whose cross-section are $2a$, $2b$ is rotating about its axis in a liquid which is otherwise at rest. It is filled with liquid of the same density. Prove that the ratio of the kinetic energy of the liquid inside to that of the liquid outside is $2ab : (a^2+b^2)$. (R.N.C.)

15. If the ellipse

$$a(x^2-y^2)+2bxy-\tfrac{1}{2}\omega(x^2+y^2)+c = 0$$

is full of liquid and is rotated round the origin with angular velocity ω, prove that the stream function is

$$\psi = a(x^2-y^2)+2bxy.$$

16. Assuming ψ of the form $C(x^3-3xy^2)$, determine C so that this will give the motion inside a rotating prism whose boundary is given by

$$x = a, \quad x+2a = \pm\sqrt{3}y,$$

and show that the time taken by a particle originally at one of the points of quadrisection of a side to move to the mid-point of the same side is $(\log_e 3)/\omega\sqrt{3}$.

Calculate the effective radius of gyration of the prism about the axis of rotation. (R.N.C.)

17. A cylindrical vessel, whose cross-section is the segment of the hyperbola $2(x^2-3y^2)+x+\alpha y = 0$ cut off by the axis $x = 0$, is filled with liquid and rotates steadily with unit angular velocity about an axis through the origin parallel to the generators of the cylinder. Prove that the stream function is given by

$$-\psi = 2(x^3-3xy^2)+\tfrac{1}{2}(x^2-y^2)+\alpha xy.$$

18. The equation

$$x^4-6x^2y^2+y^4+2a^2(x^2+y^2)-a^4 = 0$$

is the same as

$$[x^2(\sqrt{2}+1)-y^2(\sqrt{2}-1)-a^2][x^2(\sqrt{2}-1)-y^2(\sqrt{2}+1)+a^2] = 0.$$

A cylinder whose section is the closed figure formed by these two hyperbolas rotates round the origin with angular velocity ω. Prove that the motion of the contained fluid is given by

$$\psi = -(x^4-6x^2y^2+y^4)\omega/(4a^2).$$

19. A hollow cylinder of cross-section S filled with non-viscous liquid rotates with angular velocity ω about an axis parallel to its generators. Show that if χ is a function satisfying $\nabla^2\chi = -1$ within the cross-section and vanishing on the boundary, then the kinetic energy T and angular momentum G about the axis of rotation per unit length of the cylinder are given by

$$2T = \rho\omega^2(I-J), \quad G = \rho\omega(I-J),$$

where I is the second moment of the cross-section about the axis of rotation, $J = 4\int_S \chi\, dS$, and ρ is the density of the liquid.

Prove that for an elliptic cylinder rotating about a focus

$$I-J = \pi ab(a^2-b^2)(5a^2+3b^2)/4(a^2+b^2). \quad \text{(U.L.)}$$

20. A cylindrical vessel filled with incompressible fluid of density ρ is rotating about a line parallel to its generators with angular velocity ω. If the section of the vessel is bounded by a circle of radius a whose centre O is on the axis of rotation and by the radii $\theta = \pm\alpha$, verify that the stream function is given by

$$\psi = \tfrac{1}{2}\omega r^2 \frac{\cos 2\theta}{\cos 2\alpha} - 32\omega a^2\alpha^2 \sum_{n=0}^{\infty} \frac{(-1)^n (r/a)^{(2n+1)\pi/2\alpha} \cos (2n+1)\pi\theta/2\alpha}{(2n+1)\pi\{(2n+1)^2\pi^2 - 16\alpha^2\}}.$$

Calculate the kinetic energy per unit thickness of liquid.

21. A rectangular prism, the sides of whose cross-section are $2a$, $2b$, rotates with angular velocity Ω about its axis Oz and contains irrotationally moving incompressible fluid of density ρ. Show that, apart from an irrelevant constant,

$$\psi = -\frac{16\Omega}{\pi^3} \sum_{n=0}^{\infty} \frac{(-1)^n}{(2n+1)^3} \left\{ a^2 \operatorname{sech} \frac{(2n+1)\pi b}{2a} \cosh \frac{(2n+1)\pi y}{2a} \cos \frac{(2n+1)\pi x}{2a} \right.$$
$$\left. + b^2 \operatorname{sech} \frac{(2n+1)\pi a}{2b} \cosh \frac{(2n+1)\pi x}{2b} \cos \frac{(2n+1)\pi y}{2b} \right\}.$$

Write down the expression for the velocity potential ϕ and deduce an expression for the kinetic energy of the fluid per unit length of the prism. (U.L.)

22. The transverse motion of a solid cylinder in a liquid is defined by the linear velocity $Q = U + iV$ of the centroid of a section and an angular velocity ω, with respect to axes fixed in the section. Prove that on the boundary of the cylinder the value of the stream function differs by a constant from

$$\tfrac{1}{2}(i\bar{Q}z - iQ\bar{z} + \omega z\bar{z}),$$

where the bar denotes the conjugate complex.

Liquid is contained between two cylinders whose motions are defined, as above, by Q, ω, and Q', ω'. Prove that the linear momentum of the liquid is $M'Q' - MQ$ where M', M are the masses of liquid (per unit thickness) which the outer and inner cylinder could contain respectively. (U.L.)

23. A circular cylinder containing inviscid incompressible fluid is made to rotate with a gradually increasing angular velocity about an eccentric axis parallel to the axis of the cylinder. Find the motion of the fluid.

Find also the motion of the fluid if the cylinder is solid and surrounded by an infinite mass of fluid. Consider the cases where (*a*) there is initially no circulation about the cylinder; (*b*) there is initially a circulation I about the cylinder.

24. Find the lines of flow in the two-dimensional fluid motion given by

$$\phi + i\psi = -\frac{n}{2}(x+iy)^2 e^{2int}.$$

Prove or verify that the paths of the particles of the fluid (in polar coordinates) may be obtained by eliminating t from the equations

$$r\cos(nt+\theta) - x_0 = r\sin(nt+\theta) - y_0 = nt(x_0 - y_0).$$

25. If the liquid is contained between the elliptic cylinders $x^2/a^2 + y^2/b^2 = 1$ and $x^2/a^2 + y^2/b^2 = k^2$, where a, b, k are constants, and the whole rotates about Oz with angular velocity Ω, prove that the velocity potential ϕ referred to the axes Ox, Oy is given by

$$\phi = -\Omega \frac{a^2 - b^2}{a^2 + b^2} xy$$

and that the surfaces of equal pressure are the hyperbolic cylinders

$$\frac{x^2}{3a^2 + b^2} - \frac{y^2}{a^2 + 3b^2} = \text{constant}.$$

Determine also the kinetic energy and angular momentum about Oz of the liquid. (U.L.)

26. In a two-dimensional irrotational motion of an inviscid incompressible fluid of constant density ρ, the space between two cylinders whose cross-sections are the curves C_1 and C_2 is completely filled with fluid and C_1 is wholly inside C_2. Prove that

$$\int_{(C_1)} l\phi\, ds - \int_{(C_2)} l\phi\, ds = \int_{(C_1)} x\frac{\partial\phi}{\partial n}\, ds - \int_{(C_2)} x\frac{\partial\phi}{\partial n}\, ds,$$

where ϕ is the velocity potential assumed one-valued, l is the cosine of the angle between the outward normal and the axis of x, and the differentiation is along the outward normal.

An infinite solid cylinder, whose section is the curve C, moves with velocity U in the fluid along the axis of x. If for large values of $|z|$ the complex potential is given by

$$w = \phi + i\psi = \frac{U(\lambda+i\mu)}{z} + O\left(\frac{1}{|z|^2}\right),*$$

where λ and μ are real, and $z = x+iy$, prove that the kinetic energy of the fluid per unit length is equal to

$$\tfrac{1}{2}\rho U^2(2\pi\lambda - A),$$

where A is the area enclosed by the curve C.

Deduce that if the infinite right cylinder, whose section is the curve $r_1 r_2 = b^2$, where r_1 and r_2 are distances from the two points P and Q at a distance $2a$ apart, and b is greater than a, moves with velocity U along PQ in a fluid at rest at infinity, then the kinetic energy of the fluid per unit length is

$$\rho U^2 b^2 \left\{\pi - \frac{\pi a^2}{2b^2} - E\left(\frac{a^2}{b^2}\right)\right\},$$

where $$E(k) = \int_0^{\pi/2} (1 - k^2 \sin^2 x)^{\frac{1}{2}}\, dx.$$ (U.L.)

27. Two concentric cylinders, radii a, b, are moving in the line of centres with velocities U, V. Show that

$$\phi = \frac{Ua^2 - Vb^2}{b^2 - a^2} r\cos\theta + (U-V)\frac{a^2b^2}{b^2-a^2}\frac{\cos\theta}{r}.$$

Prove also that, when the direction of V is perpendicular to the direction of U,

$$\phi = -\frac{b^2V}{b^2-a^2}\left(r+\frac{a^2}{r}\right)\sin\theta + \frac{a^2U}{b^2-a^2}\left(r+\frac{b^2}{r}\right)\cos\theta,$$

where in both cases a is the radius of the inner cylinder.

28. The space between two coaxial cylindrical shells of radii a, b is filled with liquid of density ρ. The outer shell, radius a, is suddenly made to move with velocity U.

Show that the impulsive force per unit length to be applied to the inner cylinder to keep it at rest is

$$2\pi\rho a^2 b^2 U/(a^2 - b^2).$$

---

* O means "of the order of" and signifies that positive numbers K, R exist such that the absolute value of the term in question is less than K/r^2, provided that $|z| = r > R$.

Show also that the impulsive force to start the inner cylinder with velocity U, when the outer cylinder is fixed, is

$$\frac{\pi b^2}{a^2 - b^2}\{(\sigma + \rho)a^2 - (\sigma - \rho)b^2\} U,$$

where σ is the density of the cylinder.

29. Determine approximately the velocity function for two circular cylinders of radii a, a' moving with velocities V, V' in a direction perpendicular to their line of centres. Also deduce the velocity function when the cylinders are fixed in a uniform stream perpendicular to their line of centres.

If V is the velocity of a uniform stream past two fixed circular piers in a direction perpendicular to their line of centres, a the radius of each pier, and c the distance between their centres, show that, if c/a is not small, the mean velocity across the line joining the nearest points is nearly

$$V\frac{c+a}{c-a}.$$

CHAPTER X

THEOREM OF SCHWARZ AND CHRISTOFFEL

10·1. Simple closed polygons. The elementary idea of a polygon exemplified by, say, a rectangle or a regular hexagon is familiar. For hydrodynamical applications it will be necessary to extend this concept to rectilinear configurations which do not at first sight appear to resemble the polygons of elementary geometry. Let us consider two properties of the rectangle (or of the regular hexagon).

(*a*) It is possible to go from any assigned point of the boundary to any other assigned point of the boundary by following a path which never leaves the boundary. The boundary is connected.

(*b*) The boundary divides the points of the plane into two regions the points of which may be called *interior* points and *exterior* points respectively. The interior points are such that any two of them can be joined by a path which never intersects the boundary. The same holds of the exterior points. On the other hand, it is impossible to go from an interior point to an exterior point without crossing the boundary somewhere.

Any configuration of straight lines in a plane which has the properties (*a*) and (*b*) will be called a *simple closed polygon*. The adjective " simple " refers to the property that every point of the plane is either an interior point, a point of the boundary, or an exterior point, the points of each class forming a connected system.

In many problems of hydrodynamical interest the boundaries of the polygon extend to infinity.

We shall regard as the interior points of the polygon (see 5·71) those points which are in the region which is on the left of an observer who describes the boundary in a prescribed sense. Several such polygons are shown in figs. (i)-(vi). Points regarded as infinitely distant are indicated by the suffix ∞, and the exterior is denoted by hatching. In each case P denotes an interior point.

In fig. 10·1 (i) we have the case of a rectangle with two vertices at infinity. Alternatively this could be regarded as a triangle with one vertex (corresponding to A_∞ and D_∞ regarded as the same point) at infinity.

In (ii) all the vertices of the rectangle $ABCD$ can be regarded as infinitely distant.

In (iii) and (iv) we have a triangle with two vertices at infinity, the interior being regarded as inside or outside the angle $A_\infty BC_\infty$ according to the sense of description. The diagram (v) can be regarded as a rectangle in which two vertices coincide at B, C and the other two coincide at infinity, or simply as a semi-infinite straight line described twice in the senses indicated. This diagram will have several applications and we note the peculiarity that there are no exterior points. All the points of the plane belong either to the boundary or to the interior in accordance with our definition of the term interior.

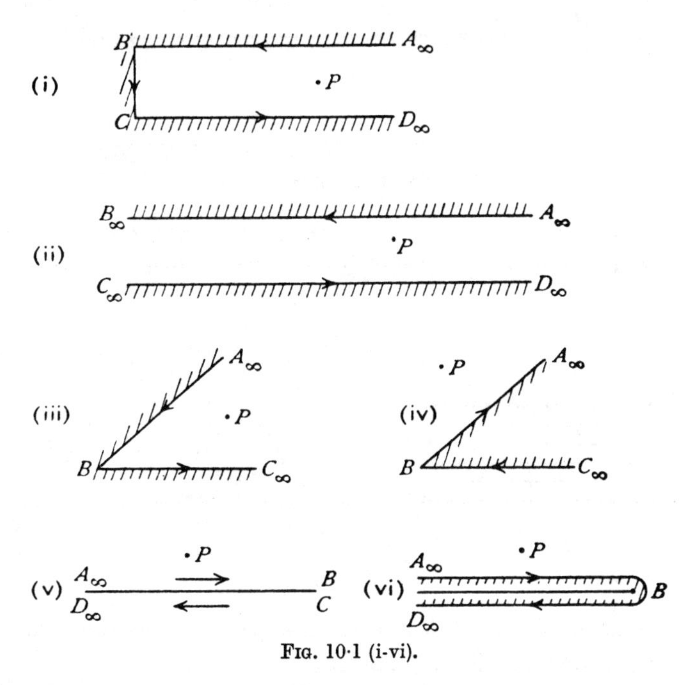

FIG. 10·1 (i-vi).

To describe more clearly in a diagram the situation envisaged in (v) we may draw the diagram as in (vi), the lines $A_\infty B$ and $D_\infty B$ being thought of as coincident.

We shall presently show that the boundary of any simple closed polygon in the z-plane can be transformed into the real axis of the ζ-plane by a conformal transformation, the interior points of the polygon then corresponding to points on one side of the real axis in the ζ-plane; the flow pattern in the polygon will then transform into a corresponding flow pattern in the half ζ-plane. Assuming this result for the moment, it is then clear that the corners of the polygon will transform into points on the real axis in the ζ-plane. We can regard the process intuitively as an opening out of the polygon until its boundary becomes an unterminated straight line accompanied by the local magnification necessary to keep the transformation conformal.

When the polygons of figs. (i)-(iii) are subjected to this process we obtain figs. (vii)-(ix). It may be noted that in case (ii) we can regard B_∞ and C_∞ as coinciding at a finite point B, C, or we can regard them as distinct, in which case we should get a result similar to the result of opening out (i).

FIG. 10·1 (vii-ix).

This intuitive method cannot of course lead to a detailed discussion of any but the simplest cases, but it does afford a useful picture of what is going on.

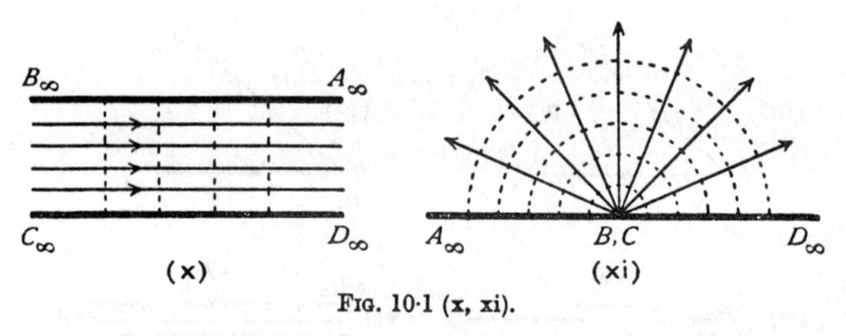

FIG. 10·1 (x, xi).

Thus if we have uniform flow in a channel with parallel sides, the stream-lines are straight and parallel to the sides, while the lines of equal velocity potential are perpendicular to the sides. The uniform flow can be regarded as due to a source at infinity on the left and an equal sink at infinity on the right. If we open up the channel regarding B_∞, C_∞ as coincident, we get the flow from a source at B, C and a sink at infinity. See the correspondence between figs. (x) and (xi). The results are trivial but illustrate very well the process of deformation involved.

10·2. Theorem of Schwarz and Christoffel. Let a, b, c, ... be n points on the real axis in the ζ-plane such that $a<b<c<\ldots$.

Let α, β, γ, ... be interior angles of a simple closed polygon of n vertices, so that

$$\alpha+\beta+\gamma+\ldots = (n-2)\pi.$$

The theorem of Schwarz and Christoffel is then as follows.

The transformation from the ζ-plane to the z-plane, defined by

$$(1) \qquad \frac{dz}{d\zeta} = K(\zeta - a)^{\frac{\alpha}{\pi}-1}(\zeta - b)^{\frac{\beta}{\pi}-1}(\zeta - c)^{\frac{\gamma}{\pi}-1}\ldots,$$

transforms the real axis in the ζ-plane into the boundary of a closed polygon in the z-plane in such a way that the vertices of the polygon correspond to the points $a, b, c, \ldots$, and the interior angles of the polygon are $\alpha, \beta, \gamma, \ldots$. Moreover, when the polygon is simple, the interior is mapped by the transformation on the upper half of the ζ-plane. K is a constant which may be complex.

Proof. The proof consists essentially in establishing the following points.

(1) As ζ increases from say a to b, z describes a straight line.

(2) As ζ passes through b this straight line turns through the angle $\pi - \beta$.

(3) That points interior to the polygon made by these lines correspond to points in the upper half of the ζ-plane.

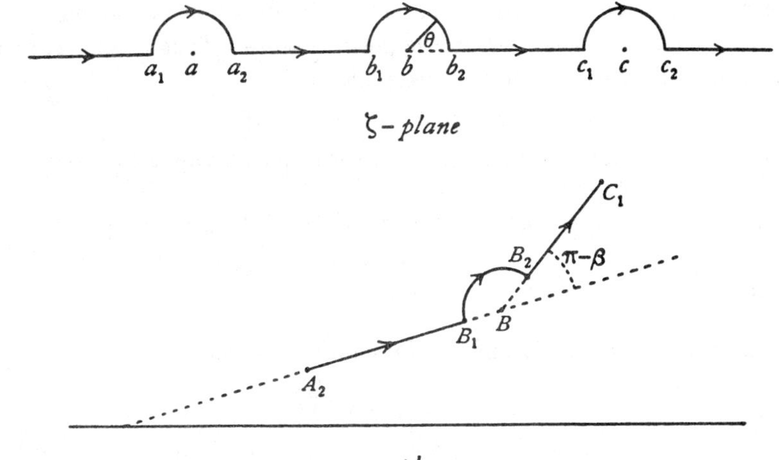

FIG. 10·2 (i).

Since $\zeta - a$ vanishes at $\zeta = a$, it follows that $dz/d\zeta$ is either zero or infinite (according as $\alpha > \pi$ or $\alpha < \pi$). We therefore avoid the points $a, b, c, \ldots$ on the real ζ-axis by drawing semicircles with these points as centres, each of small radius r and situated in the *upper half* of the ζ-plane. The semicircle, centre a, cuts the real axis in a_1, a_2, as shown in fig. 10·2 (i). We shall suppose ζ to describe the real axis in the sense of ζ increasing (so that $d\zeta$ is positive), and to avoid the points $a, b, c, \ldots$ by passing round the semicircles.

Let A_2, B_1, B_2, C_1 be the points in the z-plane which correspond to a_2, b_1, b_2, c_1.

Let $K = C\,e^{i\lambda}$, where C is a real positive constant and λ is real. Then, taking the argument of both sides of the equation of the transformation, we get

$$\arg(dz) - \arg(d\zeta) = \lambda + \left(\frac{\alpha}{\pi} - 1\right)\arg(\zeta - a)$$
$$+\left(\frac{\beta}{\pi} - 1\right)\arg(\zeta - b) + \left(\frac{\gamma}{\pi} - 1\right)\arg(\zeta - c) + \ldots.$$

As ζ moves from a_2 to b_1, $\arg(d\zeta)$ remains equal to zero; $\arg(\zeta - a) = 0$, since $\zeta - a$ is real and positive; $\arg(\zeta - b) = \arg(\zeta - c) = \ldots = \pi$, since $\zeta - b, \zeta - c, \ldots$ are all real and negative.

Thus $$\arg(dz) = \lambda + (\beta - \pi) + (\gamma - \pi) + \ldots.$$

This means that $\arg(dz)$ is constant as ζ moves from a_2 to b_1, and therefore z describes a straight line A_2B_1. The same reasoning shows that, when ζ increases from b_2 to c_1,

$$\arg(dz) = \lambda + (\gamma - \pi) + \ldots,$$

and z describes the straight line B_2C_1. Moreover, on B_2C_1, $\arg(dz)$ exceeds the value of $\arg(dz)$ on A_2B_1 by $\pi - \beta$. Thus the direction of motion of z has turned through the angle $\pi - \beta$ in the positive sense. Thus points (1) and (2) are established. Now on the semicircle b_1b_2,

$$\zeta - b = r\,e^{i\theta}, \quad d\zeta = ir\,e^{i\theta}\,d\theta.$$

Taking r to be infinitesimal, we have, with sufficient approximation,

(2) $$\frac{dz}{ir\,e^{i\theta}\,d\theta} = Ce^{i\lambda}(b-a)^{\frac{\alpha}{\pi}-1}\,r^{\frac{\beta}{\pi}-1}\,e^{i\theta\left(\frac{\beta}{\pi}-1\right)}\,(b-c)^{\frac{\gamma}{\pi}-1}\ldots;$$

so that $$\frac{dz}{d\theta} = i\,r^{\frac{\beta}{\pi}}\,e^{i\left(\lambda+\frac{\theta\beta}{\pi}\right)}\,F,$$

where F is independent of r and θ. Integrating, we get

(3) $$z = z_1 + \frac{\pi}{\beta}\,r^{\frac{\beta}{\pi}}\,e^{i\left(\lambda+\frac{\theta\beta}{\pi}\right)}\,F,$$

where z_1 is a constant. Moreover, since β is positive, we see that $z \to z_1$ when $r \to 0$, so that z_1 is the point B where the lines A_2B_1, B_2C_1 meet.

Thus the transformation makes z describe a polygon whose vertices correspond to the points $a, b, c, \ldots$ and whose internal angles are $\alpha, \beta, \gamma, \ldots$.

Again, from (3),

$$\arg(z - z_1) = \lambda + \frac{\theta\beta}{\pi} + \arg F.$$

Thus as ζ describes the semicircle, since θ decreases from π to 0, $\arg(z - z_1)$ decreases by β, and therefore z describes a circular arc, centre B, situated inside the polygon when it is a simple polygon. Thus the points in the upper half of the ζ-plane correspond to points within the polygon. This completes the essential part of the proof.

It remains to see how the polygon closes when ζ progresses from $-\infty$ to $+\infty$ along the real axis. To examine this point, consider fig. 10·2 (ii), where

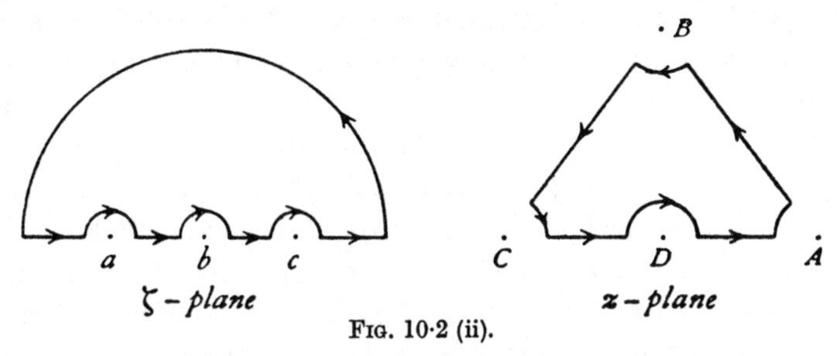

FIG. 10·2 (ii).

for simplicity we have taken the real axis in the ζ-plane indented at three points a, b, c and a large semicircle having its centre at the origin. Consider the figure so formed in the ζ-plane. As ζ goes along the portion a, b, c we get two sides AB, BC of the triangle ABC indented at A, B, C.

On the big semicircle $\zeta = Re^{i\theta}$, and if the radius R is large, we can with sufficient approximation replace $\zeta - a$, $\zeta - b$, $\zeta - c$ by $R\,e^{i\theta}$, and the equation of the transformation then gives (corresponding to (1)) the relation

$$\frac{dz}{i\,R\,e^{i\theta}\,d\theta} = C\,e^{i\lambda}\,(R\,e^{i\theta})^{\frac{\alpha+\beta+\gamma}{\pi}-3},$$

and since $\alpha+\beta+\gamma = \pi$, we get

$$\frac{dz}{d\theta} = \frac{iC}{R}\,e^{i(\lambda-\theta)},$$

which gives
$$z = z_D - \frac{C}{R}\,e^{i(\lambda-\theta)},$$

where z_D is a constant which gives the value to which z tends when $R \to \infty$.

Again, $$\arg(z - z_D) = \pi + \lambda - \theta,$$

and therefore, when ζ describes the large semicircle, θ goes from 0 to π and $\arg(z - z_D)$ goes from $\pi + \lambda$ to λ. Thus z describes the semicircle of small radius C/R about the point D, as shown in fig. 10·2 (ii). When $R \to \infty$ the semicircle in the z-plane $\to 0$, and we again see that the region within the indented triangle transforms into the upper half of the ζ-plane. Q.E.D.

If we integrate the equation of transformation, we get

$$z = C\,e^{i\lambda} f(\zeta) + L,$$

where L is an arbitrary constant, which can be removed by a proper choice of origin in the z-plane.

An alteration in the angle λ merely changes the orientation of the polygon, while an alteration in C changes the scale.

It follows that all polygons corresponding to given values of a, b, c, ..., α, β, γ, ..., are similar. In hydrodynamical applications we shall be concerned only with simple polygons generally extending to infinity. The *shape* of a polygon of n sides, when the angles are given, is determinate only when $n = 3$. Thus in the Schwarz-Christoffel transformation if $n = 3$, α, β, γ determine the shape and a, b, c are arbitrary. But if $n = 4$, $\alpha, \beta, \gamma, \delta$ do not determine the shape and so d is not arbitrary. Thus three of the numbers a, b, c may be chosen arbitrarily to correspond to three of the vertices of a given polygon, the remainder must then be arranged so as to make the polygon of the right shape. The proper choice of C and λ will then fix the scale and orientation.

When the transformation produces a simple polygon the representation is conformal, for conditions (*a*) and (*b*) of 5·62 are then satisfied for the indented real axis, and the indentations may be made infinitesimal.

Finally, it remains to discuss the situation which arises when a vertex of the polygon corresponds to a point at infinity on the real axis of the ζ-plane. If, for example $a \to -\infty$, we can by choice of K write the transformation in the form

$$\frac{dz}{d\zeta} = C\, e^{i\lambda}(-a)^{-\frac{\alpha}{\pi}+1}(\zeta-a)^{\frac{\alpha}{\pi}-1}(\zeta-b)^{\frac{\beta}{\pi}-1}\ldots .$$

When $a \to -\infty$, $\left(\frac{\zeta-a}{-a}\right)^{\frac{\alpha}{\pi}-1} \to 1$, and the transformation becomes

$$\frac{dz}{d\zeta} = C\, e^{i\lambda}(\zeta-b)^{\frac{\beta}{\pi}-1}(\zeta-c)^{\frac{\gamma}{\pi}-1}\ldots ,$$

that is to say, the factor corresponding to $a = -\infty$ is omitted from the equation of transformation, and the angle α does not appear.

10·21. Theorem of Schwarz and Christoffel for the circle.

The same formula

$$(1) \qquad \frac{dz}{d\zeta} = K(\zeta-a)^{\frac{\alpha}{\pi}-1}(\zeta-b)^{\frac{\beta}{\pi}-1}(\zeta-c)^{\frac{\gamma}{\pi}-1}\ldots ,$$

maps the inside of the polygon of 10·2 on the inside of the unit circle, $|\zeta| \leqslant 1$, where the points $a, b, c, \ldots$ are now on the circumference

$$(2) \qquad a = e^{i\mathscr{A}},\ b = e^{i\mathscr{B}}, \ldots,\ \zeta = e^{i\eta}$$

and as before

$$(3) \qquad \alpha+\beta+\gamma+\ldots = (n-2)\pi.$$

For then we get on the circumference $d\zeta/d\eta = ie^{i\eta}$ and so, using (3),

$$(4) \qquad \frac{dz}{d\eta} = K_1(\sin \tfrac{1}{2}(\eta-\mathscr{A}))^{\frac{\alpha}{\pi}-1}(\sin \tfrac{1}{2}(\eta-\mathscr{B}))^{\frac{\beta}{\pi}-1}\ldots ,$$

where $\qquad K_1 = iK(e^{\frac{1}{2}i\mathscr{A}})^{\frac{\alpha}{\pi}-1}(e^{\frac{1}{2}i\mathscr{B}})^{\frac{\beta}{\pi}-1}\ldots ,$

and as η increases, for example from $\mathscr{A}$ to $\mathscr{B}$, arg $(dz/d\eta)$ remains constant so that z describes the straight line A_2B_2 in fig. 10·2 (i) in the z-plane.

On going round the point b on the semicircle b_1b_2 in fig. 10·21 *inside* the unit circle, arg $(\zeta - b)$ increases by $\pi - \beta$ and so we turn the corner B in the z-plane.

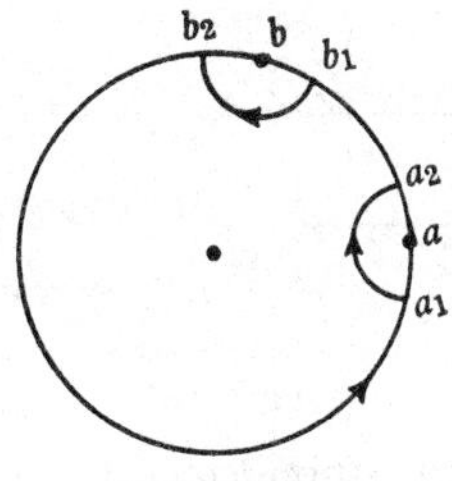

Fig. 10·21.

10·31. Mapping a semi-infinite strip. Consider a semi-infinite strip $A_\infty BCD_\infty$, of breadth a, regarded as a rectangle with two vertices at infinity.

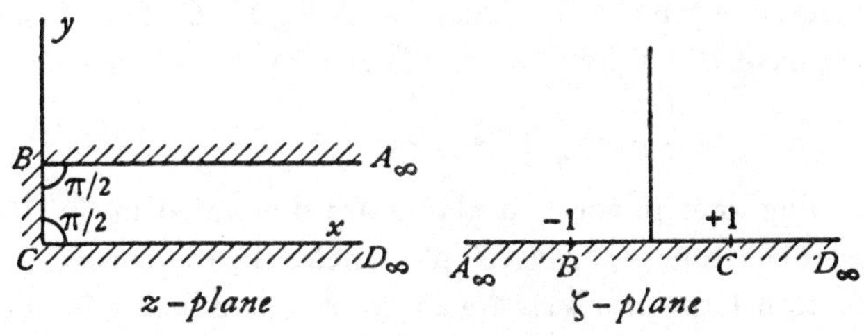

Fig. 10·31.

Let us map A_∞, B, C on the points $\zeta = -\infty$, $\zeta = -1$, $\zeta = 1$ of the real axis in the ζ-plane. If we open out the boundary and lay it along the real axis of the ζ-plane, it is then evident that the fourth vertex will lie at $\zeta = \infty$.

Thus, in accordance with the Schwarz and Christoffel theorem, the only interior angles which will appear in the transformation are those at B and C, each of which is $\pi/2$. Taking axes as shown, we get

$$\frac{dz}{d\zeta} = K(\zeta+1)^{-\frac{1}{2}}(\zeta-1)^{-\frac{1}{2}} = \frac{K}{(\zeta^2-1)^{\frac{1}{2}}},$$

which gives
$$z = K\cosh^{-1}\zeta + L.$$

Since $\cosh^{-1} x = \log(x+\sqrt{x^2-1})$, if we take $\cosh^{-1} 1 = 0$, we shall have $\cosh^{-1}(-1) = i\pi$.

Thus $L = 0$, $ai = K(i\pi)$, so that

$$z = \frac{a}{\pi}\cosh^{-1}\zeta, \quad \text{or} \quad \zeta = \cosh\frac{\pi z}{a}.$$

10·32. Mapping an infinite strip. Taking an infinite strip $A_\infty B_\infty C_\infty D_\infty$ of breadth a, let us regard B_∞, C_∞ as coincident and map the points, B_∞, C_∞ on $\zeta = 0$, the origin O on $\zeta = 1$, and $F(z = ai)$ on $\zeta = -1$.

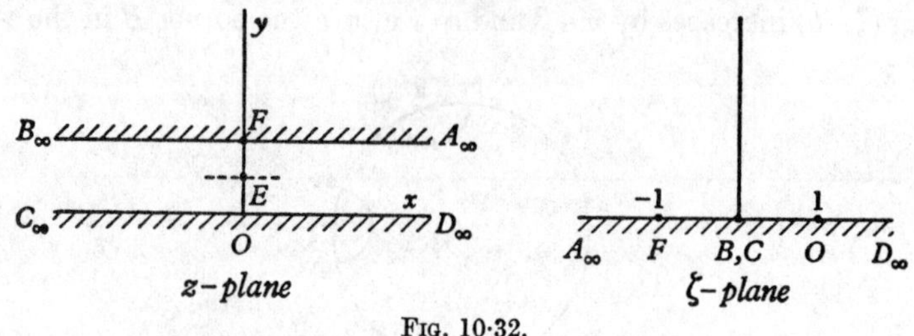

FIG. 10·32.

Then D_∞ will evidently correspond to $\zeta = \infty$.

The angle at $B_\infty C_\infty$ is zero, and we therefore get

$$\frac{dz}{d\zeta} = K\zeta^{-1}, \quad z = K\log\zeta + L.$$

Take axes as shown in fig. 10·32, and the determination of the logarithm to be that which vanishes when $\zeta = 1$. Then $0 = K\log 1 + L$, $ai = K\log(-1) + L$. Thus we must have $L = 0$, $iK\pi = ia$. Therefore

$$(1) \qquad z = \frac{a}{\pi}\log\zeta, \quad \text{or} \quad \zeta = e^{\pi z/a}.$$

Corresponding lines in the two planes are illustrated in fig. 10·1 (x), (xi). The lines $x =$ constant transform into circles $|\zeta| =$ constant; the lines $y =$ constant transform into lines $\arg\zeta =$ constant radiating from the origin in the ζ-plane.

If we map A_∞, D_∞ on $\zeta = 0$, the transformation is found to be

$$\zeta = -e^{-\pi z/a}.$$

In some cases it is convenient to take the origin in the z-plane at the point E midway between the walls. The corresponding transformation is given by writing $z + ia/2$ for z in (1), so that

$$(2) \qquad z = \frac{a}{\pi}\log\zeta - \frac{ia}{2}, \quad \text{or} \quad \zeta = i\,e^{\pi z/a}.$$

10·33. Mapping a strip on a circle.

We map the strip of breadth $2a$ shown in fig. 10·33 on the unit circle in the ζ-plane so that A_∞, B_∞ map into $\zeta = -1$, $\zeta = 1$ respectively and $z = 0$ into $\zeta = 0$. The angles at A_∞, B_∞ are zero so that using 10·21 (1)

$$\frac{dz}{d\zeta} = 2K(\zeta-1)^{-1}(\zeta+1)^{-1} = K\left(-\frac{1}{1-\zeta} - \frac{1}{1+\zeta}\right),$$

and

$$z = K\log\frac{1-\zeta}{1+\zeta} + L.$$

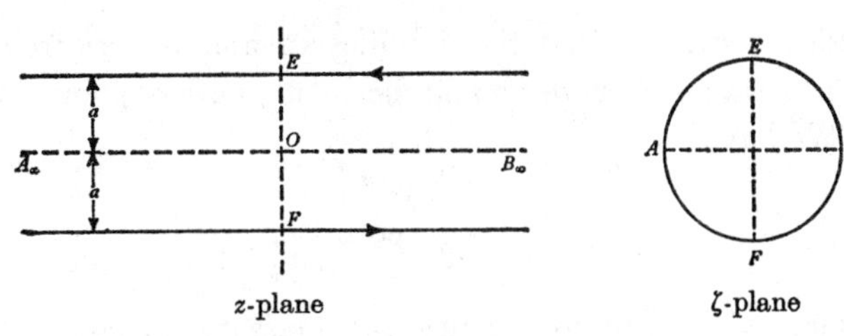

FIG. 10·33.

Since the origins correspond $L = 0$ and if $z = ai$ corresponds to $\zeta = i$, we have $ai = -Ki\pi/2$, $K = -2a/\pi$ and finally

$$z = \frac{2a}{\pi}\log\frac{1+\zeta}{1-\zeta}, \quad \text{or} \quad \zeta = \tanh\frac{\pi z}{4a}.$$

Then $z = -ai$ corresponds to $\zeta = -i$.

10·4. Flow into a channel through a narrow slit in a wall. Let the slit be at the origin, and let the real axis be taken in one side of the channel $A_\infty B_\infty C_\infty D_\infty$ of breadth a.

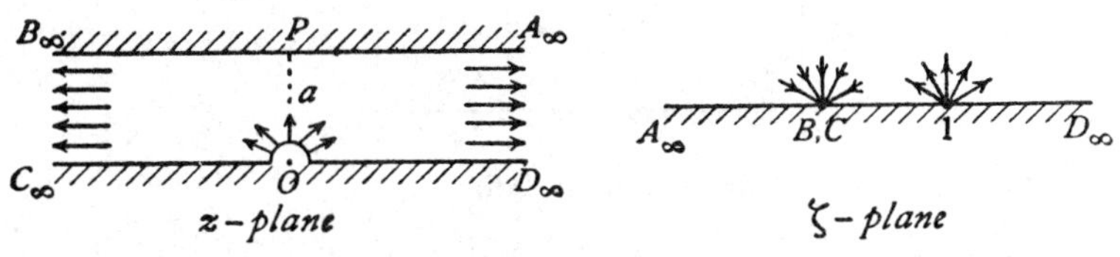

FIG. 10·4 (i).

If πm is the volume which flows in at O per unit time (per unit thickness), the flow at O will be that due to a source of output $2\pi m$, and therefore of strength m. At infinite distance from O there will be parallel flow, and therefore at C_∞ and D_∞ there will be sinks of strength $\frac{1}{2}m$.

Let us regard B_∞, C_∞ as coincident, and then open out the walls into the real axis in the ζ-plane so that B_∞, C_∞ become the origin $\zeta = 0$.

The Schwarz-Christoffel transformation then gives (10·32),

$$\zeta = e^{\pi z/a},$$

and $z = 0$ corresponds to $\zeta = 1$.

Thus in the ζ-plane we have a sink of strength $\frac{1}{2}m$ at $\zeta = 0$ and a source of strength m at $\zeta = 1$. These give rise to the complex potential

$$w = -m\log(\zeta - 1) + \tfrac{1}{2}m\log(\zeta) = -m\log(\zeta^{\frac{1}{2}} - \zeta^{-\frac{1}{2}}).$$

But
$$\zeta^{\frac{1}{2}} - \zeta^{-\frac{1}{2}} = e^{\pi z/(2a)} - e^{-\pi z/(2a)} = 2\sinh\frac{\pi z}{2a}.$$

Hence, omitting a constant, we get

$$w = -m\log\sinh\frac{\pi z}{2a}.$$

It is physically evident that the dividing streamline goes from O straight to the opposite wall, so there should be a stagnation point at this point $P(z = ai)$. We have, in fact,

$$\frac{dw}{dz} = -\frac{m\pi}{2a} \coth \frac{\pi z}{2a},$$

which vanishes when $z = ai$. Hence the pressure at points of $A_\infty B_\infty$ is maximum at P, and is therefore smaller at the remaining points of the wall. Thus the effect of the motion is to urge this wall outwards and, if unsupported, to cause an outward bulge at P. The velocity at a great distance from the origin is $m\pi/(2a)$.

Again, if we consider the streamline OP to be a rigid wall, we obtain the motion within a semi-infinite rectangular channel due to a source at one corner, fig. 10·4 (ii).

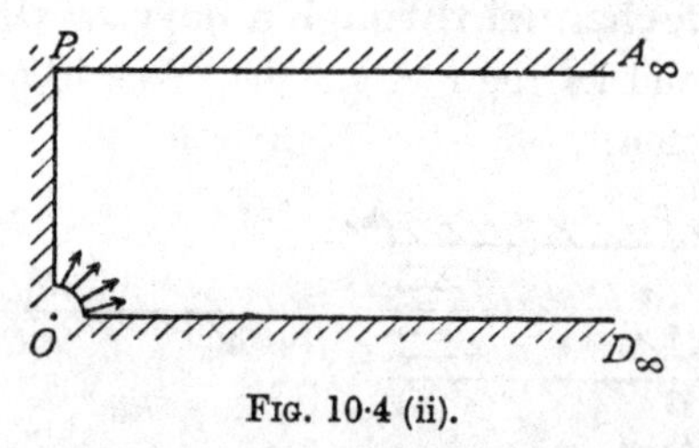

Fig. 10·4 (ii).

From another point of view we have the two-dimensional efflux from a large rectangular vessel through a small hole in the corner.

10·5. Source midway between two parallel planes. This can be obtained from 10·4 by using the principle of reflection.

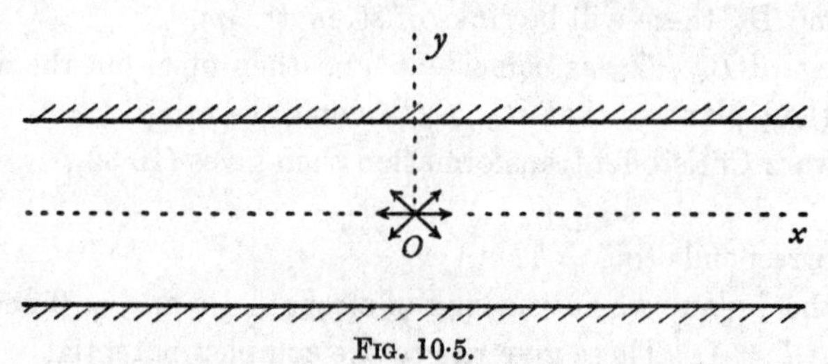

Fig. 10·5.

Taking axes as shown in fig. 10·5, let there be a source of strength m at the origin between two planes whose distance apart is $2a$. Then

$$(1) \qquad w = -m \log \sinh \frac{\pi z}{2a},$$

for this function satisfies the conditions between the upper wall and the real axis, and it is real on the real axis. The conditions of 5·53 are thus satisfied, and w can be analytically continued below the real axis by attributing to it conjugate complex values at conjugate complex points, which is precisely what (1) implies.

We may also note that (1) is the complex potential of an infinite row of sources placed on the y-axis at the distance $2a$ apart, for

$$\sinh\frac{\pi z}{2a} = 0 \quad \text{when } z = 0, \quad \pm 2ai, \quad \pm 4ai, \quad \pm 6ai, \ldots .$$

Instead we could map on the unit circle, where using fig. 10·33 with the source at the origin in the z-plane, we get sinks $m/2$ at A and B and a source m at the centre of the circle in the ζ-plane. The image system in the circle consists of sinks $m/2$ at A and B and therefore

$$w = m\log(1-\zeta^2) - m\log\zeta + \text{constant} = -m\log\sinh\frac{\pi z}{2a}$$

by choice of the constant.

10·6. A step in the bed of a deep stream. Let there be a sudden change of level at BC in the bed of a stream whose velocity at infinity is U.

The bed $A_\infty BCD_\infty$ is a simple polygon and can therefore be transformed into the real axis in the ζ-plane, B and C corresponding to $\zeta = -1$, $\zeta = 1$ respectively. By the Schwarz-Christoffel transformation,

$$\frac{dz}{d\zeta} = K(\zeta+1)^{\frac{1}{2}}(\zeta-1)^{-\frac{1}{2}} = K\frac{\zeta+1}{(\zeta^2-1)^{\frac{1}{2}}} = \frac{K\zeta}{(\zeta^2-1)^{\frac{1}{2}}} + \frac{K}{(\zeta^2-1)^{\frac{1}{2}}},$$

so that
$$z = K\{\sqrt{\zeta^2-1} + \cosh^{-1}\zeta\} + L.$$

Fig. 10·6 (i)

Since $\sqrt{(\zeta^2-1)}$ and $\cosh^{-1}\zeta$ are many-valued functions, let us consider the determinations appropriate to the different parts of the plane.

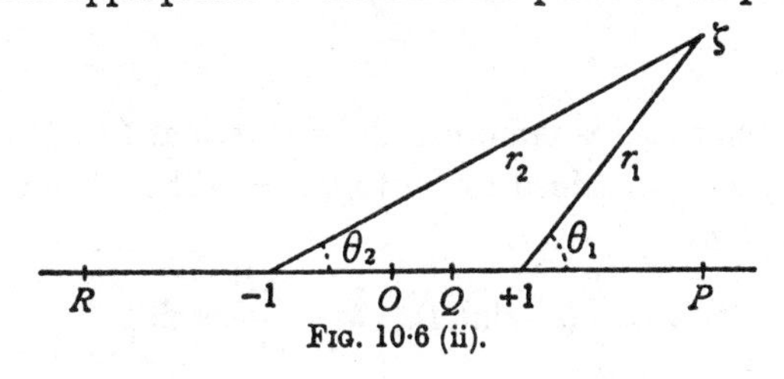

Fig. 10·6 (ii).

In fig. 10·6 (ii) we see a general point in the ζ-plane distant r_1, r_2 from $+1$ and -1, and therefore

$$\zeta-1 = r_1 e^{i\theta_1}, \quad \zeta+1 = r_2 e^{i\theta_2},$$
$$\sqrt{\zeta^2-1} = \sqrt{r_1 r_2} \,.\, e^{i(\theta_1+\theta_2)/2},$$

where $\sqrt{r_1 r_2}$ denotes the real positive square root of the product. Let us denote a point on the real axis by $\zeta = \xi$. Then when $\xi > 1$ we shall take $\theta_1 = 0$, $\theta_2 = 0$. It follows that $\theta_1 = \pi$, $\theta_2 = 0$, when $-1 < \xi < 1$, so that

$$\sqrt{\zeta^2-1} = \sqrt{r_1 r_2} \,.\, e^{i\pi/2} = i\sqrt{r_1 r_2}.$$

Again, when $\xi < -1$, $\theta_1 = \pi$, $\theta_2 = \pi$, and therefore

$$\sqrt{\zeta^2-1} = \sqrt{r_1 r_2}\, e^{i\pi} = -\sqrt{r_1 r_2}.$$

Now $$\cosh^{-1}\zeta = \log(\zeta + \sqrt{\zeta^2-1}).$$

Therefore on $A_\infty B (\xi < -1)$, we get

$$\begin{aligned} z &= \{-\surd(\xi^2-1) + \log[\xi - \surd(\xi^2-1)]\}K + L \\ &= \{-\surd(\xi^2-1) + \log[-\xi + \surd(\xi^2-1)] + i\pi\}K + L. \end{aligned}$$

On BC $\quad z = \{i\surd(1-\xi^2) + \log[\xi + i\surd(1-\xi^2)]\}K + L.$

On CD_∞ $\quad z = \{\surd(\xi^2-1) + \log[\xi + \surd(\xi^2-1)]\}K + L.$

If in the z-plane we take C to be the point $z = 0$ and B the point $z = ih$, we get $L = 0$, $ih = i\pi K$, so that $K = h/\pi$, and therefore

$$z = \frac{h}{\pi}\{\surd(\zeta^2-1) + \cosh^{-1}\zeta\}.$$

We now consider the complex potential.

A uniform stream in the z-plane may be taken to imply a source at D_∞ and an equal sink at A_∞. Thus in the ζ-plane we must also have a source and sink at the corresponding points so that there is a uniform stream, V say. Hence $w = V\zeta$, and therefore

$$\frac{dw}{dz} = V \cdot \frac{d\zeta}{dz} = \frac{V}{K}\sqrt{\frac{\zeta-1}{\zeta+1}}.$$

But at infinity $dw/dz = U$, $\zeta = \infty$.
Hence $U = V/K = V\pi/h$. Thus

$$w = \frac{hU}{\pi}\zeta.$$

Observe that the speed is infinite at B, and zero at C. A more convenient form for the solution is obtained by writing $\zeta = \cosh t$. Then in terms of the uniformising variable t,

$$z = \frac{h}{\pi}(t + \sinh t), \quad w = \frac{hU}{\pi}\cosh t.$$

The principle of reflection enables us to apply the same complex potential to a stream of infinite width flowing against a semi-infinite body of rectangular section, fig. (iii), the origin being at C and the real axis pointing upstream.

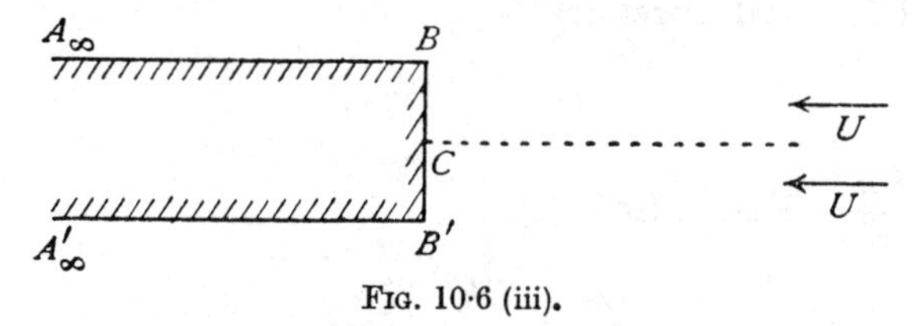

FIG. 10·6 (iii).

The reader may prove that the force on the end BB' per unit thickness is finite by integrating $\frac{1}{2}\rho q^2$ over the end.

10·7. Abrupt change in the breadth of a channel. Suppose a channel, fig. 10·7 (i), with parallel sides undergoes an abrupt change of breadth from h to k.

If the velocity at A_∞ is U, the velocity at B_∞ must by continuity be Uh/k.

We shall open out the polygonal boundary $A_\infty B_\infty C_\infty DEF_\infty$ into the real axis in the ζ-plane regarding the points B_∞, C_∞ as coincident and making them correspond to $\zeta = 0$.

We shall make D correspond to $\zeta = 1$ and E correspond to $\zeta = a$, the real number a being determined later, for we cannot arbitrarily fix a since the correspondents of B_∞, C_∞, D have already been chosen.

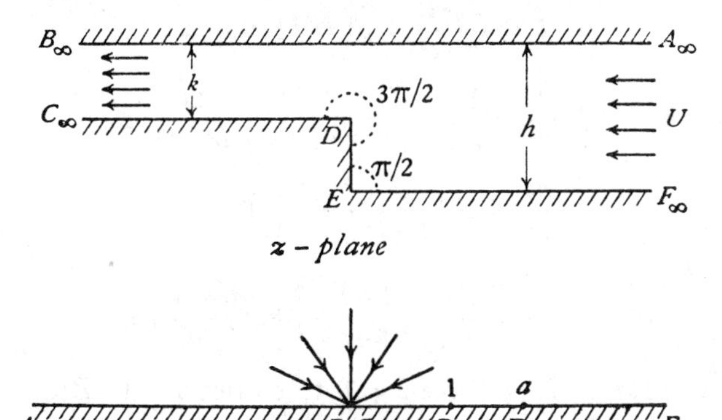

FIG. 10·7 (i).

The Schwarz-Christoffel transformation then gives

$$\frac{dz}{d\zeta} = K\zeta^{-1}(\zeta-1)^{\frac{1}{2}}(\zeta-a)^{-\frac{1}{2}}, \tag{1}$$

for the angle of the polygon at B_∞, C_∞ is zero.

Now in the z-plane the flow is from a source of *output* Uh at A_∞ to a sink of intake Uh at B_∞. Hence in the ζ-plane we have a sink at the origin which takes in the volume Uh per unit time over an angle of π. Hence the strength of the sink is Uh/π, and therefore

$$w = \frac{Uh}{\pi}\log\zeta, \tag{2}$$

so that $\dfrac{dw}{d\zeta} = \dfrac{Uh}{\pi\zeta}$. Hence, from (1),

$$\frac{dw}{dz} = \frac{Uh}{\pi K}\sqrt{\frac{\zeta - a}{\zeta - 1}}.$$

Now at $A_\infty\,(\zeta = \infty)$, $dw/dz = U$, if the real axis is parallel to $A_\infty B_\infty$. Therefore

$$U = \frac{Uh}{\pi K}, \quad K = \frac{h}{\pi}.$$

Again, at $B_\infty\,(\zeta = 0)$, $dw/dz = Uh/k$. Therefore

$$\frac{Uh}{k} = \frac{Uh}{\pi K}\sqrt{a},$$

so that

$$a = h^2/k^2.$$

To obtain an explicit relation between z and w, we must integrate (1). The integration may be simply effected by writing

$$\zeta = \frac{b^2 - t^2}{1 - t^2}, \quad b^2 = h^2/k^2 = a,$$

which gives

$$\frac{\zeta - 1}{\zeta - b^2} = \frac{1}{t^2}, \quad \frac{d\zeta}{\zeta} = \left(\frac{2t}{1 - t^2} - \frac{2t}{b^2 - t^2}\right)dt,$$

so that

$$\frac{dz}{dt} = K\left(\frac{2}{1 - t^2} - \frac{2}{b^2 - t^2}\right),$$

whence

$$z = \frac{h}{\pi}\left(\log\frac{1+t}{1-t} - \frac{1}{b}\log\frac{b+t}{b-t}\right) + L, \tag{3}$$

where L is arbitrary. If we take $z = 0$ to correspond to $E\,(\zeta = a)$, we have $t = 0$ and therefore $L = 0$.

Also substituting in (2),

$$w = \frac{Uh}{\pi}\log\frac{b^2 - t^2}{1 - t^2}, \text{ whence}$$

$$t^2 = \frac{e^{w\pi/(Uh)} - b^2}{e^{w\pi/(Uh)} - 1}. \tag{4}$$

The elimination of t between (3) and (4) gives the relation between w and z. The principle of reflection again enables us to apply the same complex potential to the streaming motion past an infinite solid of rectangular section placed symmetrically in a stream which flows between parallel banks, fig. 10·7 (ii).

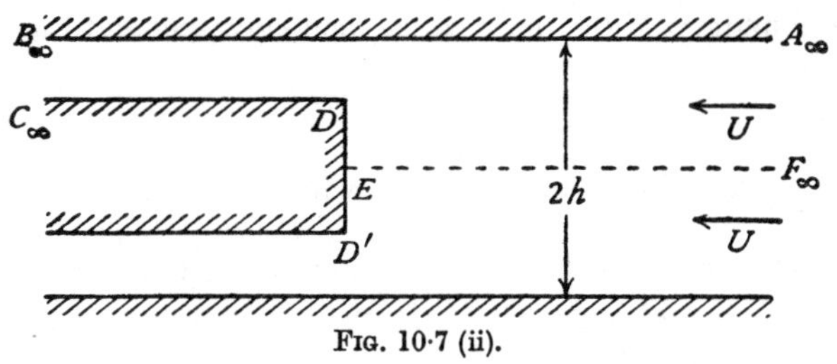

FIG. 10·7 (ii).

The reader should verify that the force on the end is finite by integrating $\frac{1}{2}\rho q^2$ over DD'. Compare with the corresponding situation at the end in fig. 10·6 (iii).

10·8. Branch in a canal. Fig. 10·8 (i) shows a branched canal with straight parallel sides in the main canal and the branch. The sides of the branch make an angle α with the sides of the main canal.

The breadths of the canal and branch are as indicated h, h_1, h_2, and the velocity at infinity upstream in the main canal is U. Our problem is to determine the downstream velocities U_1, U_2 in the main canal and the branch.

There will be a streamline l coming from infinity and dividing at C into CB_∞ and CD_∞, C being a stagnation point. The fluid to the left of l flows into the branch, that to the right of l into the main canal. On the streamline $A_\infty ED_\infty$ the stream undergoes an abrupt change of direction at E and the velocity there is consequently infinite.

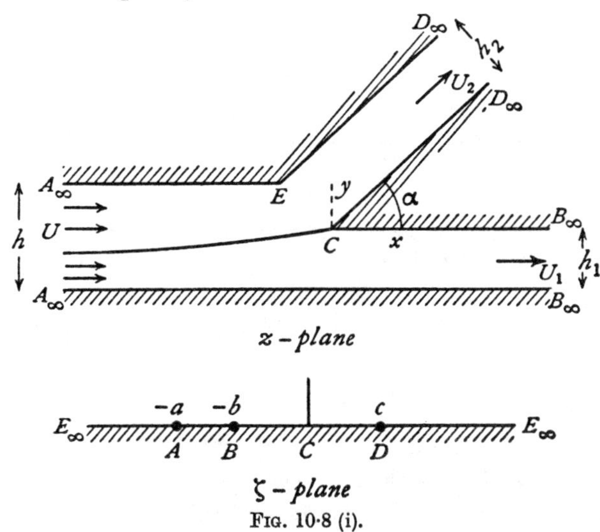

FIG. 10·8 (i).

We now map the interior of the canal on the ζ-plane in such a way that E goes to infinity on the real axis while C maps into $\zeta = 0$.

Let the points A_∞, B_∞, D_∞ correspond to $\zeta = -a, -b, c$ respectively. We note that the boundary of the main canal then corresponds to negative values of ζ.

Now, consider

$$Q = \log \frac{U}{v} = \log \frac{U}{q} + i\theta, \text{ where } v = q\, e^{-i\theta}.$$

Along the sides of the main canal, $\theta = 0$; along the sides of the branch, $\theta = \alpha$. At C, $q = 0$, and hence Q is infinite. Thus we can draw the diagram in the Q-plane, fig. 10·8 (ii).

Mapping this on the ζ-plane by means of 10·32, we get

$$(1) \qquad \zeta = -e^{-Q\pi/\alpha} = -\left(\frac{v}{U}\right)^{\frac{\pi}{\alpha}}.$$

The following values of ζ and v correspond:

$$\begin{array}{cccc} \zeta & -a & -b & c \\ v & U & U_1 & U_2\, e^{-i\alpha}. \end{array}$$

Thus

$$(2) \qquad a = 1, \quad b = \left(\frac{U_1}{U}\right)^{\frac{\pi}{\alpha}}, \quad c = \left(\frac{U_2}{U}\right)^{\frac{\pi}{\alpha}}.$$

To construct the diagram in the w-plane, let us take $\psi = 0$ on $A_\infty E D_\infty$. Then we shall have

$$\begin{aligned} &\text{on } A_\infty B_\infty, \quad \psi = Uh, \\ &\text{on } CD_\infty, \quad \psi = U_2 h_2, \\ &\text{on } CB_\infty, \quad \psi = U_2 h_2, \\ &\text{on } A_\infty B_\infty, \quad \psi = U_2 h_2 + U_1 h_1, \end{aligned}$$

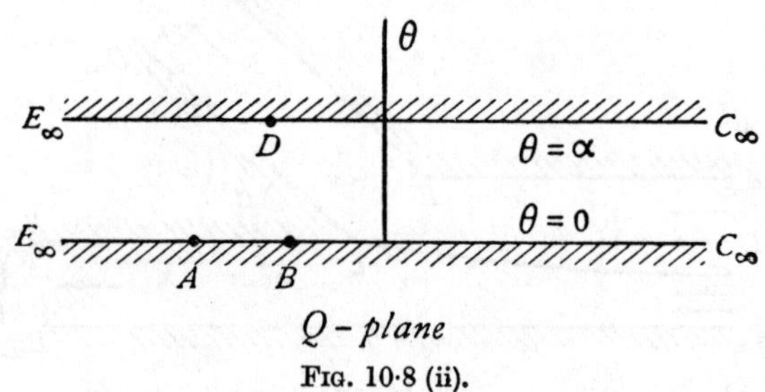

FIG. 10·8 (ii).

and therefore

$$(3) \qquad Uh = U_1 h_1 + U_2 h_2,$$

as is otherwise obvious from the equation of continuity.

Taking $\phi = \infty$ at A_∞, we shall have $\phi = -\infty$ at B_∞, D_∞.

Thus we get the required diagram, fig. 10·8 (iii).

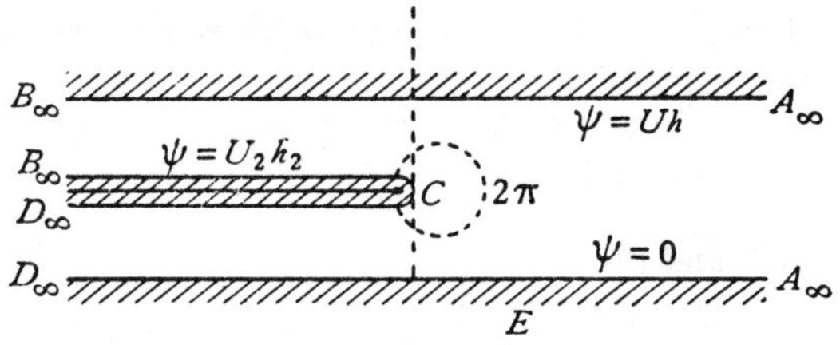

w – *plane*

FIG. 10·8 (iii).

To map the w-plane on the ζ-plane, we get

$$\frac{dw}{d\zeta} = K_1\,\zeta(\zeta+a)^{-1}(\zeta+b)^{-1}(\zeta-c)^{-1}$$

$$= -\frac{K_1\,a}{(a-b)(a+c)}\cdot\frac{1}{\zeta+a}+\frac{K_1\,b}{(a-b)(b+c)}\cdot\frac{1}{\zeta+b}+\frac{K_1\,c}{(a+c)(b+c)}\cdot\frac{1}{\zeta-c},$$

which gives, on integration and after a slight reduction,

$$w = -\frac{K_1\,a}{(a-b)(a+c)}\log\frac{\zeta+a}{\zeta-c}+\frac{K_1\,b}{(a-b)(b+c)}\log\frac{\zeta+b}{\zeta-c}+L_1.$$

Now on $D_\infty E$, $\psi = 0$, and $\zeta+a$, $\zeta+b$, $\zeta-c$ all have the same sign. Therefore the logarithms are all real and hence L_1 is real. Again at C, $\phi = 0$, $\psi = U_2\,h_2$, and $\zeta = 0$. Thus L_1 is purely imaginary. Therefore $L_1 = 0$, and putting $\zeta = 0$ we get

$$-U_2\,h_2 = \frac{K_1\,a\pi}{(a-b)(a+c)}-\frac{K_1\,b\pi}{(a-b)(b+c)}.$$

Finally, on $A_\infty B_\infty$, $\psi = Uh$, while $\zeta+a$ is positive and $\zeta+b$, $\zeta-c$ are negative. Therefore

$$-Uh = \frac{K_1\,a\pi}{(a-b)(a+c)}.$$

Hence $$\frac{K_1\,b\pi}{(a-b)(b+c)} = U_2\,h_2-Uh = -U_1\,h_1\text{, from (3),}$$

and therefore

$$w = \frac{Uh}{\pi}\log\frac{\zeta+a}{\zeta-c}-\frac{U_1\,h_1}{\pi}\log\frac{\zeta+b}{\zeta-c}. \tag{4}$$

Now from (1), we have

$$-\frac{dw}{dz} = \mathfrak{v} = U\,e^{-i\alpha}\,\zeta^{\alpha/\pi}, \tag{5}$$

while from (4),

$$\frac{dw}{d\zeta} = \frac{Uh}{\pi}\cdot\frac{1}{\zeta+a}-\frac{U_1\,h_1}{\pi}\cdot\frac{1}{\zeta+b}+\frac{U_1\,h_1-Uh}{\pi}\cdot\frac{1}{\zeta-c}, \tag{6}$$

and therefore by division we obtain $dz/d\zeta$ as a function of ζ, and thence as usual we can obtain w in terms of z, thus giving the velocity distribution.

Since $\upsilon = 0$ when $\zeta = 0$, it follows that $dw/d\zeta$ is also zero when $\zeta = 0$, and therefore, from (6),

$$\frac{Uh}{a} - \frac{U_1 h_1}{b} + \frac{Uh - U_1 h_1}{c} = 0,$$

whence by use of (3) and (2),

$$Uh = U_1 h_1 \left(\frac{U_1}{U}\right)^{-\frac{\pi}{\alpha}} - U_2 h_2 \left(\frac{U_2}{U}\right)^{-\frac{\pi}{\alpha}}, \text{ or}$$

(7) $$\left(\frac{U_1}{U}\right)^{1-\frac{\pi}{\alpha}} \frac{h_1}{h} - \left(\frac{U_2}{U}\right)^{1-\frac{\pi}{\alpha}} \frac{h_2}{h} = 1.$$

Let $\dfrac{h_1}{h} = \lambda$, $\dfrac{h_2}{h} = \mu$, $\dfrac{h_1 U_1}{hU} = \chi$. Then

$$\chi + \mu \frac{U_2}{U} = 1, \quad \frac{U_2}{U} = \frac{1-\chi}{\mu}, \quad \frac{U_1}{U} = \frac{\chi}{\lambda},$$

the first result being obtained from (3). Substitution in (7) gives

$$\lambda \left(\frac{\chi}{\lambda}\right)^{1-\frac{\pi}{\alpha}} - \mu \left(\frac{1-\chi}{\mu}\right)^{1-\frac{\pi}{\alpha}} = 1.$$

In the present problem λ, μ, α are given, and χ is determined by approximation from this transcendental equation.

The principle of reflection shows that our solution also elucidates the problem of a straight canal with two side branches, the origin still being taken

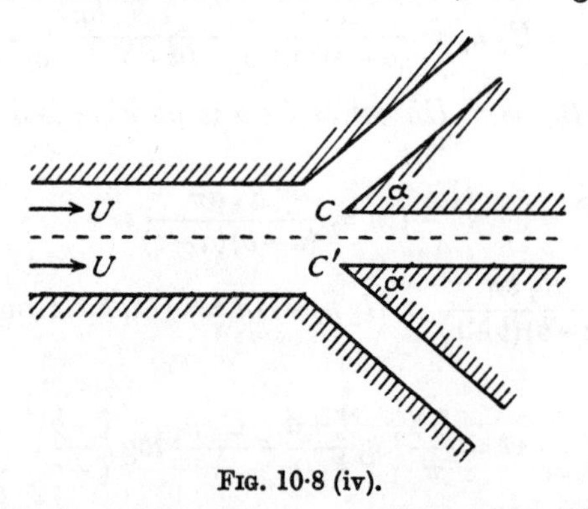

FIG. 10·8 (iv).

at C. There is, of course, no difficulty in moving the origin, say to the middle point of CC'.

EXAMPLES X

1. Apply the transformation of Schwarz and Christoffel to obtain the solution for a wide stream of velocity U flowing past a thin obstacle of length c projecting perpendicularly from a straight bank, in the form

$$w^2 = U^2(z^2 + c^2).$$

Find the pressure at any point of the obstacle and show that it becomes negative if $y > c(1+k)^{\frac{1}{2}}(1+2k)^{-\frac{1}{2}}$, where $k = \rho U^2/2p_0$, p_0 being the pressure at infinity. (U.L.

2. Prove that the complex potential

$$w = -im\pi + m \log\left(\cosh\frac{\pi z}{a} - \cosh\frac{\pi h}{a}\right)$$

gives the flow from a large vase of breadth a through a small hole in one side at height h above the bottom, the streamline $\psi = 0$ comprising the unpierced side, the base and the other side from the base to the hole. Show that at a sufficient distance from the orifice parallel flow supervenes.

3. Show that the complex potential $w = m \log \sinh \{\pi z/(2a)\}$ gives the flow from a large vase of breadth $2a$ through a small hole in the centre of its base. Trace the general form of the streamlines, and prove that at a distance from the base greater than its breadth the flow is sensibly parallel to the walls of the vessel.

4. Prove the theorem of Schwarz and Christoffel for the mapping of a polygon upon a half-plane. What happens when one of the external angles of the polygon $> 2\pi$?

Liquid flows two-dimensionally through a neck of breadth $2b$ to which converge symmetrically two wedge-shaped channels bounded by the straight lines

$$\pm y = b + x \tan\beta\pi, \quad x > 0;$$
$$\pm y = b - x \tan\beta\pi, \quad x < 0.$$

If the total flow (per unit thickness) through the neck is $2bV$, show that the motion is given by the transformations

$$w = \frac{bV}{\pi}\log\frac{t-1}{t+1}; \quad \frac{dz}{dt} = +\frac{Ct^{2\beta}}{(t^2-1)^{\alpha}},$$

where $$b = C\int_0^{\infty}\frac{u^{2\beta}}{(1+u^2)^{\alpha}}\,du,$$

and $$\alpha = 1+\beta.$$ (U.L.)

5. What problem is solved by the transformation

$$\frac{d}{dt}(x+iy) = \frac{1}{t-a}\left(\frac{\sqrt{t}+1}{\sqrt{t}-1}\right)^{\frac{1}{2}},$$
$$\phi + i\psi = \log(t-a),$$

where x and y are the cartesian coordinates of a point and ϕ and ψ are the potential and current functions respectively. (M.T.)

6. Find the transformation to give the two-dimensional flow of a stream of velocity $2U$ at infinity past a right-angled bend in a river bounded by the positive halves of the x- and y-axes and the straight lines

$$x = a,\ y > a \quad \text{and} \quad y = a,\ x > a.$$

7. Show that the transformations

$$z = \frac{a}{\pi}\{\sqrt{t^2-1} - \sec^{-1} t\}; \quad t = e^{-\frac{\pi w}{aV}},$$

where $z = x+iy$, $w = \phi + i\psi$, give the velocity potential ϕ and the stream function ψ for the flow of a straight river of breadth a, running with velocity V at right angles to the straight shore of an otherwise unlimited sheet of water into which it flows. The motion being treated as two-dimensional, show that the real axis in the t-plane corresponds to the whole boundary of the liquid.

8. Show that the problem of a stream of velocity V and infinite width going past a rectangular projection in an otherwise unlimited straight shore is given by

$$\frac{dz}{dw} = \frac{1}{V}\frac{(w^2-c^2)^{\frac{1}{2}}}{(w^2-b^2)^{\frac{1}{2}}},$$

where b and c are constants given by the equations

$$\frac{F(c/b)}{F(\sqrt{b^2-c^2}/b)} = \frac{k}{h} \quad \text{and} \quad \frac{k}{b} = \frac{1}{V}F(c/b);$$

where
$$F(\kappa) = \kappa^2 \int_0^{\pi/2} \frac{\cos^2\phi \, d\phi}{(1-\kappa^2\sin^2\phi)^{\frac{1}{2}}},$$

and h, $2k$ are the length and breadth of the rectangular projection. Obtain the complete solution in a form not involving elliptic functions, when $k = 0$.

CHAPTER XI

JETS AND CURRENTS

11·10. Free streamlines. A streamline μ in two-dimensional motion separates the fluid into two regions A and B. Neglecting external forces, we have on streamlines of the two regions

$$\frac{p_A}{\rho_A}+\tfrac{1}{2}q_A^2 = C_A, \quad \frac{p_B}{\rho_B}+\tfrac{1}{2}q_B^2 = C_B,$$

for an inviscid liquid in steady motion, the suffix denoting the region considered. Now consider a point P of the streamline μ. If we approach P from the region A, we arrive at this point with a value p_1 for the pressure and q_1 for the speed. Similarly if we approach P from the region B, we obtain p_2 and q_2. Thus

$$\frac{p_1}{\rho_A}+\tfrac{1}{2}q_1^2 = C_1, \quad \frac{p_2}{\rho_B}+\tfrac{1}{2}q_2^2 = C_2.$$

Now the pressure must be continuous (3·31); therefore $p_1 = p_2$. It follows that

$$\rho_A\, q_1^2 - \rho_B\, q_2^2 = \text{constant}.$$

In the cases which we have hitherto considered the velocity has been continuous, i.e. $q_1 = q_2$.

We now envisage a class of motions in which the velocity is *discontinuous*, for example, a layer of oil flowing over a layer of water, the speeds in the two layers being different.

To particularise the discontinuous motion still further, let us suppose that the fluid in region A is *at rest*, $q_1 = 0$.

We then see that, along μ, $q_2 =$ constant. We are thus led to frame the following definition. A streamline which separates fluid in motion from fluid at rest is called a *free streamline*.

Neglecting external forces, free streamlines have the following properties.

(i) Along a free streamline the stream function ψ is constant. This is of course a general property of all streamlines.

(ii) Along a free streamline the speed is a constant called the *skin speed*. Free streamlines are thus *isotachic* lines, or lines of constant speed.

(iii) Along a free streamline the pressure is constant. Free streamlines are thus *isobaric lines* or *isobars*, i.e. lines of constant pressure.

Proof. Since the pressure is continuous, its value on the free streamline is equal to its value in that part of the adjacent fluid which is at rest, and that value is constant when external forces are neglected.

It follows from (iii) that the fluid which is at rest could be absent.

Example. The liquid issuing as a jet from a hole in a vessel is bounded by free streamlines, the constant pressure being maintained by the atmosphere. If the atmosphere is absent, the constant pressure is zero.

11·11. Jets and currents. Neglecting external forces, suppose that we have liquid in motion (in two dimensions) bounded by the free streamlines

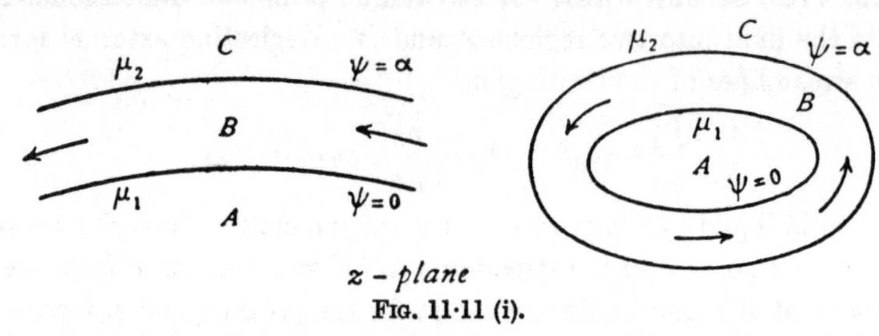

FIG. 11·11 (i).

μ_1, μ_2. These streamlines separate the plane into three regions A, B, C, the liquid in motion occupying region B. If A and C are empty of liquid, we have a *jet*; if A and C are occupied by liquid at rest, we have a *current*. Smoke issuing from a chimney or water from a hose are examples of (three-dimensional) jets. Currents are exemplified by the discharge of liquid into a pond from a submerged pipe, and by ocean currents, for example the Gulf Stream.

A jet or current may be closed or may extend to infinity, see fig. 11·11 (i).

On the free streamlines, ψ and q are both constant.

On μ_1, let $\qquad \psi = 0, \quad q = U.$

On μ_2, let $\qquad \psi = \alpha, \quad q = V.$

Then the figure in the w-plane consists of the infinite band between $\psi = 0$ and $\psi = \alpha$.

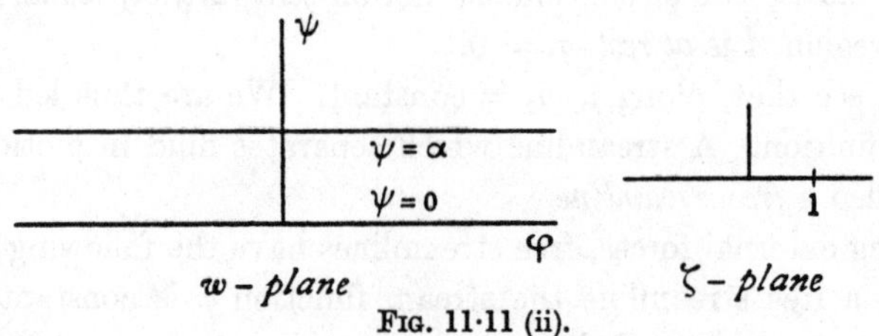

FIG. 11·11 (ii).

If we represent the w-plane on the upper half of the ζ-plane, in such a way that $w = 0$ corresponds to $\zeta = 1$, we have from 10·32

$$w = \frac{\alpha}{\pi} \log \zeta, \tag{1}$$

where the ogarithm is chosen to vanish when $\zeta = 1$.

Now consider the function ω, where

(2) $$e^{-i\omega} = -\frac{1}{U}\frac{dw}{dz} = \frac{q}{U}e^{-i\theta},$$

so that $$\omega = \theta + i \log \frac{q}{U}.$$

On μ_1, $q = U$, on μ_2, $q = V$, and therefore $\omega = \theta$ on μ_1, $\omega = \theta + i\beta$ on μ_2,

where $$\beta = \log \frac{V}{U}.$$

Thus the figure in the ω-plane, much as in the w-plane, consists of a band of breadth β and bounded on one side by the real axis.

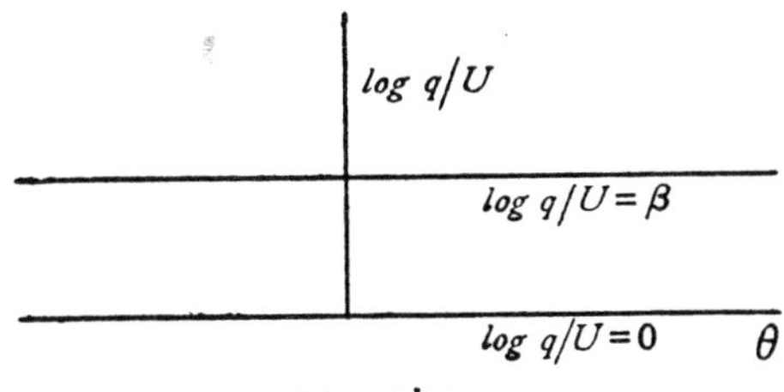

FIG. 11·11 (iii).

Representing this band on the ζ-plane we obtain (10·32).

(3) $$\omega = \frac{\beta}{\pi}\log \zeta,$$

where $\omega = 0$ corresponds to $\zeta = 1$.

Hence, from (1) and (3),

$$\omega = \frac{\beta w}{\alpha},$$

and therefore, from (2),

(4) $$-\frac{1}{U}\frac{dw}{dz} = \exp\left(-\frac{i\beta w}{\alpha}\right).$$

If $V = U$, we have $\beta = 0$, and therefore

$$w = -Uz.$$

This means that a jet in which the speed is the same on both boundaries must be straight.

If $\beta \neq 0$, (4) gives

$$z - z_0 = -\frac{\alpha}{i\beta U}\exp\left(\frac{i\beta w}{\alpha}\right)$$
$$= -\frac{\alpha}{i\beta U}\exp\left(\frac{i\beta\phi}{\alpha}\right)\exp\left(-\frac{\beta\psi}{\alpha}\right).$$

Therefore $$|z - z_0| = \frac{\alpha}{\beta U}\exp\left(-\frac{\beta\psi}{\alpha}\right).$$

So that, when ψ is constant, $|z-z_0|$ is constant and z then describes a circle whose centre is z_0. The radius of the circle is

$$\frac{\alpha}{\beta U}\exp\left(-\frac{\beta\psi}{\alpha}\right),$$

and therefore, if r_1 and r_2 are the radii of μ_1 and μ_2,

$$\frac{r_1}{r_2} = \exp(\beta) = \frac{V}{U},$$

that is to say the current speeds on the free streamlines are inversely proportional to the radii.

Thus it appears that currents bounded by free streamlines can exist and that these streamlines are either parallel lines or concentric circles. It should be observed that in the latter case, the motion being irrotational, the fluid does not rotate like a rigid ring.

11·2. Formula of Schwarz. Given a circle, centre $z = 0$, radius R, the function $f(z)$, which is holomorphic within the circle and whose real part takes the value $\phi(\theta)$ on the circumference, is given, save for an imaginary constant, by

$$f(z) = \frac{1}{2\pi}\int_0^{2\pi}\phi(\theta)\frac{R\,e^{i\theta}+z}{R\,e^{i\theta}-z}\,d\theta. \tag{1}$$

Proof. Let $\zeta = R\,e^{i\theta}$ denote a point on the circumference C of the circle. Then $\bar{\zeta} = R\,e^{-i\theta} = R^2/\zeta$. Since $\theta = -i\log(\zeta/R)$ we can write

$$\phi(\theta) = \psi(\zeta), \tag{2}$$

where $\psi(\zeta)$ is a known function of ζ.

Then on the circumference

$$f(\zeta)+\bar{f}(R^2/\zeta) = 2\phi(\theta) = 2\psi(\zeta), \tag{3}$$

and therefore

$$\frac{1}{2\pi i}\int_{(C)}\frac{f(\zeta)}{\zeta-z}\,d\zeta+\frac{1}{2\pi i}\int_{(C)}\frac{\bar{f}(R^2/\zeta)}{\zeta-z}\,d\zeta = \frac{1}{2\pi i}\int_{(C)}\frac{2\psi(\zeta)}{\zeta-z}\,d\zeta. \tag{4}$$

If z is inside C, Cauchy's formula (5·59) and the residue theorem give

$$f(z)+\bar{f}(0) = \frac{1}{2\pi i}\int_{(C)}\frac{2\psi(\zeta)}{\zeta-z}\,d\zeta. \tag{5}$$

Let $f(0) = a+ib$, then $\bar{f}(0) = a-ib$, and so putting $z = 0$ in (5),

$$2a = \frac{1}{2\pi i}\int_{(C)}\frac{2\psi(\zeta)}{\zeta}\,d\zeta, \tag{6}$$

and therefore from (5),

$$f(z) = \frac{1}{2\pi i}\int_{(C)}\psi(\zeta)\frac{\zeta+z}{\zeta(\zeta-z)}\,d\zeta+ib. \tag{7}$$

Putting $\zeta = R\,e^{i\theta}$ we have $d\zeta/\zeta = i\,d\theta$, and the required result follows. Q.E.D.

It is often advantageous to use the alternative formula (5).

11·30. Impinging jets. Fig. 11·30 shows two uniform streams, A_1, A_2, of the same speed U at infinity, meeting and branching off into two other

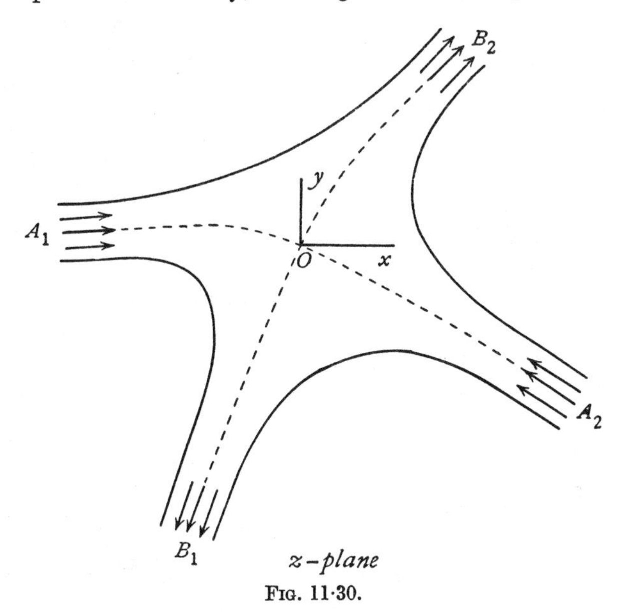

FIG. 11·30.

streams, B_1, B_2. Assuming that a steady motion of the type thus depicted is possible, the problem is to determine the streams B_1, B_2 when A_1 and A_2 are completely specified. If we imagine the streams or currents A_1, A_2 to advance from infinity, it is physically plausible that when they meet a stagnation point O will arise, and therefore that when the motion has become steady a stagnation point will continue to exist. Let us take this point as origin and the x-axis as parallel to and in the direction of flow of A_1.

The free streamlines $A_1 B_1$, $B_1 A_2$, $A_2 B_2$, $B_2 A_1$ will be lines of constant speed, and therefore the speed at infinity of all four streams must be the same, namely U. Let h_1, h_2, k_1, k_2 be the breadths at infinity of A_1, A_2, B_1, B_2. Since the inflow and outflow must balance to preserve continuity, we obtain

$$h_1 + h_2 = k_1 + k_2. \tag{1}$$

Here h_1 and h_2 are given, k_1, k_2 are unknown.

11·31. The complex velocity. Writing as usual

$$\mathfrak{v} = q e^{-i\theta} = u - iv,$$

where q is the speed and θ is the direction of the velocity, we have on the free streamlines

$$\mathfrak{v} = U e^{-i\theta}, \tag{2}$$

and therefore as we go round the free streamlines starting at A_1 and describing $A_1 B_1$, $B_1 A_2$, $A_2 B_2$, $B_2 A_1$ in turn, θ will vary from 0 to -2π, and therefore $-\theta$ will vary from 0 to 2π.

Consequently the representative point of υ drawn on the Argand diagram in the υ-plane will describe a circle whose centre is the origin and whose radius is U, fig. 11·31.

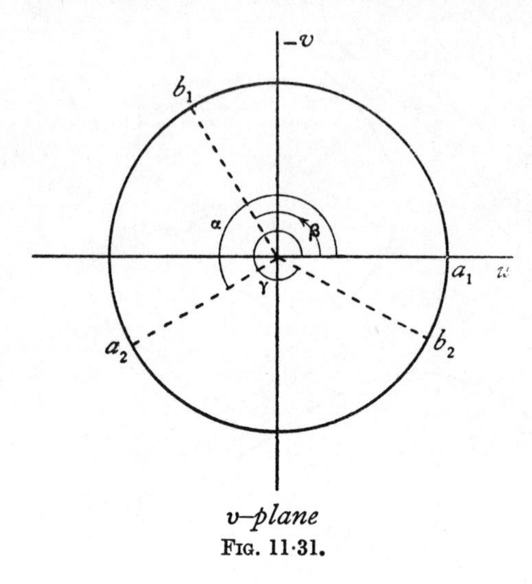

FIG. 11·31.

The points A_1, A_2, B_1, B_2 are then represented by

$$(3)\quad a_1 = U, \qquad a_2 = U\, e^{i\alpha},$$
$$b_1 = U\, e^{i\beta}, \quad b_2 = U\, e^{i\gamma},$$

where $-\alpha$, $-\beta$, $-\gamma$ are the asymptotic directions of the streams A_2, B_1, B_2. Here α is given but β and γ are unknown.

The values of the flux at A_1, A_2, B_1, B_2 are respectively

$$h_1 U, \quad h_2 U, \quad k_1 U, \quad k_2 U,$$

and therefore if we take $\psi = 0$ on $A_1 B_2$, i.e. on the arc $a_1 b_2$, we shall have

$$\psi = h_1 U \text{ on the arc } a_1 b_1,$$
$$\psi = (k_2 - h_2) U \text{ on the arc } b_1 a_2,$$
$$\psi = k_2 U \text{ on the arc } a_2 b_2.$$

11·32. Expression of the complex potential in terms of υ. To determine the complex potential $w = \phi + i\psi$ which satisfies these conditions, we observe that ψ is the real part of $-iw$, and therefore we can apply the formula of Schwarz (11·2), which gives

$$-2\pi i w = \int_0^\beta h_1 U \frac{U\, e^{i\theta} + \upsilon}{U\, e^{i\theta} - \upsilon}\, d\theta + \int_\beta^\alpha (k_2 - h_2) U \frac{U\, e^{i\theta} + \upsilon}{U\, e^{i\theta} - \upsilon}\, d\theta + \int_\alpha^\gamma k_2 U \frac{U\, e^{i\theta} + \upsilon}{U\, e^{i\theta} - \upsilon}\, d\theta.$$

Now
$$\int \frac{U\, e^{i\theta} + \upsilon}{U\, e^{i\theta} - \upsilon}\, d\theta = \int \left(-1 + \frac{2U\, e^{i\theta}}{U\, e^{i\theta} - \upsilon} \right) d\theta$$
$$= -\theta - 2i \log (U\, e^{i\theta} - \upsilon)$$
$$= -\theta - 2i \log U\, e^{i\theta} - 2i \log \left(1 - \frac{\upsilon}{U\, e^{i\theta}} \right)$$
$$= \theta - 2i \log \left(1 - \frac{\upsilon}{U\, e^{i\theta}} \right) - 2i \log U,$$

where the logarithm is determined so as to vanish when $\upsilon = 0$. Thus we find

$$\frac{-2\pi i w}{U} = h_1 \left\{\beta - 2i \log\left(1 - \frac{\upsilon}{b_1}\right) + 2i \log\left(1 - \frac{\upsilon}{U}\right)\right\}$$

$$+ (k_2 - h_2)\left\{(\alpha - \beta) - 2i \log\left(1 - \frac{\upsilon}{a_2}\right) + 2i \log\left(1 - \frac{\upsilon}{b_1}\right)\right\}$$

$$+ k_2 \left\{(\gamma - \alpha) - 2i \log\left(1 - \frac{\upsilon}{b_2}\right) + 2i \log\left(1 - \frac{\upsilon}{a_2}\right)\right\}.$$

Therefore, ignoring a constant,

$$w = -\frac{U}{\pi}\left\{h_1 \log\left(1 - \frac{\upsilon}{a_1}\right) + h_2 \log\left(1 - \frac{\upsilon}{a_2}\right)\right.$$

$$\left. - k_1 \log\left(1 - \frac{\upsilon}{b_1}\right) - k_2 \log\left(1 - \frac{\upsilon}{b_2}\right)\right\},$$

which is the required expression for the complex potential in terms of υ.

11·33. Relations between the breadths and directions of the currents.

Since momentum is conserved in the x- and y-directions, we have

$$h_1 + h_2 \cos\alpha - k_1 \cos\beta - k_2 \cos\gamma = 0,$$
$$h_2 \sin\alpha - k_1 \sin\beta - k_2 \sin\gamma = 0.$$

11·34. Expression for z in terms of υ.

Since

$$\upsilon = -\frac{dw}{dz},$$

we get

$$dz = -\frac{1}{\upsilon}\, dw = -\frac{1}{\upsilon}\frac{dw}{d\upsilon}\, d\upsilon.$$

Now, from 11·32,

$$\frac{1}{\upsilon}\frac{dw}{d\upsilon} = \frac{U}{\pi}\left\{\frac{h_1}{\upsilon(a_1 - \upsilon)} + \frac{h_2}{\upsilon(a_2 - \upsilon)} - \frac{k_1}{\upsilon(b_1 - \upsilon)} - \frac{k_2}{\upsilon(b_2 - \upsilon)}\right\}$$

$$= \frac{U}{\pi}\left\{\frac{h_1}{a_1}\frac{1}{a_1 - \upsilon} + \frac{h_2}{a_2}\frac{1}{a_2 - \upsilon} - \frac{k_1}{b_1}\frac{1}{b_1 - \upsilon} - \frac{k_2}{b_2}\frac{1}{b_2 - \upsilon}\right\}$$

$$+ \frac{U}{\pi\upsilon}\left\{\frac{h_1}{a_1} + \frac{h_2}{a_2} - \frac{k_1}{b_1} - \frac{k_2}{b_2}\right\},$$

and, from 11·33, the second term in this expression vanishes. Therefore, integrating and observing that $z = 0$ when $\upsilon = 0$, we get

$$z = \frac{U}{\pi}\left\{\frac{h_1}{a_1}\log\left(1-\frac{\upsilon}{a_1}\right)+\frac{h_2}{a_2}\log\left(1-\frac{\upsilon}{a_2}\right)-\frac{k_1}{b_1}\log\left(1-\frac{\upsilon}{b_1}\right)-\frac{k_2}{b_2}\log\left(1-\frac{\upsilon}{b_2}\right)\right\},$$

where $a_1 = U$, $a_2 = U\,e^{i\alpha}$, $b_1 = U\,e^{i\beta}$, $b_2 = U\,e^{i\gamma}$.

This result shows that the motion is *reversible*, for the above expression for z is unaltered if we change the signs of U, a_1, a_2, b_1, b_2 and υ.

11·35. The equations of the free streamlines. On a free streamline

$$\upsilon = U\,e^{-i\theta}.$$

If we substitute this in the expression for z above, we get

$$\pi z = h_1 \log(1-e^{-i\theta}) + h_2\,e^{-i\alpha}\log(1-e^{-i\theta-i\alpha})$$
$$- k_1\,e^{-i\beta}\log(1-e^{-i\theta-i\beta}) - k_2\,e^{-i\gamma}\log(1-e^{-i\theta-i\gamma}).$$

Now, $\quad 1-e^{-i\chi} = e^{-\frac{i\chi}{2}}\left(e^{\frac{i\chi}{2}}-e^{-\frac{i\chi}{2}}\right) = 2i\sin\frac{\chi}{2}\,.\,e^{-\frac{i\chi}{2}}.$

Therefore $\quad \pi z = h_1\left\{\log 2i + \log\sin\frac{\theta}{2} - \frac{i\theta}{2}\right\}$

$$+h_2\,e^{-i\alpha}\left\{\log 2i+\log\sin\frac{\theta+\alpha}{2}-i\frac{\theta+\alpha}{2}\right\}$$
$$-k_1\,e^{-i\beta}\left\{\log 2i+\log\sin\frac{\theta+\beta}{2}-i\frac{\theta+\beta}{2}\right\}$$
$$-k_2\,e^{-i\gamma}\left\{\log 2i+\log\sin\frac{\theta+\gamma}{2}-i\frac{\theta+\gamma}{2}\right\}.$$

Now, from 11·33, $h_1 + h_2\,e^{-i\alpha} - k_1\,e^{-i\beta} - k_2\,e^{-i\gamma} = 0$. Thus

$$\pi z = \frac{i}{2}(-h_2\,\alpha\,e^{-i\alpha}+k_1\,\beta\,e^{-i\beta}+k_2\,\gamma\,e^{-i\gamma})$$
$$+h_1\log\sin\frac{\theta}{2}+h_2\,e^{-i\alpha}\log\sin\frac{\theta+\alpha}{2}-k_1\,e^{-i\beta}\log\sin\frac{\theta+\beta}{2}$$
$$-k_2\,e^{-i\gamma}\log\sin\frac{\theta+\gamma}{2}.$$

If we equate the real and imaginary parts, we get the coordinates (x, y) of a point on the free streamlines expressed in terms of the parameter θ.

11·40. The indeterminateness of the problem. In solving the problem of two impinging jets we had to introduce four unknowns, namely k_1, k_2, β, γ, the asymptotic breadths and directions of the resulting branch jets. Between these constants we have found three relations, 11·30 and 11·33, so that the problem contains one undetermined constant. Thus a unique solution is, in general, not possible. The explanation of this indeterminateness no doubt lies in the fact that we are considering a steady motion already established without regard to the initial conditions from which this steady motion is supposed to arise.

Thus, for example, we could suppose the motion to be set up by starting the jets at distant points at instants separated by a time interval t. To different values of t there will no doubt correspond different steady motions, although there is no reason to suppose that they will all be stable.

11·41. Direct impact of two equal jets. In this case there is symmetry about both axes, so that we can take

$$\alpha = \pi, \quad \beta = \frac{\pi}{2}, \quad \gamma = \frac{3\pi}{2}, \quad h_1 = h_2 = k_1 = k_2 = h.$$

From 11·32,
$$w = -\frac{Uh}{\pi}\left[\log\left(1-\frac{\upsilon}{U}\right)+\log\left(1+\frac{\upsilon}{U}\right) - \log\left(1-\frac{\upsilon}{iU}\right)-\log\left(1+\frac{\upsilon}{iU}\right)\right],$$

so that

(1)
$$\exp\left(-\frac{\pi w}{Uh}\right) = \frac{U^2-\upsilon^2}{U^2+\upsilon^2},$$

and from 11·34,

(2)
$$z = \frac{h}{\pi}\left[\log\frac{U-\upsilon}{U+\upsilon}+i\log\frac{U+i\upsilon}{U-i\upsilon}\right].$$

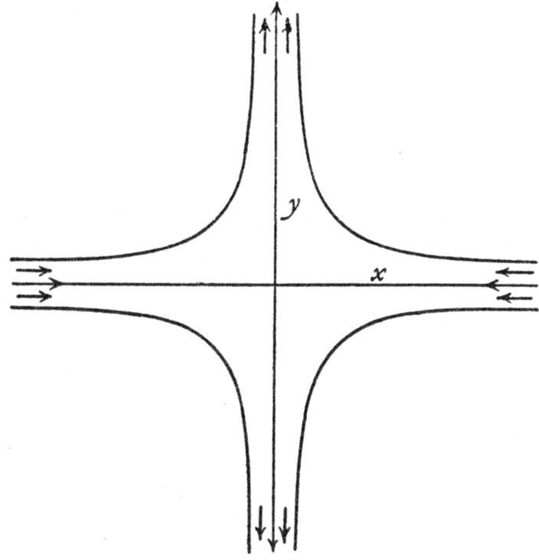

Fig. 11·41 (i).

The elimination of υ between (1) and (2) gives the relation between w and z.

On the free streamlines, $\mathfrak{v} = U\,e^{-i\theta}$. Thus

$$\frac{\pi z}{h} = \log\frac{1-e^{-i\theta}}{1+e^{-i\theta}} - i\log\frac{1-e^{i\left(\frac{\pi}{2}-\theta\right)}}{1+e^{i\left(\frac{\pi}{2}-\theta\right)}}$$

$$= \log\frac{-2i\,.\,e^{-\frac{i\theta}{2}}\sin\left(-\frac{\theta}{2}\right)}{2e^{-\frac{i\theta}{2}}\cos\left(-\frac{\theta}{2}\right)} - i\log\frac{-2i\,e^{i\left(\frac{\pi}{4}-\frac{\theta}{2}\right)}\sin\left(\frac{\pi}{4}-\frac{\theta}{2}\right)}{2e^{i\left(\frac{\pi}{4}-\frac{\theta}{2}\right)}\cos\left(\frac{\pi}{4}-\frac{\theta}{2}\right)}$$

$$= \log\left(-i\tan\frac{-\theta}{2}\right) - i\log\left(-i\tan\left(\frac{\pi}{4}-\frac{\theta}{2}\right)\right).$$

Now, on the streamline in the first quadrant,

$$-\frac{3\pi}{2} < \theta < -\pi.$$

Hence, if we put $\theta = -\pi-\chi$, then $0<\chi<\frac{\pi}{2}$, and

$$\frac{\pi z}{h} = \log\left\{-i\tan\left(\frac{\pi}{2}+\frac{\chi}{2}\right)\right\} - i\log\left\{-i\tan\left(\frac{3\pi}{4}+\frac{\chi}{2}\right)\right\}$$

$$= \log\left(i\cot\frac{\chi}{2}\right) - i\log\left\{i\cot\left(\frac{\pi}{4}+\frac{\chi}{2}\right)\right\}$$

$$= \log i - i\log i + \log\cot\frac{\chi}{2} - i\log\cot\left(\frac{\pi}{4}+\frac{\chi}{2}\right)$$

$$= \frac{i\pi}{2} - i\times\frac{i\pi}{2} + \log\cot\frac{\chi}{2} + i\log\tan\left(\frac{\pi}{4}+\frac{\chi}{2}\right).$$

Thus, if $t = \tan\frac{\chi}{2}$,

$$\frac{\pi x}{h} = \frac{\pi}{2} + \log\frac{1}{t} = \frac{\pi}{2} - \log t, \quad \frac{\pi y}{h} = \frac{\pi}{2} + \log\frac{1+t}{1-t}.$$

Hence $\quad t = \exp\left\{\frac{\pi}{2}\left(1-\frac{2x}{h}\right)\right\}, \quad \frac{1-t}{1+t} = \exp\left\{\frac{\pi}{2}\left(1-\frac{2y}{h}\right)\right\},$

and therefore eliminating t,

$$\exp\frac{\pi}{2}\left(1-\frac{2y}{h}\right) = \frac{1-\exp\frac{\pi}{2}\left(1-\frac{2x}{h}\right)}{1+\exp\frac{\pi}{2}\left(1-\frac{2x}{h}\right)} = -\tanh\frac{\pi}{4}\left(1-\frac{2x}{h}\right) = \tanh\frac{\pi}{4}\left(\frac{2x}{h}-1\right).$$

Thus $\quad \left(\frac{2y}{h}-1\right)\frac{\pi}{2} = \log\coth\frac{\pi}{4}\left(\frac{2x}{h}-1\right), \quad y = \frac{h}{2}+\frac{h}{\pi}\log\coth\frac{\pi}{4}\left(\frac{2x}{h}-1\right).$

If we regard the streamline $x = 0$ as a rigid barrier, we have also solved the problem of the direct impact of a jet on an infinite plane.

The thrust on the plane (per unit thickness of liquid) could of course be obtained by integrating the pressures, but the thrust can be at once inferred from the fact that momentum is advancing through the jet at the rate $\rho h U^2$ perpendicular to the plane and that the momentum of the fluid in contact with the plane is zero in the direction of the normal to the plane. Thus the thrust is $\rho h U^2$.

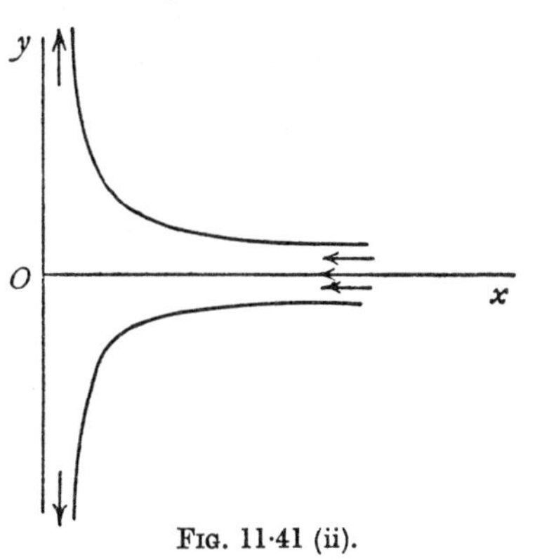

Fig. 11·41 (ii).

It should be noted that there is no indeterminateness in the above solution, since the condition of symmetry introduces a fourth relation among the unknowns.

11·42. Direct impact of two jets. When two jets with the same asymptote impinge directly as in fig. 11·42, it is clear that a symmetrical

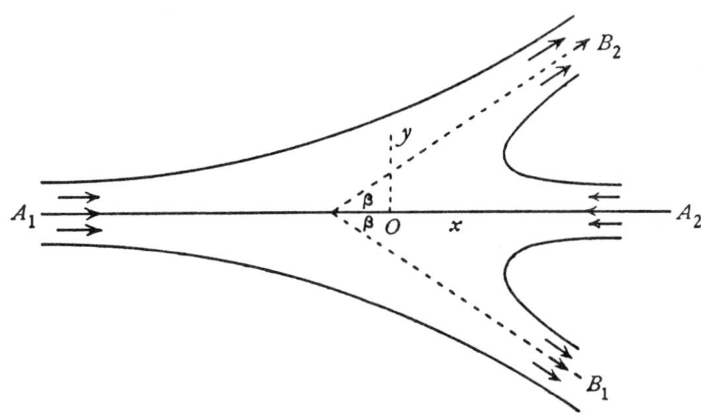

Fig. 11·42.

solution must exist. Thus $k_1 = k_2$, $\alpha = \pi$, $\gamma = 2\pi - \beta$.

Hence, from 11·33,

$$k_1 = k_2 = \tfrac{1}{2}(k_1+k_2) = \tfrac{1}{2}(h_1+h_2), \quad h_1 - h_2 - k_1 \cos\beta - k_2 \cos\beta = 0.$$

Thus
$$\cos\beta = \frac{h_1 - h_2}{k_1+k_2} = \frac{h_1 - h_2}{h_1+h_2}.$$

The parametric equations of the free streamlines can be found as before.

11·43. Oblique impact of equal jets. If two jets of the same breadth, and whose asymptotes are inclined at the angle 2β, are projected simultaneously, it is physically clear that a solution will exist which is symmetrical with respect to the bisector of the angle between the asymptotes. Further, from the principle of reversibility (11·34), it is clear that the solution will be the same as that of the preceding section if we reverse all the velocities, see fig. 11·42. In the present case, $k_1 = k_2$, β are given and h_1, h_2 are required.

We have, then,
$$h_1+h_2 = 2k_1,$$
$$h_1-h_2 = (h_1+h_2)\cos\beta = 2k_1\cos\beta.$$

Thus
$$h_1 = k_1(1+\cos\beta),$$
$$h_2 = k_1(1-\cos\beta).$$

11·50. Rigid boundaries. We shall now discuss some motions in which the moving fluid is bounded in part by free streamlines and in part by fixed rigid walls.

A rigid wall acting as a boundary is of course a streamline along which $\psi =$ constant, but it is not necessarily either an isobar or an isotachic line.

The discussion of those problems in which the rigid boundaries are *straight* may be effected by Kirchhoff's method, which depends essentially on the function
$$Q = \log\left(-U\frac{dz}{dw}\right),$$
where U is a typical speed, generally the skin speed on a free streamline.

Since
$$-\frac{dw}{dz} = u - iv = qe^{-i\theta} = \mathfrak{v},$$
we have
$$Q = \log\left(\frac{U}{\mathfrak{v}}\right) = \log\left(\frac{U}{q}\right) + i\theta.$$

Now along a free streamline the speed q is constant and therefore $\log(U/q)$ is constant.

Along a fixed straight boundary the direction of motion θ is constant, since it coincides with the direction of the boundary.

If therefore we mark the boundaries and free streamlines in the Q-plane, the diagram will consist of straight lines and will constitute a polygon whose interior can be mapped by means of the Schwarz-Christoffel transformation on the upper half of the ζ-plane. Thus a relation is obtained between Q and ζ, i.e. between dw/dz and ζ.

On the other hand, the boundaries and free streamlines when marked in the w-plane all correspond to straight lines $\psi =$ constant, and the resulting polygon can also be mapped on the upper half of the ζ-plane. This leads to a relation between w and ζ.

If we eliminate ζ between the two relations found in this manner, we obtain a relation between dw/dz and w which on integration leads to the relation between w and z, which characterises the motion.

Alternatively, we can often with advantage retain ζ as a parameter in terms of which w and z are expressed.

The precise execution of the above process will be best understood from the illustrations which follow.

11·51. Borda's mouthpiece in two dimensions. Borda's mouthpiece consists of a long straight tube projecting inwards into a large vessel.

Neglecting gravity, the ratio of the section of the escaping water at the vena contracta to the section of the tube is $\frac{1}{2}$. The two-dimensional form of this mouthpiece consists therefore of a long canal with parallel sides projecting inwards. We shall suppose the canal to be so long that the walls of the vessel do not affect the flow, in fact we consider an infinite canal.

The z-plane diagram in fig. 11·51 represents a section of the mouthpiece whose walls are $A_\infty B$, $A'_\infty B'$. The wall $A_\infty B$ is part of a streamline. The fluid flows along $A_\infty B$, turns at B, and flows out of the tube along BC_∞. The shaded area between $A_\infty B$ and BC_∞ indicates fluid at rest or absent. The lines corresponding to the wetted walls in the z-plane are indicated by special shading in all the diagrams.

At the section $C_\infty C'_\infty$ there is uniform parallel flow, with velocity U say. Let the breadth of the mouthpiece at BB' be $2a$. Then, if σ is the coefficient of contraction, the breadth of the issuing jet at $C_\infty C'_\infty$ will be $2\sigma a$ and the flux out of the mouthpiece will be $2\sigma aU$.

The central streamline $E_\infty F_\infty$ is straight. If we take $\psi = 0$ on $E_\infty F_\infty$, we shall have $\psi = -\sigma aU$ on the streamline $A_\infty BC_\infty$ and $\psi = \sigma aU$ on the streamline $A'_\infty B'C'_\infty$.

Again, let us take $\phi = 0$ at B and B', which can always be arranged since an arbitrary constant can be added to the velocity potential. Then at A'_∞, A_∞, E_∞, and at all points in the vessel at a great distance from BB', we shall have $\phi = +\infty$, while $\phi = -\infty$ at C_∞, C'_∞. Thus the w-plane is as shown in fig. 11·51.

Let us map the interior of the polygon $A'_\infty C'_\infty C_\infty A_\infty$ in the w-plane on the upper half of the ζ-plane, making B', B correspond to $\zeta = -1$, $\zeta = 1$, and C'_∞, C_∞, regarded as coincident, to $\zeta = 0$. Then from 10·32 (2), we get

$$w = \frac{2\sigma aU}{\pi} \log \zeta - i\sigma aU, \tag{1}$$

the determination of the logarithm being that which vanishes when $\zeta = 1$.

The next step is to draw the polygon described by

$$Q = \log\left(-U\frac{dz}{dw}\right) = \log\left(\frac{U\,e^{i\theta}}{q}\right) = \log\frac{U}{q} + i\theta,$$

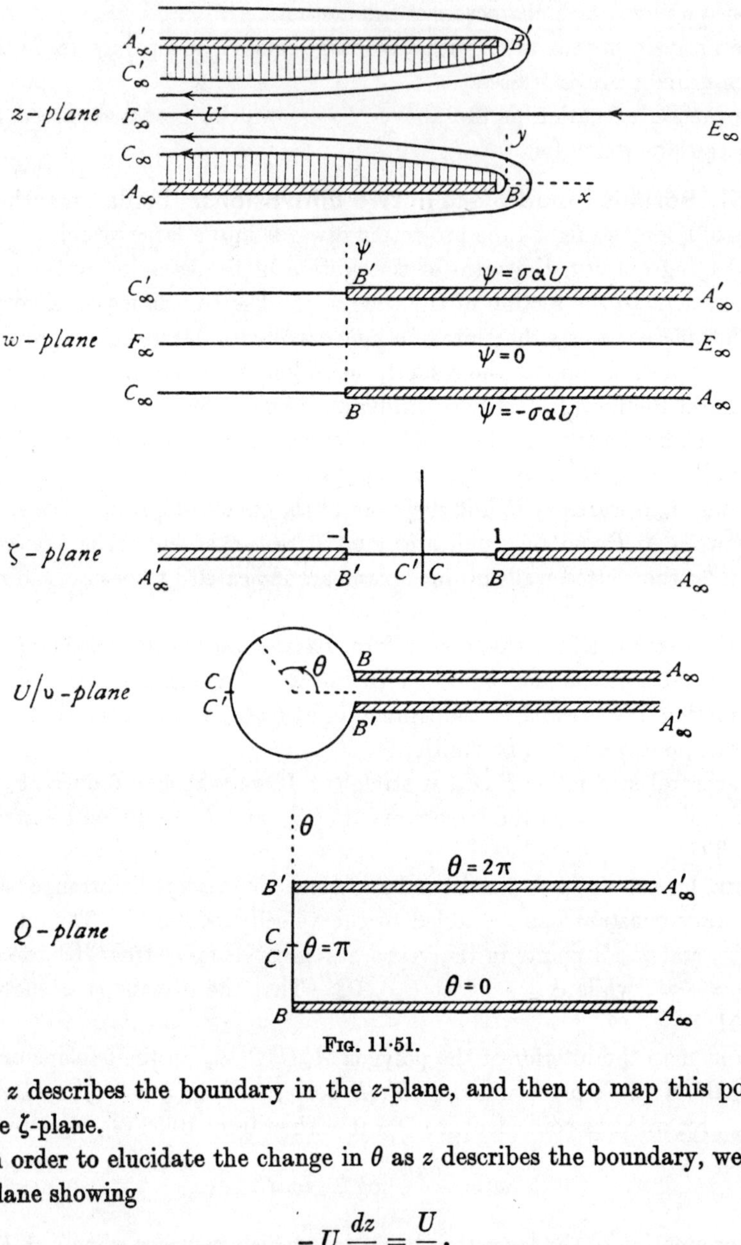

Fig. 11·51.

when z describes the boundary in the z-plane, and then to map this polygon on the ζ-plane.

In order to elucidate the change in θ as z describes the boundary, we draw the plane showing

$$-U\frac{dz}{dw} = \frac{U}{\upsilon}.$$

On the free streamlines, $q = U$, and therefore

$$\frac{U}{\upsilon} = e^{i\theta}.$$

Therefore as we move along $BC_\infty C'_\infty B'$, U/υ describes a circle of radius unity. Along $A_\infty B$ we have $\theta = 0$, while q increases from 0 at A_∞ to U at B, and hence U/υ decreases from ∞ at A_∞ to unity at B.

The diagram is shown in fig. 11·51. The lines there marked $A_\infty B$, $A'_\infty B'$ are actually coincident but are drawn separately to clarify the diagram. It now appears that $\theta = 2\pi$ along $A'_\infty B'$, and therefore

$$\begin{array}{ll} \text{on } A_\infty B & \theta = 0, \\ \text{on } BC_\infty & Q = i\theta \quad (0<\theta<\pi), \\ \text{on } C'_\infty B' & Q = i\theta \quad (\pi<\theta<2\pi), \\ \text{on } B'A'_\infty & \theta = 2\pi. \end{array}$$

The diagram in the Q-plane is therefore that shown in fig. 11·51.

We map this polygon on the ζ-plane by means of 10·31, which gives

$$(2) \qquad Q = 2\cosh^{-1}\zeta = 2\log\left(\zeta + \sqrt{(\zeta^2-1)}\right),$$

so that

$$(3) \qquad -U\frac{dz}{dw} = [\zeta + \sqrt{(\zeta^2-1)}]^2.$$

Equations (1) and (3) constitute the solution of the problem. We could eliminate ζ and then obtain the relation between w and z by integration. The interest of the problem lies, however, in determining the form of the free streamlines, which is done in the next section.

With regard to the above solution it should be observed that there are no points at which the speed becomes infinite. The fluid turns the corners at B and B' with finite speed and to this extent the solution is physically acceptable.

In an actual fluid the dead water portion is usually occupied not by fluid at rest but by fluid in eddying motion. Therefore the above investigation can only be regarded as a first approximation. On the other hand, the solution is an adequate representation of the issuing jet when the region outside the free streamlines is occupied by air or water vapour.

11·52. The equation of the free streamlines. The free streamline BC_∞ is mapped on the segment $0<\zeta<1$ of the real axis in the ζ-plane and so ζ is real. If dz is an element of this free streamline, we have

$$(1) \qquad dz = \frac{dz}{dw}\cdot\frac{dw}{d\zeta}\cdot d\zeta = \frac{-2\sigma a}{\pi}\left[2\zeta - \frac{1}{\zeta} + 2(\zeta^2-1)^{1/2}\right]$$

from 11·51 (1) and (3). Therefore in terms of the real parameter ζ we have

$$(2)\qquad \frac{\pi z}{2\sigma a} = -\int_1^{\zeta}\left[2\zeta - \frac{1}{\zeta} + 2i(1-\zeta^2)^{1/2}\right]d\zeta.$$

A more convenient parameter is the flow direction θ on BC_∞. Indeed on BC_∞ we have $-dw/dz = Ue^{-i\theta}$ and therefore from 11·51 (3) $\zeta = \cos\frac{1}{2}\theta$. Substituting in (2) and integrating we get, in terms of the uniformising variable θ,

$$(3)\qquad \frac{\pi z}{2\sigma a} = \sin^2\tfrac{1}{2}\theta - \log\sec\tfrac{1}{2}\theta + \tfrac{1}{2}i(\theta - \sin\theta).$$

Therefore

$$x = \frac{2\sigma a}{\pi}\left(\sin^2\frac{\theta}{2} - \log\sec\frac{\theta}{2}\right),\quad y = \frac{\sigma a}{\pi}(\theta - \sin\theta).$$

From these equations the free streamline BC_∞ can be plotted.
Now at C_∞, $\theta = \pi$ and therefore $y = \sigma a$.
Hence from fig. 11·51, z-plane,

$$2a = 2\sigma a + \sigma a + \sigma a.$$

whence $\sigma = \frac{1}{2}$, which is Borda's result.

11·53. Flow through an aperture. We consider fluid issuing from a very large vessel through an aperture in one of the walls. The fluid will issue as a jet bounded by free streamlines along which the speed is constant, and at infinity the flow in the jet will be uniform and parallel.

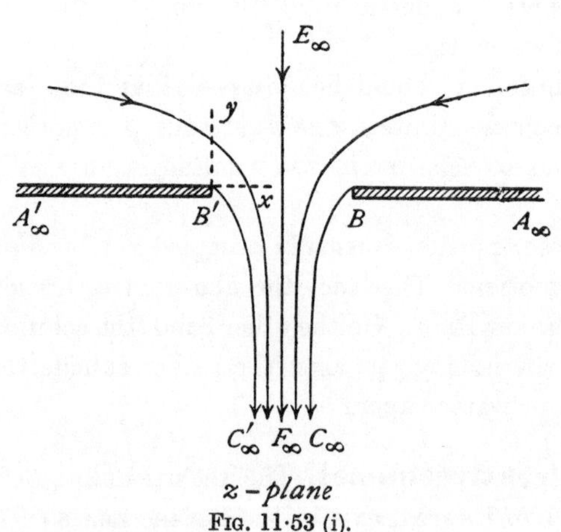

FIG. 11·53 (i).

Fig. 11·53 (i) represents the motion.

If we take $\phi = 0$ at B and B', we shall have $\phi = -\infty$ at C_∞ and $\phi = \infty$ at E_∞, where $E_\infty F_\infty$ is the central (straight) streamline taken as $\psi = 0$.

If $BB' = 2a$, and U is the uniform velocity at C_∞, the output of the jet at $C_\infty C'_\infty$ will be $2\sigma aU$, where σ is the coefficient of contraction. Hence $A_\infty BC_\infty$ will be the streamline $\psi = -\sigma aU$ and $A'_\infty B'C'_\infty$ will be $\psi = \sigma aU$.

The diagram in the w-plane will therefore be the same as in fig. 11·51, and we shall have, after mapping on the ζ-plane of the same figure,

(1) $$w = \frac{2\sigma a U}{\pi}\log\zeta - i\sigma a U.$$

The diagram in the $U/\mathfrak{v}$ plane is, however, different.

As z moves along $BC_\infty C'_\infty B'$, we obtain fig. 11·53 (ii).

Here arg $(U/\mathfrak{v})$ *decreases* by π as we go from B' to B, so that

$$\text{on } B'C'_\infty, \quad Q = i\theta \quad (0>\theta>-\tfrac{1}{2}\pi),$$
$$\text{on } BC_\infty, \quad Q = i\theta \quad (-\tfrac{1}{2}\pi>\theta>-\pi).$$

Mapping the Q-plane on the ζ-plane by means of 10·31 (the origin being moved to $Q = -i\pi$), we get

(2) $$Q = \cosh^{-1}\zeta - i\pi.$$

Equations (1) and (2) give the solution of the problem.

To find the coefficient of contraction, we use the general method of 11·52. Taking the origin at B', and considering the free streamline $B'C'_\infty$ on which ζ is real and increases from -1 to 0, we get, from (1) and (2),

(3) $$dz = \frac{dz}{dw}\cdot\frac{dw}{d\zeta}\cdot d\zeta = \frac{2\sigma a}{\pi\zeta}[\zeta+(\zeta^2-1)^{1/2}].$$

Now on $B'C'_\infty$, $Q = i\theta$ and so from (2) $\zeta = -\cos\theta$.

Substituting in (3) and integrating

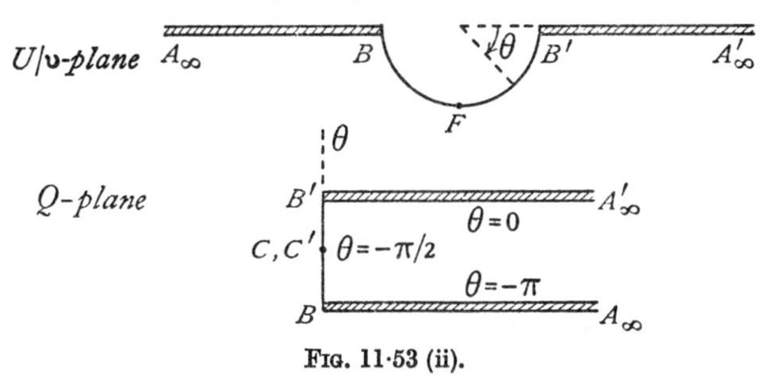

FIG. 11·53 (ii).

we have, in terms of the uniformising variable θ,

$$x = \frac{2\sigma a}{\pi}\int_0^\theta \cos\theta\tan\theta\, d\theta = \frac{2\sigma a}{\pi}(1-\cos\theta).$$

At C'_∞, $\theta = -\frac{1}{2}\pi$, $x = 2\sigma a/\pi$, which is the horizontal distance between B' and C'_∞. Therefore

$$2a = 2\sigma a + \frac{4\sigma a}{\pi}, \quad \sigma = \frac{\pi}{\pi+2} = 0{\cdot}611.$$

11·54. Curved boundaries. Looking at the problem of Borda's mouthpiece, a consideration of the method of solution of 11·51 shows that its success depends solely on the fact that the diagrams in the w-plane and the Q-plane are bounded by straight lines, thereby allowing the application of the theorem of Schwarz and Christoffel.

Professor E. B. Schieldrop has pointed out that by a slight modification of the diagram in the Q-plane, still keeping it a polygon, the solution can be

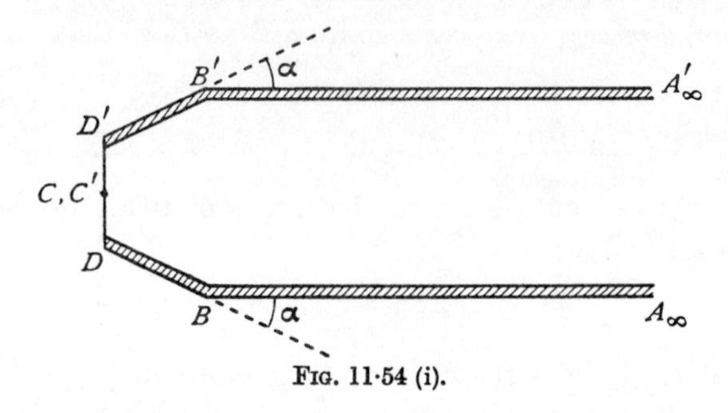

FIG. 11·54 (i).

obtained, corresponding to a rounding of Borda's mouthpiece at the entrance BB' (fig. 11·51).

Thus, if the diagram in the Q-plane is replaced by that shown in fig. 11·54 (i), got by cutting off the corners of the diagram of 11·51 by lines inclined at an angle α, on transforming back to the z-plane we get a mouthpiece with a rounded entrance, fig. 11·54 (ii).

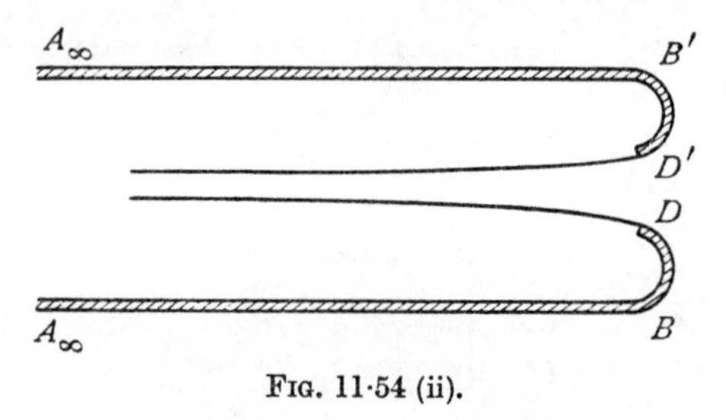

FIG. 11·54 (ii).

The actual calculations have to be performed by graphical or approximate methods, but it is remarkable that such a simple modification can produce workable designs for nozzles. The idea is, of course, applicable to all problems which can be treated by polygonal diagrams in the Q- and w-planes. For details the reader is referred to the original paper.*

EXAMPLES XI

1. Incompressible inviscid liquid in two-dimensional motion under no body forces occupies the whole space outside the region

$$-a \leqslant y \leqslant +a, \quad 0 \leqslant x \leqslant \infty,$$

* E. B. Schieldrop, *Skrifter Oslo*, No. 6, 1928.

and streams into the region through the end $x = 0$. Show that the asymptotes of the free streamlines as the liquid leaves the region at $x = \infty$, after a steady state has been reached, are $y = \pm a/2$. (U.L.)

2. In 11·53; if the breadth of the aperture is $\pi + 2$, prove that the speed q on the centre line of the jet at distance h from the aperture is given by

$$h = \frac{U}{q} - \log \frac{U+q}{U-q}.$$

3. In Borda's mouthpiece (11·51), prove that

$$x = 2\sigma a\,(\sin^2 \tfrac{1}{2}\theta - \log \sec \tfrac{1}{2}\theta)/\pi,$$

and plot the free streamlines.

4. Liquid flows in the negative direction of the axis of y between two planes defined by $x = \pm a$, $y > b$ and meets a barrier defined by $y = 0$, $l > x > -l$. The speed for large positive values of y is V. Show how to determine the ultimate velocity of the two jets and the resultant thrust on the barrier. (U.L.)

5. A jet of incompressible fluid moving irrotationally in two dimensions issues from a funnel-shaped opening of which the walls converge at an angle 2α, the width of the opening at the end being $2c$. The jet is bounded after emergence by "free streamlines" $\psi = \pm\beta$, the speed along each being V.

Prove that the motion is given by the following equations of transformation:

$$w = \frac{2\beta}{\pi} \log t - i\beta, \quad t = -\sin \frac{\pi u}{2\alpha}, \quad \frac{dz}{du} = -\frac{\beta}{\alpha V} e^{iu} \cot \frac{\pi u}{2\alpha},$$

and that
$$c = \frac{\beta}{V}\left[1 + \frac{1}{\alpha}\int_0^\alpha \cot \frac{\pi u}{2\alpha} \sin u \, du\right].$$

Find the coefficient of ultimate contraction of the jet in this case, and verify that it agrees with $\pi/(\pi + 2)$ if $\alpha = \pi/2$. (U.L.)

6. Show that the transformations

$$w = A \log (t-1) + D,$$
$$t = \cosh \{(\Omega - C)/B)\},$$

where $\Omega = \log(-dz/dw)$, and A, B, C, D are constants whose values are to be found, give the motion of a two-dimensional jet of liquid issuing symmetrically from an aperture of width $2a$ in a plane wall.

Prove that the ultimate width of the jet is $2\pi a/(\pi + 2)$, and that the equation to either of its boundaries may be put into the form

$$x = -\frac{2a}{\pi+2}\left(\log \tan \frac{\theta}{2} + \cos \theta\right), \quad y = \frac{2a}{\pi+2}(1 - \sin \theta),$$

the corresponding edge of the aperture being taken as origin, and θ being the inclination of the tangent to the ultimate velocity. (U.L.)

7. Fluid escapes from an aperture placed symmetrically in the base of a deep vessel with vertical sides. Treating the motion as two-dimensional, neglecting gravity, and regarding the region in the plane of z occupied by the fluid as bounded in the way described below, draw and explain figures showing the boundaries of the corresponding regions in the planes of w and Ω, where $\Omega = \log(-dz/dw)$, and write down equations of the form

$$\frac{dw}{dt} = f_1(t), \quad \frac{d\Omega}{dt} = f_2(t),$$

by which each of these regions can be represented conformally on the upper half-plane in the plane of an auxiliary complex variable t. Show how all the constants occurring in $f_1(t)$ and $f_2(t)$ can be determined.

[The boundary of the z region consists of (i) a semi-infinite line $x = 0$, $y \geqslant 0$; (ii) a segment $y = 0$, $a \geqslant x \geqslant 0$; (iii) a free streamline starting at $z = a$ and having an asymptote $x = b$ (not given); (iv) the infinite line $x = c$, where $c > b > a$.] (U.L.)

8. Investigate the motion given by the conformal representation

$$\frac{dw}{dt} = -\frac{m}{\pi t}, \quad t = e^{-\pi w/m},$$

$$\frac{d\Omega}{dt} = -\frac{1}{2n}\frac{\sqrt{(b-a)(b-a')}}{(t-b)\sqrt{(t-a)(t-a')}},$$

$$e^{\Omega} = -Q\frac{dz}{dw} = \frac{Q}{q}e^{i\theta} = \left[\frac{\sqrt{(b-a')(t-a)}+\sqrt{(b-a)(t-a')}}{\sqrt{(a-a')(t-b)}}\right]^{\frac{1}{n}}.$$

Calculate the breadth of the vessel in terms of the final breadth of the jet, at which the velocity is Q; determine also the intrinsic equations of the curves bounding the jet and its final direction.

Show how $n = 1$, $b = 0$, and $a = \infty$, or $a' = -\infty$ will give Helmholtz's jet with the profile a tractrix, and describe any other simple case, such as $a = \infty$, $a' = 0$. (U.L.)

9. The fixed boundaries of liquid moving in the (x, y)-plane are given by $y = x-a$ $(y < -a)$ and $y = -x+a$ $(y > a)$. The sector containing the negative part of the x-axis is completely filled with liquid which is at rest at infinity and which escapes through the opening between $(0, a)$ and $(0, -a)$. Show that the ultimate width of the jet is

$$\frac{\pi a}{\left\{\log(\sqrt{2}-1)+\sqrt{2}+\frac{\pi}{2}\right\}}.$$

Determine the form of the free streamlines. (U.L.)

10. A two-dimensional jet of liquid issues symmetrically out of two plane walls converging at an angle 2α, but terminating at equal distances from the point of convergence. Show that the equations of transformation which lead to the solution of the problem are

$$\frac{dz}{dt} = -\frac{2b}{\pi t}e^{-i\alpha}\left(t+\sqrt{(t^2-1)}\right)^{\frac{2\alpha}{\pi}},$$

$$\frac{dw}{dt} = \frac{2bV}{\pi t},$$

where $2b$ is the ultimate width of the jet and V its ultimate velocity.

Prove that, if $2c$ is the width of the opening between the walls,

$$c = b\left[1+\frac{2}{\pi}\int_0^{\frac{\pi}{2}}\tan\theta\sin\left\{\alpha\left(1-\frac{2\theta}{\pi}\right)\right\}d\theta\right]. \quad \text{(U.L.)}$$

11. A stream, whose breadth and speed at infinity are a and V respectively, flows on the side $y > 0$ of the obstacle given by

$$y = 0, \quad -\infty < x < 0, \quad x = 0, \quad 0 < y < a.$$

Show that the two-dimensional irrotational motion of the stream under no body forces is given by equations of the form

$$\frac{dw}{dt} = \frac{A}{(t-1)(t-\lambda)}, \quad \frac{d\Omega}{dt} = \frac{B}{\sqrt{(t^2-1)}}, \quad \Omega = \log\left(-V\frac{dz}{dw}\right),$$

where $-1 < \lambda < +1$. Determine A and B, and proceed to find the equation satisfied by the angle α through which the stream is deflected. (U.L.)

12. A Borda's mouthpiece of breadth a is fitted symmetrically in the base of a large rectangular vessel of breadth ka, and projects inwards to a great distance from the base. Prove that inside the vessel at a distance from the mouthpiece the flow is practically a parallel stream and that the coefficient of contraction is $k-(k^2-k)^{\frac{1}{2}}$ Deduce the result of 11·51 as a limiting case.

13. If in fig. 11·54 (i) the points D, D' correspond to $\zeta = -a$, $\zeta = a$, $(a < 1)$ respectively, show that

$$\frac{dQ}{d\zeta} = 2(\zeta^2-a^2)^{\frac{\alpha}{\pi}-\frac{1}{2}}(\zeta^2-1)^{-\frac{\alpha}{\pi}}.$$

Sketch the general shape of the corresponding diagram in the U/υ-plane. If $U = \upsilon\, re^{i\theta}$ on the part of the diagram corresponding to BD, $B'D'$, show that

$$\log r = \log r_0 + \theta \cot \alpha,$$

where r_0 is the value of r when $\theta = 0$.

14. In 11·54, show that, along $B'D'$,

$$w = (2b/\pi)\log\zeta - ib \quad \text{and} \quad -\frac{1}{U}e^{Q} = \frac{\pi\zeta}{2b}\frac{dz}{d\zeta},$$

and therefore

$$ds = -\frac{2b}{\pi}\frac{r}{\zeta}d\zeta.,$$

Deduce that

$$x = -\frac{2b}{\pi}\int_1^{\zeta} r\cos\theta\frac{d\zeta}{\zeta}, \quad y = -\frac{2b}{\pi}\int_1^{\zeta} r\sin\theta\frac{d\zeta}{\zeta},$$

where

$$\theta = 2\sin\alpha\int_{\zeta}^{1}\left(\frac{\zeta^2-a^2}{1-\zeta^2}\right)^{\frac{\alpha}{\pi}}\frac{d\zeta}{\sqrt{(\zeta^2-a^2)}}, \quad a \leqslant \zeta \leqslant 1.$$

15. In 11·54, if $\alpha = \pi/4$, prove that

$$Q = 2\int_0^{\zeta}\frac{d\zeta}{\sqrt[4]{(\zeta^2-1)(\zeta^2-a^2)}} + i\pi, \quad \theta = \sqrt{2}\int_{\zeta}^{1}\frac{d\zeta}{\sqrt[4]{(\zeta^2-1)(\zeta^2-a^2)}}.$$

Evaluate (x, y) for $a = 0{\cdot}2, 0{\cdot}4, 0{\cdot}6, 0{\cdot}8$, and hence plot the form of the mouthpiece.

16. Discuss the application of the method of 11·54 to obtain the flow through an aperture in which the edges are suitably rounded.

CHAPTER XII

HELMHOLTZ MOTIONS

12·1. Cavitation. Consider a cylinder moving from right to left with speed U totally immersed in incompressible fluid, say water, otherwise at rest. Fig. 12·1 shows the streamlines as seen by an observer moving with the cylinder. In fig. (i) the motion is just starting (cf. Plate I, fig. 1) and the cylinder is wetted all over; the points of minimum pressure are on the boundary of the cylinder at the extremities of the diameter perpendicular to the direction of motion. In fig. (ii) the cylinder has attained a high speed.* In this case it is found that

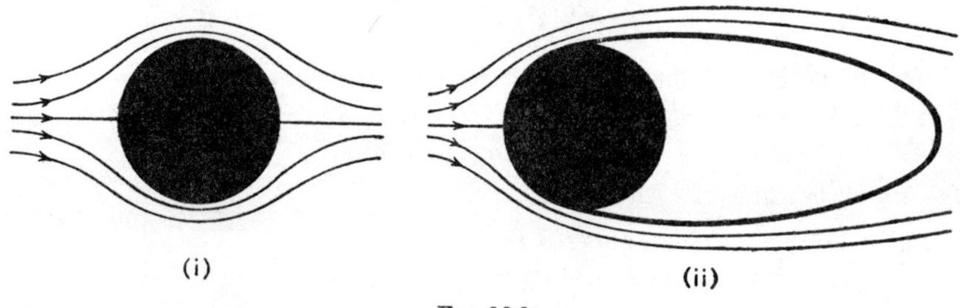

FIG. 12·1.

the water separates from the cylinder at points which are on the anterior part to form a *bubble* or *cavity*, between free streamlines, filled with water vapour.

Let Π be the pressure at infinity and p_c the pressure of the vapour within the cavity. Then Prandtl defines the *cavitation number* σ by

$$(1) \qquad \sigma = \frac{\Pi - p_c}{\frac{1}{2}\rho U^2} = \frac{V^2 - U^2}{U^2}$$

by Bernoulli's theorem if V is the fluid speed on the cavity wall.

In two-dimensional motion it is found that the width of the cavity is of order σ^{-1}, and its length of order σ^{-2}. Thus both width and length increase when σ decreases.

In cavities under water, if the external pressure is kept constant and the speed is sufficiently great, σ is positive, since p_c the vapour pressure is less than the atmospheric pressure. As the speed U increases it appears from (1)

* More properly a high Reynolds' number. See 21·31.

that σ decreases and therefore that $\sigma \to 0$ when $U \to \infty$, with a consequent indefinitely great increase in the width and length of the cavity.

In the present chapter we shall be mainly concerned with what have been called (after their discoverer) *Helmholtz motions*, characterised as follows :

(1) The motion takes place in free space, i.e. gravity is neglected.

(2) The motion is steady, i.e. $p + \frac{1}{2}\rho q^2 = \text{constant}$.

(3) The pressure along the cavity is equal to the pressure in the undisturbed stream, $p_c = \Pi$, and therefore the cavitation number vanishes.

As to (1), the difference between motion in free space (e.g. in a freely falling tank) and under gravity is slight when the object producing the cavity is moving horizontally at high speed. The effect of gravity depends essentially on the Froude number

$$\frac{U^2}{g \times (\text{length of cavity})}$$

which tends to zero as $\sigma \to 0$ and the effect of gravity is not then important.

As to (2), since the motion will be assumed steady, we may suppose the obstacle at rest and the fluid to flow past it.

12·12. Proper cavitation. This is defined to be discontinuous fluid motion in which the minimum pressure occurs on the free streamlines. In improper cavitation the minimum pressure is attained on the boundary of the obstacle.

The distinction between proper and improper cavitation derives its interest from the following facts, all applying to steady motion.

(1) In proper cavitation the free streamlines are convex seen from the fluid.

Proof. The pressure gradient along a normal to the streamline drawn into the fluid is positive. Therefore the acceleration of a particle normal to its path (the streamline) is directed into the cavity. Hence the cavity is convex seen from the fluid.

By a similar argument (applied in free space) the speed attains its maximum at a point where the boundary is convex seen from the fluid.

Corollary. The distinction between proper and improper cavitation does not arise with obstacles whose boundaries are rectilinear.

(2) At a point of a streamline where the direction of the velocity tends to two different limits at either side, the speed is either zero or infinite.

Corollary. At a point of separation from an obstacle the streamline has a continuous tangent.

(3) If a point P divides a streamline into two arcs λ_1, λ_2 such that the tangent is continuous at P but the curvature is different according as P is

approached along λ_1 or along λ_2, then the speed cannot be constant along λ_1 or λ_2.

Proof. With the notations of 12·43 consider θ and τ as functions of the complex potential w on the streamline $\psi = 0$, say. The curvature is

$$\frac{d\theta}{d\lambda} = \frac{d\theta}{dw}\frac{dw}{d\lambda} = -U e^{\tau}\frac{d\theta}{dw}.$$

Now e^{τ} is continuous near an arc on which the speed is constant, so that a discontinuity in $d\theta/d\lambda$ implies a discontinuity in $d\theta/dw$. Therefore $d\tau/dw$ has a logarithmic singularity. Therefore τ cannot remain constant on the streamline at either side of the discontinuity in curvature.

Corollary (i). At a point of separation the curvature is either continuous or infinite.

Corollary (ii). In proper cavitation the curvature at the point of separation from an obstacle of finite curvature is continuous; the streamline must not be concave, by (1), and a convex streamline of infinite curvature would cut into the obstacle.

12·20. Direct impact of a stream on a lamina. Suppose a stream of infinite breadth and velocity U to encounter a fixed lamina BB' of breadth l placed at right angles to the stream, fig. 12·20. We take the centre A of the lamina as origin and AB as axis of x. The streamline which strikes the middle of the lamina at A will divide, and, following the lamina to B and B', will then leave along the free streamlines BC_∞, $B'C'_\infty$. The vacuous region between these free streamlines constitutes the cavity. We shall suppose the dividing streamline to be $\psi = 0$ and we shall take $\phi = 0$ at A. Then $\phi = -\infty$ at C_∞, C'_∞. The w-plane is shown in fig. 12·20, where for clearness the portions $C'_\infty B'A$, $C_\infty BA$ are shown slightly separated although in fact they coincide with the negative ϕ-axis.

The diagram in the w-plane must therefore be regarded as a polygon whose boundary is $C'_\infty B'ABC_\infty$ and whose interior is the whole w-plane, the interior angle at A being 2π. By means of the Schwarz-Christoffel transformation, we map this region on the upper half of the ζ-plane, making B', A, B correspond to $\zeta = -1, 0, 1$ respectively. The transformation is therefore

$$(1) \qquad \frac{dw}{d\zeta} = K\zeta, \quad w = \tfrac{1}{2}K\zeta^2,$$

since $w = 0$ when $\zeta = 0$.

Now consider $-U\,dz/dw = U/\mathfrak{v}$. Marking this on the Argand diagram, as z describes $ABC_\infty C'_\infty B'A$, we get the figure of the $U/\mathfrak{v}$-plane, so that

$\arg(-U\,dz/dw)$ decreases by π when we go from A round $ABC_\infty C'_\infty B'A$. Therefore the Q-plane is as shown in fig. 12·20 (cf. fig. 11·53 (ii).)

We map this polygon on the upper half of the ζ-plane, making $\zeta = -1, 1, 0$ correspond to B', B, A_∞ respectively, at which points the interior angles of the polygon are $\frac{1}{2}\pi$, $\frac{1}{2}\pi$, 0, so that the Schwarz-Christoffel transformation gives

$$\frac{dQ}{d\zeta} = \frac{K'}{\zeta\sqrt{\zeta^2-1}},$$

$$Q = -iK' \cosh^{-1}\left(-\frac{1}{\zeta}\right) + L.$$

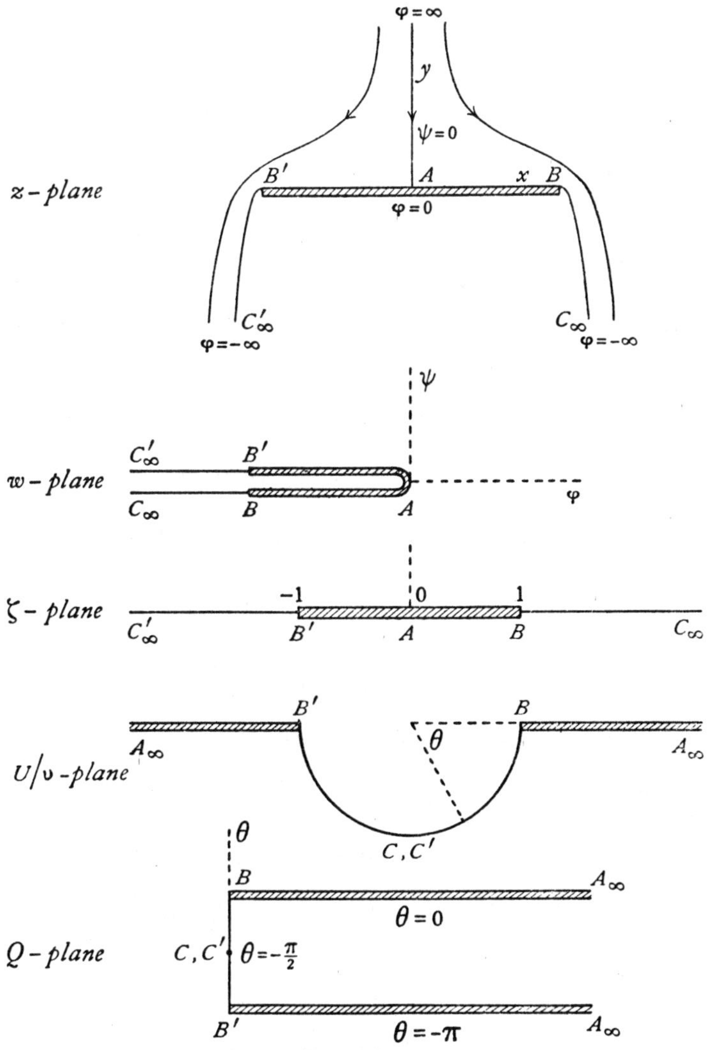

Fig. 12·20.

Now, when $\zeta = -1$, $Q = -i\pi$, and when $\zeta = 1$, $Q = 0$.
Therefore $L = -i\pi$, $-iK'(i\pi)+L = 0$, and hence

$$Q = \cosh^{-1}\left(-\frac{1}{\zeta}\right) - i\pi.$$

From this we get

$$-\frac{1}{\zeta} = \cosh(i\pi + Q) = -\cosh Q.$$

Therefore $$Q = \cosh^{-1}\frac{1}{\zeta} = \log\left(\frac{1}{\zeta} + \sqrt{\frac{1}{\zeta^2} - 1}\right).$$

But $Q = \log\left(-U\dfrac{dz}{dw}\right)$. Thus

$$-U\frac{dz}{dw} = \frac{1}{\zeta} + \sqrt{\frac{1}{\zeta^2} - 1}.$$

(2) $$-\frac{dw}{U\,dz} = \frac{1}{\zeta} - \sqrt{\frac{1}{\zeta^2} - 1}.$$

That the proper sign is here taken in front of the square root follows from the fact that there is a stagnation point at A where $\zeta = 0$, so that dw/dz must vanish when $\zeta = 0$. But for small values of ζ, the square root is $1/\zeta$, very nearly, so that $dw/dz \to 0$ when $\zeta \to 0$.

Now, from (1), $\dfrac{dw}{d\zeta} = K\zeta$. Therefore

(3) $$\frac{U\,dz}{d\zeta} = K[-1 - \sqrt{(1-\zeta^2)}].$$

Integrating from B' to B, i.e. from $\zeta = -1$ to $\zeta = 1$, we get

$$Ul = -K\int_{-1}^{1}[1+\sqrt{(1-\zeta^2)}]d\zeta = -K\int_{-\frac{\pi}{2}}^{\frac{\pi}{2}}(1+\cos\theta)\cos\theta\,d\theta.$$

(4) $$= -K(2 + \tfrac{1}{2}\pi).$$

This determines K, and therefore, from (1),

(5) $$w = -\frac{Ul\zeta^2}{\pi + 4}.$$

The solution is therefore given by (2) and (5).

12·21. The drag. To determine the thrust on the lamina or the *drag*, if p and q are the pressure and speed on the upstream face of the lamina and Π the pressure in the cavity, Bernoulli's theorem gives

$$\frac{p}{\rho} + \tfrac{1}{2}q^2 = \frac{\Pi}{\rho} + \tfrac{1}{2}U^2.$$

Hence the drag D is given by

$$D = \int (p - \Pi)\, dx = \tfrac{1}{2}\rho \int (U^2 - q^2)\, dx,$$

from B' to B along the lamina.

Since AB is the axis of x, it follows that

$$q^2 = \left(\frac{dw}{dz}\right)^2, \quad dx = dz.$$

Therefore

$$D = \tfrac{1}{2}\rho \int \left\{U^2 - \left(\frac{dw}{dz}\right)^2\right\} dz = \tfrac{1}{2}\rho l U^2 - \tfrac{1}{2}\rho \int_{-1}^{1} \left(\frac{dw}{d\zeta}\right)^2 \left(\frac{d\zeta}{dz}\right)^2 \frac{dz}{d\zeta}\, d\zeta.$$

Using (1) and (3) of 12·20, we get for the integral

$$\tfrac{1}{2}\rho U K \int_{-1}^{1} (1 - \sqrt{1-\zeta^2})\, d\zeta = -\frac{\rho l U^2}{\pi + 4}\left(2 - \frac{\pi}{2}\right).$$

Thus
$$D = \frac{\pi \rho l U^2}{\pi + 4},$$

which is the drag (per unit thickness of liquid).

12·22. Drag coefficient. In experimental work it is usual to express drag by means of a *drag coefficient* C_D, defined by the equation

$$D = C_D \cdot \tfrac{1}{2}\rho U^2 S,$$

where S is the projected area of the body perpendicular to the stream. The drag coefficient in the case just investigated of a lamina perpendicular to the stream is therefore

$$C_D = \frac{2\pi}{\pi + 4} = 0{\cdot}88.$$

This agrees with the observed value in motions with a well-defined cavity.

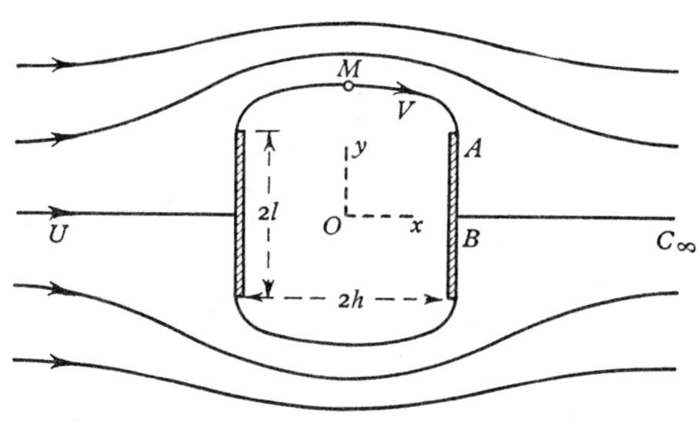

z-plane

FIG. 12·23 (i).

12·23. Riabouchinsky's problem. If a stream U is disturbed by two parallel plates, instead of one, with free streamlines joining the edges, we obtain a situation first investigated by Riabouchinsky,* and illustrated in fig. 12·23 (i) in which the plates are perpendicular to the undisturbed stream direction which is along the line joining the middle points of the plates. Here M is the mid-point of a free streamline joining two edges. The flow pattern has two lines of symmetry indicated by the axes Ox, Oy. The complex velocity v has the values V, iV, 0, U at M, A, B, C_∞ respectively where V is the constant speed on the free streamline.

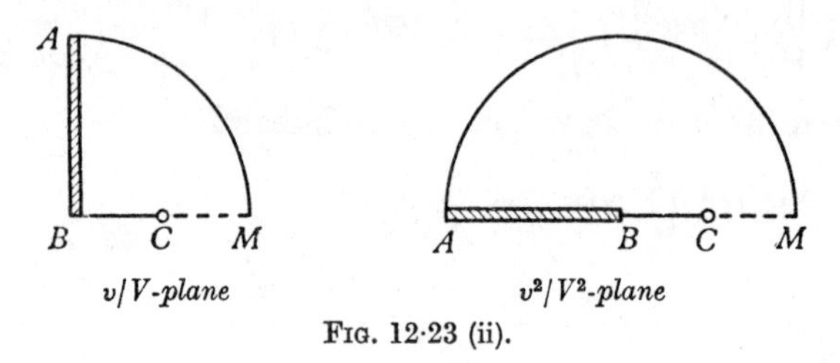

FIG. 12·23 (ii).

Fig. 12·23 (ii) shows the plane of v/V (the hodograph plane) for one quarter of the flow, and also the plane of v^2/V^2. In the former the free streamline maps into a quadrantal arc of a circle since $|v| = V$ on the free streamline and in the v^2/V^2 plane this maps on a semicircle. This semicircle we map on the upper half of the ζ-plane of fig. 12·23 (iii) by the transformation

ζ-plane

FIG. 12·23 (iii).

$$\zeta = -\tfrac{1}{2}\left(\frac{v^2}{V^2} + \frac{V^2}{v^2}\right) \tag{1}$$

which maps A into $\zeta = 1$ and M into $\zeta = -1$. To form the w-plane we take $\phi = 0$ on the y-axis, clearly allowable by the symmetry, and $\psi = 0$ on the free streamline. The part of the flow here considered maps into the third quadrant in the w-plane and therefore into the upper half of the w^2-plane, fig. 12·23 (iv).

The Möbius transformation (see Ex. V, 14).

$$w^2 = \frac{\alpha\zeta + \beta}{\gamma\zeta + \delta}, \quad \alpha\delta - \beta\gamma \neq 0 \tag{2}$$

will then map the upper half of the w^2-plane on the upper half of the ζ-plane, provided α, β, γ, δ are properly chosen real constants (real in order that the real

* *Proc. London Math. Soc.* (2), 19 (1921) 206–215.

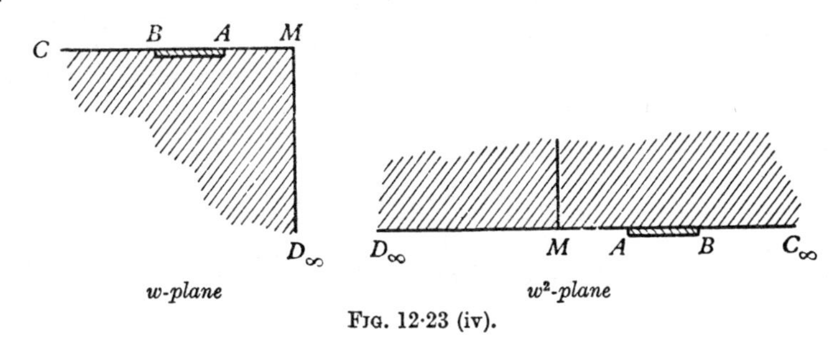

FIG. 12·23 (iv).

axes may correspond). Combining (1) and (2) the w^2-plane is mapped on the hodograph plane by

$$w^2 = G^2 \frac{v^4+\epsilon v^2 V^2+V^4}{v^4+\eta v^2 V^2+V^4},$$

where G, ϵ, η are constants. Since at M $w=0$ and $v=V$, we have $\epsilon = -2$ and the numerator is $(v^2-V^2)^2$. At C_∞ $w = -\infty$ and $v = U$ so that

$$\eta = -(U^4+V^4)/V^2U^2$$

and the denominator is $(v^2-U^2)(v^2-V^4/U^2)$. Thus we can write

$$(3) \qquad w = G\frac{t^2-1,}{(a^2-t^2)^{1/2}(t^2-b^2)^{1/2}}, \quad t = \frac{v}{V}, \quad a^2 = \frac{V^2}{U^2}, \quad ab=1,$$

and we note, from 12·1 (1), that

$$(4) \qquad a^2 = \frac{V^2}{U^2} = 1+\sigma, \quad b^2 = \frac{U^2}{V^2} = \frac{1}{1+\sigma},$$

where σ is the cavitation number.

The formula (3) is a relation between w and dw/dz and so leads to the solution of the problem by a quadrature.

In the notation of elliptic functions* write

$$(5) \qquad t = v/V = b\,\mathrm{nd}(u \mid m), \quad m = 1-\frac{b^2}{a^2}, \quad m_1 = \frac{b^2}{a^2}$$

where m is the squared modulus and m_1 is the complementary squared modulus.

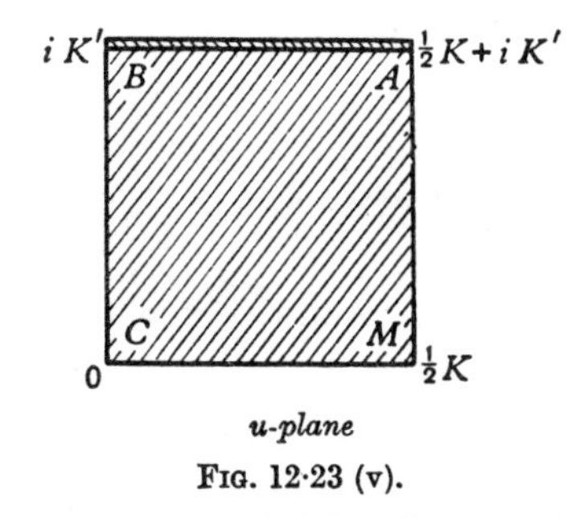

FIG. 12·23 (v).

* For the notation see Milne-Thomson, *loc. cit.*, p. 245.

The values of v at M, C, A, B are respectively $V, U, iV, 0$ and so the corresponding values of u are $\frac{1}{2}K$, 0, $\frac{1}{2}K+iK'$, iK' and the complex u-plane is therefore as shown in fig. 12·23 (v).

We then get from (3), after some reduction,

$$w = \frac{G(m_1^{1/2}\,\mathrm{nd}^2\,u - 1)\,\mathrm{dn}^2\,u}{m\,\mathrm{sn}\,u\,\mathrm{cn}\,u}. \tag{6}$$

Now $\quad dz/du = (dw/du) \div (dw/dz) = -(dw/du) \div v$

and therefore from (5) and (6)

$$\frac{dz}{du} = -\frac{G}{U(1+m_1^{1/2})}\{\mathrm{ds}^2\,u + m_1^{1/2}\,\mathrm{dc}^2\,u\}.$$

Integrating we get

$$z = -\frac{G}{U(1+m_1^{1/2})} f(u) + H, \tag{7}$$

$$f(u) = \mathrm{Ds}\,u + m_1^{1/2}\,\mathrm{Dc}\,u, \tag{8}$$

where H is an arbitrary constant and $\mathrm{Ds}\,u$, $\mathrm{Dc}\,u$ in Neville's notation* denote elliptic integrals of the second kind. The constants G and H are determined by

$$z = h + il \text{ when } u = \tfrac{1}{2}K + iK', \quad z = h \text{ when } u = iK',$$

where $2l$ is the breadth of a plate and $2h$ is the distance between the plates.

If we denote the drag coefficient on one plate by $C_D(\sigma)$ as a function of the cavitation number σ, it can be proved that when σ is small

$$C_D(\sigma) = (1+\sigma)C_D(0),$$

the cavitation number $\sigma = 0$ corresponds to an infinite value of h that is to say when one plate is infinitely distant. Thus from 12·22, $C_D(0) = 2\pi/(\pi+4)$.

In practice it is small cavitation numbers which arise, and Riabouchinsky's solution attains fundamental importance from its ability to deal with variable small cavitation numbers.

12·25. Gliding and planing. The problem derives its interest from the behaviour of seaplane floats, speed-boats, and like phenomena.

In gliding or planing on a free surface the pressure along all the free streamlines is nearly atmospheric (constant) and the cavitation number is practically zero.

In gliding near a free surface gravity can be neglected if gl/U^2, gh/U^2 are each very small compared with unity, where l is the length of the obstacle, and h is the depth of the water.

In the case of deep water, however, the effect of gravity cannot be neglected in calculating the splash formed by an object moving near the surface. There is in fact a whole complex of motions consistent with a given inclination of, say, a plate and the speed of the stream.

* E. H. Neville, *Jacobian elliptic functions*, 2nd edition, Oxford, (1951), Chapter XIV.

The problem to be considered in the next section (12·26) arises as follows. For any given depth of water, stream velocity and attitude of a plate, there is a greatest height of the trailing edge above the (upstream) water surface for which a continuous motion embracing the plate is possible; i.e. suppose a splash is established round a plate immersed in a stream, then this plate can be moved upwards above the upstream level without breaking the continuity of the splash. The work of 12·26 applies to the form of such motions in very deep water.

12·26. Gliding of a plate on the surface of a stream.* We suppose the lamina stationary and the stream to flow past it.

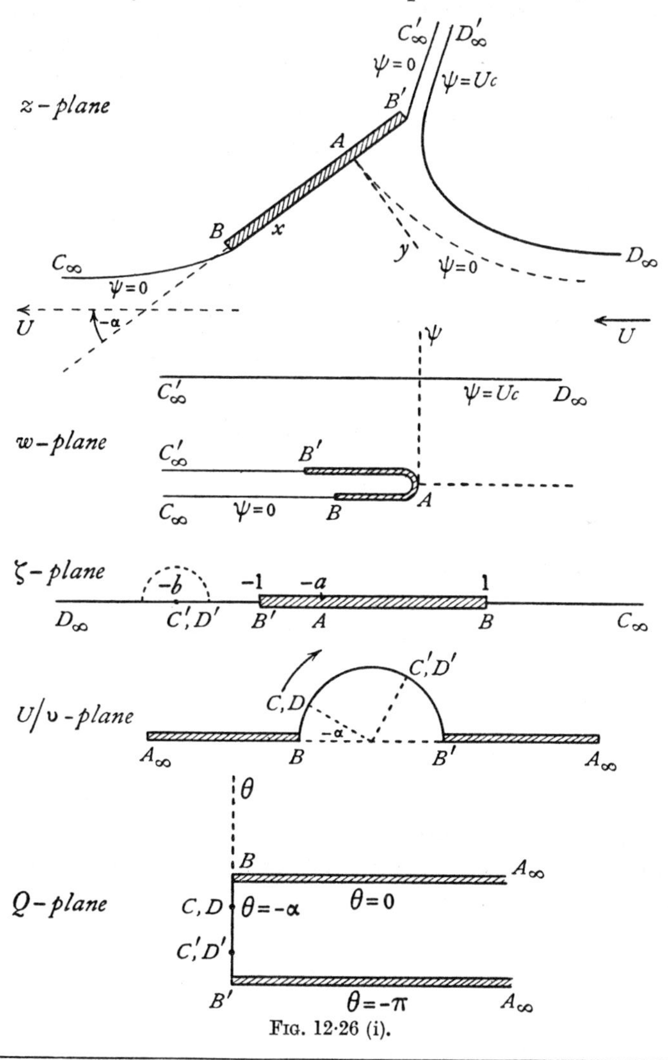

FIG. 12·26 (i).

* A. E Green, *Proc. Camb. Phil. Soc.*, 32 (1936).

Consider, fig. 12·26 (i), a stationary lamina BB' of breadth l, against which a stream of great depth and velocity U impinges. It is assumed that the stream leaves the trailing edge of the lamina at B along the free surface streamline BC_∞, while a jet or splash is formed at the leading edge B', this jet being bounded by the free streamlines $B'C'_\infty$, $D_\infty D'_\infty$. The region behind the plate between BC_∞, $B'C'_\infty$ is occupied by atmospheric air at pressure Π, and so is the region above and to the right of $D_\infty D'_\infty$. It follows that along these free streamlines the speed is constant and equal to U, the speed of the stream at D_∞.

There will be a streamline which impinges on the lamina at some point A and there divides to form the free streamlines BC_∞, $B'C'_\infty$. We shall take this dividing line to be $\psi = 0$. The origin will be taken at A and the axis of x along AB. The direction of the stream is then taken to make an angle $-\alpha$ with AB.

If c is the breadth of the jet at a great distance, we shall have along $D_\infty D'_\infty$, $\psi = Uc$.

The diagram in the w-plane is then that shown in fig. 12·26 (i), which should be compared with fig. 12·20. Transform this into the upper half of the ζ-plane, making B', B correspond to $\zeta = -1, +1$, and let A, C'_∞ then correspond to $\zeta = -a, -b$, respectively. Since the interior angles of the w-polygon are 2π at A and 0 at C'_∞, the Schwarz-Christoffel transformation gives

$$\frac{dw}{d\zeta} = K\frac{\zeta+a}{\zeta+b} = K - K\frac{b-a}{\zeta+b}, \tag{1}$$

so that
$$w = K\zeta - K(b-a)\log(\zeta+b) + L.$$

As ζ increases by passing round the point $\zeta = -b$, $\arg(\zeta+b)$ decreases from π to 0, and therefore $\log(\zeta+b)$ decreases by $i\pi$, and thus ψ, the imaginary part of w, increases by $K(b-a)\pi$. But, in passing round C', the diagram in the w-plane shows that ψ decreases from Uc to 0. Thus

$$K = -\frac{Uc}{\pi(b-a)}. \tag{2}$$

Now, consider $-U\,dz/dw$ as z describes $ABC_\infty D_\infty D'_\infty C'_\infty B'A$. Along the free streamlines the speed is constant. Thus the path described is that shown in fig. 12·26 (i) and the argument decreases from 0 to $-\pi$.

Thus the diagram in the Q-plane is as shown. To map this on the ζ-plane, the Schwarz-Christoffel transformation gives

$$\frac{dQ}{d\zeta} = \frac{K_0}{(\zeta+a)\sqrt{(\zeta^2-1)}},$$

so that
$$Q = K'\cosh^{-1}\frac{1+a\zeta}{\zeta+a} + L', \quad K_0 = K'\sqrt{(a^2-1)}.$$

Now at B', B, Q has the values $-i\pi$, 0 and ζ the values -1, 1. Therefore

$$-i\pi = K' \cosh^{-1}(-1)+L', \quad 0 = K' \cosh^{-1}(1)+L'.$$

Thus $$L' = 0, \quad K' = -1, \quad \text{and} \quad Q = -\cosh^{-1}\frac{1+a\zeta}{\zeta+a},$$

so that $$\cosh^{-1}\frac{1+a\zeta}{\zeta+a} = \log\left(-\frac{dw}{U\,dz}\right).$$

Thus $$-\frac{dw}{U\,dz} = \frac{1+a\zeta}{\zeta+a} - \sqrt{\left(\frac{1+a\zeta}{\zeta+a}\right)^2 - 1},$$

the negative sign being taken in front of the square root, since $dw/dz = 0$ at the stagnation point A, where $\zeta = -a$.

Inverting this result, we get

$$(3) \qquad -\frac{U\,dz}{dw} = \frac{1+a\zeta+\sqrt{(1-a^2)(1-\zeta^2)}}{\zeta+a}.$$

We can now obtain the value of a, for at D_∞, $-U\,dz/dw = e^{-i\alpha}$ (fig. 12·26 (i)) and $\zeta = -\infty$. Thus (3) gives

$$e^{-i\alpha} = a-\sqrt{(a^2-1)} = a-i\sqrt{1-a^2},$$

and thus $a = \cos\alpha$, since $e^{-i\alpha} = \cos\alpha - i\sin\alpha$.

Again, from (1) and (3), we get

$$\frac{dz}{d\zeta} = -\frac{K}{U}\,\frac{1+a\zeta+\sqrt{(1-a^2)(1-\zeta^2)}}{\zeta+b}.$$

Integrating this from $\zeta = -1$ to $\zeta = 1$ we get, after some reduction, the breadth of the plate,

$$(4) \qquad l = \frac{c\sin\alpha}{(b-a)}[b-\sqrt{(b^2-1)}]+\frac{c}{\pi(b-a)}\left(2a+(ab-1)\log\frac{b-1}{b+1}\right).$$

To find the thrust T on the plate, we have, as in 12·21,

$$T = \tfrac{1}{2}\rho\int_{-1}^{1}\left\{U^2-\left(\frac{dw}{dz}\right)^2\right\}\frac{dz}{d\zeta}\,d\zeta = \frac{\rho U^2 c\sin\alpha}{b-a}[b-\sqrt{(b^2-1)}],$$

after a calculation which we leave to the reader. When b is large this becomes

$$(5) \qquad T = \frac{c\rho U^2\sin\alpha}{2b(b-a)}.$$

This is of course normal to the plate, and can therefore be resolved into a drag, and a lift given respectively by

$$D = T\sin\alpha, \quad L = T\cos\alpha.$$

By division, we get from (4) and (5),

$$\frac{\rho U^2 l}{T} = 2b[b-\sqrt{(b^2-1)}]+\frac{2b}{\pi\sin\alpha}\left(2a+(ab-1)\log\frac{b-1}{b+1}\right)$$
$$= 2b^2\left\{1-\left(1-\frac{1}{2b^2}+\dots\right)\right\}+\frac{2b}{\pi\sin\alpha}\left\{2a+(ab-1)\left(-\frac{2}{b}-\dots\right)\right\},$$

taking b large and using the binomial and logarithmic series.

If we now let $b \to \infty$, we get Rayleigh's formula,

$$T = \rho U^2 l \frac{\pi \sin \alpha}{4 + \pi \sin \alpha}, \tag{6}$$

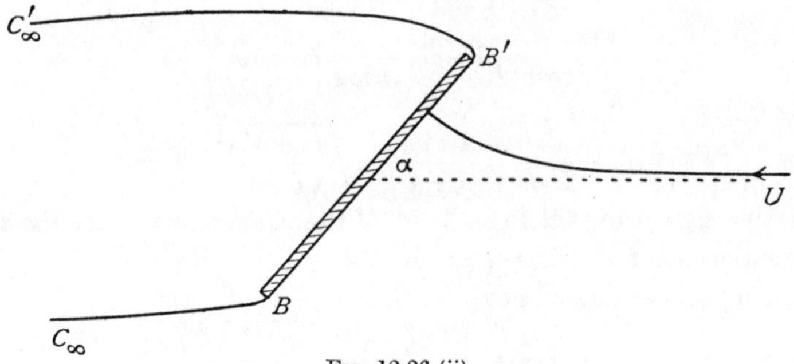

Fig. 12·26 (ii).

which gives the thrust when an infinite stream impinges on a lamina at an angle α, fig. 12·26 (ii), for when $b \to \infty$, D_∞, D'_∞ come together. We discussed the case $\alpha = \pi/2$ in 12·21.

12·30. Reflection across free streamlines. We now describe an entirely different procedure due to M. Shiffman* which consists in extending the variables which describe the flow across the free streamlines and finding the boundaries and singularities of this extension. This process is called the *principle of reflection across free streamlines* and the resulting extension of the flow is called the *image* of the actual flow. We shall indicate by a star the variables $z^\star$, $w^\star$, $v^\star$, of the image corresponding to the variables z, w, v of the actual flow.

Considering for the present flows with only one free streamline we denote by U the fluid speed on that line so that in the v-plane, or hodograph plane, the free streamline is represented by the circular arc

$$v\bar{v} = U^2. \tag{1}$$

Now consider streamlines in the z-, w-, and v-planes as pictured in fig. 12·30 (i), the free streamline being shown dotted.

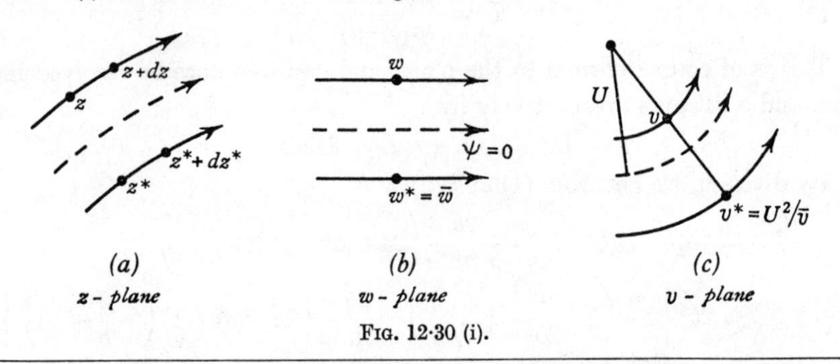

Fig. 12·30 (i).

* *Communications on Pure and Applied Mathematics*, Vol. I (1948) 89–99, Vol. II (1949) 1–11.

Without loss of generality we can take $\psi=0$ on the free streamline.

Since w and $v=-dw/dz$ are both holomorphic functions of z, it follows that the variables z, w, v are holomorphic functions of one another.

By 5·53 the function w can be continued analytically across the straight line $\psi=0$, on which it takes real values, by optical reflection and so

$$(2) \qquad w^\star = \bar{w}.$$

Since a streamline is represented in the w-plane by a line parallel to $\psi=0$, and its optical reflection in $\psi=0$ is the image streamline, it follows that *in taking the image the order of the streamlines is inverted* (see fig. 12·30 (ii)).

Again by the principle of analytic continuation (5·52), since from (1) v and $U^2/\bar{v}$ take the same value on the arc $|v| = U$, we have

$$(3) \qquad v^\star = \frac{U^2}{\bar{v}} = \frac{U^2}{q^2}v$$

and v, $v^\star$ are inverse points with respect to the circle $|v|=U$.

Therefore *the complex velocity and its image are parallel* but the speed is altered in the ratio U^2/q^2. Thus we have the following theorem :

Theorem. The image of an element of a streamline is another element of a streamline in the same direction. The order of the streamlines is inverted.

Let dz be an element of a streamline and $dz^\star$ its image. Since

$$v = -\frac{dw}{dz}, \quad v^\star = -\frac{dw^\star}{dz^\star}, \quad \bar{v} = -\frac{d\bar{w}}{d\bar{z}},$$

we have, using (2).

$$v^\star\, dz^\star = -dw^\star = -d\bar{w} = -dw = v\, dz,$$

since on a streamline $d\bar{w} = dw$ and so $\bar{v}\, d\bar{z} = v\, dz$. Combining this with (3) we get

$$(4) \qquad dz^\star = \frac{\bar{v}^2}{U^2}d\bar{z} = \frac{q^2}{U^2}dz,$$

which furnishes a second proof of the above theorem and shows that arc length is altered in the ratio q^2/U^2.

We now consider some particular situations.

Flow in a corner. Let the flow be in the angle $\alpha\pi$.

The image system is flow *outside* an equal angle. Fig. 12·30 (ii).

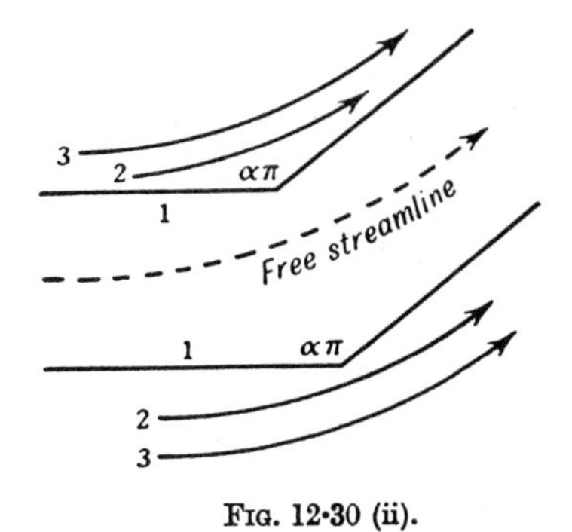

Fig. 12·30 (ii).

Consider as an application a jet running against a wall composed of two planes forming a corner ABC, fig. 12·30 (iii).

The image flow is shown shaded in fig. (iii). The whole flow, actual and image, takes place in a canal between ABC and $A^\star B^\star C^\star$. The streamline which

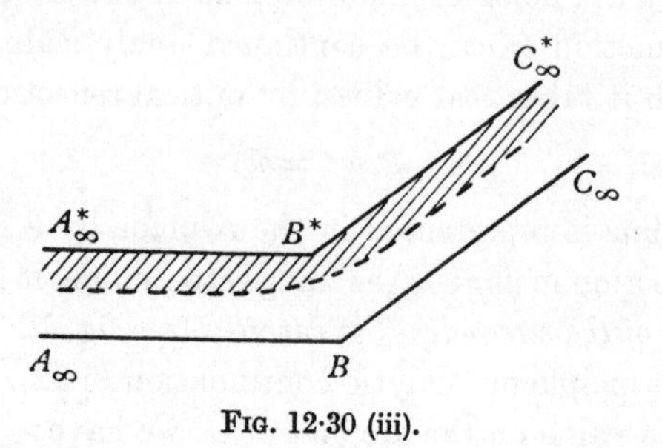

FIG. 12·30 (iii).

bisects the canal at infinity is the free streamline and has constant velocity V. Thus $A^\star B^\star$ and $B^\star C^\star$ are straight, parallel to and at distance $2h$ from AB and BC where h is the breadth of the jet at infinity; $B^\star$ is on the bisector of the angle ABC.

Stagnation point.

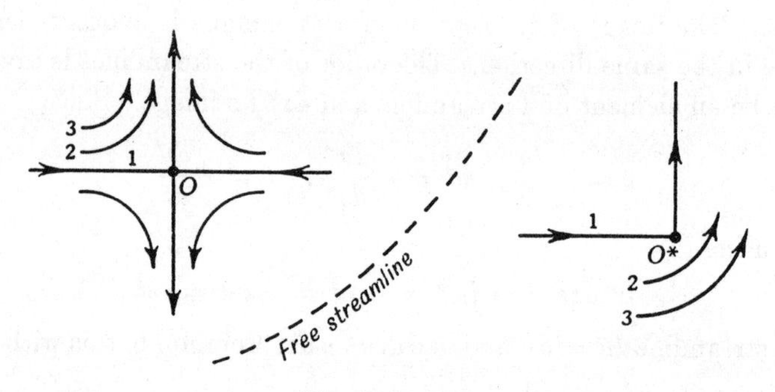

FIG. 12·30 (iv).

The image of a quadrant near the stagnation point O is flow in an angle $3\pi/2$, and so the image of the whole neighbourhood is three sheets with the branch point $O^\star$.

Uniform stream at infinity. If the stream is $v = Ve^{-i\alpha}$, (3) gives

$$v^\star = \frac{U^2}{V}\, e^{-i\alpha}$$

so that the image is a uniform stream. From (4)

$$z^\star - \frac{V^2}{U^2}\, z = \text{constant} \tag{5}$$

and therefore when z is infinite so is $z^\star$. Thus the image is a parallel uniform stream at infinity with speed altered in the ratio U^2/V^2, which is unity if $V = U$.

Free streamline. If there is a second free streamline, speed V, (4) gives (5) again so that the image is a *homothetic* free streamline. If $V=U$ the image is a translation of the original.

Image of a general point. Let the flow near the point z_0 be given by

(6) $v = \alpha(z-z_0)^\beta +$ higher powers of $(z-z_0)$, the index β being real

From (4)

$$dz^\star = \frac{\bar{\alpha}^2(\bar{z}-\bar{z}_0)^{2\beta}}{U^2}\,d\bar{z}+\ldots.$$

$$z^\star - z_0^\star = \frac{\bar{\alpha}^2}{U^2(2\beta+1)}(\bar{z}-\bar{z}_0)^{2\beta+1}+\ldots \tag{7}$$

where $z_0^\star$ is the image of z_0 and only the leading term of an expansion has been written. Combining these results with (3) we get

$$v^\star = \frac{A}{(z^\star - z_0^\star)^{\frac{\beta}{2\beta+1}}}+\ldots,\quad A = \left(\frac{U^{2\beta+2}}{(2\beta+1)^\beta\bar{\alpha}}\right)^{\frac{1}{2\beta+1}}. \tag{8}$$

From (7) we see that if $2\beta+1>0$ the image is at a finite point, while if $2\beta+1<0$ the image $z_0^\star$ is at infinity.

Simple source. In (6) put $\alpha=m$, the strength of the source, and $\beta=-1$. Hence $A=-m$ and the image is therefore an equal source at infinity (inward flow). Conversely the image of a simple source at infinity is an equal source at a finite point.

12·31. Borda's mouthpiece. This has been described in 11·51. By symmetry it is sufficient to consider only half the flow, fig. 12·31.

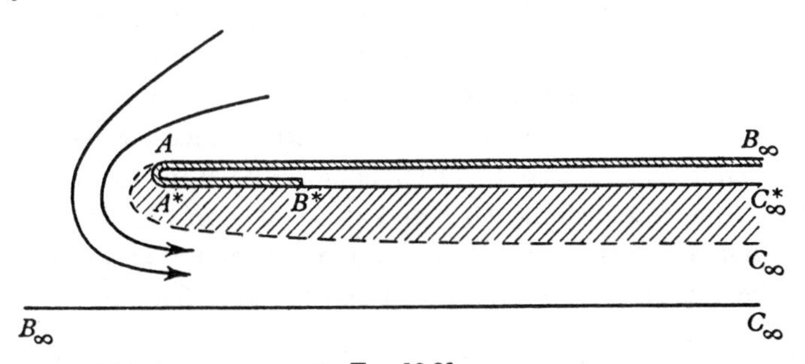

FIG. 12·31.

For simplicity of explanation the upper wall AB_∞ has been doubled by $A^\star C_\infty^\star$. If M is the influx there is a source of output M at infinity (B_∞). The image of A is $A^\star$ coincident with A and therefore the image of AB_∞ is $A^\star B^\star$, $B^\star$ being a source of output M. The image of $B_\infty C_\infty$ is the parallel line $B^\star C_\infty^\star$. Since the velocity at C_∞, $C_\infty^\star$ is the same as that on the free streamline, we see without calculation that the coefficient of contraction is 0·5.

Thus we have a simple intuitive picture of the flow. From a source at infinity the fluid enters the region between the fixed walls AB_∞, $B_\infty C_\infty$. From a source at $B^\star$ of equal strength the flow impinges on that from infinity to form the free streamline (or line of constant pressure) AC_∞.

12·32. Flow from an orifice. The problem was considered in 11·53.

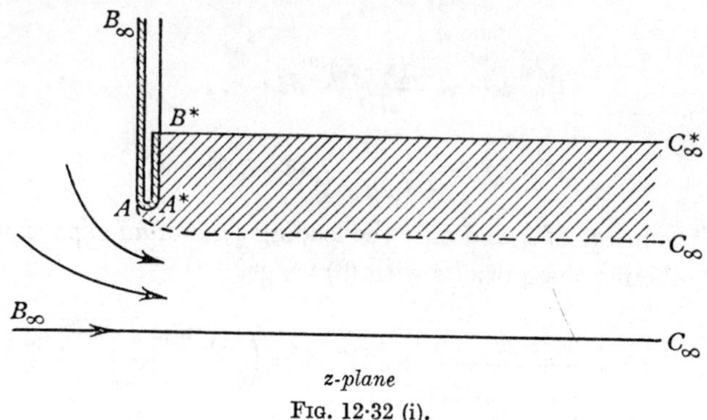

z-plane

FIG. 12·32 (i).

The line of symmetry $B_\infty C_\infty$ of the jet is a streamline so we need consider only half the flow, fig. 12·31 (i). The image of this portion of the flow is a region on the upper side of the free streamline AC_∞ bounded by this line and the images of AB_∞, $B_\infty C_\infty$. Starting from A, the point of detachment, the image of A is the coincident point $A^\star$ and therefore the image of AB_∞ is the finite straight segment $A^\star B^\star$ coincident with AB_∞ in direction, there being a source at $B^\star$ whose output is the same as that at B_∞ i.e. as the efflux from this half of the orifice. The image of the boundary $B_\infty C_\infty$ is the parallel line $B^\star C_\infty{}^\star$. Since the velocity of the jet at infinity is equal to the velocity on the free streamline, the width of the jet at infinity is preserved in the formation of the image. Thus we see, without calculation, that the coefficient of contraction exceeds 0·5. Therefore intuitively the flow can be regarded as taking place between fixed walls $B_\infty AB^\star C_\infty{}^\star$ and $B_\infty C_\infty$; the free streamline arises from the impact of the flows due to a source at infinity (B_∞) and an equal source at $B^\star$.

The position of the image point $B^\star$ is not arbitrary but is completely determined by the condition of symmetry in the w-plane, fig. 12·32 (ii), wherein

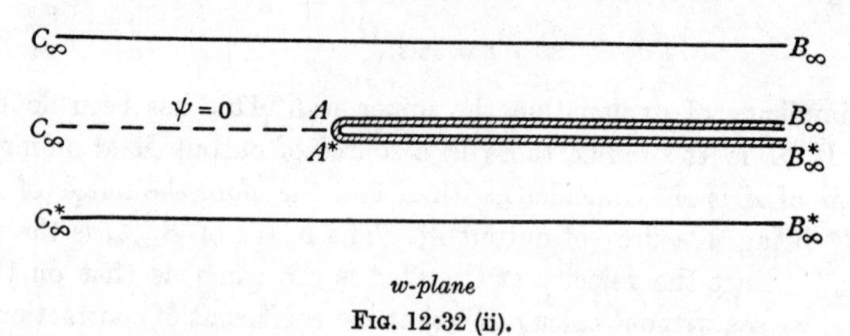

w-plane

FIG. 12·32 (ii).

$B_\infty{}^\star C_\infty{}^\star$ is the optical reflection of $B_\infty C_\infty$ in the real axis $\psi = 0$, the point of detachment of the free streamline being A.

We now proceed to the analytical expression of the flow. Let us map on the upper half of a complex ζ-plane as shown in fig. (iii)

−1 0 1

C_∞^* B^* A^*A B C_∞

ζ-plane

Fig. 12·32 (iii).

Then by the Schwarz-Christoffel transformation

$$\frac{dw}{d\zeta} = \frac{-K\zeta}{(\zeta-1)(\zeta+1)}, \quad \frac{dz}{d\zeta} = \frac{L\zeta}{(\zeta-1)^{3/2}(\zeta+1)^{1/2}},$$

where K, L are constants to be determined, whence by division

$$v = -\frac{dw}{dz} = \frac{K}{L}\left(\frac{\zeta-1}{\zeta+1}\right)^{1/2}$$

and therefore $K = LU$ since $v \to U$ when $\zeta \to \infty$.

The uniformising variable where $\zeta = -\cos\lambda$ leads on integration to

$$w = -K \log \sin \lambda + M, \text{ hence } M = 0,$$

$$z = iL\left(\tan\frac{\lambda}{2} - \lambda\right) + N, \text{ hence } N \text{ is imaginary},$$

$$v = iU \cot\frac{\lambda}{2}, \text{ hence } N = iL\pi,$$

where M, N are constants.

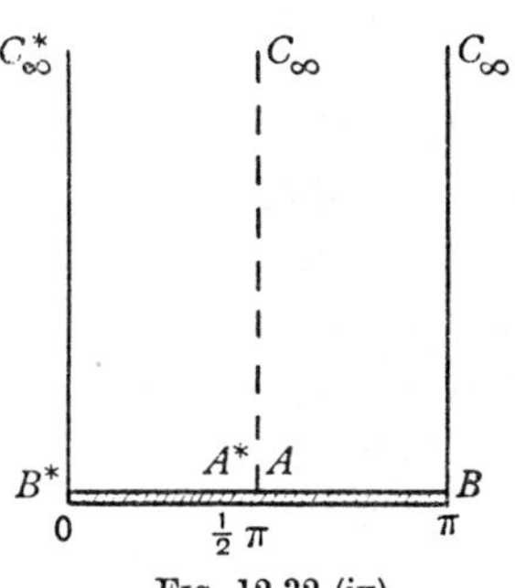

Fig. 12·32 (iv).

The domain in the complex λ-plane is shown in fig. (iv).

All the constants can be determined by the correspondence between the points A, B, C in the various planes. If $2l$ is the breadth of the orifice we have

$$\text{at } A, \quad \lambda = \pi/2, \quad z = li, \quad w = 0$$

$$\text{at } B, \quad \lambda = \pi$$

$$\text{at } C_\infty, \lambda = \pi + i\infty, \quad z = +\infty,$$

whence we get, since $K = LU$,

$$w = -\frac{2lU}{\pi+2}\log\sin\lambda, \quad z = \frac{2li}{\pi+2}\left\{\tan\frac{\lambda}{2} - \lambda + \pi\right\}, \quad v = iU\cot\frac{\lambda}{2}.$$

To determine $B^\star$ we put $\lambda = 0$ getting for z the value $2\pi li/(\pi+2)$.

The width of the jet at infinity is one half the magnitude of this i.e. $\pi l/(\pi+2)$ and the coefficient of contraction is $\pi/(\pi+2) = 0{\cdot}611$.

To get the free streamline put $\lambda = \frac{\pi}{2} + iv$. Then

$$x = \frac{2l}{\pi+2}(v - \tanh v), \quad y = \frac{2l}{\pi+2}(\text{sech } v + \pi/2), \quad 0 \leqslant v \leqslant \infty.$$

12·33. Stream impinging on a lamina. The problem was discussed in 12·20. For the reflection method we need consider only half the flow. The image system is shown in fig. 12·33 (i).

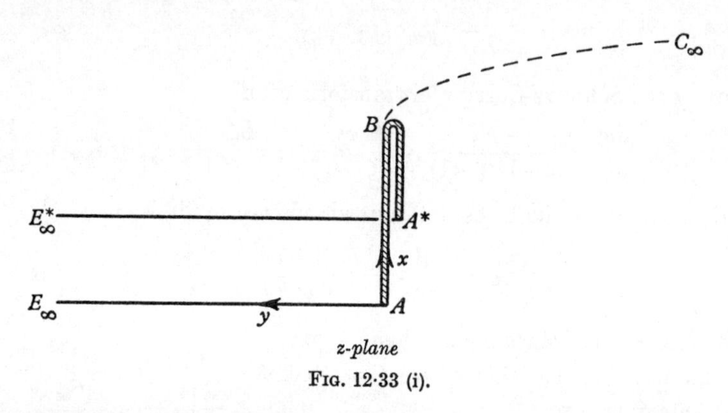

FIG. 12·33 (i).

Note that at $A^\star$ the angle is $3\pi/2$, the image of the angle $\pi/2$ at A. The image of AE_∞ is the parallel line $A^\star E_\infty{}^\star$.

If ds, $ds^\star$ are corresponding lengths of AB and its image $A^\star B^\star$ (i.e. $A^\star B$), we have from 12·30 (4) $ds^\star = ds\, q^2/U^2$; since on AB q varies from zero at A to U at B it follows that $A^\star B^\star$ is less than AB.

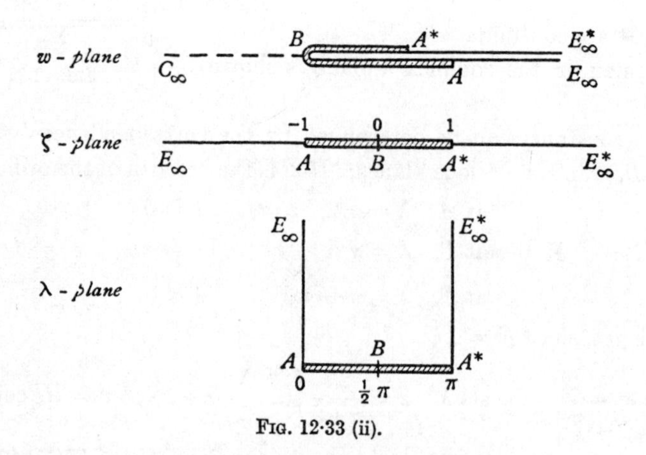

FIG. 12·33 (ii).

The image flow overlaps the actual flow to some extent, and should therefore be considered as a separate sheet of a Riemann surface.

For the mapping we have

(1) $$w = K\zeta^2, \quad \frac{dz}{d\zeta} = \frac{L\zeta(1-\zeta)^{1/2}}{(1+\zeta)^{1/2}}.$$

The uniformising variable λ, where $\zeta = -\cos\lambda$ gives

(2) $$\frac{dz}{d\lambda} = -L\cos\lambda(1+\cos\lambda)$$

so that

(3) $$z = -L(\tfrac{1}{2}\lambda + \sin\lambda + \tfrac{1}{4}\sin 2\lambda),$$

where we have taken $w = 0$ at B i.e. at $\zeta = 0$, and $z=0$ at A. We get from (1) and (2) $dw/dz = -v = 2K\sin\lambda/\{L(1+\cos\lambda)\}$.

At B, $\lambda = \pi/2$, $z = l$, $v = U$, where $2l$ is the breadth of the plate and so

$$L = -\frac{4l}{\pi+4}, \quad K = -\tfrac{1}{2}LU$$

and so finally

$$w = \frac{2lU}{\pi+4}\cos^2\lambda, \quad z = \frac{4l}{\pi+4}(\tfrac{1}{2}\lambda + \sin\lambda + \tfrac{1}{4}\sin 2\lambda).$$

We note that at A^*, where $\lambda=\pi$, the value of z is $2\pi l/(\pi+4)$. The drag coefficient was found in 12·22 to be $2\pi/(\pi+4)$. The relation between these numbers is not fortuitous as will now be proved.

12·34. Geometrical interpretation of the force.

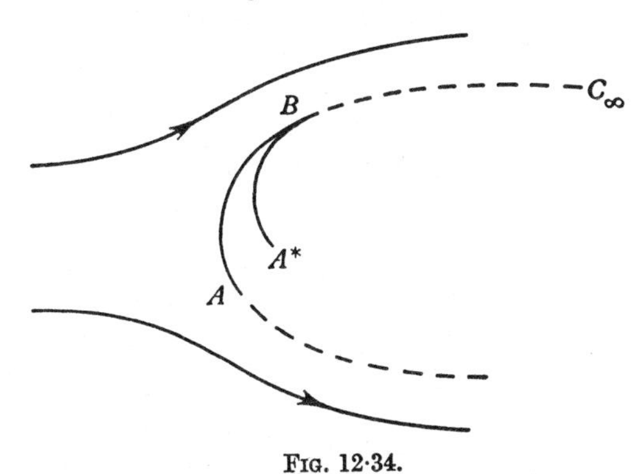

FIG. 12·34.

Consider flow past an obstacle AB. Let BC_∞ be a free streamline. By reflection in BC_∞ the obstacle has the image BA^*. Let $X+iY=F$ be the force per unit thickness of obstacle. Then from 6·41

$$F = \int_A^B -i(p-p_c)\,dz,$$

where p_c is the pressure in the cavity. By Bernoulli's theorem

$$\frac{p}{\rho}+\tfrac{1}{2}q^2=\frac{p_c}{\rho}+\tfrac{1}{2}U^2$$

and therefore

$$F=\tfrac{1}{2}\rho U^2(-i)\int_A^B\left(1-\frac{q^2}{U^2}\right)dz$$

$$=\tfrac{1}{2}\rho U^2(-i)\int_A^B(dz-dz^\star)\quad\text{from 12·30 (4)}$$

$$=-\tfrac{1}{2}\rho U^2 i(a^\star-a),$$

where $a^\star$ and a are the complex numbers for the points $A^\star$ and A.

Thus the resultant force both in magnitude and direction is the same as if the excess stagnation pressure were acting over the entire front of the line joining A to $A^\star$.

If the body and flow are symmetrical, the force is in the direction of the flow and the line $AA^\star$ is perpendicular to this direction. In that case the magnitude of the force is $\tfrac{1}{2}\rho U^2 AA^\star$ and the drag coefficient is $AA^\star/AB$ as exemplified in 12·33.

12·35. Backward jet. A type of cavity which is observed in the entry of an object into fluid is one in which a spout is formed behind but directed back

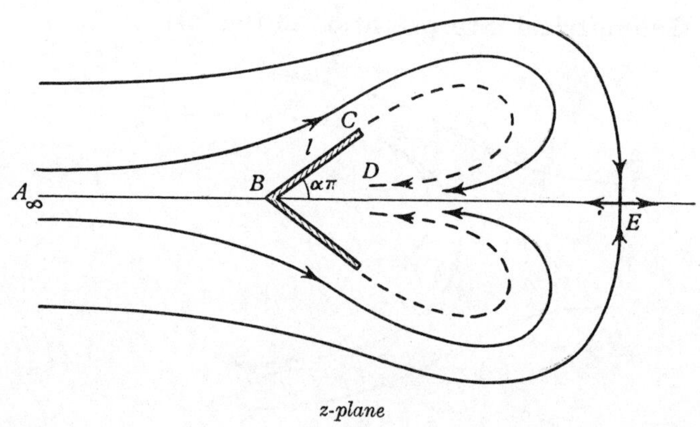

z-plane

FIG. 12·35 (i).

towards the object, and a stagnation point is formed behind the spout. Fig. 12·35 (i) depicts such a symmetrical flow past a bent lamina, the free streamlines being shown dotted as usual; E is the stagnation point. Mathematically we imagine the backward jet to continue *upstream* to infinity but on a different sheet of the flow. In actual motions the backward jet might break up before reaching the obstacle, or might first impinge on the obstacle and then disintegrate.

Consider the upper half of the flow only and map the flow and its reflection on the upper half of the ζ-plane. The w- and ζ-planes are shown in fig. (ii)

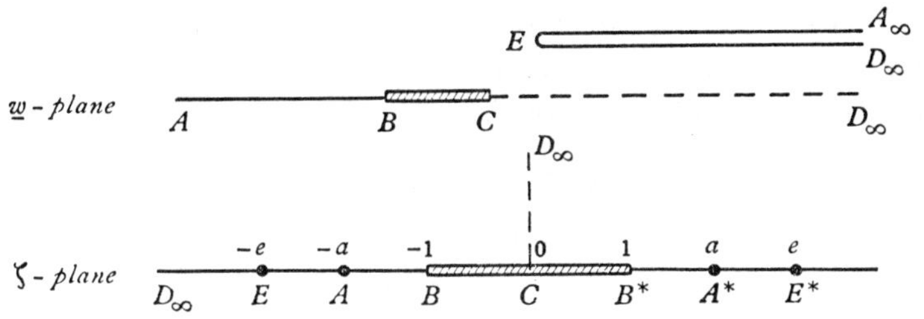

FIG. 12·35 (ii).

If $\alpha\pi$ is the inclination of the arm BC of the lamina to the asymptotic flow direction, and if B, $B^\star$ are mapped on $\zeta = -1, 1$ and $E, A, E^\star, A^\star$ on $\zeta = -e$ $-a$, e, a we have

$$(1) \qquad \frac{dz}{d\zeta} = K\zeta\left(\frac{\zeta-1}{\zeta+1}\right)^{\alpha}\frac{1}{(\zeta+a)^2(\zeta-a)^2}(\zeta-e)^2.$$

since by the principle of reflection the angle at C in the z-plane is 2π, the angle at B is $(1-\alpha)\pi$, at $B^\star$ $(1+\alpha)\pi$, at A_∞, $A_\infty{}^\star$ the angles are $-\pi$, at E and $E^\star$, π and 3π respectively.

In the w-plane the image is simply a reflection across CD_∞ and

$$(2) \qquad \frac{dw}{d\zeta} = -L\zeta\frac{(\zeta+e)(\zeta-e)}{(\zeta+a)^2(\zeta-a)^2}$$

The complex velocity is then

$$(3) \qquad v = -\frac{dw}{dz} = \frac{L}{K}\left(\frac{\zeta+1}{\zeta-1}\right)^{\alpha}\frac{\zeta+e}{\zeta-e}.$$

The constants K, L, a, e are determined by the following five conditions

(i) If $BC = l$, $\quad le^{\alpha\pi i} = z_C - z_B = K\displaystyle\int_{-1}^{0}\zeta\left(\frac{\zeta-1}{\zeta+1}\right)^{\alpha}\frac{(\zeta-e)^2}{(\zeta^2-a^2)^2}\,d\zeta.$

(ii) We denote by z_{A+} the limit of the values of z as we approach A from the right along the real axis in the ζ-plane. Similarly z_{A-} is defined by approaching A from the left. Then $Im(z_{A-}) - Im(z_{A+}) = 0$ since

$$Im(z_{A-}) - Im(z_{A+}) = Im(\pi i r)$$

where r is the residue of $dz/d\zeta$ at $\zeta = -a$.

(iii) w is real near C so that L is real.

(iv) $v_D = -V$, where V is the velocity on the free streamline

(v) $$v_{A_\infty} = U,$$

where U is the velocity of the stream.

These conditions determine the constants K, L, a, e in terms of V, U, l, all the integrals being explicitly calculable.* The quantities V and U are related by

$$\Pi + \tfrac{1}{2}\rho U^2 = p_c + \tfrac{1}{2}\rho V^2,$$

where Π is the pressure at infinity and p_c is the pressure in the cavity. By the method of 12·34 the drag coefficient is $\frac{1}{2}\pi \mid K \mid$.

12·40. Levi-Civita's method. We shall now describe a general method of determining the flow past an obstacle, on the assumption that the motion is steady, irrotational and two-dimensional, and that a cavity is formed behind the obstacle.

The method depends essentially on mapping the w-plane upon the *inside* of a semicircle in the ζ-plane in such a way that the free streamlines map into the diameter, and on the use of the function $\omega(\zeta)$ which has already been employed in the theory of jets (11·11).

12·41. Mapping the z-plane. Suppose an obstacle S to be placed in an infinite stream of velocity U, fig. 12·41. One of the streamlines v coming from infinity will meet the body normally at a stagnation point O and will there divide, following the body along arcs λ_1 and λ_2 and thereafter leaving the body at points A_1, A_2 to form two free streamlines μ_1, μ_2 between which lies the cavity. We shall take $\phi = 0$, $\psi = 0$ at O, so that the dividing streamline is $\psi = 0$. The origin is taken at O and the axis of x is drawn downstream and parallel to the velocity at infinity. The region occupied by the moving fluid is denoted by R.

It is convenient here to take $(-w)$ rather than w in the next diagram, which shows the $(-w)$-plane. The lines $\mu_1+\lambda_1$, $\mu_2+\lambda_2$, which are really coincident with the positive real axis, are drawn slightly separated as in other cases of this kind. The diagram in the $(-w)$-plane is now mapped on the upper half of an auxiliary W-plane by means of the transformation (readily deduced from the Schwarz-Christoffel theorem)

$$w = -W^2.$$

---

* The flow is discussed by another method by D. Gilbarg and D. H. Rock, Naval Ordnance Laboratory Memo. 8718 (1945).

The points corresponding to A_1 and A_2 are denoted by W_1, $-W_2$. Actually $W_1 = \surd(-\phi_1)$, $W_2 = \surd(-\phi_2)$, where ϕ_1, ϕ_2 are the velocity potentials at A_1, A_2.

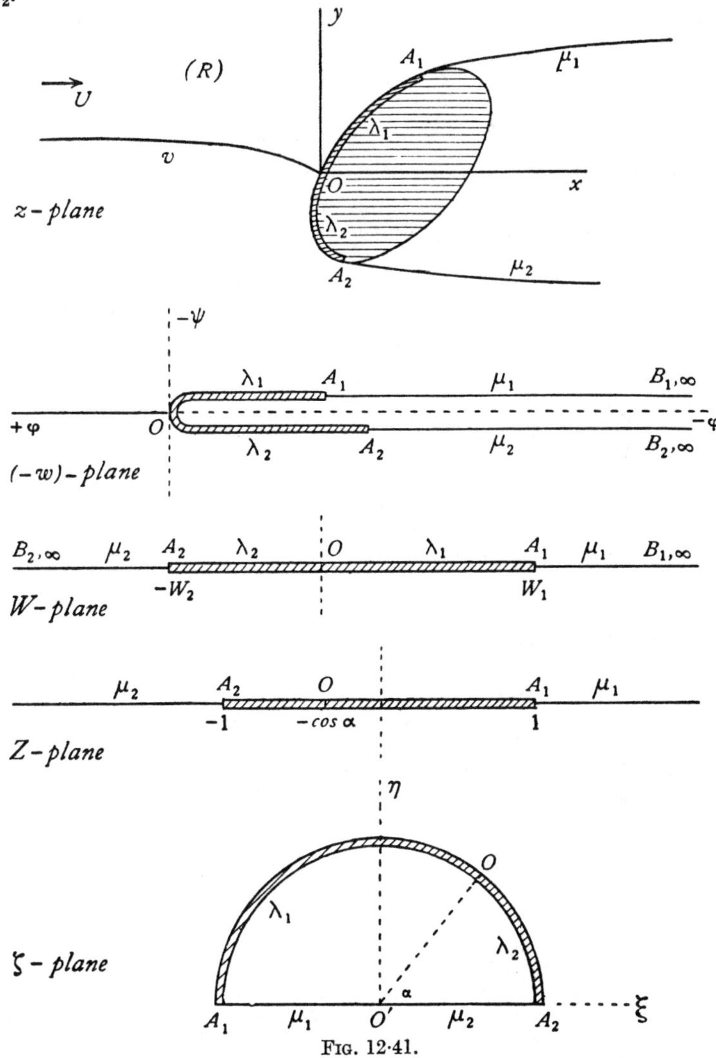

FIG. 12·41.

The upper half of the W-plane is now mapped on the upper half of the Z-plane in such a way that A_1 corresponds to $Z = 1$, and A_2 to $Z = -1$. The necessary transformation is readily seen to be

$$W = \tfrac{1}{2}Z(W_1 + W_2) + \tfrac{1}{2}(W_1 - W_2) = a(Z + \cos\alpha),$$

where
$$a = \tfrac{1}{2}(W_1 + W_2), \quad \cos\alpha = \frac{W_1 - W_2}{W_1 + W_2}.$$

Note that $W = 0$ now corresponds to $Z = -\cos\alpha$.

Finally, we map the upper half of the Z-plane on the interior of the semicircle in the ζ-plane whose radius is unity, whose centre is at the origin, and whose diameter is along the real axis, fig. 12·41.

The necessary transformation is of the Joukowski type,

$$Z = -\tfrac{1}{2}\left(\zeta + \frac{1}{\zeta}\right).$$

To verify this, we have $\zeta = e^{i\chi}$ on the semicircle, and therefore $Z = -\cos\chi$. Hence, as χ goes from 0 to π, ζ describes the semicircle, while Z goes from -1 through $-\cos\alpha$ to 1. The arc of the semicircle therefore corresponds to the segment A_1A_2 of the real axis in the Z-plane, the arc A_1O corresponding to λ_1 and OA_2 to λ_2. Again, as ζ varies from -1 through 0 to $+1$, Z goes from 1 to ∞ and then from $-\infty$ to -1. Thus the radii $O'A_1$, $O'A_2$ correspond to μ_1, μ_2. Since, in conformal transformation the senses of description also correspond, it follows that we have mapped the upper half of the Z-plane on the interior of the semicircle.

Eliminating Z and W, we thus get

$$w = -a^2\left[\cos\alpha - \tfrac{1}{2}\left(\zeta + \frac{1}{\zeta}\right)\right]^2,$$

which maps the w-plane on the interior of the semicircle. Moreover, the stagnation point O corresponds to $\zeta = e^{i\alpha}$.

12·42. The streamlines. We have

$$2i\psi = w - \bar{w} = a^2\left[\cos\alpha - \tfrac{1}{2}\left(\bar{\zeta} + \frac{1}{\bar{\zeta}}\right)\right]^2 - a^2\left[\cos\alpha - \tfrac{1}{2}\left(\zeta + \frac{1}{\zeta}\right)\right]^2$$

$$= \tfrac{1}{4}a^2\left(4\cos\alpha - \zeta - \bar{\zeta} - \frac{1}{\zeta} - \frac{1}{\bar{\zeta}}\right)\left(\zeta - \bar{\zeta} + \frac{1}{\zeta} - \frac{1}{\bar{\zeta}}\right).$$

Since $$\zeta = \xi + i\eta, \quad \frac{1}{\zeta} = \frac{\xi - i\eta}{\xi^2 + \eta^2}.$$

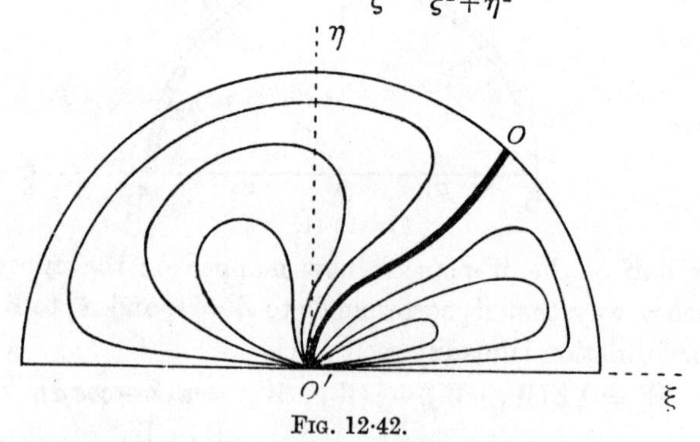

FIG. 12·42.

Therefore the equation of the streamline $\psi = 0$ is

$$\{2\cos\alpha(\xi^2 + \eta^2) - \xi(\xi^2 + \eta^2 + 1)\}\eta(\xi^2 + \eta^2 - 1) = 0.$$

Since $\eta = 0$ corresponds to the diameter of the semicircle and $\xi^2+\eta^2-1$ to the curved part, this line is the dividing streamline, the equation of the remaining part being the cubic curve

$$\xi(\xi^2+\eta^2+1)-2\cos\alpha\,(\xi^2+\eta^2) = 0,$$

which passes through the stagnation point O $(\cos\alpha, \sin\alpha)$ and touches the axis of η at the origin. We are only concerned with the part of this curve within the semicircle. This has been drawn in fig. 12·42, and the form of the streamlines indicated by the method of 6·23.

12·43. The function $\omega(\zeta)$. The function $\omega(\zeta)$ is defined by the equation

$$e^{-i\omega(\zeta)} = -\frac{1}{U}\frac{dw}{dz} = \frac{\mathfrak{v}}{U} = \frac{q\,e^{-i\theta}}{U}.$$

Thus
$$\omega(\zeta) = \theta+i\log\frac{q}{U} = \theta+i\tau,$$

and the real part of $\omega(\zeta)$ is therefore the angle which the fluid velocity makes with the axis of x in the z-plane, and the imaginary part τ is a measure of the speed, in fact

$$q = U\,e^{\tau}.$$

On the free streamlines, $q = U$ and therefore $\tau = 0$.

Thus $\omega(\zeta)$ is real on μ_1, μ_2, that is, on the real axis in the ζ-plane. Also, at infinity in the z-plane, $\theta = 0$, and therefore

$$\omega(0) = 0.$$

The function $\omega(\zeta)$ is necessarily holomorphic at all points within the semicircle, for these correspond to the region R of the z-plane where the motion is continuous. Moreover, we have seen that $\omega(\zeta)$ is real on the real ζ-axis, and therefore the function $\omega(\zeta)$ can be continued (see 5·53) over the remaining half of the unit circle by attributing to it the value $\omega(\bar{\zeta})$ at the point $\bar{\zeta}$, and hence giving θ the same value but changing the sign of τ. The function $\omega(\zeta)$ determines all the circumstances of the motion as we shall now prove.

12·44. The wetted walls. The wetted walls, fig. 12·41, or parts λ_1, λ_2 of the obstacle in contact with the fluid, are represented by the arcs A_1O, A_2O in the ζ-plane. Now from the definition of $\omega(\zeta)$, or briefly ω, and the expression of w in terms of ζ (12·41), we have

$$(1) \qquad U\,dz = -e^{i\omega}\,dw = \tfrac{1}{2}a^2e^{i\omega}\left(\zeta+\frac{1}{\zeta}-2\cos\alpha\right)\left(\zeta-\frac{1}{\zeta}\right)\frac{d\zeta}{\zeta}.$$

Now on the arc A_1O, we have

$$\zeta = e^{i\chi}.$$

Substituting and integrating from $\chi = \alpha$ to any value of χ on OA_1, and remembering that O corresponds to $z = 0$, we get

$$(2) \qquad Uz = -2a^2 \int_\alpha^\chi e^{i\omega}(\cos\chi - \cos\alpha)\sin\chi\, d\chi.$$

Equating the real and imaginary parts,

$$x = -\frac{2a^2}{U}\int_\alpha^\chi e^{-\tau}\cos\theta(\cos\chi - \cos\alpha)\sin\chi\, d\chi,$$

$$y = -\frac{2a^2}{U}\int_\alpha^\chi e^{-\tau}\sin\theta\,(\cos\chi - \cos\alpha)\sin\chi\, d\chi,$$

and these are the parametric equations of the wetted wall λ_2 if we take χ between 0 and α, and of λ_1 if we take χ between α and π.

In particular, if in (2) we take $\chi = 0$, we get the value z_2 corresponding to the point A_2.

If $d\lambda$ denotes an element of arc of λ_1 or λ_2, we have

$$U\, d\lambda = U\,|\,dz\,| = 2a^2 e^{-\tau}\,|\cos\chi - \cos\alpha\,|\sin\chi\, d\chi,$$

and therefore

$$(3) \qquad \lambda = \frac{2a^2}{U}\int_{0,\pi}^{\alpha} e^{-\tau}\,(\cos\chi - \cos\alpha)\sin\chi\, d\chi,$$

the lower limit 0 corresponding to λ_2 and the lower limit π to λ_1, since $d\chi$ is negative on going from π to α, and so is $\cos\chi - \cos\alpha$.

The radius of curvature of the wetted walls is given by

$$(4) \qquad r = \frac{d\lambda}{d\theta} = \frac{2a^2 e^{-\tau}}{U}\left|\cos\chi - \cos\alpha\right|\sin\chi\frac{d\chi}{d\theta}.$$

12·45. The free streamlines. The parametric equations of the free streamline μ_2 are got by integrating (1) of the last section from $\zeta = 1$ to $\zeta = \xi$, where ξ is a point on $O'A_2$, remembering that, at A_2, z takes the value z_2 just found. Thus

$$U(z - z_2) = \frac{a^2}{2}\int_1^\xi e^{i\omega}\left(\zeta + \frac{1}{\zeta} - 2\cos\alpha\right)\left(\zeta - \frac{1}{\zeta}\right)\frac{d\zeta}{\zeta},$$

where ζ is now a real variable. Equating the real and imaginary parts, the required result is obtained.

The velocity is given by $\mathbf{v} = U\, e^{-i\omega}$.

The pressure. Using the pressure equation, Π being the pressure in the cavity, we get

$$\frac{p}{\rho} + \tfrac{1}{2}q^2 = \frac{\Pi}{\rho} + \tfrac{1}{2}U^2.$$

Thus
$$p - \Pi = \tfrac{1}{2}\rho(U^2 - q^2) = \tfrac{1}{2}\rho U^2(1 - e^{2\tau}).$$

This is the hydrodynamic pressure.

12·46. Drag, lift, and moment. Let the resultant force due to the fluid have components X, Y along the axes at the point O in the z-plane. Then, as in the theorem of Blasius,

$$X+iY = i\int_{(A_1OA_2)} (p-\Pi)\,dz \doteq \frac{i}{2}\rho U^2 \int_{(A_1OA_2)} (1-e^{2\tau})\,dz.$$

Since $U\,dz = -e^{i\omega(\zeta)}\,dw$, we get

$$X+iY = \frac{i}{2}\rho U \int_{(A_2OA_1)} e^{i\omega(\zeta)}\frac{dw}{d\zeta}\,d\zeta - \frac{i}{2}\rho U \int_{(A_2OA_1)} e^{i\omega(\zeta)+2\tau}\frac{dw}{d\zeta}\,d\zeta,$$

the integrals now being taken round the arc of the semicircle in the ζ-plane. Now consider the analytic continuation of $\omega(\zeta)$, 5·53, in the whole circle Γ, fig. 12·46.

We have $\quad i\omega(\bar{\zeta}) = i(\theta - i\tau) = i(\theta+i\tau)+2\tau = i\omega(\zeta)+2\tau.$

Also w is real when ζ describes A_1OA_2 (see fig. 12·41) and therefore $dw = d\bar{w}$. Moreover, when ζ describes the arc A_1OA_2 of Γ in the counter-clockwise sense,

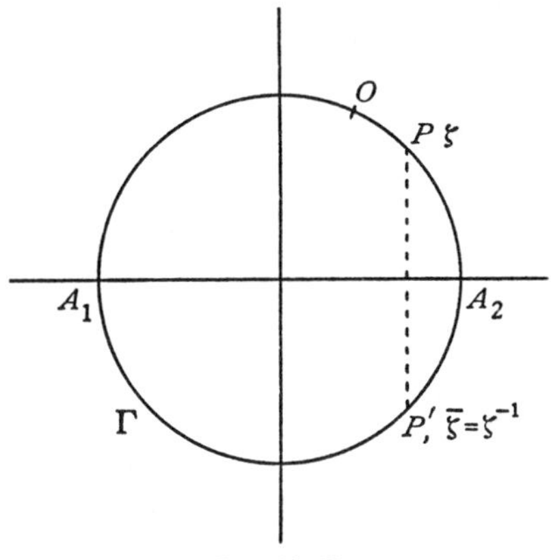

FIG. 12·46.

ζ describes the arc $A_1P'A_2$ in the clockwise sense. Thus we get

$$X+iY = \frac{i}{2}\rho U \int_{(A_2OA_1)} e^{i\omega(\zeta)}\frac{dw}{d\zeta}\,d\zeta - \frac{i}{2}\rho U \int_{(A_2P'A_1)} e^{i\omega(\bar{\zeta})}\frac{d\bar{w}}{d\bar{\zeta}}\,d\bar{\zeta}$$

$$= \frac{i}{2}\rho U \int_{(\Gamma)} e^{i\omega(\zeta)}\frac{dw}{d\zeta}\,d\zeta.$$

It therefore only remains to calculate the residue of the integrand at the only pole inside Γ, namely $\zeta = 0$.

Expanding by Maclaurin's theorem, and remembering that $\omega(0) = 0$ (12·43), we get

$$e^{i\omega(\zeta)} = 1 + i\zeta\omega'(0) + \tfrac{1}{2}\zeta^2[i\omega''(0) - (\omega'(0))^2] + \ \ldots$$

Also $$\frac{dw}{d\zeta} = \frac{a^2}{2}\left(-\zeta - \frac{2}{\zeta^2}\cos\alpha + \frac{1}{\zeta^3} + 2\cos\alpha\right).$$

Multiplying these results and picking out the coefficient of $1/\zeta$, we get the residue

$$\tfrac{1}{2}a^2\{\tfrac{1}{2}i\omega''(0)-2i\cos\alpha\,\omega'(0)-\tfrac{1}{2}[\omega'(0)]^2\}.$$

Thus, using the residue theorem, we have

$$X+iY = \tfrac{1}{4}\pi\rho U a^2(\omega'(0))^2+\tfrac{1}{4}\pi\rho U a^2 i[4\omega'(0)\cos\alpha-\omega''(0)],$$

whence $X = \tfrac{1}{4}\pi\rho U a^2[\omega'(0)]^2$, $Y = \tfrac{1}{4}\pi\rho U a^2[4\omega'(0)\cos\alpha-\omega''(0)]$,

and X is the drag and Y the lift. These are Levi-Civita's elegant results.

The moment M of the forces about the stagnation point O is found by a similar calculation to be the real part of the integral

$$M+iN = \tfrac{1}{2}\rho U\int_{(A_2OA_1)} z\left\{e^{-i\omega(\zeta)}-e^{-i\omega\left(\frac{1}{\zeta}\right)}\right\}\frac{dw}{d\zeta}\,d\zeta,$$

taken round the semicircle A_2OA_1 in the ζ-plane, and must be calculated in each particular case. The knowledge of X, Y, M allows us by the ordinary principles of Statics to find the single force which is equivalent to the action of the fluid on the obstacle. This single force will always exist unless

$$X = Y = 0.$$

12·47. Discontinuity of $\omega(\zeta)$. The function $\omega(\zeta) = \theta+i\tau$ presents a discontinuity at the stagnation point O, for its real part θ has two values there corresponding to the two directions of flow along the tangent (or tangents) at O, moreover $\tau\to-\infty$ at O, since the speed vanishes there.

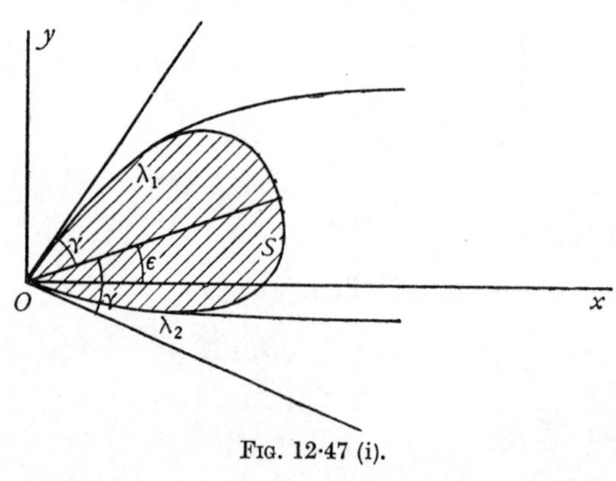

FIG. 12·47 (i).

Let the tangents to the contour of the obstacle at O make an angle 2γ with each other, fig. 12·47 (i), and let the internal bisector of the angle make an angle ϵ with the x-axis.

If O is an ordinary point of the contour $2\gamma = \pi$, and the tangents form parts of one line. If the contour is symmetrical about the x-axis, $\epsilon = 0$.

When we approach O moving along λ_1, $\theta\to\gamma+\epsilon$, but when we approach O moving along λ_2, $\theta\to-\gamma+\epsilon$.

Also, when $\zeta\to e^{i\alpha}$, $\omega(\zeta)$ becomes infinite, and the same must happen when $\zeta\to e^{-i\alpha}$.

A simple function satisfying these conditions is

$$\omega_0(\zeta) = \epsilon-\gamma+\frac{2i\gamma}{\pi}\log\frac{\zeta-e^{i\alpha}}{1-\zeta\,e^{i\alpha}}.$$

To prove this, consider the behaviour of the function

$$f(\zeta) = \frac{\zeta - e^{i\alpha}}{\zeta - e^{-i\alpha}}$$

when ζ moves inside or upon the semicircle in the ζ-plane. If we determine the logarithm so that $\log f(0) = 2i(\alpha - \pi)$, the function $\log f(\zeta)$ is one-valued and holomorphic at all points within the semicircle.

Now consider ζ to be taken at P on the arc λ_1, fig. 12·47 (ii). We get

$$\arg f(\zeta) = \arg(\zeta - e^{i\alpha}) - \arg(\zeta - e^{-i\alpha})$$
$$= -\pi + \nu_1 - \nu_2 = -\pi - (\pi - \alpha) = -2\pi + \alpha.$$

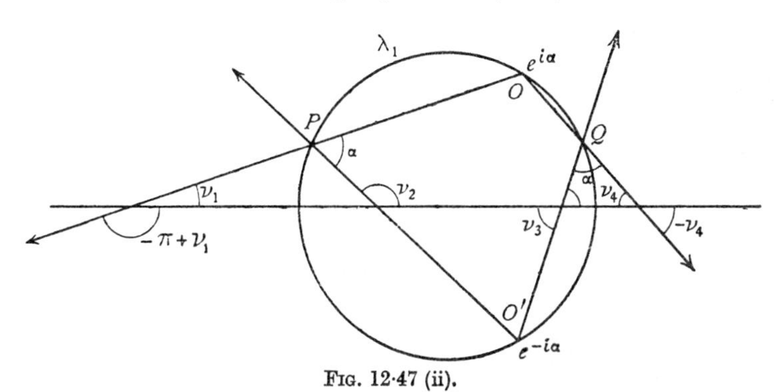

FIG. 12·47 (ii).

If ζ is taken at Q on the arc λ_2,

$$\arg f(\zeta) = -\nu_4 - \nu_3 = -\pi + \alpha.$$

Thus $\arg f(\zeta)$ has the constant value $\alpha - 2\pi$ on λ_1 and the constant value $\alpha - \pi$ on λ_2.

When ζ passes through O from λ_2 to λ_1, $\arg f(\zeta)$ decreases by π.

Now
$$\omega_0(\zeta) = \epsilon - \gamma + \frac{2i\gamma}{\pi} \log(-e^{-i\alpha}) \frac{\zeta - e^{i\alpha}}{\zeta - e^{-i\alpha}}$$
$$= \epsilon - \gamma + \frac{2i\gamma}{\pi} \{\log(-e^{-i\alpha}) + \log f(\zeta)\}.$$

But $\log(-e^{-i\alpha}) = i(\pi - \alpha)$. Therefore, when ζ is on λ_2,

$$\theta = \epsilon - \gamma + \frac{2i\gamma}{\pi} \{i(\pi - \alpha) + i(\alpha - \pi)\} = \epsilon - \gamma,$$

and when ζ is on λ_1,

$$\theta = \epsilon - \gamma + \frac{2i\gamma}{\pi} \{i(\pi - \alpha) + i(\alpha - 2\pi)\} = \epsilon + \gamma.$$

It is also evident that $\omega_0(\zeta) \to \infty$ when $\zeta \to e^{i\alpha}$ or $e^{-i\alpha}$. Thus $\omega_0(\zeta)$ has all the properties stated. Moreover, $\omega_0(\zeta)$ is real when ζ is real and can therefore be continued into the lower semicircle by the principle of reflexion. We also note that

$$\omega_0(0) = \epsilon - \frac{2\gamma}{\pi}\left(\alpha - \frac{\pi}{2}\right).$$

We have thus isolated the singular part of the function $\omega(\zeta)$ at the point O and its image O'. If we put

$$\omega(\zeta) = \omega_0(\zeta) + \Omega(\zeta),$$

where

$$\Omega(\zeta) = a_0 + a_1\zeta + a_2\zeta^2 + \dots,$$

we have the general solution for obstacles which present only one discontinuity of $\omega(\zeta)$ on the semicircle in the ζ-plane. By attributing various forms to $\Omega(\zeta)$ we can then obtain solutions for the resulting contours. The converse problem of determining $\omega(\zeta)$ for a given contour is of course more difficult and only a few cases have been completely elucidated.

12·50. Solution when $\Omega(\zeta) = 0$. In this case we have, from 12·47,

$$(1) \qquad \omega(\zeta) = \omega_0(\zeta) = \epsilon - \gamma + \frac{2i\gamma}{\pi}\log\frac{\zeta - e^{i\alpha}}{1 - \zeta e^{i\alpha}}.$$

Since $\omega(0) = 0$ from 12·43, we must have

$$(2) \qquad \epsilon = \frac{2\gamma}{\pi}\left(\alpha - \frac{\pi}{2}\right), \quad \alpha = \frac{\pi}{2}\left(1 + \frac{\epsilon}{\gamma}\right),$$

which determines α.

Also θ is constant along λ_1 and λ_2. Hence $\omega(\zeta)$ is the function corresponding to the flow past a lamina bent at an angle 2γ, fig. 12·50.

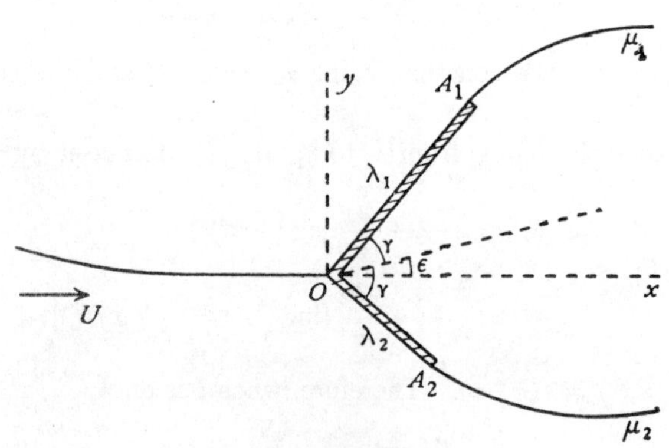

FIG. 12·50.

Since

$$e^{i\chi} - e^{i\alpha} = e^{\frac{1}{2}i(\chi+\alpha)}\left(e^{\frac{1}{2}i(\chi-\alpha)} - e^{-\frac{1}{2}i(\chi-\alpha)}\right)$$
$$= 2i\, e^{\frac{1}{2}i(\chi+\alpha)} \sin\tfrac{1}{2}(\chi - \alpha),$$

we get, from (1), the imaginary part of $\omega(\zeta)$, namely

$$\tau = \frac{\gamma}{\pi}\log\left(\frac{\sin^2\frac{1}{2}(\chi-\alpha)}{\sin^2\frac{1}{2}(\chi+\alpha)}\right).$$

Substitution of this in 12·44 (3) will give the lengths of OA_1, OA_2 and it is then obvious that the ratio of these lengths is not arbitrary. This means that the stagnation point will be at the bend, only if the lamina is correctly orientated to the stream. If not, an abrupt change of direction of the velocity will then occur at the bend; physically acceptable solutions will be found only by modifying the cavity so as to include parts of one arm or both.

It may be remarked that slight changes in γ may produce violent oscillations in the position of the stagnation point and consequent fluctuations in the moment.

To calculate the drag and lift, we have

$$\omega'(\zeta) = \frac{2i\gamma}{\pi}\left\{\frac{1}{\zeta - e^{i\alpha}} + \frac{e^{i\alpha}}{1-\zeta e^{i\alpha}}\right\},$$

$$\omega''(\zeta) = -\frac{2i\gamma}{\pi}\left\{\frac{1}{(\zeta - e^{i\alpha})^2} - \frac{e^{2i\alpha}}{(1-\zeta e^{i\alpha})^2}\right\}.$$

Thus
$$\omega'(0) = -\frac{4\gamma}{\pi}\sin\alpha, \quad \omega''(0) = -\frac{4\gamma}{\pi}\sin 2\alpha,$$

where α is given by (2).

Using Levi-Civita's formulae (12·46), we get for the drag and lift

$$X = \frac{4a^2\rho U\gamma^2}{\pi}\cos^2\frac{\pi\epsilon}{2\gamma}, \quad Y = a^2\rho U\gamma\sin\frac{\pi\epsilon}{\gamma}.$$

12·51. Stream impinging on a plate. If we put $2\gamma = \pi$, the bend in the lamina disappears and we have then the case of a stream impinging on a fixed lamina, see fig. 12·26 (ii). From 12·50 (2), we then get $\alpha = \epsilon + \frac{1}{2}\pi$, so that α is the angle at which the lamina is inclined to the asymptotic direction of the stream. The formulae for the drag and lift, 12·50, then give the results already obtained in 12·26.* When $\alpha = \frac{1}{2}\pi$, we obtain the result of 12·21.

12·52. The symmetrical case. If in the problem of 12·50 we take $\epsilon = 0$, we get the symmetrical case in which the stream impinges directly on a lamina bent in the middle, fig. 12·52, which represents an approximation to a ship with a sharp bow. In this case $\alpha = \frac{1}{2}\pi$, and the drag, 12·50, becomes $4a^2\rho U\gamma^2/\pi$, while the lift vanishes. This expression contains the constant a, whose value can be expressed in terms of $\frac{1}{2}l$, the length of OA_1, as follows. Since $\epsilon = 0$,

$$\omega(\zeta) = \theta + i\tau = \frac{2i\gamma}{\pi}\log\frac{1+i\zeta}{1-i\zeta}.$$

Putting
$$t = \tan^2(\tfrac{1}{4}\pi - \tfrac{1}{2}\chi), \quad \zeta = e^{i\chi},$$

we get
$$\tau = \frac{\gamma}{\pi}\log t,$$

* The verification of this statement involves the calculation of a^2 in terms of U and l, the length of the plate. The details of the calculation are explained in 12·52.

and therefore, from 12·44 (3),

$$\frac{l}{2} = \frac{2a^2}{U}\int_0^{\frac{1}{2}\pi} t^{-\frac{\gamma}{\pi}} \cos\chi \sin\chi \, d\chi = \frac{4a^2}{U}\int_0^1 t^{-\frac{\gamma}{\pi}} \left\{ -\frac{1}{(1+t)^2} + \frac{2}{(1+t)^3} \right\} dt.$$

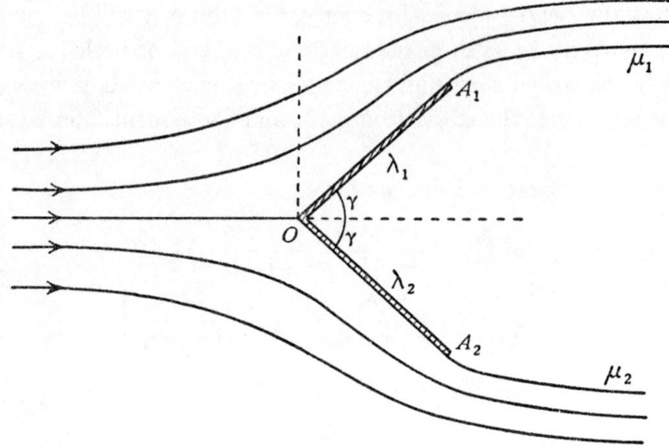

FIG. 12·52.

Denoting the value of the definite integral by f, we get

$$8a^2 = Ul/f,$$

and hence the drag is $\frac{1}{2}\rho U^2 l\gamma^2/(\pi f)$.

To evaluate f, let

$$f_n(x) = \int_0^1 \frac{t^{x-1}}{(1+t)^n} dt.$$

Then $$f_1(x) = \tfrac{1}{2}\Psi\left(\frac{x+1}{2}\right) - \tfrac{1}{2}\Psi\left(\frac{x}{2}\right),$$

where $\Psi(x)$ is the logarithmic derivative of the Gamma function.*

Then $$f = 2f_3\left(1-\frac{\gamma}{\pi}\right) - f_2\left(1-\frac{\gamma}{\pi}\right).$$

Now it is easy to establish the reduction formula

$$f_{n+1}(x) - \frac{n-x}{n} f_n(x) = \frac{1}{n2^n},$$

from which we get $$f = \tfrac{1}{4} + \frac{\gamma}{2\pi} + \frac{\gamma^2}{\pi^2} f_1\left(1-\frac{\gamma}{\pi}\right),$$

and f can then be evaluated from tables of the Ψ function.†

* Milne-Thomson, *Calculus of Finite Differences*, London (1965), 9·3.
† *British Association Tables*, Vol. I, London (1931).

EXAMPLES XII

1. Obtain the free streamline when a wide stream flowing with velocity U parallel to a straight shore impinges on a pier which projects to a distance h perpendicularly from the shore, and find the thrust on the pier due to the fluid motion.

2. A stream of finite width c whose velocity at infinity has resolutes

$$(-V \cos \alpha, \quad -V \sin \alpha)$$

impinges on the rigid plane barrier $y = 0$. Show that the two-dimensional irrotational motion of the stream under no body forces is given by equations of the form

$$\frac{dw}{dt} = \frac{A(t-\lambda)}{(t^2-1)}, \quad \frac{d\Omega}{dt} = \frac{B}{(t-\lambda)(t^2-1)^{\frac{1}{2}}}, \quad \Omega = \log\left(-V\frac{dz}{dw}\right),$$

where $-1<\lambda<+1$.

Show that the stream divides into two branches of ultimate widths $c \cos^2 \frac{1}{2}\alpha$, $c \sin^2 \frac{1}{2}\alpha$, and that the thrusts on the portions of the barrier on either side of the point of zero velocity on the barrier are in the ratio $\pi - \alpha : \alpha$, assuming the pressure on the side of the barrier opposite to the stream equal to that along the free streamlines. (U.L.)

3. When a stream impinges on a lamina to which it is inclined at the angle α, prove that the stagnation point divides the lamina in the ratio

$$\frac{2+2\cos\alpha+(\pi-\alpha)\sin\alpha+2\cos\alpha\sin^2\alpha}{2-2\cos\alpha+\alpha\sin\alpha-2\cos\alpha\sin^2\alpha}.$$

Hence show that the stagnation point is always between the centre of the lamina and the end farthest upstream.

4. Show that the transformations

$$w = -\zeta^2, \quad \frac{dQ}{d\zeta} = \frac{-i(\kappa\kappa')^{\frac{1}{4}}}{\zeta(\zeta-\sqrt{\kappa'})^{\frac{1}{2}}(\zeta+\sqrt{\kappa})^{\frac{1}{2}}}$$

lead to the solution of the problem of a plate placed obliquely in a stream with the liquid dividing along two free streamlines on either side.

If $2b$ is the breadth of the plate and it is inclined at an angle α to the stream, show that

$$\kappa = \frac{2bV}{1+\frac{\pi}{4}\sin\alpha}\cos^4\frac{\alpha}{2}, \quad \kappa' = \frac{2bV}{1+\frac{\pi}{4}\sin\alpha}\sin^4\frac{\alpha}{2},$$

where V is the stream velocity.

5. A fluid flows in two dimensions from $y = +\infty$ between the two planes $x = \pm a$, $y>b$ and impinges symmetrically on the fixed plane $y = 0$. Explain how the forms of the streamlines can be found, and show that if d is the ultimate width of the stream in contact with $y = 0$, then

$$b-d = \frac{(a^2+d^2)}{\pi a}\log\left(\frac{a+d}{a-d}\right).$$

6. A finite stream impinges on an infinite straight barrier, the motion being in two dimensions and the boundaries of the stream curves of constant speed. If the undisturbed stream makes angle $\frac{1}{2}\pi - \alpha$ with the barrier, show that the perpendicular drawn from the point on the barrier where the stream divides to the asymptote of the streamline through that point is to the breadth of the undisturbed stream as

$$\frac{\pi}{2}(1+\sin\alpha)+\alpha\cos^2\alpha+\sin\alpha\cos\alpha\log(2\cos\alpha)+2\cos\alpha\tanh^{-1}\left(\tan\frac{\alpha}{2}\right):\pi.$$

Show that the resultant thrusts on the two parts of the barrier represented by this point are in the ratio $\pi+2\alpha:\pi-2\alpha$. (U.L.)

7. A stream of incompressible fluid, whose velocity at infinity is U, impinges symmetrically upon a bent lamina whose section consists of two straight lines, each of length a, at right angles. The fluid flows over the convex side and is bounded internally on the downstream side by two free streamlines. Show that the resultant thrust on the lamina is $\sqrt{2}\,\pi\rho a U^2/\{6\sqrt{2}+\pi+2\log_e(\sqrt{2}-1)\}$, and that the intrinsic equation of either of the free streamlines may be written $s = A\cot^2 2\theta$, where A is a constant, s is measured from the edge of the lamina and θ is the inclination of the tangent to the axis of symmetry.

8. A bent plane perpendicular to the xy-plane, whose section by that plane consists of two straight lines AB, BC at right angles, is placed in incompressible liquid flowing at infinity with unit velocity in the negative direction of Ox, so that the central streamline is straight along xO, strikes the plane at B on the concave side and bisects the angle ABC. With the usual notation, verify that all the conditions can be satisfied by putting

$$\Omega = \log\frac{\sqrt{(t-a)(c-b)}+\sqrt{(c-t)(b-a)}}{\sqrt{(t-b)(c-a)}}, \quad w = (t-b)^2.$$

Choosing the scale of measurement so that $c-a = 1$, and putting $c-t = \cos^2 U$, prove that

$$L = 2\int_B^{\pi/2}\sin^{\frac{3}{2}}(U+B)\sin^{\frac{1}{2}}(U-B)\sin 2U\,dU,$$

$$P = 2\int_B^{\pi/2}\sin^{\frac{3}{2}}(U-B)\sin^{\frac{1}{2}}(U+B)\sin 2U\,dU,$$

where B is the value of U corresponding to $t = b$, whilst L is the length of either plane and P the resultant thrust on it. (U.L.)

9. In fig. 12·50, prove that the lengths of the wetted walls are given by

$$\lambda_1 = k\int_\pi^\alpha\left(\frac{\sin\dfrac{\chi-\alpha}{2}}{\sin\dfrac{\chi+\alpha}{2}}\right)^{-\frac{2\gamma}{\pi}}(\cos\chi-\cos\alpha)\sin\chi\,d\chi$$

and a second integral for λ_2 obtained by taking the lower limit to be zero. Hence prove that when λ_1 is given there is only one possible value of λ_2 to make the motion conform to the type shown in the diagram.

10, For flow into a channel with an aperture of width W in one wall, fig. 14·8 (*b*), prove that

$$\frac{W}{2h} = \frac{a}{\pi}\left(\pi+\frac{a^2+1}{a}\log\frac{1+a}{1-a}\right)$$

and that the coefficient of contraction is $\sigma = 2ah/W$ where $a = V/U$. (Conway)

CHAPTER XIII

RECTILINEAR VORTICES

13·0. In this chapter some aspects of two-dimensional vortex motion will be considered. The vorticity vector is necessarily perpendicular to the plane of the motion and we shall as usual consider unit thickness of liquid, that is to say, we suppose the liquid to be confined between two planes at unit distance apart and parallel to the plane of the motion. The vortex lines being straight and parallel, all vortex tubes are cylindrical, with generators perpendicular to the plane of the motion. Such vortices are known as *rectilinear vortices.* It is, as before, convenient to use the language of plane geometry.

13·10. Circular vortex. Let there be a single cylindrical vortex tube, whose cross-section is a circle of radius a, surrounded by unbounded fluid.

The section of the vortex by the plane of the motion is a circle and the arrangement may therefore be referred to as a *circular vortex.*

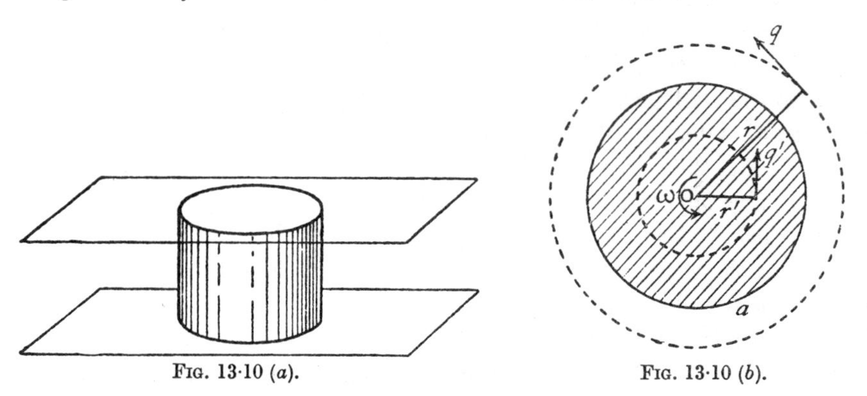

FIG. 13·10 (*a*). FIG. 13·10 (*b*).

We shall suppose that the vorticity over the area of this circle has the constant value ω. Outside the circle the vorticity is zero. Draw circles, concentric with the circle which bounds the vortex, of radii r' and r, where $r' < a < r$. Let q' and q be the speeds of fluid motion on the circles of radii r' and r respectively. It is clear from the symmetry that the speed at every point of the circle radius r' is the same, and that the velocity is tangential to this circle, for a radial component would entail a net flux across the circle and its centre O would then be a source or a sink. Similarly the velocity at any point of the circle of radius r is tangential to that circle.

Apply Stokes's circulation theorem (2·50) to these circles. Then

$$\int q'\,ds = \omega\,\pi r'^2, \quad r'<a\,; \qquad \int q\,ds = \omega\,\pi a^2, \quad r>a.$$

Since q' and q are constants on their respective circles we get

$$2\pi r'\,q' = \omega\,\pi r'^2, \quad 2\pi r\,q = \omega\,\pi a^2.$$

Thus $q' = \frac{1}{2}\omega r'$, $r'<a$; $q = \frac{1}{2}\omega a^2/r$, $r>a$. When $r' = r = a$ we have $q' = q = \frac{1}{2}a\omega$ so that the velocity is continuous as we pass through the circle.

From this it appears that the existence of a vortex such as we have described implies the co-existence of a certain distribution or *field of velocity*. This velocity field which co-exists with the vortex is known as the *induced velocity field* and the velocity at any point of it is called the *induced velocity*.

It is customary to refer to the velocity at a point of the field as the *velocity induced by the vortex*, but this must be understood merely as a convenient abbreviation of the fuller statement that were the vortex alone in the otherwise undisturbed fluid, the velocity at the point would have the value in question. In this sense, when several vortices are present, the field of each will contribute its proper amount to the velocity at a point.

Returning to the circular vortex, the induced velocity at the extremity of any radius vector r joining the centre of the vortex to a point of the fluid external to the vortex is of magnitude inversely proportional to r and is perpendicular to r. Thus the induced velocity tends to zero at great distances.

As to the fluid within the vortex, its velocity is of magnitude proportional to r and therefore the fluid composing the vortex moves like a rigid body rotating about the centre O with angular velocity $\frac{1}{2}\omega$. The velocity at the centre is zero. This important fact may be stated in the following way.

A circular vortex induces no velocity at its centre. This is to be understood to mean that the centre of a circular vortex alone in the otherwise undisturbed fluid will not tend to move.

Still considering the fluid within the vortex, the velocities at the extremities of oppositely directed radii are of the same magnitude but of opposite sense so that the mean velocity of the fluid within the vortex is zero. Thus, if a circular vortex of small radius be " placed " in a field of flow at a point where the velocity is $\mathbf{u}$ the mean velocity at its centre will still be $\mathbf{u}$ and the fluid composing the vortex will move with velocity $\mathbf{u}$; it will " swim with the stream " carrying its vorticity with it.

The circular vortex is illustrated in nature on the grand scale by the tropical cyclone (hurricane, typhoon) which attains a diameter * of from 100 to 500 miles, and travels at a speed seldom exceeding 15 miles per hour. Within

* D. Brunt, *Weather Study*, London (1942).

the area the wind can reach hurricane force, while there is a central region, " the eye of the storm ", of diameter 10 to 20 miles where conditions may be comparatively calm.

We also get from the above results

$$\frac{q}{\frac{1}{2}\omega a} = \frac{r}{a} \text{ when } r<a, \quad \frac{q}{\frac{1}{2}\omega a} = \frac{a}{r} \text{ when } r>a,$$

so that the speed q tends to zero at infinity and is greatest at the boundary.

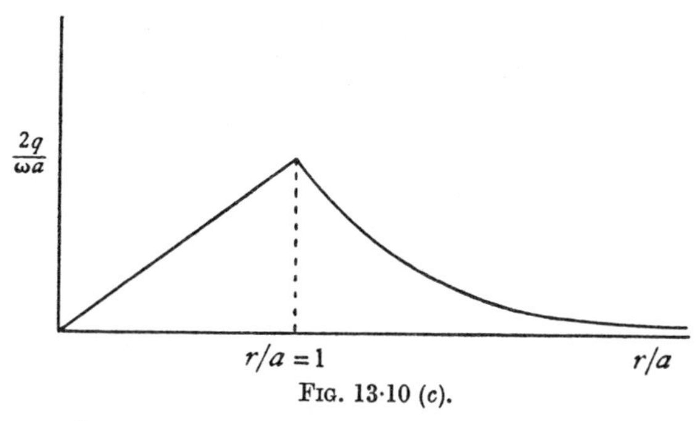

Fig. 13·10 (c).

Fig. 13·10 (c) shows the graphical relation between the above quantities, the curved part being a portion of a rectangular hyperbola.

Outside the vortex the motion is irrotational, and the velocity is $q\,e^{i(\theta+\frac{1}{2}\pi)}$ so that

$$-\frac{dw}{dz} = \frac{\frac{1}{2}a^2\omega\, e^{-i(\theta+\frac{1}{2}\pi)}}{r} = -\frac{1}{2}\frac{ia^2\omega}{z},$$

and therefore the complex potential is

$$w = \tfrac{1}{2}i\omega a^2 \log z.$$

It follows that there is a circulation of strength κ given by $\kappa = \frac{1}{2}\omega a^2$. We may therefore call κ the *strength of the vortex*,* the actual circulation being $2\pi\kappa$.

Thus for the liquid outside a circular vortex whose strength is κ and whose centre is at the point z_0, the complex potential is

$$w = i\kappa \log (z-z_0).$$

13·11. Pressure due to a circular vortex. Let p_1, p denote the pressure inside and outside the vortex. The pressure must be continuous at the boundary, and therefore

(1) $$p_1 = p \text{ when } r = a.$$

Inside the vortex, the equation of motion is

$$\frac{1}{\rho}\frac{dp_1}{dr} = \frac{r\omega^2}{4} = \frac{\kappa^2 r}{a^4},$$

* This notation avoids the recurrence of a redundant factor 2π and is analogous to the definition of the strength of a source in 8·10.

for the liquid is rotating with constant angular velocity $\omega/2$, and hence the acceleration is $r\omega^2/4$ towards O.

Integrating, we get

$$p_1 = \frac{\kappa^2 r^2 \rho}{2a^4} + p_0,$$

where p_0 is the pressure at the centre.

Outside the vortex, we may use the pressure equation in the form

$$\frac{p}{\rho} + \frac{\kappa^2}{2r^2} = \text{constant} = \frac{\Pi}{\rho},$$

where Π is the pressure at infinity. Hence, from (1),

$$(2) \qquad p_0 + \frac{\kappa^2 \rho}{2a^2} = \Pi - \frac{\kappa^2 \rho}{2a^2}, \quad \text{or} \quad p_0 = \Pi - \frac{\kappa^2 \rho}{a^2}. \quad \text{Thus}$$

$$p_1 = \Pi - \frac{\kappa^2 \rho}{a^2}\left(1 - \frac{r^2}{2a^2}\right), \quad p = \Pi - \frac{\kappa^2 \rho}{2r^2}.$$

The relation between the pressure and the radius is shown in fig. 13·11, where y = pressure/Π, $x = r/a$, and $k = \kappa^2\rho/(a^2\Pi)$.

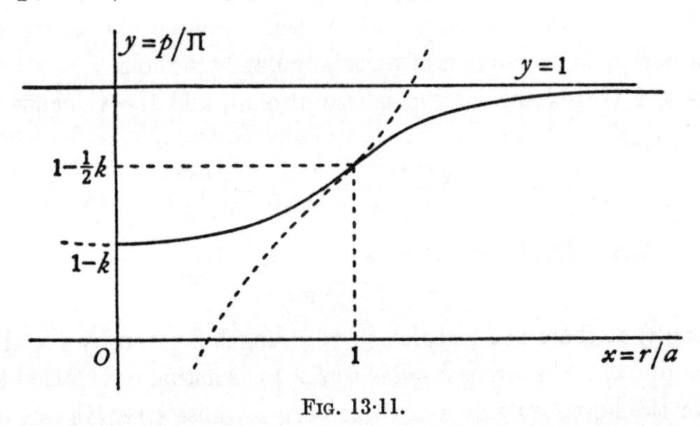

FIG. 13·11.

The curves are the parabola $y-(1-k) = \frac{1}{2}kx^2$ and $(y-1)x^2 = -\frac{1}{2}k$, the latter being asymptotic to $y = 1$. The curves touch at $x = 1$, corresponding to the boundary. The dotted portions are drawn to show how the curves lie. It appears that the pressure increases continuously from the value $\Pi(1-k)$ and tends to Π at a great distance.

13·12. Hollow circular vortex. We have just seen that the pressure is least at the centre of the vortex and has the value $\Pi(1-k)$. If $k>1$ it follows that the pressure would be negative. To avoid this, a concentric hollow containing no liquid will exist within the vortex. The pressure diagram of fig. 13·11 will now have to be modified by moving the origin to the appropriate point between $y = 1-k$ and $y = 1-\frac{1}{2}k$. As an extreme case we may suppose $k = 2$, i.e. $\kappa^2\rho = 2a^2\Pi$. We have then a completely hollow cylindrical space around which there is cyclic irrotational motion.

It also appears that when the circulation $2\pi\kappa$ and the pressure Π at infinity are given, a circular vortex, whose interior is wholly occupied by fluid, has a *minimum radius* given by $a^2 = \kappa^2\rho/\Pi$. (Cf. 13·8.)

13·13. Rankine's combined vortex. This consists of a circular cylindrical vortex with its axis vertical in a liquid which moves under the action of gravity, the upper surface being exposed to atmospheric pressure Π. This is a three-dimensional problem, but may be conveniently treated here.

Take the origin in the axis of the vortex and in the level of the liquid at infinity, fig. 13·13, and measure z vertically downwards. We see at once that the kinematical conditions at the boundary are satisfied by taking the velocity

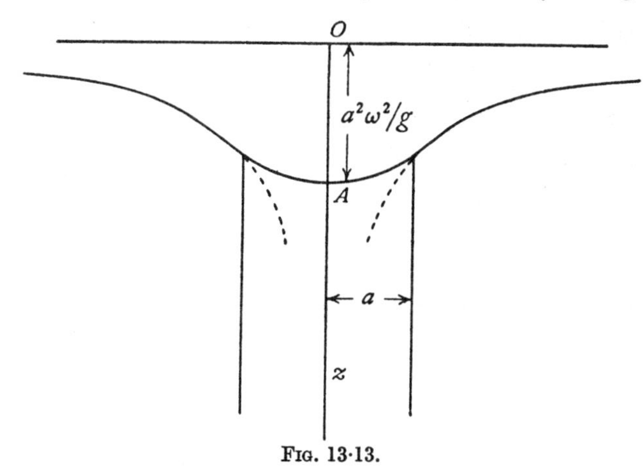

FIG. 13·13.

system found in 13·10, namely (writing 2ω for ω) ωr when $r<a$ and $\omega a^2/r$ when $r>a$, both horizontal in direction and at right angles to r. When $r>a$ the motion is evidently irrotational, for there is a velocity potential $\phi = -a^2\omega\theta$. Therefore the pressure equation gives

$$\frac{p}{\rho}+\frac{1}{2}\frac{\omega^2a^4}{r^2}-gz = \text{constant},$$

$-gz$ being the potential of the gravitational field. To determine the constant, put $r = \infty$, $z = 0$. Then $p = \Pi$, the surface pressure at infinity, and therefore

$$p = \Pi+g\rho z-\frac{\omega^2a^4\rho}{2r^2}. \tag{1}$$

To determine the pressure p_1 inside the vortex, we have the equations of motion

$$\frac{1}{\rho}\frac{\partial p_1}{\partial r} = r\omega^2, \quad \frac{1}{\rho}\frac{\partial p_1}{\partial z} = g.$$

Thus $$\frac{dp_1}{\rho} = r\omega^2 dr + g\,dz, \quad \frac{p_1}{\rho} = \frac{r^2\omega^2}{2} + gz + C.$$

To determine C, we must have $p = p_1$ when $r = a$, and therefore

$$(2) \qquad p_1 = \Pi + g\rho z - a^2\omega^2\rho\left(1 - \frac{r^2}{2a^2}\right).$$

At the free surface $p = p_1 = \Pi$, and therefore, from (1) and (2),

$$z = \frac{a^4\omega^2}{2gr^2} \quad \text{when } r > a,$$

$$(3) \qquad z = \frac{a^2\omega^2}{g}\left(1 - \frac{r^2}{2a^2}\right), \quad \text{when } r < a.$$

These equations determine the form of the free surface.

It also appears that the surfaces of constant pressure are obtained by a mere translation vertically of the free surface form which corresponds to

$$p = p_1 = \Pi.$$

To obtain the depth of the depression at A below the general level of the liquid, put $r = 0$ in (3). Then $OA = a^2\omega^2/g$.

13·20. Rectilinear vortex filament. The strength of the circular vortex was defined in 13·10 by

$$\kappa = \tfrac{1}{2}\omega a^2 = \frac{\omega}{2\pi}\cdot \pi a^2,$$

πa^2 being the area of the cross-section. If we let $a \to 0$ and $\omega \to \infty$ in such a way that the above product remains constant, we get a *rectilinear vortex filament*, that is, a two-dimensional vortex whose cross-section is an infinitesimal circle (cf. 1·12).

In view of the minimum size of a circular vortex described in 13·12, the vortex filament obtained in this way must be regarded a convenient abstraction (cf. 8·10).

A rectilinear vortex filament is represented by a point in the plane of the motion, just as a two-dimensional source is so represented. It also follows from 13·10 that the complex potential, due to a vortex filament of strength κ situated at the point z_0, is

$$w = i\kappa \log(z - z_0).$$

The strength κ is positive when the circulation round the filament is counterclockwise. We may refer to such a filament as a point vortex, or simply a vortex when there is no fear of ambiguity.

13·21. Single vortex filament. Take a vortex filament of strength κ at the point A, z_0. Then $w = i\kappa \log(z - z_0)$. The velocity of the point P, z, is therefore given by

$$\mathfrak{v} = u - iv = -\frac{dw}{dz} = \frac{-i\kappa}{z - z_0} = \frac{-i\kappa}{R\,e^{i\theta}}$$

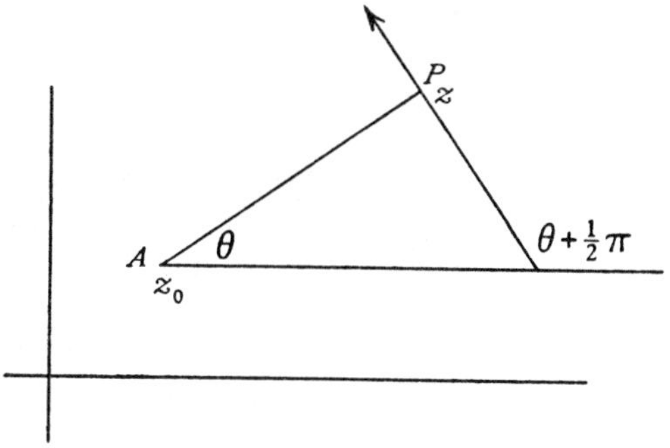

FIG. 13·21.

where $R = AP$ and $\arg(z - z_0) = \theta$. Thus

$$\bar{\mathfrak{v}} = u + iv = \frac{\kappa\, e^{i(\theta + \frac{1}{2}\pi)}}{R}.$$

Thus the direction of motion at P is at right angles to AP with speed $q = \kappa/R$ in the sense given by the rotation of the vortex at A.

It should be noted that the stream function is

$$\psi = \tfrac{1}{2}\kappa \log(z - z_0)(\bar{z} - \bar{z}_0) = \kappa \log|z - z_0| = \kappa \log R.$$

Also since $2i\psi = w(z) - \bar{w}(\bar{z})$ we have $\mathfrak{v} = -2i\,\partial\psi/\partial z$.

13·22. Motion of vortex filaments. We have seen (13·10) that a circular vortex alone in the fluid possesses no tendency to set itself in motion and the same therefore applies to a vortex filament. If therefore there are several vortex filaments, the motion of the filament at the point P is the same as the motion which would be produced at P by the remaining vortices if the vortex at P did not exist. It must be observed, however, that the complete motion of the fluid may be due not only to vortices but also to the presence of sources, streams, or other causes, and the velocity of P will then be compounded of the velocity induced by the other vortices, as just described, and the general velocity at P of the fluid due to the remaining causes.

Thus if w is the complex potential of a flow which contains a vortex of strength κ at the point z_0, the complex velocity of the vortex is

$$u_0 - iv_0 = -\left\{\frac{d}{dz}[w - i\kappa \log(z - z_0)]\right\}_0 = -2i\left\{\frac{\partial}{\partial z}[\psi - \kappa \log|z - z_0|]\right\}_0,$$

where suffix zero indicates that after the differentiation we put $z = z_0$ and $\bar{z} = \bar{z}_0$.

13·23. Two vortex filaments. Consider filaments at A_1, A_2 of strengths κ_1, κ_2. If z_1, z_2 are the affixes of A_1, A_2, then

$$w = i\kappa_1 \log(z-z_1) + i\kappa_2 \log(z-z_2).$$

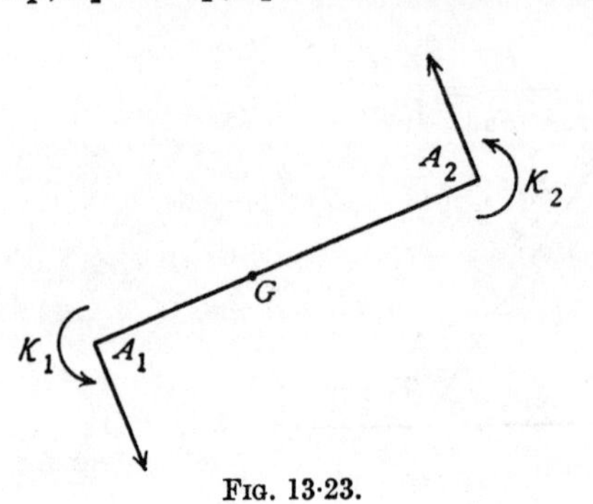

Fig. 13·23.

The velocity of A_1 is due to κ_2 alone and is therefore given by

(1) $$u_1 - iv_1 = \frac{-i\kappa_2}{z_1 - z_2}.$$

Similarly the velocity of A_2 is given by

(2) $$u_2 - iv_2 = \frac{-i\kappa_1}{z_2 - z_1}.$$ Thus,

(3) $$\kappa_1(u_1 - iv_1) + \kappa_2(u_2 - iv_2) = 0.$$

If we imagine masses κ_1, κ_2 at A_1, A_2, their centroid G will be at the point $(\kappa_1 z_1 + \kappa_2 z_2)/(\kappa_1 + \kappa_2)$, provided $\kappa_1 + \kappa_2 \neq 0$. We shall call G the *centroid of the vortices*. It follows from (3) that the centroid of the vortices remains at rest.

The velocity of A_1 is

$$\frac{\kappa_2}{A_1 A_2} = \frac{\kappa_2 A_1 A_2}{\kappa_1 + \kappa_2} \cdot \frac{\kappa_1 + \kappa_2}{A_1 A_2^2} = GA_1 \cdot \omega,$$

where $$\omega = (\kappa_1 + \kappa_2)/A_1 A_2^2,$$

and the line $A_1 A_2$ therefore rotates with this angular velocity. Since neither vortex has a component of velocity along $A_1 A_2$, it follows that $A_1 A_2$ remains constant in length. Therefore ω is constant and each vortex describes a circle with constant angular speed.

If $\kappa_1 = \kappa_2 = \kappa$ and $A_1 A_2 = a$, then each vortex describes the circle on $A_1 A_2$ as diameter with angular speed $2\kappa/a^2$.

13·24. Motion of a system of vortex filaments. If we have several filaments $\kappa_1, \kappa_2, \kappa_3, \ldots$ at points $z_1, z_2, z_3, \ldots$, the considerations of the last section show without difficulty that the function

$$W = i\, \Sigma\, \kappa_r \kappa_s \log(z_r - z_s), \quad r \neq s,$$

gives the induced velocity of any particular filament.

If for simplicity we consider three filaments, we have

$$W = i\{\kappa_1 \kappa_2 \log(z_1 - z_2) + \kappa_2 \kappa_3 \log(z_2 - z_3) + \kappa_3 \kappa_1 \log(z_3 - z_1)\},$$

and the induced velocity of the filament κ_1 is given by

$$u_1 - iv_1 = -\frac{1}{\kappa_1}\frac{\partial W}{\partial z_1} = -i\left\{\frac{\kappa_2}{z_1 - z_2} + \frac{\kappa_3}{z_1 - z_3}\right\}.$$

Writing down the corresponding velocities of κ_2 and κ_3, we get at once, after multiplication and addition,

(1) $$\kappa_1 u_1 + \kappa_2 u_2 + \kappa_3 u_3 = 0, \quad \kappa_1 v_1 + \kappa_2 v_2 + \kappa_3 v_3 = 0,$$

so that the centroid of the three filaments remains at rest. This result is easily seen to apply to any number of filaments.

If we write $W = \Phi + i\Psi$, we have

$$\kappa_1(u_1 - iv_1) = -\frac{\partial W}{\partial z_1} = -\frac{\partial \Phi}{\partial x_1} - i\frac{\partial \Psi}{\partial x_1}.$$

Hence
$$\kappa_1 u_1 = -\frac{\partial \Phi}{\partial x_1} = -\frac{\partial \Psi}{\partial y_1}, \quad \kappa_1 v_1 = +\frac{\partial \Psi}{\partial x_1}.$$

Thus Ψ is analogous to the stream function in giving the components of the velocity of the vortex. Also

$$\frac{d\Psi}{dt} = \sum_{r=1}^{3} \left\{\frac{\partial \Psi}{\partial x_r}\frac{dx_r}{dt} + \frac{\partial \Psi}{\partial y_r}\frac{dy_r}{dt}\right\}.$$

But
$$\frac{dx_r}{dt} = u_r = -\frac{1}{\kappa_r}\frac{\partial \Psi}{\partial y_r}, \quad \frac{dy_r}{dt} = v_r = \frac{1}{\kappa_r}\frac{\partial \Psi}{\partial x_r}.$$

Therefore $d\Psi/dt = 0$ and Ψ is a constant of the motion.

13·30. Vortex pair. A pair of vortices each of strength κ but of opposite rotations is called a *vortex pair*. Consider such a pair κ at A and $-\kappa$ at B where $AB = 2a$. Take the x-axis to bisect AB at right angles, fig. 13·30 (i). The vortex at A induces in B a velocity $\kappa/(2a)$ parallel to Ox, and B induces a like velocity in A. Hence the pair advances in the direction OX with the constant velocity $\kappa/(2a)$. The complex potential is

$$w = i\kappa \log \frac{z - ai}{z + ai},$$

where the origin O is the mid-point of AB.

The stream function is therefore

$$\psi = \kappa \log \frac{PA}{PB},$$

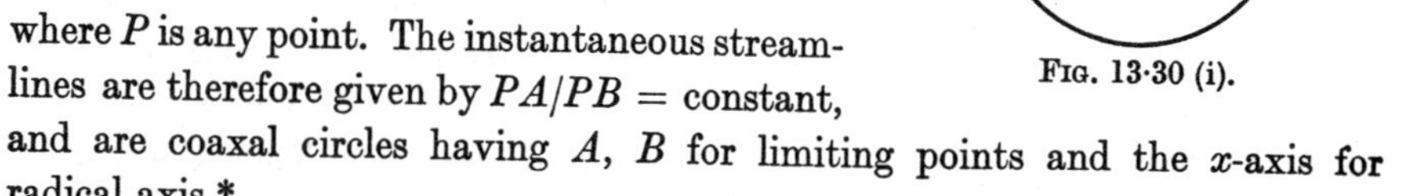

Fig. 13·30 (i).

where P is any point. The instantaneous stream-lines are therefore given by PA/PB = constant, and are coaxal circles having A, B for limiting points and the x-axis for radical axis.*

The velocity at any point of Ox is along this line and there is therefore no flow across it. The velocity at O is $2\kappa/a$, which is four times the velocity of advance of the vortex pair.

* In terms of the coaxal coordinates of 6·50 the complex potential of the vortex pair is $w = -\kappa\zeta$, provided that we take A, B as $z = \pm c$. The relation of the problem of 6·53 to the present theory now becomes evident.

To find the relative streamlines, we impose a velocity on the whole system equal and opposite to the velocity of advance of the vortex. The appropriate stream function is therefore

$$\psi = \kappa\left(\frac{y}{2a}+\log\frac{r_1}{r_2}\right),$$

where $r_1 = PA$, $r_2 = PB$, and the relative streamlines are shown in fig. 13·30 (ii).

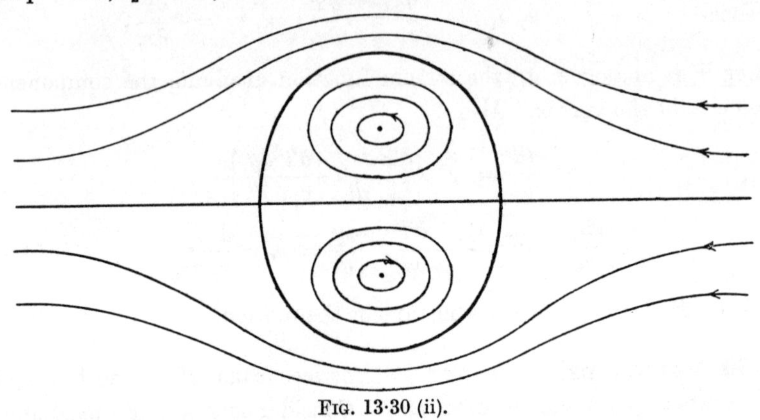

FIG. 13·30 (ii).

The semi-axes of the oval are $2{\cdot}09a$, $1{\cdot}73a$ approximately (Kelvin).

The figure also represents the streaming past a fixed cylinder whose section is that of the oval.

If U is the asymptotic velocity of the stream, we have $U = \kappa/(2a)$, and hence we can regard the motion as due to vortices at A and B of strength $\pm 2aU$ in a stream of asymptotic velocity U.

13·31. Vortex filament parallel to a plane. Let a vortex filament at A be at distance a from a plane OX, $AO = a$. If κ is the strength of the filament and we place a second filament at B where $AB = 2a$ and at the same time allow the fluid to have access also to the side of the plane on which B lies, we have a vortex pair which yields no flow across the plane. This may therefore be removed. Thus the vortex at B is the image of the vortex at A.

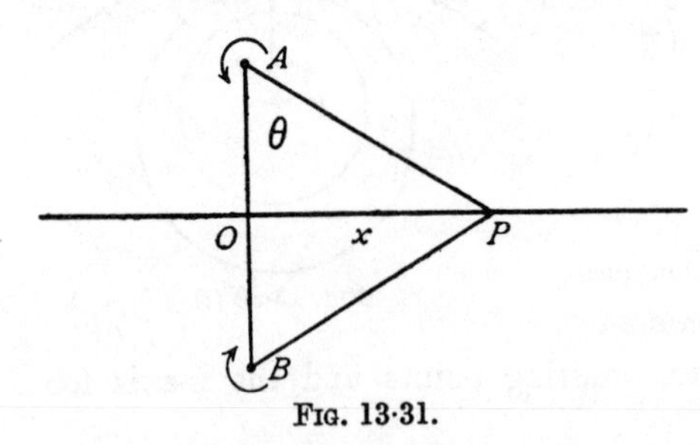

FIG. 13·31.

Since the pair moves parallel to OX with velocity $\kappa/(2a)$, it follows that the single vortex A in the presence of the plane will move parallel to the plane with the above velocity. In the position shown the complex potential is

$$w = i\kappa\log\frac{z-ai}{z+ai}.$$

Hence at time t we shall have

$$w = i\kappa \log \frac{z - ai - Vt}{z + ai - Vt},$$

where

$$V = \kappa/(2a).$$

Hence

$$\frac{\partial w}{\partial t} = -i\kappa V\left(\frac{1}{z - ai - Vt} - \frac{1}{z + ai - Vt}\right),$$

and therefore at a point on OX, taking $t = 0$,

$$\left(\frac{\partial \phi}{\partial t}\right)_{t=0} = \frac{2aV\kappa}{x^2 + a^2} = \frac{\kappa^2 \cos^2\theta}{a^2}.$$

Again when $t = 0$, the velocity at P on Ox is the resultant of κ/PA, κ/PB perpendicular to PA, PB, and so $q = 2\kappa \cos^2 \theta/a$. Thus the pressure at P is given by

$$\frac{p}{\rho} + \frac{2\kappa^2 \cos^4\theta}{a^2} - \frac{\kappa^2 \cos^2\theta}{a^2} = \frac{\Pi}{\rho},$$

where Π is the pressure at infinity ($\theta = \frac{1}{2}\pi$).

Thus

$$p = \Pi - \frac{\kappa^2 \rho}{a^2} \cos^2\theta \,.\, \cos 2\theta.$$

The force on the plane due to the motion is therefore

$$\frac{\kappa^2 \rho}{a^2} \int_{-\pi/2}^{\pi/2} \cos^2\theta \cos 2\theta \,.\, a \sec^2\theta \, d\theta = 0.$$

13·32. Vortex doublet. Consider a vortex pair, κ at $a\,e^{i\alpha}$ and $-\kappa$ at $-a\,e^{i\alpha}$. If we let $a \to 0$ and $\kappa \to \infty$ in such a way that $2a\kappa = \mu$, we get a vortex doublet inclined at an angle α to the x-axis, cf. 8·23.

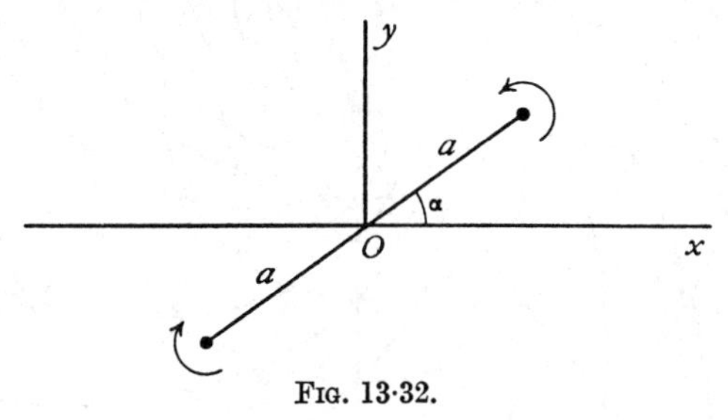

FIG. 13·32.

The direction of the doublet is reckoned from the vortex of negative rotation to that of positive rotation. The complex potential is

$$w = \lim_{a \to 0} i\kappa \left(\log (z - a\,e^{i\alpha}) - \log (z + a\,e^{i\alpha})\right)$$

$$= \lim i\kappa \left(-\frac{a\,e^{i\alpha}}{z} + \frac{a^2\,e^{2i\alpha}}{2z^2} - \ldots - \frac{a\,e^{i\alpha}}{z} - \frac{a^2\,e^{2i\alpha}}{2z^2} - \ldots\right)$$

$$= -\frac{i\mu\,e^{i\alpha}}{z}.$$

The stream function is $\psi = -\mu \cos (\alpha - \theta)/r$.

If, in particular, we take the vortex doublet to be at the origin and along the axis of y, we have $\psi = -\mu \sin\theta/r$. If we put $\mu = Ub^2$, we obtain

$$\psi = -\frac{Ub^2 \sin\theta}{r},$$

which is the stream function for a circular cylinder of radius b moving with velocity U along the x-axis.

Thus the motion due to a circular cylinder is the same as that due to a suitable vortex doublet placed at the centre, and with its axis perpendicular to the direction of motion.

We obtain a circulation round the cylinder by placing a vortex filament of suitable strength at the centre.

13·33. Source and vortex. The considerations of the previous section lead us to enquire how a source and a vortex combine. The complex potential

$$w = (-m + i\kappa)\log z,$$

FIG. 13·33.

decreases by $2\pi(im + \kappa)$ when we go once round the origin, and therefore ϕ decreases by $2\pi\kappa$ while ψ decreases by $2\pi m$, and w therefore satisfies the condition for a vortex and a source there.

The stream function is

$$\psi = -m\theta + \kappa \log r.$$

When ψ has the constant value $\kappa \log C$, we get

$$r = C e^{\frac{m\theta}{\kappa}},$$

so that the streamlines are equiangular spirals.

The streamlines can readily be drawn by the diagonal method (4·32), by combining the lines

$$m\theta = n\omega,$$

$$\kappa \log r = n\omega, \quad n = 0, 1, 2, \ldots,$$

or

$$\theta = \frac{n}{m}\omega, \quad r = e^{\frac{n\omega}{\kappa}}.$$

The above combination is known as a spiral vortex.

We could impose a longitudinal velocity perpendicular to the plane. This suggests two loose analogies : (i) the swirling flow of gas in an exhaust jet, (ii) flow towards a bath waste if a sink is substituted for the source.

13·40. Vortex filament parallel to two perpendicular planes. Take the planes for axes. Let the vortex be at (x, y). Then the image system is $-\kappa$ at $(x, -y)$, $-\kappa$ at $(-x, y)$, and κ at $(-x, -y)$. The velocity of the

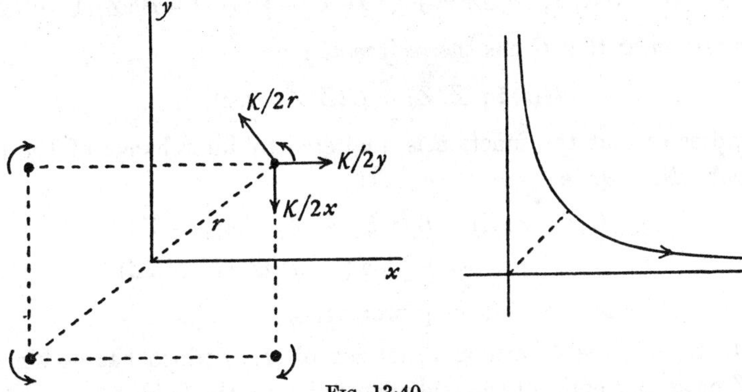

FIG. 13·40.

vortex is due solely to its images. Hence the velocity components are those shown in the figure. Since $x = r\cos\theta$, $y = r\sin\theta$, the radial and transverse components are given by

$$\frac{dr}{dt} = \frac{\kappa\cos\theta}{2r\sin\theta} - \frac{\kappa\sin\theta}{2r\cos\theta} = \frac{\kappa\cos 2\theta}{r\sin 2\theta},$$

$$r\frac{d\theta}{dt} = \frac{\kappa}{2r} - \frac{\kappa\sin\theta}{2r\sin\theta} - \frac{\kappa\cos\theta}{2r\cos\theta} = -\frac{\kappa}{2r}.$$

Therefore, by division,

$$\frac{1}{r}\frac{dr}{dt} = -\frac{2\cos 2\theta}{\sin 2\theta}\frac{d\theta}{dt}.$$

Integrating, we get $\qquad r\sin 2\theta = a,$

where a is a constant. The form of the path of the vortex and its sense of description are shown, $x = \frac{1}{2}a$, $y = \frac{1}{2}a$ being asymptotes.

13·50. Vortex in or outside a circular cylinder. Let there be a vortex of strength κ at the point $Z=X+iY$ outside the cylinder $|z|=a$. If the motion is due solely to the vortex the circle theorem gives the complex potential

(1) $$i\kappa \log (z-Z) - i\kappa \log (a^2/z - \bar{Z}),$$

which, omitting an irrelevant constant, is equivalent to

$$i\kappa \log (z-Z) - i\kappa \log (z - a^2/\bar{Z}) + i\kappa \log z.$$

This shows (cf. 8·61) that the image system consists of a vortex of strength $-\kappa$ at the inverse point and a vortex of strength κ at the centre.

The addition to (1) of the term $i\kappa \log (-\bar{Z})$, which is independent of z gives the complex potential in the form

(2) $$w = i\kappa \log (z-Z) - i\kappa \log (1 - a^2/z\bar{Z}) = \phi + i\kappa G,$$

where the stream function κG is constant on the boundary of the cylinder and is given by $2i\kappa G = w - \bar{w}$, so that

(3) $$G = G(z, \bar{z}\,;\ Z, \bar{Z}) = \log |z-Z| - \tfrac{1}{2} \log (1 - a^2/z\bar{Z})(1 - a^2/\bar{z}Z),$$

and it is now clear that G has the *reciprocal property*

(4) $$G(z, \bar{z}\,;\ Z, \bar{Z}) = G(Z, \bar{Z}\,;\ z, \bar{z}),$$

which expresses that the function is unaltered by interchange of the pairs of variables.* Now write

(5) $$g(z, \bar{z}\,;\ Z, \bar{Z}) = G(z, \bar{z}\,;\ Z, \bar{Z}) - \log |z-Z|$$
$$= -\tfrac{1}{2} \log (1 - a^2/z\bar{Z})(1 - a^2/\bar{z}Z).$$

The function g has the following properties.

(i) $g(z, \bar{z}\,;\ Z, \bar{Z})$ is a harmonic function of (x, y) which has no singularity at $z = Z$ or at any point of the region occupied by the fluid.

(ii) The function g has the reciprocal property

(6) $$g(z, \bar{z}\,;\ Z, \bar{Z}) = g(Z, \bar{Z}\,;\ z, \bar{z}).$$

Now

$$\frac{\partial}{\partial Z} g(Z, \bar{Z}\,;\ Z, \bar{Z}) = \left(\frac{\partial}{\partial z} g(z, \bar{z}\,;\ Z, \bar{Z}) + \frac{\partial}{\partial z} g(Z, \bar{Z}\,;\ z, \bar{z})\right)_1,$$

where suffix 1 indicates that after differentiation we must put $z = Z$, $\bar{z} = \bar{Z}$. It then follows from (ii) that

(7) $$\left(\frac{\partial}{\partial z} g(z, \bar{z}\,;\ Z, \bar{Z})\right)_1 = \tfrac{1}{2} \frac{\partial}{\partial Z} g(Z, \bar{Z}\,;\ Z, \bar{Z}).$$

* G is in fact a Green's function.

This result is fundamental for the present investigation. It is a direct consequence of the property (ii). In the present case (5) shows that

$$g(Z, \overline{Z}\,;\; Z, \overline{Z}) = -\log\left(1-\frac{a^2}{Z\overline{Z}}\right), \tag{8}$$

and it is easily shown by direct differentiation of (5) and (8) that this particular function satisfies (7).

Let us now superpose on the flow due to the vortex alone in the presence of the cylinder any other field of flow (e.g. a uniform stream or a circulation round the cylinder) whose stream function $\psi_0(z, \bar{z})$ is constant on the boundary of the cylinder and has no singularity at $z = Z$. The stream function for the combined flow is

$$\psi(z, \bar{z}) = \psi_0(z, \bar{z}) + \kappa G(z, \bar{z}\,;\; Z, \overline{Z}). \tag{9}$$

To find the complex velocity of the vortex we use the principle stated in 13·22 by forming the function

$$\chi = \psi - \kappa \log |\, z - Z \,|,$$

i.e. by subtracting the stream function of the vortex itself alone in unbounded fluid. Then from 13·22 the complex velocity of the vortex is

$$u_1 - iv_1 = -2i\left(\frac{\partial \chi}{\partial z}\right)_1.$$

Now from (5) and (9)

$$\chi = \psi_0(z, \bar{z}) + \kappa g(z, \bar{z}\,;\; Z, \overline{Z}). \tag{10}$$

Therefore the complex velocity of the vortex is

$$u_1 - iv_1 = -2i\frac{\partial \chi_1}{\partial Z}, \text{ where} \tag{11}$$

$$\chi_1 = \psi_0(Z, \overline{Z}) + \tfrac{1}{2}\kappa g(Z, \overline{Z}\,;\; Z, \overline{Z}). \tag{12}$$

Comparing (10) with (12) we see that the factor κ in the last term of the expression for χ becomes the factor $\frac{1}{2}\kappa$ in the last term of the expression for χ_1 on account of (7). Observe that $\chi_1 = \chi_1(Z, \overline{Z})$ is a function of Z, $\overline{Z}$ only and is independent of z. From (11) we get

$$\frac{dX}{dt} = u_1 = -\frac{\partial \chi_1}{\partial Y}, \quad \frac{dY}{dt} = v_1 = \frac{\partial \chi_1}{\partial X}. \tag{13}$$

The function χ_1 is called Routh's stream function.

In particular, if the vortex at (X, Y) is the only mobile singularity in the flow and if $\psi_0(Z, \overline{Z})$ depends on the time t only through (X, Y), we have

$$\frac{d\chi_1}{dt} = \frac{\partial \chi_1}{\partial X}\cdot\frac{dX}{dt} + \frac{\partial \chi_1}{\partial Y}\cdot\frac{dY}{dt} = 0$$

from (13). Therefore

$$\chi_1(Z, \bar{Z}) = \text{constant},$$

and this is the equation of the path of the vortex.

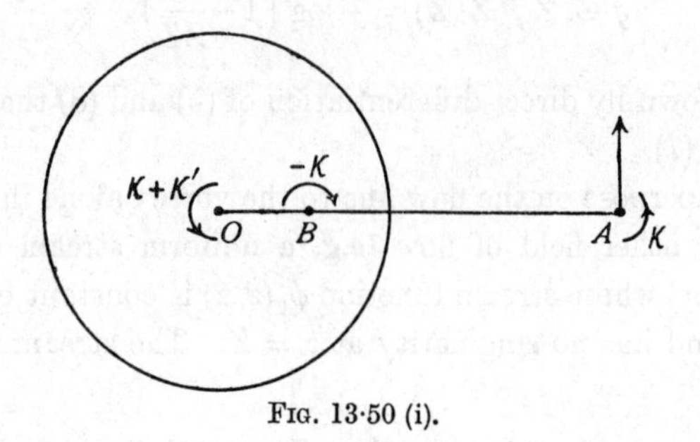

FIG. 13·50 (i).

In the case of a single vortex of strength κ outside the cylinder and a circulation of strength κ' about the cylinder we have $\psi_0 = \kappa' \log r$, where $r = |z|$, while from (8) $g(Z, \bar{Z}\,;\; Z, \bar{Z}) = -\log(1-a^2/R^2)$, where $R = |Z|$. Therefore the path of the vortex is given by

$$\text{(14)} \qquad \chi_1 = \kappa' \log R - \tfrac{1}{2}\kappa \log(1-a^2/R^2) = \text{constant},$$

so that R remains constant and the vortex describes a circle concentric with the cylinder and with speed

$$\frac{\kappa+\kappa'}{R} - \frac{\kappa R}{R^2-a^2}.$$

To deal with the case of a vortex *inside* the cylinder we see that the function g of (8) is unsuitable, for it has the singularity $Z = 0$ in the region of flow.

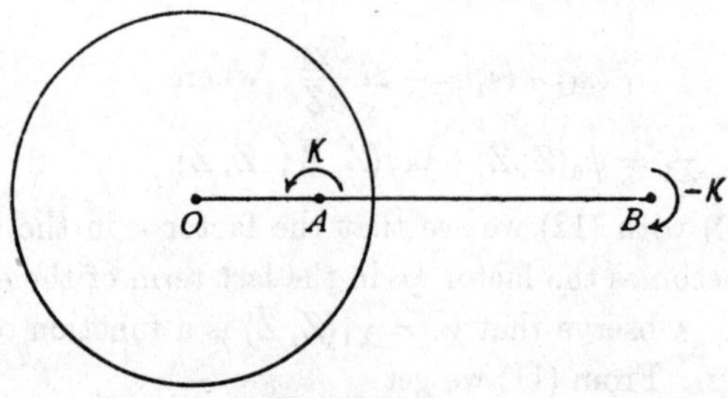

FIG. 13·50 (ii).

The complex potential, however, is still obtainable from the circle theorem in the form

$$i\kappa' \log z - i\kappa \log\left(\frac{a^2}{z} - \bar{Z}\right) + i\kappa \log(z-Z).$$

Take $\kappa+\kappa' = 0$. Then we have the situation of fig. 13·50 (ii), where

$$w = i\kappa \log(z-Z) - i\kappa \log(a^2 - z\bar{Z}).$$

and we now see that the appropriate form for g, having the reciprocal property, is

$$g(z, \bar{z}\,;\; Z, \bar{Z}) = -\tfrac{1}{2}\log(a^2 - z\bar{Z})(a^2 - \bar{z}Z) \quad \text{leading to}$$

$$(15) \qquad g(Z, \bar{Z}\,;\; Z, \bar{Z}) = -\log(a^2 - R^2).$$

Thus $\psi_0 = 0$ and the path is given by

$$\chi_1 = -\tfrac{1}{2}\kappa \log(a^2 - R^2) = \text{constant},$$

so that the vortex again describes a concentric circle with the speed $\kappa R/(a^2 - R^2)$. It appears from this discussion that the image of an *internal* vortex of strength κ is a vortex of strength $-\kappa$ at the inverse point and a circulation of strength $-\kappa$ about the cylinder.

As a final illustration * consider a vortex of strength κ at the point Z outside the cylinder together with a uniform stream (of complex potential $-Uze^{-i\alpha}$) and circulation of strength κ' about the cylinder. Then if (R, Θ) are the polar coordinates of the vortex, we have

$$\psi_0(Z, \bar{Z}) = -U(R - a^2/R)\sin(\Theta - \alpha) + \kappa' \log R,$$

and the path of the vortex is given by

$$(16) \quad \chi_1 = -U\left(R - \frac{a^2}{R}\right)\sin(\Theta - \alpha) + \kappa' \log R - \tfrac{1}{2}\kappa \log\left(1 - \frac{a^2}{R^2}\right) = \text{constant}.$$

If the flow whose stream function is $\psi_0(z, \bar{z})$ is superposed on the field which contains n vortices, κ_r at the point z_r, $r = 1, 2, \ldots, n$, the stream function of the combined flow is, from (9),

$$(18) \qquad \psi(z, \bar{z}) = \psi_0(z, \bar{z}) + \Sigma \kappa_r G(z, \bar{z}\,;\; z_r, \bar{z}_r),$$

the summation being from $r = 1$ to $r = n$. Write

$$(19) \quad \psi_s = \psi_0(z_s, \bar{z}_s) + \tfrac{1}{2}\sum_{r \neq s} \kappa_r G(z_r, \bar{z}_r\,;\; z_s, \bar{z}_s), \quad g_s = g(z_s, \bar{z}_s\,;\; z_s, \bar{z}_s),$$

$$(20) \qquad \Psi = \Sigma \kappa_s(\psi_s + \tfrac{1}{2}\kappa_s g_s) \quad \text{summed from } s = 1 \text{ to } s = n.$$

It then follows from (11) that the complex velocity of the vortex at z_s is v_s where

$$(21) \qquad \kappa_s v_s = -2i\, \partial\Psi/\partial z_s.$$

Thus the function (20) is entirely analogous to the function Ψ of 13·24 and is a constant of the motion if ψ_0 does not depend on time explicitly.

The function g has been found in (5) for the region exterior to a circle and in (15) for the region interior to a circle. When the boundary is other than circular, the conformal mapping of the region occupied by the fluid on the region exterior or interior to a circle, as the case may be, will then reduce the

* For a detailed study of the ideas here outlined, consult C. C. Lin, *On the Motion of Vortices in Two-Dimensions*, University of Toronto Press (1943). This paper discusses the most general problem.

problem to the one considered here. The relation between the function g and its transform under the mapping is given by 13·60 (4), (5).

13·51. Vortices in the presence of a circular cylinder. Consider vortices κ at A, z_1, and $-\kappa$ at B, $\bar{z}_1$, outside the circular cylinder $|z| = a$. The complex potential in the absence of the cylinder is $i\kappa \log (z-z_1)/(z-\bar{z}_1)$, and if we insert the cylinder $|z| = a$ the circle theorem gives

$$w = i\kappa \log (z-z_1)/(z-\bar{z}_1) - i\kappa \log \left(\frac{a^2}{z} - \bar{z}_1\right) \bigg/ \left(\frac{a^2}{z} - z_1\right),$$

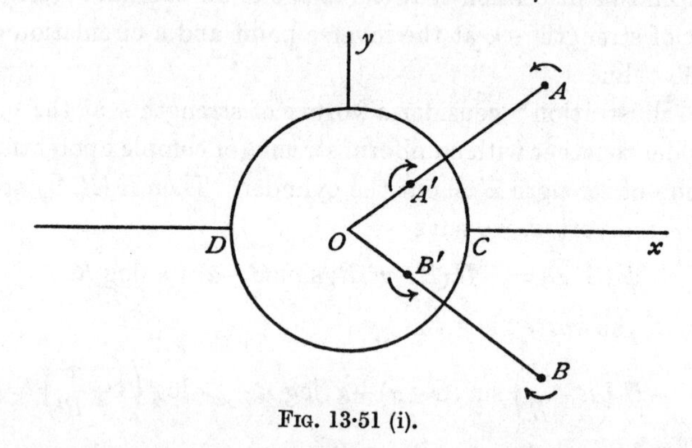

FIG. 13·51 (i).

so that the image system consists of opposite vortices at the inverse points and $w = i\kappa \log (z-z_1) + w_z$, where, ignoring a constant,

$$(1) \qquad w_z = -i\kappa \log \frac{(z-\bar{z}_1)\left(z - \dfrac{a^2}{\bar{z}_1}\right)}{z - \dfrac{a^2}{z_1}},$$

and the complex velocity of the vortex at A is the value of $-dw_z/dz$ when $z = z_1$.

If we write, as in 13·50,

$$w = \phi + i\kappa G = \phi + i\kappa (g + \log |z - z_1|),$$

so that

$$g(z, \bar{z}\,;\, z_1, \bar{z}_1) = -\log |z - \bar{z}_1| - \log \left|\frac{a^2}{z} - \bar{z}_1\right| + \log \left|\frac{a^2}{z} - z_1\right|,$$

the path of the vortex at A is given by

$$\text{constant} = \tfrac{1}{2}\kappa g(z_1, \bar{z}_1\,;\, z_1, \bar{z}_1) = \tfrac{1}{2}\kappa \log \frac{AB'}{AB\,.\,AA'}, \text{ so that}$$

$$AB\,.\,AA' = 2k\,.\,AB',$$

where k is a constant whose value depends on the initial conditions.

If A is the point (x, y) or (r, θ), then

$$r \sin\theta \left(r - \frac{a^2}{r}\right) = k\left[r^2 + \frac{a^4}{r^2} - 2a^2 \cos 2\theta\right]^{\frac{1}{2}},$$

which reduces to $(x^2+y^2-a^2)^2\,y^2 = k^2[(x^2+y^2-a^2)^2+4a^2y^2]$.

Taking $k = 0$, this gives the cylinder and the x-axis, which is the "dividing streamline."

We can therefore draw the form of the curves, fig. 13·51 (ii), which consist of two loops within the cylinder and branches outside which are asymptotic

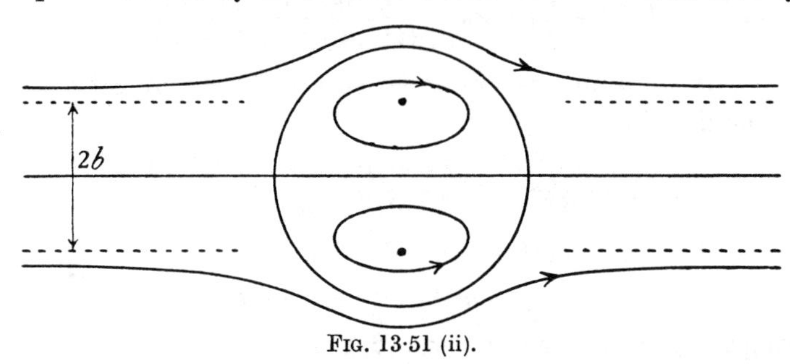

FIG. 13·51 (ii).

to $y = \pm k$, for when $x \to \infty$, $y^2 \to k^2$. The outside curves are described by a vortex pair whose distance apart at infinity is $2b$, where b is the value of the constant k. The inner loops are described by a vortex pair within the cylinder. The motions inside and outside can co-exist but the line joining corresponding vortices within and without does not continue to pass through the centre of the circle.

The loops may degenerate into points. In that case there will be a stationary vortex pair within the cylinder. To get the condition for this, take all four vortices on the y-axis and let r be the distance of A' from the centre. Then A' will be stationary if its induced velocity is zero, i.e. if

$$-\frac{1}{2r} + \frac{1}{-r+a^2/r} + \frac{1}{r+a^2/r} = 0,$$

which gives $r^4+4a^2r^2-a^4 = 0$, whence

$$r^2/a^2 = \sqrt{5}-2 = 0{\cdot}236067, \text{ and } r/a = 0{\cdot}486,$$

nearly. A vortex pair thus situated within the cylinder will remain at rest.

13·52. Stationary vortex filaments in the presence of a cylinder.

If in fig. 13·51 (i) we reverse the sense of rotation of all the vortices, the motion of the vortex A is given by (1) of the last section, with the sign of κ changed.

Let us impose on this system a stream in the direction of OX whose velocity at infinity is U. The complex potential for the cylinder alone in the stream is

$$-U\left(z + \frac{a^2}{z}\right)$$

Hence the motion of the vortex A is obtained from the function

$$(1) \qquad w_z = -U\left(z+\frac{a^2}{z}\right)+i\kappa\log\frac{(z-\bar{z}_1)\left(z-\dfrac{a^2}{\bar{z}_1}\right)}{z-\dfrac{a^2}{z_1}}.$$

Hence the vortex A will be at rest if $dw_z/dz = 0$, when $z = z_1$. Performing the differentiation and dropping the suffix 1 for simplicity we get

$$(2) \qquad U\left(1-\frac{a^2}{z^2}\right) = i\kappa\,\frac{(z^2-a^2)(z\bar{z}-a^2)+a^2(z-\bar{z})^2}{(z\bar{z}-a^2)(z-\bar{z})(z^2-a^2)}.$$

If two complex numbers are equal so are their conjugates. Expressing this fact and dividing, we get

$$\frac{(z^2-a^2)^2\bar{z}^2}{(\bar{z}^2-a^2)^2z^2} = \frac{(z^2-a^2)(z\bar{z}-a^2)+a^2(z-\bar{z})^2}{(\bar{z}^2-a^2)(z\bar{z}-a^2)+a^2(z-\bar{z})^2},$$

which leads, after a straightforward reduction, to

$$(3) \qquad (z\bar{z}-a^2)^2+z\bar{z}(z-\bar{z})^2 = 0.$$

Putting $z = r\,e^{i\theta}$, where $0 < \theta < \pi/2$, this gives

$$(r^2-a^2)^2 = 4r^4\sin^2\theta, \quad \text{or,} \quad \left(r-\frac{a^2}{r}\right) = 2r\sin\theta.$$

Hence $AA' = AB$, and if this condition is satisfied the vortices can be at rest behind the cylinder.

From (2) and (3) we deduce that $\kappa = U(r^2-a^2)^2(r^2+a^2)/r^5$.

It is clear from the symmetry that if A is at rest, so is B. Thus it appears that equal but opposite vortices of the strength given above can remain at rest behind a circular cylinder in a uniform stream U, provided that they are placed at image points A and B, such that $AA' = AB$.

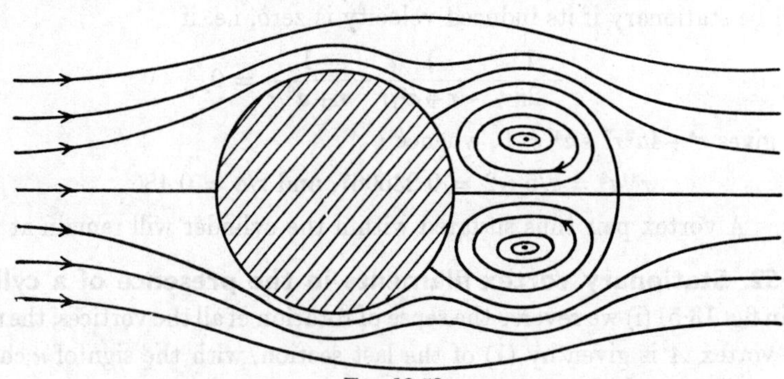

FIG. 13·52.

This result is of great interest since such vortices have frequently been photographed (see Plate 1) in slow streaming past a cylinder. The general form of the streamlines is shown in fig. 13·52.

There are four stagnation points, three on the cylinder and one on the axis of the stream.

We deduce the complex potential of the fluid motion from (1) by adding the term $-i\kappa \log(z-z_1)$ for the vortex at A.

13·60. Conformal transformation. Let there be a vortex at Π in the ζ-plane, and let P in the z-plane correspond to Π in the ζ-plane by means of the conformal transformation

(1) $$z = f(\zeta).$$

Let γ be a small curve surrounding Π, and c a small curve consisting of points corresponding to the points of γ and therefore surrounding P. If

(2) $$w = \phi + i\psi$$

be the complex potential giving an irrotational fluid motion in the ζ-plane, there will be a corresponding fluid motion in the z-plane, obtained by eliminating ζ from (1) and (2), and the values of ϕ, ψ and w will be equal at corresponding points. Hence the circulations round γ and c will be equal, i.e.

$$\int_{(\gamma)} -d\phi = \int_{(c)} -d\phi.$$

Hence, if there is a vortex filament at Π of strength κ, there will be a vortex filament of strength κ at the corresponding point P.* It does not,

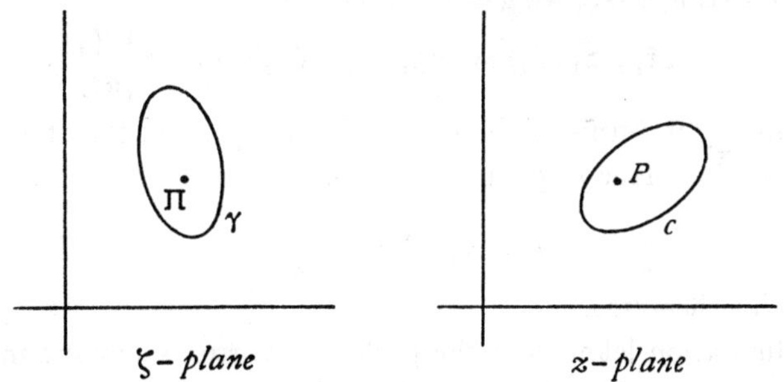

FIG. 13·60.

however, follow that these vortices will move so as to continue to occupy corresponding points. If we know the motion of one, we can nevertheless determine the motion of the other by means of a theorem due to Routh. The theorem may be obtained as follows.

Let Π be the point ζ_1, P the point z_1.

Suppose the transformation (1) maps the region exterior to the profile A in the z-plane on the region exterior to the cylinder C or $|\zeta| = a$ in the ζ-plane (cf. fig. 6·29). The principles of conformal mapping allow us

* Cf. 8·50. It is assumed that c encircles P once only.

to relate corresponding flows in the two planes through their stream functions, say

$$\text{(3)} \qquad \psi(z, \bar{z}) = \omega(\zeta, \bar{\zeta}).$$

If either of these is given the other is determined.

If the only mobile singularity is a vortex of strength κ at z_1 and therefore a vortex of strength κ at ζ_1, the path of the vortex in the ζ-plane is determined by the function χ_1 of 13·50 (12) where, with an obvious modification of notation,

$$\chi_1 = \omega_0(\zeta_1, \bar{\zeta}_1) + \tfrac{1}{2}\kappa\gamma(\zeta_1, \bar{\zeta}_1;\ \zeta_1, \bar{\zeta}_1),$$

where γ is given by 13·50 (5) in the form

$$\gamma(\zeta, \bar{\zeta};\ \zeta_1, \bar{\zeta}_1) = \Gamma(\zeta, \bar{\zeta};\ \zeta_1, \bar{\zeta}_1) - \log|\zeta - \zeta_1|.$$

Now in the z-plane we have

$$g(z, \bar{z};\ z_1, \bar{z}_1) = G(z, \bar{z};\ z_1, \bar{z}_1) - \log|z - z_1|,$$

where G is a function (at present unknown) which has no singularities in the fluid except at $z = z_1$, and has the reciprocal property 13·50 (4). On the other hand, Γ and G are both stream functions, one the transform of the other under (1). Therefore Γ and G take the same value at corresponding points, and so by subtraction

$$\text{(4)} \qquad g(z, \bar{z};\ z_1, \bar{z}_1) = \gamma(\zeta, \bar{\zeta};\ \zeta_1, \bar{\zeta}_1) + \log\left|\frac{\zeta - \zeta_1}{z - z_1}\right|.$$

Therefore letting $z \to z_1$ we get

$$\text{(5)} \qquad g(z_1, \bar{z}_1;\ z_1, \bar{z}_1) = \gamma(\zeta_1, \bar{\zeta}_1;\ \zeta_1, \bar{\zeta}_1) + \log\left|\frac{d\zeta_1}{dz_1}\right|.$$

This determines g in terms of the known function γ, and therefore the path of the vortex in the z-plane is given by $\chi =$ constant, where

$$\chi = \chi_1 + \tfrac{1}{2}\kappa \log\left|\frac{d\zeta_1}{dz_1}\right|.$$

This is Routh's theorem.

As an illustration let us find the path of a vortex κ moving in the z-plane in the presence of a flat plate stretching from $z = -2a$ to $z = 2a$. Such a plate is mapped on the circle $|\zeta| = a$ by the Joukowski transformation

$$z = \zeta + a^2/\zeta.$$

The problem for the circle is solved by 13·50 (14). For brevity take $\kappa' = 0$, so that if $\zeta_1 = r(\cos\theta + i\sin\theta)$

$$\chi_1 = -\tfrac{1}{2}\kappa\log\left(1 - \frac{a^2}{r^2}\right), \qquad \frac{dz_1}{d\zeta_1} = 1 - \frac{a^2}{r^2}(\cos 2\theta - i\sin 2\theta).$$

Thus in the z-plane

$$\chi = -\tfrac{1}{2}\kappa\log\left(1 - \frac{a^2}{r^2}\right) - \tfrac{1}{4}\kappa\log\left(1 - \frac{2a^2}{r^2}\cos 2\theta + \frac{a^4}{r^4}\right),$$

with $$x_1 = \left(r + \frac{a^2}{r}\right) \cos\theta, \quad y_1 = \left(r - \frac{a^2}{r}\right) \sin\theta,$$

and the path is χ = constant. The problem which includes streaming and circulation round the plate, 13·50 (16), offers no additional difficulties.

13·61. Vortex outside a cylinder. Just as in the case of a source described in 8·71, the complex potential can be written down in terms of a mapping function,

$$(1) \qquad z = f(Z),$$

which maps the region exterior to the contour C of the cylinder in the z-plane on the region exterior to the unit circle $|Z| = 1$ in the Z-plane. Thus if there is a vortex of strength κ at z_0 outside C, there is also a vortex of strength κ at the corresponding point Z_0 outside the unit circle in the Z-plane, and by the circle theorem

$$(2) \qquad w = i\kappa \log (Z - Z_0) - i\kappa \log \left(\frac{1}{Z} - \bar{Z}_0\right),$$

which with (1) determines w as a function of z. Any distribution of vortices can be treated in this manner (see 8·70, 8·71).

13·64. Green's equivalent stratum of sources and vortices. We use the description and notation of 8·24. Since $u - iv = -dw/dz$ is a holomorphic function of z in the region L of the flow bounded by the contour C, Cauchy's formula (5·59) gives

$$(1) \qquad \frac{1}{2\pi i}\int_{(C)} \frac{(u-iv)_C \, d\zeta}{\zeta - z} = u - iv \quad \text{or} \quad 0,$$

according as z is in L or outside. Here $(u - iv)_C$ denotes the complex velocity $u - iv$ at points of the contour C. Let q_s be the tangential component of velocity in the direction of positive description of C, and let q_n be the normal component directed into L. Then

$$(2) \qquad (u - iv)_C \, d\zeta = (u - iv)_C \, ds \, e^{i\theta} = (q_s - iq_n) \, ds,$$

and therefore at a point z in the field of flow (1) gives

$$u - iv = \int_{(C)} \frac{(q_n/2\pi) \, ds}{z - \zeta} + i \int_{(C)} \frac{(q_s/2\pi) \, ds}{z - \zeta},$$

and this is the complex velocity at z of a distribution of sources of strength $q_n/2\pi$ and of vortices of strength $q_s/2\pi$, per unit length ranged on the boundary C. By (1) this distribution produces zero velocity outside L.

13·70. Vortex sheet. In 13·20 we defined a rectilinear vortex as the idealised limit of a cylindrical region of vorticity whose cross-section shrinks to a point while the amount of vorticity remains unaltered. We use an analogous process in defining a vortex sheet.

In fig. 13·70 **n** is the unit normal vector at the point P of a surface Σ. Let ϵ be an infinitesimal positive scalar and consider the points P_1, P_0 whose position vectors referred to P are $\frac{1}{2}\epsilon\mathbf{n}$, $-\frac{1}{2}\epsilon\mathbf{n}$ respectively. As P describes the surface Σ the points P_1, P_0 describe surfaces S_1, S_0 parallel to Σ which is halfway between them. Take an infinitesimal area of Σ, say dS, whose centroid is P.

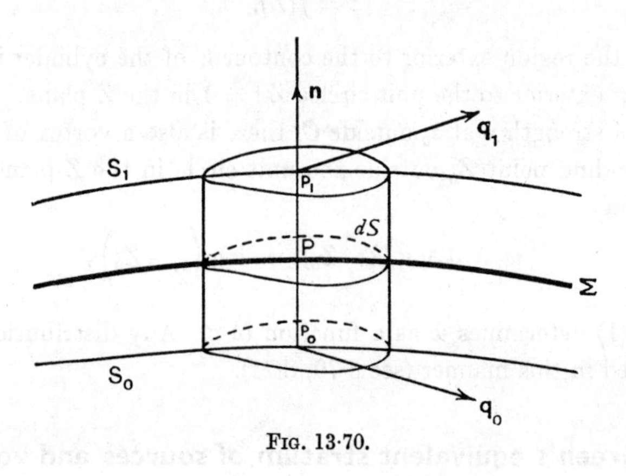

FIG. 13·70.

The normals to Σ at the boundary of dS together with the surface S_1, S_0 will delimit a cylindrical element of volume $d\tau = \epsilon\, dS$.

Now imagine the above surfaces to be drawn in fluid which is moving irrotationally everywhere except in that part which lies between S_1 and S_0. Let $\boldsymbol{\zeta}$ be the vorticity vector at P. Then we can write $\boldsymbol{\zeta}\, d\tau = \boldsymbol{\zeta}\epsilon\, dS = \boldsymbol{\omega}\, dS$, where

$$\text{(1)} \qquad \boldsymbol{\omega} = \boldsymbol{\zeta}\epsilon.$$

If we now let $\epsilon \to 0$, $\zeta \to \infty$ in such a way that $\boldsymbol{\omega}$ remains unaltered, the surface Σ is called a *vortex sheet* of vorticity $\boldsymbol{\omega}$ per unit *area*.

Before the passage to the limit, the velocity will be continuous throughout the fluid, and if $\mathbf{q}$, $\mathbf{q}_1$, $\mathbf{q}_0$ are the velocities at P, P_1, P_0, we have

$$\text{(2)} \qquad \mathbf{q}_1 = \mathbf{q} + \tfrac{1}{2}\epsilon(\mathbf{n}\nabla)\mathbf{q}, \quad \mathbf{q}_0 = \mathbf{q} - \tfrac{1}{2}\epsilon(\mathbf{n}\nabla)\mathbf{q},$$

whence by addition

$$\text{(3)} \qquad \mathbf{q} = \tfrac{1}{2}(\mathbf{q}_0 + \mathbf{q}_1).$$

This result is true however small ϵ may be. Thus the velocity of a point P *of* a vortex sheet is the arithmetic mean of the velocities just above and just below P on the normal at P.

If we apply Gauss's theorem, 2·61 (2) with $\mathbf{a} = \mathbf{q}$, to the elementary cylinder of volume $d\tau$ in fig. 13·70, we get approximately

$$\boldsymbol{\zeta}\epsilon\, dS = \mathbf{n}_\wedge(\mathbf{q}_1 - \mathbf{q}_0)\, dS,$$

neglecting a contribution of higher order of smallness from the curved surface of the cylinder. Dividing by dS and letting $\epsilon \to 0$ as before, (1) gives the exact result

$$\boldsymbol{\omega} = \mathbf{n}_\wedge(\mathbf{q}_1 - \mathbf{q}_0) \tag{4}$$

for the surface vorticity $\boldsymbol{\omega}$ of the sheet.

It is clear that a non-zero value of $\boldsymbol{\omega}$ is associated with a discontinuity of the components of $\mathbf{q}_0$, $\mathbf{q}_1$ perpendicular to $\mathbf{n}$. It follows that *a surface across which the tangential velocity changes abruptly is a vortex sheet.*

It also appears from (4) that $\boldsymbol{\omega}$ is perpendicular to $\mathbf{n}$ and is therefore tangential to the vortex sheet. A two-dimensional *vortex sheet* is represented by a line AB in the plane of the motion, such that there is an abrupt change in the tangential velocity, but no change in the normal velocity, on crossing the line AB.

For example, in rowing, the blade of an immersed oar separates fluid moving in opposite directions (cf. fig. 6·34) along the face of the blade. When the oarsman suddenly removes the blade from the water, the hollow space left quickly fills up with fluid presenting an abrupt change of tangential velocity on two sides, in fact a vortex sheet. This sheet is unstable and rolls up to form the vortex so frequently observed. A similar explanation may be offered for the vortices which follow the tip of a spoon moved across the surface of a cup of tea.

It is important to notice that the formation of vortex sheets, in the wake of a moving aerofoil for example, is in no way contradictory to the theorem that motion started by impulses must be irrotational.

In some motions (see Plate 4) a vortex trail consisting of two rows follows the body. This may be regarded as a rolling up of portions of a vortex sheet into concentrated vortices. We shall therefore develop the theory of two rows of vortices.

13·71. Single infinite row. Consider an infinite row of vortices each of strength κ at the points

$$0,\ \pm a,\ \pm 2a,\ \ldots,\ \pm na,\ \ldots.$$

The complex potential of the $2n+1$ vortices nearest the origin is

$$
\begin{aligned}
w_n &= i\kappa \log z + i\kappa \log (z-a) + \ldots + i\kappa \log (z-na) \\
&\qquad + i\kappa \log (z+a) + \ldots + i\kappa \log (z+na) \\
&= i\kappa \log \{z(z^2-a^2)(z^2-2^2a^2)\ldots(z^2-n^2a^2)\} \\
&= i\kappa \log \left\{\frac{\pi z}{a}\left(1-\frac{z^2}{a^2}\right)\left(1-\frac{z^2}{2^2a^2}\right)\ldots\left(1-\frac{z^2}{n^2a^2}\right)\right\} \\
&\qquad + i\kappa \log \left\{\frac{a}{\pi}\,.\,a^2\,.\,2^2a^2\ldots n^2a^2\right\}.
\end{aligned}
$$

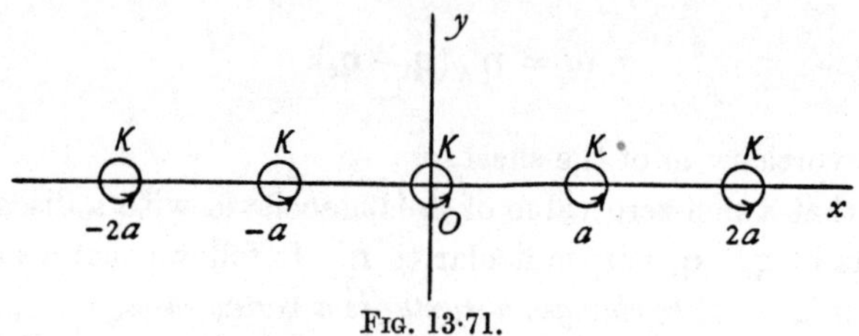

FIG. 13·71.

The constant term may be omitted, so that we write

$$w_n = i\kappa \log \left\{\frac{\pi z}{a}\left(1-\frac{z^2}{a^2}\right)\left(1-\frac{z^2}{2^2a^2}\right)\ldots\left(1-\frac{z^2}{n^2a^2}\right)\right\}.$$

Now, $\sin x$ can be expressed as an infinite product in the form *

$$\sin x = x\left(1-\frac{x^2}{\pi^2}\right)\left(1-\frac{x^2}{2^2\pi^2}\right)\ldots\left(1-\frac{x^2}{n^2\pi^2}\right)\ldots.$$

If we let $n \to \infty$, we get for the row

$$w = i\kappa \log \sin \frac{\pi z}{a}.$$

Consider the vortex at $z = 0$. Its complex velocity is given by

$$-\frac{d}{dz}\left\{i\kappa \log \sin \frac{\pi z}{a} - i\kappa \log z\right\}_{z=0} = -i\kappa\left(\frac{\pi}{a}\cot\frac{\pi z}{a} - \frac{1}{z}\right)_{z=0} = 0.$$

Thus the vortex at the origin is at rest and so therefore are all the vortices of the row. Thus the row induces no velocity in itself.

The stream function is given by

$$2i\psi = w(z) - \bar{w}(\bar{z}) = i\kappa \log \left(\sin\frac{\pi z}{a}\sin\frac{\pi \bar{z}}{a}\right),$$

$$\psi = \tfrac{1}{2}\kappa \log \tfrac{1}{2}\left(\cosh\frac{2\pi y}{a} - \cos\frac{2\pi x}{a}\right).$$

* See e.g. Hobson, *Plane Trigonometry*, § 282.

For large values of y/a we can neglect the term $\cos 2\pi x/a$, for its modulus never exceeds unity, and therefore along the streamlines ψ = constant we have y = constant. Thus at a great distance from the row the streamlines are parallel to the row.

Again, if $\mathfrak{v}_1$, $\mathfrak{v}_2$ are the complex velocities at the points z, $\bar{z}$, respectively, we have

$$\mathfrak{v}_1+\mathfrak{v}_2 = -\frac{i\kappa\pi}{a}\cot\frac{\pi z}{a} - \frac{i\kappa\pi}{a}\cot\frac{\pi\bar{z}}{a}$$

$$= -\frac{i\kappa\pi}{a}\frac{2\sin\dfrac{2\pi x}{a}}{\cosh\dfrac{2\pi y}{a}-\cos\dfrac{2\pi x}{a}},$$

which is purely imaginary and tends to zero when y tends to infinity. Thus the velocities along the distant streamlines are parallel to the row but in opposite directions. The row therefore behaves like a vortex sheet as regards distant points.

13·72. Kármán vortex street. This consists of two parallel infinite rows of the same spacing, say a, but of opposite vorticities κ and $-\kappa$, so arranged that each vortex of the upper row is directly above the mid-point of the line joining two vortices of the lower row, fig. 13·72. Taking the con-

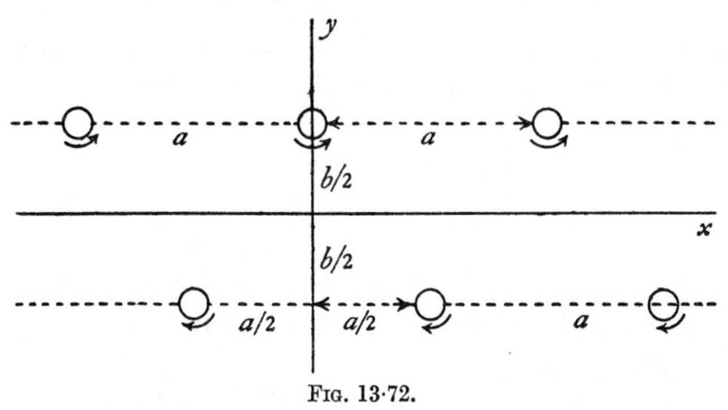

FIG. 13·72.

figuration at time $t = 0$, we take the axes as shown in the figure, the x-axis being midway between and parallel to the rows which are at the distance b apart. At this instant the vortices in the upper row are at the points $ma+\frac{1}{2}ib$, and those in the lower at the points $(m+\frac{1}{2})a-\frac{1}{2}ib$, where $m = 0$, ± 1, ± 2, ...

The complex potential at the instant $t = 0$ is therefore, by the preceding section,

$$w = i\kappa\log\sin\frac{\pi}{a}\left(z-\frac{ib}{2}\right) - i\kappa\log\sin\frac{\pi}{a}\left(z-\frac{a}{2}+\frac{ib}{2}\right).$$

Since neither row induces any velocity in itself, the velocity of the vortex at $z = \frac{1}{2}a - \frac{1}{2}ib$ will be given by

$$-u_1 + iv_1 = \left[\frac{d}{dz}\, i\kappa \log \sin \frac{\pi}{a}(z - \tfrac{1}{2}ib)\right]_{z = \frac{1}{2}a - \frac{1}{2}ib}$$

$$= \frac{i\kappa\pi}{a} \cot\left(\frac{\pi}{2} - i\frac{\pi b}{a}\right) = -\frac{\kappa\pi}{a}\tanh\frac{\pi b}{a}.$$

Thus the lower row advances with velocity

$$V = \frac{\kappa\pi}{a}\tanh\frac{\pi b}{a},$$

and similarly the upper row advances with the same velocity. The rows will advance the distance a in time $\tau = a/V$ and the configuration will be the same after this interval as at the initial instant.

To examine the stability of the arrangement, we observe that at time t the vortices of the upper row will be at the points $ma + Vt + \frac{1}{2}ib$ and those of the lower at the points $(n+\frac{1}{2})a + Vt - \frac{1}{2}ib$, where m and n take all integral values including zero from $-\infty$ to $+\infty$. If we displace each vortex slightly, those in the upper row will move to $ma + Vt + \frac{1}{2}ib + z_m$, and those in the lower row to $(n+\frac{1}{2})a + Vt - \frac{1}{2}ib + z_n'$, where $|z_m|$, $|z_n'|$ are all small initially. The system will be stable if these quantities remain small. Now the complex velocity of the vortex for which $m = 0$ will be

$$V + \frac{d\bar{z}_0}{dt}. \tag{1}$$

The contributions (13·21) to this velocity from the vortices corresponding to $\pm m$ in the upper row and $-n-1$, n in the lower will be

$$-i\kappa\left(\frac{1}{z_0 - z_m - ma} + \frac{1}{z_0 - z_{-m} + ma}\right)$$

$$+ i\kappa\left(\frac{1}{z_0 - z_n' - (n+\frac{1}{2})a + ib} + \frac{1}{z_0 - z'_{-n-1} + (n+\frac{1}{2})a + ib}\right).$$

Expanding by the binomial theorem and retaining only the first powers of z_0, z_m, z_{-m}, z'_{-n-1}, z_n', whose moduli are all small, we get

$$i\kappa\left\{\frac{z_0 - z_m + z_0 - z_{-m}}{m^2a^2} - \frac{z_0 - z'_{-n-1}}{[(n+\frac{1}{2})a + ib]^2} - \frac{z_0 - z_n'}{[(n+\frac{1}{2})a - ib]^2}\right\}$$

$$- i\kappa\left\{\frac{1}{(n+\frac{1}{2})a - ib} - \frac{1}{(n+\frac{1}{2})a + ib}\right\}.$$

If we put * $z_m = \gamma \cos m\theta$, $z_n' = \gamma' \cos(n+\frac{1}{2})\theta$, where γ and γ' are small complex numbers, the above contribution becomes

$$\frac{2\kappa i}{a^2}\frac{\gamma(1 - \cos m\theta)}{m^2} - \frac{2\kappa i(\gamma - \gamma'\cos(n+\frac{1}{2})\theta)((n+\frac{1}{2})^2 - k^2)}{a^2((n+\frac{1}{2})^2 + k^2)^2}$$

$$+ \frac{\kappa}{a^2}\cdot\frac{2ka}{(n+\frac{1}{2})^2 + k^2}, \quad k = \frac{b}{a}.$$

---

* These correspond to a displacement of an undulatory character of the rows.

Now it is known * that

$$(2) \qquad \sum_{0}^{\infty} \frac{\kappa}{a^2} \cdot \frac{2ka}{(n+\frac{1}{2})^2+k^2} = \frac{\kappa\pi}{a} \tanh k\pi = V,$$

and $d\bar{z}_0/dt = d\bar{\gamma}/dt$. Thus, summing and using (1), the disturbing effect on the vortex for which $m = 0$ is given by

$$(3) \qquad \frac{d\bar{\gamma}}{dt} = \frac{2i\kappa}{a^2}(A\gamma + C\gamma'),$$

where
$$A = \sum_{m=1}^{\infty} \frac{1-\cos m\theta}{m^2} - \sum_{n=0}^{\infty} \frac{(n+\frac{1}{2})^2 - k^2}{[(n+\frac{1}{2})^2+k^2]^2},$$

$$C = \sum_{n=0}^{\infty} \frac{[(n+\frac{1}{2})^2 - k^2]\cos(n+\frac{1}{2})\theta}{[(n+\frac{1}{2})^2+k^2]^2}.$$

For a vortex in the lower row we put $-\kappa$ for κ and interchange γ and γ', which gives

$$(4) \qquad \frac{d\bar{\gamma}'}{dt} = \frac{-2i\kappa}{a^2}(A\gamma' + C\gamma).$$

To solve these equations, differentiation of the conjugate complex of (3) gives

$$\frac{d^2\gamma}{dt^2} = -\frac{2i\kappa}{a^2}\left(A\frac{d\bar{\gamma}}{dt} + C\frac{d\bar{\gamma}'}{dt}\right) = \frac{4\kappa^2}{a^4}(A^2 - C^2)\gamma,$$

on using (3) and (4) again. Substitution of

$$\gamma = h \exp\left(\frac{2\kappa\lambda t}{a^2}\right)$$

then gives

$$\lambda^2 + C^2 = A^2.$$

Therefore λ is real and the motion is unstable if $A^2 > .C^2$

On the other hand, λ is purely imaginary and the motion is periodic and therefore stable if $C^2 > A^2$.

But when $\theta = \pi$, we get $C = 0$, for every term vanishes.

Thus we must have $A = 0$ when $\theta = \pi$ as a *necessary* condition of stability for this type of displacement.

To find A, we get from (2), by differentiation with respect to k,

$$\sum_{n=0}^{\infty} \frac{(n+\frac{1}{2})^2 - k^2}{[(n+\frac{1}{2})^2+k^2]^2} = \frac{\pi^2}{2\cosh^2 k\pi},$$

and it is easily verified by applying the rule for expansion in Fourier series that

$$\sum_{m=1}^{\infty} \frac{1-\cos m\theta}{m^2} = \tfrac{1}{4}\theta(2\pi - \theta),$$

* Hobson's *Trigonometry*, § 294.

and therefore when $\theta = \pi$,

$$A = \frac{\pi^2}{4} - \frac{\pi^2}{2\cosh^2 k\pi}.$$

Thus $A = 0$ if $\cosh k\pi = \sqrt{2}$, so that

$$k\pi = 0{\cdot}8814, \quad \text{or} \quad b = 0{\cdot}281a,$$

and the vortex street cannot be stable unless this condition is satisfied. For a further discussion of this question reference may be made to Lamb's *Hydrodynamics.**

13·73. The drag due to a vortex wake. When a cylindrical body is placed in a stream, at fairly low Reynolds' numbers 21·31 it is found that vortices leave the opposite edges alternately,† with a definite period

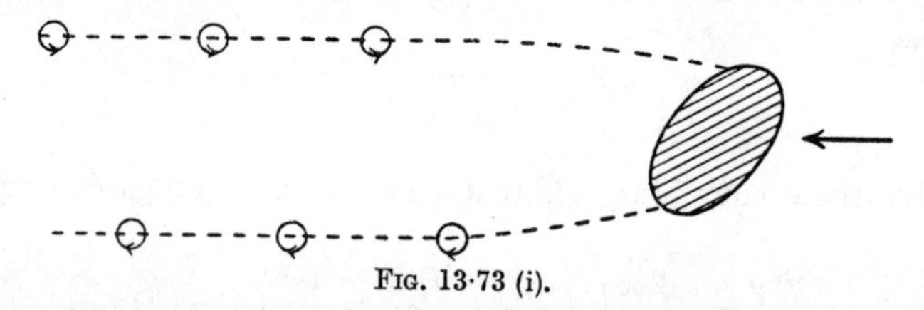

FIG. 13·73 (i).

between the formation of successive vortices, and at a distance behind the body a fully developed vortex street exists. In the immediate neighbourhood of the body the form of the vortex trail is obscure. At a great distance downstream the vortices are damped out by viscosity. In the intermediate part the vortex street exists in the form already described. We shall now investigate an approximate expression for the drag due to this form of wake.

We shall make the following assumptions:

(i) The wake can be represented by point vortices.

(ii) The origin being taken in the midst of the regular portion of the wake, the complex potential will be nearly the same as that for the infinite vortex street discussed above.

(iii) If we surround the cylinder by a contour which advances with the same velocity as the wake and whose dimensions are large compared with the cylinder and the distances between successive vortices and the rows, the motion on the boundaries of the contour will be steady.

(iv) That the formation of the vortices is truly periodic.

We shall consider the cylinder to be in motion with velocity U in liquid otherwise at rest. Take the vortices at distance a apart in the rows and let

* See also L. Rosenhead, *Proc. Roy. Soc.* (A). 127, (1930), where the stability is discussed when the vortices have finite cross-sections.

† See Plate 4.

b be the distance between the rows, fig. 13·73 (ii). Then the vortex street will advance with velocity

$$(1) \qquad V = \frac{\kappa\pi}{a} \tanh \frac{\pi b}{a},$$

where κ is the strength of each vortex. Since vortices are continually shed from the body at interval τ say, the period of the motion will be τ, and we shall have $V < U$ and $(U - V)\tau = a$.

We take the x-axis midway between the rows and in the direction of advance. By imposing a constant velocity $-V$ on the whole system the

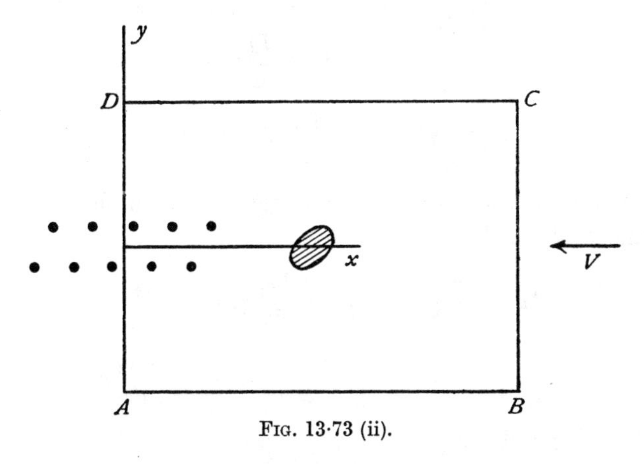

Fig. 13·73 (ii).

vortex street will be brought to rest, the cylinder will advance with velocity $U - V$, and the fluid will have a general streaming velocity $-V$ (except near the wake). The dynamical conditions will be unaltered.

We now draw a rectangle $ABCD$ of dimensions large compared with those of the cylinder and a, b, the side AD being taken coincident with the y axis which is so chosen that the origin is at the centre of the parallelogram whose vertices are the four vortices nearest the origin. Thus no vortex occurs on the boundary of this rectangle. The complex potential will then be

$$(2) \qquad w_1 = Vz + i\kappa \log \frac{\sin \dfrac{\pi(z - z_0)}{a}}{\sin \dfrac{\pi(z + z_0)}{a}} = Vz + w, \text{ say, where}$$

$$(3) \qquad z_0 = \tfrac{1}{4}a + \tfrac{1}{2}ib.$$

The term Vz represents the stream which has been superposed. Let $-X - iY$ be the action of the liquid on the cylinder. Then the *liquid* within $ABCD$ is acted upon by the thrust $X + iY$ on the internal boundary, i.e. by the cylinder, and by the pressure over the external boundary of the liquid outside.

If $H_x + iH_y$ denotes the momentum of the liquid within $ABCD$, and $F_x + iF_y$ denotes the flux of momentum *outwards* across the boundary $ABCD$, Euler's momentum theorem (3·42) gives

$$(4) \qquad X - iY - i\int_{(c)} p\, d\bar{z} = \frac{\partial}{\partial t}(H_x - iH_y) + F_x - iF_y,$$

where c denotes the contour of the rectangle $ABCD$.

Now the normal velocity outwards across the arc ds of the boundary is $-\partial\psi/\partial s$, and therefore

$$F_x = -\int_{(c)} \rho u \frac{\partial \psi}{\partial s}\, ds, \quad F_y = -\int_{(c)} \rho v \frac{\partial \psi}{\partial s}\, ds,$$

$$(5) \qquad F_x - iF_y = \int_{(c)} \rho \frac{dw_1}{dz}\, d\psi.$$

Now
$$-i\int_{(c)} p\, d\bar{z} = -i\int_{(c)} (C - \tfrac{1}{2}\rho q^2)\, d\bar{z},$$

from the pressure equation, the régime being steady on the boundary. The integral of the constant C vanishes and

$$q^2\, d\bar{z} = \frac{dw_1}{dz} \cdot \frac{d\bar{w}_1}{d\bar{z}}\, d\bar{z} = \frac{dw_1}{dz}(dw_1 - 2i\, d\psi).$$

Thus
$$-i\int_{(c)} p\, d\bar{z} = \tfrac{1}{2} i\rho \int_{(c)} \left(\frac{dw_1}{dz}\right)^2 dz + \rho\int_{(c)} \frac{dw_1}{dz}\, d\psi.$$

Substitute in (4) and use (5). Then

$$X - iY = -\tfrac{1}{2} i\rho \int_{(c)} \left(\frac{dw_1}{dz}\right)^2 dz + \frac{\partial}{\partial t}(H_x - iH_y).$$

Now, from (2),
$$\left(\frac{dw_1}{dz}\right)^2 = V^2 + 2V\frac{dw}{dz} + \left(\frac{dw}{dz}\right)^2,$$

and the integral of the first term taken round c vanishes, while that of the second is real. Therefore X is the real part of

$$-\tfrac{1}{2} i\rho \int_{(c)} \left(\frac{dw}{dz}\right)^2 dz + \frac{\partial}{\partial t}(H_x - iH_y),$$

and X, the drag, is a function of the time. We shall calculate the mean value D of the drag.

Since dw/dz is independent of the time on the contour, see assumption (iii) above, and since the motion is periodic with period τ, we have, on integrating from 0 to τ and taking the real part,

$$(6) \qquad \tau D = (H_x)_\tau - (H_x)_0 - \text{real part of } \left\{\frac{i\rho\tau}{2}\int_{(c)} \left(\frac{dw}{dz}\right)^2 dz\right\},$$

the first two terms on the right indicating the increase in x-momentum due to the appearance of two new vortices within the contour in the interval τ.

To find the first terms of (6) it is therefore necessary to calculate the increase of momentum due to the entrance of a vortex pair into the rectangle $ABCD$, considered as infinitely large. Consider

$$H'_x - iH'_y = \int \rho(u-iv)\,dS = -\rho \int \frac{dw'}{dz}\,dS,$$

where $w' = i\kappa \log(z-z_0) - i\kappa \log(z-\bar{z}_0)$. Thus, if we take for convenience of calculation, $z_0 = ih$, $\bar{z}_0 = -ih$, we get the components H'_x, H'_y given by

$$H'_x - iH'_y = -\rho \int_{-\infty}^{+\infty} dy \int_{-\infty}^{+\infty} dx \left(\frac{i\kappa}{z-ih} - \frac{i\kappa}{z+ih}\right)$$

$$= -i\rho\kappa \int_{-\infty}^{+\infty} dy \Big[\log(z-ih) - \log(z+ih)\Big]_{x=-\infty}^{x=\infty}.$$

As x increases from $-\infty$ to $+\infty$ the increase in $\log(z-ih)$ is equal to $-i\pi$ or $i\pi$ according as $y>h$ or $y<h$.

Thus the integrated part is $2\pi i$ or 0 according as y does or does not lie between $-h$ and $+h$. Hence

$$H'_x - iH'_y = -i\rho\kappa \int_{-h}^{h} dy\,(2\pi i) = 4\pi\rho\kappa h, \quad H'_x = 4\pi\rho\kappa h.$$

Resolving this along the x-axis of our problem, we get for the first pair of terms in (6) the value

$$(7) \qquad 2\pi\rho\kappa b.$$

To evaluate the integral in (6), we have, from (2),

$$-\left(\frac{dw}{dz}\right)^2 = \frac{\kappa^2\pi^2}{a^2}\left\{\cot^2\frac{\pi(z-z_0)}{a} + \cot^2\frac{\pi(z+z_0)}{a} - 2\cot\frac{\pi(z-z_0)}{a}\cot\frac{\pi(z+z_0)}{a}\right\},$$

and therefore

$$\int \left(\frac{dw}{dz}\right)^2 dz = \frac{\kappa^2\pi}{a}\left\{\cot\frac{\pi(z-z_0)}{a} + \cot\frac{\pi(z+z_0)}{a} + 2\cot\frac{2\pi z_0}{a}\log\frac{\sin\frac{\pi(z-z_0)}{a}}{\sin\frac{\pi(z+z_0)}{a}}\right\}.$$

Now $dw/dz = 0$ except on DA. Hence the value of the integral is got by finding the change in the above expression for the values $z = iy = +i\infty$ and $z = iy = -i\infty$.

Put
$$\log\zeta_1 = \frac{i\pi(z-z_0)}{a} = -\frac{\pi}{a}\left(y-\frac{b}{2}\right) - \frac{i\pi}{4},$$

$$\log\zeta_2 = \frac{i\pi(z+z_0)}{a} = -\frac{\pi}{a}\left(y+\frac{b}{2}\right) + \frac{i\pi}{4}.$$

Then

$$\int_{(DA)} \left(\frac{dw}{dz}\right)^2 dz = \frac{\kappa^2 \pi}{a}\left[\frac{i(\zeta_1^2+1)}{\zeta_1^2-1}+\frac{i(\zeta_2^2+1)}{\zeta_2^2-1}-2i\tanh\frac{\pi b}{a}\log\frac{\zeta_2(\zeta_1^2-1)}{\zeta_1(\zeta_2^2-1)}\right]_{y=\infty}^{y=-\infty}$$

$$= \frac{\kappa^2\pi}{a}\left\{4i-4i\tanh\frac{\pi b}{a}\left(-\frac{\pi i}{2}+\frac{\pi b}{a}\right)\right\}.$$

Substituting in (6) and using (7), we get

$$D = \frac{2\pi\kappa\rho b}{\tau}+\frac{2\pi\kappa^2\rho}{a}\left(1-\frac{\pi b}{a}\tanh\frac{\pi b}{a}\right),$$

which is Kármán's formula for the drag. In terms of V, this can be written, since $a = \tau(U-V)$,

$$D = \frac{2\pi\kappa\rho b}{a}\{U-2V\}+\frac{2\pi\kappa^2\rho}{a}.$$

It must be emphasised that the above calculation depends on the assumptions stated and can be regarded only as an approximation.

13·8. Vortex in compressible flow. Suppose the streamlines are circles and that in each such circle there is the same circulation $2\pi\kappa$.

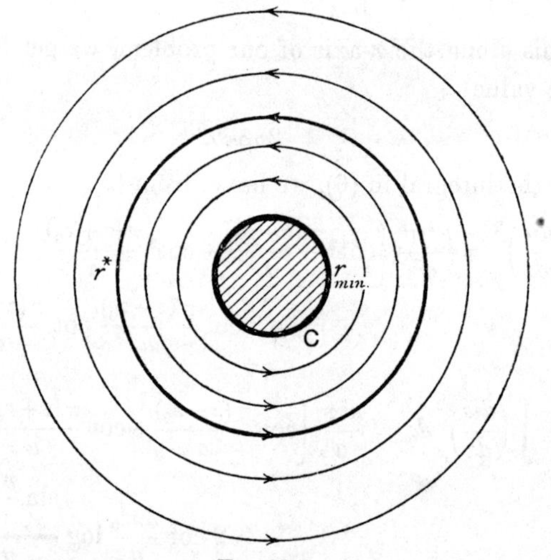

FIG. 13·8.

Then if q is the speed at radius r we have the circulation $2\pi r q = 2\pi\kappa$, so that

$$r = \frac{\kappa}{q} = \frac{\kappa}{Mc_0}\cdot\frac{c_0}{c} = \frac{\kappa}{c_0}\left\{\frac{1}{M^2}+\tfrac{1}{2}(\gamma-1)\right\}^{\frac{1}{2}},$$

using the notations of 1·63 and 1·63 (7).

Clearly then r has a minimum value $r_{\min}$ when $M = \infty$, and so

$$\frac{r}{r_{\min}} = \left\{1+\frac{2}{(\gamma-1)M^2}\right\}^{\frac{1}{2}}.$$

As r increases M must steadily decrease. M attains the critical value unity when

$$r = r^\star = r_{\min}\sqrt{\frac{\gamma+1}{\gamma-1}}.$$

Thus in Fig. 13·8 if the circle C has the radius $r_{\min}$ the motion here contemplated cannot be continued within C, so that this region must be empty of fluid or perhaps occupied by a solid core. In the region between C and the circle of radius $r^\star$ we have $M>1$ and the flow is supersonic. When $r>r^\star$ the flow is subsonic.

EXAMPLES XIII

1. If a rectilinear vortex moves (in two dimensions) in fluid bounded by a fixed plane, prove that a streamline can never coincide with a line of constant pressure.

2. Prove that the pressure due to a spiral vortex is the same as that due to a source of suitable strength.

3. A region in the plane of x, y is bounded by the lines $y = \pm c$. Two-dimensional fluid motion in the region is due to a vortex at the origin. Prove that the stream function is

$$\frac{\kappa}{2}\log\frac{1-t}{1+t},$$

where
$$t = \cos(\pi y/2c)\,\text{sech}\,(\pi x/2c),$$

and $2\pi\kappa$ is the circulation round the vortex. (U.L.)

4. Investigate the motion of two infinitely long parallel straight line vortices of the same strength, in infinite liquid.

Prove that the equation of the streamlines of the liquid relative to moving axes, so chosen that the coordinates of the vortices are $(\pm c, 0)$, is

$$\log\{[(x-c)^2+y^2][(x+c)^2+y^2]\}-(x^2+y^2)/2c^2 = \text{constant}.$$

5. Three parallel rectilinear vortices of the same strength κ and in the same sense meet any plane perpendicular to them in an equilateral triangle of side a. Show that the vortices all move round the same cylinder with uniform speed in time $2\pi a^2/(3\kappa)$.

6. A two-dimensional vortex filament of strength m is near a corner of a large rectangular tank filled with perfect fluid, the filament being parallel to the edge of the corner.

Show that the filament will trace out in plan the curve $r\sin 2\theta = $ constant, and that the motion will be regulated by the equation $r^2\dot{\theta} = m/2$.

7. Determine the motion of a rectilinear vortex filament of strength κ in infinite liquid bounded by two perpendicular infinite plane walls whose line of intersection is parallel to the filament. Show that the time taken by the vortex in moving from the position midway between the two planes to another position is proportional to $\cot 2\theta$, where θ is the angle between one of the planes and the plane containing the filament and the common line of the two planes.

Find the effect of the presence of the vortex on the pressure at any point P on one of the walls at the instant when the plane through P containing the vortex is perpendicular to the wall. (U.L.)

8. Two parallel rectilinear vortices of strengths k_1, k_2 move in a perfect fluid of infinite extent, and cross a plane perpendicular to their length at A and B respectively. G is the centre of mass for masses k_1 and k_2 at A and B, and C when the masses are interchanged. Show that the vortices rotate in circles about G with angular velocity $(k_1+k_2)/AB^2$, and the speed of a particle P of fluid in the plane is $(k_1+k_2)CP/AP \,.\, BP$.

Prove that, when ABP is an equilateral triangle, the particle P moves as if rigidly attached to the vortices; also that the same is true if P is a point on the line AB such that

$$-k_2/k_1 = (x+\tfrac{1}{2})^2(x-\tfrac{3}{2})/(x-\tfrac{1}{2})^2(x+\tfrac{3}{2}),$$

where $x = OP/AB$ and O is the mid-point of AB. (U.L.)

9. A vortex of strength m is inside a fixed circular cylinder of radius a, filled with liquid moving irrotationally, at a distance $b(<a)$ from the centre of this cylinder. Find how this vortex moves and compare with the case of the vortex which lies in an infinite liquid outside the same cylinder, $(b>a)$, there being no circulation round the cylinder alone.

Determine $\partial\phi/\partial t$ in both cases, ϕ being the velocity potential.

10. A thin rectilinear vortex exists inside and parallel to the generators of a cylindrical vessel, whose normal cross-section is bounded by a semicircle of radius a and the diameter joining its ends. Find the velocity of the vortex in any position and prove that there is a point of equilibrium on the radius bisecting the semicircle at a distance from the centre nearly equal to $0{\cdot}49a$. (U.L.)

11. A rectilinear vortex of strength κ is situated in an infinite fluid surrounding a fixed circular cylinder of radius a. The vortex is parallel to and at a distance f from the axis of the cylinder and there is no circulation in any circuit which does not enclose the vortex. Show that the vortex moves about the axis of the cylinder with a constant angular velocity equal to

$$\kappa a^2/f^2(f^2-a^2).$$

Find the velocity of the fluid at a point on the cylinder such that the axial plane through the point makes an angle θ with the axial plane through the vortex, and proceed to show how the resultant thrust on the cylinder may be calculated.

12. A long fixed cylinder of radius a is surrounded by infinite frictionless incompressible liquid, and there is in the liquid a vortex filament of strength m, which is parallel to the axis of the cylinder at a distance $c(c>a)$ from this axis. Given that there is no circulation round any circuit enclosing the cylinder but not the filament, show that the speed q of the fluid at the surface of the cylinder is

$$\frac{m}{a}\left(1-\frac{c^2-a^2}{r^2}\right),$$

r being the distance of the point considered from the filament.

Show further that, at the surface of the cylinder,

$$p = p_0 - \frac{m^2\rho}{2a^2}\left\{1-\frac{2(a^4+c^4-a^2c^2)}{c^2r^2}+\frac{(c^2-a^2)^2}{r^4}\right\},$$

where p_0 is the pressure at infinity. (U.L.)

13. Determine the stream function of the motion (of homogeneous incompressible frictionless fluid) due to a rectilinear vortex in the region bounded by two right circular cylinders with a common axis, which is parallel to the line of the vortex.

The radii of the cylinders being r_0 and r_1, and the distance of the vortex from the axis being c, prove that, if $c^2 = r_0 r_1$, the vortex remains stationary, otherwise, the path of any point on it is a circle. (U.L.)

14. Inviscid incompressible fluid is flowing past a fixed circular cylinder of radius a, its undisturbed velocity at a great distance from the cylinder being V parallel to the axis of x. The motion is two-dimensional, and the origin of coordinates is taken at the centre O of the section of the cylinder. Behind the cylinder is a vortex pair symmetrically situated with respect to the axis of x. Prove that the vortices can maintain their positions relative to the cylinder if they lie on the curve

$$2ry = r^2 - a^2,$$

and that the strengths of the vortices corresponding to a given position on this curve are

$$\pm 4Vy^2\left(\frac{1}{r} + \frac{a^2}{r^3}\right),$$

where r is the distance from O in the plane. State, without proof, whether the arrangement is stable or unstable; and explain briefly the connection of these theoretical results with the actual observed flow of a fluid of small viscosity past a circular cylinder. (U.L.)

15. Vortex filaments, all parallel to Oz, of strengths $\kappa_1, \kappa_2, \ldots$, cut the plane $z = 0$ at points $(x_1, y_1), (x_2, y_2), \ldots$. Prove that

$$\Sigma\kappa_1 x_1 = A, \quad \Sigma\kappa_1 y_1 = B, \quad \Sigma\kappa_1 r_1{}^2 = C, \quad \Sigma\kappa_1 r_1{}^2 \dot{\theta}_1 = 2\pi\Sigma\kappa_1\kappa_2.$$

If a pair of equal and opposite vortex filaments are situated inside, or outside, a circular cylinder of radius a at equal distances from its axis, prove that the equation of the cylinder described by each vortex is

$$(r^2 - a^2)^2(r^2 \sin^2\theta - b^2) = 4a^2b^2r^2 \sin^2\theta,$$

where b is a constant. (U.L.)

16. A vortex of strength κ is placed at the point $\zeta = id$ outside the circle $|\zeta| = c$. Apply the conformal transformation $iz = \zeta + c^2/\zeta$ to find the complex potential due to a vortex at $z = f$ behind a flat plate of length $4c$ about which there is circulation $2\pi\kappa(\lambda - 1)$. Show that for the velocity of the vortex to vanish $\lambda = (d^4 + c^4)/(d^4 - c^4)$, but for the velocity at the edges of the plate to be finite $\lambda = (d^2 - c^2)/(d^2 + c^2)$, and hence that the velocity at the ends cannot be finite if the vortex is at rest. Assume d and f to be real. (U.L.)

17. Three vortex filaments, each of strength m, are symmetrically placed inside a fixed circular cylinder of radius a, and pass through the corners of an equilateral triangle of side $\sqrt{3}\,.\,b$. If there is no circulation in the fluid other than that due to the vortices, show that they will revolve about the axis of the cylinder with angular velocity

$$\frac{m}{b^2}\left[\frac{a^6 + 2b^6}{a^6 - b^6}\right]. \quad \text{(U.L.)}$$

18. Show that an infinite cylinder of liquid whose cross-section is an ellipse inside which the vorticity vector $\boldsymbol{\zeta}$ is constant and parallel to the generators of the cylinder can maintain its form when rotating as a rigid body (the centres of the cross-sections being at rest) with an angular velocity $\boldsymbol{\omega} = \lambda\boldsymbol{\zeta}$, where λ depends only upon the eccentricity of the cross-section.

Find the paths of fluid particles inside the cylinder (i) relative to the rotating cross-section, and (ii) relative to a fixed frame. (U.L.)

19. Prove that a cylindrical vortex of uniform vorticity, whose normal section is bounded by an ellipse of semi-axes a, b, can exist in incompressible non-viscous liquid of uniform density ρ, at rest at infinity under pressure P, provided that it rotates about its axis with a suitable constant angular velocity n.

Show that cavitation will occur at the ends of the minor axis, unless $P > \rho n^2(a+b)a$, and that, when there is no cavitation, the relative streamlines are lines of constant pressure inside the vortex. (U.L.)

20. Prove that in the steady two-dimensional motion of a liquid of uniform vorticity ζ,

$$p/\rho = C - \tfrac{1}{2}q^2 + \zeta\psi.$$

Prove that, if $\zeta = 0$, the resultant force exerted by a uniform stream of velocity V on a fixed cylinder of any form of section is $k\rho V$ at right angles to the direction of flow, where k is the circulation in any circuit embracing the cylinder.

If $\zeta \neq 0$ and the cylinder is circular, find the form of ψ, and if the motion at infinity is a shearing motion parallel to Ox, prove that the preceding result holds, provided k is replaced by $k' + \pi a^2 \zeta$, where k' is the circulation immediately surrounding the cylinder, radius a, and V is the velocity at infinity on that streamline which would, if produced, pass through the centre of the cylinder. (U.L.)

21. An infinite row of equidistant rectilinear vortices of equal numerical strengths κ, but alternately of opposite signs, are spaced at distances a apart in infinite fluid. Show that the complex potential is

$$w = i\kappa \log \tan \frac{\pi z}{2a},$$

the origin of coordinates being at one of the vortices of positive sign, and hence show that the row remains at rest in this configuration.

Show further that if the very small radius of cross-section of each vortex filament is ϵa, then the amount of flow between two consecutive vortices is approximately $2\kappa \log 2/\pi\epsilon$.

22. Show that the complex potential w for a very long row of equidistant parallel line vortices, each of strength κ, whose traces on a plane perpendicular to them have coordinates

$$z_r = ra, \quad (r = 0, \ \pm 1, \ \pm 2, \ldots),$$

is given by

$$w = i\kappa \log \sin \frac{\pi z}{a}.$$

If the vortices suffer a *small* disturbance so that

$$z_r = ra + (\xi + i\eta)e^{ir\alpha}, \quad (0 < \alpha < 2\pi),$$

show that the disturbance increases as $e^{\lambda t}$, where

$$\lambda = \kappa\alpha(2\pi - \alpha)/2a^2. \qquad \text{(U.L.)}$$

23. Show that, at any point z, the velocity (u, v) due to an infinite row of vortices, each of strength m, at the points $z = z_0 + na$, where n is any positive or negative integer or zero and a is real and positive, is given by

$$-u + iv = \frac{im\pi}{a} \cot \frac{\pi(z - z_0)}{a}.$$

If now the vortices undergo small displacements

$$\zeta_n = \zeta_0 \cos n\chi,$$

where $0<\chi<2\pi$, obtain the equation of motion in the form

$$\frac{d\bar{\zeta}_0}{dt} = \frac{im\pi^2}{2a^2}\zeta_0\left\{1-\left(1-\frac{\chi}{\pi}\right)^2\right\},$$

where $\bar{\zeta}_0$ is the conjugate imaginary to ζ_0.

Hence show that such a single row of vortices is fundamentally unstable. (U.L.)

24. Show that the stream function for a row of an infinite number of rectilinear vortices of equal strength κ, evenly spaced at intervals a along the x-axis in infinite fluid, is

$$\psi = \frac{\kappa}{2}\log\left[\cosh\frac{2\pi y}{a} - \cos\frac{2\pi x}{a}\right],$$

all the vortices being parallel to the z-axis.

The position of a second row of such vortices of strength $-\kappa$ would be obtained by a rigid body displacement of the first set, defined by $x = \lambda a$, $y = -\mu a$. Show that such a double row or vortex " street " advances with speed

$$\frac{\pi\kappa}{a}\left(\frac{\cosh 2\pi\mu + \cos 2\pi\lambda}{\cosh 2\pi\mu - \cos 2\pi\lambda}\right)^{\frac{1}{2}}$$

in a direction θ with the street, given by

$$\tan\theta = \sin 2\pi\lambda/\sinh 2\pi\mu. \qquad \text{(U.L.)}$$

25. Two parallel rows of rectilinear vortices, evenly spaced at intervals a, are situated at distance b apart. All the vortices in one row are of equal strength K and those in the other of equal strength $-K$. Find for what arrangements the system will move forward with uniform speed, and determine these speeds. Show that if each vortex of the one row is exactly opposite one of the other the arrangement is an unstable one. (U.L.)

26. An infinite " street " of linear parallel vortices is given by the following: $x = ra$, $y = b$, strength $= m$; $x = ra$, $y = -b$, strength $= -m$, where r is any positive or negative integer, or zero. Prove that, if the fluid at infinity is at rest, the street moves as a whole, in the direction of its length, with speed

$$\frac{\pi m}{a}\coth\frac{2\pi b}{a}.$$

Show that such a street of vortices is necessarily unstable for a displacement limited to a single vortex. (U.L.)

27. Calculate the velocity of the vortices in a Kármán vortex street, the strength of each vortex in one row being m, and that of each vortex in the other row being $-m$, and the vortices of one row alternating with those of the other.

If all the vortices but one are supposed constrained to retain their relative positions, investigate the stability of the motion of the remaining vortex, showing that it cannot be stable unless a certain relation is satisfied between the mutual distances of the vortices, and find this relation. (U.L.)

28. In the notation of 13·70 prove that

$$\text{(i)}\quad \mathbf{n}(\mathbf{q}_1-\mathbf{q}_0) = 0,$$

$$\text{(ii)}\quad \boldsymbol{\omega}\wedge\mathbf{n} = \mathbf{q}_1-\mathbf{q}_0,$$

$$\text{(iii)}\quad \mathbf{q}_1 = \mathbf{q}+\tfrac{1}{2}\boldsymbol{\omega}\wedge\mathbf{n}, \quad \mathbf{q}_0 = \mathbf{q}-\tfrac{1}{2}\boldsymbol{\omega}\wedge\mathbf{n}.$$

CHAPTER XIV

FLOWS UNDER GRAVITY WITH A FREE SURFACE

In this chapter we consider some flows in which the force of gravity is taken into account.

14·10. Flow under gravity with a free surface. A free surface is a surface which always consists of the same fluid particles and along which the pressure is constant.

In the case of two-dimensional motion such a free surface is cylindrical, and we consider the curve which is the section of this cylinder by the plane of the motion.

Let then the free surface be typified by the curve C. The form of C will depend on the time t, and will have a parametric representation.

$$(1) \qquad z = f(\alpha, t) \quad \text{on } C,$$

where α is a real-valued Lagrangian coordinate (3·44) for the particles of C, such that the total t-derivatives of z agree with the partial t-derivatives of f, i.e.

$$(2) \qquad \frac{dz}{dt} = \frac{\partial f}{\partial t} = f_t \quad , \quad \frac{d^2 z}{dt^2} = \frac{\partial^2 f}{\partial t^2} = f_{tt}.$$

If $\mathbf{g}$ is the acceleration due to gravity, the equation of motion is

$$(3) \qquad d\mathbf{q}/dt - \mathbf{g} = -(\nabla p)/\rho,$$

and since ∇p is normal to a surface of constant pressure, the condition of constant pressure at the free surface is that the vector $d\mathbf{q}/dt - \mathbf{g}$ is normal to the free surface.*

In the two-dimensional case, with the y-axis vertically upwards, this condition states that $d^2z/dt^2 + ig$ is normal to C when z is on C. Since f_α is in the direction of the tangent to C, we see from (2) that

$$(4) \qquad f_{tt} + ig = ir(\alpha, t) f_\alpha \quad \text{on } C,$$

where $r(\alpha, t)$ is real valued when α is real. Thus the most general two-dimensional free surface can be represented in the form (1), where $f(\alpha, t)$, for real values of α, is a solution of (4), which is a partial differential equation of parabolic type.

---

* This can be made the basis of a three-dimensional treatment.

We note that $ir(\alpha, t)$ points in the direction of decreasing pressure.

The determination of a suitable form for the function $r(\alpha, t)$ constitutes the central difficulty of the general problem. (Cf. 14·20.)

When the function $r(\alpha, t)$ is *given*, the problem reduces to solving equation (4) subject to boundary conditions. Observe also that when $r(\alpha, t)$ is given, (4) is a linear equation for $f(\alpha, t)$ and therefore solutions can be added, i.e. the principle of superposition applies.

14·12. Potential flow with a free surface. Referring to 14·10 (1), (4), let us determine the complex potential $w(z, t)$ on the assumption that the motion is to be irrotational.

We have $u+iv = dz/dt$ and therefore

$$\frac{dz}{dt} = -\frac{\partial \bar{w}(\bar{z}, t)}{\partial \bar{z}} = -\bar{w}_{\bar{z}} = -\overline{w_z}, \tag{1}$$

and therefore from 14·10 (2)

$$f_t(\alpha, t) = -\overline{w_z(z, t)} \quad \text{on } C. \tag{2}$$

By means of 14·10 (1) we can express $w(z, t)$ as a function of α and t and therefore

$$w_\alpha(\alpha, t) = w_z(z, t)\, z_\alpha = -\overline{f_t(\alpha, t)}\, f_\alpha(\alpha, t) \quad \text{on } C.$$

Since α is real on C, this relation can be written in the form

$$w_\alpha(\alpha, t) = -\overline{f_t(\bar{\alpha}, t)}\, f_\alpha(\alpha, t). \tag{3}$$

Assuming that $f(\alpha, t)$ and $f_t(\alpha, t)$ are analytic functions of α, the right-hand side of (3) is also an analytic function of α. Thus we can use (3) to define w as an analytic function of α for complex values of the parameter α, and hence as an analytic function of z. The complex potential $w(z, t)$ so defined will be the complex potential of a flow which is consistent with the free surface given by $z = f(\alpha, t)$.

Therefore any analytic solution $f(\alpha, t)$ of 14·10 (4), wherein the coefficient $r(\alpha, t)$ is a real valued function for real values of α, represents a possible free surface motion, for which the corresponding free surface is $z = f(\alpha, t)$, and for which the complex potential is determined from (3) by a quadrature.*

14·14. Steady potential flow with a free surface. We take axes x_0, y_0. Let $z_0 = x_0 + iy_0$ and let the complex potential be $w_0 = w_0(z_0) = \phi_0 + i\psi_0$. We take the free surface to be the streamline $\psi_0 = 0$, so that

$$\psi_0 = 0, \; w_0(z_0) = \bar{w}_0(\bar{z}_0), \quad \text{at the free surface.} \tag{1}$$

* The method here described is due to Fritz John, *Communications on Pure and Applied Mathematics*, VI (1953), 497–503.

Therefore at the free surface

(2) $$q_0^2 = \frac{dw_0}{dz_0}\frac{d\bar{w}_0}{d\bar{z}_0} = \frac{d\phi_0}{dz_0}\frac{d\phi_0}{d\bar{z}_0}.$$

Therefore by Bernoulli's theorem, since the pressure is constant at the free surface,

(3) $$\tfrac{1}{2}\frac{d\phi_0}{dz_0}\frac{d\phi_0}{d\bar{z}_0} + \Omega_0 = \tfrac{1}{2}V^2,$$

where Ω_0 is the gravitational potential and V is a constant of the dimensions of velocity. By proper choice of the height of the origin we can always arrange that V^2 is positive. From (3) we get

(4) $$\frac{dz_0}{d\phi_0}\frac{d\bar{z}_0}{d\phi_0} = \frac{1}{V^2 - 2\Omega_0}, \quad \text{at the free surface.}$$

Let the components of gravity be $(g \sin \alpha, \ -g \cos \alpha)$. Then

(5) $$\Omega_0 = -\tfrac{1}{2}ig\,(z_0 e^{-i\alpha} - \bar{z}_0 e^{i\alpha}),$$

which becomes gy_0 when $\alpha = 0$, and $-gx_0$ when $\alpha = \frac{1}{2}\pi$.

Let a be a fixed length and introduce dimensionless quantities z, w, Ω, F^2 defined by

(6) $$z_0 = az, \ w_0 = aVw(z), \ \Omega_0 = ag\Omega, \ F^2 = V^2/(ag).$$

Then (4) can be written

(7) $$\frac{dz}{d\phi}\frac{d\bar{z}}{d\phi} = \frac{F^2}{F^2 - 2\Omega\,(\phi)}, \quad \text{at the free surface.}$$

Here F^2 is a Froude number (cf. 12·1) which is infinite when $g = 0$. Also (5) becomes

(8) $$\Omega = -\tfrac{1}{2}i\,(ze^{-i\alpha} - \bar{z}e^{i\alpha}).$$

Differentiate this with respect to ϕ. Then

(9) $$\frac{d\bar{z}}{d\phi}e^{i\alpha} = \frac{dz}{d\phi}e^{-i\alpha} - 2i\frac{d\Omega(\phi)}{d\phi}, \quad \text{at the free surface.}$$

Observe in this connection that at the free surface z and $\bar{z}$ are functionally related and therefore Ω as defined by (8) is a function of ϕ,

(10) $$\Omega = \Omega(\phi), \quad \text{at the free surface.}$$

Now eliminate $d\bar{z}/d\phi$ between (7) and (9). Then

(11) $$\frac{dz}{d\phi}\left(\frac{dz}{d\phi}e^{-i\alpha} - 2i\frac{d\Omega(\phi)}{d\phi}\right) = \frac{F^2 e^{i\alpha}}{F^2 - 2\Omega(\phi)}, \quad \text{at the free surface.}$$

This equation defines z as an analytic function of ϕ, and therefore since $w = \phi$ at the free surface where (11) is satisfied we have the equation

$$\frac{dz}{dw}\left(\frac{dz}{dw}e^{-i\alpha}-2i\frac{d\Omega(w)}{dw}\right)=\frac{F^2e^{i\alpha}}{F^2-2\Omega(w)} \tag{12}$$

to determine z as a function of w in terms of the arbitrary analytic function $\Omega(w)$ which reduces to the real valued function $\Omega(\phi)$ when $\psi = 0$.

Observe that $\Omega(\phi)$ is not known in advance, for it will be determined from (8) only when the functional relation between z and $\bar{z}$ has been determined.

If we assign functional forms to $\Omega(\phi)$ or $\Omega(w)$, we can find the corresponding flow. The idea is due to M. C. Sautreaux.*

If we denote by a dash differentiation with respect to w, (12) can be written

$$z'^2-2i\Omega' e^{i\alpha}z'+\frac{F^2e^{2i\alpha}}{2\Omega-F^2}=0, \tag{13}$$

which likewise represents (11) at the free surface.

If $\alpha = 0$, i.e. if the y-axis is vertically upwards,

$$z'^2-2i\Omega'z'+\frac{F^2}{2\Omega-F^2}=0. \tag{14}$$

If $g = 0$, F^2 is infinite and (14) becomes

$$z'^2-2i\Omega'z'-1=0. \tag{15}$$

This equation therefore is apt to represent all flows with free streamlines when the gravitational field is absent.

Solving (14) for z' we get

$$\frac{dz}{dw}=i\Omega'(w)\pm\sqrt{\left(-\Omega'^2(w)-\frac{F^2}{2\Omega(w)-F^2}\right)}. \tag{16}$$

Since $dz/dw = -1/(u-iv) = -u/q^2-iv/q^2$, we find u and v from the real and imaginary parts of the right-hand side of (16).

Integrating (16) we get

$$x+iy=z=i\Omega(w)\pm\int\sqrt{\left(-\Omega'^2(w)-\frac{F^2}{2\Omega(w)-F^2}\right)}\,dw. \tag{17}$$

If in (17) we put $\psi = 0$, we find the equation of the streamline $\psi = 0$, x and y being expressed as functions of the parameter ϕ. If the radicand is negative for a range of ϕ, we get x = constant for this range, so that part of the streamline $\psi = 0$ will consist of a vertical line which could be replaced by a rigid wall or boundary. The free streamline will therefore correspond to the case where the radicand is positive. Combining this with (7), for which the left-hand side is necessarily positive, we see that on the free streamline

* "Sur une question d'hydrodynamique", *Ann. Scient. de l'École Normale Supérieure,* 10 (1893), 95–182.

$$(18) \qquad -\frac{1}{\Omega'^2(\phi)} < \frac{2\Omega(\phi)-F^2}{F^2} < 0.$$

This inequality therefore delimits the range of values of ϕ which correspond to points on the free streamline.

Example (i). $F^2 = 1$, $2\Omega(w)-1=2w$, $\alpha = 0$. Then (17) gives

$$(19) \qquad x+iy = z = i(w+\tfrac{1}{2}) \pm \int \sqrt{\left(-1-\frac{1}{2w}\right)}\, dw.$$

The streamline $\psi = 0$ is given by

$$(20) \qquad x+iy = i(\phi+\tfrac{1}{2}) - \int \sqrt{\left(-1-\frac{1}{2\phi}\right)}\, d\phi,$$

where we have chosen the negative sign for the radical.

The radicand is negative unless ϕ lies in the interval $(-\frac{1}{2}, 0)$. For values of ϕ outside this interval x = constant. Thus $\psi = 0$ consists in part of a vertical wall. If $-\frac{1}{2} < \phi < 0$ we put $\phi = -\frac{1}{4}(1+\cos 2\theta)$, and then

$$(21) \qquad x+iy = \tfrac{1}{4}i(1-\cos 2\theta) - \int \sin^2\theta\, d\theta$$
$$= \tfrac{1}{4}i(1-\cos 2\theta) - \tfrac{1}{2}\theta + \tfrac{1}{4}\sin 2\theta + C.$$

The constant C is arbitrary, but it will be found convenient (not essential) to give it the value $\frac{1}{4}\pi$. We then get

$$(22) \qquad x = \tfrac{1}{4}\pi - \tfrac{1}{2}\theta + \tfrac{1}{4}\sin 2\theta, \quad y = \tfrac{1}{4}(1-\cos 2\theta).$$

This is the equation of a cycloid whose cusps correspond to $\theta = 0$, $(\frac{1}{4}\pi, 0)$, $\theta = \pi$, $(-\frac{1}{4}\pi, 0)$ and whose vertex is at $\theta = \frac{1}{2}\pi$, $(0, \frac{1}{2})$.

From (16) $-\dfrac{u}{q^2} - \dfrac{iv}{q^2} = i - \tan\theta$ when $\psi = 0$. Thus $u = q^2 \tan\theta$, $v = -q^2$.

When $\theta = \frac{1}{2}\pi$, $u/q^2 = \infty$ and therefore $u = v = q = 0$, so that the vertex of the cycloid is a stagnation point. As θ goes from $\frac{1}{2}\pi$ to 0, u is positive, while as θ goes from $\frac{1}{2}\pi$ to π, u is negative. Thus the cycloid is described opposite ways when a particle moves from the vertex to the cusps, in fact there is symmetry of the flow about $x = 0$.

Therefore using the principle of the dividing streamline we have the flow shown in fig. 14·14.

If we wish to discuss the interior streamlines we write

$$(23) \qquad w = -\tfrac{1}{4}(1+\cos(2\theta+2i\eta)),$$

which reduces to $\phi = -\frac{1}{4}(1+\cos 2\theta)$ when $\eta = 0$.

From (23)

$$(24) \qquad \psi = \tfrac{1}{4}\sin 2\theta \sinh 2\eta,$$

which gives the streamlines ψ = constant. In particular $\psi = 0$ when $\eta = 0$, the case just discussed, or when $\theta = 0$, $\frac{1}{2}\pi$, π, and so on. Thus the flow pattern is a periodic reproduction of fig. 14·14.

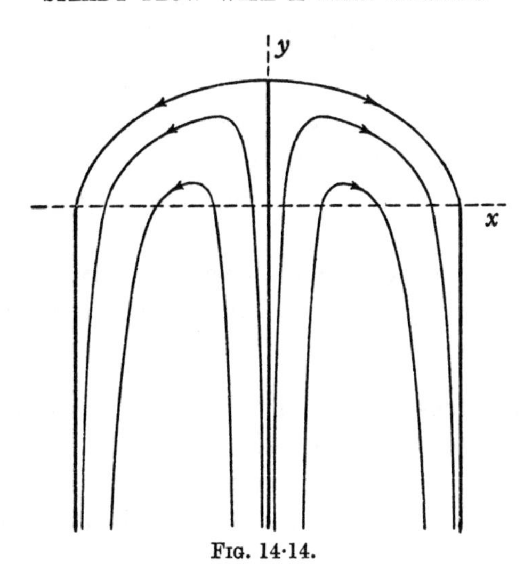

FIG. 14·14.

Example (*ii*). $F^2 = \infty$, $\Omega'(w) = e^w$, $\alpha = -\frac{1}{2}\pi$.

This is a case in which gravity is absent and which leads to flow through an orifice as discussed in 11·53.

Equation (13) becomes

(25) $$z'^2 - 2e^w z' + 1 = 0.$$

Putting $\psi = 0$ and then solving we get

(26) $$\frac{dz}{d\phi} = e^\phi + \sqrt{(e^{2\phi} - 1)},$$

where we have taken the positive sign for the radical.

If $\phi > 0$, the right-hand side is real, and since $-d\phi/dz = u - iv$, we see that $v = 0$ so that there is a rigid wall parallel to the x-axis, with which, by choice of origin, it will coincide.

When $\phi < 0$ we put

(27) $$e^\phi = \sin\theta,$$

and then (26) becomes

$$\frac{dz}{d\theta} = \cot\theta\,(\sin\theta + i\cos\theta) = \cos\theta + i\left(\frac{1}{\sin\theta} - \sin\theta\right).$$

Integration, taking the arbitrary constant to be $\frac{1}{2}\pi$, gives

(28) $$z = \tfrac{1}{2}\pi + \sin\theta + i(\cos\theta + \log\tan\tfrac{1}{2}\theta),$$

which gives the free streamline

$$x = \tfrac{1}{2}\pi + \sin\theta, \quad y = \cos\theta + \log\tan\tfrac{1}{2}\theta.$$

If in (25) we write $\psi = \pi$ instead of $\psi = 0$, we get a second free streamline for which (27) becomes $e^\phi = -\sin\theta$. To get the equation we simply change the sign of θ in (28). Since $\log[-\tan\frac{1}{2}\theta] = i\pi + \log\tan\frac{1}{2}\theta$, this gives for the second free streamline

$$z = -\tfrac{1}{2}\pi - \sin\theta + i(\cos\theta + \log\tan\tfrac{1}{2}\theta)$$

which is the mirror image of (27) in the y-axis.

The flow is shown in fig. 11·53 (i), but here the origin is the middle point of BB'.

The width of the orifice, in our dimensionless coordinates is $2(\frac{1}{2}\pi + \sin\frac{1}{2}\pi)$ and of the jet at infinity is $2(\frac{1}{2}\pi + \sin 0)$. Therefore the coefficient of contraction is $\pi/(\pi+2)$.

14·20. Tangent flows. Referring to 14·10 (1) let z_0 denote a point of the free surface, the motion being assumed steady. We define the Lagrangian coordinate α of a surface particle by the condition that $-\alpha$ is the time at which the particle occupies the position z_0. In steady motion all particles take the same time β to travel from position z_0 to position z. Therefore the function $z = f(\alpha, \beta - \alpha)$ must be independent of α for every β. Hence

$$(1) \qquad z = f(\alpha, t) = z(\beta), \quad \beta = \alpha + t.$$

From 14·10 (4) it then follows that $r(\alpha, t) = (f_{tt} + ig)/if_\alpha$ must be a function of β alone, say

$$(2) \qquad r(\alpha, t) = S'(\beta).$$

Then 11·60 (4) reduces to the ordinary differential equation

$$(3) \qquad z''(\beta) + ig = iS'(\beta)z'(\beta),$$

where $S'(\beta)$ and therefore $S(\beta)$ is real for real β. This linear equation can be solved by two quadratures to give $z(\beta)$.

If the motion is irrotational, it follows from 14·12 (3) that w is also a function of $\beta = \alpha + t$ which is determined by a quadrature from

$$(4) \qquad \frac{dw}{d\beta} = -\overline{z'(\bar{\beta})}\, z'(\beta).$$

We can now make an important remark. If in (3) we put $g = 0$, we have a problem in which gravity is absent, the type of problem already considered in chapter XII. Having found $z_0(\beta)$, the solution of this gravity-free problem, we are in possession of the complementary function of the linear equation (3) when $g = 0$, namely

$$(5) \qquad iS'(\beta) = z_0''(\beta)/z_0'(\beta) \quad \text{when } g = 0.$$

If we insert this in (3), we obtain the equation

$$(6) \qquad z''(\beta) + ig = \frac{z_0''(\beta)}{z_0'(\beta)}\, z'(\beta),$$

which can be solved in the form

$$(7) \qquad z_1(\beta) = z_0(\beta) + g\{F(\beta) - F(0)\},$$

where $F(\beta)$ is a particular integral of (6). Since (6) is of type (3) the solution

$z_1(\beta)$ gives a free surface of constant pressure, which reduces to $z_0(\beta)$ when $g = 0$.

Also the solution $z_1(\beta)$ contains a free parameter, namely the skin speed in the solution $z_0(\beta)$.

We shall call $z_1(\beta)$ the *tangent solution** to the given problem, for it reduces to the gravity-free problem when $g = 0$.

14·21. The tangent solution to the vertically downwards jet.

Consider the vertically downwards jet which issues from an orifice of breadth $2a$ in a flat horizontal infinite plate, fig. 11·53 (i). We shall take the origin not at B, but at the mid-point of BB' and we shall consider, when $g = 0$, the free streamline BC_∞.

For this free streamline, we have from 14·14 (28),

$$(1) \quad x_0 = k(\tfrac{1}{2}\pi + \sin\theta), \quad y_0 = k(-\log\cot\tfrac{1}{2}\theta + \cos\theta), \quad k = 2a/(\pi+2),$$

where θ here denotes the acute angle between the direction of motion and the vertical so that at B, $\theta = \frac{1}{2}\pi$ and at C_∞, $\theta = 0$.

If U is the skin speed of the jet, and β is the time taken by a particle to move from B to P on the free streamline, we have

$$(2) \qquad U^2 = z_0'(\beta)\bar{z}_0'(\beta),$$

and from (1)

$$(3) \qquad z_0(\theta) = k\{\tfrac{1}{2}\pi + \sin\theta + i\cos\theta - i\log\cot\tfrac{1}{2}\theta\}.$$

The differential equation of the tangent solution 14·20 (6) can be written

$$\frac{d}{d\beta}\left(\frac{z'(\beta)}{z_0'(\beta)}\right) = -\frac{ig}{z_0'(\beta)}\cdot = \frac{-ig}{U^2}\bar{z}_0'(\beta) \quad \text{from (2).}$$

Changing the variable from β to θ we have

$$(4) \qquad \frac{d}{d\theta}\left(\frac{z'(\theta)}{z_0'(\theta)}\right) = -\frac{ig}{U^2}\bar{z}_0'(\theta)$$

whence

$$\frac{z'(\theta)}{z_0'(\theta)} = A - \frac{ig}{U^2}\bar{z}_0(\theta)$$

and so

$$(5) \qquad z(\theta) = Az_0(\theta) + B - \frac{ig}{U^2}\int_{\frac{1}{2}\pi}^{\theta} z_0'(\theta)\,\bar{z}_0(\theta)\,d\theta,$$

We must take $A = 1$, $B = 0$, and so

$$(6) \qquad z_1(\theta) = z_0(\theta) - ig[F(\theta) - F(\tfrac{1}{2}\pi)].$$

* L. M. Milne-Thomson, *Proc. Midwestern Conference on Solid and Fluid Mechanics* 1959.

Performing the integration we get

$$(7) \qquad z_1(\theta) = z_0(\theta) + \frac{gk^2}{U^2}\left\{\theta - \frac{\pi}{2} - \frac{1}{2}i\left(\log\cot\frac{\theta}{2}\right)^2 + ie^{-i\theta}\log\cot\frac{\theta}{2}\right\}.$$

Observe that when $\theta = 0$, $\log\cot\frac{1}{2}\theta = \infty$ but that

$$(8) \qquad \lim_{\theta\to 0}(\sin\theta\log\cot\tfrac{1}{2}\theta) = 0.$$

From (7)

$$x = x_0 + \frac{gk^2}{U^2}\left\{\theta - \frac{\pi}{2} + \sin\theta\log\cot\frac{\theta}{2}\right\}$$

$$= k\left\{\frac{\pi}{2} + \sin\theta\right\} + \frac{gk^2}{U^2}\left(\theta - \frac{\pi}{2} + \sin\theta\log\cot\frac{\theta}{2}\right),$$

and therefore

$$(9) \qquad \lim_{\theta\to 0} x = \frac{\pi}{2}\left\{k - \frac{gk^2}{U^2}\right\}$$

and this gives the asymptote of the free streamline of the tangent flow.

Now by the equation of continuity the same quantity of liquid must flow through each cross-section of the jet in the same time, and, as the velocity clearly tends to infinity as $\theta \to 0$, the ultimate form of the jet will coincide with this asymptote.

In particular, let us choose the skin speed such that (9) vanishes; that is to say

$$(10) \qquad U^2 = gk.$$

Introduce this into (7) and substitute for $z_0(\theta)$ from (3). Then the free streamline of the tangent solution is

$$(11) \qquad z_1 = \frac{U^2}{g}\{\theta + \lambda\sin\theta + i[\lambda\cos\theta + \tfrac{1}{2}(1-\lambda^2)]\},$$

where $\lambda = 1 + \log\cot\frac{1}{2}\theta$.

When $\theta = \frac{1}{2}\pi$ we have $\lambda = 1$ and

$$z_1\left(\frac{\pi}{2}\right) = \frac{U^2}{g}\left(\frac{\pi}{2} + 1\right) = k\left(\frac{\pi}{2} + 1\right) = a.$$

At infinity $\theta = 0$, $\lambda = \infty$ and, since $\lambda\sin\theta \to 0$,

$$z_1(0) = \frac{U^2}{g}\{i(-\infty)\} = -i\infty.$$

We can easily verify that the pressure is constant on the free streamline, for from (11)

$$(12) \qquad 2gy_1 = U^2\{2\lambda\cos\theta + 1 - \lambda^2\}.$$

Also from 14·20 (4)

$$\frac{dw}{d\beta} = -\overline{z'(\beta)}\, z'(\beta).$$

Therefore

$$\frac{dw}{dz} = -\overline{z'(\beta)} = \frac{U}{k}\frac{\overline{z'(\theta)}}{\cot\theta}.$$

Therefore

$$q^2 = \frac{U^2}{k^2}\frac{z_1'(\theta)\,.\,\bar{z}_1'(\theta)}{\cot^2\theta} = U^2(\lambda^2 - 2\lambda\cos\theta + 1).$$

Adding this to (12) we get

$$q^2 + 2gy_1 = 2U^2,$$

which shows, by Bernoulli's theorem, that the pressure is constant on the surface of the jet.

14·30. Waves. A *wave motion* of a liquid, acted upon by gravity and having a free surface, is a motion in which the elevation of the free surface above some chosen fixed horizontal plane varies.

A *progressive wave* is one in which the surface pattern moves forwards.

If the surface pattern repeats when the wave has moved forwards a distance λ, then λ is the wave-length. (See figs. 15·10 (i), (ii), (iii).)

A *stationary or standing wave* is one in which the surface changes shape by moving vertically without translation. (See figs. 15·32, 15·33.)

In two-dimensional waves the curve in which the free surface meets the plane of the motion is called a *wave profile* or simply a *profile*.

14·40. Gerstner's trochoidal wave. In 1802 Gerstner, Professor of Mathematics at Prague, showed that a trochoidal profile properly chosen would make the pressure constant without approximation at the free surface of deep water. This is the only known exact solution of the problem of wave motion. The motion is, however, not irrotational.

Take the axis of x horizontal and the axis of y vertically upwards. Let a, b be Lagrangian parameters which fix the position of a particular fluid particle when there is no wave. Then Gerstner's wave is obtained by supposing that the position of this particular particle at time t is given by

$$(1) \qquad x = a + \frac{1}{m} e^{mb} \sin m(a+ct), \quad y = b - \frac{1}{m} e^{mb} \cos m(a+ct).$$

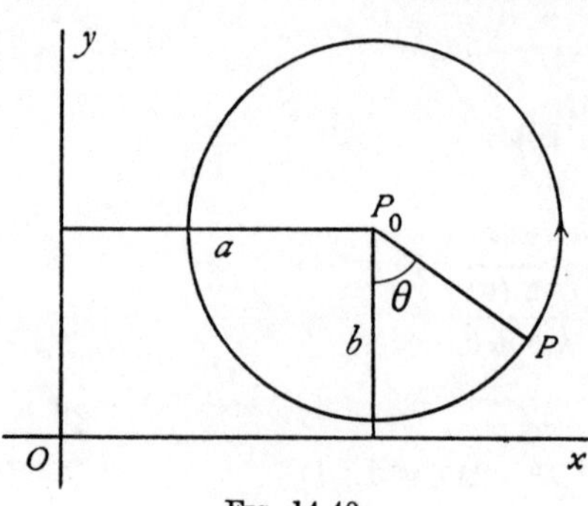

FIG. 14·40.

From this it is evident that the path of this particle is the circle whose centre is (a, b), and whose radius is e^{mb}/m, fig. 14·40.

The angular velocity of the radius joining the particle P to the centre P_0 is mc.

If we fix our attention on another particle we merely change the values of a, b in (1).

To show that (1) represents a possible fluid motion, we must prove that the equation of continuity is satisfied. We have, from (1),

$$z = a + ib - \frac{i}{m} \exp\,[m(b + ia + ict)]. \tag{2}$$

From 3·44 the equation of continuity is

$$\text{constant} = \frac{\partial(x, y)}{\partial(a, b)} = \frac{\partial x}{\partial a}\frac{\partial y}{\partial b} - \frac{\partial x}{\partial b}\frac{\partial y}{\partial a} = Re\left(i\,\frac{\partial z}{\partial a}\frac{\partial \bar{z}}{\partial b}\right) = 1 - e^{2mb},$$

which is constant, so that the motion is possible.

We must now obtain the surface condition. The accelerations of the particle are $\partial^2 x/\partial t^2$, $\partial^2 y/\partial t^2$, and therefore the equations of motion are

$$\frac{\partial^2 x}{\partial t^2} = -\frac{1}{\rho}\frac{\partial p}{\partial x}, \quad \frac{\partial^2 y}{\partial t^2} = -g - \frac{1}{\rho}\frac{\partial p}{\partial y},$$

or

$$\frac{\partial}{\partial x}\left(\frac{p}{\rho} + gy\right) = mc^2 e^{mb} \sin m(a + ct),$$

$$\frac{\partial}{\partial y}\left(\frac{p}{\rho} + gy\right) = -\,mc^2 e^{mb} \cos m(a + ct).$$

Multiply these respectively by

$$\frac{\partial x}{\partial a} = 1 + e^{mb} \cos m(a + ct), \quad \frac{\partial y}{\partial a} = e^{mb} \sin m(a + ct),$$

and then add. This gives

$$\frac{\partial}{\partial a}\left(\frac{p}{\rho} + gy\right) = mc^2 e^{mb} \sin m(a + ct). \tag{3}$$

Similarly, by adding after multiplication by $\partial x/\partial b$, $\partial y/\partial b$, we get

$$\frac{\partial}{\partial b}\left(\frac{p}{\rho} + gy\right) = -\,mc^2 e^{mb} \cos m(a + ct) + mc^2 e^{2mb}. \tag{4}$$

Multiply (3) and (4) by da, db respectively and add, giving

$$d\left(\frac{p}{\rho}+gy\right) = d[-c^2 e^{mb}\cos m(a+ct)+\tfrac{1}{2}c^2 e^{2mb}],$$

and therefore

$$\frac{p}{\rho} = \text{constant} - g\left(b-\frac{1}{m}e^{mb}\cos m(a+ct)\right) - c^2 e^{mb}\cos m(a+ct)+\tfrac{1}{2}c^2 e^{2mb}.$$

For a particle at the free surface p must be constant if surface tension is ignored, and therefore the coefficient of $\cos m(a+ct)$ must vanish, which gives

$$c^2 = g/m = g\lambda/2\pi \tag{5}$$

Thus the free surface condition is satisfied without approximation, and the pressure at *any fluid particle* whose parameters are (a, b) is given by

$$\frac{p}{\rho} = \text{constant} - bg + \tfrac{1}{2}c^2 e^{2mb},$$

and therefore the pressure is constant if b is constant.

This means that the pressure has the same value for any given particle as it moves about.

In particular, the pressure is constant for all particles for which the parameter b is the same irrespectively of the value of a.

If we take $b = \beta$ for particles in the free surface and Π to be the pressure there, we get

$$\frac{p-\Pi}{\rho} = g(\beta-b)+\tfrac{1}{2}c^2(e^{2mb}-e^{2m\beta}),$$

which determines the pressure at any other particle.

The group velocity, 15·22, is $c_g = c-\lambda\dfrac{dc}{d\lambda} = \tfrac{1}{2}c$ from (5).

14·41. Form of the free surface. To examine the form of the curves of constant pressure, equations (1) show that x and y are periodic functions of t, the period being $2\pi/(mc)$.

Keeping b and t fixed, the values of y recur when a is increased by $2\pi/m$, while the value of x undergoes a linear shift of amount $2\pi/m$. Thus the greatest values of y recur at points separated by the distance $2\pi/m$. Fixing attention on one of these greatest values of y, we see that an increase in t will cause that value to occur for a smaller value of a in order to keep the phase angle $m(a+ct)$ the same. Thus the profiles of the surfaces of equal pressure move in the *negative* direction of the axis of x with velocity equal to the wavelength $2\pi/m$ divided by the period $2\pi/(mc)$, i.e. with velocity c. If we impose on every particle a forward velocity c, the motion will become steady and the profiles will remain fixed. Writing $\theta = m(a+ct)$, we have then for the equation of the profiles of the surfaces of equal pressure

$$x = \frac{\theta}{m} + \frac{1}{m} e^{mb} \sin\theta, \quad y = b - \frac{1}{m} e^{mb} \cos\theta.$$

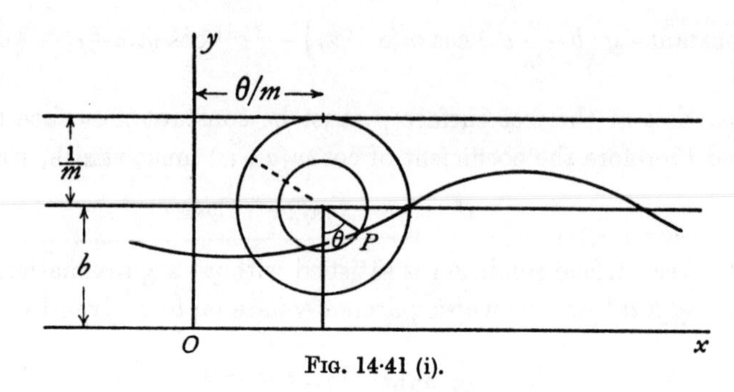

FIG. 14·41 (i).

These curves are trochoids generated by a point carried at the distance e^{mb}/m from the centre of a circle of radius $1/m$ which rolls on the underside of the line $y = b+1/m$, fig. 14·41 (i).

If we take $b = 0$ in the free surface, the corresponding profile is a cycloid. The curves of equal pressure are shown in fig. 14·41 (ii).

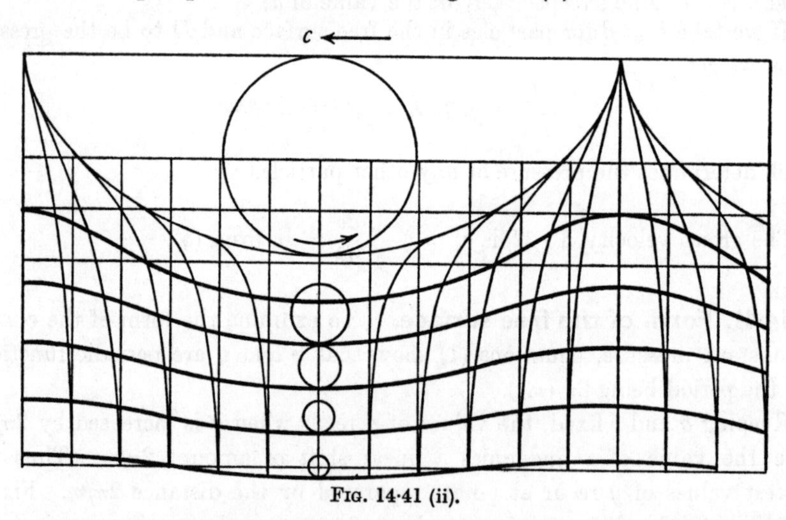

FIG. 14·41 (ii).

Any one of these may be taken as the free surface. The extreme form is the cycloid with cusps upwards. The vertical lines show the undisturbed positions of the water columns.

To find the mean level $y = k$ corresponding to any trochoid, that is to say, the level with respect to which the same amount of water is elevated as depressed, we observe that $\int (y-k)\,dx = 0$ taken over a wave-length. Thus

$$\int_0^{2\pi} \left(b-k-\frac{1}{m}\,e^{mb}\cos\theta\right)(1+e^{mb}\cos\theta)\,d\theta = 0,$$

which gives

$$b-k = \frac{e^{2mb}}{2m}. \tag{1}$$

Thus the mean level is below the path of the centre of the generating circle by this amount.

As we go down into the liquid, the distance of the tracing point from the centre of the generating circle decreases.

For the progressive wave, the kinetic energy (per unit thickness) is found by integrating over a wave-length the kinetic energy of the elementary mass $\rho(1-e^{2mb})\,da\,db$, defined by a fluid particle (a, b). We have from 14·40 (2)

$$q^2 = \frac{\partial z}{\partial t}\frac{\partial \bar{z}}{\partial t} = c^2 e^{2mb} \quad \text{from 14·40 (1).}$$

Hence, if we take $b = \beta$ to define the free surface, the kinetic energy is

$$T = \tfrac{1}{2}\rho c^2 \int_{-\infty}^{\beta}\int_a^{a+\frac{2\pi}{m}} (e^{2mb}-e^{4mb})\,da\,db = \tfrac{1}{2}\rho c^2 \cdot \frac{2\pi}{m} \times \left(\frac{e^{2m\beta}}{2m} - \frac{e^{4m\beta}}{4m}\right).$$

Put $\lambda = 2\pi/m$, $h = 2\,e^{m\beta}/m$, so that h is the height of the crest above the trough. Then

$$T = \frac{g\rho\lambda h^2}{16}\left(1-\frac{\pi^2 h^2}{2\lambda^2}\right),$$

which is the kinetic energy per wave-length.

For the potential energy, whether of the progressive wave or the steady profile, we have, taking the mean level as datum and using (1),

$$V = \iint g\rho\,\frac{e^{2mb}}{2m}\,(1-e^{2mb})\,da\,db.$$

But, from 14·40 (5), $g/m = c^2$, and therefore

$$T = V.$$

We can use the fact that $T = V$ to give an intuitive interpretation of group velocity.

The fluid particles describe circles with constant speed, and the pressure at a particle is the same at every position in its orbit. Now consider any particle whose orbit meets a fixed vertical plane at A and B, fig. 14·41 (iii). No kinetic energy or pressure thrust work crosses this plane during a period, for what crosses from left to right at A goes back from right to left at B. On the other hand potential energy does cross, for the potential energy per unit mass at A exceeds that at B by $g \,.\, AB$. Clearly the potential energy moves with the wave, that is with velocity c. But the potential energy is half the total energy. Therefore the total energy is transported with velocity $\frac{1}{2}c$, the group velocity.

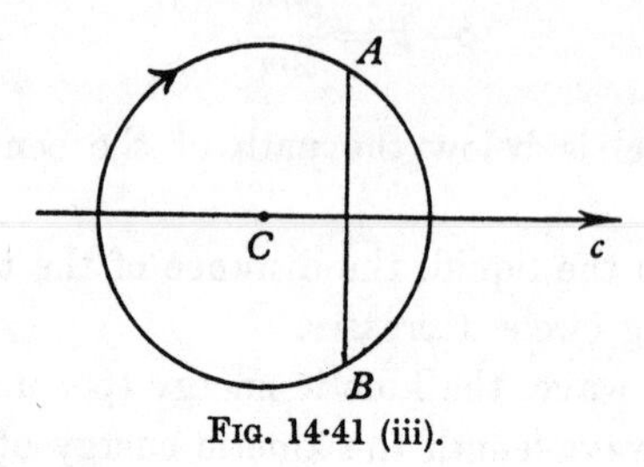

FIG. 14·41 (iii).

To prove that the motion is rotational, we observe that

$$u\,dx + v\,dy = \text{scalar product of the velocity and position vector}$$

$$= \text{real part of } (\partial z/\partial t)\,d\bar{z}.$$

From 14·40 (2),

$$\frac{\partial z}{\partial t}\,d\bar{z} = \{c \exp m(i\bar{l} + ict)\}\{d\bar{l} + dl \exp m(-il - ict)\},$$

where $l = a + ib$. Thus

$$u\,dx + v\,dy = d\left(\frac{c}{m}\,e^{mb} \sin m(a + ct)\right) + c\,e^{2mb}\,da.$$

This is not an exact differential and the motion is therefore rotational. The circulation in an elementary parallelogram of the liquid is obtained from the second of the above terms (since the first is an exact differential) and is therefore equal to

$$-\frac{\partial}{\partial b}(c\,e^{2mb}\,\delta a)\,\delta b.$$

Dividing by the area of the parallelogram, we get, for the vorticity,

$$\omega = -\frac{2mc\,e^{2mb}}{1 - e^{2mb}},$$

the negative sign indicating that the vorticity is in the opposite sense to the revolution of the particles in their circular orbits.

The vorticity is infinite at the cusp of the cycloid ($b = 0$) and decreases rapidly as we descend into the liquid.

14·50. The limiting form of a progressive wave at the crest. It is well established by observation that the limiting form at the crest of a progressive wave is a sharp point with two distinct tangents. The wave is then about to break at the crest. At such a point the velocity of the fluid relative to axes moving with the wave is necessarily zero, otherwise there would be infinite acceleration. Taking then axes moving with the wave, the x-axis opposite to the direction of propagation, the y-axis vertically downwards and the origin at the crest we shall have steady motion.

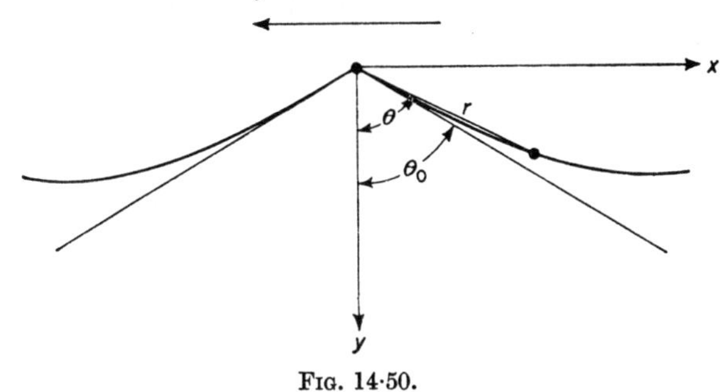

Fig. 14·50.

We take, fig. 14·50, polar coordinates given by

$$x = r\sin\theta, \quad y = r\cos\theta$$

and a stream function $\psi(r, \theta)$.

The vorticity is then

$$\omega = \nabla^2\psi = \frac{\partial^2\psi}{\partial r^2} + \frac{1}{r}\frac{\partial\psi}{\partial r} + \frac{1}{r^2}\frac{\partial^2\psi}{\partial\theta^2}. \tag{1}$$

Taking the free streamline to be $\psi = 0$, Bernoulli's equation gives, at the surface,

$$\left(\frac{\partial\psi}{\partial r}\right)^2 + \frac{1}{r^2}\left(\frac{\partial\psi}{\partial\theta}\right)^2 = 2gy = 2rg\cos\theta, \quad \psi = 0. \tag{2}$$

Confining attention to a small neighbourhood of the angular crest, we have to find a relation

$$\theta = \theta_0 + \epsilon(r), \; \epsilon(0) = 0 \tag{3}$$

to satisfy the simultaneous non-linear equations (2) where θ_0 is the angle which a tangent at the crest makes with the vertical.

We note that (2) shows that y is positive and therefore that

$$-\tfrac{1}{2}\pi < \theta_0 < \tfrac{1}{2}\pi. \tag{4}$$

Taking ω to be constant (1) is satisfied by

$$\psi = \tfrac{1}{4}\omega r^2 + Ar^n \cos n(\theta - \alpha) + \ldots \tag{5}$$

and on the right we shall suppose n to be the smallest exponent in a series of powers of r. It is clear that $n > 1$, since $q = 0$ when $r = 0$.

From (2) and (3)

(6) $$n^2A^2r^{2n-2}+\omega nAr^n\cos n(\theta-\alpha)+\tfrac{1}{4}\omega^2r^2+\ldots = 2gr\cos\theta$$

and equating the indices of the lowest power of r on the two sides we have $2n-2=1$ or $n=3/2$. Substituting for n in (6) and equating to zero the coefficients of the lowest powers of r, we get

(7) $$\tfrac{9}{4}A^2 = 2g\cos\theta_0\,,\quad A\cos\tfrac{3}{2}(\theta_0-\alpha) = 0$$

whence $$\theta_0 = \alpha+\tfrac{2}{3}(2s+1)\frac{\pi}{2}\,,\quad A = \tfrac{2}{3}(2g\cos\theta_0)^{\frac{1}{2}}\,.$$

The first gives

$$\theta_0 = -\frac{\pi}{3}+\alpha\,,\ \frac{\pi}{3}+\alpha\,,\ \pi+\alpha\ldots$$

and the only solutions which fit the condition $\cos(-\theta_0) = \cos\theta_0$ derived from (2) entail

$$\alpha = 0,\quad \theta_0 = -\frac{\pi}{3}\,,\frac{\pi}{3}\,,\quad A = \tfrac{2}{3}\sqrt{g}.$$

Thus the tangents to the two branches of the wave at the crest are equally inclined to the vertical and enclose an angle of 120°.

This result is true whether the motion is rotational or irrotational, provided the wave is progressive and the bottom is horizontal and fixed.

The above reasoning remains valid even if ω is not constant but a bounded function of r, for in the term ωr^2 we can put the leading value of ω. See also Ex. XIV, 5.

14·51. The Highest Irrotational wave. This has been investigated by J. H. Michell (*Phil. Mag.* (5) 36 (1893) 430). Taking the free surface as $\psi = 0$ and supposing consecutive crests to correspond to $\phi = 0$ and $\phi = \pi$, Michell finds for the complex potential w

$$\frac{dw}{dz} = (-i\sin w)^{\frac{1}{3}}\,e^{\frac{1}{3}iw}(1+\cdot 0397e^{2iw}+\cdot 0094e^{4iw}+\cdot 002e^{6iw}+\ldots)$$

from which he concludes that the *camber*, that is ratio of the height (crest above trough) of the highest wave to the wave length, is 0·142, i.e. the height is nearly one-seventh of the length. As was to be expected the actual record of high waves does not attain this. For example Michell cites a wave 46 feet high and 765 feet long observed by Abercromby.

14·60. An exact irrotational wave. John's equation 14·2 (3) can also be applied to wave motion.

Consider the steady motion obtained by taking

(1) $$S(\beta) = \omega\beta,$$

where ω is a constant of dimensions $[T^{-1}]$. Then

$$z(\beta) = B + ae^{i\omega\beta} + g\beta/\omega, \tag{2}$$

where, without loss of generality, we can take $B=0$ and a real and positive so that

$$z = \frac{g\beta}{\omega} + ae^{i\omega\beta}. \tag{3}$$

From 14·20 (4) we then get

$$-\frac{dw}{d\beta} = \left(\frac{g}{\omega} - ai\omega\, e^{-i\omega\beta}\right)\left(\frac{g}{\omega} + ai\omega\, e^{i\omega\beta}\right), \text{ whence}$$

$$w = -\left(\frac{g^2}{\omega^2} + a^2\omega^2\right)\beta - \frac{2ag}{\omega}\cos\omega\beta. \tag{4}$$

The free surface given by (3) is a trochoid without double points if the amplitude $a < g/\omega^2$, the wave-length being

$$\lambda = 2\pi g/\omega^2, \tag{5}$$

so that the condition $a < g/\omega^2$ implies $2\pi a < \lambda$.

The velocity $-dw/dz$ becomes infinite if $dz/d\beta=0$ which gives

$$\omega\beta = 2\pi(n+\tfrac{1}{4}) - i\log\frac{\lambda}{2\pi a}, \tag{6}$$

where n is any integer. The corresponding values of z are the singular points

$$z = \lambda\left\{(n+\tfrac{1}{4}) + \frac{i}{2\pi}\left(1 - \log\frac{\lambda}{2\pi a}\right)\right\}. \tag{7}$$

Such singularities must be excluded from the flow. In order to do this we can take as bottom any streamline above or through the singular points given by (7). Fig. 14·60 (taken from John's paper) shows the free surface and the bottom surface formed by the streamline through the singular points, for various values of the ratio $A=2\pi a/\lambda$. In the diagram the units have been adjusted so that $\omega=g=1$, and $\lambda=2\pi$. For small values of a/λ the depth of the liquid is large compared with a, λ of the order $\lambda\log(\lambda/a)$, and the amplitude of the bottom surface is infinitesimal compared with that of the free surface of order a^2/λ. On the other hand for a/λ close to $1/2\pi$, the depth of the liquid is small and the bottom follows the surface closely. In the motion given by (3) and (4), the free surface remains unchanged in time and each particle has a horizontal velocity varying between $g/\omega - a\omega$ and $g/\omega + a\omega$.

Introduce a new coordinate system moving to the right with constant horizontal velocity g/ω relative to the old one. We then obtain a motion of the type of a progressive wave. To do this we write in (3) and (4)

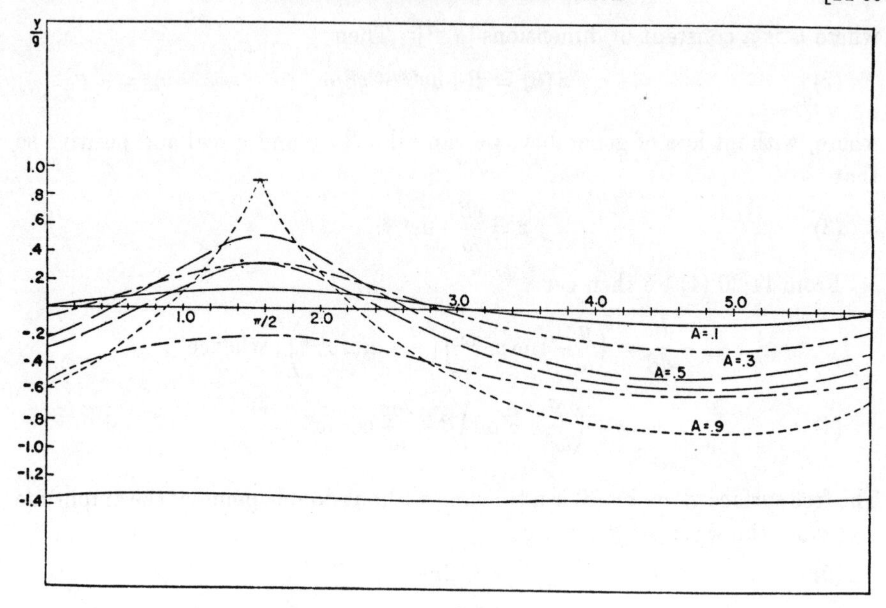

FIG. 14·60.

$$z = \frac{gt}{\omega} + Z, \quad w = W - \frac{gz}{\omega}, \quad \beta = \alpha + t,$$

which lead to

$$Z = \frac{g\alpha}{\omega} + a \exp\{i\omega(\alpha+t)\}, \tag{8}$$

$$W = \omega^2 a^2(\alpha+t) + \frac{ag}{\omega}\exp\{-i\omega(\alpha+t)\}. \tag{9}$$

Elimination of $(\alpha+t)$ shows that W is a function only of $Z+gt/\omega$, so that the wave progresses with velocity g/ω to the left. As the singularities are no longer fixed we have to associate a moving bottom surface with the wave. For small values of a/λ the bottom surface can be taken so far down that the motion reduces essentially to the infinitesimal motion of a liquid of infinite depth. The phase rate g/ω is $(g\lambda/2\pi)^{1/2}$ as in the classical approximation.

In the motion given by (3) and (4), α is a Lagrangian coordinate only for the real values which correspond to particles in the free surface. The motion of the surface particles is identical with that of a Gerstner's trochoidal wave (14·41), given by

$$z = \frac{g\alpha}{\omega} + a \exp\{i\omega(\bar{\alpha}+t)\},$$

in which α is a Lagrangian coordinate even for complex values, and the resulting fluid motion is rotational.

14·65. Levi-Civita's surface condition. A wave of constant profile shape can be reduced to steady motion. Consider such a wave profile, progressing from right to left with velocity c, to be reduced to rest by superposing a velocity c from left to right on the whole system, as for example in fig. 15·58.

In the notation of 12·43 we write

$$(1) \qquad -\frac{dw}{dz} = v = qe^{-i\theta} = ce^{-i\omega},$$

$$(2) \qquad \omega = \theta + i\tau, \quad q = ce^{\tau},$$

and we take the free surface to be the streamline $\psi=0$.

Since in steady motion the time variable does not enter, the complex potential w is a holomorphic function of z alone, and we may take w for independent variable instead of z. At the free surface $\psi=0$ and so $w=\phi$; therefore z, q, ω are functions of the real variable ϕ alone. Again at the free surface, by Bernoulli's theorem, $\frac{1}{2}q^2+gy$ is constant and therefore

$$(3) \qquad q\frac{\partial q}{\partial \phi}+g\frac{\partial y}{\partial \phi} = 0 \quad \text{when } \psi=0.$$

But from (1), when $\psi=0$, $\partial z/\partial\phi = -e^{i\theta}/q$ and therefore

$$\frac{\partial y}{\partial \phi} = -\frac{1}{q}\sin\theta, \quad \text{while from (2)} \quad \frac{\partial q}{\partial \phi} = q\frac{\partial \tau}{\partial \phi}.$$

Thus (3) can be written

$$\frac{\partial \tau}{\partial \phi} = \frac{g}{c^3}e^{-3\tau}\sin\theta, \quad \text{or} \quad \frac{\partial q^3}{\partial \phi} = 3g\sin\theta.$$

But ω is a holomorphic function of w and therefore $\partial\tau/\partial\phi = -\partial\theta/\partial\psi$. Thus finally

$$(4) \qquad \frac{\partial \theta}{\partial \psi} = -\frac{g}{c^3}e^{-3\tau}\sin\theta, \quad \psi = 0.$$

This form of the surface condition is due to Levi-Civita.*

14·70. An exact non-linear theory of waves of constant form. Consider a wave of constant form to be moving from right to left with velocity c on the surface of water of infinite depth.

We assume the wave to have a vertical axis of symmetry through a crest C. The wave length λ is the distance between, say, two consecutive troughs T_1, T_2 one on each side of C, fig. 14·70.

We reduce the wave form to rest by superposing a velocity c from left to right on the whole system, so that the liquid now flows under the fixed form with general velocity c from left to right. We take the x-axis to be in the direction of this general velocity c and the y-axis to be vertically upwards through

* *Math. Ann.* 93 (1925) 264.

the crest C. We call H the height of the wave, that is the vertical distance of crest above trough.

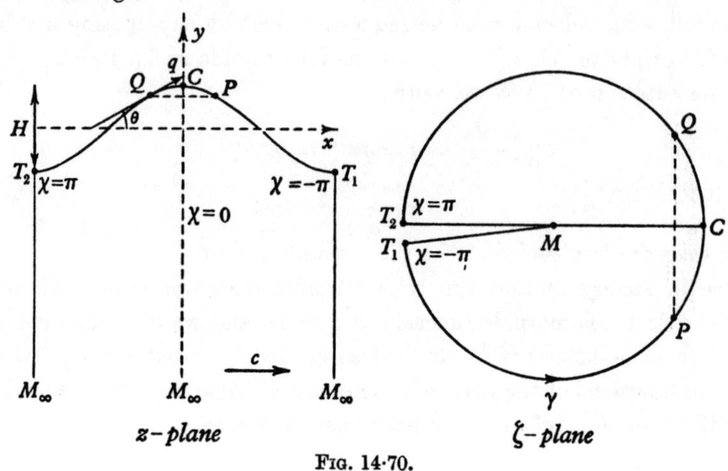

FIG. 14·70.

The form of the free surface profile is unknown. We proceed to map this unknown profile T_1CT_2 on a known curve, the circumference γ of the unit circle in a ζ-plane. If T_1M_∞, T_2M_∞ are verticals through the troughs, we can conveniently call *one wave* the region bounded by these verticals and the profile T_1CT_2.

We map the region defining one wave on the interior of the unit circle cut along a certain radius. We postulate that the point M_∞ shall map into the centre M of the unit circle and the line $M_\infty C$ into the radius MC, which we take to lie along the real axis in the ζ-plane. Thus if

$$\zeta = re^{i\chi}, \tag{1}$$

we shall have $\chi = 0$ on MC.

The cut MT will then be placed along the radius of γ opposite to MC. We shall think of the edges of the cut as being slightly separated to form the radius MT_1 on which $\chi = -\pi$ and the radius MT_2 on which $\chi = \pi$.

Then as we describe the circumference γ of the unit circle by varying χ from $-\pi$ to π, i.e. by following the path T_1CT_2 in the ζ-plane, the point z will describe the wave profile T_1CT_2, and on going from T_1 to T_2, x will decrease by the wave length λ. This decrease can be accomplished by the mapping function

$$z = \frac{i\lambda}{2\pi}\log\zeta,$$

and the same function maps $\zeta = 0$ into M_∞. Thus the mapping may be effected by

$$z = \frac{i\lambda}{2\pi}(\log\zeta + a_1\zeta + \tfrac{1}{2}a_2\zeta^2 + \tfrac{1}{3}a_3\zeta^3 + \ldots), \tag{2}$$

where, as will appear, to obtain a symmetrical profile the coefficients a_n, $n=1, 2, \dots$ must be real. We then get

$$\frac{dz}{d\zeta} = \frac{i\lambda}{2\pi}\left(\frac{1}{\zeta}+a_1+a_2\zeta+a_3\zeta^2+\dots\right) = \frac{i\lambda}{2\pi}\frac{f(\zeta)}{\zeta}, \tag{3}$$

where

$$f(\zeta) = 1+a_1\zeta+a_2\zeta^2+a_3\zeta^3+\dots . \tag{4}$$

On the free surface we put

$$\zeta = \sigma = e^{i\chi}, \quad \bar{\sigma} = 1/\sigma. \tag{5}$$

Then from (2), for a point (x, y) on the free surface,

$$\begin{cases} x = -\dfrac{\lambda}{2\pi}(\chi+a_1 \sin\chi+\tfrac{1}{2}a_2 \sin 2\chi+\tfrac{1}{3}a_3 \sin 3\chi+\dots), \\ y = \dfrac{\lambda}{2\pi}(a_1 \cos\chi+\tfrac{1}{2}a_2 \cos 2\chi+\tfrac{1}{3}a_3 \cos 3\chi+\dots), \end{cases} \tag{6}$$

and this verifies the symmetry about CM or $\chi = 0$, for x changes sign with χ, and y does not. This can happen only if all the a_n are real.

Then from (3) and (5), on the free surface,

$$\frac{dz}{d\chi} = \frac{dz}{d\sigma}\frac{d\sigma}{d\chi} = \frac{i\lambda}{2\pi}\frac{f(\sigma)}{\sigma}i\sigma = -\frac{\lambda}{2\pi}f(\sigma). \tag{7}$$

Further, if we put

$$f(\sigma) = Re^{i\theta}, \tag{8}$$

where R and θ are real we note, on account of (5), that R and θ are functions of χ so that

$$R = R(\chi), \quad \theta = \theta(\chi). \tag{9}$$

Also taking logarithms in (8) we get at the point $\sigma = e^{i\epsilon}$

$$\begin{aligned} \log R(\epsilon)+i\theta(\epsilon) &= \log f(e^{i\epsilon}) = \log(1+a_1 e^{i\epsilon}+a_2 e^{2i\epsilon}+\dots) \\ &= b_1 e^{i\epsilon}+b_2 e^{2i\epsilon}+b_3 e^{3i\epsilon}+b_4 e^{4i\epsilon}+\dots , \end{aligned}$$

where

$$b_1 = a_1, \quad b_2 = a_2-\tfrac{1}{2}a_1^2, \quad b_3 = a_3-a_1a_2+\tfrac{1}{3}a_1^3, \dots , \tag{10}$$

so that the b_n are all real and are known when the a_n are known. Thus

$$\log R(\epsilon) = b_1 \cos\epsilon+b_2 \cos 2\epsilon+b_3 \cos 3\epsilon+\dots , \tag{11}$$

$$\theta(\epsilon) = b_1 \sin\epsilon+b_2 \sin 2\epsilon+b_3 \sin 3\epsilon+\dots . \tag{12}$$

We can use (11) and (12) to obtain a relation between θ and R as follows. From (11)

$$\frac{d}{d\epsilon}\log R(\epsilon) = -b_1 \sin\epsilon-2b_2 \sin 2\epsilon-3b_3 \sin 3\epsilon-\dots .$$

But $\displaystyle\int_{-\pi}^{\pi} \frac{\sin n\epsilon \sin n\chi}{n}(b_1 \sin\epsilon+2b_2 \sin 2\epsilon+3b_3 \sin 3\epsilon+\dots)\,d\epsilon = \pi b_n \sin n\chi.$

Therefore

$$(13)\quad -\frac{1}{\pi}\int_{-\pi}^{\pi}\left(\sum_{n=1}^{\infty}\frac{\sin n\epsilon \sin n\chi}{n}\right)\left(\frac{d}{d\epsilon}\log R(\epsilon)\right)d\epsilon$$

$$= b_1 \sin\chi + b_2 \sin 2\chi + b_3 \sin 3\chi + \ldots = \theta(\chi).$$

It will be found that (13) is the key to the solution of our problem. Moreover up to this point we have studied only the properties of the mapping. Let us now consider the fluid motion.

If we take $\psi = 0$ at the surface, the boundary conditions are

$$(14)\qquad \psi = 0 \quad \text{at the surface,} \quad r = 1.$$

$$(15)\qquad \psi = \infty \ \text{at } M_\infty, \quad r = 0.$$

$$(16)\qquad u - iv = c \quad \text{at } M_\infty, \quad \zeta = 0.$$

They can all be satisfied by the complex potential

$$(17)\qquad w = -\frac{ic\lambda}{2\pi}\log\zeta,$$

which gives $\psi = -(c\lambda/2\pi)\log r$ and this satisfies (14) and (15). Also, using (3),

$$(18)\qquad u - iv = -\frac{dw}{dz} = -\frac{dw}{d\zeta}\frac{d\zeta}{dz} = \frac{c}{f(\zeta)},$$

and when $\zeta = 0, f(\zeta) = 1$, so that (16) is also satisfied. Again from (18) and (8), on the surface we have

$$u - iv = \frac{c}{f(\sigma)} = \frac{c}{R}e^{-i\theta},$$

and therefore

(19) $q = c/R$; θ is the inclination of the velocity vector to the horizontal.

At the free surface the pressure is constant and therefore Bernoulli's theorem gives $q^2 + 2gy$ = constant or, from (19), $(c^2/R^2) + 2gy$ = constant, whence, by differentiation and the use of (7),

$$\frac{d}{d\chi}\left(\frac{1}{R^2(\chi)}\right) = -\frac{2g}{c^2}\frac{dy}{d\chi} = \frac{g\lambda}{\pi c^2}R(\chi)\sin\theta(\chi),$$

which can be written

$$\frac{d}{d\chi}\left(\frac{1}{R^3(\chi)}\right) = \frac{3}{2}\frac{g\lambda}{\pi c^2}\sin\theta(\chi).$$

If we integrate this equation from 0 to ϵ, we get

$$(20)\qquad \frac{1}{R^3(\epsilon)} = \frac{3g\lambda}{2\pi c^2}\left[\int_0^\epsilon \sin\theta(\omega)\,d\omega + \frac{1}{\mu}\right],$$

where μ is an arbitrary constant.

From (19) and (20) if Q is the speed at the crest ($\epsilon = 0$), we have

$$\mu = \frac{3g\lambda c}{2\pi Q^3}.$$

Comparing (13) and (20) we now see that it is possible to eliminate $R(\epsilon)$ and so obtain an equation to give $\theta(\chi)$. To do this, take logarithms of both sides of (20) and differentiate with respect to ϵ. Then

$$-3\frac{d}{d\epsilon}\log R(\epsilon) = \frac{\mu \sin\theta(\epsilon)}{1+\mu\int_0^\epsilon \sin\theta(\omega)\,d\omega}$$

Substitute this in (13). Then

$$(21) \qquad \theta(\chi) = \frac{\mu}{3\pi}\int_{-\pi}^{\pi} \frac{\sin\theta(\epsilon)}{1+\mu\int_0^\epsilon \sin\theta(\omega)\,d\omega}\left[\sum_{n=1}^{\infty}\frac{\sin n\epsilon \sin n\chi}{n}\right]d\epsilon.$$

This is a non-linear integral equation for the slope $\theta(\chi)$ of the wave as a function of χ. After solving this, $b_1, b_2, b_3, \ldots$ are found from (12), then $a_1, a_2, a_3, \ldots$ from (10), and finally the wave profile from (6).

Moreover when $a_1, a_2, a_3, \ldots$ are known, so is the function $f(\zeta)$ by (4), and therefore the velocity at any point by (18).

Thus it appears that the whole exact theory of waves of constant form follows from the solution of the non-linear integral equation * (21).

To find the kinetic energy we must remove the superposed velocity c, so that now

$$q^2 = (u-c)^2+v^2 = (u-iv-c)(u+iv-c).$$

But from (18) $u-iv = c/f(\zeta)$. Therefore

$$q^2 = c^2[1-f(\zeta)][1-\bar{f}(\bar{\zeta})]/[f(\zeta)\bar{f}(\bar{\zeta})].$$

Now if dS and dA are corresponding elements of area of one wave and the unit circle, we have from (3) and 6·29

$$\frac{dS}{dA} = \left|\frac{dz}{d\zeta}\right|^2 = \frac{\lambda^2}{4\pi^2\zeta\bar{\zeta}}\, f(\zeta)\bar{f}(\bar{\zeta}).$$

Therefore the kinetic energy of one wave is given by

$$T = \tfrac{1}{2}\rho\int_{(\text{one wave})} q^2\,dS = \tfrac{1}{2}\rho\int_\gamma \frac{\lambda^2}{4\pi^2\zeta\bar{\zeta}} f(\zeta)\bar{f}(\bar{\zeta})\,q^2\,dA.$$

$$(22) \qquad T = \frac{1}{8}\frac{\rho c^2\lambda^2}{\pi^2}\int_\gamma \frac{[f(\zeta)-1][\bar{f}(\bar{\zeta})-1]\,dA.}{\zeta\bar{\zeta}}$$

Notes

(i) In 14·65 (4) Levi-Civita's non-linear surface condition poses the problem of solving a partial differential equation. The problem presented by the non-linear integral equation (21) is quite different in that it is one which can be tackled numerically with modern high-speed computers.†

---

* This equation is due to A. I. Nekrasov, *Izv. Ivanovo-Vosnosonk. Politehn. Inst.*, 3 (1921), 52–65 ; 6 (1922), 155–71.

† J. W. Thomas, on the exact form of gravity waves on the surface of the inviscid liquid, *Dissertation*, University of Arizona, 1967.

(ii) The problem is an eigenvalue one; indeed it can be shown that no solution different from $\theta \equiv 0$ exists when $\mu < 3$.

(iii) The kernel of (21) is

$$K(\epsilon, \chi) = \sum_{n=1}^{\infty} \frac{\sin n\epsilon \sin n\chi}{n} = \tfrac{1}{4} \log \frac{1-\cos(\epsilon+\chi)}{1-\cos(\epsilon-\chi)}. \tag{23}$$

To see this, the identity

$$\log(1-e^{i\omega}) = -\sum_{n=1}^{\infty} \left(\frac{\cos n\omega + i \sin n\omega}{n}\right) \text{ gives } \log 2 \sin \frac{\omega}{2} = -\sum_{n=1}^{\infty} \frac{\cos n\omega}{n}.$$

In this put $\omega = \epsilon+\chi$ and $\omega = \epsilon-\chi$ in turn and subtract, thus getting

$$K(\epsilon, \chi) = \tfrac{1}{2} \log \left| \frac{\sin \frac{1}{2}(\epsilon+\chi)}{\sin \frac{1}{2}(\epsilon-\chi)} \right| = \tfrac{1}{4} \log \frac{1-\cos(\epsilon+\chi)}{1-\cos(\epsilon-\chi)}.$$

(iv) In terms of the kernel $K(\epsilon, \chi)$ the equation (21) can be written

$$\theta(\chi) = \frac{\mu}{3\pi} \int_{-\pi}^{\pi} \frac{\sin \theta(\epsilon)}{1+\mu \int_0^{\epsilon} \sin \theta(\omega)\, d\omega} K(\epsilon, \chi)\, d\epsilon. \tag{24}$$

(v) If in (21) we write

$$\mu = 3+\nu, \quad 0 < \nu < 1,$$

and put

$$\theta(\chi) = \theta(\chi, \nu) = \nu\, \theta_1(\chi) + \nu^2\, \theta_2(\chi) + \nu^3\, \theta_3(\nu) + \dots,$$

equating like powers of ν on the two sides leads to an infinite series of integral equations for $\theta_1(\chi)$, $\vartheta_2(\chi)$, ..., which can be solved with increasing labour.

Correct to ν^3 the process, which is convergent for $\nu < 1$, leads to

$$\theta(\chi, \nu) = \left(\frac{1}{9}\nu - \frac{8}{243}\nu^2 + \frac{185}{52488}\nu^3\right) \sin \chi + \left(\frac{1}{54}\nu^2 - \frac{8}{729}\nu^3\right) \sin 2\chi + \frac{17}{4374}\nu^3 \sin 3\chi. \tag{25}$$

(vi) It appears from (25) that if ν^2 and higher powers of ν are negligible, the solution of (21) is of the form $\theta(\chi) = \beta \sin \chi$, where β is a small constant.

(vii) Combining (11) and (20) we see that

$$1+\mu \int_0^{\epsilon} \sin \theta(\omega)\, d\omega = \frac{2\pi c^2 \mu}{3g\lambda} \exp\left[-3(b_1 \cos \epsilon + b_2 \cos 2\epsilon + \dots)\right] \tag{26}$$

and the right-hand side never vanishes. Therefore the expression on the left is never zero, being in fact positive for $\mu > 0$. Observe also that c^2 can be obtained from (26).

(viii) That μ is necessarily positive follows from (26) by putting $\epsilon = 0$.

(ix) From (21) we see that

$$\theta(\chi) = -\theta(-\chi), \quad \theta(-\pi) = \theta(0) = \theta(\pi) = 0.$$

Therefore it is sufficient to know the values of $\theta(\chi)$ in the interval $0 < \chi < \pi$, and therefore we can replace (21) by

$$\theta(\chi) = \frac{2\mu}{3\pi} \int^{\pi} \frac{\sin \theta(\epsilon)}{1 + \mu \int_0^{\epsilon} \sin \theta(\omega)\, d\omega} K(\epsilon, \chi)\, d\epsilon, \tag{27}$$

which simplifies numerical computations.

(x) In a sine wave the form of the profile near the crest and the form near a trough are the same. Since, from (21), $\theta(\pi - \chi) \neq \theta(\chi)$, that property does not hold in the exact wave.

(xi) Equation (21) being non-linear we are not able to superpose solutions by addition.

14·71. The case of finite depth. For simplicity in 14·70 the case of infinite depth was treated. When the depth is finite, we map on a concentric circular annulus in the ζ-plane, the outer circumference of radius 1 corresponding to the

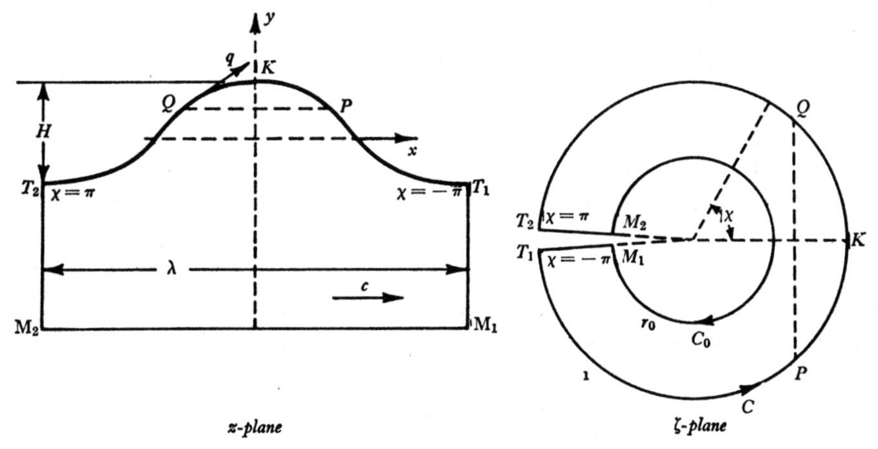

Fig. 14·71.

free surface of one wave, fig. 14·71, the inner circumference corresponding to the bottom, $y = -h$, and the part of a radius between these circumferences corresponding to T_1M_1 and T_2M_2 (analogues of T_1H_∞, T_2M_∞). The equation corresponding to 14·70 (24) is got by replacing the kernel by

$$(1) \qquad L(\epsilon, \chi) = \tfrac{1}{2} \log \left| \operatorname{sn} \frac{K}{\pi} (\epsilon + \chi) \Big/ \operatorname{sn} \frac{K}{\pi} (\epsilon - \chi) \right|,$$

where sn denotes a Jacobian elliptic function* whose quarter periods K, K' are defined by

$$(2) \qquad \frac{K'}{K} = \frac{4h}{\lambda}$$

so that

$$(3) \qquad \theta(\chi) = \frac{\mu}{3\pi} \int_{-\pi}^{\pi} \frac{\sin \theta(\epsilon)}{1 + \mu \int_0^\epsilon \sin \theta(\omega)\, d\omega} L(\epsilon, \chi) d\epsilon.$$

See J. W. Thomas footnote on p. 413.

14·75. An exact integral equation for the solitary wave.† Consider a solitary wave, i.e. a single elevation of invariable form, moving with constant velocity c from right to left on the surface of inviscid incompressible fluid which is at rest at infinity. The motion is assumed to be irrotational and two-dimensional. The bottom is horizontal and the depth of the undisturbed fluid is h.

Since the motion is two-dimensional, it is sufficient to consider only the motion in a vertical plane parallel to the direction of c. Such a vertical plane will intersect the free surface in a curve which will be called the wave profile.

If a constant velocity c from left to right is superposed on the fluid, the wave form is reduced to rest in space, the motion becomes steady and the velocity at infinity becomes c from left to right. This steady motion will now be studied.

Using the suffix ∞ to indicate an infinitely distant point the x-axis will be taken along the horizontal bottom $A_\infty B_\infty$ where $x = -\infty$ at A_∞ on the left and $x = \infty$ at B_∞ on the right.

Levi-Cività has shown that a wave of invariable form must have a vertical axis of symmetry, while the experiments of Scott Russell show that a solitary wave consists of a single elevation. The y-axis will be taken vertically upwards to coincide with this axis of symmetry and will therefore meet the free surface in C, the highest point or crest of the wave profile. Thus the origin O in the z-plane, ($z = x + iy$), will be on the bottom vertically below the crest C. The wave profile $A'_\infty C B'_\infty$ will extend from the point A'_∞ $(-\infty + ih)$ on the left to B'_∞ $(\infty + ih)$ on the right.

---

* Milne-Thomson, *Theoretical Aerodynamics*, London (1966), 14·7.

† Milne-Thomson, *Rev. Roum. Sci. Techn. — Méc. Appl.*, *Tome 9*, pp. 1189–1194, Bucarest, 1964.

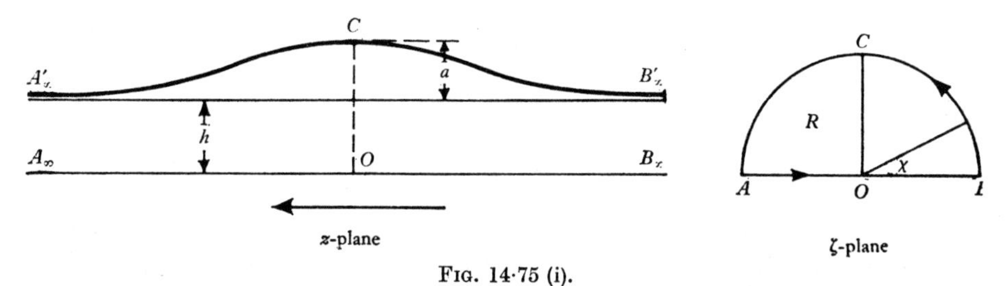

Fig. 14·75 (i).

In a ζ-plane, ($\zeta = \xi + i\eta = re^{i\chi}$), take a semicircular region R bounded by the semicircular arc ACB or γ and the diameter AB or b, defined by

$$R : |\zeta| < 1, \eta > 0, \quad \gamma : |\zeta| = 1, \eta > 0, \quad b : |\xi| \leqslant 1, \eta = 0.$$

Map the region occupied by the fluid in the z-plane on to the region $R+\gamma$ and the bottom $A_\infty B_\infty$ on to the diameter b. Let the origin O in the z-plane be mapped on the centre O of the diameter AB in the ζ-plane. In this mapping corresponding points in the two planes are denoted (without confusion) by the same letter. In particular A'_∞ and A_∞ both map on to $A\,(\zeta = -1)$ and B'_∞ and B_∞ on to $B(\zeta = 1)$. The general point P on the semicircular arc ACB will be denoted by $\sigma = e^{i\chi}$, and therefore $\bar{\sigma} = e^{-i\chi} = 1/\sigma$ where, as usual, the bar denotes the complex conjugate.

In the z-plane the motion may be considered as due to a source of output ch at A_∞ and a sink of intake ch at B_∞. Therefore in the ζ-plane there is a source of output ch at A in the semicircle and a sink of intake ch at B. The flow in the ζ-plane due to this arrangement is clearly the same as that due to a source of output $2\,ch$ at A and a sink of intake $2\,ch$ at B in the region within the disc $|\zeta| \leqslant 1$ consisting of the semicircle and its image in the real axis. The circle theorem then gives for the complex potential $w = \phi + i\psi$

$$(1) \qquad w = \frac{2ch}{\pi}\log(\zeta - 1) - \frac{2ch}{\pi}\log(\zeta + 1).$$

This gives for the stream function ψ_0 on the wave profile

$$2i\psi_0 = w(\sigma) - \bar{w}(\bar{\sigma}) = -\frac{2ch}{\pi}\log(-1), \quad \psi_0 = -ch.$$

Similarly on the bottom where $\zeta = \xi = \bar{\zeta}$, the stream function has the value zero.

Now the complex velocity $u - iv$ in the z-plane is given by

$$u - iv = -\frac{dw}{dz} = -\frac{dw}{d\zeta}\Big/\frac{dz}{d\zeta} = \frac{4ch}{\pi(1-\zeta^2)}\Big/\frac{dz}{d\zeta} \text{ from (1).}$$

Also at A_∞, B_∞ i.e. at $\zeta = -1$, $\zeta = 1$, $u - iv = c$. Therefore putting

(2) $$\frac{dz}{d\zeta} = \frac{4h}{\pi(1-\zeta^2)}\{1+(1-\zeta^2)f(\zeta)\},$$

gives

(3) $$u - iv = \frac{c}{1+(1-\zeta^2)f(\zeta)},$$

where $f(\zeta)$ is holomorphic in the region R and continuous on the boundary. Moreover $u-iv$ is nowhere infinite in the region of the flow and therefore $1+(1-\zeta^2)f(\zeta)$ can have no zeros in or upon the boundary of the semicircle. Therefore we can write

(4) $$1+(1-\zeta^2)f(\zeta) = e^{\Omega(\zeta)},$$

where $\Omega(\zeta)$ is also holomorphic in the region R and continuous on the boundary.

Let

(5) $$f(\zeta) = a_0 + a_1\zeta + a_2\zeta^2 + \ldots$$

Since $u-iv$ is real along the bottom i.e. when ζ is real, it follows that $f(\zeta)$ is real when ζ is real and therefore the coefficients a_1, a_2, $a($, ... in (5) are all real.

Integration of (2) gives

$$z = \frac{2h}{\pi}\left\{\log\frac{\zeta+1}{\zeta-1} + 2F(\zeta) + K\right\},$$

where K is a constant and

$$F(\zeta) = a_0\zeta + \tfrac{1}{2}a_1\zeta^2 + \tfrac{1}{3}a_2\zeta^3 + \ldots$$

Since $z = 0$ maps into $\zeta = 0$, it follows that $K + \log(-1) = 0$ and therefore

(6) $$z = \frac{2h}{\pi}\left\{\log\frac{1+\zeta}{1-\zeta} + 2F(\zeta)\right\}.$$

From the symmetry of the wave profile with respect to the y-axis, if the point z lies on the profile, so does the point $-\bar{z}$. But on the semicircle the points z and $\bar{z}$ correspond to the points $\sigma = e^{i\chi}$ and $e^{i(\pi-\chi)} = -e^{-i\chi} = -1/\sigma$. Substitution in (6) gives $F(-1/\sigma) = -F(1/\sigma)$ and $F(\zeta)$ is an odd function of ζ so that the odd coefficients a_1, a_3, a_5, ... all vanish. Therefore

(7) $$f(\zeta) = a_0 + a_2\zeta^2 + a_4\zeta^4 + \ldots$$

(8) $$z = \frac{4h}{\pi}\left\{\tfrac{1}{2}\log\frac{1+\zeta}{1-\zeta} + a_0\zeta + \tfrac{1}{3}a_2\zeta^3 + \tfrac{1}{5}a_4\zeta^5 + \ldots\right\},$$

and (8) gives the mapping of the z- on the ζ-plane. In (8) put $\zeta = \sigma = e^{i\chi}$ and equate real and imaginary parts. Then the wave-profile is given in terms of the parameter χ by

$$(9)\qquad x = \frac{4h}{\pi}\{\tfrac{1}{2}\log\cot\tfrac{1}{2}\chi + a_0\cos\chi + \tfrac{1}{3}a_2\cos 3\chi + \tfrac{1}{5}a_4\cos 5\chi + \ldots\}$$

$$y = \frac{4h}{\pi}\left\{\frac{\pi}{4} + a_0\sin\chi + \tfrac{1}{3}a_2\sin 3\chi + \tfrac{1}{5}a_4\sin 5\chi + \ldots\right\}.$$

Thus if $a_0, a_2, a_4, \ldots$ can be found, (1) and (8) give the complex potential, while (9) gives the form of the wave-profile. In particular if $h+a$ is the height of the crest ($\chi = \frac{1}{2}\pi$) above the bottom, (9) gives

$$(10)\qquad a = \frac{4h}{\pi}\{a_0 - \tfrac{1}{3}a_2 + \tfrac{1}{5}a_4 - \ldots\}.$$

It now follows from (7) and (4) that $\Omega(\zeta)$ is an even function and therefore has an expansion

$$(11)\qquad \Omega(\zeta) = b_0 + b_2\zeta^2 + b_4\zeta^4 + \ldots$$

while by taking logarithms of (4)

$$\Omega(\zeta) = \log\,[1 + (1-\zeta^2)f(\zeta)].$$

Substitution of $f(\zeta)$ from (7), expansion of the right hand side in powers of ζ^2, and equating coefficients lead to equations, linear in the b_{2n}, to determine the b_{2n} in terms of the a_{2n}. Since the a_{2n} are real, so are the coefficients b_{2n}.

Substituting (7) and (11) in (4), expanding, and equating coefficients lead to equations, linear in the a_{2n}, to determine the a_{2n} in terms of the b_{2n}. Therefore to solve the problem it is sufficient to find the b_{2n}.

In (11) put $\zeta = re^{i\chi}$ and separate the real and imaginary parts, say

$$(12)\qquad \Omega(\zeta) = \tau(r,\chi) + i\theta(r,\chi).$$

Then from (3) and (4)

$$(13)\qquad u - iv = ce^{-\tau(r,\chi)}\,e^{-i\theta(r,\chi)}.$$

Therefore $ce^{-\tau(r,\chi)}$ is the fluid speed and $\theta(r,\chi)$ is the inclination to the horizontal of the velocity of the fluid particle at the point which maps into $\zeta = re^{i\chi}$.

At the point $\sigma = e^{i\chi}$ on the wave profile we shall write

$$(14)\qquad \Omega(\sigma) = \tau(\chi) + i\theta(\chi)$$

and then on the wave profile the velocity is given by

$$(15)\qquad u - iv = qe^{-i\theta(\chi)},\quad q = ce^{-\tau(\chi)},$$

where q is the fluid speed and $\theta(\chi)$ is the inclination of the velocity to the horizontal.

Also since the coefficients b_{2n} are real (11) and (14) give

(16) $$\theta(\chi) = b_2 \sin 2\chi + b_4 \sin 4\chi + b_6 \sin 6\chi + \ldots$$

(17) $$\tau(\chi) = b_0 + b_2 \cos 2\chi + b_4 \cos 4\chi + \ldots$$

The function $\Omega(\zeta)$ can be expressed in terms of $\tau(\chi)$ as follows. From (17)

(18) $$b_0 = \frac{1}{\pi}\int_0^{\pi} \tau(\chi)\, d\chi, \quad b_{2n} = \frac{2}{\pi}\int_0^{\pi} \tau(\chi) \cos 2n\chi\, d\chi.$$

Therefore from (11)

(19) $$\Omega(\zeta) = \frac{1}{\pi}\int_0^{\pi} \tau(\epsilon)\, d\epsilon + \frac{2}{\pi}\int_0^{\pi} \tau(\epsilon) \sum_{n=1}^{\infty} \zeta^{2n} \cos 2n\epsilon\, d\epsilon.$$

Now, if t is real,

$$\sum_{n=1}^{\infty} t^{2n} \cos 2n\epsilon = \text{Real part of } \frac{t^2 e^{2i\epsilon}}{1-t^2 e^{2i\epsilon}} = -1 + \frac{1-t^2 \cos 2\epsilon}{1-2t^2 \cos 2\epsilon + t^4}.$$

Substitution in (19) gives

(20) $$\Omega(\zeta) = \frac{1}{\pi}\int_0^{\pi} \tau(\epsilon) \frac{1-\zeta^4}{1-2\zeta^2 \cos 2\epsilon + \zeta^4}\, d\epsilon.$$

On the wave profile $\zeta = e^{i\chi} = \sigma$ and therefore

$$\frac{1-\zeta^4}{1-2\zeta^2 \cos 2\epsilon + \zeta^4} = \frac{-(\sigma^2 - 1/\sigma^2)}{\sigma^2 + 1/\sigma^2 - 2\cos 2\epsilon} = \frac{-2i \sin 2\chi}{2\cos 2\chi - 2\cos 2\epsilon}$$

$$= \tfrac{1}{2}i[\cot(\chi+\epsilon) + \cot(\chi-\epsilon)] = \tfrac{1}{2}i \frac{\partial}{\partial\epsilon} \log \frac{\sin(\epsilon+\chi)}{\sin(\epsilon-\chi)}.$$

and therefore on the wave profile

(21) $$\tau(\chi) + i\theta(\chi) = \Omega(\sigma) = \frac{i}{2\pi}\int_0^{\pi} \tau(\epsilon) \frac{\partial}{\partial\epsilon} \log \frac{\sin(\epsilon+\chi)}{\sin(\epsilon-\chi)}\, d\epsilon$$

where the integral is to be interpreted as a Cauchy principal value. Integration by parts gives

$$\tau(\chi) + i\theta(\chi) = \frac{i}{2\pi}\left[\tau(\epsilon) \log \frac{\sin(\epsilon+\chi)}{\sin(\epsilon-\chi)}\right]_0^{\pi} - \frac{i}{2\pi}\int_0^{\pi} \tau'(\epsilon) \log \frac{\sin(\epsilon+\chi)}{\sin(\epsilon-\chi)}\, d\epsilon.$$

Now $\log \sin(\epsilon-\chi)$ has a jump discontinuity $i\pi$ when ϵ increases through the value χ and $\log \sin(\epsilon+\chi)$ has a jump discontinuity $i\pi$ when ϵ increases through the value $\pi-\chi$. Therefore equating the imaginary parts gives

(22) $$\theta(\chi) = -\frac{1}{2\pi}\int_0^{\pi} \tau'(\epsilon) \log \left|\frac{\sin(\epsilon+\chi)}{\sin(\epsilon-\chi)}\right| d\epsilon.$$

We now proceed to eliminate $\tau(\epsilon)$. By Bernoulli's theorem, on the wave profile, where the pressure is constant

$$q^2 + 2gy = \text{constant}$$

and therefore

$$\frac{dq^2}{d\chi} = -2g\frac{dy}{d\chi}, \quad q^2 = ce^{-2\tau(\chi)}, \tag{23}$$

from (15). But on the wave profile $\zeta = \sigma = e^{i\chi}$ so that

$$\frac{dz}{d\chi} = \frac{dz}{d\sigma}\frac{d\sigma}{d\chi} = i\sigma\frac{dz}{d\sigma}.$$

Therefore from (4) and (6)

$$\frac{dz}{d\chi} = -i\frac{4h}{\pi}\frac{e^{\Omega(\sigma)}}{\sigma - 1/\sigma}$$

so that

$$\frac{dy}{d\chi} = \frac{-2h}{\pi \sin\chi} e^{\tau(\chi)} \sin\theta(\chi). \tag{24}$$

Combining this with (23) we get

$$\frac{d}{d\chi}(e^{-3\tau(\chi)}) = \frac{6gh}{\pi c^2}\frac{\sin\theta(\chi)}{\sin\chi}$$

which gives on integration

$$e^{-3\tau(\chi)} = \frac{6gh}{\pi\mu c^2}\left\{1+\mu\int_0^\chi \frac{\sin\theta(\epsilon)}{\sin\epsilon}d\epsilon\right\} \tag{25}$$

where μ is a constant of integration. Logarithmic differentiation gives

$$\tau'(\chi) = -\tfrac{1}{3}\frac{\mu\sin\theta(\chi)\operatorname{cosec}\chi}{1+\mu\int_0^\chi \sin\theta(\epsilon)\operatorname{cosec}\epsilon\, d\epsilon}. \tag{26}$$

Substitution in (22) then gives the non-linear integral equation for $\theta(\chi)$

$$\theta(\chi) = \frac{\mu}{6\pi}\int_0^\pi \frac{\sin\theta(\epsilon)\operatorname{cosec}\epsilon}{1+\mu\int_0^\epsilon \sin\theta(u)\operatorname{cosec}u\, du}\log\left|\frac{\sin(\epsilon+\chi)}{\sin(\epsilon-\chi)}\right| d\epsilon. \tag{27}$$

The solution of this equation, by comparison with (16) would yield the coefficients $b_2, b_4, b_6, \ldots$

Since $q = c$ at B_∞ where $\sigma = 0$, it follows from (15) that $\tau(0) = 0$, and therefore from (17) that

$$b_0 = -b_2 - b_4 - b_6 - \ldots \tag{28}$$

Therefore as previously explained the coefficients a_{2n} can be obtained and then the equations (9) give the form of the wave profile.

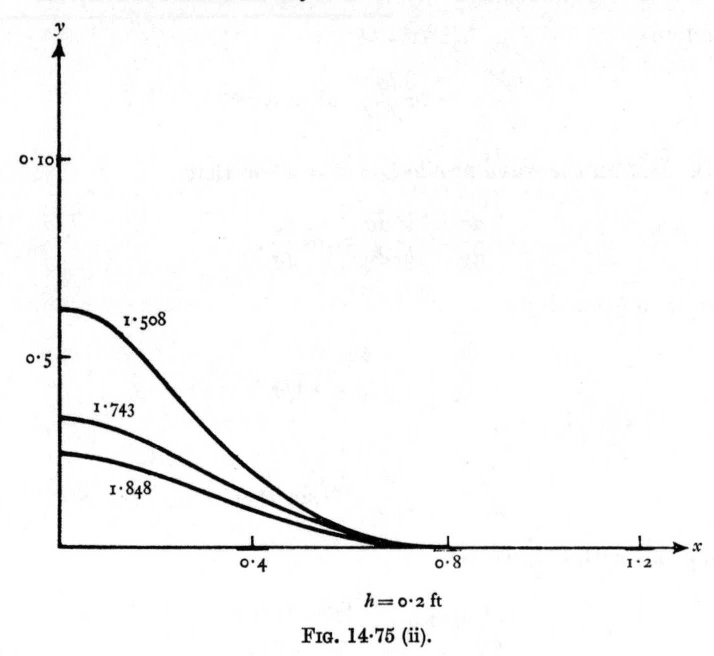

$h = 0{\cdot}2$ ft

FIG. 14·75 (ii).

The form of the profile has been calculated by J. D. Schwitters,* for $h=0{\cdot}2$ ft., and is shown for three values of μ in fig. 14·75 (ii) (scales in ft.). Schwitters finds that the trivial solution $\theta(\chi)=0$ is found for μ near $6/\pi$ and that the wave is higher and more sharply peaked for values of μ near 1·4.

14·8. Other problems. The method illustrated in 14·70, 14·71, 14·75 of setting up a non-linear integral equation by mapping on a circle or semicircle is clearly of a general character. Other problems whose solution has been

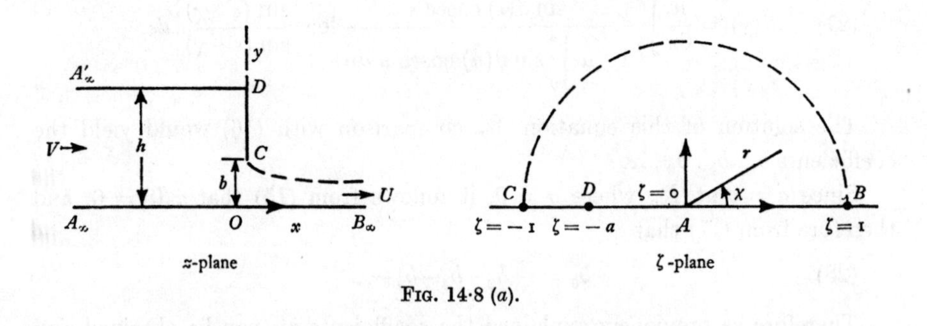

FIG. 14·8 (a).

* The exact profile of the solitary wave, *Dissertation*, University of Arizona, 1966

carried to a numerical conclusion are that of the sluice gate,* fig. 14·8 (*a*), and the vertical jet,† fig. 14·8 (*b*).

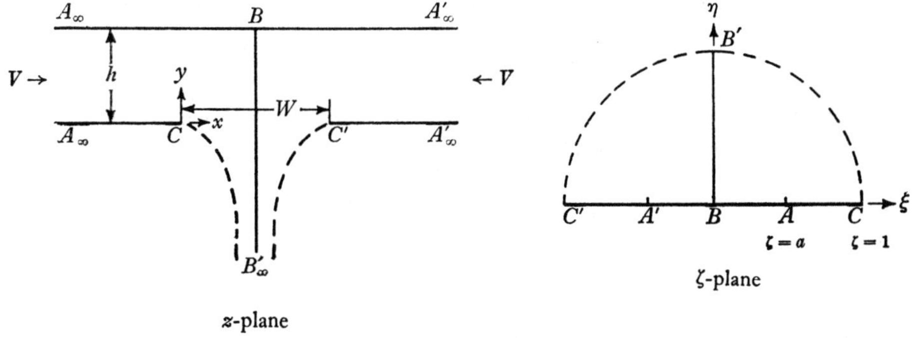

FIG. 14·8 (*b*).

EXAMPLES XIV

1. In the case of a wave about to break at the crest show that the curvature of one branch is

$$[-\omega/(6\sqrt{g})]r^{-\frac{1}{2}}.$$

2. If the vorticity has the same sense as that in which the fluid particles of a wave describe their orbits, the two branches at an angular crest are above their tangents at the crest; their concavity is turned upwards.

Show also that the reverse is true of the senses if the vorticity and description of orbits are opposite; the branches have their concavity downwards.

3. In the case of a wave about to break at the crest show that the curvature is infinite and the radius of curvature is zero if the vorticity is not zero.

If the vorticity is zero at the crest (irrotational motion at the crest), the curvature is zero at the crest.

4. In the case of a wave presenting a pointed crest, show that, to the second approximation, the stream function and the trajectories of the liquid particles are symmetrical about the vertical through the crest.

5. With the notation of 14·50 prove that the second approximation to θ is

$$\theta = \frac{\pi}{3} - \frac{\omega r^{\frac{1}{2}}}{3\sqrt{g}},$$

where θ is as in fig. 14·50.

6. Use the result of ex. 5 to show that the two branches at the crest are symmetrical with respect to the vertical through the crest in its immediate neighbourhood.

7. Prove that the radius of curvature of the limiting crest of a progressive wave is zero if the vorticity is different from zero.

Discuss the case of Gerstner's wave at the cusp of the cycloid.

8. Prove that at the pointed crest of the highest irrotational wave the curvature is zero.

---

* V. J. Klassen, A free boundary problem for the flow of a liquid under gravity through a partially obstructed orifice, *Dissertation*, University of Arizona, 1965.

† W. E. Conway, The two-dimensional jet under gravity from an aperture in the lower of two horizontal planes which bound a liquid, *Dissertation*, University of Arizona, 1965.

9. In the notation of 14·70 prove that the height of the wave (crest above trough) is

$$\frac{\lambda}{2\pi}\int_{-1}^{1}\frac{f(\zeta)-1}{\zeta}\,d\zeta.$$

10. Prove that the free surface of John's exact irrotational wave (14·60) is a cycloid if $a = g/\omega^2$.

Explain the relation between this result and Gerstner's trochoidal wave.

11. In the notation of 14·71 show that

$$L(\epsilon,\chi) = \sum_{n=1}^{\infty}\frac{\phi_n(\epsilon)\,\phi_n(\chi)}{\mu_n},$$

where

$$\phi_n(\epsilon) = \frac{\sin n\epsilon}{\sqrt{\pi}}, \quad \mu_n = \frac{n}{\pi}\coth\frac{2\pi hn}{\lambda}.$$

12. Find the tangent solution corresponding to the flow through the sluice gate depicted in fig. 14·8 (*a*).

13. Using fig. 14·8 (*b*) show that for the flow depicted the complex potential is

$$w = -2m\log(\zeta^2-a^2)-2m\log(\zeta^2-1/a^2)+4m\log(\zeta^2+1)$$

where $a = BA = BA'$ and $m = Vh/(2\pi)$. (Conway)

14. For the vertical jet of 14·8 show that the form of the mapping function must be given by

$$\frac{dz}{d\zeta} = \frac{(1-\zeta^2)\,f(\zeta)\,[\log\{4/(\zeta^2+1)\}]^{\alpha}}{(\zeta^2-a^2)(\zeta^2+1)},$$

where $f(\zeta)$ is holomorphic inside and upon the unit circle. (Conway)

15. In the preceding example, by considering the flow at B'_∞ in the z-plane, show that $\alpha = -1/3$. (Conway)

16. Use the preceding examples to show that for the vertical jet of 14·8

$$h = \frac{\pi(1-a^2)\,f(a)\,[\log(4/(a^2+1))]^{-1/3}}{2a\,(a^2+1)},$$

which determines a when $f(\zeta)$ is known. (Conway)

17. In the vertical jet of 14·8 put

$$\exp\Omega(\zeta) = (1/a^2-\zeta^2)\,f(\zeta)/h, \quad \Omega(re^{i\chi}) = \omega(r,\chi)+i\psi(r,\chi)$$

and hence obtain a non-linear integral equation for ψ of the form

$$\psi(1,\chi) = \beta(\chi)+\frac{2\mu}{3\pi}\int_0^{\pi}L(\epsilon,\chi)\,\frac{A(\epsilon)\sin\{B(\epsilon)+\psi(1,\epsilon)\}}{1+\mu\displaystyle\int_0^{\epsilon}[A(t)\sin\{B(t)+\psi(1,t)\}]\,dt}\,d\epsilon,$$

where $A(\epsilon)$, $B(\epsilon)$, $\beta(\chi)$ are known functions and the kernel is

$$L(\epsilon,\chi) = \sum_{n=1}^{\infty}\frac{\sin 2n\epsilon\,\sin 2n\chi}{2n}.$$ (Conway)

18. Using fig. 14·8 (*a*) show that for the flow depicted

$$w = -m\log\zeta+2m\log(\zeta-1), \quad m = Uh/\pi.$$ (Klassen)

19. For the sluice gate of 14·8 show that the mapping function must be given by an equation of the form

$$\frac{dz}{d\zeta} = \frac{(\zeta+1)\,hf(\zeta)}{\zeta(\zeta-1)\,(\zeta+a)^d}.$$ (Klassen)

20. In the preceding example by considering the velocity on DC, prove that

$$d = 1/2. \qquad \text{(Klassen)}$$

21. Show that the sluice gate problem of 14·8 leads to a non-linear integral equation of the form

$$\theta(\chi) = A(\chi) + \frac{2\mu}{3\pi}\int_0^\pi N(\epsilon, \chi) \frac{\cot \frac{1}{2}\epsilon \sin[\theta(\epsilon) - B(\epsilon)]}{1 + \mu \displaystyle\int_0^\epsilon \cot \tfrac{1}{2}t \sin[\theta(t) - B(t)]dt} d\epsilon,$$

where $A(\chi)$ and $B(\chi)$ are known functions and the kernel is

$$N(\epsilon, \chi) = \sum_{n=1}^{\infty} \frac{\sin n\epsilon \sin n\chi}{n}. \qquad \text{(Klassen)}$$

22. In the case of waves on water of finite depth (14·71) show that the appropriate postulations are

$$w = -\frac{i\lambda c}{2\pi}\log\zeta, \quad \frac{dz}{d\zeta} = \frac{i\lambda}{2\pi}\left(\frac{1}{\zeta} + f(\zeta)\right),$$

where $f(\zeta)$ is a Laurent series about $\zeta = 0$.

23. In fig. 14·71 if the circle C has unit radius and the circle C_0 has radius r_0, prove that $r_0 = \exp(-2\pi h/\lambda)$ where λ is the wave-length and h is the depth.

24. For waves on water of finite depth (14·71) show that

$$qe^{-i\theta} = \frac{c}{1 + \zeta f(\zeta)} = \frac{c}{e^{\Omega(\zeta)}},$$

where

$$\Omega(\zeta) = a_0 + \sum_{n=1}^{\infty} a_n \left[\left(\frac{\zeta}{r_0}\right)^n + \left(\frac{r_0}{\zeta}\right)^n\right],$$

the coefficients a_n, $n = 0, 1, 2, \ldots$ all being real.

25. In the wave of 14·71 with the notation of the preceding example, if

$$\Omega(e^{i\chi}) = \Psi(\chi) + i\theta(\chi),$$

show that θ is the inclination of the velocity on the wave profile to the horizontal and that $q = ce^{-\Psi(\chi)}$.

Prove also that

$$\frac{d}{d\chi}(e^{-3\Psi(\chi)}) = \frac{3g\lambda}{2\pi c^2}\sin\theta(\chi).$$

26. Use the result of the preceding example to show that for the wave of 14·71

$$e^{-3\Psi(\chi)} = \frac{3g\lambda}{2\pi c^2}\frac{1}{\mu}\left[1 + \mu\int_0^\chi \sin\theta(\epsilon)\, d\epsilon\right],$$

where μ is a constant such that

$$\mu = \frac{3g\lambda}{2\pi c^2} e^{3\Psi(0)}.$$

27. In the wave of 14·71 show that c is the mean speed along the bottom

$$c = \frac{1}{\lambda}\int_0^\lambda u\, dx,$$

where u is the speed on the bottom.

CHAPTER XV

LINEARISED GRAVITY WAVES

15·10. Wave motion. A *wave motion* of a liquid acted upon by gravity and having a free surface is a motion in which the elevation of the free surface above some chosen fixed horizontal plane varies.

Taking the axis of x to be horizontal and the axis of y to be vertically upwards, a motion in which the equation of the vertical section of the free surface at time t is of the form

$$(1) \qquad y = a \sin (mx - nt),$$

where a, m, n are constants, is called a *simple harmonic progressive* wave.

If we draw the *profile* of the free surface (1) at time $t = 0$, we get the sine curve $y = a \sin mx$, fig. 15·10 (i). Since (1) can be written in the form

$$(2) \qquad y = a \sin m \left(x - \frac{nt}{m}\right),$$

we see that the profile at time t is exactly the same shape as at time $t = 0$, for we have simply to move the origin to O' where $OO' = nt/m$ to recover the original form of the profile, fig. 15·10 (ii).

Equation (1) therefore represents a motion in which the curve

$$y = a \sin mx$$

moves in the direction of the axis of x with the velocity $c = n/m$, which is called the *speed of propagation* of the wave. When $a = 0$ the profile of the liquid is $y = 0$, which is the *mean level.*

The quantity a is called the *amplitude* of the wave and measures the maximum departure of the actual free surface from the mean level. The points $C_1, C_2, \ldots,$ of maximum elevation are called *crests*, the points T_1, $T_2, \ldots,$ of maximum depression are the *troughs* of the wave. The distance between successive crests is called the *wave-length* λ. Thus

$$\lambda = \frac{2\pi}{m}.$$

The aspect of the free surface is exactly the same at times t and $t + 2\pi/n$. The time

$$\tau = \frac{2\pi}{n}$$

is called the *period* of the wave. The reciprocal of the period is the *frequency* $n/2\pi$. The angle $mx - nt$ is called the *phase angle*, and the number n may be conveniently called the *phase rate*.

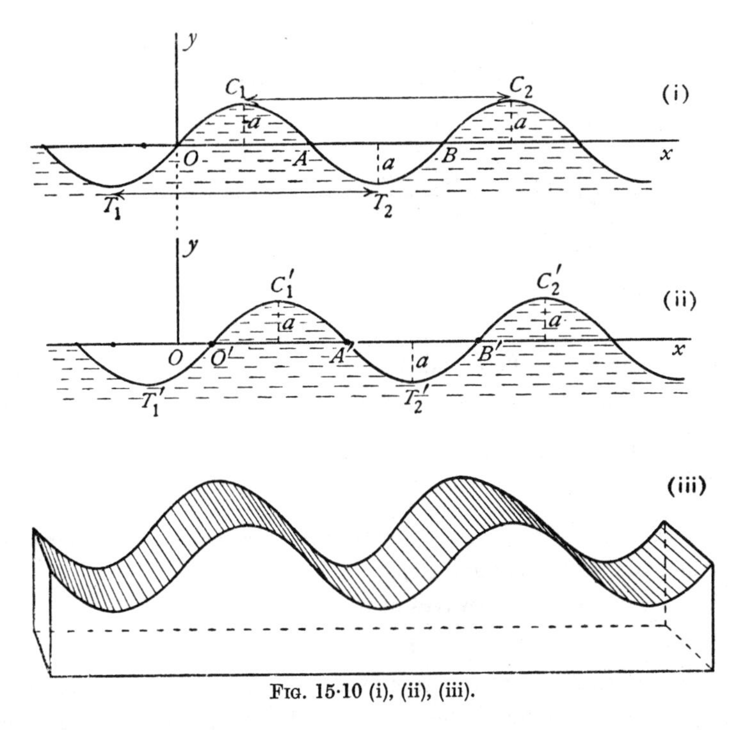

FIG. 15·10 (i), (ii), (iii).

From the above numbers we obtain the relation $\lambda = c\tau$. The equation of the profile can be also written in the form

$$y = a \sin \frac{2\pi}{\lambda}(x - ct).$$

It will be observed that the motion represented by (1) is a two-dimensional motion. In this chapter we shall be concerned only with two-dimensional wave motions which may therefore be supposed to take place between two vertical planes at unit distance apart, fig. 15·10 (iii), and this convention will be assumed in the absence of explicit statement to the contrary.

15·11. Kinematical condition at the free surface. Consider water of depth h in which waves of height $\eta = \eta(x, t)$ above the mean level are propagated, the height being measured from the undisturbed level, and the axis of x being taken along the bottom * in the direction of propagation. The equa-

* We consider only the case of constant depth. For variable depth the reader may consult J. J. Stoker, *Water waves. The mathematical theory with applications*, New York, 1957.

tion of the free surface is then $y-\eta-h = 0$, and since the surface moves with the fluid $d(y-\eta-h)/dt = 0$, so that

$$\frac{\partial \eta}{\partial t}+u\frac{\partial \eta}{\partial x} = v.$$

Unless the contrary is explicitly stated, we shall confine our attention to the *linearised theory* in which the squares and products of the variable parts of

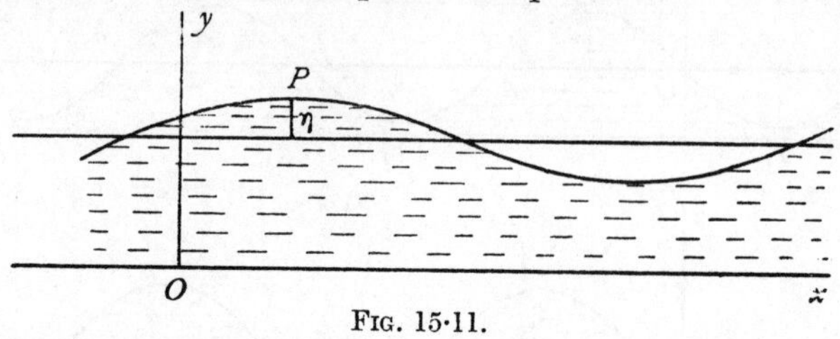

FIG. 15·11.

all quantities and their differential coefficients may be neglected. In particular $\partial\eta/\partial x$, which measures the slope of the profile, will be taken as small. We then get at the free surface

$$(1) \qquad \frac{\partial \eta}{\partial t} = \frac{\partial \psi}{\partial x},$$

where ψ is the stream function. This is the *kinematical surface condition* for wave profiles of *small height and slope.*

In the case of *irrotational* waves of profile

$$(2) \qquad \eta = a \sin (mx-nt),$$

we see from (1) that, when $y = h$, the stream function ψ is proportional to $\sin (mx-nt)$. We therefore attempt to satisfy (1) by the complex potential $w = b \cos (mz-nt)$, which gives $\psi = -b \sin (mx-nt) \sinh mh$ at the free surface to our degree of approximation. Substitution in (1) gives $bm \sinh mh = an$, so that

$$(3) \qquad w = \frac{ac}{\sinh mh} \cos (mz-nt),$$

where $c = n/m$ is the speed of propagation.

It should be noted that no hypothesis has been made as to conditions above the wave profile and therefore (3) will continue to hold if the profile is an *interface* between two fluids.

15·12. Pressure condition at the free surface. Let p_i be the pressure just inside the liquid at P in fig. 15·11 and p_0 the pressure just outside. We shall again assume the motion to be irrotational, as will indeed be the case for waves started in an inviscid liquid by natural forces. The pressure equation (neglecting the term $\frac{1}{2}q^2$) then gives

$$p_i = \rho\left(\frac{\partial \phi}{\partial t}-g\eta\right)+C(t),$$

and $C(t)$ may be taken to be independent of t by incorporating any time variable terms in $\partial\phi/\partial t$. Thus

$$p_i - p_0 = \rho\left(\frac{\partial\phi}{\partial t} - g\eta\right) - p_0 + \text{constant}.$$

In the case of constant (atmospheric) pressure, we may write

$$(1) \qquad p_i - p_0 = \rho\left(\frac{\partial\phi}{\partial t} - g\eta\right),$$

by a suitable adjustment of ϕ, and in this connection we may observe that p_i and p_0 can only differ by a small quantity and therefore $\partial\phi/\partial t$ must be small. Thus (1) is the pressure condition at the free surface of irrotational waves of small height.

If we neglect surface tension (see 15·50), we have $p_i - p_0 = 0$ and therefore $g\eta - \partial\phi/\partial t = 0$.

Now at the free surface

$$\phi = \phi(x, h+\eta, t) = \phi(x, h, t) + \eta\left(\frac{\partial\phi(x, y, t)}{\partial y}\right)_{y=h} + \ldots,$$

and therefore in the above surface condition we may suppose η to be put equal to zero in the second term, in other words,

$$(2) \qquad g\eta = \frac{\partial\phi(x, h, t)}{\partial t},$$

and this is the surface condition. Notice that (2) gives the surface elevation when ϕ is known.

15·13. Surface waves. If we combine the kinematical and pressure boundary conditions 15·11 (1), 15·12 (2), we get

$$(1) \qquad \frac{\partial^2\phi}{\partial t^2} - g\frac{\partial\psi}{\partial x} = 0, \quad y = h.$$

Now from 15·11 in the case of a simple harmonic progressive wave, fig. 15·11, we have

$$(2) \qquad w = \frac{ac}{\sinh mh}\cos(mz - nt), \quad \eta = a\sin(mx - nt).$$

Thus (1) gives, after a simple reduction,

$$(3) \qquad c^2 = \frac{g}{m}\tanh mh,$$

the equation giving the speed of propagation of waves of length $2\pi/m$.

Equations (2) and (3) completely characterise these waves at the surface of water of depth h, and it is worth observing that, while (2) is deducible from kinematic considerations alone, (3) gives the relation which must subsist between

n and m ($c = n/m$) in order that the solution may be physically satisfactory. The speed of propagation is in fact a function of the wave-length.

More generally, taking axes as in fig. 15·11, the conditions to be satisfied are (1) and

$$(4) \qquad \psi = 0, \quad \text{when} \quad y = 0,$$

since the bottom has to be a line across which there is no flow.

The complex potential is

$$w = w(x+iy, t) = \phi(x, y, t) + i\psi(x, y, t).$$

Condition (4) therefore states that w is real when $y = 0$, and therefore the holomorphic function w can be continued by the principle of reflection (5·53) into the region for which y is negative, more precisely $-h \leqslant y < 0$. Thus

$$w(x-iy, t) = \phi(x, y, t) - i\psi(x, y, t),$$

and therefore
$$\phi(x, y, t) = \tfrac{1}{2}[w(x+iy, t) + w(x-iy, t)],$$
$$\psi(x, y, t) = -\tfrac{1}{2}i[w(x+iy, t) - w(x-iy, t)].$$

Putting $y = h$ and substituting in (1), we get

$$\frac{\partial^2}{\partial t^2}[w(x+ih, t) + w(x-ih, t)] + ig\frac{\partial}{\partial x}[w(x+ih, t) - w(x-ih, t)] = 0.$$

Since w is a holomorphic function this relation must hold for any point in the region of its existence. We can therefore write z for x and so obtain

$$(5) \quad \frac{\partial^2}{\partial t^2}[w(z+ih, t) + w(z-ih, t)] + ig\frac{\partial}{\partial z}[w(z+ih, t) - w(z-ih, t)] = 0.$$

This equation is due to Cisotti.*

It therefore follows that any holomorphic function $w(z, t)$ which is real on the real axis, i.e. which is real when z is real, and which satisfies equation (5), will be the complex potential of an infinitesimal motion of water of depth h. The boundary conditions (1) and (4) are automatically satisfied.

The reader should verify that substitution of the complex potential (2) in (5) leads to (3). Thus Cisotti's equation contains the whole theory of waves of the type described.

15·14. Speed of propagation. The speed of propagation, in terms of the wave-length $\lambda = 2\pi/m$, is given by 15·13 (3) in the form

$$\frac{c^2}{gh} = \frac{\lambda}{2\pi h}\tanh\frac{2\pi h}{\lambda}.$$

When λ/h is small $2\pi h/\lambda$ is large, and therefore $\dfrac{c^2}{gh} = \dfrac{\lambda}{2\pi h}$ nearly, since $\tanh\theta \to 1$ when $\theta \to \infty$. Thus for small values of λ/h, c is proportional to

* *Rend. Lincei* (6), 29 (1920).

$\sqrt{\lambda}$. Again, when λ/h is large, h/λ is small, and therefore $c^2 = gh$ nearly, so that the speed of propagation tends to the constant value $\sqrt{(gh)}$ which it cannot exceed. The results are exhibited graphically in fig. 15·14, from which it is

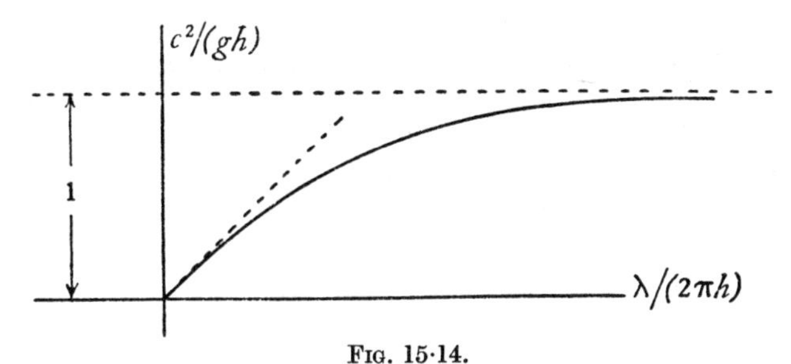

FIG. 15·14.

clear that there is only one wave-length for a given value of $c < \sqrt{(gh)}$ and that every such value is the speed of some wave. The results are considerably modified by surface tension, 15·54.

15·15. The paths of the particles. Let z be a *fixed* reference point and $z + z'$ the position of a water particle at time t, where $|z'|$ is assumed to be small. Then for a wave of small height the fluid velocity at $z + z'$ will be equal to the fluid velocity at z, neglecting second order quantities. Thus

$$\frac{d\bar{z}'}{dt} = -\frac{dw}{dz} = \frac{acm}{\sinh mh} \sin(mz - nt),$$

from 15·13 (2). Integrating, and supposing the fixed value z to be so adjusted that the added constant of integration is zero, we get

$$\bar{z}' = \frac{a}{\sinh mh} \cos(mz - nt),$$

since $c = n/m$. Equating the real and imaginary parts, we get

$$x' = a \cos(mx - nt) \cosh my / \sinh mh,$$
$$y' = a \sin(mx - nt) \sinh my / \sinh mh,$$

and therefore

$$\frac{x'^2}{\alpha^2} + \frac{y'^2}{\beta^2} = 1, \quad \alpha = \frac{a \cosh my}{\sinh mh}, \quad \beta = \frac{a \sinh my}{\sinh mh}.$$

The path of the particle is therefore an ellipse whose semi-axes are α, β, horizontal and vertical respectively, and whose centre is at the mean position z.

Since $\alpha^2 - \beta^2 = a^2/\sinh^2 mh$, all the ellipses have the same distance between their foci, but the lengths of their axes decrease as we go downwards into the liquid. At the bottom, $y = 0$ and therefore $\beta = 0$, so that the ellipse degenerates into a straight line, and the particles on the bottom simply move to and fro. The general nature of the paths of particles whose mean positions are in the same vertical line is shown in fig. 15·15.

We observe that the phase angle $(mx-nt)$ of the wave is also the eccentric angle in the ellipse, so that each particle describes its ellipse in the periodic time of the wave and all are in the same phase. The motion of a line of

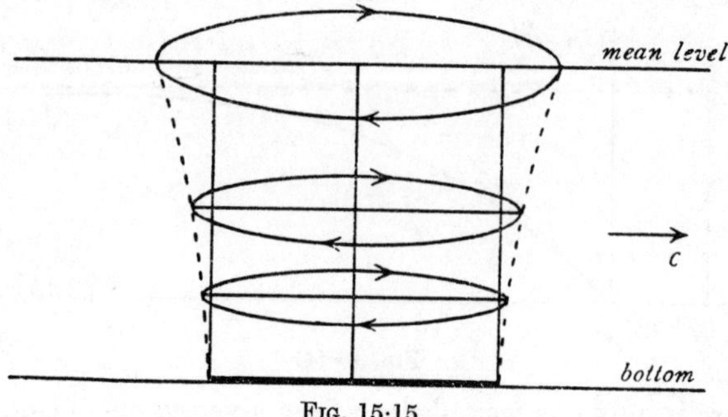

FIG. 15·15.

particles originally vertical is therefore a bending of the line, illustrated by the dotted lines in the figure, much as a blade of grass bends in the wind, but here the line also suffers a translation.

It should also be noted that the particles below a crest or trough are all moving horizontally in the same vertical line. In particular, the particle at the crest is moving forward, at the highest point of its ellipse, while at the trough it is moving backwards, at the lowest point of its ellipse. This observation has a bearing on the phenomena of tides and tidal currents.

15·17. Progressive waves on deep water. For a wave whose surface elevation is given by

(1) $$\eta = a \sin(mx-nt),$$

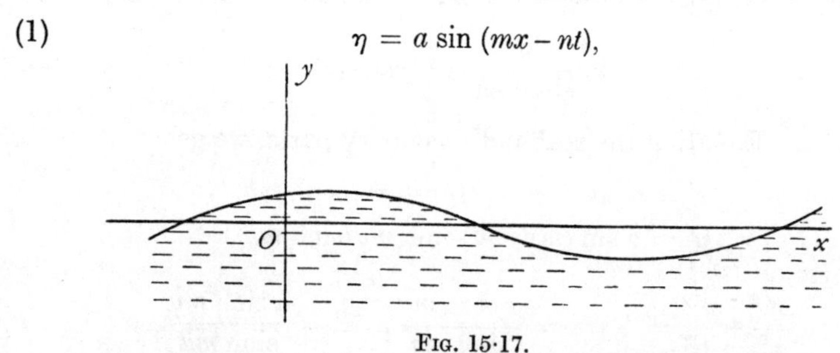

FIG. 15·17.

the complex potential is given by 15·13 (2). Move the origin to the undisturbed surface. Then

$$w = \frac{ac \cos(mz+mih-nt)}{\sinh mh}$$
$$= ac\,[\cos(mz-nt)\coth mh - i \sin(mz-nt)].$$

If we let $h \to \infty$, we get $\coth mh \to 1$, and therefore, for waves in deep water,

(2) $$w = ac\, e^{-i(mz-nt)}.$$

Also the speed of propagation, 15·13 (3), is given by

$$c^2 = \frac{g}{m} = \frac{g\lambda}{2\pi},$$

so that the speed is proportional to the square root of the wave-length.

For the paths of the particles, the method and notation of 15·15 give

$$\frac{d\bar{z}'}{dt} = -\frac{dw}{dz} = iacm\, e^{-i(mz-nt)},$$

$$\bar{z}' = a\, e^{-i(mz-nt)} = a\, e^{-i(mx-nt)} e^{my},$$

so that $|z'| = a\, e^{my}$, and the paths are circles of this radius. As $y \to -\infty$ the radii of the circles $\to 0$.

In order that the water may be regarded as deep in the above investigation it is only necessary for us to be able to take $\coth mh = 1 = \tanh mh$. Now $\tanh 2{\cdot}65 = 0{\cdot}99$, and this condition is amply satisfied if

$$mh = \frac{2\pi h}{\lambda} > 2{\cdot}65,$$

so that the water can certainly be regarded as deep if the depth exceeds half a wave-length.

A submarine whose depth is half a wave-length would hardly notice the motion due to surface waves.

15·18. Pressure due to a deep water wave. If p is the pressure at the particle whose mean position is z, the pressure equation is

$$\frac{p}{\rho} + g(y+y') - \frac{\partial\phi}{\partial t} = \text{constant}.$$

Now
$$gy' = \text{imaginary part of } -ag\, e^{-i(mz-nt)}$$
$$= \text{real part of } iag\, e^{-i(mz-nt)},$$
$$\frac{\partial\phi}{\partial t} = \text{real part of } iacn\, e^{-i(mz-nt)},$$

and therefore $\dfrac{\partial\phi}{\partial t} = gy'$ since $c^2 = g/m$.

Thus $p/\rho + gy = \text{constant}$, in other words, the pressure at any particle is equal to the pressure at the mean position of that particle.

15·20. Kinetic energy of progressive waves. By the kinetic energy of a progressive wave we shall mean the kinetic energy of the liquid (per unit thickness) between two vertical planes placed at a wave-length's distance apart and perpendicular to the direction of propagation, fig. 15·20.

Taking one of the planes through the y-axis, we have from 9·10, since the liquid is inside the contour,

$$T = +\tfrac{1}{4} i\rho \int_{(OABC)} w\, d\bar{w}.$$

From the periodic character of the motion it is clear that w has the same values at corresponding points of AB, OC, and therefore the contribution of

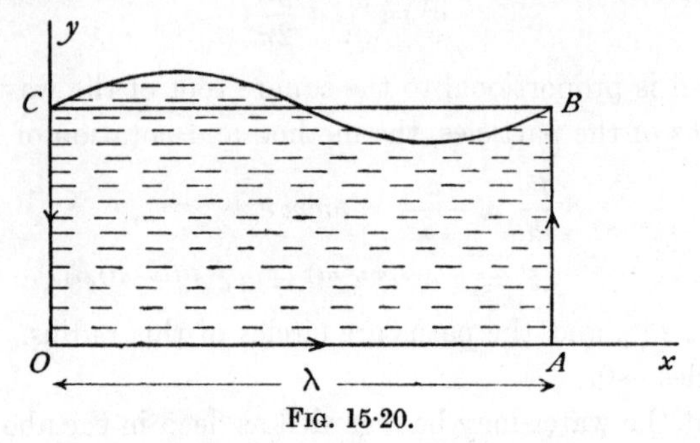

Fig. 15·20.

these lines to the integral is zero. Thus for the progressive wave of 15·13 we have

$$T = -\tfrac{1}{4} i\rho \frac{a^2 c^2 m}{\sinh^2 mh} \int_{(OA)+(BC)} \cos(mz - nt) \sin(m\bar{z} - nt)\, d\bar{z}$$

$$= -\tfrac{1}{8} i\rho \frac{a^2 c^2 m}{\sinh^2 mh} \int_{(OA)+(BC)} (\sin 2(mx - nt) - i \sinh 2my)\, d\bar{z}.$$

Now instead of integrating along the wave profile BC, we can integrate along the straight line BC, since the elevation is a small quantity. Thus

$$T = -\tfrac{1}{8} i\rho \frac{a^2 c^2 m}{\sinh^2 mh} \int_0^\lambda i \sinh 2mh\, dx = \tfrac{1}{4} a^2 g\rho\lambda,$$

using the propagation equation, 15·13 (3).

15·21. Potential energy. The potential energy (per unit thickness) is simply due to the elevated water in a wave-length and is therefore, reckoned from the undisturbed level,

$$\int_0^\lambda \tfrac{1}{2}\eta \times g\rho\eta\, dx = \tfrac{1}{4} a^2 g\rho \int_0^\lambda 2 \sin^2(mx - nt)\, dx = \tfrac{1}{4} a^2 g\rho\lambda,$$

and this is equal to the kinetic energy.

Thus the total energy of a progressive sine wave is $\tfrac{1}{2} a^2 g\rho\lambda$ per wave-length, and half of this is kinetic energy, and half potential energy due to elevation above the undisturbed level.

It may be noted also that the average energy *per unit length* of wave is $\tfrac{1}{2} a^2 g\rho$.

15·22. Group velocity. A local disturbance of the surface of still water will give rise to a wave which can be analysed into a set of simple harmonic components each of different wave-length. We have seen that the velocity of propagation depends upon the wave-length and so the waves of different wave-lengths will be gradually sorted out into groups of waves of approximately the same wave-length. In the case of water waves, the velocity of the group is, in general, less than the velocity of the individual waves composing it. What happens in this case is that the waves in front pass out of the group and new waves enter the group from behind. The energy within the group remains the same.

To study the properties of such a group, consider first the particular case of the disturbance due to the superposition of two waves of the same amplitude

$$\eta_1 = a\sin(mx-nt), \quad \eta_2 = a\sin\{(m+\delta m)x-(n+\delta n)t\},$$

where δm, δn are infinitesimal. The resulting disturbance will be

$$(1) \qquad \eta = 2a\cos\tfrac{1}{2}(x\,\delta m - t\,\delta n)\sin(mx-nt) = A\sin(mx-nt),$$

where $A = 2a\cos(x\,\delta m - t\,\delta n)$. We can therefore look upon (1) as a progressive sine wave whose amplitude A is not constant but is itself varying as a wave of velocity $c_g = \delta n/\delta m$. This velocity is called the *group velocity* and, in the case of waves of length λ, is given by

$$c_g = \frac{dn}{dm} = \frac{d(mc)}{dm} = c+m\frac{dc}{dm} = c-\lambda\frac{dc}{d\lambda}.$$

Using the value of the wave velocity c, given in 15·13 (3), we get for a single group of simple harmonic waves

$$c_g = c\left(1+\frac{m}{2c^2}\frac{dc^2}{dm}\right) = \tfrac{1}{2}c\left(1+\frac{2mh}{\sinh 2mh}\right).$$

When $mh = 2\pi h/\lambda$ is large, the group velocity is $\frac{1}{2}c$. Thus for waves on deep water the group velocity is half the wave velocity. If the water is very shallow (h/λ small), the group velocity is equal to the wave velocity.

More generally from a given local disturbance, such as a splash, waves of a variety of wave-lengths and of microscopic amplitudes a, a_1, a_2, ... will travel out. Considering only those waves of approximately the same length $2\pi/m$ the elevation at distance x at time t will be due to the sum of a large number of infinitesimal terms ; thus

$$\begin{aligned}\eta &= a\sin[mx-nt]+a_1\sin[(m+\delta m_1)x-(n+\delta n_1)t+\epsilon_1]+\ldots\\ &= A\sin(mx-nt)+B\cos(mx-nt) = C\sin(mx-nt+\epsilon),\end{aligned}$$

where

$$\begin{aligned}A &= a+a_1\cos(x\,\delta m_1-t\,\delta n_1+\epsilon_1)+a_2\cos(x\,\delta m_2-t\,\delta n_2+\epsilon_2)+\ldots,\\ B &= \phantom{a+{}} a_1\sin(x\,\delta m_1-t\,\delta n_1+\epsilon_1)+a_2\sin(x\,\delta m_2-t\,\delta n_2+\epsilon_2)+\ldots,\\ C^2 &= A^2+B^2, \quad \tan\epsilon = B/A.\end{aligned}$$

Now

$$x\,\delta m_1-t\,\delta n_1+\epsilon_1 = \delta m_1\left(x-\frac{\delta n_1}{\delta m_1}t\right)+\epsilon_1 = \delta m_1(x-c_g t)+\epsilon_1,$$

$$x\,\delta m_2-t\,\delta n_2+\epsilon_2 = \delta m_2(x-c_g t)+\epsilon_2, \ldots .$$

Thus A, B and therefore C and ϵ are functions of $(x - c_g t)$. Therefore the amplitude graph moves as a wave with velocity c_g.

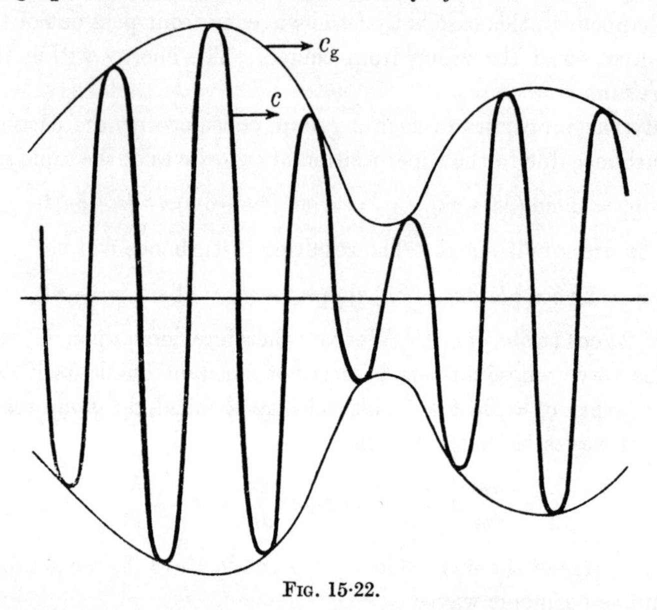

FIG. 15·22.

15·23. Dynamical significance of group velocity. In a simple harmonic train of waves, energy crosses a fixed vertical plane perpendicular to the direction of propagation at an average rate equal to the group velocity.

Proof. If p denotes the variable part of the pressure and u the horizontal velocity for a fixed value of x, the rate at which work is being done on the fluid to the right of x is

$$\frac{\partial W}{\partial t} = \int_0^h pu\, dy.$$

Now $\quad p = \rho \dfrac{\partial \phi}{\partial t} = \text{real part of } \rho \dfrac{\partial w}{\partial t} = \dfrac{\rho nac}{\sinh mh} \sin(mx - nt) \cosh my,$

$$u = \text{real part of } -\frac{dw}{dz} = \frac{mac}{\sinh mh} \sin(mx - nt) \cosh my.$$

Thus $\quad \dfrac{\partial W}{\partial t} = \dfrac{a^2 c^2 mn\rho}{\sinh^2 mh} \sin^2(mx - nt) \tfrac{1}{2}\left(h + \dfrac{\sinh 2mh}{2m}\right).$

The average value of $\sin^2(mx - nt)$ over a period is $\tfrac{1}{2}$. Thus the average rate of working is

$$\frac{a^2c^2mn\rho}{4\sinh^2 mh}\,\frac{\sinh 2mh}{2m}\left(1 + \frac{2mh}{\sinh 2mh}\right) = \frac{a^2c^2m\rho}{2} \coth mh\, c_g = \tfrac{1}{2} g\rho a^2 c_g,$$

using 15·13 (3). Now $\tfrac{1}{2}g\rho a^2$ is the energy per unit length of wave. Thus energy crosses the plane at the average rate c_g (cf. 14·41). Q.E.D.

15·24. Wave resistance. A body such as a ship travelling over the surface of water leaves behind it a train of waves. These waves possess energy which is carried away and dissipated. This energy must have been produced at the expense of the energy of the moving body which therefore experiences from this cause a resistance R. If c is the velocity of the body, and therefore also the velocity of the wave train, the rate at which work is being done to overcome R is Rc. If we consider a fixed plane drawn on the downstream side (the motion being treated as two-dimensional) perpendicular to the direction of motion of the body, the rate at which the length of the wave train is increasing ahead of this plane is c, and therefore the rate of increase of energy ahead of the plane is $c \times \frac{1}{2} g\rho a^2$, where a is the amplitude. But energy is crossing the fixed plane at a rate equal to the group velocity. Thus

$$c \times \tfrac{1}{2} g\rho a^2 = c_g \times \tfrac{1}{2} g\rho a^2 + Rc,$$

and therefore $$R = \frac{c - c_g}{c} \times \tfrac{1}{2} g\rho a^2 = \tfrac{1}{4} g\rho a^2 \left(1 - \frac{2mh}{\sinh 2mh}\right),$$

if $2\pi/m$ is the wave-length and h the depth (15·22).

Since the speed of propagation of waves cannot exceed $\surd(gh)$, it follows that when the body has a speed greater than this no wave train can accompany it and the resistance from this cause vanishes, a fact well supported by observation.

15·30. Stationary waves. Two simple harmonic wave trains of equal amplitude travelling in opposite directions are given by the surface elevations

$$\eta_1 = \tfrac{1}{2} a \sin (mx - nt), \quad \eta_2 = \tfrac{1}{2} a \sin (mx + nt).$$

The result of superposing these is the elevation

$$\eta = a \sin mx \cos nt.$$

A motion of this type is called a *stationary* or *standing* wave. At a given value of x the surface of the water moves up and down. For a given value of t the form of the surface is a sine curve of amplitude $a \cos nt$, which therefore varies between 0 and a. A wave of this type is not propagated.

The points for which $mx = s\pi$, $s = \ldots, -2, -1, 0, 1, 2, \ldots$ are always at rest in the mean surface and are called *nodes*. The points for which $mx = (2s+1)\pi/2$ are points of maximum displacement for a given value of t and are called *loops*. When $\cos nt = \pm 1$ the surface is in the form of the sine curve $\eta = \pm a \sin mx$, which represents the maximum departure from the mean level. When $\cos nt = 0$ the free surface coincides with the mean level.

When a progressive train of waves represented by η_1 impinges on a fixed vertical barrier and is there reflected (η_2), the resulting disturbance when a steady state is reached consists of stationary waves.

Such waves can, for example, be generated by tilting slightly a rectangular vessel containing water and then restoring it to the level position. The water level at each end of the vessel then moves up and down the vertical faces which are loops.

Conversely a progressive wave can be regarded as due to the superposition of two standing waves.

15·31. Complex potential of stationary waves. To obtain oscillatory waves of stationary type, we can substitute in 15·13 (5) a suitable harmonic function for w. Taking

$$w(z, t) = A \sin mz \sin nt,$$

we get

$$w(z+ih, t)+w(z-ih, t) = 2A \sin mz \cosh mh \sin nt,$$

$$w(z+ih, t)-w(z-ih, t) = 2iA \cos mz \sinh mh \sin nt,$$

and on substitution we get $n^2 = mg \tanh mh$. This equation connects the frequency with the wave-length.

The surface elevation is given by 15·12 (2):

$$g\eta = \text{real part of } \frac{\partial w(x+ih, t)}{\partial t} = An \sin mx \cosh mh \cos nt.$$

Hence if the surface elevation is $\eta = a \sin mx \cos nt$, we get

$$a = An \cosh mh/g = Am \sinh mh/n,$$

so that

$$w = \frac{ac}{\sinh mh} \sin mz \sin nt, \quad c = \frac{n}{m}.$$

This result could have been deduced at once by superposing two solutions of the form found in 15·13, in this case

$$w = \tfrac{1}{2} ac \operatorname{cosech} mh \cos(mz-nt) - \tfrac{1}{2} ac \operatorname{cosech} mh \cos(mz+nt).$$

15·32. Paths of the particles in a stationary wave. Using the method of 15·15, if $z+z'$ is the displaced position at time t of the particle whose mean position is z, we get

$$\frac{d\bar{z}'}{dt} = -\frac{dw}{dz} = -\frac{acm}{\sinh mh} \cos mz \sin nt,$$

$$\bar{z}' = \frac{a}{\sinh mh} \cos mz \cos nt.$$

Thus $\arg \bar{z}'$ is constant, and therefore the particles describe straight lines with simple harmonic motion of period $2\pi/n$, the period of the wave. The amplitude is

$$\frac{a}{\sinh mh} \,|\cos mz\,| = \frac{a}{\sinh mh} \{\tfrac{1}{2}(\cos 2mx + \cosh 2my)\}^{\frac{1}{2}}.$$

Thus the amplitude decreases as we move downwards from the surface. The inclination of the line corresponding to the mean position z is

$$\arg z' = \arg \cos m\bar{z} = \tan^{-1}\{\tan mx \tanh my\}.$$

This inclination is therefore zero below the nodes ($mx = s\pi$) and $\frac{1}{2}\pi$ below the loops. Thus the particles below the nodes move horizontally, those below the loops move vertically, fig. 15·32.

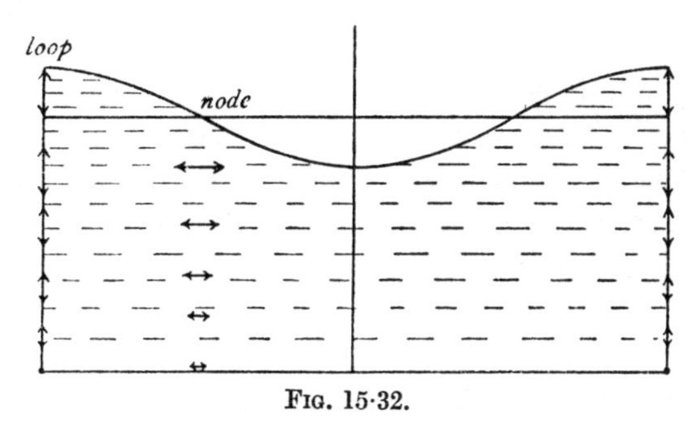

FIG. 15·32.

15·33. Stationary waves in a rectangular tank. Since the motion under the loops in a standing wave is vertical, the motion between any two given loops would be unaltered if fixed rigid vertical planes were inserted there. We should then have the case of liquid oscillating in a tank of finite dimensions.

Consider a rectangular tank of length l. Take the origin at the bottom at an end and the y-axis vertically upwards. Since $x = 0$ has to be a loop, the surface disturbance must be of the form

$$\eta = a \cos mx \cos nt,$$

and hence

$$w = ac \cos mz \sin nt/\sinh mh.$$

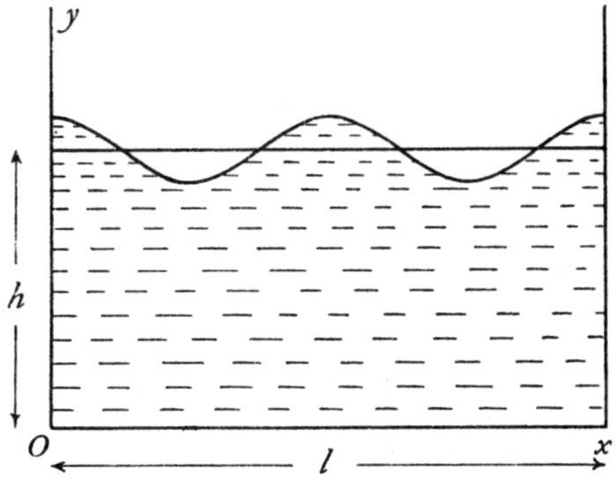

FIG. 15·33.

Since there must also be a loop at $x = l$, we must have

$$ml = s\pi,$$

where s is any integer. Thus the possible wave-lengths of the oscillations are given by

$$\lambda = 2l,\ l,\ 2l/3,\ 2l/4, \ldots .$$

When a photographic plate is developed by rocking a dish containing the developing solution, care must be taken to vary the frequency of the oscillations, otherwise the portions of the plate which are below the loops of the wave will be underdeveloped, since the solution in the neighbourhood of these points has very little motion and its chemical action soon ceases, causing a streaky effect on the negative.

15·34. Energy of stationary waves. Considering unit thickness, the potential energy in a wave-length is given by

$$V = \int_0^\lambda \tfrac{1}{2}\eta \,.\, g\rho\eta \, dx.$$

Taking $\eta = a \sin mx \cos nt$, we get $V = \tfrac{1}{4} g\rho a^2 \cos^2 nt \,.\, \lambda$, and when $\cos nt = 1$, the potential energy is $\tfrac{1}{4} g\rho a^2 \lambda$.

Now at this instant the energy is wholly potential, for the kinetic energy depends on the normal velocity of the boundary, 3·72, which is instantaneously zero. Thus the kinetic energy at time t is

$$T = \tfrac{1}{4} g\rho a^2 \sin^2 nt \,.\, \lambda.$$

15·40. Steady motion. The complex potential for a simple sine wave moving forward was obtained in 15·11 (3). If we take axes of reference moving with the wave, the complex potential will be deduced by writing $z' + ct$ for z and therefore the complex potential becomes

$$\frac{ac \cos mz'}{\sinh mh}.$$

If we superpose on the whole system a velocity c in the direction of the negative axis of x, the axes and the wave profile will be reduced to rest and the fluid will have the general velocity c from right to left, the complex potential now being

$$w = cz' + \frac{ac \cos mz'}{\sinh mh}.$$

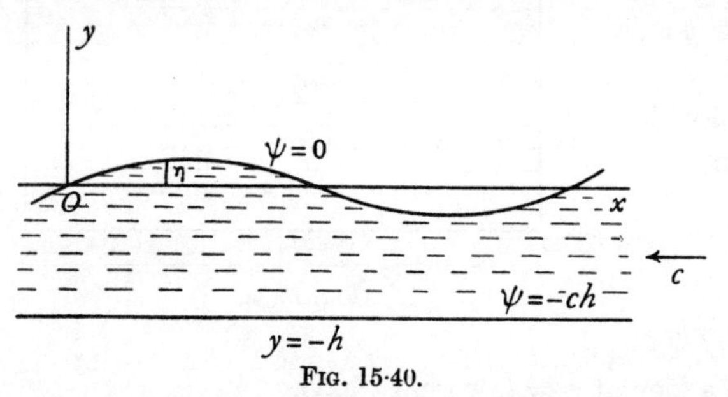

Fig. 15·40.

This represents a steady motion in which the force on any particle is unaltered, for the addition of a constant velocity has no dynamical effect. But there is an advantage in this reduction inasmuch as the profile is now a streamline corresponding to a constant value of ψ. For the applications we shall make, it is more convenient to take the origin in the undisturbed surface, which means writing $z + ih$ for z'. Thus, finally, dropping a constant cih,

$$(1) \qquad w = cz + \frac{ac \cos m(z+ih)}{\sinh mh}.$$

The bottom is now the streamline

$$\psi = -ch,$$

while, at the free surface,

$$\psi = c\eta - \frac{ac}{\sinh mh}\sinh m(\eta+h)\sin mx$$
$$= c\eta - ac\,(\cosh m\eta + \sinh m\eta\coth mh)\sin mx.$$

But, to the order of approximation adopted, the surface is the streamline $\psi = 0$, so that neglecting η^2,

(2) $$\eta = a\sin mx.$$

In the case of deep water, the complex potential becomes

(3) $$w = cz + ac\,e^{-imz},$$

and the surface streamline $\psi = 0$ gives

(4) $$\eta = a\,e^{m\eta}\sin mx,$$

which agrees with (2) to the same order of approximation.

15·41. Second approximation to the wave speed. Taking for simplicity the case of deep water, the complex potential and the surface profile are given by 15·40 (3) and (4). We note here that (4) is the exact result of putting $\psi = 0$ in (3), but it is not to be assumed that therefore (4) is necessarily a surface of constant pressure. We have

$$q^2 = \frac{dw}{dz}\frac{d\bar{w}}{d\bar{z}} = c^2(1-ima\,e^{-miz})(1+ima\,e^{im\bar{z}})$$
$$= c^2(1-2am\,e^{my}\sin mx + a^2m^2\,e^{2my}).$$

Hence at the free surface

$$q^2 = c^2\{1-2m\eta + a^2m^2\,e^{2m\eta}\} = c^2\{1-2m\eta+a^2m^2(1+2m\eta+2m^2\eta^2+\ldots)\}.$$

If p is the pressure in the liquid at the free surface and Π the external pressure, the pressure equation now gives

$$p-\Pi = \rho\{-g\eta + m\eta c^2 - c^2a^2m^3\eta - c^2a^2m^4\eta^2 - \ldots\} + \text{constant}.$$

(1) $$= \rho\eta\{-g+c^2m-c^2a^2m^3\} - \rho c^2a^2m^4\eta^2 - \ldots + \text{constant}.$$

If in this result we neglect terms containing a^2, we get $p = \Pi$ if $c^2 = g/m$, the result already obtained. A much closer agreement between p and Π will be obtained if we neglect the terms containing $a^2m^4\eta^2$, which are of the *fourth* and higher order of small quantities. This will make the free surface one of constant pressure if $-g+c^2m-c^2a^2m^3 = 0$, which gives

$$c^2 = \frac{g}{m}(1-a^2m^2)^{-1} = \frac{g}{m}(1+a^2m^2),$$

neglecting terms of the fourth order, and this is a closer approximation to the wave speed. It will be observed that the speed thus found depends not only on the wave-length but also on the amplitude.

The maximum value of the first term neglected in (1), namely $-\rho c^2 a^2 m^4 \eta^2$, is $-2g\rho a(a^3m^3/2)$, which is the fraction $a^3m^3/2$ of the difference in pressure between the crest and the trough. Thus, for a wave of amplitude 4 feet and 80 feet long, this fraction is

$$\tfrac{1}{2}\left(\frac{2\pi}{80}\times 4\right)^3 = 0{\cdot}015,$$

and the pressure neglected is therefore at most that of

$$0{\cdot}015 \times 8 \text{ ft. of water} = 1{\cdot}4 \text{ in. of water.}$$

15·42. Waves at an interface. Consider liquid of density ρ' and depth h' flowing with constant velocity V' over a layer of liquid of density ρ and depth h which flows with constant velocity V, the fluids being bounded above and below by rigid horizontal planes.

Take the axis of x in the (geometrical) interface which separates the fluids and which constitutes a vortex sheet. To investigate the condition that a wave of small elevation $\eta = a \sin(mx - nt)$ may be propagated at the interface

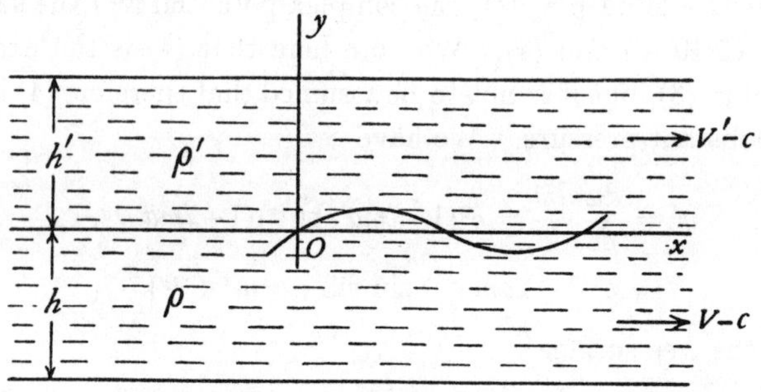

FIG. 15·42.

with velocity $c = n/m$, we impose on the whole mass of fluid a velocity c opposite to the direction of propagation, thus reducing the profile to rest and changing the velocities of the streams to $V'-c$, $V-c$. From 15·40, it is evident that the complex potential for the lower fluid is

$$(1) \qquad w = -(V-c)z - \frac{a(V-c)}{\sinh mh}\cos m(z+ih),$$

for the streamline $\psi = 0$ is then $\eta = a \sin mx$.

We deduce at once the complex potential for the upper liquid by writing $-h'$ for h, thus giving

$$(2) \qquad w' = -(V'-c)z + \frac{a(V'-c)}{\sinh mh'}\cos m(z-ih').$$

The speed in the lower liquid, neglecting a^2, is then given by

$$q^2 = \frac{dw}{dz}\cdot\frac{d\bar{w}}{d\bar{z}} = (V-c)^2 - \frac{2ma(V-c)^2}{\sinh mh}\cosh m(y+h)\sin mx,$$

and therefore at the interface the speed is given by

$$q_0^2 = (V-c)^2\{1-2m\eta \coth mh\},$$

and for the upper liquid by

$$q_0'^2 = (V'-c)^2\{1+2m\eta \coth mh'\}.$$

Now at the interface the pressure equation gives for the two liquids

$$(3) \qquad p'+\tfrac{1}{2}\rho' q_0'^2+\rho' g\eta = \text{constant},$$

$$(4) \qquad p+\tfrac{1}{2}\rho q_0^2+\rho\, g\eta = \text{constant}.$$

But the pressure must be continuous, and therefore $p = p'$. By subtraction,

$$\tfrac{1}{2}\rho' q_0'^2-\tfrac{1}{2}\rho q_0^2+g\eta(\rho'-\rho) = \text{constant}.$$

Thus the coefficient of η must vanish, and therefore

$$m\rho(V-c)^2 \coth mh+m\rho'(V'-c)^2 \coth mh' = g(\rho-\rho').$$

This equation determines the velocity of propagation. We note that

(i) If $\rho' = 0$, $V = 0$, then the equation reduces to 15·13 (3).

(ii) If the liquids are of infinite depth the equation simplifies to

$$m\rho(V-c)^2+m\rho'(V'-c)^2 = g(\rho-\rho').$$

(iii) The condition of stability is the condition that waves of the prescribed type can be propagated, i.e. that c shall be real.

(iv) There are, in general, two values of c for which the equation is satisfied.

(v) If $c = 0$, there is a stationary wave.

(vi) If the liquids are at rest, save for the wave motion $V = V' = 0$, and then, if the depths are infinite,

$$c^2 = \frac{g(\rho-\rho')}{m(\rho+\rho')}.$$

It follows that we must have $\rho>\rho'$, i.e. the heavier liquid must be underneath, but see 15·54.

As a particular deduction, consider the upper fluid to be air of specific gravity s and of infinite depth. Then putting $V = V' = 0$, we get

$$c^2 = \frac{g}{m}\,\frac{\rho(1-s)}{\rho(\coth mh+s)} = \frac{g}{m}\tanh mh\{1-s(1+\tanh mh)\},$$

approximately, since s is small. Comparing this with 15·13 (3), we see that the presence of the atmosphere tends to diminish the wave velocity.

This result is of general application, as is seen from (vi) above, which also shows that, if ρ and ρ' are nearly equal, the periods of the oscillations of the common surface will be long, compared with period of the oscillations of a free surface of liquid.

15·43. Steady flow over a sinuous bottom. Let a stream of mean depth h flow with general speed U over a bottom at which the elevation is

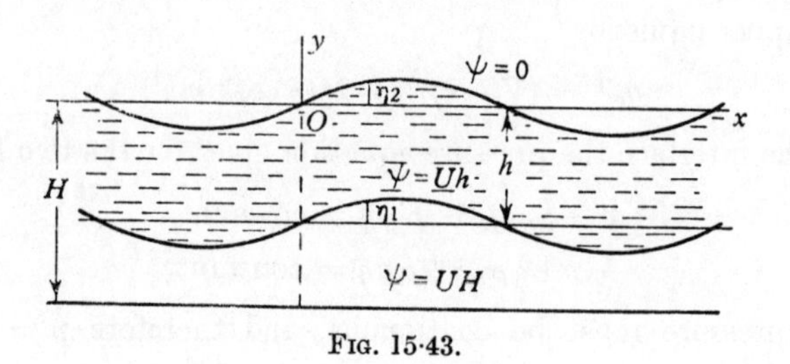

FIG. 15·43.

given by $\eta_1 = a \sin mx$, a being small and the axis of x horizontal.

Take the origin in the free surface. Then the complex potential (15·40)

$$w = -U\left\{z + \frac{b}{\sinh mH} \cos m(z+iH)\right\}, \tag{1}$$

where H is determined by

$$U^2 = \frac{g}{m} \tanh mH, \tag{2}$$

gives the steady wave motion with surface elevation $\eta_2 = b \sin mx$ on a steady stream of depth H, The free surface is the streamline $\psi = 0$ and the bottom is $\psi = UH$. We determine b so that $\psi = Uh$ is the streamline $y = -h + \eta_1$. Putting $\psi = Uh$ in the stream function gives

$$\eta_1 = \frac{b}{\sinh mH} \sin mx \sinh m(H-h)$$

which corresponds to $\eta_1 = a \sin mx$ provided that

$$a = \frac{b \sinh m(H-h)}{\sinh mH} \quad \text{or}$$

$$\frac{b}{a} = \frac{1}{\cosh mh - \dfrac{g}{mU^2} \sinh mh}. \tag{3}$$

This gives the ratio $\eta_2 : \eta_1$ for a given value of x.

Thus the crests and troughs of the free surface and the bottom correspond or are opposite according as

$$U^2 \gtrless \frac{g}{m} \tanh mh, \quad \text{or} \quad U^2 \gtrless c^2,$$

where c is the speed of propagation of waves of length $2\pi/m$ in water of depth h.

If $U = c$, the ratio $\eta_2 : \eta_1$ becomes infinite. This means that the free surface cannot then be represented by a simple sine curve, and the assumption on which the solution was obtained then breaks down.

15·44. Waves at an interface when the upper surface is free. The problem considered in 15·42 admits of interesting generalisation, if we consider the upper surface to be free instead of being bounded by a fixed horizontal plane. Taking the case of liquid of depth h and density ρ lying on liquid of density ρ', we consider the propagation of waves at the interface. This resembles the problem of 15·43, if instead of a fixed sinuous bottom we consider liquid to be present below the sinuosities. Taking the figure and notations of that section, we get in the upper liquid the same complex potential (1) and the same ratio (3) of the elevations at the surface and interface, U now denoting the wave velocity. The additional condition to be satisfied is the continuity of the pressure at the interface, which means that

$$\tfrac{1}{2}\rho q^2 + g\rho\eta - \tfrac{1}{2}\rho' q'^2 - g\rho'\eta = \text{constant},$$

where q, η refer to the upper liquid at the interface and q' to the lower. If the latter is of great depth, we can take $w' = -U(z + ae^{-im(z+ih)})$, as in 15·40 (3), which leads to

$$ag(\rho' - \rho) + \rho bmU^2 \cosh m(H-h) \operatorname{cosech} mH - \rho' amU^2 = 0.$$

Eliminating the ratio $b : a$ by means of the relation 15·43 (3), we get, after some reduction,

$$\{mU^2(\rho + \rho' \coth mh) - g(\rho' - \rho)\}\{mU^2 - g\} = 0,$$

which gives

$$(1) \qquad U^2 = \frac{g}{m}, \quad U^2 = \frac{g(\rho' - \rho)}{m(\rho + \rho' \coth mh)}.$$

Thus, corresponding to a given velocity U, there are two possible wavelengths of which the first is the same as if the upper liquid were absent.

To find the values of m in the second case, put

$$\frac{g(\rho' - \rho)h}{\rho' U^2} = l, \quad \frac{\rho}{\rho'} = s, \quad mh = x.$$

Then we have to solve the equation

$$f(x) = s + \coth x - \frac{l}{x} = 0. \quad \text{Now } f(x) = s + \left(\coth x - \frac{1}{x}\right) - \frac{l-1}{x}.$$

But $\coth x - 1/x$ is positive if x is positive, and $(l-1)/x$ is negative if $l<1$. Hence $f(x)$ is positive if $l<1$ and there are no positive roots to the equation. On the other hand, if $l>1$, $f(0) = -\infty$, $f(\infty) = 1+s$, and therefore the equation has a real positive root; that there is only one such root follows from the fact that $f'(x)$ is positive when $l>1$. Thus if

$$U^2 > \frac{g(\rho' - \rho)h}{\rho'},$$

only one type of wave is generated, but, if U^2 is less than this value, a second type of wave exists for which the ratio of the elevation at the interface to that at the free surface is given by 15·43 (3), where U^2 has the second of the values given by (1). This ratio is

$$\frac{a}{b} = -\frac{\rho\, e^{mh}}{\rho' - \rho},$$

so that if $\rho' - \rho$ is small the elevation of the waves at the interface will be very large compared with the surface elevation. This result has been used to explain the abnormal resistance sometimes experienced by ships near the mouths of some of the Norwegian fiords where there is a layer of fresh water over salt, the enhanced resistance being ascribed to the generation of waves of large amplitude at the interface.

15·50. Surface tension. An interface between two fluids which do not mix behaves as if it were in a state of uniform tension. This tension is called the *surface tension* and depends on the nature of the two fluids and on the temperature.

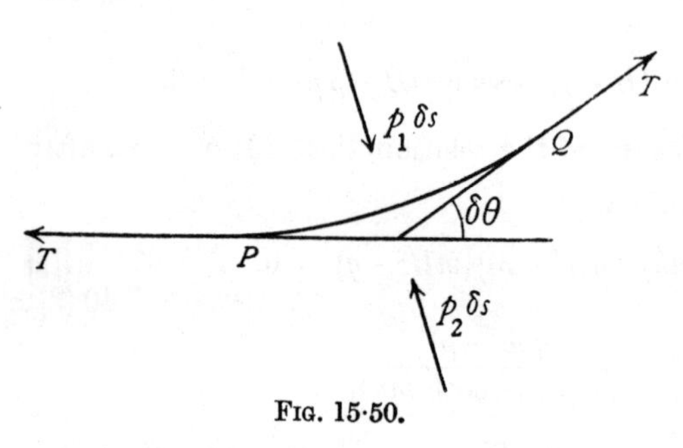

Fig. 15·50.

Let $PQ = \delta s$ be an element of arc of a cross-section of a cylindrical surface forming the interface between two fluids whose surface tension is T. If p_1, p_2 are the pressures on either side, $\delta\theta$ the angle between the tangents at P and Q, resolution along the normal at P gives the approximate equation

$$-p_1\,\delta s + p_2\,\delta s + T\,\delta\theta = 0$$

and therefore

$$p_1 - p_2 = T/R,$$

where R is the radius of curvature.

Thus at an interface there is a discontinuity of pressure. Phenomena involving surface tension are generally described by the adjective *capillary*. Referring to fig. 15·11 and the notations of 15·12, the difference between the internal and external pressures at the point P of the interface is

$$\rho\left(\frac{\partial\phi}{\partial t} - g\eta\right).$$

Also the curvature, since the slope is small, is $\partial^2\eta/\partial x^2$. Thus the pressure boundary condition at the interface is

$$\rho\left(\frac{\partial\phi}{\partial t} - g\eta\right) = -T\frac{\partial^2\eta}{\partial x^2},$$

the negative sign being due to the fact that the slope decreases as x increases.

Differentiate with respect to t, and note that $\partial\eta/\partial t = \partial\psi/\partial x$. Then

$$\frac{\partial^2\phi}{\partial t^2} - g\frac{\partial\psi}{\partial x} + \frac{T}{\rho}\frac{\partial^3\psi}{\partial x^3} = 0.$$

This now replaces the surface condition 15·13 (1).

15·51. Equation satisfied by the complex potential. The argument of 15·13 still applies, and we get Cisotti's equation in the form

$$\frac{\partial^2}{\partial t^2}[w(z+ih,t)+w(z-ih,t)]+ig\frac{\partial}{\partial z}[w(z+ih,t)-w(z-ih,t)]$$
$$-i\frac{T}{\rho}\frac{\partial^3}{\partial z^3}[w(z+ih,t)-w(z-ih,t)] = 0.$$

15·52. Surface waves. To obtain the surface waves in water of depth h, take the periodic solution

$$w(z,t) = A\cos(mz-nt),$$

which is real when $y = 0$. Substituting in the equation satisfied by w, 15·51, we get

$$-n^2\cosh mh + mg\sinh mh + \frac{m^3T}{\rho}\sinh mh = 0,$$

so that

$$n^2 = m\left(g+\frac{Tm^2}{\rho}\right)\tanh mh,$$

$$c^2 = \left(\frac{g}{m}+\frac{Tm}{\rho}\right)\tanh mh,$$

which gives the speed of propagation of waves of length $2\pi/m$.

15·53. Effect of capillarity on waves at an interface. Using the figure and notations of 15·42, we obtain the pressure equations (3) and (4) by exactly the same steps. The effect of the surface tension at the interface is to replace the condition $p = p'$ by the condition

$$p-p' = -T\frac{\partial^2\eta}{\partial x^2} = T\,am^2\sin mx,$$

and the propagation equation becomes

$$m\rho\,(V-c)^2\coth mh + m\rho'\,(V'-c)^2\coth mh' = g(\rho-\rho')+Tm^2.$$

15·54. Speed of propagation. Consider waves propagated at the interface between two layers of liquid of great depth, and otherwise at rest. We have, from 15·53, if ρ' is the density of the upper fluid,

$$c^2 = \frac{g}{m}\frac{\rho-\rho'}{\rho+\rho'}+\frac{Tm}{\rho+\rho'}.$$

If the wave-length $2\pi/m$ is large, the first term on the right is large compared with the second, and the effect of capillarity is inconsiderable. On the other hand, for small wave-lengths the second term predominates and gravity can be neglected.

Put
$$s = \frac{\rho'}{\rho}, \quad T = g\rho(1-s)l^2,$$

then s is the specific gravity of the upper liquid and l is a length which may be regarded as a measure of the surface tension. In terms of s and l,

$$c^2 = \frac{g(1-s)}{1+s}\left(\frac{1}{m}+ml^2\right).$$

By differentiation we see that c^2 has a minimum value when $m = 1/l$, so that the velocity of propagation is least for waves of length

$$\lambda_0 = 2\pi l,$$

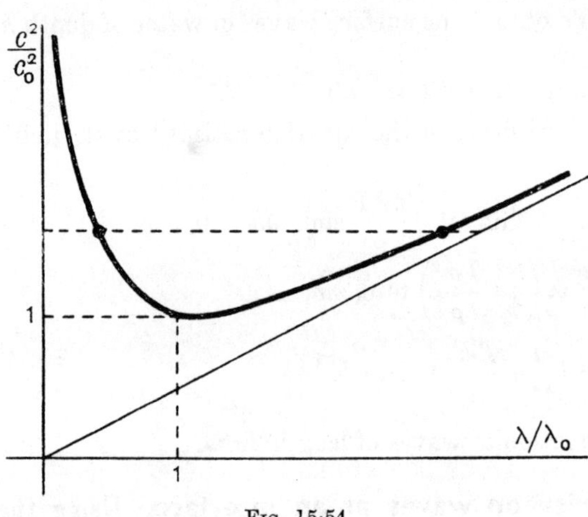

FIG. 15·54.

and the least value of c is given by

$$c_0{}^2 = \frac{2gl(1-s)}{1+s}.$$

Thus

$$\frac{c^2}{c_0{}^2} = \tfrac{1}{2}\left(\frac{\lambda}{\lambda_0}+\frac{\lambda_0}{\lambda}\right).$$

This shows that when $c > c_0$ there are two admissible values of λ/λ_0, and these values are reciprocals.

Waves of length less than λ_0 are called *ripples*, so that ripples are waves in whose propagation capillarity plays the predominating part.

The group velocity is given by

$$U = c - \lambda\frac{dc}{d\lambda} = c\left(1 - \tfrac{1}{2}\frac{\lambda^2-\lambda_0{}^2}{\lambda^2+\lambda_0{}^2}\right).$$

Thus for ripples the group velocity tends to the value $3c/2$, which is greater than the wave velocity, while for waves in which λ is much greater than λ_0 (*gravity waves*) the group velocity tends to $c/2$, as already found in 15·22.

The condition for stability is that c^2 shall be positive, so that c is real. This condition is always satisfied if $\rho > \rho'$. But it is worthy of note that it is also satisfied if $\rho < \rho'$, provided that

$$\left(\frac{\lambda}{2\pi}\right)^2 < \frac{T}{g(\rho'-\rho)}.$$

This result is illustrated by the experiment in which water is retained by atmospheric pressure in an inverted tumbler whose mouth is closed by gauze of fine mesh.

15·55. Effect of wind on deep water. If the water is deep and at rest except for the wave motion, we get, from 15·53,

$$(1) \qquad c^2 - \frac{2V's}{1+s}c + \frac{V'^2 s}{1+s} = \frac{g}{m}\frac{1-s}{1+s} + \frac{Tm}{\rho(1+s)} = c_1^2,$$

where $s = \rho'/\rho$, and c_1 is the wave velocity when there is no wind. For a given wave-length the wave velocity c will be greatest when $dc/dV' = 0$, i.e. when $c = V'$, and the maximum velocity is then

$$c_m = c_1\surd(1+s).$$

If the wind has any other velocity greater or less than c_m, the wave velocity is less than c_m.

Again, the values of c are imaginary if

$$V'^2 > c_1^2\frac{(1+s)^2}{s}.$$

Remembering that c_1 depends on the wave-length $2\pi/m$, and has the minimum value c_0 (15·54), it follows also that

$$V'^2 > \tfrac{1}{2}c_0^2\frac{(1+s)^2}{s}\left(\frac{\lambda}{\lambda_0}+\frac{\lambda_0}{\lambda}\right).$$

This last inequality implies that waves within a certain range of wave-lengths cannot then be propagated. They are blown into spindrift.

This means that the water surface is then unstable, even if a flat calm prevails previously to the starting of the wind. Lamb gives the minimum value of V' in this case as about 12·5 knots.

The two values of c given by (1) are

$$c = \frac{V's}{1+s} \pm \sqrt{\left(c_1^2 - \frac{sV'^2}{(1+s)^2}\right)},$$

and if

$$V' < c_1(1+s^{-1})^{\frac{1}{2}},$$

these velocities have opposite signs. Hence the waves can travel either with the wind or against it, but they travel faster with the wind than against it. If V' exceeds the value just given, the waves cannot travel against the wind.

It must be remembered that the above conclusions are based on arguments which take no account of viscosity.

15·58. The linearised form of Levi-Civita's surface condition. The condition given in 14·65 is non-linear. The linearised approximation is obtained by assuming $|\omega|$ to be small of the first order. This means that θ and τ are small so that $\sin\theta = \theta$ and $q = c$ nearly. Thus the linearised form is

(1) $$\frac{\partial \theta}{\partial \psi} = -\frac{g}{c^3}\theta, \quad \psi = 0.$$

The theory based on this surface condition is completely equivalent to that given in the preceding sections of this chapter.

To see that this is so, consider a symmetrical wave profile of wave-length λ and take the origin at a crest, fig. 15·58. For simplicity we take the depth of the water to be infinite.

From the symmetry ϕ is constant and $\theta=0$ on the vertical through a crest or trough, and since at a great depth $-\partial\phi/\partial x = c$, $-\partial\phi/\partial y = 0$, we may take $\phi = -cx$ and therefore $\phi=0$ on the y-axis, and $\phi = \pm\frac{1}{2}c\lambda$ on the verticals through the adjacent troughs on the left and right.

Thus the boundary conditions to be satisfied are

(2) $\theta = 0$ when $\phi = \pm\frac{1}{2}c\lambda$,

(3) $\omega \to 0$ when $\psi \to \infty$,

(4) the surface condition.

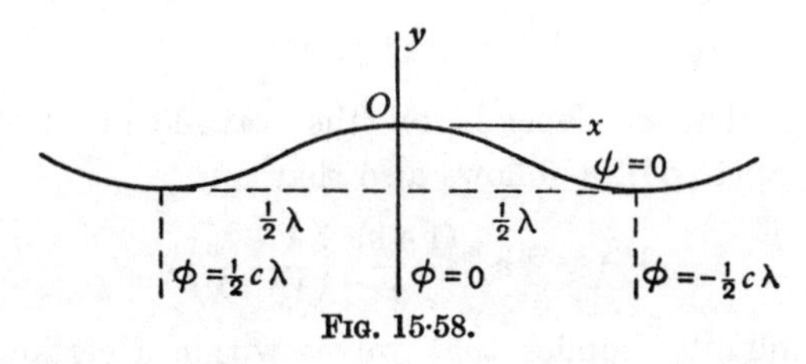

Fig. 15·58.

In the present case the surface condition is given by (1).

It is readily verified that all these conditions are satisfied by

(5) $$\omega = -iAe^{igw/c^3}, \quad c^2 = g\lambda/2\pi,$$

where A is a real constant, and the supposition that ω is small entails A being small. If then we write $m = 2\pi/\lambda$, $A = ma$, and expand the exponential in 14·65 (1), we get

$$-\frac{dz}{dw} = \frac{1}{c}e^{i\omega} = \frac{1}{c}(1+i\omega) = \frac{1}{c}(1+mae^{imw/c}),$$

whence by integration, observing that $z=0$ when $w=0$,

$$w = -c\{(z+ai) - iae^{imw/c}\}.$$

Since a is small a first approximation is $w = -c(z+ai)$ and therefore substituting this in the index of the exponential we get

(6) $$w = -c\{(z+ai) + ae^{-im(z+ai+\lambda/4)}\}.$$

This agrees with 15·40 (3) if in the latter we move the origin to the crest, that is to say write $z+ai+\lambda/4$ for z. The result then differs from (6) only by a constant. Thus the linearised approximation (1) agrees with the previous theory, and indeed gives precision to the assumptions of that theory.

15·59. The method of T. V. Davies. There is, however, a serious limitation to the use of the linearised approximation. A wave will break at the crest when the fluid velocity there exceeds the velocity of the wave. The critical case is clearly when the fluid velocity at the crest is exactly equal to the velocity of the wave, that is to say in the steady motion case $v = 0$. From 14·65 (2) this means that at the crest $e^{\tau} = 0$ and so $\tau = -\infty$. It follows that no approximation based upon τ being small can throw any light on this case. T. V. Davies* has proposed an approximation to the boundary condition 14·65 (4) which preserves its essential features and allows τ to be large. The approximation is

$$(1) \qquad \frac{\partial\theta}{\partial\psi} = -\frac{g}{3c^3}e^{-3\tau}\sin 3\theta, \quad \psi = 0,$$

which differs from 14·65 (4) only in putting $\frac{1}{3}\sin 3\theta$ instead of $\sin\theta$.

It is left as an exercise to verify that the boundary conditions 15·58 (2) and (3), and the surface condition (1) are satisfied by

$$(2) \qquad e^{-3i\omega} = 1 - 3Ae^{2\pi i w/c\lambda}, \quad c^2 = g\lambda/2\pi,$$

where $3A$ is an arbitrary real constant. When $|\omega|$ is small (2) reduces to 15·58 (6).

Since $w = 0$ at the crest it follows from 14·65 (1) and (2) that $v = 0$ there when $3A = 1$, which is the condition for breaking at the crest.

When this condition is satisfied we have near the crest, where w is small,

$$e^{-3i\omega} = 1 - \left(1 + \frac{2\pi i w}{c\lambda}\right) = -\frac{2\pi i w}{c\lambda}$$

and therefore from 14·65 (1)

$$\frac{dz}{dw} \propto w^{-1/3} \quad \text{and} \quad w \propto z^{3/2}.$$

This expresses that when breaking is about to occur, in the neighbourhood of the crest the wave is in the form of a wedge of angle 120°. This agrees with observations of waves just before breaking and with a theoretical result obtained by Stokes. See 14·50.

15·60. Long waves. The surface waves which have been considered in the preceding sections were not restricted as to wave-length. We shall now consider waves whose length is large compared with the depth of the water. Thus for water of depth h, contained say in a horizontal canal, the hypothesis

---

* *Proc. Roy. Soc.*, *A*, 208 (1951) 475; *Quart. Appld. Math.* 10 (1952) 57. My colleague B. A. Packham (to whom I am indebted for this section), *Proc. Roy. Soc.*, A, 213 (1952) 238, has successfully applied the method to the solitary wave.

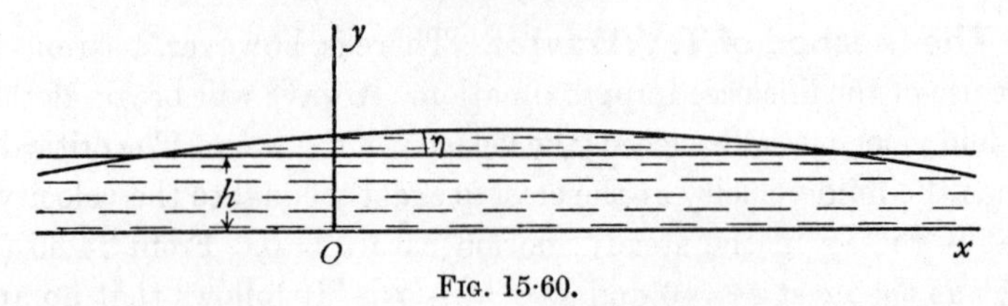

Fig. 15·60.

is that h/λ is small where λ is a typical wave-length. The previous limitation that the surface elevation and slope of the waves is small will of course be retained. In the present instance this implies that η/h and $d\eta/dx$ are small.

On the hypothesis of long waves, the propagation equation, 15·13, simplifies and the general solution can be readily obtained.

We have, in fact, if $w(z, t)$ is the complex potential,

$$w(z+ih, t) = w(z, t)+ih\frac{\partial w(z, t)}{\partial z}+\dots,$$

$$w(z-ih, t) = w(z, t)-ih\frac{\partial w(z, t)}{\partial z}+\dots,$$

and therefore, neglecting terms containing h^2, the equation for w becomes

$$\text{(1)} \qquad \frac{\partial^2 w(z, t)}{\partial t^2} - gh\frac{\partial^2 w(z, t)}{\partial z^2} = 0.$$

As before, w must be real on the real axis $y = 0$.

To solve this equation, let $c^2 = gh$, and put $z_1 = z+ct$, $z_2 = z-ct$.

Then $$\frac{\partial w}{\partial z} = \frac{\partial w}{\partial z_1}+\frac{\partial w}{\partial z_2}, \qquad \frac{\partial w}{\partial t} = c\frac{\partial w}{\partial z_1}-c\frac{\partial w}{\partial z_2},$$

and therefore (1) reduces to

$$4\frac{\partial^2 w}{\partial z_1\,\partial z_2} = 0. \text{ Integrating, we get } \frac{\partial w}{\partial z_1} = w_1'(z_1),$$

where $w_1(z_1)$ is an arbitrary function of z_1.

Integrating once more, $w = w_1(z_1)+w_2(z_2)$, where $w_2(z_2)$ is an arbitrary function of z_2, and therefore the general solution of (1) is

$$\text{(2)} \qquad w = w_1(z+ct)+w_2(z-ct),$$

where the (holomorphic) functions w_1, w_2 can be arbitrarily chosen, subject to the sole condition that w is real when $y = 0$.

Equating the real and imaginary parts, we have for the velocity potential and stream function,

$$\text{(3)} \qquad \phi = \phi(x, y, t) = \phi_1(x+ct, y)+\phi_2(x-ct, y),$$

$$\text{(4)} \qquad \psi = \psi(x, y, t) = \psi_1(x+ct, y)+\psi_2(x-ct, y),$$

and since w is real when $y = 0$,

(5) $$\psi(x, 0, t) = 0.$$

By Maclaurin's theorem,

(6) $$\phi = \phi(x, 0, t) + y\left(\frac{\partial\phi(x, y, t)}{\partial y}\right)_{y=0} + \ldots,$$

and since y lies between 0 and h the second term is infinitesimal compared with the first. Therefore we can put $y = 0$ throughout, and hence

(7) $$\phi = \phi_1(x+ct) + \phi_2(x-ct).$$

The same argument applied to (4) shows, in view of (5), that $\psi = 0$.

Thus (7) is the complete solution of the problem of long waves.

It follows from (7) that all particles which are in the same vertical plane have the same horizontal velocity, $-\partial\phi/\partial x$, and therefore remain in a vertical plane.

Again, from (6),

$$-\frac{\partial\phi}{\partial y} = -\left(\frac{\partial\phi(x, y, t)}{\partial y}\right)_{y=0} - y\left(\frac{\partial^2\phi(x, y, t)}{\partial y^2}\right)_{y=0} - \ldots .$$

Since the first term on the right is the vertical velocity at the bottom, which is zero, it follows that the vertical velocity is of the second order and is proportional to the height above the bottom.

15·61. The pressure. If Π is the pressure at the free surface and η the surface elevation for given values of x and t, the pressure equation becomes

$$\frac{p}{\rho} + gy - \frac{\partial\phi}{\partial t} = \frac{\Pi}{\rho} + g(h+\eta) - \frac{\partial\phi}{\partial t},$$

since q^2 is neglected and $\partial\phi/\partial t$ is independent of y. Thus

$$p = \Pi + g\rho(h+\eta-y),$$

which shows that the pressure at the depth $h+\eta-y$ is the same as that calculated by the laws of hydrostatics. This is sometimes expressed by saying that the vertical acceleration is negligible.*

15·62. The surface elevation. From 15·60 (7), the velocity potential is

(1) $$\phi = \phi_1(x+ct) + \phi_2(x-ct).$$

Hence, from 15·12, the surface elevation is given by

(2) $$\eta = \frac{c}{g}(\phi_1'(x+ct) - \phi_2'(x-ct)).$$

Thus the surface elevation is the sum of the elevations due to two progressive waves whose initial forms are

$$\frac{c}{g}\phi_1'(x), \quad -\frac{c}{g}\phi_2'(x),$$

* An alternative treatment of long waves based on assuming this result is indicated in Ex. XV, 68, 69, 70. See also 15·70.

advancing in opposite directions with speed c given by

$$c^2 = gh.$$

This is the characteristic property of long waves, that the wave velocity depends only on the depth of the water and not on the wave-length.

It also follows from (1) that the velocity u (necessarily horizontal on our assumptions) is given by

$$u = -\phi_1'(x+ct) - \phi_2'(x-ct). \qquad (3)$$

15·63. Wave progressing in one direction only. Consider a wave progressing in the positive direction of the x-axis. If u is the velocity of the liquid, η the corresponding surface elevation, we have, from 15·62 (2) and (3),

$$u = -\phi'(x-ct), \quad \eta = -\frac{c}{g}\phi'(x-ct),$$

and therefore $u = g\eta/c = c\eta/h$, since $c^2 = gh$. Thus for a wave progressing in the positive direction of the x-axis

$$\frac{u}{c} = \frac{\eta}{h}. \qquad (1)$$

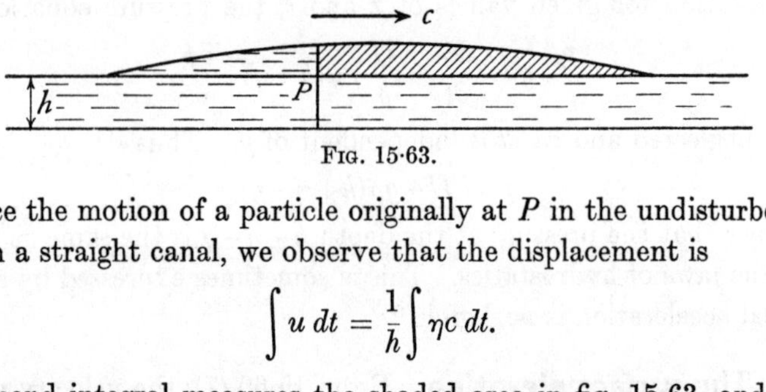

FIG. 15·63.

To trace the motion of a particle originally at P in the undisturbed surface of water in a straight canal, we observe that the displacement is

$$\int u\,dt = \frac{1}{h}\int \eta c\,dt.$$

The second integral measures the shaded area in fig. 15·63, and therefore the displacement of the particle is obtained by dividing the area of the profile which has passed P by the depth of the undisturbed water. When the wave has finally passed, the particle is left ahead of its initial position by the volume of the elevated water divided by the cross-sectional area of the water.

It also follows, from (1), that $u^2 h = g\eta^2$, and so

$$\tfrac{1}{2}\rho u^2 h\,dx = \tfrac{1}{2}g\rho\eta^2 dx,$$

which expresses the fact that the kinetic energy in a length dx of the wave is equal to the potential energy (measured from the undisturbed level) in the same length. This result is true only for a wave progressing in one direction.

15·64. Change of profile in long waves. The case of a long wave travelling in one direction without change of profile can be reduced to steady

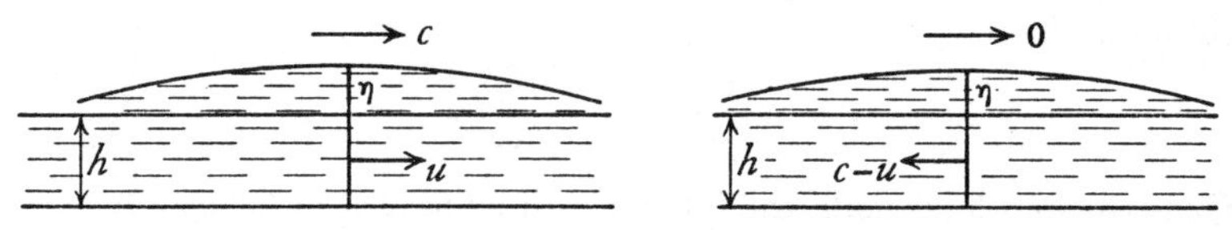

FIG. 15·64.

motion by impressing on the whole system a velocity equal but opposite to the velocity c of propagation.

The wave form then remains stationary in space and the fluid flows under it with the local velocity $-c+u$, where u is the (small) forward velocity in the progressive wave when the elevation is η.

The equation of continuity is then

$$(c-u)(h+\eta) = ch,$$

whence $u = c\eta/h$ approximately, a result already obtained.

For the reason already explained (15·60), the vertical velocity is small compared with u and therefore its square is of the fourth order of small quantities when u is taken as of the first order, so that the pressure equation at the free surface becomes

$$\frac{p}{\rho}+\tfrac{1}{2}(c-u)^2+g\eta = \frac{\Pi}{\rho}+\tfrac{1}{2}c^2.$$

Eliminating u by means of the equation of continuity, we get

$$\frac{p-\Pi}{\rho} = \frac{c^2(2h\eta+\eta^2)-2g\eta(h+\eta)^2}{2(h+\eta)^2} = \left(\frac{c^2}{h}-g\right)\eta-\frac{3\eta^2c^2}{2h^2},$$

neglecting the cube of η/h. The condition for the existence of the steady motion is that $p = \Pi$ at the free surface. Thus, unless η^2/h^2 is negligible, a free surface cannot exist when $c^2 = gh$. Thus a long wave of finite height cannot be propagated without change of profile.

It also appears that, when η^2/h^2 cannot be neglected, the condition $p = \Pi$ can be more nearly satisfied by taking a somewhat larger value of c when η is positive, and a somewhat smaller value when η is negative. Thus an elevation tends to travel faster than a depression, the wave tends to get steeper in front of a crest and observation shows that it curls over and ultimately breaks.*

15·70. Effect of small disturbing forces. Let X, Y be horizontal and vertical components of a *small* disturbing force acting on water in a horizontal

* See J. J. Stoker, "The formation of breakers and bores", *Communications on Applied Mathematics*, New York (1948) I, pp. 1–87.

canal of small depth h, X being in the direction of the canal. The equation of motion is then

$$\frac{\partial u}{\partial t} = X - \frac{1}{\rho}\frac{\partial p}{\partial x}.$$

Since the depth is small, the force Y will be practically constant as y varies from 0 to h, and Y will therefore merely operate to change slightly the value of g and the effect will be of the second order. Y can therefore be neglected.

The pressure is then given by

$$p = \Pi + g\rho\,(h+\eta-y),$$

and therefore

$$\frac{\partial u}{\partial t} = X - g\frac{\partial \eta}{\partial x}.$$

If ξ is the horizontal displacement of a particle from its undisturbed position, we have

$$u = \frac{\partial \xi}{\partial t}.$$

The equation of continuity is

$$(h+\eta)\left(dx + \frac{\partial \xi}{\partial x}dx\right) = h\,dx.$$

which expresses the fact that the same volume of fluid lies between the planes x, and $x+dx$ in the disturbed and undisturbed positions. Thus

$$(1) \qquad \eta = -h\frac{\partial \xi}{\partial x},$$

and the equation of motion is

$$(2) \qquad \frac{\partial^2 \xi}{\partial t^2} = X + c^2\frac{\partial^2 \xi}{\partial x^2}.$$

Multiply by $-h$, differentiate with respect to x and use (1). We then get

$$(3) \qquad \frac{\partial^2 \eta}{\partial t^2} = c^2\frac{\partial^2 \eta}{\partial x^2} - h\frac{\partial X}{\partial x},$$

which is the equation determining the changes in elevation.

15·71. Tides in an equatorial canal. We consider a shallow canal of uniform depth coincident with the earth's equator, and we suppose the only tide-raising force to be due to the Moon moving in the equatorial plane.

If $\mathbf{F}$ is the force of the Moon's gravitational attraction at the earth's centre, the force at two diametrically opposite points of the equator will be $\mathbf{F}+\mathbf{f}$, $\mathbf{F}-\mathbf{f}$, where $\mathbf{f}$ is the small variation in $\mathbf{F}$ as we move outwards along the radii, whose length is small compared with the Moon's distance. The force $\mathbf{f}$ is the *tide-raising force*, and the above explanation shows why tides are generated simultaneously at antipodal points of the earth.

Fig. 15·71 shows a diagram of the equatorial canal, O is the earth's centre, M the point directly under the Moon, G the point where the zero meridian of Greenwich meets the equator, and P the point, which we shall consider, in longitude α.

The Moon moves westward relatively to the earth with angular velocity n (supposed constant), and at the instant t the angle GOM will be $nt+\epsilon$. The tide-raising force of the Moon per unit mass has at P the horizontal component $f \sin(2\angle POM)$,* in the sense shown in the diagram.

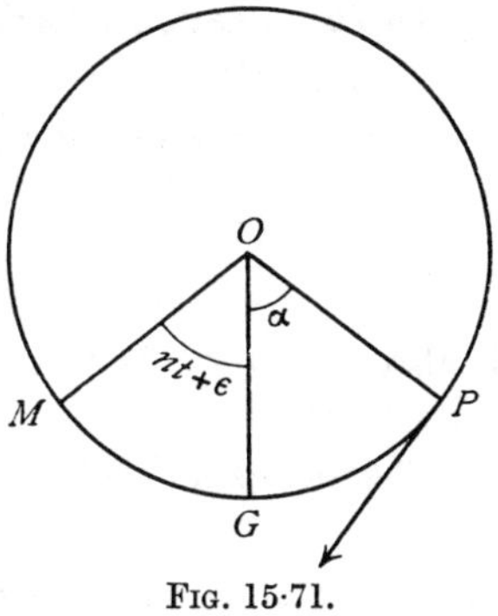

FIG. 15·71.

In the case of the Moon,

$$f/g = 8{\cdot}57 \times 10^{-8};$$

for the Sun,

$$f/g = 3{\cdot}78 \times 10^{-8}.$$

If we take at P the axis of x horizontal, i.e. perpendicular to OP, and x and α to increase in the same sense, we shall have $dx = a\, d\alpha$, where a is the radius of the earth. Thus the elevation η is given by 15·70 (3),

$$\frac{\partial^2 \eta}{\partial t^2} = \frac{c^2}{a^2}\frac{\partial^2 \eta}{\partial x^2} + \frac{2hf}{a}\cos 2(nt+\epsilon+\alpha).$$

The complete solution of this equation may be regarded as the sum of a complementary function containing arbitrary functions, and a particular integral. The complementary function represents free vibrations of the water of small amplitude which are quickly damped out by friction. The particular integral gives the forced oscillation which is the tide. To find this, we assume $\eta = A\cos 2(nt+\epsilon+\alpha)$, which gives on substitution in the equation

$$\eta = \frac{c^2}{2(c^2-n^2a^2)}\frac{af}{g}\cos 2(nt+\epsilon+\alpha).$$

Taking $a = 21 \times 10^6$ ft., the value of af/g is 1·80 ft. in the case of the Moon, and 0·79 ft. for the Sun. It follows that the tide is *semi-diurnal*, i.e. high water and low water each occur twice in a lunar day. Moreover,

$$\frac{c^2}{n^2a^2} = \frac{g}{n^2a}\cdot\frac{h}{a} = 311\,\frac{h}{a},$$

and, since h/a in the case of the actual ocean is a small fraction, $c^2 - n^2a^2$ is negative and therefore on this theory the tides are *inverted*. This means that low water occurs at the point which has the Moon in the zenith, and also at the antipodal point.

* For details the reader is referred to Lamb's *Hydrodynamics*, from which the numerical data are quoted. We shall assume the result.

15·75. Exact linearised theory. We use this term for the theory of waves of small slope treated by the exact method of 14·70. If $\theta(\chi)$ is small of the first order, we have $\sin\theta(\chi) = \theta(\chi)$ and therefore

$$\int_0^\epsilon \sin\theta(\omega)\, d\omega$$

is also a small quantity of the first order.

Therefore $\sin\theta(\epsilon) \Big/ \left[1+\mu\int_0^\epsilon \sin\theta(\omega)\, d\omega\right] = \theta(\epsilon)$ to the first order, and so the non-linear integral equation 14·70 (21) reduces to the homogeneous linear integral equation

$$\text{(1)} \qquad \theta(\chi) = \frac{\mu}{3\pi}\int_{-\pi}^{\pi} \theta(\epsilon) \sum_{n=1}^{\infty} \frac{\sin n\epsilon \sin n\chi}{n}\, d\epsilon.$$

If in this case we put $\theta(\epsilon) = \sin s\epsilon$, we get

$$\theta(\chi) = \frac{\mu}{3\pi s}\, \pi \sin s\chi,$$

and thus $\theta(\chi) = \sin s\chi$ is a solution if and only if $\mu = 3s$. Thus (1) has the eigen values

$$\mu = 3, 6, 9, \ldots, 3s, \ldots$$

and corresponding eigen functions

$$\sin\chi,\ \sin 2\chi,\ \sin 3\chi, \ldots,\ \sin s\chi, \ldots .$$

Inasmuch as the complete circuit of γ in fig. 14·70 corresponds to one wave, we must take $\mu = 3$ (cf. 14·70, Note (vi)) and

$$\text{(2)} \qquad \theta(\chi) = \beta \sin\chi,$$

where β is of the first order. Then from 14·70 (12) we find that, to the first order, all the b_s vanish except $b_1 = \beta$, and therefore all the a_s vanish except $a_1 = b_1 = \beta$.

Therefore from 14·70 (6) $y = (\lambda/2\pi)\,\beta\cos\chi$ at the free surface.

Taking the difference of the values of y at $\chi = 0$ (a crest) and $\chi = \pi$ (a trough), we find that the height of the wave is given by $H = (\lambda/\pi)\,\beta$ and therefore

$$\text{(3)} \qquad \beta = \pi H/\lambda.$$

Returning to 14·70 (6) we then find for the wave profile

$$\text{(4)} \qquad x = -(\lambda/2\pi)\,\chi - \tfrac{1}{2}H\sin\chi, \quad y = \tfrac{1}{2}H\cos\chi.$$

This is a trochoid, not a sine curve as in the ordinary linearised theory of the earlier sections of this chapter (cf. Gerstner's wave, John's wave).

We can introduce an amplitude a by writing

$$H = 2a. \tag{5}$$

To find the speed of propagation, in 14·70 (26) put

$$\mu = 3, \quad b_1 = \beta = 2\pi a/\lambda, \quad \epsilon = 0.$$

Then

$$c^2 = \frac{g\lambda}{2\pi} e^{6\pi a/\lambda} \tag{6}$$

Comparing this with $c^2 = g\lambda/2\pi$ obtained from the ordinary theory, we find agreement when a/λ is negligible, and we note that the speed of surface waves on deep water increases with increase of the ratio amplitude/wave length.

To find the kinetic energy we use 14·70 (22). Here from 14·70 (4),

$$f(\zeta) - 1 = a_1\zeta = \beta\zeta$$

and the area of the unit circle is π. Therefore using (3), (5), (6),

$$T = \tfrac{1}{8}\frac{\rho\lambda^2 c^2 \beta^2}{\pi} = \tfrac{1}{4}a^2 g\rho\lambda\, e^{6\pi a/\lambda}, \tag{7}$$

which differs from the value found in 15·20 by the presence of the exponential factor.

For the potential energy we have from (4)

$$V = \tfrac{1}{2}g\rho\int_0^\lambda y^2\,dx = \tfrac{1}{4}a^2 g\rho\lambda = Te^{-6\pi a/\lambda}. \tag{8}$$

Thus measured for the datum here used $V \neq T$.

15·86. Sound waves. Although sound waves are not gravity waves it is convenient to discuss them in connexion with the linearised theory. We shall suppose that sound waves are propagated in a gas by small to-and-fro motions of the medium whereby the disturbance passes rapidly from place to place without causing a transference of the medium itself. The basic assumptions are the following:

(i) The variations of the pressure, density, and velocity from their equilibrium values p_0, ρ_0, 0 are infinitesimal quantities of the first order whose powers and products may be neglected.

(ii) The motion is irrotational.

(iii) The pressure is a function of the density; in particular, the adiabatic law $p = \kappa\rho^\gamma$ will be assumed.

From (i), the quadratic terms in the equation of motion are negligible, and therefore omitting the external forces,

$$\frac{\partial \mathbf{q}}{\partial t} = -\frac{1}{\rho}\nabla p.$$

Also, from (ii), $\mathbf{q} = -\nabla\phi,$

and therefore $$\frac{\partial}{\partial t}\nabla\phi = \frac{1}{\rho}\nabla p.$$

Taking the scalar product by $d\mathbf{r}$, we get

$$\frac{\partial}{\partial t}(d\phi) = \frac{1}{\rho}\,dp.$$

Thus $$\frac{\partial\phi}{\partial t} = \int_{p_0}^{p}\frac{dp}{\rho} = \frac{p-p_0}{\rho_0},$$

since the difference $p-p_0$ is infinitesimal. Therefore

(1) $$p-p_0 = \rho_0\frac{\partial\phi}{\partial t}.$$

We can write for the density

(2) $$\rho = \rho_0(1+s),$$

where s, the *condensation*, is infinitesimal.

With this notation the equation of continuity (3·20 (5)) will assume the form

(3) $$\frac{\partial s}{\partial t} + \nabla\mathbf{q} = 0, \quad \text{or} \quad \frac{\partial s}{\partial t} = \nabla^2\phi.$$

From (iii), $p = \kappa\rho^\gamma = \kappa\rho_0{}^\gamma(1+s)^\gamma = p_0(1+\gamma s),$

ignoring the higher powers of s. Thus (1) gives

(4) $$\frac{\partial\phi}{\partial t} = c^2 s, \quad c^2 = \frac{\gamma p_0}{\rho_0}.$$

Eliminating s between (3) and (4), we get

(5) $$\frac{\partial^2\phi}{\partial t^2} = c^2\nabla^2\phi,$$

which is the equation satisfied by the velocity potential in the propagation of sound waves.

15·87. Plane waves. If the sound waves are propagated in one dimension only, say, parallel to the x-axis, the equation becomes

$$\frac{\partial^2\phi}{\partial t^2} = c^2\frac{\partial^2\phi}{\partial x^2},$$

the solution of which (15·60) is

$$\phi = \phi_1(x-ct)+\phi_2(x+ct),$$

where ϕ_1 and ϕ_2 are arbitrary functions. This represents a motion in which the velocity potential $\phi_1(x)$ is propagated with speed c in the positive direction of the x-axis, while the velocity potential ϕ_2 is propagated in the opposite direction also with speed c.

Thus c is the speed of sound. Since from (4),

$$c = \sqrt{\frac{\gamma p_0}{\rho_0}},$$

the speed of sound in any gas can be calculated. The result for air at 0° C., about 330 metres per second, agrees closely with the observed figure and justifies the choice of the adiabatic law.

On the above hypothesis of one-dimensional propagation, the velocity potential has the same value over any plane for which the value of x is given. Such waves are therefore called plane waves.

The velocity potential of a plane simple harmonic progressive wave is of the form

$$\phi = A \cos \frac{2\pi}{\lambda}(x - ct),$$

where λ is the wave-length. The period is $\tau = \lambda/c$.

Sound waves travel with a velocity independent of the wave-length and are in this respect analogous to long water waves.

If the particle whose equilibrium position is x is at time t in the position $x + \xi$, we have

$$\frac{\partial \xi}{\partial t} = u = -\frac{\partial \phi}{\partial x} = \frac{2\pi A}{\lambda} \sin \frac{2\pi}{\lambda}(x - ct), \quad \xi = \frac{A}{c} \cos \frac{2\pi}{\lambda}(x - ct).$$

Thus $\phi = c\xi$, and the actual amplitude of the displacement namely A/c is proportional to the amplitude of the velocity potential.

To obtain a measure of the *intensity* of sound we may take this as proportional to the mean rate at which energy is transmitted across a unit area of the wave front. The rate at which the pressure works is given by

$$I = pu = \left(p_0 + \rho_0 \frac{\partial \phi}{\partial t}\right)\left(-\frac{\partial \phi}{\partial x}\right)$$

$$= \frac{2\pi A p_0}{\lambda} \sin \frac{2\pi}{\lambda}(x - ct) + \frac{4\pi^2 A^2 c\, \rho_0}{\lambda^2} \sin^2 \frac{2\pi}{\lambda}(x - ct),$$

the mean value of which over a period is

$$I_m = \frac{2\pi^2 A^2 c\, \rho_0}{\lambda^2} = 2\pi^2 c\, \rho_0 \left(\frac{A}{c}\right)^2 \frac{1}{\tau^2}.$$

Thus the intensity is proportional to the square of the amplitude and inversely proportional to the square of the period.

15·88. Plane waves in a cylindrical pipe. Let l be the length of the pipe whose cross-section may be any plane curve and whose generators are parallel to the axis of x. We shall seek periodic solutions to represent stationary waves. To do this, assume $\phi = f(x)\cos nt$. Then the equation

$$\frac{\partial^2\phi}{\partial t^2} = c^2\frac{\partial^2\phi}{\partial x^2} \text{ gives } \frac{d^2 f}{dx^2}+\frac{n^2}{c^2}f = 0. \text{ Thus}$$

$$(1) \qquad \phi = \left(A\cos\frac{nx}{c}+B\sin\frac{nx}{c}\right)\cos nt.$$

The ends of the pipe may be open or closed. At a closed end the velocity vanishes, i.e. $\partial\phi/\partial x = 0$.

At an open end which communicates with the outside air whose pressure is p_0, the condition $p = p_0$ must be approximately satisfied, provided that the diameter of the pipe is small compared with the wave-length. Thus at an open end $\partial\phi/\partial t = 0$.

If the pipe is closed at $x = 0$ and $x = l$, we get, from (1),

$$B = 0, \quad \sin(nl/c) = 0.$$

This latter condition gives

$$\frac{nl}{c} = \pi,\ 2\pi,\ 3\pi,\ \ldots,$$

and therefore the periods $2\pi/n$ are

$$\frac{2l}{c},\ \frac{2l}{2c},\ \frac{2l}{3c},\ \ldots, \quad \text{and} \quad \phi = A\cos\frac{nx}{c}\cos nt,$$

where n has any of the above values.

These solutions can be superposed so that

$$\phi = A_1\cos\frac{\pi x}{l}\cos\frac{\pi ct}{l}+A_2\cos\frac{2\pi x}{l}\cos\frac{2\pi ct}{l}+\ldots.$$

Of these terms the first is called the *gravest* or *fundamental note*, the others *overtones*. The *frequency* of the gravest note is $c/(2l)$. The velocity vanishes at each end when the pipe is emitting the gravest note, and in addition at other points when emitting an overtone. Such points are *nodes*, while points of maximum speed for a given value of t are *loops*, using the same terminology as in the case of water waves. At a loop the pressure is constant, while at a node it is stationary for a given value of t.

For a *stopped pipe*, i.e. closed at one end, say $x = 0$, and open at the other, we have again $B = 0$, but since $\partial\phi/\partial t$ vanishes when $x = l$, we get $\cos(nl/c) = 0$, Thus

$$\frac{nl}{c} = (2s+1)\frac{\pi}{2}, \quad s = 0, 1, 2, \ldots,$$

and the frequency of the gravest note is $n/(2\pi) = c/(4l)$. The open end is a loop.

For a pipe open at both ends we get $A = 0$ and $\sin(nl/c) = 0$, so that the frequencies are the same as if both ends were closed but the open ends are now loops.

15·89. Spherical waves. When the disturbance is symmetrical with respect to the origin, ϕ will be a function of the distance r only and of the time. We then get, from 15·86 (5) and 2·72,

$$\frac{\partial^2\phi}{\partial t^2} = \frac{c^2}{r^2}\frac{\partial}{\partial r}\left(r^2\frac{\partial\phi}{\partial r}\right), \quad \text{or} \quad \frac{\partial^2(r\phi)}{\partial t^2} = c^2\frac{\partial^2(r\phi)}{\partial r^2}.$$

Thus, exactly as in 15·60, we get

$$r\phi = f_1(r-ct)+f_2(r+ct),$$

representing the sum of a diverging and a converging disturbance.

In the case of a wave diverging from the origin we can write

$$r\phi = f\left(t-\frac{r}{c}\right),$$

and the motion can be regarded as due to a source of strength $f(t)$ at the origin. If the source is in action for a limited time and then ceases, by integration over an interval of time which includes the whole time of transit of the disturbance past a given point we get, from 15·86 (4),

$$\int s\,dt = 0,$$

since the value of ϕ is zero before and after the passage of the wave. This result means that s is sometimes positive and sometimes negative, or that a diverging wave must necessarily contain both condensed and rarefied portions. This remark is due to Stokes. Thus a diverging spherical wave of condensation cannot exist alone.

EXAMPLES XV

1. The crests of rollers which are directly following a ship 220 ft. long are observed to overtake it at intervals of 16·5 sec., and it takes a crest 6 sec. to run along the ship. Find the length of the waves and the speed of the ship. (M.T.)

2. Prove that

$$w = A\cos\frac{2\pi}{\lambda}(z+ih-Vt)$$

is the complex potential for the propagation of simple harmonic surface waves of small height on water of depth h, the origin being in the undisturbed free surface. Express A in terms of the amplitude a of the surface oscillations.

Prove that

$$V^2 = \frac{g\lambda}{2\pi}\tanh\frac{2\pi h}{\lambda},$$

and deduce that every value of V less than $\sqrt{(gh)}$ is the velocity of some wave.

Prove that each particle describes an ellipse about its equilibrium position. Obtain the corresponding result when the water is infinitely deep. (U.L.)

3. Liquid of uniform depth h, contained in a vessel with vertical sides parallel to Oz, is slightly disturbed; find the equations determining the motion.

Show that the velocity potential is of the form

$$\phi = f(x, y) \cosh k(z+h) \cos(\sigma t + \epsilon)$$

and explain how f, k and σ are to be found. Illustrate your answer by the case where the horizontal section of the vessel is a rectangle of sides a, b. (U.L.)

4. Calculate the kinetic and potential energies associated with a single train of progressive waves on deep water, and from the condition that these energies are equal obtain the formula

$$V^2 = \frac{g\lambda}{2\pi}.$$

Show how this result is modified when the wave-length is so small that the potential energy due to surface tension is not negligible. (R.N.C.)

5. A train of simple harmonic waves of length λ passes over the surface of water of great depth. Prove that, at a point whose depth below the undisturbed surface is h, the pressure at the instants when the disturbed depth of the point is $h+\eta$ bears to the undisturbed pressure at the same point the ratio

$$1 + \frac{\eta}{h} \exp\left(-\frac{2\pi h}{\lambda}\right) : 1.$$

(M.T.)

6. Show that the wave-length λ of stationary waves on a river of depth h, flowing with velocity v, is given by

$$v^2 = \frac{g\lambda}{2\pi} \tanh \frac{2\pi h}{\lambda}.$$

Deduce that, if the velocity of the stream exceeds $\sqrt{(gh)}$, such stationary waves cannot exist.

7. In a train of waves on deep water given by

$$\phi = \tfrac{1}{2} V h e^{-\frac{2\pi z}{l}} \cos \frac{2\pi}{l}(x - Vt),$$

show that, if $(h/l)^2$ is negligible, the fluid particles describe circles with uniform speed.

Prove that, to a second approximation, the surface particles have a slight mean drift in the direction of propagation. (R.N.C.)

8. Plane progressive waves, in water of depth h, whose velocity potential is

$$\frac{ga}{n} \frac{\cosh m(z+h)}{\cosh mh} \cos\{m(x \cos\alpha + y \sin\alpha) - nt\}$$

are reflected at a rigid vertical wall occupying the plane $x = 0$, the axis of z being vertically upwards and the origin of coordinates in the undisturbed surface. Find the velocity potential of the reflected waves and show that the paths of the particles are ellipses the planes of which are vertical only in the plane $x = 0$. (U.L.)

9. Investigate the wave motion occurring at a horizontal interface between two fluids, of which the upper one of density ρ_2 has a general stream velocity U, and the lower one of density ρ_1 is at rest except for the small motion, the fluids being otherwise unlimited.

Show that the wave velocity c of waves of length λ is given by the equation

$$g(\rho_1 - \rho_2) = \frac{2\pi}{\lambda}(\rho_1 c^2 + \rho_2 (c - U)^2),$$

and prove that, for a given value of U, waves below a certain wave-length cannot be propagated. (U.L.)

10. An infinite liquid of density σ lies above an infinite liquid of density ρ, the two liquids being separated by a horizontal plane interface. Show that the velocity v of propagation of waves of length λ along the interface is given by

$$v^2 = \frac{g\lambda}{2\pi}\frac{\rho-\sigma}{\rho+\sigma}.$$

Prove that, for any group of such waves, the group velocity is equal to one-half of the wave velocity. (R.N.C.)

11. A layer of liquid of density ρ and depth h lies over liquid of infinite depth and density $\sigma(>\rho)$. Neglecting surface tension, show that two possible types of waves of length $2\pi/m$ can be propagated along the layer, with velocities given by

$$V^2 = \frac{g}{m} \quad \text{and} \quad V^2 = \frac{g}{m}\frac{\sigma-\rho}{\sigma\coth mh+\rho}.$$ (R.N.C.)

12. Two incompressible fluids of densities ρ_1, $\rho_2(\rho_1>\rho_2)$ are superposed. The upper fluid is moving as a whole with velocity U_2, and the lower with velocity U_1, in the direction of the axis of x, which is horizontal, that of y being vertically upwards. Show that the height η of a wave disturbance, whose velocity potentials in the two fluids are ϕ_1, ϕ_2 respectively, satisfies the following equations at the boundary:

$$\rho_1\left(\frac{\partial\phi_1}{\partial t}+U_1\frac{\partial\phi_1}{\partial x}\right)-\rho_2\left(\frac{\partial\phi_2}{\partial t}+U_2\frac{\partial\phi_2}{\partial x}\right) = (\rho_1-\rho_2)g\eta$$

$$-\frac{\partial\phi_1}{\partial y} = \frac{\partial\eta}{\partial t}+U_1\frac{\partial\eta}{\partial x}, \quad -\frac{\partial\phi_2}{\partial y} = \frac{\partial\eta}{\partial t}+U_2\frac{\partial\eta}{\partial x}.$$

Obtain the velocity of propagation of waves of length λ at such an interface if $U_1 = U_2 = 0$, and both fluids are of infinite depth. (R.N.C.)

13. The fluid in the region $0<z<h$, of density ρ_2, separates two fluids of densities ρ_1 and ρ_3, occupying the regions $h<z<\infty$ and $-\infty<z<0$, respectively, when at rest under gravity; and $\rho_1<\rho_2<\rho_3$. If waves of length λ, large compared with h, are set up in the middle layer, find the two possible velocities V_1, V_2 of propagation, showing that one value V_1 is independent of ρ_2 and such that a group of such waves of sensibly the same length advances with a velocity $\frac{1}{2}V_1$, whilst the other value V_2 is independent of λ. [The axis of z is taken vertically upwards.]

14. If the plane $z = 0$ is the horizontal interface of two otherwise unlimited incompressible fluids, of which the upper one, of density ρ_1, is moving as a whole with velocity U in the direction of the axis of x, while the other one, of density ρ_2, is at rest, show that the conditions of continuity satisfied at the interface by the velocity potentials ϕ_1, ϕ_2 of small disturbances from the steady state in the two fluids can be written in the form

$$\frac{\partial^2}{\partial z\,\partial t}(\phi_1-\phi_2) = U\frac{\partial^2\phi_2}{\partial x\,\partial z},$$

$$\rho_1\left(\frac{\partial^2\phi_1}{\partial t^2}+\frac{U\,\partial^2\phi_1}{\partial x\,\partial t}\right)-\rho_2\frac{\partial^2\phi_2}{\partial t^2} = (\rho_2-\rho_1)\frac{g\,\partial\phi_2}{\partial z}.$$

Prove that a disturbance of wave-length λ will be propagated along the surface of separation with a real velocity only if

$$\lambda>\frac{2\pi U^2}{g}\frac{\rho_1\rho_2}{\rho_2^2-\rho_1^2}.$$

15. Obtain the conditions to be satisfied for small oscillations at the horizontal interface of two semi-infinite liquids of densities ρ, ρ' $(\rho > \rho')$ moving with general stream velocities U, U' in the same horizontal direction, the surface tension T being taken into account.

Show that there are two possible wave velocities for a wave of length λ, namely

$$V = \frac{\rho U + \rho' U'}{\rho + \rho'} \pm \sqrt{\frac{\lambda g}{2\pi}\frac{\rho - \rho'}{\rho + \rho'} + \frac{2\pi T}{\lambda(\rho + \rho')} - \frac{\rho\rho'(U - U')^2}{(\rho + \rho')^2}}.$$

(U.L.)

16. Two liquids, which do not mix, occupy the region between two fixed horizontal planes. The upper, of density ρ' and mean depth h', is flowing with the general velocity U over the lower, which is of density ρ and mean depth h, and is at rest except for wave motions. Prove, neglecting viscosity, that the velocity V of waves of length $2\pi/k$, travelling over the common surface in the direction of U, is given by

$$\rho V^2 \coth kh + \rho' (U - V)^2 \coth kh' = T_1 k + g(\rho - \rho')/k,$$

where T_1 is the surface tension.

Apply the result to discuss the stability of the surface of deep water over which a wind is blowing with a given velocity. [For numerical purposes g may be taken as 980, and T_1 as 74 in C.G.S. units, and ρ'/ρ may be taken to be 0·0013.] (U.L.)

17. Liquid of density ρ and depth h lies over a fixed horizontal bed; above it is a layer of liquid of density ρ' $(<\rho)$ and thickness h', and the upper surface is a fixed horizontal plane. Obtain an equation to determine the velocity V of waves of length $2\pi/m$ at the common surface, the surface tension between the two liquids being T_1.

Prove that, if h, h' are both small compared with $2\pi/m$,

$$V^2 = hh' \frac{(\rho - \rho')g + T_1 m^2}{\rho h' + \rho' h}$$

approximately. (U.L.)

18. Two portions of a large uniform stream of liquid of density ρ, flowing with velocity U, are separated by a plane boundary of perfectly flexible fabric, of mass m per unit area, and subject to a tension T, the boundary being parallel to the stream. Show that waves of length λ can be propagated along the fabric, in the direction of the stream, with a velocity V given by

$$mV^2 - T + \frac{\lambda\rho}{\pi}(U - V)^2 = 0,$$

provided that

$$T\left(1 + \frac{m\pi}{\lambda\rho}\right) > mU^2.$$

(R.N.C.)

19. Explain, giving the necessary theory, why a flag flaps in a breeze.

20. Find the wave velocity of a train of simple harmonic waves, of wave-length λ, moving under the influence of gravity and capillarity on the common surface of two fluids of densities ρ and ρ', when T is the surface tension. Show that there is a minimum wave velocity; find its value and that of the corresponding wave-length. Prove that the group velocity of a group of waves of nearly the same amplitude, wave-length. and phase is greater or less than the wave velocity according as the wave-length is less or greater than that corresponding to the minimum wave velocity. Mention any phenomenon which is explained by this result. (U.L.)

21. A layer of liquid of density ρ_1 and height h rests upon the horizontal surface of unlimited liquid of density $\rho_2 (\rho_2 > \rho_1)$. If T_1, T_2 are the surface tensions at the upper and lower boundaries of the layer, prove that the velocity V with which waves are propagated along the layer satisfies the equation

$$\begin{aligned} &V^4 k^2 \rho_1 (\rho_2 + \rho_1 \tanh kh) \\ &\quad - V^2 k [k^2 \{\rho_1 (T_1 + T_2) + \rho_2 T_1 \tanh kh\} + \rho_1 \rho_2 g (1 + \tanh kh)] \\ &\quad + \{k^2 T_1 + \rho_1 g\}\{k^2 T_2 + (\rho_2 - \rho_1) g\} \tanh kh = 0, \end{aligned}$$

where $2\pi/k$ = wave-length. (U.L.)

22. An impulsive pressure $\varpi_0 + \varpi_1 \sin mx$ is applied to the free surface of deep water at rest. Find the impulsive pressure at any point in the water. Show that the initial kinetic energy of the water is $m\varpi_1{}^2/4\rho$ per unit area of the free surface. (R.N.C.)

23. An impulsive pressure $\varpi \sin mx$ is applied to the surface of deep water at rest, the origin being in the free surface and the axis of z downwards. Determine the velocity function of the initial motion, and show that the fluid velocity at a depth z is $m\varpi\, e^{-mz}/\rho$.

Work out the corresponding results for shallow water of depth d. (R.N.C.)

24. Sketch the two-dimensional, approximate theory of the propagation of surface waves of small height on a horizontal sheet of liquid of uniform depth.

Show that the velocity potential ϕ and the stream function ψ of a solitary wave are given approximately by

$$\phi + i\psi = -c(x+iy) + c\alpha \tanh \tfrac{1}{2} m (x+iy),$$

where the x-axis is taken along the bottom of the liquid and the y-axis vertically upwards, and where

$$mc^2 = g \tanh mh, \quad 3m\alpha = 2 \sinh^2 mh,$$

h being the depth of the liquid.

Verify that the height of the wave at a distance x from the point of maximum height is approximately

$$\eta = \eta_0 \operatorname{sech}^2 \tfrac{1}{2} mx,$$

and that, to the same degree of approximation,

$$c^2 = g(h + \eta_0). \qquad \text{(U.L.)}$$

25. A volume $4lhb$ of water is in a tank bounded by the vertical planes $x = \pm l$, $y = \pm b$ and the horizontal plane $z = -h$. Initially the water is at rest under external pressure at its upper surface equal to $p_0 + p_1 x/l$, where p_0 and p_1 are constants and p_1 is small. Suddenly this external pressure alters to a uniform pressure p_0. Determine the form of the upper surface at any subsequent time. (U.L.)

26. A rectangular trough, of length $2a$, is filled with liquid to a depth h, and made to oscillate in the direction of its length with velocity $u_0 \cos pt$. Show that the velocity potential of the forced oscillations is given by

$$\phi = \left\{ -xu_0 + \sum_{n=0}^{\infty} A_n \sin \frac{(2n+1)\pi x}{2a} \cosh \frac{(2n+1)\pi (y+h)}{2a} \right\} \cos pt,$$

where
$$A_n = 8au_0(-)^n \operatorname{sech} \frac{(2n+1)\pi h}{2a} \Big/ (2n+1)^2 \pi^2 (1 - p_n{}^2/p^2),$$

p_n denoting the period of free waves of length $4a/(2n+1)$ in liquid of depth h. (R.N.C.)

27. A long rectangular tank of length $2a$, filled with water up to a small height h, is initially at rest, and is then given a small longitudinal velocity $V \sin nt$. Show that the height η of the free surface above the equilibrium level at time t and at a distance x from that end of the tank which is initially rearmost, is given by

$$c\eta/Vh = -\sin n\frac{(x-a)}{c}\cdot \cos nt + \frac{2n}{ca}\sum_{s=0}^{\infty}\frac{\cos\{(s+\frac{1}{2})\pi x/a\}\cos\{(s+\frac{1}{2})\pi ct/a\}}{n^2/c^2+(s+\frac{1}{2})^2\pi^2/a^2},$$

where $c^2 = gh$ and s is an integer. (R.N.C.)

28. Prove that, if a canal of rectangular section is terminated by two rigid vertical walls whose distance apart is $2a$, and if the water is initially at rest and has its surface plane and inclined at a small angle β to the length of the canal, the altitude η of the wave at any time t is given by

$$\eta = \frac{8\alpha\beta}{\pi^2}\sum_0^{\infty}\frac{(-1)^n}{(2n+1)^2}\cdot \sin(2n+1)\frac{\pi x}{2a}\cdot\cos(2n+1)\frac{\pi ct}{2a},$$

where c is the velocity of a wave of length $4a/(2n+1)$ in an infinitely long canal.

29. A rectangular box, with four very long edges and two of the faces bounded by them horizontal, is completely filled with three non-miscible liquids, whose densities and depths are σ_1, σ_2, σ_3 and l_1, l_2, l_3 in downward order respectively when in equilibrium. Show that c, the velocity of propagation of waves of small amplitude along the common surfaces, is given by

$$[c^2m(\sigma_1 \coth ml_1+\sigma_2\coth ml_2)-g(\sigma_2-\sigma_1)]$$
$$[c^2m(\sigma_2\coth ml_2+\sigma_3\coth ml_3)-g(\sigma_3-\sigma_2)] = c^4m^2\sigma_2{}^2\,\text{cosech}^2\, ml_2,$$

where the wave-length is $2\pi/m$. (U.L.)

30. Using cylindrical coordinates (z, ϖ, θ), show that the differential equation for ϕ is satisfied by

$$z\varpi^n \sin n\theta \cos\sigma t,$$

the free surface of the undisturbed fluid being given by $z = h$. Oz being drawn upwards, find σ and show that the solution can represent standing waves of small amplitude on the surface of fluid bounded by one of a family of surfaces of revolution and by two suitable meridian planes, which are to be determined.

Find the path described to-and-fro by a particle of the fluid which passes through the point $(0, \varpi_0, 0)$. (U.L.)

31. Find the velocity of straight-crested simple harmonic irrotational waves of wave-length λ, over the surface of deep water. Supposing the waves to be due to an initial elevation on a very narrow strip of the surface containing the line $x = 0$, $z = 0$, prove that, at time t, the form of the surface is given by the equation

$$z = \frac{b^2}{x}\sum_0^{\infty}\left\{(-1)^n\frac{1}{1.3.5\ldots(4n+1)}\left(\frac{gt^2}{2x}\right)^{2n+1}\right\},$$

where b is a constant depending upon the initial elevation.

32. The axes of x and y being horizontal and the axis of z vertically downwards, verify that

$$\phi = A\exp\left(\frac{-2\pi z}{b\cos\alpha}\right)\sin\left(\frac{2\pi y}{b\cot\alpha}\right)\cos\frac{2\pi(x-vt)}{b}$$

is a possible velocity potential for a wave motion in deep water bounded by the vertical planes $y = \pm\frac{1}{4}b\cot\alpha$, and determine the velocity v of propagation.

33. Give the characteristics of long waves in a canal, and determine the velocity of propagation. Show that, for a propagation in one direction, the fluid velocity at any section of the canal is proportional to the height of the free surface above the equilibrium level.

34. Obtain the equation of motion of long waves in a shallow canal of depth h, under gravity, and find the possible disturbances of harmonic type in such a canal of length $2l$ and closed at both ends by a vertical boundary.

35. The cross-section of a canal is a semicircle of radius a. Prove that the velocity of propagation of long waves is $\frac{1}{2}(\pi a g)^{\frac{1}{2}}$, the banks of the canal being supposed vertical.

36. The bottom of a straight uniform canal of rectangular cross-section has its vertical longitudinal section in the form $y = a \sin mx$, where a is small compared with the mean depth h of the liquid in the canal. If the liquid is moving horizontally with a mean velocity u in the direction of the axis of x, show that the free surface has the form

$$\eta = a \frac{\sinh mh'}{\sinh m(h'-h)} \sin mx$$

where h' is given by $mu^2 = g \tanh mh'$. (U.L.)

37. If the bottom of a canal is slightly corrugated, so that the depth is given by $h + c \sin Kx$, c and Kh being small, prove that, if a stream of velocity U flows along the canal, there will be standing waves in the latter, of height η given by

$$\eta = c \sin Kx \Big/ \left(\frac{gh}{U^2} - 1\right).$$

Do the corrugations affect the velocity of progressive waves along the canal? (R.N.C.)

38. If the breadth at the free surface and the quantity of water per unit length in a canal of uniform cross-section are given, prove that the velocity of propagation of long waves is the same for all shapes of the cross-section.

A straight horizontal tube of length l, closed at both ends, whose cross-section is a circle of radius a, is half-filled with water. The tube is slightly tilted and again made horizontal. Find the period of the free oscillations of the water. Prove also that the amplitude of the forced oscillations of the free surface due to the prescribed motion of a diaphragm at one end of the tube, whose displacement at time t is $b \sin nt$, where b is small, is

$$\frac{nb}{2}\sqrt{\frac{\pi a}{g}} \operatorname{cosec} \frac{2nl}{\sqrt{\pi a g}}.$$

(U.L.)

39. Obtain the equation of motion for long waves in a shallow trough of depth h. Such a trough is closed at one end ($x = 0$) by a fixed vertical wall, and at the other end ($x = l$) by a piston subject to a prescribed simple harmonic displacement $\xi = a \cos pt$. Find the forced oscillation in the trough and show that, at the piston, the rise η of the water above the equilibrium level is given by

$$\eta = -\frac{ph\xi}{c} \cot \frac{pl}{c},$$

where $c^2 = gh$. (R.N.C.)

40. Prove that, for long waves in a horizontal canal of uniform depth h, and uniform rectangular cross-section, the following equations hold:

$$\frac{\partial \eta}{\partial t} = -h \frac{\partial u}{\partial x}, \quad \frac{\partial u}{\partial t} = -g \frac{\partial \eta}{\partial x},$$

where u is the horizontal velocity and η is the height of the wave above the equilibrium level.

Such a canal is unlimited in the direction x increasing, and is closed at the end $x = 0$ by a cross-section movable longitudinally. At $t = 0$ the water in the canal is at rest; the boundary is then given a small velocity $u = \psi(t)$, the function ψ being such that the total displacement of the boundary is always small. Show that this generates in the canal a disturbance which is purely progressive, and that $\eta = 0$ if $t < x/c$, but $\eta = (c/g)\psi\left(t - \dfrac{x}{c}\right)$ if $t > x/c$, where $c^2 = gh$. (R.N.C.)

41. A shallow trough of length $2l$ is filled with water up to a height h and is closed by two pistons with vertical walls, which are constrained to move horizontally with simple harmonic motions

$$a \sin(nt - \epsilon) \text{ (when } x = -l) \quad \text{and} \quad a \sin(nt + \epsilon) \text{ (when } x = +l).$$

Find the resulting forced oscillation, and show that the amplitude of η is

$$amh\,(\cos^2\epsilon \sin^2 mx \sec^2 ml + \sin^2\epsilon \cos^2 mx \operatorname{cosec}^2 ml)^{\frac{1}{2}},$$

where $m = n/\sqrt{(gh)}$. (U.L.)

42. Obtain the equation of motion of long waves of small amplitude in a uniform canal of depth h.

An isolated wave, of any form, travelling in such a canal, strikes a vertical wall which forms a cross-section. Show that the wave is reflected without change of type, and that, during the impact of the wave on the wall, the water rises to twice the normal height of the isolated wave.

Show also that the horizontal momentum of such a wave is equal to the total excess mass of water above the equilibrium level multiplied by the velocity of the wave, and deduce the time integral of the additional pressure on the wall, due to the impact of the wave. (U.L.)

43. Two-dimensional long waves are travelling parallel to the axis of x in water of variable depth h. Prove that, if η is the height of the free surface above the equilibrium level, η satisfies the equation

$$\frac{\partial^2 \eta}{\partial t^2} = \frac{\partial}{\partial x}\left(gh\frac{\partial \eta}{\partial x}\right).$$

If $h = x^2/2b$, prove that

$$\eta = ax^{-\frac{1}{2}} \cos\left[p\left\{\sqrt{\frac{2b}{g} - \frac{1}{4p^2}}\log x + t\right\} + \alpha\right]$$

is a typical solution of period $2\pi/p$; and use this result to illustrate the variations of amplitude and wave-length to be expected in the case of waves moving in from deep water up a gradually shelving beach. (U.L.)

44. Obtain the differential equation of motion of long waves in a canal of variable depth h in the form

$$\frac{\partial^2 \eta}{\partial t^2} = \frac{\partial}{\partial x}\left(gh\frac{\partial \eta}{\partial x}\right),$$

where η is the height of the wave above the equilibrium free surface.

The depths of a canal for $x < 0$ and $x > 0$ are h_1 and h_2 respectively. A progressive wave $\eta = a \sin m(x - V_1 t)$, where $V_1{}^2 = gh_1$, travels along the portion of depth h_1. Obtain the amplitudes of the reflected and transmitted waves and discuss its bearing upon the magnitudes of tides in a river separated from the sea by a " bar " of shallow depth.

45. In a canal both breadth and depth change suddenly, the variation of the breadth being from b_1 to b_2. A progressive wave travels with velocity c_1 along the part of breadth b_1 and is partially reflected and partially transmitted at the discontinuous section, the velocity of the transmitted part being c_2. Prove that the ratio of the elevations at the discontinuous section of the reflected and incident waves is

$$\frac{b_1 c_1 - b_2 c_2}{b_1 c_1 + b_2 c_2}. \qquad \text{(R.N.C.)}$$

46. A straight-crested earthquake wave passes along the bed of an ocean of uniform depth h, so that the elevation of the bed is given by $a \cos 2\pi(x - ct)/\lambda$, where a is small. Show that the amplitude of the consequent surface waves is

$$a\left\{\left(1 - \frac{V^2}{c^2}\right) \cosh \frac{2\pi h}{\lambda}\right\}^{-1},$$

where V is the velocity of surface waves of length λ.

47. Give the theory of "long waves" in a canal of uniform width and of depth h, proving that the velocity of free waves is $\sqrt{(gh)}$. An earthquake wave $\eta_0 = C \cos k(ct - x)$ travels along the bottom. Prove that the consequent wave on the free surface is

$$\eta = \frac{Cc^2}{c^2 - gh} \cos k(ct - x).$$

48. Obtain the equations satisfied by the elevation η and horizontal displacement ζ in tidal waves in straight canals of uniform depth.

Neglecting the rotation and curvature of the earth, and assuming a tide-producing celestial body to move uniformly once around the plane of the earth's equator in a day, show that an equatorial canal would contain a progressive wave, giving a "direct" or "inverted" tide according as its depth were greater or less than about 13 miles. (U.L.)

49. If tidal waves are due to a body assumed to move in the plane of the equator at a constant angular rate (once per diem) relative to the point Ω at which a great circle canal crosses the equator at inclination α, show that this results in a permanent change of level proportional to $\sin^2 \alpha \cos 2x/a$, and the addition of two semi-diurnal tides of standing waves, with amplitudes proportional to $(1 + \cos^2 \alpha) \cos 2x/a$, and $2 \cos \alpha \sin 2x/a$ respectively, x being the distance along the canal measured from Ω, and a the radius of the earth. (U.L.)

50. Establish the equation

$$b \frac{\partial^2 \eta}{\partial t^2} = g \frac{\partial}{\partial x}\left(S \frac{\partial \eta}{\partial x}\right),$$

for the elevation η of the surface in tidal wave motion in a canal of variable section, where b denotes the breadth at the surface, and S is the area of the section.

Prove from this equation that the amplitude of a progressive wave is nearly proportional to $b^{-\frac{1}{2}} h^{-\frac{1}{4}}$, where h is the mean depth across a section, if b and h and their rates of change along the canal vary only by small fractions of themselves in distances of the order of a wave-length; and verify that this corresponds to assuming continuous propagation of energy without reflection. (U.L.)

51. Give an account of the approximate theory of long or tidal waves, explaining the assumptions made.

A harmonic train of such waves, proceeding with velocity c_1, meets a "shelf" over which the wave velocity is c_2; show that it gives rise to a reflected and a transmitted wave, and compare the amplitudes of these with that of the incident wave.

If, after passing over the shelf, the original depth is again restored, show that the ratio of the amplitudes after and before passing over the shelf (neglecting the effects of multiple reflections) is given by

$$4c_1 c_2/(c_1+c_2)^2,$$

and that the amplitude is always reduced, whether the waves have crossed a shelf or a deep. (U.L.)

52. Obtain, stating any assumption made, the equation of tidal motion in a canal of varying section, the breadth at the surface being b, and the mean depth over this width being h, in the form

$$\frac{\partial^2 \eta}{\partial t^2} = \frac{g}{b}\cdot\frac{\partial}{\partial x}\left(hb\,\frac{\partial \eta}{\partial x}\right).$$

An estuary for which $b = \beta x/a$, $h = \gamma x/a$, where $0<x<a$, and β and γ are constants, communicates with the open sea at $x = a$, in which a tidal oscillation $\eta = C\cos(nt+\epsilon)$ is maintained.

Show that the tidal waves of the estuary are given by

$$\eta = C\frac{J_1(2\kappa^{\frac{1}{2}}x^{\frac{1}{2}})}{J_1(2\kappa^{\frac{1}{2}}a^{\frac{1}{2}})}\cdot\frac{a^{\frac{1}{2}}}{x^{\frac{1}{2}}}\cdot\cos(nt+\epsilon),$$

where $\kappa = n^2 a/g\gamma$. (U.L.)

53. Give the theory of long waves in a canal of uniform depth h due to a disturbing potential $\Omega = H\exp i(\sigma t - Kx)$. If the bottom yields to the disturbing force so that its elevation is $\eta_0 = \alpha\exp i(\sigma t - Kx)$, prove that the *relative* height of the waves is the same as if the potential had been diminished in the ratio $1-\mu$, where μ denotes the ratio of α to the "equilibrium height", $-H/g$, due to the disturbance. Prove that this conclusion is not confined to simple harmonic waves. (M.T.)

54. Taking $c = 1100$ ft./sec., calculate the length of an organ pipe, open at both ends, whose fundamental note has the frequency 128. Prove that the fundamental frequency will be unaltered if a rigid diaphragm is placed at the middle of the pipe. Explain the physical reason for this phenomenon. (R.N.C.)

55. Show that the possible periods of the air vibrations in a pipe, open at both ends and of length $2l$, are

$$T,\ \frac{T}{2},\ \frac{T}{3},\ \ldots,$$

where $T = 4l/c$, and c is the velocity of sound in air.

If a thin frictionless piston, of mass M, is placed at the middle of the pipe, show that the periods $(2\pi/n)$ of modes other than the symmetrical ones are now given by

$$\cot\frac{nl}{c} + \frac{M'c}{Mnl} = 0,$$

where M' is the total mass of air in the pipe.

Hence show that, if M'/M is small, these periods are

$$T\left(1-\frac{4M'}{\pi^2 M}\right),\quad \frac{T}{3}\left(1-\frac{4M'}{3^2\pi^2 M}\right),\quad \frac{T}{5}\left(1-\frac{4M'}{5^2\pi^2 M}\right),\ \ldots$$

very nearly. (R.N.C.)

56. A straight tube of length l is open to the atmosphere at one end. The other end opens into a large vessel in which the pressure at time t is

$$\Pi(1-\alpha \sin nt),$$

where Π is the atmospheric pressure and α is a small constant. Find the velocity potential of the air within the tube. (R.N.C.)

57. A horizontal pipe, of length l, is rigidly closed at one end and open at the other. Show that the periods of the air vibrations in the pipe are $4l/cN$, where N is an odd integer, and c is the velocity of sound in air.

If there is a thin frictionless piston, of mass M, at the middle section of the pipe, show that the periods $(2\pi/n)$ of the free vibrations are given by

$$\frac{Mnl}{c} = 2M' \cot\frac{nl}{c},$$

where M' is the mass of the air within the pipe. (R.N.C.)

58. A straight tube of length l rigidly closed at one end has the other end stopped by a plug of mass M which can move without friction in the tube and is controlled by a spring. If no air is present the plug can perform small oscillations of frequency $n/2\pi$. If the tube be now filled by a mass M' of air at atmospheric pressure, and the other side of the plug be also exposed to the atmosphere, prove that the frequency $\sigma/2\pi$ of the free vibrations is given by

$$(\sigma^2-n^2)\tan\frac{\sigma l}{c} = \frac{\sigma c}{l}\frac{M'}{M}.$$

(R.N.C.)

59. A straight pipe, of length l, is rigidly closed at one end, and at the other end is an air-tight piston which is caused to oscillate, its displacement at time t being $a\cos nt$, where a is small. Find the velocity function for the air vibration set up in the pipe, and show that the kinetic energy of the enclosed air is

$$\frac{ma^2}{4}\left\{n^2 \operatorname{cosec}^2\frac{nl}{c} - \frac{cn}{l}\cot\frac{nl}{c}\right\}\sin^2 nt,$$

where m is the mass of the air within the pipe, and c is the velocity of sound. (R.N.C.)

60. A thin piston of mass M, placed at the middle of a straight tube open at both ends, is controlled by a spring such that the natural period *in vacuo* is $2\pi/m$. Show that if the presence of the air is taken into account the natural period is $2\pi/n$ given by

$$M(m^2-n^2) = 2\rho\, cnS\tan\frac{nl}{c},$$

where $2l$ is the length of the tube, and S is the sectional area. (R.N.C.)

61. A pipe, of length $2l$, is stopped at one end and open at the other, and is divided into two parts by a thin close-fitting piston which slides in the pipe without friction, but is controlled by a spring of such strength that its natural period of vibration is $2\pi/m$. In equilibrium the piston is at the middle point of the pipe, and the enclosed air is at atmospheric pressure. Show that the period $2\pi/n$ of a normal mode of vibration is given by

$$a(n^2-m^2) = 2cn\cot\frac{2nl}{c},$$

where a is the length of pipe required to contain air equal in mass to the piston, and c is the velocity of propagation of sound in air at atmospheric pressure. (U.L.)

62. Determine what happens when a train of plane waves of sound impinges directly on the surface of separation of two gases in which the speeds of sound are c and c'. Show that the fraction $(c'-c)^2/(c'+c)^2$ of the incident energy is reflected. (R.N.C.)

63. If $n/2\pi$ is the frequency of waves symmetrical about the origin within a rigid spherical case, of radius a, show that

$$\tan\frac{na}{c} = \frac{na}{c}.$$

(R.N.C.)

64. Prove that in sound waves of small amplitude the velocity potential ϕ satisfies the equation

$$\frac{\partial^2\phi}{\partial t^2} = c^2\,\nabla^2\phi,$$

and prove further that the value of ϕ at any time t at any point P of an unbounded medium is given by

$$4\pi\phi = \frac{d}{dt}\left[t\int F(ct)\,d\omega\right] + t\int G(ct)\,d\omega,$$

where the integrations are taken with respect to solid angle ($d\omega$) over a sphere of centre P and radius ct, and F, G denote the initial values of ϕ, $\dfrac{\partial\phi}{\partial t}$, respectively.

Prove that, at a point where there is initially no disturbance, the time integral of the condensation over the whole interval during which the waves are passing the point is, in general, zero. (U.L.)

65. The centre of a rigid sphere of radius a at time t is at the point $x = b\sin nt$, where b is small. Verify that all the conditions for the surrounding gas are satisfied by

$$\phi = \text{real part of } A\,\frac{\partial}{\partial r}\left\{\frac{e^{in(t-r/c)}}{r}\right\}\cos\theta,$$

where c is the velocity of sound, and A is constant. Find the value of A, the mechanical force needed to maintain the prescribed motion of the sphere and the work done by it in one vibration. (U.L.)

66. A point source of sound gives rise to a vibratory motion for which

$$\phi = \frac{a\cos k(ct-r)}{r}.$$

Show that the mean rate of transmission of energy across the surface of a concentric spherical surface is $2\pi\rho ck^2a^2$.

An organ pipe has one end open and one end closed. Discuss the effect of the open end on the periods, and show that for vibrations of fundamental mode the modulus of decay is $16l^3/\pi\omega c$, where l is the length and ω the area of cross-section of the pipe. (U.L.)

67. If the velocity potential (in spherical polar coordinates) for sound waves is of the form $f(r)\,e^{int}\cos\theta$, show that

$$f(r) = \frac{d}{dr}\left\{\frac{1}{r}\left(A\,e^{ikr} + B\,e^{-ikr}\right)\right\},$$

where $k = n/c$ and A and B are arbitrary constants.

A rigid spherical envelope of radius a containing air executes small oscillations, so that its centre at any instant is at the point $r = b\sin nt$, $\theta = 0$.

Prove that the velocity potential of the air inside the sphere is

$$C\left\{\frac{\cos kr}{kr}-\frac{\sin kr}{k^2r^2}\right\}\cos\theta\cos nt,$$

where $C = nk^2a^3b/\{(2-k^2a^2)\sin ka - 2ka\cos ka\}$. (M.T.)

68. For long waves in a canal, assuming the result of 15·61, prove that

$$\frac{\partial p}{\partial x} = g\rho\frac{\partial \eta}{\partial x},$$

and use the fact that the right-hand side is independent of y to infer that particles in a vertical plane perpendicular to the direction of propagation remain in such a vertical plane.

69. If (u, v) are the small components of velocity in a long wave, use the equation of motion and Ex. 68 to show that

$$\frac{\partial u}{\partial t} = -\frac{1}{\rho}\frac{\partial p}{\partial x} = -g\frac{\partial \eta}{\partial x} \text{ and that}$$

$$\frac{\partial^2 \xi}{\partial t^2} = -g\frac{\partial \eta}{\partial x}, \text{ where } \xi = \int_0^t u\,dt.$$

Obtain the equation of continuity in the form

$$\eta = -h\frac{\partial \xi}{\partial x},$$

where h is the mean depth.

70. Use Ex. 69 to show that

$$\frac{\partial^2 \eta}{\partial t^2} = c^2\frac{\partial^2 \eta}{\partial x^2}, \quad c^2 = gh,$$

and hence prove that

$$\eta = f_1(x+ct) + f_2(x-ct).$$

CHAPTER XVI

STOKES' STREAM FUNCTION

16·0. Axisymmetrical motions. In the preceding chapters we have been able to discuss two-dimensional motions in terms of a single complex variable and a complex potential. In proceeding to consider motion in three dimensions, we can no longer have recourse to the complex potential. The simplest case is that in which the motion is the same in every plane through a certain line called the *axis*. Such a motion occurs, for example, when a solid of revolution moves in the direction of its axis of revolution in a liquid otherwise at rest.

This type of motion, which is called *axisymmetrical*, presents some analogies with the two-dimensional case; in particular, a stream function can be defined, and when the motion is irrotational a velocity potential of course always exists.

The axis of symmetry will be taken as x-axis and the motions are most conveniently discussed in terms of spherical polar coordinates (r, θ, ω), or cylindrical coordinates (x, ϖ, ω), fig. 2·72 (i), (ii).

16·1. Stokes' stream function. Consider a fixed point A on the axis of symmetry and an arbitrary point P. Join P to A by curves AQ_1P, AQ_2P both lying in the same plane through the axis, which for convenience may be called a *meridian plane*. The position of a point in this plane can be fixed by the cylindrical coordinates (x, ϖ). If we rotate the *meridian curves* AQ_1P, AQ_2P about the axis of symmetry, a closed surface will be formed into which as much liquid flows from right to left across the surface generated by AQ_2P as flows out in the same time across the surface generated by AQ_1P, assuming that no liquid is created or destroyed within the surface.

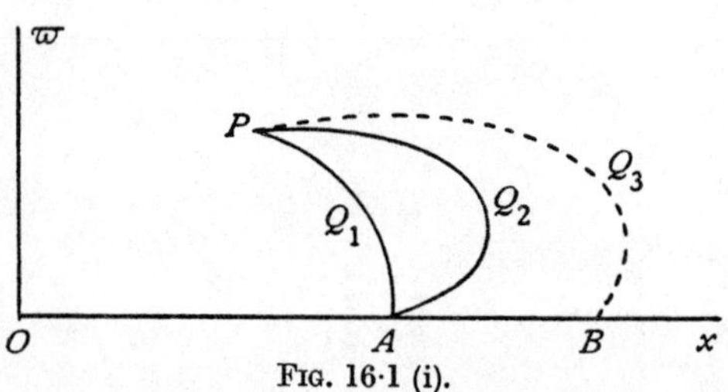

Fig. 16·1 (i).

If we denote the flux across either of these surfaces by $2\pi\psi$, the function ψ is Stokes' *stream function*, also known as the *current function*. If we keep AQ_1P fixed and replace AQ_2P by any other meridian curve joining A to P, the argument shows that the value of ψ is unaltered. The stream function ψ

depends therefore on the position of P, and perhaps on the fixed point A. If we take another fixed point B on the axis and draw the curve BQ_3P, the flux across the surface generated by BQ_3P will be the same as that across AQ_1P, for from the symmetry there is no flow across AB. It follows that the value of ψ does not depend on the particular fixed point chosen for the definition, provided that this lies on the axis. On this understanding the value of the stream function at P depends solely on the position of P, and when P is on the axis we have $\psi = 0$.

If ψ_P, $\psi_{P'}$ denote the values of the stream function at P and P', the flux from right to left across the surface generated by the revolution about the axis of any line joining P and P' is $2\pi\psi_{P'} - 2\pi\psi_P$. Taking P, P' at infinitesimal distance δs apart, the normal velocity from *right to left* across PP' is therefore given by

$$2\pi\varpi\, \delta s\, q_n = 2\pi(\psi_{P'} - \psi_P),$$

whence by proceeding to the limit

$$q_n = \frac{1}{\varpi}\frac{\partial\psi}{\partial s}.$$

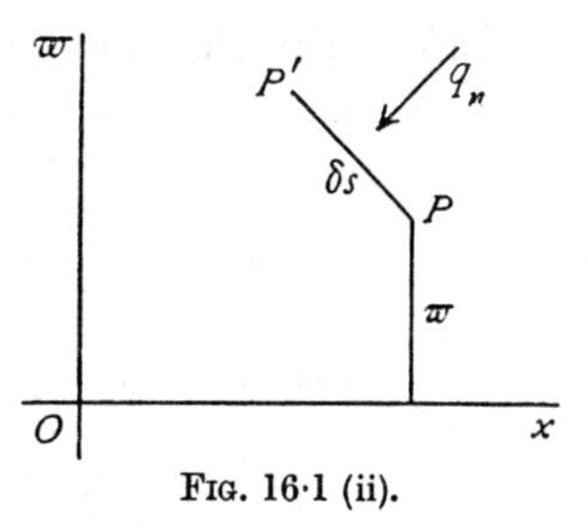

FIG. 16·1 (ii).

As particular applications of this important result, by taking ds in turn equal to $d\varpi$, dx, $r\, d\theta$, dr, we have

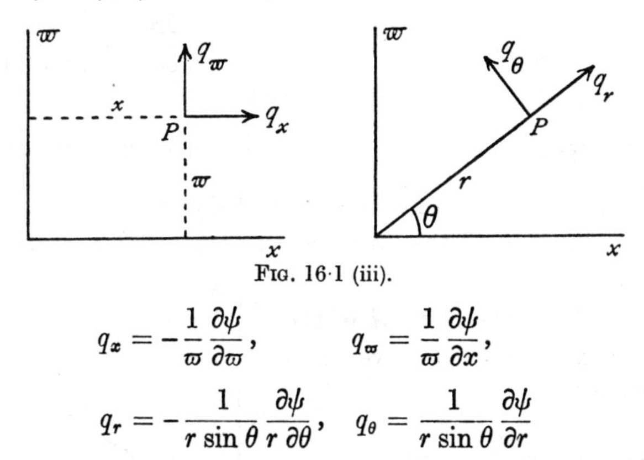

FIG. 16·1 (iii).

$$q_x = -\frac{1}{\varpi}\frac{\partial\psi}{\partial\varpi}, \qquad q_\varpi = \frac{1}{\varpi}\frac{\partial\psi}{\partial x},$$

$$q_r = -\frac{1}{r\sin\theta}\frac{\partial\psi}{r\,\partial\theta}, \quad q_\theta = \frac{1}{r\sin\theta}\frac{\partial\psi}{\partial r}$$

which give the velocity components in cylindrical and spherical polar coordinates. There is no component perpendicular to the meridian plane.

The streamlines are given by the equation

$$\psi = \text{constant},$$

for across such a line there is no flow.

The dimensions of ψ are L^3T^{-1}, but the dimensions of the velocity potential ϕ are L^2T^{-1}.

It should be observed that ψ exists in virtue of the continuity of the motion, and therefore the equation of continuity is automatically satisfied. We also note that from the above values of the velocity components

$$\frac{\partial(\varpi q_x)}{\partial x} + \frac{\partial(\varpi q_\varpi)}{\partial \varpi} = 0,$$

which is another form of the equation of continuity.

The stream function has been defined with reference to a base point on the axis. To take the base point elsewhere merely changes ψ by a constant (cf. 4·30). Since differences and derivatives of ψ are alone involved we may, if convenient, regard ψ as containing an arbitrary constant.

16·20. Simple source. A *simple source* is a point of outward radial flow. If the source emits the volume $4\pi m$ per unit time, m is the *strength* of the source.*

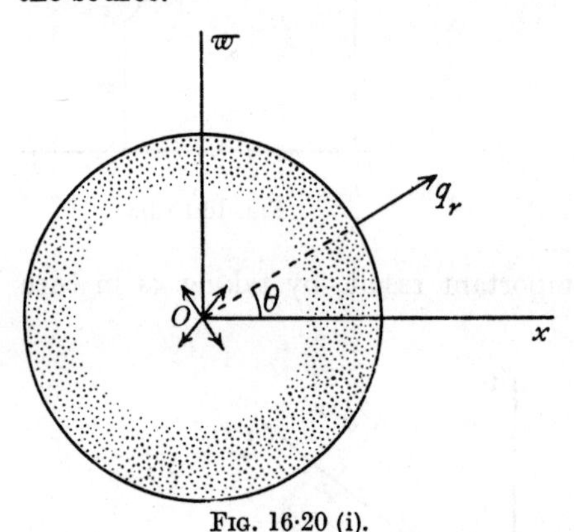

Fig. 16·20 (i).

A *sink* is a point of inward radial flow.

If there is a source of strength m at the origin, the outward flux across a sphere of radius r, whose centre is at the source, is related to the radial velocity by the formula $4\pi m = 4\pi r^2 q_r$. Thus

$$q_r = -\frac{\partial\phi}{\partial r} = -\frac{1}{r\sin\theta}\frac{\partial\psi}{r\,\partial\theta} = \frac{m}{r^2},$$

whence

$$\phi = \frac{m}{r}, \quad \psi = m\cos\theta = \frac{mx}{r}.$$

The stream function can also be obtained directly from the definition by considering the flux across the spherical cap cut off by a plane through P perpendicular to Ox.

If the source is at the point A of the axis instead of at the origin, we shall have, fig. 16·20 (ii),

$$\phi = \frac{m}{AP} = \frac{m}{\sqrt{r^2+c^2-2cr\cos\theta}},$$

$$\psi = m\cos\theta_1 = \frac{m(r\cos\theta - c)}{\sqrt{r^2+c^2-2cr\cos\theta}}.$$

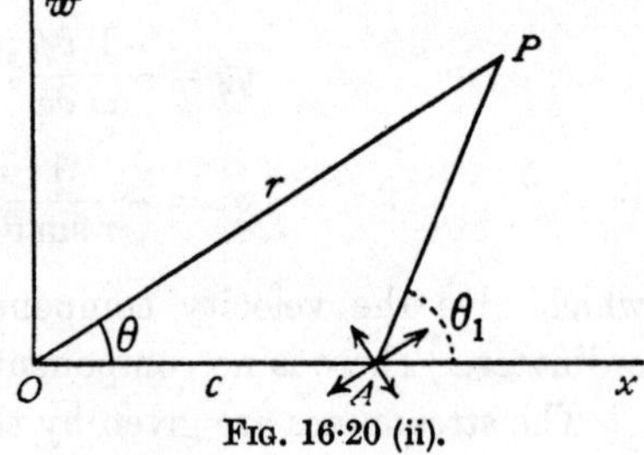

Fig. 16·20 (ii).

In terms of x and ϖ, we have

$$\phi = \frac{m}{\sqrt{(x-c)^2+\varpi^2}}, \quad \psi = \frac{m(x-c)}{\sqrt{(x-c)^2+\varpi^2}}.$$

* Thus the *output* is $M = 4\pi m$. Some writers call M the strength (cf. p. 210, footnote).

We note that these functions involve x and c only through the combination $x-c$. Hence

$$\frac{\partial\phi}{\partial c} = -\frac{\partial\phi}{\partial x}, \quad \frac{\partial\psi}{\partial c} = -\frac{\partial\psi}{\partial x}.$$

The image of a source in a plane is clearly an equal source at the optical image in the plane, cf. 8·40.

16·21. Submarine explosion. If a spherical cavity of radius R_0 containing gas at pressure p_0 begins to expand rapidly in surrounding unbounded liquid, we have a state of affairs closely imitating the effect of a submarine explosion. Let R be the radius of the cavity at time t, p_1 the pressure of the gas which is assumed to expand adiabatically, and let the inertia of the gas be neglected. Then, by the law of adiabatic expansion,

$$\frac{p_1}{p_0} = \left(\frac{R_0{}^3}{R^3}\right)^{\gamma}.$$

Gravity being neglected, the motion of the liquid is radial, the velocity at the boundary of the cavity being $dR/dt = R'$.

Thus the motion will resemble that due to a source, and we can put

$$\phi = \frac{m}{r}, \quad -\frac{\partial\phi}{\partial r} = \frac{m}{r^2}.$$

Therefore when $r = R$, $m/R^2 = R'$. Hence

$$\phi = \frac{R^2R'}{r}, \quad \frac{\partial\phi}{\partial t} = \frac{R^2R''+2RR'^2}{r}.$$

The pressure equation then gives

$$\frac{p}{\rho}+\tfrac{1}{2}\left(\frac{R^2R'}{r^2}\right)^2 - \frac{R^2R''+2RR'^2}{r} = F(t).$$

If the pressure at infinity is negligible, we see that $F(t)$ is equal to zero, for that is the value taken by the left side when $r = \infty$. Putting $r = R$, we have $p = p_1$, and therefore

$$RR''+\tfrac{3}{2}R'^2 = \frac{R_0{}^{3\gamma}}{R^{3\gamma}}\cdot\frac{p_0}{\rho}.$$

Multiply by $2R^2R'$ and introduce a constant $c^2 = p_0/\rho$. Then

$$\frac{d}{dt}(R^3R'^2) = \frac{2c^2R_0{}^{3\gamma}}{R^{3\gamma-2}}R'.$$

Integration, observing that $R' = 0$ when $R = R_0$, gives

$$\frac{R'^2}{c^2} = \frac{2}{3(\gamma-1)}\left[\left(\frac{R_0}{R}\right)^3 - \left(\frac{R_0}{R}\right)^{3\gamma}\right].$$

If $\gamma = 4/3$, the solution can be completed by writing $R = (1+n)R_0$, which gives

$$\frac{ct}{R_0} = (1+\tfrac{2}{3}n+\tfrac{1}{5}n^2)\sqrt{(2n)}.$$

As an illustration, if $p_0 = 1000$ atmospheres and $R_0 = 50$ cm., then $c = 3{\cdot}16\times 10^4$ cm./sec., the radius of the cavity is doubled in 0·004 seconds, and the initial acceleration of the radius is $2{\cdot}00\times 10^7$ cm./sec^2, which justifies the neglect of gravity.

16·22. Uniform stream. For a uniform stream U, parallel to Ox, by taking the flux from right to left across a circle of radius ϖ whose centre is

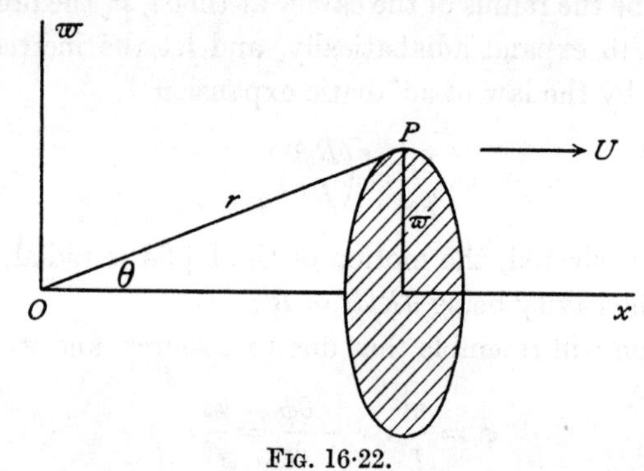

FIG. 16·22.

on Ox and whose plane is perpendicular to Ox, we have $2\pi\psi = -\pi\varpi^2 U$, and therefore

(1) $$\psi = -\tfrac{1}{2}\varpi^2 U = -\tfrac{1}{2}Ur^2\sin^2\theta.$$

This result could also be obtained by integrating the equation

$$-\frac{1}{\varpi}\frac{\partial\psi}{\partial\varpi} = U.$$

The velocity potential is clearly

(2) $$\phi = -Ux = -Ur\cos\theta.$$

16·23. Source in a uniform stream. If we combine a source and a uniform stream, we get

(1) $$\psi = -\tfrac{1}{2}Ur^2\sin^2\theta + m\cos\theta.$$

The stagnation point is such that $q_r = 0$, $q_\theta = 0$, or

$$U\cos\theta + \frac{m}{r^2} = 0, \quad -U\sin\theta = 0,$$

which give $\theta = \pi$, $r^2 = m/U = a^2$, say.

The streamline which passes through the stagnation point is therefore

$$-\tfrac{1}{2}Ur^2\sin^2\theta + m\cos\theta = -Ua^2.$$

This is the dividing streamline, whose equation may be written in the form

$$\varpi^2 = 2a^2(1+\cos\theta),$$

and therefore when $\theta \to 0$, $\varpi \to 2a$, which gives the asymptotes.

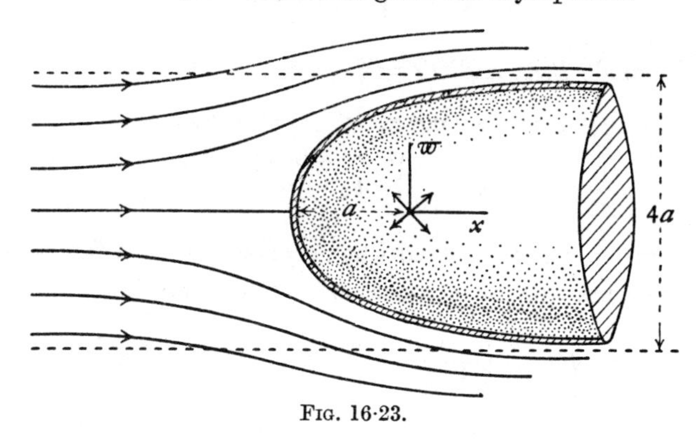

FIG. 16·23.

The dividing line is shown in fig. 16·23 and can be easily traced by Rankine's method or from the equation

$$r = a \operatorname{cosec} \frac{\theta}{2}.$$

Equation (1) therefore gives the streaming motion past a blunt-nosed cylindrical body whose diameter is ultimately $4a$.

The pressure equation gives

$$(2) \qquad \frac{p}{\rho} + \tfrac{1}{2}U^2\left(1 + \frac{2a^2}{r^2}\cos\theta + \frac{a^4}{r^4}\right) = \frac{\Pi}{\rho} + \tfrac{1}{2}U^2,$$

which shows that $p \to \Pi$ as r increases.

This result may be used for calibrating a Pitot tube for different positions of the side openings, the opening at the nose measuring $\Pi + \frac{1}{2}\rho U^2$, while the side opening measures p.

Equation (2) may also be used for calculating the pressure distribution near the nose of an airship.

16·24. Finite line source. Consider a line source stretching along the axis from O to A, the strength at the distance ξ from O being m_ξ/a per unit length where $OA = a$.

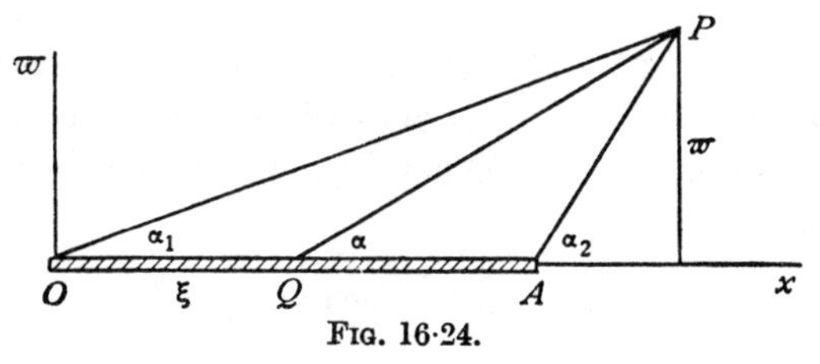

FIG. 16·24.

The stream function is got by superposing the stream functions of a series of elementary point sources of strengths $m_\xi \delta\xi/a$, and is therefore

$$\psi = \frac{1}{a}\int_0^a m_\xi \cos\alpha \, d\xi,$$

where α is the angle PQx, and $OQ = \xi$.

Since $$\xi = x - \varpi \cot\alpha, \quad d\xi = \varpi \operatorname{cosec}^2\alpha \, d\alpha,$$

we have $$\psi = \frac{1}{a}\int_{\alpha_1}^{\alpha_2} m_\xi \frac{\varpi \cos\alpha}{\sin^2\alpha} \, d\alpha,$$

and the integration can be effected when m_ξ is known as a function of ξ. The simplest case occurs when m_ξ = constant = m say, and then

$$\psi = \frac{m}{a}\left(\frac{\varpi}{\sin\alpha_1} - \frac{\varpi}{\sin\alpha_2}\right) = \frac{m}{a}(PO - PA).$$

The streamlines are hyperbolas with foci at O and A.

If we superpose a uniform stream U, we get

$$\psi = -\tfrac{1}{2}Ur^2 \sin^2\theta + \frac{m}{a}(PO - PA).$$

The dividing streamline must contain the negative x-axis as part of itself and therefore corresponds to $\psi = -m$, and has for equation

$$\varpi^2 = \frac{2m}{U}\left\{\frac{PO - PA}{a} + 1\right\}.$$

Now $$PA^2 = r^2 + a^2 - 2ar\cos\alpha_1, \quad PO = r.$$

Hence for large values of r,

$$PO - PA = r - r\left(1 - \frac{a\cos\alpha_1}{r} - \ldots\right)$$
$$= a\cos\alpha_1 + \text{powers of } r^{-1}.$$

When P recedes to infinity, $\alpha_1 \to 0$.

Thus the dividing streamlines are asymptotic to

$$\varpi^2 = 4b^2, \quad b^2 = m/U.$$

We have therefore again the case of streaming past a cylinder with a nose, but in this case more pointed than that illustrated in fig. 16·23.

16·25. Airship forms. If we combine a uniform stream in the positive direction of the x-axis with a point source m at the origin and a uniform line sink of total strength $-m$ stretching from the origin to $x = a$, we get the stream function

$$\psi = -\tfrac{1}{2}U\varpi^2 - \frac{m}{a}(PO - PA) + \frac{mx}{r}.$$

When P is on the positive x-axis, $x = r$, $PO - PA = a$, and when P is on the negative x-axis, $x = -r$, $PO - PA = -a$.

Thus $\psi = 0$ contains the whole of the x-axis, and the dividing streamline consists of this and a closed portion of airship form, fig. 16·25.

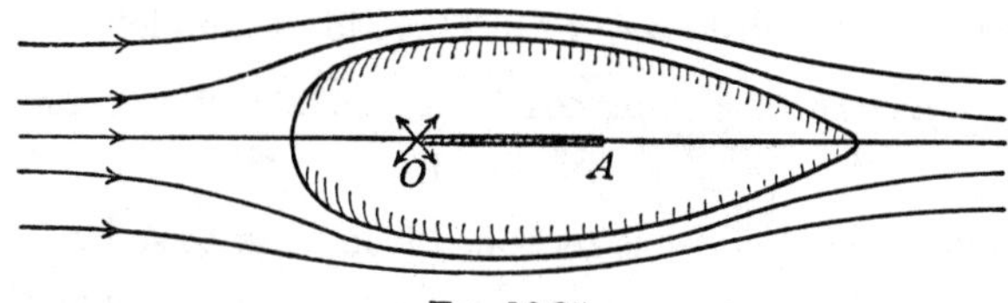

Fig. 16·25.

By assuming other laws of variation of the line sink, provided the total intake remains equal to the output of the source, a variety of such forms can be produced.

16·26. Source and equal sink. Doublet. Another simple combination consists of a source of strength m at the point $(a, 0)$ and a sink of strength $-m$ at the point $(-a, 0)$.

With the notations of fig. 16·26 (i), we get

$$\phi = m\left(\frac{1}{r_2} - \frac{1}{r_1}\right), \quad \psi = m(\cos\theta_2 - \cos\theta_1),$$

from which the streamlines can easily be drawn.

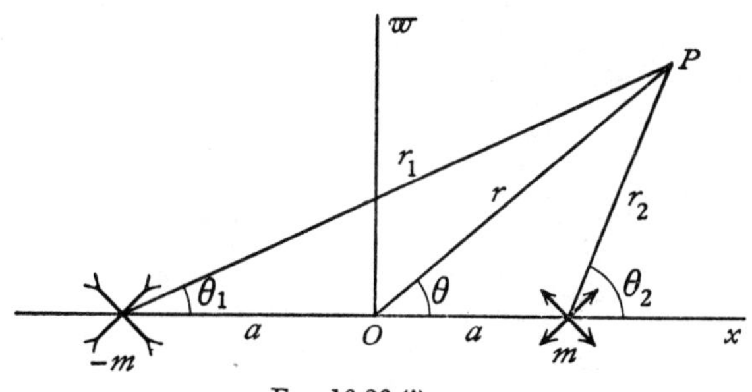

Fig. 16·26 (i).

If the product $2ma = \mu$ remains constant when $m \to \infty$ and $2a \to 0$, the combination becomes a *double source* or *doublet*. The corresponding values of ϕ and ψ may be obtained quite simply as follows.

We have, by the sine rule,

$$\frac{r_1}{\sin\theta_2} = \frac{r_2}{\sin\theta_1} = \frac{2a}{\sin(\theta_2 - \theta_1)} = \frac{2a}{2\sin\frac{1}{2}(\theta_2 - \theta_1)\cos\frac{1}{2}(\theta_2 - \theta_1)}.$$

Hence $$r_1 - r_2 = \frac{a(\sin\theta_2 - \sin\theta_1)}{\sin\frac{1}{2}(\theta_2 - \theta_1)\cos\frac{1}{2}(\theta_2 - \theta_1)} = \frac{2a\cos\frac{1}{2}(\theta_2 + \theta_1)}{\cos\frac{1}{2}(\theta_2 - \theta_1)}.$$

Therefore
$$\phi = \frac{\mu \cos \frac{1}{2}(\theta_2+\theta_1)}{r_1 r_2 \cos \frac{1}{2}(\theta_2-\theta_1)},$$

$$\psi = \frac{m(x-a)}{r_2} - \frac{m(x+a)}{r_1} = \frac{\mu x \cos \frac{1}{2}(\theta_2+\theta_1)}{r_1 r_2 \cos \frac{1}{2}(\theta_2-\theta_1)} - \frac{\mu}{2}\left(\frac{1}{r_2}+\frac{1}{r_1}\right).$$

When $a \to 0$, $\theta_2 \to \theta_1 \to \theta$, $r_2 \to r_1 \to r$. Thus, for a doublet,

$$\phi = \frac{\mu \cos \theta}{r^2}, \quad \psi = \frac{\mu x \cos \theta}{r^2} - \frac{\mu}{r} = \frac{\mu(x^2-r^2)}{r^3} = -\frac{\mu \varpi^2}{r^3}.$$

The direction of the doublet is reckoned from sink to source.

These results also follow from Maclaurin's theorem, using the remark at the end of section 16·20. Thus if $\phi_1 = 1/r$, when a is small,

$$\phi = \frac{m}{r} + am\left(\frac{\partial \phi_1}{\partial a}\right)_{a=0} - \frac{m}{r} + am\left(\frac{\partial \phi_1}{\partial a}\right)_{a=0} = -\mu \frac{\partial}{\partial x}\left(\frac{1}{r}\right) = \frac{\mu x}{r^3}.$$

Similarly, $$\psi = -\mu \frac{\partial}{\partial x}\left(\frac{x}{r}\right) = -\mu\left(\frac{1}{r} - \frac{x^2}{r^3}\right) = -\frac{\mu \varpi^2}{r^3}.$$

The streamlines for a doublet are shown in fig. 16·26 (ii). The method of drawing them is exhibited in (a) of the same figure. Taking $\psi = -n\omega$,

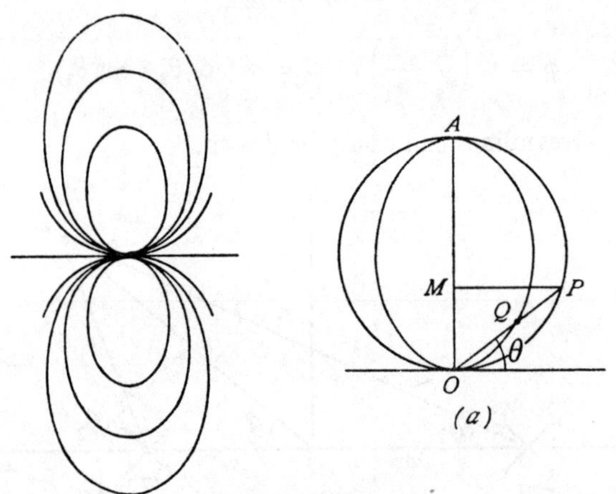

FIG. 16·26 (ii).

$n = 1, 2, 3, \ldots$, we draw the circle diameter $OA = \mu/(n\omega)$ touching the axis of the doublet at O. Draw PM perpendicular to OA and mark

$$OQ = OM = OP \sin \theta = OA \sin^2 \theta.$$

Then, if $OQ = r$,

$$\frac{\mu \sin^2 \theta}{r} = n\omega,$$

and so Q is a point on the streamline.

The image of a doublet in a plane is an equal but anti-parallel doublet at the optical image in the plane, cf. 8·42.

16·27. Rankine's solids. If we combine the source and equal sink of section 16·26 with a uniform stream U in the negative direction of the x-axis the stream function is

$$(1) \qquad \psi = \tfrac{1}{2}Ur^2\sin^2\theta + m(\cos\theta_2 - \cos\theta_1).$$

When P is on the axis, $\theta = 0$ or π, while $\theta_2 - \theta_1 = 0$, except for points between source and sink, where $\theta_2 - \theta_1 = \pi$.

Thus $\psi = 0$ contains the whole of the axis except the part between source and sink and therefore gives the dividing streamline, whose equation is

$$(2) \qquad \varpi^2 + b^2(\cos\theta_2 - \cos\theta_1) = 0, \quad b^2 = 2m/U.$$

Since $\cos\theta_1$, $\cos\theta_2$ are each numerically less than unity, it follows that ϖ^2 cannot exceed $2b^2$, and therefore the dividing streamline is closed.

The dividing line generates by rotation about the axis a dividing stream surface which is clearly symmetrical with respect to the plane $x = 0$, since the equation is unaltered if the signs of m and U are reversed. We have thus the streaming motion past a closed solid of revolution of oval section, fig. 16·27, where A is the sink, B is the source, called a Rankine's solid.

The points C and D where the stream divides are stagnation points. To determine them we may differentiate the stream function, or more simply

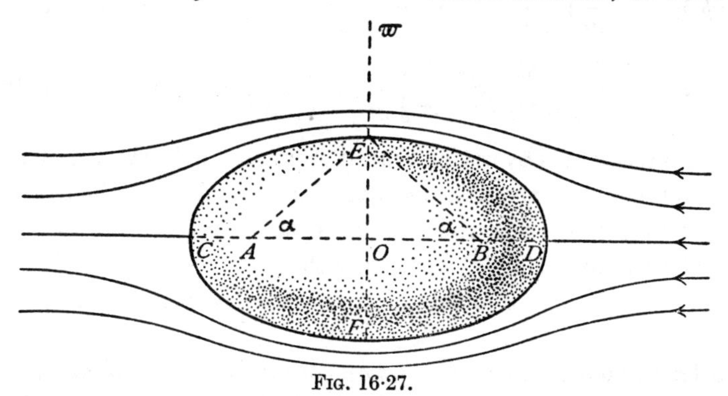

FIG. 16·27.

observe that the stream at D neutralises the velocity due to source and sink, so that if $OD = l$, $OB = a$,

$$(3) \qquad -\frac{m}{(l+a)^2} + \frac{m}{(l-a)^2} = U, \quad \text{or} \quad (l^2 - a^2)^2 = 2ab^2 l,$$

which determines l and therefore the length of the solid. To determine the breadth, if $OE = h$, equation (2) gives $2b^2\cos\alpha = h^2$, where α is the angle EAO. Hence

$$(4) \qquad \frac{2a}{\sqrt{h^2 + a^2}} = \frac{h^2}{b^2},$$

which determines the breadth.

Solids constructed in the above manner by a suitable distribution of sources and sinks have a practical value as well as a theoretical interest, for, the source distribution being known, it is easy to calculate the velocity and hence the pressure. The comparison of the calculated with observed results shows that the pressure follows the theoretical distribution closely on the anterior part, and for moderate streams shows a departure only near the rear where there is a sudden drop below the theoretical value. It is this drop which causes the drag actually found in practice.

16·28. Green's equivalent strata.

A connected closed surface S separates space into two regions R_1 and R_2. Let dn_1, dn_2 denote elements of normal to S drawn into R_1 and R_2 respectively. Then

$$(1) \qquad \frac{\partial}{\partial n_2}\left(\frac{1}{r}\right) = -\frac{\partial}{\partial n_1}\left(\frac{1}{r}\right).$$

Let ϕ_1 and ϕ_2 denote the velocity potential of acyclic irrotational motions in the regions R_1 and R_2 respectively.

Consider the motion given by ϕ_1. From 2·63 (2) we have

$$(2) \quad (\phi_1)_P = -\frac{1}{4\pi}\int_{(S)} \frac{1}{r}\frac{\partial \phi_1}{\partial n_1}\,dS + \frac{1}{4\pi}\int_{(S)} \phi_1 \frac{\partial}{\partial n_1}\left(\frac{1}{r}\right) dS \quad \text{when } P \text{ is in } R_1,$$

$$(3) \quad 0 = -\frac{1}{4\pi}\int_{(S)} \frac{1}{r}\frac{\partial \phi_1}{\partial n_1}\,dS + \frac{1}{4\pi}\int_{(S)} \phi_1 \frac{\partial}{\partial n_1}\left(\frac{1}{r}\right) dS \quad \text{when } P \text{ is in } R_2.$$

We can interpret (1) by saying that at any point of R_1 the velocity potential of the actual motion is the same as that which would be produced by

(i) A distribution of sources of strength $(-\partial\phi_1/\partial n_1)/4\pi$ per unit area distributed over the surface S together with

(ii) A distribution of doublets of strength $\phi_1/4\pi$ per unit area distributed over S.

These distributions constitute *Green's equivalent stratum of sources and doublets.*

The distributions will give the actual velocity at any point of R_1 and zero velocity at any point of R_2.

Turn now to ϕ_2. A point P in R_1 is external to R_2 and therefore (3) gives

$$(4) \qquad 0 = -\frac{1}{4\pi}\int_{(S)} \frac{1}{r}\frac{\partial \phi_2}{\partial n_2}\,dS + \frac{1}{4\pi}\int_{(S)} \phi_2 \frac{\partial}{\partial n_2}\left(\frac{1}{r}\right) dS, \qquad P \text{ in } R_1.$$

Adding (2) and (4) we get

$$(5) \quad \phi_P = -\frac{1}{4\pi}\int_{(S)} \frac{1}{r}\left(\frac{\partial \phi_1}{\partial n_1} + \frac{\partial \phi_2}{\partial n_2}\right) dS + \frac{1}{4\pi}\int_{(S)} (\phi_1 - \phi_2)\,\frac{\partial}{\partial n_1}\left(\frac{1}{r}\right) dS.$$

This again can be interpreted as a distribution of sources and doublets. The equivalent stratum already found is therefore not unique. If, however, we take $\phi_2 = \phi_1$ over S, the second integral in (5) vanishes. Also if we replace S in thought only, by a membrane, we have on S, $-\partial\phi_1/\partial s = -\partial\phi_2/\partial s$, so that the tangential velocity is continuous but the normal velocity is discontinuous. In this case we have a *unique source distribution* of strength

$$-(\partial\phi_1/\partial n_1 + \partial\phi_2/\partial n_2)/4\pi$$

per unit area which would produce the given motion.

Alternatively let us choose ϕ_2 so that $(\partial\phi_1/\partial n_1 + \partial\phi_2/\partial n_2) = 0$ over S. Then the first integral in (5) vanishes and we have continuous normal velocity but discontinuous tangential velocity over S, which is therefore a vortex sheet. In this case we have a *unique doublet distribution* of strength $(\phi_1 - \phi_2)/4\pi$ per unit area which could produce the given motion.

It emerges from this result that a vortex sheet can be replaced by a distribution of doublets.

If the motion in R_1 is cyclic, with cyclic constants $\kappa_1, \kappa_2, \ldots$, we can use doublets, but not sources, to give

$$(\phi_1)_P = \frac{1}{4\pi}\int_{(S)} (\phi_1 - \phi_2)\frac{\partial}{\partial n_1}\left(\frac{1}{r}\right) dS + \frac{\kappa_1}{4\pi}\int_{\sigma_1} \frac{\partial}{\partial n_1}\left(\frac{1}{r}\right) d\sigma_1 + \ldots.$$

Here ϕ_1 is one-valued in the region R_1 modified to become simply connected by introducing barriers $\sigma_1, \sigma_2, \ldots$, and ϕ_2 is the velocity potential of acyclic motion generated in R_2, when proper normal velocities are applied to each element dS of an imagined membrane coincident in position with the original boundary.

16·29. Butler's sphere theorem.

The circle theorem of 6·21 has an analogue which applies to axisymmetrical motions. Let $f(r, \theta)$ be a given function of the two spherical polar coordinates r and θ and let a be a given positive constant. Define

$$(1) \qquad f^* = f^*(r, \theta) = \frac{r}{a} f\left(\frac{a^2}{r}, \ \theta\right).$$

We can then state the following theorem.†

Butler's sphere theorem. Let there be axisymmetrical irrotational flow, in incompressible inviscid fluid with no rigid boundaries, characterised by the stream function $\psi_0 = \psi_0(r, \theta)$ all of whose singularities are at a distance greater than a from the origin, and let $\psi_0 = O(r^2)$ at the origin. If the rigid sphere $r = a$ be introduced into the flow, the stream function becomes

$$(2) \qquad \psi = \psi_0 - \psi_0^* = \psi_0(r, \theta) - \frac{r}{a}\psi_0\left(\frac{a^2}{r}, \theta\right).$$

† S. F. J. Butler. *Proc. Camb. Phil. Soc.* 49 (1953) 169–174.

Proof. The conditions to be satisfied are

(i) the flow given by ψ must be irrotational,

(ii) ψ=constant, when $r=a$,

(iii) $\psi_0{}^\star$ has no singularities outside $r=a$.

(iv) The velocity due to $\psi_0{}^\star$ must tend to zero as r tends to infinity, and $\psi_0{}^\star$ must introduce no net flux over the sphere at infinity.

From 16·1 and 2·72 (4) the condition of zero vorticity derived from the stream function ψ is

$$(3) \qquad r^2 \frac{\partial^2 \psi}{\partial r^2} + \sin\theta \frac{\partial}{\partial \theta}\left(\frac{1}{\sin\theta}\frac{\partial \psi}{\partial \theta}\right) = 0.$$

By direct differentiation we readily verify that if ψ_0 satisfies (3) so does

$$(4) \qquad \psi_0{}^\star = \frac{r}{a}\psi_0\left(\frac{a^2}{r}, \theta\right).$$

This disposes of (i), and (ii) is clearly satisfied since $\psi=0$ when $r=a$.

Since r and a^2/r are inverse points with respect to the sphere $r=a$, if one point is inside the sphere, the other is outside. Thus the singularities of ψ_0 being all outside the sphere, those of $\psi_0{}^\star$ are all inside. Thus (iii) is satisfied.

As to (iv), ψ_0 is regular inside the sphere $r = a$ and near the origin $\psi_0 = O(r^2)$. Therefore at infinity $\psi_0{}^\star = O(1/r)$. From 16·1 it then follows that the velocity at infinity due to $\psi_0{}^\star$ is $O(1/r^3)$ which tends to zero as r tends to infinity. For the flux we have $\int q_r\, dS = O(1/r)$ which also tends to zero. Q.E.D.

The same method of proof shows that if all the singularities of $\psi_0(r, \theta)$ are *inside* the sphere $r=a$, and if $\psi_0=O(1/r)$ for large r, then (2) gives the flow inside the sphere when $r=a$ is made a rigid boundary. Here (iv) is replaced by the condition that $\psi_0{}^\star$ gives finite velocity at the origin. The proof is left as an exercise.

16·30. Sphere in a stream. The stream function for a uniform stream from right to left is $\frac{1}{2}Ur^2 \sin^2\theta$. Therefore from Butler's sphere theorem, when the sphere $r = a$ is inserted,

$$(1) \qquad \psi = \tfrac{1}{2}Ur^2 \sin^2\theta \left(1 - \frac{a^3}{r^3}\right).$$

We note that this is the stream function due to the combination of a stream $-U$ and a doublet of strength $\frac{1}{2}Ua^3$ at the origin. Thus the velocity potential is

$$(2) \qquad \phi = U\left(r\cos\theta + \frac{a^3\cos\theta}{2r^2}\right).$$

The streamlines can be drawn directly from (1), or more easily by first drawing the streamlines of the doublet as in 16·26 and then applying Rankine's diagonal method to a superposed uniform stream.

The velocity at any point of the sphere is tangential and therefore from (2) its value is $-\partial\phi/r\,\partial\theta = 3U\sin\theta/2$. The stagnation points occur on the

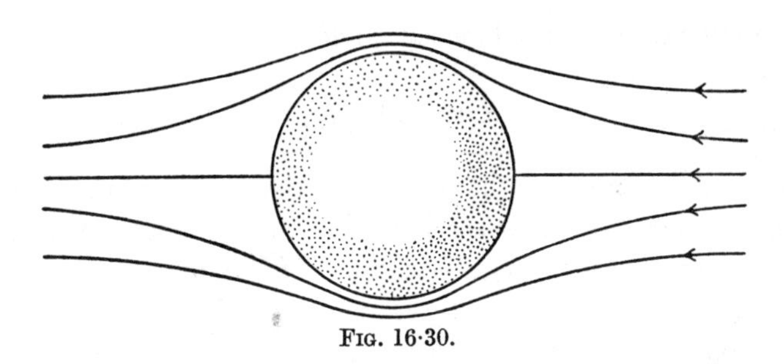

FIG. 16·30.

axis when $\theta = 0$ or π, and the maximum velocity of slip is $3U/2$ round the equatorial belt which is perpendicular to the direction of the stream.

The pressure at any point of the sphere is given by

$$\frac{p}{\rho}+\frac{9U^2\sin^2\theta}{8} = \frac{\Pi}{\rho}+\tfrac{1}{2}U^2,$$

where Π is the pressure at infinity. The points of minimum pressure occur on the equatorial belt mentioned above and the pressure there is p_0, where

$$p_0 = \Pi - \tfrac{5}{8}\rho U^2,$$

and therefore the condition that there shall be no cavitation is that $p_0 > 0$, i.e.

$$\Pi > \tfrac{5}{8}\rho U^2.$$

In accordance with d'Alembert's paradox, the resultant thrust on the sphere is zero. The thrust on the anterior hemisphere is given by

$$F = \int_0^{\pi/2} p\cos\theta\,.\,2\pi a^2\sin\theta\,d\theta = \pi a^2\left\{\Pi - \frac{\rho U^2}{16}\right\}.$$

The thrust on the rear hemisphere is equal but opposite to this.

16·31. Kinetic energy.

When the motion is irrotational, the kinetic energy of the liquid contained in any region bounded by surfaces of revolution about the axis is given (3·72) by

$$T = -\tfrac{1}{2}\rho\int \phi\frac{\partial\phi}{\partial n}\,dS,$$

where dn is an element of normal drawn into the liquid at the element dS of area of the bounding surface. In the present case, $dS = 2\pi\varpi\, ds$, where ds is an element of arc of the meridian curve of the boundary. Also

$$-\frac{\partial\phi}{\partial n} = \frac{1}{\varpi}\frac{\partial\psi}{\partial s},$$

since each represents the normal velocity, and therefore

$$(1) \qquad T = \pi\rho \int \phi\, d\psi,$$

the integral being taken round the portions of the meridian curves on one side of the axis, fig. 16·31, in the sense indicated by the arrows, the fluid being comprised between the surfaces generated by ABC, DEO.

If the outer boundary is absent, the integral is then taken round ABC in the clockwise sense. Changing the sense of description, we get

$$(2) \qquad T = -\pi\rho \int_{(CBA)} \phi\, d\psi,$$

the sense now being anticlockwise.

Another expression for T in terms of the stream function only is obtained by observing that integration of (1) by parts gives

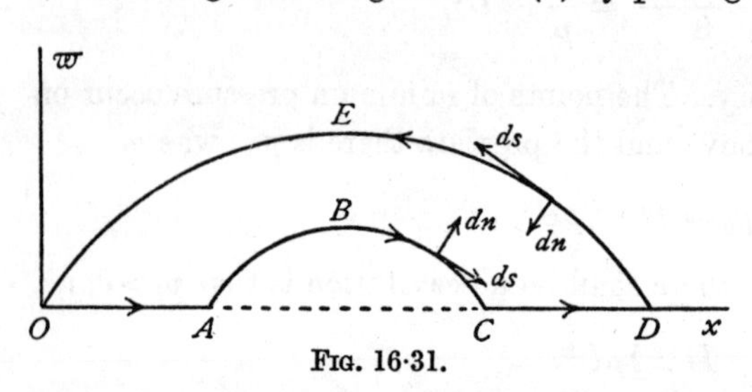

FIG. 16·31.

$$T = -\pi\rho \int \psi\, d\phi,$$

since the integrated part vanishes. But

$$\frac{\partial\phi}{\partial s} = \frac{1}{\varpi}\frac{\partial\psi}{\partial n}. \quad \text{Hence}$$

$$(3) \quad T = -\pi\rho \int \frac{\psi}{\varpi}\frac{\partial\psi}{\partial n}\, ds,$$

taken round the boundary in the sense indicated in fig. 16·31.

16·32. Moving sphere. When a sphere moves with velocity U in a liquid at rest at infinity, the velocity potential and stream function are at once deduced from 16·30 (2) and (1) by superposing a uniform velocity U in the positive direction of the x-axis, so that

$$\phi = \tfrac{1}{2}Ua^3\frac{\cos\theta}{r^2}, \quad \psi = -\tfrac{1}{2}Ua^3\frac{\sin^2\theta}{r}.$$

It is important to observe that these results now refer to an origin moving with the sphere, so that even when U is constant the motion is not steady.*

The kinetic energy of the liquid is given by, 16·31,

* It is then relatively steady, 1·11.

$$T = -\pi\rho\int \phi\, d\psi = \tfrac{1}{2}\pi\rho U^2 a^3 \int_0^{\pi} \cos^2\theta \sin\theta\, d\theta = \tfrac{1}{3}\pi\rho U^2 a^3 = \tfrac{1}{4}M'U^2,$$

where M' is the mass of liquid which the sphere could displace.

Thus the total kinetic energy of the system, solid plus liquid, is

$$T_1 = \tfrac{1}{2}(M+\tfrac{1}{2}M')U^2,$$

where M is the mass of the sphere. Thus the virtual mass is $(M+\frac{1}{2}M')$ cf. 9·221.

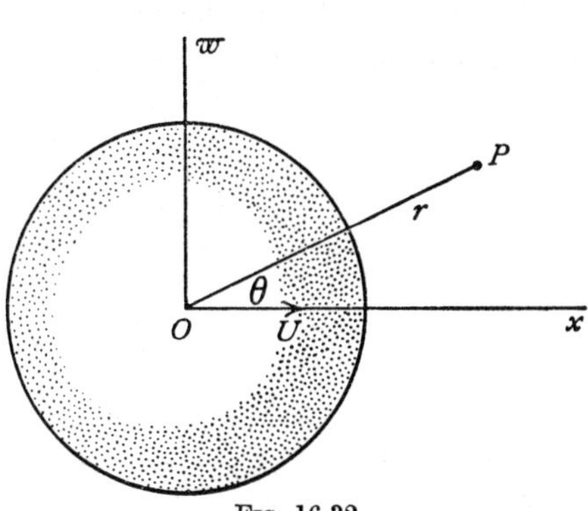

FIG. 16·32.

If F is the *resistance* of the liquid, by equating the rates of working, we have

$$FU = \frac{dT}{dt} = \tfrac{1}{2}M'U\frac{dU}{dt},$$

and therefore

$$F = \tfrac{1}{2}M'\frac{dU}{dt},$$

which vanishes when U is constant.

If the sphere falls under gravity in an infinite liquid, the forces acting upon it are the weight Mg vertically downwards, the buoyancy $M'g$ vertically upwards, and the resistance $\frac{1}{2}M'\,dU/dt$ also vertically upwards. Thus

$$Mg - M'g - \tfrac{1}{2}M'\frac{dU}{dt} = M\frac{dU}{dt},$$

so that the acceleration is

$$\frac{dU}{dt} = \frac{M-M'}{M+\frac{1}{2}M'}g = \frac{s-1}{s+\frac{1}{2}}g,$$

where s is the specific gravity of the sphere compared with the liquid.

This result implies that the effect of the liquid is to reduce the acceleration due to gravity in the ratio $s-1 : s+\frac{1}{2}$. In particular, if $s<1$, the sphere rises with the acceleration given by the above formula. This has an obvious application to the motion of a balloon.

Darwin has shown* that the type of investigation leading to 9·222 (9) can be applied to a three-dimensional body moving in the x-direction to give for the drift volume

$$D = \frac{1}{U}\iiint (-f_x)\, dx\, dy\, dz,$$

wherein the x-integration must be done first. The hydrodynamic mass is then ρD.

In the case of the sphere $D=\frac{2}{3}\pi a^3$ and the hydrodynamic mass is therefore $\frac{1}{2}M'$ as obtained above.

* *Loc. cit.* 9·21.

16·33. Pressure on a moving sphere. The pressure equation is

$$(1) \qquad \frac{p}{\rho}+\tfrac{1}{2}q^2-\frac{\partial\phi}{\partial t} = F(t),$$

$$(2) \qquad q^2 = \left(\frac{\partial\phi}{\partial r}\right)^2+\left(\frac{\partial\phi}{r\partial\theta}\right)^2 = \frac{U^2a^6}{r^6}\left(\cos^2\theta+\frac{\sin^2\theta}{4}\right).$$

Let $\mathbf{r} = \overrightarrow{OP}$ be the position vector of the point P (fixed in space) with regard to O the centre of the sphere. Then

$$\phi = \tfrac{1}{2}\frac{a^3}{r^3}\mathbf{U}\mathbf{r},$$

$$(3) \qquad \frac{\partial\phi}{\partial t} = \tfrac{1}{2}\frac{a^3}{r^3}\mathbf{r}\frac{\partial\mathbf{U}}{\partial t}+\tfrac{1}{2}\frac{a^3}{r^3}\mathbf{U}\frac{\partial\mathbf{r}}{\partial t}-\frac{3a^3}{2r^4}\frac{\partial r}{\partial t}\mathbf{U}\mathbf{r},$$

$$\mathbf{U} = \frac{\partial}{\partial t}(\overrightarrow{PO}) = -\frac{\partial\mathbf{r}}{\partial t},$$

and therefore $U\cos\theta$ = velocity of O along $OP = -\partial r/\partial t$.

Let $\mathbf{f} = \partial\mathbf{U}/dt$ be the acceleration of the centre of the sphere.

Substituting in (3), we get

$$\frac{\partial\phi}{\partial t} = \tfrac{1}{2}\frac{a^3}{r^3}\mathbf{r}\mathbf{f}-\tfrac{1}{2}\frac{a^3}{r^3}U^2+\frac{3a^3}{2r^3}U^2\cos^2\theta.$$

Substituting in (1), we get

$$\frac{p}{\rho}-\tfrac{1}{2}\frac{a^3}{r^3}\mathbf{r}\mathbf{f}+\tfrac{1}{8}\frac{U^2a^6}{r^6}(3\cos^2\theta+1)-\tfrac{1}{2}\frac{U^2a^3}{r^3}(3\cos^2\theta-1) = \frac{\Pi}{\rho},$$

since all the terms on the left except the first vanish when $r = \infty$.

Thus the pressure on the surface of the sphere is given by

$$\frac{p-\Pi}{\rho} = \tfrac{1}{2}\mathbf{a}\mathbf{f}+\tfrac{1}{8}U^2(9\cos^2\theta-5),$$

where $\mathbf{a}$ is the point $(\mathbf{r})_{r=a}$ of the sphere.

These results can also be obtained at once from 3·61.

16·40. Image of a source in a sphere. Consider a sphere, centre O radius a.

Let there be a source of strength m at the point A $(f, 0, 0)$, and let P be any point. If AP makes angle θ_1 with the positive direction of the axis, the stream function for the source alone, adjusted to vanish at the origin is

$$\psi_0 = m(1+\cos\theta_1)$$

and therefore, by Butler's sphere theorem, when the sphere is placed in the field,

$$(1) \qquad \psi = m(1+\cos\theta_1)-m(1+\cos\theta_1)^*$$

$$= m+m\cos\theta_1-\frac{mr}{a}-m(\cos\theta_1)^*$$

The evaluation of $(\cos\theta_1)^\star$ offers no special difficulties but the solution of this and similar problems connected with the sphere are facilitated by some simple geometrical considerations. In fig. 16·40, B is the inverse of A so that

(2) $$OB = \frac{a^2}{f} = f', \quad ff' = a^2$$

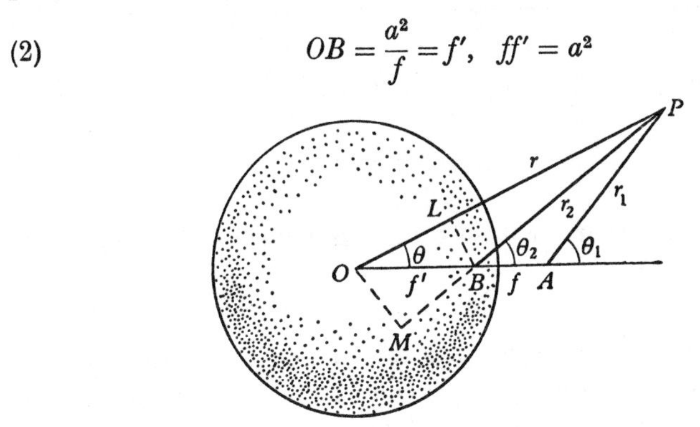

FIG. 16·40.

Draw BL, OM perpendicular to OP, PB. Then O, M, B, L are concyclic so that

(3) $$PO \,.\, PL = PB \,.\, PM.$$

Now $r_1^2 = r^2 + f^2 - 2fr\cos\theta$, $r_2^2 = r^2 + f'^2 - 2f'r\cos\theta$, and therefore from 16·29 (1)

(4) $$(r_1)^\star = \frac{r}{a}\left(\frac{a^4}{r^2} + f^2 - \frac{2fa^2}{r}\cos\theta\right)^{1/2} = \frac{r}{a}\cdot\frac{fr_2}{r} = \frac{fr_2}{a}.$$

Again $\cos\theta_1 = (r\cos\theta - f)/r_1$ and therefore

$$(\cos\theta_1)^\star = \frac{r}{a}\left(\frac{a^2}{r}\cos\theta - f\right)\bigg/\frac{fr_2}{r} = -\frac{r}{a}\cdot\frac{PL}{PB} = -\frac{r}{a}\cdot\frac{PM}{PO}$$

from (3) and so

(5) $$(\cos\theta_1)^\star = -\frac{PM}{a} = \frac{-f'\cos\theta_2 - r_2}{a}.$$

Therefore from (1)

$$\psi = m + m\cos\theta_1 - \frac{m}{a}(r - r_2) + \frac{ma}{f}\cos\theta_2.$$

The last two terms constitute the image of the source in the sphere, which thus consists of a source of strength ma/f at the inverse point and a line sink of strength m/a per unit length stretching from the inverse point to the centre.

16·41. Image of a radial doublet in a sphere. Consider a doublet of strength μ placed at A along a radius of a sphere of centre O, radius a. Taking OA as axis of μ, the stream function due to μ alone is

$$\psi_0 = -\mu\sin^2\theta_1/r_1 = -\mu(1 - \cos^2\theta_1)/r_1,$$

using the diagram and notation of 16·40. By Butler's sphere theorem, in the presence of the sphere $r = a$, we have

$$\psi = -\frac{\mu \sin^2 \theta_1}{r_1} + \frac{\mu r}{a}\frac{1 - PM^2/PO^2}{fr_2/r} = -\frac{\mu \sin^2 \theta_1}{r_1} + \frac{\mu OM^2}{afr_2}.$$

But $OM = f' \sin \theta_2 = (a^2/f) \sin \theta_2$. Therefore

$$\psi = -\frac{\mu \varpi^2}{r_1{}^3} + \left(\frac{a^3}{f^3}\mu\right)\frac{\varpi^2}{r_2{}^3}.$$

Thus the required image is an oppositely directed doublet of strength $\mu a^3/f^3$ at the inverse point (cf. 8·81).

16·42. Force on an obstacle. Let the motion be steady and irrotational. Let there be n singularities of the flow each at a finite distance from the obstacle. Let S_0 be the surface of the obstacle, and let S_i ($i = 1, 2, \ldots, n$) be spheres of infinitesimal radius, one round each singularity. Let S_{n+1} be a sphere of large radius R conceived as enclosing S_i ($i = 0, 1, 2, \ldots, n$), and let V be the volume exterior to these S_i but interior to S_{n+1}. Then by Gauss's theorem

$$\sum_{i=0}^{n+1} \int_{(S_i)} [\mathbf{n}\, q^2 - 2\mathbf{q}(\mathbf{n}\,\mathbf{q})]dS = -\int_{(V)} [\nabla\, q^2 - 2\mathbf{q}(\nabla\,\mathbf{q}) - 2(\mathbf{q}\,\nabla)\mathbf{q}]d\tau$$

$$= -\int_{(V)} [2\mathbf{q}_\wedge(\nabla_\wedge \mathbf{q}) - 2\mathbf{q}(\nabla\mathbf{q})]d\tau = 0,$$

since $\nabla\,\mathbf{q} = 0$ and $\nabla_\wedge \mathbf{q} = 0$, and therefore

$$\sum_{i=0}^{n} \int_{(S_i)} [\mathbf{n}\, q^2 - 2\mathbf{q}(\mathbf{n}\,\mathbf{q})]dS = -\int_{(S_{n+1})} [\mathbf{n}\, q^2 - 2\mathbf{q}(\mathbf{n}\,\mathbf{q})]dS.$$

The integral on the left is independent of S_{n+1} and therefore so also is the integral on the right, and if $\mathbf{q} = O(1/R^2)$, it is clear that the integral on the right has its integrand $O(1/R^2)$ and therefore must anyway tend to zero as $R \to \infty$. Thus the integral is identically zero and therefore, writing

(1) $$\mathbf{u} = -\tfrac{1}{2}\mathbf{n}\, q^2 + \mathbf{q}(\mathbf{n}\,\mathbf{q}),$$

we have

(2) $$-\int_{(S_0)} \mathbf{u}\, dS = \sum_{i=1}^{n} \int_{(S_i)} \mathbf{u}\, dS.$$

In exactly the same manner we prove that

(3) $$-\int_{(S_0)} \mathbf{r}_\wedge \mathbf{u}\, dS = \sum_{i=1}^{n} \int_{(S_i)} \mathbf{r}_\wedge \mathbf{u}\, dS.$$

Thus if $\mathbf{F}$, $\mathbf{L}$ are the force and moment on the obstacle, we see from 3·62 that

$$(4) \qquad \mathbf{F} = \sum_{i=1}^{n} \rho \int_{(S_i)} \mathbf{u}\, dS, \quad \mathbf{L} = \sum_{i=1}^{n} \rho \int_{(S_i)} \mathbf{r}_\wedge \mathbf{u}\, dS.$$

Thus the action on the obstacle may be regarded as the resultant of a system of forces and moments

$$(5) \qquad \mathbf{F}_i = \rho \int_{(S_i)} \mathbf{u}\, dS, \quad \mathbf{L}_i = \rho \int_{(S_i)} \mathbf{r}_\wedge \mathbf{u}\, dS.$$

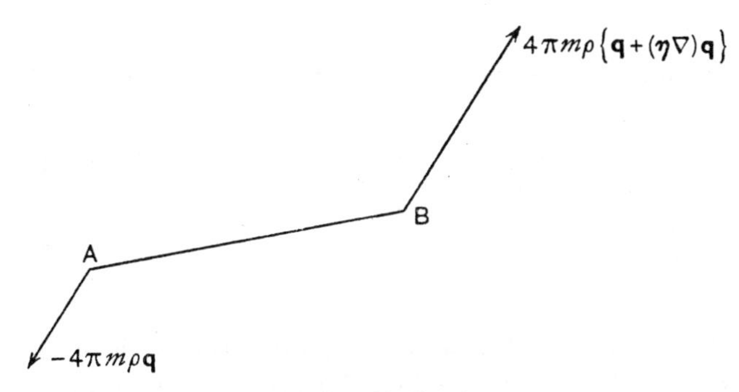

FIG. 16·42.

Suppose that the ith singularity is a source of strength m_i situated at the position $\mathbf{r}_i$. Then, if r is the radius of the infinitesimal sphere S_i, we can write for points on the surface of this sphere

$$\mathbf{q} = \frac{m_i\, \mathbf{n}}{r^2} + \mathbf{q}_i\,,$$

where $\mathbf{q}_i$ is the velocity at the point $\mathbf{r}_i$ induced by all causes except the singularity there. Substituting in (5) and remembering that $\int \mathbf{n}\, dS = 0$ over a closed surface, we get

$$(6) \qquad \mathbf{F}_i = 4\pi m_i\, \rho\, \mathbf{q}_i\,, \quad \mathbf{L}_i = \mathbf{r}_{i\,\wedge}\, 4\pi m_i\, \rho\, \mathbf{q}_i.$$

This shows that in the case of sources, we can suppose the fluid action on the obstacle to be due simply to a force $\mathbf{F}_i$ localised at the ith source ($i = 1, 2, \ldots, n$).

To find the effect of a doublet, consider a sink $-m$ at A and a source m at B, where $\overrightarrow{AB} = \boldsymbol{\eta}$.

If $\mathbf{q}$ is the velocity induced at the sink by all causes except the *source and the sink*, the velocity at the source due to all causes except the source is

$$\mathbf{q}_+ = \mathbf{q} + (\boldsymbol{\eta}\, \nabla)\, \mathbf{q} - m\boldsymbol{\eta}/\eta^3,$$

while the velocity at the sink due to all causes except the sink is

$$\mathbf{q}_- = \mathbf{q} - m\boldsymbol{\eta}/\eta^3.$$

Thus from (6) we see that at the source there is a force $4\pi m\,\rho\,\mathbf{q}_+$ and at the sink a force $-4\pi m\,\rho\,\mathbf{q}_-$. The forces in the line AB cancel and we are left with the forces shown in fig. 16·42, i.e. in the limit when we have a doublet of strength $\boldsymbol{\mu} = \boldsymbol{\eta}\, m$, there is a force and couple

$$\text{(7)} \qquad \mathbf{F} = 4\pi\rho\,(\boldsymbol{\mu}\,\nabla)\,\mathbf{q}, \quad \mathbf{L} = 4\pi\rho\,\boldsymbol{\mu}\wedge\mathbf{q},$$

where $\mathbf{q}$ is the velocity induced at the doublet.

16·43. Action of a source on a sphere. From 16·40 we find that the image system of a source m at distance f from the centre of a sphere of radius a, induces a radial velocity $ma^3 f^{-1}(f^2-a^2)^{-2}$ and therefore the sphere is urged towards the source by a force

$$\frac{4\pi\rho m^2 a^3}{f(f^2-a^2)^2}.$$

16·44. Action of a radial doublet on a sphere. From 16·41 we find the velocity induced by the image doublet at the point (r, θ) to be

$$-\frac{2\mu\cos\theta}{r^3}\frac{a^3}{f^3},$$

the pole of coordinates being the inverse of the position of the doublet. From 16·42 (7) the sphere is urged towards the doublet by a force

$$4\pi\rho\left\{\mu\frac{\partial}{\partial r}\left(-\frac{2\mu\cos\theta}{r^3}\frac{a^3}{f^3}\right)\right\}_{\theta=0,\, r=f-a^2/f} = \frac{24\pi\rho\mu^2 a^3 f}{(f^2-a^2)^4}.$$

16·50. The equation satisfied by the stream function when the motion is irrotational. If the flow is symmetrical about the x-axis, the vorticity is (from 2·72 (8))

$$\frac{\partial q_\varpi}{\partial x} - \frac{\partial q_x}{\partial \varpi} = \frac{\partial}{\partial x}\left(\frac{1}{\varpi}\frac{\partial\psi}{\partial x}\right) + \frac{\partial}{\partial\varpi}\left(\frac{1}{\varpi}\frac{\partial\psi}{\partial\varpi}\right) = \frac{1}{\sin\omega}\nabla^2\left(\frac{\psi\sin\omega}{\varpi}\right).$$

If the motion is irrotational, it follows that

$$\text{(1)} \qquad \frac{\partial}{\partial x}\left(\frac{1}{\varpi}\frac{\partial\psi}{\partial x}\right) + \frac{\partial}{\partial\varpi}\left(\frac{1}{\varpi}\frac{\partial\psi}{\partial\varpi}\right) = 0,$$

which is the required equation.

We shall now prove that the conformal transformation

$$z_1 = x + i\varpi = f(\xi + i\eta) = f(\zeta)$$

transforms the above equation into

$$(2) \qquad \frac{\partial}{\partial \xi}\left(\frac{1}{\varpi}\frac{\partial \psi}{\partial \xi}\right)+\frac{\partial}{\partial \eta}\left(\frac{1}{\varpi}\frac{\partial \psi}{\partial \eta}\right)=0,$$

where ϖ is considered as the function of ξ, η given by the transformation.

Proof. Since

$$(3) \qquad 2\frac{\partial}{\partial z_1}=\frac{\partial}{\partial x}-i\frac{\partial}{\partial \varpi}, \quad 2\frac{\partial}{\partial \zeta}=\frac{\partial}{\partial \xi}-i\frac{\partial}{\partial \eta},$$

we see * that (1) is equivalent to the vanishing of the real part of

$$\frac{\partial}{\partial z_1}\left(\frac{1}{\varpi}\frac{\partial \psi}{\partial \bar{z}_1}\right)=\frac{\partial}{\partial z_1}\left(\frac{1}{\varpi}\frac{\partial \bar{\zeta}}{\partial \bar{z}_1}\frac{\partial \psi}{\partial \bar{\zeta}}\right)=\frac{\partial \bar{\zeta}}{\partial \bar{z}_1}\frac{\partial}{\partial z_1}\left(\frac{1}{\varpi}\frac{\partial \psi}{\partial \bar{\zeta}}\right)=\frac{\partial \bar{\zeta}}{\partial \bar{z}_1}\frac{\partial \zeta}{\partial z_1}\frac{\partial}{\partial \zeta}\left(\frac{1}{\varpi}\frac{\partial \psi}{\partial \bar{\zeta}}\right).$$

The first two factors on the right are conjugate imaginaries whose product is therefore real, and consequently the real part of

$$\frac{\partial}{\partial \zeta}\left(\frac{1}{\varpi}\frac{\partial \psi}{\partial \bar{\zeta}}\right)$$

vanishes and this is equation (2). Q.E.D.

16·51. The velocity.

We have

$$q_\varpi+iq_x=\frac{1}{\varpi}\frac{\partial \psi}{\partial x}-\frac{i}{\varpi}\frac{\partial \psi}{\partial \varpi}=2\frac{1}{\varpi}\frac{\partial \psi}{\partial z_1},$$

and therefore

$$q^2=\frac{4}{\varpi^2}\frac{\partial \psi}{\partial z_1}\frac{\partial \psi}{\partial \bar{z}_1}.$$

The conformal transformation $z_1=f(\zeta)$ then gives

$$q^2=\frac{4}{\varpi^2}\frac{\partial \zeta}{\partial z_1}\frac{\partial \psi}{\partial \zeta}\frac{\partial \bar{\zeta}}{\partial \bar{z}_1}\frac{\partial \psi}{\partial \bar{\zeta}},$$

and therefore

$$q^2\varpi^2 f'(\zeta)\bar{f}'(\bar{\zeta})=4\frac{\partial \psi}{\partial \zeta}\frac{\partial \psi}{\partial \bar{\zeta}}=\left(\frac{\partial \psi}{\partial \xi}\right)^2+\left(\frac{\partial \psi}{\partial \eta}\right)^2.$$

Again, if ds_ξ, ds_η are elements of arc in the directions in which ξ and η increase respectively, we have

$$(ds_\xi)^2+(ds_\eta)^2=(ds)^2=(dx)^2+(d\varpi)^2=f'(\zeta)d\zeta\times\bar{f}'(\bar{\zeta})d\bar{\zeta}$$
$$=J^2[(d\xi)^2+(d\eta)^2],$$

where $J^2=f'(\zeta)\bar{f}'(\bar{\zeta})$. Thus the velocity components in the directions in which ξ, η increase are given by

$$q_\xi=-\frac{\partial \phi}{\partial s_\xi}=-\frac{1}{J}\frac{\partial \phi}{\partial \xi}, \quad q_\eta=-\frac{\partial \phi}{\partial s_\eta}=-\frac{1}{J}\frac{\partial \phi}{\partial \eta},$$

* Note that ψ may be regarded as a function of the independent variables z_1, $\bar{z}_1$ or of ζ, $\bar{\zeta}$. Also ζ is a function of z_1 only and therefore $\bar{\zeta}$ is a function of $\bar{z}_1$ only.

or in terms of the stream function by

$$q_\xi = -\frac{1}{\varpi}\frac{\partial\psi}{\partial s_\eta} = -\frac{1}{J\varpi}\frac{\partial\psi}{\partial\eta}, \quad q_\eta = \frac{1}{\varpi}\frac{\partial\psi}{\partial s_\xi} = \frac{1}{J\varpi}\frac{\partial\psi}{\partial\xi}.$$

From these results we obtain the equations (cf. 5·30),

$$\frac{\partial\phi}{\partial\xi} = \frac{1}{\varpi}\frac{\partial\psi}{\partial\eta}, \quad \frac{\partial\phi}{\partial\eta} = -\frac{1}{\varpi}\frac{\partial\psi}{\partial\xi}.$$

16·52. Boundary condition satisfied by the stream function. When a solid of revolution moves with velocity U in the direction of its axis in a liquid, the normal velocities of the solid and the liquid in contact with it are the same. Thus

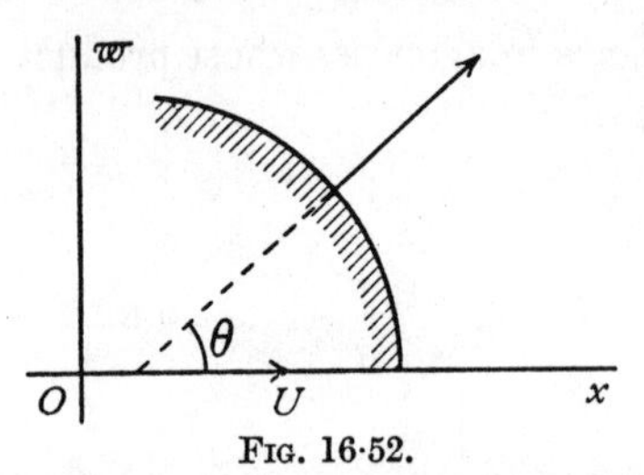

FIG. 16·52.

$$-\frac{1}{\varpi}\frac{\partial\psi}{\partial s} = U\cos\theta = U\frac{d\varpi}{ds}.$$

Integrating round the boundary, we get

$$\psi = -\tfrac{1}{2}U\varpi^2 + \text{constant}.$$

If the liquid is at rest at infinity, the motion there must be unaffected by the presence of the solid and therefore ψ must tend to a constant value at infinity. Without loss of generality, the constant may be assumed to be zero.

16·53. The sphere. One of the simplest applications of the foregoing results is to the motion of a sphere. The transformation

$$z_1 = x + i\varpi = c\,e^\zeta$$

gives $x = c\,e^\xi \cos\eta$, $\varpi = c\,e^\xi \sin\eta$, so that the surfaces ξ = constant are spheres. For a sphere of radius a we have $a = c\,e^{\xi_0}$. The equation satisfied by the stream function is

$$(1) \qquad \frac{\partial}{\partial\xi}\left(\frac{1}{e^\xi \sin\eta}\frac{\partial\psi}{\partial\xi}\right) + \frac{\partial}{\partial\eta}\left(\frac{1}{e^\xi \sin\eta}\frac{\partial\psi}{\partial\eta}\right) = 0,$$

while if the sphere moves forward with velocity U in the direction of the x-axis,

$$(2) \qquad \psi = -\tfrac{1}{2}Uc^2 e^{2\xi_0}\sin^2\eta \text{ at the surface;}$$

(3) $\qquad \psi \to 0$ at infinity, where the liquid is undisturbed.

Equation (2) suggests the trial solution $\psi = f(\xi)\sin^2\eta$. Substitution in (1) then gives successively

$$\frac{\partial}{\partial \xi}[e^{-\xi}f'(\xi)] - 2e^{-\xi}f(\xi) = 0,$$

$$f''(\xi) - f'(\xi) - 2f(\xi) = 0, \quad f(\xi) = B\,e^{2\xi} + C\,e^{-\xi}.$$

From (3), we see that $B = 0$ and (2) then gives

$$C\,e^{-\xi_0} = -\tfrac{1}{2}c^2 U\,e^{2\xi_0},$$

whence
$$\psi = -\tfrac{1}{2}c^2 U\,e^{3\xi_0}\sin^2\eta/e^{\xi} = -\tfrac{1}{2}a^3 U\sin^2\theta/r,$$

the result already obtained in 16·32.

16·54. Stream function for a planetary ellipsoid. A *planetary* (or disc-shaped) ellipsoid is the figure obtained by rotating an ellipse about its minor axis. This figure is also known as an oblate spheroid. The figures of the earth and of the planet Jupiter are approximately of this form.

The transformation

$$z_1 = x + i\varpi = c\sinh\zeta \tag{1}$$

gives $x = c\sinh\xi\cos\eta$, $\varpi = c\cosh\xi\sin\eta$, and therefore the curve $\xi = \xi_0$ is an ellipse in the meridian plane whose semi-axes are

$$a = c\cosh\xi_0, \quad b = c\sinh\xi_0,$$

and so $\xi = \xi_0$ gives a planetary ellipsoid.

The stream function satisfies the equation (16·50)

$$\frac{\partial}{\partial\xi}\left(\frac{1}{\cosh\xi\sin\eta}\frac{\partial\psi}{\partial\xi}\right) + \frac{\partial}{\partial\eta}\left(\frac{1}{\cosh\xi\sin\eta}\frac{\partial\psi}{\partial\eta}\right) = 0. \tag{2}$$

When the ellipsoid moves forward with velocity U in liquid at rest at infinity the conditions to be satisfied by ψ are

(3) $\psi = -\tfrac{1}{2}U c^2\cosh^2\xi_0\sin^2\eta$ at the solid surface,

(4) $\psi \to 0$ at infinity.

Condition (3) suggests the trial solution, $\psi = f(\xi)\sin^2\eta$.

Substitution in (2) gives successively

$$f''(\xi)\cosh\xi - f'(\xi)\sinh\xi - 2f(\xi)\cosh\xi = 0,$$

$$f'(\xi)\cosh\xi - 2f(\xi)\sinh\xi = B,$$

$$\frac{d}{d\xi}\left(\frac{f(\xi)}{\cosh^2\xi}\right) = \frac{B}{\cosh^3\xi},$$

$$f(\xi) = B\cosh^2\xi\int\frac{d\xi}{\cosh^3\xi} + C\cosh^2\xi,$$

where B and C are constants and (4) shows that $C = 0$.

Now, by integration by parts or by direct verification,

$$\int \frac{d\xi}{\cosh^3\xi} = \tfrac{1}{2}\left(\frac{\sinh\xi}{\cosh^2\xi} - \cot^{-1}\sinh\xi\right) + D,$$

and we take $D = 0$, since the other terms tend to zero when $\xi \to \infty$. Thus

$$f(\xi) = \tfrac{1}{2}B\cosh^2\xi\left(\frac{\sinh\xi}{\cosh^2\xi} - \cot^{-1}\sinh\xi\right).$$

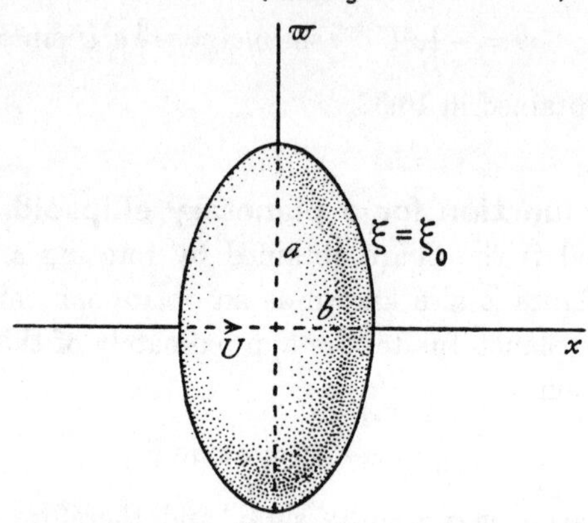

FIG. 16·54.

To verify that $f(\xi) \to 0$ when $\xi \to \infty$, we have, for large values of ξ,

$$\sinh\xi - \cosh^2\xi\cot^{-1}\sinh\xi = \sinh\xi - \frac{\cosh^2\xi}{\sinh\xi} = -\frac{1}{\sinh\xi},$$

which clearly tends to zero. Thus (4) is satisfied.

To determine B, we have from (3)

$$B\cosh^2\xi_0\left(\frac{\sinh\xi_0}{\cosh^2\xi_0} - \cot^{-1}\sinh\xi_0\right) = -Uc^2\cosh^2\xi_0.$$

Now $\qquad a = c\cosh\xi_0, \quad b = a\sqrt{1-e^2} = c\sinh\xi_0.$

Therefore $\qquad B = -Uc^2/(e\sqrt{1-e^2} - \sin^{-1}e).$

Thus, finally,

$$\psi = \frac{-\tfrac{1}{2}Uc^2(\sinh\xi - \cosh^2\xi\cot^{-1}\sinh\xi)}{e\sqrt{1-e^2} - \sin^{-1}e}\sin^2\eta. \tag{5}$$

To find the velocity potential, we have from 16·51

$$\frac{\partial\phi}{\partial\eta} = -\frac{1}{c\cosh\xi\sin\eta}\frac{\partial\psi}{\partial\xi}.$$

Therefore, from (5),

$$\frac{\partial\phi}{\partial\eta} = \frac{k}{c}(2 - 2\sinh\xi\cot^{-1}\sinh\xi)\sin\eta, \quad k = \frac{\tfrac{1}{2}Uc^2}{e\sqrt{1-e^2} - \sin^{-1}e}.$$

$$\phi = -\frac{Uc(1 - \sinh\xi\cot^{-1}\sinh\xi)}{e\sqrt{1-e^2} - \sin^{-1}e}\cos\eta.$$

Note that ϕ is of the form $Uf(\xi)x$.

The kinetic energy and hence the virtual mass are then easily calculated from the formula

$$T = -\pi\rho \int_{\eta=0}^{\eta=\pi} \phi \, d\psi = \tfrac{2}{3}\pi\rho a^3 U^2 \frac{e - (\sin^{-1} e)\sqrt{(1-e^2)}}{\sin^{-1} e - e\sqrt{(1-e^2)}}.$$

The streaming motion past a planetary ellipsoid is easily deduced by superposing a stream $-U$ on the solution found above.

16·55. For a circular disc moving perpendicularly to its plane, we put $e = 1$, $c = a$ in the formulae of 16·54.

Thus on the face of the disc ($\xi_0 = 0$), we have

$$\phi = \frac{2Ua}{\pi} \cos \eta, \quad \psi = -\tfrac{1}{2} U a^2 \sin^2 \eta,$$

and the kinetic energy is

$$T = -\pi\rho \int_{\eta=0}^{\eta=\pi} \phi \, d\psi = \tfrac{4}{3}\rho a^3 U^2.$$

16·56. Venturi tube. To find solutions of 16·54 (2) which are independent of ξ we put $\partial\psi/\partial\xi = 0$ which leads at once to

$$\psi = Ac \cos \eta,$$

where A is an arbitrary constant. The streamlines, η = constant, are hyperbolas and the stream surfaces are therefore the hyperboloids (of one sheet) generated by the revolution of these hyperbolas about the x-axis.

If we take a particular constant value η_0, we get the flow of liquid through a tube whose wall is the corresponding hyperboloid, the smallest section, or *throat*, of the tube being a circle of radius $c \sin \eta_0$. The taper of the tube in the neighbourhood of the throat can be made of any degree of fineness by taking η_0 sufficiently small.

We have thus an idealised representation of the flow through the throat of a Venturi tube (1·7), or the working part of a high-speed wind tunnel.

From the definition of ψ the flux through the throat is $2\pi Ac(1 - \cos \eta_0)$ which determines A in terms of the flux.

As an extreme case, taking $\eta_0 = \pi/2$, we get the flow through a circular aperture of radius c in an infinite plane wall ($x = 0$). As usual in such cases the speed at the edge of the aperture is infinite (cf. 6·10).

16·57. Stream function for an ovary ellipsoid. The *ovary* (or egg-shaped) ellipsoid, also called a prolate spheroid, is generated by the rotation of an ellipse about its major axis. The method of 16·54 can be applied to this case by means of the transformation

$$x + i\varpi = c \cosh \zeta.$$

Pursuing the same steps, we obtain

$$\psi = -\frac{\frac{1}{2}Ub^2\left(\cosh\xi + \sinh^2\xi \log\tanh\dfrac{\xi}{2}\right)\sin^2\eta}{\dfrac{a}{c}+\dfrac{b^2}{c^2}\log\dfrac{a+b-c}{a+b+c}},$$

for the spheroid, defined by $\xi = \xi_0$, $a = c\cosh\xi_0$, $b = c\sinh\xi_0$, moving forward with velocity U in the direction of the x-axis.

16·58. Paraboloid of revolution.

The transformation

$$x + i\varpi = c(\xi + i\eta)^2$$

gives

$$x = c(\xi^2 - \eta^2), \quad \varpi = 2c\xi\eta,$$

and therefore

$$x - c\xi^2 = -\frac{\varpi^2}{4c\xi^2}.$$

Thus the surfaces ξ = constant are paraboloids of revolution whose foci are at the origin. To discuss the motion of such a paraboloid progressing with velocity U in liquid otherwise at rest, we observe that, at the boundary of the paraboloid $\xi = \xi_0$, we must have

$$\psi = -\tfrac{1}{2}U\,4c^2\xi_0^{\,2}\eta^2, \tag{1}$$

while the stream function must satisfy

$$\frac{\partial}{\partial\xi}\left(\frac{1}{\xi\eta}\frac{\partial\psi}{\partial\xi}\right) + \frac{\partial}{\partial\eta}\left(\frac{1}{\xi\eta}\frac{\partial\psi}{\partial\eta}\right) = 0. \tag{2}$$

We therefore put $\psi = f(\xi)\eta^2$ in (2), which gives successively

$$\frac{f'(\xi)}{\xi} = B, \quad f(\xi) = \tfrac{1}{2}B\xi^2 + C,$$

and therefore $\psi = (\frac{1}{2}B\xi^2 + C)\eta^2$. The condition at infinity is no longer $\psi = 0$, for the paraboloid itself is of infinite extent and disturbs the fluid. This must therefore be replaced by the condition that the velocity vanishes at infinity for points not near the paraboloid. From 16·51, we have

$$q^2\varpi^2c^2 \,.\, 2\zeta \,.\, 2\bar{\zeta} = (B\xi\eta^2)^2 + \eta^2(B\xi^2 + 2C)^2,$$

$$q^2 = \frac{B^2}{16c^4} + \frac{BC\,\xi^2 + C^2}{4c^4\xi^2(\xi^2+\eta^2)}.$$

The first term on the right does not vanish unless $B = 0$. Therefore we must have $B = 0$.

Hence

$$\psi = C\eta^2.$$

Comparing this with (1), we get

$$\psi = -2c^2U\xi_0^{\,2}\eta^2. \tag{3}$$

In the case of the streaming motion past the paraboloid we have, by superposing a stream U from right to left,

$$\psi = 2c^2U(\xi^2-\xi_0^2)\eta^2.$$

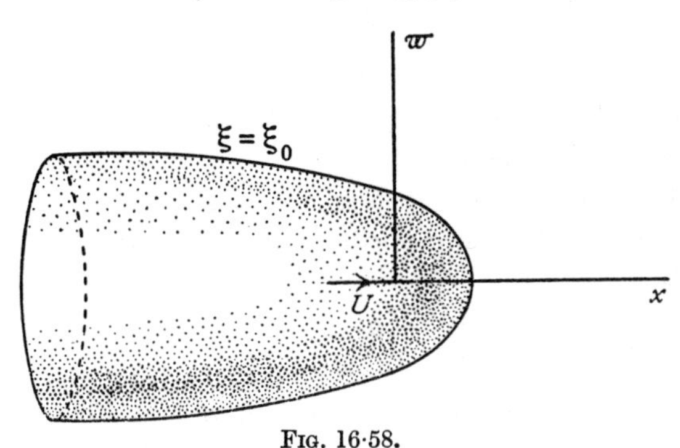

FIG. 16·58.

The result (3) can also be deduced as a limiting case of the motion of an ovary ellipsoid as follows. Moving the origin to the focus, the transformation of 16·57 can be written

$$x+i\varpi = c\cosh\zeta - c = 2c\sinh^2\tfrac{1}{2}\zeta.$$

If we write $2k^2c$ for c, and ζ/k for ζ, this becomes

$$x+i\varpi = 4ck^2\sinh^2\frac{\zeta}{2k},$$

and when $k\to\infty$ this goes over into $x+i\varpi = c\zeta^2$.

The corresponding changes in the stream function of 16·57 are as follows a, b, c become respectively

$$2k^2c\cosh\frac{\xi_0}{k},\quad 2k^2c\sinh\frac{\xi_0}{k},\quad 2k^2c,$$

while ξ, η become ξ/k, η/k. Making $k\to\infty$ then gives (3).

16·60 Comparison theorems. We consider irrotational flow of an inviscid incompressible fluid, bounded by streamlines, in a region R of the xy plane. There are no sources or sinks inside R.

The plane may be that of two-dimensional flow or a meridian plane of axisymmetrical flow, the x-axis being the axis of symmetry.

The velocity from right to left across an element dn of normal to a streamline ψ = constant will be $\epsilon\,\partial\psi/\partial n$ where $\epsilon = 1$ or $1/y$ according as the flow is plane or axisymmetrical so that ϵ is always positive.

A *strip domain* is the region bounded by two non-intersecting streamlines each having its end-points at $x = \pm\infty$.

A point P of the boundary will be called a *regular point* if it is not on the axis (in the case of axisymmetrical flow) and is on the circumference of some circle which touches the boundary at P and whose interior lies entirely in R.

The stream function ψ satisfies the differential equation

$$\psi_{xx}+\psi_{yy} = \epsilon\psi_y \tag{1}$$

This is an equation of elliptic type and its solutions obey a *maximum principle*, namely that a non-constant ψ can achieve neither a maximum nor minimum inside its region of definition.

Physically this means that the presence of such an interior maximum or minimum would demand interior vorticity which contradicts the hypothesis of irrotational flow.

If therefore ψ is zero on one boundary and equal to a positive constant on the other, it follows that $\psi \geqslant 0$ in the whole region between the boundaries.

Comparison Theorem 1. Let D, $D^\star$ be strip domains bounded respectively by streamlines γ, Γ and $\gamma^\star$, Γ and let D be contained in $D^\star$.

Let two distinct axisymmetric flows through D and $D^\star$ be defined by stream functions ψ and $\psi^\star$ such that

$$\psi = 0 \text{ on } \gamma, \quad \psi^\star = 0 \text{ on } \gamma^\star, \quad \psi = Q = \psi^\star \text{ on } \Gamma,$$

where Q is a positive constant; that is the two flows have the same flux. These flows are not to be thought of as superposed.

If P is a regular point common to γ and $\gamma^\star$, and if R is any regular point on Γ, then, for the speeds at P and R of the two flows,

$$q(P) \leqslant q^\star(P), \quad q(R) \geqslant q^\star(R). \tag{2}$$

For either equality to hold it is necessary and sufficient that $D = D^\star, \psi = \psi^\star$.

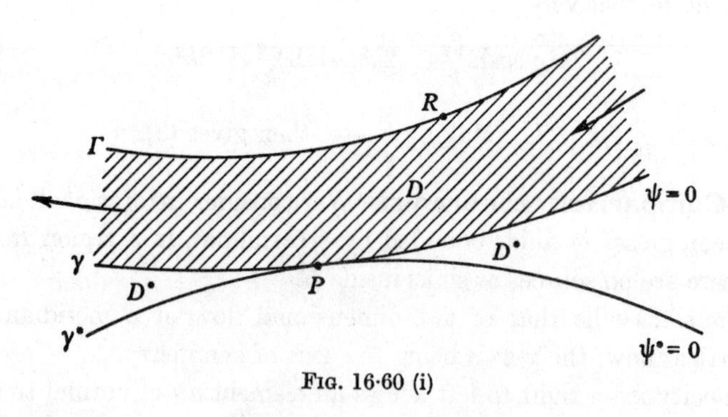

FIG. 16·60 (i)

Proof. Let $\Omega = \psi^\star - \psi$ and suppose the general flow is from right to left, fig. 16·60 (i).

On γ, $\qquad \psi = 0 \quad \text{and } \Omega = \psi^\star \geqslant 0;$

on Γ, $\qquad \psi = \psi^\star \text{ and } \Omega = 0.$

By the maximum principle if $\Omega = 0$ on γ, $\Omega \equiv 0$ throughout D.

If $\Omega > 0$ on γ, $\Omega \geqslant 0$ throughout D.

At P, $\Omega_P = 0$. Therefore $\epsilon\, \partial\Omega/\partial n \geqslant 0$,

$$\text{i.e. } q^\star(P) - q(P) \geqslant 0 \text{ or } q^\star(P) \geqslant q(P).$$

At R, $\Omega_R = 0$. Therefore $\epsilon\, \partial\Omega/\partial n \geqslant 0$,

$$\text{i.e. } -q^\star(R) + q(R) \geqslant 0 \text{ or } q(R) \geqslant q^\star(R).$$

This proves the theorem for plane or axisymmetrical flow for which $\epsilon = 1$ or $1/y$, both of which are positive.

Clearly for either equality to hold we must have $D = D^\star, \psi = \psi^\star$. Q.E.D.

If we let Γ recede to infinity so that Q increases indefinitely we still have $\Omega = 0$ on Γ at every stage so that finally $\Omega = 0$ at ∞. The second conclusion concerning the velocity of the flows at R ceases to be meaningful, but the first subsists in the form of the following theorem.

Comparison Theorem 2. Let D and $D^\star$ be flow regions for two plane (or axisymmetrical) flows having the same non-zero uniform velocity at infinity. Let D and $D^\star$ be bounded by the single streamlines γ and $\gamma^\star$ extending to $x = \pm\infty$. If D is a part of $D^\star$ and if γ and $\gamma^\star$ have a regular point P in common, then the speeds at P follow the inequality

$$q(P) \leqslant q^\star(P). \tag{3}$$

The equality holds only if $D = D^\star$ and the two flows are identical.

These comparison theorems seem to be originally due to Lavrentieff, but have been given a sharper form by Gilbarg,*

Serrin's under-over comparison theorem.† Let R_1 and R_2 be two regions occupied by plane or axisymmetrical irrotational flows and let S_1 and S_2 denote the respective streamlines $\psi = 0$. We assume $\psi \geqslant 0$ in each flow. Let S_1 and S_2 have an arc MN in common such that the direction of each flow on MN is from M to N. Further, suppose that the arcs QM of S_1 and NQ of S_2, having only the point Q in common, together with MN bound a region

$$MNQ = R_3$$

interior both to R_1 and R_2. Let M and N be regular boundary points, and let $q(M, R_1)$ denote the boundary velocity at M for the flow R_1, and so on. Then

$$\frac{q(M, R_1)}{q(N, R_1)} \leqslant \frac{q(M, R_2)}{q(N, R_2)}, \tag{4}$$

the equality holding if and only if the flows are geometrically similar.

* *Jour. Rational Mechanics and Analysis*. 1 (1952), 309–20.

† J. Serrin, *Jour. Rational Mechanics and Analysis*, 1 (1952), 563–72. Comparison theorems have been applied by Serrin and Gilbarg to prove various uniqueness theorems for plane and axisymmetrical flows.

Proof. Fig. 16·60 (ii) shows the case where Q is a finite point and the case where Q is at infinity.

If the flows are similar, the theorem is obvious with equality.

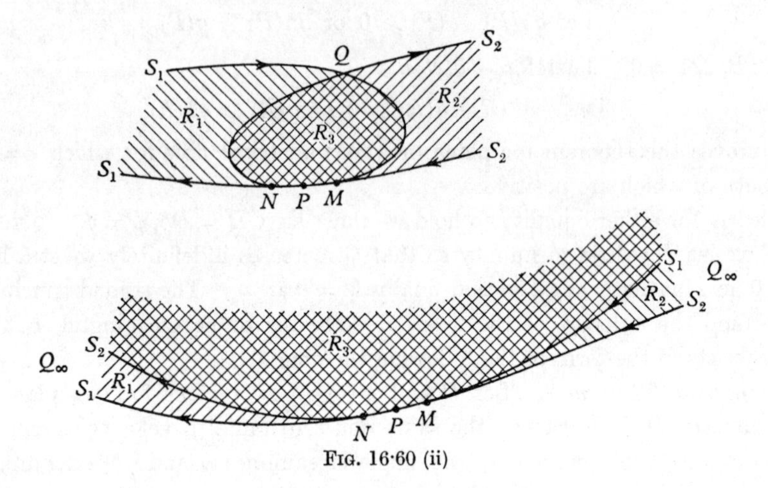

FIG. 16·60 (ii)

Let ψ_1, ψ_2 be the respective stream functions and let P be a point of MN. Let

$$\Omega_P = q(P, R_1)\,\psi_2 - q(P, R_2)\,\psi_1. \tag{5}$$

On QM, $\qquad \Omega_P = q(P, R_1)\,\psi_2 \geqslant 0\,;$

on MN, $\qquad \Omega_P = 0\,;$

on NQ, $\qquad \Omega_P = -q(P, R_2)\,\psi_1 \leqslant 0.$

If dn is an element of normal at P drawn into the flows,

$$\epsilon\frac{\partial \Omega_P}{\partial n} = q(P, R_1)\,\epsilon\frac{\partial \psi_2}{\partial n} - q(P, R_2)\,\epsilon\frac{\partial \psi_1}{\partial n}$$

$$= q(P, R_1)\,q(P, R_2) - q(P, R_2)\,q(P, R_1) = 0.$$

Hence a line C_P issues from P on which $\Omega_P = 0$, fig. 16·60 (iii).

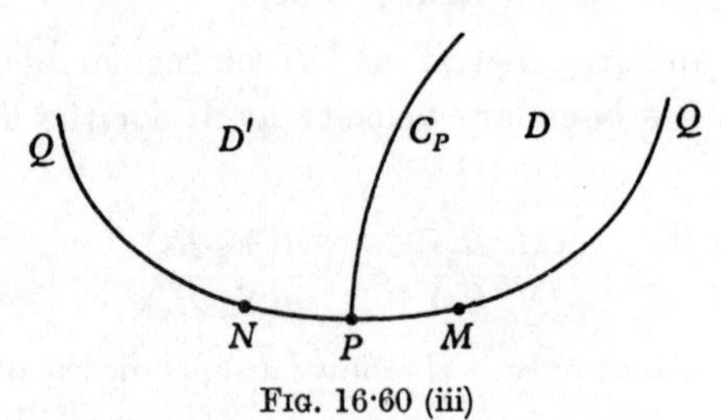

FIG. 16·60 (iii)

Consider the region D bounded by QM, MP, C_P.

On QM, $\qquad \Omega_P \geqslant 0\,;$

on MP, $\qquad \Omega_P = 0,$

on C_P, $\qquad \Omega_P = 0.$

By the maximum principle either $\Omega_P \equiv 0$ in D or $\Omega_P > 0$ in D. The latter must be the case.

Hence $(\epsilon\, \partial\Omega_P/\partial n)_M > 0$, i.e.

$$(6) \qquad q(P, R_1)\, q(M, R_2) - q(P, R_2)\, q(M, R_1) > 0.$$

Similarly in the region D' bounded by C_P, PN, NQ, we find that $\Omega_P < 0$ and $(\partial\Omega_P/\partial n)_N < 0$ or

$$(7) \qquad q(P, R_1)\, q(N, R_2) - q(P, R_2)\, q(N, R_1) < 0,$$

and from these inequalities, (6) and (7), the theorem follows at once. Q.E.D.

We leave the reader to prove that the level line C_P passes through Q.

EXAMPLES XVI

1. Construct graphically the streamlines for a source and equal sink in three dimensions.

2. A source of strength m is placed at the origin in a stream of incompressible fluid moving with velocity U in the direction of negative x. Find the equation of the surfaces of constant pressure, and trace roughly the shape of the meridian section of three such surfaces, corresponding to $p \gtreqless p_0$ respectively, where p_0 is the pressure at infinity.

In the case where $p = p_0 + \rho\kappa^4$ and $\kappa^4 < \frac{3}{8}U^2$, prove that the plane $x = 4\kappa m^{\frac{1}{2}}/6^{\frac{3}{4}} \,.\, U$ touches the surface of constant pressure along a circle, and find the radius of this circle. (U.L.)

3. If AB be a uniform line source, and A, B equal sinks of such strength that there is no total gain or loss of fluid, show that

$$\psi = C\{(r_1 - r_2)^2 - c^2\}\left(\frac{1}{r_1} - \frac{1}{r_2}\right),$$

where $c = AB$, r_1 and r_2 are distances from A, B respectively, and C is a constant depending on the strengths of the sources. (R.N.C.)

4. Two sources of strengths m, m' are placed at two points, A, B respectively, in an infinite stream of velocity V parallel to AB.

Obtain the equations of the streamlines in the form

$$m \cos\theta + m' \cos\theta' - V\varpi^2/2 = \text{constant},$$

where (r, θ), (r', θ') are bipolar coordinates referred to A, B as poles, and AB as initial line, and ϖ is the perpendicular distance of any point from AB.

Show that the main stream, the stream issuing from A, and that issuing from B, are separated by the loci

$$m\Big/\left(r \sin\frac{\theta}{2}\right)^2 + m'\Big/\left(r' \sin\frac{\theta'}{2}\right)^2 = V,$$

and

$$-m\Big/\left(r \cos\frac{\theta}{2}\right)^2 + m'\Big/\left(r' \sin\frac{\theta'}{2}\right)^2 = V.$$

(R.N.C.)

5. A and B are a simple source and sink of strengths μ and μ' respectively, in an infinite liquid. Show that the equation of the streamlines is

$$\mu \cos\theta - \mu' \cos\theta' = \text{constant},$$

where θ, θ' are the angles which AP, BP make with AB, P being any point.

Prove also that, if $\mu > \mu'$, the cone defined by the equation

$$\cos\theta = 1 - 2\mu'/\mu,$$

divides the streamlines issuing from A into two sets, one extending to infinity, and the other terminating at B. (R.N.C.)

6. Prove that, if O, C_1, C_2 are points on the axis of x, such that $OC_1 = c_1$, $OC_2 = c_2$, and $c_1 c_2 = a^2$, the function

$$\psi = m\left[\frac{r_2 - r}{a} + \frac{a}{c_1}\frac{x - c_2}{r_2} + \frac{x - c_1}{r_1}\right],$$

where r, r_1, r_2 are the distances of any point from O, C_1, C_2 respectively, and O is the origin, gives the motion of liquid due to a simple source of strength m at C_1, in the presence of a fixed sphere $r = a$. (R.N.C.)

7. Find an expression for the potential due to a continuous distribution of sources and sinks along the axis of x in a perfect fluid.

If the distribution is of constant strength s from $x = 0$ to $x = a$, show that the equipotential surfaces are ellipsoids of revolution with foci at the two ends of the line.

If, in addition to the above, there is a sink of total equal strength at the origin, and a steady streaming with velocity V at infinity parallel to the axis of x, show that there is a closed stream surface of revolution of airship shape whose total length is the difference of the roots of

$$x^3 \pm x^2 a = sa^2/V. \quad \text{(U.L.)}$$

8. Interpret the motions for which

$$\text{(i)}\ \psi = C\varpi^2\left(\frac{a^3}{r^3} - 1\right),$$

$$\text{(ii)}\ \psi = C\varpi^2\left(\frac{a^3}{r^3} - \frac{b^3}{r'^3}\right),$$

where r and r' are measured from two fixed points O, O' in Ox.

9. If there is a doublet at the origin of strength μ in the direction of the unit vector $\mathbf{a}_1$, prove that its velocity potential is

$$\phi = \mu(\mathbf{a}_1\, \nabla)\left(\frac{1}{r}\right),$$

where $\mathbf{r}$ is the position vector of the point at which ϕ is calculated. Interpret the expression

$$\mu_1 \mu_2 (\mathbf{a}_1\, \nabla)(\mathbf{b}_1\, \nabla)\left(\frac{1}{r}\right).$$

10. Determine ψ when the velocity function is

$$x\left\{2 + \frac{1}{(x^2 + \varpi^2)^{\frac{3}{2}}}\right\}. \quad \text{(R.N.C.)}$$

11. Verify that

$$\psi = \left(\frac{A}{r^2}\cos\theta + Br^2\right)\sin^2\theta$$

is a possible form of Stokes' stream function, and find the corresponding velocity potential. (U.L.)

12. A spherical mass of liquid, of radius b and density ρ, has a concentric spherical cavity of radius a, which contains gas at pressure p whose mass may be neglected. The liquid is at rest when an impulsive pressure ϖ is applied to the external boundary. Show that the initial kinetic energy generated is

$$2\pi\varpi^2 ab/\rho\,(b-a).$$

If, during the subsequent motion, the gas obeys Boyle's law and there is no pressure on the external boundary, find the radius of the cavity when the liquid first comes to rest. (R.N.C.)

13. A mass of fluid of density ρ is bounded by two concentric spherical free surfaces of radii r_1 and r_2, and, the fluid being at rest, impulsive pressures ϖ_1 and ϖ_2 are applied to these surfaces. Show that the surfaces begin to move with velocities

$$\frac{\varpi_1-\varpi_2}{\rho(r_2-r_1)}\frac{r_2}{r_1}, \quad \text{and} \quad \frac{\varpi_1-\varpi_2}{\rho(r_2-r_1)}\frac{r_1}{r_2}. \qquad \text{(R.N.C.)}$$

14. A mass of fluid of density ρ and volume $4\pi c^3/3$ is in the form of a spherical shell. There is a constant pressure p on the external surface, and zero pressure on the internal surface. Initially the fluid is at rest, and the external radius is $2nc$. Show that when the external radius becomes nc, the velocity U of the external surface is given by

$$U^2 = \frac{14p}{3\rho}\frac{(n^3-1)^{\frac{1}{3}}}{n-(n^3-1)^{\frac{1}{3}}}. \qquad \text{(R.N.C.)}$$

15. A mass of fluid of density ρ and volume $4\pi c^3/3$ is in the form of a spherical shell. A constant pressure p_0 is exerted on the external surface of the shell. There is no pressure on the internal surface and no other forces act on the liquid. Initially the liquid is at rest and the internal radius of the shell is $2c$. Prove that the velocity of the internal surface when its radius is c is

$$\left(\frac{14p_0}{3\rho}\frac{2^{\frac{1}{3}}}{2^{\frac{1}{3}}-1}\right)^{\frac{1}{2}}.$$

16. An infinite mass of liquid is at rest subject to a uniform pressure p_0 and contains a spherical cavity of radius a filled with gas at pressure mp_0. Prove that if the inertia of the gas be neglected and Boyle's law be supposed to hold throughout the motion, the radius of the sphere will oscillate between a and na, where n is determined by

$$1+3m\log n-n^3 = 0.$$

If ρ be the density of the fluid, and m be nearly equal to unity, show that the time of oscillation will be $2\pi(a^2\rho/(3p_0))^{\frac{1}{2}}$.

17. A quantity of fluid, self-attracting according to the law of gravitation, surrounds a solid sphere of radius a, the radius of the external free surface being b. The solid sphere is suddenly annihilated. Show that, when the radius of the inner surface is r, the square of the velocity of any point of it is

$$\frac{kR[3(r^5-a^5)-5(r^3R^2-a^3b^2)+2(R^5-b^5)]}{r^3(R-r)},$$

where $R^3 = r^3+b^3-a^3$ and k is a constant.

18. A volume of gravitating liquid is initially at rest in the form of a spherical shell of very great radius and contracts under its own attraction, there being no pressure on either surface of the shell. Prove that, when the inner radius is x,

$$x^3\left(\frac{dx}{dt}\right)^2 = \frac{4\pi\gamma\rho}{15}\,y(y-x)(3x^3+6x^2y+4xy^2+2y^3),$$

where $y^3 = x^3+c^3$, γ is the constant of gravitation and ρ and $4\pi c^3/3$ are the density and volume of the liquid.

19. A sphere is moving forward in a straight line with velocity U. Find the force required, by direct calculation of the resultant thrust of the fluid.

20. A sphere is projected under gravity at a great depth with velocity U at inclination 45° to the horizontal. If the density of the sphere be twice the density of the liquid, prove that the greatest height above the point of projection attained by the sphere is $5U^2/(8g)$. (R.N.C.)

21. A sphere of radius a is placed in an infinite stream of liquid flowing with uniform velocity V. Show that the streamlines are given by the equation

$$(a^3-r^3)\sin^2\theta/r = \text{constant}.$$

If the sphere is divided into two parts by a diametral plane perpendicular to the direction of motion of the stream, show that the resultant force between the two parts is less than it would be if the liquid were at rest, the pressure at infinity remaining the same, by an amount $\pi\rho a^2V^2/16$.

22. A sphere of radius a is moving with constant velocity V through an infinite liquid at rest at infinity. If p_0 is the pressure at infinity, prove that the pressure p at any point P distant r from the centre O of the sphere, and such that OP makes an angle θ with the velocity of the sphere, is given by

$$p = p_0 - \rho\frac{V^2a^3}{2r^3}\left[\left(1+\frac{a^3}{4r^3}\right) - \frac{3x^2}{r^2}\left(1-\frac{a^3}{4r^3}\right)\right].$$

Show further that if V exceed $\sqrt{8p_0/5\rho}$, a hollow ring is formed in the liquid round the equator of the sphere. (R.N.C.)

23. A sphere of radius a is moving in an infinite liquid with variable speed V in the direction of the axis of x. Show that the pressure at the surface of the sphere is least over the small circle

$$x = -\frac{2a^2}{9V^2}\frac{dV}{dt},$$

the centre of the sphere being the origin. (R.N.C.)

24. Obtain the solution for the irrotational motion of incompressible liquid in which a sphere of radius a is moving with velocity U.

Find the equation of the streamlines in this motion, and prove that the equation to the path of a particle relative to the centre of the moving sphere is

$$r^2\sin^2\theta\left(1-\frac{a^3}{r^3}\right) = b^2,$$

where b is a constant depending on the particle.

Explain why this equation is not identical with that of a streamline; and show that the position of the particle on its path is expressed in terms of the time by the equation

$$Ut = -\int\left\{\left(1-\frac{a^3}{r^3}\right)\left(1-\frac{a^3}{r^3}-\frac{b^2}{r^2}\right)\right\}^{-\frac{1}{2}}dr.$$

(R.N.C.)

25. A sphere of radius a moves with uniform velocity V in an infinite liquid. Find the velocity potential and show that the equation to the path of a particle in the fluid is obtained by eliminating r, θ between the equations

$$x = r\cos\theta - \int \frac{r^3\,dr}{\sqrt{(r^3-a^3)(r^3-a^3-rc^2)}},$$

$$y = r\sin\theta, \quad r^3 - a^3 = rc^2 \operatorname{cosec}^2\theta,$$

where c is an arbitrary constant. (R.N.C.)

26. A sphere of radius a is fixed in a liquid which is flowing past it in such a manner that at a great distance from the sphere the velocity is constant. A coloured particle of fluid is inserted upstream at a point which lies on the axis of the system and its motion is observed. If, while the particle is upstream, its distance from the centre changes from z_1 to z_2 in time T, show that the maximum velocity of slip on the sphere is

$$\frac{3}{2T}\left[z_1 - z_2 - \frac{a}{6}\log\frac{(z_1^3-a^3)(z_2-a)^3}{(z_2^3-a^3)(z_1-a)^3} - \frac{a}{\sqrt{3}}\tan^{-1}\frac{a(z_1-z_2)\sqrt{3}}{2(z_1 z_2+a^2)+a(z_1+z_2)}\right].$$

(M.T.)

27. A stream of water of great depth is flowing with uniform speed V over a plane level bottom. A hemisphere of weight w in water and of radius a, rests with its base on the bottom. Prove that the average pressure between the base of the hemisphere and the bottom is less than the fluid pressure at any point of the bottom at a great distance from the hemisphere, if $V^2 > 32w/11\pi a^2\rho$.

28. A uniform sphere of mass M floats half-immersed in liquid of unlimited extent and depth, under gravity. If a velocity U vertically downward is suddenly impressed upon the sphere, show that the required impulse $= 3MU/2$. Prove that the upward velocity of the fluid in contact with the sphere at the free surface $= \frac{1}{2}U$. (R.N.C.)

29. A sphere of radius a is moving with constant velocity U through an infinite liquid at rest at infinity. If p_0 is the pressure at infinity, show that the pressure p at any point of the surface of the sphere, the radius to which point makes an angle θ with the direction of motion, is given by

$$p = p_0 + \tfrac{1}{2}\rho U^2(1 - \tfrac{9}{4}\sin^2\theta).$$

If the sphere be divided into two hemispheres by a plane inclined at an angle α to the direction of motion, show that the normal and tangential components of the reaction between the two hemispheres, due to the fluid pressures, are

$$\pi a^2\left[p_0 - \rho U^2\frac{(11 - 9\sin^2\alpha)}{32}\right] \text{ and } \tfrac{9}{16}\pi\rho a^2 U^2 \sin\alpha\cos\alpha \text{ respectively.}$$ (U.L.)

30. If two doublets of strengths μ, μ' have a common axis, show that one of the stream sheets is a sphere.

31. Find the stream function ψ for a double source at O inside a fixed sphere of radius a whose centre is on the axis of the double source at a distance c from O.
Calculate the pressure at any point of the sphere in this case. (R.N.C.)

32. A double source of unit strength with its axis parallel to the axis of x is placed at the point $(0, 0, c)$ outside a fixed sphere of radius a, having its centre at the origin, and immersed in liquid which is otherwise unbounded. Prove that near the sphere the velocity potential Φ due to the double source and its image is

$$\sum_{n=2}^{\infty} \frac{1}{c^{n+1}}\left(r^{n-1}+\frac{n-1}{n}\frac{a^{2n-1}}{r^n}\right)\sin\theta\frac{dP_{n-1}(\mu)}{d\mu}\cos\phi,$$

where r, θ, ϕ are spherical polar coordinates, $\mu = \cos\theta$, and $P_{n-1}(\mu)$ is the zonal surface harmonic of degree $(n-1)$.

[The theorem $\frac{\mu dP_n(\mu)}{d\mu} - nP_n(\mu) = \frac{dP_{n-1}(\mu)}{d\mu}$ may be assumed, if necessary.]

Verity the result that the x-component of the velocity of the liquid at $(0, 0, c)$, due to the sphere moving with a given velocity (in the absence of the double source), is

$$-\frac{1}{4\pi}\iint \Phi q_\nu dS,$$

where q_ν is the normal velocity at an element dS of the surface of the sphere, and the integral is taken over the surface of the sphere. (U.L.)

33. If a double source S is placed in the presence of a fixed sphere of radius a whose centre O is distant c from S, find the stream function and show that the speed at a point P on the surface of the sphere is

$$\{3m(c^2-a^2)\sin\Theta\}/r^5,$$

where $r = SP$ and Θ = angle SOP.

Prove that the pressures on the sphere have a resultant

$$24m^2\rho\pi a^3 c/(c^2-a^2)^4$$

towards the double source. (U.L.)

34. Determine the hydrodynamical image with respect to a sphere of a doublet whose axis passes through the centre of the sphere.

Prove that, if the distance of the doublet from the centre is great compared with the radius, the resultant thrust on the sphere is approximately proportional to the inverse seventh power of the distance.

35. A double source of strength μ is placed at the centre of a fixed hollow sphere of radius a, which is filled with incompressible inviscid fluid. Show how to obtain the pressure at any point, given the pressure p_0 at the point A of the sphere which lies on the axis of the double source, and show that the equation to one of the surfaces of equal pressure is

$$(r/a)^3 = (1+\tfrac{1}{4}\tan^2\theta)/(2-\tan^2\theta). \qquad \text{(R.N.C.)}$$

36. A solid is bounded by the exterior portions of two equal spheres of radius a which cut one another orthogonally and is surrounded by infinite fluid. If the solid is set in motion with velocity u in the direction of the line of centres, show that the velocity potential of the resulting motion is

$$\tfrac{1}{2}a^3u\left(\frac{\cos\theta}{r^2}+\frac{\cos\theta'}{r'^2}-\frac{\cos\Theta}{2\sqrt{2}R^2}\right),$$

where r, r', R are the radii vectores of a point measured respectively from the centres of the two spheres and the point midway between them, and θ, θ', Θ are the angles which these radii vectores make with the direction of motion of the solid.

37. Find the velocity potential due to a simple source outside a fixed sphere, in an unlimited frictionless liquid.

Prove that the sphere is apparently attracted towards the source and that, when the radius is small compared with the distance of the source, the attraction varies, to the first approximation, inversely as the fifth power of the distance.

38. Prove that the velocity potential due to the image of a source of strength m in a sphere of radius a is the same as that due to a distribution of doublets over the surface of the sphere, the axes being normal to the surface and the strength per unit area being

$$m\left(\frac{2a}{cR}-\frac{1}{a}\log\frac{a+R+a^2/c}{a+R-a^2/c}\right),$$

where c is the distance of the source from the centre and R is the distance from the inverse point.

39. A source of strength m is situated in fluid, bounded internally by a fixed sphere of radius a, at a distance c from the centre of the sphere. Prove that the velocity potential at a point on the surface is

$$\frac{2m}{r}-\frac{m}{a}\log\frac{r+c+a}{r+c-a},$$

r being the distance of the point from the source.

Find the magnitude of the velocity at any point on the surface. (U.L.)

40. Define Stokes' stream function for motion of incompressible fluid symmetrical about an axis; show that the following are possible Stokes' functions, and give their interpretation; $r-r'$ and $\cos\theta$, where $r = OP$, $r' = O'P$, $\theta = POO'$; O, O' being any two fixed points on the axis of symmetry.

Prove that $$\psi = \mu\left\{\cos\theta+\frac{a}{c}\cos\theta'+\frac{r'-R}{a}\right\}$$

gives the motion due to a simple source S of strength μ placed at a distance c from the centre of a fixed sphere of radius a, R being measured from the centre of the sphere, (r, θ) from S, and (r', θ') from the inverse point of S with respect to the sphere. (U.L.)

41. A source and a sink of equal strengths are placed at the points $(0, 0, \pm c)$ *inside* a sphere of radius a with its centre at the point $(0, 0, 0)$. Find an expression for the velocity potential at points within the sphere. (U.L.)

42. Find the image of a source with regard to a sphere. O is the centre, P, Q are points outside the sphere on the same radius, Q being nearer the sphere, and P', Q' are their inverse points. Prove that a source of strength μ at Q and one of strength $\mu a/OQ$ at Q' produce the same radial flow at every part of the surface of the sphere as a line source uniformly distributed along QP of total strength μ, together with a line source uniformly distributed along $P'Q'$ of total strength $\mu a/OQ$.

43. A solid sphere of radius a oscillates in an infinite liquid with simple harmonic displacement $c\cos pt$, where c is small. Determine the direction and magnitude of the resulting oscillation at any given point of the fluid. (R.N.C.)

44. The centre of a sphere of radius a, in an unbounded liquid, performs small linear oscillations, the displacement at time t being $c\sin nt$. Prove that the mean kinetic energy per unit volume of the fluid at the point (r, θ), referred to the centre of the sphere and its line of motion, is

$$\frac{\rho c^2 n^2 a^6}{32r^6}(5+3\cos 2\theta).$$

Calculate the periodic force necessary to maintain the motion. (R.N.C.)

45. A sphere of variable radius a moves through an infinite fluid with a variable velocity v in a fixed direction. Find the pressure at any point on its surface, and show that the resultant thrust of the fluid on the sphere is

$$\tfrac{2}{3}\pi\rho a^2(a\dot{v}+3v\dot{a}).$$

46. Fluid, extending to infinity, surrounds a spherical boundary whose radius at time t is $a+b\sin nt$, the centre being fixed. If there are no external forces, show that the pressure at the boundary is

$$p_0+\tfrac{1}{4}\rho\, bn^2(5b\cos 2nt-4a\sin nt+b),$$

where p_0 is the pressure at infinity. (R.N.C.)

47. A sphere of radius a is surrounded by infinite liquid which is at rest at a very great distance from the sphere, under a pressure p_0. If the sphere is made to vibrate radially so that the radius at any time is $a+b\cos nt$, and there are no body forces, find the pressure on the surface of the sphere at any time, and show that its least value is

$$p_0-n^2\rho b(a+b).$$

48. Show that the motion set up by impulsive pressures applied to the boundary of a liquid is irrotational.

A spherical bubble of steam in a large mass of water of density ρ under no body forces suddenly loses practically all its internal pressure by condensation of the steam. If at this instant its radius is a, show that the bubble collapses in a time

$$a\sqrt{\frac{\rho}{6p_0}}\,\frac{\Gamma(\frac{5}{6})\Gamma(\frac{1}{2})}{\Gamma(\frac{4}{3})},$$

p_0 being the pressure at a great distance, and that the energy dissipated is p_0 (original volume of bubble). (U.L.)

49. An infinite mass of liquid fills up the region outside a fixed sphere of radius a and is attracted towards the centre of the sphere by a force μ/r^2 per unit mass. If the pressure at infinity is ϖ and the sphere is suddenly annihilated, show that the instantaneous change in the pressure at a distance r is $(\varpi a+\mu\rho)/r$.

Find the velocity of the inner boundary of the liquid at any subsequent time and in the particular case where the pressure at infinity is zero, find the time taken to fill up the cavity. (U.L.)

50. An infinite liquid of uniform density ρ at rest under uniform pressure P contains a spherieal bubble of radius a_0, full of vapour, which carries an electric charge e uniformly distributed over the surface. Assuming that this charge always remains the same and produces an outward thrust on unit area of the surface of amount $e^2/8\pi a^4$ when the radius is a, and that the vapour suddenly condenses so that the internal pressure falls to zero, find the pressure at a distance r from the centre of the bubble when its radius is a and prove that then

$$a^4\dot{a}^2 = \frac{2P}{3\rho}(a_0^3 - a^3)a - \frac{e^2}{4\pi\rho a_0}(a_0 - a).$$

Hence show that, if $3e^2/8\pi a_0 P = a_1^3 + a_1^2 a_0 + a_1 a_0^2$, the radius of the bubble will oscillate between the limits a_0 and a_1 and find an expression for the period. (U.L.)

51. Obtain the differential equation of the surfaces which move so as always to be made up of the same fluid particles.

At time $t = 0$ a spherical bubble of gas of radius a is at rest within a large surrounding mass of heavy liquid of density ρ which is also at rest. The pressure of the gas is p_0 and the pressure of the liquid in the horizontal plane through the centre of the bubble is p. Prove that, in the initial motion, the radius of the bubble begins to increase with acceleration $(p_0 - p)/(\rho a)$, the centre of bubble begins to move upwards with acceleration $2g$, and the bubble remains approximately spherical, the inertia of the gas and the surface tension being neglected. (U.L.)

52. The equation of the meridian section of a surface of revolution is $r = a \sec \frac{1}{2}\theta$, where $0 \leqslant \theta \leqslant \pi$. The surface is placed in a steady stream of velocity U. Show that the stream function is

$$U\{\tfrac{1}{2}r^2 \sin^2\theta - a^2(1 - \cos\theta)\}$$

and find the velocity potential. (U.L.)

53. A light thin circular disc of radius c is at rest on the surface of still liquid of density ρ, of infinite extent and depth. A vertical downward impulse I is applied to the centre of the disc. Show that the velocity communicated to the disc is $3I/(4\rho c^3)$. (U.L.)

54. (Oblate) spheroidal coordinates ζ, μ, ω are obtained from cylindrical coordinates ϖ, x, ω by the transformations

$$\varpi + ix = k\sin(\theta + i\eta),$$
$$\zeta = \sinh\eta, \quad \mu = \cos\theta.$$

Obtain Laplace's equation in these coordinates in the form

$$\frac{\partial}{\partial\mu}\left\{(1-\mu^2)\frac{\partial\phi}{\partial\mu}\right\} + \frac{1}{1-\mu^2}\frac{\partial^2\phi}{\partial\omega^2} = -\frac{\partial}{\partial\zeta}\left\{(\zeta^2+1)\frac{\partial\phi}{\partial\zeta}\right\} + \frac{1}{\zeta^2+1}\frac{\partial^2\phi}{\partial\omega^2}.$$

Find the boundary conditions in terms of these coordinates when an oblate spheroid is moving along its axis with velocity U in an unbounded fluid otherwise at rest.

55. An airship in the shape of a prolate spheroid, of polar semi-axis a, equatorial semi-axis b, is moving with speed U parallel to its axis of revolution in air, which may be treated as an incompressible fluid. Find an expression for the velocity potential at any point of the fluid, and also for the pressure at any point of the envelope of the airship, the pressure at infinity being π_0.

56. Prove that when a spheroidal disc in which $a = b = 100c$ moves through water, with a velocity of 1 ft./sec., in the direction of its smallest axis, the speed at the rim is about 63 ft./sec.

57. Obtain the formula
$$-\tfrac{1}{2}\rho\iint \phi\frac{\partial\phi}{\partial n}\,dS$$

for the kinetic energy of a fluid, which is in a region bounded internally by a moving surface S, and is at rest at infinite distances; n being drawn into the fluid.

Prove that, if S is a prolate spheroid, of eccentricity $\tanh\alpha$, moving parallel to its axis of symmetry with velocity V, the kinetic energy of the fluid is

$$\tfrac{1}{2}M'V^2 \frac{\alpha - \tanh\alpha}{\sinh\alpha\cosh\alpha - \alpha},$$

where M' is the mass of the displaced fluid. (U.L.)

58. Show that in spheroidal coordinates (μ, ζ, ω) defined by the equations

$$x = a\mu\zeta, \quad y = \varpi\cos\omega, \quad z = \varpi\sin\omega,$$

$$\varpi = a(1-\mu^2)^{\frac{1}{2}}(\zeta^2+1)^{\frac{1}{2}},$$

the equation of continuity becomes

$$\frac{\partial}{\partial\mu}\left\{(1-\mu^2)\frac{\partial\phi}{\partial\mu}\right\} + \frac{1}{1-\mu^2}\frac{\partial^2\phi}{\partial\omega^2} = -\frac{\partial}{\partial\zeta}\left\{(\zeta^2+1)\frac{\partial\phi}{\partial\zeta}\right\} + \frac{1}{1+\zeta^2}\frac{\partial^2\phi}{\partial\omega^2}.$$

If a thin circular disc of radius a is moving with velocity U parallel to its axis in an infinite mass of liquid, prove that the velocity potential is

$$\phi = (2aU/\pi)\mu(1-\zeta\cot^{-1}\zeta),$$

and show that the kinetic energy of the liquid is $\frac{4}{3}\rho a^3 U^2$. (U.L.)

59. The space bounded by the paraboloids $x^2+y^2 = az$, $x^2+y^2 = b(z-c)$ [where a, b, c are positive and $b>a$] outside the former and inside the latter contains liquid at rest. Suddenly the bounding surfaces are made to move with speeds U, V respectively in the direction of the z-axis. Prove that in the motion instantaneously set up the surfaces over which the current function is constant are paraboloids of latus rectum $ab(U-V)/(aU-bV)$.

60. If $\varpi + ix = f(\xi + i\eta)$, show that the equation

$$\frac{\partial}{\partial\xi}\left(\frac{1}{\varpi}\frac{\partial\psi}{\partial\xi}\right) + \frac{\partial}{\partial\eta}\left(\frac{1}{\varpi}\frac{\partial\psi}{\partial\eta}\right) = 0$$

has a solution of the type $\psi = \varpi^{\frac{1}{2}}UV$ where U, V are respectively functions of ξ, η provided that

$$\left(\frac{\partial\varpi}{\partial\xi}\right)^2 + \left(\frac{\partial\varpi}{\partial\eta}\right)^2 = \varpi^2[f_1(\xi) + f_2(\eta)].$$

If $(\xi + i\eta)(\varpi + ix) = a$, show that there is a solution of the type

$$\psi = \xi(\xi^2+\eta^2)^{-\frac{1}{2}} e^{\pm n\eta} J_1(n\eta).$$

61. Use the definition in 16·29 to prove that for two functions $f(r, \theta)$, $g(r, \theta)$

$$(fg)^\star = \frac{a}{r}f^\star g^\star, \quad \left(\frac{f}{g}\right)^\star = \frac{r}{a}\frac{f^\star}{g^\star}, \quad \left(\frac{1}{f}\right)^\star = \frac{r^2}{a^2}\frac{1}{f^\star}.$$

CHAPTER XVII

SPHERES AND ELLIPSOIDS

17·0. The discussion of irrotational motion of a liquid in space when symmetry about an axis no longer exists, resolves itself into the determination of the velocity potential which satisfies given boundary conditions.

Apart from the boundary conditions the equation of continuity has to be satisfied, in other words the velocity potential must satisfy Laplace's equation $\nabla^2\phi = 0$. Solutions of this equation are called *harmonic functions*, concerning which a vast literature exists which it would be impossible even to summarise here. We shall merely investigate certain special types of solution of which immediate application will be made to the motion of two spheres and the ellipsoid.

17·1. Spherical harmonics. Laplace's equation in cartesian coordinates is

$$\frac{\partial^2\phi}{\partial x^2}+\frac{\partial^2\phi}{\partial y^2}+\frac{\partial^2\phi}{\partial z^2} = 0. \tag{1}$$

Any homogeneous solution of this equation is called a *spherical harmonic.* Obvious examples of solutions are 1, x, y, z, yz, x^2-y^2. If ϕ is a harmonic function such that $\phi = \phi_m+\phi_n$, where ϕ_m and ϕ_n are each homogeneous functions of x, y, z of different degrees m and n respectively, it is obvious that ϕ_m and ϕ_n are also spherical harmonics, for the results of operating on them with ∇^2 are also homogeneous functions of different degrees and therefore cannot cancel one another identically when added together. Expressed in polar coordinates r, θ, ω, the equation becomes, 2·72,

$$\frac{\partial}{\partial r}\left(r^2\frac{\partial\phi}{\partial r}\right)+\frac{1}{\sin\theta}\frac{\partial}{\partial\theta}\left(\sin\theta\frac{\partial\phi}{\partial\theta}\right)+\frac{1}{\sin^2\theta}\frac{\partial^2\phi}{\partial\omega^2} = 0. \tag{2}$$

The velocity potential of a simple source $\phi = m/r$ is a spherical harmonic, as is also immediately obvious by substitution in (2). If the source is at a point A on the axis of x distant c from the origin, fig. 17·1, we have $\phi = m/R$, where $R = AP$, and this, being a velocity potential, must satisfy Laplace's equation, for it was derived in 16·20 from the equation of continuity. Now

$$\begin{aligned} R^2 &= r^2+c^2-2cr\cos\theta \\ &= r^2\left(1-\frac{2c}{r}\cos\theta+\frac{c^2}{r^2}\right) = c^2\left(1-\frac{2r}{c}\cos\theta+\frac{r^2}{c^2}\right) \end{aligned}$$

If $|\lambda|<1$, we have an expansion of the form

$$(1-2\lambda\cos\theta+\lambda^2)^{-\frac{1}{2}} = 1+\lambda P_1(\cos\theta)+\lambda^2 P_2(\cos\theta)+\ldots,$$

where the coefficients $P_1(\cos\theta)$, $P_2(\cos\theta)$, ..., are independent of λ.

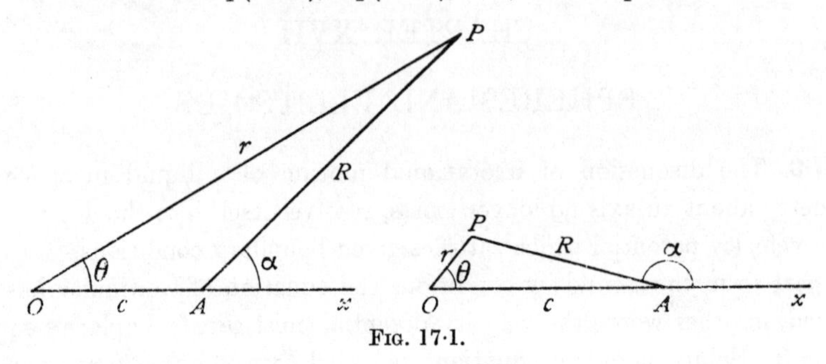

Fig. 17·1.

Thus, if $r<c$, putting $\lambda = r/c$, we have

$$\frac{1}{R} = \frac{1}{c}+\frac{r}{c^2}P_1(\cos\theta)+\frac{r^2}{c^3}P_2(\cos\theta)+\ldots,$$

while, if $r>c$,

$$\frac{1}{R} = \frac{1}{r}+\frac{c}{r^2}P_1(\cos\theta)+\frac{c^2}{r^3}P_2(\cos\theta)+\ldots.$$

Since the terms in $r, r^2, \ldots, r^{-1}, r^{-2}, \ldots$ of these expansions are homogeneous but of different degrees, as remarked above, each must be a spherical harmonic. Thus we have the two sets of spherical harmonics (ignoring the constant c),

$$1,\quad rP_1(\cos\theta),\quad r^2P_2(\cos\theta),\ldots,$$

$$\frac{1}{r},\quad \frac{P_1(\cos\theta)}{r^2},\quad \frac{P_2(\cos\theta)}{r^3},\ldots,$$

each of which satisfies Laplace's equation identically. It is easily proved by expanding by the binomial theorem that

$$P_1(\cos\theta) = \cos\theta,\quad P_2(\cos\theta) = \tfrac{1}{2}(3\cos^2\theta-1),$$

and so on. The functions $P_n(\cos\theta)$, $n = 1, 2, 3, \ldots$, are known as Legendre's functions or *zonal harmonics* (of the first kind). These functions are appropriate to problems dealing with spherical boundaries. Thus for streaming past a sphere, we have

$$\phi = U\left(r\cos\theta+\frac{a^3}{2r^2}\cos\theta\right)$$

$$= UrP_1(\cos\theta)+\tfrac{1}{2}Ua^3\frac{P_1(\cos\theta)}{r^2},$$

which involves the two spherical harmonics belonging to the zonal harmonic $P_1(\cos\theta)$.

In the case of a doublet of strength μ at A, we have, from 16·26,

$$\phi = \frac{\mu \cos \alpha}{R^2} = \frac{\mu(r \cos \theta - c)}{(r^2+c^2-2cr \cos \theta)^{\frac{3}{2}}} = \mu \frac{\partial}{\partial c} \frac{1}{(r^2+c^2-2cr \cos \theta)^{\frac{1}{2}}}.$$

Thus, if $r<c$,

$$\frac{\cos \alpha}{R^2} = -\left(\frac{1}{c^2}+\frac{2rP_1(\cos \theta)}{c^3}+\frac{3r^2P_2(\cos \theta)}{c^4}+\ldots\right),$$

while if $r>c$,

$$\frac{\cos \alpha}{R^2} = \frac{1}{r^2} P_1(\cos \theta)+\frac{2c}{r^3} P_2(\cos \theta)+\frac{3c^2}{r^4} P_3(\cos \theta)+\ldots .$$

These expansions give the velocity potential of a doublet in terms of the zonal harmonics.

We may add the following observation. If ϕ is a spherical harmonic, so are all its partial differential coefficients of any order with respect to x, y, z. Thus, for example, $\partial\phi/\partial x$ is a spherical harmonic, as is at once obvious by substitution in (1). From the spherical harmonic $1/r$, we derive in this manner the further harmonics

$$\frac{x}{r^3}, \quad \frac{y}{r^3}, \quad \frac{z}{r^3}.$$

17·12 Kelvin's inversion theorem. If $\phi = \phi(r, \theta, \omega)$ is a harmonic function, then $\phi^\star = \dfrac{a^2}{r} \phi\left(\dfrac{a^2}{r}, \theta, \omega\right)$ is also harmonic, where a is any constant.

Proof. Let $R = a^2/r$. Then $\phi^\star = R\phi(R, \theta, \omega)$. By hypothesis $\phi(r, \theta, \omega)$ satisfies Laplace's equation 17·1 (2), and therefore $\phi(R, \theta, \omega)$ satisfies the same equation with R written for r, namely

$$(1) \qquad \frac{\partial}{\partial R}\left(R^2 \frac{\partial \phi(R, \theta, \omega)}{\partial R}\right)+\frac{1}{\sin \theta}\frac{\partial}{\partial \theta}\left(\sin \theta \frac{\partial \phi(R, \theta, \omega)}{\partial \theta}\right) + \frac{1}{\sin^2 \theta}\frac{\partial^2 \phi(R, \theta, \omega)}{\partial \omega^2} = 0.$$

Now

$$r^2 \frac{\partial \phi^\star}{\partial r} = \left(-\frac{a^2}{r^2}\phi(R, \theta, \omega)-\frac{a^4}{r^3}\frac{\partial \phi(R, \theta, \omega)}{\partial R}\right) r^2,$$

and therefore

$$\frac{\partial}{\partial r}\left(r^2 \frac{\partial \phi^\star}{\partial r}\right) = \frac{2a^4}{r^2}\frac{\partial \phi(R, \theta, \omega)}{\partial R}+\frac{a^6}{r^3}\frac{\partial^2 \phi(R, \theta, \omega)}{\partial R^2} = R\frac{\partial}{\partial R}\left(R^2 \frac{\partial \phi(R, \theta, \omega)}{\partial R}\right).$$

Therefore

$$\frac{\partial}{\partial r}\left(r^2\frac{\partial\phi^\star}{\partial r}\right)+\frac{1}{\sin\theta}\frac{\partial}{\partial\theta}\left(\sin\theta\frac{\partial\phi^\star}{\partial\theta}\right)+\frac{1}{\sin^2\theta}\frac{\partial^2\phi^\star}{\partial\omega^2}$$

$$=R\left\{\frac{\partial}{\partial R}\left(R^2\frac{\partial\phi(R,\theta,\omega)}{\partial R}\right)+\frac{1}{\sin\theta}\frac{\partial}{\partial\theta}\left(\sin\theta\frac{\partial\phi(R,\theta,\omega)}{\partial\theta}\right)+\frac{1}{\sin^2\theta}\frac{\partial^2\phi(R,\theta,\omega)}{\partial\omega^2}\right\}=0.$$

Q.E.D.

Note that (r, θ, ω), $(a^2/r, \theta, \omega)$ are inverse points with respect to the sphere $r = a$, so if one is inside, the other is outside the sphere.

17·13. Weiss's sphere theorem.

The circle theorem of 6·21 has a general three-dimensional analogue, not confined to axisymmetrical motion, as follows.*

Weiss's sphere theorem. Let there be irrotational flow, in incompressible inviscid fluid with no rigid boundaries, characterised by the velocity potential $\phi(r, \theta, \omega)$, all of whose singularities are at a distance greater than a from the origin. If the sphere $r = a$ be introduced into the flow, the velocity potential becomes

$$(1) \qquad \Phi(r,\theta,\omega)=\phi(r,\theta,\omega)+\frac{1}{a}\int_0^{a^2/r}R\,\frac{\partial\phi(R,\theta,\omega)}{\partial R}\,dR.$$

Proof. Let the velocity potential after the introduction of the sphere become $\phi(r, \theta, \omega)+\chi(r, \theta, \omega)$, so that $\chi(r, \theta, \omega)$ is the perturbation velocity potential due to the introduction of the sphere. The conditions to be satisfied are, satisfaction of Laplace's equation, no perturbation at infinity, zero normal velocity at the sphere. More precisely

(i) $\nabla^2\chi = 0$, and χ has no singularities outside the sphere $r = a$.

(ii) $\chi(r, \theta, \omega) = O\left(\dfrac{1}{r^2}\right)$ for large r.

(iii) $\left(\dfrac{\partial\Phi}{\partial r}\right)_{r=a} = 0.$

Taking χ to be defined by (1) it follows from 17·12 that $\nabla^2\chi = 0$, and also since all the singularities of ϕ are external to the sphere, all those of χ are internal to it, since the exterior inverts into the interior. Thus (i) is satisfied.

Again, since ϕ is by hypothesis regular near the origin, we have there a series expansion of the form

$$\phi(r,\theta,\omega)=A_0+A_1r+A_2r^2+\dots,$$

where A_0, A_1, A_2 are independent of r

* P. Weiss, *Proc. Camb. Phil. Soc.*, 40 (1945).

Inserting this in the expression for χ we find easily that the terms of the lowest order is

$$\tfrac{1}{2} A_1 \frac{a^3}{r^2}$$

which shows that (ii) is satisfied.

To verify condition (iii) we have

$$\frac{\partial \Phi}{\partial r} = \frac{\partial \phi(r, \theta, \omega)}{\partial r} + \frac{1}{a}\left(-\frac{a^2}{r^2}\right)\left(R \frac{\partial \phi(R, \theta, \omega)}{\partial R}\right)_{R=a^2/r}$$

which vanishes when $r = a$, for then $R = a$, so that condition (iii) is satisfied.

Q.E.D.

We observe that the application of this theorem is not restricted to axisymmetrical motions.

17·20. Concentric spheres. The region between a solid sphere of radius a and a concentric spherical envelope of internal radius b is filled with liquid.

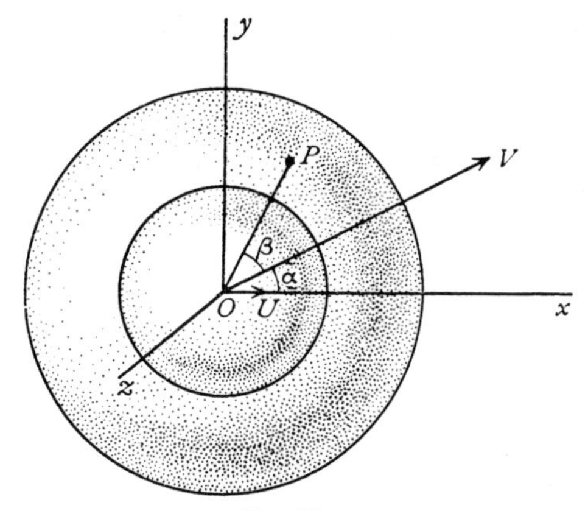

FIG. 17·20.

Impulses are applied to the sphere and envelope, thereby causing the sphere to start into motion with velocity U and the envelope with velocity V at an angle α to the direction of U. To discuss the *initial* motion (the spheres will only be concentric initially) take the direction of U as x-axis, the origin being at the common centre. The boundary conditions are then as follows:

(1) $$-\frac{\partial \phi}{\partial r} = U \cos \theta, \text{ when } r = a;$$

(2) $$-\frac{\partial \phi}{\partial r} = V \cos \beta, \text{ when } r = b,$$

where β is the angle between the direction of V and OP, P being any point on the envelope.

The cartesian coordinates of P will be $b\cos\theta$, $b\sin\theta\cos\omega$, $b\sin\theta\sin\omega$, and therefore the unit vector in the direction of OP is

$$\mathbf{i}\cos\theta+\mathbf{j}\sin\theta\cos\omega+\mathbf{k}\sin\theta\sin\omega.$$

If we take the direction of V to be in the x, y plane, the unit vector in the direction of V will be

$$\mathbf{i}\cos\alpha+\mathbf{j}\sin\alpha,$$

where $\mathbf{i}$, $\mathbf{j}$, $\mathbf{k}$ are unit vectors along the axes.

The scalar product of these vectors gives (2·11)

$$\cos\beta = \cos\alpha\cos\theta+\sin\alpha\sin\theta\cos\omega = \frac{x}{r}\cos\alpha+\frac{y}{r}\sin\alpha.$$

Boundary condition (1) therefore suggests that ϕ will involve the harmonics x, x/r^3, while (2) suggests further the harmonics y, y/r^3.

Hence we assume

$$\phi = Ax+\frac{Bx}{r^3}+Cy+\frac{Dy}{r^3},$$

or, reverting to polar coordinates,

$$\phi = \left(Ar+\frac{B}{r^2}\right)\cos\theta+\left(Cr+\frac{D}{r^2}\right)\sin\theta\cos\omega,$$

which gives

$$-\frac{\partial\phi}{\partial r} = \left(-A+\frac{2B}{r^3}\right)\cos\theta+\left(-C+\frac{2D}{r^3}\right)\sin\theta\cos\omega.$$

Equations (1) and (2) then give

$$\left(-A+\frac{2B}{a^3}\right)\cos\theta+\left(-C+\frac{2D}{a^3}\right)\sin\theta\cos\omega = U\cos\theta,$$

$$\left(-A+\frac{2B}{b^3}\right)\cos\theta+\left(-C+\frac{2D}{b^3}\right)\sin\theta\cos\omega = V(\cos\alpha\cos\theta+\sin\alpha\sin\theta\cos\omega).$$

These equations must be satisfied for all values of θ and ω and therefore *

$$-A+\frac{2B}{a^3} = U, \quad -C+\frac{2D}{a^3} = 0,$$

$$-A+\frac{2B}{b^3} = V\cos\alpha, \quad -C+\frac{2D}{b^3} = V\sin\alpha.$$

---

* These equations can be deduced by putting θ in turn equal to 0 and $\pi/2$.

Writing, for brevity, $c^3 = b^3 - a^3$, we get at once

$$B = \frac{a^3 b^3}{2c^3}(U - V\cos\alpha), \quad A = \frac{a^3 U - b^3 V\cos\alpha}{c^3},$$

$$D = -\frac{a^3 b^3}{2c^3} V\sin\alpha, \qquad C = -\frac{b^3 V\sin\alpha}{c^3},$$

and therefore

$$\phi = \frac{\cos\theta}{c^3}\left[(a^3 U - b^3 V\cos\alpha)r + \frac{a^3 b^3}{2r^2}(U - V\cos\alpha)\right]$$
$$- \frac{b^3 V\sin\alpha}{c^3}\left[r + \frac{a^3}{2r^2}\right]\sin\theta\cos\omega.$$

It must be emphasised that this velocity potential only represents the motion at the instant the boundaries are concentric.

17·21. Concentric spheres moving in the same direction. In the result of 17·20, if $\alpha = 0$, U and V are in the same direction, and then

$$\phi = \frac{\cos\theta}{c^3}\left[(a^3 U - b^3 V)r + \frac{a^3 b^3 (U - V)}{2r^2}\right].$$

In this case the impulsive pressures on the boundaries when the motion is started from rest, namely $\rho\phi$ (see 3·64), are

$$\varpi_1 = \frac{a\cos\theta}{c^3}\left[\left(a^3 + \frac{b^3}{2}\right)U - \frac{3b^3}{2}V\right]\rho \text{ on the inner,}$$

$$\varpi_2 = \frac{b\cos\theta}{c^3}\left[\frac{3a^3}{2}U - \left(\frac{a^3}{2} + b^3\right)V\right]\rho \text{ on the outer.}$$

The impulsive thrust on the inner boundary is therefore

$$I_1 = \int_0^\pi \varpi_1 \cos\theta\, 2\pi a^2 \sin\theta\, d\theta$$
$$= \tfrac{4}{3}\pi a^3 \rho\left[\left(a^3 + \frac{b^3}{2}\right)U - \frac{3b^3}{2}V\right]\Big/ c^3.$$

Similarly, on the outer boundary the impulsive thrust is

$$I_2 = \tfrac{4}{3}\pi b^3 \rho\left[\frac{3a^3}{2}U - \left(\frac{a^3}{2} + b^3\right)V\right]\Big/ c^3,$$

the directions of these thrusts being shown in fig. 17·21.

If we reverse the impulsive thrusts I_1 and I_2, we get the impulsive thrusts exerted *on* the liquid *by* the boundaries, and their resultant is equal to the momentum set up in the liquid, which is therefore, in the direction of Ox,

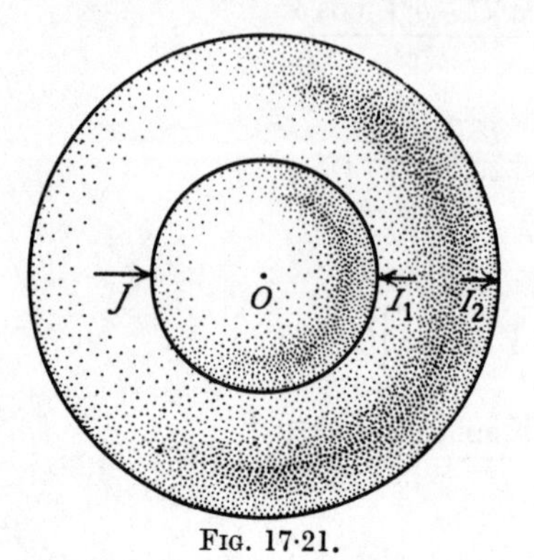

FIG. 17·21.

$$I_1 - I_2 = \tfrac{4}{3}\pi\rho(b^3 V - a^3 U) = M_2' V - M_1' U,$$

where M_1', M_2' are respectively the masses of the liquid which the inner and outer boundaries could contain. This result is true not only of spheres but of any two surfaces moving in any manner. For the momentum of the liquid is independent of the density of the inner body. Thus if we imagine the inner body to have the same density as the liquid, the centre of mass O of this body and the liquid is fixed with respect to the envelope and moves with the velocity $\mathbf{V}$ of the envelope. Therefore the total momentum of the liquid and inner body is $M_2'\mathbf{V}$. Thus the momentum of the liquid alone is $M_2'\mathbf{V} - M_1'\mathbf{U}$.

To find the impulse J required to start the inner sphere, we have, by the principle of momentum,

$$J - I_1 = M_1 U,$$

where M_1 is the mass of the sphere.

17·22. If the outer envelope is at rest, $V = 0$, and then

$$J = \left(M_1 + M_1' \frac{2a^3 + b^3}{2b^3 - 2a^3}\right) U,$$

so that the apparent added mass of the sphere, when a fixed outer boundary is present, is

$$M_1' \frac{2a^3 + b^3}{2b^3 - 2a^3},$$

which tends to $\tfrac{1}{2}M_1'$ when $b \to \infty$ (cf. 16·32).

The kinetic energy of the liquid can be found by integration or inferred directly from the added mass in the form

$$T = \tfrac{1}{2}M_1' \frac{2a^3 + b^3}{2b^3 - 2a^3} U^2.$$

If the outer envelope is at rest and the inner sphere is accelerated from rest by a force F which produces the acceleration f, in time δt the impulse is $J = F\,\delta t$, and the velocity is $U = f\,\delta t$, and therefore

$$F = \left(M_1 + M_1' \frac{2a^3+b^3}{2b^3-2a^3}\right) f.$$

17·30. Two spheres moving in the line of centres. Consider two spheres, centres A, B, radii a, b, moving towards one another with velocities U, V, fig. 17·30.

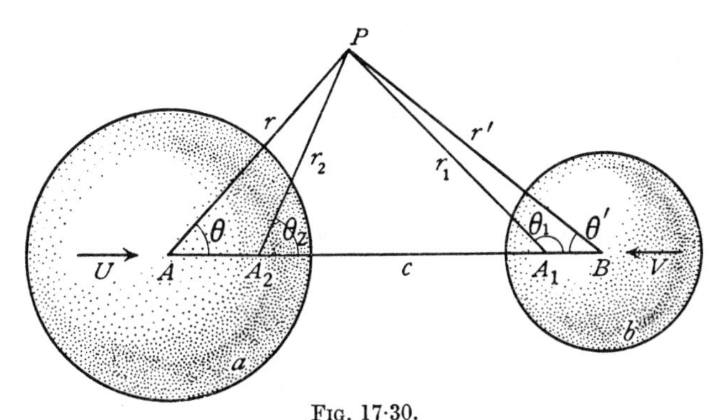

FIG. 17·30.

The position of a point P is fixed in a meridian plane by its polar coordinates (r, θ) referred to A and (r', θ') referred to B. The velocity potential ϕ must satisfy the boundary conditions

$$-\left(\frac{\partial \phi}{\partial r}\right)_{r=a} = U \cos \theta, \quad -\left(\frac{\partial \phi}{\partial r'}\right)_{r'=b} = V \cos \theta',$$

and therefore we can write

$$(1) \qquad \phi = U\phi_1 + V\phi_2,$$

where ϕ_1 and ϕ_2 each satisfy Laplace's equation and the boundary conditions

$$(2) \qquad -\left(\frac{\partial \phi_1}{\partial r}\right)_{r=a} = \cos \theta, \quad -\left(\frac{\partial \phi_2}{\partial r}\right)_{r=a} = 0.$$

$$(3) \qquad -\left(\frac{\partial \phi_1}{\partial r'}\right)_{r'=b} = 0, \quad -\left(\frac{\partial \phi_2}{\partial r'}\right)_{r'=b} = \cos \theta'.$$

Thus ϕ_1 is the velocity potential when the sphere A moves with unit velocity towards B, the latter being at rest.

If B were absent, ϕ_1 would be the velocity potential due to a doublet at A

in the direction AB of strength $\mu_0 = \frac{1}{2}a^3$. The presence of B, however, causes the first boundary condition of (3) to be violated.

To satisfy this, we introduce the image of μ_0 in B, which is a doublet, μ_1, directed along BA at A_1, the inverse point of A with respect to B. This image requires an image μ_2 at A_2, the inverse point of A_1 with respect to A, and so on. Thus we have an infinite series of images at points $A_1, A_2, A_3, \ldots$, of strengths $\mu_1, \mu_2, \mu_3, \ldots$, where the odd suffixes refer to points within B and the even suffixes to points within A. Let $f_n = AA_n$. Then, if $AB = c$,

$$(4) \qquad f_1 = c - \frac{b^2}{c}, \quad f_2 = \frac{a^2}{f_1}, \quad f_3 = c - \frac{b^2}{c - f_2}, \ldots,$$

$$(5) \qquad \mu_1 = \mu_0\left(-\frac{b^3}{c^3}\right), \quad \mu_2 = \mu_1\left(-\frac{a^3}{f_1^{\,3}}\right), \quad \mu_3 = \mu_2\left(-\frac{b^3}{(c-f_2)^3}\right), \ldots .$$

The equations for the f_n lead to a difference equation of Riccati's form,* which can be in this case completely solved and the value of μ_n can then be written down.

With the notations of the figure, we then have

$$\phi_1 = \frac{\mu_0 \cos\theta}{r^2} + \frac{\mu_1 \cos\theta_1}{r_1^{\,2}} + \frac{\mu_2 \cos\theta_2}{r_2^{\,2}} + \ldots .$$

This is an exact solution of the problem, but in an unwieldy form.

To obtain an approximate solution correct to the term in c^{-3}, we observe that if B were absent, ϕ_1 would be

$$\frac{1}{2}\frac{a^3 \cos\theta}{r^2}.$$

Using the expansion of 17·1, we get, near the sphere B, when B is taken as origin,

$$\frac{1}{2}\frac{a^3 \cos\theta}{r^2} = \frac{1}{2}\frac{a^3}{c^2} + \frac{a^3 r' P_1(\cos\theta')}{c^3} + \ldots,$$

which gives over B the normal velocity

$$-\frac{a^3 \cos\theta'}{c^3} - \ldots .$$

This can be cancelled by adding a term to the first approximation, which gives the second approximation,

$$(6) \qquad \phi_1 = \frac{1}{2}\frac{a^3 \cos\theta}{r^2} + \frac{1}{2}\frac{a^3 b^3}{c^3}\frac{\cos\theta'}{r'^2},$$

* Milne-Thomson, *Calculus of Finite Differences*, 11·8.

and the normal velocity over B now vanishes to the order c^{-3} at least. Similarly,

$$\phi_2 = \frac{1}{2}\frac{b^3 \cos\theta'}{r'^2} + \frac{1}{2}\frac{a^3 b^3}{c^3}\frac{\cos\theta}{r^2}, \tag{7}$$

to the same order of approximation.

Near A, the same expansion gives

$$\frac{\cos\theta'}{r'^2} = \frac{1}{c^2} + \frac{2r\cos\theta}{c^3} + \ldots,$$

and therefore, when $r = a$,

$$\phi_1 = \tfrac{1}{2}a\cos\theta, \quad \phi_2 = \frac{1}{2}\frac{b^3}{c^2} + \frac{3}{2}\frac{ab^3}{c^3}\cos\theta. \tag{8}$$

To find the kinetic energy of the liquid, we have

$$T = -\tfrac{1}{2}\rho\int \phi\frac{\partial\phi}{\partial n}\,dS_A - \tfrac{1}{2}\rho\int\phi\frac{\partial\phi}{\partial n}\,dS_B,$$

taken over the spheres A and B. Thus, using (2) and (3),

$$T = \tfrac{1}{2}\rho[P_1 U^2 + (Q_1 + Q_2)UV + P_2 V^2],$$

where

$$P_1 = -\int\phi_1\frac{\partial\phi_1}{\partial n}\,dS_A, \quad P_2 = -\int\phi_2\frac{\partial\phi_2}{\partial n}\,dS_B,$$

$$Q_1 = -\int\phi_2\frac{\partial\phi_1}{\partial n}\,dS_A, \quad Q_2 = -\int\phi_1\frac{\partial\phi_2}{\partial n}\,dS_B.$$

From Green's theorem (or by direct calculation) we have $Q_1 = Q_2$. Also on A, $\partial\phi_1/\partial n = -\cos\theta$, $dS_A = 2\pi a^2 \sin\theta\, d\theta$ and

$$\int_0^\pi \cos^2\theta\sin\theta\,d\theta = \tfrac{2}{3}.$$

Therefore, correct to terms in c^{-3},

$$P_1 = \tfrac{2}{3}\pi a^3, \quad Q_1 = Q_2 = \frac{2\pi a^3 b^3}{c^3}, \quad P_2 = \tfrac{2}{3}\pi b^3,$$

and therefore

$$T = \tfrac{1}{4}M_1' U^2 + \frac{2\pi a^3 b^3\rho}{c^3}UV + \tfrac{1}{4}M_2' V^2,$$

where M_1', M_2' are the masses of liquid displaced by the respective spheres.

17·31. Sphere moving perpendicularly to a wall. If in the problem considered in 17·30 we put $V = U$, $b = a$, it is evident that the plane which bisects AB at right angles is a plane of symmetry across which there is no flow. We can therefore replace this plane by an infinite rigid wall and we thus get the case of a sphere moving with velocity U towards the wall. Putting $c = 2h$, where h is the distance of the centre of the sphere from the wall, the kinetic energy of the liquid is

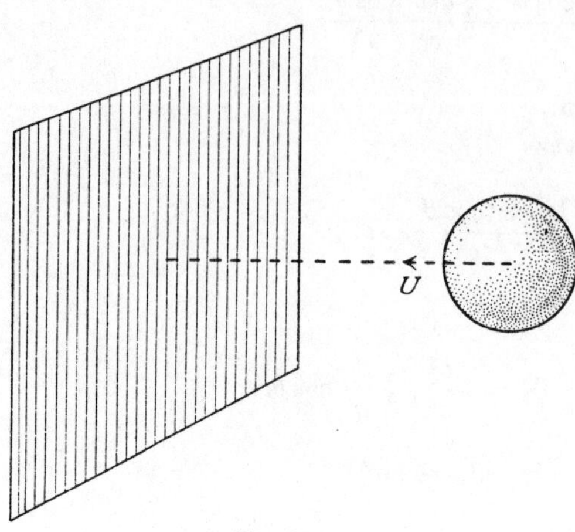

FIG. 17·31.

$$T = \tfrac{1}{4} M' \left(1 + \tfrac{3}{8}\frac{a^3}{h^3}\right) U^2,$$

where M' is the mass of liquid displaced by the sphere.

The sphere moves as if the fluid were unbounded and another sphere were moving with the optical image of the first in the wall.

If the sphere is moving towards the wall and there are no extraneous forces, the total energy remains constant, i.e.

$$\tfrac{1}{4}\left(2M + M' + \tfrac{3}{8}\frac{M'a^3}{h^3}\right) U^2 = \text{constant}.$$

As the sphere approaches the wall h decreases, and therefore $1/h^3$ increases. Hence U must decrease and the sphere is repelled from the wall. Similarly, if the velocity is away from the wall, as h increases $1/h^3$ decreases and therefore U increases. Thus in either case the sphere is repelled from the wall.

It follows that two equal spheres moving with the same speed in opposite directions along the line of centres will appear to repel one another whether the distance between them is increasing or decreasing. Observe that only the relative velocity of approach is concerned in this result, so that the spheres may have any velocities along the line of centres. This phenomenon minimises the prospects of head-on collision between floating bodies.

17·40. Two spheres moving at right angles to the line of centres. If spheres, centres A, B, radii a, b, move with velocities U, V parallel in direction and at right angles to AB, the velocity potential will be of the form

$$\phi = U\phi_1 + V\phi_2.$$

Subject to the boundary conditions, fig. 17·40,

$$(1) \qquad -\left(\frac{\partial \phi_1}{\partial r}\right)_{r=a} = \cos\theta, \quad -\left(\frac{\partial \phi_1}{\partial r'}\right)_{r'=b} = 0,$$

$$-\left(\frac{\partial \phi_2}{\partial r}\right)_{r=a} = 0, \quad -\left(\frac{\partial \phi_2}{\partial r'}\right)_{r'=b} = \cos\theta'.$$

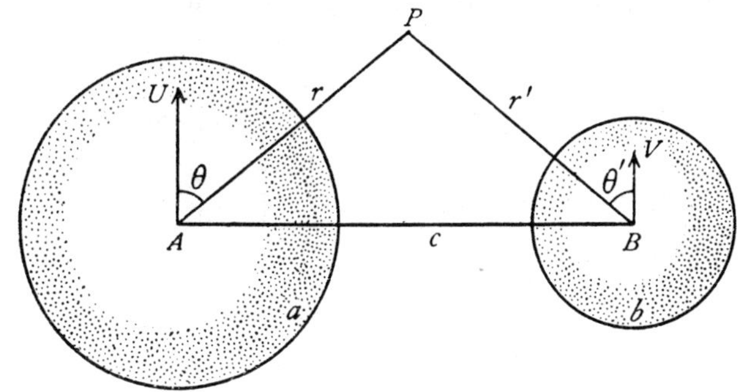

FIG. 17·40.

If the distance c between the centres is very great, each sphere will be almost unaffected by the presence of the other and we shall have, as a first approximation to ϕ_1, the potential

$$\frac{a^3}{2}\frac{\cos\theta}{r^2}.$$

Now when c is large, at points near B we shall have approximately $r = c$, and therefore

$$\frac{a^3}{2}\frac{\cos\theta}{r^2} = \frac{a^3 r\cos\theta}{2r^3} = \frac{a^3 r'\cos\theta'}{2c^3}, \text{ nearly.}$$

This gives over B the normal velocity

$$-\frac{a^3\cos\theta'}{2c^3}$$

instead of zero, as demanded by (1). This normal velocity will be cancelled if we take

$$(2) \qquad \phi_1 = \frac{a^3}{2}\frac{\cos\theta}{r^2} + \frac{a^3b^3}{4c^3}\frac{\cos\theta'}{r'^2}.$$

On A, the same method of approximation gives

$$-\frac{\partial \phi_1}{\partial r} = \cos\theta - \frac{a^3b^3}{4c^6}\cos\theta,$$

so that, if c^{-6} is neglected, (2) gives an approximation to the required velocity potential.

Near B, we have

$$\phi_1 = \frac{a^3r'\cos\theta'}{2c^3} + \frac{a^3b^3}{4c^3}\frac{\cos\theta'}{r'^2},$$

and therefore, when $r = a$,

(3) $$\phi_1 = \tfrac{1}{2}a\cos\theta,$$

and, when $r' = b$,

(4) $$\phi_1 = \frac{3a^3b}{4c^3}\cos\theta',$$

provided terms containing c^{-6} and higher powers are negligible.

The kinetic energy of the liquid is then given, as in 17·30, by

$$T = \tfrac{1}{2}\rho[p_1U^2+(q_1+q_2)UV+p_2V^2],$$

where
$$p_1 = -\int \phi_1\frac{\partial\phi_1}{\partial n}\,dS_A,\quad p_2 = -\int\phi_2\frac{\partial\phi_2}{\partial n}\,dS_B,$$
$$q_1 = -\int\phi_2\frac{\partial\phi_1}{\partial n}\,dS_A,\quad q_2 = -\int\phi_1\frac{\partial\phi_2}{\partial n}\,dS_B,$$

and, from Green's theorem, $q_1 = q_2$.

On the surface of A, $\quad \partial\phi_1/\partial n = -\cos\theta,$

and on the surface of B, $\quad \partial\phi_2/\partial n = -\cos\theta'.$

We therefore get, as in 17·30,

$$p_1 = \tfrac{2}{3}\pi a^3,\quad q_1 = q_2 = \pi\frac{a^3b^3}{c^3},\quad p_2 = \tfrac{2}{3}\pi b^3,$$

and therefore
$$T = \tfrac{1}{4}M_1'U^2+\frac{\pi a^3b^3\rho}{c^3}UV+\tfrac{1}{4}M_2'V^2,$$

where M_1', M_2' are the masses of liquid displaced by the spheres.

17·41. Sphere moving parallel to a wall. Putting $V = U$, $b = a$ in the results of 17·40, we get the case of a sphere moving parallel to a fixed rigid plane wall, for the plane bisecting AB at right angles being a plane across which there is no flow may be taken as a boundary. If $c = 2h$, so that h is the distance of the centre from the wall, we have

$$T = \tfrac{1}{4}M'U^2\left(1+\tfrac{3}{16}\frac{a^3}{h^3}\right).$$

The sphere moves as if the fluid were unbounded, and another equal sphere moved with the optical image of the first in the wall.

17·50. Ellipsoidal coordinates. The equation

(1) $$\frac{x^2}{a^2+\theta}+\frac{y^2}{b^2+\theta}+\frac{z^2}{c^2+\theta} = 1,$$

where a, b, c are fixed and θ is a parameter, represents for any constant value of θ a central quadric of a confocal system.* In particular, if $\theta = 0$, we have an ellipsoid. Equation (1) leads to

(2) $$f(\theta) = x^2(b^2+\theta)(c^2+\theta)+y^2(c^2+\theta)(a^2+\theta)+z^2(a^2+\theta)(b^2+\theta) - (a^2+\theta)(b^2+\theta)(c^2+\theta) = 0,$$

* See for example, R. J. T. Bell, *Coordinate Geometry of Three Dimensions* (1926), Chapter X.

which is a cubic equation in θ, and has therefore three roots, say λ, μ, ν. This shows that, given the point P (x, y, z), there are three central quadrics which pass through P. These are in fact an ellipsoid, a hyperboloid of one sheet, and a hyperboloid of two sheets. Moreover, these quadrics cut orthogonally at P. For a proof of these statements the reader is referred to works on solid geometry. We shall assume their truth.

Since λ, μ, ν are the roots of the equation (2), the identity

$$f(\theta) = (\lambda-\theta)(\mu-\theta)(\nu-\theta) \tag{3}$$

follows, for the function on the right vanishes when $\theta = \lambda$, μ, ν, and the coefficients of θ^3 agree.

If we put in turn $\theta = -a^2$, $-b^2$, $-c^2$, we get, from this identity,

$$x^2 = \frac{(a^2+\lambda)(a^2+\mu)(a^2+\nu)}{(a^2-b^2)(a^2-c^2)},$$

$$y^2 = \frac{(b^2+\lambda)(b^2+\mu)(b^2+\nu)}{(b^2-c^2)(b^2-a^2)}, \tag{4}$$

$$z^2 = \frac{(c^2+\lambda)(c^2+\mu)(c^2+\nu)}{(c^2-a^2)(c^2-b^2)},$$

which give the values of x, y, z when λ, μ, ν are known. Thus λ, μ, ν can be used to fix the position of a point in space, and we take them as a system of orthogonal curvilinear coordinates called *ellipsoidal coordinates*. The surfaces $\lambda =$ constant, $\mu =$ constant, $\nu =$ constant are the confocal quadrics, and in particular we shall always suppose that $\lambda =$ constant gives the ellipsoids.

In order to make hydrodynamical applications of these coordinates we must find the appropriate expression for $\nabla^2\phi$. Reference to 2·72 shows that we must first calculate h_1, h_2, h_3 where

$$(ds)^2 = (dx)^2+(dy)^2+(dz)^2 = h_1^2(d\lambda)^2+h_2^2(d\mu)^2+h_3^2(d\nu)^2.$$

Since
$$dx = \frac{\partial x}{\partial \lambda}d\lambda+\frac{\partial x}{\partial \mu}d\mu+\frac{\partial x}{\partial \nu}d\nu,$$

putting $d\mu = 0$, $d\nu = 0$, we get

$$h_1^2 = \left(\frac{\partial x}{\partial \lambda}\right)^2+\left(\frac{\partial y}{\partial \lambda}\right)^2+\left(\frac{\partial z}{\partial \lambda}\right)^2. \tag{5}$$

From equations (4) we get by logarithmic differentiation

$$\frac{\partial x}{\partial \lambda} = \frac{1}{2}\frac{x}{a^2+\lambda}, \quad \frac{\partial y}{\partial \lambda} = \frac{1}{2}\frac{y}{b^2+\lambda}, \quad \frac{\partial z}{\partial \lambda} = \frac{1}{2}\frac{z}{c^2+\lambda}, \tag{6}$$

and therefore, from (5),

$$h_1^2 = \tfrac{1}{4}\left\{\frac{x^2}{(a^2+\lambda)^2}+\frac{y^2}{(b^2+\lambda)^2}+\frac{z^2}{(c^2+\lambda)^2}\right\}$$

$$= \tfrac{1}{4}\left\{\frac{(a^2+\mu)(a^2+\nu)}{(a^2+\lambda)(a^2-b^2)(a^2-c^2)}+\frac{(b^2+\mu)(b^2+\nu)}{(b^2+\lambda)(b^2-a^2)(b^2-c^2)}+\frac{(c^2+\mu)(c^2+\nu)}{(c^2+\lambda)(c^2-a^2)(c^2-b^2)}\right\}.$$

$$(7) \qquad = \tfrac{1}{4}\frac{(\lambda-\mu)(\lambda-\nu)}{(a^2+\lambda)(b^2+\lambda)(c^2+\lambda)},$$

the second line being obtained by using equations (4) and then observing that this results from putting the third line into partial fractions by the usual method of substituting $-a^2$, $-b^2$, $-c^2$ for λ. The values of h_2, h_3 can be written down from symmetry.

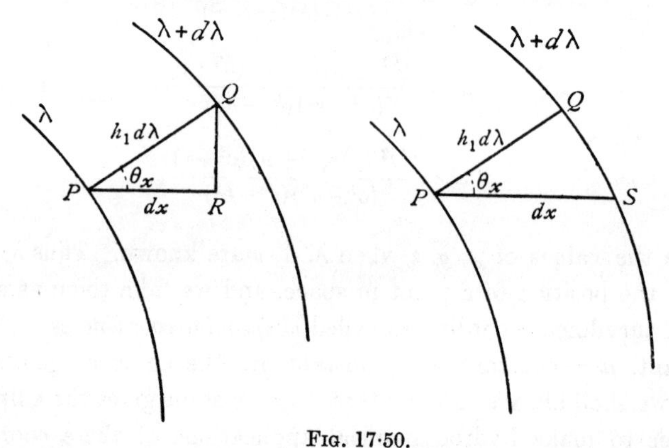

FIG. 17·50.

If we regard (x, y, z) as functions of λ and proceed from a point P on the surface $\lambda =$ constant along the normal at P to a point Q on the surface $\lambda+d\lambda =$ constant, we shall have $PQ = h_1\, d\lambda$ and

$$(8) \qquad \frac{1}{h_1}\frac{\partial x}{\partial \lambda} = \cos\theta_x,$$

where θ_x is the angle between PQ and the axis of x. If, on the other hand, we regard λ as a function of x, y, z, and, if we proceed in the x-direction a distance dx keeping y, z constant, we shall arrive at the point S of the surface $\lambda+d\lambda =$ constant, and then

$$\frac{h_1\,\partial\lambda}{\partial x} = \cos\theta_x$$

Thus

$$h_1\frac{\partial\lambda}{\partial x} = \frac{1}{h_1}\frac{\partial x}{\partial\lambda}.$$

If we put

$$(9) \qquad (k_\lambda)^2 = (a^2+\lambda)(b^2+\lambda)(c^2+\lambda),$$

we get from (7),

$$(2h_1 k_\lambda)^2 = (\lambda-\mu)(\lambda-\nu), \quad (2h_2 k_\mu)^2 = (\mu-\nu)(\mu-\lambda), \quad (2h_3 k_\nu)^2 = (\nu-\lambda)(\nu-\mu).$$

Therefore
$$\left(\frac{h_2 h_3}{h_1}\right)^2 = -\tfrac{1}{4}(\mu-\nu)^2\left(\frac{k_\lambda}{k_\mu k_\nu}\right)^2,$$

and
$$(h_1 h_2 h_3)^2 = -\frac{(\lambda-\mu)^2(\mu-\nu)^2(\nu-\lambda)^2}{64k_\lambda^2 k_\mu^2 k_\nu^2}.$$

Observing that k_μ, k_ν are independent of λ, the operation ∇^2 gives, from 2·72 (3),

$$(10) \quad \nabla^2\phi = -\frac{4}{(\lambda-\mu)(\mu-\nu)(\nu-\lambda)}\left\{(\mu-\nu)k_\lambda\frac{\partial}{\partial\lambda}\left(k_\lambda\frac{\partial\phi}{\partial\lambda}\right)\right.$$
$$\left. +(\nu-\lambda)k_\mu\frac{\partial}{\partial\mu}\left(k_\mu\frac{\partial\phi}{\partial\mu}\right)+(\lambda-\mu)k_\nu\frac{\partial}{\partial\nu}\left(k_\nu\frac{\partial\phi}{\partial\nu}\right)\right\}.$$

Equating this to zero we have Laplace's equation expressed in ellipsoidal coordinates. Solutions of this equation are called *ellipsoidal harmonics*.

17·51. Ellipsoidal harmonics. With the notations of 17·50, Laplace's equation in ellipsoidal coordinates can be written in the form

$$(1) \quad (\mu-\nu)\left(k_\lambda\frac{\partial}{\partial\lambda}\right)^2\phi+(\nu-\lambda)\left(k_\mu\frac{\partial}{\partial\mu}\right)^2\phi+(\lambda-\mu)\left(k_\nu\frac{\partial}{\partial\nu}\right)^2\phi = 0.$$

Let α be a value of ϕ which satisfies this equation, and let us seek solutions of the form

$$\phi = \alpha\chi,$$

where χ is a function of λ *only*. We get at once

$$k_\lambda\frac{\partial(\alpha\chi)}{\partial\lambda} = k_\lambda\frac{\partial\alpha}{\partial\lambda}\chi+k_\lambda\alpha\frac{\partial\chi}{\partial\lambda}, \quad \frac{\partial}{\partial\lambda}\left\{k_\lambda\frac{\partial(\alpha\chi)}{\partial\lambda}\right\} = \chi\frac{\partial}{\partial\lambda}\left(k_\lambda\frac{\partial\alpha}{\partial\lambda}\right)+R,$$

$$(2) \qquad R = 2k_\lambda\frac{\partial\alpha}{\partial\lambda}\frac{\partial\chi}{\partial\lambda}+\alpha\frac{\partial k_\lambda}{\partial\lambda}\frac{\partial\chi}{\partial\lambda}+\alpha k_\lambda\frac{\partial^2\chi}{\partial\lambda^2}.$$

Substitution in (1), remembering that α is a solution of (1), gives $R = 0$. which can be written in the form

$$(3) \qquad \frac{\partial}{\partial\lambda}\log\left(k_\lambda\frac{\partial\chi}{\partial\lambda}\right) = -\frac{2}{\alpha}\frac{\partial\alpha}{\partial\lambda}.$$

Since the left side is a function of λ only, the right side must be independent of μ, ν, and therefore a solution of the proposed form is only possible if α be such that this condition is satisfied. This means that α must be of the form

$$(4) \qquad \alpha = \alpha_\lambda f(\mu,\nu),$$

where α_λ is independent of μ, ν and $f(\mu, \nu)$ is independent of λ. In this case (3) becomes

$$\frac{d}{d\lambda} \log \left(k_\lambda \frac{d\chi}{d\lambda}\right) = \frac{d}{d\lambda} \log \frac{1}{{\alpha_\lambda}^2},$$

which, on integration, gives

$$\chi = A \int \frac{d\lambda}{{\alpha_\lambda}^2 k_\lambda} + B,$$

where A and B are arbitrary constants.

Thus if α is an ellipsoidal harmonic having the postulated properties, so also are

$$(5) \qquad \alpha \int \frac{d\lambda}{{\alpha_\lambda}^2 k_\lambda}, \quad \int \frac{d\lambda}{k_\lambda},$$

the second being obtained by taking $\alpha = 1$, which is obviously a solution of (1).

Now (1) is merely Laplace's equation $\nabla^2 \phi = 0$ expressed in a particular system of coordinates.

Therefore x, y, z, xy, yz, zx, and in fact any spherical harmonic, are solutions of (1).

The values of x, y, z are given in 17·50 (4), so that we may take

$$\alpha = (a^2+\lambda)^{\frac{1}{2}} (a^2+\mu)^{\frac{1}{2}} (a^2+\nu)^{\frac{1}{2}}, \text{ corresponding to } x,$$

or

$$\alpha = (b^2+\lambda)^{\frac{1}{2}} (b^2+\mu)^{\frac{1}{2}} (b^2+\nu)^{\frac{1}{2}} (c^2+\lambda)^{\frac{1}{2}} (c^2+\mu)^{\frac{1}{2}} (c^2+\nu)^{\frac{1}{2}},$$

corresponding to yz, which both obviously satisfy (4).

Therefore the ellipsoidal harmonics given by the first function of (5) are

$$(6) \qquad \phi_x = Cx \int_\lambda^\infty \frac{d\lambda}{(a^2+\lambda) k_\lambda}, \quad \phi_{yz} = Cyz \int_\lambda^\infty \frac{d\lambda}{(b^2+\lambda)(c^2+\lambda) k_\lambda},$$

where C is an arbitrary constant, and x, y, z are supposed expressed in terms of λ, μ, ν by means of 17·50 (4). The limits have been adjusted to make the integrals vanish at $\lambda = \infty$. These are the only forms of which we shall make applications.

All functions of the type (6) are included in the forms ∇V, $\mathbf{r}_\wedge (\nabla V)$, where

$$\mathbf{r} = \mathbf{i}x + \mathbf{j}y + \mathbf{k}z,$$

$$V = \int_\lambda^\infty \left(\frac{x^2}{a^2+\lambda} + \frac{y^2}{b^2+\lambda} + \frac{z^2}{c^2+\lambda} - 1\right) \frac{d\lambda}{k_\lambda}.$$

17·52. Translatory motion of an ellipsoid. Consider the ellipsoid

$$(1) \qquad \frac{x^2}{a^2} + \frac{y^2}{b^2} + \frac{z^2}{c^2} = 1,$$

which corresponds to $\lambda = 0$, moving in the direction of the x-axis with velocity U. The boundary condition is

$$(2) \qquad -\frac{\partial \phi}{\partial n} = U \cos \theta_x \quad \text{or} \quad \frac{\partial \phi}{\partial \lambda} = -U \frac{\partial x}{\partial \lambda}, \quad \lambda = 0,$$

since $dn = h_1 d\lambda$, $\cos \theta_x = \partial x / h_1 \partial \lambda$, where θ_x is the angle between the normal and the x-axis (17·50 (8)).

Thus when $\lambda = 0$, $\phi = -Ux$, and when $\lambda \to \infty$, $\phi \to 0$. These conditions are satisfied by the function ϕ_x of 17·51 (6). We therefore take

$$\phi = Cx \int_\lambda^\infty \frac{d\lambda}{(a^2+\lambda) k_\lambda}.$$

Condition (2) then gives

$$-U \frac{\partial x}{\partial \lambda} = C \frac{\partial x}{\partial \lambda} \int_0^\infty \frac{d\lambda}{(a^2+\lambda) k_\lambda} - \frac{Cx}{a^2 . abc}, \quad \text{when } \lambda = 0.$$

From 17·50 (6), $\partial x / \partial \lambda = \frac{1}{2} x / a^2$, when $\lambda = 0$, and therefore

$$(3) \qquad C = \frac{abcU}{2-\alpha_0}, \quad \text{where } \alpha_0 = abc \int_0^\infty \frac{d\lambda}{(a^2+\lambda) k_\lambda}.$$

The constant α_0 depends solely on the semi-axes a, b, c of the ellipsoid. Its numerical evaluation requires the use of elliptic integrals.

Thus, finally,

$$(4) \qquad \phi = \frac{abcUx}{2-\alpha_0} \int_\lambda^\infty \frac{d\lambda}{(a^2+\lambda)^{\frac{3}{2}} (b^2+\lambda)^{\frac{1}{2}} (c^2+\lambda)^{\frac{1}{2}}},$$

and on the ellipsoid we have, from (3),

$$\phi = \frac{\alpha_0 x U}{2-\alpha_0}.$$

The kinetic energy of the liquid is

$$T = -\tfrac{1}{2} \rho \int \phi \frac{\partial \phi}{\partial n} dS = \frac{\alpha_0}{2(2-\alpha_0)} \rho U^2 \int x \cos \theta_x \, dS.$$

Since $\cos \theta_x dS$ is the projection on the plane $x = 0$ of the area dS of the surface, fig. 17·52, the last integral gives the volume of the ellipsoid $4\pi abc/3$ and

$$T = \tfrac{1}{2} M' \frac{\alpha_0}{2-\alpha_0} U^2,$$

where M' is the mass of liquid displaced by the ellipsoid.

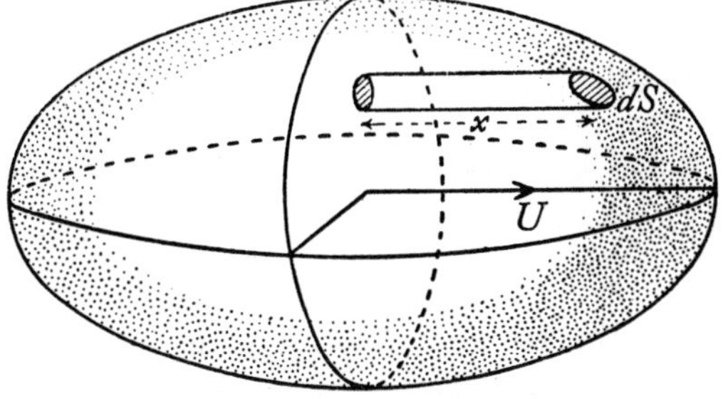

Fig. 17·52.

The case of the sphere is obtained by putting $a = b = c$, when all the integrals can easily be evaluated.

When the ellipsoid has in

addition velocity components V, W parallel to the y- and z-axes, we get, by superposing the results analogous to (4), the velocity potential

$$\frac{abcUx}{2-\alpha_0}\int_\lambda^\infty \frac{d\lambda}{(a^2+\lambda)k_\lambda}+\frac{abcVy}{2-\beta_0}\int_\lambda^\infty \frac{d\lambda}{(b^2+\lambda)k_\lambda}+\frac{abcWz}{2-\gamma_0}\int_\lambda^\infty \frac{d\lambda}{(c^2+\lambda)k_\lambda},$$

where β_0, γ_0 are defined by writing $b^2+\lambda$, $c^2+\lambda$ for $a^2+\lambda$ in the integrand of (3).

The ovary and planetary ellipsoids can be regarded as special cases of the above.

17·53. Rotating ellipsoid. When the ellipsoid rotates with angular velocity $\boldsymbol{\omega} = \omega_x \mathbf{i}+\omega_y \mathbf{j}+\omega_z \mathbf{k}$, the velocity of the point $\mathbf{r} = x\mathbf{i}+y\mathbf{j}+z\mathbf{k}$ on the boundary is $\boldsymbol{\omega}_\wedge \mathbf{r}$. If $\omega_y = \omega_z = 0$, the velocity is therefore $-\mathbf{j}\omega_x z+\mathbf{k}\omega_x y$.

If θ_y, θ_z are the angles between the normal to the ellipsoid and the y- and z-axes, the boundary condition is

$$-\frac{\partial\phi}{\partial n} = (-z\cos\theta_y+y\cos\theta_z)\,\omega_x,$$

or

$$\frac{\partial\phi}{\partial\lambda} = \left(z\frac{\partial y}{\partial\lambda}-y\frac{\partial z}{\partial\lambda}\right)\omega_x.$$

The function ϕ_{yz} of 17·51 (6) can be adapted to this form of boundary condition. Thus taking

$$(1)\qquad \phi = Cyz\int_\lambda^\infty \frac{d\lambda}{(b^2+\lambda)(c^2+\lambda)k_\lambda},$$

we get, when $\lambda = 0$,

$$\left(\frac{zy}{2b^2}-\frac{yz}{2c^2}\right)\omega_x = -\frac{Cyz}{b^2c^2\,.\,abc}+C\left(\frac{zy}{2b^2}+\frac{yz}{2c^2}\right)I,$$

where we have put $\partial y/\partial\lambda = \frac{1}{2}y/b^2$, $\partial z/\partial\lambda = \frac{1}{2}z/c^2$, and

$$I = \int_0^\infty \frac{d\lambda}{(b^2+\lambda)(c^2+\lambda)k_\lambda}.$$

Since

$$\frac{1}{(b^2+\lambda)(c^2+\lambda)} = \frac{-1}{(b^2-c^2)}\left(\frac{1}{b^2+\lambda}-\frac{1}{c^2+\lambda}\right),$$

we get

$$I = \frac{-(\beta_0-\gamma_0)}{(b^2-c^2)abc},$$

where β_0, γ_0 have the same meanings as in the last section. Thus

$$C = \frac{(b^2-c^2)^2}{2(b^2-c^2)+(b^2+c^2)(\beta_0-\gamma_0)}\,abc\,\omega_x,$$

and the required velocity potential is obtained by substituting this value of C in (1). The kinetic energy of the liquid can be calculated by the same method as before.

When the ellipsoid has angular velocity components ω_y, ω_z in addition, the complete velocity potential is found by superposing the results obtained by symmetry from the above.

17·54. Rotating ellipsoidal shell. If the interior of the ellipsoid

$$\frac{x^2}{a^2}+\frac{y^2}{b^2}+\frac{z^2}{c^2} = 1$$

is filled with liquid and rotates about the x-axis with angular velocity ω_x, the boundary condition is

$$-\frac{x}{a^2}\frac{\partial\phi}{\partial x}-\frac{y}{b^2}\frac{\partial\phi}{\partial y}-\frac{z}{c^2}\frac{\partial\phi}{\partial z} = -\frac{y}{b^2}\,\omega_x z+\frac{z}{c^2}\,\omega_x y.$$

We can satisfy this by taking $\phi = Ayz$, for this is a spherical harmonic. We then get

$$A\left(\frac{1}{b^2}+\frac{1}{c^2}\right) = \omega_x\left(\frac{1}{b^2}-\frac{1}{c^2}\right),$$

which determines A. If, in addition, the shell has a velocity u_x along the x-axis, we must have $\phi = -xu_x$. Thus if the shell moves in any manner,

$$\phi = -xu_x-yu_y-zu_z-\frac{b^2-c^2}{b^2+c^2}\,\omega_x yz-\frac{c^2-a^2}{c^2+a^2}\,\omega_y zx-\frac{a^2-b^2}{a^2+b^2}\,\omega_z xy.$$

EXAMPLES XVII

1. If $\phi = r^n S$ is a spherical harmonic, prove that, S being independent of r,

$$\frac{1}{\sin\theta}\frac{\partial}{\partial\theta}\left(\sin\theta\frac{\partial S}{\partial\theta}\right)+\frac{1}{\sin^2\theta}\frac{\partial^2 S}{\partial\omega^2}+S(n+1)n = 0,$$

and deduce that S/r^{n+1} is also a spherical harmonic.

2. If $\phi = r^n S$ is a spherical harmonic, symmetrical about the axis of x, and S is independent of r, show that

$$\frac{d}{d\mu}\left[(1-\mu^2)\frac{dS}{d\mu}\right]+n(n+1)\,S = 0,$$

where $\mu = \cos\theta$. Show that solutions of this equation corresponding to $n = 0$, $n = 1$ are $P_0(\mu)$, $P_1(\mu)$, and also

$$Q_0(\mu) = \tfrac{1}{2}\log\frac{1+\mu}{1-\mu},\quad Q_1(\mu) = \tfrac{1}{2}\mu\log\frac{1+\mu}{1-\mu}-1.$$

Show that the velocity potential of a line source along the axis from 0 to a is $m(Q_0(\mu)-Q_0(\mu'))/a$, where m is the total strength and $\mu' = \cos\theta'$, where θ' is the polar angle at the end a.

3. The motion of fluid is given by the velocity potential

$$\phi = C\left\{\left(1+\frac{1}{n}\right)\frac{r^n}{a^{n-1}}+\frac{a^{n+2}}{r^{n+1}}\right\}P_n(\cos\theta),$$

in which C is constant, and r and θ are spherical polar coordinates. Determine the stream function. (U.L.)

4. A sphere of radius a is surrounded by a concentric spherical shell of radius b, and the space between is filled with liquid. If the sphere be moving with velocity V, show that

$$\phi = \frac{Va^3}{b^3-a^3}\left\{r+\frac{b^3}{2r^2}\right\}\cos\theta,$$

and find the current function.

5. A thin spherical shell whose mass may be neglected surrounds a concentric sphere of mass m and density σ, the intervening space being filled with a mass m' of liquid of density ρ. Prove that, if the outer sphere be given a normal impulse, the momentum is divided between the sphere and the liquid in the ratio $3m\rho/[m'(2\sigma+\rho)]$.

6. The space between two concentric spheres, radii a, b, is filled with liquid. The spheres have velocities U, V in the same direction. Find the velocity potential. Prove that the kinetic energy of the liquid is

$$\frac{\pi\rho}{3(b^3-a^3)}[a^3b^3(U-V)^2+2(Ua^3-Vb^3)^2].$$

Deduce the impulse required to set the outer sphere in motion with velocity V, the masses of the spheres being M_1, M_2.

7. The space between a solid sphere, of radius a, and a concentric spherical shell, of radius $2a$, is filled with homogeneous liquid, and, the system being at rest, an impulse is applied to the shell, causing it to start with velocity V; given that the velocity function of the initial fluid motion is of the form $(Ar+B/r^2)\cos\theta$, show that the sphere starts with velocity

$$\frac{12\rho V}{7\sigma+5\rho},$$

where σ, ρ are, respectively, the densities of the sphere and the liquid.

Show also that, if the mass of the shell is negligible, the magnitude of the applied impulse is

$$\frac{4}{7}\,\frac{17\sigma+7\rho}{7\sigma+5\rho}MV,$$

where M is the mass of the liquid. (R.N.C.)

8. A hollow spherical shell of inner radius a contains a concentric solid uniform sphere of radius b and density σ, and the space between the two is filled with liquid of density ρ. If the shell is suddenly made to move with speed u, prove that speed v is imparted to the inner sphere, where

$$v = \frac{3ua^3}{2\frac{\sigma}{\rho}(a^3-b^3)+a^3+2b^3}.$$

(R.N.C.)

9. Find the values of A and B for which

$$\left(Ar+\frac{B}{r^2}\right)\cos\theta$$

is the velocity function of the motion of an incompressible fluid which fills the space between a solid sphere of radius a and a concentric spherical shell of radius $2a$, when the sphere has a velocity U and the shell is at rest. Prove that the kinetic energy of the fluid of density ρ is $10\pi\rho a^3U^2/21$.

If the sphere, of density σ, is initially at rest in contact with the shell at the highest point and falls down under gravity, show that the velocity in the concentric position is given by

$$U^2 = \frac{14ga(\sigma-\rho)}{7\sigma+5\rho}.$$

(R.N.C.)

10. The space between a solid sphere, of mass M and radius a, and a fixed concentric spherical shell of inner radius b is filled with liquid. An impulse I acts directly on the sphere. Prove that the sphere starts with velocity

$$I\bigg/\left(M+M'\frac{b^3+2a^3}{2b^3-2a^3}\right),$$

where M' is the mass of liquid displaced by the sphere.

Find the initial value of Stokes' current function for the motion. (R.N.C.)

11. Liquid of density ρ fills the space between a solid sphere of radius a and density σ and a fixed concentric spherical envelope of radius b. Prove that the work done by an impulse which starts the solid sphere with velocity V is

$$\tfrac{1}{3}\pi a^3V^2\left(2\sigma+\frac{2a^3+b^3}{b^3-a^3}\rho\right).$$

Calculate the initial momentum of the liquid. (R.N.C.)

12. A sphere, of radius a and density σ, is surrounded by a concentric spherical shell of radius b, and the space between the sphere and shell is filled with fluid of density ρ. The whole system is moving with a velocity v when the shell is suddenly stopped. Find the velocity of the sphere immediately after the impact. (R.N.C.)

13. The space between two concentric spheres, radii a, b, of which the outer is fixed, is filled with fluid of density ρ. Show that, if the inner starts from rest with acceleration f, the initial resultant thrust on the outer is

$$2\pi\rho fa^3b^3/(b^3-a^3).$$

14. Homogeneous liquid occupies the simply connected region bounded internally by a surface S_1 and externally by a fixed surface S_0. Irrotational motion is set up by moving S_1 in any way without change of volume. Prove that the kinetic energy of the fluid is greater than it would be if there were no external boundary.

Verify the theorem by calculating the kinetic energy when S_1 and S_0 are instantaneously concentric spheres and S_1 is set in motion as a rigid boundary.

15. A sphere of radius a is moving with speed v along a diameter of a fixed sphere of radius b, the space between the two surfaces being filled with fluid. Prove that, when the distance between the centres is x, the kinetic energy of the fluid motion is

$$2\pi\rho v\left(\tfrac{1}{3}\mu_0+\sum_{n=1}^{\infty}\mu_n\right),$$

where $\mu_{n+1}(b^2-xc_n)^3 = \mu_n a^3b^3,\quad \mu_0 = \tfrac{1}{2}va^3,$

and $c_{n+1}(b^2-xc_n) = xb^2-c_n(x^2-a^2),\quad c_0 = x.$ (U.L.)

16. A sphere, of mass M and radius a, is at rest with its centre at a distance h from a plane boundary. Show that the magnitude of the impulse necessary to start the sphere with a velocity V directly towards the boundary is

$$V\left\{M+\tfrac{1}{2}M'\left(1+\frac{3a^3}{8h^3}\right)\right\},$$

very nearly, where M' is the mass of the displaced fluid. Find also the impulse on the plane boundary. (R.N.C.)

17. A sphere of radius a moves in a semi-infinite liquid of density ρ bounded by a plane wall, its centre being at a great distance h from the wall. Show that the approximate kinetic energy of the fluid is

$$\tfrac{1}{3}\pi\rho a^3V^2\left\{1+\frac{3}{16}\frac{a^3}{h^3}(1+\sin^2\alpha)\right\},$$

the sphere moving at an angle α with the wall at a speed V. (U.L.)

18. An infinite rigid plane separates liquid otherwise unbounded into two parts. A sphere moves in a direction perpendicular to the plane. Explain by general reasoning the effect of making a circular opening in the plane with its centre on the line along which the sphere is moving, (a) when its velocity is towards the plane; (b) when its velocity is away from the plane.

19. Two equal spheres, radius a, distance between centres d, are fixed in a stream U perpendicular to the line of centres. Show that the velocity halfway between them is approximately

$$U\left\{1+\frac{8a^3}{d^3}\right\}.$$

Find the velocity when the stream is parallel to the line of centres.

20. Two spheres, radii a, b, distance c apart, are surrounded by fluid. The first is made to move with velocity U towards the second. Show that the second starts with velocity A/B approximately where

$$A=\frac{3a^3}{2c^3}UM',\quad B=M+\tfrac{1}{2}M'\left\{1+\frac{3a^3b^3}{c^6}\right\}.$$

21. A sphere, of radius a, immersed in liquid of density ρ, whose only boundary is an infinite plane wall, is moving with velocity U directly towards this wall, which is at a distance c from the centre of the sphere, c being large compared with a. Neglecting a^4/c^4, prove that the velocity potential in the immediate neighbourhood of the sphere is given by

$$\phi=\tfrac{1}{2}Ua^3\left[\left(1+\frac{a^3}{8c^3}\right)\Big/r^2+\frac{1}{4}\frac{r}{c^3}\right]\cos\theta,$$

where r is the radius vector of a point, measured from the centre of the sphere, and θ is the angle r makes with the direction of motion of the sphere.

Calculate approximately the kinetic energy of the liquid. (R.N.C.)

22. A sphere of radius a is moving with speed V parallel to a fixed plane wall, the wall being at a distance c from the centre of the sphere. Show that, in the neighbourhood of the sphere, the velocity potential is approximately given by

$$\phi=V\frac{a^3}{2}y[(1+a^3/16c^3)/r^3+1/r'^3],$$

where r and r' are distances from the centre of the sphere and its image in the wall respectively, and y is measured parallel to the direction of motion.

Calculate to the same approximation the pressure on the sphere.

23. A mine at a distance a from a plane infinite wall and at a depth b below the surface of still water, which extends to infinity in depth and away from the wall, explodes symmetrically. If E is the total energy generated by the mine, calculate the normal velocity at any point of the free surface immediately after the explosion, and also the normal impulsive pressure at any point of the wall.

24. A sphere of radius a moves with velocity u directly towards a fixed plane, which bounds a region occupied by homogeneous frictionless liquid. Show how to determine the velocity potential of the motion when the centre of the sphere is at a distance c from the plane.

Prove that the kinetic energy of the liquid is

$$\tfrac{1}{4}M'u^2\left\{1+3\sum_{n=1}^{\infty}(\mu_n/\mu_0)\right\},$$

where M' is the mass of liquid displaced by the sphere, and

$$\mu_0 = \tfrac{1}{2}ua^3, \quad \mu_n = \mu_{n-1}(p_n/q_n)^3,$$

p_n/q_n being the nth convergent to the continued fraction

$$\frac{a}{2c-}\,\frac{a^2}{2c-}\,\frac{a^2}{2c-\dots},$$

in which all the partial quotients after the first are equal to $-a^2/2c$. (U.L.)

25. Two equal circular cylinders, of radius a, with their centres d apart, are fixed in a uniform stream of velocity V perpendicular to the line of centres. Obtain an approximate value for the velocity function, assuming that a/d is small. Show that the velocity midway between the centres is $V\left(1+8\dfrac{a^2}{d^2}\right)$ very nearly. (R.N.C.)

26. The space between a long solid cylinder, radius a, and a concentric cylindrical shell, radius b, is filled with homogeneous liquid. Find the velocity function for the fluid motion when the cylinder and shell have velocities U and V, respectively, perpendicular to their common axis and in the same direction.

If, when the system is at rest, an impulse applied to the shell causes it to start with a velocity V, find the initial velocity of the cylinder, and show that the velocity function for the initial fluid motion is

$$\frac{Vb^2}{r}\,\frac{\rho(a^2-r^2)-\sigma(a^2+r^2)}{\rho(b^2+a^2)+\sigma(b^2-a^2)}\cos\theta,$$

where ρ, σ are, respectively, the densities of the liquid and cylinder. (R.N.C.)

27. A circular aperture of radius a in the wall of a large tank filled with liquid of density ρ is closed by a piston with a plane end flush with the wall. The piston is suddenly pushed inwards with velocity U. Show that the impulsive pressure P on the wall is given by

$$P = \rho U\left(\frac{1}{2}\frac{a^2}{r}+\frac{1^2}{2^2\,.\,4}\frac{a^4}{r^3}+\dots+\frac{1^2\,.\,3^2\dots(2k-3)^2}{2^2\,.\,4^2\dots(2k-2)^2}\frac{a^{2k}}{2kr^{2k-1}}+\dots\right),$$

where r is the distance of the point considered from the centre of the aperture. (U.L.)

28. An ellipsoid of semi-axes a, b, c moves with velocity **V** through an infinite liquid at rest at infinity in the direction of the axis of length $2a$. Find the velocity potential of the motion and show that at a great distance the motion approximates to that due to a doublet at the centre of the ellipsoid, of axis and strength given by

$$\frac{8}{3}\frac{\pi}{2-\alpha_0}abc\,\mathbf{V},$$

where
$$\alpha_0 = abc\int_0^\infty \frac{du}{(a^2+u)^{\frac{3}{2}}(b^2+u)^{\frac{1}{2}}(c^2+u)^{\frac{1}{2}}}.$$
(U.L.)

29. The ellipsoid $x^2/a^2+y^2/b^2+z^2/c^2 = 1$ is placed in a uniform stream parallel to the x-axis. Prove that the lines of equal pressure on the ellipsoid are its curves of intersection with the cones $y^2/b^2+z^2/c^2 = x^2/h^2$, where h is an arbitrary constant.

30. A stream of infinite depth, whose bed is the plane $z = 0$, flows with velocity U parallel to the x-axis and is disturbed only by an obstacle in the shape of the upper half of an ellipsoid. If λ is the positive root of

$$\frac{x^2}{a^2+\lambda}+\frac{y^2}{b^2+\lambda}+\frac{z^2}{c^2+\lambda} = 1,$$

and the ellipsoid is given by $\lambda = 0$, show that the velocity potential of the motion of the stream is

$$\phi = -Ux\left\{1+\frac{\alpha_\lambda}{2-\alpha_0}\right\},$$

where
$$\alpha_\lambda = abc\int_\lambda^\infty \frac{du}{(a^2+u)^{\frac{3}{2}}(b^2+u)^{\frac{1}{2}}(c^2+u)^{\frac{1}{2}}}.$$

Prove also that the slip velocity at all points over the section by $x = 0$ is $2U/(2-\alpha_0)$. (U.L.)

31. A rigid ellipsoid, semi-axes a, b, c, is moving with velocities U, V parallel to the axes a, b respectively. Show that to maintain the motion a couple is required about the axis c of moment

$$\frac{8\pi\rho abc(\beta_0-\alpha_0)\,UV}{3(2-\alpha_0)(2-\beta_0)},$$

reckoned positive from the axis a to axis b. (U.L.)

32. The region outside the ellipsoid

$$x^2/a^2+y^2/b^2+z^2/c^2 = 1$$

is occupied by liquid which is at rest at infinity. The ellipsoid rotates with angular velocity ω about the axis of x. Find the velocity potential and show that the kinetic energy of the liquid is

$$\frac{1}{10}\frac{(b^2-c^2)^2(\gamma_0-\beta_0)}{2(b^2-c^2)-(b^2+c^2)(\gamma_0-\beta_0)}M\omega^2,$$

where
$$\beta_0 = abc\int_0^\infty (a^2+\lambda)^{-\frac{1}{2}}(b^2+\lambda)^{-\frac{3}{2}}(c^2+\lambda)^{-\frac{1}{2}}\,d\lambda,$$

$$\gamma_0 = abc\int_0^\infty (a^2+\lambda)^{-\frac{1}{2}}(b^2+\lambda)^{-\frac{1}{2}}(c^2+\lambda)^{-\frac{3}{2}}\,d\lambda,$$

and M is the mass of liquid displaced by the ellipsoid. Hence find the effective moment of inertia of the ellipsoid. (U.L.)

33. Show that when a circular disc of radius a rotates about a diameter in liquid at rest at infinity, the kinetic energy of the liquid is $8\rho a^5\omega^2/45$, ω being the angular velocity of the disc and ρ the density of the liquid.

34. Find the only solutions of Laplace's equation in ellipsoidal coordinates λ, μ, ν which are independent of μ and ν.

The axes of an ellipsoid which is surrounded by an unlimited liquid vary with the time in such a manner that the ellipsoid always remains similar to itself. Prove that

$$\phi = -\tfrac{1}{6}abc(\dot{a}/a+\dot{b}/b+\dot{c}/c)\int_\lambda^\infty \frac{d\lambda}{\sqrt{(a^2+\lambda)(b^2+\lambda)(c^2+\lambda)}}.$$

35. Prove that if a solution of Laplace's equations in ellipsoidal coordinates λ, μ, ν is of the product form $L\,.\,M\,.\,N$, then a possible value of L satisfying the equation is of the form

$$(a^2+\lambda)^{\frac{1}{2}}(b^2+\lambda)^{\frac{1}{2}}(c^2+\lambda)^{\frac{1}{2}}.$$

Find a second solution in λ which satisfies the differential equation for λ in this case, and hence obtain three solutions of Laplace's equations in the form

$$xyz\,.\,F,$$

where F is a function of either λ alone, or μ alone, or ν alone.

36. Show that if λ is a root of the equation

$$x^2/(a^2+\lambda)+y^2/(b^2+\lambda)+z^2/(c^2+\lambda) = 1,$$

then $L = (a^2+\lambda)^{\frac{1}{2}}$ is a solution of the equation

$$\frac{1}{L}\frac{d^2L}{d\alpha^2} = A\lambda+B,$$

where $d\alpha = d\lambda/\sqrt{(a^2+\lambda)(b^2+\lambda)(c^2+\lambda)}$ and A and B have certain values.

Prove also that for certain other values of A and B, $(b^2+\lambda)^{\frac{1}{2}}$, $(c^2+\lambda)^{\frac{1}{2}}$ are also solutions, but that it is impossible to obtain a solution of the form

$$L = p(a^2+\lambda)^{\frac{1}{2}}+q(b^2+\lambda)^{\frac{1}{2}}+r(c^2+\lambda)^{\frac{1}{2}},$$

where no two of p, q, and r are zero, if a, b, c are all different.

37. An ellipsoidal vessel of semi-axes a, b, c is filled with frictionless liquid of density ρ, and is rotating about the axis of x with angular velocity ω. Prove that the velocities at any point of the liquid are given by

$$u = 0, \quad v = Cz, \quad w = Cy,$$

and determine the constant C.

38. An ellipsoid is filled with fluid and has velocity components U, V, W, ω_1, ω_2, ω_3, the axes of reference being principal axes. Show that relative to the ellipsoid the paths of the particles are ellipses, the period being $2\pi/m$, where

$$m = 2abc\left[\frac{\omega_1^2}{a^2(b^2+c^2)^2}+\frac{\omega_2^2}{b^2(c^2+a^2)^2}+\frac{\omega_3^2}{c^2(a^2+b^2)^2}\right]^{\frac{1}{2}}.$$

39. An ellipsoidal thin shell of semi-axes a, b, c is filled with liquid of density ρ and rotates about the axis c with spin ω. Find the velocity potential of the motion, and show that the kinetic energy is

$$\frac{2}{15}\frac{\pi\rho abc(a^2-b^2)^2}{a^2+b^2}\omega^2.$$

40. An ellipsoidal shell is filled with liquid and rotates uniformly about a given diameter. Prove that the path of every particle of liquid relative to the ellipsoid will be an ellipse whose plane is conjugate to the given diameter, and that every particle will sweep out, about the centre of its elliptic path, equal areas in equal times.

41. The axes of an ellipsoid which is filled with liquid vary with the time in such a manner that the volume of the ellipsoid remains constant. Prove that the velocity potential of the liquid is

$$\phi = (\dot{a}x^2/a + \dot{b}y^2/b + \dot{c}z^2/c)/2.$$

42. Given that $x = a(\cosh\alpha + \cos\beta - \cosh\gamma)$, $y = 4a\cosh\frac{1}{2}\alpha\cos\frac{1}{2}\beta\sinh\frac{1}{2}\gamma$, $z = 4a\sinh\frac{1}{2}\alpha\sin\frac{1}{2}\beta\cosh\frac{1}{2}\gamma$, transform the equation of continuity into

$$(\cos\beta + \cosh\gamma)\frac{\partial^2\phi}{\partial\alpha^2} + (\cosh\gamma + \cosh\alpha)\frac{\partial^2\phi}{\partial\beta^2} + (\cosh\alpha - \cos\beta)\frac{\partial^2\phi}{\partial\gamma^2} = 0,$$

and show that the surfaces for which α, β, γ are constant are confocal paraboloids.

Hence show that the velocity potential for infinite liquid streaming past the fixed hyperbolic paraboloid $\beta = \beta_0$, with speed V, parallel at infinity to the axis of x, is given by $\phi = V(x - \alpha\beta\sin\beta_0)$, and write down the corresponding values of ϕ when the fixed surface is the elliptic paraboloid $\alpha = \alpha_0$, or $\gamma = \gamma_0$.

43. An infinite mass of liquid has the plane $z = 0$ for free surface. If on the surface an impulsive pressure $\varpi = \varpi_0 \sin mx \sin my$ is applied, show that the initial motion is given by $\rho\phi = \varpi_0 \exp(-z(m^2+n^2)^{\frac{1}{2}})$, z being the position in the fluid.

44. A right circular cone of height h has a base radius of $h\sqrt{2}$. A mass of fluid of this form is moving parallel to the axis and base first with velocity V, when the base strikes against a fixed plane. Taking the fixed plane as the plane xy and the centre of the base as origin, prove that the velocity function just after impact is $V(2z^2 - x^2 - y^2)/(4h)$, that the impulsive pressure of the fluid is $V\rho(2(z-h)^2 - x^2 - y^2)/(4h)$, and that the impulse on the plane is 3/4 of what it would have been had the cone been solid and of the same mass.

45. Show that any irrotational motion of homogeneous liquid moving in a simply connected region bounded internally by a closed surface, and at rest at infinite distances, can be regarded as due to sources and doublets distributed over the surface. Proceed to explain what must be done in order to dispense with either the sources or the doublets.

In the case of a sphere of radius a, which is being deformed so that after a small interval of time t the equation of the surface is $r = a + Ut\,P_2(\cos\theta)$, determine a distribution (i) of sources, (ii) of doublets, on the surface which will give rise to the same velocity potential.

46. Irrotational motion of homogeneous incompressible frictionless fluid outside a closed surface S is due to the motion of S with outward normal velocity q_ν of given magnitude at any point of S. The velocity potential due to a double source, of unit strength, and having its axis parallel to the axis of x, situated at a point P outside S (supposed fixed), is denoted by Φ. Prove that the x-component of the velocity at P, due to the motion q_ν, is

$$-\frac{1}{4\pi}\iint \Phi q_\nu\, dS.$$

(U.L.)

CHAPTER XVIII

SOLID MOVING THROUGH A LIQUID

18·10. Motion of a solid through a liquid. Consider a solid S immersed and at rest within an unbounded liquid which is also at rest. If the solid be set in motion in any manner, the resulting motion of the liquid will be irrotational and acyclic. Moreover, such a motion once established will instantly cease (3·77 (VI)) if the solid be brought to rest. We shall consider only motions of the liquid which are due solely to the motion of the solid in the sense just described. In such a motion the pressure of the liquid at the surface of the solid is finite, and therefore to generate a given motion of the solid requires only a finite amount of energy, which is shared between the solid and the liquid. The kinetic energy of the liquid is therefore finite, and so the velocity at infinity must be zero. Thus the velocity potential ϕ must satisfy the conditions

$$(1) \qquad \nabla^2 \phi = 0 \text{ everywhere}, \quad \nabla \phi = 0 \text{ at infinity}.$$

To describe the conditions* to be satisfied at the boundary of the solid, take a frame of reference R' fixed relatively to the solid, say an origin O' and three cartesian axes $O'x$, $O'y$, $O'z$. The motion of the solid is then specified by a velocity $\mathbf{u}$ of the origin and an angular velocity $\boldsymbol{\omega}$. Thus at the point $\mathbf{r}$ of the boundary the velocity is $\mathbf{u}+\boldsymbol{\omega}_\wedge \mathbf{r}$, and if $\mathbf{n}$ is the unit *outward* normal vector at this point to the surface of the solid, the boundary condition is

$$(2) \qquad -\frac{\partial \phi}{\partial n} = \mathbf{n}(\mathbf{u}+\boldsymbol{\omega}_\wedge \mathbf{r}) = \mathbf{u}\mathbf{n}+\boldsymbol{\omega}(\mathbf{r}_\wedge \mathbf{n}),$$

using the triple scalar product (2·13). This condition may be satisfied by writing

$$(3) \qquad \phi = \mathbf{u}\boldsymbol{\varphi}+\boldsymbol{\omega}\boldsymbol{\chi},$$

where $\boldsymbol{\varphi}$, $\boldsymbol{\chi}$ are vectors whose components along cartesian axes, say, are solutions of Laplace's equation whose gradient tends to zero at infinity, and which satisfy the boundary conditions

$$(4) \qquad -\frac{\partial \boldsymbol{\varphi}}{\partial n} = \mathbf{n}, \quad -\frac{\partial \boldsymbol{\chi}}{\partial n} = (\mathbf{r}_\wedge \mathbf{n}),$$

so that $\boldsymbol{\varphi}$, $\boldsymbol{\chi}$ depend solely on the shape of the solid and not on its motion.

---

* L. M. Milne-Thomson *Vth International Congress for Applied Mechanics (1938)*

Several special cases of the determination of ϕ have already been discussed, for example, the sphere and ellipsoid. We now proceed to investigate the motion of the solid by a method which depends essentially on considering the solid and liquid to form a single system. The pressure thrusts at the boundary then become internal forces and their evaluation is unnecessary.

18·20. Kinetic energy of the liquid. This is given by

$$(1) \quad T_L = -\tfrac{1}{2}\rho \int_{(S)} \phi \frac{\partial \phi}{\partial n}\, dS = \tfrac{1}{2}\rho \int_{(S)} (\mathbf{u}\boldsymbol{\varphi} + \boldsymbol{\omega}\boldsymbol{\chi})(\mathbf{n}[\mathbf{u} + \boldsymbol{\omega}_\wedge \mathbf{r}])\, dS,$$

using the boundary conditions 18·10 (2), (3), the integral being taken over the boundary of the solid.

This expression shows that T_L is a homogeneous quadratic function of the vectors $\mathbf{u}$, $\boldsymbol{\omega}$. For, if λ is any scalar, the effect of changing $\mathbf{u}$, $\boldsymbol{\omega}$ into $\lambda\mathbf{u}$, $\lambda\boldsymbol{\omega}$ is merely to change T_L into $\lambda^2 T_L$. Therefore by Euler's theorem on homogeneous functions (2·71) we have

$$(2) \qquad \mathbf{u}\frac{\partial T_L}{\partial \mathbf{u}} + \boldsymbol{\omega}\frac{\partial T_L}{\partial \boldsymbol{\omega}} = 2T_L.$$

Again, from 18·10 (2), (3), we get

$$\frac{\partial}{\partial \mathbf{u}}\left(\phi \frac{\partial \phi}{\partial n}\right) = -\mathbf{n}\phi + \boldsymbol{\varphi}\frac{\partial \phi}{\partial n}, \text{ and therefore}$$

$$\frac{\partial T_L}{\partial \mathbf{u}} = \tfrac{1}{2}\rho \int_{(S)} \mathbf{n}\phi\, dS - \tfrac{1}{2}\rho \int_{(S)} \boldsymbol{\varphi}\frac{\partial \phi}{\partial n}\, dS.$$

But since the components of $\boldsymbol{\varphi}$ satisfy Laplace's equation, Green's theorem, 2·62 (2), gives

$$\int_{(S)} \boldsymbol{\varphi}\frac{\partial \phi}{\partial n}\, dS = \int_{(S)} \phi \frac{\partial \boldsymbol{\varphi}}{\partial n}\, dS = -\int \mathbf{n}\phi\, dS$$

from 18·10 (4). Therefore we get the first of the following results, and the second by a similar argument,

$$(3) \qquad \frac{\partial T_L}{\partial \mathbf{u}} = \rho \int_{(S)} \mathbf{n}\phi\, dS, \quad \frac{\partial T_L}{\partial \boldsymbol{\omega}} = \rho \int_{(S)} (\mathbf{r}_\wedge \mathbf{n})\phi\, dS.$$

Were the motion started by impulses (see 18·31), the integrals on the right of (3) would be the linear impulse and impulsive moment applied by the solid S to the boundary of the liquid in contact with it.

When written in full for, say, cartesian coordinates, the above expression (1) for the kinetic energy will be found to consist of 21 terms containing the quadratic combinations two at a time of the six components of $\mathbf{u}$ and $\boldsymbol{\omega}$.

If in (3) we write $\lambda\mathbf{u}$, $\lambda\boldsymbol{\omega}$ for $\mathbf{u}$ and $\boldsymbol{\omega}$, these expressions become multiplied by λ. Thus the partial derivatives of the kinetic energy are homogeneous linear functions of the vectors $\mathbf{u}$, $\boldsymbol{\omega}$.

In cartesian coordinates, we have

$$\mathbf{u} = \mathbf{i}u_x + \mathbf{j}u_y + \mathbf{k}u_z\,,$$

$$\boldsymbol{\omega} = \mathbf{i}\omega_x + \mathbf{j}\omega_y + \mathbf{k}\omega_z\,,$$

and therefore (2·71)

$$\frac{\partial T_L}{\partial \mathbf{u}} = \mathbf{i}\,\frac{\partial T_L}{\partial u_x} + \mathbf{j}\,\frac{\partial T_L}{\partial u_y} + \mathbf{k}\,\frac{\partial T_L}{\partial u_z}\,,$$

$$\frac{\partial T_L}{\partial \boldsymbol{\omega}} = \mathbf{i}\,\frac{\partial T_L}{\partial \omega_x} + \mathbf{j}\,\frac{\partial T_L}{\partial \omega_y} + \mathbf{k}\,\frac{\partial T_L}{\partial \omega_z}\,.$$

18·21. The kinetic energy of the solid is given by

(1) $$T_S = \tfrac{1}{2}\int_{(V)} \sigma[\mathbf{u} + (\boldsymbol{\omega}_{\wedge}\mathbf{r})]^2 d\tau,$$

taken throughout the volume V enclosed by the solid, σ being the density which may be constant or variable. In cartesian coordinates this expression contains 10 independent coefficients. By Euler's theorem

(2) $$\mathbf{u}\,\frac{\partial T_S}{\partial \mathbf{u}} + \boldsymbol{\omega}\,\frac{\partial T_S}{\partial \boldsymbol{\omega}} = 2T_S.$$

From (1) we get at once

(3) $$\frac{\partial T_S}{\partial \mathbf{u}} = \int_{(V)} \sigma(\mathbf{u} + \boldsymbol{\omega}_{\wedge}\mathbf{r})\,d\tau = \mathbf{M}_S\,,$$

(4) $$\frac{\partial T_S}{\partial \boldsymbol{\omega}} = \int_{(V)} \sigma\mathbf{r}_{\wedge}(\mathbf{u} + \boldsymbol{\omega}_{\wedge}\mathbf{r})\,d\tau = \mathbf{H}_S\,,$$

the linear and angular momentum of the solid.

If $T = T_L + T_S$ is the total kinetic energy of the system, combining (2) with 18·20 (2) gives

(5) $$\mathbf{u}\,\frac{\partial T}{\partial \mathbf{u}} + \boldsymbol{\omega}\,\frac{\partial T}{\partial \boldsymbol{\omega}} = 2T.$$

18·30. The wrench. A system of forces represented by *localised* vectors, acting at given points, has for resultant a single force $\mathbf{F}$ acting through any chosen *base-point* O and a couple $\mathbf{L}$. The force $\mathbf{F}$ is then the vector, localised in a line through O, obtained by drawing through O vectors equal and parallel to the given forces and taking their vector sum. Thus the magnitude and direction of $\mathbf{F}$ do not depend on the position of O. On the other hand, $\mathbf{L}$ is the sum of the moments about O of the given localised forces, and consequently its magnitude and direction depend on the position of O, but $\mathbf{L}$ is a *free* vector. The vector pair ($\mathbf{F}$, $\mathbf{L}$) is called a *force wrench*. For two force wrenches to be equal, both force and couple components must be equal when referred to the same base-point. By suitable choice of O it is possible to arrange that the axis of the couple shall be parallel to $\mathbf{F}$. The line in which $\mathbf{F}$ is then localised is called the *central axis*. If the corresponding couple is $\mathbf{\Gamma}$, this reduction is unique and $\mathbf{F}_\wedge \mathbf{\Gamma} = 0$.

In exactly the same way a system of localised impulses gives rise to an *impulse wrench* $(\boldsymbol{\xi}, \boldsymbol{\lambda})$, where $\boldsymbol{\xi}$ is the vector sum of the impulses and $\boldsymbol{\lambda}$ is the sum of their vector moments about the base-point. Again, a *momentum wrench* $(\mathbf{M}, \mathbf{H})$ arises when we compound localised linear momenta into a single localised linear momentum vector $\mathbf{M}$, and an angular momentum vector $\mathbf{H}$.

18·31. The impulse. When a solid S moves in a given manner in unbounded liquid, the motion of the liquid being due solely to the motion of the solid, the motion of the liquid is uniquely determinate and the velocity potential ϕ is likewise determinate, see 3·77 (VII), (save for an irrelevant added constant). The motion of the liquid which actually exists at time t could be generated *instantaneously* from rest by applying to the solid a suitable impulse wrench. This impulse wrench must be so adjusted as to produce instantaneously in the solid the momentum wrench which actually exists at time t and also to overcome the perfectly definite impulse wrench exerted on the boundary of the solid by the impulsive pressures $\rho\phi$ of the liquid, see 3·64.

The impulse wrench on the solid which would thus generate the motion from rest is called the *impulse* of the system at the instant considered.

18·32. Rate of change of the impulse. Instead of the moving frame R' of origin O' fixed relatively to the solid S, we consider in this section a frame of reference R, origin O, fixed in space (see 3·55). We shall denote time rate of change with respect to this frame R by $\partial/\partial t$. We shall prove that if $(\boldsymbol{\xi}, \boldsymbol{\lambda})$ is the impulse wrench, as defined in 18·31, and $(\mathbf{F}, \mathbf{L})$ the external force wrench on the solid, referred to the same base-point O, then

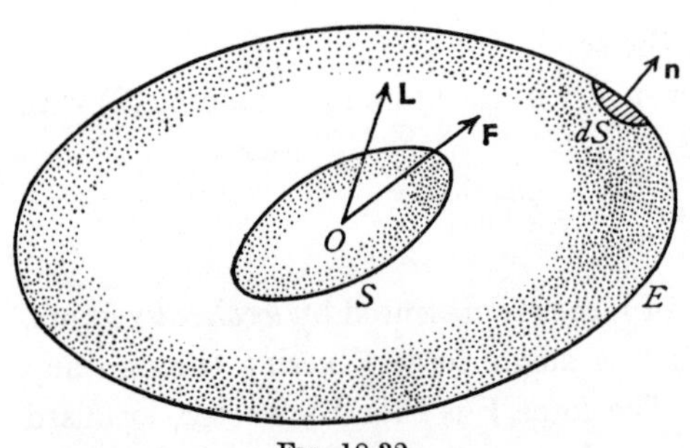

Fig. 18·32.

$$(1) \qquad \frac{\partial \boldsymbol{\xi}}{\partial t} = \mathbf{F}, \quad \frac{\partial \boldsymbol{\lambda}}{\partial t} = \mathbf{L}.$$

Proof. Let us imagine a closed surface E, fixed in space, to enclose the solid S. This surface is merely geometrical and is not a material barrier to the motion. Let $(\mathbf{M}_E, \mathbf{H}_E)$ be the momentum of the system Σ_E consisting of the solid and the liquid within E at time t. If we suppose the motion of the solid and unbounded liquid which actually exists at time t to be generated instantaneously from rest as described in 18·31 by the impulse $(\boldsymbol{\xi}, \boldsymbol{\lambda})$ applied to the *solid*, there will be an impulsive pressure $\rho\phi$ throughout the liquid, and therefore the external impulse acting on the system Σ_E will consist solely and entirely of $(\boldsymbol{\xi}, \boldsymbol{\lambda})$ and an impulsive pressure $\rho\phi$ over the surface E. These impulses therefore generate the momentum $(\mathbf{M}_E, \mathbf{H}_E)$. Thus, if $\mathbf{n}$ is the unit outward normal to dS,

$$(2) \qquad \boldsymbol{\xi} - \int_{(E)} \mathbf{n}\rho\phi \, dS = \mathbf{M}_E, \quad \boldsymbol{\lambda} - \int_{(E)} (\mathbf{r}_\wedge \mathbf{n}\rho\phi)\, dS = \mathbf{H}_E,$$

the second integral being the moment about O of the impulsive pressure thrust.

Again, the pressure equation gives

$$p = -\tfrac{1}{2}\rho q^2 + \rho\frac{\partial\phi}{\partial t} + C,$$

and a pressure C, uniform over the boundary, has no resultant. Therefore the equations of motion of the system Σ_E are

$$\mathbf{F} - \int_{(E)} \mathbf{n}\left(\rho\frac{\partial\phi}{\partial t} - \tfrac{1}{2}\rho q^2\right) dS = \frac{\partial \mathbf{M}_E}{\partial t} + \int_{(E)} \rho\mathbf{q}(\mathbf{n}\mathbf{q})\, dS,$$

$$\mathbf{L} - \int_{(E)} (\mathbf{r}_\wedge \mathbf{n})\left(\rho\frac{\partial\phi}{\partial t} - \tfrac{1}{2}\rho q^2\right) dS = \frac{\partial \mathbf{H}_E}{\partial t} + \int_{(E)} \rho\mathbf{r}_\wedge \mathbf{q}(\mathbf{n}\mathbf{q})\, dS,$$

the integrals on the right expressing the flux of linear and angular momentum through E (see 3·40, 3·42). Eliminating $\mathbf{M}_E$, $\mathbf{H}_E$ by means of equations (2), we get

$$(3) \qquad \frac{\partial\boldsymbol{\xi}}{\partial t} - \mathbf{F} = \rho\int_{(E)} [\tfrac{1}{2}\mathbf{n}q^2 - (\mathbf{n}\mathbf{q})\mathbf{q}]\, dS,$$

$$(4) \qquad \frac{\partial\boldsymbol{\lambda}}{\partial t} - \mathbf{L} = \rho\int_{(E)} [\tfrac{1}{2}(\mathbf{r}_\wedge \mathbf{n})\, q^2 - (\mathbf{n}\mathbf{q})(\mathbf{r}_\wedge \mathbf{q})]\, dS.$$

Since the left sides of these equations are independent of E, these equations show that the integrals on the right are independent of our particular choice of the envelope E.* We shall prove, by taking all points of the envelope to be infinitely distant from the solid, that these integrals are zero.† The result (1) will then follow.

Now, from 3·75 (3), we have at any point P of the liquid

$$\phi_P = C - \int_{(S)} \frac{1}{r}\left(\frac{1}{4\pi}\frac{\partial\phi}{\partial n}\right) dS + \int_{(S)} \frac{\phi}{4\pi}\frac{\partial}{\partial n}\left(\frac{1}{r}\right) dS,$$

where r is the distance of P from the element dS of the surface of the solid over which the integrals are taken.

For points P at a great distance R from O we can write $r = R + s$, where s/R is infinitesimal and therefore approximately

$$\frac{1}{r} = \frac{1}{R} - \frac{s}{R^2}, \quad \frac{1}{r^2} = \frac{1}{R^2} - \frac{2s}{R^3}.$$

---

* This independence also follows from 3·63.

† Observe that we are not stating that these integrals are zero in the limit but that they have a constant value. That this constant value is zero is then inferred from their limiting behaviour at infinity.

Noting that $\int_{(S)} \frac{\partial \phi}{\partial n} dS = 0$ by the equation of continuity, we see that to the above order of approximation

$$\phi_P = C + \frac{A}{R^2} + \dots ,$$

where A is independent of R. It follows that the speed q is of order R^{-3}. Also for points on the envelope E, $dS = R^2 d\omega$, where $d\omega$ is an elementary solid angle. Thus in (3) and (4) the magnitudes are of the order

$$\int_{(E)} \frac{d\omega}{R^4}, \quad \int_{(E)} \frac{d\omega}{R^3},$$

and these clearly tend to zero when $R \to \infty$. Q.E.D.

18·40. Moving origin. It is convenient to refer the motion not to the frame R, origin O, fixed in space, but to the frame R', origin O', fixed in the solid (18·10). At time t the frame R' occupies a certain position in space. We choose the frame R so that it coincides with R' at this instant. Let the motion of R' be described by the velocity $\mathbf{u}$ of the origin O' and an angular velocity $\boldsymbol{\omega}$, both relative to the fixed frame R. We consider the changes in $\boldsymbol{\xi}$, $\boldsymbol{\lambda}$ in a short interval dt. Since the interval is infinitesimal we can consider separately the effects of the translation $\mathbf{u}\, dt$ of the origin, the rotation $\boldsymbol{\omega}\, dt$ of the frame, and the changes in the vectors in the interval dt as estimated by an observer moving with R', and then add the results.

To consider the effect of the translation we ignore the rotation, and suppose the vectors $\boldsymbol{\xi}$, $\boldsymbol{\lambda}$ to remain unchanged with respect to an observer moving with the frame R'.

Since the localised vector $\boldsymbol{\xi}$ is merely moved parallel to itself it undergoes no change. On the other hand, the moment of the impulse with respect to

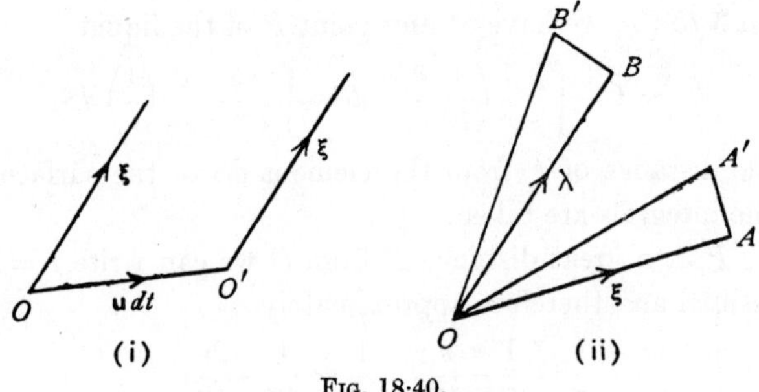

FIG. 18·40.

the fixed origin O is increased by the moment about O of $\boldsymbol{\xi}$ in its new position at O', fig. 18·40 (i), namely by $\mathbf{u}\, dt \wedge \boldsymbol{\xi}$. Thus, due to the translation of the origin, $\boldsymbol{\lambda}$ increases at the rate $\mathbf{u} \wedge \boldsymbol{\xi}$.

Now consider the rotation $\boldsymbol{\omega}\, dt$, the origin remaining fixed and the impulses remaining unchanged with respect to an observer moving with the frame R'.

If OA, OB, fig. 18·40 (ii), represent $\boldsymbol{\xi}$, $\boldsymbol{\lambda}$ at time t they will be represented by OA', OB' at time $t+dt$, where

$$\overrightarrow{AA'} = \boldsymbol{\omega}\, dt \wedge \overrightarrow{OA}, \quad \overrightarrow{BB'} = \boldsymbol{\omega}\, dt \wedge \overrightarrow{OB},$$

and therefore with regard to the fixed frame R the rates of increase are $\boldsymbol{\omega} \wedge \boldsymbol{\xi}$ and $\boldsymbol{\omega} \wedge \boldsymbol{\lambda}$ respectively.

Lastly, to an observer moving with the frame R', $\boldsymbol{\xi}$, $\boldsymbol{\lambda}$ appear to vary with the time at the rates which we shall denote by $d\boldsymbol{\xi}/dt$, $d\boldsymbol{\lambda}/dt$.

Thus the rates of change of $\boldsymbol{\xi}$, $\boldsymbol{\lambda}$ with respect to the fixed frame R with which the moving frame R' instantaneously coincides, are respectively

$$\frac{\partial \boldsymbol{\xi}}{\partial t} = \frac{d\boldsymbol{\xi}}{dt} + \boldsymbol{\omega} \wedge \boldsymbol{\xi}, \quad \frac{\partial \boldsymbol{\lambda}}{\partial t} = \frac{d\boldsymbol{\lambda}}{dt} + \boldsymbol{\omega} \wedge \boldsymbol{\lambda} + \mathbf{u} \wedge \boldsymbol{\xi}.$$

18·41. Equations of motion. Since the rate of change of the impulse is equal to the external force, 18·32, we have

$$\frac{d\boldsymbol{\xi}}{dt} + \boldsymbol{\omega} \wedge \boldsymbol{\xi} = \mathbf{F},$$

$$\frac{d\boldsymbol{\lambda}}{dt} + \boldsymbol{\omega} \wedge \boldsymbol{\lambda} + \mathbf{u} \wedge \boldsymbol{\xi} = \mathbf{L}.$$

These are the equations of motion in the form suitable for resolution along axes fixed in the moving solid.

18·42. Impulse derived from the kinetic energy. If $(\boldsymbol{\xi}, \boldsymbol{\lambda})$ are the components of the impulse at the base-point O, a force wrench applied to the body for an infinitesimal time, δt, will increase the velocity from $(\mathbf{u}, \boldsymbol{\omega})$ to $(\mathbf{u}+\delta\mathbf{u}, \boldsymbol{\omega}+\delta\boldsymbol{\omega})$, and the corresponding impulse will then be $(\boldsymbol{\xi}+\delta\boldsymbol{\xi}, \boldsymbol{\lambda}+\delta\boldsymbol{\lambda})$, all the increments being infinitesimal. The work done is, by the definition of an impulse,* $\mathbf{u}\,\delta\boldsymbol{\xi} + \boldsymbol{\omega}\,\delta\boldsymbol{\lambda}$, and this must be equal to the increase in total kinetic energy

$$T = T_S + T_L.$$

Therefore

$$\text{(1)} \qquad \mathbf{u}\,\delta\boldsymbol{\xi} + \boldsymbol{\omega}\,\delta\boldsymbol{\lambda} = \frac{\partial T}{\partial \mathbf{u}}\,\delta\mathbf{u} + \frac{\partial T}{\partial \boldsymbol{\omega}}\,\delta\boldsymbol{\omega} = \delta T.$$

If we take $\delta\mathbf{u} = h\mathbf{u}$, $\delta\boldsymbol{\omega} = h\boldsymbol{\omega}$, where h is an infinitesimal scalar constant,

* If $(\mathbf{F}, \mathbf{L})$ is the force wrench in question the work done is

$$\mathbf{F}(\mathbf{u}\,\delta t) + \mathbf{L}(\boldsymbol{\omega}\,\delta t) = \mathbf{u}(\mathbf{F}\,\delta t) + \boldsymbol{\omega}(\mathbf{L}\,\delta t) = \mathbf{u}\,\delta\boldsymbol{\xi} + \boldsymbol{\omega}\,\delta\boldsymbol{\lambda}.$$

since the impulse is a homogeneous linear function of the velocities, we must have also $\delta\boldsymbol{\xi} = h\boldsymbol{\xi}$, $\delta\boldsymbol{\lambda} = h\boldsymbol{\lambda}$, and therefore

$$\mathbf{u}\boldsymbol{\xi}+\boldsymbol{\omega}\boldsymbol{\lambda} = \frac{\partial T}{\partial \mathbf{u}}\mathbf{u}+\frac{\partial T}{\partial \boldsymbol{\omega}}\boldsymbol{\omega} = 2T,$$

from 18·21 (5). Taking an infinitesimal variation of this equation, we get

$$\mathbf{u}\,\delta\boldsymbol{\xi}+\boldsymbol{\xi}\,\delta\mathbf{u}+\boldsymbol{\omega}\,\delta\boldsymbol{\lambda}+\boldsymbol{\lambda}\,\delta\boldsymbol{\omega} = 2\,\delta T,$$

and therefore, from (1),

$$\boldsymbol{\xi}\,\delta\mathbf{u}+\boldsymbol{\lambda}\,\delta\boldsymbol{\omega} = \frac{\partial T}{\partial \mathbf{u}}\delta\mathbf{u}+\frac{\partial T}{\partial \boldsymbol{\omega}}\delta\boldsymbol{\omega}.$$

Since $\delta\mathbf{u}$, $\delta\boldsymbol{\omega}$ are independent, this gives

$$\boldsymbol{\xi} = \frac{\partial T}{\partial \mathbf{u}}, \quad \boldsymbol{\lambda} = \frac{\partial T}{\partial \boldsymbol{\omega}},$$

which expresses the impulse in terms of the partial derivatives of the total kinetic energy.*

18·43. Equations in terms of kinetic energy. The equations of motion, 18·41, now become

$$\frac{d}{dt}\left(\frac{\partial T}{\partial \mathbf{u}}\right)+\boldsymbol{\omega}\wedge\frac{\partial T}{\partial \mathbf{u}} = \mathbf{F},$$

$$\frac{d}{dt}\left(\frac{\partial T}{\partial \boldsymbol{\omega}}\right)+\boldsymbol{\omega}\wedge\frac{\partial T}{\partial \boldsymbol{\omega}}+\mathbf{u}\wedge\frac{\partial T}{\partial \mathbf{u}} = \mathbf{L}.$$

These are Kirchhoff's equations in vector form.

If we observe that $T = T_S+T_L$, these equations can be written

$$\frac{d}{dt}\left(\frac{\partial T_S}{\partial \mathbf{u}}\right)+\boldsymbol{\omega}\wedge\frac{\partial T_S}{\partial \mathbf{u}} = \mathbf{F}-\frac{d}{dt}\left(\frac{\partial T_L}{\partial \mathbf{u}}\right)-\boldsymbol{\omega}\wedge\frac{\partial T_L}{\partial \mathbf{u}},$$

$$\frac{d}{dt}\left(\frac{\partial T_S}{\partial \boldsymbol{\omega}}\right)+\boldsymbol{\omega}\wedge\frac{\partial T_S}{\partial \boldsymbol{\omega}}+\mathbf{u}\wedge\frac{\partial T_S}{\partial \mathbf{u}} = \mathbf{L}-\frac{d}{dt}\left(\frac{\partial T_L}{\partial \boldsymbol{\omega}}\right)-\boldsymbol{\omega}\wedge\frac{\partial T_L}{\partial \boldsymbol{\omega}}-\mathbf{u}\wedge\frac{\partial T_L}{\partial \mathbf{u}}.$$

Now, if the liquid had been absent ($T_L = 0$), the right side of these equations would have contained only **F** and **L**. The action of the liquid pressures on the body must therefore be represented by the remaining terms on the right. Thus the action of the liquid is represented by the force and couple

$$\mathbf{F}_L = -\frac{d}{dt}\left(\frac{\partial T_L}{\partial \mathbf{u}}\right)-\boldsymbol{\omega}\wedge\frac{\partial T_L}{\partial \mathbf{u}},$$

$$\mathbf{L}_L = -\frac{d}{dt}\left(\frac{\partial T_L}{\partial \boldsymbol{\omega}}\right)-\boldsymbol{\omega}\wedge\frac{\partial T_L}{\partial \boldsymbol{\omega}}-\mathbf{u}\wedge\frac{\partial T_L}{\partial \mathbf{u}}.$$

---

* This result can also be inferred by combining 18·20 (3) with 18·21 (3), (4). This is left as an exercise.

18·50. Permanent translation. If the motion is steady and the solid does not rotate, the action of the liquid on the solid reduces to zero force, d'Alembert's paradox, and a couple

$$(1) \qquad -\mathbf{u}_{\wedge}\frac{\partial T_L}{\partial \mathbf{u}}.$$

This couple (cf. 6·42) will tend to rotate the solid. The couple, however, vanishes if, and only if, the above vector product vanishes, which means that the vectors $\mathbf{u}$ and $\partial T_L/\partial \mathbf{u}$ are parallel.

Therefore in this case the velocity $\mathbf{u}$ is in the direction of the normal to the ellipsoid

$$\mathbf{u}\frac{\partial T_L}{\partial \mathbf{u}} = c,$$

where c is a constant.* From 18·20 (2), since $\boldsymbol{\omega} = 0$, the equation of this ellipsoid is also

$$(2) \qquad T_L = \tfrac{1}{2}c,$$

and is analogous to the momental ellipsoid of a rigid body. The direction of $\mathbf{u}$ can only be parallel to the normal if $\mathbf{u}$ is along one of the principal axes of the ellipsoid.

Since there are three principal axes, it follows that there are three directions in space, mutually perpendicular, such that, if the solid is set moving along one of them without rotation, it will continue so to move. These directions are known as *directions of permanent translation.*

When the solid is moving in a direction of permanent translation with velocity $\mathbf{u}$, a small disturbance is effected by changing $\mathbf{u}$ to $\mathbf{u}+\mathbf{v}$ and giving the body an angular velocity $\boldsymbol{\omega}$, where $\mathbf{v}$, $\boldsymbol{\omega}$ are initially infinitesimal so that, when their quadratic terms are neglected, Kirchhoff's equations become linear. The discussion of the stability entails the solution of these equations and is, except in certain cases of symmetry, rather complicated. We can however obtain a general idea of the stability from the following argument in which the effect of $\boldsymbol{\omega}$ is ignored. Taking our solid to be itself an ellipsoid, the expression for T_L is obtained from 17·52. It appears from numerical calculation that the greatest axis of the ellipsoid (2) is in the direction of the greatest axis of the solid, fig. 18·50.

The sense of the couple (1) is also shown in the figure. Thus, if S moves in the direction OB of its least axis, the couple (1) tends to cancel any slight deviation therefrom. On the other hand, if the direction of motion is that of the greatest axis OA of S, any deviation is increased by the couple. If the direction of motion be that of the intermediate axis, the couple is restorative

* See Ex. II, 27. The ellipsoid here mentioned is an ellipsoid in the velocity-space in which the velocity components (u, v, w) take the place of the cartesian coordinates (x, y, z).

or not according to the direction of the deviated velocity. Thus when a body of this general shape moves through a liquid, the motion is stable only when it moves broadside on.

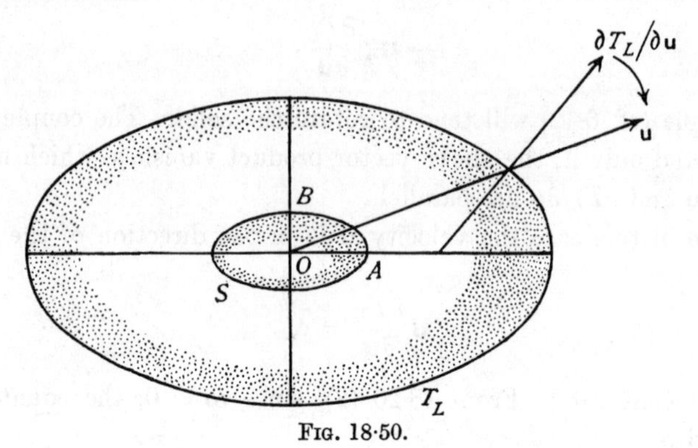

FIG. 18·50.

There are many well-known phenomena which are explained in principle by this remark. Thus a ship has to be kept on her course by the helmsman; an elongated airship requires similar attention. A sailing ship will not sail permanently before the wind with the helm lashed, but tends to set itself at right angles to the wind. A body sinking in liquid tends to sink with its longest dimension horizontal.

Lastly, we may remark that, to hold a body at rest in a uniform stream $\mathbf{u}$, requires a couple

$$\mathbf{u} \wedge \frac{\partial T_L}{\partial \mathbf{u}},$$

where T_L is the kinetic energy of the fluid when it is reduced to rest at infinity and the solid moves with velocity $\mathbf{u}$.

This may be regarded as supplementing d'Alembert's paradox by the statement that a solid in a uniform stream is acted upon by a couple except when presented to the stream in any of the three orientations corresponding to the directions of permanent translation.

18·51. Permanent rotation. When the solid rotates steadily without translation, we have $\mathbf{u} = 0$, and the body is acted upon by the couple

$$-\boldsymbol{\omega} \wedge \frac{\partial T_L}{\partial \boldsymbol{\omega}}.$$

This couple vanishes when the vectors $\boldsymbol{\omega}$ and $\partial T_L/\partial \boldsymbol{\omega}$ are parallel, that to say, when the axis of rotation is parallel to the normal to the ellipsoid

$$\boldsymbol{\omega} \frac{\partial T_L}{\partial \boldsymbol{\omega}} = c \quad \text{or} \quad T_L = \tfrac{1}{2}c.$$

Thus there are also three axes of permanent rotation mutually perpendicular, but not necessarily intersecting, for the ellipsoid determines merely the directions and not the position of the axes.

18·52. Solid of revolution. When the solid has three perpendicular planes of symmetry, the total kinetic energy referred to their intersections as axes must assume the form

$$2T = Pu_x^2 + Qu_y^2 + Ru_z^2 + A\omega_x^2 + B\omega_y^2 + C\omega_z^2,$$

for the reversal of any velocity component must leave the kinetic energy unaltered, and therefore no product terms can occur. When the solid is one of revolution about the x-axis, T will be unaltered when u_y, u_z or ω_y, ω_z are interchanged. Therefore $Q = R$, $B = C$. If, in addition, the axis of revolution moves always in the xy-plane and there is no rotation about that axis, we must have $u_z = 0$, $\omega_y = \omega_x = 0$. Thus, in this case,

$$T = \tfrac{1}{2}(Pu_x^2 + Qu_y^2 + C\omega_z^2).$$

The equations of motion are then, if there are no external forces,

$$\mathbf{i}P\dot{u}_x + \mathbf{j}Q\dot{u}_y + P\omega_z u_x\mathbf{j} - Q\omega_z u_y\mathbf{i} = 0.$$

$$(1) \qquad C\dot{\omega}_z\mathbf{k} + (Q-P)\,\mathbf{k}u_x u_y = 0,$$

where **i**, **j**, **k** are unit vectors along the axes.

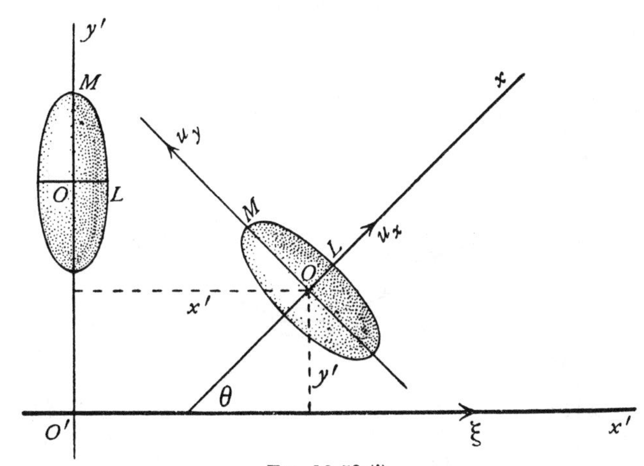

Fig. 18·52 (i).

Since there are no external forces, 18·32 (1) shows that the impulse components are constant. In the present case the angular component $\boldsymbol{\lambda}$ is perpendicular to the plane of the motion and therefore the impulse reduces to a single linear resultant $\boldsymbol{\xi}$ localised in the line $O'x'$ say. Then

$$(2) \qquad Pu_x = \xi\cos\theta, \quad Qu_y = -\xi\sin\theta, \quad \omega_z = \dot{\theta},$$

where θ is the inclination to the x'-axis of the line OL fixed in the solid and coinciding with the x-axis.

From (1),

$$(3) \qquad C\ddot{\theta}+\frac{\xi^2(P-Q)}{PQ}\sin\theta\cos\theta = 0.$$

Writing $\chi = 2\theta$, this becomes

$$(4) \qquad C\ddot{\chi}+\frac{\xi^2(P-Q)}{PQ}\sin\chi = 0.$$

If $P>Q$, this is the equation of motion of a pendulum. The value of χ given by (4) is therefore periodic, and hence so also is the value of θ given by (3).

If (x', y') are the coordinates of the centre of the solid, we get, from (2),

$$(5) \qquad \dot{x}' = u_x\cos\theta - u_y\sin\theta = \xi\left(\frac{\cos^2\theta}{P}+\frac{\sin^2\theta}{Q}\right),$$

$$(6) \qquad \dot{y}' = u_x\sin\theta + u_y\cos\theta = \xi\left(\frac{1}{P}-\frac{1}{Q}\right)\sin\theta\cos\theta = \frac{C\ddot{\theta}}{\xi},$$

from (3).

Equation (5) shows that $\dot{x}'$ is never negative, so that the centre always moves forwards, the path having no loops. From (6), we get *

$$y' = C\dot{\theta}/\xi,$$

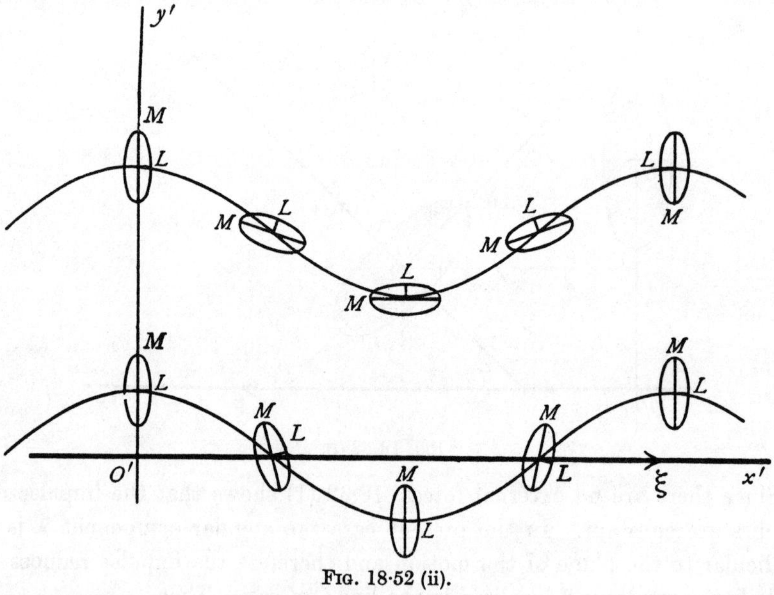

FIG. 18·52 (ii).

* No arbitrary constant is added since the moment of the impulse about the centre vanishes with y'.

so that y' is periodic, since θ is periodic, and therefore the path of the centre is a sinuous curve. This equation shows that y' is proportional to $\dot{\theta}$.

Two main cases can arise according as the solid performs complete revolutions or oscillates between the positions given by $\theta = \alpha$, $\theta = -\alpha$. These are illustrated in fig. 18·52 (ii). In the first case $\dot{\theta}$ has a fixed sign, so that the path does not intersect the line of the impulse. When the solid oscillates, however, $\dot{\theta}$ (and therefore y') vanishes in the extreme positions and the path lies symmetrically about the line of the impulse.

18·53. Stability due to rotation. In the case of a solid of revolution the kinetic energy can be put in the form

$$T = \tfrac{1}{2}[Au_x^2 + Bu_y^2 + Bu_z^2 + P\omega_x^2 + Q\omega_y^2 + Q\omega_z^2].$$

If the solid moves with velocity $\mathbf{i}u$, $\mathbf{i}\omega$, we shall have $u_x = u$, $\omega_x = \omega$ and, in a slight disturbance, u_y, ω_y, u_z, ω_z will be infinitesimal. Now

$$\frac{\partial T}{\partial \mathbf{u}} = \mathbf{i}Au_x + \mathbf{j}Bu_y + \mathbf{k}Bu_z,$$

$$\frac{\partial T}{\partial \boldsymbol{\omega}} = \mathbf{i}P\omega_x + \mathbf{j}Q\omega_y + \mathbf{k}Q\omega_z.$$

Therefore, neglecting products of the infinitesimals, the equations of motion parallel to the x-axis become

$$\frac{du_x}{dt} = 0, \quad \frac{d\omega_x}{dt} = 0.$$

Thus, to the first order, $u_x = u$, $\omega_x = \omega$.
To the same order of approximation, the remaining equations are

$$B\frac{du_y}{dt} - B\omega u_z + Au\omega_z = 0, \quad B\frac{du_z}{dt} + B\omega u_y - Au\omega_y = 0,$$

$$Q\frac{d\omega_y}{dt} + (P-Q)\,\omega\omega_z + (A-B)\,uu_z = 0,$$

$$Q\frac{d\omega_z}{dt} - (P-Q)\,\omega\omega_y - (A-B)\,uu_y = 0.$$

To solve these put $u_y = a\,e^{i\lambda t}$, $u_z = b\,e^{i\lambda t}$, $\omega_y = \alpha\,e^{i\lambda t}$, $\omega_z = \beta\,e^{i\lambda t}$. This gives the four equations:

$$\begin{array}{lllll}
iB\lambda a & -B\omega b & & +Au\beta & = 0,\\
B\omega a & +iB\lambda b & -Au\alpha & & = 0,\\
 & (A-B)ub & +Qi\lambda\alpha & +(P-Q)\omega\beta & = 0,\\
-(A-B)ua & & -(P-Q)\omega\alpha & +Qi\lambda\beta & = 0.
\end{array}$$

The elimination of a, b, α, β leads to the determinant

$$\begin{vmatrix} iB\lambda & -B\omega & 0 & Au \\ -B\omega & -iB\lambda & Au & 0 \\ 0 & (A-B)u & iQ\lambda & (P-Q)\omega \\ (A-B)u & 0 & (P-Q)\omega & -iQ\lambda \end{vmatrix} = 0$$

or, on expanding,

$$[BQ\lambda^2-(P-Q)B\omega^2-A(A-B)u^2]^2-[B(P-2Q)\lambda\omega]^2 = 0.$$

This gives for λ the two quadratic equations

$$BQ\lambda^2-B(P-2Q)\lambda\omega-B(P-Q)\omega^2-A(A-B)u^2 = 0,$$
$$BQ\lambda^2+B(P-2Q)\lambda\omega-B(P-Q)\omega^2-A(A-B)u^2 = 0.$$

Now the condition for stability is that λ should be real, for then $e^{i\lambda t}$ is periodic and therefore a disturbance, if once small, remains small.

The roots of both the above quadratics are real if

$$B^2(P-2Q)^2\omega^2 > -4BQ[B(P-Q)\omega^2+A(A-B)u^2],$$

that is to say, if $$B^2P^2\omega^2+4ABQ(A-B)u^2>0.$$

This condition is always satisfied if $A \geqslant B$, and can be satisfied in any case by taking ω large enough.

A familiar example of this principle is the stability given to a projectile by rifling the gun barrel.

18·54. Solid containing a cavity. When a solid has a cavity filled with liquid in acyclic motion, the total energy of the system is equal to that of the solid plus that of the liquid. The kinetic energy of the latter is clearly a homogeneous quadratic function of the velocity $(\mathbf{u}, \boldsymbol{\omega})$ of the solid, for the previous argument shows that the velocity potential is a homogeneous linear function of $(\mathbf{u}, \boldsymbol{\omega})$. Thus the effect of the liquid in the cavity is merely to alter the apparent constants of inertia of the solid, and the motion of the system is the same as that of the solid with these altered constants of inertia.

18·60. Lagrange's equations. The configuration of a dynamical system is known when the coordinates of every point in the system are known, or at least ascertainable from known quantities. These coordinates may be the ordinary cartesian x, y, z, or any other quantities in terms of which these may be expressed. Thus, in the case of a top spinning under gravity about a fixed point on its axis of revolution, it is sufficient to know the inclination θ of the axis to the vertical, and the angle ω which the vertical plane through this axis makes with a fixed vertical plane. Given θ and ω as functions of the time and the initial configuration and motion of the top, the position at time t of any point of the top can be ascertained. These quantities θ, ω are called *generalised coordinates.* Extending this idea, we can conceive the position of any specified dynamical system to be determinable in terms of a certain number of generalised coordinates $q_1, q_2, \ldots, q_n$.

When the position vector $\mathbf{r}$ of every point of the system is given explicitly by a relation of the type

$$(1) \qquad \mathbf{r} = \mathbf{r}(q_1, q_2, \ldots, q_n),$$

the system is said to be *holonomic.* As a direct consequence, the velocity is

$$(2) \qquad \mathbf{v} = \dot{\mathbf{r}} = \Sigma_i \boldsymbol{\alpha}_i \dot{q}_i,$$

where the subscript i in Σ_i denotes summation from $i = 1$ to $i = n$, and where

$$(3) \qquad \boldsymbol{\alpha}_i = \partial \mathbf{r}/\partial q_i, \quad \text{so that from (2)}$$

$$(4) \qquad \frac{\partial \dot{\mathbf{r}}}{\partial \dot{q}_i} = \boldsymbol{\alpha}_i = \frac{\partial \mathbf{r}}{\partial q_i}.$$

For a non-holonomic system, we have (2) without however (3), so that the non-integrable equation $d\mathbf{r} = \Sigma_i \boldsymbol{\alpha}_i \, dq_i$ takes the place of (1); and equation (4) no longer holds.

Consider now a system of solids S moving in inviscid liquid L, which may be either unbounded or contained within a fixed enveloping surface E. We shall assume that the solids form a holonomic system, and that the motion of the liquid is due entirely to the motion of the solids and would instantly cease, were the solids brought simultaneously to rest. The motion will then be irrotational and acyclic.

We cannot assume that the liquid is a holonomic system, for if the solids are moved through a cycle of positions returning each to its original position, we have no evidence that the fluid particles are then each in their original positions, indeed examples can be constructed which seem to point to the contrary conclusion. Therefore we cannot assume (1) to hold for the fluid particles.

At the boundary of a solid we have the condition

$$(5) \qquad -\frac{\partial \phi}{\partial n} = V_n,$$

where ϕ is the velocity potential and V_n is the normal component of the velocity of the solid. At the envelope E we have $V_n = 0$. But by hypothesis V_n is a *linear* function of the generalised velocities $\dot{q}_1, \dot{q}_2, \ldots, \dot{q}_n$, so that the boundary conditions (5) together with Laplace's equation determine ϕ uniquely as a *linear* function of the generalised velocities. Thus we can write

$$(6) \qquad \phi = \Sigma_i \phi_i \dot{q}_i,$$

where the ϕ_i are functions of the generalised coordinates, but not of the velocities, and satisfy Laplace's equation. Therefore taking the gradient it appears that (2) holds also for the liquid.

Now consider the rate at which work is being done by all the forces on the system in an imagined or *virtual motion* in which the generalised virtual velocities, which we shall denote by Dq_i/Dt, are geometrically possible so that for a point of a solid

$$\frac{D\mathbf{r}}{Dt} = \Sigma_i \frac{\partial \mathbf{r}}{\partial q_i}\frac{Dq_i}{Dt} \tag{7}$$

and for the motion of the liquid

$$\phi = \Sigma_i \phi_i \frac{Dq_i}{Dt}. \tag{8}$$

For brevity we shall term the rate of doing work by any system of forces their *power*, and denote it by DW/Dt for the virtual displacements.

Consider the solids only. If $\mathbf{F}_S$ is the total force, internal and external, on the typical particle of mass m at the point P, the virtual power of the forces acting on the solids is

$$\frac{DW_S}{Dt} = \Sigma_P \left(\mathbf{F}_S \frac{D\mathbf{r}}{Dt}\right) = \Sigma_{P,i}\left(\mathbf{F}_S \frac{\partial \mathbf{r}}{\partial q_i}\right)\frac{Dq_i}{Dt} = \Sigma_i Q_{S,i}\frac{Dq_i}{Dt} \tag{9}$$

on using (7), where

$$Q_{S,i} = \Sigma_P \left(\mathbf{F}_S \frac{\partial \mathbf{r}}{\partial q_i}\right) \tag{10}$$

is the *generalised force* corresponding with the generalised coordinate q_i. The equation of motion of the particle at P is $\mathbf{F}_S = m\ddot{\mathbf{r}}$, and therefore (10) gives

$$Q_{S,i} = \Sigma_P\, m\ddot{\mathbf{r}}\frac{\partial \mathbf{r}}{\partial q_i} = \frac{d}{dt}\left(\Sigma_P m\dot{\mathbf{r}}\frac{\partial \mathbf{r}}{\partial q_i}\right) - \Sigma_P m\dot{\mathbf{r}}\frac{\partial \dot{\mathbf{r}}}{\partial q_i}.$$

At this stage we use the holonomic property (4). Replacing $\partial\mathbf{r}/\partial q_i$ by $\partial\dot{\mathbf{r}}/\partial\dot{q}_i$ and noting that the kinetic energy of the solid is $T_S = \frac{1}{2}\Sigma_P m\dot{\mathbf{r}}\,\dot{\mathbf{r}}$, we get *Lagrange's equations* for the solids, namely

$$Q_{S,i} = \frac{d}{dt}\left(\frac{\partial T_S}{\partial \dot{q}_i}\right) - \frac{\partial T_S}{\partial q_i}, \quad i = 1, 2, \ldots, n. \tag{11}$$

Consider the liquid. To avoid confusion with the generalised coordinates we shall denote the fluid velocity by $\mathbf{v}$ instead of the customary $\mathbf{q}$. Then

$$\mathbf{v} = \frac{d\mathbf{r}}{dt} = -\frac{\partial \phi}{\partial \mathbf{r}} = -\Sigma_i \frac{\partial \phi_i}{\partial \mathbf{r}}\dot{q}_i, \tag{12}$$

on using (6), and therefore

$$\frac{\partial \mathbf{v}}{\partial \dot{q}_i} = -\frac{\partial \phi_i}{\partial \mathbf{r}}. \tag{13}$$

The kinetic energy of the liquid is $T_L = \frac{1}{2}\int \rho\mathbf{v}\,\mathbf{v}\,d\tau$, and therefore

$$\frac{\partial T_L}{\partial \dot{q}_i} = \rho\int \mathbf{v}\frac{\partial \mathbf{v}}{\partial \dot{q}_i}d\tau = -\rho\int \mathbf{v}\frac{\partial \phi_i}{\partial \mathbf{r}}d\tau \quad \text{from (13)}, \tag{14}$$

where the integrals are extended to the whole volume of the liquid. In a virtual motion of the liquid we write for the virtual velocity:

$$(15)\qquad \mathbf{V} = \frac{D\mathbf{r}}{Dt} = -\Sigma_i \frac{\partial \phi_i}{\partial \mathbf{r}} \frac{Dq_i}{Dt} \quad \text{from (8).}$$

Since d/dt and D/Dt are independent operators, (12) and (15) give

$$(16)\qquad \frac{D\mathbf{v}}{Dt} = \frac{d\mathbf{V}}{dt}.$$

Let $\mathbf{F}_L$ be the total force, including pressure thrust, per unit mass on a fluid particle. Then the equation of motion is

$$(17)\qquad \mathbf{F}_L = \dot{\mathbf{v}},$$

and the virtual power of the forces on the liquid is

$$(18)\qquad \frac{DW_L}{Dt} = \int \rho \mathbf{F}_L \mathbf{V}\, d\tau = -\Sigma_i \int \rho \mathbf{F}_L \frac{\partial \phi_i}{\partial \mathbf{r}} \frac{Dq_i}{Dt}\, d\tau = \Sigma_i Q_{L,i} \frac{Dq_i}{Dt},$$

$$(19)\qquad Q_{L,i} = -\int \rho \mathbf{F}_L \frac{\partial \phi_i}{\partial \mathbf{r}}\, d\tau.$$

Now consider

$$\int \rho \mathbf{v}\, \mathbf{V}\, d\tau = -\Sigma_i \int \rho \mathbf{v} \frac{\partial \phi_i}{\partial \mathbf{r}} \frac{Dq_i}{Dt}\, d\tau = \Sigma_i \frac{\partial T_L}{\partial \dot{q}_i} \frac{Dq_i}{Dt}$$

from (14). Operate on this equation with d/dt. Then

$$(20)\qquad \int \rho \dot{\mathbf{v}}\, \mathbf{V}\, d\tau + \int \rho \mathbf{v} \frac{d\mathbf{V}}{dt}\, d\tau = \Sigma_i \frac{d}{dt}\left(\frac{\partial T_L}{\partial \dot{q}_i}\right) \frac{Dq_i}{Dt} + \Sigma_i \frac{\partial T_L}{\partial \dot{q}_i} \frac{D\dot{q}_i}{Dt}.$$

But from (16)

$$\int \rho \mathbf{v} \frac{d\mathbf{V}}{dt}\, d\tau = \frac{D}{Dt} \int \tfrac{1}{2} \rho \mathbf{v}\, \mathbf{v}\, d\tau = \frac{DT_L}{Dt} = \Sigma_i \frac{\partial T_L}{\partial q_i} \frac{Dq_i}{Dt} + \Sigma_i \frac{\partial T_L}{\partial \dot{q}_i} \frac{D\dot{q}_i}{Dt}.$$

Combining this with (17), (18), (20) we get

$$\Sigma_i \left\{ Q_{L,i} - \frac{d}{dt}\left(\frac{\partial T_L}{\partial \dot{q}_i}\right) + \frac{\partial T_L}{\partial q_i} \right\} \frac{Dq_i}{Dt} = 0.$$

Since the Dq_i/Dt are independent we may put all but one equal to zero, and so we get Lagrange's equations for the liquid, namely,

$$(21)\qquad Q_{L,i} = \frac{d}{dt}\left(\frac{\partial T_L}{\partial \dot{q}_i}\right) - \frac{\partial T_L}{\partial q_i}, \quad i = 1, 2, \ldots, n.$$

If we write $T = T_S + T_L$ and add (11) and (21), we get

$$(22)\qquad Q_i = \frac{d}{dt}\left(\frac{\partial T}{\partial \dot{q}_i}\right) - \frac{\partial T}{\partial q_i},$$

$$(23)\qquad Q_i = \Sigma_P \mathbf{F}_S \frac{\partial \mathbf{r}}{\partial q_i} - \int \rho \mathbf{F}_L \frac{\partial \phi_i}{\partial \mathbf{r}}\, d\tau,$$

which constitute Lagrange's equations for the whole system of solids and liquid.

The generalised forces Q_i are the coefficients of Dq_i/Dt in the combination of virtual powers arising from (9) and (18), namely $D(W_S+W_L)/Dt$. In this expression the only forces which contribute to the virtual power are the *external forces* on the system (third law of motion), the only possible exception being pressure thrusts on the fixed envelope E. These, however, do no work, since the normal component of the velocity at E vanishes. In calculating the Q_i from (23) $\mathbf{F}_S$ and $\mathbf{F}_L$ may therefore be taken as the external forces on the solids and liquid. In the absence of external forces the Q_i are zero.

18·61. Sphere in the presence of a wall. When a sphere moves in liquid bounded by an infinite fixed rigid wall, the kinetic energy is given to a first approximation from the results of 17·31, 17·41 in the form

$$T = \tfrac{1}{2}(A\dot{x}^2+B\dot{y}^2),$$

where (x, y) are the coordinates of the centre referred to axes perpendicular to and along the wall, fig. 18·61, and

$$A = M+\tfrac{1}{2}M'\left(1+\tfrac{3}{8}\frac{a^3}{x^3}\right), \quad B = M+\tfrac{1}{2}M'\left(1+\tfrac{3}{16}\frac{a^3}{x^3}\right),$$

M being the mass of the sphere and M' the mass of the displaced liquid.

Then, by Lagrange's equations, if X, Y are the components of the external force acting on the sphere,

$$X = \frac{d}{dt}(A\dot{x})-\tfrac{1}{2}\dot{x}^2\frac{\partial A}{\partial x}-\tfrac{1}{2}\dot{y}^2\frac{\partial B}{\partial x},$$

$$Y = \frac{d}{dt}(B\dot{y}).$$

If the forces are so adjusted that $\dot{x}$, $\dot{y}$ remain constant, these equations give

$$X = \frac{9M'a^3}{64x^4}(-2\dot{x}^2+\dot{y}^2), \quad Y = -\frac{9M'a^3}{32x^4}\dot{x}\dot{y}.$$

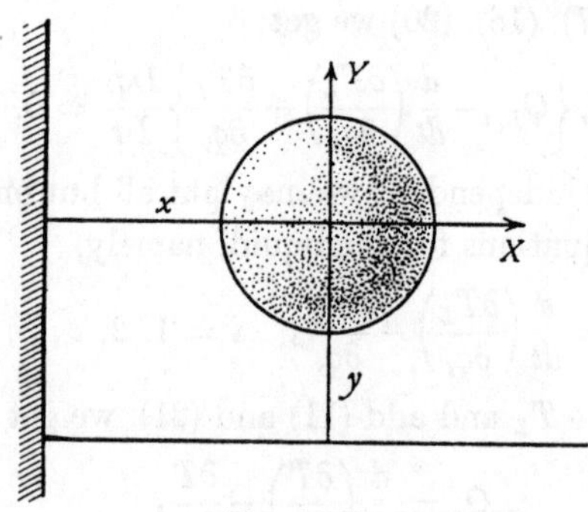

FIG. 18·61.

Thus if the sphere is moving directly towards or away from the wall ($\dot{y} = 0$), X is negative, and therefore a force towards the wall is required to maintain $\dot{x}$ constant. The sphere is therefore apparently repelled from the wall. On the other hand, when moving parallel to the wall ($\dot{x} = 0$), X is positive, and therefore a force away from the wall is required to maintain $\dot{y}$ constant. Thus the sphere is apparently attracted towards the wall.

Analogous results in the case of the motion of two spheres may easily be obtained by the application of Lagrange's equations.

18·70. Solid of revolution athwart an inviscid stream. Consider a stream U disturbed by a solid of revolution held with its axis perpendicular to the stream.

Let Π be the plane which contains the axis of the solid and the direction of the stream. Let γ be the circumference of the cross-section of the solid by a plane at distance x from a fixed point on the axis of the solid. Then any point P of the surface S of the solid is defined by coordinates (x, ω) where ω is the azimuth of the meridian plane through P measured from the plane Π.

The fluid velocity at P may be regarded as compounded of a component q_ω tangential to γ and a component q_m tangential to the meridian curve through P. It is then evident that we can write

$$(1) \qquad q_m(x, \omega) = Uf(x, \omega), \quad q_\omega(x, \omega) = Ug(x, \omega),$$

where the functions f and g are independent of U. We shall prove that *

$$(2) \qquad q_m(x, \omega) = q_m(x, 0) \cos \omega, \quad q_\omega(x, \omega) = q_\omega(x, \pi/2) \sin \omega.$$

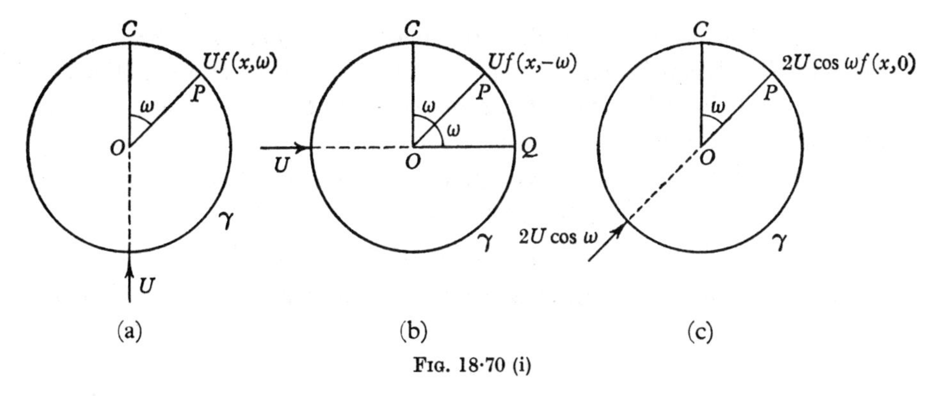

FIG. 18·70 (i)

Proof. Consider fig. 18·70 (i), which shows the point P on the circle γ, its centre O, and the point C of γ which in (a) is the point where the radius in the direction of the stream U meets γ. In (b) the radius in the direction of the stream U meets γ in Q, where OQ and OC are equally inclined to OP. In (c) the intensity of the stream is considered to be $2U \cos \omega$, and its direction is along OP.

* These elegant results are due to I. J. Campbell, *Q.J.M. and A.M.*, ix (1956), 140–2.

From (1) the meridional components of velocity are $Uf(x, \omega)$, $Uf(x, -\omega)$ and $2Uf(x, 0) \cos \omega$ in (a), (b), (c) respectively. Now the stream in (c) is the resultant of the superposition of those in (a) and (b). Therefore

$$Uf(x, \omega) + Uf(x, -\omega) = 2U \cos \omega f(x, 0).$$

But from the circular symmetry $f(x, -\omega) = f(x, \omega)$. Therefore

$$Uf(x, \omega) = Uf(x, 0) \cos \omega,$$

and this proves that

$$q_m(x, \omega) = q_m(x, 0) \cos \omega.$$

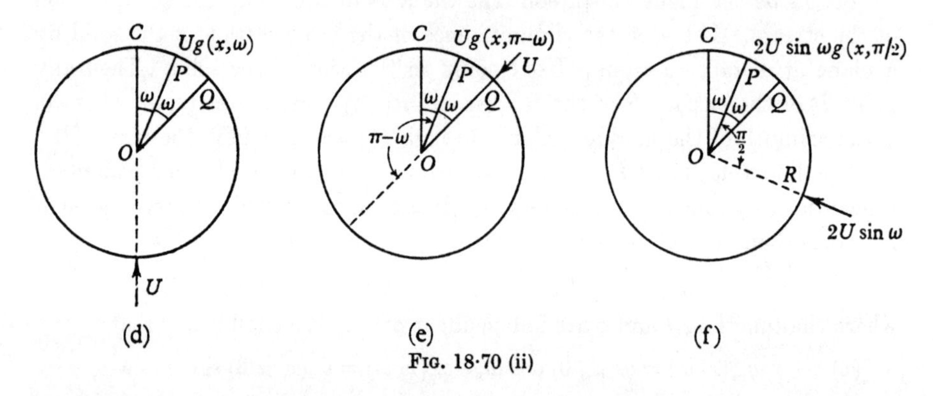

Fig. 18·70 (ii)

Now consider fig. 18·70 (ii) where the points C, P, Q are as before. In (d) the stream U is directed along OC, in (e) along QO. In (f) the stream is $2U \sin \omega$, and is directed along RO, where OR is derived from OP by clockwise rotation through a right angle. From (1) the components of velocity tangential to γ are $Ug(x, \omega)$, $Ug(x, \pi - \omega)$, $2U \sin \omega\, g(x, \pi/2)$ in (d), (e), (f) respectively, and since by superposition of (d) and (e) we get (f) it follows that

$$Ug(x, \omega) + Ug(x, \pi - \omega) = 2U \sin \omega\, g(x, \pi/2).$$

Now $g(x, \pi - \omega) = g(x, \omega)$, as is seen by reversing the stream, and therefore we have proved that

$$q_\omega(x, \omega) = q_\omega(x, \pi/2) \sin \omega.$$

Q.E.D.

EXAMPLES XVIII

1. If ϕ, ϕ' are the velocity potentials of two possible motions of an incompressible fluid in a simply connected region, prove that

$$\int \phi \frac{\partial \phi'}{\partial n} dS = \int \phi' \frac{\partial \phi}{\partial n} dS,$$

taken over the boundary.

A solid is surrounded by liquid enclosed in a fixed envelope. If the solid be set in motion with velocity $\mathbf{v}$, prove that the momentum given to the fluid is $-M\mathbf{v}$, where M is the mass of fluid displaced by the solid.

2. Define the impulse wrench $\mathbf{I}$ of a rigid body on a surrounding infinite liquid, where there are no irreducible circuits in the liquid, and show that it is not in general identical with the wrench of momentum of the fluid.

Prove that the force wrench applied by the solid to the liquid is equivalent to

$$\frac{d\mathbf{I}}{dt} + \mathcal{W}\!\!\int \rho\{\mathbf{V}(\boldsymbol{\nu}\mathbf{V}) - \tfrac{1}{2}\boldsymbol{\nu}V^2\}d\Sigma,$$

where $\mathcal{W}\!\!\int$ denotes a "wrench-integral", $\mathbf{V}$ is the velocity at any point P of a large fixed surface Σ enclosing the solid, and $\boldsymbol{\nu}$ is the unit vector in the direction of the outwards normal to Σ.

What conditions must be satisfied if the last wrench-integral is to be null in the limit, when all the points of the surface Σ are removed to infinity?

[A wrench-integral is the limit of the sum of infinitesimal localised vectors.]

(U.L.)

3. Obtain the velocity potential due to a sphere of centre O and radius a, moving with velocity U in the direction Ox, in an infinite liquid of density ρ.

Show that the x-momentum of the fluid contained between this sphere and any concentric sphere is zero, but that the x-momentum of the fluid contained between the sphere and any infinitely long circular cylinder of axis Ox is $\frac{1}{2}mU$, where m is the mass of fluid displaced by the sphere. (R.N.C.)

4. A rigid body is moving without rotation in an infinite liquid, the resolutes of its velocity parallel to the axes being (U, V, W) and its volume being v. Assuming the velocity potential ϕ of the flow produced to be given, at large distances from the body, by the development

$$\phi = \frac{ax+by+cz}{r^3} + \frac{S_2}{r^3} + \frac{S_3}{r^4} + \dots + \frac{S_m}{r^{m+1}} + \dots,$$

where the origin is some point of the body and S_m is a surface spherical harmonic of degree m, prove that the kinetic energy T of the moving liquid is given by

$$2T/\rho = 4\pi(aU+bV+cW) - v(U^2+V^2+W^2). \qquad \text{(U.L.)}$$

5. Obtain the equations of motion of a body, moving through unbounded liquid, in the form

$$\frac{d}{dt}\left(\frac{\partial T}{\partial u}\right) + \omega_2\frac{\partial T}{\partial w} - \omega_3\frac{\partial T}{\partial v} = X,$$

$$\frac{d}{dt}\left(\frac{\partial T}{\partial \omega_1}\right) + \omega_2\frac{\partial T}{\partial \omega_3} - \omega_3\frac{\partial T}{\partial \omega_2} + v\frac{\partial T}{\partial w} - w\frac{\partial T}{\partial v} = L,$$

where, in the notation of 18·43, $\mathbf{u} = (u, v, w)$, $\boldsymbol{\omega} = (\omega_1, \omega_2, \omega_3)$, $\mathbf{F} = (X, Y, Z)$, $\mathbf{L} = (L, M, N)$.

6. Assuming that the velocity potential of the motion set up in a liquid by the motion of an ellipsoid of semi-axes a, b, c with velocity u parallel to the direction of the axis of length $2a$ is of the form

$$Cux \int_{\lambda}^{\infty} \frac{d\psi}{(a^2+\psi)^{\frac{3}{2}}(b^2+\psi)^{\frac{1}{2}}(c^2+\psi)^{\frac{1}{2}}},$$

where C is a constant, determine C.

Find the kinetic energy of the fluid and the " impulse " of the motion. (U.L.

7. A prolate ellipsoid of revolution, of semi-axes a, b, is held in a stream of general velocity V, flowing in a direction which makes an angle θ with the major axis. Determine the couple, arising from the fluid pressure, which tends to set the major axis transverse to the stream.

[The fluid is homogeneous, incompressible, and frictionless, and has no external boundary.] (U.L.)

8. Show that if a solid of revolution moves through a liquid, then the form of the kinetic energy T is given by

$$2T = A(u^2+v^2)+Cw^2+P(\omega_1{}^2+\omega_2{}^2)+R\omega_3{}^2,$$

and prove that the steady motion given by

$$u = v = 0, \quad w = V; \quad \omega_1 = \omega_2 = 0, \quad \omega_3 = \Omega$$

is stable, provided $\Omega^2 > 4V^2PC(A-C)/AR^2$. (U.L.)

9. A solid of revolution, of uniform density and free from holes, immersed in an infinite liquid, is such that, when its motion is given by the velocity (u, v, w) of its centre of mass and the angular velocity $(\omega_1, \omega_2, \omega_3)$, the kinetic energy of the system is

$$\tfrac{1}{2}\{Au^2+B(v^2+w^2)+C\omega_1{}^2+D(\omega_2{}^2+\omega_3{}^2)\}.$$

The solid is initially at rest under gravity in an infinite liquid. Show that the equation determining the inclination θ of its axis to the vertical at any time is

$$D\ddot{\theta}+\frac{M^2g^2(A-B)}{AB}t^2\sin\theta\cos\theta = 0,$$

where M is the mass of the solid less the mass of the fluid displaced. (U.L.)

10. A solid of revolution with a plane of symmetry perpendicular to its axis moves through a fluid with the velocity $(\boldsymbol{\omega}, \mathbf{u})$. Show that for such a solid there is a possible steady motion in which $u_y\omega_z - u_z\omega_y = 0$, where the x-axis is the axis of revolution, and determine the character of the motion.

11. The kinetic energy of a solid moving two-dimensionally in an infinite liquid is given by the expression

$$2T = Au^2+Bv^2+C\omega^2,$$

where (u, v) is the velocity of the centre of mass referred to two axes, Ox, Oy fixed in the body, and ω is its angular velocity about the perpendicular axis Oz. Show that if the solid is initially moving with velocity U in the direction Ox, and without rotation, the motion will be stable if slightly disturbed, provided $A > B$. (U.L.)

12. If A, or B, and G are the force components and couple required to act per unit time in order to generate unit velocity perpendicular, or parallel, to the axis of a spheroid and unit angular velocity about a perpendicular axis, and C is the effective moment of inertia about the axis when the body moves in infinite liquid at rest at infinity, prove that the total kinetic energy T with the usual notation is given by

$$2T = A(u^2+v^2)+Bw^2+G(p^2+q^2)+Cr^2.$$

Express T in terms of Lagrange's coordinates $x, y, z, \theta, \phi, \psi$, and show that, if the impressed impulse is F parallel to Oz, then

$$G(\dot\theta^2+\dot\psi^2\sin^2\theta)+Cr^2+F^2\left(\frac{\sin^2\theta}{A}+\frac{\cos^2\theta}{B}\right) = 2T,$$

$$G\dot\psi\sin^2\theta+Cr\cos\theta = E, \text{ a constant}, \dot\phi+\dot\psi\cos\theta = r,$$

$$\dot x = -F\left(\frac{1}{A}-\frac{1}{B}\right)\sin\theta\cos\theta\cos\psi,$$

$$\dot y = -F\left(\frac{1}{A}-\frac{1}{B}\right)\sin\theta\cos\theta\sin\psi, \quad \dot z = F\left(\frac{\sin^2\theta}{A}+\frac{\cos^2\theta}{B}\right). \qquad \text{(U.L.)}$$

13. A pendulum consists of a rigid bar, free to turn about a fixed horizontal axis at its upper end, and a bob, in the form of a thin elliptic cylindrical shell filled with liquid. The generators of the cylinder are parallel to the fixed axis, the cylinder has plane ends at right angles to its generators, the central line of the bar (produced) lies along the minor axis of the middle cross-section, the whole mass, including the liquid, of the pendulum is M, its centre of mass is at a distance h from the fixed axis, the mass of the liquid is m, the major and minor semi-axes of a cross-section are a and b, the length of the simple equivalent pendulum is L, and this would become L' if the liquid solidified. Prove that

$$(L'-L)Mh(a^2+b^2) = ma^2b^2.$$

14. An anchor ring is immersed in fluid which is moving so that the circulation in any circuit which threads the ring once is constant. Prove that the motion is necessarily irrotational and that the circulation in any reducible circuit is zero. (U.L.)

15. A cylinder of negligible mass whose cross-section is an ellipse of axes $2a$, $2b$ is filled with water and placed at rest on a table with the major axis, $2a$, of the sections vertical and allowed to roll over. Find the angular speed when the major axis is horizontal in the cases (i) where the table is perfectly rough; (ii) where it is perfectly smooth; and show that the squares of these angular speeds are in the ratio

$$(a^2-b^2)^2 : (a^2-b^2)^2+4b^2(a^2+b^2).$$

16. A simple closed surface contains liquid and a solid. The surface is set in motion in any given manner, T_1 is the kinetic energy of the fluid when the solid is free, and T_2 when the solid is fixed; while T' is the kinetic energy of the fluid when, the boundary being held fixed, the solid is moved as in the first case; show that

$$T_2 < T_1 + T'. \qquad \text{(U.L.)}$$

17. Any number of spheres are moving in infinite liquid. Show that the "impulse" is compounded of impulses through the centre of each sphere and, if T is the total kinetic energy of all the solids and the liquid, then the impulse at the centre of the sphere whose position vector is $\mathbf{r}_0$ is

$$\partial T/\partial \dot{\mathbf{r}}_0 .$$

18. Two circular cylinders of unit length are placed between two parallel planes at unit distance apart. The cylinders can slide without friction between the planes and the intervening space is filled with liquid. If the cylinders are simultaneously projected at right angles to the plane of their axes, prove that they experience a mutual repulsion or attraction according as the directions of projection are in the opposite or the same sense.

19. Two spheres are moving in their line of centres at distance c apart, great compared with their radii a, b. Calculate the approximate value of the kinetic energy of the motion and write down the equations of motion.

If the spheres perform small oscillations about fixed positions, show that the mean value of the force acting on each is $3\pi\rho(a^3b^3/c^4)\,kk'p^2 \cos\epsilon$, where k, k' are the amplitudes of the oscillations, $2\pi/p$ the period, ϵ the phase difference. (U.L.)

CHAPTER XIX

VORTEX MOTION

19·10. Poisson's equation. Let f_Q be a continuous function whose value is defined at every point Q of a certain volume V. Let

$$\phi_P = \int_{(V)} \frac{f_Q}{PQ}\, d\tau_Q ,$$

where P is a point of V, and $d\tau_Q$ is an element of volume at Q. Then ϕ_P satisfies the equation

$$\text{(1)} \qquad \frac{\partial^2 \phi_P}{\partial \mathbf{P}^2} = \nabla^2 \phi_P = -4\pi f_P ,$$

which is known as Poisson's equation.

Proof. Consider a closed surface S enclosing P, containing the volume γ, and situated entirely within V. We can regard ϕ_P as the velocity potential due to a continuous distribution of sources of strength f_Q per unit volume, so that the element of volume $d\tau_Q$ behaves as a source of strength $f_Q\, d\tau_Q$ and therefore of output $4\pi f_Q\, d\tau_Q$. Then the outward normal flux across S is simply the sum of the outputs of all the sources within S and is therefore

$$4\pi \int_{(\gamma)} f_Q\, d\tau_Q = 4\pi \int_{(\gamma)} f_P\, d\tau_P .$$

Again, by Gauss's theorem, the outward normal flux is

$$-\int_{(\gamma)} \nabla^2 \phi_P\, d\tau_P .$$

Thus

$$\int_{(\gamma)} (\nabla^2 \phi_P + 4\pi f_P)\, d\tau_P = 0,$$

and, since (γ) is arbitrary, we have (1). Q.E.D.

Poisson's equation is also applicable, when ϕ and f are replaced by vectors, in the form

$$\partial^2 \boldsymbol{\varphi}_P / \partial \mathbf{P}^2 = \nabla^2 \boldsymbol{\varphi}_P = -4\pi \mathbf{f}_P ,$$

for the vectors can be resolved along three fixed vectors and the formula (1) is then applicable to each resolved part.

19·20. Velocity expressed in terms of vorticity. Consider liquid enclosed within a *fixed* envelope E, and suppose the vorticity $\boldsymbol{\zeta}$ to be given at every point. In those parts of the fluid, if any, where the motion is irrotational we shall have $\boldsymbol{\zeta} = 0$.

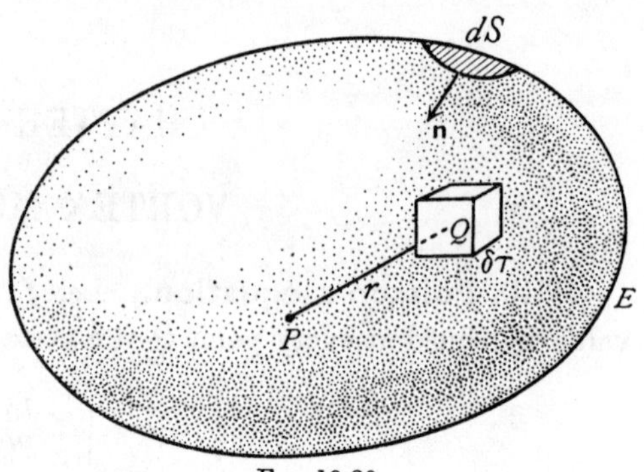

FIG. 19·20.

If $\mathbf{n}$ is the unit inward normal at the element dS of E, the boundary condition is

(1) $\mathbf{nq} = 0$ on E.

Take a point P within the fluid and regard P as fixed. The velocity at P will be denoted by $\mathbf{q}_P$, the velocity at Q by $\mathbf{q}_Q$, where Q is any other point in the fluid. Let us consider the vector

$$(2) \qquad \mathbf{A}_P = \frac{1}{4\pi}\int_{(V)} \frac{\mathbf{q}_Q}{PQ}\, d\tau,$$

where the integral is taken through the volume V enclosed by E, the point P remaining fixed.

As we shall have to differentiate sometimes regarding P as fixed and Q as variable and sometimes Q as fixed and P as variable, we use temporarily $\partial/\partial\mathbf{Q}$ or $\partial/\partial\mathbf{P}$ for ∇ according to which case is considered. The volume element $d\tau$ is $d\tau_Q$ throughout.

Then by Poisson's equation (19·10),

$$\frac{\partial^2 \mathbf{A}_P}{\partial \mathbf{P}^2} = -\mathbf{q}_P,$$

and therefore, from 2·32 (V),

$$(3) \qquad \mathbf{q}_P = \frac{\partial}{\partial\mathbf{P}} \wedge \left\{\frac{\partial}{\partial\mathbf{P}} \wedge \mathbf{A}_P\right\} - \frac{\partial}{\partial\mathbf{P}}\left(\frac{\partial\mathbf{A}_P}{\partial\mathbf{P}}\right).$$

Now $\mathbf{q}_Q$ is independent of the position of P, and therefore

$$\frac{\partial\mathbf{A}_P}{\partial\mathbf{P}} = \frac{1}{4\pi}\int_{(V)} \mathbf{q}_Q \frac{\partial}{\partial\mathbf{P}}\left(\frac{1}{PQ}\right) d\tau = -\frac{1}{4\pi}\int_{(V)} \mathbf{q}_Q \frac{\partial}{\partial\mathbf{Q}}\left(\frac{1}{PQ}\right) d\tau,$$

since $\dfrac{\partial}{\partial\mathbf{Q}}\left(\dfrac{1}{PQ}\right)$ is the velocity at Q due to a unit sink at P, and $\dfrac{\partial}{\partial\mathbf{P}}\left(\dfrac{1}{PQ}\right)$ is the velocity at P due to a unit sink at Q. These are equal but opposite vectors. Also, from 2·34 (VI),

$$\frac{\partial}{\partial\mathbf{Q}}\left(\frac{\mathbf{q}_Q}{PQ}\right) = \frac{1}{PQ}\frac{\partial\mathbf{q}_Q}{\partial\mathbf{Q}} + \mathbf{q}_Q \frac{\partial}{\partial\mathbf{Q}}\left(\frac{1}{PQ}\right),$$

and $\partial \mathbf{q}_Q/\partial \mathbf{Q} = 0$, from the equation of continuity. Thus

$$\frac{\partial \mathbf{A}_P}{\partial \mathbf{P}} = -\frac{1}{4\pi}\int_{(V)} \frac{\partial}{\partial \mathbf{Q}}\left(\frac{\mathbf{q}_Q}{PQ}\right) d\tau = \frac{1}{4\pi}\int_{(E)} \frac{\mathbf{n}\mathbf{q}_Q}{PQ}\, dS = 0,$$

from (1). Therefore (3) gives

$$\text{(4)} \qquad \mathbf{q}_P = \frac{\partial}{\partial \mathbf{P}} \wedge \left(\frac{\partial}{\partial \mathbf{P}} \wedge \mathbf{A}_P\right) = \frac{\partial}{\partial \mathbf{P}} \wedge \mathbf{B}_P,$$

where $\mathbf{B}_P$ is the *vector potential* of the velocity defined by

$$\mathbf{B}_P = \frac{\partial}{\partial \mathbf{P}} \wedge \mathbf{A}_P.$$

The velocity is obtained as the curl of the vector potential, just as in irrotational motion it is obtained as the gradient of the scalar velocity potential.

To find $\mathbf{B}_P$, we have from its definition and 2·34 (VII),

$$\mathbf{B}_P = \frac{1}{4\pi}\frac{\partial}{\partial \mathbf{P}} \wedge \int_{(V)} \frac{\mathbf{q}_Q}{PQ}\, d\tau = -\frac{1}{4\pi}\int_{(V)} \mathbf{q}_Q \wedge \frac{\partial}{\partial \mathbf{P}}\left(\frac{1}{PQ}\right) d\tau$$

$$= \frac{1}{4\pi}\int_{(V)} \mathbf{q}_Q \wedge \frac{\partial}{\partial \mathbf{Q}}\left(\frac{1}{PQ}\right) d\tau$$

$$= \frac{1}{4\pi}\int_{(V)} \frac{\partial}{\partial \mathbf{Q}} \wedge \left(\frac{\mathbf{q}_Q}{PQ}\right) d\tau + \frac{1}{4\pi}\int_{(V)} \frac{1}{PQ}\left(\frac{\partial}{\partial \mathbf{Q}} \wedge \mathbf{q}_Q\right) d\tau,$$

$$\text{(5)} \qquad \mathbf{B}_P = \frac{1}{4\pi}\int_{(E)} \frac{\mathbf{n} \wedge \mathbf{q}_Q}{PQ}\, dS + \frac{1}{4\pi}\int_{(V)} \frac{\boldsymbol{\zeta}_Q}{PQ}\, d\tau,$$

the fourth statement being obtained by a second application of 2·34 (VII) and the last by Gauss's theorem in the form 2·61 (2). This result expresses the vector potential in terms of the vorticity and the velocity at the boundary E.

19·21. Flux through a circuit. The flux through a circuit C can be expressed in terms of the vector potential as follows. If we close the circuit by a diaphragm S, the flux is

$$\int_{(S)} \mathbf{n}\mathbf{q}\, dS = \int_{(S)} \mathbf{n}(\nabla \wedge \mathbf{B})\, dS.$$

By the triple scalar product $\mathbf{n}(\nabla \wedge \mathbf{B}) = (\mathbf{n} \wedge \nabla)\mathbf{B}$, and therefore, by Stokes' theorem, the flux through C is

$$\int_{(C)} \mathbf{B}\, d\mathbf{s},$$

taken round the circuit. The direction of the flux is related to the sense of description by the right-handed screw rule.

19·22. Unbounded fluid. When the fluid is unbounded and the speed q at a great distance is of order $1/r^2$ at least, where $r = PQ$, the surface integral in 19·20 (5) tends to zero, since $dS = r^2\,d\omega$, where $d\omega$ is an elementary solid angle, and therefore

$$\mathbf{B}_P = \frac{1}{4\pi}\int \frac{\boldsymbol{\zeta}_Q}{r}\,d\tau, \quad \mathbf{q}_P = \nabla_\wedge \mathbf{B}_P$$

and the velocity is therefore a function of the vorticity only.

Thus we have, using 2·34 (VII),

$$\mathbf{q}_P = -\frac{1}{4\pi}\int \boldsymbol{\zeta}_{Q\wedge}\nabla\left(\frac{1}{r}\right)d\tau = \frac{1}{4\pi}\int \frac{\boldsymbol{\zeta}_{Q\wedge}\mathbf{r}}{r^3}\,d\tau,$$

where $\mathbf{r}$ is the position vector of P with respect to Q (not vice versa).

The above result means that the velocity at P can be regarded as the vector sum of elementary velocities, that corresponding to the vorticity in the volume element $d\tau$ at Q being

$$d\mathbf{q}_P = \frac{d\tau}{4\pi r^3}\,\boldsymbol{\zeta}_{Q\wedge}\overrightarrow{QP}.$$

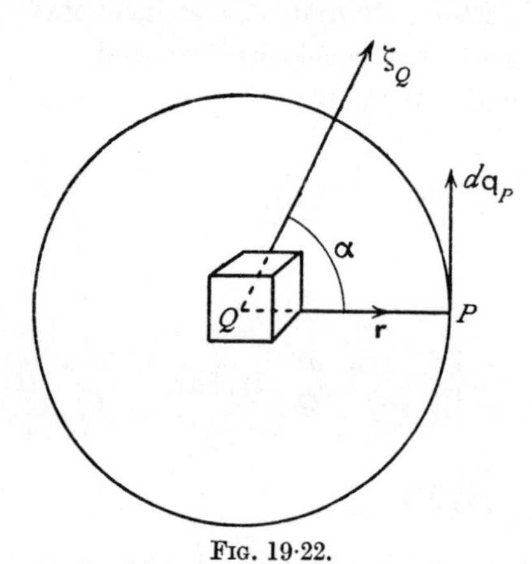

FIG. 19·22.

The relation of the vectors is shown in fig. 19·22. The magnitude of this velocity is

$$dq_P = \frac{d\tau}{4\pi r^2}\,\zeta_Q \sin\alpha,$$

where α is the angle between $\boldsymbol{\zeta}_Q$ and $\mathbf{r}$. This fictitious elementary velocity may be referred to as the velocity *induced* at P by the element at Q.

19·23. Vortex filament. Let the vorticity be concentrated in a single vortex filament. It has been proved, 3·52, that the product of the magnitude of the vorticity and the (infinitesimal) area of the cross-section of such a filament is constant. Calling this product κ, the strength of the filament, the velocity induced at P by the length ds of the filament, fig. 19·23 (i), will be

$$\frac{\kappa\,ds}{4\pi r^3}\,(\mathbf{s}_{1\wedge}\mathbf{r}),$$

where $\mathbf{s}_1$ is a unit vector in the direction of the tangent to the filament.

In the case of a re-entrant or closed filament C (vortex ring of infinitesimal cross-section), we shall have

$$\mathbf{q}_P = \frac{\kappa}{4\pi}\int_{(C)} \frac{\mathbf{s}_{1\wedge}\mathbf{r}}{r^3}\,ds = \frac{\kappa}{4\pi}\int_{(C)} \mathbf{s}_{1\wedge}\frac{\partial}{\partial\mathbf{Q}}\left(\frac{1}{r}\right)ds.$$

Applying Stokes' theorem in the form 2·51 (3), we get

$$\mathbf{q}_P = \frac{\kappa}{4\pi}\int_{(S)} \left(\mathbf{n} \wedge \frac{\partial}{\partial \mathbf{Q}}\right) \wedge \left[\frac{\partial}{\partial \mathbf{Q}}\left(\frac{1}{r}\right)\right] dS$$

over any diaphragm S which has C for boundary.

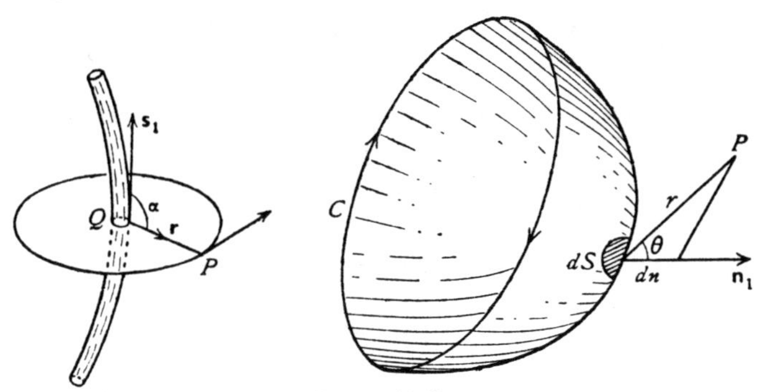

FIG. 19·23 (i).

Now, by the triple vector product,

$$\left(\mathbf{n} \wedge \frac{\partial}{\partial \mathbf{Q}}\right) \wedge \frac{\partial}{\partial \mathbf{Q}}\left(\frac{1}{r}\right) = \frac{\partial}{\partial \mathbf{Q}}\left\{\mathbf{n}\frac{\partial}{\partial \mathbf{Q}}\left(\frac{1}{r}\right)\right\} - \mathbf{n}\left\{\frac{\partial^2}{\partial \mathbf{Q}^2}\left(\frac{1}{r}\right)\right\},$$

and the last term vanishes, since $1/r$ is a spherical harmonic.

Hence, since $\mathbf{n}\partial/\partial\mathbf{Q} = \partial/\partial n$,

$$\mathbf{q}_P = \frac{\kappa}{4\pi}\int_{(S)} \frac{\partial}{\partial \mathbf{Q}}\left\{\frac{\partial}{\partial n}\left(\frac{1}{r}\right)\right\} dS = -\frac{\partial}{\partial \mathbf{P}}\left\{\frac{\kappa}{4\pi}\int_{(S)} \frac{\partial}{\partial n}\left(\frac{1}{r}\right) dS\right\}.$$

It follows that the velocity at P is derived from the velocity potential *

$$\phi = \frac{\kappa}{4\pi}\int_{(S)} \frac{\partial}{\partial n}\left(\frac{1}{r}\right) dS.$$

Now $\partial(1/r)/\partial n = \cos\theta/r^2$, where θ is the angle between dn and the line joining dS to P. This is shown in fig. 19·23 (i). (Observe that in the figure as drawn dr is negative.) Also, $dS\cos\theta$ is the projection of the area dS on the plane perpendicular to r, and therefore $dS\cos\theta/r^2 = d\omega$, the elementary solid angle subtended at P by the area dS. Thus, finally, we get

$$\phi = \frac{\kappa\omega_P}{4\pi},$$

where ω_P is the solid angle subtended at P by any diaphragm which closes the filament C.

This is illustrated in fig. 19·23 (ii), which shows a sphere of unit radius, centre P, on whose surface the solid angle is measured. It may be observed

* Comparing with 16·26 we see that this is also the velocity potential of a sheet of doublets, normally over S, of strength $-\kappa$ per unit area cf. 16·28.

that the value of ϕ just found is equal to the flux through the aperture presented by the vortex ring C, due to a point source at P of strength $\kappa/(4\pi)$. If we take P round a closed circuit which threads the ring once, the solid angle

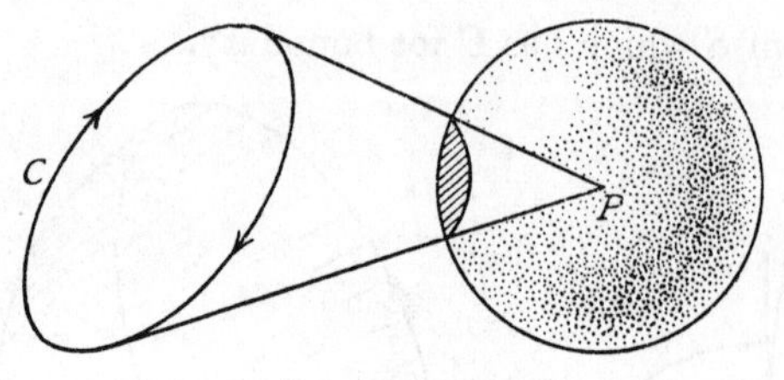

FIG. 19·23 (ii).

increases or diminishes by 4π, according to the sense of description. Thus ϕ is many valued. This is, of course, in agreement with the fact that the presence of the vortex ring renders the space doubly connected.

The momentum $\mathbf{M}$ of the fluid being assumed to be $\int \rho\phi\, d\mathbf{S}$ over both sides of S (cf. 18·20 (3)), we have

$$\mathbf{M} = \kappa\rho \int_{(S)} d\mathbf{S}$$

taken over one side of S, which is the same for all diaphragms, since $\int d\mathbf{S}$ over a closed surface vanishes. If the vortex ring is a *plane* curve of area A and normal $\mathbf{n}$, we have $\mathbf{M} = \kappa\rho A\mathbf{n}$.

19·24. Electrical analogy. There is an exact correspondence between the formulae concerning vortex motion and those concerning certain electromagnetic phenomena. In this analogy a vortex filament corresponds to an electric circuit, the strength to the electric current, and the fluid velocity to magnetic force. Thus the formula for induced velocity corresponds exactly to the formula of Biot and Savart for the magnetic effect of a current. The analogy is still further extended by observing that sources and sinks correspond to positive and negative magnetic poles.

19·30. Kinetic energy. This is given by

$$T = \tfrac{1}{2}\rho \int q^2\, d\tau.$$

If $\mathbf{B}$ is the vector potential,

$$\mathbf{q} = \nabla_\wedge \mathbf{B}, \quad q^2 = \mathbf{q}(\nabla_\wedge \mathbf{B}) = \nabla(\mathbf{B}_\wedge \mathbf{q}) + \mathbf{B}\boldsymbol{\zeta},$$

from 2·34 (I). Applying Gauss's theorem,

$$T = \tfrac{1}{2}\rho \int_{(V)} \mathbf{B}\boldsymbol{\zeta}\, d\tau + \tfrac{1}{2}\rho \int_{(S)} \mathbf{B}(\mathbf{n}_\wedge \mathbf{q})\, dS,$$

taken throughout the volume V enclosed by the bounding surface S. If the liquid is unbounded and the first integral converges, we have

$$T = \tfrac{1}{2}\rho \int \mathbf{B}\boldsymbol{\zeta}\, d\tau = \frac{\rho}{8\pi}\int\int \frac{\zeta_P \zeta_Q}{PQ}\, d\tau_P d\tau_Q\,,$$

where $\boldsymbol{\zeta}_P$, $\boldsymbol{\zeta}_Q$ are the vorticities at P and Q and $d\tau_P$, $d\tau_Q$ elements of volume at these points.

Another expression for the kinetic energy is

$$T = \rho \int_{(V)} \mathbf{q}(\mathbf{r}_\wedge \boldsymbol{\zeta}) d\tau + \rho \int_{(S)} \{(\mathbf{nq})(\mathbf{qr}) - \tfrac{1}{2}(\mathbf{nr}) q^2\} dS,$$

taken throughout the volume V contained by the boundary S.

To prove this result we have, by the triple scalar product and 2·34 (IV),

$$\begin{aligned}\mathbf{q}(\mathbf{r}_\wedge \boldsymbol{\zeta}) &= -\mathbf{r}[\mathbf{q}_\wedge(\nabla_\wedge \mathbf{q})] = \mathbf{r}[(\mathbf{q}\,\nabla)\mathbf{q}] - \tfrac{1}{2}\mathbf{r}\,\nabla\, q^2 \\ &= (\mathbf{q}\,\nabla)(\mathbf{qr}) - \mathbf{q}[(\mathbf{q}\,\nabla)\,\mathbf{r}] + \tfrac{1}{2}q^2\,\nabla\,\mathbf{r} - \tfrac{1}{2}\nabla(\mathbf{r}q^2).\end{aligned}$$

Now, $$\nabla\,\mathbf{r} = 3, \quad (\mathbf{q}\,\nabla)\,\mathbf{r} = \mathbf{q}.$$

Therefore $$\mathbf{q}(\mathbf{r}_\wedge \boldsymbol{\zeta}) = \tfrac{1}{2}q^2 + \nabla[\mathbf{q}(\mathbf{qr})] - (\mathbf{qr})\,\nabla\,\mathbf{q} - \tfrac{1}{2}\nabla\,(\mathbf{r}q^2).$$

Integrating and applying Gauss's theorem, the result follows, since $\nabla\,\mathbf{q} = 0$. In the case of a fixed boundary, $\mathbf{nq} = 0$. If the liquid extends to infinity and the velocity at a great distance is of order r^{-2}, the energy is represented by the first integral alone.

19·40. Axisymmetrical motions. When the motion is symmetrical about the x-axis, the vortex lines must be circles whose centres are upon this axis and whose planes are perpendicular thereto. Such motions are conveniently discussed with the aid of Stokes' stream function, whose existence does not depend on the motion being irrotational.

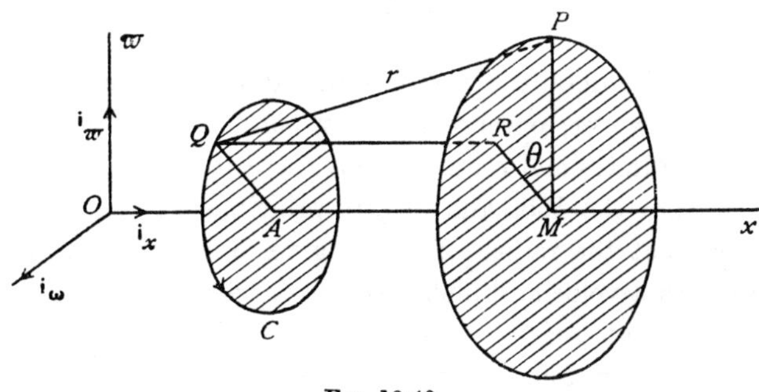

FIG. 19·40.

To obtain the form of the stream function, consider a point P, coordinates (x, ϖ), in a meridian plane. Draw the circle, centre M, perpendicular to the x-axis, on which P lies.

Let $\mathbf{B}$ be the vector potential at P. Since $\mathbf{q} = \nabla_\wedge \mathbf{B}$, and since the components of $\mathbf{q}$ lie in the meridian plane, it is at once evident that $\mathbf{B}$ must be perpendicular to the meridian plane. It also follows from the symmetry that $\mathbf{B}$ has the same magnitude B at every point of the circle. Since the flux through the circle is measured by the circulation of $\mathbf{B}$ round it (19·21), it follows that this flux is $2\pi\varpi B$. If we take the vorticity on a vortex line C to be related by the right-handed rule to the direction of the axis, this flux is from left to right. Hence, if ψ is the stream function,

$$2\pi\psi = -2\pi\varpi B, \quad \psi = -\varpi B.$$

This gives the stream function in terms of the magnitude of the vector potential.

19·41. Circular vortex filament. Consider the circular vortex filament C, fig. 19·40, of very small cross-section σ. Then the strength of the filament is $\zeta\sigma = 4\pi\kappa$, say. If Q is any point of C, whose centre is A, where $OA = \xi$, draw MR equal and parallel to AQ. Let the angle PMR be θ, and let $AQ = \eta$. Then the element of arc at Q is $\eta\, d\theta$, and the vorticity vector at Q is a tangent to C. Thus the vorticity at Q is $\zeta \cos\theta \,.\, \mathbf{i}_\omega - \zeta \sin\theta \,.\, \mathbf{i}_\varpi$, where $\mathbf{i}_\varpi$ and $\mathbf{i}_\omega$ are unit vectors parallel to the axis of ϖ and perpendicular to the meridian plane respectively. Thus, from 19·22

$$\mathbf{B} = \kappa \int_0^{2\pi} \frac{\mathbf{i}_\omega \cos\theta - \mathbf{i}_\varpi \sin\theta}{PQ}\, \eta\, d\theta,$$

$$PQ^2 = r^2 = (x-\xi)^2 + \eta^2 + \varpi^2 - 2\eta\varpi\cos\theta.$$

The coefficient of $\mathbf{i}_\varpi$ vanishes for the reason already explained, and the fact is indeed obvious in this case on performing the integration. The coefficient of $\mathbf{i}_\omega$ is the magnitude of $\mathbf{B}$, and therefore the stream function is

$$\psi = -\kappa\varpi\eta \int_0^{2\pi} \frac{\cos\theta\, d\theta}{r}.$$

A discussion of the details of the motion requires the use of elliptic functions. We may, however, observe that for points in the plane of the ring (considered as of infinitesimal cross-section) there is no radial velocity. This follows at once from the Biot and Savart principle, explained in 19·23. It therefore follows that the radius of the ring remains constant, and the ring moves forward with a velocity which must be constant since the motion must be steady relatively to the ring.

When two such rings follow one another with the same axis and sense of rotation, the effect of the induced velocity will be to enlarge the diameter of the leading ring and diminish that of the other, which may eventually pass right through the leader when the rôles become interchanged.

If two equal rings of opposite rotations approach one another, the induced velocity will tend to enlarge each, and on the plane midway between them the velocity will be perpendicular to the axis. Thus as a ring moves towards a wall to which its plane is parallel, the diameter will continually increase, and its velocity will continually diminish.

19·50. Equation satisfied by the stream function. Taking the curl of 3·43 (3) we have

$$\frac{\partial \boldsymbol{\zeta}}{\partial t} - \nabla_\wedge(\mathbf{q}_\wedge \boldsymbol{\zeta}) = 0. \tag{1}$$

In the case of axisymmetrical motion

$$\mathbf{q} = \mathbf{i}_x q_x + \mathbf{i}_\varpi q_\varpi, \quad \boldsymbol{\zeta} = \mathbf{i}_\omega \zeta,$$

where $\mathbf{i}_x$, $\mathbf{i}_\varpi$, $\mathbf{i}_\omega$ are unit vectors in the meridian plane and perpendicular to that plane. Thus $\mathbf{q}_\wedge \boldsymbol{\zeta} = \mathbf{i}_x q_\varpi \zeta - \mathbf{i}_\varpi q_x \zeta$, and therefore, from 2·72 (4), taking $h_1 = h_2 = 1$, $h_3 = \varpi$, we get

$$-\nabla_\wedge(\mathbf{q}_\wedge \boldsymbol{\zeta}) = \mathbf{i}_\omega \left(\frac{\partial(q_x \zeta)}{\partial x} + \frac{\partial(q_\varpi \zeta)}{\partial \varpi} \right),$$

and therefore (1) gives

$$\frac{\partial \zeta}{\partial t} + \frac{\partial(q_x \zeta)}{\partial x} + \frac{\partial(q_\varpi \zeta)}{\partial \varpi} = 0.$$

Using the equation of continuity (16·1),

$$\frac{\partial(q_x \varpi)}{\partial x} + \frac{\partial(q_\varpi \varpi)}{\partial \varpi} = 0,$$

this becomes

$$\frac{\partial \zeta}{\partial t} + \varpi \left\{ q_x \frac{\partial}{\partial x}\left(\frac{\zeta}{\varpi}\right) + q_\varpi \frac{\partial}{\partial \varpi}\left(\frac{\zeta}{\varpi}\right) \right\} = 0.$$

Thus, in terms of the stream function,

$$\frac{\partial}{\partial t}\left(\frac{\zeta}{\varpi}\right) - \frac{1}{\varpi}\frac{\partial \psi}{\partial \varpi}\frac{\partial}{\partial x}\left(\frac{\zeta}{\varpi}\right) + \frac{1}{\varpi}\frac{\partial \psi}{\partial x}\frac{\partial}{\partial \varpi}\left(\frac{\zeta}{\varpi}\right) = 0. \tag{2}$$

When the motion is steady this gives

$$\begin{vmatrix} \dfrac{\partial \psi}{\partial x} & \dfrac{\partial \psi}{\partial \varpi} \\ \dfrac{\partial}{\partial x}\left(\dfrac{\zeta}{\varpi}\right) & \dfrac{\partial}{\partial \varpi}\left(\dfrac{\zeta}{\varpi}\right) \end{vmatrix} = 0,$$

which implies that ζ/ϖ is a function of ψ, say,

$$\zeta = \varpi f(\psi). \tag{3}$$

Equation (3) shows the relation which must be satisfied by the vorticity for the motion to be steady (cf. 4·41). Now

$$\zeta = \frac{\partial q_\varpi}{\partial x} - \frac{\partial q_x}{\partial \varpi} = \frac{1}{\varpi}\left(\frac{\partial^2 \psi}{\partial x^2} + \frac{\partial^2 \psi}{\partial \varpi^2} - \frac{1}{\varpi}\frac{\partial \psi}{\partial \varpi}\right) = \frac{1}{\sin\omega}\nabla^2\left(\frac{\psi \sin\omega}{\varpi}\right) \tag{4}$$

$$= \frac{1}{\varpi} E^2 \psi, \text{ where } E^2 = \frac{\partial^2}{\partial x^2} + \frac{\partial^2}{\partial \varpi^2} - \frac{1}{\varpi}\frac{\partial}{\partial \varpi}.$$

Therefore the equation satisfied by the stream function is, from (2),

$$(5)\qquad \frac{\partial}{\partial t}(E^2\psi)+\varpi\begin{vmatrix}\dfrac{\partial\psi}{\partial x} & \dfrac{\partial\psi}{\partial\varpi}\\ \dfrac{\partial}{\partial x}\left(\dfrac{1}{\varpi^2}E^2\psi\right) & \dfrac{\partial}{\partial\varpi}\left(\dfrac{1}{\varpi^2}E^2\psi\right)\end{vmatrix}=0.$$

When the motion is steady, (3) and (4) yield the simpler equation

$$(6)\qquad E^2\psi=\varpi^2 f(\psi).$$

Taking the value of $\zeta(=\varpi^{-1}E^2\psi)$ in polar coordinates, from 2·72 (4), we get

$$\frac{\partial}{\partial r}\left(\frac{1}{\sin\theta}\frac{\partial\psi}{\partial r}\right)+\frac{\partial}{\partial\theta}\left(\frac{1}{r^2\sin\theta}\frac{\partial\psi}{\partial\theta}\right)=r^2\sin\theta f(\psi).$$

By attributing forms to $f(\psi)$ we have a differential equation to determine ψ. The simplest assumption of this nature is to take $f(\psi)=A$, a constant. We may then seek for solutions of the type

$$\psi=F(r)\sin^2\theta,$$

which gives

$$r^2F''(r)-2F(r)=Ar^4.$$

To find the complementary function, put $A=0$, $F(r)=Kr^n$, which gives $n=2$, or -1. The same substitution gives for the particular integral $n=4$, $K=A/10$.

Thus

$$\psi=\left(\frac{B}{r}+Cr^2+\frac{A}{10}r^4\right)\sin^2\theta.$$

19·51. Hill's spherical vortex. The stream function just found will represent the motion within a fixed sphere of radius a, if the value of ψ is finite at all points within the sphere and the normal velocity vanishes at the boundary. These conditions give $B=0$, and

$$-\left(\frac{1}{r\sin\theta}\frac{\partial\psi}{r\,\partial\theta}\right)_{r=a}=-\left(C+\frac{Aa^2}{10}\right)2\cos\theta=0,$$

whence $C=-Aa^2/10$. Thus

$$(1)\qquad \psi=-\frac{A}{10}(a^2-r^2)r^2\sin^2\theta,\quad r<a,$$

fulfils the required conditions, whatever the value of A.

The vorticity, given by direct calculation or by 19·50 (3), is $\zeta=Ar\sin\theta$. The vortex lines are circles perpendicular to the axis of symmetry. On all such circles of the same radius the vorticity has the same value.

There are stagnation points in the meridian plane given by the solutions of the simultaneous equation $q_r=0$, $q_\theta=0$, i.e. by

$$(2a^2-4r^2)\sin\theta=0,\quad 2(a^2-r^2)\cos\theta=0,$$

whence $\theta = \pm\pi/2$, $r = a/\sqrt{2}$. Thus there is a ring of stagnation points of radius $r = a/\sqrt{2}$.

The stream surfaces are given by

$$(a^2 - r^2) r^2 \sin^2\theta = c^4,$$

where c is a constant. These include the sphere and the axis of symmetry on which the stream divides. The principle of the dividing streamline then enables us to draw the form of the streamlines in the meridian plane, fig. 19·51, which shrink to zero at the stagnation points.

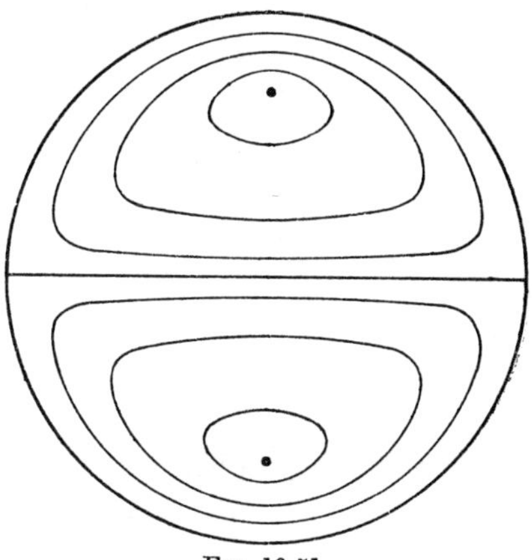

FIG. 19·51.

Taking advantage of the arbitrary constant A, the remarkable fact emerges that such a vortex can exist at rest in surrounding fluid which streams past it. The stream function for streaming past a sphere is, 16·30,

(2)

$$\psi = \tfrac{1}{2} U r^2 \sin^2\theta \left(1 - \frac{a^3}{r^3}\right), \quad r > a.$$

When $r = a$, (1) and (2) give $\psi = 0$, and the normal velocity is zero at the boundary. In order that the motion may exist, we must also have continuity of the tangential velocity, which gives, on equating the values of $\partial\psi/\partial r$,

$$A = \frac{15U}{2a^2},$$

and therefore the stream function (1) for the internal motion becomes

$$\psi = -\frac{3U}{4a^2} (a^2 - r^2) r^2 \sin^2\theta.$$

If we impress on the whole system a velocity U from left to right, we have a spherical vortex of radius a moving forward with velocity U in fluid at rest at infinity. The motion of the fluid external to the vortex is irrotational and the same as that produced by the motion of a solid sphere of the same radius.

19·60. Aerofoil of finite span. The Joukowski aerofoil considered in Chapter VII was a cylinder of infinite length of which we merely considered a unit segment. The aerofoils actually in use being of finite length or span, the motion cannot be considered as entirely two-dimensional.

Consider an aerofoil of span $2b$ symmetrical with regard to the central section perpendicular to the span, fig. 19·60 (a). In this figure the aerofoil is considered to be at rest and the wind stream to impinge on the leading edge, the wind direction at infinity upstream being that of the z-axis. The axis of y is taken vertically upwards and the axis of x along the span, the origin

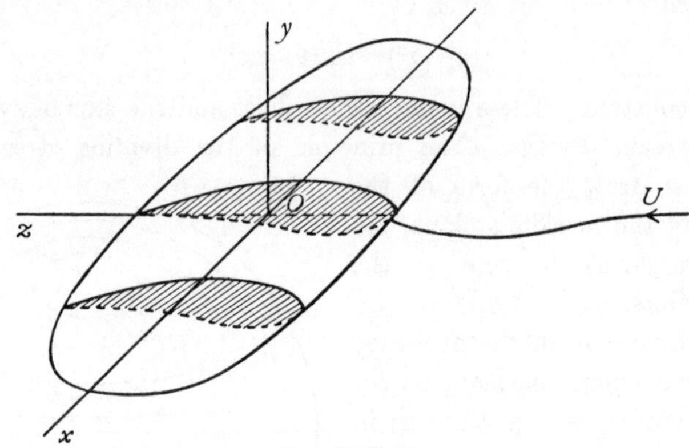

Fig. 19·60 (a).

being in the central section. In fig. 19·60 (b), which is purely schematic to show the principle, each streamline which impinges on the leading edge divides into two streamlines, the upper s going over the top of the aerofoil and the other s'

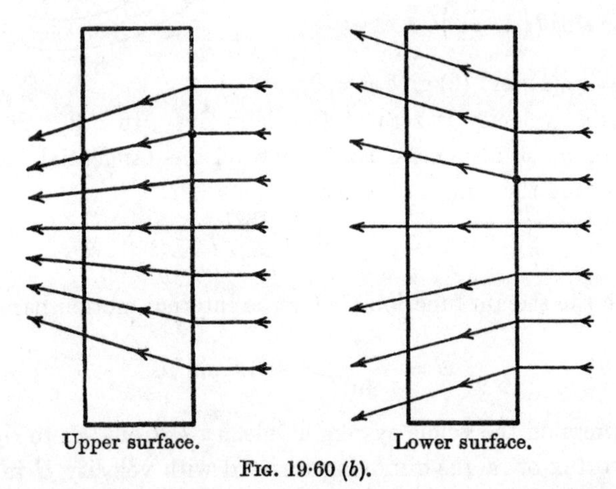

Fig. 19·60 (b).

underneath. These lines s, s' do not necessarily follow the transverse section of the aerofoil, and therefore do not leave it at the same point of the trailing edge.

The locus of the lines s will be a surface S, and the locus of the lines s' will be a second surface S'. We shall assume that immediately behind the trailing edge these surfaces coincide and form a single surface Σ across which the tangential velocity is discontinuous in direction but has the same magnitude. Since the pressure equation contains only the square of the magnitude of the velocity, the pressure will then be continuous. The surface Σ is a vortex sheet of the kind described in 13·70, and can be considered as consisting of vortices spread over it. Since the speeds above and below are equal at any point of Σ, the vortex lines will bisect the angles between the directions of the velocities.

In order to obtain a simple problem we shall suppose that these vortex lines are all straight and parallel to Oz. As a further simplification we shall take the trailing edge to be straight and the surface Σ to begin at that edge. These assumptions are not so restrictive as might appear at first view.

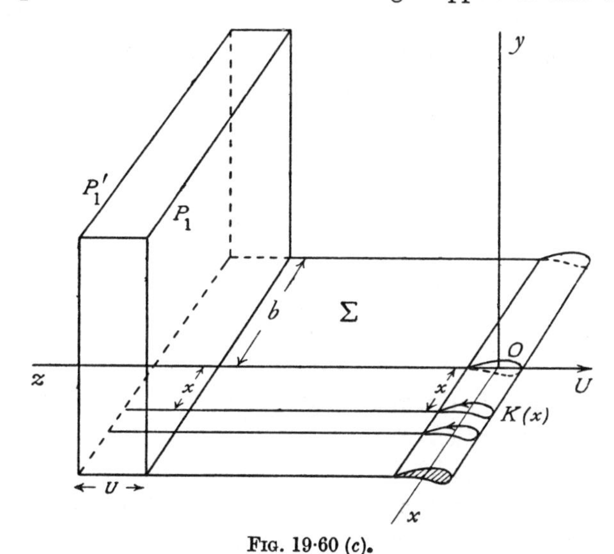

FIG. 19·60 (c).

To evaluate the resistance or drag,* it is more convenient to consider the aerofoil as in motion with velocity U and the air to be otherwise at rest. We consider two fixed infinite planes P, P_1 drawn perpendicular to the direction of motion, P a great distance ahead and P_1 a great distance astern. See fig. 19·60 (c), where P is not shown. If we draw a second plane P_1' parallel to P_1 and at a distance U behind it, the increase in energy per unit time of the fluid between P and P_1 will be due to the transference of that part of the vortex sheet Σ which lies between P_1' and P_1 into the region between P and P_1, for the irrotational parts of the motion ahead and astern will make no contribution, on account of the quasi-steady character of the motion between P and P_1. Thus, if ϕ is the velocity potential and R is the drag, by equating the rate of working of R to the rate of increase in kinetic energy, we get

$$(1) \qquad RU = \tfrac{1}{2}\rho \iint (\nabla \phi)^2 dx\, dy\, U.$$

* The drag here considered is the *induced drag* caused by the induced velocity of the vortex wake. It is less than the observed drag which includes skin friction and other effects.

Transforming this by Green's theorem, we get

$$R = \tfrac{1}{2}\rho \int_{-b}^{b} \phi \frac{\partial \phi}{\partial y}\, dx - \tfrac{1}{2}\rho \int_{-b}^{b} \phi' \frac{\partial \phi'}{\partial y}\, dx,$$

where ϕ refers to the upper side of Σ and ϕ' to the lower side. Since the normal velocity $-\partial\phi/\partial y$ is continuous, this gives

$$(2) \qquad R = \tfrac{1}{2}\rho \int_{-b}^{b} (\phi - \phi') \frac{\partial \phi}{\partial y}\, dx.$$

Now consider the section of the aerofoil at distance x from O. Let $K(x)$ be the circulation round this section. As we go through the vortex sheet Σ from above to below, the velocity potential decreases by the circulation. Thus $\phi - \phi' = K(x)$. Therefore, finally,

$$(3) \qquad R = \tfrac{1}{2}\rho \int_{-b}^{b} K(x) \frac{\partial \phi}{\partial y}\, dx.$$

To calculate the lift Y, we have by the theorem of Kutta and Joukowski for the section of the aerofoil between x and $x+dx$ the lift $\rho U K(x)\, dx$. Thus

$$(4) \qquad Y = \rho U \int_{-b}^{b} K(x)\, dx.$$

19·61. Aerofoil of minimum induced drag. We are now in a position to inquire what distribution of circulation $K(x)$ along the aerofoil will give the least resistance when the lift is given. With the notations of the preceding section we have to make R a minimum, subject to the condition that Y is given. Using the method of undetermined multipliers,* we must have

$$\delta R - \lambda\, \delta Y = 0$$

for any variation in $K(x)$. Now, from 19·60 (1),

$$\delta R = \rho \iint \nabla \phi \,.\, \nabla\, \delta\phi\, dx\, dy = \rho \int_{-b}^{b} (\delta\phi - \delta\phi') \frac{\partial \phi}{\partial y}\, dx,$$

using the same transformation as before. Also $\delta\phi - \delta\phi' = \delta K(x)$. Therefore

$$\int_{-b}^{b} \delta K(x) \frac{\partial \phi}{\partial y}\, dx - \lambda U \int_{-b}^{b} \delta K(x)\, dx = 0,$$

and if this is to hold for any arbitrary variation $\delta K(x)$, we must have

$$\frac{\partial \phi}{\partial y} = \lambda U = \text{constant} = V, \text{ say.}$$

The wake therefore behaves like a flat plate of breadth $2b$, moving with velocity V in a direction perpendicular to its length. Superposing a velocity $-V$ on the solution given in 6·34, we have the complex potential

$$w = -iV[z - \sqrt{(z^2 - b^2)}], \quad z = x + iy,$$

* See for example, Edwards's *Differential Calculus*.

and therefore on the plane $y = 0$, we have

$$\phi = \pm V\sqrt{(b^2 - x^2)},$$

the upper sign being taken on the upper side.

Thus, the circulation being given by the decrease in ϕ on passing round the plate,

$$K(x) = 2V\sqrt{(b^2 - x^2)}.$$

The circulation at the middle section ($x = 0$) is

$$K_0 = 2Vb,$$

and therefore

$$K(x) = \frac{K_0}{b}\sqrt{(b^2 - x^2)}.$$

This can also be written in the form

$$\frac{[K(x)]^2}{K_0^2} + \frac{x^2}{b^2} = 1,$$

which is the equation of an ellipse described by the point of coordinates $(x, K(x))$.†

EXAMPLES XIX

1. If S is a surface bounded by a curve C, prove that

$$\int_C [\mathbf{F}_\wedge \mathbf{ds}] = \int_S \{\mathbf{n} \operatorname{div} \mathbf{F} - \operatorname{grad} (\mathbf{Fn})\} dS,$$

n denoting the normal to S. (Stokes' theorem may be assumed, if necessary.)

An infinite liquid is at rest at infinity and the motion is due to a closed vortex filament of boundary C and strength κ; show that the velocity at a point P is

$$\mathbf{q} = -\frac{\kappa}{4\pi}\int_C [\operatorname{grad} (1/r)_\wedge \mathbf{ds}],$$

where r is the distance between P and the element **ds**.

Hence show that $\mathbf{q} = -\kappa \operatorname{grad} \Omega/4\pi$, where Ω is the solid angle subtended by the closed filament at P. (U.L.)

2. If the vorticity $\boldsymbol{\omega}$ is given at all points within a fluid, prove that the correct values of the vorticity are given if

$$\mathbf{v} = \operatorname{curl} \mathbf{A},$$

where

$$\mathbf{A} = \frac{1}{4\pi}\int \frac{\boldsymbol{\omega}\, d\tau}{r},$$

and the integrals extend through the fluid.

If the velocity has also a known divergence θ, show that this can be allowed for by adding to **v** a portion

$$-\nabla \frac{1}{4\pi}\int \frac{\theta\, d\tau}{r}.$$

† For a detailed discussion of the vortex sheets connected with aerofoils see Milne-Thomson's *Theoretical Aerodynamics*, London (1966).

If the circulation k is zero about all paths in the fluid except such as enclose a thin vortex, prove that the circulation is the same about all circuits that enclose this vortex and that the vortex cannot end within the fluid. Prove also that

$$\mathbf{A} = \frac{k}{4\pi}\int \frac{d\mathbf{s}}{r},$$

where the integral is taken vectorially along the vortex. (U.L.)

3. If the components of rate of pure strain are (a, b, c, f, g, h), show that

$$\frac{\omega}{\rho} = \frac{\omega_0}{\rho_0}\exp\int_0^t (a\lambda^2 + b\mu^2 + c\nu^2 + 2f\mu\nu + 2g\nu\lambda + 2h\lambda\mu)\,dt,$$

where λ, μ, ν are the direction cosines of the element ω of a vortex filament.

Interpret this result physically and discuss its connection with Kelvin's theorem as to the permanence of the circulation in a circuit moving with the fluid. (U.L.)

4. Show that the velocity due to a rectilinear segment AB of a vortex filament is perpendicular to the plane PAB and equal to

$$\frac{\kappa}{4\pi p}(\cos PAB + \cos PBA),$$

where p is the perpendicular from P on AB.

Calculate the velocity at any point due to a rectangular vortex filament, the sides of the rectangle being given by $z = 0$ and $x = \pm a$, $y = \pm b$. (U.L.)

5. A cylinder of any cross-section containing fluid rotates with given angular velocity about its axis and the fluid possesses constant vorticity ζ. Show that the kinetic energy per unit length of the cylinder of this motion exceeds the kinetic energy of the irrotational motion by

$$\tfrac{1}{2}\rho\zeta^2 \iint [(\partial V/\partial x)^2 + (\partial V/\partial y)^2]\,dx\,dy,$$

where V is the solution of $\nabla^2 V = 1$, which is finite and continuous at all internal points and is constant on the boundary.

6. Liquid moves in two-dimensions within an elliptic cylinder whose axes are $2a$, $2b$. If the vorticity has the constant value ω at every point, prove that the streamlines are similar ellipses described in the periodic time $2\pi(a^2+b^2)/(ab\omega)$.

7. Prove that a stream function of type $\psi = Ax^2 + By^2$ can represent steady motion of perfect fluid, with uniform vorticity ζ_0, taking place inside a cylinder bounded by an ellipse of semi-axes (a, b) which is rotating round its axis with uniform angular velocity ω_0, to be determined in terms of ζ_0. Show that the paths of particles of liquid relative to the boundary are similar ellipses.

By transforming to elliptic coordinates given by

$$x = c\cosh\xi\cos\eta, \quad y = c\sinh\xi\sin\eta,$$

show that if the very thin rigid cylindrical boundary has infinite liquid outside it moving irrotationally, then, provided it is of the same density as the liquid outside, this rigid interface between them may be supposed dissolved without disturbing the steady state of motion.

8. Prove that in a steady two-dimensional motion of a liquid of uniform vorticity 2ζ, under no body-force,

$$\frac{p}{\rho} = \text{constant} - \tfrac{1}{2}q^2 + 2\zeta\psi,$$

where q is the speed and ψ the stream function.

Liquid is flowing past a fixed circular cylinder of radius a. The vorticity is constant and equal to 2ζ, and, if the origin is at the centre of a section of the cylinder, the motion at infinity is the shearing motion

$$u = U - 2\zeta y, \quad v = 0.$$

The circulation immediately around the cylinder is K. Find the form of ψ, and prove that the resultant force on the cylinder exerted by the fluid pressure is $\rho U(K + 2\pi a^2 \zeta)$ along the axis of y. (U.L.)

9. The motion of an incompressible fluid in two dimensions is such that the vorticity 2ζ is uniform; show that the stream function ψ is given by

$$\psi = \tfrac{1}{2}\zeta(x^2 + y^2) + f(x + iy) + f(x - iy),$$

where f is an arbitrary function.

The space between two confocal elliptic cylinders, with semi-axes $c \cosh \alpha$, $c \sinh \alpha$, and $c \cosh \beta$, $c \sinh \beta$ respectively, where $\alpha > \beta$, is filled with liquid of uniform vorticity ζ. Determine the stream function, and prove that the kinetic energy per unit length is equal to

$$\tfrac{1}{16}\pi\rho\zeta^2 c^4 \{\sinh 4\alpha - \sinh 4\beta - 4 \tanh(\alpha - \beta)\}. \qquad \text{(U.L.)}$$

10. A cylindrical vortex sheet is such that the vortex lines are generators of the cylinder and the vorticity at any point is $2U \sin \theta$, where θ is the angle measured from a fixed plane through the axis of the cylinder. Prove that the vortex sheet moves through the liquid with velocity U parallel to the fixed plane. (M.T.)

11. Homogeneous liquid is circulating irrotationally in two dimensions round a hollow cylindrical vortex of radius a and circulation $2\pi\kappa$. Prove that the pressure at a great distance must be $\rho\kappa^2/(2a^2)$.

Prove that the system can oscillate freely in a mode in which the boundary of the cross-section of the vortex becomes a sinuous curve with n wave-lengths to the circumference and that the period has one or other of the values

$$\frac{2\pi a^2}{n^{\frac{1}{2}}(n^{\frac{1}{2}} \pm 1)\kappa}.$$

12. A mass of liquid, whose outer boundary is an infinitely long cylinder of radius b, is in a state of cyclic irrotational motion and is under the action of a uniform pressure P over the external surface. Prove that there must be a concentric cylindrical hollow whose radius a is determined by the equation

$$8\pi^3 a^2 b^2 P = M\kappa^2,$$

where M is the mass of unit length of the liquid and κ is the circulation.

If the liquid receive a small symmetrical displacement, prove that the time of a small oscillation is

$$(4\pi^2/\kappa)a^2 b^2 [\log(b/a)/(b^4 - a^4)]^{\frac{1}{2}}.$$

13. The motion of fluid in an unbounded region is due to a thin vortex ring, the circulation through which is k. Prove that the velocity at any point, not in the substance of the ring, can be expressed by either of the formulae

$$-\operatorname{grad} \phi \quad \text{and} \quad \operatorname{curl} \vec{A}.$$

Obtain expressions for ϕ and the components of $\vec{A}$, and verify that the values of the components of velocity are the same whether they are derived from the one formula or the other. (U.L.)

14. Show that the velocity $\mathbf{q}$ at a point P in an incompressible non-viscous fluid, extending to infinity, where it is at rest, and containing a closed vortex filament of boundary C and strength κ, is given by $\mathbf{q} = \text{curl}\,\mathbf{\Psi}$, where $\mathbf{\Psi} = \dfrac{\kappa}{4\pi}\displaystyle\int \dfrac{d\mathbf{s}}{r}$.

If the boundary C is a circle, find the relation between $\mathbf{\Psi}$ and Stokes' stream function for the problem. Hence, or otherwise, show that the velocity resolutes parallel and perpendicular to the axis of a circular vortex filament, at a point P near the axis, are respectively given by

$$u = \kappa a^2/2r^3, \quad v = 3\kappa a^2 \varpi (r^2 - a^2)^{\frac{1}{2}}/4r^5,$$

where ϖ and r are the distances of P from the nearest points of the axis and the vortex filament respectively. (U.L.)

15. Prove that the velocity at the centre of a circular vortex ring of strength m and radius a is $m/2a$, and find the velocity at any point on the axis of the ring.

16. Prove that the effect of a circular vortex ring at a great distance from itself is approximately the same as that of a double source of strength $ma^2/4$, where m is the strength of the vortex and a its radius.

17. Obtain the approximate formula $(K/4\pi b)\{\log(8b/a) - 1/4\}$ for the velocity of advance of a thin circular vortex ring, b being the radius of the line of centres of the cross-sections, a the radius of any cross-section, and K the circulation.

18. If q, q' are the velocities of the liquid due to a thin circular vortex ring of strength m and radius a at two points in the plane of the ring at distances r, r' from its centre, where $rr' = a^2$ and $r > r'$, prove that

$$qr^{\frac{1}{2}} + q'r'^{\frac{1}{2}} = \frac{m}{\pi}\int_0^{\pi/2} \frac{d\theta}{(r - r'\sin^2\theta)^{\frac{1}{2}}}.$$

19. Prove that for a single thin vortex ring of radius a the stream function at a point near the ring and distant x from its plane is approximately equal to

$$-\frac{\kappa x^2 a^2}{4(a^2 + x^2)^{\frac{3}{2}}},$$

where κ is the circulation through the ring.

20. Prove that the velocity due to a thin circular vortex ring of radius a and strength K, at a point P of its plane distant r from the centre of the ring, is

$$\frac{K}{2a}\left[1 + \frac{3r^2}{a^2}\left(\frac{1}{2}\right)^2 + \frac{5r^4}{a^4}\left(\frac{1\,.\,3}{2\,.\,4}\right)^2 + \frac{7r^6}{a^6}\left(\frac{1\,.\,3\,.\,5}{2\,.\,4\,.\,6}\right)^2 + \ldots\right],$$

where $r < a$, and calculate the velocity when $r > a$.

21. A circular line vortex of strength κ lies on a sphere of radius f and centre O; prove that the vortex has an image in a concentric sphere of radius a, and that it lies on a concentric sphere of radius f', its strength is κ', and its radius and that of the given vortex subtend the same angle α at O, provided that

$$ff' = a^2, \quad \kappa\sqrt{f} = -\kappa'\sqrt{f'}.$$

Prove that, at a point P on Ox, the axis of the first vortex, outside a rigid sphere, centre O and radius $a\,(<f)$, the velocity is along Ox and is equal to

$$\frac{\kappa}{2x}\sum_{i=1}^{\infty}\frac{f^{2i+1} - a^{2i+1}}{f^i x^{i+1}} P_i^1(\cos\alpha),$$

where $P_i^1(\cos\alpha) = \sin^2\alpha\, dP_i(\cos\alpha)/d(\cos\alpha)$, P_i being the zonal harmonic of order i. (U.L.)

22. If the vorticity is given at all points within an incompressible fluid, prove that a possible value of the velocity $\mathbf{v}$ is given by

$$\mathbf{v} = \operatorname{curl} \mathbf{A},$$

where, if (ξ, η, ζ) are the components of the vorticity, the components of $\mathbf{A}$ are

$$\frac{1}{4\pi}\iiint \xi \frac{dx\,dy\,dz}{r}, \quad \frac{1}{4\pi}\iiint \eta \frac{dx\,dy\,dz}{r}, \quad \frac{1}{4\pi}\iiint \zeta \frac{dx\,dy\,dz}{r},$$

and the integrals extend throughout the fluid.

For a single circular vortex filament of radius a and strength K, with the axis of x as axis of symmetry, prove that, at any point P, $\mathbf{A}$ is at right angles to the axis of x and to the perpendicular from P on to the axis of x, and that its magnitude is

$$\frac{K}{4\pi}\left(\frac{a}{r}\right)^{\frac{1}{2}} \left\{\left(\frac{2}{k} - k\right) F(k) - \frac{2}{k} E(k)\right\},$$

where

$$k = \frac{4ar}{x^2 + (r+a)^2},$$

$$F(k) = \int_0^{\frac{1}{2}\pi} \frac{du}{(1 - k^2 \sin^2 u)^{\frac{1}{2}}}, \quad E(k) = \int_0^{\frac{1}{2}\pi} (1 - k^2 \sin^2 u)^{\frac{1}{2}}\, du,$$

r is the distance of P from the axis of x, and x the distance of P from the plane of the vortex ring. (U.L.)

23. Prove that the force and couple components of the impulse $(\mathbf{F}, \mathbf{L})$ and the kinetic energy of a vortex system are given by

$$\mathbf{F} = \rho \int \mathbf{r}_\wedge \boldsymbol{\zeta}\, d\tau, \quad L_x = \rho \int (y^2 + z^2) \zeta_x d\tau, \quad T = \tfrac{1}{2}\rho \int \mathbf{r}(\boldsymbol{\zeta}_\wedge \mathbf{q})\, d\tau,$$

subject to certain conditions which should be stated.

Deduce that for a circular vortex filament of strength κ and radius ϖ, with its axis along Ox,

$$T = -\pi\rho\kappa\psi = 2\pi\rho\kappa\varpi(\varpi u - xv), \quad F_x = \pi\rho\kappa\varpi^2,$$

where u, v are the velocities along and perpendicular to Ox and ψ is Stokes' current function. Deduce also that for a circular vortex ring, whose section is a circle of radius a small compared with the radius ϖ_0 of its circular axis, at a distance s from the centre of the normal section

$$\psi = -\kappa\varpi_0 a^2 \log\left(\frac{8\varpi_0}{a} - \frac{3}{2} - \frac{s^2}{2a^2}\right), \quad T = \tfrac{1}{2}\rho\kappa^2\varpi_0\left(\log\frac{8\varpi_0}{a} - \frac{7}{4}\right). \qquad \text{(U.L.)}$$

24. Assuming the law of vorticity $\omega/\varpi = \omega_0/\varpi_0$ in a thin vortex ring in steady translatory motion, prove that, if the axial section of the ring is a circle of radius c with its centre at a distance ϖ_0 from the axis of symmetry, the velocity of advance of the ring is

$$\frac{\omega_0}{4\pi\varpi_0}\int_0^{2\pi}\left[\iint \frac{(\varpi - \varpi_0 \cos\alpha)\varpi^2}{\{\varpi^2 + \varpi_0{}^2 - 2\varpi\varpi_0\cos\alpha + (c - x)^2\}^{\frac{3}{2}}}\, d\varpi\, dx\right] d\alpha,$$

the integration with respect to ϖ and x being taken over an axial section, where ω is the resultant vorticity and the ring is supposed to move with constant speed in the direction of the x-axis. (U.L.)

25. Show that, for an aerofoil of finite span, the induced drag is a minimum for a given lift, when the distribution of lift across the span is elliptical.

If V is the speed of the aerofoil relative to the air, L the lift, D the induced drag, ρ the density of the air, and $2s$ the span of the aerofoil, prove that when D is a minimum

$$D = L^2/2\pi\rho s^2 V^2. \qquad \text{(U.L.)}$$

CHAPTER XX

SUBSONIC AND SUPERSONIC FLOW

20·0. With a few exceptions, the preceding investigations have been concerned with liquids or incompressible fluid, typically water. The Mach number (1·63) has been taken equal to zero.

In this chapter we shall be concerned with compressible fluids, typically air. The fluid will be assumed inviscid. The most important consequence of viscosity is probably the drag due to the skin friction in the boundary layer. External forces will be neglected, which implies, as explained in 1·44, that we are concerned only with hydrodynamic, or here more appropriately *aerodynamic*, pressure.

20·01. Thermodynamical considerations. Consider a *unit mass* of gas, volume v, density ρ, so that

$$v\rho = 1. \tag{1}$$

Let T be the absolute temperature (temperature measured from the absolute zero, about $-273°$ C.) of the gas. The gas is said to be *perfect* if it obeys the law,

$$pv = RT, \quad \text{or} \quad p = R\rho T, \tag{2}$$

where p is the pressure and R is a constant. Thus of the four quantities p, v, ρ, T only two are independent.

Logarithmic differentiation of (2) gives the relations

$$\frac{dp}{p}+\frac{dv}{v}=\frac{dT}{T}, \quad \frac{dp}{p}=\frac{d\rho}{\rho}+\frac{dT}{T}. \tag{3}$$

We shall consider only a perfect gas.

The first law of thermodynamics asserts that heat is a form of energy.

Let us imagine our unit mass of gas to receive a small quantity q of heat.

Hypothesis. For all gases, in mean motion or not, there exists an internal energy function E, independent of the mean motion and dependent only on the variables of state p, ρ, T, such that, when a small quantity of heat q is communicated to the gas,

$$q = dE+p\,dv. \tag{4}$$

The quantity dE is the excess of the energy supplied over the mechanical work done by the pressure.

Hypothesis. In a perfect gas the internal energy E is a function of the absolute temperature T alone.

This hypothesis is a generalisation from the results of experiment. It is also known as *Joule's law.* It follows that

$$dE = k\,dT \tag{5}$$

and (4) now becomes

$$q = k\,dT + p\,dv. \tag{6}$$

If, in communicating the small quantity q of heat to the gas, the expansion is prevented ($dv = 0$), the temperature of the gas will rise, say dT, and we can write

$$q = c_v dT.$$

The quantity c_v is called the *specific heat at constant volume.* It is the quantity of heat required to raise the temperature one unit when the volume is kept constant. Putting $dv = 0$ in (6) therefore gives

$$k = c_v. \tag{7}$$

We similarly define c_p, the *specific heat at constant pressure,* as the quantity of heat required to raise the temperature one unit when the pressure is kept constant. Now, if p is constant, (3) gives $dv/v = dT/T$, and therefore from (6)

$$q = \left(k + \frac{pv}{T}\right) dT,$$

and therefore

$$c_p = k + R = c_v + R$$

from (7).

We therefore conclude that

$$R = (c_p - c_v). \tag{8}$$

Hypothesis. In a perfect gas c_p, c_v are constant.

This is also based on the results of experiment.

In the above we have denoted the small quantity of heat by q and not by what would seem the more natural notation dQ. The reason for this is that there is, in general, no function Q of which q is an exact differential. We can, however, write

$$q = T\,dS, \tag{9}$$

where dS is the differential of a function S called the *entropy.*

To justify (9), observe that (6) and (7) give

$$dS = c_v \frac{dT}{T} + \frac{p}{T} dv = c_v \frac{dp}{p} + c_p \frac{dv}{v},$$

using (2) and (8) which proves that dS is an exact differential. Now write

$$(10) \qquad \gamma = c_p/c_v,$$

and we get at once

$$dS = c_v d \log (pv^\gamma).$$

If the state changes from (p_1, v_1) to (p_2, v_2), the increase of entropy is therefore

$$(11) \qquad S_2 - S_1 = c_v \log (p_2 v_2{}^\gamma) - c_v \log (p_1 v_1{}^\gamma).$$

The second law of thermodynamics asserts that the entropy of an isolated system can never decrease, i.e. $dS \geqslant 0$.

If the entropy retains the same constant value throughout the fluid, the flow is said to be *homentropic*. The condition for homentropic flow is therefore $dS = 0$. It follows from (11) that, if the flow is homentropic,

$$(12) \qquad pv^\gamma = \kappa, \quad \text{or} \quad p = \kappa\rho^\gamma,$$

where κ is a constant which depends on the entropy. This is the adiabatic law (cf. 1·62).

The steady flow of gas is governed by the equations of motion and continuity in the form

$$(13) \qquad -\frac{1}{\rho}\nabla p = (\mathbf{q}\nabla)\mathbf{q}, \quad \nabla(\rho\,\mathbf{q}) = 0,$$

and, as there are three unknowns p, ρ, $\mathbf{q}$, these equations are insufficient to determine the motion. In the case of homentropic flow, however, we can adjoin the adiabatic relation (12) and so obtain a determinate system of equations.

To calculate the internal energy we have

$$dE = c_v dT = \frac{c_v d(pv)}{R} = \frac{d(pv)}{\gamma - 1},$$

and thus, save for an added constant, we have the alternative forms

$$(14) \qquad E = \frac{pv}{\gamma - 1} = \frac{p}{(\gamma - 1)\rho} = c_v T.$$

The *enthalpy* or *total heat* I is the heat which communicated to a unit mass of a perfect gas will raise the temperature, at constant pressure, from absolute zero to the present temperature.

Thus from (4), since p is constant,

$$(15) \qquad I = E + pv = \frac{\gamma pv}{\gamma - 1} = \frac{\gamma p}{(\gamma - 1)\rho} = c_p T,$$

and therefore from (4) and (9)

$$(16) \qquad dI = v\,dp + T\,dS.$$

In the *isentropic* case, where S is constant along a streamline but not necessarily the same constant on different streamlines, we have

$$(17) \qquad dI = \frac{dp}{\rho} \text{ along a streamline.}$$

20·1. Crocco's equation. From 1·62 Bernoulli's equation in terms of the enthalpy, 20·01, can be written

(1) $$I+\tfrac{1}{2}q^2 = H, \text{ constant along a streamline.}$$

The function H is the *total energy* or *stagnation enthalpy* (i.e. the enthalpy at $q = 0$) of the streamline. In general H has different values on different streamlines. Flow in which H has the same value everywhere is called *homenergic*.

Now from 3·43 (3) the equation of steady motion under no forces is

(2) $$\mathbf{q}_\wedge \boldsymbol{\zeta} = \nabla(\tfrac{1}{2}q^2)+\frac{1}{\rho}\nabla p.$$

Eliminating p and q with the aid of 20·01 (16) we get Crocco's equation

(3) $$\mathbf{q}_\wedge \boldsymbol{\zeta} = \nabla H - T\ \nabla S.$$

Thus, neglecting viscosity and heat conduction, we shall find vorticity in the field of flow whenever the distribution of the total energy H or the entropy S is not uniform. This can happen, for example, when the fluid starts from a state of rest but of non-uniform temperature, or downstream of a curved shock line (20·6).

From (3) it follows that steady irrotational flow if homenergic, is also homentropic, and if homentropic, also homenergic.

20·12. Addition of a constant velocity. Let F be a given two-dimensional flow in the x, y plane. If this flow is referred to cartesian axes x, y, z which are moving uniformly with velocity $-V$ in the direction of the z-axis, the flow F' as viewed from the moving axes differs from F by the addition of a constant velocity V normal to the plane of the motion at every point. The velocity components u, v and the pressure, temperature, and density are the same functions of x, y, and the time as for the flow F. The addition of this constant velocity has no effect on the acceleration of the fluid particles or the vorticity.*

Thus, for example, the above addition to the flow pattern for the compressible vortex of 13·8 leads to a spiral flow about an axis. The streamlines are helices on coaxial cylinders, any pair of which may be taken as boundaries. This instance has some interest in connection with the flow of gases in an exhaust pipe. The method has also been applied to a side-slipping or swept-back supersonic aerofoil and to an oblique shock wave.

20·13. Steady motion. Neglecting viscosity, heat conduction and heat radiation, we have for the steady flow of a gas the following set of equations:

(1) $$\nabla(\rho\mathbf{q}) = 0, \qquad \text{equation of continuity,}$$

* H. Poritsky, *Journal of Applied Mechanics*, 13 (1946), pp. 53–60.

(2) $$(\mathbf{q}\,\nabla)\,\mathbf{q} = -\frac{1}{\rho}\,\nabla\,p,$$ equation of motion,

(3) $$p = f(\rho, S),$$ equation of state,

where S is the entropy.

These are three equations for ρ, p, $\mathbf{q}$, S. To obtain a determinate problem a fourth condition is required. Such a condition can be obtained by supposing the flow to be isentropic. Then

(4) $$(\mathbf{q}\,\nabla)\,S = 0,$$ entropy constant along a streamline.

It is convenient to write

(5) $$c^2 = \frac{\partial p}{\partial \rho} \quad (S \text{ constant}),$$

where c is the local sound speed (15·86). We then get from (3)

(6) $$\nabla\,p = c^2\nabla\,\rho + \frac{\partial p}{\partial S}\,\nabla\,S.$$

Taking the scalar product by $\mathbf{q}$, and using (4) and (1),

(7) $$(\mathbf{q}\,\nabla)p = \mathbf{q}\,c^2\nabla\,\rho = -c^2\rho\,\nabla\,\mathbf{q}.$$

Now $(\mathbf{q}\,\nabla)\mathbf{q} = \nabla(\frac{1}{2}q^2) - \mathbf{q}_\wedge\boldsymbol{\zeta}$ from 2·34 (IV). Take the scalar product of (2) by $\mathbf{q}$ and use (7). Then since $\mathbf{q}$ and $\mathbf{q}_\wedge\boldsymbol{\zeta}$ are perpendicular,

(8) $$\mathbf{q}\,\nabla(\tfrac{1}{2}q^2) = c^2\nabla\,\mathbf{q}.$$

This is the equation satisfied by the velocity, and we may regard c^2 as defined by Bernoulli's equation, 1·63 (4),

(9) $$c^2 = \tfrac{1}{2}(\gamma-1)(q_{\max}^2 - q^2)$$

along a streamline, which holds on account of (4).

20·2. Steady irrotational motion. Here we have 20·13 (8) together with $\mathbf{q} = -\nabla\,\phi$. Thus in cartesian coordinates

(1) $$(u^2-c^2)\frac{\partial^2\phi}{\partial x^2} + (v^2-c^2)\frac{\partial^2\phi}{\partial y^2} + (w^2-c^2)\frac{\partial^2\phi}{\partial z^2}$$
$$+\,2uv\frac{\partial^2\phi}{\partial x\,\partial y} + 2vw\frac{\partial^2\phi}{\partial y\,\partial z} + 2wu\frac{\partial^2\phi}{\partial z\,\partial x} = 0,$$

where

(2) $$u = -\frac{\partial\phi}{\partial x}, \quad v = -\frac{\partial\phi}{\partial y}, \quad w = -\frac{\partial\phi}{\partial z};$$

(3) $$c^2 = \tfrac{1}{2}(\gamma-1)\left[q_{\max}^2 - \left(\frac{\partial\phi}{\partial x}\right)^2 - \left(\frac{\partial\phi}{\partial y}\right)^2 - \left(\frac{\partial\phi}{\partial z}\right)^2\right].$$

If the values given by (2) and (3) are substituted in (1), we get the non-linear equation satisfied by the velocity potential for compressible flow.

In the case of incompressible fluid ($c = \infty$) this equation reduces to Laplace's equation.

Simple examples of steady irrotational motion have already been given for the source (8·9) and the vortex (13·8).

Considerable progress is possible with the *linearised theory*, in which small perturbations of a uniform stream by an immersed slender obstacle are considered.*

20·3. The hodograph method. Consider two-dimensional steady motion. Let PQR be an arc of a curve in the plane of the flow, the (x, y) plane, which may be conveniently called the *physical plane*. From the points P, Q, R, $\ldots$, draw vectors $\overrightarrow{PP_1}, \overrightarrow{QQ_1}, \overrightarrow{RR_1} \ldots$, to represent the fluid velocity at these points. From a fixed point H draw vectors $\overrightarrow{HP'}, \overrightarrow{HQ'}, \overrightarrow{HR'}, \ldots$, equal and parallel to these velocity vectors. The points $P', Q', R', \ldots$, describe the

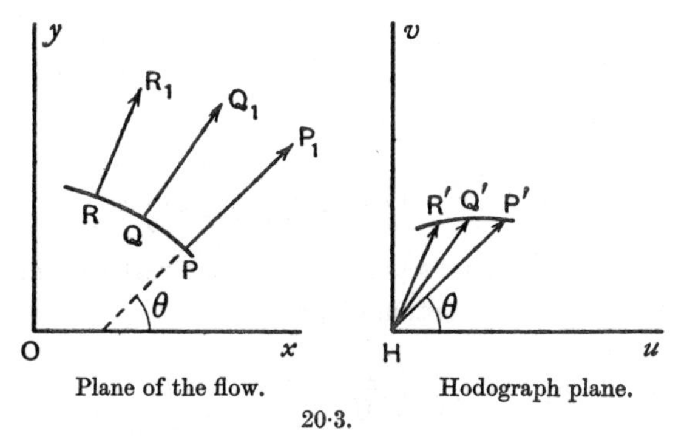

Plane of the flow. Hodograph plane.

20·3.

hodograph of the given curve PQR, and the plane of this curve is the *hodograph plane* of the given motion. If we take the axis Hu in the hodograph plane parallel to Ox in the plane of the flow, the velocity at P will be

$$u+iv = q\, e^{i\theta},$$

and P' will have cartesian coordinates (u, v) or polar coordinates (q, θ).

We have seen in 20·2 that the velocity potential of an irrotational compressible flow satisfies a non-linear differential equation. We shall show that if (q, θ) or (u, v) are taken as variables, the equation becomes linear.

It is useful to introduce the stream function ψ. The equation of continuity is, in the case of steady motion,

$$\frac{\partial(\rho u)}{\partial x}+\frac{\partial(\rho v)}{\partial y} = 0,$$

and we can satisfy this by taking

$$(1) \qquad \rho u = -\rho_0 \frac{\partial \psi}{\partial y}, \quad \rho v = \rho_0 \frac{\partial \psi}{\partial x},$$

* Milne-Thomson, *Theoretical Aerodynamics*, 4th Edition, London (1966), Chapters XVI, XVII.

where ρ_0 is any constant, which may be conveniently identified with the density, say, in the main stream, when we consider flow past an aerofoil. The function ψ is the *stream function*. Thus if ϕ is the velocity potential, we have

$$-d\phi = u\,dx+v\,dy, \quad -\frac{\rho_0}{\rho}\,d\psi = -v\,dx+u\,dy,$$

and therefore, as is easily verified,

$$-\left(d\phi+\frac{i\rho_0}{\rho}\,d\psi\right) = (u-iv)\,dz = q\,e^{-i\theta}dz,$$

so that

$$(2) \qquad dz = -\frac{e^{i\theta}}{q}\left(d\phi+\frac{i\rho_0}{\rho}\,d\psi\right).$$

If suffixes denote partial differentiation, $z_q = \partial z/\partial q$, we have at once

$$z_q = -\frac{e^{i\theta}}{q}\left(\phi_q+\frac{i\rho_0}{\rho}\,\psi_q\right), \quad z_\theta = -\frac{e^{i\theta}}{q}\left(\phi_\theta+\frac{i\rho_0}{\rho}\,\psi_\theta\right),$$

and since $z_{q\theta} = z_{\theta q}$, we get

$$\frac{\partial}{\partial\theta}\left\{\frac{e^{i\theta}}{q}\left(\phi_q+\frac{i\rho_0}{\rho}\,\psi_q\right)\right\} = \frac{\partial}{\partial q}\left\{\frac{e^{i\theta}}{q}\left(\phi_\theta+\frac{i\rho_0}{\rho}\,\psi_\theta\right)\right\}.$$

Performing the differentiations and equating the real and imaginary parts, we get, observing that ρ is a function of q only,

$$(3) \qquad \phi_q = q\psi_\theta\,\frac{\partial}{\partial q}\left(\frac{\rho_0}{q\rho}\right), \quad \phi_\theta = \frac{\rho_0 q}{\rho}\,\psi_q.$$

These are the *hodograph equations*. To get the equation satisfied by the stream function since $\phi_{q\theta} = \phi_{\theta q}$, we have

$$\frac{\partial}{\partial q}\left(\frac{\rho_0 q}{\rho}\,\psi_q\right) = \frac{\partial}{\partial\theta}\left\{q\psi_\theta\,\frac{\partial}{\partial q}\left(\frac{\rho_0}{q\rho}\right)\right\}, \text{ or}$$

$$(4) \qquad \frac{\rho_0 q}{\rho}\,\psi_{qq}+\psi_q\,\frac{\partial}{\partial q}\left(\frac{\rho_0 q}{\rho}\right) - q\psi_{\theta\theta}\,\frac{\partial}{\partial q}\left(\frac{\rho_0}{q\rho}\right) = 0,$$

since ρ is independent of θ.

Now
$$\frac{d}{dq}\left(\frac{\rho_0}{\rho}\right) = -\frac{\rho_0}{\rho^2}\frac{d\rho}{dp}\frac{dp}{dq} = \frac{\rho_0}{\rho^2}\frac{1}{c^2}\,q\rho = \frac{\rho_0}{\rho}\frac{q}{c^2},$$

using Bernoulli's theorem 1·61 (4) and $c^2 = dp/d\rho$.

We then find that (4) becomes finally

$$(5) \qquad q^2\,\frac{\partial^2\psi}{\partial q^2}+q(1+M^2)\,\frac{\partial\psi}{\partial q}+(1-M^2)\,\frac{\partial^2\psi}{\partial\theta^2} = 0, \quad M = \frac{q}{c}.$$

This is the linear equation satisfied by the stream function. The equation is due to Chaplygin.*

---

* See also R. Sauer, *Theoretische Einführung in die Gasdynamik*, Berlin (1943), p. 94.

20·31. The hodograph equations for homentropic flow. Assuming the adiabatic relation $p/p_0 = (\rho/\rho_0)^\gamma$, introduce the non-dimensional speed variable

$$(1) \qquad \tau = \frac{q^2}{q_{max}^2}, \quad \beta = \frac{1}{\gamma - 1} \, (= 2{\cdot}5 \text{ for air}).$$

Observe that $0 \leqslant \tau \leqslant 1$, and that $M^2 = 2\beta\tau/(1-\tau)$.

It is then easy to show that Bernoulli's equation can be exhibited in the form

$$(2) \qquad \rho = \rho_0(1-\tau)^\beta,$$

and that the hodograph equations become

$$(3) \qquad 2\tau\,(1-\tau)^{\beta+1}\phi_\tau = -\{1-(2\beta+1)\tau\}\psi_\theta\,, \quad (1-\tau)^\beta\phi_\theta = 2\tau\psi_\tau.$$

The elimination of ϕ leads now to Chaplygin's equation, 20·3 (5), in the new variables, namely,

$$(4) \qquad 2\tau(1-\tau)^{\beta+1}\frac{\partial}{\partial\tau}\{2\tau(1-\tau)^{-\beta}\psi_\tau\} + \{1-(2\beta+1)\tau\}\psi_{\theta\theta} = 0.$$

Since this is a linear equation for ψ, we can seek to build up solutions by superposition by addition of elementary solutions of the type

$$(5) \qquad \psi = B_m\tau^{\frac{1}{2}m}F_m(\tau)\sin(m\theta + \epsilon_m),$$

where B_m, ϵ_m are arbitrary constants. Substitution in (4) leads to the hypergeometric equation

$$(6) \quad \tau(1-\tau)F_m''(\tau) + \{m+1-(m+1-\beta)\tau\}F_m'(\tau) + \tfrac{1}{2}m(m+1)\beta F_m(\tau) = 0,$$

which is satisfied by the hypergeometric function *

$$F_m(\tau) = F(a, b\,; c\,; \tau) = 1 + \frac{a\,.\,b}{1\,.\,c}\tau + \frac{a(a+1)b(b+1)}{1\,.\,2\,.\,c(c+1)}\tau^2 + \ldots,$$

where $a+b = m-\beta$, $c = m+1$, $ab = -\frac{1}{2}\beta m(m+1)$.

The corresponding value of ϕ is then found from (3) to be

$$(7) \qquad \phi = -B_m\tau^{\frac{1}{2}m}(1-\tau)^{-\beta}\left\{F_m(\tau) + \frac{2\tau}{m}F_m'(\tau)\right\}\cos(m\theta + \epsilon_m).$$

There are solutions in compact form when $m = 0$ or -1. The case $m = 0$ is exceptional. To solve (3) we can then assume

$$(8) \qquad \psi = A\theta \quad \text{or} \quad \phi = B\theta,$$

which lead respectively to ϕ and ψ as functions of τ alone. In the physical plane the source (8·9) and the vortex (13·8) are comprised in the solution (8), and more generally the type of spiral flow obtained by combining a source and vortex, which was discussed in the case of a liquid in 13·33.†

---

* Milne-Thomson, *Calculus of Finite Differences*, London (1966), 9·8.

† Considerable progress has been made by S. Bergman in the elucidation of steady isentropic flows. See in particular, *N.A.C.A. Technical Notes*, Nos. 972, 973, 1018, 1096. See also M. J. Lighthill, the hodograph transformation in trans-sonic flow, *Proc. Roy. Soc.* (*A*) 191 (1947), pp. 323–369.

20·32. The case $m = -1$. In this case 20·31 (6) becomes

$$(1-\tau)\,F''_{-1}(\tau)+\beta F'_{-1}(\tau) = 0,$$

whence
$$F_{-1}(\tau) = A(1-\tau)^{\beta+1}+B,$$

so that there is a pair of fundamental solutions

(1) $$F^{(1)}_{-1}(\tau) = 1, \quad F^{(2)}_{-1}(\tau) = (1-\tau)^{\beta+1}.$$

Corresponding with the first, we get from 20·31 (5), (7),

(2) $$\psi = \frac{-A}{2q_{\max}}\,\tau^{-\frac{1}{2}}\sin\theta, \quad \phi = \frac{-A}{2q_{\max}}\,\tau^{-\frac{1}{2}}(1-\tau)^{-\beta}\cos\theta,$$

where we have put $\epsilon_{-1} = 0$ as is clearly permissible and written the constant B_{-1} as $A/2q_{\max}$.

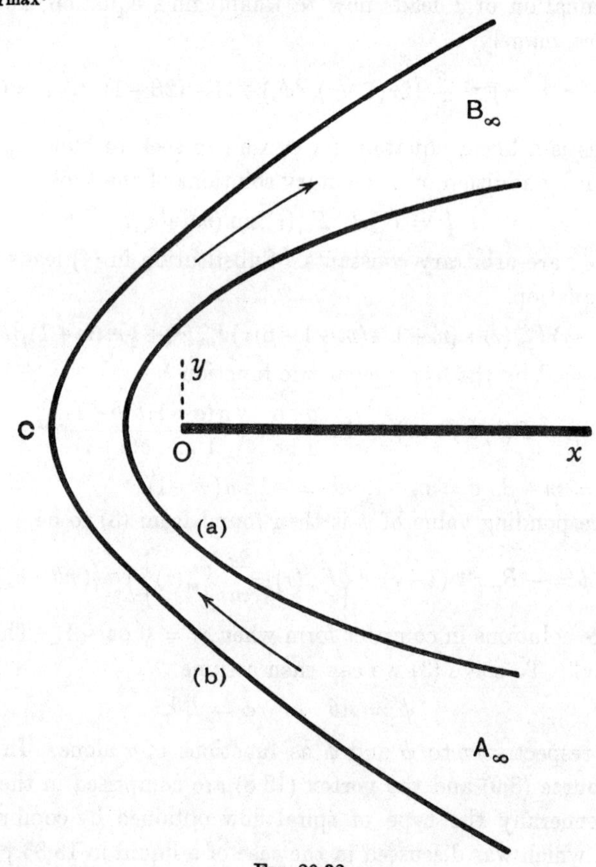

Fig. 20·32.

The second solution of (1) above leads similarly to

(3) $$\psi = \frac{-A}{2q_{\max}}\,\tau^{-\frac{1}{2}}(1-\tau)^{\beta+1}\sin\theta, \quad \phi = \frac{-A}{2q_{\max}}\,\tau^{-\frac{1}{2}}\{1+(2\beta+1)\tau\}\cos\theta.$$

The flows given by (2) and (3) degenerate into an incompressible flow when $q_{\max} \to \infty$ (and therefore $\tau \to 0$), given by

$$\text{(4)} \qquad \psi = -\frac{A}{2q}\sin\theta, \quad \phi = -\frac{A}{2q}\cos\theta.$$

If w is the complex potential of (4), we get

$$w = \phi + i\psi = \frac{-A}{2q\,e^{-i\theta}} = \frac{A}{2}\frac{dz}{dw},$$

and therefore $w^2 = Az$, whence

$$Ax = \phi^2 - \psi^2, \quad Ay = 2\phi\psi, \quad y^2 = \frac{4\psi^2}{A}\left(x + \frac{\psi^2}{A}\right).$$

Thus the curves ψ = constant are confocal parabolas (cf. Ex. VI, 20).

If we take two of these parabolas, say (a) and (b) in fig. 20·32, as boundaries we get the flow of a liquid in a curved two-dimensional channel or nozzle. The nozzle converges from A_∞, where the speed is zero, to its narrowest part at C and then diverges to B_∞, where the speed is again zero. We might therefore expect (2) and (3) to give a somewhat similar type of flow. The flow (2) has been discussed by Ringleb,* and the flow (3) by Temple.

20·33. Compressible flow in a convergent-divergent nozzle. We shall discuss 20·32 (2). It is convenient to replace A by $4aq_{\max}^2$, so that

$$\text{(1)} \qquad \psi = -2aq_{\max}\tau^{-\frac{1}{2}}\sin\theta, \quad \phi = -2aq_{\max}\tau^{-\frac{1}{2}}(1-\tau)^{-\beta}\cos\theta.$$

From Bernoulli's equation in the form 20·31 (2) we have

$$d\phi + \frac{i\rho_0}{\rho}\,d\psi = d\phi + i(1-\tau)^{-\beta}d\psi,$$

and therefore from 20·3 (2), after some reduction,

$$dz = 2a\{i\tau^{-1}(1-\tau)^{-\beta}e^{2i\theta}d\theta + \tfrac{1}{2}\beta\tau^{-1}(1-\tau)^{-\beta-1}(1+e^{2i\theta})d\tau - \tfrac{1}{2}\tau^{-2}(1-\tau)^{-\beta}e^{2i\theta}d\tau\},$$

and therefore on integration

$$\text{(2)} \qquad \frac{z}{a} = \tau^{-1}(1-\tau)^{-\beta}e^{2i\theta} + \beta\int_\alpha^\tau \tau^{-1}(1-\tau)^{-\beta-1}d\tau,$$

where α is an arbitrary constant lying between 0 and 1. The choice of this constant merely determines the position of the origin in the physical plane.

The streamlines ψ = constant are now obtained by eliminating τ and θ between (2) and the first equation of (1). If we write

$$\text{(3)} \qquad X = X(\tau) = a\beta\int_\alpha^\tau \tau^{-1}(1-\tau)^{-\beta-1}d\tau, \quad R = R(\tau) = a\tau^{-1}(1-\tau)^{-\beta},$$

* F. Ringleb, "Exakte Lösungen der Differentialgleichung einer adiabatischen Gasströmung" *Z a.M.M.*, 20 (1940), pp. 185–198.

equation (2) gives

$$(4) \qquad z = X + R\, e^{2i\theta},$$

and therefore on elimination of θ

$$(5) \qquad (z - X)(\bar{z} - X) = R^2,$$

so that the curves of constant speed (τ = constant) are circles whose centres $z = X(\tau)$ are on the real axis and whose radii are $R(\tau)$. Also

$$(6) \qquad \frac{dX}{d\tau} = \frac{a\beta}{\tau(1-\tau)^{\beta+1}}, \quad \frac{dR}{d\tau} = \frac{a\tau(\beta+1) - a}{\tau^2(1-\tau)^{\beta+1}}.$$

Thus as τ increases from zero X always increases, while R decreases to a minimum value when $\tau = 1/(1+\beta)$ and thereafter increases.

The condition that consecutive circles of the system (5) corresponding to values τ and $\tau + \delta\tau$ shall intersect is easily seen from a diagram to be

$$-\delta X < \delta R < \delta X.$$

The values (6) show that this implies

$$(2\beta+1)^{-1} < \tau < 1, \quad \text{i.e. } c^{\star 2} < q^2 < q^2_{max},$$

on using 20·31 (1) and 1·63 (3). Thus in the supersonic region consecutive circles of constant speed always intersect, in the subsonic region never. The critical case occurs where consecutive circles touch, and therefore the envelope of the system (5) separates the z-plane into two regions, one in which consecutive circles of constant speed intersect and one in which they do not.

To find the envelope, differentiate (5) with respect to τ. This gives

$$2R\frac{dR}{d\tau} = -\frac{dX}{d\tau}(z + \bar{z} - 2X) = -2R\frac{dX}{d\tau}\cos 2\theta$$

from (4), and therefore using (6),

$$(7) \qquad \frac{1}{\tau} = 1 + \beta(1 + \cos 2\theta),$$

and the envelope is then given by (7) and the two equations implied by (4).

If in (4) we regard τ as the function of θ given by (7), at a singular point of the envelope we must have $dz/d\theta = 0$, which gives after a simple reduction $\cos 2\theta = -1/(2\beta)$.

The corresponding value of τ from (7) is

$$(8) \qquad \tau = \frac{2}{2\beta+1}, \text{ whence } q = c^\star\sqrt{2}.$$

At this point of the envelope two consecutive circles of constant speed touch, and therefore the singularity is a cusp. By the symmetry there are two such cusps, optical images in the x-axis. The envelope is indicated by —·—·— in fig. 20·33 (i).

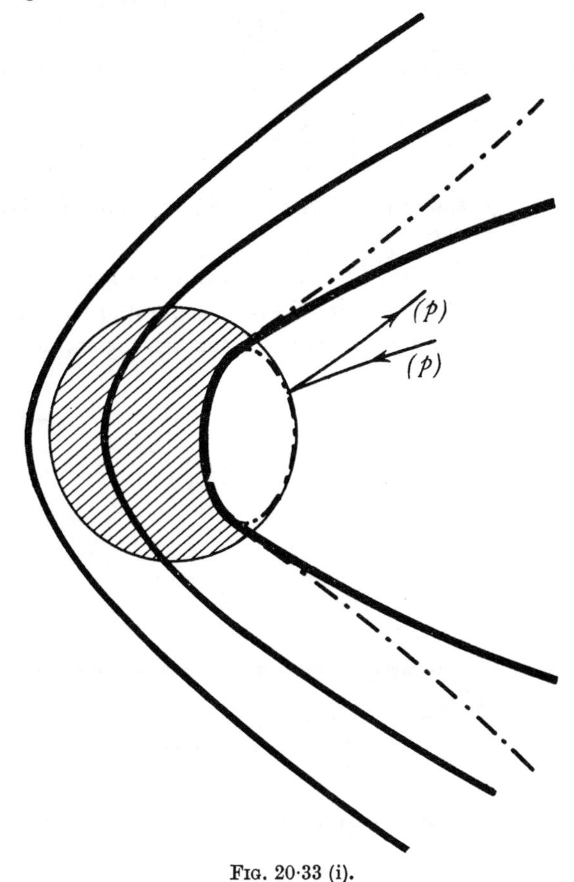

FIG. 20·33 (i).

In the hodograph plane if we take (τ, θ) as polar coordinates, the envelope is the ellipse given by (7), while a streamline ψ = constant is given by the first of equations (1). Eliminating θ between (1) and (7), we get for τ the quadratic

(9) $$\frac{\beta\psi^2}{2a^2 q_{\max}^2}\tau^2 - (2\beta+1)\tau + 1 = 0,$$

and so to each value of ψ there correspond two values of τ, a physically impossible régime characteristic of the region where consecutive circles of constant speed intersect; a streamline such as (p) in fig. 20·33 (i) turns back at the envelope.

These values are, however, imaginary if

(10) $$\psi^2 > \frac{(2\beta+1)^2}{2\beta} a^2 q_{\max}^2 .$$

The critical case arises when inequality is replaced by equality or

$$(11) \qquad \frac{\psi}{-2aq_{\max}} = \frac{2\beta+1}{\sqrt{(8\beta)}} = 1{\cdot}342,$$

taking $\beta = 2{\cdot}5$, the case of air. This streamline in the hodograph plane is

$$(12) \qquad \frac{2\beta+1}{\sqrt{(8\beta)}} = \frac{\sin\theta}{\sqrt{\tau}},$$

which passes through the cusps of the envelope where τ is given by (8) and $\cos 2\theta = -1/(2\beta)$.

The heavy line in fig. 20·33 (i) shows this streamline, which touches the envelope at the cusps and passes into the region behind it. The region to the right (in the figure) of this streamline is a forbidden region, in which no flow is

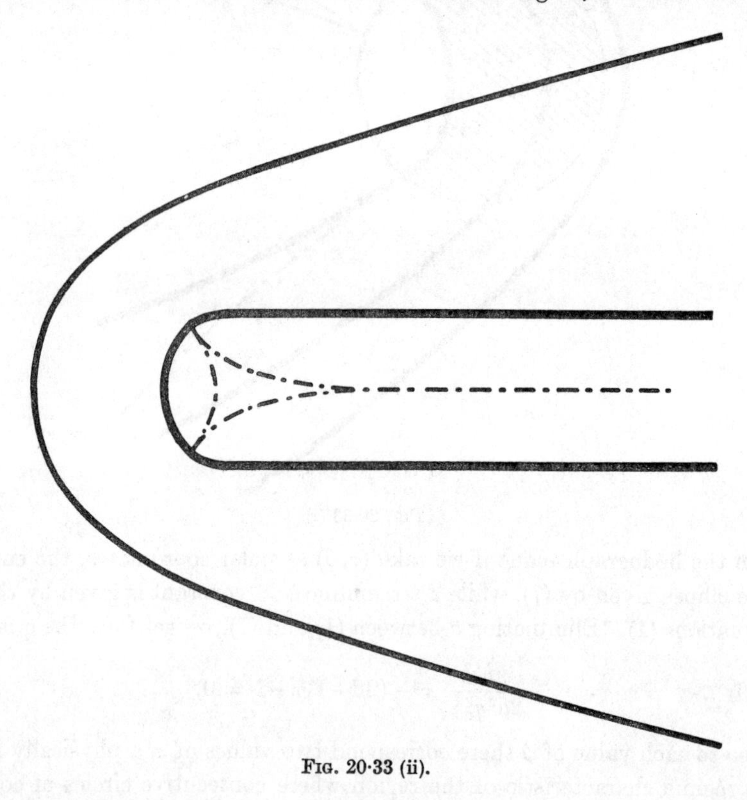

FIG. 20·33 (ii).

physically possible. To obtain a nozzle, we can take as rigid boundaries any two streamlines to the left of this critical streamline.

We also observe that the circle of constant speed on which $q = c^{*}$ is given by the value $\tau = 1/(2\beta+1)$, and the flow in the part of the nozzle (shaded in fig. 20·33 (i)) interior to this circle is supersonic. Thus Ringleb's solution, besides satisfying the hodograph equations exactly, gives an example of a compressible flow in which the régime can pass from subsonic to supersonic and back without shock.

It also appears from (12) that the maximum speed attainable on the critical streamline occurs when $\theta = \pi/2$ and

$$\tau = \tau_{\max} = \frac{8\beta}{(2\beta+1)^2} = \tfrac{5}{9} \text{ for air.}$$

Thus the maximum local Mach number attainable in this type of flow is

$$M_{\max} = \left(\frac{2\beta\tau_{\max}}{1-\tau_{\max}}\right)^{1/2} = 2{\cdot}5 \text{ for air.}$$

A similar discussion can be made of the solution 20·32 (3). See fig. 20·33 (ii).

Here the curves of constant speed are found to be trochoids which have a two-cusped envelope. A critical streamline passes through the cusps and separates a forbidden from a permitted region of flow. In this case also the flow passes from subsonic to supersonic and back without shock. The maximum Mach number attainable is about 2 for air.

Comparing the figures of this section with fig. 20·32, we see that in the incompressible case the forbidden region degenerates into a straight barrier.

20·4. Moving disturbance. Before considering supersonic flow let us examine a special problem. Let a feeble instantaneous disturbance such as a cry originate at a point P in air otherwise at rest.

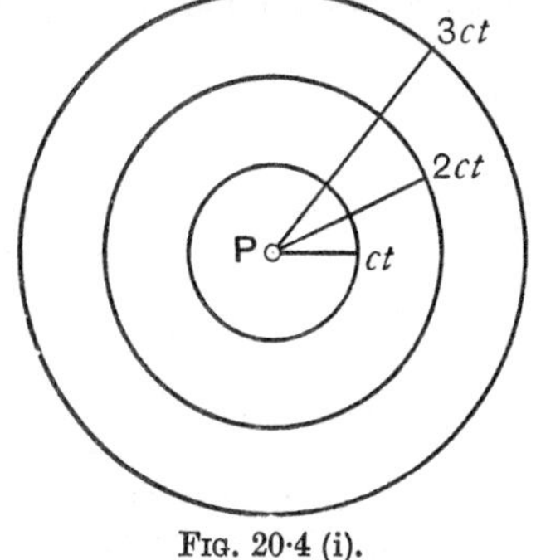

FIG. 20·4 (i).

Such a disturbance will spread in a spherical wave, with P as centre, with the speed of sound c, so that at times t, $2t$, $3t$, ... the disturbance will have reached points which lie on concentric spheres, centre P, radii ct, $2ct$, $3ct$, ... If, however, the air is in motion with velocity V from right to left, the points reached by the disturbance at time nt will lie on a sphere of radius nct whose centre is at distance Vnt from P. If $V<c$ these spheres will not intersect, and it is clear from fig. 20·4 (ii) that the disturbance will ultimately reach any pre-assigned point of space.

But when $V>c$ the state of affairs is different, fig. 20·4 (iii), for then the disturbance never reaches points which lie outside a cone whose vertex is P, whose axis is in the direction of V, and whose angle is 2μ, where $\sin\mu = c/V = 1/M$. The angle μ is called the *Mach angle* and the cone is the *Mach cone.*

In two-dimensional motion the Mach cone is replaced by a wedge and the lines in which the plane of the motion cuts the wedge are *Mach lines.*

A similar phenomenon is observed when uniform flow $V(>c)$ takes place parallel to a wall which is smooth save for a small roughness (such as a projecting seam) at P. Here a disturbance originates at P and is continually renewed as the oncoming air reaches P. The waves continually generated at P give a

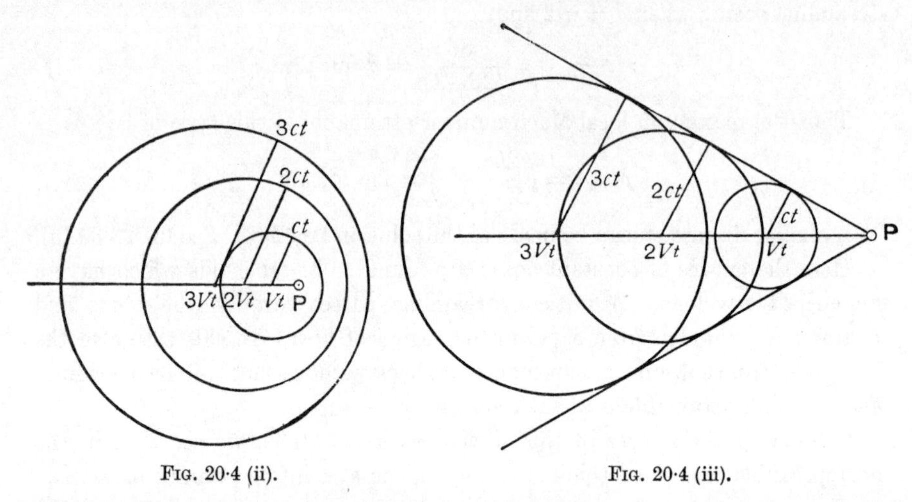

Fig. 20·4 (ii). Fig. 20·4 (iii).

noticeable disturbance only where they lie most densely, i.e. on m, the Mach line which issues from P. In the steady state the disturbance at every point of m is the same; the disturbance is not damped, at least in theory, as we recede from the wall along m. If there are several such roughnesses, each will give rise

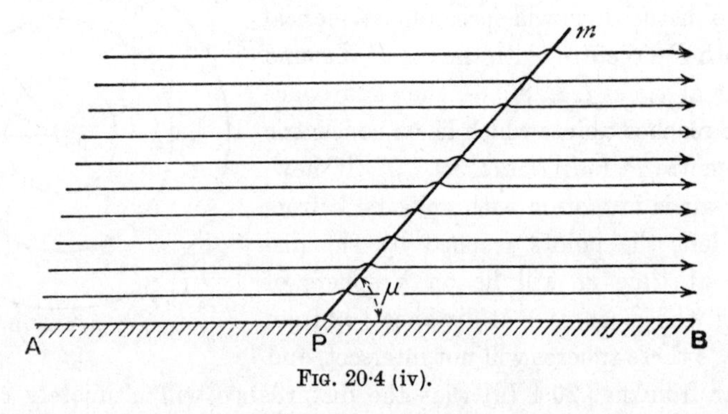

Fig. 20·4 (iv).

to a Mach line. Along such a line there is air density slightly different from the density of the smooth flow, and this circumstance renders it possible to photograph the lines whose existence is thus well attested.

From this it appears that supersonic flow, in which the airspeed exceeds the critical value, is physically different from subsonic flow. This manifests itself mathematically by the change of the differential equations from the elliptic to the hyperbolic type.

20·41. Characteristics. Consider a geometrical surface C conceived to be moving through the fluid. Let the point P belonging to the surface have the velocity $\mathbf{q}_C$, and let $\mathbf{q}$ be the velocity of the fluid particle with which P instantaneously coincides. The velocity of the point P of the surface relative to the fluid is then $\mathbf{q}_C - \mathbf{q}$.

Def. A *characteristic* is a surface which moves through the fluid in such a way that the magnitude of the component of the velocity of each point P of the surface relative to the fluid in the direction of the normal to the surface at P is equal to the local speed of sound at P.

In symbols

$$\mathbf{n}(\mathbf{q}_C - \mathbf{q}) = \pm c,$$

where c is the speed of sound at P, and $\mathbf{n}$ is the unit normal to the surface at P.

Since small disturbances are propagated with the speed of sound (15·86), it follows that the wave front of such a small disturbance is a characteristic.*

20·42. Characteristics for steady motion. In the case of two-dimensional steady motion, with which alone we shall be concerned,† the character-

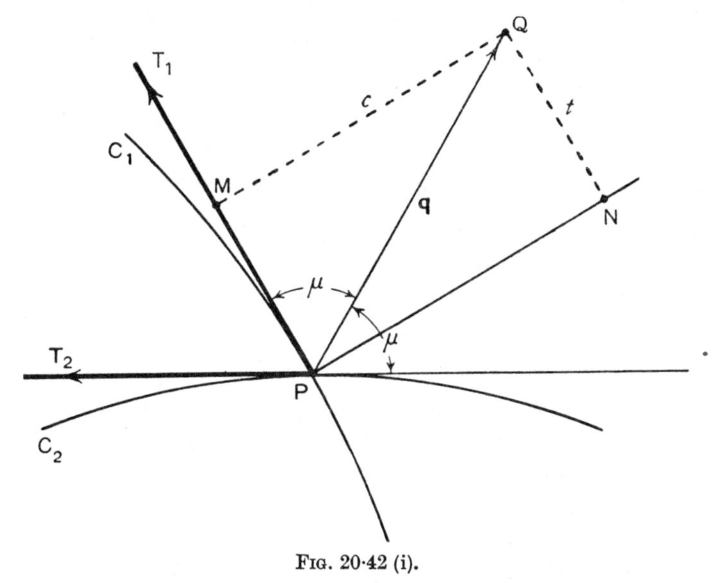

FIG. 20·42 (i).

istics will be cylindrical surfaces represented by a curve in the plane of the motion and will be *at rest*. Thus

$$(1) \qquad \mathbf{nq} = \pm c, \quad \text{or} \quad q_n = \pm c,$$

where q_n is the normal component of the *fluid velocity*.

* T. Levi-Civita, *Caratteristiche e propagazione ondosa*, Bologna (1931).

† The theory of characteristics can be applied also to steady axisymmetrical motions. For simplicity of exposition the two-dimensional case only is discussed.

Thus if PT_1 is the tangent to a characteristic C_1 and if $\overrightarrow{PQ}$ is the fluid velocity vector at P, the projection of PQ on the normal at P is equal to c. If then μ is the acute angle between the tangent to the characteristic and the fluid velocity,

$$\sin \mu = \frac{c}{q} = \frac{1}{M}, \tag{2}$$

and the angle μ is called the Mach angle at P (cf. 20·4).

It is clear from (2) that the Mach angle can exist only if the Mach number $M \geqslant 1$. Thus real characteristics in the sense of the definition exist only where the flow is supersonic. It is also clear from the figure and from (2) that there are two possible directions for the tangent to the characteristic at P, namely, PT_1 and T_2P, each making the angle μ with PQ. Thus exactly two characteristics pass through each point of fluid in supersonic motion.

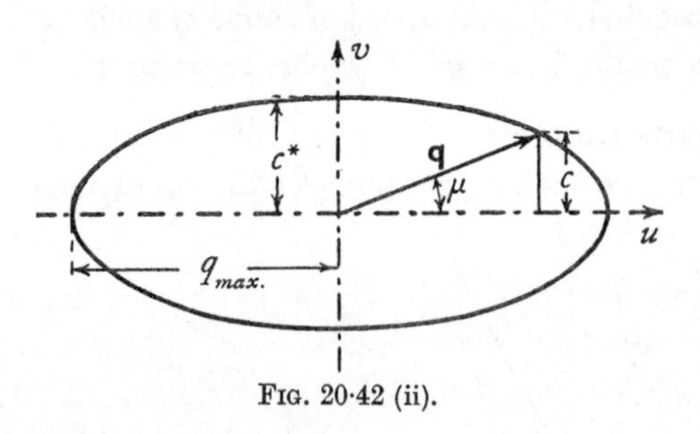

Fig. 20·42 (ii).

The directions of the characteristics are readily found by means of the *adiabatic ellipse* whose equation in the hodograph plane (cartesian coordinates u, v) is

$$\frac{u^2}{q_{\max}^2} + \frac{v^2}{c^{\star 2}} = 1.$$

To use this ellipse * to determine the directions of the tangents to the characteristics at a point P, draw the velocity vector $\overrightarrow{PQ} = \mathbf{q}$ at P and place the ellipse with its centre at P and so that it passes through Q. This can be done in two ways, and in each case the major axis is along the tangent to the corresponding characteristic at P.

As to the sense of the tangents to the characteristics, we can conveniently take the positive sense of the normal to be that which makes an acute angle with the velocity vector, and the positive sense on the tangents PT_1 and PT_2 to be that obtained by giving the normal a counterclockwise rotation of one right angle. With this convention, PT_1 in fig. 20·42 (iii) makes the angle μ with $\mathbf{q}$ and PT_2 the angle $\pi - \mu$.

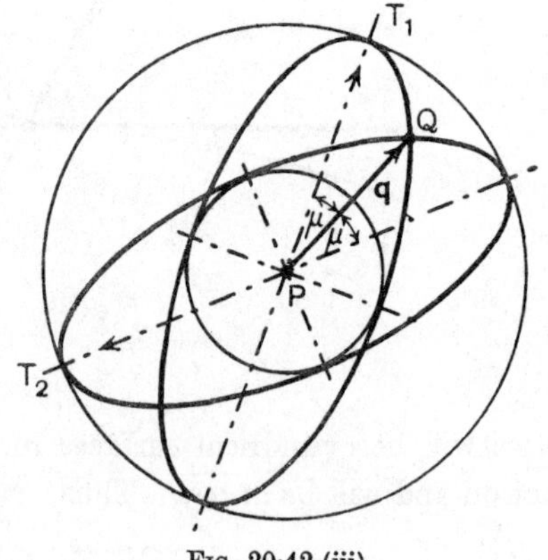

Fig. 20·42 (iii).

* Since 1·63 shows that $c^\star/q_{\max} = \sqrt{[(\gamma-1)/(\gamma+1)]}$, when γ is given all adiabatic ellipses are similar.

The corresponding tangential components of the velocity are $q \cos \mu = t$, and $-q \cos \mu = -t$.

The normal components by definition are then both c, and therefore Bernoulli's theorem, 1·63 (4), gives

$$c^2 = k^2 (q_{\max}^2 - q^2 \cos^2 \mu) = k^2 (q_{\max}^2 - t^2), \tag{3}$$

$$k^2 = \frac{\gamma - 1}{\gamma + 1} = \frac{c^{\star 2}}{q_{\max}^2}. \tag{4}$$

The Mach lines of 20·4 are the same as the characteristics of this section, and indeed for steady flow the terms Mach line and characteristic are interchangeable.

20·43. Variation of speed along a characteristic. For steady two-dimensional flow the equation of continuity, 21·39 (3), can be written

$$\frac{\partial (\rho q)}{\partial s} + \kappa_n \rho q = 0, \tag{1}$$

where ds is an element of arc of a streamline and $\kappa_n = \partial\theta/\partial n$ is the curvature at this element of the orthogonal trajectory. Here θ is the angle which the tangent to the streamline makes with some arbitrarily chosen fixed direction.

If in addition the flow is irrotational, $\zeta = 0$ and therefore, 21·39 (4),

$$\frac{\partial q}{\partial n} = \kappa_s q = q \frac{\partial \theta}{\partial s}, \tag{2}$$

where $\kappa_s = \partial\theta/\partial s$ is the curvature of the streamline. We now assume that the flow is also homenergic. It then follows from Crocco's equation (20·1) that the flow is also homentropic. Thus 1·61 (3)

$$q\, dq = -\frac{dp}{\rho} = -c^2 \frac{d\rho}{\rho} \tag{3}$$

holds not only along a streamline, but for variation in any direction.

From (1) and (3) we get successively

$$\kappa_n q = -\frac{\partial q}{\partial s} - \frac{q}{\rho}\frac{\partial \rho}{\partial s} = -\frac{\partial q}{\partial s} + \frac{q^2}{c^2}\frac{\partial q}{\partial s},$$

$$\frac{\partial q}{\partial s} = \kappa_n q \tan^2 \mu = q \tan^2 \mu \frac{\partial \theta}{\partial n}. \tag{4}$$

Now consider the variation of the speed q as we proceed from P to an adjacent point R of the characteristic whose tangent PT_1 makes the acute angle μ with the tangent PS to the streamline through P. From (2) and (4)

$$dq = \frac{\partial q}{\partial s} ds + \frac{\partial q}{\partial n} dn = q\left(\tan^2 \mu \frac{\partial \theta}{\partial n} ds + q \frac{\partial \theta}{\partial s} dn\right).$$

Now from fig. 20·43, $dn = ds \tan\mu$, and therefore

$$dq = q \tan\mu \left(\frac{\partial\theta}{\partial n}\,dn + \frac{\partial\theta}{\partial s}\,ds\right) = q \tan\mu\, d\theta,$$

that is to say,

$$\text{(5)} \qquad \frac{q\,d\theta}{dq} = \cot\mu \text{ along } PT_1.$$

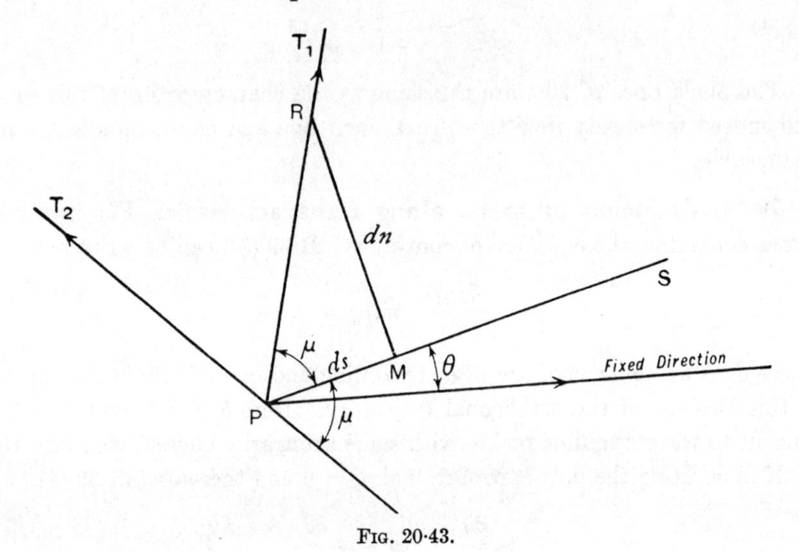

FIG. 20·43.

For the other characteristic through P we get, by writing $\pi - \mu$ for μ,

$$\text{(6)} \qquad \frac{q\,d\theta}{dq} = -\cot\mu \text{ along } PT_2.$$

20·44. Characteristic coordinates. Consider the characteristic whose tangent PT_1 makes the angle μ with the velocity at P. If t is the component of $\mathbf{q}$ along PT_1, we have $t = q\cos\mu$, and therefore

$$dt = \cos\mu\, dq - q \sin\mu\, d\mu = q\sin\mu(d\theta - d\mu)$$

from 20·43 (5). Using 20·42 (3) and observing that $q \sin\mu = c$, we get

$$d\theta - d\mu = \frac{dt}{c} = \frac{dt}{k\sqrt{(q_{max}^2 - t^2)}},$$

$$\theta - \mu = \frac{1}{k}\sin^{-1}\frac{t}{q_{max}} - \tfrac{1}{2}\pi + \alpha,$$

where $\alpha - \frac{1}{2}\pi$ is an arbitrary constant of integration. When $q = c^\star$, the critical sound speed, $\mu = \frac{1}{2}\pi$ and therefore $\theta = \alpha$. Using 20·42 (3) again we get

$$\sin^{-1}\frac{t}{q_{max}} = \tan^{-1}\frac{kt}{c} = \tan^{-1}(k\cot\mu),$$

so that

$$(1) \qquad \theta = \mu + \frac{1}{k}\tan^{-1}(k \cot \mu) - \tfrac{1}{2}\pi + \alpha.$$

Again, from Bernoulli's theorem 1·63 (4), putting $c = q \sin \mu$ we get

$$(2) \qquad q^2 = \frac{q^2_{\max}}{1 + \dfrac{2 \sin^2 \mu}{\gamma - 1}}.$$

It follows that (1) and (2) give the polar coordinates (q, θ) of points on the hodograph of a characteristic which makes an acute angle μ with $\mathbf{q}$, the different characteristics of this system being obtained by varying α. Writing

$$(3) \qquad f(\mu) = \mu + \frac{1}{k}\tan^{-1}(k \cot \mu) - \tfrac{1}{2}\pi,$$

we observe that $f(\pi - \mu) = -f(\mu)$, and (1) becomes

$$(4) \qquad \theta - f(\mu) = \alpha,$$

while the corresponding equation for the system of characteristics which make the angle $\pi - \mu$ with $\mathbf{q}$ is

$$(5) \qquad \theta + f(\mu) = \beta.$$

Thus taken in conjunction (4) and (5) are the equations of the hodographs of the two families of characteristics. On a member of the first family α is constant; on a member of the second family β is constant. Thus α, β are curvilinear coordinates. Given α and β, a point in the field of flow is determined by the intersection of the two corresponding characteristics; and to each point in the field there corresponds a pair of numbers α, β. Thus if α, β are known at each point of the field, the flow is thereby completely determined, for the characteristics can be plotted and the streamlines obtained as explained below.

The practical application of the method is facilitated by a change of notation. Let

$$D = \text{number of degrees in } \theta \text{ radians},$$

$$P = 1000 - [\text{number of degrees in } f(\mu) \text{ radians}].$$

Then (4) and (5) may be replaced by

$$(6) \qquad P + D = 2A, \quad P - D = 2B,$$

where A and B are new constants of integration. Thus

$$(7) \qquad P = A + B, \quad D = A - B.$$

Observe that when P is given, μ can be determined from (3), and therefore q^2 from (2), and hence the pressure. For this reason Busemann calls P the *pressure number*. The term *direction number* is applied to D by Temple *; it is the angle between the local direction of flow and some arbitrarily fixed line.

To carry out the method for an actual flow of air ($\gamma = 1{\cdot}405$) all that is required is contained in the following table of corresponding values of P, p/p_0, and μ.

TABLE

Pressure Number P	p/p_0	Mach Angle μ°	Pressure Number P	p/p_0	Mach Angle μ°	Pressure Number P	p/p_0	Mach Angle μ°
1000	0·527	90°·00	980	0·179	34°·26	960	0·054	23°·12
999	0·479	67°·70	979	0·170	33°·50	959	0·051	22°·70
998	0·449	61°·96	978	0·161	32°·80	958	0·047	22°·29
997	0·424	58°·18	977	0·153	32°·10	957	0·044	21°·89
996	0·401	55°·15	976	0·145	31°·41	956	0·041	21°·49
995	0·381	52°·66	975	0·137	30°·80	955	0·038	21°·11
994	0·363	50°·58	974	0·130	30°·19	954	0·036	20°·73
993	0·345	48°·70	973	0·123	29°·58	953	0·033	20°·37
992	0·329	47°·07	972	0·116	28°·98	952	0·031	20°·00
991	0·313	45°·54	971	0·110	28°·42	951	0·029	19°·64
990	0·298	44°·16	970	0·104	27°·88	950	0·027	19°·31
989	0·284	42°·84	969	0·097	27°·34	949	0·025	18°·93
988	0·270	41°·62	968	0·092	26°·82	948	0·023	18°·59
987	0·257	40°·51	967	0·086	26°·32	947	0·021	18°·26
986	0·245	39°·48	966	0·080	25°·80	946	0·019	17°·97
985	0·233	38°·47	965	0·075	25°·33			
984	0·221	37°·53	964	0·071	24°·87	870·68	0·000	0°·00
983	0·210	36°·67	963	0·066	24°·42			
982	0·199	35°·82	962	0·062	23°·98			
981	0·189	35°·02	961	0·058	23°·54			
980	0·179	34°·26	960	0·054	23°·12			

20·45. Straight-walled nozzle. Fig. 20·45 (i) shows some of the characteristics and streamlines for supersonic flow through a divergent two-dimensional nozzle with straight walls. Such a flow could be regarded as due to a source (8·9) placed at the intersection of the walls. The (slightly curved) characteristics divide the field into diamond-shaped cells.

Through a vertex of each cell there pass two characteristics. Along one A is constant, along the other B is constant.

An approximate representation of the field of flow will be obtained if the curved sides of each cell are replaced by straight lines. A curved side, which is an arc of a characteristic of the system A = constant, joins two vertices whose coordinates are, say, A, B and $A, B-\epsilon$.

* G. Temple, "The method of characteristics in supersonic flow", *R. and M.*, No. 2091 (1944). The present account of the method, which was originated by Busemann, is based on this paper.

To a degree of approximation which depends on the smallness of ϵ, the straight line joining these vertices will be parallel to that characteristic of the system A = constant which passes through the point whose coordinates are

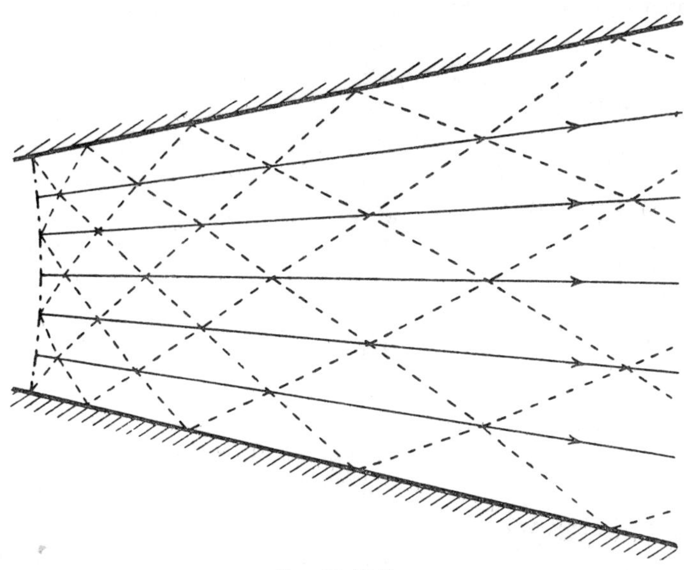

FIG. 20·45 (i).

A, $B-\frac{1}{2}\epsilon$. The angle between this line and the local direction of flow is the Mach angle μ which corresponds to the pressure number $P = A+B-\frac{1}{2}\epsilon$. The local direction of flow is obtained from the direction number $D = A-B+\frac{1}{2}\epsilon$.

Similarly the straight line which joins the points A, B and $A-\epsilon$, B is approximately parallel to the characteristic of the system B = constant which passes through $A-\frac{1}{2}\epsilon$, B. The angle it makes with the local direction of flow is the value of μ which corresponds to $P = A+B-\frac{1}{2}\epsilon$, and the local direction of flow is obtained from $D = A-B-\frac{1}{2}\epsilon$.

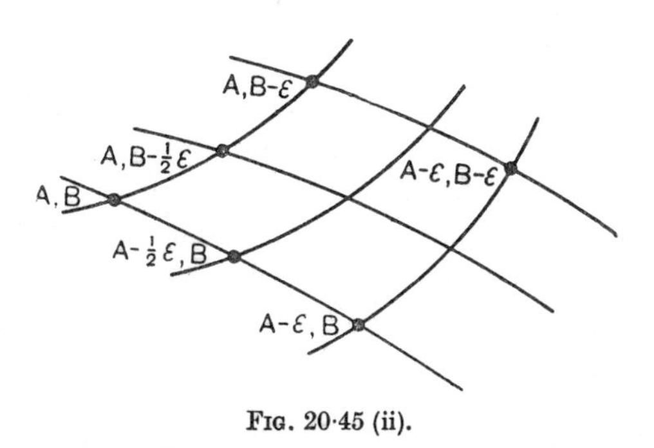

FIG. 20·45 (ii).

To draw a diagram such as fig. 20·45 (i), let us follow Temple in supposing that we are *given* the value of the pressure at the section VZ, fig. 20·45 (iii), of the nozzle to be $p = 0{\cdot}449\, p_0$. Then from the table the corresponding pressure number is 998. Let us further suppose that the walls are inclined at 16°. Divide the arc VZ into four equal parts at the points W, X, Y.

Then if we take the line of the nozzle wall through V as the initial line from which directions are measured, the direction numbers at V, W, X, Y, Z are 0, 4, 8, 12, 16. Thus the coordinates of these points are respectively (499, 499), (501, 497), (503, 495), (505, 493), (507, 491).

The next step is to draw the lines VW_1, W_1W, WX_1, X_1X, and so on. It will be sufficient to show how to determine a typical point, say X_1.

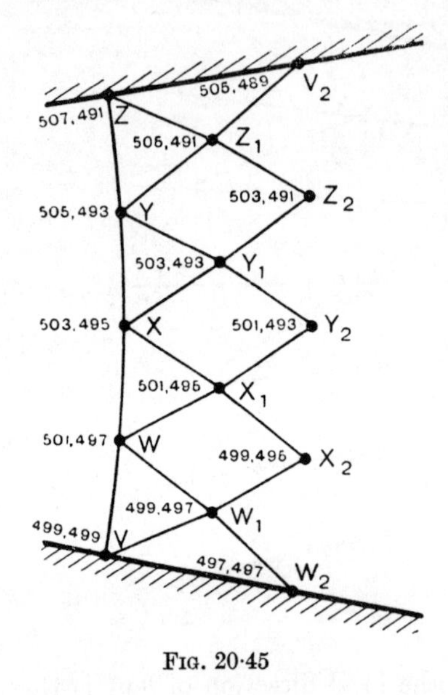

FIG. 20·45

The point X_1 lies on a characteristic A = constant through W (501, 497) and on a characteristic B = constant through X (503, 495). Therefore X_1 is the point (501, 495). Hence WX_1 has the same direction as the characteristic of the system A = constant which passes through the point (501, 496).

The pressure number for this point is $P = 501+496 = 997$ and the direction number is $D = 501-496 = 5$. From the table the corresponding Mach angle is $\mu = 58°{\cdot}18$. Using a protractor, we draw through W a line which makes the angle $D+\mu = 5°+58°{\cdot}18$ with the direction of the wall through V. Similarly through X we draw the line in the direction of the characteristic of the system B = constant which passes through (502, 495). These lines determine X_1 by their intersection. When V_1, W_1, X_1, Y_1, Z_1 are determined, we proceed to find W_2, X_2, Y_2, Z_2, and so on.

This illustrates the method for straight walls. If the walls are curved, we replace them by an approximating polygon in which the directions of successive sides differ by the chosen standard amount ϵ.

The application of the method is limited to continuous flow free from shock waves, whose presence will be indicated by the intersection of neighbouring characteristics of the same system and the appearance of an envelope of these characteristics.

20·5. Flow round a corner. Consider fluid streaming with constant supersonic speed V_0 parallel to a straight wall AB, which bends away from the stream into a second straight part BC at the corner B. In the uniform stream the Mach angle is given by $\sin \mu_0 = c_0/V_0$ and is therefore known. Thus the flow will begin to turn the corner along a straight characteristic or Mach line, m_0 in fig. 20·5. Assuming for the moment that the final state is uniform flow of speed V_1 parallel to BC, the turn will be completed at a second straight

Mach line m_1. The method of characteristics then shows at once that all the Mach lines issuing from B are straight and that the velocity at each point of any one of them, say m, is the same. If ϕ is the velocity potential, it follows that

$$(1) \qquad q_r = -\frac{\partial \phi}{\partial r}, \quad q_\theta = -\frac{1}{r}\frac{\partial \phi}{\partial \theta},$$

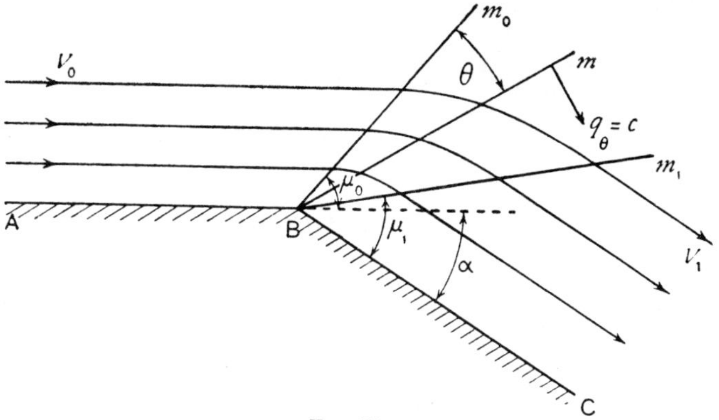

FIG. 20·5.

are independent of r. Moreover, since m is a characteristic $q_\theta = c$, and therefore Bernoulli's theorem, 1·63 (4) gives

$$(2) \qquad \left(\frac{1}{r}\frac{\partial \phi}{\partial \theta}\right)^2 = c^2 = \tfrac{1}{2}(\gamma-1)\left\{q_{\max}^2 - \left(\frac{\partial \phi}{\partial r}\right)^2 - \left(\frac{1}{r}\frac{\partial \phi}{\partial \theta}\right)^2\right\}.$$

Since q_r, q_θ are independent of r, we try to satisfy (1) and (2) by assuming

$$(3) \qquad \phi = rf(\theta),$$

where $f(\theta)$ is independent of r. Substitution in (2) then gives, using 20·42 (4),

$$\frac{1}{k^2}[f'(\theta)]^2 + [f(\theta)]^2 = q_{\max}^2.$$

This equation has the obvious solution

$$f(\theta) = -q_{\max} \sin(k\theta+\epsilon),$$

where ϵ is an arbitrary constant, and so

$$(4) \qquad q_r = q_{\max} \sin(k\theta+\epsilon), \quad q_\theta = c^\star \cos(k\theta+\epsilon),$$

since $c^\star = kq_{\max}$.

Let us measure θ from the initial Mach line m_0. Then when $\theta = 0$,

$$q_{\max} \sin \epsilon = V_0 \cos \mu_0, \quad c^\star \cos \epsilon = V_0 \sin \mu_0,$$

so that

$$(5) \qquad \tan \epsilon = k \cot \mu_0 = k\sqrt{(M_0{}^2 - 1)},$$

where M_0 is the Mach number of the oncoming stream V_0.

To find the position of m_1 we have, on this line, $\theta = \theta_1 = \mu_0 + \alpha - \mu_1$, where α is the angle BC makes with AB, i.e. the angle through which the oncoming stream has been deflected. Then

(6) $V_1 \cos \mu_1 = q_{max} \sin (k\theta_1 + \epsilon), \quad V_1 \sin \mu_1 = c^\star \cos (k\theta_1 + \epsilon).$

By division an equation is obtained to give μ_1, and V_1 is then found from (6). To determine the pressure, we have

$$\frac{\gamma p}{\rho} = c^2 = q_\theta^2 = c^{\star 2} \cos^2 (k\theta + \epsilon).$$

Now $\dfrac{p}{p_0} = \left(\dfrac{\rho}{\rho_0}\right)^\gamma$ and $c_0^2 = \dfrac{\gamma p_0}{\rho_0} = \frac{1}{2}(\gamma+1)c^{\star 2}$. Therefore

(7) $$\left(\frac{p}{p_0}\right)^{\frac{\gamma-1}{\gamma}} = \frac{2 \cos^2 (k\theta + \epsilon)}{\gamma+1}.$$

The maximum value of θ which is physically possible is that which makes $p = 0$, i.e.

(8) $$k\theta_{max} + \epsilon = \tfrac{1}{2}\pi.$$

It follows that if $\alpha + \mu_0 > \theta_{max}$, i.e. if

(9) $$\alpha > \frac{1}{k}(\tfrac{1}{2}\pi - \epsilon) - \mu_0,$$

the fluid will not be in contact with the wall BC, but will be separated from it by a vacuum bounded by BC and the line $\theta = \theta_{max}$ which is simultaneously a streamline and a characteristic.

If instead of presenting a sharp corner, the wall has a continuous bend, the bend may be replaced by an approximating polygon and the solution obtained by a limiting process. It is, however, simpler in this case to use the method of characteristics.

Lastly, we observe that the flow is irrotational and homentropic, and is therefore reversible. In fig. 20·5 the flow is expansive, i.e. pressure and density decrease in the direction of the flow, and the Mach line m_0 " leans away " from the oncoming stream. If we reverse the direction of motion on all the streamlines, the characteristic m_1 will lean towards the oncoming stream V_1 and the flow will be compressive, entailing increase of pressure and density.

20·6. Shock waves. If we try to apply the method of investigation for flow round a convex bend (20·5) to a concave bend, the Mach lines will be found to develop an envelope, E.

This would entail a mathematically ambiguous state of flow behind the envelope, where the fluid velocity would not be uniquely determined. Such a state is not physically possible. Experimental observations indicate that this situation gives rise to a *shock line* S which starts at the cusp of the envelope and runs between the two branches. In crossing this line the normal velocity component decreases suddenly,* the density, pressure, temperature and entropy

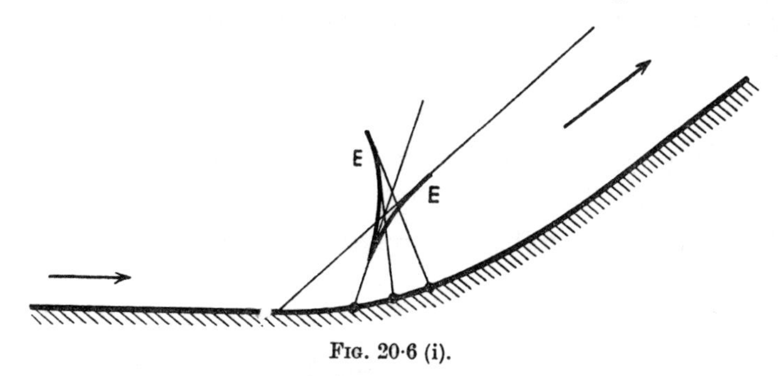

FIG. 20·6 (i).

suddenly increase. Fig. 20·6 (ii) shows the application of the shock and expansive flow (20·5) to a flat aerofoil BC.

The straight streamline AB impinges at B and a straight streamline CD departs from C.

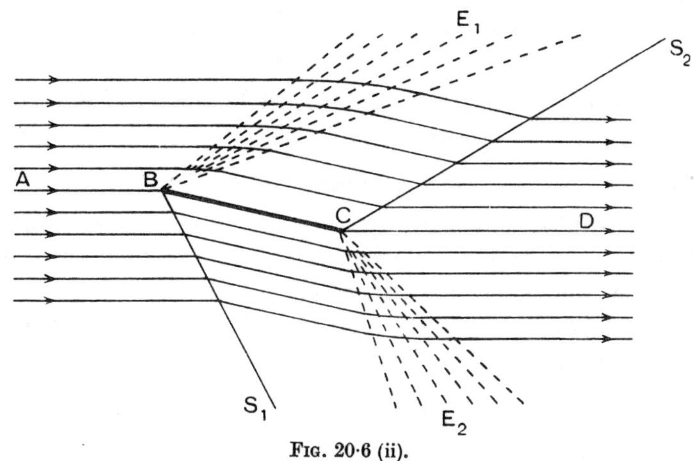

FIG. 20·6 (ii).

On the *upper surface* we have therefore expansive flow E_1 round the bend ABC which turns the oncoming stream into one parallel to BC. This then flows into the concave bend BCD and a shock line S_2 passes through C. Similarly on the *lower surface* we get a shock line S_1 at B and an expansion E_2 at C.

Consider the straight stationary shock wave occurring at an obtuse angle $\pi - \theta$.

Let suffix 0 refer to conditions in front of the shock line S and suffix 1 to conditions behind that line, so that V_0 is the speed of the oncoming, V_1 that of

---

* It will appear shortly that the tangential component is unaltered, so that the speed is always reduced in passing through the shock line.

the deflected flow. Let S make the angle σ with the direction of V_0 and let w_0, w_1 denote the components of V_0 and V_1 perpendicular to S. If we consider the conditions in front and behind a small line element dl of S, the oncoming flux of matter must be the same as the departing flux (equation of continuity) so that

$$\rho_0 w_0 = \rho_1 w_1. \tag{1}$$

Since the pressure thrust acts normally to dl there is no change of the momentum flux parallel to S, therefore

$$\rho_0 w_0 V_0 \cos\sigma = \rho_1 w_1 V_1 \cos(\sigma - \theta). \tag{2}$$

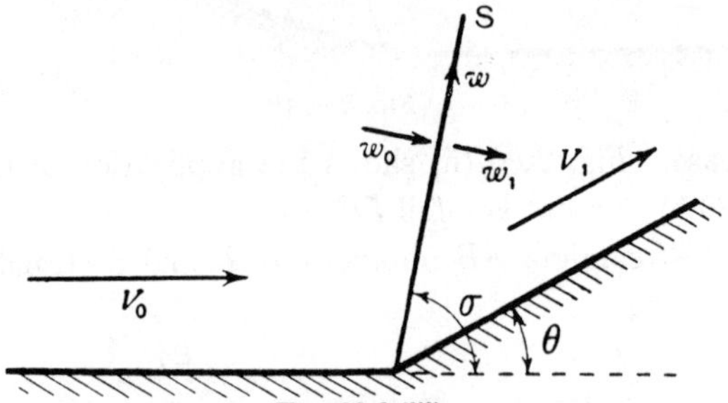

Fig. 20·6 (iii).

The difference in pressure thrusts on dl must be equal to the normal flux of momentum through dl. Therefore

$$p_1 - p_0 = \rho_0 w_0^2 - \rho_1 w_1^2. \tag{3}$$

These are the equations of ordinary mechanics. We obtain a fourth relation by applying the principle of conservation of energy including thermal energy.

If E is the internal energy per unit mass of air the total energy is $E + \frac{1}{2}V^2$ per unit mass. We equate the flux of energy to the rate at which work is done by the pressure thrusts. Thus

$$p_0 w_0 - p_1 w_1 = \rho_1 w_1 (E_1 + \tfrac{1}{2}V_1^2) - \rho_0 w_0 (E_0 + \tfrac{1}{2}V_0^2),$$

and therefore from (1)

$$\frac{p_0}{\rho_0} + E_0 + \tfrac{1}{2}V_0^2 = \frac{p_1}{\rho_1} + E_1 + \tfrac{1}{2}V_1^2.$$

Using 20·01 (14), we get

$$\frac{\gamma p_0}{(\gamma - 1)\rho_0} + \tfrac{1}{2}V_0^2 = \frac{\gamma p_1}{(\gamma - 1)\rho_1} + \tfrac{1}{2}V_1^2 = \tfrac{1}{2}q_{max}^2. \tag{4}$$

By Bernoulli's theorem each of the first two expressions in (4) is equal to the appropriate value of $\frac{1}{2}q_{max}^2$. The equality of these first two expressions shows that q_{max} is unaltered by the shock.

This equation is of the same form as Bernoulli's equation but in fact the states (p_0, ρ_0) and (p_1, ρ_1) here correspond with different values of the entropy, so that (4) cannot be written down from the principles of isentropic flow on which Bernoulli's equation is based. The increase of entropy from 20·01 (1) is

$$S_1 - S_0 = c_v \log \frac{p_1 \rho_0^{\gamma}}{p_0 \rho_1^{\gamma}}.$$

From (1) and (2) we see that the tangential component of the velocity parallel to the shock front is unaltered. Calling this component w, we get from (4)

$$\frac{2\gamma}{\gamma-1}\frac{p_0}{\rho_0} = q_{max}^2 - w^2 - w_0^2, \quad \frac{2\gamma}{\gamma-1}\frac{p_1}{\rho_1} = q_{max}^2 - w^2 - w_1^2.$$

Substitute for p_0, p_1 in (3) and eliminate ρ_0, ρ_1 by means of (1). Then, after a simple reduction, we get the important relation, due to Prandtl,

(5) $$w_0 w_1 = \frac{\gamma-1}{\gamma+1}(q_{max}^2 - w^2) = c^{\star 2} - k^2 w^2,$$

with the notation 20·42 (4), and observing that $c^{\star} = kq_{max}$.

Observing that $w_0 = V_0 \sin \sigma$, $w_1 = V_1 \sin(\sigma - \theta)$, equations (2) to (4) reduce easily to the following set.

(6) $$V_0 \cos \sigma = V_1 \cos(\sigma - \theta),$$

(7) $$p_0 + \rho_0 V_0^2 \sin^2 \sigma = p_1 + \rho_1 V_1^2 \sin^2(\sigma - \theta),$$

(8) $$\frac{\gamma p_0}{(\gamma-1)\rho_0} + \tfrac{1}{2} V_0^2 \sin^2 \sigma = \frac{\gamma p_1}{(\gamma-1)\rho_1} + \tfrac{1}{2} V_1^2 \sin^2(\sigma - \theta),$$

the last being got by squaring both sides of (6) and subtracting half the result from (4).

Putting $$\Delta\rho_0 = \rho_1 - \rho_0, \quad \Delta p_0 = p_1 - p_0,$$

we get, after some easy reductions,

(9) $$1 + \frac{\Delta\rho_0}{\rho_0} = \frac{\tan \sigma}{\tan(\sigma - \theta)} \quad \text{from (1) and (6),}$$

(10) $$\Delta p_0 = \rho_0 V_0^2 \sin^2 \sigma \frac{\Delta\rho_0}{\rho_0 + \Delta\rho_0} \quad \text{from (1) and (3),}$$

(11) $$\frac{\Delta p_0}{\Delta\rho_0} = \gamma \frac{2p_0 + \Delta p_0}{2\rho_0 + \Delta\rho_0} \quad \text{from (8) and (10).}$$

From these equations Δp_0, $\Delta\rho_0$ and σ may be calculated when θ, p_0, ρ_0, V_0 are given.

Also from (11)

$$\frac{p_1}{p_0} - \frac{\rho_1}{\rho_0} = \frac{\gamma-1}{2}\left(1 + \frac{p_1}{p_0}\right)\left(\frac{\rho_1}{\rho_0} - 1\right)$$

which determines the Hugoniot curve of p_1/p_0 against ρ_1/ρ_0, a rectangular hyperbola. When $p_1/p_0 \to \infty$, we get

$$\frac{\rho_1}{\rho_0} \to \frac{\gamma+1}{\gamma-1} = 6 \text{ approximately}$$

for air. Thus a shock wave can compress air at most to six times its original density. The dotted curve is the adiabatic $p_1/p_0 = (\rho_1/\rho_0)^\gamma$. When $\Delta p \to 0$, $\Delta \rho \to 0$, (11) goes over into the differential equation $dp/d\rho = \gamma p/\rho$ of the adiabatic. The two curves therefore touch at their starting point $\rho_1/\rho_0 = 1$. The ratio p/ρ and therefore the temperature rise more steeply in the Hugoniot curve than in the adiabatic.

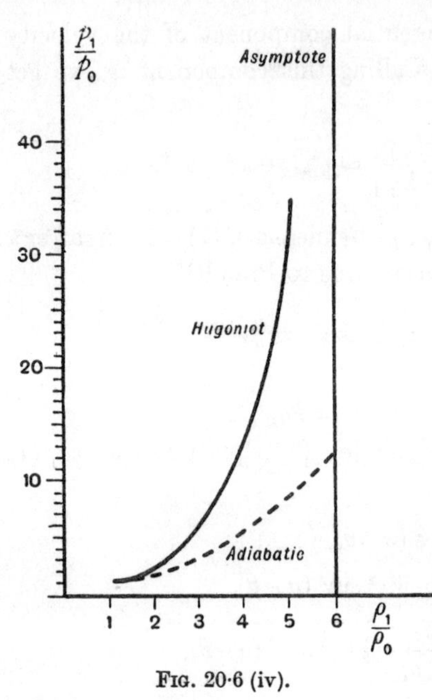

Fig. 20·6 (iv).

Finally, we may note that the conditions in front of the shock line here discussed must be supersonic. The conditions behind may be either supersonic or subsonic. It is the normal component of velocity which is reduced, the component tangential to the shock front is unaltered. Thus the velocity is refracted towards the shock front in passing from front to back. If the shock front is sufficiently oblique to the oncoming air, the conditions behind may still be supersonic.

20·61. The shock polar. In the hodograph plane represent the velocity $\mathbf{V}_0$ of the oncoming flow by the segment OA of the u-axis. From O draw the vector OP to represent the velocity $\mathbf{V}_1$ (components u, v) of the stream deflected by the shock through the angle θ. The locus of P is the *shock polar* belonging to V_0.

With the notations of fig. 20·61 (i) we have

$$w = V_0 \cos\sigma, \quad w_0 = V_0 \sin\sigma, \quad w_1 = V_0 \sin\sigma - \frac{v}{\cos\sigma}.$$

Substituting in 20·6 (5) these give

$$V_0^2 \sin^2\sigma - V_0 v \tan\sigma = k^2(q_{\max}^2 - V_0^2 \cos^2\sigma), \tag{1}$$

which together with

$$\tan\sigma = \frac{V_0 - u}{v}, \tag{2}$$

determines the locus of $P(u, v)$, i.e. the shock polar.

The elimination of σ between (1) and (2) leads directly to the equation

(3) $$v^2\{k^2(q_{\max}^2 - V_0^2) + V_0(V_0 - u)\} = (V_0 - u)^2(V_0 u - k^2 q_{\max}^2).$$

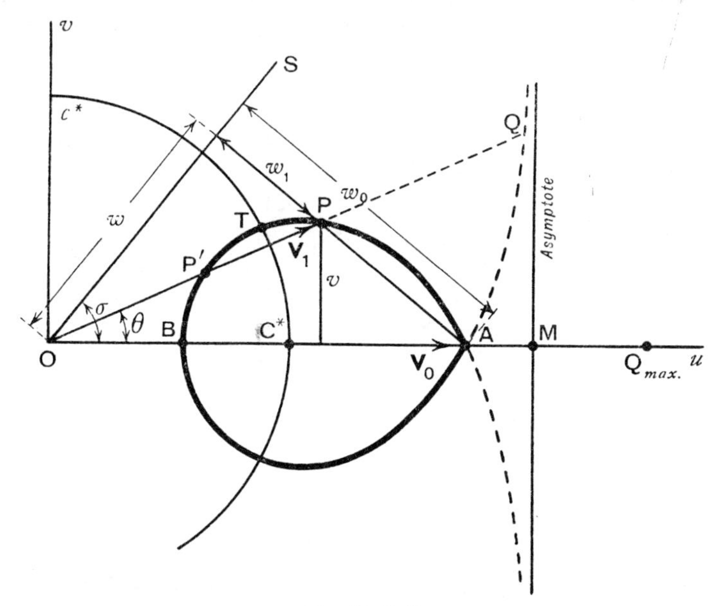

FIG. 20·61 (i).

The shock polar is therefore a cubic curve * symmetrical with respect to the u-axis which it meets at the points A, B in fig. (i),

(4) $$u = V_0, \quad u = k^2 q_{\max}^2 / V_0,$$

so that

(5) $$OA \,.\, OB = k^2 q_{\max}^2 = c^{\star 2} = OC^{\star 2}.$$

Thus A and B are inverse points with respect to the sonic circle $u^2 + v^2 = c^{\star 2}$. Points on the polar within this circle correspond with a subsonic régime after the shock. When

(6) $$OM = u = \frac{k^2(q_{\max}^2 - V_0^2) + V_0^2}{V_0},$$

v is infinite and the real asymptote is therefore $u = OM$. If we produce OP to meet the curve again at Q, an initial velocity represented by OQ will, after the shock, be reduced to the velocity represented by OA.

The shock polar corresponding to given values V_0 and $q_{\max}$ can be constructed point by point as follows.

Mark the points A, B, M given by (4) and (6), and on AB, MB as diameters draw circles C_1, C_2. Join any point Q on C_2 to B, and let QB intersect C_1 in R.

* In fact the Folium of Descartes.

Then the point P where AR meets QN, the perpendicular to AB, is a point on the shock polar. The proof is left as an exercise.

One application of the shock polar is obvious from fig. (i). Supposing the polar to be given, the direction of the shock line which deflects the stream

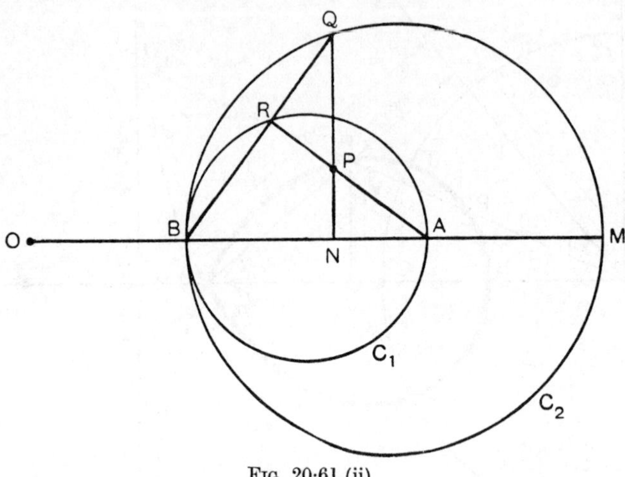

FIG. 20·61 (ii).

through the angle θ, is obtained by drawing a normal to AP, where P is the point where the line through O, which makes the angle θ with the oncoming stream, cuts the polar. Also we get from this construction $V_1 = OP$. Since OP cuts the polar at a second point P' there is a second possible shock line perpendicular to AP'; but experimental results seem to indicate that for compressive flow at a bend the one corresponding with P actually occurs. The tangent OT from O (the point T lies on the circle $u^2+v^2 = c^{\star 2}$) gives the

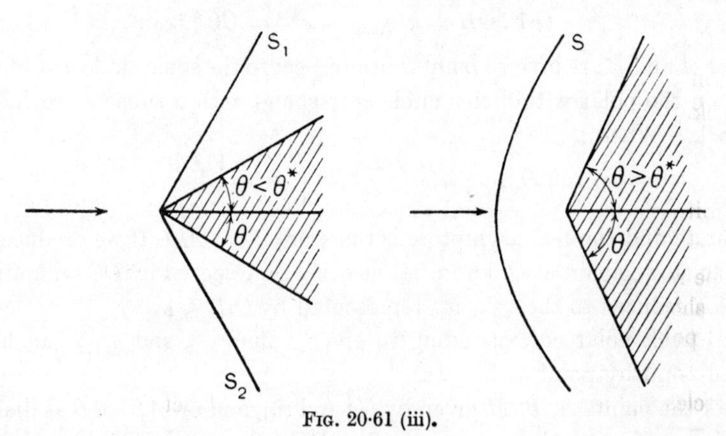

FIG. 20·61 (iii).

critical angle $\theta^\star$ where the two possible shock lines coincide. If $\theta > \theta^\star$, the above construction fails and there is a curved shock line in front of the corner.

We observe, further, that as $\theta \to 0$, i.e. as P approaches the double point A of the shock polar the shock becomes weaker and the conditions of no shock are being approached, so that the direction of the shock line must tend to coincidence with a Mach line. Therefore the angle between the tangents to the shock polar at the double point A must be $\pi - 2\mu$, where μ is the Mach angle.

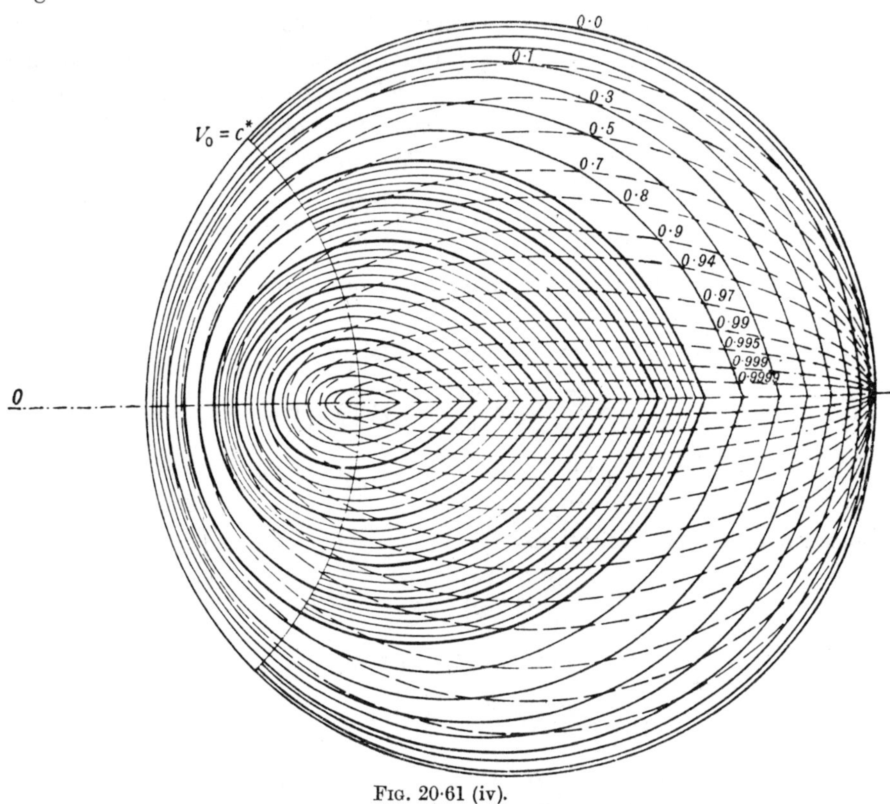

FIG. 20·61 (iv).

Fig. (iv), due to Busemann, shows a family of shock polars for $c^\star < V_0 \leqslant q_{\max}$. They all enclose the point $C^\star$, and lie within the circle to which they tend when $V_0 \to q_{\max}$. On the dotted curves the ratio of the stagnation pressure behind the shock to that before it has the constant value shown.

20·7. Characteristics in isentropic flow. The steady flow considered in 20·43 was homentropic. In this section we envisage the more general case of isentropic flow, in which the entropy S remains constant along each streamline but not necessarily the same constant on different streamlines.

Imagine the field of flow to be covered with a geometric network consisting of a family of curves C and their orthogonal trajectories, the family of curves N.

Consider the C and N curves at the point P. For the curve C denote by ds the element of arc, by $\mathbf{t}$ the unit tangent vector drawn in the direction of increasing s and by κ_s the curvature at P. For the orthogonal curve N the corresponding quantities will be denoted by dn, $\mathbf{n}$, κ_n, and we shall take as our standard disposition that shown in fig. 20·7, in which the direction of $\mathbf{n}$ is obtained by a counterclockwise rotation of $\mathbf{t}$ through one right angle, thereby determining the positive sense of dn.

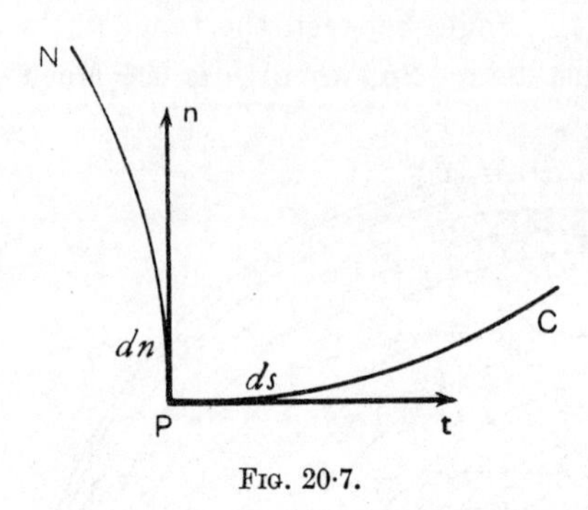

Fig. 20·7.

Then by Frenet's formula *

$$(1) \qquad \frac{\partial \mathbf{t}}{\partial s} = \kappa_s \mathbf{n}, \quad \frac{\partial \mathbf{n}}{\partial n} = -\kappa_n \mathbf{t},$$

since the positive direction of the normal vector of the curve N is, by our convention, that of $-\mathbf{t}$.

By differentiating the formula $\mathbf{nt} = 0$, we get

$$\mathbf{n}\frac{\partial \mathbf{t}}{\partial n} = -\mathbf{t}\frac{\partial \mathbf{n}}{\partial n} = \kappa_n \mathbf{tt} = \kappa_n \mathbf{nn},$$

$$\mathbf{t}\frac{\partial \mathbf{n}}{\partial s} = -\mathbf{n}\frac{\partial \mathbf{t}}{\partial s} = -\kappa_s \mathbf{nn} = -\kappa_s \mathbf{tt},$$

since $\mathbf{nn} = 1 = \mathbf{tt}$. Therefore

$$(2) \qquad \frac{\partial \mathbf{t}}{\partial n} = \kappa_n \mathbf{n}, \quad \frac{\partial \mathbf{n}}{\partial s} = -\kappa_s \mathbf{t}.$$

With these notations we have, if $\mathbf{q} = q_s \mathbf{t} + q_n \mathbf{n}$,

$$(3) \qquad \nabla = \mathbf{t}\frac{\partial}{\partial s} + \mathbf{n}\frac{\partial}{\partial n}, \quad \mathbf{q}\nabla = q_s \frac{\partial}{\partial s} + q_n \frac{\partial}{\partial n}.$$

If we now apply (3) to the equations of motion and continuity, 20·13 (2), (1), and take account of (1) and (2), we get

$$(4) \qquad q_s \frac{\partial q_s}{\partial s} + q_n \frac{\partial q_s}{\partial n} - q_s q_n \kappa_s - q_n^2 \kappa_n = -\frac{1}{\rho}\frac{\partial p}{\partial s}.$$

$$(5) \qquad q_s \frac{\partial q_n}{\partial s} + q_n \frac{\partial q_n}{\partial n} + q_s^2 \kappa_s + q_s q_n \kappa_n = -\frac{1}{\rho}\frac{\partial p}{\partial n}.$$

$$(6) \qquad q_s \frac{\partial \rho}{\partial s} + q_n \frac{\partial \rho}{\partial n} + \rho \frac{\partial q_s}{\partial s} + \rho \frac{\partial q_n}{\partial n} - \rho q_n \kappa_s + \rho q_s \kappa_n = 0.$$

The constancy of entropy along a streamline, 20·13 (4), is expressed by

$$(7) \qquad q_s \frac{\partial S}{\partial s} + q_n \frac{\partial S}{\partial n} = 0.$$

* C. E. Weatherburn, *Elementary Vector Analysis*, London (1926), p. 85.

Lastly, 20·13 (6), which is derived from the equation of state, yields

$$\frac{\partial p}{\partial s} = c^2 \frac{\partial \rho}{\partial s} + \frac{\partial p}{\partial S}\frac{\partial S}{\partial s}, \tag{8}$$

$$\frac{\partial p}{\partial n} = c^2 \frac{\partial \rho}{\partial n} + \frac{\partial p}{\partial S}\frac{\partial S}{\partial n}. \tag{9}$$

We can use (8) and (9) to eliminate $\partial p/\partial n$ and $\partial p/dS$ from (5). Then (4) to (7) are four ordinary simultaneous linear algebraic equations to determine the four quantities

$$\frac{\partial q_s}{\partial n}, \quad \frac{\partial q_n}{\partial n}, \quad \frac{\partial \rho}{\partial n}, \quad \frac{\partial S}{\partial n}, \tag{10}$$

which are the derivates of q_s, q_n, ρ, S in the direction normal to the C-curves.

If we use the determinantal method of solution we get

$$\frac{\partial q_s/\partial n}{\Delta_5} = \frac{\partial q_n/\partial n}{-\Delta_4} = \frac{\partial \rho/\partial n}{\Delta_3} = \frac{\partial S/\partial n}{-\Delta_2} = \frac{1}{\Delta_1}, \tag{11}$$

where on reduction of the determinants Δ_1, Δ_2, etc., we find

$$\Delta_1 = -q_n^2(q_n^2 - c^2), \quad \Delta_2 = -q_s q_n (q_n^2 - c^2)\frac{\partial S}{\partial s},$$

$$\Delta_3 = \rho q_n \left\{ \frac{q_s}{\rho}\frac{\partial p}{\partial s} + q_n^2 \frac{\partial q_s}{\partial s} - q_n q_s \frac{\partial q_n}{\partial s} - q_n \kappa_s (q_s^2 + q_n^2) + \frac{q_s}{\rho}\frac{\partial \rho}{\partial s}(q_n^2 - c^2)\right\},$$

$$\Delta_4 = q_n^2 \left\{ \frac{q_s}{\rho}\frac{\partial p}{\partial s} + c^2 \frac{\partial q_s}{\partial s} - q_n q_s \frac{\partial q_n}{\partial s} - q_n \kappa_s (q_s^2 + c^2) + q_s \kappa_n (c^2 - q_n^2)\right\},$$

$$\Delta_5 = q_n (q_n^2 - c^2)\left\{ q_s \frac{\partial q_s}{\partial s} - q_s q_n \kappa_s - q_n^2 \kappa_n + \frac{1}{\rho}\frac{\partial p}{\partial s}\right\}.$$

Following Meyer * we now enquire " are there any lines (C-curves) along which the equations of motion (4) to (7) will say nothing about the normal derivatives (10) and therefore also about $\partial p/\partial n$? "

Clearly this will be the case if, and only if, the values given by (11) are indeterminate, that is, if all the determinants $\Delta_k (k = 1, 2, 3, 4, 5)$ vanish. A necessary condition is the vanishing of Δ_1 (and therefore also of Δ_5), which gives $q_n = 0$ or $q_n = \pm c$. If $q_n = 0$ all the determinants vanish and the C-curves are then the streamlines. We shall return to this case later. If

$$q_n = \pm c, \tag{12}$$

* R. E. Meyer, " The method of characteristics for problems of compressible flow involving two independent variables ", Sixth International Congress for Applied Mechanics, Paris, 1946, *Quarterly Journal of Mechanics and Applied Mathematics*, I, (1948), pp. 196–219.

we have $\Delta_1 = \Delta_2 = \Delta_5 = 0$, while the conditions $\Delta_3 = \Delta_4 = 0$, when combined with (12), yield the single additional equation

$$q_n \left\{ q_n \frac{\partial q_s}{\partial s} - q_s \frac{\partial q_n}{\partial s} - \kappa_s (c^2 + q_s^2) \right\} + \frac{q_s}{\rho} \frac{\partial p}{\partial s} = 0. \tag{13}$$

This equation, together with (12), is independent of all the derivatives normal to a C-curve and also of the curvature κ_n of the orthogonal N-curve.

In particular, a flow which has a discontinuity of the normal derivatives of pressure, density, velocity and entropy along a curve on which (13) holds is compatible with the equations of motion. Such curves are called *characteristics* or *Mach lines*, and it is the possibility of these discontinuities and their confinement to the Mach lines which distinguishes steady supersonic from steady subsonic flow.

It should be noted that the characteristics as here defined on which (12) holds are the *same curves* as those defined in 20·41.

The alternative condition for the vanishing of the Δ_k is, as observed above, $q_n = 0$. Inserting this in equations (4) to (7), we get

$$q_s \frac{\partial q_s}{\partial s} = -\frac{1}{\rho} \frac{\partial p}{\partial s}, \quad \kappa_s q_s^2 = -\frac{1}{\rho} \frac{\partial p}{\partial n}, \tag{14}$$

$$\frac{\partial (\rho q_s)}{\partial s} + \rho q_s \kappa_n = 0, \tag{15}$$

$$\frac{\partial S}{\partial s} = 0, \tag{16}$$

which express the equations of motion along a streamline, the equation of continuity, and the original assumption of constant entropy along a streamline.

These equations show that the streamlines too have certain properties of characteristics. Since the above equations, (14) to (16), say nothing concerning $\partial q_s / \partial n$ and $\partial S / \partial n$, discontinuities of these quantities may occur on the streamlines and are propagated with the velocity of the fluid. Such discontinuities correspond with the presence of vorticity in the flow (cf. 20·1). Usually, however, the entropy and total energy are known on every streamline, so that the vorticity is determined by Crocco's equation, 20·1 (3), and the streamlines then have no properties of characteristics.

Thus it appears that the only lines on which the equations allow discontinuities in normal derivatives are the Mach lines and the streamlines. The indeterminateness of normal derivatives holds for derivatives of any order, as can be proved by differentiating the equations any number of times with respect to n.

We also observe that (12) means that the Mach lines make with the streamlines the Mach angle $\mu = \sin^{-1}(c/q)$. Thus if θ is the inclination of a streamline to the x-axis, we have for the slope of the Mach lines

$$\frac{dy}{dx} = \tan(\theta \mp \mu).$$

This shows that a Mach line along which the state of flow (i.e. $\mathbf{q}$, p, S) is constant must be straight (cf. 20·5).

Meyer (see p. 621, footnote) has shown that (13) to (16) are adapted to a step-by-step method of integration devised by Massau.* This method anticipates those since devised by Busemann and others. For details the reader is referred to Meyer's paper.

In the case of axisymmetrical motion we consider the C and N curves which lie in a meridian plane. Let the tangent at P make the angle θ with the axis and let ϖ be the distance of P from the axis. Then the only modification of the foregoing consists in the addition to the left sides of (6) and (13) the respective terms ρR and $c^2 R$, where

$$\varpi R = q_s \sin\theta + q_n \cos\theta.$$

20·71. Uniqueness theorem. We consider steady adiabatic two-dimensional supersonic flow.

Def. A curvilinear polygon which consists of arcs each of which is either a Mach line, a streamline, or a sonic line † is called a *characteristic polygon.*

The boundary between a region where the state of motion (defined by velocity, density or pressure, and entropy) is uniform, and a region where it is not must consist of lines on which the normal derivatives of some order are discontinuous. The boundary must therefore be a characteristic polygon.

Again, if two flow patterns which show no finite discontinuity of the velocity, density, and entropy are identical in one region but not in others, the boundary of this region must be a characteristic polygon.

Def. An *ordinary line* is a curve which does not meet any one Mach line or streamline at more than one point.

On an ordinary line the equations of motion admit and determine one set of values of the normal derivatives of any order of $\mathbf{q}$, ρ, S for any arbitrary continuous distribution of these variables themselves on the line.

Uniqueness theorem. The state of supersonic flow (i.e. the values of the variables $\mathbf{q}$, ρ, S) along an arc AB of an ordinary line, determines uniquely the field of flow in the smallest characteristic polygon which contains AB.

---

* J. Massau, *Mémoire sur l'intégration graphique des équations aux dérivées partielles*, Ghent (1900–1903). See also *Enzykl. d. Math. Wiss.*, II, 3_1, p. 159.

† A line on which the fluid speed is equal to the critical sound speed $c^\star$. Where such a line occurs it usually separates sub- from super-sonic flow; cf. the critical circle in Ringleb's nozzle (20·33), and the vortex of 13·8.

Proof. Consider two flow patterns F_1 and F_2 which have the same state of motion (i.e. the same values of $\mathbf{q}$, ρ, S) along the arc AB of an ordinary line.

It follows from the definition of an ordinary line that for F_1 and F_2, the variables $\mathbf{q}$, ρ, S can be expanded into the same Taylor series in the neighbourhood of AB, and therefore the flow patterns F_1 and F_2 are identical in a small finite region containing AB. Any such region must, however, extend to some characteristic polygon. Q.E.D.

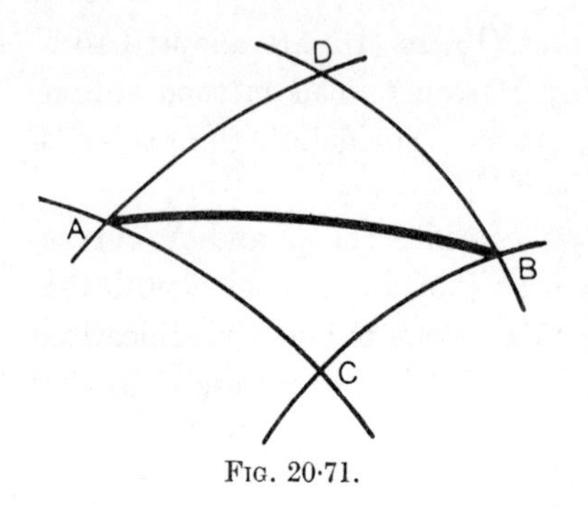

FIG. 20·71.

Usually the entropy and total energy are known in every streamline. The streamlines then cease to have the properties of characteristics. Except for the possible intervention of sonic lines, the uniqueness theorem then asserts that the state of motion on AB determines the flow uniquely in the Mach quadrilateral $ACBD$ bounded by the pair of Mach lines through A and the pair through B.

20·8. Flows dependent on time. Consider a flow which depends on the time t and on one space coordinate r, the distance from a fixed origin. Let q denote the velocity in the direction in which r increases. Then the equations of motion and continuity are

$$\text{(1)} \qquad \frac{\partial q}{\partial t}+q\frac{\partial q}{\partial r}=-\frac{1}{\rho}\frac{\partial p}{\partial r},$$

$$\text{(2)} \qquad \frac{\partial \rho}{\partial t}+\rho\frac{\partial q}{\partial r}+q\frac{\partial \rho}{\partial r}+\frac{(n-1)\rho q}{r}=0,$$

where $n = 1$ for one-dimensional flow, say along a straight tube, $n = 2$ for two-dimensional flow with circular symmetry as for a two-dimensional source, and $n = 3$ for three-dimensional flow with spherical symmetry as in the case of a three-dimensional source. Making the change of variable (due to Riemann), and regarding all variables as dimensionless,

$$\text{(3)} \qquad \omega=\int_{\rho_0}^{\rho}\frac{c\,d\rho}{\rho}, \quad c^2=\frac{\partial p}{\partial \rho},$$

we get

$$c\frac{\partial \omega}{\partial r}=\frac{c^2}{\rho}\frac{\partial \rho}{\partial r}=\frac{1}{\rho}\frac{\partial p}{\partial r}, \quad \frac{\partial \omega}{\partial t}=\frac{c}{\rho}\frac{\partial \rho}{\partial t}.$$

Equations (1) and (2) can now be written

$$\frac{\partial q}{\partial t}+q\frac{\partial q}{\partial r}+c\frac{\partial \omega}{\partial r}=0, \quad \frac{\partial \omega}{\partial t}+c\frac{\partial q}{\partial r}+q\frac{\partial \omega}{\partial r}+\frac{(n-1)cq}{r}=0,$$

which combine by addition and subtraction to give

$$(4) \qquad \frac{\partial}{\partial t}(\omega+q)+(q+c)\frac{\partial}{\partial r}(\omega+q) = -\frac{(n-1)cq}{r}.$$

$$(5) \qquad \frac{\partial}{\partial t}(\omega-q)+(q-c)\frac{\partial}{\partial r}(\omega-q) = -\frac{(n-1)cq}{r}.$$

As in 20·7 imagine the (r, t) plane to be covered with two orthogonal families of curves C and N. Let ds, dn be elements of arc of C and N respectively at P, and let the tangent at P make the angle α with the r-axis. Then

$$ds + i\,dn = e^{-i\alpha}(dr + i\,dt),$$

and therefore

$$\frac{\partial}{\partial t} = \sin\alpha\frac{\partial}{\partial s} + \cos\alpha\frac{\partial}{\partial n},$$

$$\frac{\partial}{\partial r} = \cos\alpha\frac{\partial}{\partial s} - \sin\alpha\frac{\partial}{\partial n},$$

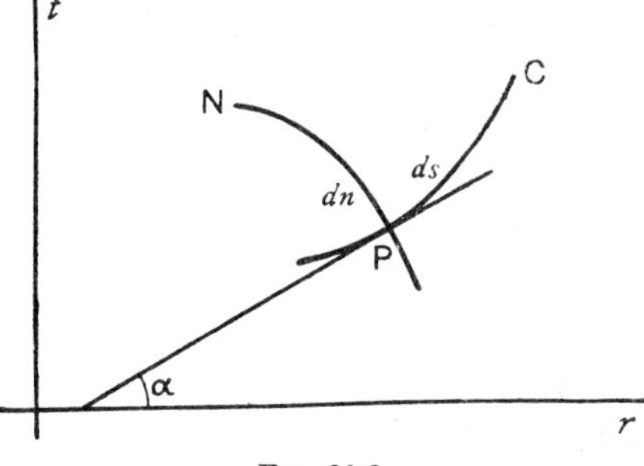

FIG. 20·8.

so that (4) is equivalent * to

$$[\sin\alpha+(q+c)\cos\alpha]\frac{\partial(\omega+q)}{\partial s}+[\cos\alpha-(q+c)\sin\alpha]\frac{\partial(\omega+q)}{\partial n} = -\frac{(n-1)cq}{r}.$$

This equation will have nothing to say concerning the normal derivative $\partial(\omega+q)/\partial n$ if we so choose the direction of the C curve that

$$\cos\alpha-(q+c)\sin\alpha = 0$$

along C, i.e. so that

$$(6) \qquad \cot\alpha = \frac{dr}{dt} = q+c \quad \text{along } C,$$

and our equation then becomes

$$(7) \qquad \frac{\partial(\omega+q)}{\partial s} = -\frac{(n-1)cq}{r}\sin\alpha.$$

Thus discontinuities in the normal derivatives of $\omega+q$ in the direction in the (r, t) plane normal to the C curves travel with the velocity $q+c$ given by (6). This means that in the physical plane these discontinuities are propagated with the velocity $q+c$, i.e. with velocity c relative to the fluid. Thus in accordance with the definition of 20·41 the characteristics are points,† circles, or spheres according as $n = 1$, 2, or 3.

A similar investigation applied to (5) shows that the discontinuities of the

* Observe that on the right side n is a number 1, 2, or 3 and has no connection with dn.

† Or, more generally, planes perpendicular to the direction of the r-axis.

normal derivates of $\omega - q$ travel with the velocity $q-c$, thus defining a second set of characteristics.

It appears from (3) that so long as $c^2 = \partial p/\partial \rho$ is positive the foregoing considerations are valid, so that in this case characteristics exist for subsonic as well as for supersonic flow.

In the case of one-dimensional flow we obtain at once from (4), (5) Riemann's results, namely that $\omega + q$ is constant for a geometrical point moving with the velocity $q+c$, and $\omega - q$ is constant for a geometrical point moving with the velocity $q-c$. For subsonic flow these velocities have opposite and for supersonic flow the same directions.

EXAMPLES XX

1. If E is the internal energy of a gas, S the entropy, show that

$$p = -\frac{\partial E}{\partial v}, \quad T = \frac{\partial E}{\partial S}.$$

2. Draw a graph to show the relation between the pressure p and the speed q in isentropic flow along a streamline, and show that the curve has a point of inflexion where $q = c^\star$.

3. In isentropic flow along a streamline show that

$$\frac{\rho q}{\rho^\star q^\star} = M^\star\{\tfrac{1}{2}(\gamma+1) - \tfrac{1}{2}(\gamma-1)M^{\star 2}\}^{\frac{1}{\gamma-1}},$$

where $\rho^\star$ is the density where $q = q^\star = c^\star$ and $M^\star = q/c^\star$.

4. Obtain Bernoulli's equation in the form

$$k^2 q^2 + (1-k^2)c^2 = c^{\star 2},$$

where $k^2 = (\gamma-1)/(\gamma+1)$.

5. In two-dimensional steady irrotational flow prove that

$$\frac{\partial u}{\partial y} - \frac{\partial v}{\partial x} = 0, \quad \frac{\partial(\rho u)}{\partial x} + \frac{\partial(\rho v)}{\partial y} = 0.$$

6. Use the velocity potential

$$\phi = -Vx + \frac{K\theta}{2\pi}$$

to prove that in compressible subsonic flow with the same circulation the radial and transverse components of velocity are

$$U_r = V\cos\theta, \quad U_\theta = -V\sin\theta - \frac{K}{2\pi r}\frac{\sqrt{1-M_0^2}}{1-M_0^2\sin^2\theta},$$

and hence prove the Kutta-Joukowski theorem for lift on an aerofoil.

Prove that the drag is zero.

7. Show that the change of variable given by

$$\frac{d\lambda}{\lambda} = \tfrac{1}{2}(1-\tau)^\beta\frac{d\tau}{\tau}, \quad \frac{\lambda^2}{\tau} \to 1, \quad \text{when } \tau \to 0$$

leads to the hodograph equations

$$\lambda\,\partial\phi/\partial\lambda = -F\,\partial\psi/\partial\theta, \quad \lambda\,\partial\psi/\partial\lambda = \partial\phi/\partial\theta,$$

where

$$F = \{1-(2\beta+1)\tau\}(1-\tau)^{-2\beta-1}.$$

8. Referring to the second solution of the hodograph equation, when $m = -1$, given by 20·32 (3), show that

$$\frac{z}{a} = \left(\beta+\frac{1}{\tau}\right)e^{2i\theta}+(2\beta+1)(\log\tau-2i\theta)+2i\beta\theta.$$

Prove that the curves of constant speed are trochoids generated by rolling a circle of radius $a(1+\beta)$ along the line $x = a(1+\beta)+a(2\beta+1)\log\tau$, the trochoid being described by the extremity of a radial line of length $a(\beta+\tau^{-1})$. (Temple.)

9. Prove that the hodograph of the envelope of the trochoids of Ex. 8 is

$$\tan^2\theta = \frac{(1-\tau)\{(2\beta+1)\tau-1\}}{\{(2\beta+1)\tau+1\}^2}.$$

(Temple.)

10. Prove the following construction for the normals to the characteristics at a point P in steady two-dimensional supersonic flow where the local sound speed is c. Draw PQ to represent the velocity vector. Let the circle whose centre is P and whose radius is c intersect the circle on PQ as diameter at N_1 and N_2. Then PN_1, PN_2 are the normals to the characteristics.

11. Use 20·43 (5) and (6) to prove that, if C_1, C_2 are the characteristics at P, the tangent to the hodograph of C_1 at P is parallel to the normal to C_2 at P, and a similar result with C_1 and C_2 interchanged.

12. Considering steady, irrotational, two-dimensional supersonic flow, a triangle OAD is drawn at the origin O in the hodograph plane and P is a point on AD between A and D. The triangle is so drawn that OP represents the velocity at P' in the field of flow, PA represents the local sound speed at P', A is a right angle, and OD represents q_{max}. The line PC is drawn parallel to AO to meet OD at C. Prove that (i) $AP/AD = k$ (see 20.42 (4)), (ii) OC represents $c^\star$, (iii) APD is parallel to the normal to a characteristic at P', (iv) APD is the tangent at P to the hodograph of the other characteristic through P'. (Temple.)

13. Use Ex. 12 to show that the point P describes an epicycloid obtained by rolling the circle on CD as diameter on the fixed circle centre O, radius OC, and infer that in steady irrotational two-dimensional supersonic flow the hodograph of any characteristic is the epicycloid generated by rolling a circle of diameter $q_{max}-c^\star$ on a fixed circle of radius $c^\star$.

14. In the case of flow round a corner, show that in the notation of 20·5 the Mach number is given by

$$M^2 = 1+\frac{1}{k^2}\tan^2(k\theta+\epsilon).$$

15. Discuss flow round a corner (20·5) when the oncoming stream has the critical speed $c^\star$, and prove that the equation of the streamlines is then

$$r = r_0[\cos k\theta]^{-1/k^2},$$

where r_0 is a constant, and hence show that the streamlines are homothetic curves

16. Draw by the method of characteristics the complete flow diagram for flow round a corner when the oncoming stream has the critical speed $c^\star$ and the angle α of the second wall with the first is so great that the flow does not reach the second wall.

17. In expansive supersonic flow round a polygonal bend the air stream is deflected through the small angle θ_n at the nth corner ($n = 1, 2, 3, \ldots,$). If p_n is the pressure and μ_n the local Mach angle after the nth corner is passed, prove that, approximately,

$$\frac{p_n}{p_{n-1}} = 1 - 2\gamma\theta_n \operatorname{cosec} 2\mu_n, \quad \mu_n = \mu_{n-1} - \tfrac{1}{2}\theta_n[(\gamma+1)\sec^2\mu_{n-1} - 2].$$

(Lighthill.)

18. In the preceding example, show that, if the bend is continuous,

$$\frac{d\mu}{d\theta} = -\tfrac{1}{2}(\gamma+1)\sec^2\mu + 1,$$

and hence prove, or verify, that

$$\theta = f(\mu_0) - f(\mu)$$

where

$$f(\mu) = \sqrt{\left(\frac{\gamma+1}{\gamma-1}\right)} \tan^{-1} \sqrt{\left(\frac{\gamma+1}{\gamma-1}\right)} \tan\mu - \mu.$$

(Lighthill.)

19. In Ex. 18, prove that if the bend is continuous

$$\frac{1}{p}\frac{dp}{d\theta} = -2\gamma \operatorname{cosec} 2\mu,$$

and hence prove, or verify, that

$$\frac{p}{p_0} = \frac{g(\mu)}{g(\mu_0)}, \text{ where}$$

$$g(\mu) = \left(\frac{\sin^2\mu}{\gamma - \cos 2\mu}\right)^{\gamma/(\gamma-1)}$$

(Lighthill.)

20. In Ex. 19, prove that the velocity V at the deflection θ is given by

$$\frac{V}{V_0} = \sqrt{\frac{\gamma - \cos 2\mu_0}{\gamma - \cos 2\mu}},$$

where V_0 is the velocity of the undeflected stream. (Lighthill.)

21. Show that the pressure behind a plane shock front which deflects through the angle θ is, approximately,

$$p_1 = p_0[1 - 2\gamma\theta \operatorname{cosec} 2\mu_0].$$

22. Show that 20·6 does in fact yield an increase of entropy in the case of a shock wave, and that the increase is approximately

$$c_v \frac{\gamma^3 - \gamma}{12}\left(\frac{\Delta\rho}{\rho_0}\right)^3,$$

where $\Delta\rho/\rho_0 = 2\theta \operatorname{cosec} 2\mu_0$ nearly.

23. With the notations of 20·6 prove that

$$\frac{\Delta p}{\Delta\rho} = w_0 w_1.$$

24. Gas is flowing steadily in a parallel supersonic stream along a straight pipe. If the pressure downstream is greater than the pressure upstream, show that a shock front perpendicular to the stream must exist and that the flow downstream of the shock front is subsonic. Prove that

$$\Delta p = \rho_0 w_0^2\left(1 - \frac{c^{\star 2}}{w_0^2}\right), \quad \Delta\rho = \rho_0\left(\frac{w_0^2}{c^{\star 2}} - 1\right), \quad w_0 w_1 = c^{\star 2}.$$

25. For the perpendicular shock front of Ex. 24 show that

$$\frac{p_1}{p_{00}} = \frac{\dfrac{2\gamma}{\gamma+1} M^2 - \dfrac{\gamma-1}{\gamma+1}}{\{\frac{1}{2}(\gamma-1) M^2+1\}^{\gamma/(\gamma-1)}},$$

$$\frac{\rho_1}{\rho_{00}} = \frac{\gamma+1}{\gamma-1}\left\{1-\left(\frac{\rho_1}{\rho_{00}}\right)^{\gamma-1}\right\}\frac{\rho_1}{\rho_{00}},$$

$$\frac{T_1}{T_{00}} = \frac{\dfrac{4\gamma}{(\gamma+1)^2} - \dfrac{T_0}{T_{00}}}{1-\dfrac{T_0}{T_{00}}},$$

where p_{00}, ρ_{00}, T_{00} are the stagnation values for the oncoming flow.

26. Use the method of the addition of a constant velocity to deduce the equations of a straight oblique shock front from those for the perpendicular shock front described in Ex. 24.

27. In one-dimensional flow of a gas the pressure is a function of the density. Obtain the equations of flow in the form

$$\frac{\partial u}{\partial t} + u\frac{\partial u}{\partial x} + \frac{c^2}{\rho}\frac{\partial \rho}{\partial x} = 0,$$

$$\frac{\partial \rho}{\partial t} + \rho\frac{\partial u}{\partial x} + u\frac{\partial \rho}{\partial x} = 0.$$

28. A gas flows parallel to the x-axis. A particle is at x at time t and at x_0 at time $t = 0$ and

$$\int_{x_0}^{x} \rho\, dx = m.$$

Prove that $$\frac{1}{\rho} = \frac{\partial x}{\partial m}, \quad \frac{\partial^2 x}{\partial t^2} = -\frac{\partial p}{\partial m}.$$

29. Show that the linear equation satisfied by the velocity potential ϕ in 20·3 is

$$(1-M^2)\, q^2\frac{\partial^2\phi}{\partial q^2} + (1+\gamma M^4)\, q\frac{\partial\phi}{\partial q} + (1-M^2)^2\frac{\partial^2\phi}{\partial\theta^2} = 0.$$

30. The pressure, density and normal velocity resolute on either side of a straight line stationary compression shock are p_0, ρ_0, w_0 and p_1, ρ_1, w_1. Establish the equations

$$\rho_0 w_0 = \rho_1 w_1 = m,$$

$$p_0 - p_1 = m(w_1 - w_0), \qquad \gamma(p_0+p_1) = m(w_0+w_1),$$

where γ is the adiabatic index.

If c_0 and c_1 are the velocities of sound on either side of the shock, prove that

$$c_0^2 c_1^2 = w_0^2 w_1^2 - \tfrac{1}{4}(\gamma^2-1)\, w_0 w_1 (w_0 - w_1)^2. \qquad \text{(U.L.)}$$

CHAPTER XXI

VISCOSITY

21·01. The existence of a stress tensor. We consider continuous material, not necessarily fluid, as described in 1·0.

Consider a tetrahedron, centroid P, of the material isolated in thought from the material which surrounds it. Let the lengths of the edges be infinitesimal of order l.

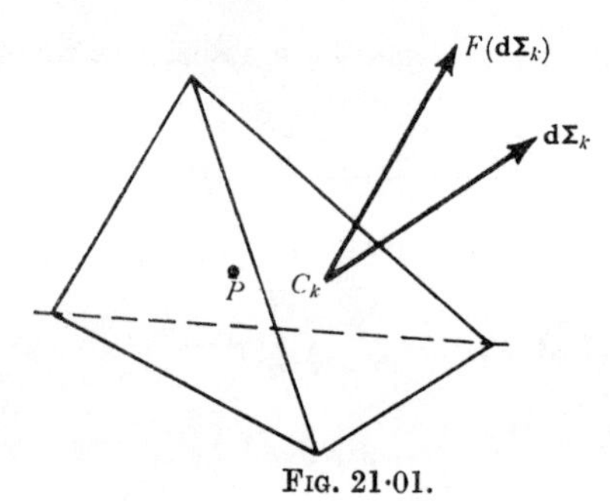

FIG. 21·01.

Let the vector areas of the four faces directed along the outward normals be $\mathbf{d\Sigma}_k$, $k = 1, 2, 3, 4$.

Then by projection on an arbitrary plane we see that

(1) $$\mathbf{d\Sigma}_1+\mathbf{d\Sigma}_2+\mathbf{d\Sigma}_3+\mathbf{d\Sigma}_4 = 0.$$

Hypothesis. The force exerted on any face k by the material on that side of the face into which $\mathbf{d\Sigma}_k$ points is a vector $F(\mathbf{d\Sigma}_k)$ which is a function of the vector area $\mathbf{d\Sigma}_k$ of the face, and acts at the centroid C_k of the face.

From this it follows by the law of action and reaction that

(2) $$F(-\mathbf{d\Sigma}_k) = -F(\mathbf{d\Sigma}_k)$$

so that F is an odd function.

We can neglect the inertial forces which are proportional to the volume (of order l^3) in comparison with the surface forces which are proportional to the area (of order l^2). Therefore *the surface forces form a system in equilibrium* (cf. 1·3), that is to say

(3) $$F(\mathbf{d\Sigma}_1)+F(\mathbf{d\Sigma}_2)+F(\mathbf{d\Sigma}_3)+F(\mathbf{d\Sigma}_4) = 0$$

and therefore

(4) $$F(\mathbf{d\Sigma}_1)+F(\mathbf{d\Sigma}_2)+F(\mathbf{d\Sigma}_3) = -F(\mathbf{d\Sigma}_4) = F(-\mathbf{d\Sigma}_4) = F(\mathbf{d\Sigma}_1+\mathbf{d\Sigma}_2+\mathbf{d\Sigma}_3)$$

on using (2) and then (1).

Thus the vector $F(\mathbf{d\Sigma})$ is a *linear function* (2·16) of $\mathbf{d\Sigma}$ so that we can write

(5) $$F(\mathbf{d\Sigma}) = \mathbf{d\Sigma}\,\mathbf{\Phi}$$

where $\mathbf{\Phi}$ is a tensor of the second rank, called the *stress tensor* at P.

The vector $\mathbf{d\Sigma}\,\mathbf{\Phi}$ is the force exerted on the directed element of area $\mathbf{d\Sigma}$. Let $\mathbf{d\Sigma} = \mathbf{n}\,d\Sigma$ where $\mathbf{n}$ is the unit normal vector. The force per unit area, namely $\mathbf{n\Phi}$, is called the *stress vector* for an infinitesimal area whose unit normal is $\mathbf{n}$.

21·02. The equation of motion for continuous material. Consider a material body τ bounded by a closed surface S. Let $\mathbf{n}$ be the unit *outward* normal at the element of area dS. Then (21·01) the force exerted on this area by the material outside the body is $dS\,\mathbf{n\Phi}$ where $\mathbf{\Phi}$ is the stress tensor. Therefore if $\mathbf{F}$ is the body force per unit mass, ρ is the density, and $\mathbf{a}$ is the acceleration, the second law of motion gives (cf. 3·41)

(1) $$\int_{(\tau)} \mathbf{a}\,\rho d\tau = \int_{(\tau)} \mathbf{F}\,\rho d\tau + \int_{(S)} \mathbf{n\Phi}\,dS$$

or using Gauss's theorem

$$\int_{(\tau)} (\mathbf{a}\rho - \mathbf{F}\rho - \nabla\mathbf{\Phi})\,d\tau = 0.$$

Since the volume of integration is arbitrary the integrand must vanish and we have the equation of motion

(2) $$\rho\mathbf{a} = \rho\mathbf{F} + \nabla\mathbf{\Phi}.$$

This equation applies to all continuous materials for which a stress tensor exists.

We now prove that if the stress tensor is symmetric

(3) $$\mathbf{\Phi} = \mathbf{\Phi}_c \quad \text{so that} \quad \mathbf{\Phi}^{\times} = 0,$$

then the equation of angular momentum rate also holds namely

(4) $$\frac{d}{dt}\int_{(\tau)} \rho(\mathbf{r}\wedge\mathbf{q})\,d\tau = \int_{(\tau)} \rho(\mathbf{r}\wedge\mathbf{F})\,d\tau + \int_{(S)} \mathbf{r}\wedge\mathbf{n\Phi}\,dS$$

which expresses that the rate of change of moment of momentum is equal to the moment of the forces acting on the material body.

Proof. From the transport theorem which depends solely on conservation of mass, 3·20 (2), we have

$$(5)\quad \frac{d}{dt}\int_{(\tau)} \rho(\mathbf{r}\wedge\mathbf{q})\,d\tau = \int_{(\tau)}\rho\left[\frac{d\mathbf{r}}{dt}\wedge\mathbf{q}+\mathbf{r}\wedge\frac{d\mathbf{q}}{dt}\right]d\tau = \int_{(\tau)}\rho\mathbf{r}\wedge\frac{d\mathbf{q}}{dt}\,d\tau$$

since $d\mathbf{r}/dt = \mathbf{q}$. Substitute for $\rho\, d\mathbf{q}/dt(= \rho\mathbf{a})$ from (2). Then the right-hand side of (5) is

$$(6)\quad \int_{(\tau)}\rho(\mathbf{r}\wedge\mathbf{F})\,d\tau+\int_{(\tau)}\mathbf{r}\wedge\nabla\mathbf{\Phi}\,d\tau = \int_{(\tau)}\rho(\mathbf{r}\wedge\mathbf{F})\,d\tau - \int_{(\tau)}\nabla(\mathbf{\Phi}\wedge\mathbf{r})\,d\tau,$$

on using 2·34 (XI) with $\mathbf{\Phi}^{\times} = 0$ from (3). Finally using Gauss's theorem (5) and (6) give (4). Q.E.D.

Thus (2) and (6) give the laws of motion for the material body. Our application will be to the fluid body.

21·021. General solution of the equation of motion.* In the case of conservative forces derivable from a potential Ω per unit volume we have the equations of motion and continuity which can be written, cf. 3·43 (4), 3·20 (4),

$$(1)\quad \frac{\partial}{\partial t}(\rho\mathbf{q}) = \nabla(\mathbf{\Phi}-\rho\mathbf{q}\,;\mathbf{q}-\Omega\mathbf{I}).$$

$$(2)\quad \frac{\partial\rho}{\partial t}+\nabla(\rho\mathbf{q}) = 0.$$

From (1) we see that $\rho\mathbf{q}$ is obtained by pre-multiplication of a symmetric tensor of the second rank by ∇. Thus we write

$$(3)\quad \rho\mathbf{q} = -\nabla\left(\frac{\partial\mathbf{F}}{\partial t}\right) = -\frac{\partial}{\partial t}(\nabla\mathbf{F}) = -\frac{\partial}{\partial t}(\mathbf{F}\nabla),$$

where $\mathbf{F}$ is a symmetric 2-tensor. Substitution in (2) gives

$$(4)\quad \rho = \nabla\mathbf{F}\nabla.$$

Thus from (1) and (4) we get

$$(5)\quad \nabla\left(\mathbf{\Phi}-\rho\mathbf{q}\,;\mathbf{q}-\Omega\mathbf{I}+\frac{\partial^2\mathbf{F}}{\partial t^2}\right) = 0.$$

This equation is satisfied identically by

$$(6)\quad \mathbf{\Phi} = \Omega\mathbf{I}+\rho\mathbf{q}\,;\mathbf{q}-\frac{\partial^2\mathbf{F}}{\partial t^2}+\nabla\wedge\mathbf{\Psi}\wedge\nabla,$$

where $\mathbf{\Psi}$ is a symmetric 2-tensor so that

$$(7)\quad \nabla(\nabla\wedge\mathbf{\Psi}\wedge\nabla) = 0.$$

* G. D. Nigam, A note on Milne-Thomson's general solution of the Navier-Stokes equations, *Proc. Camb. Phil. Soc.* 61 (1965), 915.

Equations (3), (4), (6) furnish the velocity, density and stress in terms of two symmetric 2-tensors $\mathbf{F}$ and $\mathbf{\Psi}$.

That the velocity field given by (3) is the most general solution can be proved as follows.

Equation (3) can be written

$$\rho\mathbf{q} = \nabla\mathbf{M}, \quad \mathbf{M} = -\partial\mathbf{F}/\partial t. \tag{8}$$

The most general representation of the vector field $\rho\mathbf{q}$ is

$$\rho\mathbf{q} = \nabla\phi + \nabla \wedge \boldsymbol{\beta} = \nabla\mathbf{N}, \tag{9}$$

where ϕ is a scalar, $\boldsymbol{\beta}$ is a vector and therefore

$$\mathbf{N} = \phi\mathbf{I} + \mathbf{I} \wedge \boldsymbol{\beta}. \tag{10}$$

The representation (8) will be the most general one if

$$\nabla(\mathbf{M} - \mathbf{N}) = 0, \tag{11}$$

and therefore

$$\mathbf{M} - \mathbf{N} = \nabla \wedge \mathbf{T} \wedge \nabla, \tag{12}$$

where $\mathbf{T}$ is an arbitrary 2-tensor.

Thus for a given $\mathbf{N}$ we can always find $\mathbf{M}$ from (12) which is indeterminate up to an arbitrary 2-tensor $\mathbf{T}$.

21·03. The stress tensor of a perfect fluid. In an ideal or perfect fluid the force exerted by the surrounding fluid on an element dS of the surface of a fluid particle is a normal thrust $-p\mathbf{n}\, dS$, where $\mathbf{n}$ is the unit outward normal and p is the pressure. We can therefore regard the stress (or force per unit area) as obtained from the stress tensor

$$\mathbf{\Psi} = -p\mathbf{I}, \tag{1}$$

where $\mathbf{I}$ is the idemfactor (2·16), by scalar multiplication by $\mathbf{n}$, that is to say,

$$\text{stress} = \mathbf{n\Psi} = -p\mathbf{nI} = -p\mathbf{n}.$$

In terms of three mutually perpendicular unit vectors $\mathbf{i}$, $\mathbf{j}$, $\mathbf{k}$ we have $\mathbf{I} = \mathbf{i} ; \mathbf{i} + \mathbf{j} ; \mathbf{j} + \mathbf{k} ; \mathbf{k}$. If therefore in (1) we replace dyadic by scalar multiplication we obtain the *first scalar invariant* of the stress tensor denoted by

$$\mathbf{\Psi}_I = -p(\mathbf{ii} + \mathbf{jj} + \mathbf{kk}) = -3p, \tag{2}$$

and we could use (2) to define the pressure p when $\mathbf{\Psi}$ is given.

The stress tensor (1) possesses all-round or spherical symmetry ; that is to say, the direction of the stress is normal to, and its intensity is independent of the orientation of, dS.

21·04. The viscosity hypothesis. In the case of a viscous fluid, that is to say, a fluid which is subject to internal friction, the stress on the element dS of the surface of a fluid particle is not necessarily normal to dS, and so the stress tensor will be of the form

$$\boldsymbol{\Phi} = -p'\mathbf{I} + \Xi, \tag{1}$$

where the tensor $-p'\mathbf{I}$ has all-round symmetry as in the case of no viscosity, while the tensor Ξ depends directly on the viscosity. The stress on dS will then be

$$\mathbf{n}\boldsymbol{\Phi} = -p'\mathbf{n} + \mathbf{n}\Xi. \tag{2}$$

In 2·40 we analysed the motion of a fluid particle into a movement of the particle as a whole, like a rigid body, compounded with a rate of pure strain in which the direction of motion of each point of the particle is normal to a certain quadric. If we regard viscosity as manifesting itself through action of a frictional character on the surface of our fluid particle by the surrounding fluid, it is clear that rigid body movements, since they cause no relative motion, can have no effect in producing forces of a frictional character. The natural hypothesis is to attribute the stress $\mathbf{n}\Xi$ of (2) solely to the pure strain.

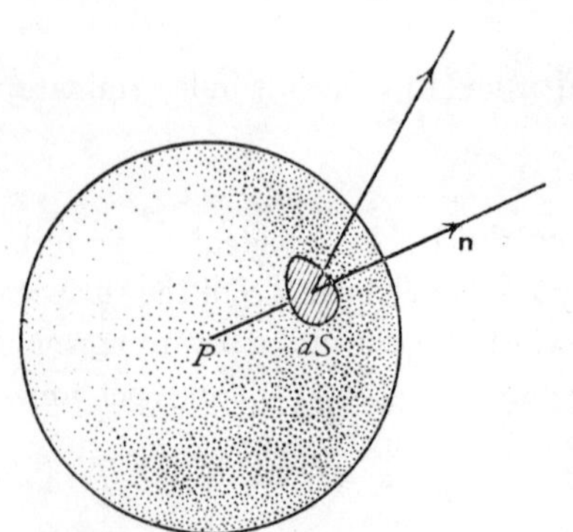

FIG. 21·04.

Consider a spherical particle, centre P, of infinitesimal radius h, fig. 21·04.

If $\mathbf{n}$ is the unit outward normal at the area dS of this particle, the pure strain is causing dS to move relatively to the centre of the sphere with velocity 2·40 (iii),

$$f(h\mathbf{n}) = h\mathbf{n}\mathbf{D}$$

The classical viscosity hypothesis is that $\mathbf{n}\Xi$ is proportional to $f(\mathbf{n})$, more precisely that

$$\mathbf{n}\Xi = 2\mu f(\mathbf{n}) = 2\mu\mathbf{n}\mathbf{D}, \tag{3}$$

where μ is called the *coefficient of viscosity*. The physical dimensions of μ are expressed by $ML^{-1}T^{-1}$ in terms of measure-ratios of mass length and time.

It then follows from (3) that

$$\Xi = \mu(\nabla\,;\,\mathbf{q} + \mathbf{q}\,;\,\nabla), \tag{4}$$

and therefore (1) becomes

$$\boldsymbol{\Phi} = -p'\mathbf{I} + \mu(\nabla\,;\,\mathbf{q} + \mathbf{q}\,;\,\nabla).$$

The *pressure* p is now *defined* by (cf. 21·03) the scalar invariant of this, namely,

$$-3p = \boldsymbol{\Phi}_I = -3p' + 2\mu(\nabla\,\mathbf{q}),$$

and so finally the stress tensor is the symmetric tensor given by the *constitutive equation*

$$\Phi = -p\mathbf{I} - \tfrac{2}{3}\mu(\nabla\, \mathbf{q})\mathbf{I} + \mu(\nabla\, ;\, \mathbf{q} + \mathbf{q}\, ;\, \nabla). \tag{5}$$

Thus the stress on dS is

$$\mathbf{n}\Phi = -p\mathbf{n} - \tfrac{2}{3}\mathbf{n}\mu(\nabla\, \mathbf{q}) + (\mathbf{n}\, 2\mu\, \nabla)\mathbf{q} + \mathbf{n}_\wedge \mu\zeta, \tag{6}$$

where $\zeta = \nabla_\wedge \mathbf{q}$ is the vorticity.

The stress on this hypothesis is a linear function of the direction of the normal to the area across which it is supposed to act. By choosing different elements on the surface of the sphere, we obtain the corresponding viscous stress. For an inviscid fluid, $\mu = 0$. Also when the fluid is at rest, $\mathbf{q} = 0$. In both these cases the viscous stress vanishes. That the above hypothesis applies generally is an assumption whose justification requires an investigation into the transfer of momentum due to random motions of the molecules to which the stress must ultimately owe its existence. To enter into these considerations is beyond the scope of this work, and it will therefore be assumed that (5) represents the effect of internal friction in the fluids which we shall consider. Such fluids which obey the classical constitutive equation (5) have been variously described as Newtonian, Stokesian, or Navier–Stokesian.

21·05. Boundary conditions in a viscous fluid. The kinematical condition that the normal velocity of the fluid in contact with a moving boundary is equal to the normal velocity of the boundary holds for fluids whether viscous or not.

When a viscous fluid is in contact with a solid, the tangential velocity is the same for both. This assertion is of a physical character founded on experiment. It is known as the *adherence condition* or *no slip condition.*

The surface of a solid boundary is often conveniently called the *wall.*

Thus there is no relative motion of the wall and the fluid in contact with it.

Further *the vorticity on the wall is tangential.* For if C is closed circuit enclosing the area S of a fixed wall, since $\mathbf{q} = 0$ on the wall it follows from 2·51 (1) that $\mathbf{n}\zeta = 0$ on the wall. Thus ζ is at right angles to $\mathbf{n}$ and is therefore tangential to the wall.

Moreover for a moving wall if $\mathbf{q}_r$ and $\zeta_r = \nabla_\wedge \mathbf{q}_r$ are the velocity and vorticity relative to the wall, the same argument shows that ζ_r is tangential to the wall.

At an interface separating two fluids the normal pressure and the viscous stress are continuous, provided surface tension is neglected.

21·06. Action of the fluid on the wall.

*Theorem.** Draw a unit normal $\mathbf{n}$ into the wall at P. Then the action of the fluid on the wall is given by the stress vector

(1) $$\mathbf{A} = \mathbf{n}(p - \tfrac{4}{3}\mu\, \nabla\mathbf{q}) + \mu\mathbf{n} \wedge \boldsymbol{\zeta}.$$

FIG. 21·06 (i).

Proof. Let C be a closed circuit drawn round the point P of the wall to enclose the area S of the wall. Since $\mathbf{q} = 0$ on the wall 2·51 (3) shows that

(2) $$(\mathbf{n}\,\nabla)\mathbf{q} - \mathbf{n}(\nabla\mathbf{q}) + \mathbf{n} \wedge \boldsymbol{\zeta} = 0 \text{ on the wall.}$$

But from 21·04 (6)

(3) $$\mathbf{A} = -\mathbf{n}\boldsymbol{\Phi} = \mathbf{n}[p + \tfrac{2}{3}\mu(\nabla\mathbf{q})] - 2\mu(\mathbf{n}\,\nabla)\mathbf{q} - \mathbf{n} \wedge \mu\boldsymbol{\zeta}.$$

Substitution for $(\mathbf{n}\nabla)\mathbf{q}$ from (2) gives (1). Q.E.D.

It follows that the action consists of a normal pressure $p - \frac{4}{3}\mu\nabla\mathbf{q}$ and a shear of magnitude $\mu\zeta$ in the sense of the vorticity (which is tangential to the wall) turned through a right angle in the positive sense of rotation about $\mathbf{n}$, where $\mathbf{n}$ is the normal directed into the wall.

For a liquid $\nabla\mathbf{q} = 0$ and the normal pressure reduces to p.

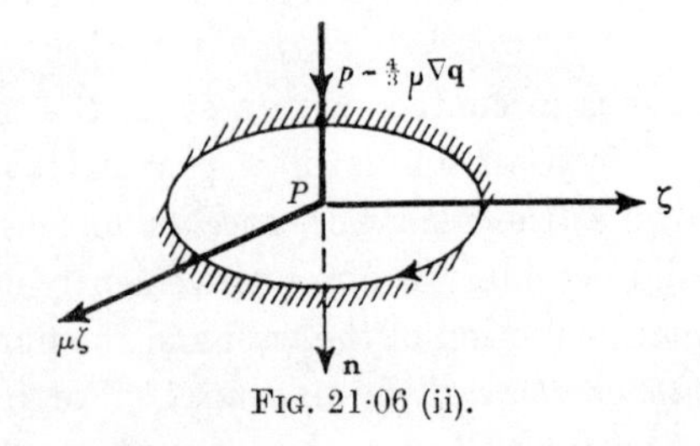

FIG. 21·06 (ii).

21·10. Dissipation of energy.

Let Σ be the surface of a fluid body which therefore always encloses the same fluid particles. The kinetic and internal energies are

$$T_k = \int_{(V)} \tfrac{1}{2}\rho\mathbf{q}\mathbf{q}\, d\tau, \quad J = \int_{(V)} \rho E\, d\tau,$$

* R. Berker, *C. R. Acad. Sc. Paris*, 258 (1964), 5144–5147.

taken through the volume V enclosed by Σ. Here E is the internal energy per unit mass (cf. 1·6 and 20·01).

The time rates of increase of T_k and J are, from 3·20 (2),

$$\frac{dT_k}{dt} = \int_{(V)} \rho \mathbf{aq}\, d\tau, \quad \frac{dJ}{dt} = \int_{(V)} \rho \frac{dE}{dt}\, d\tau, \tag{1}$$

where $\mathbf{a}$ is the acceleration given by the equation of motion, 21·02 (1),

$$\rho \mathbf{a} = \rho \mathbf{F} + \nabla \mathbf{\Phi}. \tag{2}$$

We then have the following energy balance.

The rate of increase of kinetic and internal energy = rate of working of the stress forces on the boundary Σ + rate of working of the body forces + the rate at which heat is supplied ; in symbols

$$\frac{dT_k}{dt} + \frac{dJ}{dt} = \int_{(\Sigma)} -(\mathbf{n\Phi})\, \mathbf{q}\, d\Sigma + \int_{(V)} \rho \mathbf{Fq}\, d\tau + \int_{(V)} Q\, d\tau. \tag{3}$$

Here Q is the rate per unit volume at which heat is supplied, for example by conduction through Σ, or by radiation from sources external to V.

Using (1) and Gauss's theorem, (3) gives

$$\int_{(V)} \left\{ \rho \mathbf{aq} - \rho \mathbf{Fq} - \nabla(\mathbf{\Phi q}) + \rho \frac{dE}{dt} - Q \right\} d\tau = 0. \tag{4}$$

Now $$\nabla(\mathbf{\Phi q}) = (\nabla \mathbf{\Phi})\, \mathbf{q} + (\mathbf{\Phi} \nabla)\, \mathbf{q},$$

and if we put this in (4) and remember (2), we get

$$\int_{(V)} \left\{ \rho \frac{dE}{dt} - (\mathbf{\Phi} \nabla)\, \mathbf{q} - Q \right\} d\tau = 0. \tag{5}$$

Since the volume of integration is arbitrary, the integrand vanishes and we have, for any deformable material,

$$\rho \frac{dE}{dt} = Q + (\mathbf{\Phi} \nabla)\, \mathbf{q}. \tag{6}$$

In order to keep the discussion general we introduce the *second coefficient of viscosity* λ, related to the " bulk modulus " κ (cf. elasticity theory) by

$$\kappa = \lambda + \tfrac{2}{3}\mu. \tag{7}$$

If $\kappa = 0$ we have the case discussed in 21·04. Further, we introduce the rate of deformation tensor $\mathbf{D}$ defined by

$$\mathbf{D} = \tfrac{1}{2}(\nabla\, ;\, \mathbf{q} + \mathbf{q}\, ;\, \nabla), \tag{8}$$

and we note that the first scalar invariant (2·16) is

$$\mathbf{D}_I = \nabla \mathbf{q}. \tag{9}$$

With these notations the stress tensor of 21·04 (5) in its generalised form becomes

$$(10) \qquad \mathbf{\Phi} = -p\mathbf{I} + (\kappa - \tfrac{2}{3}\mu)\,\mathbf{I}\mathbf{D}_I + 2\mu\mathbf{D}.$$

Now from (8) $\nabla \,;\, \mathbf{q} = \mathbf{D} + \frac{1}{2}(\nabla \,;\, \mathbf{q} - \mathbf{q} \,;\, \nabla)$ and the tensor in the brackets is skew symmetric, whereas $\mathbf{\Phi}$ and $\mathbf{D}$ are symmetric. Therefore from 2·16

$$(\mathbf{\Phi}\nabla)\,\mathbf{q} = \mathbf{\Phi}\cdot\cdot(\nabla \,;\, \mathbf{q}) = \mathbf{\Phi}\cdot\cdot\mathbf{D}$$

and therefore (6) becomes

$$(11) \qquad \rho\frac{dE}{dt} = \mathbf{\Phi}\cdot\cdot\mathbf{D} + Q.$$

Now if T is the absolute temperature and S is the entropy, we have from 20·01 (4), (9),

$$(12) \qquad T\,dS = dE + pd(1/\rho).$$

Therefore $T\,dS/dt$ is the rate of gain of heat per unit *mass*, so that the rate of gain of heat per unit *volume* is

$$(13) \qquad \rho T\frac{dS}{dt} = \rho\frac{dE}{dt} + p\rho\frac{d}{dt}\left(\frac{1}{\rho}\right) = \rho\frac{dE}{dt} + p\mathbf{D}_I,$$

since, from 3·20 (5), $\mathbf{D}_I = \nabla\mathbf{q} = -(1/\rho)\,d\rho/dt$. Therefore (11) and (13) give for the rate of gain of heat per unit volume

$$(14) \qquad \rho T\frac{dS}{dt} = \mathbf{\Phi}\cdot\cdot\mathbf{D} + p\mathbf{D}_I + Q.$$

Now Q is the rate at which heat is supplied by conduction and other external causes. Therefore

$$(15) \qquad w_i = \mathbf{\Phi}\cdot\cdot\mathbf{D} + p\mathbf{D}_I$$

is the rate per unit volume at which a fluid element gains heat at the expense of other forms of energy. Therefore w_i is the rate of dissipation of energy due to internal friction and for that reason is known as the *dissipation function*.

Now use (10) and note that $\mathbf{D}\cdot\cdot\mathbf{I} = \mathbf{D}_I$. Then

$$(16) \qquad w_i = 2\mu\,\{(\mathbf{D}\cdot\cdot\mathbf{D}) - \tfrac{1}{3}\mathbf{D}_I{}^2)\} + \kappa\mathbf{D}_I{}^2.$$

For a spherically symmetrical expansion or contraction the term in curled brackets vanishes. The last term will vanish if either

$$\kappa = 0 \quad \text{or} \quad \mathbf{D}_I = 0.$$

For a liquid $\mathbf{D}_I = 0$ in any case so that κ does not enter here or in (10). For a gas $\mathbf{D}_I \neq 0$ and the question whether $\kappa = 0$ remains open.

In cartesian coordinates we can write

$$\mathbf{D} = \sum_{r,s=1,2,3} e_{rs}(\mathbf{i}_r\,;\,\mathbf{i}_s), \quad e_{rs} = \tfrac{1}{2}\left(\frac{\partial q_r}{\partial x_s} + \frac{\partial q_s}{\partial x_r}\right) = e_{sr}\,.$$

Then

$$w_i = 2\mu\,\{2(e_{23}{}^2 + e_{31}{}^2 + e_{12}{}^2) + \tfrac{1}{3}(e_{22} - e_{33})^2 + \tfrac{1}{3}(e_{33} - e_{11})^2 + \tfrac{1}{3}(e_{11} - e_{22})^2\} + \kappa(e_{11} + e_{22} + e_{33})^2,$$

which is essentially non-negative and can vanish only if the fluid moves like a rigid body implying that $e_{11} = e_{22} = e_{33} = e_{23} = e_{31} = e_{12} = 0$.

Observe that for a liquid

$$w_i = 2\mu(\mathbf{D}\cdot\cdot\,\mathbf{D}) = 2\mu(e_{11}{}^2 + e_{22}{}^2 + e_{33}{}^2 + 2e_{23}{}^2 + 2e_{31}{}^2 + 2e_{12}{}^2).$$

In the case of liquid within a fixed closed envelope S we find by similar steps that the rate of dissipation of energy is

$$W_i = \int_{(V)} \mu\zeta^2\,d\tau + 2\mu\int_{(S)}\left\{\mathbf{n}\,(\mathbf{q}\wedge\boldsymbol{\zeta}) - \tfrac{1}{2}\frac{\partial q^2}{\partial n}\right\}dS.$$

But $\mathbf{q} = 0$ at the fixed surface. Therefore $W_i = \int_V \mu\zeta^2\,d\tau$, and we can suppose energy to be dissipated at the rate $\mu\zeta^2$ per unit volume.

21·11. The flow of heat in a fluid. The question of dissipation of energy is bound up with the rate of flow of heat. Treating the same fluid body as in 21·10 and using the same notations equations (11) and (15) give

$$(1) \qquad \rho\frac{dE}{dt} = Q + w_i - p\mathbf{D}_I\,,\ \mathbf{D}_I = \nabla\,\mathbf{q}.$$

If K is the thermal conductivity and T is the absolute temperature, the rate at which the fluid body gains heat by conduction through the bounding surface S is

$$\int_{(S)} -K(\mathbf{n}\,\nabla)T\,dS = \int_{(V)} \nabla(K\nabla T)\,d\tau,$$

i.e. at the rate $\nabla(K\nabla T)$ per unit volume. This is included in Q so if we write

$$(2) \qquad Q = Q_1 + \nabla(K\nabla T),$$

Q_1 will be the rate at which heat is supplied from all other causes, e.g. by radiation.

Also from 20·01, for a perfect gas,

$$(3) \qquad E = c_v T$$

where c_v is the specific heat at constant volume. Therefore the rate of flow of heat is governed by the equation

(4) $$\rho \frac{d}{dt}(c_v T) = w_i + \nabla(K\nabla T) + Q_1 - p\nabla \mathbf{q}.$$

In the case of an incompressible fluid and if $Q_1 = 0$, we have

(5) $$\rho \frac{d}{dt}(cT) = w_i + \nabla(K\nabla T),$$

where c is the specific heat of the liquid.

This equation together with the equation of motion and the equation of continuity serve to determine the three quantities p, $\mathbf{q}$, T which characterise the general motion of a viscous liquid. In the case of a gas it would also be necessary to take account of the equation of state connecting the pressure, density, and entropy.

21·14. Components of stress. If u_1, u_2, u_3 are orthogonal coordinates, we denote the components of stress in the directions u_1, u_2, u_3 across a plane perpendicular to h by the notations

$$\widehat{hu_1},\ \widehat{hu_2},\ \widehat{hu_3}.$$

Thus in cartesian coordinates we have nine components across planes perpendicular to x, y, z, namely

$$\widehat{xx},\ \widehat{xy},\ \widehat{xz};\ \widehat{yx},\ \widehat{yy},\ \widehat{yz};\ \widehat{zx},\ \widehat{zy},\ \widehat{zz}.$$

Taking $\mathbf{n} = \mathbf{i}$ in the formula giving the stress, 21·04 (6), we have

$$\mathbf{i}\,\widehat{xx} + \mathbf{j}\,\widehat{xy} + \mathbf{k}\,\widehat{xz} = 2\mu(\mathbf{i}\,\nabla)\mathbf{q} + \mu\mathbf{i}_\wedge \boldsymbol{\zeta} - p'\mathbf{i}$$

$$= 2\mu\frac{\partial}{\partial x}(\mathbf{i}q_x + \mathbf{j}q_y + \mathbf{k}q_z) + \mu(-\mathbf{j}\zeta_z + \mathbf{k}\zeta_y) - p'\mathbf{i},$$

where from 21·04, $$p' = p + \tfrac{2}{3}\mu\,\nabla\mathbf{q},$$

and so $$\widehat{xx} = -p' + 2\mu\frac{\partial q_x}{\partial x},\quad \widehat{xy} = \mu\left(\frac{\partial q_y}{\partial x} + \frac{\partial q_x}{\partial y}\right),\quad \widehat{xz} = \mu\left(\frac{\partial q_x}{\partial z} + \frac{\partial q_z}{\partial x}\right).$$

It follows from this that $\widehat{xy} = \widehat{yx}$, $\widehat{xz} = \widehat{zx}$, $\widehat{yz} = \widehat{zy}$, so that the nine components actually reduce to six, namely $\widehat{xx}$, $\widehat{yy}$, $\widehat{zz}$; $\widehat{xy}$, $\widehat{yz}$, $\widehat{zx}$.

This result can also be obtained by equating to zero the moments about lines parallel to the edges through the centre of an infinitesimal parallelepiped, and the same method shows that the result is true for any system of coordinates.

More generally, for any orthogonal system of coordinates (2·72), we have

$$\mathbf{i}_1\,\widehat{u_1u_1} + \mathbf{i}_2\,\widehat{u_1u_2} + \mathbf{i}_3\,\widehat{u_1u_3} = -p'\mathbf{i}_1 + 2\mu(\mathbf{i}_1\,\nabla)\mathbf{q} + \mu(\mathbf{i}_{1\wedge}\boldsymbol{\zeta}).$$

Now, from 2·34 (IV),

$$2(\mathbf{i}_1\,\nabla)\mathbf{q} + \mathbf{i}_{1\wedge}\boldsymbol{\zeta} = \nabla(\mathbf{i}_1\mathbf{q}) - \nabla_\wedge(\mathbf{i}_{1\wedge}\mathbf{q}) - \mathbf{q}_\wedge(\nabla_\wedge \mathbf{i}_1) - \mathbf{q}(\nabla\,\mathbf{i}_1) + \mathbf{i}_1(\nabla\,\mathbf{q}),$$

and $\mathbf{i}_1\mathbf{q} = q_1$, $\mathbf{i}_{1\wedge}\mathbf{q} = -\mathbf{i}_2q_3 + \mathbf{i}_3q_2$. Hence using the method of 2·72, we get successively

$$\nabla(\mathbf{i}_1\mathbf{q}) = \nabla q_1 = \frac{\mathbf{i}_1}{h_1}\frac{\partial q_1}{\partial u_1}+\frac{\mathbf{i}_2}{h_2}\frac{\partial q_1}{\partial u_2}+\frac{\mathbf{i}_3}{h_3}\frac{\partial q_1}{\partial u_3},$$

$$\nabla_\wedge(\mathbf{i}_1{}_\wedge\mathbf{q}) = \frac{1}{h_1h_2h_3}\left\{\mathbf{i}_1h_1\left(\frac{\partial(q_2h_3)}{\partial u_2}+\frac{\partial(q_3h_2)}{\partial u_3}\right)\right.$$
$$\left. -\mathbf{i}_2h_2\frac{\partial(q_2h_3)}{\partial u_1}-\mathbf{i}_3h_3\frac{\partial(q_3h_2)}{\partial u_1}\right\},$$

$$\mathbf{q}_\wedge(\nabla_\wedge\mathbf{i}_1) = -\mathbf{i}_1\left(\frac{q_2}{h_1h_2}\frac{\partial h_1}{\partial u_2}+\frac{q_3}{h_1h_3}\frac{\partial h_1}{\partial u_3}\right)+\frac{\mathbf{i}_2q_1}{h_1h_2}\frac{\partial h_1}{\partial u_2}+\frac{\mathbf{i}_3q_1}{h_1h_3}\frac{\partial h_1}{\partial u_3},$$

$$\mathbf{q}(\nabla\,\mathbf{i}_1) = \frac{\mathbf{i}_1q_1}{h_1h_2h_3}\frac{\partial(h_2h_3)}{\partial u_1}+\frac{\mathbf{i}_2q_2}{h_1h_2h_3}\frac{\partial(h_2h_3)}{\partial u_1}+\frac{\mathbf{i}_3q_3}{h_1h_2h_3}\frac{\partial(h_2h_3)}{\partial u_1},$$

$$\mathbf{i}_1(\nabla\,\mathbf{q}) = \frac{\mathbf{i}_1}{h_1h_2h_3}\left\{\frac{\partial(q_1h_2h_3)}{\partial u_1}+\frac{\partial(q_2h_3h_1)}{\partial u_2}+\frac{\partial(q_3h_1h_2)}{\partial u_3}\right\}.$$

Thus, omitting terms which cancel, we get

$$\widehat{u_1u_1} = -p'+\frac{2\mu}{h_1}\left\{\frac{\partial q_1}{\partial u_1}+\frac{q_2}{h_2}\frac{\partial h_1}{\partial u_2}+\frac{q_3}{h_3}\frac{\partial h_1}{\partial u_3}\right\},$$

$$\widehat{u_1u_2} = \mu\left\{\frac{1}{h_1}\frac{\partial q_2}{\partial u_1}+\frac{1}{h_2}\frac{\partial q_1}{\partial u_2}-\frac{q_2}{h_1h_2}\frac{\partial h_2}{\partial u_1}-\frac{q_1}{h_1h_2}\frac{\partial h_1}{\partial u_2}\right\},$$

$$\widehat{u_1u_3} = \mu\left\{\frac{1}{h_1}\frac{\partial q_3}{\partial u_1}+\frac{1}{h_3}\frac{\partial q_1}{\partial u_3}-\frac{q_3}{h_1h_3}\frac{\partial h_3}{\partial u_1}-\frac{q_1}{h_1h_3}\frac{\partial h_1}{\partial u_3}\right\}.$$

The remaining stress components can be written down at once. It is evident from the above that $\widehat{u_1u_2} = \widehat{u_2u_1}$, $\widehat{u_1u_3} = \widehat{u_3u_1}$, for the relations are not altered by an interchange of suffixes.

In the case of cylindrical coordinates (2·72), $u_1 = x$, $u_2 = \varpi$, $u_3 = \omega$, we get $h_1 = 1$, $h_2 = 1$, $h_3 = \varpi$,

$$\widehat{xx} = -p'+2\mu\frac{\partial q_x}{\partial x}, \qquad \widehat{\varpi\omega} = \mu\left\{\frac{\partial q_\omega}{\partial \varpi}+\frac{1}{\varpi}\frac{\partial q_\varpi}{\partial \omega}-\frac{q_\omega}{\varpi}\right\},$$

$$\widehat{\varpi\varpi} = -p'+2\mu\frac{\partial q_\varpi}{\partial \varpi}, \qquad \widehat{\omega x} = \mu\left\{\frac{\partial q_\omega}{\partial x}+\frac{1}{\varpi}\frac{\partial q_x}{\partial \omega}\right\},$$

$$\widehat{\omega\omega} = -p'+\frac{2\mu}{\varpi}\left(\frac{\partial q_\omega}{\partial \omega}+q_\varpi\right), \qquad \widehat{x\varpi} = \mu\left\{\frac{\partial q_\varpi}{\partial x}+\frac{\partial q_x}{\partial \varpi}\right\}.$$

The stress components for spherical polar coordinates are given in Ex. XXI, 20.

For a liquid $\nabla\mathbf{q} = 0$ so that $p' \doteq p$.

21·20. The equation of motion of a viscous fluid. In the case of a viscous fluid the equation of motion is derived at once from 21·02 by attributing to $\mathbf{\Phi}$ the value given in 21·04 (5). To calculate $\nabla\mathbf{\Phi}$ we merely write ∇ for $\mathbf{n}$, i.e. we use Gauss's theorem. The resulting equation of motion is

(1) $$\rho \frac{d\mathbf{q}}{dt} = \rho\mathbf{F} - \nabla p - \tfrac{2}{3}\nabla[\mu(\nabla\mathbf{q})] + [\nabla(2\mu\nabla)]\mathbf{q} + \nabla \wedge (\mu\boldsymbol{\zeta}).$$

Writing $\boldsymbol{\alpha} = \nabla\mu$ this readily reduces to

(2) $$\rho \frac{d\mathbf{q}}{dt} = \rho\mathbf{F} - \nabla p - \mu\nabla \wedge \boldsymbol{\zeta} + \tfrac{4}{3}\mu\nabla(\nabla\mathbf{q}) + 2(\boldsymbol{\alpha}\nabla)\mathbf{q} + \boldsymbol{\alpha}_\wedge\boldsymbol{\zeta} - \tfrac{2}{3}\boldsymbol{\alpha}(\nabla\mathbf{q}).$$

In the case where the viscosity μ is independent of position, $\boldsymbol{\alpha} = 0$ and the equation of motion becomes

(3) $$\rho \frac{d\mathbf{q}}{dt} = \rho\mathbf{F} - \nabla p - \mu\nabla \wedge \boldsymbol{\zeta} + \tfrac{4}{3}\mu\nabla(\nabla\mathbf{q}).$$

To this we must adjoin the equation of continuity, 3·20 (5),

(4) $$\nabla\mathbf{q} = \frac{d}{dt}\log\left(\frac{1}{\rho}\right).$$

21·22. Steady motion ; no external forces. In this case 21·20 (1) becomes

$$\nabla\boldsymbol{\Phi} = (\rho\mathbf{q}\,\nabla)\mathbf{q} = \nabla(\rho\mathbf{q}\,;\,\mathbf{q}) - \mathbf{q}[\nabla(\rho\mathbf{q})]$$

from 2·34 (X). But the equation of continuity is $\nabla(\rho\mathbf{q}) = 0$, and therefore

(1) $$\nabla(\boldsymbol{\Phi} - \rho\mathbf{q}\,;\,\mathbf{q}) = 0.$$

This equation is valid even when ρ and μ are functions of position.

21·30. Equation of motion of a viscous liquid. We now proceed to study the flow of incompressible fluids. *It will be assumed throughout that the coefficient of viscosity is independent of position and that the density is constant.*

It is important to remember that the *stability* of the flows whose exact solutions will be obtained depends strongly on the Reynolds number, 21·31, in general the smaller it is the more stable is the flow.

Breakdown of stability leads to turbulent flow concerning which we make only the general observation that a satisfactory mathematical theory still awaits discovery.

Since the equation of continuity for an incompressible fluid is $\nabla\mathbf{q} = 0$, it follows from 21·20 (3) that the equation of motion is

(1) $$\rho \frac{d\mathbf{q}}{dt} = \rho\mathbf{F} - \nabla p - \mu\nabla_\wedge\boldsymbol{\zeta}.$$

Alternatively, using 2·32 (V), we can write this as

(2) $$\rho \frac{d\mathbf{q}}{dt} = \rho\mathbf{F} - \nabla p + \mu\nabla^2\mathbf{q}.$$

These equations are known as the Navier-Stokes equations.

For some purposes it is convenient to use the *kinematic coefficient of viscosity* $\nu = \mu/\rho$ of dimensions L^2T^{-1}, whose value for water at 15° C. is $1\cdot 23\times 10^{-5}$ ft.²/sec. and for air $1\cdot 59\times 10^{-4}$. Judged by this standard, air is more viscous than water. With rising temperature ν decreases for water and increases for air.

In the case of conservative forces we can write the equation of motion (2) in the following forms.

$$\frac{d\mathbf{q}}{dt} = -\nabla\left(\frac{p}{\rho}+\Omega\right)+\nu\nabla^2\mathbf{q}, \tag{3}$$

$$\frac{\partial\mathbf{q}}{\partial t}-\mathbf{q}\wedge\boldsymbol{\zeta} = -\nabla\left(\frac{p}{\rho}+\tfrac{1}{2}q^2+\Omega\right)+\nu\nabla^2\mathbf{q}. \tag{4}$$

$$\frac{\partial\mathbf{q}}{\partial t}-\mathbf{q}\wedge\boldsymbol{\zeta} = -\nabla\chi-\nu\nabla\wedge\boldsymbol{\zeta}, \tag{5}$$

where

$$\chi = p/\rho+\tfrac{1}{2}q^2+\Omega.$$

From this equation it appears that for steady irrotational motion $\nabla\chi = 0$, and therefore χ has the same constant value throughout the fluid.

The form (5) is convenient for transformation to any system of orthogonal curvilinear coordinates by the methods of 2·72.

In particular, for two-dimensional motion in terms of cartesian coordinates, we have, from (5),

$$\frac{\partial q_x}{\partial t}-q_y\zeta = -\frac{\partial\chi}{\partial x}-\nu\frac{\partial\zeta}{\partial y},$$

$$\frac{\partial q_y}{\partial t}+q_x\zeta = -\frac{\partial\chi}{\partial y}+\nu\frac{\partial\zeta}{\partial x}.$$

In these equations ζ and χ are unaffected by change of coordinates, and we may therefore regard x, y as any orthogonal curvilinear coordinates. Thus, in the case of plane polar coordinates (r, θ),

$$\frac{\partial q_r}{\partial t}-q_\theta\zeta = -\frac{\partial\chi}{\partial r}-\nu\frac{\partial\zeta}{r\,\partial\theta},$$

$$\frac{\partial q_\theta}{\partial t}+q_r\zeta = -\frac{\partial\chi}{r\,\partial\theta}+\nu\frac{\partial\zeta}{\partial r}.$$

The expression for ζ in terms of the velocity components is

$$\zeta = \frac{\partial q_\theta}{\partial r}+\frac{q_\theta}{r}-\frac{\partial q_r}{r\,\partial\theta}. \tag{6}$$

21·31. Similarity; Reynolds number. Consider the equations of motion and continuity of a liquid and the stress tensor $\boldsymbol{\Phi}$

$$\frac{\partial\mathbf{q}}{\partial t}+(\mathbf{q}\nabla)\mathbf{q} = -\nabla P+\nu\nabla^2\mathbf{q}, \tag{1}$$

(2) $$\nabla \mathbf{q} = 0.$$

(3) $$\mathbf{\Phi} = -p\mathbf{I} + \mu(\nabla\,;\mathbf{q} + \mathbf{q}\,;\nabla).$$

(4) $$P = \frac{p}{\rho} + \Omega,$$

where the body-force is assumed to be conservative.

Let us arbitrarily choose a constant speed U and a constant length l as typical. Make, for example, the change of variables

(5) $$\mathbf{q} = U\mathbf{q}_1\,,\quad \mathbf{r} = l\mathbf{r}_1\,,\quad t = l^2 t_1/\nu,\quad P = \nu U\, P_1/l.$$

Then the variables with suffix 1 are dimensionless. As a derived result we have

(6) $$\nabla = \frac{1}{l}\nabla_1,\quad \nabla_1 = \frac{\partial}{\partial \mathbf{r}_1}.$$

Substitution in (1) and (2) gives

(7) $$\frac{1}{\mathscr{R}}\frac{\partial \mathbf{q}_1}{\partial t_1} + (\mathbf{q}_1\nabla_1)\mathbf{q}_1 = -\frac{1}{\mathscr{R}}\nabla_1 P_1 + \frac{1}{\mathscr{R}}\nabla_1{}^2\,\mathbf{q}_1\,;\ \nabla_1\mathbf{q}_1 = 0.$$

(8) $$\mathscr{R} = \frac{Ul}{\nu}.$$

The number $\mathscr{R}$ is called the *Reynolds number* and it is clear from the foregoing that two geometrically similar flows for which the Reynolds number is the same are *dynamically similar* in the sense that the change of variables (5) will reduce each to the same form (7).

In the same order of ideas we can write

(9) $$\mathbf{\Phi} = \rho U^2 \mathbf{\Phi}_1\,,$$

where $\mathbf{\Phi}_1$ is the dimensionless stress tensor

(10) $$\mathbf{\Phi}_1 = -p_1\mathbf{I} + \frac{1}{\mathscr{R}}(\nabla_1\,;\mathbf{q}_1 + \mathbf{q}_1\,;\nabla_1).$$

Thus equality of Reynolds numbers is the necessary and sufficient condition that two flows shall be dynamically similar.

In experiments made with a model in a wind tunnel, speeds and lengths are less than the same quantities for the full-scale machine, while ν is the same for both. This has led to the introduction of compressed-air wind tunnels, for $\nu = \mu/\rho$ can be made smaller by increasing ρ. This often makes possible the equalizing of the Reynolds number for the model and full-scale machine.

In comparing forces of drag and lift, we observe that any force F can be written in the form

(11) $$F = \tfrac{1}{2}\rho U^2 l^2 f(\mathscr{R}).$$

The dimensionless number $f(\mathscr{R})$ is the *drag or lift* coefficient corresponding to our choice of U and l.

Sometimes it is convenient to replace l^2 in (11) by some suitably chosen area S and the drag or lift coefficient will be then $F/(\frac{1}{2}\rho U^2 S)$.

21·34. The equation of compatibility. Taking the curl of equation 21·30 (5) and observing that the curl of a gradient is zero we get

$$\frac{\partial \boldsymbol{\zeta}}{\partial t} = \nabla_\wedge(\mathbf{q}_\wedge \boldsymbol{\zeta}) - \nu \nabla_\wedge(\nabla_\wedge \boldsymbol{\zeta}) \tag{1}$$

since $\nabla \boldsymbol{\zeta} = 0$ (3·52) and therefore from 2·32 (V)

$$\nabla_\wedge(\nabla^2 \mathbf{q}) = -\nabla_\wedge(\nabla_\wedge \boldsymbol{\zeta}) = \nabla^2 \boldsymbol{\zeta}. \tag{2}$$

Equation (1) is the condition the velocity field $\mathbf{q}$ is compatible with the flow of a viscous liquid under conservative forces when the coefficient of viscosity μ is independent of position.

Equation (1) is called the *compatibility equation.**

The compatibility equation can also be written in the forms

$$\frac{\partial \boldsymbol{\zeta}}{\partial t} = (\boldsymbol{\zeta} \nabla)\mathbf{q} - (\mathbf{q} \nabla)\boldsymbol{\zeta} + \nu \nabla^2 \boldsymbol{\zeta}, \tag{3}$$

$$\frac{d\boldsymbol{\zeta}}{dt} = (\boldsymbol{\zeta} \nabla)\mathbf{q} + \nu \nabla^2 \boldsymbol{\zeta}. \tag{4}$$

If we start a viscous liquid into motion from rest, initially $\boldsymbol{\zeta} = 0$, and therefore (4) becomes initially

$$\frac{d\boldsymbol{\zeta}}{dt} = \nu \nabla^2 \boldsymbol{\zeta}. \tag{5}$$

Since $\boldsymbol{\zeta}$ does not in general vanish at the boundaries, it follows that vorticity may ultimately be generated by diffusing inwards from the boundaries in accordance with the above equation.

In two-dimensional motion the vorticity is always perpendicular to the plane of the motion and therefore (5) applies at all times.

That in actual fluids the vorticity exists to any marked extent only in those parts of the fluid which have passed near to rigid boundaries is a fact well supported by observation, and is strikingly exhibited in the case of the wake behind a sailing vessel which arises solely from the water which has passed near to the ship's hull. The same observation shows that the eddy disturbance in the wake is damped out by friction.

Another illustration is of some interest. A discussion arises from time to time as to whether the sense of rotation of the vortex which is often seen when water runs out of a bath is different in the Northern and Southern

* This convenient term is due to R. Berker.

hemispheres. It is not difficult to prove by experiment that either sense of rotation can be obtained according as the bath is filled with the hot or the cold tap, the fluid from one or the other acquiring opposite vorticities as it moves near the boundary.

21·35. Equation satisfied by the stream function in two-dimensional motion. In two-dimensional motion we have, in terms of the stream function,

$$u = -\frac{\partial \psi}{\partial y}, \quad v = \frac{\partial \psi}{\partial x}, \quad \boldsymbol{\zeta} = \mathbf{k}\, \nabla^2 \psi, \tag{1}$$

while from 21·34 (3) the equation of compatibility is

$$\frac{\partial \boldsymbol{\zeta}}{\partial t} + (\mathbf{q}\, \nabla)\boldsymbol{\zeta} = \nu\, \nabla^2 \boldsymbol{\zeta}, \tag{2}$$

since $(\boldsymbol{\zeta}\, \nabla)\mathbf{q} = 0$.

Also $(\mathbf{q}\, \nabla) = u\dfrac{\partial}{\partial x} + v\dfrac{\partial}{\partial y} = -\dfrac{\partial \psi}{\partial y}\dfrac{\partial}{\partial x} + \dfrac{\partial \psi}{\partial x}\dfrac{\partial}{\partial y}$ and $(\mathbf{q}\, \nabla)\lambda = \dfrac{\partial(\psi, \lambda)}{\partial(x, y)}$,

the notation referring to the Jacobian (see 21·36).

Therefore

$$\frac{\partial}{\partial t}(\nabla^2 \psi) + \frac{\partial(\psi, \nabla^2 \psi)}{\partial(x, y)} - \nu\, \nabla^4 \psi = 0. \tag{3}$$

For plane polar coordinates $x = r \cos\theta$, $y = r \sin\theta$

$$\frac{\partial(\psi, \nabla^2 \psi)}{\partial(r, \theta)} = \frac{\partial(\psi, \nabla^2 \psi)}{\partial(x, y)}\frac{\partial(x, y)}{\partial(r, \theta)} = r\frac{\partial(\psi, \nabla^2 \psi)}{\partial(x, y)}. \tag{4}$$

Thus

$$\frac{\partial}{\partial t}(\nabla^2 \psi) + \frac{1}{r}\frac{\partial(\psi, \nabla^2 \psi)}{\partial(r, \theta)} - \nu\, \nabla^4 \psi = 0. \tag{5}$$

At a fixed boundary to which the fluid adheres

$$0 = q_s = -\partial\psi/\partial n, \quad 0 = q_n = \partial\psi/\partial s$$

and therefore

(6) ψ=constant, and $\partial\psi/\partial n = 0$ on the boundary

21·36. Equation satisfied by the stream function in axisymmetrical motion. Resuming the argument and notations of 19·50, we see that, to allow for viscosity, the left side of equation (5) must be modified by the addition of the term corresponding to $\varpi\nu\, \nabla_\wedge(\nabla_\wedge \boldsymbol{\zeta})$.

From 2·72 (4), we have successively

$$\nabla_\wedge \boldsymbol{\zeta} = \frac{\mathbf{i}_x}{\varpi}\frac{\partial}{\partial \varpi}(\varpi\zeta) - \frac{\mathbf{i}_\varpi}{\varpi}\frac{\partial}{\partial x}(\varpi\zeta),$$

$$\nabla_\wedge(\nabla_\wedge \boldsymbol{\zeta}) = -\mathbf{i}_\omega\left[\frac{\partial}{\partial x}\left(\frac{1}{\varpi}\frac{\partial(\varpi\zeta)}{\partial x}\right) + \frac{\partial}{\partial \varpi}\left(\frac{1}{\varpi}\frac{\partial(\varpi\zeta)}{\partial \varpi}\right)\right] = -\frac{\mathbf{i}_\omega}{\varpi}E^2(\varpi\zeta) = -\frac{\mathbf{i}_\omega}{\varpi}E^4\psi,$$

since
$$\zeta = \frac{1}{\varpi}\left(\frac{\partial^2\psi}{\partial x^2} + \frac{\partial^2\psi}{\partial \varpi^2} - \frac{1}{\varpi}\frac{\partial\psi}{\partial\varpi}\right) = \frac{1}{\varpi}E^2\psi.$$

Thus the equation satisfied by the stream function becomes

$$(1) \qquad \frac{\partial}{\partial t}(E^2\psi) + \varpi\frac{\partial(\psi, \varpi^{-2}E^2\psi)}{\partial(x, \varpi)} - \nu E^4\psi = 0,$$

the second term being an alternative notation for the determinant of 19·50 (5). This determinant is known as the Jacobian or functional determinant, and its vanishing implies a functional relation between ψ and $\varpi^{-2}E^2\psi$.

If we change to spherical polar coordinates (r, θ, ω), we have $x = r\cos\theta$, $\varpi = r\sin\theta$ and so $\partial(r, \theta)/\partial(x, \varpi) = 1/r$ and therefore the compatibility equation (1) becomes

$$(2) \qquad \frac{\partial}{\partial t}(E^2\psi) + \sin\theta\,\frac{\partial(\psi, (E^2\psi)/(r^2\sin^2\theta))}{\partial(r, \theta)} - \nu E^4\psi = 0$$

$$E^2 = \frac{\partial^2}{\partial r^2} + \frac{\sin\theta}{r^2}\frac{\partial}{\partial\theta}\left(\frac{1}{\sin\theta}\frac{\partial}{\partial\theta}\right).$$

We note that the motion represented by (1) is not reversible, for a change of sign of ψ alters the signs of the first and last terms, but not that of the middle term.

21·38. Circulation in a viscous liquid. If C is the circulation in a closed circuit moving with the fluid, then

$$C = \int \mathbf{q}\, d\mathbf{r},$$

taken round the circuit.

From 3·51 (2) and 21·30 (3) we get, for a liquid,

$$\frac{dC}{dt} = \int\left[-\nabla\left(\frac{p}{\rho} + \Omega\right) + \nu\nabla^2\mathbf{q}\right]d\mathbf{r} = \nu\int\nabla^2\mathbf{q}\,.\,d\mathbf{r} = -\nu\int(\nabla_\wedge\boldsymbol{\zeta})d\mathbf{r}.$$

Thus as the circuit moves with the liquid the rate of change of circulation depends only on the vorticity in the neighbourhood of the circuit. Hence, if the liquid is originally at rest ($\zeta = 0$), circulation can only arise by the diffusion of vorticity inwards from the boundary (cf. 21·34).

21·39. Intrinsic equations. Consider a streamline OP and its orthogonal trajectory ON, the motion being steady and two-dimensional, fig. 21·39.

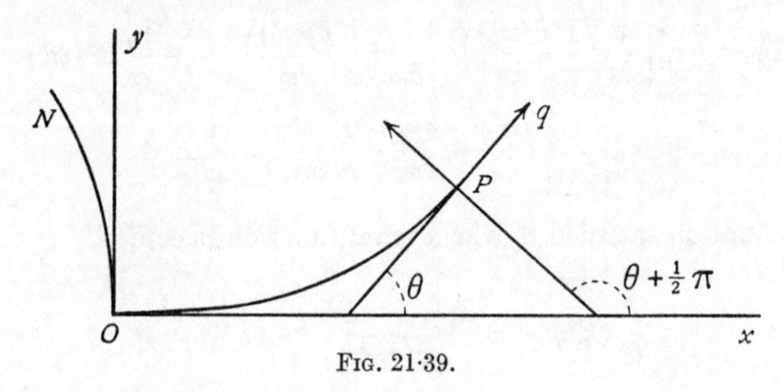

FIG. 21·39.

Take as axes at O the tangent and normal to the streamline. The intrinsic equations for an inviscid liquid have been given in 4·25. It remains to complete them by adding on the right the terms corresponding to $\nu \nabla^2 \bar{v}$, where $\bar{v} = u+iv = qe^{i\theta}$, θ being the inclination to Ox of the tangent at the point P of the streamline. We require the values at O, where $\theta = 0$. Let ds, dn be elements of arc of OP and ON, κ_s, κ_n the corresponding curvatures at O. Then when $\theta = 0$,

$$\kappa_s = \frac{\partial \theta}{\partial s}, \quad \kappa_n = \frac{\partial \theta}{\partial n}. \quad \text{Also } \frac{\partial x}{\partial s} = \cos\theta, \quad \frac{\partial y}{\partial s} = \sin\theta,$$

$$\frac{\partial x}{\partial n} = -\sin\theta, \quad \frac{\partial y}{\partial n} = \cos\theta.$$

Therefore, differentiating and putting $\theta = 0$,

$$\frac{\partial^2 x}{\partial s^2} = 0, \quad \frac{\partial^2 y}{\partial s^2} = \kappa_s, \quad \frac{\partial^2 x}{\partial n^2} = -\kappa_n, \quad \frac{\partial^2 y}{\partial n^2} = 0.$$

Now, if f is any function of x and y,

$$\frac{\partial f}{\partial s} = \frac{\partial f}{\partial x}\frac{\partial x}{\partial s} + \frac{\partial f}{\partial y}\frac{\partial y}{\partial s}, \quad \frac{\partial f}{\partial n} = \frac{\partial f}{\partial x}\frac{\partial x}{\partial n} + \frac{\partial f}{\partial y}\frac{\partial y}{\partial n},$$

$$\frac{\partial^2 f}{\partial s^2} = \frac{\partial}{\partial s}\left(\frac{\partial f}{\partial x}\right)\frac{\partial x}{\partial s} + \frac{\partial f}{\partial x}\frac{\partial^2 x}{\partial s^2} + \frac{\partial}{\partial s}\left(\frac{\partial f}{\partial y}\right)\frac{\partial y}{\partial s} + \frac{\partial f}{\partial y}\frac{\partial^2 y}{\partial s^2},$$

$$\frac{\partial^2 f}{\partial n^2} = \frac{\partial}{\partial n}\left(\frac{\partial f}{\partial x}\right)\frac{\partial x}{\partial n} + \frac{\partial f}{\partial x}\frac{\partial^2 x}{\partial n^2} + \frac{\partial}{\partial n}\left(\frac{\partial f}{\partial y}\right)\frac{\partial y}{\partial n} + \frac{\partial f}{\partial y}\frac{\partial^2 y}{\partial n^2}.$$

Therefore at O we have

$$\frac{\partial f}{\partial s} = \frac{\partial f}{\partial x}, \quad \frac{\partial f}{\partial n} = \frac{\partial f}{\partial y},$$

$$\frac{\partial^2 f}{\partial s^2} = \frac{\partial}{\partial s}\left(\frac{\partial f}{\partial x}\right) + \kappa_s \frac{\partial f}{\partial y} = \frac{\partial^2 f}{\partial x^2} + \kappa_s \frac{\partial f}{\partial n},$$

$$\frac{\partial^2 f}{\partial n^2} = \frac{\partial}{\partial n}\left(\frac{\partial f}{\partial y}\right) - \kappa_n \frac{\partial f}{\partial x} = \frac{\partial^2 f}{\partial y^2} - \kappa_n \frac{\partial f}{\partial s}.$$

Hence, on the understanding that θ is to be made zero after the differentiation, we have

$$\nabla^2 (q\, e^{i\theta}) = \frac{\partial^2}{\partial s^2}(q\, e^{i\theta}) + \frac{\partial^2}{\partial n^2}(q\, e^{i\theta}) - \kappa_s \frac{\partial}{\partial n}(q\, e^{i\theta}) + \kappa_n \frac{\partial}{\partial s}(q\, e^{i\theta}).$$

The real and imaginary parts are the required components and we thus get the intrinsic equations of motion, namely

$$(1) \qquad \frac{1}{\rho}\frac{\partial p}{\partial s} + q\frac{\partial q}{\partial s} + \frac{\partial \Omega}{\partial s} = \nu\left\{\frac{\partial^2 q}{\partial s^2} + \frac{\partial^2 q}{\partial n^2} - \kappa_s \frac{\partial q}{\partial n} + \kappa_n \frac{\partial q}{\partial s} - q(\kappa_s^2 + \kappa_n^2)\right\}.$$

$$(2) \qquad \frac{1}{\rho}\frac{\partial p}{\partial n} + \kappa_s q^2 + \frac{\partial \Omega}{\partial n} = \nu\left\{2\kappa_s \frac{\partial q}{\partial s} + 2\kappa_n \frac{\partial q}{\partial n} + q\left(\frac{\partial \kappa_s}{\partial s} + \frac{\partial \kappa_n}{\partial n}\right)\right\},$$

where Ω is the potential of the external force.*

To these we must adjoin the equation of continuity, which is (when $\theta = 0$)

$$\frac{\partial}{\partial x}(\rho q \cos\theta) + \frac{\partial}{\partial y}(\rho q \sin\theta) = 0, \quad \text{or}$$

$$(3) \qquad \frac{\partial(\rho q)}{\partial s} + \kappa_n \rho q = 0, \quad \text{i.e.} \quad \frac{\partial q}{\partial s} + \kappa_n q = 0 \text{ for a liquid.}$$

From 4·20 the vorticity is

$$(4) \qquad \zeta = -\frac{\partial q}{\partial n} + \kappa_s q.$$

Equation (1) can be written with the aid of (3) and (4) in the form

$$(5) \qquad \frac{\partial}{\partial s}\left(\frac{p}{\rho} + \tfrac{1}{2}q^2 + \Omega\right) = \nu\left\{-\frac{\partial \zeta}{\partial n} + q\left(\frac{\partial \kappa_s}{\partial n} - \frac{\partial \kappa_n}{\partial s}\right) - q(\kappa_s^2 + \kappa_n^2)\right\},$$

and therefore, integrating along a streamline from 0 to s, we get

$$\left[\frac{p}{\rho} + \tfrac{1}{2}q^2 + \Omega\right]_0^s = \nu F, \quad F = \int_0^s \left\{-\frac{\partial \zeta}{\partial n} + q\left(\frac{\partial \kappa_s}{\partial n} - \frac{\partial \kappa_n}{\partial s}\right) - q(\kappa_s^2 + \kappa_n^2)\right\} ds.$$

Ω is the potential per unit mass of the external forces

In a fluid of small viscosity ν is small, and therefore the magnitude of F gives a measure of the range of applicability of Bernoulli's theorem as a first approximation. In particular, at the boundary of a solid $q = 0$, and then

$$F = \int_0^s \left(-\frac{\partial \zeta}{\partial n}\right) ds.$$

This last result also applies when the streamlines are straight, for then κ_s, κ_n vanish.

* The equations of this section are readily obtained by specialising the general intrinsic method explained in 20·7.

21·40. Flow between parallel plates. Consider liquid forced under pressure to move between two fixed parallel plates at the distance h apart, fig. 21·40.

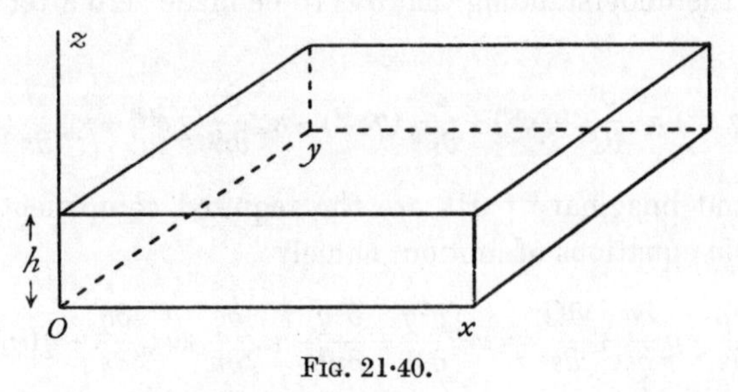

FIG. 21·40.

Take one plate to be in the x, y plane and the other to be $z = h$.

Suppose, first, that the motion is in the x-direction only, so that if

$$\mathbf{q} = \mathbf{i}\, u + \mathbf{j}\, v + \mathbf{k}\, w,$$

then $v = 0$, $w = 0$. The equation of continuity is $\partial u/\partial x = 0$, so that u is independent of x. When the motion is steady, u will therefore be a function of z only and independent of the time. Thus the equations of motion are

$$0 = -\frac{\partial p}{\partial x} + \mu \frac{\partial^2 u}{\partial z^2}, \quad 0 = -\frac{\partial p}{\partial y}, \quad 0 = -\frac{\partial p}{\partial z}.$$

Therefore $-\partial p/\partial x = P$ is independent of x, y, z, and hence

$$u = A + Bz - \frac{P}{2\mu} z^2.$$

Since $u = 0$ when $z = 0$ and $z = h$, we get

$$u = \frac{1}{2\mu} z(h-z) P.$$

The average value of u across a section perpendicular to x is

$$u_0 = \frac{1}{h}\int_0^h u\, dz = \frac{Ph^2}{12\mu},$$

and $u = 6u_0 z(h-z)/h^2$, the velocity midway between the plates being $3u_0/2$.

The velocity across a section follows the " parabolic law ", namely, if at each point of a line parallel to Oz we draw the velocity vector, the extremities of these vectors will lie on a parabola, fig. 1·0.

The case in which the plates are not fixed but slide along themselves with given velocities in a fixed direction (Couette flow) is easily treated by the same method.

The motion is not irrotational, for the vorticity vector is

$$\boldsymbol{\zeta} = \mathbf{j}\,\frac{\partial u}{\partial z} = \mathbf{j}\,\frac{6u_0(h-2z)}{h^2}.$$

The viscous traction exerted on the upper plate by the fluid is, from 21·06,

$$\mu\mathbf{k}_\wedge\boldsymbol{\zeta} = -\mathbf{i}\mu\frac{\partial u}{\partial z} = 6\mathbf{i}\mu u_0/h.$$

Thus there is a traction of amount $6\mu u_0/h$ per unit area in the direction of flow exerted by the fluid on each plate.

The rate of dissipation of energy per unit volume is

$$\mu\zeta^2 = \mu\times 36u_0^2(h-2z)^2/h^4,$$

and therefore, considering a column of height h, the rate of dissipation per unit area of plate is

$$\int_0^h \mu\zeta^2 dz = 12\mu u_0^2/h.$$

To discuss the rate of flow of heat, let us make the hypothesis that each plate is maintained at the same constant temperature T_0. Then $\partial T/\partial x = 0$, and we get from 21·11

$$0 = \mu\zeta^2 + K\,\nabla^2 T = \mu\zeta^2 + K\frac{\partial^2 T}{\partial z^2}.$$

If we further assume the viscosity μ to be independent of the temperature distribution, which will be nearly true if the plates are close together, this gives

$$KT = -\frac{3\mu u_0^2}{4h^4}(h-2z)^4 + Az + B.$$

Since $T = T_0$ when $z = 0$ and when $z = h$, we get

$$K(T-T_0) = \frac{3\mu u_0^2}{4h^4}[h^4-(h-2z)^4].$$

Secondly, let us suppose that the flow is two-dimensional, i.e. $w = 0$ everywhere. We shall suppose the plates to be very close together. Then u and v vary from their maximum values midway between the plates to zero in the short distance $h/2$. Thus the rate of variation of these components in the z-direction must be very great compared with the rates in the x- and y-directions. Neglecting these latter in comparison, the equation of motion becomes

$$-\mathbf{i}\frac{\partial p}{\partial x} - \mathbf{j}\frac{\partial p}{\partial y} - \mathbf{k}\frac{\partial p}{\partial z} + \mu\left(\mathbf{i}\frac{\partial^2 u}{\partial z^2} + \mathbf{j}\frac{\partial^2 v}{dz^2}\right) = 0.$$

Hence $\partial p/\partial z = 0$, and p is a function of x, y only.

Thus
$$\mu \frac{\partial^2 u}{\partial z^2} = \frac{\partial p}{\partial x}, \quad \mu \frac{\partial^2 v}{\partial z^2} = \frac{\partial p}{\partial y}.$$

Therefore we get, as above,

$$u = -\frac{6u_0 z(z-h)}{h^2}, \quad v = -\frac{6v_0 z(z-h)}{h^2},$$

where u_0, v_0 are the mean values of u, v as before.

Consequently,
$$\frac{\partial p}{\partial x} = -\frac{12\mu}{h^2} u_0, \quad \frac{\partial p}{\partial y} = -\frac{12\mu}{h^2} v_0,$$

and therefore u_0, v_0 are the components of a two-dimensional motion of an inviscid liquid in which the velocity potential is

$$\phi = \frac{ph^2}{12\mu}.$$

Thus when a portion of the region between the plates is obstructed by a cylinder of thickness h, the mean flow will be the same as that of an inviscid fluid flowing past a cylinder of the same cross-section, with the reservation that the analogy must break down at distances from the obstacle comparable with h. Since h can be made as small as we please this restriction is insignificant. This has enabled Hele-Shaw and others to make very beautiful experimental models of two-dimensional flow of an ideal liquid by injecting colouring matter to display the streamlines.

21·42. Flow down an incline.

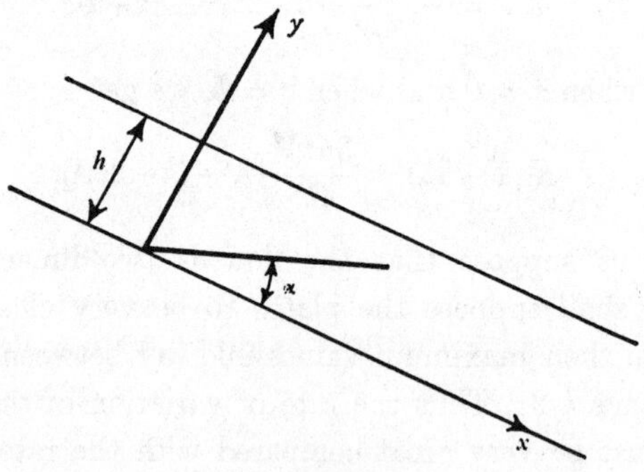

Fig. 21·42.

Consider two-dimensional steady flow of depth h down a plane inclined at angle α to the horizontal, the flow velocity being in straight lines parallel to the lines of greatest slope, one of which is taken as x-axis. Take the y-axis in the same vertical plane as the x-axis, fig. 21·42. Then $v = 0$ $w = 0$ and the equation continuity is $\partial u/\partial x = 0$, so that u is a function of y only. The compatibility equation then gives

(1) $$\partial^3 u/\partial y^3 = 0$$

so that

(2) $$u = Ay^2 + By$$

since $u = 0$ when $y = 0$.

Neglecting surface tension the conditions at the free surface $y = h$ are

(3) $$\widehat{yx} = 0, \quad \widehat{yy} = -\Pi, \quad \widehat{yz} = 0$$

where Π is the atmosphere pressure.

Now from 21·14

$$\widehat{yx} = \mu\left(\frac{\partial u}{\partial y}+\frac{\partial v}{\partial x}\right), \quad \widehat{yy} = -p+2\mu\frac{\partial v}{\partial y}, \quad \widehat{yz} = \mu\left(\frac{\partial v}{\partial z}+\frac{\partial w}{\partial y}\right),$$

and therefore the boundary stress conditions are satisfied by

(4) $$p = \Pi, \quad \partial u/\partial y = 0 \text{ when } y = h.$$

Therefore from (2)

(5) $$u = A(y^2 - 2hy), \quad \mathbf{q} = \mathbf{i}u.$$

Since $(\mathbf{q}\,\nabla)\mathbf{q} = \partial(\mathbf{i}u)/\partial x = 0$ the equation of steady motion is

(6) $$-(\nabla p)/\rho - \nabla\Omega + 2\nu A\mathbf{i} = 0, \quad \Omega = g(y\cos\alpha - x\sin\alpha).$$

Multiply by $d\mathbf{r}$ and integrate. Then

(7) $$\frac{p}{\rho} = -gy\cos\alpha + (2\nu A + g\sin\alpha)\,x + \frac{p_0}{\rho}$$

where p_0 is a constant. Now $p = \Pi =$ constant when $y = h$ for all x and therefore $2\nu A + g\sin\alpha = 0$. Thus the problem is solved by

(8) $$u = g\sin\alpha\, y(2h-y)/(2\nu), \quad p = \Pi + g\rho\cos\alpha(h-y).$$

21·44. Flow through a pipe. When viscous liquid flows steadily through a cylindrical pipe of any cross-section whose axis is in the direction of the z-axis, the equation of continuity shows that the velocity is independent of z if there are no components of velocity at right angles to the axis. Then we can put

$$\mathbf{q} = \mathbf{k}q,$$

where q is a function of x and y only. The equation of motion is then

$$\left(q\frac{\partial}{\partial z}\right)\mathbf{q} = -\frac{1}{\rho}\nabla p + \nu\nabla^2(\mathbf{k}q),$$

whence $$-\frac{\partial p}{\partial x}=0, \quad -\frac{\partial p}{\partial y}=0, \quad -\frac{\partial p}{\partial z}+\mu\left(\frac{\partial^2 q}{\partial x^2}+\frac{\partial^2 q}{\partial y^2}\right)=0.$$

Let $P = -\partial p/\partial z$ denote the pressure gradient along the pipe in the direction of flow. This is constant since $\partial^2 p/\partial z^2 = 0$. Writing

(1) $$q=\psi-\frac{P}{4\mu}(x^2+y^2),$$

the last equation gives

(2) $$\nabla^2\psi=0.$$

Since $q = 0$ at the boundary, (1) and (2) show that ψ is the stream function for inviscid liquid filling a cylinder of the same cross-section as the pipe, and rotating about its axis with angular velocity $P/(2\mu)$. Thus the problem becomes the same as that of 9·70.

Taking the case of a pipe whose cross-section is the ellipse

(3) $$\frac{x^2}{a^2}+\frac{y^2}{b^2}=1,$$

we have, from 9·71,

$$\psi=\frac{P}{4\mu}\frac{a^2-b^2}{a^2+b^2}(x^2-y^2)+\text{constant}.$$

So that $$q=A-\frac{P}{2\mu}\left(\frac{x^2}{a^2}+\frac{y^2}{b^2}\right)\frac{a^2b^2}{a^2+b^2}.$$

Therefore $q = 0$ on (3), if we take

$$A=\frac{P}{2\mu}\frac{a^2b^2}{a^2+b^2}, \quad \text{so that } q=\frac{P}{2\mu}\frac{a^2b^2}{a^2+b^2}\left(1-\frac{x^2}{a^2}-\frac{y^2}{b^2}\right).$$

The rate of discharge is $R = \iint q\,dx\,dy$, over the cross-section of the pipe.

To evaluate the integral, observe that on the ellipse given by $x = \lambda a \cos\theta$, $y = \lambda b \sin\theta$, the integrand is $A(1-\lambda^2)$, and the area between this ellipse and that corresponding to $\lambda + d\lambda$ is $2\pi ab\lambda\, d\lambda$. Thus

$$\iint\left(1-\frac{x^2}{a^2}-\frac{y^2}{b^2}\right)dx\,dy=\int_0^1 2\pi ab\lambda(1-\lambda^2)\,d\lambda=\frac{\pi ab}{2},$$

and $$R=\frac{P}{4\mu}\frac{a^2b^2}{a^2+b^2}\pi ab=\frac{P}{4\mu}\frac{a^2b^2}{a^2+b^2}S$$

where S is the area of the section.

Hence the mean velocity across the section is

$$q_0=\frac{P}{4\mu}\frac{a^2b^2}{a^2+b^2}, \quad \text{and therefore } q=2q_0\left(1-\frac{x^2}{a^2}-\frac{y^2}{b^2}\right).$$

For a pipe whose cross-section is a circle of radius c, Poiseuille flow, we have

$$q = 2q_0\left(1-\frac{r^2}{c^2}\right), \quad q_0 = \frac{Pc^2}{8\mu}, \quad R = \frac{\pi c^4 P}{8\mu},$$

where r is the distance from the axis.

If we take $c^2 = ab$ and write $b = a\alpha$, so that the elliptical and circular sections have the same area, the ratio of the rates of discharge is $2\alpha : (1+\alpha^2)$, which is less than unity. Thus a circular pipe discharges at a greater rate than an elliptic one of the same cross-sectional area.

Measurements of the rate of discharge from circular pipes provide evidence that the assumption of no slip at the wall is justified, for slip would increase the discharge by an amount which would destroy the validity of the law, found above, that the rate varies as the fourth power of the diameter.

The above results also furnish a method of measuring μ.

21·50. Radial plane flow. We consider steady two-dimensional flow in straight streamlines which radiate from a fixed point which will be taken as the origin. The radial and transverse components of the flow in terms of the stream function are

$$(1) \qquad q_r = -\frac{1}{r}\frac{\partial\psi}{\partial\theta}, \quad q_\theta = \frac{\partial\psi}{\partial r},$$

and since $q_\theta = 0$, the stream function is independent of r, $\psi = \psi(\theta)$. Therefore there is a source (or sink) at the origin the absolute value of whose output will be denoted by M. The vorticity is

$$(2) \qquad \nabla^2\psi = \frac{1}{r^2}\frac{\partial^2\psi}{\partial\theta^2} = \frac{\psi''}{r^2},$$

where dashes denote differentiation with respect to θ. Therefore

$$\nabla^4\psi = \frac{1}{r}\left\{\frac{\partial}{\partial r}\left[r\frac{\partial}{\partial r}\left(\frac{\psi''}{r^2}\right)\right]+\frac{\partial}{\partial\theta}\left[\frac{1}{r}\frac{\partial}{\partial\theta}\left(\frac{\psi''}{r^2}\right)\right]\right\} = \frac{1}{r^4}(4\psi''+\psi''''),$$

$$\frac{\partial(\psi, \nabla^2\psi)}{\partial(r, \theta)} = \frac{2\psi'\psi''}{r^3},$$

and consequently the compatibility equation is

$$(3) \qquad \nu(\psi''''+4\psi'')-2\psi'\psi'' = 0$$

which integrates once to give, in terms of a constant A,

$$(4) \qquad \psi'''+4\psi'-\frac{1}{v}(\psi')^2+A = 0.$$

The equation of steady motion is

$$\nabla\left(\frac{p}{\rho}+\Omega\right) = -(\mathbf{q}\,\nabla)\,\mathbf{q}-\nu\,\nabla\wedge\boldsymbol{\zeta} = \left(\frac{-\nu}{r^3}\,\psi'''+\frac{\psi'^2}{r^3}\right)\mathbf{i}_r - \frac{2\nu}{r^3}\,\psi''\,\mathbf{i}_\theta.$$

Take the scalar product by $d\mathbf{r}$. Then, using (4),

$$d\left(\frac{p}{\rho}+\Omega\right) = \frac{\nu}{r^3}\,(4\psi'+A)\,dr - \frac{2\nu}{r^2}\,\psi''\,d\theta.$$

Integrating we get, in terms of a constant p_0,

(5) $$\frac{p}{\rho}+\Omega = -\,\frac{\nu}{2r^2}\,(A+4\psi')+\frac{p_0}{\rho}.$$

Now write

(6) $$\psi' = -rq_r = -Mv(\theta) = -Mv.$$

Then $v(\theta)$ is a dimensionless measure of the radial velocity and (5) and (4) give respectively

(7) $$p+\rho\Omega = \frac{\mu}{2r^2}\,(4Mv-A)+p_0.$$

(8) $$v''+4v+\lambda v^2-B = 0, \quad \lambda = M/\nu, \quad B = A/M.$$

The dimensionless parameter λ is effectively a Reynolds number.

Observe that the appropriate solution v of (8) gives the velocity field from (6), and the pressure field from (7).

Also we have

$$M = \pm\int_{-\pi}^{\pi} q_r\,r d\theta \quad\text{or}\quad M = \pm\int_{-\alpha/2}^{\alpha/2} q_r\,r d\theta,$$

according as there is a source or a sink and the fluid is unbounded or confined between fixed walls $\theta = -\alpha/2$, $\theta = \alpha/2$. Thus using (6) according as the fluid is unbounded or bounded,

(9) $$\int_{-\pi}^{\pi} v\,d\theta = \pm 1 \quad\text{or}\quad \int_{-\alpha/2}^{\alpha/2} v\,d\theta = \pm 1,$$

the upper sign for a source.

Multiply (8) by $2v'$ and integrate. Then

(10) $$v'^2 = \frac{2\lambda}{3}\,F(v),$$

where $F(v)$ is the cubic polynomial

(11) $$F(v) = -v^3-\frac{6}{\lambda}\,v^2+Cv+D = -(v-e_1)(v-e_2)(v-e_3),$$

C, D being real constants of integration and e_1, e_2, e_3 being the zeros of $F(v)$. Thus

$$\theta\sqrt{\frac{2\lambda}{3}} = \pm\int\frac{dv}{\sqrt{F(v)}}. \tag{12}$$

Note

Equation (10) can be solved in terms of the $\wp$ function of Weierstrass* $\wp(u)$ or $\wp(u\,;g_2\,,g_3)$ defined by

$$u = \int_{\wp(u)}^{\infty}\frac{dt}{[4t^3-g_2t-g_3]^{1/2}},$$

and which therefore satisfies the differential equation

$$[\wp'(u)]^2 = 4[\wp(u)]^3 - g_2\wp(u) - g_3.$$

Equation (10) reduces to this if we write

$$v(\theta) = \chi(\alpha) - \frac{2}{\lambda}, \quad \alpha = i\theta\sqrt{\frac{\lambda}{6}},$$

and therefore

$$v(\theta) = -\frac{2}{\lambda} + \wp\left[i(\theta-\theta_0)\sqrt{\frac{\lambda}{6}}\,;g_2\,,g_3\right],$$

where θ_0, g_2, g_3 are arbitrary constants.

21·51. The zeros of F(v) for flow between fixed walls. From 21·50 (11) we see that the equation $F(v) = 0$ has real coefficients and therefore of the three roots e_1, e_2, e_3 one at least, say e_1, is real and the other two e_2, e_3 are either real or conjugate complex.

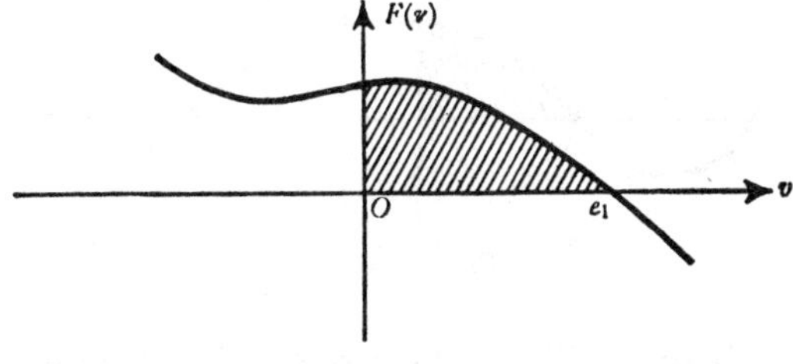

FIG. 21·51 (i).

Case of only one real root. Here e_2, e_3 are conjugate complex and therefore for real v, $(v-e_2)(v-e_3)>0$. But θ is real and therefore $F(v)$ must be positive so that $e_1 - v \geqslant 0$ or $v \leqslant e_1$. Thus e_1 is an upper bound for v.

But on the walls the adherence condition gives $v = 0$ and therefore $0 \leqslant e_1$, so that e_1 is non-negative.

* Whittaker and Watson, *Modern Analysis*, 4th edition Ch. 20.

Thus *if v does not change sign*, either $v \leqslant 0$ so that v does not remain finite and this case must be excluded, or $0 \leqslant v \leqslant e_1$. This indicates outward radial flow.

Therefore the case of only one real root corresponds to a source at O.

Case of three real roots. Here we can number them so that

$$e_1 \geqslant e_2 \geqslant e_3, \quad \text{while} \quad e_1 + e_2 + e_3 = -6/\lambda \tag{1}$$

from 21·50 (11). Therefore e_3 is negative.

Then $F(v) > 0$ if $-\infty \leqslant v \leqslant e_3$ (which is impossible for $v = 0$ on the walls while e_3 is negative) or

$$e_2 \leqslant v \leqslant e_1. \tag{2}$$

Since the range of v must include $v = 0$ we have

$$e_2 \leqslant 0 \leqslant e_1. \tag{3}$$

Thus e_2 is non-positive and e_1 is non-negative. If $v > 0$ we have a source, fig. 21·51 (ii). If $v < 0$ we have a sink, fig. 2·51 (iii).

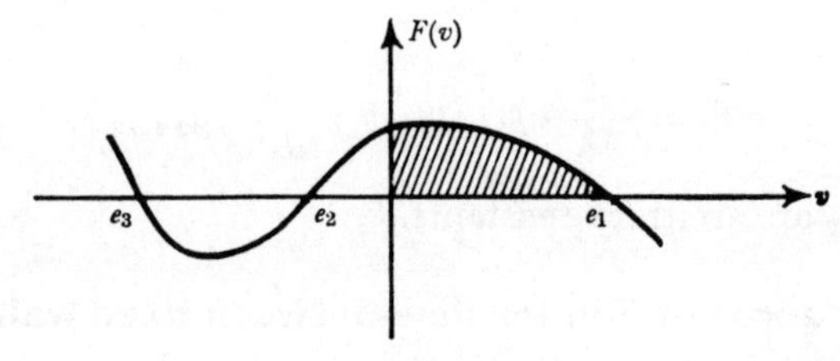

Fig. 21·51 (ii).

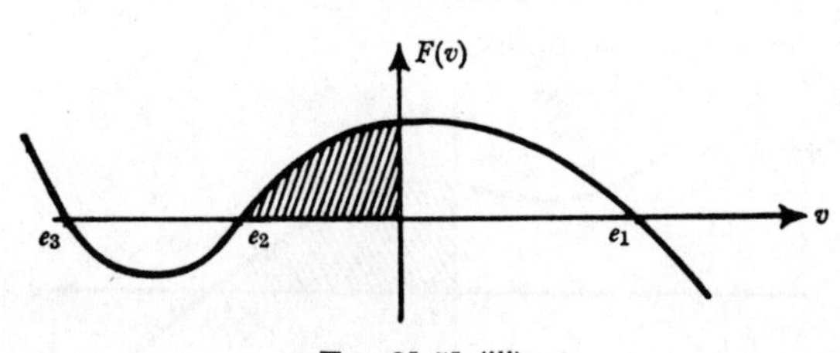

Fig. 21·51 (iii).

21·52. The diffuser. This is the case of outward radial flow, $v \geqslant 0$, from a source at the junction of two fixed walls which are taken to be $\theta = -\alpha/2$, $\theta = \alpha/2$.

From the symmetry it is clear that the maximum value of the dimensionless velocity v is e_1 attained on the axis $\theta = 0$. With increase of θ, v decreases and therefore in 21·50 (12) the negative sign is to be taken so that

$$\theta \sqrt{\frac{2\lambda}{3}} = -\int_{e_1}^{v} \frac{dv}{\sqrt{F(v)}}, \quad 0 \leqslant \theta \leqslant \alpha/2. \tag{1}$$

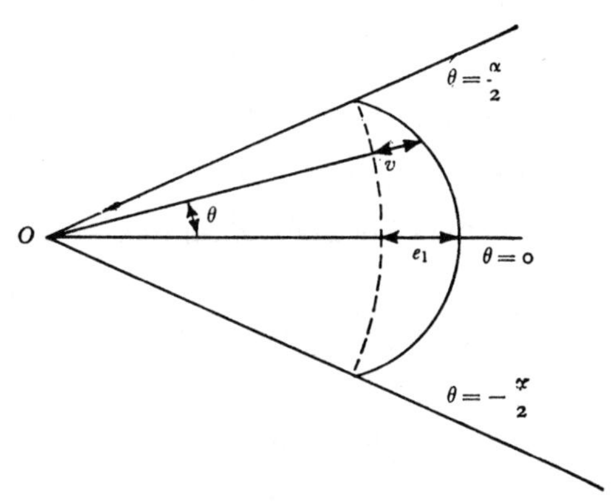

FIG. 21·52.

Since $\theta = \alpha/2$ corresponds to $v = 0$

(2) $$\alpha\sqrt{\frac{\lambda}{6}} = \int_0^{e_1} \frac{dv}{\sqrt{F(v)}}.$$

Also from 21·50 (9)

(3) $$\int_0^{\alpha/2} v(\theta)\, d\theta = \tfrac{1}{2},$$

which is equivalent to

(4) $$\sqrt{\frac{\lambda}{6}} = \int_0^{e_1} \frac{v\, dv}{\sqrt{F(v)}}$$

while from 21·51 (1)

(5) $$e_1 + e_2 + e_3 = -6/\lambda \quad \text{where} \quad F(v) = (e_1 - v)(v - e_2)(v - e_3).$$

The three equations (2), (4), (5) in principle serve to determine e_1, e_2, e_3 when λ and α are given. We proceed to show that a solution does not always exist.

From (5)

$$(v - e_2)(v - e_3) = v^2 + (e_1 + 6/\lambda)v + e_2 e_3,$$

and since e_2, e_3 are both conjugate complex or both negative

$$(v - e_2)(v - e_3) > (e_1 + 6/\lambda)v.$$

Therefore from (2)

$$\alpha\sqrt{\frac{\lambda}{6}} < \int_0^{e_1} \frac{dv}{[(e_1 - v)\, v(e_1 + 6/\lambda)]^{1/2}} = \frac{\pi}{(e_1 + 6/\lambda)^{1/2}}.$$

Therefore

(6) $$\alpha < \frac{\pi}{(1+e_1\lambda/6)^{1/2}},$$

and therefore α must be less than π.

Now suppose α to be given beforehand. Then from (2) and (4)

$$\sqrt{\frac{\lambda}{6}} < e_1 \int_0^{e_1} \frac{dv}{[(e_1-v)(v-e_2)(v-e_3)]^{1/2}} = e_1\alpha\sqrt{\frac{\lambda}{6}},$$

from which it follows that $e_1\alpha > 1$.

But from (6) $\alpha^2 e_1\lambda < 6(\pi^2-\alpha^2)$ and therefore

(7) $$\lambda < 6(\pi^2/\alpha - \alpha).$$

Therefore if λ is sufficiently large when α is given, the above type of flow is impossible. Thus for small λ we have flow of the above type but as λ increases there comes a critical case in which not only v but $dv/d\theta$ vanishes on the wall and beyond this purely outward radial flow does not exist.

This limitation is not present in the case of a sink.

21·56. Steady rotatory motion. When the motion is two-dimensional, consisting of rotation about the axis of x with an angular velocity n which is a function only of the distance ϖ from the axis of rotation, it appears that the only velocity component is ϖn perpendicular to the radius vector. Thus the viscous stress components (21·14) all vanish except $\widehat{\varpi\omega}$, which is equal to

$$\mu[\partial(\varpi n)/\partial\varpi - n] = \mu\varpi\, \partial n/\partial\varpi.$$

Therefore the moment about the axis of the viscous drag on a circular cylindrical surface of radius ϖ and of unit length is $\varpi\, \mu\varpi(\partial n/\partial\varpi)2\pi\varpi$.

When the motion is steady, there is no change in the angular momentum of the fluid contained between two such cylinders, and therefore the above moment has the same value (but opposite signs) at the inner and outer surfaces. Hence

$$\varpi^3 \frac{dn}{d\varpi} = A,$$

where A is independent of ϖ. Thus

$$n = -\frac{A}{2\varpi^2} + B.$$

If the fluid is bounded internally by a cylindrical surface of radius a moving with angular velocity n_1 and externally by a concentric cylinder of radius b moving with angular velocity n_2, we must have

$$n_1 = -\frac{A}{2a^2} + B, \quad n_2 = -\frac{A}{2b^2} + B,$$

and therefore

$$n = \frac{n_2 b^2 - n_1 a^2}{b^2 - a^2} + \frac{a^2 b^2 (n_1 - n_2)}{\varpi^2 (b^2 - a^2)} \tag{1}$$

In this argument n_1 and n_2 are not restricted to have the same sign. If we suppose $n_2 = -n_3$, where n_1 and n_3 have the same sign, the angular velocity n will vanish when

$$\varpi^2 = \frac{a^2 b^2 (n_1 + n_3)}{a^2 n_1 + b^2 n_3}$$

and the fluid on the two sides of the " stagnation " cylinder so defined will be rotating in opposite senses.

Again, if in (1) we put $b = \infty$, $n_2 = 0$, we get $n/n_1 = a^2/\varpi^2$, which gives the velocity distribution when the fluid is bounded internally only.

If the fluid is bounded externally but not internally, we have $a = 0$ and therefore $n = n_2$, so that the whole system rotates (in steady motion) like a rigid body.

If the inner cylinder is at rest, we get

$$n = \frac{b^2}{\varpi^2} \frac{\varpi^2 - a^2}{b^2 - a^2} n_2.$$

This steady motion has been shown by G. I. Taylor to be stable for all values of n_2. The friction couple on the outer cylinder is then

$$\left(\mu \,.\, 2\pi\varpi^3 \frac{dn}{d\varpi}\right) = 4\pi\mu \frac{a^2 b^2 n_2}{b^2 - a^2}.$$

When the outer cylinder is fixed and the inner one rotates, Taylor * has shown that the motion is stable only for sufficiently small angular velocities of the inner.

In a later paper Taylor † has shown that, while the motion remains stable in the above sense when the inner cylinder is at rest, turbulence sets in when the angular velocity n_2 is sufficiently large.

21·58. Effect of viscosity on water waves. When waves of small height

$$\eta = a \sin (mx - nt)$$

are propagated on deep water in the absence of viscosity, the complex potential (15·17) is $w = ac\, e^{-i(mz - nt)}$, so that the velocity is $u - iv = imac\, e^{-i(mz-nt)}$, giving $u = mac\, e^{my} \sin (mx - nt)$, $v = -mac\, e^{my} \cos (mx - nt)$.

If the liquid is viscous, the surface stresses due to these velocity components are (21·14), when $y = 0$,

* G. I. Taylor, *Phil Trans.* (A), 223 (1922). † *Proc. Roy. Soc.* (A), 157 (1936).

$$\widehat{yy} = -p + 2\mu\frac{\partial v}{\partial y} = -p - 2\mu m^2 ac\cos(mx - nt),$$

$$\widehat{yx} = \mu\left(\frac{\partial v}{\partial x} + \frac{\partial u}{\partial y}\right) = 2\mu m^2 ac\sin(mx - nt),$$

and if these forces are applied to the surface by an external agency the wave as given above will persist even when the fluid is viscous.

The rate at which the forces do work is

$$\widehat{yy}\, v + \widehat{yx}\, u = pmac\cos(mx - nt) + 2\mu m^3 a^2 c^2,$$

and the mean value of this is $2\mu m^3 a^2 c^2$.

Now the total energy of the wave (per unit surface area) is (15·21)

$$\tfrac{1}{2}a^2 g\rho = \tfrac{1}{2}a^2 mc^2\rho,$$

and, in the absence of the external agency mentioned above, the rate of dissipation of energy in the wave must be equal to the mean rate of working of the viscous forces. Thus

$$\frac{d}{dt}\left(\tfrac{1}{2}a^2 mc^2\rho\right) = -2\mu m^3 a^2 c^2, \quad \text{or} \quad \frac{da}{dt} = -2\nu m^2 a.$$

Hence $a = a_0\exp(-2\nu m^2 t)$, where a_0 is the initial value of a, and so the wave at time t has the profile given by

$$\eta = a_0\exp(-2\nu m^2 t)\sin(mx - nt),$$

the amplitude of which is continually diminishing with the time. The time taken for the index of the exponential to attain the value -1 is

$$t_1 = \frac{1}{2\nu m^2} = \frac{\lambda^2}{8\pi^2\nu}.$$

and after the lapse of this time the amplitude of the wave will be

$$a_0 e^{-1} = 0{\cdot}37 \times a_0.$$

Taking $\nu = 0{\cdot}0178$ cm.²/sec. for water, we get $t_1 = 0{\cdot}711\lambda^2$ sec., when λ is measured in centimetres.

Thus when $\lambda = 1$ cm., t_1 is less than 1 sec., while if $\lambda = 100$ cm., t_1 is about two hours. Thus capillary waves are suppressed by viscosity almost immediately while gravity waves are affected very little.

When waves travel in the direction of the wind, but with less velocity, the crest shelters the leeward face, while the windward face from trough to crest receives the full force of the wind. The part of the wave on which the wind thus impinges directly is, owing to the propagation, receding from the wind which therefore pushes in the direction in which the water is already moving. On the leeward face the water is, owing to the propagation, rising, and the sheltering prevents the wind from opposing this motion; it may even be helped by the back eddy which often exists on the lee side of an obstacle. Thus the wind always urges in the direction in which the water is moving and so tends to increase the energy which the viscosity tends to dissipate.

21·60. Time-dependent plane flow in parallel lines. Consider two-dimensional flow in the xy plane, the streamlines being straight and parallel to the x-axis. Then

$$(1)\qquad \mathbf{q} = \mathbf{i}u, \quad (\mathbf{q}\,\nabla)\mathbf{q} = u\frac{\partial u}{\partial x}\mathbf{i}, \quad \nabla\mathbf{q} = \frac{\partial u}{\partial x}.$$

The equation of continuity is therefore

$$(2)\qquad \frac{\partial u}{\partial x} = 0,$$

and so $u = u(y, t)$ is independent of x, while $(\mathbf{q}\nabla)\mathbf{q} = 0$.

The equation of motion is therefore

$$\mathbf{i}\frac{\partial u}{\partial t} = -\nabla\left(\frac{p}{\rho}+\Omega\right)+\nu\mathbf{i}\left(\frac{\partial^2 u}{\partial y^2}\right)$$

or

$$(3)\qquad \mathbf{i}\left(\frac{\partial u}{\partial t}-\nu\frac{\partial^2 u}{\partial y^2}\right) = -\nabla\left(\frac{p}{\rho}+\Omega\right).$$

Taking the curl of (3) we get the compatibility equation

$$\frac{\partial^2 u}{\partial y\,\partial t}-\nu\frac{\partial^3 u}{\partial y^3} = 0,$$

which integrates to give

$$(4)\qquad \nu\frac{\partial^2 u}{\partial y^2}-\frac{\partial u}{\partial t} = f_1(t).$$

Take the scalar product of (3) with $d\mathbf{r}$ and integrate to obtain

$$(5)\qquad \frac{p}{\rho}+\Omega = xf_1(t)+f(t),$$

where $f_1(t)$, $f(t)$ are arbitary functions of time. Of these $f_1(t)$ must vanish otherwise the pressure would tend to infinity and therefore

$$(6)\qquad \frac{p}{\rho}+\Omega = f(t).$$

$$(7)\qquad \nu\frac{\partial^2 u}{\partial y^2}-\frac{\partial u}{\partial t} = 0, \quad u = u(y, t).$$

Equation (7) has the same form as the equation for the flow of heat in one dimension, and is in fact the particular case of 21·11 (5) obtained by putting

$$T = u, \quad w_i = 0, \quad \nu = K/(\rho c).$$

We state three problems relative to this equation and give their solutions without proof, leaving verification to the reader. (See Courant and Hilbert, *the methods of mathematical physics.*)

Problem I. The solution of (7) for $-\infty<y<\infty$, $t>0$ defined by

$$u(y, 0) = f(y), \quad -\infty<y<\infty \tag{8}$$

where $f(y)$ is a given function is

$$u(y, t) = \frac{1}{2(\pi\nu t)^{1/2}}\int_{-\infty}^{\infty} \exp\left[\frac{-(y-\alpha)^2}{4\nu t}\right] f(\alpha)\, d\alpha, \tag{9}$$

provided that we take at a point of discontinuity

$$f(y) = \tfrac{1}{2}[f(y-0)+f(y+0)]. \tag{10}$$

The length $(\nu t)^{1/2}$ is sometimes called the *viscous thickness.*

Problem II. The solution of (7) for $0<y<\infty$, $0<t<\infty$, defined by

$$u(y, 0) = f(y) \text{ for } 0<y<\infty, \quad u(0, t) = 0 \text{ for } 0<t<\infty \tag{11}$$

is

$$u(y, t) = \frac{1}{2(\pi\nu t)^{1/2}}\int_0^{\infty} \left\{\exp\left[-\frac{(y-\alpha)^2}{4\nu t}\right] - \exp\left[-\frac{(y+\alpha)^2}{4\nu t}\right]\right\} f(\alpha)\, d\alpha. \tag{12}$$

Problem III. The solution of (7) for $0<y<\infty$, $0<t<\infty$ defined by

$$u(y, 0) = 0 \quad \text{for} \quad 0<y<\infty, \quad u(0, t) = g(t) \quad \text{for} \quad 0<t<\infty \tag{13}$$

is

$$u(y, t) = \frac{2}{\sqrt{\pi}}\int_{y/[2(\nu t)^{1/2}]}^{\infty} \exp(-\alpha^2)\, g[t - y^2/(4\nu\alpha^2)]\, d\alpha. \tag{14}$$

If we add (12) and (14) we get the solution of (7) for

$$\begin{cases} u(y, 0) = f(y) & \text{for} \quad 0<y<\infty \\ u(0, t) = g(t) & \text{for} \quad 0<t<\infty. \end{cases} \tag{15}$$

21·61. On the generation of vorticity by conservative forces. The following problem was considered by Boussinesq. Liquid occupies a half-space bounded by a vertical wall and is at rest. At time $t = 0$ the fluid is started into motion by gravity, the streamlines being assumed to be straight and vertical and the motion two-dimensional. We shall take the x-axis to be vertically downwards and the y-axis to be horizontal. This idealised problem can be thought of as arising when the fluid occupies a two-dimensional tank bounded by $y = 0$ and $y = h$ as vertical sides and $x = -k$ and $x = k$ as top and bottom. If, at time

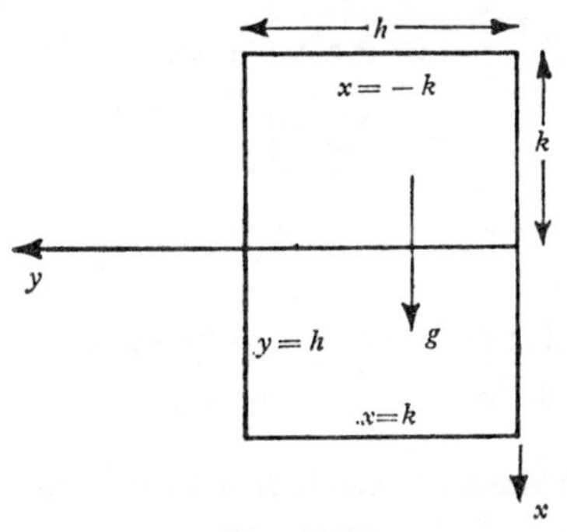

FIG. 21·61.

$t = 0$, we remove the bottom the fluid starts into motion. The Boussinesq's problem is the limit of this state of affairs when h and k both tend to infinity. Assuming that $\partial p/\partial x = 0$, 21·60 (3) gives the equation of motion

(1) $$\nu \frac{\partial^2 u}{\partial y^2} - \frac{\partial u}{\partial t} = -g,$$

where g is the acceleration due to gravity so that $\Omega = -gx$.

The boundary conditions are

(2) $$u(y, 0) = 0 \quad \text{for} \quad 0<y<\infty; \quad u(0, t) = 0 \quad \text{for} \quad 0<t<\infty.$$

The problem so stated can be reduced to Problem III of 21·60 by writing

(3) $$w = u - gt$$

so that (1) and (2) become

(4) $$\nu \frac{\partial^2 w}{\partial y^2} - \frac{\partial w}{\partial t} = 0.$$

(5) $$w(y, 0) = 0 \quad \text{for} \quad 0<y<\infty; \; w(0, t) = -gt \quad \text{for} \quad 0<t<\infty.$$

Using the solution 21·60 (14) we get w, and so

(6) $$u(y, t) = gt\left\{1 - \frac{2}{\sqrt{\pi}}\int_\alpha^\infty e^{-\beta^2}\left(1 - \frac{\alpha^2}{\beta^2}\right) d\beta\right\}, \; \alpha = \frac{y}{2\sqrt{(\nu t)}}.$$

The vorticity is given by

(7) $$\zeta = -\frac{\partial u}{\partial y} = -\frac{2g\sqrt{t}}{\sqrt{(\pi\nu)}} F(\alpha), \; F(\alpha) = \alpha\int_\alpha^\infty \frac{e^{-\beta^2}}{\beta^2}\, d\beta.$$

Now when α is small the integral in (7) is dominated by the term $1/\alpha$ and therefore

(8) $$\lim_{\alpha\to 0} F(\alpha) = 1.$$

Again $\alpha F(\alpha) < \int_\alpha^\infty \beta^2 \frac{e^{-\beta^2}}{\beta^2} d\beta \to 0$ when $\alpha \to \infty$. Therefore

$$\lim_{\alpha\to\infty} F(\alpha) = 0. \tag{9}$$

Thus the vorticity ζ grows from $-2g\sqrt{t}/\sqrt{(\pi\nu)}$ to 0 as $y/\sqrt{(\nu t)}$ grows from 0 to ∞.

Thus for a given small value of t, the vorticity is appreciable only in a layer whose thickness varies as $\sqrt{\nu}$.

21·62. Disappearance of a surface of discontinuity. Liquid fills all space and flows in lines parallel to the x-axis in the xy plane, the velocity being $u(y, t)$.

At time $t = 0$, $u(y, 0) = u_0$ for $y > 0$ and $u(y, 0) = -u_0$ for $y < 0$ so that the surface $y = 0$ is a surface of discontinuity for u, in fact a vortex sheet.

Then from 21·60, Problem I, we have

$$u(y, t) = \frac{-u_0}{2\sqrt{(\pi\nu t)}} \int_{-\infty}^0 \exp\left(\frac{-(y-\alpha)^2}{4\nu t}\right) d\alpha + \frac{u_0}{2\sqrt{(\pi\nu t)}} \int_0^\infty \exp\left(\frac{-(y-\alpha)^2}{4\nu t}\right) d\alpha. \tag{1}$$

After a simple reduction this becomes

$$u(y, t) = u_0 \operatorname{erf}\left(\frac{y}{2\sqrt{(\nu t)}}\right), \quad \operatorname{erf} x = \frac{2}{\sqrt{\pi}} \int_0^x \exp(-\beta^2)\, d\beta \tag{2}$$

in terms of the error function erf x.

The vorticity is given by

$$\zeta = -\frac{\partial u}{\partial y} = \frac{-u_0}{\sqrt{(\pi\nu t)}} \exp\left(\frac{-y^2}{4\nu t}\right). \tag{3}$$

For $t = 0$, $\zeta = 0$ everywhere except on $y = 0$.

For $t > 0$, ζ is different from zero everywhere.

Thus the vortex sheet diffuses instantly into the whole fluid, so that the surface of discontinuity instantly disappears.

21·63. Decay of vorticity. When the motion is in circles about the z-axis, the velocity being a function of the distance r from that axis, we get, from 21·34,

$$\frac{\partial \zeta}{\partial t} = \nu \left(\frac{\partial^2 \zeta}{\partial r^2} + \frac{1}{r}\frac{\partial \zeta}{\partial r}\right). \tag{1}$$

This equation is identical with that for the radial flow of heat in two dimensions. Thus, in the case of an isolated rectilinear vortex of strength κ, initially concentrated along the axis of z, we get the solution

$$\zeta = \frac{\kappa}{2\nu t} \exp\left(-\frac{r^2}{4\nu t}\right),$$

which is easily verified by differentiation to satisfy (1). The circulation in a circle of radius r is then

$$\int_0^r \zeta \, 2\pi r \, dr = 2\pi\kappa\left\{1 - \exp\left(-\frac{r^2}{4\nu t}\right)\right\}.$$

When $t \to 0$ this is $2\pi\kappa$, while when $t \to \infty$ it tends to zero. This shows how rapidly vorticity is damped out by the viscosity to which it owes its origin.*

21·70. Vector circulation. Let C be a curve in the plane of a two-dimensional motion, and let $\mathbf{k}$ be the unit vector normal to the plane. Let

$$K = \int_{(C)} \mathbf{q} \, d\mathbf{r}. \tag{1}$$

When C is a *closed curve*, enclosing the area Σ, the scalar K is the circulation (2·42) in this curve, and by Stokes's theorem

$$K = \int_{(\Sigma)} \mathbf{k}(\nabla_\wedge \mathbf{q}) dS = \mathbf{k} \int_{(\Sigma)} \boldsymbol{\zeta} \, dS.$$

If, as is usual in two-dimensional motion, we conceive the fluid to have unit thickness, we can call $\boldsymbol{\zeta} \, dS$ the (vector) amount of vorticity in the volume dS of a cylinder of unit thickness, and then, since $\boldsymbol{\zeta} = \mathbf{k}\zeta$, the circulation K is the (scalar) amount of vorticity in the cylinder Σ of unit thickness.

More generally, we can extend the definition (1) to an open curve C (plane or twisted) by defining the circulation in C as the scalar K.

Now consider the vector

$$\boldsymbol{\Gamma} = \int_{(S)} \mathbf{n}_\wedge \mathbf{q} \, dS, \tag{2}$$

where the integral is taken over a surface S. If S is a *closed surface* which encloses the volume V, Gauss's theorem gives, taking $\mathbf{n}$ as the outward normal,

$$\boldsymbol{\Gamma} = \int_{(V)} \nabla_\wedge \mathbf{q} \, d\tau = \int_{(V)} \boldsymbol{\zeta} \, d\tau. \tag{3}$$

Thus $\boldsymbol{\Gamma}$ measures the (vector) amount of vorticity in the volume V. It is left as a simple exercise to show that for the two-dimensional motion just considered, in which S will now denote the whole surface of the cylinder Σ, we have $K = \mathbf{k}\boldsymbol{\Gamma}$.

Definition. The vector $\boldsymbol{\Gamma}$ defined by (2) is called the *vector circulation* over the closed or open surface S.

---

*"Big whirls have little whirls which feed on their velocity;
Little whirls have smaller whirls and so on to viscosity."

Attributed to L. F. Richardson.

There is a useful alternative expression for the vector circulation over a *closed* surface S, namely

$$(4) \qquad \mathbf{\Gamma} = \int_{(S)} \mathbf{r}\,(\mathbf{n}\boldsymbol{\zeta})\,dS.$$

Proof. If X is any differentiable function of position, it follows from Stokes' theorem that

$$(5) \qquad \int_{(S)} (\mathbf{n}_{\wedge}\nabla)X\,dS = 0,$$

for any closed curve C drawn on S divides it into two diaphragms S_1 and S_2, each closing C and the surface integrals over these diaphragms are equal line integrals taken in opposite senses round C, and therefore cancel when added.

Using the dyadic notation and the fact, 2·71 (2), that $\nabla\,;\,\mathbf{r} = \mathbf{I}$, the idemfactor, we have

$$\begin{aligned}(\mathbf{n}_{\wedge}\nabla)(\mathbf{q}\,;\,\mathbf{r}) &= (\mathbf{n}_{\wedge}\nabla)(\mathbf{q}_0\,;\,\mathbf{r})+(\mathbf{n}_{\wedge}\nabla)(\mathbf{q}\,;\,\mathbf{r}_0) = \{\mathbf{q}(\mathbf{n}_{\wedge}\nabla)\}\mathbf{r}+\{(\mathbf{n}_{\wedge}\nabla)\mathbf{q}\}\mathbf{r}\\ &= \mathbf{q}\{\mathbf{n}_{\wedge}(\nabla\,;\,\mathbf{r})\}+\{\mathbf{n}(\nabla_{\wedge}\mathbf{q})\}\mathbf{r} = \mathbf{q}(\mathbf{n}_{\wedge}\mathbf{I})+\mathbf{r}(\mathbf{n}\boldsymbol{\zeta}) = -(\mathbf{n}_{\wedge}\mathbf{q})+\mathbf{r}(\mathbf{n}\boldsymbol{\zeta}).\end{aligned}$$

The result (4) follows by integrating over S and using (5) with $X = \mathbf{q}\,;\,\mathbf{r}$.

Q.E.D.

Corollary. For irrotational motion $\boldsymbol{\zeta} = 0$ and therefore $\mathbf{\Gamma} = 0$.

It is important to observe that the above proof has been so framed as to avoid volume integrals and (4) therefore takes no account of the circumstances inside S. The only restriction on (4) is that implied by (5) which demands that X shall be finite, one-valued, and continuous.

In the case of a closed surface S moving with the fluid, we have from (3) and the equation of continuity in the form $d(\rho\,d\tau)/dt = 0$, the rate of change of the circulation in the form

$$\frac{d\mathbf{\Gamma}}{dt} = \frac{d}{dt}\int_{(V)} \frac{\boldsymbol{\zeta}}{\rho}(\rho\,d\tau) = \int_{(V)} \frac{d}{dt}\left(\frac{\boldsymbol{\zeta}}{\rho}\right)\rho\,d\tau = \int_{(V)} (\boldsymbol{\zeta}\nabla)\,\mathbf{q}\,d\tau$$

from 3·53 (2). Now from 2·34 (X)

$$\nabla\,(\boldsymbol{\zeta}\,;\,\mathbf{q}) = (\boldsymbol{\zeta}\nabla)\,\mathbf{q}+\mathbf{q}(\nabla\boldsymbol{\zeta}) = (\boldsymbol{\zeta}\nabla)\,\mathbf{q},$$

since $\nabla\boldsymbol{\zeta} = 0$. Therefore

$$\frac{d\mathbf{\Gamma}}{dt} = \int_{(V)} \nabla(\boldsymbol{\zeta};\mathbf{q})\,d\tau = -\int_{(S)} (\mathbf{n}\,\boldsymbol{\zeta})\,\mathbf{q}\,dS.$$

From this it follows that the circulation $\mathbf{\Gamma}$ remains constant for a closed surface generated by vortex lines, as, for example, in the case of Hill's spherical vortex, 19·51.

21·80. The wake. When a body, typically an aerofoil, moves through fluid or when the fluid streams past an aerofoil at rest, a *wake* is formed which consists of fluid which has passed near to the surface of the aerofoil, and, as remarked in 21·34, the vorticity is largely confined to the fluid which constitutes the wake.

FIG. 21·80.

We propose to develop some consequences which flow from two particular hypotheses.

(i) The wake consists of fluid in regular motion which can be described by streamlines and vortex lines.

(ii) Outside the wake the vorticity is negligible, i.e. we can assume $\boldsymbol{\zeta} = 0$.

Hypothesis (ii) may be regarded as a definition of the wake.

21·81. The net vorticity in the wake.

Theorem. Let S be a closed surface every point of which is in contact with the fluid and which cuts the wake in vortex lines. Then, if the fluid velocity is finite and continuous over S, the vector circulation over S is zero.

Proof. From 21·70 (4), $$\boldsymbol{\Gamma} = \int_{(S)} \mathbf{r}(\mathbf{n}\boldsymbol{\zeta})\,dS.$$

Outside the wake $\boldsymbol{\zeta} = 0$, inside it $\mathbf{n}\boldsymbol{\zeta} = 0$ since the vortex lines lie on S. Therefore $\boldsymbol{\Gamma} = 0$. Q.E.D.

Corollary. The net vorticity in any section of the wake cut off by a closed surface which intersects the wake in vortex lines is zero, for from 21·70 (3)

$$\text{Net vorticity} = \int \boldsymbol{\zeta}\,d\tau = \boldsymbol{\Gamma} = 0,$$

where the volume integral is taken through the section in question.

This corollary assumes that the whole of the interior of the surface is occupied by fluid.

In the case of a closed surface S_1 which surrounds an aerofoil A and cuts the wake in vortex lines, consider the fluid between A and S_1. Then the circulation over S_1 is zero by the theorem, and the circulation over A is zero, since $\mathbf{q} = 0$ on A in the case of a viscous fluid, where $\mathbf{q}$ is the fluid velocity relative to A.

The net vorticity in the boundary layer and that portion of the wake which lies inside S_1 is therefore zero.

These results are purely kinematical. They hold for compressible viscous fluids and do not assume steady motion or constant viscosity.

21·82. Vorticity transport. Referring to fig. 2·50 (iii), let

$$(1) \qquad \mathbf{T}_S = \int_{(S)} (\mathbf{nq})\boldsymbol{\zeta}\, dS - \int_{(S)} \nu \frac{\partial \boldsymbol{\zeta}}{\partial n}\, dS.$$

The first integral represents the rate of transport of vorticity through the open surface S due to convection, the second the rate of transport due to diffusion.

If in fig. 2·50 (iii) we take C to be a vortex line and the diaphragm S, which closes C, to be a surface consisting of vortex lines, we shall call S a *vortex diaphragm* closing the vortex line C.

$\mathbf{T}_S$ then represents the rate of transport of vorticity, through a vortex diaphragm, due to convection and diffusion.

*Preston's theorem.** In the steady motion of a homogeneous liquid of uniform viscosity the rate of transport of vorticity through a vortex diaphragm which closes the vortex line C is

$$(2) \qquad \mathbf{T}_S = -\int_{(C)} \chi\, d\mathbf{s}, \quad \chi = \frac{p}{\rho} + \tfrac{1}{2}q^2 + \Omega.$$

Proof. Multiply the equation of motion 21·30 (5) by $\mathbf{n}_\wedge$, put $\partial\mathbf{q}/\partial t = 0$, and integrate over S. Then

$$(3) \qquad -\int_{(S)} \mathbf{n}_\wedge(\mathbf{q}_\wedge\boldsymbol{\zeta})dS + \nu\int_{(S)} \mathbf{n}_\wedge(\nabla_\wedge\boldsymbol{\zeta})dS = -\int_{(S)} (\mathbf{n}_\wedge\nabla)\chi\, dS.$$

Now $\nabla\boldsymbol{\zeta} = \nabla(\nabla_\wedge\mathbf{q}) = 0$ from 2·32 (II), and since S is a vortex diaphragm $\mathbf{n}\boldsymbol{\zeta} = 0$. Therefore using the triple vector product

$$\mathbf{n}_\wedge(\nabla_\wedge\boldsymbol{\zeta}) = (\mathbf{n}_\wedge\nabla)_\wedge\boldsymbol{\zeta} - (\mathbf{n}\,\nabla)\boldsymbol{\zeta} + \mathbf{n}(\nabla\,\boldsymbol{\zeta}) = (\mathbf{n}_\wedge\nabla)_\wedge\boldsymbol{\zeta} - \frac{\partial\boldsymbol{\zeta}}{\partial n},$$

$$\mathbf{n}_\wedge(\mathbf{q}_\wedge\boldsymbol{\zeta}) = -(\mathbf{nq})\boldsymbol{\zeta} + (\mathbf{n}\boldsymbol{\zeta})\mathbf{q} = -(\mathbf{nq})\boldsymbol{\zeta}.$$

Substitute in (3) and use (1). Then by Stokes' theorem

$$\mathbf{T}_S = -\nu\int_{(S)} (\mathbf{n}_\wedge\nabla)_\wedge\boldsymbol{\zeta}\, ds - \int_{(S)} (\mathbf{n}_\wedge\nabla)\chi\, dS = -\nu\int_{(C)} d\mathbf{s}_\wedge\boldsymbol{\zeta} - \int_{(C)} d\mathbf{s}\chi.$$

But C is a vortex line and therefore on C the vectors $d\mathbf{s}$ and $\boldsymbol{\zeta}$ are parallel so that $d\mathbf{s}_\wedge\boldsymbol{\zeta} = 0$. Q.E.D.

21·83. The force on an aerofoil. Consider a three-dimensional aerofoil A at rest in a steady stream $\mathbf{V} = \mathbf{i}V$ and ignore body force.

Let Σ be an imagined fixed closed surface (i.e. not a physical boundary) entirely surrounding the aerofoil A. The equation of steady motion, 21·22 (1), is

$$\nabla[\boldsymbol{\Phi} - \rho(\mathbf{q}\,;\mathbf{q})] = 0.$$

* The two-dimensional form of this theorem is due to J. H. Preston, *A.R.C. Report*, No. 6372.

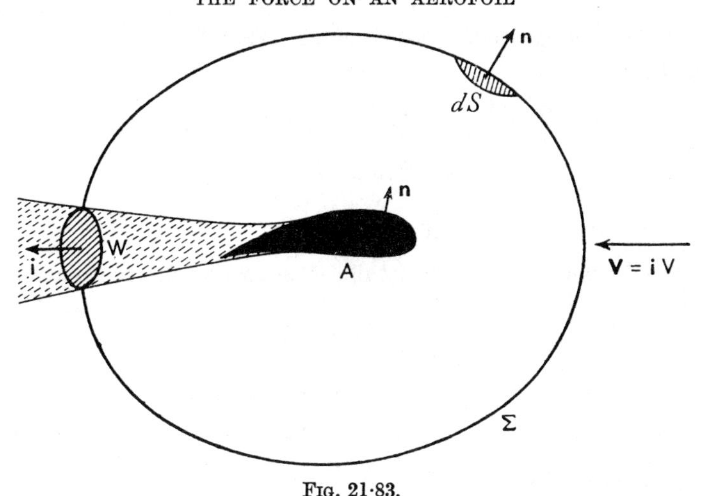

FIG. 21·83.

By Gauss's theorem integration over the volume between Σ and A gives

$$0 = -\int_{(A)} \mathbf{n}[\mathbf{\Phi} - \rho(\mathbf{q}\,;\,\mathbf{q})]\,dS + \int_{(\Sigma)} \mathbf{n}[\mathbf{\Phi} - \rho(\mathbf{q}\,;\,\mathbf{q})]\,dS.$$

Now $\mathbf{n}(\mathbf{q}\,;\,\mathbf{q}) = (\mathbf{nq})\,\mathbf{q}$, and this vanishes on A since $\mathbf{q} = 0$ on A if the fluid is viscous, and $\mathbf{nq} = 0$ on A if the fluid is inviscid. Therefore the force on the aerofoil is

$$(1) \qquad \mathbf{F} = \int_{(A)} \mathbf{n\Phi}\,dS = \int_{(\Sigma)} [\mathbf{n\Phi} - \rho(\mathbf{nq})\,\mathbf{q}]\,dS.$$

Thus, using 21·04 (6), we get

$$(2) \qquad \mathbf{F} = \int_{(\Sigma)} [-p\mathbf{n} - \rho(\mathbf{nq})\,\mathbf{q} + \tfrac{4}{3}\mu\mathbf{n}(\nabla\,\mathbf{q}) - \mu\mathbf{n}_{\wedge}\boldsymbol{\zeta}]\,dS$$
$$+\,2\int_{(\Sigma)} \mu[(\mathbf{n}\,\nabla)\,\mathbf{q} - \mathbf{n}(\nabla\,\mathbf{q}) + \mathbf{n}_{\wedge}\boldsymbol{\zeta}]\,dS.$$

This is a general result which applies to any steady motion, whether ρ and μ are constants or functions of position.

We now show that if μ is constant, the second integral vanishes. For by the triple vector product, since $\boldsymbol{\zeta} = \nabla_{\wedge}\mathbf{q}$,

$$\int_{(\Sigma)} [(\mathbf{n}\,\nabla)\,\mathbf{q} - \mathbf{n}(\nabla\,\mathbf{q}) + \mathbf{n}_{\wedge}\boldsymbol{\zeta}]\,dS = \int_{(\Sigma)} (\mathbf{n}_{\wedge}\nabla)_{\wedge}\mathbf{q}\,dS = 0$$

from 21·70 (5).

Thus when μ is constant the force on the aerofoil is given by the first integral of (2).

Let us now assume that not only μ but also ρ is constant, so that we are dealing with a homogeneous viscous liquid. Then $\nabla \mathbf{q} = 0$ and the force on the aerofoil is

$$(3) \qquad \mathbf{F} = \int_{(\Sigma)} [-p\mathbf{n} - \mu \mathbf{n}_\wedge \boldsymbol{\zeta} - \rho(\mathbf{nq})\mathbf{q}]dS.$$

This investigation is continued in 22·75.

EXAMPLES XXI

1. Water flows along a pipe of circular section with velocity q under a pressure gradient P. Prove that

$$\frac{\partial}{\partial r}\left(r\frac{\partial q}{\partial r}\right) = -\frac{Pr}{\mu},$$

where r is the distance from the axis, and find the rate of discharge.

2. Viscous fluid flows steadily parallel to the axis in the annular space between two coaxial cylinders of radii a, na $(n > 1)$.

Show that the rate of discharge is

$$\frac{\pi P a^4}{8\mu}\left\{n^4 - 1 - \frac{(n^2-1)^2}{\log n}\right\},$$

where P is the pressure gradient. Find the average velocity.

3. If water flows along a cylindrical pipe of circular section, inclined at an angle α to the horizontal, prove that the rate of discharge is

$$\frac{\pi a^4}{8\mu}(P + g\rho \sin \alpha),$$

where P is the pressure gradient.

4. Show that if a viscous incompressible fluid is flowing steadily in straight lines along a cylinder whose generators are parallel to the axis of z, the speed w at any point satisfies the equation

$$\frac{\partial^2 w}{\partial x^2} + \frac{\partial^2 w}{\partial y^2} = \text{constant}.$$

Fluid is flowing steadily along a channel of rectangular section of sides $2a$ and $2b$, under a pressure gradient P per unit length. Show that the total flow per second is

$$\frac{4ab^3}{3\mu} P\left\{1 - \frac{192b}{\pi^5 a}\left(\tanh\frac{\pi a}{2b} + \frac{1}{3^5}\tanh\frac{3\pi a}{2b} + \ldots\right)\right\}.$$

If the mean velocity across a section when $a = b$ is V, and R is the drag on the walls per unit area, deduce that

$$R/\rho V^2 = 3{\cdot}8\bigg/\frac{Va}{\nu}. \qquad \text{(U.L.)}$$

5. In the transformation $z = c\cosh(\xi + i\eta)$, $\xi = \xi_0$ is the cross-section of a solid cylinder which is being dragged longitudinally with constant velocity U through a pipe whose cross-section is $\xi = \xi_1$, the intervening space being filled with liquid, at constant pressure, moving parallel to the axis with velocity u.

Show that $\nabla^2 u = 0$ and that all conditions are satisfied by

$$u = U(\xi_1 - \xi)/(\xi_1 - \xi_0).$$

Prove that the drag on the solid cylinder is $2\pi\mu U/(\xi_1 - \xi_0)$ per unit length.

6. Prove that the stream function $\psi = C(b^2y - \frac{1}{3}y^3)$ corresponds to a steady stream of liquid in a straight channel of breadth $2b$, there being no slip at the boundary.

Show that this stream function satisfies the differential equation of viscous liquid motion, and calculate the pressure at any point, the coefficient of kinematic viscosity being ν and the density ρ.

7. Show that in the two-dimensional motion of a viscous liquid the stream function satisfies the equation

$$\left(\nu\nabla^2 - \frac{\partial}{\partial t}\right)\nabla^2\psi = \frac{\partial(\psi, \nabla^2\psi)}{\partial(x, y)}.$$

Hence show that any steady motion for which the streamlines are independent of the degree of viscosity must be *either* (1) a motion for which the resultant velocity and vorticity are each constant along every streamline or (2) a rigid body rotation superposed upon an irrotational motion.

8. A viscous liquid is bounded by parallel planes at distance h apart. One plane is fixed and the other moves parallel to itself with simple harmonic motion $a\cos nt$. Show that the tangential drag per unit area on the fixed plane has the maximum value

$$\frac{an\mu\omega}{h(\cosh\omega - \cos\omega)^{\frac{1}{2}}}, \quad \omega^2 = \frac{2nh^2}{\mu}.$$

9. Show that the circulation $I = \int(\mathbf{q} \cdot d\mathbf{s})$ round a closed circuit always composed of the same particles of fluid remains invariable if, and only if,

$$d\mathbf{q}/dt = -\nabla Q,$$

where Q is a scalar function of position and time t, and that, if Q does not contain t, then $Q + \frac{1}{2}q^2$ is constant along the path of a particle.

Prove that, when the body forces are conservative and p is a function of ρ, the accelerations are certainly derivable from such a function Q if (i) $\mu = 0$, or if (ii) μ/ρ = constant and $\nabla^2\mathbf{q} = 0$, and give Q in each case. (U.L.)

10. A viscous incompressible fluid is flowing steadily along a cylinder in straight lines parallel to the generators and to the axis of z. Show that the velocity at any point is given by

$$w = Ax^2 + Bxy + Cy^2 + \psi,$$

where ψ satisfies the equation $\dfrac{\partial^2\psi}{\partial x^2} + \dfrac{\partial^2\psi}{\partial y^2} = 0$, and A, B, and C are constants.

If the cross-section of the cylinder be a semicircle and its diameter, and the fluid be flowing under a constant pressure gradient P per unit length, find the mean velocity across any section. (U.L.)

11. Incompressible viscous liquid under no body forces moves in a *thin* film between the fixed plane $z = 0$ and the rigid moving surface $z = h(x, y)$. If (U, V, W) are the velocity resolutes of the point (x, y, h) of the moving surface, show that the differential equation for the pressure at points (x, y, z) is

$$\frac{\partial}{\partial x}\left(h^3\frac{\partial p}{\partial x}\right) + \frac{\partial}{\partial y}\left(h^3\frac{\partial p}{\partial y}\right) + 6\mu\left\{\frac{\partial}{\partial x}(hU) + \frac{\partial}{\partial y}(hV)\right\} = 12\mu W,$$

stating any assumptions made.

The moving surface is plane, unlimited in the y-direction, inclined at a small angle α to the plane $z = 0$, and the leading and trailing edges are at heights h_1, h_2 respectively. Show that, if π is the pressure at these edges, then, if $V = W = 0$,

$$p-\pi = \frac{6\mu U}{\alpha(h_1+h_2)h^2}(h_1-h)(h-h_2)$$

gives the pressure at points in the section whose thickness is h. (U.L.)

12. Show that in the steady motion of a viscous liquid of kinematic viscosity ν

$$\left(\nabla^2 - \frac{1}{\nu}q\frac{\partial}{\partial s}\right)\left(\frac{p}{\rho}+\Omega+\tfrac{1}{2}q^2\right) = \zeta^2,$$

where s is taken along a streamline.

13. Find an expression for F, the total rate of dissipation of energy in a viscous fluid. Show that if the boundaries be at rest and there is no slip, the rate of dissipation is given by

$$F = \mu\iiint(\xi^2+\eta^2+\zeta^2)\,dx\,dy\,dz.$$

If the motion is two-dimensional and due entirely to the steady motion of a cylinder travelling with velocity V at right angles to its generators, find the appropriate form of F. (U.L.)

14. Obtain the dynamical equations of motion for a fluid, taking viscosity and compressibility into account.

Show that the rate of doing work of the internal reactions in the fluid is $-F$, where

$$F = \widehat{xx}\frac{\partial u}{\partial x}+\widehat{yy}\frac{\partial v}{\partial y}+\widehat{zz}\frac{\partial w}{\partial z}+\widehat{yz}\left(\frac{\partial w}{\partial y}+\frac{\partial v}{\partial z}\right)+\widehat{zx}\left(\frac{\partial u}{\partial z}+\frac{\partial w}{\partial x}\right)+\widehat{xy}\left(\frac{\partial v}{\partial x}+\frac{\partial u}{\partial y}\right),$$

and that the equation of flow of heat in the fluid takes the form

$$\frac{\partial}{\partial x}\left(K\frac{\partial\theta}{\partial x}\right)+\frac{\partial}{\partial y}\left(K\frac{\partial\theta}{\partial y}\right)+\frac{\partial}{\partial z}\left(K\frac{\partial\theta}{\partial z}\right)+F = \rho\frac{D(c\theta)}{Dt},$$

where θ is the absolute temperature, K is the thermal conductivity, ρ the density, and c is the specific heat at constant volume.

What other relations are required to obtain a theoretically sufficient set of equations? (U.L.)

15. A viscous fluid is in two-dimensional motion such that at any instant the streamlines are circles about the axis of x. Show that the stream function ψ satisfies the equation

$$\frac{\partial\psi}{\partial t} = \nu\left(\frac{\partial^2\psi}{\partial r^2}+\frac{1}{r}\frac{\partial\psi}{\partial r}\right).$$

Examine the form of the solutions corresponding to ψ a function of r^2/t only.

A simple rectilinear vortex of strength k is generated along the z-axis at time $t=0$. Find the velocity of the fluid at any position distant r from the axis at time t; and show that if a circle with its centre on the axis is to spread outwards so as to enclose a constant amount of vorticity, the area of the circle must increase steadily. (U.L.)

16. Viscous incompressible fluid is in steady two-dimensioned radial motion between two non-parallel plane walls; r and ϕ are polar coordinates, r being the distance from the line of intersection of the planes of the walls, which are $\phi = \pm\alpha$. Show that the velocity is given by

$$u = f(\phi)/r,$$

where

$$\left(\frac{df}{d\phi}\right)^2 = \frac{2}{3\nu}(h-3\nu kf-6\nu f^2-f^3),$$

h and k being constants and ν the kinematic viscosity.

If $\mathscr{R} = ru_{\max}/\nu$, show that for purely divergent flow and a given value of $\mathscr{R}$ the greatest value of α is given by

$$(\mathscr{R}+3)^{\frac{1}{2}}\alpha = \sqrt{3}\int_0^{\frac{1}{2}\pi} \frac{d\psi}{\left\{1-\frac{1}{2}\left(1+\frac{3}{\mathscr{R}}\right)^{-1}\sin^2\psi\right\}^{\frac{1}{2}}}. \qquad \text{(M.T.)}$$

17. Prove that for viscous liquid filling a closed vessel which is at rest the rate of dissipation of energy is

$$\mu\int \zeta^2\,d\tau.$$

If the vessel has the form of a solid of revolution and is rotating about its axis (which is the axis of z) with angular velocity ω, prove that the rate of dissipation of energy has an additional term

$$\mu\omega\int (l\,Du+m\,Dv)\,dS, \quad D \equiv y\frac{\partial}{\partial x}-x\frac{\partial}{\partial y},$$

and l, m, n are the direction cosines of the inward normal at the element dS of the surface of the vessel. (U.L.)

18. Incompressible viscous liquid under no body force completely fills the space between an infinite circular cylindrical axle rotating with angular velocity ω in an eccentric circular cylindrical bearing. If O and O' are the centres of the cross-sections, which have radii a, $a+\epsilon$ respectively, where ϵ is small, and $OO' = \lambda\epsilon$, $(0<\lambda<1)$, show that the pressure p at a point P in the liquid satisfies the approximate equation

$$\frac{dp}{d\theta} = \frac{6\mu\omega a^2\lambda(\cos\theta+C)}{\epsilon^2(1+\lambda\cos\theta)^3},$$

where θ is the angle POO', μ is the viscosity, and C is a constant. Neglect the curvature of the lubricating channel. Find p and show that

$$C = 3\lambda/(2+\lambda^2). \qquad \text{(U.L.)}$$

19. Transform the equations of motion of a viscous incompressible fluid, and the equation of continuity, to cylindrical polar coordinates r, θ, z on the assumption that the pressure p and the components u, v, w of the fluid velocity in the directions of r, θ and z increasing, respectively, are all independent of θ.

The fluid fills the space $z>0$, being bounded by the plane $z = 0$ only, and this plane is rotating with constant angular velocity ω about the axis $r = 0$. Verify that the steady motion is given by

$$u = \omega rF(\zeta), \quad v = \omega rG(\zeta), \quad w = (\nu\omega)^{\frac{1}{2}}H(\zeta), \quad p=\rho\nu\omega P(\zeta),$$

where

$$z = (\nu/\omega)^{\frac{1}{2}}\zeta,$$

ρ is the density and ν the kinetic viscosity of the fluid; F, G, H, and P are independent of ρ, ν and ω and satisfy certain ordinary differential equations and

$$F(0) = 0, \quad G(0) = 1, \quad H(0) = 0, \quad F(\infty) = 0, \quad G(\infty) = 0.$$

The boundary conditions in the physical problem may be taken as $u = 0$, $v = \omega r$, $w = 0$ at $z = 0$, and $u = 0$, $v = 0$ at $z = \infty$; w must not be taken to vanish when $z = \infty$. (U.L.)

20. Prove that in polar coordinates the stress components are

$$\widehat{rr} = -p+2\mu\frac{\partial q_r}{\partial r}, \quad \widehat{\theta\theta} = -p+\frac{2\mu}{r}\left(\frac{\partial q_\theta}{\partial \theta}+q_r\right),$$

$$\widehat{\omega\omega} = -p+\frac{2\mu}{r\sin\theta}\left\{\frac{\partial q_\omega}{\partial \omega}+q_r\sin\theta+q_\theta\cos\theta\right\},$$

$$\widehat{r\theta} = \mu\left(\frac{\partial q_r}{r\,\partial\theta}+\frac{\partial q_\theta}{\partial r}-\frac{q_\theta}{r}\right),$$

$$\widehat{\theta\omega} = \mu\left\{\frac{\partial q_\omega}{r\,\partial\theta}+\frac{\partial q_\theta}{r\sin\theta\,\partial\omega}-\frac{q_\omega\cot\theta}{r}\right\},$$

$$\widehat{r\omega} = \mu\left\{\frac{\partial q_r}{r\sin\theta\,\partial\omega}+\frac{\partial q_\omega}{\partial r}-\frac{q_\omega}{r}\right\}.$$

21. Obtain the equations for the steady motion of a liquid under pressure only, in the non-dimensional form

$$\left(u\frac{\partial u}{\partial x}+v\frac{\partial u}{\partial y}+w\frac{\partial u}{\partial z}+\frac{\partial p}{\partial x}\right) = \frac{1}{\mathscr{R}}\nabla^2 u,$$

where $\mathscr{R}$ is the Reynolds number. (U.L.)

22. Obtain the transformation formulae relating the stress components and slide velocities referred to two different sets of rectangular axes.

Assuming the stresses in a fluid of viscosity μ to be given by

$$\widehat{xx}-\alpha = \widehat{yy}-\beta = \widehat{zz}-\gamma = -p, \quad \text{where} \quad -p = \tfrac{1}{3}(\widehat{xx}+\widehat{yy}+\widehat{zz}),$$

and three equations of the type

$$\widehat{yz} = \mu\left(\frac{\partial w}{\partial y}+\frac{\partial v}{\partial z}\right),$$

and α, β, γ are linear functions of the slide velocities, show that, if the form of these equations is invariant for a change of rectangular axes, then

$$\alpha-2\mu\frac{\partial u}{\partial x} = \beta-2\mu\frac{\partial v}{\partial y} = \gamma-2\mu\frac{\partial w}{\partial z} = -\tfrac{2}{3}\mu\delta,$$

where

$$\delta = \frac{\partial u}{\partial x}+\frac{\partial v}{\partial y}+\frac{\partial w}{\partial z}.$$

(U.L.)

23. Assuming the stresses in a viscous fluid to be given by equations of the type

$$\widehat{xx} = -p-\tfrac{2}{3}\mu\,\nabla\,\mathbf{q}+2\mu\frac{\partial u}{\partial x}, \quad \widehat{xy} = \mu\left(\frac{\partial v}{\partial x}+\frac{\partial u}{\partial y}\right), \quad \widehat{xz} = \mu\left(\frac{\partial u}{\partial z}+\frac{\partial w}{\partial x}\right),$$

find the equations of motion parallel to the coordinate axes.

24. Prove that the equation of motion of a viscous compressible fluid can be put in the form

$$\frac{d\mathbf{q}}{dt} = -\nabla\left(\Omega+\int\frac{dp}{\rho}\right)+\tfrac{1}{3}\nu\,\nabla(\nabla\,\mathbf{q})+\nu\,\nabla^2\mathbf{q},$$

on the assumption that the surface traction contains a term proportional to $\nabla \mathbf{q}$ in addition to the terms arising when the fluid is incompressible.

25. Assuming that the coefficient of kinematic viscosity in a compressible viscous fluid is constant, prove that the equation of motion has an integral of the form

$$\int \frac{dp}{\rho} + \tfrac{1}{2}q^2 + \Omega - \frac{\partial \phi}{\partial t} + \tfrac{4}{3}\nu\, \nabla^2 \phi = F(t),$$

the motion being assumed irrotational.

26. A sphere is in steady motion with velocity V along the axis of Z in an infinite perfect fluid, while the fluid rotates with constant angular velocity $\boldsymbol{\Omega}$ about that axis. Show that the stream function is of the form $f(r)\sin^2\theta$ in polar coordinates where $f(r)$ satisfies the equation

$$r^3 f''' - 2r^2 f'' - rf' + k^2 r^2 (rf' - 2f) = 0,$$

where $k = 2\Omega/V$. Find the integral of this equation that satisfies the " no slip " condition at the boundary, and discuss the state of flow in the neighbourhood of the sphere. (U.L.)

27. For steady motion of a liquid in the xy plane where the streamlines are parallel to the x-axis, show that the compatibility equation is $\partial^3 u/\partial y^3 = 0$ and that the stream function is of the form

$$\Psi = -\tfrac{1}{3}Ay^3 - \tfrac{1}{2}By^2 - Cy - D.$$

Prove further that the pressure is given by

$$\frac{p}{\rho} + \Omega = 2A\nu x + \frac{p_0}{\rho}.$$

28. In the case of steady motion in which the streamlines are circles whose centre is on the z-axis prove that the only component of the velocity is u tangential to the circles.

Prove also that

$$u = \frac{A}{r} + Br + Cr \log r,$$

where r is the radius of the circle, and A, B, C are constants of which C is in general zero.

29. If two-dimensional steady motions have the same streamlines whether the liquid is viscous or not, prove that the stream function must satisfy simultaneously the equations

$$\nabla^4 \psi = 0, \quad \partial(\psi, \nabla^2 \psi)/\partial(x, y) = 0,$$

and that these equations are satisfied when the vorticity is constant.

30. Prove that in the two-dimensional flow of a viscous liquid in which the vorticity is constant

$$\frac{p}{\rho} + \tfrac{1}{2}q^2 + \Omega$$

is constant along a streamline.

31. Show that the flow whose stream function is

$$\psi = Uy[1-\exp(-Ux/\nu)], \quad U>0, \quad x>0$$

and its reflection in the y-axis gives the impact of streams U and $-U$ impinging at the y-axis. (Riabouchinsky)

32. Prove that the compatibility equation can be written

$$\frac{\partial \boldsymbol{\zeta}}{\partial t}+(\mathbf{q}\nabla)\boldsymbol{\zeta} = (\boldsymbol{\zeta}\nabla)\mathbf{q}+\nu\nabla^2\boldsymbol{\zeta}.$$

33. Prove that the steady flow for which the stream function is $\psi = xF(y)+G(y)$, i.e. is linear in x, satisfies the boundary layer equations 23·20.

CHAPTER XXII

STOKES AND OSEEN FLOWS

22·1. General remarks. The flows considered in this chapter are mostly those with small Reynolds number, generally the case $\mathscr{R} = o(1)$.

The Reynolds number Ul/ν can be small by reason of the typical velocity U or the typical length l being small or by the kinematic viscosity ν being large. When U is small we have *slow motion* or *creeping motion*; when l is small we have the motion of minute objects, for example *Brownian motion.* The case of large ν is illustrated by the flow of pitch past a peg fixed on an inclined plane, any visible change taking months.

We proceed to consider the effect of small $\mathscr{R}$ on the Navier-Stokes equations.

22·11. Stokes flow. Referring to 21·31 suppose that the Reynolds number $\mathscr{R}$ is small, $\mathscr{R} = o(1)$, on account of the velocity being small. It then appears that the inertia term $(\mathbf{q}\,\nabla)\mathbf{q}$, which is quadratic in the velocity, should be smaller than the other terms in the equation of motion. The consequence of neglecting this term entirely is to render the equation linear and we get the equations of *Stokes flow*, namely

$$(1) \qquad \partial\mathbf{q}/\partial t = -\nabla P + \nu\nabla^2\mathbf{q}, \quad \nabla\mathbf{q} = 0, \quad P = \Omega + p/\rho.$$

Noting that $(\mathbf{q}\,\nabla)\mathbf{q} = \nabla(\tfrac{1}{2}q^2) - \mathbf{q}\wedge\boldsymbol{\zeta}$ we could replace P in (1) by

$$(2) \qquad P' = \Omega + p/\rho + \tfrac{1}{2}q^2$$

without changing essentially the form of the equation. Since q^2 is already quadratic, little is gained, but we notice that the neglect of the term $\mathbf{q}\wedge\boldsymbol{\zeta}$ amounts to neglect of the convection of vorticity.

The change of variables 21·31 (5) may be written as a change to *Stokes non-dimensional variables* (suffix s) in the form

$$(3) \qquad \mathbf{q} = U\mathbf{q}_S, \quad \mathbf{r} = l\mathbf{r}_S, \quad t = \mathscr{R}lt_S/U, \quad P = U^2P_S/\mathscr{R}, \quad \nabla_S = l\nabla$$

which lead to 21·31 (7) in form

$$(4) \qquad \partial\mathbf{q}_S/\partial t_S + \mathscr{R}(\mathbf{q}_S\nabla_S)\mathbf{q}_S = -\nabla_S P_S + \nabla_S^2\mathbf{q}_S, \quad \nabla_S\mathbf{q}_S = 0.$$

When $\mathscr{R}\to 0$, the Stokes variables remaining fixed, (4) becomes

$$(5) \qquad \partial\mathbf{q}_S/\partial t_S = -\nabla_S P_S + \nabla_S^2\mathbf{q}_S, \quad \nabla_S\mathbf{q}_S = 0$$

which is the non-dimensional form of a flow similar to that given by (1). We may therefore regard (1) as the standard form for Stokes flow, whatever the reason for supposing the Reynolds number to be small.

In the case of streaming past a fixed obstacle, however large the Reynolds number (provided turbulence does not occur), the adherence condition ensures small velocity and therefore Stokes flow in the immediate neighbourhood of the boundary.

For consideration of the approximation involved see 22·24.

22·20. Slow streaming past a sphere. Let a solid sphere of radius a be held fixed in a uniform stream U flowing steadily in the positive direction of the axis of x. If we neglect the quadratic terms in the equation of motion, the stream function satisfies the equation 21·36 (2), $E^4\psi = 0$ or

$$(1) \qquad \left[\frac{\partial^2}{\partial r^2}+\frac{\sin\theta}{r^2}\frac{\partial}{\partial\theta}\left(\frac{1}{\sin\theta}\frac{\partial}{\partial\theta}\right)\right]^2\psi = 0.$$

At infinity we have a uniform stream

$$(2) \qquad \psi = -\tfrac{1}{2}Ur^2\sin^2\theta$$

and this suggests the trial solution

$$\psi = f(r)\sin^2\theta.$$

Substitution in (1) gives successively

$$\left[\frac{\partial^2}{\partial r^2}+\frac{\sin\theta}{r^2}\frac{\partial}{\partial\theta}\left(\frac{1}{\sin\theta}\frac{\partial}{\partial\theta}\right)\right]\left[\left(\frac{d^2f(r)}{dr^2}-\frac{2f(r)}{r^2}\right)\sin^2\theta\right] = 0,$$

$$\left(\frac{d^2}{dr^2}-\frac{2}{r^2}\right)\left(\frac{d^2}{dr^2}-\frac{2}{r^2}\right)f(r) = 0.$$

We can satisfy this linear homogeneous equation of the fourth order by a sum of terms of the form Ar^n, provided that

$$[(n-2)(n-3)-2][n(n-1)-2] = 0,$$

whence $n = -1, 1, 2, 4$, and therefore

$$f(r) = \frac{A}{r}+Br+Cr^2+Dr^4.$$

Condition (2) shows that $C = -\tfrac{1}{2}U$, $D = 0$, and therefore

$$\psi = \left(\frac{A}{r}+Br-\tfrac{1}{2}Ur^2\right)\sin^2\theta.$$

The velocity components, which must vanish on the surface of the sphere, are

$$q_r = -\frac{1}{r\sin\theta}\frac{\partial\psi}{r\,\partial\theta} = U\cos\theta-2\left(\frac{A}{r^3}+\frac{B}{r}\right)\cos\theta,$$

$$q_\theta = \frac{1}{r\sin\theta}\frac{\partial\psi}{\partial r} = -U\sin\theta-\left(\frac{A}{r^3}-\frac{B}{r}\right)\sin\theta.$$

Putting $r = a$, gives $A = -\frac{1}{4}Ua^3$, $B = \frac{3}{4}Ua$, and so

$$(3) \qquad \psi = -\tfrac{1}{2}U\left(r^2 - \tfrac{3}{2}ar + \tfrac{1}{2}\frac{a^3}{r}\right)\sin^2\theta = -\tfrac{1}{4}U\,\frac{(r-a)^2(2r+a)}{r}\sin^2\theta.$$

The streamlines are shown in fig. 22·20.

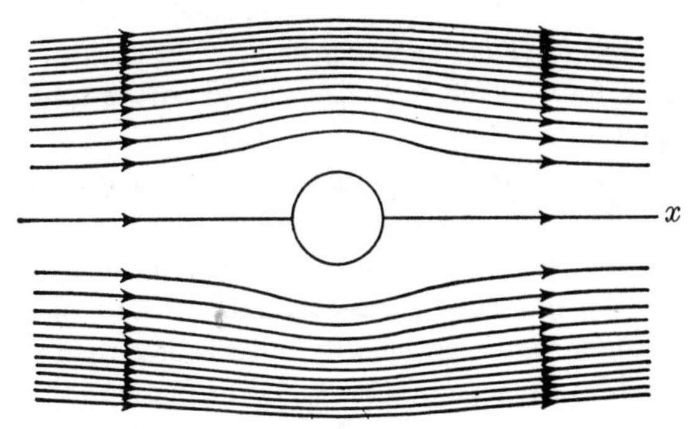

FIG. 22·20.

Observe that the stream function ψ given by (3) is unaltered when θ is replaced by $\pi-\theta$ and so is symmetrical about $\theta = \pi/2$. Thus the Stokes flow about a sphere shows no wake.

From 2·72 (4) and (7), the vorticity is

$$(4) \qquad \zeta = \frac{1}{r}\frac{\partial(rq_\theta)}{\partial r} - \frac{\partial q_r}{r\,\partial\theta} = -\frac{3a}{2r^2}U\sin\theta.$$

22·21. Drag on a slowly moving sphere. In the problem just discussed the liquid is reduced to rest and the sphere moves forward with velocity U, if we impress on the system a velocity U in the direction of x decreasing. The corresponding stream function is then

$$\psi = -\tfrac{1}{4}U\left(-3ar + \frac{a^3}{r}\right)\sin^2\theta.$$

The streamlines are shown in fig. 22·21.

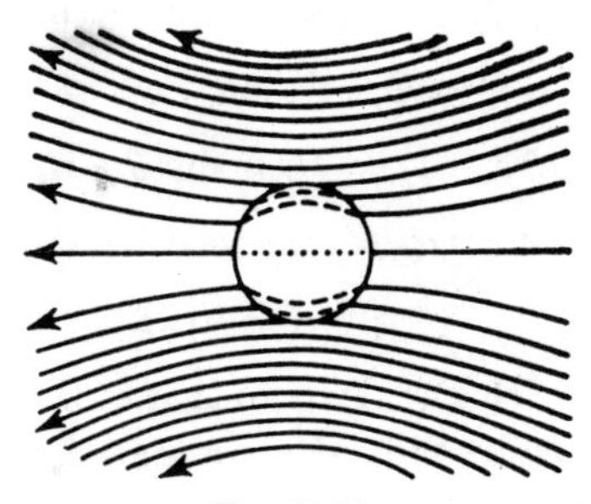

FIG. 22·21.

If P is the drag, the rate of doing work is PU, and this must just balance the rate of dissipation of energy given by 21·10.

The vorticity is still given by 22·20 (4), and therefore

$$PU = \mu \int_a^\infty dr \int_0^\pi \frac{9a^2}{4r^4} U^2 \sin^2\theta \,.\, 2\pi r^2 \sin\theta \, d\theta = 6\pi\mu U^2 a,$$

and so $P = 6\pi\mu Ua$, a formula due to Stokes.

This is also the force which must be applied to the sphere to hold it at rest in a steady stream U.

Take $\mathscr{R} = 2aU/\nu$ and divide by the frontal area πa^2. Then the drag coefficient is $C_D = 6\pi\mu Ua/(\frac{1}{2}\rho U^2 \pi a^2) = 24/\mathscr{R}$.

It must be remembered that the foregoing analysis applies only to motions in which the Reynolds number Ua/ν is small. Thus for a sphere of one millimetre radius moving in water the velocity must be less than 0·2 cm./sec. One application of this formula occurs in studying the motion of minute particles.

To find the terminal velocity (or velocity when the resultant force is zero) of a sphere of density σ falling in a liquid of density ρ, we have, on equating the weight to the buoyancy plus the resistance,

$$\tfrac{4}{3}\pi\sigma a^3 g = \tfrac{4}{3}\pi\rho a^3 g + 6\pi\mu Ua, \quad U = \tfrac{2}{9}\,\frac{\sigma-\rho}{\mu}\, a^2 g.$$

22·24. Stokes flow at a distance. In obtaining the equation of Stokes flow, 22·11 (1), the term $(\mathbf{q}\,\nabla)\mathbf{q}$ of order U^2/r at distance r from the centre was neglected while the viscous term $\nu\,\nabla^2\,\mathbf{q}$ of order $\nu U/r^2$ was retained. The ratio of the first of these to the second is

$$(1) \qquad Ur/\nu = \mathscr{R}r/l.$$

This ratio is small when $\mathscr{R}$ is small provided that r/l is not too large. Nevertheless, however small $\mathscr{R}$ may be it is always possible to make $\mathscr{R}r/l$ as large as desired by choice of r. Thus if $r = O(l/\mathscr{R})$, the neglected inertia terms are comparable with the retained viscous terms and the basis of Stokes' approximation is destroyed. Nevertheless in the case of a sphere of radius a disturbing a stream $U\mathbf{i}$ we have from 22·20 the stream function of the Stokes flow

$$\psi = \psi_1 + \psi_2 + \psi_3 \quad \text{where}$$

$$(2) \qquad \psi_1 = -\tfrac{1}{2}Ur^2\sin^2\theta, \quad \psi_2 = \tfrac{3}{4}Uar\sin^2\theta, \quad \psi_3 = -\tfrac{1}{4}U(a^3/r)\sin^2\theta.$$

Thus from 16·1 when $r = O(l/\mathscr{R})$ the velocity $\mathbf{q}$ calculated from ψ is

$$(3) \qquad \mathbf{q} = U\mathbf{i} + O(\mathscr{R}).$$

Thus when $\mathscr{R} \to 0$ the velocity tends to that of the uniform stream $U\mathbf{i}$ *before* r attains values which make Stokes' approximation break down. Therefore the Stokes solution provides a uniformly valid approximation to the *total* velocity distribution and therefore a valid approximation to many bulk properties such as the resistance.

We also observe that in (2) ψ_1 is the stream function for the stream $U\mathbf{i}$ and ψ_3 is the stream function of a doublet both of which correspond to irrotational flows and contribute nothing to the drag or vorticity. The stream function ψ_2 is that of a singularity called a *Stokeslet* (Hancock) which may be interpreted physically as a force applied to the fluid. Thus if the force $-D\mathbf{i}$ is applied to the fluid at the origin the corresponding stream function is $Dr\sin^2\theta/(8\pi\mu)$ and if $D = 6\pi\mu Ua$ we get ψ_2.

22·311. Steady Stokes flow in two-dimensions. From 21·34 (1), (2) we have

$$(1) \qquad \frac{\partial \boldsymbol{\zeta}}{\partial t} = \nabla \wedge (\mathbf{q} \wedge \boldsymbol{\zeta}) + \nu \nabla^2 \boldsymbol{\zeta}.$$

In the case of Stokes flow we neglect $\mathbf{q} \wedge \boldsymbol{\zeta}$ and therefore for steady motion

$$(2) \qquad \nabla^2 \boldsymbol{\zeta} = 0.$$

Now in two-dimensional flow $\boldsymbol{\zeta} = \mathbf{k}\nabla^2\psi$ in terms of the stream function ψ (4·40 (2)). Therefore ψ satisfies the *biharmonic equation*

$$(3) \qquad \nabla^4 \psi = 0.$$

If the motion takes place in the xy plane, the coordinate perpendicular to this plane will not be required and therefore we shall be free to write, as in Chapter 5,

$$(4) \qquad z = x + iy, \quad \bar{z} = x - iy$$

and then (5·33) we have the operational equivalence

$$(5) \qquad \nabla^2 = \frac{4\partial^2}{\partial z\, \partial \bar{z}}.$$

Thus finally the equation (3) satisfied by the stream function can be written

$$(6) \qquad \frac{\partial^4 (2i\psi)}{\partial z^2\, \partial \bar{z}^2} = 0$$

where the factor $2i$ has been introduced for subsequent convenience. Integrating we get successively

$$(7) \qquad \frac{\partial^3 (2i\psi)}{\partial z^2\, \partial \bar{z}} = W''(z), \quad \frac{\partial^2 (2i\psi)}{\partial z\, \partial \bar{z}} = W'(z) - \overline{W}'(\bar{z})$$

where $W''(z)$ denotes an arbitrary function of z only. Since $2i\psi$ is imaginary, so is $\partial^2(2i\psi)/(\partial z\, \partial \bar{z})$ and therefore the arbitrary function introduced by the second integration must be $-\overline{W}'(\bar{z})$. Now from 6·41 (4), $u + iv = 2i\partial\psi/\partial\bar{z}$. Therefore integrating once more with respect to z we get from (7) the fundamental result

$$(8)\qquad u+iv = 2i\partial\psi/\partial\bar{z} = W(z)-z\overline{W}'(\bar{z})-\bar{w}(\bar{z})$$

where $w(z)$ is another arbitrary function.

We shall call $W(z)$ and $w(z)$ the *generalised complex velocities.*

A further integration of (8) gives the stream function

$$(9)\qquad 2i\psi = \bar{z}W(z)-z\overline{W}(\bar{z})-{}'\bar{w}(\bar{z})+{}'w(z),$$

where we have used the notation for indefinite integration*

$$(10)\qquad {}'f(\alpha) = \int f(\alpha)\,d\alpha.$$

It should be observed that (9) gives the most general solution of the two-dimensional biharmonic equation (3).

22·312. Determinateness of the generalised complex velocities. When the flow is given u and v are known and from 22·311 (8)

$$(1)\qquad u+iv = W(z)-z\overline{W}'(\bar{z})-\bar{w}(\bar{z}).$$

Let, if possible, two other generalised complex velocities $W_1(z)$, $w_1(z)$ lead to the same value of $u+iv$. Then

$$(2)\qquad W_1(z)-z\overline{W}_1'(\bar{z})-\bar{w}_1(\bar{z}) = W(z)-z\overline{W}'(\bar{z})-\bar{w}(\bar{z}).$$

Differentiate partially with respect to z and rearrange. Then

$$W_1'(z)-W'(z) = \overline{W}_1'(\bar{z})-\overline{W}'(\bar{z}).$$

Since these equal quantities are complex conjugate each must be a real constant, say a. Then

$$(3)\qquad W_1'(z)-W'(z) = a$$

and therefore

$$(4)\qquad W_1(z) = W(z)+az+b, \quad a \text{ real.}$$

Substituting in (2) we get

$$(5)\qquad w_1(z) = w(z)+\bar{b}.$$

From (4) and (5) we see that $W(z)$ is determinate save for an added arbitrary term $az+b$, a real, and that $w(z)$ is determinate save for an arbitrary added constant $\bar{b}$.

22·313. Vorticity and pressure. The vorticity is given by

$$\boldsymbol{\zeta} = \mathbf{k}\zeta = \mathbf{k}\nabla^2\psi$$

and therefore from 22·311 (5) and (7)

* Milne-Thomson, *Plane Elastic Systems*, Springer, Berlin, 1968.

(1) $$\zeta = -2i[W'(z) - \overline{W}'(\bar{z})].$$

The complex variable form of the equation of steady Stokes flow, 22·11 (1), is

(2) $$2\frac{\partial}{\partial \bar{z}}\left(\frac{p}{\rho}+\Omega\right) = 4\nu \frac{\partial^2}{\partial z\, \partial \bar{z}}(u+iv) = -4\nu \overline{W}''(\bar{z}),$$

whence by integration

(3) $$p+\rho\Omega = -2\mu[\overline{W}'(\bar{z})+W'(z)]+p_0$$

where the arbitrary function $W'(z)$ has been adjusted to make the right-hand side real and p_0 is an arbitrary real constant, so that (3) gives the pressure.

Combining (1) and (3) we get

(4) $$p+\rho\Omega - i\mu\zeta = -4\mu W'(z)+p_0.$$

Now when $u+iv$ is given $\zeta = \partial v/\partial x - \partial u/\partial y$ is one-valued and p is necessarily one-valued. Therefore $W'(z)$ is one-valued, while from (2) its derivative $W''(z)$ exists. We therefore get the important conclusion

(5) *$W'(z)$ is holomorphic in the region of flow.*

22·314. The Cyclic function. By means of the relations $x = (\bar{z}+z)/2$, $y = i(\bar{z}-z)/2$ any function of x and y can be expressed as a function of z and $\bar{z}$. Let

(1) $$f = f(z, \bar{z})$$

be such a function defined in the region D.

Let C be a simple closed curve lying in the region D and let z be a point of C which is not a singular point of $f(z, \bar{z})$. The function f may be one-valued as, for example, $z^4\bar{z}^2+z^3$, or many-valued as $(z-1)^{1/2}$ or $\bar{z}\log z$.

Let us start from z with a definite determination of the function $f(z, \bar{z})$ say $f_1(z, \bar{z})$ and follow the variation of this as z describes the circuit C in the positive sense (5·40). After describing the circuit we shall arrive back at z with the determination $f_2(z, \bar{z})$ of the function. We write

(2) $$[f(z, \bar{z})]_C = f_2(z, \bar{z}) - f_1(z, \bar{z}).$$

We call the function $[f(z, \bar{z})]_C$ *the cyclic function** of $f(z, \bar{z})$ relative to the circuit C and the initial determination of $f(z, \bar{z})$.

It follows from (2) that

(3) $$\frac{\partial}{\partial z}[f(z, \bar{z})]_C = \frac{\partial f_2(z, \bar{z})}{\partial z} - \frac{\partial f_1(z, \bar{z})}{\partial z} = \left[\frac{\partial f(z, \bar{z})}{\partial z}\right]_C.$$

* The term is due to Filon. See Milne-Thomson, *Plane Elastic Systems.*

From (3) it follows that if $[\partial f/\partial \bar{z}]_C = 0$, then $[f(z, \bar{z})]_C$ must be a function of $\bar{z}$ only.

If therefore f is independent of $\bar{z}$, we see with the notation of 22·311 (10) that

(4) $$[f(z)]_C = 0 \text{ implies } [{}^{\backprime}f(z)]_C = \alpha, \quad [{}^{\backprime\backprime}f(z)]_C = \alpha z + \beta$$

where α and β are constants.

We also note that for a function of z only

(5) $$[f(z)]_C = \int_C \frac{df}{dz}\, dz = 2\pi i \times \text{sum of residues of } \frac{df}{dz} \text{ inside } C.$$

This gives a simple and rapid method of evaluation.

We give two theorems which are easily proved from the definition

(i) If f and g are any two functions, then $[f+g]_C = [f]_C + [g]_C$.

(ii) If $[g]_C = 0$, then $[fg]_C = g[f]_C$.

22·315. Cyclic properties of the generalised complex velocities. Consider a closed contour C enclosing a finite number of obstacles whose contours are $C_1, C_2, \ldots, C_n$, the intervening region being occupied by viscous liquid. Fig. 5·55 illustrates the case $n = 3$. Since the pressure and vorticity are one-valued it follows from 22·313 (4) that

(1) $$[W'(z)]_{C_k} = 0, \quad k = 1, 2, \ldots, n.$$

Therefore from 22·314 (4)

(2) $$[W(z)]_{C_k} = 2\pi i \gamma_k,$$

where $2\pi i \gamma_k$ is a constant. Now let z_k be any point inside the contour C_k. Then

(3) $$[\gamma_k \log (z - z_k)]_{C_k} = 2\pi i \gamma_k.$$

Therefore

(4) $$[W(z) - \gamma_k \log (z - z_k)]_{C_k} = 0.$$

Therefore for any contour reconcilable with C_k without enclosing other contours

(5) $$W(z) = \gamma_k \log (z - z_k) + h_k(z)$$

where $$[h_k(z)]_{C_k} = 0.$$

Observing that $[\gamma_k \log (z - z_k)]_{C_s} = 0$ for $s \neq k$ it follows that for the whole region D bounded by C and all the C_k

(6) $$W(z) = \sum_{k=1}^{n} \gamma_k \log (z - z_k) + W^0(z),$$

where $W^0(z)$ is a function holomorphic throughout D.

Since $u+iv$ is necessarily one-valued $[u+iv]_C = 0$ and therefore from 22·311 (8), since $W'(z)$ is holomorphic in D,

$$[W(z) - \overline{w}(\bar{z})]_C = 0$$

and therefore
$$[w(z)]_C = \overline{[W(z)]_C} = -2\pi i \sum_{k=1}^{n} \overline{\gamma_k}$$

so that

$$w(z) = -\sum_{k=1}^{n} \overline{\gamma_k} \log(z - z_k) + w^0(z) \tag{7}$$

while $w^0(z)$ is holomorphic throughout D.

Equations (6) and (7) give the forms of the generalised complex velocities in a finite region. When the region extends to infinity and the *pressure and vorticity remain bounded*, we easily show that for sufficiently large $|z|$

$$W(z) = \sum_{k=1}^{n} \gamma_k \log z + W^0(z) + Az + E \tag{8}$$

$$w(z) = -\sum_{k=1}^{n} \overline{\gamma_k} \log z + w^0(z) + Bz \tag{9}$$

where $W^0(z)$, $w^0(z)$ are holomorphic at infinity, and vanish there, while A, E, B are constants.

We make some comparisons with 22·311 (8).

(i) For a uniform stream $U+iV$, $W(z) = U+iV$, $w(z) = 0$.

(ii) For rotation with angular speed n about the origin $u = -yn$, $v = xn$ and $u+iv = inz$. Therefore

$$W(z) = \tfrac{1}{2}inz, \quad w(z) = 0.$$

(iii) For uniform shear flow ω parallel to $y \cos\beta - x \sin\beta = 0$ from 7·15 (3)

$$W(z) = \tfrac{1}{4}i\omega z, \quad w(z) = -\tfrac{1}{2}i\omega z e^{-2i\beta}.$$

All these cases are comprised in

$$W(z) = Az + E, \quad w(z) = Bz \tag{10}$$

where
$$A = \tfrac{1}{2}in + \tfrac{1}{4}i\omega, \quad E = U+iV, \quad B = -\tfrac{1}{2}i\omega e^{-2i\beta}.$$

On account of Stokes' paradox (22·319) it will appear that in (8) and (10) above we must take $E = 0$. This means that Stokes flow in two-dimensions does not exist for an obstacle disturbing a uniform stream.

22·316. Fundamental stress combinations. The analogy between linear elasticity and the present situation leads to the introduction of the *fundamental stress combinations.**

* Milne-Thomson, *loc. cit.* p. 684.

(1) $$\Theta = \widehat{xx}+\widehat{yy}, \quad \Phi = \widehat{yy}-\widehat{xx}+2i\widehat{xy}$$

so that from 21·14

(2) $$\Theta = -2p+2\mu\left(\frac{\partial u}{\partial x}+\frac{\partial v}{\partial y}\right) = 4\mu[W'(z)+\overline{W'(\bar{z})}]-2p_0+2\rho\Omega$$

from 22·313 (3).

(3) $$\Phi = 2\mu\left[\frac{\partial v}{\partial y}-\frac{\partial u}{\partial x}+i\left(\frac{\partial v}{\partial x}+\frac{\partial u}{\partial y}\right)\right] = -4\mu\frac{\partial}{\partial \bar{z}}(u-iv)$$
$$= 4\mu[\bar{z}W''(z)+w'(z)].$$

We note that

(4) $$\widehat{yy}-i\widehat{xy} = \tfrac{1}{2}(\Theta+\bar{\Phi}), \quad \widehat{xx}+i\widehat{xy} = \tfrac{1}{2}(\Theta-\bar{\Phi}).$$

22·317. Force and moment on an obstacle. Referring to 6·41 let the force be (X, Y) and the moment about the origin M. Then the contribution to the force on the element of arc ds at the point P of the contour C is

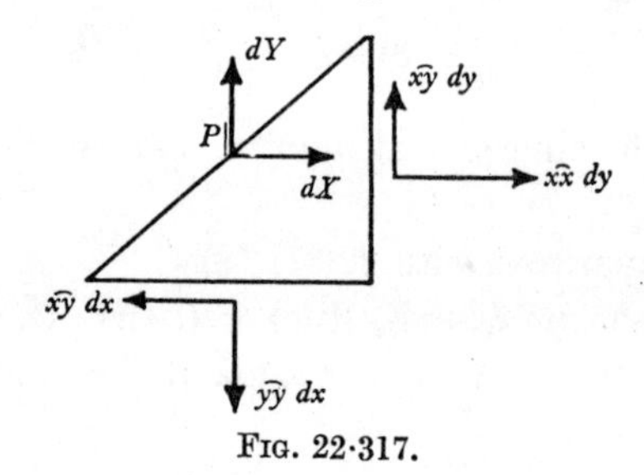

FIG. 22·317.

(1) $$dX = \widehat{xx}\,dy-\widehat{xy}\,dx, \quad dY = \widehat{xy}\,dy-\widehat{yy}\,dx$$

and therefore from 23·316

$$dX+idY = -idx(\widehat{yy}-i\widehat{xy})+dy(\widehat{xx}+i\widehat{xy}) = -\tfrac{1}{2}i\Theta dz-\tfrac{1}{2}i\bar{\Phi}d\bar{z}$$
$$= -2i\mu\{W'(z)dz+\overline{W}'(\bar{z})dz+z\overline{W}''(\bar{z})d\bar{z}+\bar{w}'(\bar{z})d\bar{z}\}+idz(p_0-\rho\Omega).$$

Integrating round the contour C we have

(2) $$X+iY = -2i\mu[W(z)+z\overline{W}'(\bar{z})+\bar{w}(\bar{z})]_C-i\int\rho\Omega dz,$$

since p_0 being a constant pressure makes no contribution to the resultant force. Now $W(z)+z\overline{W}'(\bar{z})+\bar{w}(\bar{z}) = 2W(z)-(u+iv)$ and $[u+iv]_C = 0$. Writing (X_B, Y_B), M_B for the force and moment due to reversed body force we have

(3) $$X_B+iY_B = -i\int\rho\Omega dz$$

and therefore

$$(4) \quad X+iY-(X_B+iY_B) = -4i\mu[W(z)]_C$$
$$= 8\pi\mu \times \text{sum of residues of } W'(z) \text{ inside } C.$$

Observe that $W'(z)$ has only been proved holomorphic outside C.

By a similar calculation it is easily proved that

$$(5) \quad M-M_B = 2\mu \times \text{Real part of } [{}^{\backprime}w(z)-zw(z)]_C$$
$$= -4\pi\mu \times \text{Real part of sum of residues of } izdw/dz \text{ inside } C.$$

22·318. Uniqueness. We shall consider the exterior problem in which the liquid extends to infinity in all directions and the velocity is given on all boundaries and also at infinity. The boundaries will be, for example, two fixed closed contours C_1 and C_2 at a finite distance.

Then the generalised complex velocities $W(z)$, $w(z)$ are uniquely determined by the boundary conditions.

Proof. Suppose that there are possibly two solutions W_1, w_1; W_2, w_2 and let

$$(1) \qquad W(z) = W_1-W_2, \quad w(z) = w_1-w_2.$$

If there are any singularities (say a source) in the fluid or on the boundaries, or any streams, rotation or shear flow, these will not appear in W, w, for they will disappear on taking the difference. Then

$$(2) \qquad G(z, \bar{z}) = W(z)-z\overline{W}'(\bar{z})-\overline{w}(\bar{z}),$$

which gives the velocity difference of flows 1 and 2, will be bounded, and will in fact tend to zero at infinity.

Describe a circle K_r about the origin whose radius r is large enough for K_r to contain all the boundaries C_1, C_2 and consider the integral

$$(3) \qquad J = \int_{C_1+C_2+K_r} (W'\overline{G}\,dz - G\overline{W}'\,d\bar{z}) = \int_{K_r} (W'\overline{G}\,dz - \overline{W}'G\,d\bar{z})$$

since G vanishes on C_1 and C_2. Therefore

$$(4) \qquad |J| \leqslant 4\pi r\,|W'|_{\max}\,|G|_{\max}$$

where the maxima are taken on K_r.

But for large values of $|z|$

$$(5) \qquad W(z) = W^0(z)$$

where $W^0(z) \to 0$ as $z \to \infty$.

Therefore

$$|W'| = O\left(\frac{1}{r^2}\right) < \frac{M_1}{r^2} \text{ on } K_r,\ M_1 \text{ a constant.}$$

But G is bounded, $|G| < M_2$ say. Therefore

(6) $$|J| < \frac{4\pi M_1 M_2}{r}.$$

Therefore $|J| \to 0$ when $r \to \infty$.

But by the Area theorem (5·43)

(7) $J = -2i \int (W' - \overline{W}')^2 \, dS$ over the area bounded by C_1, C_2, K_r

and $-2(W' - \overline{W}')^2$ being a non-negative number, $|J|$ can not diminish as r increases. Therefore $W' = \overline{W}' = a$, a real constant. But $W' = O(1/z^2)$. Therefore $a = 0$.

Therefore

(8) $$W(z) = \text{constant} = 0, \text{ using (5).}$$

Therefore $w(z) = 0$ on C_1 and C_2, and $w(\infty) = 0$. Therefore $W(z) = w(z) = 0$ in the whole domain occupied by the fluid and the solution is unique. Q.E.D.

Proof of uniqueness for the interior problem follows the same lines.

22·319. Stokes' paradox. We have seen (22·20) that Stokes flow past a sphere can be calculated. The corresponding plane problem of a uniform stream disturbed by a circular cylinder has no solution. This result is known as *Stokes' paradox.* The reason is given in the following theorem.

Theorem. Viscous liquid occupies the region R exterior to fixed closed finite contours $C_1, C_2, \ldots, C_n$ on which the velocity is zero. If the velocity, pressure and vorticity are all bounded in R, then the velocity is identically zero in R.

Proof. Since the pressure depends on $W' + \overline{W}'$ and the vorticity depends on $W' - \overline{W}'$ it follows that W' is bounded. Since the velocity is also bounded we have from 22·315 (8)

(1) $$W(z) = E + W^0(z), \quad w = w^0(z)$$

where $W^0(z)$, $w^0(z)$ are holomorphic and tend to zero at infinity.

As in 22·318 we have

(2) $$W' = \overline{W}' = a, \text{ a real constant.}$$

But a must be zero since from (1) $W'(z) = O(1/z^2)$.

Therefore $W(z) = E$ throughout R while $w(\infty) = w^0(\infty) = 0$.

Now from 22·318 (2), $G(z, \bar{z}) = E - \overline{w}(\bar{z})$ and therefore since $G = 0$ on all boundaries, we have

(3) $$w(z) = \overline{E} \text{ on all boundaries.}$$

Therefore since $w(\infty) = 0$, we have by Cauchy's formula for infinite regions, 5·591,

$$(4) \qquad w(z) = \overline{E} \text{ in } R.$$

Therefore† $u+iv = G = 0$ in R. Q.E.D.

Note. From another point of view since $w(\infty) = 0$, (4) can be held to show that $\overline{E}$ and therefore $E = 0$, or alternatively that there is a contradiction in the region R, supposing $E \neq 0$. The conclusion is the same.

22·320. Functional equation satisfied by the generalised complex velocities. Consider first the interior problem in which the liquid occupies the region R interior to a simple closed contour C in the z-plane.

Let the region R be mapped onto the interior R^* of the unit circle $|\zeta|<1$ and the contour C onto the circumference C^*: $|\zeta| = 1$ by the mapping

$$(1) \qquad z = f(\zeta), \quad z = 0 \text{ corresponding to } \zeta = 0,$$

and let $W_1(z)$, $w_1(z)$ be the complex velocities. Write

$$(2) \qquad W_1(z) = W_1[f(\zeta)] = W(\zeta), \quad w_1(z) = w(\zeta)$$

since $W_1'(z) = W'(\zeta)\,d\zeta/dz = W'(\zeta)/f'(\zeta)$ we have from 22·311 (8)

$$(3) \qquad u+iv = W(\zeta) - f(\zeta)\overline{W}'(\overline{\zeta})/\overline{f}'(\overline{\zeta}) - \overline{w}(\overline{\zeta}).$$

Let the velocity on the boundary C be $B_1(z)$ and let $\sigma = e^{i\theta}$ denote the point of C^* corresponding to the point z of C. Then if $B(\sigma) = B_1(z) = B_1[f(\sigma)]$, the function $B(\sigma)$ will be called the *boundary function* and the boundary condition is

$$(4) \qquad W(\sigma) - f(\sigma)\overline{W}'(\overline{\sigma})/\overline{f}'(\overline{\sigma}) - \overline{w}(\overline{\sigma}) = B(\sigma)$$

or, since $\overline{\sigma} = e^{-i\theta} = 1/\sigma$,

$$(5) \qquad W(\sigma) - f(\sigma)\overline{W}'(1/\sigma)/\overline{f}'(1/\sigma) - \overline{w}(1/\sigma) = B(\sigma).$$

On account of the indeterminancy of W and w we can arrange the constants in 22·312 (4), (5) so that

$$(6) \qquad w(0) = 0, \quad \text{Real part of } \{W'(0)/f'(0)\} = 0.$$

Multiply (5) by $d\sigma/[2\pi i(\sigma-\zeta)]$ where ζ is a point inside C^* and then integrate round C^*, using Cauchy's formula, 5·591. Then

$$(7) \qquad W(\zeta) - \frac{1}{2\pi i}\int_{C^*} \frac{f(\sigma)}{\overline{f}'(1/\sigma)} \frac{\overline{W}'(1/\sigma)}{\sigma-\zeta}\,d\sigma = \frac{1}{2\pi i}\int_{C^*} \frac{B(\sigma)\,d\sigma}{\sigma-\zeta},$$

† The above statement and proof of Stokes' Paradox is given by D. Gh. Ionescu, "La méthode des fonctions analytiques dans l'hydrodynamique des liquides visqueux", *Revue de Mécanique Appliquée*, VIII (1963) 676.

for $\overline{w}(1/\sigma)$ is holomorphic when σ is outside C^* and by (6) vanishes when $\sigma = \infty$. Therefore the corresponding Cauchy integral is zero.

Thus (7) is the functional equation which determines $W(\zeta)$ and therefore $W_1(z)$ in terms of the boundary function.

Apply the same method to the complex conjugate of (5). Then

$$(8) \qquad w(\zeta) = -\frac{1}{2\pi i}\int_{C^*}\frac{\overline{f}(1/\sigma)}{f'(\sigma)}\frac{W'(\sigma)}{\sigma-\zeta}\,d\sigma - \frac{1}{2\pi i}\int_{C^*}\frac{\overline{B}(1/\sigma)}{\sigma-\zeta}\,d\sigma + \overline{W}(0),$$

since here $\overline{W}(1/\sigma)$ is holomorphic outside C^* but takes the value $\overline{W}(0)$ when $\sigma = \infty$.

Notice that (8) does not present a functional equation to be solved, but a formula which gives $w(\zeta)$ after $W(\zeta)$ has been determined from (7).

Thus in principle the interior problem is solved.

For the *exterior* problem in which the liquid occupies the region R_2 outside C, we still map this region onto R^* the region inside the unit circle but now we make $z = \infty$ map into $\zeta = 0$. The mapping then takes the form

$$(9) \qquad z = f(\zeta) = \frac{\lambda}{\zeta} + f^0(\zeta), \quad z = \infty \text{ corresponds to } \zeta = 0,$$

where λ is a constant and $f^0(\zeta)$ is holomorphic inside C^*. In this case, 22·315 (10),

$$(10) \qquad W(\zeta) = \frac{\lambda A}{\zeta} + W^0(\zeta), \quad w(\zeta) = \frac{\lambda B}{\zeta} + w^0(\zeta),$$

where $W^0(\zeta)$, $w^0(\zeta)$ are holomorphic in R^*.

Substituting in (5) we get the relation

$$(11) \qquad W^0(\sigma) - f(\sigma)\overline{W}^{0\prime}(1/\sigma)/\overline{f}'(1/\sigma) - \overline{w}^0(1/\sigma) = F(\sigma)$$

where

$$(12) \qquad F(\sigma) = B(\sigma) - \frac{A\lambda}{\sigma} - \frac{f(\sigma)}{\overline{f}'(1/\sigma)}\,\bar{A}\bar{\lambda}\sigma^2 + \sigma\bar{\lambda}\bar{B}.$$

Thus $W^0(\zeta)$, $w^0(\zeta)$ are determined by (7) and (8) with $F(\sigma)$ replacing $B(\sigma)$.

22·321. Solution of the interior problem for the circle. Suppose the liquid occupies the region interior to a circumference C of radius R about the origin. This is mapped onto the unit circle by

$$(1) \qquad z = f(\zeta) = R\zeta, \quad z = 0 \text{ corresponds to } \zeta = 0.$$

Let

$$(2) \qquad W(\zeta) = \sum_{n=0}^{\infty} a_n \zeta^n.$$

Then $W'(\zeta) = a_1 + 2a_2\zeta + 3a_3\zeta^2 + \ldots$ and therefore

$$f(\sigma)\overline{W}'(1/\sigma)/\bar{f}'(1/\sigma) = \bar{a}_1\sigma + 2\bar{a}_2 + \frac{b_1}{\sigma} + \frac{b_2}{\sigma^2} + \ldots$$

where b_1, b_2, ... are constants. Therefore by 5·591 (3)

$$\frac{1}{2\pi i}\int_{C^*} \frac{f(\sigma)\overline{W}'(1/\sigma)}{\bar{f}'(1/\sigma)(\sigma-\zeta)}\,d\sigma = \bar{a}_1\zeta + 2\bar{a}_2.$$

Therefore from 22·320 (7)

$$W(\zeta) = \bar{a}_1\zeta + 2\bar{a}_2 + \frac{1}{2\pi i}\int_{C^*} \frac{B(\sigma)}{\sigma-\zeta}\,d\sigma. \tag{3}$$

Now from (2) $a_1 = W'(0)$. Therefore from (3)

$$a_1 = \bar{a}_1 + \frac{1}{2\pi i}\int_{C^*} \frac{B(\sigma)\,d\sigma}{\sigma^2}.$$

But from 22·320 (6), Real part of $W'(0) = 0$. Therefore

$$a_1 = \frac{1}{4\pi i}\int_{C^*} \frac{B(\sigma)}{\sigma^2}\,d\sigma. \tag{4}$$

Also $\qquad a_2 = \frac{1}{2}W''(0)$. Therefore

$$a_2 = \frac{1}{2\pi i}\int_{C^*} \frac{B(\sigma)}{\sigma^3}\,d\sigma, \tag{5}$$

so that $W(\zeta)$ is completely determined by (3), (4), (5). From 22·320 (8) we then find that

$$w(\zeta) = \frac{a_1}{\zeta} - \frac{1}{\zeta}W'(\zeta) + \overline{W}(0) - \frac{1}{2\pi i}\int_{C^*} \frac{\overline{B}(1/\sigma)d\sigma}{\sigma-\zeta}. \tag{6}$$

For the *exterior problem* we write $z = f(\zeta) = R/\zeta$ and then proceed as in 22·320. For an example see 22·40.

As an example of the interior problem consider liquid in a cylinder of radius R rotating steadily about its axis with angular speed n.

Then on the boundary the adherence condition gives

$$u + iv = Rne^{i(\theta+\frac{1}{2}\pi)} = iRn\sigma = B(\sigma).$$

Therefore $\quad a_1 = \frac{1}{2}iRn, \quad a_2 = 0 \quad$ and therefore

$$W(\zeta) = \tfrac{1}{2}iRn\zeta, \quad w(\zeta) = 0.$$

22·40. Circular cylinder disturbing uniform shear flow. Let the cylinder whose cross-section is $|z| \leqslant R$ disturb the shear flow of 22·315 with $\beta = 0$. Then

$$A = \tfrac{1}{4}i\omega, \quad B = -\tfrac{1}{2}i\omega, \quad z = f(\zeta) = R/\zeta \tag{1}$$

and therefore we write, 22·320 (10), (12)

(2) $$W(\zeta) = AR/\zeta + W^0(\zeta), \quad w(\zeta) = BR/\zeta + w^0(\zeta)$$

(3) $$F(\sigma) = \tfrac{1}{2}iR\omega(\sigma - 1/\sigma).$$

Since $B(\sigma) = 0$ on the boundary.

Observing that $f(\sigma)\overline{W}^{0\prime}(1/\sigma)/\overline{f}'(1/\sigma) = -\sigma^{-3}\overline{W}^{0\prime}(1/\sigma)$ is holomorphic *outside* the unit circle and therefore contributes nothing to the integral in 22·320 (7), and writing $F(\sigma)$ for $B(\sigma)$ we get

(4) $$W^0(\zeta) = \tfrac{1}{2}iR\omega\zeta.$$

Similarly from 22·320 (8) with $\overline{F}(1/\sigma)$ for $\overline{B}(1/\sigma)$ we get

(5) $$w^0(\zeta) = \tfrac{1}{2}iR\omega\zeta^3 - \tfrac{1}{2}iR\omega\zeta.$$

Thus (2) and (5) determine the generalised complex velocities in terms of ζ. Since $\zeta = R/z$ we get, in terms of z,

(6) $$W_1(z) = \tfrac{1}{4}i\omega z + \tfrac{1}{2}i\omega R^2/z$$

(7) $$w_1(z) = -\tfrac{1}{2}i\omega z - \tfrac{1}{2}i\omega R^2/z + \tfrac{1}{2}i\omega R^4/z^3.$$

To determine the stream function we use 22·311 (9) which gives

(8) $$\psi = \tfrac{1}{4}\omega\{r^2 - R^2 \log r^2 + (2R^2 - r^2 - R^4/r^2)\cos 2\theta\}, \quad z = re^{i\theta}.$$

Some of the streamlines are shown (not to scale) in fig. 22·40.

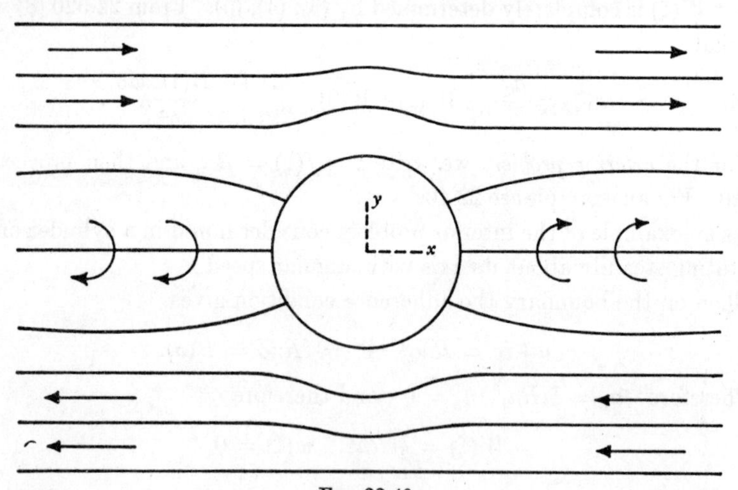

FIG. 22·40.

Since (8) is unchanged when either $\pi - \theta$ or $-\theta$ is written for θ, the streamlines are symmetrical about both axes of reference. The *direction of flow*, however, has rotational symmetry with respect to rotation of the diagram through two right angles about the centre of the circle.

22·50. Oseen's method. We have shown that Stokes flow becomes almost a uniform stream before the approximation fails. Oseen suggested that the appropriate approximation at large distances is that obtained by regarding the flow as a small departure from uniform flow under inertial and viscous forces of comparable magnitude. This he achieved by writing, in the case of an obstacle which disturbs the stream $U\mathbf{i}$

$$(1) \qquad \frac{\partial \mathbf{q}}{\partial t}+(U\mathbf{i}\nabla)\mathbf{q} = -\nabla P+\nu\nabla^2\mathbf{q}, \quad \nabla\mathbf{q} = 0, \quad P = \frac{p}{\rho}+\Omega,$$

or in the case of steady motion, with which alone we shall be concerned,

$$(2) \qquad U(\mathbf{i}\nabla)\mathbf{q} = -\nabla P+\nu\nabla^2\mathbf{q}.$$

This equation is valid at *sufficiently great distances* for any Reynolds number $\mathscr{R}$, *however large*. See 22·75.

When, however, $\mathscr{R} = o(1)$, the equation is uniformly valid in the whole region of flow, for it is valid at a distance while in the finite part it differs from the exact Navier-Stokes equation only by the order of the negligible inertia terms.

Notes. (i) When $\mathscr{R} = o(1)$ the term $U(\mathbf{i}\nabla)\mathbf{q}$ is negligible near the obstacle in comparison with the viscous term and Oseen's equation then coincides with that of Stokes.

(ii) When $\mathscr{R} = o(1)$ the boundary condition $\mathbf{q} = 0$ at the obstacle is satisfied to $O(\mathscr{R})$ and Oseen's equation is not any less accurate than that of Stokes in the neighbourhood of the obstacle.

(iii) Oseen's equation is not a second approximation which improves on Stokes' equation.

(iv) In general Oseen's equation will not furnish a second approximation to the drag.

(v) Since Oseen's equation is only a first order approximation it is unnecessary to satisfy the boundary conditions to more than the first order in $\mathscr{R}$.

Referring to 21·31 let us change to a flow similar to that given by the Navier-Stokes equations but now in *Oseen non-dimensional variables* defined as follows, with suffix O;

$$(3) \qquad \mathbf{q} = U\mathbf{q}_O, \quad \mathbf{r} = l\mathbf{r}_O/\mathscr{R}, \quad t = lt_O/(\mathscr{R}U), \quad P = U^2P_O$$

so that

$$(4) \qquad \nabla_O = \partial/\partial\mathbf{r}_O = (l/\mathscr{R})\nabla.$$

Then the Navier-Stokes equations became

$$(5) \qquad \partial\mathbf{q}_O/\partial t_O+(\mathbf{q}_O\nabla_O)\mathbf{q}_O = -\nabla_O P_O+\nabla^2{}_O\mathbf{q}_O, \quad \nabla_O\mathbf{q}_O = 0.$$

Since $\mathbf{r}/l = \mathbf{r}_O/\mathscr{R}$ it follows that *geometrically* the transformation is a simple homothety and for a fixed value of $\mathbf{r}$, as $\mathscr{R}\to 0$ so does $\mathbf{r}_O$. That is to say the obstacle in Oseen coordinates appears to shrink in size, the limit being a *point obstacle.* Since a point can not produce a finite disturbance, the limit of a steady Oseen flow, in Oseen coordinates, is a flow in which the velocity is everywhere that of the uniform stream $\mathbf{i}$ except at the point obstacle.

This result in the case $\mathscr{R} = o(1)$ enables us to regard (5) as arising from the perturbation of a stream $\mathbf{i}$ by a small obstacle, the small perturbation velocity $\mathbf{v}_O$ being given by $\mathbf{q}_O = \mathbf{i}+\mathbf{v}_O$ and then (5) can be written, to this order, as

$$\partial \mathbf{v}_O/\partial t_O + (\mathbf{i}\nabla_O)\mathbf{v}_O = -\nabla_O P_O + \nabla^2{}_O \mathbf{v}_O$$

which is the Oseen flow (1) in its non-dimensional form.

22·51. Oseen's equation for steady motion. The equation of motion is

$$U\frac{\partial \mathbf{q}}{\partial x} = -\nabla P + \nu\nabla^2 \mathbf{q}, \quad P = \frac{p}{\rho}+\Omega, \tag{1}$$

and the equation of continuity is

$$\nabla \mathbf{q} = 0. \tag{2}$$

Taking the divergence of (1) we get

$$\nabla^2 P = 0. \tag{3}$$

So that P is a harmonic function.

Take the curl of (1). Then

$$U\frac{\partial}{\partial x}\boldsymbol{\zeta} = \nu\nabla^2\boldsymbol{\zeta} \tag{4}$$

is the *compatibility equation* for steady Oseen flow.

Since $\nabla\boldsymbol{\zeta} = 0$ it follows (see Ex. III, 25) that

$$\nabla \wedge \mathbf{q} = \boldsymbol{\zeta} = \nabla\alpha \wedge \nabla\beta \tag{5}$$

where α and β are two suitably chosen scalar functions and the vectors $\boldsymbol{\zeta}$ are tangent to the curves whose equations are

$$\alpha(x, y, z) = \text{constant}, \quad \beta(x, y, z) = \text{constant}. \tag{6}$$

22·52. Case where the vorticity is perpendicular to the stream at infinity. This case arises in particular when the motion is two-dimensional in the xy-plane, or when the motion is axisymmetrical with respect to the x-axis. We can then prove the following theorem.

Theorem. Every solution of Oseen's equation for steady flow in which the vorticity is everywhere perpendicular to the velocity at infinity $U\mathbf{i}$ can be put in the form

(1) $$\mathbf{q} = -\nabla\phi + \frac{1}{2k}\nabla\chi - \mathbf{i}\chi, \quad k = \frac{U}{2\nu}$$

where ϕ is a harmonic function and χ satisfies

(2) $$\nabla^2\chi - 2k\frac{\partial\chi}{\partial x} = 0.$$

The pressure is given by

(3) $$P = \frac{p}{\rho} + \Omega = \frac{p_0}{\rho} + U\frac{\partial\phi}{\partial x}$$

where p_0 is an arbitrary constant.

Proof. Since the vorticity is perpendicular to the x-axis, in 22·51 (5) we can take $\alpha = x$ and so we can write

(4) $$\nabla \wedge \mathbf{q} = \boldsymbol{\zeta} = \mathbf{i} \wedge \nabla\chi$$

where $\chi = \chi(x, y, z)$ is determinate save for an added function of x. A particular solution of equation (4) is

(5) $$\mathbf{q} = -\mathbf{i}\chi$$

and therefore the general solution is

(6) $$\mathbf{q} = -\nabla A - \mathbf{i}\chi$$

where A is an arbitrary scalar function.

Substituting (4) in the compatibility equation 22·51 (4) we get

$$\mathbf{i} \wedge \nabla\{2k\,\partial\chi/\partial x - \nabla^2\chi\} = 0$$

and therefore

$$\nabla^2\chi - 2k\,\partial\chi/\partial x = f(x),$$

where $f(x)$ is an arbitrary function. Since χ already contains an added arbitrary function we can take $f(x)$ to be zero and therefore χ satisfies the equation

(7) $$\nabla^2\chi - 2k\,\partial\chi/\partial x = 0.$$

Also since $\mathbf{q}$ must satisfy the equation of continuity, (6) and (7) give

(8) $$0 = \nabla^2 A + \partial\chi/\partial x = \nabla^2(A + \chi/(2k))$$

so that $A + \chi/(2k)$ is a harmonic function, say ϕ, and $A = -\chi/(2k) + \phi$. Therefore from (6) we get (1).

Substitute the value of $\mathbf{q}$ given by (1) in Oseen's equation 22·51 (1) and we get (3). Q.E.D.

Now, as is easily proved, $(\nabla^2\chi - 2k\,\partial\chi/\partial x) = e^{kx}(\nabla^2 - k^2)(e^{-kx}\chi)$ and therefore equation (7) is equivalent to

(9) $$(\nabla^2 - k^2)(e^{-kx}\chi) = 0.$$

Let $\eta = e^{-kx}\chi$. Then if η is independent of θ and ω where (r, θ, ω) are spherical polar coordinates, (9) becomes

$$\frac{\partial^2(r\eta)}{\partial r^2} - k^2 r\eta = 0 \tag{10}$$

with the solution $r\eta = Ae^{-kr} + Be^{kr}$. Therefore the solution for χ which vanishes at infinity is a constant multiple of $e^{-k(r-x)}/r$. Thus if we write

$$\lambda = \lambda(r) = e^{-kr}/r \tag{11}$$

the general solution for χ which is independent of ω (the axisymmetrical case) and vanishes at infinity is

$$\chi = e^{kx}\{a_0\lambda + a_1\,\partial\lambda/\partial x + a_2\,\partial^2\lambda/\partial x^2 + \ldots\}, \tag{12}$$

where $a_0, a_1, a_2, \ldots$ are arbitrary constants.

The value of ϕ which is such that $-\nabla\phi = \mathbf{i}U +$ terms which vanish at infinity and is independent of ω is (17·1).

$$\phi = -Ux + b_0/r + b_1\,\partial(1/r)/\partial x + b_2\partial^2(1/r)/\partial x^2 + \ldots \tag{13}$$

where $b_0, b_1, b_2, \ldots$ are arbitrary constants.

On account of (1) equations (12) and (13) are appropriate to the axisymmetrical case in which a fixed obstacle disturbs the uniform stream $U\mathbf{i}$. If the same obstacle moves with velocity $U\mathbf{i}$ in liquid otherwise at rest, we omit the term $-Ux$ from ϕ. The two cases are in fact dynamically equivalent if the axes of reference are fixed in the body.

22·60. Oseen streaming past a sphere. Let the sphere, $r = a$ in spherical polar coordinates (r, θ, ω), disturb the uniform stream $U\mathbf{i}$. We shall base the Reynolds number on the diameter so that

$$\mathscr{R} = 2aU/\nu. \tag{1}$$

The boundary conditions at the surface of the sphere are

$$q_r = 0, \quad q_\theta = 0 \quad \text{when} \quad r = a. \tag{2}$$

From 22·52 (1) we have

$$q_r = -\frac{\partial\phi}{\partial r} + \frac{1}{2k}\frac{\partial\chi}{\partial r} - \chi\cos\theta, \quad q_\theta = -\frac{1}{r}\frac{\partial\phi}{\partial\theta} + \frac{1}{2kr}\frac{\partial\chi}{\partial\theta} + \chi\sin\theta. \tag{3}$$

Now use 22·52 (12) observing that

$$\frac{e^{-k(r-x)}}{r} = \frac{e^{-kr(1-\cos\theta)}}{r} = \frac{1}{r} - k(1-\cos\theta) + \frac{k^2 r}{2}(1-\cos\theta)^2 + \ldots$$

$$e^{kx}\frac{\partial}{\partial x}\left(\frac{e^{-kr}}{r}\right) = -\frac{\cos\theta}{r^2} - \frac{k\cos^2\theta}{r} - \ldots, \quad e^{kx}\frac{\partial^2}{\partial x^2}\left(\frac{e^{-kr}}{r}\right) = \frac{3\cos^2\theta - 1}{r^3} + \ldots$$

so that we have the approximation

$$(4)\qquad \chi = a_0\left\{\frac{1}{r} - k(1-\cos\theta) + \frac{k^2 r}{2}(1-\cos\theta)^2\right\} - a_1\left(\frac{\cos\theta}{r^2} + \frac{k\cos^2\theta}{r}\right) + a_2\,\frac{3\cos^2\theta - 1}{r^3}.$$

For ϕ we have the approximation, 22·52 (13), 17·1

$$(5)\qquad \phi = -Ur\cos\theta + \frac{b_0}{r} - \frac{b_1\cos\theta}{r^2} + \frac{b_2(3\cos^2\theta - 1)}{r^3}.$$

Substitute (4) and (5) in (3) and then put $r = a$ in accordance with (2). Equating to zero the coefficients of the powers of $\cos\theta$ we find equations connecting a_0, a_1, a_2, b_0, b_1, b_2. Solving these and neglecting in the result powers of $ka\,(=\mathscr{R}/4)$ greater than the first we find that

$$(6)\qquad \begin{aligned} a_0 &= \frac{3Ua}{2}\left(1+\tfrac{3}{16}\mathscr{R}\right), & a_1 &= \tfrac{3}{8}Ua^2\mathscr{R} \\ b_0 &= \frac{3Ua^2}{\mathscr{R}}\left(1+\tfrac{3}{16}\mathscr{R}\right), & b_1 &= \tfrac{1}{2}Ua^2. \end{aligned}$$

To get the stream function ψ near the sphere correct to $O(\mathscr{R})$ it is necessary to take a_2 into account and doing so Goldstein has found*

$$(7)\qquad \psi = -\tfrac{1}{2}Ua^2\sin^2\theta\left(\frac{r}{a}-1\right)^2\left[\left(1+\tfrac{3}{16}\mathscr{R}\right)\left(1+\frac{a}{2r}\right) - \tfrac{3}{16}\mathscr{R}\left(1+\frac{a}{r}\right)^2\cos\theta\right]$$

which agrees with the result for Stokes flow (22·20 (3)) when the terms in $\mathscr{R}$ are omitted.

22·62. Moving sphere. By superposing a velocity $-U\mathbf{i}$ on the stationary

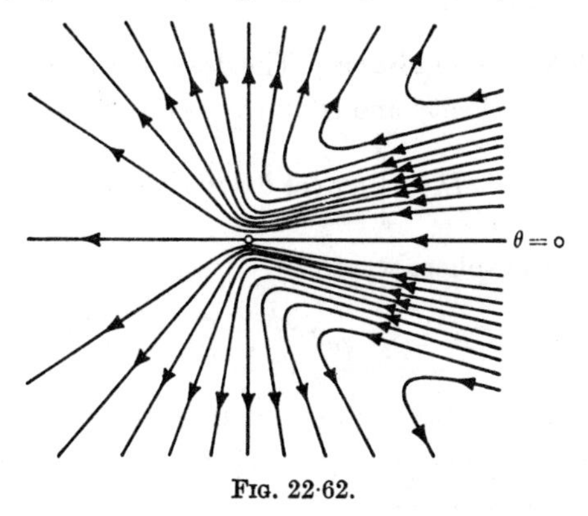

FIG. 22·62.

* S. Goldstein, *Proc. Royal Soc. A.*, **123** (1929), 216–235.

sphere of 22·60 we obtain the case of the sphere moving from right to left in fluid otherwise at rest. The form of the streamlines when $\mathcal{R} = o(1)$ is shown in fig. 22·62.

The flow pattern is not symmetrical about $\theta = \pi/2$ and except for a fairly narrow wake behind the sphere the flow is ultimately radially outwards as from a source of strength, see 22·60 (5) and (6),

$$b_0 = \frac{3Ua^2}{\mathcal{R}}(1+\tfrac{3}{16}\mathcal{R}) \tag{1}$$

placed at the centre of the sphere. In the wake the flow is inwards towards the sphere thus compensating for the outwards source flow.

On the x-axis to the right, $\theta = 0$, we find that

$$u = -\tfrac{3}{2}\frac{Ua}{r}+\ldots \tag{2}$$

so that in the wake the inwards flow has a velocity varying inversely as the first power instead of the square of the distance from the centre which explains the compensation.

As to the vorticity, in cylindrical coordinates, (ϖ, ω, x) we have

$$\zeta = \tfrac{3}{2}\,Ua\,(1+kr)\frac{\varpi}{r^3}\,e^{-k(r-x)}. \tag{3}$$

On account of the exponential factor, ζ is practically insensible except in the region bounded by a paraboloidal surface of revolution for which $k(r-x)$ is $O(1)$. Thus the motion may be described as nearly irrotational except within the wake as vaguely defined above.

The Oseen method can also be applied to find the flow past a circular cylinder by the same method. The Stokes flow for this problem does not exist on account of Stokes' paradox (22·319).

22·63. The drag on a sphere. The drag on the fixed sphere is clearly in the direction $\mathbf{i}$ of the velocity and so in the notation of 21·14 we have

$$D = \int (\widehat{rr}\cos\theta - \widehat{r\theta}\sin\theta)\,dS$$

over the surface of the sphere where

$$\widehat{rr} = -p = -p_0 - \rho U\frac{\partial\phi}{\partial x}, \quad \widehat{r\theta} = \mu\left(\frac{\partial q_\theta}{\partial r}+\frac{1}{r}\frac{\partial q_r}{\partial\theta}\right)$$

external forces being omitted. But $q_r = 0$ on the surface and so $\partial q_r/\partial\theta = 0$. Moreover the constant pressure p_0 gives no resultant force and therefore

$$D = \int_S \left(-\rho U\frac{\partial\phi}{\partial x}\cos\theta - \mu\frac{\partial q_\theta}{\partial r}\sin\theta\right)dS, \quad r = a. \tag{1}$$

Now the vorticity is $\mathbf{i} \wedge \nabla \chi$.

But on the surface of the sphere the vorticity is $\mathbf{i}_\omega\, \partial q_\theta / \partial r$ from 2·72. Therefore

$$\mu \mathbf{i}_\omega\, \partial q_\theta/\partial r = \mu \mathbf{i} \wedge \nabla \chi = \rho U \mathbf{i} \wedge \nabla \phi \tag{2}$$

from 22·52 (1) since at the surface of the sphere $\mathbf{q} = 0$.

Thus $\mu\, \partial q_\theta/\partial r = \rho U(\partial \phi/\partial y) \cos \omega \sin \theta + \rho U (\partial \phi/\partial z) \sin \omega \sin \theta$.

Substituting in (1) we get

$$D = -\rho U \int \frac{\partial \phi}{\partial r}\, dS\,, \quad r = a. \tag{3}$$

Using the expression 22·60 (5) for ϕ and noticing that by the properties of zonal harmonics the only surviving term after integration over the whole surface of the sphere arises from b_0/r, we get

$$\frac{D}{\rho U} = \frac{b_0}{a^2} \times 4\pi a^2 = 4\pi b_0 = \frac{12\pi U a^2}{\mathscr{R}}\,(1 + \tfrac{3}{16}\mathscr{R}) \tag{4}$$

and the drag coefficient is

$$C_D = \frac{24}{\mathscr{R}}\,(1 + \tfrac{3}{16}\mathscr{R}) \tag{5}$$

which may be compared with the value $24/\mathscr{R}$ obtained in 22·20 for Stokes flow.

Since the Stokes and Oseen flows are equivalent in the neighbourhood of the sphere, the results of calculating the drag from either flow should be the same and the result (5) should be illusory. The reason why the result is correct has been shown* to be that reversal of the uniform flow at infinity reverses the force without change of magnitude, which is obviously true for a sphere.

22·66. Expansions at small Reynolds numbers. The problem of obtaining higher approximations to the flow past a sphere and cylinder than those of Stokes and Oseen has been treated by Proudman and Pearson.† The technique adopted is as follows. It is assumed that "Stokes" and "Oseen" expansions of the stream function are respectively of the forms

$$\Sigma f_n(\mathscr{R})\psi_n(r, \theta) \quad \text{and} \quad \Sigma F_n(\mathscr{R})\Psi_n(\mathscr{R}r, \theta)$$

where (r, θ) are spherical (or cylindrical) polar coordinates and f_{n+1}/f_n and F_{n+1}/F_n vanish with $\mathscr{R}$.

Substitution of these expansions in the Navier-Stokes equation then yields a set of differential equations for the coefficients ψ_n and Ψ_n, but only one set of physical boundary conditions is applicable to each expansion (the adherence condition for the Stokes expansion and the uniform-stream condition for the Oseen expansion) so that unique solutions can not be derived immediately.

However, the fact that the two expansions are (in principle) both derived

* W. Chester, "On Oseen's approximation", *Journal of fluid mechanics*, **13** (1962), 557–569.
† *Journal of fluid mechanics*, **2** (1957), 237–262.

from the same exact solution leads to a "matching" procedure which yields further boundary conditions for each expansion. It is thus possible to determine alternately successive terms in each expansion. Two expansions are said to match when they agree to any prescribed order of accuracy.

For details the reader is referred to the original paper.

22·75. Oseen's approximation at a distance. We resume the discussion of the force on an aerofoil started in 21·83. Here some approximation will be involved. In fig. 21·83 the closed surface Σ is arbitrary. Let us take it to be a sphere of radius so large that we can write

$$(1) \qquad \mathbf{q} = \mathbf{V}+\mathbf{v}, \quad p = \Pi+p',$$

where $\mathbf{v}$ and p' are small deviations of the first order from the uniform state of velocity $\mathbf{V}$ and pressure Π. Then the equation of motion assumes Oseen's form (22·51):

$$V\frac{\partial \mathbf{v}}{\partial x} = -\frac{1}{\rho}\nabla p' + \nu\nabla^2\mathbf{v}.$$

For convenience introduce the parameter k, defined by

$$(2) \qquad V = 2k\nu.$$

The equations of motion and continuity then become

$$(3) \qquad \frac{1}{\rho}\nabla p' = \nu\left(\nabla^2 - 2k\frac{\partial}{\partial x}\right)\mathbf{v}, \quad \nabla\mathbf{v} = 0.$$

It follows from (3) that $\nabla^2 p' = 0$, so that p' is a harmonic function, and if we write

$$(4) \qquad p' = \rho V\frac{\partial\phi}{\partial x}, \quad \nabla^2\phi = 0,$$

we get a particular solution of (3), namely

$$(5) \qquad \mathbf{v} = \mathbf{q}_1 = -\nabla\phi.$$

The complete solution will be of the form

$$(6) \qquad \mathbf{v} = \mathbf{q}_1 + \mathbf{v}_2,$$

where $\mathbf{v}_2$ satisfies the equations

$$(7) \qquad \left(\nabla^2 - 2k\frac{\partial}{\partial x}\right)\mathbf{v}_2 = 0, \quad \nabla\mathbf{v}_2 = 0.$$

Let us examine the solution (5) in more detail. Since ultimate vanishing over the distant sphere Σ is required, we should expect the appropriate form of the velocity potential ϕ expressed in spherical polar coordinates r, θ, ω to be a sum of harmonic terms of the type $S_n(\theta, \omega)/r^{n+1}$, and of such terms the dominant one for large r should be

$$\phi_0 = \frac{S_0(\theta, \omega)}{r}.$$

Substitution in 17·1 (2) gives the equation satisfied by S_0, namely

$$\frac{1}{\sin\theta}\frac{\partial}{\partial\theta}\left(\sin\theta\,\frac{\partial S_0}{\partial\theta}\right)+\frac{1}{\sin^2\theta}\frac{\partial^2 S_0}{\partial\omega^2}=0.$$

To solve this equation write

$$u = \log\tan\tfrac{1}{2}\theta, \quad \text{and} \quad S_0 = f_m(u)\cos m\omega \quad \text{or} \quad f_m(u)\sin m\omega.$$

We then get

$$\frac{d^2 f_m}{du^2}-m^2 f_m = 0,$$

whence $\quad f_m = A_m e^{-mu}+B_m e^{mu} = A_m(\cot\tfrac{1}{2}\theta)^m+B_m(\tan\tfrac{1}{2}\theta)^m.$

The second term $\to\infty$ when $\theta\to\pi$, i.e. upstream, and this is clearly unsuitable. We therefore take $B_m = 0$.

On the other hand, the first term $\to\infty$ when $\theta\to 0$, i.e. in the wake. At first sight this seems to demand $A_m = 0$ also, but it will appear from the calculations which follow that if $A_m = 0$, there will be no lift.

Confining our attention to the case $m = 1$, we then get the particular potential

$$(8) \qquad \phi_{10} = -\frac{1}{r}\cot\tfrac{1}{2}\theta(\alpha\cos\omega+\beta\sin\omega),$$

and the velocity $\mathbf{q}_{10}$ derived from this becomes infinite when $\theta = 0$. To remove this infinity we write $\mathbf{v}_2 = \mathbf{q}_2+\mathbf{v}_3$, where $\mathbf{q}_2$ is chosen to satisfy

$$(9) \qquad \left(\nabla^2-2k\frac{\partial}{\partial x}\right)\mathbf{q}_2 = 0, \quad \mathbf{q}_2 = -\nabla\psi.$$

We then seek to determine the potential ψ to satisfy

$$\left(\nabla^2-2k\frac{\partial}{\partial x}\right)\psi = 0 \quad \text{or} \quad (\nabla^2-k^2)(e^{-kx}\psi) = 0.$$

It is readily verified that this equation has the particular solution

$$(10) \qquad e^{-kx}\psi = \frac{e^{-kr}}{r}\cot\tfrac{1}{2}\theta(\alpha\cos\omega+\beta\sin\omega), \quad x = r\cos\theta,$$

and the combination

$$(11) \qquad \phi_{10}+\psi = -\frac{1-e^{-kr(1-\cos\theta)}}{r}\cot\tfrac{1}{2}\theta(\alpha\cos\omega+\beta\sin\omega)$$

has no infinity when $\theta = 0$, since the term $\operatorname{cosec}\tfrac{1}{2}\theta$ which causes it can be cancelled. But $\mathbf{q}_2$ determined from (9) does not satisfy the equation of continuity, for

$$\nabla\,\mathbf{q}_2 = -\nabla^2\psi = -2k\frac{\partial\psi}{\partial x} \quad \text{from (9).}$$

We therefore add a further velocity $\mathbf{q}_3$ which satisfies

$$\left(\nabla^2 - 2k\frac{\partial}{\partial x}\right)\mathbf{q}_3 = 0, \quad \text{and} \quad \nabla\,\mathbf{q}_3 = 2k\frac{\partial\psi}{\partial x},$$

so that $\nabla(\mathbf{q}_2+\mathbf{q}_3) = 0$. Now the assumption

$$\mathbf{q}_3 = \frac{e^{-k(r-x)}}{r}(\alpha'\mathbf{j}+\beta'\mathbf{k}) \quad \text{gives} \quad \nabla\,\mathbf{q}_3 = -\frac{e^{-k(r-x)}}{r^3}(1+kr)(\alpha' y+\beta' z).$$

Comparing this with $2k\partial\psi/\partial x$ derived from (10) we get $\alpha' = -2k\alpha$, $\beta' = -2k\beta$, and the appropriate velocity is

$$\mathbf{q}_3 = -\frac{2k}{r}e^{-k(r-x)}(\alpha\mathbf{j}+\beta\mathbf{k}), \tag{12}$$

and we note that $\mathbf{q}_3$ is perpendicular to $\mathbf{V}$, so that $\mathbf{V}\mathbf{q}_3 = 0$.

The complete solution built up on these lines is of the form

$$\mathbf{q} = \mathbf{V}+\mathbf{q}_1+\mathbf{q}_2+\mathbf{q}_3+\mathbf{q}_4. \tag{13}$$

Here $\mathbf{q}_1$ is the irrotational solution associated with the pressure (equation (4)), and includes the particular term $\mathbf{q}_{10}$ calculated from ϕ_{10} (equation (8)), which becomes infinite when $\theta = 0$; $\mathbf{q}_2$ is the special solution $-\nabla\psi$ where ψ is given by (10), so chosen that along the wake ($\theta = 0$) the infinities of ϕ_{10} and ψ cancel one another; $\mathbf{q}_3$ is a further special solution chosen to ensure that the equation of continuity $\nabla(\mathbf{q}_2+\mathbf{q}_3) = 0$ is satisfied; $\mathbf{q}_4$ is the complementary solution which satisfies the Oseen equation (7) and the equation of continuity.

The velocity $\mathbf{q}$ given by (13) is finite and continuous over the whole surface of the sphere Σ.

The presence of the exponential factor shows that $\mathbf{q}_3$ is negligible unless $r-x$ is small, that is to say in the wake, which at a distance is bounded rather vaguely by a paraboloidal surface $r-x = \epsilon$, where ϵ is a small constant. The vorticity arises only from $\nabla_\wedge\mathbf{q}_3$ and $\nabla_\wedge\mathbf{q}_4$ since $\nabla_\wedge\mathbf{q}_1 = \nabla_\wedge\mathbf{q}_2 = 0$. It will appear presently that the value of $\nabla_\wedge\mathbf{q}_4$ does not affect the force on the aerofoil, so that the vorticity is effectively confined to the wake.

It should be emphasised that the above method of approximation concerns only the distant sphere. It has nothing to say concerning the flow in the neighbourhood of the aerofoil.

22·76. Lift and drag.

From 21·83 (3) the force on the aerofoil is

$$\mathbf{F} = \mathbf{P}+\mathbf{Q}+\mathbf{R},$$

where

$$\mathbf{P} = -\int_{(\Sigma)} p\mathbf{n}\,dS, \quad \mathbf{Q} = -\int_{(\Sigma)} \mu\mathbf{n}_\wedge\boldsymbol{\zeta}\,dS, \quad \mathbf{R} = -\int_{(\Sigma)} \rho\mathbf{q}(\mathbf{n}\mathbf{q})\,dS. \tag{1}$$

From 22·75 (1), (4), (13), we have

$$p = \Pi + p' = \Pi + \rho V \frac{\partial \phi}{\partial x} = \Pi - \rho \mathbf{V}\mathbf{q}_1, \quad -p = -\Pi + \rho \mathbf{V}\mathbf{v} - \rho \mathbf{V}\mathbf{q}_2 - \rho \mathbf{V}\mathbf{q}_4,$$

where $\mathbf{v} = \mathbf{q}_1 + \mathbf{q}_2 + \mathbf{q}_3 + \mathbf{q}_4$ and $\mathbf{V}\mathbf{q}_3 = 0.$

Since $\int_{(\Sigma)} \mathbf{n}\, dS = 0,$ we get

$$\mathbf{P} = \rho \int_{(\Sigma)} \mathbf{n}(\mathbf{V}\mathbf{v}) - \rho \int_{(\Sigma)} \mathbf{n}(\mathbf{V}\mathbf{q}_2) dS - \rho \int_{(\Sigma)} \mathbf{n}(\mathbf{V}\mathbf{q}_4) dS.$$

Also $\mathbf{q}(\mathbf{n}\mathbf{q}) = \mathbf{V}(\mathbf{n}\mathbf{V}) + \mathbf{V}(\mathbf{n}\mathbf{v}) + \mathbf{v}(\mathbf{n}\mathbf{V})$

to the first order and $\int_{(\Sigma)} \mathbf{n}\mathbf{v}\, dS = 0$ by the equation of continuity. Therefore

$$(2) \quad \mathbf{P} + \mathbf{R} = \rho \int_{(\Sigma)} \mathbf{V}_\wedge (\mathbf{n}_\wedge \mathbf{v}) dS - \rho \mathbf{V} \int_{(\Sigma)} (\mathbf{n}\mathbf{q}_4) dS + \rho \int_{(\Sigma)} \mathbf{q}_{4\wedge} (\mathbf{V}_\wedge \mathbf{n}) dS - \rho \int_{(\Sigma)} \mathbf{n}(\mathbf{V}\mathbf{q}_2) dS.$$

Let

$$(3) \qquad I = -\int_{(\Sigma)} \rho(\mathbf{n}\mathbf{q}_4) dS.$$

Then I is the inflow into Σ from the complementary solution ; predominantly an inflow into the wake.

Again the vector circulation over Σ is

$$\boldsymbol{\Gamma} = \int_{(\Sigma)} \mathbf{n}_\wedge \mathbf{v}\, dS = \int_{(\Sigma)} \mathbf{n}_\wedge \mathbf{q}_1 dS + \int_{(\Sigma)} \mathbf{n}_\wedge \mathbf{q}_2 dS + \int_{(\Sigma)} \mathbf{n}_\wedge \mathbf{q}_3 dS + \int_{(\Sigma)} \mathbf{n}_\wedge \mathbf{q}_4 dS.$$

The first two integrals on the right are vector circulations due to the irrotational velocities $\mathbf{q}_1$ and $\mathbf{q}_2$ and must therefore vanish (21·70), and so we can write $\boldsymbol{\Gamma} = \boldsymbol{\Gamma}_3 + \boldsymbol{\Gamma}_4$, where $\boldsymbol{\Gamma}_3$ and $\boldsymbol{\Gamma}_4$ are the vector circulations due to $\mathbf{q}_3$ and $\mathbf{q}_4$. Lastly, put $\boldsymbol{\zeta}_3 = \nabla_\wedge \mathbf{q}_3$, $\boldsymbol{\zeta}_4 = \nabla_\wedge \mathbf{q}_4$, then from (1), (2) and (3)

$$(4) \qquad \mathbf{F} = \rho \mathbf{V}_\wedge \boldsymbol{\Gamma}_3 + \mathbf{V} I + \mathbf{F}' + \mathbf{F}'', \text{ where}$$

$$(5) \qquad \mathbf{F}' = -\rho \int_{(\Sigma)} \mathbf{n}(\mathbf{V}\mathbf{q}_2) dS - \int_{(\Sigma)} \mu \mathbf{n}_\wedge \boldsymbol{\zeta}_3 dS,$$

$$\mathbf{F}'' = \rho \mathbf{V}_\wedge \boldsymbol{\Gamma}_4 + \rho \int_{(\Sigma)} \mathbf{q}_{4\wedge} (\mathbf{V}_\wedge \mathbf{n}) dS - \int_{(\Sigma)} \mu \mathbf{n}_\wedge \boldsymbol{\zeta}_4 dS.$$

We now prove that $\mathbf{F}'' = 0$. Since $\nabla \mathbf{q}_4 = 0$, 2·32 (IV) gives

$$\nabla_\wedge \boldsymbol{\zeta}_4 = \nabla_\wedge (\nabla_\wedge \mathbf{q}_4) = -\nabla^2 \mathbf{q}_4, \quad \text{and} \quad \nabla_\wedge (\mathbf{q}_{4\wedge} \mathbf{i}) = (\mathbf{i}\, \nabla) \mathbf{q}_4 = \frac{\partial \mathbf{q}_4}{\partial x},$$

and therefore, since $\mathbf{q}_4$ satisfies 22·75 (7),

$$\nabla_\wedge \{\boldsymbol{\zeta}_4 + 2k \mathbf{q}_{4\wedge} \mathbf{i}\} = 0.$$

Therefore there exists a scalar function Z such that

$$\boldsymbol{\zeta}_4 + 2k \mathbf{q}_{4\wedge} \mathbf{i} = \nabla Z.$$

Therefore

$$\int_{(\Sigma)} \mu \mathbf{n}_\wedge \boldsymbol{\zeta}_4 dS = \int_{(\Sigma)} \mu(\mathbf{n}_\wedge \nabla) Z\, dS - 2k\mu \int_{(\Sigma)} \mathbf{n}_\wedge (\mathbf{q}_{4\wedge}\mathbf{i}) dS.$$

The first integral on the right vanishes by 21·70 (5) and $2k\mu = \rho V$ by 22·75 (2). Therefore

$$\int_{(\Sigma)} \mu \mathbf{n}_\wedge \boldsymbol{\zeta}_4 dS = -\rho \int_{(\Sigma)} \mathbf{n}_\wedge (\mathbf{q}_{4\wedge}\mathbf{V}) dS, \text{ and so}$$

$$\rho \int_{(\Sigma)} \mathbf{q}_{4\wedge}(\mathbf{V}_\wedge \mathbf{n}) dS - \int_{(\Sigma)} \mu \mathbf{n}_\wedge \boldsymbol{\zeta}_4\, dS = -\rho \mathbf{V}_\wedge \int_{(\Sigma)} \mathbf{n}_\wedge \mathbf{q}_4 dS = -\rho \mathbf{V}_\wedge \boldsymbol{\Gamma}_4,$$

and therefore $\mathbf{F}'' = 0$.

Returning to (5) we can show that $\mathbf{F}' \to 0$ as the radius of Σ tends to infinity. This is a simple consequence of the expressions for $\mathbf{q}_2$ and $\mathbf{q}_3$ given in 22·75 and is left as an exercise.

We now get from (4) the asymptotic result $\mathbf{F} = \mathbf{L} + \mathbf{D}$ where

$$(6) \qquad \mathbf{L} = \rho \mathbf{V}_\wedge \boldsymbol{\Gamma}_3, \quad \mathbf{D} = \mathbf{V} I.$$

Thus $\mathbf{L}$ is perpendicular to $\mathbf{V}$ and is therefore a lift, whilst $\mathbf{D}$ is the drag. These results improve in accuracy the greater the radius of the sphere Σ and constitute the generalisation of the Kutta-Joukowski theorem for an inviscid and Filon's formula * for a viscous liquid in two-dimensional motion. Here $\boldsymbol{\Gamma}_3$ is the vector circulation over Σ due to $\mathbf{q}_3$, and I is the inflow into the wake due to $\mathbf{q}_4$.

To simplify $\boldsymbol{\Gamma}_3$ write $\mathbf{q}_3 = v_3 \mathbf{j} + w_3 \mathbf{k}$. Then

$$\begin{aligned} \boldsymbol{\Gamma}_3 &= 2\pi r^2 \int_0^\pi (v_3 \mathbf{k} - w_3 \mathbf{j}) \sin\theta \cos\theta \, d\theta \\ &= 4\pi k r (\beta \mathbf{j} - \alpha \mathbf{k}) e^{-kr} \int_{-1}^{1} e^{kru} u\, du, \quad u = \cos\theta \\ &= 4\pi(\beta \mathbf{j} - \alpha \mathbf{k})\{1 + e^{-2kr} - (1 - e^{-2kr})/(kr)\}. \end{aligned}$$

Hence as $r \to \infty$, $\boldsymbol{\Gamma}_3 \to 4\pi(\beta\mathbf{j} - \alpha\mathbf{k})$, which gives rise to the lift

$$\mathbf{L} = \rho \mathbf{V}_\wedge \boldsymbol{\Gamma}_3 = 4\pi\rho V(\alpha \mathbf{j} + \beta \mathbf{k}).$$

This result justifies the statement made in 22·75 that there can be no lift when α and β are both zero.

It can be shown from the above results that the components of the lift are associated with circulations in the distant circuits in which the sphere Σ is cut by the diametral planes $\omega = 0$ and $\omega = \frac{1}{2}\pi$. There will be zero circulation in any circuit surrounding but not threading the wake.†

---

* L. N. G. Filon, " Forces on a cylinder ", *Proc. Roy. Soc.* (A), 113 (1926).

† Cf. T. E. Garstang, *Phil. Trans. Roy. Soc.* (*A*) 236 (1936), p. 25.

EXAMPLES XXII

1. Verify that the velocity

$$\vec{q} = A \operatorname{grad} \frac{\partial}{\partial x}(1/r) + Bx \operatorname{grad}(1/r) + \{U - B/r, 0, 0\}$$

satisfies the equations of slow steady motion of an incompressible viscous liquid (neglecting the so-called inertia terms).

Determine the constants A and B in order that the solution is applicable to the streaming of infinite liquid past the fixed solid sphere $x^2+y^2+z^2 = a^2$, the velocity of the stream at a distance being $(U, 0, 0)$, and show that the stream exerts a force $6\pi\mu aU$ upon the sphere in the direction of the stream. (U.L.)

2. A sphere of radius a is fixed with its centre at the origin O in viscous liquid, whose velocity at infinity is U parallel to Ox. Verify that when the "inertia terms" are neglected altogether, the equations of motion and boundary conditions are satisfied by the velocity components

$$u = U\left\{1 - \frac{a}{r} + \frac{a}{4r^3}\left(1 - \frac{a^2}{r^2}\right)(r^2 - 3x^2)\right\},$$

$$v = -\frac{3Ua}{4r^3}\left(1 - \frac{a^2}{r^2}\right)xy, \quad w = -\frac{3Ua}{4r^3}\left(1 - \frac{a^2}{r^2}\right)xz,$$

provided that the pressure p is properly determined. Find the resultant force on the sphere. (U.L.)

3. Two infinite circular cylinders, of radii a, a', are rotating with uniform angular velocities ω, ω', so as always to be in contact along the z-axis, and are surrounded by viscous incompressible liquid of density ρ and viscosity coefficient μ. Neglecting inertia terms, prove that all the necessary conditions can be satisfied by a stream function of the form

$$\psi = A\sin^2\theta + Br^2 + Cr\sin\theta + \frac{D\sin^3\theta}{r},$$

and determine the constants A, B, C, D, when the equations of the norma lsections of the cylinders are

$$r = 2a\sin\theta, \quad r = -2a'\sin\theta,$$

where (z, r, θ) are cylindrical coordinates.

Examine the case $a' = a$, $\omega' = -\omega$ more particularly, showing that then

$$\psi = \tfrac{1}{2}\omega a\left(r - \frac{4a^2\sin^2\theta}{r}\right)\sin\theta;$$

determine the stress components p_{rr}, $p_{\theta\theta}$, $p_{r\theta}$, so far as they depend on μ, and deduce the tangential stress on one of the cylinders, noting any peculiarity in your result and discussing its bearing on the validity of the solution. (U.L.)

4. Discuss the approximations made by Oseen in the mathematical discussion of the flow of viscous fluid past a fixed obstacle at small Reynolds numbers. Find the equation satisfied by the vorticity on Oseen's theory and explain its physical significance.

Verify that, for two-dimensional flow past a cylinder of any section, the equations of motion and continuity are satisfied by

$$u = \frac{\partial\phi}{\partial x} + \frac{1}{2k}\frac{\partial\chi}{\partial x} - \chi, \quad v = \frac{\partial\phi}{\partial y} + \frac{1}{2k}\frac{\partial\chi}{\partial y}, \quad p = -\rho U\frac{\partial\phi}{\partial x},$$

where $k = U/(2\nu)$, U is the undisturbed velocity of the stream and is in the direction of the axis of x, ν is the kinematic viscosity and

$$\nabla^2\phi = 0, \quad \left(\nabla^2 - 2k\frac{\partial}{\partial x}\right)\chi = 0, \quad \nabla^2 \equiv \frac{\partial^2}{\partial x^2} + \frac{\partial^2}{\partial y^2}.$$

Assuming that solutions for ϕ, χ can be found that make u, v vanish at the surface of the cylinder, prove that the drag force on the cylinder per unit length is

$$\rho U \int \frac{\partial\phi}{\partial n}\, ds,$$

taken round the boundary of the cylinder and dn is an element of outward normal. (U.L.)

5. State the arguments by which Oseen reduced the equations of motion of a viscous liquid, moving at a great distance from a fixed solid body with velocity U parallel to Ox, to the form

$$U\frac{\partial}{\partial x}(u, v, w) = -\frac{1}{\rho}\nabla p + \nu\nabla^2(u, v, w).$$

Writing $U = 2\nu k$, verify that these equations are satisfied by

$$u = \frac{\partial\phi}{\partial x} + \frac{1}{2k}\frac{\partial\chi}{\partial x} - \chi, \quad v = \frac{\partial\phi}{\partial y} + \frac{1}{2k}\frac{\partial\chi}{\partial y}, \quad w = \frac{\partial\phi}{\partial z} + \frac{1}{2k}\frac{\partial\chi}{\partial z},$$

where
$$\nabla^2\phi = 0, \quad \left(\nabla^2 - 2k\frac{\partial}{\partial x}\right)\chi = 0.$$

Discuss the solution

$$\phi = \sum_{n=0}^{\infty} A_n S_n r^{-n-1},$$

$$= e^{kx}\sum_{n=0}^{\infty}(2n+1)B_n S_n (\pi/2kr)^{\frac{1}{2}} K_{n+\frac{1}{2}}(kr),$$

where S_n is a spherical surface harmonic of order n and $K_{n+\frac{1}{2}}$ is the Bessel function of the second kind of half integral order.

Explain how the terms involving χ account for the wake behind the solid. (U.L.)

6. In Oseen's approximation, if ϕ is the velocity potential of the irrotational motion outside the wake, prove that the drag on any body of revolution, with its axis parallel to a steady stream of liquid of density ρ, is

$$\rho U \int \frac{\partial\phi}{\partial n}\, dS,$$

where the integral is over the surface of the body and U is the undisturbed velocity of the stream. (M.T.)

7. A sphere of radius a moves with constant velocity U along the x-axis through a viscous liquid at rest at infinity. Verify that, on Oseen's hypothesis, the stream function is

$$\psi = \frac{Ua^3\sin^2\theta}{4r} - \frac{3\nu a}{2}(1+\cos\theta)(1-e^{-k(r-x)}), \quad k = \frac{U}{2\nu}.$$

CHAPTER XXIII

BOUNDARY LAYERS

23·10. Introduction. In this chapter we take up and amplify some of the questions adumbrated in 1·93. We shall confine our attention to the two-dimensional flow of a homogeneous viscous incompressible fluid.

The problems of the uniform rectilinear motion of a body through liquid at rest at infinity and that of a stream uniform at infinity disturbed by the introduction of the same body held at rest are dynamically equivalent. The force experienced by the body in the first case is equal in magnitude and opposite in direction to the force required to hold the body at rest in the second case. To fix our ideas we shall therefore consider the case where the fixed body disturbs the uniform stream. The force exerted on the body by the stream will be referred to as the *drag*.

The theorem known as d'Alembert's paradox, 1·9, asserts that when the liquid is inviscid, the drag is zero. As inviscid liquids exist only in thought, we must consider the effect of the viscosity exhibited by real liquids. We shall consider the case where the kinematic viscosity ν is small, or more exactly where the Reynolds number $\mathscr{R} = Ul/\nu$ is large, but not so large as to cause turbulence.

It is a matter of common observation that in the case of a real fluid the drag is never zero and, for a given body, will be greater or less according to the body's presentation to the stream. In practice the problem of minimizing the drag is important. A simple illustration is afforded by a properly designed aerofoil which is so shaped as to present a small drag in a certain limited range of incidence and stream velocity. Photographs show that the flow pattern contains streamlines which hug the body closely and a narrow wake behind the trailing edge. The aerofoil is then said to be *streamlined* or to be of *easy shape*. When the incidence is increased well beyond the designer's range, " broadside on " for example, the streamlines cease to hug the body at some point upstream of the trailing edge. The flow is then said to have *separated* (from the body) and a broad wake of turbulent flow ensues, accompanied, in general, by increased drag.

These commonly observed facts led, even before the beginning of this century, to the view that classical hydrodynamics was a useless subject in so far as it was unable to yield physical predictions of drag due to viscosity. Relief was however at hand. In 1904 Prandtl* enunciated a theory which qualitatively settled the whole question.

Applied to a body of easy shape, at large Reynolds number, Prandtl's idea was substantially as follows. The flow past the body is practically the potential flow of classical hydrodynamics except that between this flow and the surface of the body there exists a *thin layer*, the *boundary layer*, in which the fluid velocity rapidly adjusts from zero at the surface of the body (viscous adherence) to the velocity of the potential flow just outside the layer. Within the layer the gradient normal to the surface of the body of the tangential component of velocity will be large and the viscous shearing stress will exert an appreciable effect on the flow.

In this statement there is an implied approximation process, namely that in the flow outside the thin boundary layer we can neglect the viscosity, while in the boundary layer certain quantities are so large that others may be neglected in comparison.

Thus at one blow Prandtl rehabilitated the classical theory of hydrodynamics by replacing the problem of flow past a body of easy shape by two problems (i) the viscous flow in the thin boundary layer (ii) the classical inviscid flow about a new body consisting of the original body enhanced by the thickness of a boundary layer (and possibly a wake). Although this statement is an oversimplification, the new role of the classical theory is evident. It is concerning the first of the above problems that this chapter gives a brief and limited account.

23·11. Flow into a stagnation point. Consider the inviscid flow whose complex potential is $w = -\frac{1}{2}Uz^2/a$. The velocity in this potential flow is

$$(1) \qquad u_p = Ux/a, \quad v_p = -Uy/a$$

which represents flow in the negative direction of the y-axis, against a flat plate, $y = 0$, towards the stagnation point $(0, 0)$, fig. 4·70.

We consider the problem of finding a viscous steady flow with the same general structure. If, guided by (1), we put

$$(2) \qquad v = -f(y),$$

the equation of continuity $\partial u/\partial x + \partial v/\partial y = 0$ demands that

$$(3) \qquad u = xf'(y).$$

The Navier-Stokes equations for the steady flow are

* L. Prandtl, Ueber Flüssigkeitsbewegung bei sehr kleiner Reibung. *Proc. third int. math. Congress* Heidelberg, 1904.

$$u\frac{\partial u}{\partial x}+v\frac{\partial u}{\partial y} = -\frac{1}{\rho}\frac{\partial p}{\partial x}+\nu\left(\frac{\partial^2 u}{\partial x^2}+\frac{\partial^2 u}{\partial y^2}\right),$$

$$u\frac{\partial v}{\partial x}+v\frac{\partial v}{\partial y} = -\frac{1}{\rho}\frac{\partial p}{\partial y}+\nu\left(\frac{\partial^2 v}{\partial x^2}+\frac{\partial^2 v}{\partial y^2}\right),$$

which give on substituting (2) and (3)

$$(4) \qquad xf'^2 - xff'' = -\frac{1}{\rho}\frac{\partial p}{\partial x}+\nu xf''',$$

$$(5) \qquad ff' = -\frac{1}{\rho}\frac{\partial p}{\partial y} - \nu f''.$$

Integrating (5) we get

$$(6) \qquad \frac{p}{\rho} = -\nu f' - \tfrac{1}{2}f^2 + g(x)$$

where $g(x)$ is independent of y. Substitution in (4) gives

$$(7) \qquad \frac{g'(x)}{x} = \nu f''' + ff'' - f'^2 = \text{constant} = \frac{-U^2}{a^2},$$

since the first member is independent of y and the second is independent of x. The choice of constant is guided by the fact that in the inviscid flow (1),

$$f''' = f'' = 0, \quad f' = U/a.$$

The result is consistent with the following boundary conditions for the viscous flow

$$(8) \qquad f(0) = f'(0) = 0, \quad f'(\infty) = U/a.$$

Equation (7) then gives $g(x) = -\tfrac{1}{2}x^2 U^2/a^2 + \text{constant}$, and therefore from (6)

$$(9) \qquad \frac{p-p_0}{\rho} = -\nu f' - \tfrac{1}{2}f^2 - \tfrac{1}{2}x^2\frac{U^2}{a^2},$$

where p_0 is the pressure at the stagnation point (0, 0).

Thus (9) determines the pressure when $f(y)$ is known. Since ν has the dimensions of a velocity times a length we can introduce dimensionless numbers η and $F(\eta)$ defined by

$$(10) \qquad \eta = y\sqrt{\frac{U}{a\nu}}, \quad f(y) = F(\eta)\sqrt{\frac{\nu U}{a}},$$

and then (7) gives

$$(11) \qquad F''' + FF'' + 1 - F'^2 = 0$$

with boundary conditions

$$(12) \qquad F(0) = F'(0) = 0, \quad F'(\infty) = 1.$$

This ordinary differential equation can be solved numerically. We give a short 2-decimal table

η	0·0	0·4	0·8	1·2	1·6	2·0	2·4	2·8	3·2
F	0·00	·09	·31	·62	·98	1·36	1·76	2·15	2·55
F'	0·00	·41	·69	·85	·93	·97	·99	1·00	1·00
F''	1·23	·85	·53	·29	·15	·07	·03	·01	·00
$\eta - F$	0·00	·31	·49	·58	·62	·64	·64	·65	·65

From the numerical calculations it appears that

(13) $F'(\eta) > 0{\cdot}99$ when $\eta > 2{\cdot}4$, and $\eta - F \to 0{\cdot}65$ when $\eta \to \infty$.

Comparing then the inviscid flow given by (1) with the viscous flow given by (2) and (3) we have

$$\frac{u_p - u}{u_p} = 1 - F', \quad \frac{v - v_p}{U} = \frac{\eta - F}{\sqrt{\mathscr{R}}} \text{ so that}$$

(14) $$\frac{u_p - u}{u_p} < 0{\cdot}01 \text{ when } y > 2{\cdot}4a/\sqrt{\mathscr{R}}$$

$$v < v_p + 0{\cdot}65U/\sqrt{\mathscr{R}} \text{ for all } y.$$

These inequalities show that, for large $\mathscr{R}$, u attains more than 99 per cent. of the value of u_p outside the layer bounded above by

(15) $$y = \delta = 2{\cdot}4a/\sqrt{\mathscr{R}} = k\sqrt{\nu}, \quad k \text{ a constant,}$$

while outside this layer v and v_p scarcely differ if $\mathscr{R}$ is large.

Thus the thickness of the layer is proportional to $\sqrt{\nu}$.

As a numerical example take a = 25 cm., U = 400 cm./sec. and

$$\nu = 0{\cdot}01 \text{ cm.}^2/\text{sec.}$$

Then $\mathscr{R} = Ua/\nu = 10^6$ and $\delta = 0{\cdot}6$ mm.

Thus when $\mathscr{R}$ is large the potential flow accurately represents the motion at quite a short distance from the wall in accordance with Prandtl's assumption.

Also from (1), (2), (10), (14),

(16) $$v \leqslant -\frac{U}{a}(y - 0{\cdot}65a/\sqrt{\mathscr{R}}),$$

while from (1) the velocity on the right of (16) occurs in the potential flow for

(17) $$y_p = y - 0{\cdot}65a/\sqrt{\mathscr{R}}.$$

Thus when $\mathscr{R}$ is sufficiently large, the thin boundary layer just found causes a general shift upwards of the potential flow streamlines by the distance

$$0{\cdot}65a/\sqrt{\mathscr{R}}.$$

This shift is called the *displacement thickness* (see 23·24). Assuming that $F(\eta)$ is known (as indeed it is to any prescribed degree of accuracy) the foregoing is the exact solution for flow into a stagnation point and the existence of the boundary layer is established in this case.

The existence of the boundary layer just proved refers to a case of steady motion. A boundary layer can also exist, but with thickness increasing with time, in a case of time dependent flow as will be exemplified in section 23·16.

23·16. Stokes' problem. An infinite rigid plate immersed in unbounded liquid at rest is instantaneously given a velocity U parallel to itself.

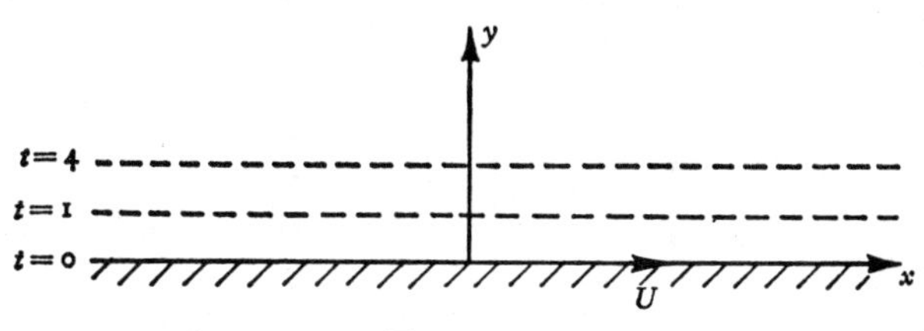

FIG. 23·16.

Consider two-dimensional motion in which the plate coincides with $y = 0$ at time $t = 0$. Clearly it is only necessary to consider the flow for $y \geqslant 0$, for the motion in the lower half-plane would be the reflection, in the plane $y = 0$, of the motion in the upper half-plane.

Let (u, v, w) be the velocity of the liquid at time t. Then the boundary conditions may be taken to be

$$(1) \qquad u = 0 \text{ for } t = 0,\ y>0\,;\ v = 0 \text{ and } w = 0 \text{ for all } y \text{ and } t$$
$$u = U \text{ for } t>0,\ y = 0.$$

This is Problem III of 21·60 and in terms of the *error function* erf α the solution from 21·60 (14) is

$$(2) \qquad u = U\left(1 - \operatorname{erf}\frac{y}{2\sqrt{(\nu t)}}\right), \quad \operatorname{erf}\alpha = \frac{2}{\sqrt{\pi}}\int_0^\alpha e^{-\beta^2}\,d\beta.$$

If we define boundary layer thickness by δ_t as the normal distance from the plate at which u falls to within one per cent. of its value at the plate at time t we have

$$\tfrac{1}{100} = 1 - \operatorname{erf}\left(\tfrac{1}{2}\delta_t/(\nu t)^{1/2}\right)$$

whence from a table of erf x*

$$(3) \qquad \delta_t = 4(\nu t)^{1/2}.$$

Thus the thickness at time $t = 4$ is double that at time $t = 1$ and the

* Milne-Thomson and Comrie, *Standard four-figure mathematical tables* (Macmillan).

thickness grows indefinitely as t increases according to the parabolic law

$$\delta_t^2 = 16\nu t.$$

This investigation proves the existence of a boundary layer in this case of time dependent flow.

Observe that in this and the problem of the preceding section we have solved the full equations of motion and so the solutions are not only exact but apply to all values of ν small or large.

23·20. The boundary layer equations for flow past a straight wall. We consider two-dimensional flow of a liquid in the x, y plane, the wall coinciding with $y = 0$.

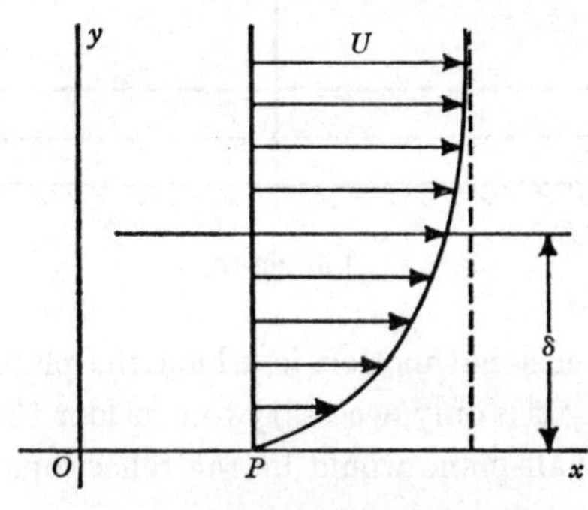

Fig. 23·20.

If we draw an ordinate at the point P of the wall and mark vectors to show the x-component u of the velocity at points of the ordinate, the extremities of these vectors will lie on a curve called the *u-profile*, fig. 23·20. The boundary layer hypothesis demands that the u-profile shall have an asymptote parallel to the ordinate at P. We call U the *slip velocity*, i.e. the stream velocity which would obtain at P were there no viscosity. Let δ be the thickness of the boundary layer which is here again temporarily defined as the distance from the wall at which u has attained a certain percentage, say 99 per cent. of U. Various definitions of this type may be given each leading to a different measure δ of the thickness but all of the same order of magnitude (cf. the numerical illustration in 23·11).

Since $u = 0$ at the wall and rises nearly to the value U in the small distance δ, U/δ and therefore $\partial u/\partial y$ will be great as y increases from 0 to δ within the layer. On the other hand the transverse component v will be small throughout.

Introduce the variable η defined by $y = \eta\delta$.

Then η will be a dimensionless variable comparable as regards order of magnitude with the variable x. The equation of continuity $\partial u/\partial x + \partial v/\partial y = 0$ then gives

$$v = -\left[\int_0^\eta \frac{\partial u}{\partial x}\, d\eta\right]\delta$$

and therefore we can write

$$(1) \qquad v = v_0 \delta$$

where v_0 is comparable in magnitude with u.

The equations of motion then become

$$(2) \qquad \frac{\partial u}{\partial t} + u\frac{\partial u}{\partial x} + v_0\frac{\partial u}{\partial \eta} = -\frac{1}{\rho}\frac{\partial p}{\partial x} + \nu\frac{\partial^2 u}{\partial x^2} + \frac{\nu}{\delta^2}\frac{\partial^2 u}{\partial \eta^2}.$$

$$(3) \qquad \frac{\partial v_0}{\partial t}\delta + \frac{\partial v_0}{\partial x}u\delta + \frac{\partial v_0}{\partial \eta}v_0\delta = -\frac{1}{\rho\delta}\frac{\partial p}{\partial \eta} + (\nu\delta)\frac{\partial^2 v_0}{\partial x^2} + \frac{\nu}{\delta}\frac{\partial^2 v_0}{\partial \eta^2}.$$

Prandtl's assumption. In the boundary layer the inertia term $(\mathbf{q}\nabla)\mathbf{q}$ is of the same order of magnitude as the friction term $\nu\nabla^2\mathbf{q}$.

Apply this to equation (2). Here u, v_0, are of the same order of magnitude and so are x and η. Therefore on the right hand side $\nu\partial^2 u/\partial x^2$ is negligible compared with $\nu(\partial^2 u/\partial\eta^2)/\delta^2$ which is therefore of the same order of magnitude as $u\,\partial u/\partial x$. Thus ν varies as δ^2. Therefore δ varies as $\sqrt{\nu}$ that is to say the thickness of the boundary layer is proportional to the square root of the kinematic viscosity. From equation (3) it now follows that

$$(4) \qquad \frac{\partial p}{\partial y} = 0 \text{ in the boundary layer.}$$

Thus the pressure does not vary in passing normally through the boundary layer.

The equations therefore reduce to (4), and

$$(5) \qquad \frac{\partial u}{\partial t} + u\frac{\partial u}{\partial x} + v\frac{\partial u}{\partial y} = -\frac{1}{\rho}\frac{\partial p}{\partial x} + \nu\frac{\partial^2 u}{\partial y^2},$$

together with the equation of continuity

$$(6) \qquad \frac{\partial u}{\partial x} + \frac{\partial v}{\partial y} = 0.$$

Equations (4), (5), (6), together with the appropriate boundary conditions, determine the flow in the boundary layer.

23·22. Flow along a curved wall. We take as coordinates distance x measured along the wall from a fixed point on it, and distance y measured along the normal to the wall.

Let the normal to the wall which passes through $P(x, y)$ meet the wall at the point M where the curvature is κ. Let C be the centre of curvature and let PN be perpendicular to the adjacent radius CM' through $P'(x+dx, y+dy)$.

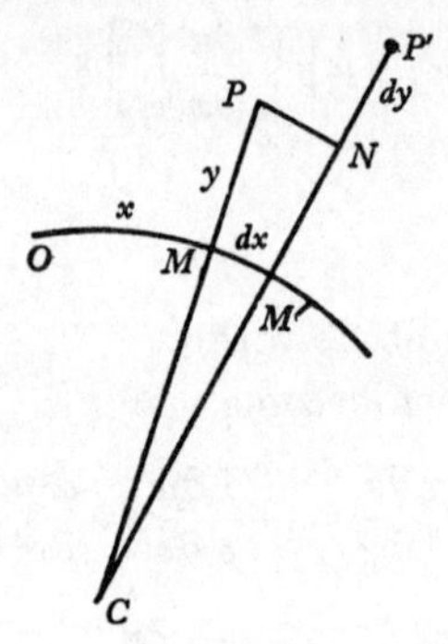

FIG. 23·22.

Then $OM = x, \quad MP = y, \quad MM' = dx, \quad NP' = dy$ and

$$\frac{PN}{dx} = \frac{1/\kappa+y}{1/\kappa} = 1+\kappa y, \quad PN = (1+\kappa y)\,dx.$$

We can now use the results of 2·72 with

(1) $$u_1 = x,\ u_2 = y,\ h_1 = (1+\kappa y),\ h_2 = 1.$$

Thus if (u, v) is the velocity at P, from 2·72 (2) we have the equation of continuity

(2) $$\frac{\partial u}{\partial x}+\frac{\partial(h_1 v)}{\partial y} = \frac{\partial u}{\partial x}+(1+\kappa y)\frac{\partial v}{\partial y}+\kappa v = 0.$$

Observe that κ is a function x only and is independent of y.

Omitting external force, using 2·32 (V), we get from 21·30 the component equations

(3) $$\frac{du}{dt} = -\frac{1}{h_1}\frac{\partial}{\partial x}\left(\frac{p}{\rho}\right)-\nu\frac{\partial\zeta}{\partial y},$$

(4) $$\frac{dv}{dt} = -\frac{\partial}{\partial y}\left(\frac{p}{\rho}\right)+\nu\frac{\partial\zeta}{h_1\partial x}.$$

From 2·72 (4) the vorticity is

(5) $$\zeta = \frac{1}{h_1}\frac{\partial v}{\partial x}-\frac{\partial u}{\partial y}-\frac{\kappa u}{h_1}.$$

From 2·72 (6) we have the accelerations

(6) $$\frac{du}{dt} = \frac{\partial u}{\partial t}+\frac{1}{1+\kappa y}u\frac{\partial u}{\partial x}+v\frac{\partial u}{\partial y}+\frac{\kappa}{1+\kappa y}uv,$$

(7) $$\frac{dv}{dt} = \frac{\partial v}{\partial t}+\frac{1}{1+\kappa y}u\frac{\partial v}{\partial x}+\frac{v\,\partial v}{\partial y}-\frac{\kappa u^2}{1+\kappa y}.$$

Forming $\partial\zeta/\partial y$ and eliminating $\partial^2 v/\partial x\partial y$ with the aid of the equation of continuity we get from (3) and (6) (and a corresponding result from (4) and (7)), the equations of motion

$$(8)\quad \frac{\partial u}{\partial t}+\frac{u}{h_1}\frac{\partial u}{\partial x}+v\frac{\partial u}{\partial y}+\frac{\kappa uv}{h_1}$$

$$= -\frac{1}{\rho}\frac{\partial p}{h_1\partial x}+\nu\left(\frac{1}{h_1^2}\frac{\partial^2 u}{\partial x^2}+\frac{\partial^2 u}{\partial y^2}+\frac{2\kappa}{h_1^2}\frac{\partial v}{\partial x}+\frac{\kappa' y}{h_1^2}\frac{\partial v}{\partial y}+\frac{\kappa}{h_1}\frac{\partial u}{\partial y}-\frac{\kappa^2}{h_1^2}u+\frac{\kappa'}{h_1^2}v\right),$$

$$(9)\quad \frac{\partial v}{\partial t}+\frac{u}{h_1}\frac{\partial v}{\partial x}+v\frac{\partial v}{\partial y}-\frac{\kappa u^2}{h_1}$$

$$= -\frac{1}{\rho}\frac{\partial p}{\partial y}+\nu\left(\frac{1}{h_1^2}\frac{\partial^2 v}{\partial x^2}+\frac{\partial^2 v}{\partial y^2}+\frac{2\kappa}{h_1}\frac{\partial v}{\partial y}-\frac{\kappa}{h_1^2}\frac{\partial u}{\partial x}-\frac{\kappa' y}{h_1^3}\frac{\partial v}{\partial x}-\frac{\kappa' u}{h_1^3}\right),$$

wherein $\kappa' = d\kappa/dx, \quad h_1 = 1+\kappa y.$

We now make the boundary layer approximation exactly as before by putting $y = \eta\delta$, $v = v_0\delta$ and making the additional geometrical assumption that κ does not change abruptly or become infinite, more precisely that $\kappa\delta$ and $\kappa'\delta^2$ are small. The resulting equations are then

$$(10)\quad \frac{\partial u}{\partial t}+u\frac{\partial u}{\partial x}+v\frac{\partial u}{\partial y} = -\frac{1}{\rho}\frac{\partial p}{\partial x}+\nu\frac{\partial^2 u}{\partial y^2},$$

$$(11)\quad \frac{\partial p}{\partial y} = \kappa\rho u^2,$$

$$(12)\quad \frac{\partial u}{\partial x}+\frac{\partial v}{\partial y} = 0.$$

Observe that except for (11), which gives the pressure gradient required to balance the centrifugal effect of flow round a curved wall, these equations are the same as for a straight wall for which $\kappa = 0$.

In so far as δ is small (11) shows that the pressure varies but little in passing normally through the boundary layer.

Since in the main stream

$$(13)\quad \frac{\partial U}{\partial t}+U\frac{\partial U}{\partial x} = -\frac{1}{\rho}\frac{\partial p}{\partial x}$$

we can replace (10) by

$$(14)\quad \frac{\partial u}{\partial t}+u\frac{\partial u}{\partial x}+v\frac{\partial u}{\partial y} = \frac{\partial U}{\partial t}+U\frac{\partial U}{\partial x}+\nu\frac{\partial^2 u}{\partial y^2}.$$

23·23. Boundary conditions. For flow along a straight wall the boundary layer equations are

(1) $$\frac{\partial u}{\partial t}+u\frac{\partial u}{\partial x}+v\frac{\partial u}{\partial y} = -\frac{1}{\rho}\frac{\partial p}{\partial x}+\nu\frac{\partial^2 u}{\partial y^2},$$

(2) $$\frac{\partial u}{\partial x}+\frac{\partial v}{\partial y} = 0, \quad \frac{\partial p}{\partial y} = 0.$$

The last equation of (2) states that as we proceed normally to the wall, the pressure does not change and is equal to the pressure just outside the boundary layer

(3) $$p = p(x, t)$$

and is a given function. Therefore the pressure $p(x, t)$ is the same as the pressure would be at the wall were there no viscosity. If however the boundary layer separates from the wall, p must be determined otherwise, generally by experiment.

Let $U = U(x, t)$ be the slip velocity, and let

(4) $$u = u_B(x, y, t), \quad v = v_B(x, y, t)$$

where suffix B indicates that these are solutions of the boundary layer equations (1) and (2), not, in general, solutions of the complete Navier-Stokes equations.

Let us again temporarily define boundary layer thickness δ as in 23·20, namely as the distance from the wall at which $u_B(x, y, t)$ attains the value $99U(x, t)/100$ or more generally as the distance at which

(5) $$u_B(x, y, t) = aU(x, t), \quad y = \delta$$

where a is some agreed fraction. Now the approach of $u_B(x, y, t)$ to $U(x, t)$ is asymptotic (see fig. 23·20) and therefore as $a \to 1$, $\delta \to \infty$.

Therefore the condition (5) becomes in the limit

(6) $$u_B(x, y, t) = U(x, t), \quad y = \infty.$$

This is the condition on $u_B(x, y, t)$ at $y = \infty$. It is usually referred to as the condition at the *outer edge of the boundary layer* on the principle that the boundary layer is thin, and so (6) is practically satisfied for quite moderate values of $\mathscr{R}^{1/2}y$.

The conditions at the wall are those of adherence,

(7) $$u_B(x, y, t) = 0, \quad v_B(x, y, t) = 0 \text{ when } y = 0.$$

Thus we must regard (4) as the solution of (1) and (2) subject to the conditions (3), (6), (7). As to the *existence* in general of such a solution we shall not enquire.

The foregoing heuristic treatment may be cast in a more mathematical form by the introduction of *Prandtl non-dimensional variables* distinguished by suffix P and, in the two-dimensional case, defined by

(8) $x = lx_P,\ y = ly_P/\mathscr{R}^{1/2},\ u = Vu_P,\ v = Vv_P/\mathscr{R}^{1/2},\ t = lt_P/V,\ P = V^2P_P,$

where $\mathscr{R} = Vl/\nu$ and V is a constant velocity, and where, $u = u(x, y, t)$, $u_P = u_P(x_P, y_P, t_P)$ and so on. This is an affine transformation which stretches y more rapidly than x. In particular when $\mathscr{R}$ is very large small values of y correspond to large values of y_P. In this sense the outer edge $y = \delta$ of the boundary layer will correspond to a large value of y_P and in the limit, when $\mathscr{R}\to\infty$, $y = 0$ corresponds to $y_P = \infty$.

With the above transformations the Navier-Stokes equations become

$$(9)\quad \frac{du_P}{dt_P} = -\frac{\partial P_P}{\partial x_P} + \frac{\partial^2 u_P}{\partial y^2_P} + \frac{1}{\mathscr{R}}\frac{\partial^2 u_P}{\partial x^2_P},\quad \frac{1}{\mathscr{R}}\frac{dv_P}{dt_P} = -\frac{\partial P_P}{\partial y_P} + \frac{1}{\mathscr{R}}\frac{\partial^2 v_P}{\partial y^2_P} + \frac{1}{\mathscr{R}^2}\frac{\partial^2 v_P}{\partial x^2_P}$$

$$\frac{\partial u_P}{\partial x_P} + \frac{\partial v_P}{\partial y_P} = 0,\quad \frac{d}{dt_P} = \frac{\partial}{\partial t_P} + u_P\frac{\partial}{\partial x_P} + v_P\frac{\partial}{\partial y_P},\quad P_P = \frac{p}{\rho V^2}$$

where we have omitted external force.

Now let $\mathscr{R}\to\infty$ while the Prandtl variables remain fixed. Then (9) go over into the boundary layer equations

$$(10)\quad \frac{\partial u_P}{\partial t_P} + u_P\frac{\partial u_P}{\partial x_P} + v_P\frac{\partial u_P}{\partial y_P} = -\frac{\partial P_P}{\partial x_P} + \frac{\partial^2 u_P}{\partial y^2_P},\quad \frac{\partial P_P}{\partial y_P} = 0,\quad \frac{\partial u_P}{\partial x_P} + \frac{\partial v_P}{\partial y_P} = 0$$

which are the equations (1), (2) in Prandtl variables and are to be solved under the boundary conditions

$$(11)\quad u_P(x_P, 0, t_P) = 0,\ v_P(x_P, 0, t_P) = 0,\ u_P(x_P, \infty, t_P) = U/V$$
$$P_P(x_P, 0, t_P) = p(x, t)/\rho V^2.$$

By the introduction of the Prandtl variables the general problem has thus been reduced to the definite problem posed by the solution of (10) subject to the boundary conditions (11). Moreover we can return to the physical variables by the use of (8). Thus the boundary conditions (11) agree with the conditions (3), (6), (7) previously found.

23·24. Boundary layer thicknesses. Hitherto the thickness of the boundary layer has tentatively been regarded as the distance from the wall at which u has attained a given percentage, e.g. 99 per cent., of the *slip velocity* U which would obtain *at the wall* were the viscosity zero. Since the velocity U is attained only asymptotically we are led to substitute the more precise definition given below. As usual we consider the liquid between two planes parallel to the flow and at unit distance apart.

We then define *displacement thickness* δ_1 by

$$(1)\quad U\delta_1 = \int_0^\infty (U - u)dy.$$

The upper limit infinity allows for the asymptotic approach of u to U (see

23·23), but in practice the upper limit is the value of y beyond which the integrand is negligible. The quantity $U\delta_1$ thus defined measures the diminution due to the boundary layer of the volume flux across a normal to the wall. Thus the streamlines of the flow, were the viscosity zero, are displaced from the wall through the distance δ_1 as obtained in 23·11 for flow into a stagnation point.

The *momentum thickness* δ_2 is defined by

(2) $$U^2\delta_2 = \int_0^\infty u(U-u)dy,$$

where $\rho U^2\delta_2$ measures the flux of defect of momentum of the actual from the inviscid flow.

Similarly we define *kinetic energy thickness* δ_3 by

(3) $$U^3\delta_3 = \int_0^\infty u(U^2-u^2)dy.$$

More generally if χ is any property of the flow and χ_i the corresponding value in the inviscid flow the χ *thickness* δ_χ can be defined by

(4) $$U\chi_i\delta_\chi = \int_0^\infty u(\chi_i-\chi)dy.$$

Observe that, in general, all these thicknesses are small (cf. 23·11).

23·26. The stream function near the wall. Let $\psi(x, y) = \psi$ be the stream function at the point $P(x, y)$ near to the wall $y = 0$.

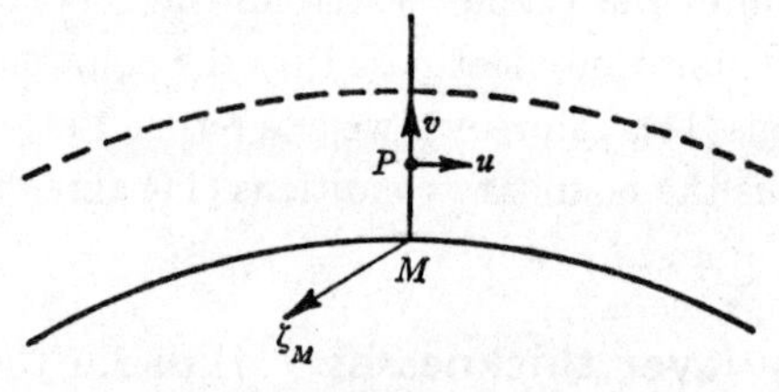

FIG. 23·26.

Then the vorticity at M, the foot of the perpendicular from P to M, is

(1) $$\zeta_M = \zeta(x, 0) = -(\partial u/\partial y)_{y=0}$$

and therefore near the wall

(2) $$u = -\zeta_M y.$$

Also from the equation of continuity

$$\frac{\partial v}{\partial y} = -\frac{\partial u}{\partial x} = \zeta'_M y$$

where the dash refers to differentiation with respect to x.

Therefore near the wall

(3) $$v = \tfrac{1}{2}\zeta'_M y^2$$

and (2) and (3) yield the first approximation $\frac{1}{2}\zeta_M y^2$ to the stream function. To get a second approximation write $\psi = \frac{1}{2}\zeta_M y^2 + Ay^3$, substitute in the equation of motion 23·22 (8) and then put $y = 0$. This gives A and we find

(4) $$\psi = \tfrac{1}{2}y^2\left\{\zeta_M - \tfrac{1}{3}\left(\frac{1}{\mu}\frac{\partial p}{\partial x} + \kappa\zeta_M\right)y\right\}.$$

23·28. Separation and attachment. We have from 23·26 (4)

(1) $$\psi = \tfrac{1}{2}y^2\{\zeta_M + Ay\}, \quad A = -\tfrac{1}{3}\left(\frac{p'}{\mu} + \kappa\zeta_M\right).$$

The streamline $\psi = k$ is therefore

(2) $$k = \tfrac{1}{2}y^2(\zeta_M + Ay).$$

Also the wall $y = 0$ is part of the streamline $\psi = 0$ and therefore the streamline $\psi = 0$ near M on the wall consists of

(3) $$y^2 = 0, \quad \zeta_M + Ay = 0,$$

that is to say the wall and a line nearly parallel to the wall. If, however, $\zeta_M = 0$ the second of (3) gives for the slope at M

(4) $$y' = -\zeta'_M/A = 3\mu\zeta'_M/p'$$

and the two loci of (3) are no longer nearly parallel but intersecting and therefore constitute a dividing streamline for the flow which from 23·26 (3) actually leaves the wall if ζ'_M and therefore v is positive. Since $\zeta_M = (-\partial u/\partial y)_M$, $\zeta_M = 0$ implies that $\partial u/\partial y$ changes sign at M and so the flow is forwards on one side of M and backwards on the other.

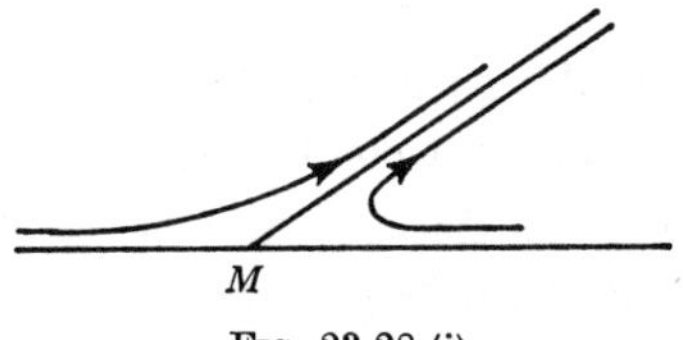

FIG. 23·28 (i).

If we consider a thin layer of fluid adjacent to the wall and wholly inside the boundary layer, this thin layer is urged forward by the viscous pull of the superincumbent fluid, and is retarded by the friction at the wall. If the pressure gradient is favourable (pressure decreasing in the direction of flow), the thin layer will continue to move forward. Near the wall as we have seen the forward velocity is small and therefore the momentum of the fluid may be insufficient for the fluid to force its way for very long against an unfavourable pressure gradient (pressure increasing in the direction of flow). This circumstance would bring the fluid to rest and a slow back-flow may set in. The forward going stream then leaves the surface.

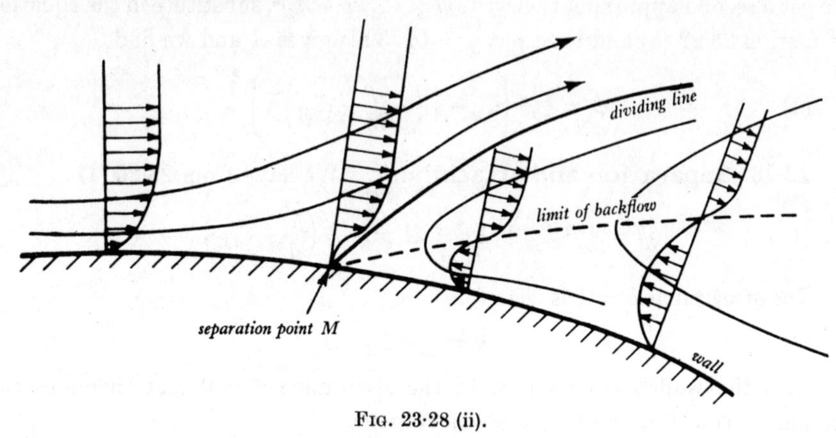

FIG. 23·28 (ii).

Fig. (ii) shows some velocity profiles and the dividing streamline springing from the point M at which $\partial u/\partial y = -\zeta_M$ vanishes. Beyond M a thin layer of vorticity leaves the wall and enters the interior of the main flow and the boundary layer approximation ceases to apply. The dotted line delimits the region of back-flow.

Definition. A point M on the wall at which $\zeta_M = 0$ and $\zeta'_M > 0$ is called a *point of separation.*

At a point of separation we see from the equation of motion and the profile graph of fig. (ii) that when $y = 0$

$$(5) \qquad p' = -\mu \left(\frac{\partial \zeta}{\partial y}\right)_{y=0} = \nu \frac{\partial^2 u}{\partial y^2} > 0.$$

It then follows from (4) that y' is positive and the situation is as shown in fig. (i).

We observe that to determine a point of separation it is also necessary to know the pressure field outside the boundary layer and if separation indeed takes place, there is vorticity elsewhere in the fluid and in general the pressure distribution can be found only from experiment.

Definition. A point M on the wall at which $\zeta_M = 0$ and $\zeta'_M < 0$ is called a *point of attachment.*

Here in fig. 23·26 v is negative and the fluid does not leave the wall.

At a point of attachment there is a favourable pressure gradient, $p' < 0$. The forward stagnation point on a body which obstructs a stream is a point of attachment.

It is of course possible for the fluid to separate and then attach again. See e.g. Plate 1, figs. 4 and 5.

It may be remarked that from a practical point of view prediction of the point of separation does not fall neatly into the above theory, although the general notion seems to be substantially satisfactory.

23·30. Boundary layer along a flat plate at zero incidence. Consider a semi-infinite flat plate which disturbs a uniform steady stream U from left to right and parallel to the plate.

Take the plate to occupy the region $y = 0$, $x>0$ so that the origin is at the leading edge.

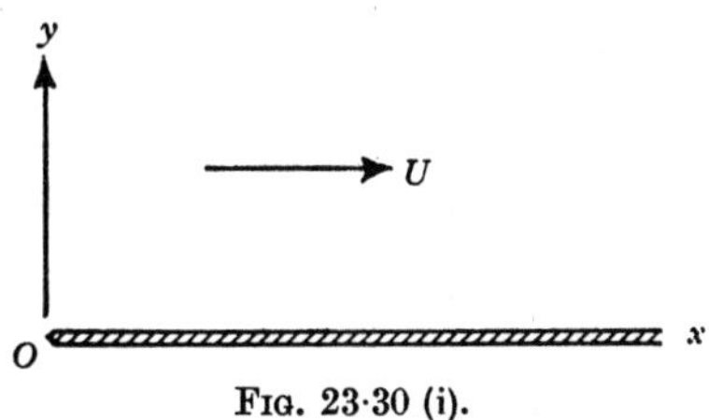

FIG. 23·30 (i).

We assume that the stream is negligibly affected by the presence of the plate except in the boundary layer.

The pressure in the boundary layer is a function of x only, $p = p(x)$. Therefore at the edge of the boundary layer we have

$$-\frac{1}{\rho}\frac{\partial p}{\partial x} = \frac{\partial U}{\partial t} + U\frac{\partial U}{\partial x} = 0$$

and so there is no pressure gradient along the plate. Thus the boundary layer equations are

(1) $$u\frac{\partial u}{\partial x} + v\frac{\partial u}{\partial y} = \nu\frac{\partial^2 u}{\partial y^2},$$

(2) $$\frac{\partial u}{\partial x} + \frac{\partial v}{\partial y} = 0.$$

The boundary conditions will be taken as

(3) $u = v = 0$ at $y = 0$, $x>0$ (adherence),

(4) $u = U$ at $y = \infty$, for all x (edge of boundary layer),

(5) $u = U$ at $x = 0$ (leading edge).

Since the leading edge is also a stagnation point, (5) implies an infinite gradient in speed at $x = 0$ and therefore a singularity in the mathematical solution. Since no linear scale is preferred in this problem we define the *local Reynolds number* $\mathcal{R}_x$ by

(6) $$\mathcal{R}_x = Ux/\nu$$

and this vanishes at $x = 0$. Thus the boundary layer hypothesis of large Reynolds number breaks down at the leading edge and so the solution which we shall derive from (1) and (2) must be held to begin to apply not at $x = 0$ but at a short distance downstream from this point. The consequent conclusions agree well with experiment.

Following Blasius we look for a solution of the form

$$(7) \qquad u = u(x, y) = Uf'(\eta), \text{ where } \eta = \frac{y}{g(x)}, \quad \frac{\partial}{\partial y} = \frac{1}{g}\frac{\partial}{\partial \eta}.$$

Here the dash refers to differentiation with respect to η.

Substituting in the equation of continuity (2) we get

$$\frac{\partial v}{\partial y} = -\frac{\partial u}{\partial x} = \frac{U}{g}\frac{dg}{dx}\eta f''$$

which integrates to give

$$(8) \qquad v = U\frac{dg}{dx}(\eta f' - f), \text{ assuming } f(0) = 0.$$

Substituting for u and v in (1) we have

$$(9) \qquad \nu f''' + Ug\frac{dg}{dx}ff'' = 0.$$

If this is to be satisfied for all values of x, we must have $g\, dg/dx =$ constant, whence by integration $g = k(x-c)^{1/2}$, where k and c are constants. The constant c merely fixes the origin so with $c = 0$ we can take, by choice of k,

$$(10) \qquad g = \left(\frac{2\nu x}{U}\right)^{1/2},$$

which implies zero thickness for the boundary layer at the leading edge. Substitution in (9) now gives the equation of Blasius

$$(11) \qquad f''' + ff'' = 0, \quad f = f(\eta)$$

with the boundary conditions

$$(12) \qquad f = f' = 0 \text{ at } \eta = 0, \quad f' = 1 \text{ at } \eta = \infty.$$

This equation with the given boundary conditions may be solved to any required degree of accuracy by computing machines. The following short table gives some values to two decimal places.

η	0	0·5	1·0	2·0	3·0	4·0	5·0	6·0
f	0	·06	·23	·80	1·80	2·78	3·78	4·78
f'	0	·23	·46	·80	·97	1·00	1·00	1·00
f''	0·47	·46	·43	·28	·07	·01	·00	·00

The displacement thickness is, 23·14,

$$(13) \qquad \delta_1 = g(x)\int_0^\infty [1-f'(\eta)]d\eta = 1{\cdot}72\left(\frac{\nu x}{U}\right)^{1/2} = \frac{1{\cdot}72x}{\sqrt{\mathscr{R}_x}}$$

using the above table, and equation (6). Thus the graph of δ_1 against x is a parabola of latus rectum $2{\cdot}96\nu/U$.

For the normal velocity we have from (8) and (10)

$$(14) \qquad \frac{v}{U} = \frac{1}{(2\mathscr{R}_x)^{1/2}}(\eta f'-f) \text{ which tends to } \frac{0{\cdot}865}{(\mathscr{R}_x)^{1/2}}$$

when $\eta \to \infty$.

Thus v exhibits no boundary layer in the sense that it does not tend to zero for large η. On the other hand $v \to 0$ when $x \to \infty$ and therefore $\mathscr{R}_x$ increases.

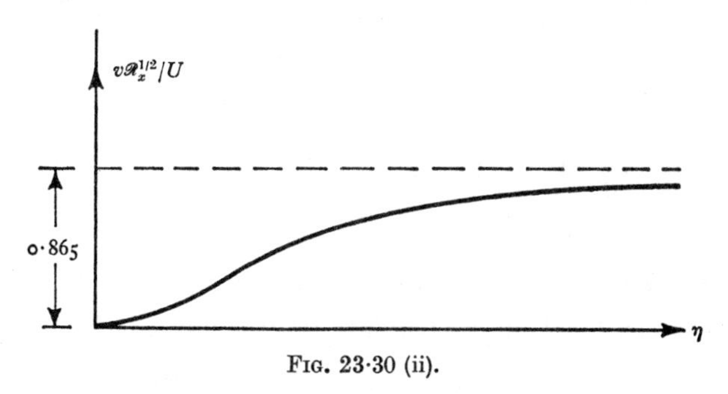

Fig. 23·30 (ii).

This result means that the main flow is slightly distorted by the presence of the plate.

The shearing stress on the wall is, by 21·14,

$$(15) \qquad (\widehat{xy})_{y=0} = \left(\mu\frac{\partial u}{\partial y}\right)_{y=0} = \rho U^2 f''(0)/(2\mathscr{R}_x)^{1/2}.$$

Hence the drag on a width b and length l (from the leading edge) of the plate, reckoning both sides of the plate is

$$(16) \qquad 2b\int_0^l (\widehat{xy})_{y=0}\,dx = 4blf''(0)\rho U^2/\sqrt{2\mathscr{R}_l}, \; \mathscr{R}_l = \frac{Ul}{\nu}$$

whence, dividing by $bl\rho U^2$ we get the drag coefficient

$$(17) \qquad C_D = 1{\cdot}33/\mathscr{R}_l^{1/2},$$

where we have taken $f''(0) = 0{\cdot}47$.

23·40. Similar solutions. Two solutions are said to be *similar* if at every given x-location, the velocity profiles differ only by scale factor in u and y.

Without loss of generality we may assume that x is not negative.

Let $U(x)$ be the slip velocity. Then similar solutions must be expressible in the form

(1) $\quad u(x, y) = U(x)f'(\eta), \quad y = \eta g(x)$ whence $\partial\eta/\partial x = -\eta g'(x)/g(x)$,

where $f(\eta)$ is a suitable function of η and $g(x)$ is a suitable function of x. From (1) $f'(\infty) = 1$.

It is evident that if we plot

(2) $$\frac{u(x, y)}{U(x)} \quad \text{against} \quad \frac{y}{g(x)}$$

the resulting profile depends only on the variable η once the function $f(\eta)$ has been found.

From the equation of continuity we have

$$\frac{1}{g(x)}\frac{\partial v}{\partial \eta} = \frac{\partial v}{\partial y} = -\frac{\partial u}{\partial x} = -U'(x)f'(\eta) + U(x)f''(\eta)\frac{\eta g'(x)}{g(x)}.$$

Therefore

$$\frac{\partial v}{\partial \eta} = \frac{\partial}{\partial \eta}[\eta U(x)g'(x)f'(\eta) - f(\eta)\{g(x)U'(x) + g'(x)U(x)\}]$$

(3) $$v = U(x)g'(x)\eta f'(\eta) - f(\eta)[g(x)U(x)]'$$

where we must have $f(0) = 0$ since $v = 0$ when $y = 0$.

For steady motion the first boundary layer equation, 23·22 (14), is

$$u\frac{\partial u}{\partial x} + v\frac{\partial u}{\partial y} = U(x)U'(x) + \nu\frac{\partial^2 u}{\partial y^2}$$

so substituting for u and v from (1) and (3) we get the Falkner-Skan equation*

(4) $$f''' + \alpha f f'' + \beta(1 - f'^2) = 0$$

(5) $$\alpha = g(x)[U(x)g(x)]'/\nu, \quad \beta = g^2(x)U'(x)/\nu$$

with the boundary conditions

$$f(0) = f'(0) = 0, \quad f'(\infty) = 1.$$

The Blasius equation 23·30 (11) is a particular case of (4) with $\alpha = 1, \beta = 0$. Putting $\eta = 0$ in (4) we see that

(6) $$\beta = -f'''(0)$$

so that β (which is independent of η) must be a constant. We then see from (4) that α must be a function of η only and since, from (5), α is independent of η, α must also be a constant.

It therefore follows that the functions $U(x)$, $g(x)$ can not be chosen arbitrarily but must satisfy the equations

* V. M. Falkner and S. W. Skan. *Phil. Mag.* (7) 12 (1931) 865–896.

(7) $$g^2U'+gg'U = \nu\alpha, \quad g^2U' = \nu\beta$$

where α, β are constants.

From these equations we get

(8) $$(2\alpha-\beta)\nu = g^2U'+2gg'U = (g^2U)'.$$

If $2\alpha-\beta \neq 0$ we have

(9) $$g^2U = (2\alpha-\beta)\nu x+\gamma.$$

The constant γ merely locates the origin and putting $\gamma = 0$

(10) $$g^2(x) = \frac{(2\alpha-\beta)\nu x}{U(x)}, \quad 2\alpha-\beta \neq 0.$$

Combining this with $g^2U' = \nu\beta$ from (7) we get

(11) $$\frac{U'(x)}{U(x)} = \frac{\beta}{(2\alpha-\beta)x}$$

and therefore, C being a constant,

(12) $$U(x) = Cx^m, \quad x>0, \quad 2\alpha-\beta \neq 0, \quad m = \beta/(2\alpha-\beta).$$

On the other hand if $2\alpha-\beta = 0$, (9) gives

(13) $$g^2U = \gamma.$$

Combined with the second of (7) this gives

(14) $$\frac{U'(x)}{U(x)} = \frac{\nu\beta}{\gamma} = p \quad \text{say}$$

and therefore

(15) $$U(x) = Ce^{px}, \quad 2\alpha-\beta = 0, \quad x>0$$

where C and p are constants.

Thus we have

Goldstein's theorem. There exist similar solutions of the boundary layer equations only if

$$U(x) = Cx^m \text{ or } U(x) = Ce^{px}$$

where $U(x)$ is the slip velocity and C, m, p are constants.

The condition that the above flows shall be possible is that $g^2(x)>0$. This will happen in the following cases

$$(a) \quad U = Cx^m, \quad C>0, \quad m>-1$$

$$(b) \quad U = Cx^m, \quad C<0, \quad m<0$$

$$(c) \quad U = Ce^{px}, \quad C>0, \quad p>0.$$

In all these cases except (*a*) when $-1<m<0$, the pressure is favourable.

Examples of (*a*) and (*b*) are furnished by flow on one or both faces of a wedge as is shown by 6·04 (3).

23·60. Kármán's momentum integral. The boundary layer equation 23·22 (14) and the equation of continuity can be written in the form

$$(1) \qquad -\nu \frac{\partial^2 u}{\partial y^2} = \frac{\partial}{\partial t}(U-u) + U\frac{\partial U}{\partial x} - u\frac{\partial u}{\partial x} - v\frac{\partial u}{\partial y},$$

$$(2) \qquad 0 = (U-u)\frac{\partial u}{\partial x} + (U-u)\frac{\partial v}{\partial y},$$

where $U = U(x, t)$ is the slip velocity.

Adding (1) and (2) we get

$$(3) \qquad -\nu\frac{\partial^2 u}{\partial y^2} = \frac{\partial}{\partial t}(U-u) + \frac{\partial}{\partial x}[u(U-u)] + (U-u)\frac{\partial U}{\partial x} + \frac{\partial}{\partial y}[v(U-u)].$$

We first make the following observations. At $y = \infty$, $u = U$ and

$$\zeta = -\frac{\partial u}{\partial y} = -\frac{\partial U(x,t)}{\partial y} = 0. \quad \text{Further } v(U-u) = 0.$$

If we suppose the wall to be porous we can assume a *suction velocity* there so that at $y = 0$, $u = 0$, $v = -v_S$ where v_S is the suction velocity. We shall denote by ζ_0 the vorticity at the wall.

Now integrate (3) through the boundary layer, i.e. from $y = 0$ to $y = \infty$. Then

$$(4) \qquad -\nu\zeta_0 = \frac{\partial}{\partial t}\int_0^\infty (U-u)\,dy + \frac{\partial}{\partial x}\int_0^\infty u(U-u)\,dy + \frac{\partial U}{\partial x}\int_0^\infty (U-u)\,dy + Uv_S.$$

This is Kármán's momentum integral.

If we introduce the displacement and momentum thickness δ_1 and δ_2 defined in 23·24, and observe that the shearing stress on the wall is $(\widehat{xy})_0 = -\mu\zeta_0$ (21·06) equation (4) becomes

$$(5) \qquad \frac{1}{\rho}(\widehat{xy})_0 = \frac{\partial}{\partial t}(U\delta_1) + U^2\frac{\partial \delta_2}{\partial x} + (\delta_1 + 2\delta_2)U\frac{\partial U}{\partial x} + Uv_S.$$

Kármán's integral is the basis of numerous suggested methods of approximation to flow in the boundary layer. The classical method is due to Polhausen. In the case of steady motion with constant U he writes δ in (4) instead of the upper limit ∞ where δ is a thickness of the boundary layer and takes

$$(6) \qquad u = a_1(x)y + a_2(x)y^2 + a_3(x)y^3 + \ldots$$

He then seeks to satisfy the conditions

(7) $u = 0,\ \partial^2 u/\partial y^2 = 0$ when $y = 0$; $u = U,\ \partial u/\partial y = 0$ when $y = \delta$.

The condition on $\partial^2 u/\partial y^2$ is obtained from (1) and $\partial u/\partial y = 0$ when $y = \delta$ expresses that the vorticity vanishes at the edge of the boundary layer.

Thus in the case of steady flow along a flat plate $\partial U/\partial x = 0$ and (4) becomes

$$-\nu\left(\frac{\partial u}{\partial y}\right)_0 = \frac{\partial}{\partial x}\int_0^\delta u^2 dy - U\frac{\partial}{\partial x}\int_0^\delta u\, dy. \tag{8}$$

Assuming three terms in the expansion (6) we find after considerable computation

$$\delta = \sqrt{\left(\frac{280\nu x}{13U}\right)} = x\sqrt{\frac{280}{13\mathscr{R}_x}},\ (\widehat{xy})_0 = \frac{0{\cdot}323\rho U^2}{\sqrt{\mathscr{R}_x}},\ \mathscr{R}_x = \frac{Ux}{\nu}. \tag{9}$$

Another approximation due to Lamb is the direct assumption

$$u = U\sin\frac{\pi y}{2\delta} \tag{10}$$

which satisfies all the conditions (7). Substitution in (8) then gives

$$U^2\frac{\partial\delta}{\partial x} = \frac{\pi^2}{4-\pi}\frac{\nu U}{\delta} \quad \text{whence}$$

$$\delta^2 = \frac{2\pi^2}{4-\pi}\frac{\nu x}{U} \tag{11}$$

which gives the thickness of the boundary layer.

The traction on the wall is (21·06)

$$\mu\zeta = \rho U^2\sqrt{\frac{4-\pi}{8}}\cdot\sqrt{\frac{\nu}{Ux}}.$$

Thus the drag on a length l (from the leading edge) reckoning both sides of the plate is obtained by doubling the above and integrating from 0 to l along the plate. This gives

$$\rho U^2 l\sqrt{(8-2\pi)}\sqrt{\frac{\nu}{Ul}} = 1{\cdot}310\rho U^2 l/\mathscr{R}^{1/2},\ \mathscr{R} = \frac{Ul}{\nu}. \tag{12}$$

The coefficient 1·310 thus obtained agrees well with the coefficient 1·328 obtained by Blasius without assumption (10) or the use of Kármán's integral.

23·61. Intrinsic equations. Continuing 21·39 if we make the approximations of the boundary layer theory, 21·39 (1) reduces near the wall to

$$\frac{1}{\rho}\frac{\partial p}{\partial s} + q\frac{\partial q}{\partial s} + \frac{\partial\Omega}{\partial s} = \nu\frac{\partial^2 q}{\partial n^2}, \tag{1}$$

provided that the curvatures are not large, and then

$$F = \int_0^s \frac{\partial^2 q}{\partial n^2}\, ds.$$

With the same approximations, 21·39 (2) gives

$$\frac{1}{\rho}\frac{\partial p}{\partial n} + \frac{\partial \Omega}{\partial n} = 0. \tag{2}$$

Eliminating $p/\rho + \Omega$ from (1) and (2) and using 21·39 (3), we get

$$\frac{\partial}{\partial n}\left(\nu \frac{\partial^2 q}{\partial n^2} + \kappa_n q^2\right) = 0.$$

Thus, in the boundary layer,

$$\nu \frac{\partial^2 q}{\partial n^2} + \kappa_n q^2 = A,$$

where A is independent of n and is therefore the value of $\nu\, \partial^2 q/\partial n^2$ on the boundary.

If κ_n is *constant*, this equation gives

$$\nu \left(\frac{\partial q}{\partial n}\right)^2 + \tfrac{2}{3}\kappa_n q^3 = 2Aq + B,$$

and this admits of further integration in terms of elliptic functions.

EXAMPLES XXIII

1. Taking the velocity at infinity to be zero discuss the velocity field

$$u = \nu^{-1} x\, U^2 \exp(-Uy/\nu), \quad v = -U(1 - \exp(-Uy/\nu)).$$

Show that there is a stagnation point at the origin and find the pressure.

2. Solve Stokes's problem when the plate at $y = 0$ oscillates with velocity $U \cos nt$, assuming the solution to be of the form u = real part of $f(y)e^{int}$.

3. For the Blasius equation draw a graph to show the dependence of $f'(\eta)$ on $f(\eta)$.

4. For flow along a flat plate show that the value of η corresponding to $y = \delta_1$ is 1·22 approximately.

5. A flat plate is placed at zero incidence in a steady uniform stream U. If δ is the " thickness " of the boundary layer show that the velocity distribution

$$u = U \sin \frac{\pi y}{2\delta}$$

makes $u = U$, $\partial u/\partial y = 0$, when $y = \delta$ and $u = 0$, $\partial^2 u/\partial y^2 = 0$ when $y = 0$. Hence show that

$$\delta^2 = \frac{2\pi^2 \nu x}{(4-\pi)\, U}.$$

6. In the previous example show that the drag, reckoning both sides, on a length l of the plate taken from the leading edge is

$$1{\cdot}310\rho U^2 l/\mathscr{R}^{1/2},$$

where $\mathscr{R} = Ul/\nu$ is taken as the Reynolds number.

7. For flow over a flat plate (23·30) show that the graph of $v\mathscr{R}^{1/2}/U$ as a function of η has a point of inflexion and find its approximate position.

8. Prove that for steady two-dimensional motion of a liquid the von Mises transformation

$$\xi = x, \quad \eta = -\psi(x, y) = \int_0^y u(x, y)\, dy$$

enables the boundary layer equation to be put in the form

$$\frac{\partial Z}{\partial \xi} = \nu u \frac{\partial^2 Z}{\partial \eta^2}, \quad Z = -U^2 + u^2.$$

9. For the Blasius solution show that v is a monotone increasing function of y which tends to a finite limit

$$v_\infty = kU/\mathscr{R}^{1/2}$$

where k is a constant, when $y \to \infty$.

Show that $v = 0$ when $y = 0$ and the graph of v has a single point of inflexion. (Serrin)

10. In the Falkner-Skan equation 23·40 (4) when $\alpha = 1, \beta > 0$ prove that, if $\beta_2 > \beta_1 \geqslant 0$, then $f'_2(\eta) > f'_1(\eta)$ where $f_1(\eta), f_2(\eta)$ correspond to $\beta = \beta_1, \beta_2$.

11. For the similar solutions corresponding to $U = Cx^m$ prove that the shear stress at the wall is

$$[\tfrac{1}{2}(1+m)]^{1/2} f''(0)\rho U^2/\sqrt{\mathscr{R}_x}, \quad \mathscr{R}_x = |\, Ux/\mu \,|.$$

Using this result in the case $C > 0$, $m > -\frac{1}{3}$ show that the drag coefficient for a portion of the surface from $x = 0$ to $x = l$, of width b is

$$C_D = 4f''(0)(3m+1)^{-1}[\tfrac{1}{2}(1+m)]^{1/2}\, bl/\mathscr{R}_x^{1/2}.$$

12. In the Blasius solution prove the momentum thickness is $2f''(0)x\mathscr{R}^{-1/2}$.

13. In the Blasius solution, calculate the kinetic energy thickness.

14. For the boundary layer in compressible flow show that the enthalpy thickness can be defined by

$$\delta_I = \int_0^\infty \frac{\rho u}{\rho_i U}\left(1 - \frac{I}{I_i}\right) dy$$

where suffix i refers to the corresponding inviscid flow at the wall and U is the slip velocity.

15. Obtain the approximate equations of motion for the two-dimensional steady flow of an incompressible, slightly viscous fluid in a boundary layer along a plane wall in the form

$$u\frac{\partial u}{\partial x}+v\frac{\partial u}{\partial y} = U\frac{\partial U}{\partial x}+\nu\frac{\partial^2 u}{\partial y^2}, \quad u = -\frac{\partial \psi}{\partial y}, \quad v = \frac{\partial \psi}{\partial x},$$

where U is the velocity in the main stream just outside the boundary layer.

Fluid is flowing between two non-parallel plane walls, towards the intersection of the planes, so that, if x is measured along a wall from the intersection of the planes, U is negative and inversely proportional to x. Verify that a solution of the differential equations may be obtained in which ψ is a function of y/x only and hence obtain the solution

$$\frac{u}{U} = 3\tanh^2\left\{\alpha+\left(\frac{|U|}{2\nu x}\right)^{\frac{1}{2}} y\right\}-2,$$

where $\tanh^2\alpha = \frac{2}{3}$, for the velocity u in the boundary layer along one of the walls. (U.L.)

16. A jet of air issues from a small hole in a wall, and mixes with the surrounding air. Write down equations to determine approximately the velocity in the jet at some distance from the hole, on the assumptions that the compressibility of the air may be neglected and that the motion is laminar and symmetrical about an axis. If M is the rate at which momentum flows across a section of the jet, μ is the viscosity and ρ the density of the air, and if the axis of x is along the jet and y is the distance from that axis, prove that the velocity in the jet parallel to the axis is given by

$$u = \frac{3M}{8\pi\mu}\frac{1}{1+\xi^2/4}, \quad \text{where} \quad \xi = \frac{\rho}{4\mu}\sqrt{\left(\frac{3M}{\pi\rho}\right)}\frac{y}{x}.$$ (M.T.)

17. A jet of air issues from a straight narrow slit in a wall, and mixes with the surrounding air. On the assumptions that the compressibility of the air may be neglected, that the motion is steady (non-turbulent) and two-dimensional, and that the approximations of the boundary layer theory may be applied, show that at some distance from the slit the velocity along the axis of the jet is

$$\left(\frac{3M^2}{32\rho^2\nu x}\right)^{\frac{1}{3}}\operatorname{sech}^2\left[\left(\frac{M}{48\rho\nu^2x^2}\right)^{\frac{1}{3}} y\right],$$

where M denotes the rate at which momentum flows across unit length of a section of the jet. The axis of x is along, and the axis of y perpendicular to, the axis of the jet.

Find the flux across a section of the jet. (M.T.)

INDEX

The references are to pages.